Student Solutions Manual

Algebra and Trigonometry

FOURTH EDITION

James Stewart
McMaster University and University of Toronto

Lothar Redlin
The Pennsylvania State University

Saleem Watson
California State University, Long Beach

Prepared by

Andy Bulman-Fleming

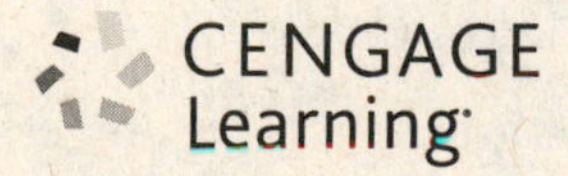

Australia • Brazil • Mexico • Singapore • United Kingdom • United States

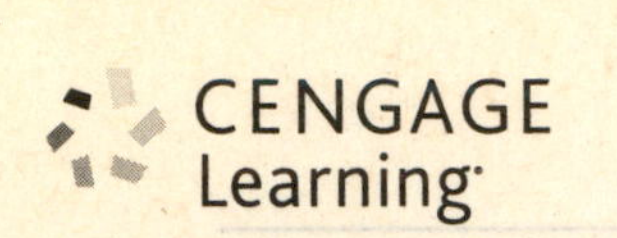

WCN: 01-100-101

For product information and technology assistance, contact us at **Cengage Learning Customer & Sales Support, 1-800-354-9706**.

For permission to use material from this text or product, submit all requests online at **www.cengage.com/permissions**
Further permissions questions can be emailed to **permissionrequest@cengage.com**.

ISBN: 978-1-305-11815-7

Cengage Learning
20 Channel Center Street
Boston, MA 02210
USA

Cengage Learning is a leading provider of customized learning solutions with office locations around the globe, including Singapore, the United Kingdom, Australia, Mexico, Brazil, and Japan. Locate your local office at: **www.cengage.com/global**.

Cengage Learning products are represented in Canada by Nelson Education, Ltd.

To learn more about Cengage Learning Solutions, visit **www.cengage.com**.

Purchase any of our products at your local college store or at our preferred online store **www.cengagebrain.com**.

Printed in the United States of America
Print Number: 01 Print Year: 2014

9/20

CONTENTS

■ **PROLOGUE:** Principles of Problem Solving 1

CHAPTER P **PREREQUISITES** 3

P.1 Modeling the Real World with Algebra 3
P.2 Real Numbers 3
P.3 Integer Exponents and Scientific Notation 6
P.4 Rational Exponents and Radicals 8
P.5 Algebraic Expressions 11
P.6 Factoring 13
P.7 Rational Expressions 16
P.8 Solving Basic Equations 19
P.9 Modeling with Equations 22
Chapter P Review 25
Chapter P Test 28
■ **FOCUS ON MODELING:** Making the Best Decisions 29

CHAPTER 1 **EQUATIONS AND GRAPHS** 33

1.1 The Coordinate Plane 33
1.2 Graphs of Equations in Two Variables; Circles 38
1.3 Lines 45
1.4 Solving Quadratic Equations 50
1.5 Complex Numbers 55
1.6 Solving Other Types of Equations 56
1.7 Solving Inequalities 60
1.8 Solving Absolute Value Equations and Inequalities 70
1.9 Solving Equations and Inequalities Graphically 71
1.10 Modeling Variation 74
Chapter 1 Review 77
Chapter 1 Test 86

■ **FOCUS ON MODELING:** Fitting Lines to Data 89

CHAPTER 2 **FUNCTIONS** 91

2.1 Functions 91
2.2 Graphs of Functions 95
2.3 Getting Information from the Graph of a Function 102
2.4 Average Rate of Change of a Function 108
2.5 Linear Functions and Models 111
2.6 Transformations of Functions 114
2.7 Combining Functions 121
2.8 One-to-One Functions and Their Inverses 125
Chapter 2 Review 130
Chapter 2 Test 137
■ **FOCUS ON MODELING:** Modeling with Functions 138

CHAPTER 3 **POLYNOMIAL AND RATIONAL FUNCTIONS** 143

3.1 Quadratic Functions and Models 143
3.2 Polynomial Functions and Their Graphs 148
3.3 Dividing Polynomials 156
3.4 Real Zeros of Polynomials 160
3.5 Complex Zeros and the Fundamental Theorem of Algebra 176
3.6 Rational Functions 181
Chapter 3 Review 198
Chapter 3 Test 208
■ **FOCUS ON MODELING:** Fitting Polynomial Curves to Data 210

CHAPTER 4 **EXPONENTIAL AND LOGARITHMIC FUNCTIONS** 213

4.1 Exponential Functions 213
4.2 The Natural Exponential Function 217
4.3 Logarithmic Functions 220
4.4 Laws of Logarithms 225
4.5 Exponential and Logarithmic Equations 227

4.6 Modeling with Exponential Functions 231

4.7 Logarithmic Scales 233

Chapter 4 Review 234

Chapter 4 Test 239

■ **FOCUS ON MODELING:** Fitting Exponential and Power Curves to Data 240

CHAPTER **5** **TRIGONOMETRIC FUNCTIONS: RIGHT TRIANGLE APPROACH** **243**

5.1 Angle Measure 243

5.2 Trigonometry of Right Triangles 245

5.3 Trigonometric Functions of Angles 247

5.4 Inverse Trigonometric Functions and Right Triangles 250

5.5 The Law of Sines 252

5.6 The Law of Cosines 254

Chapter 5 Review 257

Chapter 5 Test 259

■ **FOCUS ON MODELING:** Surveying 287

CHAPTER **6** **TRIGONOMETRIC FUNCTIONS: UNIT CIRCLE APPROACH** **263**

6.1 The Unit Circle 263

6.2 Trigonometric Functions of Real Numbers 265

6.3 Trigonometric Graphs 267

6.4 More Trigonometric Graphs 274

6.5 Inverse Trigonometric Functions and Their Graphs 277

6.6 Modeling Harmonic Motion 279

Chapter 6 Review 282

Chapter 6 Test 286

■ **FOCUS ON MODELING:** Fitting Sinusoidal Curves to Data 260

CHAPTER **7** **ANALYTIC TRIGONOMETRY** **291**

7.1 Trigonometric Identities 291

7.2 Addition and Subtraction Formulas 295

7.3 Double-Angle, Half-Angle, and Product-Sum Formulas 298

7.4 Basic Trigonometric Equations 304

7.5 More Trigonometric Equations 306

Chapter 7 Review 309

Chapter 7 Test 312

■ **FOCUS ON MODELING:** Traveling and Standing Waves 313

CHAPTER **8** **POLAR COORDINATES AND PARAMETRIC EQUATIONS** **315**

8.1 Polar Coordinates 315

8.2 Graphs of Polar Equations 317

8.3 Polar Form of Complex Numbers; De Moivre's Theorem 321

8.4 Plane Curves and Parametric Equations 327

Chapter 8 Review 334

Chapter 8 Test 337

■ **FOCUS ON MODELING:** The Path of a Projectile 338

CHAPTER **9** **VECTORS IN TWO AND THREE DIMENSIONS** **341**

9.1 Vectors in Two Dimensions 341

9.2 The Dot Product 344

9.3 Three-Dimensional Coordinate Geometry 346

9.4 Vectors in Three Dimensions 347

9.5 The Cross Product 349

9.6 Equations of Lines and Planes 350

Chapter 9 Review 352

Chapter 9 Test 354

■ **FOCUS ON MODELING:** Vector Fields 355

CHAPTER **10** **SYSTEMS OF EQUATIONS AND INEQUALITIES** **357**

10.1 Systems of Linear Equations in Two Variables 357

10.2 Systems of Linear Equations in Several Variables 361

10.3 Partial Fractions 365

10.4 Systems of Nonlinear Equations 370

10.5 Systems of Inequalities 374

Chapter 10 Review 380

Chapter 10 Test 384

■ **FOCUS ON MODELING:** Linear Programming 386

CHAPTER 11 **MATRICES AND DETERMINANTS** **391**

11.1 Matrices and Systems of Linear Equations 391

11.2 The Algebra of Matrices 397

11.3 Inverses of Matrices and Matrix Equations 401

11.4 Determinants and Cramer's Rule 406

Chapter 11 Review 413

Chapter 11 Test 418

■ **FOCUS ON MODELING:** Computer Graphics 420

CHAPTER 12 **CONIC SECTIONS** **423**

12.1 Parabolas 423

12.2 Ellipses 425

12.3 Hyperbolas 430

12.4 Shifted Conics 433

12.5 Rotation of Axes 439

12.6 Polar Equations of Conics 445

Chapter 12 Review 450

Chapter 12 Test 457

■ **FOCUS ON MODELING:** Conics in Architecture 458

CHAPTER 13 **SEQUENCES AND SERIES** **459**

13.1 Sequences and Summation Notation 459

13.2 Arithmetic Sequences 461

13.3 Geometric Sequences 464

13.4 Mathematics of Finance 468

13.5 Mathematical Induction 470

13.6 The Binomial Theorem 475

Chapter 13 Review 477

Chapter 13 Test 481

■ **FOCUS ON MODELING:** Modeling with Recursive Sequences 482

CHAPTER 14 COUNTING AND PROBABILITY 483

14.1 Counting 483

14.2 Probability 486

14.3 Binomial Probability 490

14.4 Expected Value 493

Chapter 14 Review 494

Chapter 14 Test 497

■ **FOCUS ON MODELING:** The Monte Carlo Method 497

APPENDIXES 499

A Geometry Review 499

B Calculations and Significant Figures 499

C Graphing with a Graphing Calculator 500

PROLOGUE: Principles of Problem Solving

1. Let r be the rate of the descent. We use the formula time $= \dfrac{\text{distance}}{\text{rate}}$; the ascent takes $\dfrac{1}{15}$ h, the descent takes $\dfrac{1}{r}$ h, and the total trip should take $\dfrac{2}{30} = \dfrac{1}{15}$ h. Thus we have $\dfrac{1}{15} + \dfrac{1}{r} = \dfrac{1}{15} \quad\Leftrightarrow\quad \dfrac{1}{r} = 0$, which is impossible. So the car cannot go fast enough to average 30 mi/h for the 2-mile trip.

2. Let us start with a given price P. After a discount of 40%, the price decreases to $0.6P$. After a discount of 20%, the price decreases to $0.8P$, and after another 20% discount, it becomes $0.8\,(0.8P) = 0.64P$. Since $0.6P < 0.64P$, a 40% discount is better.

3. We continue the pattern. Three parallel cuts produce 10 pieces. Thus, each new cut produces an additional 3 pieces. Since the first cut produces 4 pieces, we get the formula $f(n) = 4 + 3(n - 1)$, $n \geq 1$. Since $f(142) = 4 + 3(141) = 427$, we see that 142 parallel cuts produce 427 pieces.

4. By placing two amoebas into the vessel, we skip the first simple division which took 3 minutes. Thus when we place two amoebas into the vessel, it will take $60 - 3 = 57$ minutes for the vessel to be full of amoebas.

5. The statement is false. Here is one particular counterexample:

	Player A	Player B
First half	1 hit in 99 at-bats: average $= \frac{1}{99}$	0 hit in 1 at-bat: average $= \frac{0}{1}$
Second half	1 hit in 1 at-bat: average $= \frac{1}{1}$	98 hits in 99 at-bats: average $= \frac{98}{99}$
Entire season	2 hits in 100 at-bats: average $= \frac{2}{100}$	99 hits in 100 at-bats: average $= \frac{99}{100}$

6. *Method 1:* After the exchanges, the volume of liquid in the pitcher and in the cup is the same as it was to begin with. Thus, any coffee in the pitcher of cream must be replacing an equal amount of cream that has ended up in the coffee cup.

Method 2: Alternatively, look at the drawing of the spoonful of coffee and cream mixture being returned to the pitcher of cream. Suppose it is possible to separate the cream and the coffee, as shown. Then you can see that the coffee going into the cream occupies the same volume as the cream that was left in the coffee.

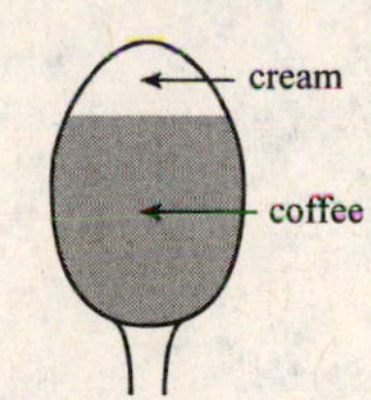

Method 3 (an algebraic approach): Suppose the cup of coffee has y spoonfuls of coffee. When one spoonful of cream is added to the coffee cup, the resulting mixture has the following ratios: $\dfrac{\text{cream}}{\text{mixture}} = \dfrac{1}{y+1}$ and $\dfrac{\text{coffee}}{\text{mixture}} = \dfrac{y}{y+1}$.

So, when we remove a spoonful of the mixture and put it into the pitcher of cream, we are really removing $\dfrac{1}{y+1}$ of a spoonful of cream and $\dfrac{y}{y+1}$ spoonful of coffee. Thus the amount of cream left in the mixture (cream in the coffee) is $1 - \dfrac{1}{y+1} = \dfrac{y}{y+1}$ of a spoonful. This is the same as the amount of coffee we added to the cream.

7. Let r be the radius of the earth in feet. Then the circumference (length of the ribbon) is $2\pi r$. When we increase the radius by 1 foot, the new radius is $r + 1$, so the new circumference is $2\pi(r + 1)$. Thus you need $2\pi(r + 1) - 2\pi r = 2\pi$ extra feet of ribbon.

8. The north pole is such a point. And there are others: Consider a point a_1 near the south pole such that the parallel passing through a_1 forms a circle C_1 with circumference exactly one mile. Any point P_1 exactly one mile north of the circle C_1 along a meridian is a point satisfying the conditions in the problem: starting at P_1 she walks one mile south to the point a_1 on the circle C_1, then one mile east along C_1 returning to the point a_1, then north for one mile to P_1. That's not all. If a point a_2 (or a_3, a_4, a_5, ...) is chosen near the south pole so that the parallel passing through it forms a circle C_2 (C_3, C_4, C_5, ...) with a circumference of exactly $\frac{1}{2}$ mile ($\frac{1}{3}$ mi, $\frac{1}{4}$ mi, $\frac{1}{5}$ mi, ...), then the point P_2 (P_3, P_4, P_5, ...) one mile north of a_2 (a_3, a_4, a_5, ...) along a meridian satisfies the conditions of the problem: she walks one mile south from P_2 (P_3, P_4, P_5, ...) arriving at a_2 (a_3, a_4, a_5, ...) along the circle C_2 (C_3, C_4, C_5, ...), walks east along the circle for one mile thus traversing the circle twice (three times, four times, five times, ...) returning to a_2 (a_3, a_4, a_5, ...), and then walks north one mile to P_2 (P_3, P_4, P_5, ...).

P PREREQUISITES

P.1 MODELING THE REAL WORLD WITH ALGEBRA

1. Using this model, we find that if $S = 12$, $L = 4S = 4(12) = 48$. Thus, 12 sheep have 48 legs.

3. If $x = \$120$ and $T = 0.06x$, then $T = 0.06(120) = 7.2$. The sales tax is \$7.20.

5. If $v = 70$, $t = 3.5$, and $d = vt$, then $d = 70 \cdot 3.5 = 245$. The car has traveled 245 miles.

7. (a) $M = \dfrac{N}{G} = \dfrac{240}{8} = 30$ miles/gallon

(b) $25 = \dfrac{175}{G} \Leftrightarrow G = \dfrac{175}{25} = 7$ gallons

9. (a) $V = 9.5S = 9.5\left(4\text{ km}^3\right) = 38\text{ km}^3$

(b) $19\text{ km}^3 = 9.5S \Leftrightarrow S = 2\text{ km}^3$

11. (a)

Depth (ft)	Pressure (lb/in^2)
0	$0.45(0) + 14.7 = 14.7$
10	$0.45(10) + 14.7 = 19.2$
20	$0.45(20) + 14.7 = 23.7$
30	$0.45(30) + 14.7 = 28.2$
40	$0.45(40) + 14.7 = 32.7$
50	$0.45(50) + 14.7 = 37.2$
60	$0.45(60) + 14.7 = 41.7$

(b) We know that $P = 30$ and we want to find d, so we solve the equation $30 = 14.7 + 0.45d \Leftrightarrow 15.3 = 0.45d \Leftrightarrow$ $d = \dfrac{15.3}{0.45} = 34.0$. Thus, if the pressure is 30 lb/in^2, the depth is 34 ft.

13. The number N of cents in q quarters is $N = 25q$.

15. The cost C of purchasing x gallons of gas at \$3.50 a gallon is $C = 3.5x$.

17. The distance d in miles that a car travels in t hours at 60 mi/h is $d = 60t$.

19. (a) $\$12 + 3(\$1) = \$12 + \$3 = \$15$

(b) The cost C, in dollars, of a pizza with n toppings is $C = 12 + n$.

(c) Using the model $C = 12 + n$ with $C = 16$, we get $16 = 12 + n \Leftrightarrow n = 4$. So the pizza has four toppings.

21. (a) (i) For an all-electric car, the energy cost of driving x miles is $C_e = 0.04x$.

(ii) For an average gasoline powered car, the energy cost of driving x miles is $C_g = 0.12x$.

(b) (i) The cost of driving 10,000 miles with an all-electric car is $C_e = 0.04(10{,}000) = \$400$.

(ii) The cost of driving 10,000 miles with a gasoline powered car is $C_g = 0.12(10{,}000) = \$1200$.

23. (a) The GPA is $\dfrac{4a + 3b + 2c + 1d + 0f}{a + b + c + d + f} = \dfrac{4a + 3b + 2c + d}{a + b + c + d + f}$.

(b) Using $a = 2 \cdot 3 = 6$, $b = 4$, $c = 3 \cdot 3 = 9$, and $d = f = 0$ in the formula from part (a), we find the GPA to be $\dfrac{4 \cdot 6 + 3 \cdot 4 + 2 \cdot 9}{6 + 4 + 9} = \dfrac{54}{19} \approx 2.84$.

P.2 THE REAL NUMBERS

1. (a) The natural numbers are $\{1, 2, 3, \ldots\}$.

(b) The numbers $\{\ldots, -3, -2, -1, 0\}$ are integers but not natural numbers.

(c) Any irreducible fraction $\frac{p}{q}$ with $q \neq 1$ is rational but is not an integer. Examples: $\frac{3}{2}, -\frac{5}{12}, \frac{1729}{23}$.

(d) Any number which cannot be expressed as a ratio $\frac{p}{q}$ of two integers is irrational. Examples are $\sqrt{2}, \sqrt{3}, \pi$, and e.

3. The set of numbers between but not including 2 and 7 can be written as **(a)** $\{x \mid 2 < x < 7\}$ in interval notation, or **(b)** $(2, 7)$ in interval notation.

5. The distance between a and b on the real line is $d(a, b) = |b - a|$. So the distance between -5 and 2 is $|2 - (-5)| = 7$.

7. **(a)** No: $a - b = -(b - a) \neq b - a$ in general.

(b) No; by the Distributive Property, $-2(a - 5) = -2a + -2(-5) = -2a + 10 \neq -2a - 10$.

9. **(a)** Natural number: 100

(b) Integers: 0, 100, -8

(c) Rational numbers: $-1.5, 0, \frac{5}{2}, 2.71, 3.1\overline{4}, 100, -8$

(d) Irrational numbers: $\sqrt{7}, -\pi$

11. Commutative Property of addition

13. Associative Property of addition

15. Distributive Property

17. Commutative Property of multiplication

19. $x + 3 = 3 + x$

21. $4(A + B) = 4A + 4B$

23. $3(x + y) = 3x + 3y$

25. $4(2m) = (4 \cdot 2)m = 8m$

27. $-\frac{5}{2}(2x - 4y) = -\frac{5}{2}(2x) + \frac{5}{2}(4y) = -5x + 10y$

29. **(a)** $\frac{3}{10} + \frac{4}{15} = \frac{9}{30} + \frac{8}{30} = \frac{17}{30}$

(b) $\frac{1}{4} + \frac{1}{5} = \frac{5}{20} + \frac{4}{20} = \frac{9}{20}$

31. **(a)** $\frac{2}{3}\left(6 - \frac{3}{2}\right) = \frac{2}{3} \cdot 6 - \frac{2}{3} \cdot \frac{3}{2} = 4 - 1 = 3$

(b) $\left(3 + \frac{1}{4}\right)\left(1 - \frac{4}{5}\right) = \left(\frac{12}{4} + \frac{1}{4}\right)\left(\frac{5}{5} - \frac{4}{5}\right) = \frac{13}{4} \cdot \frac{1}{5} = \frac{13}{20}$

33. **(a)** $2 \cdot 3 = 6$ and $2 \cdot \frac{7}{2} = 7$, so $3 < \frac{7}{2}$

(b) $-6 > -7$

(c) $3.5 = \frac{7}{2}$

35. **(a)** False

(b) True

37. **(a)** True

(b) False

39. **(a)** $x > 0$ **(b)** $t < 4$

(c) $a \geq \pi$ **(d)** $-5 < x < \frac{1}{3}$

(e) $|p - 3| \leq 5$

41. **(a)** $A \cup B = \{1, 2, 3, 4, 5, 6, 7, 8\}$

(b) $A \cap B = \{2, 4, 6\}$

43. **(a)** $A \cup C = \{1, 2, 3, 4, 5, 6, 7, 8, 9, 10\}$

(b) $A \cap C = \{7\}$

45. **(a)** $B \cup C = \{x \mid x \leq 5\}$

(b) $B \cap C = \{x \mid -1 < x < 4\}$

47. $(-3, 0) = \{x \mid -3 < x < 0\}$

49. $[2, 8) = \{x \mid 2 \leq x < 8\}$

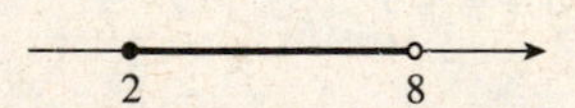

51. $[2, \infty) = \{x \mid x \geq 2\}$

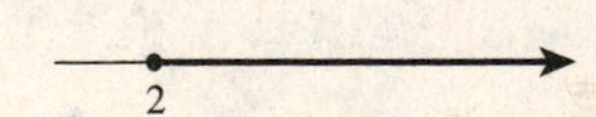

53. $x \leq 1 \Leftrightarrow x \in (-\infty, 1]$

1

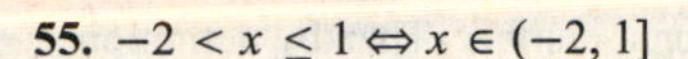

55. $-2 < x \le 1 \Leftrightarrow x \in (-2, 1]$

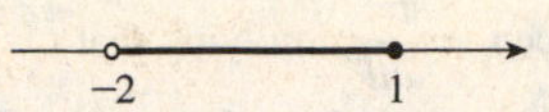

57. $x > -1 \Leftrightarrow x \in (-1, \infty)$

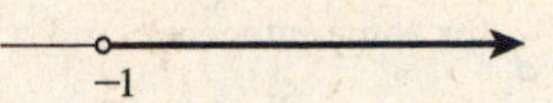

59. (a) $[-3, 5]$ (b) $(-3, 5]$

61. $(-2, 0) \cup (-1, 1) = (-2, 1)$

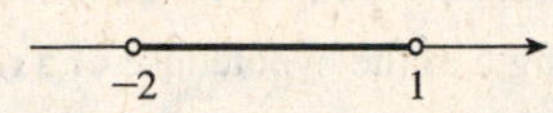

63. $[-4, 6] \cap [0, 8) = [0, 6]$

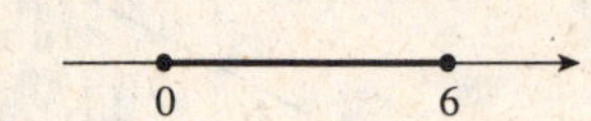

65. $(-\infty, -4) \cup (4, \infty)$

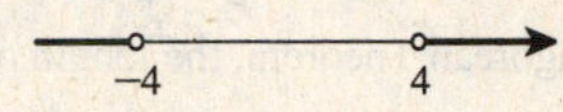

67. **(a)** $|100| = 100$

(b) $|-73| = 73$

69. **(a)** $||-6| - |-4|| = |6 - 4| = |2| = 2$

(b) $\frac{-1}{|-1|} = \frac{-1}{1} = -1$

71. **(a)** $|(-2) \cdot 6| = |-12| = 12$

(b) $\left|\left(-\frac{1}{3}\right)(-15)\right| = |5| = 5$

73. $|(-2) - 3| = |-5| = 5$

75. **(a)** $|17 - 2| = 15$

(b) $|21 - (-3)| = |21 + 3| = |24| = 24$

(c) $\left|-\frac{3}{10} - \frac{11}{8}\right| = \left|-\frac{12}{40} - \frac{55}{40}\right| = \left|-\frac{67}{40}\right| = \frac{67}{40}$

77. **(a)** Let $x = 0.777\ldots$. So $10x = 7.7777\ldots \Leftrightarrow x = 0.7777\ldots \Leftrightarrow 9x = 7$. Thus, $x = \frac{7}{9}$.

(b) Let $x = 0.2888\ldots$. So $100x = 28.8888\ldots \Leftrightarrow 10x = 2.8888\ldots \Leftrightarrow 90x = 26$. Thus, $x = \frac{26}{90} = \frac{13}{45}$.

(c) Let $x = 0.575757\ldots$. So $100x = 57.5757\ldots \Leftrightarrow x = 0.5757\ldots \Leftrightarrow 99x = 57$. Thus, $x = \frac{57}{99} = \frac{19}{33}$.

79. $\pi > 3$, so $|\pi - 3| = \pi - 3$.

81. $a < b$, so $|a - b| = -(a - b) = b - a$.

83. **(a)** $-a$ is negative because a is positive.

(b) bc is positive because the product of two negative numbers is positive.

(c) $a - ba + (-b)$ is positive because it is the sum of two positive numbers.

(d) $ab + ac$ is negative: each summand is the product of a positive number and a negative number, and the sum of two negative numbers is negative.

85. Distributive Property

87. **(a)** When $L = 60$, $x = 8$, and $y = 6$, we have $L + 2(x + y) = 60 + 2(8 + 6) = 60 + 28 = 88$. Because $88 \le 108$ the post office will accept this package.

When $L = 48$, $x = 24$, and $y = 24$, we have $L + 2(x + y) = 48 + 2(24 + 24) = 48 + 96 = 144$, and since $144 \not\le 108$, the post office will *not* accept this package.

(b) If $x = y = 9$, then $L + 2(9 + 9) \le 108 \Leftrightarrow L + 36 \le 108 \Leftrightarrow L \le 72$. So the length can be as long as 72 in. = 6 ft.

89. $\frac{1}{2} + \sqrt{2}$ is irrational. If it were rational, then by Exercise 6(a), the sum $\left(\frac{1}{2} + \sqrt{2}\right) + \left(-\frac{1}{2}\right) = \sqrt{2}$ would be rational, but this is not the case.

Similarly, $\frac{1}{2} \cdot \sqrt{2}$ is irrational.

(a) Following the hint, suppose that $r + t = q$, a rational number. Then by Exercise 6(a), the sum of the two rational numbers $r + t$ and $-r$ is rational. But $(r + t) + (-r) = t$, which we know to be irrational. This is a contradiction, and hence our original premise—that $r + t$ is rational—was false.

(b) r is rational, so $r = \frac{a}{b}$ for some integers a and b. Let us assume that $rt = q$, a rational number. Then by definition, $q = \frac{c}{d}$ for some integers c and d. But then $rt = q \Leftrightarrow \frac{a}{b}t = \frac{c}{d}$, whence $t = \frac{bc}{ad}$, implying that t is rational. Once again we have arrived at a contradiction, and we conclude that the product of a rational number and an irrational number is irrational.

91. (a) Construct the number $\sqrt{2}$ on the number line by transferring the length of the hypotenuse of a right triangle with legs of length 1 and 1.

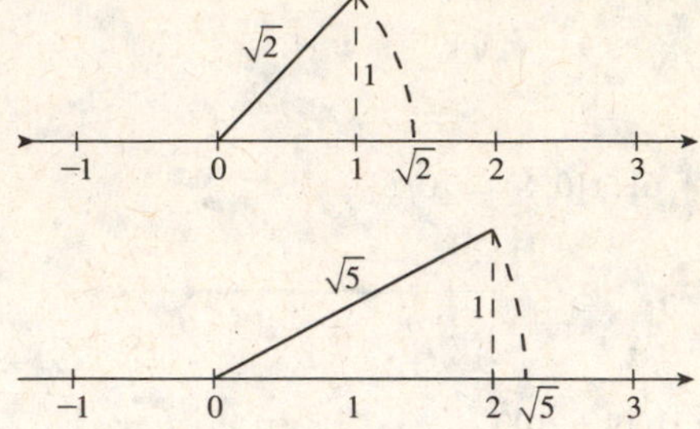

(b) Construct a right triangle with legs of length 1 and 2. By the Pythagorean Theorem, the length of the hypotenuse is $\sqrt{1^2 + 2^2} = \sqrt{5}$. Then transfer the length of the hypotenuse to the number line.

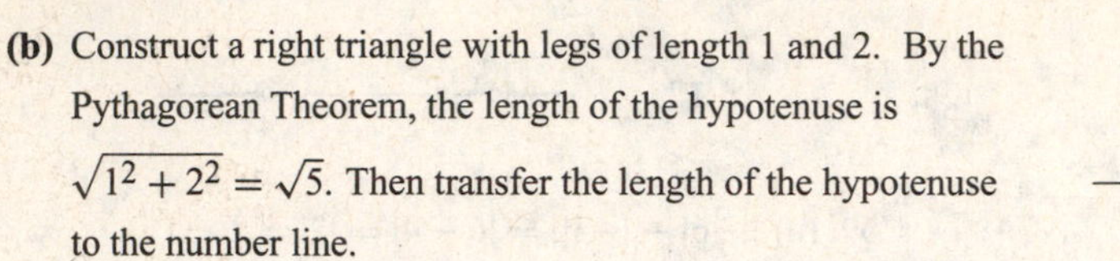

(c) Construct a right triangle with legs of length $\sqrt{2}$ and 2 [construct $\sqrt{2}$ as in part (a)]. By the Pythagorean Theorem, the length of the hypotenuse is $\sqrt{\left(\sqrt{2}\right)^2 + 2^2} = \sqrt{6}$. Then transfer the length of the hypotenuse to the number line.

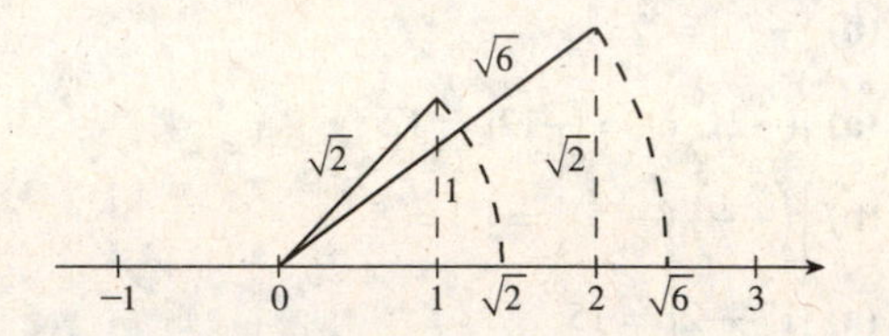

93. Answers will vary.

P.3 INTEGER EXPONENTS AND SCIENTIFIC NOTATION

1. Using exponential notation we can write the product $5 \cdot 5 \cdot 5 \cdot 5 \cdot 5 \cdot 5$ as 5^6.

3. In the expression 3^4, the number 3 is called the *base* and the number 4 is called the *exponent*.

5. When we divide two powers with the same base, we *subtract* the exponents. So $\frac{3^5}{3^2} = 3^3$.

7. (a) $2^{-1} = \frac{1}{2}$ **(b)** $2^{-3} = \frac{1}{8}$ **(c)** $\left(\frac{1}{2}\right)^{-1} = 2$ **(d)** $\frac{1}{2^{-3}} = 2^3 = 8$

9. (a) No, $\left(\frac{2}{3}\right)^{-2} = \left(\frac{3}{2}\right)^2 = \frac{9}{4}$. **(b)** Yes, $(-5)^4 = 625$ and $-5^4 = -\left(5^4\right) = -625$.

11. (a) $-2^6 = -64$ **(b)** $(-2)^6 = 64$ **(c)** $\left(\frac{1}{5}\right)^2 \cdot (-3)^3 = \frac{1^2(-3)^3}{5^2} = -\frac{27}{25}$

13. (a) $\left(\frac{5}{3}\right)^0 \cdot 2^{-1} = \frac{1}{2}$ **(b)** $\frac{2^{-3}}{3^0} = \frac{1}{2^3} = \frac{1}{8}$ **(c)** $\left(\frac{1}{4}\right)^{-2} = 4^2 = 16$

15. (a) $5^3 \cdot 5 = 5^4 = 625$ **(b)** $3^2 \cdot 3^0 = 3^2 = 9$ **(c)** $\left(2^2\right)^3 = 2^6 = 64$

17. (a) $5^4 \cdot 5^{-2} = 5^2 = 25$ **(b)** $\frac{10^7}{10^4} = 10^3 = 1000$ **(c)** $\frac{3^2}{3^4} = \frac{1}{3^2} = \frac{1}{9}$

19. (a) $x^2x^3 = x^{2+3} = x^5$ **(b)** $\left(-x^2\right)^3 = (-1)^3 x^{2 \cdot 3} = -x^6$ **(c)** $t^{-3}t^5 = t^{-3+5} = t^2$

21. (a) $x^{-5} \cdot x^3 = x^{-5+3} = x^{-2} = \frac{1}{x^2}$ **(b)** $w^{-2}w^{-4}w^5 = w^{-2-4+5} = w^{-1} = \frac{1}{w}$

(c) $\dfrac{y^{10}y^0}{y^7} = y^{10+0-7} = y^3$

23. (a) $\dfrac{a^9a^{-2}}{a} = a^{9-2-1} = a^6$ **(b)** $\left(a^2a^4\right)^3 = \left(a^{2+4}\right)^3 = \left(a^6\right)^3 = a^{6\cdot 3} = a^{18}$

(c) $(2x)^2\left(5x^6\right) = 2^2x^2 \cdot 5x^6 = 20x^{2+6} = 20x^8$

25. (a) $\left(3x^2y\right)\left(2x^3\right) = 3 \cdot 2x^{2+3}y = 6x^5y$

(b) $\left(2a^2b^{-1}\right)\left(3a^{-2}b^2\right) = 2 \cdot 3a^{2-2}b^{-1+2} = 6b$

(c) $\left(4y^2\right)\left(x^4y\right)^2 = 4y^2x^{4\cdot 2}y^2 = 4x^8y^{2+2} = 4x^8y^4$

27. (a) $\left(2x^2y^3\right)^2(3y) = 2^2x^{2\cdot 2}y^{3\cdot 2} \cdot 3y = 12x^4y^7$

(b) $\dfrac{x^2y^{-1}}{x^{-5}} = x^{2-(-5)}y^{-1} = x^7y^{-1} = \dfrac{x^7}{y}$

(c) $\left(\dfrac{x^2y}{3}\right)^3 = \dfrac{x^{2\cdot 3}y^3}{3^3} = \dfrac{x^6y^3}{27}$

29. (a) $\left(x^3y^3\right)^{-1} = \dfrac{1}{x^3y^3}$

(b) $\left(a^2b^{-2}\right)^{-3}\left(a^3\right)^{-2} = a^{2(-3)}b^{-2(-3)}a^{3(-2)} = a^{-6}b^6a^{-6} = \dfrac{b^6}{a^{12}}$

(c) $\left(\dfrac{x^2}{y^{-2}}\right)^{-2}\left(\dfrac{2y^{-3}}{x^2}\right)^3 = x^{2(-2)}y^{-(-2)(-2)} \cdot 2^3y^{-3(3)}x^{-2(3)} = x^{-4}y^{-4} \cdot 8y^{-9}x^{-6} = 8x^{-4-6}y^{-4-9} = \dfrac{8}{x^{10}y^{13}}$

31. (a) $\dfrac{3x^{-2}y^5}{9x^{-3}y^2} = \dfrac{xy^3}{3}$

(b) $\left(\dfrac{2x^3y^{-1}}{y^2}\right)^{-2} = \left(\dfrac{2x^3}{y^3}\right)^{-2} = \dfrac{y^{3\cdot 2}}{2^2x^{3\cdot 2}} = \dfrac{y^6}{4x^6}$

(c) $\left(\dfrac{y^{-1}}{x^{-2}}\right)^{-1}\left(\dfrac{3x^{-3}}{y^2}\right)^{-2} = y^{-1(-1)}x^{-(-2)(-1)}3^{-2}x^{-3(-2)}y^{-2(-2)} = \dfrac{x^4y^5}{9}$

33. (a) $\left(\dfrac{3a}{b^3}\right)^{-1} = 3^{-1}a^{-1}b^{-3(-1)} = \dfrac{b^3}{3a}$

(b) $\left(\dfrac{q^{-1}r^{-1}s^{-2}}{r^{-5}sq^{-8}}\right)^{-1} = \dfrac{r^{-5}sq^{-8}}{q^{-1}r^{-1}s^{-2}} = q^{-8-(-1)}r^{-5-(-1)}s^{1-(-2)} = \dfrac{s^3}{q^7r^4}$

35. (a) $69{,}300{,}000 = 6.93 \times 10^7$

(b) $7{,}200{,}000{,}000{,}000 = 7.2 \times 10^{12}$

(c) $0.000028536 = 2.8536 \times 10^{-5}$

(d) $0.0001213 = 1.213 \times 10^{-4}$

37. (a) $3.19 \times 10^5 = 319{,}000$

(b) $2.721 \times 10^8 = 272{,}100{,}000$

(c) $2.670 \times 10^{-8} = 0.00000002670$

(d) $9.999 \times 10^{-9} = 0.000000009999$

39. (a) $5{,}900{,}000{,}000{,}000$ mi$= 5.9 \times 10^{12}$ mi

(b) 0.0000000000004 cm $= 4 \times 10^{-13}$ cm

(c) 33 billion billion molecules $= 33 \times 10^9 \times 10^9 = 3.3 \times 10^{19}$ molecules

41. $\left(7.2 \times 10^{-9}\right)\left(1.806 \times 10^{-12}\right) = 7.2 \times 1.806 \times 10^{-9} \times 10^{-12} \approx 13.0 \times 10^{-21} = 1.3 \times 10^{-20}$

43. $\dfrac{1.295643 \times 10^{9}}{\left(3.610 \times 10^{-17}\right)\left(2.511 \times 10^{6}\right)} = \dfrac{1.295643}{3.610 \times 2.511} \times 10^{9+17-6} \approx 0.1429 \times 10^{19} = 1.429 \times 10^{19}$

45. $\dfrac{(0.0000162)(0.01582)}{(594621000)(0.0058)} = \dfrac{\left(1.62 \times 10^{-5}\right)\left(1.582 \times 10^{-2}\right)}{\left(5.94621 \times 10^{8}\right)\left(5.8 \times 10^{-3}\right)} = \dfrac{1.62 \times 1.582}{5.94621 \times 5.8} \times 10^{-5-2-8+3} = 0.074 \times 10^{-12}$
$= 7.4 \times 10^{-14}$

47. $\left|10^{50} - 10^{10}\right| < 10^{50}$, whereas $\left|10^{101} - 10^{100}\right| = 10^{100}\left|10 - 1\right| = 9 \times 10^{100} > 10^{50}$. So 10^{10} is closer to 10^{50} than 10^{100} is to 10^{101}.

49. Since one light year is 5.9×10^{12} miles, Centauri is about $4.3 \times 5.9 \times 10^{12} \approx 2.54 \times 10^{13}$ miles away or 25,400,000,000,000 miles away.

51. Volume $=$ (average depth) (area) $= \left(3.7 \times 10^{3}\text{ m}\right)\left(3.6 \times 10^{14}\text{ m}^2\right)\left(\dfrac{10^3\text{ liters}}{\text{m}^3}\right) \approx 1.33 \times 10^{21}$ liters

53. The number of molecules is equal to

$$(\text{volume}) \cdot \left(\frac{\text{liters}}{\text{m}^3}\right) \cdot \left(\frac{\text{molecules}}{22.4\text{ liters}}\right) = (5 \cdot 10 \cdot 3) \cdot \left(10^3\right) \cdot \left(\frac{6.02 \times 10^{23}}{22.4}\right) \approx 4.03 \times 10^{27}$$

55.

Year	Total interest
1	\$152.08
2	308.79
3	470.26
4	636.64
5	808.08

57. **(a)** $\dfrac{18^5}{9^5} = \left(\dfrac{18}{9}\right)^5 = 2^5 = 32$

(b) $20^6 \cdot (0.5)^6 = (20 \cdot 0.5)^6 = 10^6 = 1{,}000{,}000$

59. **(a)** We wish to prove that $\left(\dfrac{a}{b}\right)^{-n} = \dfrac{b^n}{a^n}$. By definition, and using the result from Exercise 58(b),
$\left(\dfrac{a}{b}\right)^{-n} = \dfrac{1}{\left(\dfrac{a}{b}\right)^n} = \dfrac{1}{\dfrac{a^n}{b^n}} = \dfrac{b^n}{a^n}$.

(b) We wish to prove that $\dfrac{a^{-n}}{b^{-m}} = \dfrac{b^m}{a^n}$. By definition, $\dfrac{a^{-n}}{b^{-m}} = \dfrac{\frac{1}{a^n}}{\frac{1}{b^m}} = \dfrac{1}{a^n} \cdot \dfrac{b^m}{1} = \dfrac{b^m}{a^n}$.

P.4 RATIONAL EXPONENTS AND RADICALS

1. Using exponential notation we can write $\sqrt[3]{5}$ as $5^{1/3}$.

3. No. $\sqrt{5^2} = \left(5^2\right)^{1/2} = 5^{2(1/2)} = 5$ and $\left(\sqrt{5}\right)^2 = \left(5^{1/2}\right)^2 = 5^{(1/2)2} = 5$.

5. Because the denominator is of the form $\sqrt{a}$, we multiply numerator and denominator by $\sqrt{a}$: $\frac{1}{\sqrt{3}} = \frac{1}{\sqrt{3}} \cdot \frac{\sqrt{3}}{\sqrt{3}} = \frac{\sqrt{3}}{3}$.

7. No. If a is negative, then $\sqrt{4a^2} = -2a$.

9. $\dfrac{1}{\sqrt{3}} = 3^{-1/2}$

11. $4^{2/3} = \sqrt[3]{4^2} = \sqrt[3]{16}$

13. $\sqrt[5]{5^3} = 5^{3/5}$

15. $a^{2/5} = \sqrt[5]{a^2}$

17. $\sqrt[3]{y^4} = y^{4/3}$

19. (a) $\sqrt{16} = \sqrt{4^2} = 4$

(b) $\sqrt[4]{16} = \sqrt[4]{2^4} = 2$

(c) $\sqrt[4]{\dfrac{1}{16}} = \sqrt[4]{\left(\dfrac{1}{2}\right)^4} = \dfrac{1}{2}$

21. (a) $3\sqrt[3]{16} = 3\sqrt[3]{2 \cdot 2^3} = 6\sqrt[3]{2}$

(b) $\dfrac{\sqrt{18}}{\sqrt{81}} = \sqrt{\dfrac{18}{81}} = \sqrt{\dfrac{2}{9}} = \dfrac{\sqrt{2}}{3}$

(c) $\sqrt{\dfrac{27}{4}} = \sqrt{\dfrac{3 \cdot 3^2}{2^2}} = \dfrac{3\sqrt{3}}{2}$

23. (a) $\sqrt{7}\sqrt{28} = \sqrt{7 \cdot 28} = \sqrt{196} = 14$

(b) $\dfrac{\sqrt{48}}{\sqrt{3}} = \sqrt{\dfrac{48}{3}} = \sqrt{16} = 4$

(c) $\sqrt[4]{24}\sqrt[4]{54} = \sqrt[4]{24 \cdot 54} = \sqrt[4]{1296} = 6$

25. (a) $\dfrac{\sqrt{216}}{\sqrt{6}} = \sqrt{\dfrac{216}{6}} = \sqrt{36} = 6$

(b) $\sqrt[3]{2}\sqrt[3]{32} = \sqrt[3]{64} = 4$

(c) $\sqrt[4]{\dfrac{1}{4}}\sqrt[4]{\dfrac{1}{64}} = \sqrt[4]{\dfrac{1}{256}} = \dfrac{1}{\sqrt[4]{256}} = \dfrac{1}{4}$

27. $\sqrt[4]{x^4} = |x|$

29. $\sqrt[5]{32y^6} = \sqrt[5]{2^5y^6} = 2\sqrt[5]{y^6} = 2y\sqrt[5]{y}$

31. $\sqrt[4]{16x^8} = \sqrt[4]{2^4x^8} = 2x^2$

33. $\sqrt[3]{x^3y} = \left(x^3\right)^{1/3} y^{1/3} = x\sqrt[3]{y}$

35. $\sqrt{36r^2t^4} = \sqrt{\left(6rt^2\right)^2} = 6|r|t^2$

37. $\sqrt[3]{\sqrt{64x^6}} = \left(8\left|x^3\right|\right)^{1/3} = 2|x|$

39. $\sqrt{32} + \sqrt{18} = \sqrt{16 \cdot 2} + \sqrt{9 \cdot 2} = \sqrt{4^2 \cdot 2} + \sqrt{3^2 \cdot 2} = 4\sqrt{2} + 3\sqrt{2} = 7\sqrt{2}$

41. $\sqrt{125} - \sqrt{45} = \sqrt{25 \cdot 5} - \sqrt{9 \cdot 5} = \sqrt{5^2 \cdot 5} - \sqrt{3^2 \cdot 5} = 5\sqrt{5} - 3\sqrt{5} = 2\sqrt{5}$

43. $\sqrt{9a^3} - \sqrt{a} = \sqrt{3^2a^2 \cdot a} - \sqrt{a} = 3a\sqrt{a} - \sqrt{a} = (3a - 1)\sqrt{a}$

45. $\sqrt[3]{x^4} + \sqrt[3]{8x} = \sqrt[3]{x^3x} + \sqrt[3]{2^3x} = x\sqrt[3]{x} + 2\sqrt[3]{x} = (x + 2)\sqrt[3]{x}$

47. $\sqrt{81x^2 + 81} = \sqrt{81\left(x^2 + 1\right)} = \sqrt{81}\sqrt{x^2 + 1} = 9\sqrt{x^2 + 1}$

49. (a) $16^{1/4} = 2$ (b) $-125^{1/3} = -5$ (c) $9^{-1/2} = \dfrac{1}{9^{1/2}} = \dfrac{1}{3}$

51. (a) $32^{2/5} = \left(32^{1/5}\right)^2 = 2^2 = 4$ (b) $\left(\dfrac{4}{9}\right)^{-1/2} = \left(\dfrac{9}{4}\right)^{1/2} = \dfrac{3}{2}$ (c) $\left(\dfrac{16}{81}\right)^{3/4} = \left(\dfrac{2}{3}\right)^3 = \dfrac{8}{27}$

53. (a) $5^{2/3} \cdot 5^{1/3} = 5^{2/3+1/3} = 5^1 = 5$ (b) $\dfrac{3^{3/5}}{3^{2/5}} = 3^{3/5-2/5} = \sqrt[5]{3}$ (c) $\left(\sqrt[3]{4}\right)^3 = 4^{(1/3)3} = 4$

55. When $x = 3$, $y = 4$, $z = -1$ we have $\sqrt{x^2 + y^2} = \sqrt{3^2 + 4^2} = \sqrt{9 + 16} = \sqrt{25} = 5$.

57. When $x = 3$, $y = 4$, $z = -1$ we have

$$(9x)^{2/3} + (2y)^{2/3} + z^{2/3} = (9 \cdot 3)^{2/3} + (2 \cdot 4)^{2/3} + (-1)^{2/3} = \left(3^3\right)^{2/3} + \left(2^3\right)^{2/3} + (1)^{1/3}$$
$$= 3^2 + 2^2 + 1 = 9 + 4 + 1 = 14.$$

59. (a) $x^{3/4}x^{5/4} = x^{3/4+5/4} = x^2$ (b) $y^{2/3}y^{4/3} = y^{2/3+4/3} = y^2$

61. (a) $\dfrac{w^{4/3}w^{2/3}}{w^{1/3}} = w^{4/3+2/3-1/3} = w^{5/3}$

(b) $\dfrac{a^{5/4}\left(2a^{3/4}\right)^3}{a^{1/4}} = 2^3 a^{(5/2)+(3/4)3-1/4} = 8a^{13/4}$

63. (a) $\left(8a^6b^{3/2}\right)^{2/3} = 8^{2/3}a^{6(2/3)}b^{(3/2)(2/3)} = 4a^4b$

(b) $(4a^6b^8)^{3/2} = 4^{3/2}a^{6(3/2)}b^{8(3/2)} = 8a^9b^{12}$

65. (a) $\left(8y^3\right)^{-2/3} = 8^{-2/3}y^{3(-2/3)} = \dfrac{1}{4y^2}$

(b) $\left(u^4v^6\right)^{-1/3} = u^{4(-1/3)}v^{6(-1/3)} = \dfrac{1}{u^{4/3}v^2}$

67. (a) $\left(\dfrac{x^{-2/3}}{y^{1/2}}\right)\left(\dfrac{x^{-2}}{y^{-3}}\right)^{1/6} = x^{-2/3+(-2)(1/6)}y^{-1/2-(-3)(1/6)} = \dfrac{1}{x}$

(b) $\left(\dfrac{x^{1/2}y^2}{2y^{1/4}}\right)^4\left(\dfrac{4x^{-2}y^{-4}}{y^2}\right)^{1/2} = x^{(1/2)(4)}y^{2(4)}2^{-1(4)}y^{(-1/4)(4)}4^{1/2}x^{-2(1/2)}y^{-4(1/2)}y^{-2(1/2)}$

$= x^2y^8 2^{-4}y^{-1}2x^{-1}y^{-2}y^{-1} = 2^{-4+1}x^{2-1}y^{8-1-2-1} = \dfrac{xy^4}{8}$

69. $\sqrt{x^3} = x^{3/2}$

71. $\sqrt[9]{x^5} = x^{5/9}$

73. $\left(\sqrt[6]{y^5}\right)\left(\sqrt[3]{y^2}\right) = y^{5/6}\cdot y^{2/3} = y^{5/6+2/3} = y^{3/2}$

75. $\left(5\sqrt[3]{x}\right)\left(2\sqrt[4]{x}\right) = 5\cdot 2x^{1/3+1/4} = 10x^{7/12}$

77. $\dfrac{\sqrt[4]{x^7}}{\sqrt[4]{x^3}} = \sqrt[4]{x^4} = x$

79. $\sqrt{\dfrac{16u^3v}{uv^5}} = \sqrt{\dfrac{16u^2}{v^4}} = \dfrac{4u}{v^2}$

81. $\dfrac{\sqrt{xy}}{\sqrt[4]{16xy}} = 16^{-1/4}x^{1/2-1/4}y^{1/2-1/4} = \dfrac{x^{1/4}y^{1/4}}{2}$

83. $\sqrt[3]{y\sqrt{y}} = \left(y^{1+1/2}\right)^{1/3} = y^{(3/2)(1/3)} = y^{1/2}$

85. (a) $\dfrac{1}{\sqrt{6}} = \dfrac{1}{\sqrt{6}}\cdot\dfrac{\sqrt{6}}{\sqrt{6}} = \dfrac{\sqrt{6}}{6}$

(b) $\sqrt{\dfrac{3}{2}} = \dfrac{\sqrt{3}}{\sqrt{2}}\cdot\dfrac{\sqrt{2}}{\sqrt{2}} = \dfrac{\sqrt{6}}{2}$

(c) $\dfrac{9}{\sqrt[4]{2}} = \dfrac{9}{2^{1/4}}\cdot\dfrac{2^{3/4}}{2^{3/2}} = \dfrac{9\sqrt[4]{8}}{2}$

87. (a) $\dfrac{1}{\sqrt{5x}} = \dfrac{1}{\sqrt{5x}}\cdot\dfrac{\sqrt{5x}}{\sqrt{5x}} = \dfrac{\sqrt{5x}}{5x}$

(b) $\sqrt{\dfrac{x}{5}} = \dfrac{\sqrt{x}}{\sqrt{5}}\cdot\dfrac{\sqrt{5}}{\sqrt{5}} = \dfrac{\sqrt{5x}}{5}$

(c) $\sqrt[5]{\dfrac{1}{x^3}} = \dfrac{1}{x^{3/5}}\cdot\dfrac{x^{2/5}}{x^{2/5}} = \dfrac{x^{2/5}}{x}$

89. (a) $\dfrac{1}{\sqrt[3]{x}} = \dfrac{1}{\sqrt[3]{x}}\cdot\dfrac{\sqrt[3]{x^2}}{\sqrt[3]{x^2}} = \dfrac{\sqrt[3]{x^2}}{x}$

(b) $\dfrac{1}{\sqrt[6]{x^5}} = \dfrac{1}{\sqrt[6]{x^5}}\cdot\dfrac{\sqrt[6]{x}}{\sqrt[6]{x}} = \dfrac{\sqrt[6]{x}}{x}$

(c) $\dfrac{1}{\sqrt[7]{x^3}} = \dfrac{1}{\sqrt[7]{x^3}}\cdot\dfrac{\sqrt[7]{x^4}}{\sqrt[7]{x^4}} = \dfrac{\sqrt[7]{x^4}}{x}$

91. (a) Since $\frac{1}{2} > \frac{1}{3}$, $2^{1/2} > 2^{1/3}$.

(b) $\left(\frac{1}{2}\right)^{1/2} = 2^{-1/2}$ and $\left(\frac{1}{2}\right)^{1/3} = 2^{-1/3}$. Since $-\frac{1}{2} < -\frac{1}{3}$, we have $\left(\frac{1}{2}\right)^{1/2} < \left(\frac{1}{2}\right)^{1/3}$.

93. First convert 1135 feet to miles. This gives $1135\text{ ft} = 1135\cdot\dfrac{1\text{ mile}}{5280\text{ feet}} = 0.215$ mi. Thus the distance you can see is given by $D = \sqrt{2rh+h^2} = \sqrt{2(3960)(0.215)+(0.215)^2} \approx \sqrt{1702.8} \approx 41.3$ miles.

95. (a) Substituting, we get $0.30(60)+0.38(3400)^{1/2}-3(650)^{1/3} \approx 18+0.38(58.31)-3(8.66) \approx 18+22.16-25.98 \approx 14.18$. Since this value is less than 16, the sailboat qualifies for the race.

(b) Solve for A when $L = 65$and $V = 600$. Substituting, we get $0.30(65)+0.38A^{1/2}-3(600)^{1/3} \le 16 \Leftrightarrow 19.5+0.38A^{1/2}-25.30 \le 16 \Leftrightarrow 0.38A^{1/2}-5.80 \le 16 \Leftrightarrow 0.38A^{1/2} \le 21.80 \Leftrightarrow A^{1/2} \le 57.38 \Leftrightarrow A \le 3292.0$. Thus, the largest possible sail is 3292 ft^2.

97. (a)

n	1	2	5	10	100
$2^{1/n}$	$2^{1/1}=2$	$2^{1/2}=1.414$	$2^{1/5}=1.149$	$2^{1/10}=1.072$	$2^{1/100}=1.007$

So when n gets large, $2^{1/n}$ decreases toward 1.

(b)

n	1	2	5	10	100
$\left(\frac{1}{2}\right)^{1/n}$	$\left(\frac{1}{2}\right)^{1/1}=0.5$	$\left(\frac{1}{2}\right)^{1/2}=0.707$	$\left(\frac{1}{2}\right)^{1/5}=0.871$	$\left(\frac{1}{2}\right)^{1/10}=0.933$	$\left(\frac{1}{2}\right)^{1/100}=0.993$

So when n gets large, $\left(\frac{1}{2}\right)^{1/n}$ increases toward 1.

P.5 ALGEBRAIC EXPRESSIONS

1. (a) $2x^3-\frac{1}{2}x+\sqrt{3}$ is a polynomial. (The constant term is not an integer, but all exponents are integers.)

(b) $x^2-\frac{1}{2}-3\sqrt{x}=x^2-\frac{1}{2}-3x^{1/2}$ is not a polynomial because the exponent $\frac{1}{2}$ is not an integer.

(c) $\dfrac{1}{x^2+4x+7}$ is not a polynomial. (It is the reciprocal of the polynomial x^2+4x+7.)

(d) $x^5+7x^2-x+100$ is a polynomial.

(e) $\sqrt[3]{8x^6-5x^3+7x-3}$ is not a polynomial. (It is the cube root of the polynomial $8x^6-5x^3+7x-3$.)

(f) $\sqrt{3}x^4+\sqrt{5}x^2-15x$ is a polynomial. (Some coefficients are not integers, but all exponents are integers.)

3. To subtract polynomials we subtract *like* terms. So
$\left(2x^3+9x^2+x+10\right)-\left(x^3+x^2+6x+8\right)=(2-1)x^3+(9-1)x^2+(1-6)x+(10-8)=x^3+8x^2-5x+2.$

5. The Special Product Formula for the "square of a sum" is $(A+B)^2=A^2+2AB+B^2$. So
$(2x+3)^2=(2x)^2+2(2x)(3)+3^2=4x^2+12x+9.$

7. (a) No, $(x+5)^2=x^2+10x+25\neq x^2+25$.

(b) Yes, if $a\neq 0$, then $(x+a)^2=x^2+2ax+a^2$.

9. Binomial, terms $5x^3$ and 6, degree 3

11. Monomial, term -8, degree 0

13. Four terms, terms x, $-x^2$, x^3, and $-x^4$, degree 4

15. $(6x-3)+(3x+7)=(6x+3x)+(-3+7)=9x+4$

17. $\left(2x^2-5x\right)-\left(x^2-8x+3\right)=\left(2x^2-x^2\right)+[-5x-(-8x)]+(-3)=x^2+3x-3$

19. $3(x-1)+4(x+2)=3x-3+4x+8=7x+5$

21. $\left(5x^3+4x^2-3x\right)-\left(x^2+7x+2\right)=5x^3+\left(4x^2-x^2\right)+(-3x-7x)-2=5x^3+3x^2-10x-2$

23. $2x(x-1)=2x^2-2x$

25. $x^2(x+3)=x^3+3x^2$

27. $2(2-5t)+t(t+10)=4-10t+t^2+10t=t^2+4$

29. $r\left(r^2-9\right)+3r^2(2r-1)=r^3-9r+6r^3-3r^2$
$=7r^3-3r^2-9r$

31. $x^2\left(2x^2-x+1\right)=2x^4-x^3+x^2$

33. $(x-3)(x+5)=x^2+5x-3x-15=x^2+2x-15$

35. $(s+6)(2s+3)=2s^2+3s+12s+18=2s^2+15s+18$

37. $(3t-2)(7t-4)=21t^2-12t-14t+8=21t^2-26t+8$

39. $(3x+5)(2x-1)=6x^2+10x-3x-5=6x^2+7x-5$

41. $(x+3y)(2x-y)=2x^2+5xy-3y^2$

43. $(2r - 5s)(3r - 2s) = 6r^2 - 19rs + 10s^2$

45. $(5x + 1)^2 = 25x^2 + 10x + 1$

47. $(3y - 1)^2 = (3y)^2 - 2(3y)(1) + 1^2 = 9y^2 - 6y + 1$

49. $(2u + v)^2 = 4u^2 + 4uv + v^2$

51. $(2x + 3y)^2 = 4x^2 + 12xy + 9y^2$

53. $\left(x^2 + 1\right)^2 = x^4 + 2x^2 + 1$

55. $(x + 6)(x - 6) = x^2 - 36$

57. $(3x - 4)(3x + 4) = (3x)^2 - 4^2 = 9x^2 - 16$

59. $(x + 3y)(x - 3y) = x^2 - (3y)^2 = x^2 - 9y^2$

59. $(\sqrt{x} + 2)(\sqrt{x} - 2) = x - 4$

63. $(y + 2)^3 = y^3 + 3y^2(2) + 3y\left(2^2\right) + 2^3 = y^3 + 6y^2 + 12y + 8$

65. $(1 - 2r)^3 = 1^3 - 3\left(1^2\right)(2r) + 3(1)(2r)^2 - (2r)^3 = -8r^3 + 12r^2 - 6r + 1$

67. $(x + 2)\left(x^2 + 2x + 3\right) = x^3 + 2x^2 + 3x + 2x^2 + 4x + 6 = x^3 + 4x^2 + 7x + 6$

69. $(2x - 5)\left(x^2 - x + 1\right) = 2x^3 - 2x^2 + 2x - 5x^2 + 5x - 5 = 2x^3 - 7x^2 + 7x - 5$

71. $\sqrt{x}(x - \sqrt{x}) = x\sqrt{x} - (\sqrt{x})^2 = x\sqrt{x} - x$

73. $y^{1/3}\left(y^{2/3} + y^{5/3}\right) = y^{1/3+2/3} + y^{1/3+5/3} = y^2 + y$

75. $\left(x^2 + y^2\right)^2 = \left(x^2\right)^2 + \left(y^2\right)^2 + 2x^2y^2 = x^4 + y^4 + 2x^2y^2$

77. $\left(x^2 - a^2\right)\left(x^2 + a^2\right) = x^4 - a^4$

79. $(\sqrt{a} - b)(\sqrt{a} + b) = a - b^2$

81. $\left(1 + x^{2/3}\right)\left(1 - x^{2/3}\right) = 1 - x^{4/3}$

83. $\left((x - 1) + x^2\right)\left((x - 1) - x^2\right) = (x - 1)^2 - \left(x^2\right)^2 = x^2 - 2x + 1 - x^4 = -x^4 + x^2 - 2x + 1$

85. $(2x + y - 3)(2x + y + 3) = (2x + y)^2 - 3^2 = 4x^2 + 4xy + y^2 - 9$

87. **(a)** $\text{RHS} = \frac{1}{2}\left[(a + b)^2 - \left(a^2 + b^2\right)\right] = \frac{1}{2}\left[\left(a^2 + b^2 + 2ab\right) - a^2 + b^2\right] = \frac{1}{2}(2ab) = ab = \text{LHS}$

(b) $\text{LHS} = \left(a^2 + b^2\right)^2 - \left(a^2 - b^2\right)^2 = \left(a^2\right)^2 + \left(b^2\right)^2 + 2a^2b^2 - \left[\left(a^2\right)^2 + \left(b^2\right)^2 - 2a^2b^2\right] = 4a^2b^2 = \text{RHS}$

89. **(a)** The height of the box is x, its width is $6 - 2x$, and its length is $10 - 2x$. Since Volume = height × width × length, we have $V = x(6 - 2x)(10 - 2x)$.

(b) $V = x\left(60 - 32x + 4x^2\right) = 60x - 32x^2 + 4x^3$, degree 3.

(c) When $x = 1$, the volume is $V = 60(1) - 32\left(1^2\right) + 4\left(1^3\right) = 32$, and when $x = 2$, the volume is $V = 60(2) - 32\left(2^2\right) + 4\left(2^3\right) = 24$.

91. **(a)** $A = 2000(1 + r)^3 = 2000\left(1 + 3r + 3r^2 + r^3\right) = 2000 + 6000r + 6000r^2 + 2000r^3$, degree 3.

(b) Remember that % means divide by 100, so 2% = 0.02.

Interest rate r	2%	3%	4.5%	6%	10%
Amount A	\$2122.42	\$2185.45	\$2282.33	\$2382.03	\$2662.00

93. **(a)** When $x = 1$, $(x + 5)^2 = (1 + 5)^2 = 36$ and $x^2 + 25 = 1^2 + 25 = 26$.

(b) $(x + 5)^2 = x^2 + 10x + 25$

P.6 FACTORING

1. The polynomial $2x^5 + 6x^4 + 4x^3$ has three terms: $2x^5$, $6x^4$, and $4x^3$.

3. To factor the trinomial $x^2 + 7x + 10$ we look for two integers whose product is 10 and whose sum is 7. These integers are 5 and 2, so the trinomial factors as $(x + 5)(x + 2)$.

5. The Special Factoring Formula for the "difference of squares" is $A^2 - B^2 = (A - B)(A + B)$. So $4x^2 - 25 = (2x - 5)(2x + 5)$.

7. $5a - 20 = 5(a - 4)$

9. $-2x^3 + x = -x\left(2x^2 - 1\right)$

11. $2x^2y - 6xy^2 + 3xy = xy(2x - 6y + 3)$

13. $y(y - 6) + 9(y - 6) = (y - 6)(y + 9)$

15. $x^2 + 8x + 7 = (x + 7)(x + 1)$

17. $x^2 + 2x - 15 = (x + 5)(x - 3)$

19. $3x^2 - 16x + 5 = (3x - 1)(x - 5)$

21. $(3x + 2)^2 + 8(3x + 2) + 12 = [(3x + 2) + 2][(3x + 2) + 6] = (3x + 4)(3x + 8)$

23. $x^2 - 25 = (x - 5)(x + 5)$

25. $49 - 4z^2 = (7 - 2z)(7 + 2z)$

27. $16y^2 - z^2 = (4y - z)(4y + z)$

29. $(x + 3)^2 - y^2 = [(x + 3) - y][(x + 3) + y] = (x - y + 3)(x + y + 3)$

31. $x^2 + 10x + 25 = (x + 5)^2$

33. $z^2 - 12z + 36 = (z - 6)^2$

35. $4t^2 - 20t + 25 = (2t - 5)^2$

37. $9u^2 - 6uv + v^2 = (3u - v)^2$

39. $x^3 + 27 = (x + 3)\left(x^2 - 3x + 9\right)$

41. $8a^3 - 1 = (2a - 1)\left(4a^2 + 2a + 1\right)$

43. $27x^3 + y^3 = (3x + y)\left(9x^2 - 3xy + y^2\right)$

45. $u^3 - v^6 = u^3 - \left(v^2\right)^3 = \left(u - v^2\right)\left(u^2 + uv^2 + v^4\right)$

47. $x^3 + 4x^2 + x + 4 = x^2(x + 4) + 1(x + 4) = (x + 4)\left(x^2 + 1\right)$

49. $5x^3 + x^2 + 5x + 1 = x^2(5x + 1) + (5x + 1) = \left(x^2 + 1\right)(5x + 1)$

51. $x^3 + x^2 + x + 1 = x^2(x + 1) + 1(x + 1) = (x + 1)\left(x^2 + 1\right)$

53. $x^{5/2} - x^{1/2} = x^{1/2}\left(x^2 - 1\right) = \sqrt{x}(x - 1)(x + 1)$

55. Start by factoring out the power of x with the smallest exponent, that is, $x^{-3/2}$. So

$$x^{-3/2} + 2x^{-1/2} + x^{1/2} = x^{-3/2}\left(1 + 2x + x^2\right) = \frac{(1 + x)^2}{x^{3/2}}.$$

57. Start by factoring out the power of $\left(x^2 + 1\right)$ with the smallest exponent, that is, $\left(x^2 + 1\right)^{-1/2}$. So

$$\left(x^2 + 1\right)^{1/2} + 2\left(x^2 + 1\right)^{-1/2} = \left(x^2 + 1\right)^{-1/2}\left[\left(x^2 + 1\right) + 2\right] = \frac{x^2 + 3}{\sqrt{x^2 + 1}}.$$

59. $2x^{1/3}(x - 2)^{2/3} - 5x^{4/3}(x - 2)^{-1/3} = x^{1/3}(x - 2)^{-1/3}[2(x - 2) - 5x] = x^{1/3}(x - 2)^{-1/3}(2x - 4 - 5x)$

$$= x^{1/3}(x - 2)^{-1/3}(-3x - 4) = \frac{(-3x - 4)\sqrt[3]{x}}{\sqrt[3]{x - 2}}$$

61. $12x^3 + 18x = 6x\left(2x^2 + 3\right)$

63. $6y^4 - 15y^3 = 3y^3(2y - 5)$

65. $x^2 - 2x - 8 = (x - 4)(x + 2)$

67. $y^2 - 8y + 15 = (y - 3)(y - 5)$

69. $2x^2 + 5x + 3 = (2x + 3)(x + 1)$

71. $9x^2 - 36x - 45 = 9\left(x^2 - 4x - 5\right) = 9(x - 5)(x + 1)$

73. $6x^2 - 5x - 6 = (3x + 2)(2x - 3)$

75. $x^2 - 36 = (x - 6)(x + 6)$

77. $49 - 4y^2 = (7 - 2y)(7 + 2y)$

79. $t^2 - 6t + 9 = (t - 3)^2$

81. $4x^2 + 4xy + y^2 = (2x + y)^2$

83. $t^3 + 1 = (t + 1)\left(t^2 - t + 1\right)$

85. $8x^3 - 125 = (2x)^3 - 5^3 = (2x - 5)\left[(2x)^2 + (2x)(5) + 5^2\right] = (2x - 5)\left(4x^2 + 10x + 25\right)$

87. $x^3 + 2x^2 + x = x\left(x^2 + 2x + 1\right) = x(x + 1)^2$

89. $x^4 + 2x^3 - 3x^2 = x^2\left(x^2 + 2x - 3\right) = x^2(x - 1)(x + 3)$

91. $x^4y^3 - x^2y^5 = x^2y^3\left(x^2 - y^2\right) = x^2y^3(x + y)(x - y)$

93. $x^6 - 8y^3 = \left(x^2\right)^3 - (2y)^3 = \left(x^2 - 2y\right)\left[\left(x^2\right)^2 + \left(x^2\right)(2y) + (2y)^2\right] = \left(x^2 - 2y\right)\left(x^4 + 2x^2y + 4y^2\right)$

95. $y^3 - 3y^2 - 4y + 12 = \left(y^3 - 3y^2\right) + (-4y + 12) = y^2(y - 3) + (-4)(y - 3) = (y - 3)\left(y^2 - 4\right)$
$= (y - 3)(y - 2)(y + 2)$ (factor by grouping)

97. $3x^3 - x^2 - 12x + 4 = 3x^3 - 12x - x^2 + 4 = 3x\left(x^2 - 4\right) - \left(x^2 - 4\right) = (3x - 1)\left(x^2 - 4\right) = (3x - 1)(x - 2)(x + 2)$
(factor by grouping)

99. $(a + b)^2 - (a - b)^2 = [(a + b) - (a - b)][(a + b) + (a - b)] = (2b)(2a) = 4ab$

101. $x^2\left(x^2 - 1\right) - 9\left(x^2 - 1\right) = \left(x^2 - 1\right)\left(x^2 - 9\right) = (x - 1)(x + 1)(x - 3)(x + 3)$

103. $(x - 1)(x + 2)^2 - (x - 1)^2(x + 2) = (x - 1)(x + 2)[(x + 2) - (x - 1)] = 3(x - 1)(x + 2)$

105. $y^4(y + 2)^3 + y^5(y + 2)^4 = y^4(y + 2)^3[(1) + y(y + 2)] = y^4(y + 2)^3\left(y^2 + 2y + 1\right) = y^4(y + 2)^3(y + 1)^2$

107. Start by factoring $y^2 - 7y + 10$, and then substitute $a^2 + 1$ for y. This gives

$$\left(a^2 + 1\right)^2 - 7\left(a^2 + 1\right) + 10 = \left[\left(a^2 + 1\right) - 2\right]\left[\left(a^2 + 1\right) - 5\right] = \left(a^2 - 1\right)\left(a^2 - 4\right) = (a - 1)(a + 1)(a - 2)(a + 2)$$

109. $3x^2(4x - 12)^2 + x^3(2)(4x - 12)(4) = x^2(4x - 12)[3(4x - 12) + x(2)(4)] = 4x^2(x - 3)(12x - 36 + 8x)$
$= 4x^2(x - 3)(20x - 36) = 16x^2(x - 3)(5x - 9)$

111. $3(2x - 1)^2(2)(x + 3)^{1/2} + (2x - 1)^3\left(\frac{1}{2}\right)(x + 3)^{-1/2} = (2x - 1)^2(x + 3)^{-1/2}\left[6(x + 3) + (2x - 1)\left(\frac{1}{2}\right)\right]$
$= (2x - 1)^2(x + 3)^{-1/2}\left(6x + 18 + x - \frac{1}{2}\right) = (2x - 1)^2(x + 3)^{-1/2}\left(7x + \frac{35}{2}\right)$

113. $\left(x^2 + 3\right)^{-1/3} - \frac{2}{3}x^2\left(x^2 + 3\right)^{-4/3} = \left(x^2 + 3\right)^{-4/3}\left[\left(x^2 + 3\right) - \frac{2}{3}x^2\right] = \left(x^2 + 3\right)^{-4/3}\left(\frac{1}{3}x^2 + 3\right) = \dfrac{\frac{1}{3}x^2 + 3}{\left(x^2 + 3\right)^{4/3}}$

115. The volume of the shell is the difference between the volumes of the outside cylinder (with radius R) and the inside cylinder (with radius r). Thus $V = \pi R^2 h - \pi r^2 h = \pi\left(R^2 - r^2\right)h = \pi(R-r)(R+r)h = 2\pi \cdot \dfrac{R+r}{2} \cdot h \cdot (R-r)$. The average radius is $\dfrac{R+r}{2}$ and $2\pi \cdot \dfrac{R+r}{2}$ is the average circumference (length of the rectangular box), h is the height, and $R-r$ is the thickness of the rectangular box. Thus $V = \pi R^2 h - \pi r^2 h = 2\pi \cdot \dfrac{R+r}{2} \cdot h \cdot (R-r) = 2\pi \cdot$ (average radius) $\cdot$ (height) $\cdot$ (thickness)

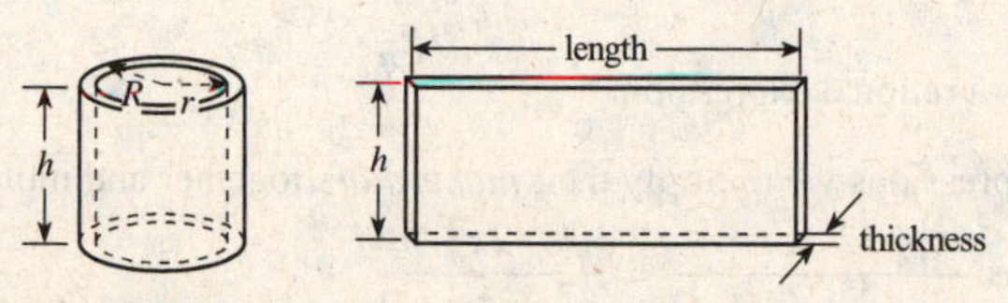

117. **(a)** $528^2 - 527^2 = (528-527)(528+527) = 1(1055) = 1055$

(b) $122^2 - 120^2 = (122-120)(122+120) = 2(242) = 484$

(c) $1020^2 - 1010^2 = (1020-1010)(1020+1010) = 10(2030) = 20{,}300$

119. **(a)** $A^4 - B^4 = \left(A^2 - B^2\right)\left(A^2 + B^2\right) = (A-B)(A+B)\left(A^2+B^2\right)$

$$\begin{aligned} A^6 - B^6 &= \left(A^3 - B^3\right)\left(A^3 + B^3\right) \text{ (difference of squares)} \\ &= (A-B)\left(A^2 + AB + B^2\right)(A+B)\left(A^2 - AB + B^2\right) \text{ (difference and sum of cubes)} \end{aligned}$$

(b) $12^4 - 7^4 = 20{,}736 - 2{,}401 = 18{,}335$; $12^6 - 7^6 = 2{,}985{,}984 - 117{,}649 = 2{,}868{,}335$

(c) $18{,}335 = 12^4 - 7^4 = (12-7)(12+7)\left(12^2 + 7^2\right) = 5(19)(144+49) = 5(19)(193)$

$$\begin{aligned} 2{,}868{,}335 &= 12^6 - 7^6 = (12-7)(12+7)\left[12^2 + 12(7) + 7^2\right]\left[12^2 - 12(7) + 7^2\right] \\ &= 5(19)(144+84+49)(144-84+49) = 5(19)(277)(109) \end{aligned}$$

121. **(a)**

$$\begin{array}{rr} & A+1 \\ \times & A-1 \\ \hline & -A-1 \\ A^2+A & \\ \hline A^2 & -1 \end{array} \qquad \begin{array}{rr} & A^2+A+1 \\ \times & A-1 \\ \hline & -A^2-A-1 \\ A^3+A^2+A & \\ \hline A^3 & -1 \end{array} \qquad \begin{array}{rr} & A^3+A^2+A+1 \\ \times & A-1 \\ \hline & -A^3-A^2-A-1 \\ A^4+A^3+A^2+A & \\ \hline A^4 & -1 \end{array}$$

(b) Based on the pattern in part (a), we suspect that $A^5 - 1 = (A-1)\left(A^4 + A^3 + A^2 + A + 1\right)$. Check:

$$\begin{array}{rr} & A^4+A^3+A^2+A+1 \\ \times & A-1 \\ \hline & -A^4-A^3-A^2-A-1 \\ A^5+A^4+A^3+A^2+A & \\ \hline A^5 & -1 \end{array}$$

The general pattern is $A^n - 1 = (A-1)\left(A^{n-1} + A^{n-2} + \cdots + A^2 + A + 1\right)$, where n is a positive integer.

P.7 RATIONAL EXPRESSIONS

1. (a) $\dfrac{3x}{x^2-1}$ is a rational expression.

(b) $\dfrac{\sqrt{x+1}}{2x+3}$ is not a rational expression. A rational expression must be a polynomial divided by a polynomial, and the numerator of the expression is $\sqrt{x+1}$, which is not a polynomial.

(c) $\dfrac{x(x^2-1)}{x+3} = \dfrac{x^3-x}{x+3}$ is a rational expression.

3. To multiply two rational expressions we multiply their *numerators* together and multiply their *denominators* together. So $\dfrac{2}{x+1} \cdot \dfrac{x}{x+3}$ is the same as $\dfrac{2 \cdot x}{(x+1)\cdot(x+3)} = \dfrac{2x}{x^2+4x+3}$.

5. (a) Yes. Cancelling $x+1$, we have $\dfrac{x(x+1)}{(x+1)^2} = \dfrac{x}{x+1}$.

(b) No; $(x+5)^2 = x^2+10x+25 \neq x^2+25$, so $x+5 = \sqrt{x^2+10x+25} \neq \sqrt{x^2+25}$.

7. The domain of $4x^2-10x+3$ is all real numbers.

9. Since $x-3 \neq 0$ we have $x \neq 3$. Domain: $\{x \mid x \neq 3\}$

11. Since $x+3 \geq 0$, $x \geq -3$. Domain; $\{x \mid x \geq -3\}$

13. $x^2-x-2 = (x+1)(x-2) \neq 0 \Leftrightarrow x \neq -1$ or 2, so the domain is $\{x \mid x \neq -1, 2\}$.

15. $\dfrac{5(x-3)(2x+1)}{10(x-3)^2} = \dfrac{5(x-3)(2x+1)}{5(x-3)\cdot 2(x-3)} = \dfrac{2x+1}{2(x-3)}$

17. $\dfrac{x-2}{x^2-4} = \dfrac{x-2}{(x-2)(x+2)} = \dfrac{1}{x+2}$

19. $\dfrac{x^2+5x+6}{x^2+8x+15} = \dfrac{(x+2)(x+3)}{(x+5)(x+3)} = \dfrac{x+2}{x+5}$

21. $\dfrac{y^2+y}{y^2-1} = \dfrac{y(y+1)}{(y-1)(y+1)} = \dfrac{y}{y-1}$

23. $\dfrac{2x^3-x^2-6x}{2x^2-7x+6} = \dfrac{x\left(2x^2-x-6\right)}{(2x-3)(x-2)} = \dfrac{x(2x+3)(x-2)}{(2x-3)(x-2)} = \dfrac{x(2x+3)}{2x-3}$

25. $\dfrac{4x}{x^2-4} \cdot \dfrac{x+2}{16x} = \dfrac{4x}{(x-2)(x+2)} \cdot \dfrac{x+2}{16x} = \dfrac{1}{4(x-2)}$

27. $\dfrac{x^2+2x-15}{x^2-25} \cdot \dfrac{x-5}{x+2} = \dfrac{(x+5)(x-3)(x-5)}{(x+5)(x-5)(x+2)} = \dfrac{x-3}{x+2}$

29. $\dfrac{t-3}{t^2+9} \cdot \dfrac{t+3}{t^2-9} = \dfrac{(t-3)(t+3)}{\left(t^2+9\right)(t-3)(t+3)} = \dfrac{1}{t^2+9}$

31. $\dfrac{x^2+7x+12}{x^2+3x+2} \cdot \dfrac{x^2+5x+6}{x^2+6x+9} = \dfrac{(x+3)(x+4)}{(x+1)(x+2)} \cdot \dfrac{(x+2)(x+3)}{(x+3)(x+3)} = \dfrac{x+4}{x+1}$

33. $\dfrac{x+3}{4x^2-9} \div \dfrac{x^2+7x+12}{2x^2+7x-15} = \dfrac{x+3}{4x^2-9} \cdot \dfrac{2x^2+7x-15}{x^2+7x+12} = \dfrac{x+3}{(2x-3)(2x+3)} \cdot \dfrac{(x+5)(2x-3)}{(x+3)(x+4)} = \dfrac{x+5}{(2x+3)(x+4)}$

35. $\dfrac{\dfrac{x^3}{x+1}}{\dfrac{x}{x^2+2x+1}} = \dfrac{x^3}{x+1} \cdot \dfrac{x^2+2x+1}{x} = \dfrac{x^3(x+1)(x+1)}{(x+1)x} = x^2(x+1)$

37. $\dfrac{x/y}{z} = \dfrac{x}{y} \cdot \dfrac{1}{z} = \dfrac{x}{yz}$

39. $1 + \dfrac{1}{x+3} = \dfrac{x+3}{x+3} + \dfrac{1}{x+3} = \dfrac{x+4}{x+3}$

41. $\dfrac{1}{x+5} + \dfrac{2}{x-3} = \dfrac{x-3}{(x+5)(x-3)} + \dfrac{2(x+5)}{(x+5)(x-3)} = \dfrac{x-3+2x+10}{(x+5)(x-3)} = \dfrac{3x+7}{(x+5)(x-3)}$

43. $\dfrac{3}{x+1}-\dfrac{1}{x+2}=\dfrac{3(x+2)}{(x+1)(x+2)}-\dfrac{x+1}{(x+1)(x+2)}=\dfrac{3x+6-x-1}{(x+1)(x+2)}=\dfrac{2x+5}{(x+1)(x+2)}$

45. $\dfrac{5}{2x-3}-\dfrac{3}{(2x-3)^2}=\dfrac{5(2x-3)}{(2x-3)^2}-\dfrac{3}{(2x-3)^2}=\dfrac{10x-15-3}{(2x-3)^2}=\dfrac{10x-18}{(2x-3)^2}=\dfrac{2(5x-9)}{(2x-3)^2}$

47. $u+1+\dfrac{u}{u+1}=\dfrac{(u+1)(u+1)}{u+1}+\dfrac{u}{u+1}=\dfrac{u^2+2u+1+u}{u+1}=\dfrac{u^2+3u+1}{u+1}$

49. $\dfrac{1}{x^2}+\dfrac{1}{x^2+x}=\dfrac{1}{x^2}+\dfrac{1}{x(x+1)}=\dfrac{x+1}{x^2(x+1)}+\dfrac{x}{x^2(x+1)}=\dfrac{2x+1}{x^2(x+1)}$

51. $\dfrac{2}{x+3}-\dfrac{1}{x^2+7x+12}=\dfrac{2}{x+3}-\dfrac{1}{(x+3)(x+4)}=\dfrac{2(x+4)}{(x+3)(x+4)}+\dfrac{-1}{(x+3)(x+4)}$
$=\dfrac{2x+8-1}{(x+3)(x+4)}=\dfrac{2x+7}{(x+3)(x+4)}$

53. $\dfrac{1}{x+3}+\dfrac{1}{x^2-9}=\dfrac{1}{x+3}+\dfrac{1}{(x-3)(x+3)}=\dfrac{x-3}{(x-3)(x+3)}+\dfrac{1}{(x-3)(x+3)}=\dfrac{x-2}{(x-3)(x+3)}$

55. $\dfrac{2}{x}+\dfrac{3}{x-1}-\dfrac{4}{x^2-x}=\dfrac{2}{x}+\dfrac{3}{x-1}-\dfrac{4}{x(x-1)}=\dfrac{2(x-1)}{x(x-1)}+\dfrac{3x}{x(x-1)}+\dfrac{-4}{x(x-1)}=\dfrac{2x-2+3x-4}{x(x-1)}=\dfrac{5x-6}{x(x-1)}$

57. $\dfrac{1}{x^2+3x+2}-\dfrac{1}{x^2-2x-3}=\dfrac{1}{(x+2)(x+1)}-\dfrac{1}{(x-3)(x+1)}$
$=\dfrac{x-3}{(x-3)(x+2)(x+1)}+\dfrac{-(x+2)}{(x-3)(x+2)(x+1)}=\dfrac{x-3-x-2}{(x-3)(x+2)(x+1)}=\dfrac{-5}{(x-3)(x+2)(x+1)}$

59. $\dfrac{1+\frac{1}{x}}{\frac{1}{x}-2}=\dfrac{x\left(1+\frac{1}{x}\right)}{x\left(\frac{1}{x}-2\right)}=\dfrac{x+1}{1-2x}$

61. $\dfrac{1+\dfrac{1}{x+2}}{1-\dfrac{1}{x+2}}=\dfrac{(x+2)\left(1+\dfrac{1}{x+2}\right)}{(x+2)\left(1-\dfrac{1}{x+2}\right)}=\dfrac{(x+2)+1}{(x+2)-1}=\dfrac{x+3}{x+1}$

63. $\dfrac{\dfrac{1}{x-1}+\dfrac{1}{x+3}}{x+1}=\dfrac{(x-1)(x+3)\left(\dfrac{1}{x-1}+\dfrac{1}{x+3}\right)}{(x-1)(x+3)(x+1)}=\dfrac{(x+3)+(x-1)}{(x-1)(x+1)(x+3)}=\dfrac{2(x+1)}{(x-1)(x+1)(x+3)}$
$=\dfrac{2}{(x-1)(x+3)}$

65. $\dfrac{x-\dfrac{x}{y}}{y-\dfrac{y}{x}}=\dfrac{xy\left(x-\dfrac{x}{y}\right)}{xy\left(y-\dfrac{y}{x}\right)}=\dfrac{x^2y-x^2}{xy^2-y^2}=\dfrac{x^2(y-1)}{y^2(x-1)}$

67. $\dfrac{\dfrac{x}{y}-\dfrac{y}{x}}{\dfrac{1}{x^2}-\dfrac{1}{y^2}}=\dfrac{\dfrac{x^2-y^2}{xy}}{\dfrac{y^2-x^2}{x^2y^2}}=\dfrac{x^2-y^2}{xy}\cdot\dfrac{x^2y^2}{y^2-x^2}=\dfrac{xy}{-1}=-xy$. An alternative method is to multiply the numerator and denominator by the common denominator of both the numerator and denominator, in this case x^2y^2:

$\dfrac{\dfrac{x}{y}-\dfrac{y}{x}}{\dfrac{1}{x^2}-\dfrac{1}{y^2}}=\dfrac{\left(\dfrac{x}{y}-\dfrac{y}{x}\right)}{\left(\dfrac{1}{x^2}-\dfrac{1}{y^2}\right)}\cdot\dfrac{x^2y^2}{x^2y^2}=\dfrac{x^3y-xy^3}{y^2-x^2}=\dfrac{xy\left(x^2-y^2\right)}{y^2-x^2}=-xy.$

69. $\dfrac{x^{-2}-y^{-2}}{x^{-1}+y^{-1}}=\dfrac{\dfrac{1}{x^2}-\dfrac{1}{y^2}}{\dfrac{1}{x}+\dfrac{1}{y}}=\dfrac{\dfrac{y^2}{x^2y^2}-\dfrac{x^2}{x^2y^2}}{\dfrac{y}{xy}+\dfrac{x}{xy}}=\dfrac{y^2-x^2}{x^2y^2}\cdot\dfrac{xy}{y+x}=\dfrac{(y-x)(y+x)xy}{x^2y^2(y+x)}=\dfrac{y-x}{xy}$

Alternatively, $\dfrac{x^{-2}-y^{-2}}{x^{-1}+y^{-1}}=\dfrac{\left(\dfrac{1}{x^2}-\dfrac{1}{y^2}\right)}{\left(\dfrac{1}{x}+\dfrac{1}{y}\right)}\cdot\dfrac{x^2y^2}{x^2y^2}=\dfrac{y^2-x^2}{xy^2+x^2y}=\dfrac{(y-x)(y+x)}{xy(y+x)}=\dfrac{y-x}{xy}$.

71. $1-\dfrac{1}{1-\dfrac{1}{x}}=1-\dfrac{x}{x-1}=\dfrac{x-1-x}{x-1}=\dfrac{1}{1-x}$

73. $\dfrac{\dfrac{1}{1+x+h}-\dfrac{1}{1+x}}{h}=\dfrac{(1+x)-(1+x+h)}{h(1+x)(1+x+h)}=-\dfrac{1}{(1+x)(1+x+h)}$

75. $\dfrac{\dfrac{1}{(x+h)^2}-\dfrac{1}{x^2}}{h}=\dfrac{x^2-(x+h)^2}{hx^2(x+h)^2}=\dfrac{x^2-\left(x^2+2xh+h^2\right)}{hx^2(x+h)^2}=-\dfrac{2x+h}{x^2(x+h)^2}$

77. $\sqrt{1+\left(\dfrac{x}{\sqrt{1-x^2}}\right)^2}=\sqrt{1+\dfrac{x^2}{1-x^2}}=\sqrt{\dfrac{1-x^2}{1-x^2}+\dfrac{x^2}{1-x^2}}=\sqrt{\dfrac{1}{1-x^2}}=\dfrac{1}{\sqrt{1-x^2}}$

79. $\dfrac{3(x+2)^2(x-3)^2-(x+2)^3(2)(x-3)}{(x-3)^4}=\dfrac{(x+2)^2(x-3)[3(x-3)-(x+2)(2)]}{(x-3)^4}$

$=\dfrac{(x+2)^2(3x-9-2x-4)}{(x-3)^3}=\dfrac{(x+2)^2(x-13)}{(x-3)^3}$

81. $\dfrac{2(1+x)^{1/2}-x(1+x)^{-1/2}}{1+x}=\dfrac{(1+x)^{-1/2}[2(1+x)-x]}{1+x}=\dfrac{x+2}{(1+x)^{3/2}}$

83. $\dfrac{3(1+x)^{1/3}-x(1+x)^{-2/3}}{(1+x)^{2/3}}=\dfrac{(1+x)^{-2/3}[3(1+x)-x]}{(1+x)^{2/3}}=\dfrac{2x+3}{(1+x)^{4/3}}$

85. $\dfrac{1}{5-\sqrt{3}}=\dfrac{1}{5-\sqrt{3}}\cdot\dfrac{5+\sqrt{3}}{5+\sqrt{3}}=\dfrac{5+\sqrt{3}}{25-3}=\dfrac{5+\sqrt{3}}{22}$

87. $\dfrac{2}{\sqrt{2}+\sqrt{7}}=\dfrac{2}{\sqrt{2}+\sqrt{7}}\cdot\dfrac{\sqrt{2}-\sqrt{7}}{\sqrt{2}-\sqrt{7}}=\dfrac{2\left(\sqrt{2}-\sqrt{7}\right)}{2-7}=\dfrac{2\left(\sqrt{2}-\sqrt{7}\right)}{-5}=\dfrac{2\left(\sqrt{7}-\sqrt{2}\right)}{5}$

89. $\dfrac{y}{\sqrt{3}+\sqrt{y}}=\dfrac{y}{\sqrt{3}+\sqrt{y}}\cdot\dfrac{\sqrt{3}-\sqrt{y}}{\sqrt{3}-\sqrt{y}}=\dfrac{y\left(\sqrt{3}-\sqrt{y}\right)}{3-y}=\dfrac{y\sqrt{3}-y\sqrt{y}}{3-y}$

91. $\dfrac{1-\sqrt{5}}{3}=\dfrac{1-\sqrt{5}}{3}\cdot\dfrac{1+\sqrt{5}}{1+\sqrt{5}}=\dfrac{1-5}{3\left(1+\sqrt{5}\right)}=\dfrac{-4}{3\left(1+\sqrt{5}\right)}$

93. $\dfrac{\sqrt{r}+\sqrt{2}}{5}=\dfrac{\sqrt{r}+\sqrt{2}}{5}\cdot\dfrac{\sqrt{r}-\sqrt{2}}{\sqrt{r}-\sqrt{2}}=\dfrac{r-2}{5\left(\sqrt{r}-\sqrt{2}\right)}$

95. $\sqrt{x^2+1}-x=\dfrac{\sqrt{x^2+1}-x}{1}\cdot\dfrac{\sqrt{x^2+1}+x}{\sqrt{x^2+1}+x}=\dfrac{x^2+1-x^2}{\sqrt{x^2+1}+x}=\dfrac{1}{\sqrt{x^2+1}+x}$

97. (a) $R=\dfrac{1}{\dfrac{1}{R_1}+\dfrac{1}{R_2}}=\dfrac{1}{\dfrac{1}{R_1}+\dfrac{1}{R_2}}\cdot\dfrac{R_1R_2}{R_1R_2}=\dfrac{R_1R_2}{R_2+R_1}$

(b) Substituting $R_1=10$ ohms and $R_2=20$ ohms gives $R=\dfrac{(10)(20)}{(20)+(10)}=\dfrac{200}{30}\approx 6.7$ ohms.

99.

x	2.80	2.90	2.95	2.99	2.999	3	3.001	3.01	3.05	3.10	3.20
$\frac{x^2-9}{x-3}$	5.80	5.90	5.95	5.99	5.999	?	6.001	6.01	6.05	6.10	6.20

From the table, we see that the expression $\frac{x^2-9}{x-3}$ approaches 6 as x approaches 3. We simplify the expression: $\frac{x^2-9}{x-3} = \frac{(x-3)(x+3)}{x-3} = x+3, x \neq 3$. Clearly as x approaches 3, $x+3$ approaches 6. This explains the result in the table.

101. Answers will vary.

Algebraic Error	Counterexample
$\frac{1}{a}+\frac{1}{b} \neq \frac{1}{a+b}$	$\frac{1}{2}+\frac{1}{2} \neq \frac{1}{2+2}$
$(a+b)^2 \neq a^2+b^2$	$(1+3)^2 \neq 1^2+3^2$
$\sqrt{a^2+b^2} \neq a+b$	$\sqrt{5^2+12^2} \neq 5+12$
$\frac{a+b}{a} \neq b$	$\frac{2+6}{2} \neq 6$
$\frac{a}{a+b} \neq \frac{1}{b}$	$\frac{1}{1+1} \neq 1$
$\frac{a^m}{a^n} \neq a^{m/n}$	$\frac{3^5}{3^2} \neq 3^{5/2}$

103. (a)

x	1	3	$\frac{1}{2}$	$\frac{9}{10}$	$\frac{99}{100}$	$\frac{999}{1000}$	$\frac{9999}{10,000}$
$x+\frac{1}{x}$	2	3.333	2.5	2.011	2.0001	2.000001	2.00000001

It appears that the smallest possible value of $x+\frac{1}{x}$ is 2.

(b) Because $x > 0$, we can multiply both sides by x and preserve the inequality: $x+\frac{1}{x} \geq 2 \Leftrightarrow x\left(x+\frac{1}{x}\right) \geq 2x \Leftrightarrow x^2+1 \geq 2x \Leftrightarrow x^2-2x+1 \geq 0 \Leftrightarrow (x-1)^2 \geq 0$. The last statement is true for all $x > 0$, and because each step is reversible, we have shown that $x+\frac{1}{x} \geq 2$ for all $x > 0$.

P.8 SOLVING BASIC EQUATIONS

1. Substituting $x = 3$ in the equation $4x - 2 = 10$ makes the equation true, so the number 3 is a *solution* of the equation.

3. (a) $\frac{x}{2}+2x = 10$ is equivalent to $\frac{5}{2}x - 10 = 0$, so it is a linear equation.

(b) $\frac{2}{x} - 2x = 1$ is not linear because it contains the term $\frac{2}{x}$, a multiple of the reciprocal of the variable.

(c) $x+7 = 5-3x \Leftrightarrow 4x-2 = 0$, so it is linear.

5. (a) This is true: If $a = b$, then $a+x = b+x$.

(b) This is false, because the number could be zero. However, it is true that multiplying each side of an equation by a *nonzero* number always gives an equivalent equation.

(c) This is false. For example, $-5 = 5$ is false, but $(-5)^2 = 5^2$ is true.

7. (a) When $x = -2$, LHS $= 4(-2) + 7 = -8 + 7 = -1$ and RHS $= 9(-2) - 3 = -18 - 3 = -21$. Since LHS $\neq$ RHS, $x = -2$ is not a solution.

(b) When $x = 2$, LHS $= 4(-2) + 7 = 8 + 7 = 15$ and RHS $= 9(2) - 3 = 18 - 3 = 15$. Since LHS $=$ RHS, $x = 2$ is a solution.

9. (a) When $x = 2$, LHS $= 1 - [2 - (3 - (2))] = 1 - [2 - 1] = 1 - 1 = 0$ and RHS $= 4(2) - (6 + (2)) = 8 - 8 = 0$. Since LHS $=$ RHS, $x = 2$ is a solution.

(b) When $x = 4$ LHS $= 1 - [2 - (3 - (4))] = 1 - [2 - (-1)] = 1 - 3 = -2$ and RHS $= 4(4) - (6 + (4)) = 16 - 10 = 6$. Since LHS $\neq$ RHS, $x = 4$ is not a solution.

11. (a) When $x = -1$, LHS $= 2(-1)^{1/3} - 3 = 2(-1) - 3 = -2 - 3 = -5$. Since LHS $\neq 1$, $x = -1$ is not a solution.

(b) When $x = 8$ LHS $= 2(8)^{1/3} - 3 = 2(2) - 3 = 4 - 3 = 1 =$ RHS. So $x = 8$ is a solution.

13. (a) When $x = 0$, LHS $= \dfrac{0 - a}{0 - b} = \dfrac{-a}{-b} = \dfrac{a}{b} =$ RHS. So $x = 0$ is a solution.

(b) When $x = b$, LHS $= \dfrac{b - a}{b - b} = \dfrac{b - a}{0}$ is not defined, so $x = b$ is not a solution.

15. $5x - 6 = 14 \Leftrightarrow 5x = 20 \Leftrightarrow x = 4$

17. $7 - 2x = 15 \Leftrightarrow 2x = -8 \Leftrightarrow x = -4$

19. $\frac{1}{2}x + 7 = 3 \Leftrightarrow \frac{1}{2}x = -4 \Leftrightarrow x = -8$

21. $-3x - 3 = 5x - 3 \Leftrightarrow 0 = 8x \Leftrightarrow x = 0$

23. $7x + 1 = 4 - 2x \Leftrightarrow 9x = 3 \Leftrightarrow x = \frac{1}{3}$

25. $-x + 3 = 4x \Leftrightarrow 3 = 5x \Leftrightarrow x = \frac{3}{5}$

27. $\frac{x}{3} - 1 = \frac{5}{3}x + 7 \Leftrightarrow x - 3 = 5x + 21 \Leftrightarrow 4x = -24 \Leftrightarrow x = -6$

29. $2(1 - x) = 3(1 + 2x) + 5 \Leftrightarrow 2 - 2x = 3 + 6x + 5 \Leftrightarrow 2 - 2x = 8 + 6x \Leftrightarrow -6 = 8x \Leftrightarrow x = -\frac{3}{4}$

31. $4\left(y - \frac{1}{2}\right) - y = 6(5 - y) \Leftrightarrow 4y - 2 - y = 30 - 6y \Leftrightarrow 3y - 2 = 30 - 6y \Leftrightarrow 9y = 32 \Leftrightarrow y = \frac{32}{9}$

33. $x - \frac{1}{3}x - \frac{1}{2}x - 5 = 0 \Leftrightarrow 6x - 2x - 3x - 30 = 0$ (multiply both sides by 6) $\Leftrightarrow x = 30$

35. $2x - \dfrac{x}{2} + \dfrac{x + 1}{4} = 6x \Leftrightarrow 8x - 2x + x + 1 = 24x \Leftrightarrow 7x + 1 = 24x \Leftrightarrow 1 = 17x \Leftrightarrow x = \frac{1}{17}$

37. $(x - 1)(x + 2) = (x - 2)(x - 3) \Leftrightarrow x^2 + x - 2 = x^2 - 5x + 6 \Leftrightarrow x - 2 = -5x + 6 \Leftrightarrow 6x = 8 \Leftrightarrow x = \frac{4}{3}$

39. $(x - 1)(4x + 5) = (2x - 3)^2 \Leftrightarrow 4x^2 + x - 5 = 4x^2 - 12x + 9 \Leftrightarrow x - 5 = -12x + 9 \Leftrightarrow 13x = 14 \Leftrightarrow x = \frac{14}{13}$

41. $\dfrac{1}{x} = \dfrac{4}{3x} + 1 \Rightarrow 3 = 4 + 3x$ (multiply both sides by the LCD, $3x$) $\Leftrightarrow -1 = 3x \Leftrightarrow x = -\frac{1}{3}$

43. $\dfrac{2x - 1}{x + 2} = \dfrac{4}{5} \Rightarrow 5(2x - 1) = 4(x + 2) \Leftrightarrow 10x - 5 = 4x + 8 \Leftrightarrow 6x = 13 \Leftrightarrow x = \frac{13}{6}$

45. $\dfrac{2}{t + 6} = \dfrac{3}{t - 1} \Rightarrow 2(t - 1) = 3(t + 6)$ [multiply both sides by the LCD, $(t - 1)(t + 6)$] $\Leftrightarrow 2t - 2 = 3t + 18 \Leftrightarrow -20 = t$

47. $\dfrac{3}{x + 1} - \dfrac{1}{2} = \dfrac{1}{3x + 3} \Rightarrow 3(6) - (3x + 3) = 2$ [multiply both sides by $6(x + 1)$] $\Leftrightarrow 18 - 3x - 3 = 2 \Leftrightarrow -3x + 15 = 2 \Leftrightarrow -3x = -13 \Leftrightarrow x = \frac{13}{3}$

49. $\dfrac{1}{z} - \dfrac{1}{2z} - \dfrac{1}{5z} = \dfrac{10}{z + 1} \Rightarrow 10(z + 1) - 5(z + 1) - 2(z + 1) = 10(10z)$ [multiply both sides by $10z(z + 1)$] $\Leftrightarrow 3(z + 1) = 100z \Leftrightarrow 3z + 3 = 100z \Leftrightarrow 3 = 97z \Leftrightarrow \frac{3}{97} = z$

51. $\dfrac{x}{2x - 4} - 2 = \dfrac{1}{x - 2} \Rightarrow x - 2(2x - 4) = 2$ [multiply both sides by $2(x - 2)$] $\Leftrightarrow x - 4x + 8 = 2 \Leftrightarrow -3x = -6 \Leftrightarrow x = 2$. But substituting $x = 2$ into the original equation does not work, since we cannot divide by 0. Thus there is no solution.

53. $\dfrac{3}{x + 4} = \dfrac{1}{x} + \dfrac{6x + 12}{x^2 + 4x} \Rightarrow 3(x) = (x + 4) + 6x + 12$ (multiply both sides by $x(x + 4)$) $\Leftrightarrow 3x = 7x + 16 \Leftrightarrow -4x = 16 \Leftrightarrow x = -4$. But substituting $x = -4$ into the original equation does not work, since we cannot divide by 0. Thus, there is no solution.

55. $x^2 = 25 \Leftrightarrow x = \pm 5$

57. $5x^2 = 15 \Leftrightarrow x^2 = 3 \Leftrightarrow x = \pm\sqrt{3}$

59. $8x^2 - 64 = 0 \Leftrightarrow x^2 - 8 = 0 \Leftrightarrow x^2 = 8 \Leftrightarrow x = \pm\sqrt{8} = \pm 2\sqrt{2}$

61. $x^2 + 16 = 0 \Leftrightarrow x^2 = -16$ which has no real solution.

63. $(x-3)^2 = 5 \Leftrightarrow x - 3 = \pm\sqrt{5} \Leftrightarrow x = 3 \pm \sqrt{5}$

65. $x^3 = 27 \Leftrightarrow x = 27^{1/3} = 3$

67. $0 = x^4 - 16 = \left(x^2+4\right)\left(x^2-4\right) = \left(x^2+4\right)(x-2)(x+2)$. $x^2 + 4 = 0$ has no real solution. If $x - 2 = 0$, then $x = 2$. If $x + 2 = 0$, then $x = -2$. The solutions are ± 2.

69. $x^4 + 64 = 0 \Leftrightarrow x^4 = -64$ which has no real solution.

71. $(x+2)^4 - 81 = 0 \Leftrightarrow (x+2)^4 = 81 \Leftrightarrow \left[(x+2)^4\right]^{1/4} = \pm 81^{1/4} \Leftrightarrow x + 2 = \pm 3$. So $x + 2 = 3$, then $x = 1$. If $x + 2 = -3$, then $x = -5$. The solutions are -5 and 1.

73. $3(x-3)^3 = 375 \Leftrightarrow (x-3)^3 = 125 \Leftrightarrow (x-3) = 125^{1/3} = 5 \Leftrightarrow x = 3 + 5 = 8$

75. $\sqrt[3]{x} = 5 \Leftrightarrow x = 5^3 = 125$

77. $2x^{5/3} + 64 = 0 \Leftrightarrow 2x^{5/3} = -64 \Leftrightarrow x^{5/3} = -32 \Leftrightarrow x = (-32)^{3/5} = \left(-2^5\right)^{1/5} = (-2)^3 = -8$

79. $3.02x + 1.48 = 10.92 \Leftrightarrow 3.02x = 9.44 \Leftrightarrow x = \dfrac{9.44}{3.02} \approx 3.13$

81. $2.15x - 4.63 = x + 1.19 \Leftrightarrow 1.15x = 5.82 \Leftrightarrow x = \dfrac{5.82}{1.19} \approx 5.06$

83. $3.16(x + 4.63) = 4.19(x - 7.24) \Leftrightarrow 3.16x + 14.63 = 4.19x - 30.34 \Leftrightarrow 44.97 = 1.03x \Leftrightarrow x = \dfrac{44.97}{1.03} \approx 43.66$

85. $\dfrac{0.26x - 1.94}{3.03 - 2.44x} = 1.76 \Rightarrow 0.26x - 1.94 = 1.76(3.03 - 2.44x) \Leftrightarrow 0.26x - 1.94 = 5.33 - 4.29x \Leftrightarrow 4.55x = 7.27 \Leftrightarrow x = \dfrac{7.27}{4.55} \approx 1.60$

87. $r = \dfrac{12}{M} \Leftrightarrow M = \dfrac{12}{r}$

89. $PV = nRT \Leftrightarrow R = \dfrac{PV}{nT}$

91. $P = 2l + 2w \Leftrightarrow 2w = P - 2l \Leftrightarrow w = \dfrac{P - 2l}{2}$

93. $V = \frac{1}{3}\pi r^2 h \Leftrightarrow r^2 = \dfrac{3V}{\pi h} \Rightarrow r = \pm\sqrt{\dfrac{3V}{\pi h}}$

95. $V = \frac{4}{3}\pi r^3 \Leftrightarrow r^3 = \dfrac{3V}{4\pi} \Leftrightarrow r = \sqrt[3]{\dfrac{3V}{4\pi}}$

97. $A = P\left(1 + \dfrac{i}{100}\right)^2 \Leftrightarrow \dfrac{A}{P} = \left(1 + \dfrac{i}{100}\right)^2 \Rightarrow 1 + \dfrac{i}{100} = \pm\sqrt{\dfrac{A}{P}} \Leftrightarrow \dfrac{i}{100} = -1 \pm \sqrt{\dfrac{A}{P}} \Leftrightarrow i = -100 \pm 100\sqrt{\dfrac{A}{P}}$

99. $\dfrac{ax+b}{cx+d} = 2 \Leftrightarrow ax + b = 2(cx+d) \Leftrightarrow ax + b = 2cx + 2d \Leftrightarrow ax - 2cx = 2d - b \Leftrightarrow (a - 2c)x = 2d - b \Leftrightarrow x = \dfrac{2d - b}{a - 2c}$

101. **(a)** The shrinkage factor when $w = 250$ is $S = \dfrac{0.032(250) - 2.5}{10{,}000} = \dfrac{8 - 2.5}{10{,}000} = 0.00055$. So the beam shrinks $0.00055 \times 12.025 \approx 0.007$ m, so when it dries it will be $12.025 - 0.007 = 12.018$ m long.

(b) Substituting $S = 0.00050$ we get $0.00050 = \dfrac{0.032w - 2.5}{10{,}000} \Leftrightarrow 5 = 0.032w - 2.5 \Leftrightarrow 7.5 = 0.032w \Leftrightarrow w = \dfrac{7.5}{0.032} \approx 234.375$. So the water content should be $234.375\ \text{kg/m}^3$.

103. **(a)** Solving for v when $P = 10{,}000$ we get $10{,}000 = 15.6v^3 \Leftrightarrow v^3 \approx 641.02 \Leftrightarrow v \approx 8.6$ km/h.

(b) Solving for v when $P = 50{,}000$ we get $50{,}000 = 15.6v^3 \Leftrightarrow v^3 \approx 3205.13 \Leftrightarrow v \approx 14.7$ km/h.

105. **(a)** $3(0)+k-5=k(0)-k+1 \Leftrightarrow k-5=-k+1 \Leftrightarrow 2k=6 \Leftrightarrow k=3$

(b) $3(1)+k-5=k(1)-k+1 \Leftrightarrow 3+k-5=k-k+1 \Leftrightarrow k-2=1 \Leftrightarrow k=3$

(c) $3(2)+k-5=k(2)-k+1 \Leftrightarrow 6+k-5=2k-k+1 \Leftrightarrow k+1=k+1$. $x=2$ is a solution for every value of k. That is, $x=2$ is a solution to every member of this family of equations.

P.9 MODELING WITH EQUATIONS

1. An equation modeling a real-world situation can be used to help us understand a real-world problem using mathematical methods. We translate real-world ideas into the language of algebra to construct our model, and translate our mathematical results back into real-world ideas in order to interpret our findings.

3. **(a)** A square of side x has area $A=x^2$.

(b) A rectangle of length l and width w has area $A=lw$.

(c) A circle of radius r has area $A=\pi r^2$.

5. A painter paints a wall in x hours, so the fraction of the wall she paints in one hour is $\frac{1 \text{ wall}}{x \text{ hours}}=\frac{1}{x}$.

7. If n is the first integer, then $n+1$ is the middle integer, and $n+2$ is the third integer. So the sum of the three consecutive integers is $n+(n+1)+(n+2)=3n+3$.

9. If n is the first even integer, then $n+2$ is the second even integer and $n+4$ is the third. So the sum of three consecutive even integers is $n+(n+2)+(n+4)=3n+6$.

11. If s is the third test score, then since the other test scores are 78 and 82, the average of the three test scores is $\frac{78+82+s}{3}=\frac{160+s}{3}$.

13. If x dollars are invested at $2\frac{1}{2}\%$ simple interest, then the first year you will receive $0.025x$ dollars in interest.

15. Since w is the width of the rectangle, the length is four times the width, or $4w$. Then

$$\text{area} = \text{length} \times \text{width} = 4w \times w = 4w^2 \text{ ft}^2$$

17. If d is the given distance, in miles, and distance $=$ rate $\times$ time, we have time $=\frac{\text{distance}}{\text{rate}}=\frac{d}{55}$.

19. If x is the quantity of pure water added, the mixture will contain 25 oz of salt and $3+x$ gallons of water. Thus the concentration is $\frac{25}{3+x}$.

21. If d is the number of days and m the number of miles, then the cost of a rental is $C=65d+0.20m$. In this case, $d=3$ and $C=275$, so we solve for m: $275=65\cdot 3+0.20m \Leftrightarrow 275=195+0.2m \Leftrightarrow 0.2m=80 \Leftrightarrow m=\frac{80}{0.2}=400$. Thus, Michael drove 400 miles.

23. If x is Linh's score on her final exam, then because the final counts twice as much as each midterm, her average score is $\frac{82+75+71+2x}{3(100)+200}=\frac{228+2x}{500}=\frac{114+x}{250}$. For her to average 80%, we must have $\frac{114+x}{250}=80\%=0.8 \Leftrightarrow$ $114+x=250(0.8)=200 \Leftrightarrow x=86$. So Linh scored 86% on her final exam.

25. Let m be the amount invested at $4\frac{1}{2}\%$. Then $12{,}000-m$ is the amount invested at 4%.

Since the total interest is equal to the interest earned at $4\frac{1}{2}\%$ plus the interest earned at 4%, we have

$525=0.045m+0.04(12{,}000-m) \Leftrightarrow 525=0.045m+480-0.04m \Leftrightarrow 45=0.005m \Leftrightarrow m=\frac{45}{0.005}=9000$. Thus $9000 is invested at $4\frac{1}{2}\%$, and $12,000 − 9000 = $3000 is invested at 4%.

27. Using the formula $I=Prt$ and solving for r, we get $262.50=3500\cdot r\cdot 1 \Leftrightarrow r=\frac{262.5}{3500}=0.075$ or 7.5%.

29. Let x be her monthly salary. Since her annual salary $=$ $12 \times$ (monthly salary) $+$ (Christmas bonus) we have $97{,}300 = 12x + 8{,}500 \Leftrightarrow 88{,}800 = 12x \Leftrightarrow x \approx 7{,}400$. Her monthly salary is \$7,400.

31. Let x be the overtime hours Helen works. Since gross pay $=$ regular salary $+$ overtime pay, we obtain the equation $352.50 = 7.50 \times 35 + 7.50 \times 1.5 \times x \Leftrightarrow 352.50 = 262.50 + 11.25x \Leftrightarrow 90 = 11.25x \Leftrightarrow x = \dfrac{90}{11.25} = 8$. Thus Helen worked 8 hours of overtime.

33. All ages are in terms of the daughter's age 7 years ago. Let y be age of the daughter 7 years ago. Then $11y$ is the age of the movie star 7 years ago. Today, the daughter is $y + 7$, and the movie star is $11y + 7$. But the movie star is also 4 times his daughter's age today. So $4(y + 7) = 11y + 7 \Leftrightarrow 4y + 28 = 11y + 7 \Leftrightarrow 21 = 7y \Leftrightarrow y = 3$. Thus the movie star's age today is $11(3) + 7 = 40$ years.

35. Let p be the number of pennies. Then p is the number of nickels and p is the number of dimes. So the value of the coins in the purse is the value of the pennies plus the value of the nickels plus the value of the dimes. Thus $1.44 = 0.01p + 0.05p + 0.10p \Leftrightarrow 1.44 = 0.16p \Leftrightarrow p = \frac{1.44}{0.16} = 9$. So the purse contains 9 pennies, 9 nickels, and 9 dimes.

37. Let l be the length of the garden. Since area $=$ width $\cdot$ length, we obtain the equation $1125 = 25l \Leftrightarrow l = \frac{1125}{25} = 45$ ft. So the garden is 45 feet long.

39. Let x be the length of a side of the square plot. As shown in the figure, area of the plot $=$ area of the building $+$ area of the parking lot. Thus, $x^2 = 60(40) + 12{,}000 = 2{,}400 + 12{,}000 = 14{,}400 \Rightarrow x = \pm 120$. So the plot of land measures 120 feet by 120 feet.

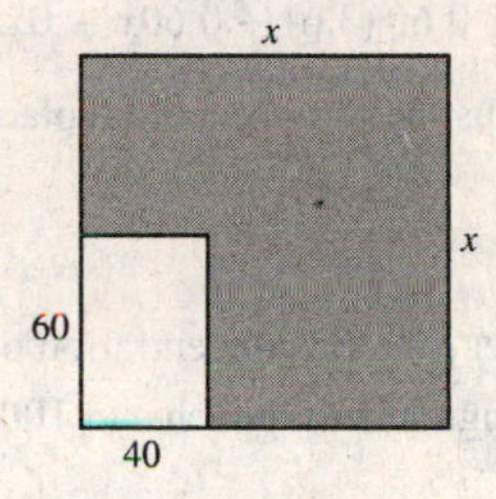

41. The figure is a trapezoid, so its area is $\dfrac{\text{base}_1 + \text{base}_2}{2}$ (height). Putting in the known quantities, we have $120 = \dfrac{y + 2y}{2}(y) = \frac{3}{2}y^2 \Leftrightarrow y^2 = 80 \Rightarrow y = \pm\sqrt{80} = \pm 4\sqrt{5}$. Since length is positive, $y = 4\sqrt{5} \approx 8.94$ inches.

43. Let x be the width of the strip. Then the length of the mat is $20 + 2x$, and the width of the mat is $15 + 2x$. Now the perimeter is twice the length plus twice the width, so $102 = 2(20 + 2x) + 2(15 + 2x) \Leftrightarrow 102 = 40 + 4x + 30 + 4x \Leftrightarrow 102 = 70 + 8x \Leftrightarrow 32 = 8x \Leftrightarrow x = 4$. Thus the strip of mat is 4 inches wide.

45. Let x be the length of the man's shadow, in meters. Using similar triangles, $\dfrac{10 + x}{6} = \dfrac{x}{2} \Leftrightarrow 20 + 2x = 6x \Leftrightarrow 4x = 20 \Leftrightarrow x = 5$. Thus the man's shadow is 5 meters long.

47. Let x be the amount (in mL) of 60% acid solution to be used. Then $300 - x$ mL of 30% solution would have to be used to yield a total of 300 mL of solution.

	60% acid	30% acid	Mixture
mL	x	$300 - x$	300
Rate (% acid)	0.60	0.30	0.50
Value	$0.60x$	$0.30(300 - x)$	$0.50(300)$

Thus the total amount of pure acid used is $0.60x + 0.30(300 - x) = 0.50(300) \Leftrightarrow 0.3x + 90 = 150 \Leftrightarrow x = \dfrac{60}{0.3} = 200$. So 200 mL of 60% acid solution must be mixed with 100 mL of 30% solution to get 300 mL of 50% acid solution.

49. Let x be the number of grams of silver added. The weight of the rings is $5 \times 18\text{ g} = 90\text{ g}$.

	5 rings	Pure silver	Mixture
Grams	90	x	$90 + x$
Rate (% gold)	0.90	0	0.75
Value	0.90 (90)	$0x$	$0.75\,(90 + x)$

So $0.90\,(90) + 0x = 0.75\,(90 + x) \Leftrightarrow 81 = 67.5 + 0.75x \Leftrightarrow 0.75x = 13.5 \Leftrightarrow x = \frac{13.5}{0.75} = 18$. Thus 18 grams of silver must be added to get the required mixture.

51. Let x be the number of liters of coolant removed and replaced by water.

	60% antifreeze	60% antifreeze (removed)	Water	Mixture
Liters	3.6	x	x	3.6
Rate (% antifreeze)	0.60	0.60	0	0.50
Value	0.60 (3.6)	$-0.60x$	$0x$	0.50 (3.6)

So $0.60\,(3.6) - 0.60x + 0x = 0.50\,(3.6) \Leftrightarrow 2.16 - 0.6x = 1.8 \Leftrightarrow -0.6x = -0.36 \Leftrightarrow x = \dfrac{-0.36}{-0.6} = 0.6$. Thus 0.6 liters must be removed and replaced by water.

53. Let c be the concentration of fruit juice in the cheaper brand. The new mixture that Jill makes will consist of 650 mL of the original fruit punch and 100 mL of the cheaper fruit punch.

	Original Fruit Punch	Cheaper Fruit Punch	Mixture
mL	650	100	750
Concentration	0.50	c	0.48
Juice	$0.50 \cdot 650$	$100c$	$0.48 \cdot 750$

So $0.50 \cdot 650 + 100c = 0.48 \cdot 750 \Leftrightarrow 325 + 100c = 360 \Leftrightarrow 100c = 35 \Leftrightarrow c = 0.35$. Thus the cheaper brand is only 35% fruit juice.

55. Let t be the time in minutes it would take Candy and Tim if they work together. Candy delivers the papers at a rate of $\frac{1}{70}$ of the job per minute, while Tim delivers the paper at a rate of $\frac{1}{80}$ of the job per minute. The sum of the fractions of the job that each can do individually in one minute equals the fraction of the job they can do working together. So we have $\dfrac{1}{t} = \frac{1}{70} + \frac{1}{80} \Leftrightarrow 560 = 8t + 7t \Leftrightarrow 560 = 15t \Leftrightarrow t = 37\frac{1}{3}$ minutes. Since $\frac{1}{3}$ of a minute is 20 seconds, it would take them 37 minutes 20 seconds if they worked together.

57. Let t be the time, in hours, it takes Karen to paint a house alone. Then working together, Karen and Betty can paint a house in $\frac{2}{3}t$ hours. The sum of their individual rates equals their rate working together, so $\dfrac{1}{t} + \frac{1}{6} = \dfrac{1}{\frac{2}{3}t} \Leftrightarrow \dfrac{1}{t} + \frac{1}{6} = \dfrac{3}{2t} \Leftrightarrow$ $6 + t = 9 \Leftrightarrow t = 3$. Thus it would take Karen 3 hours to paint a house alone.

59. Let t be the time in hours that Wendy spent on the train. Then $\frac{11}{2}-t$ is the time in hours that Wendy spent on the bus. We construct a table:

	Rate	Time	Distance
By train	40	t	$40t$
By bus	60	$\frac{11}{2}-t$	$60\left(\frac{11}{2}-t\right)$

The total distance traveled is the sum of the distances traveled by bus and by train, so $300=40t+60\left(\frac{11}{2}-t\right)\Leftrightarrow$ $300=40t+330-60t\Leftrightarrow -30=-20t\Leftrightarrow t=\frac{-30}{-20}=1.5$ hours. So the time spent on the train is $5.5-1.5=4$ hours.

61. Let r be the speed of the plane from Montreal to Los Angeles. Then $r+0.20r=1.20r$ is the speed of the plane from Los Angeles to Montreal.

	Rate	Time	Distance
Montreal to L.A.	r	$\dfrac{2500}{r}$	2500
L.A. to Montreal	$1.2r$	$\dfrac{2500}{1.2r}$	2500

The total time is the sum of the times each way, so $9\frac{1}{6}=\dfrac{2500}{r}+\dfrac{2500}{1.2r}\Leftrightarrow\dfrac{55}{6}=\dfrac{2500}{r}+\dfrac{2500}{1.2r}\Leftrightarrow$ $55\cdot 1.2r=2500\cdot 6\cdot 1.2+2500\cdot 6\Leftrightarrow 66r=18{,}000+15{,}000\Leftrightarrow 66r=33{,}000\Leftrightarrow r=\frac{33{,}000}{66}=500$. Thus the plane flew at a speed of 500 mi/h on the trip from Montreal to Los Angeles.

63. Let x be the distance from the fulcrum to where the mother sits. Then substituting the known values into the formula given, we have $100\,(8)=125x\quad\Leftrightarrow\quad 800=125x\Leftrightarrow x=6.4$. So the mother should sit 6.4 feet from the fulcrum.

65. Let l be the length of the lot in feet. Then the length of the diagonal is $l+10$. We apply the Pythagorean Theorem with the hypotenuse as the diagonal. So $l^2+50^2=(l+10)^2\Leftrightarrow l^2+2500=l^2+20l+100\Leftrightarrow 20l=2400\Leftrightarrow l=120$. Thus the length of the lot is 120 feet.

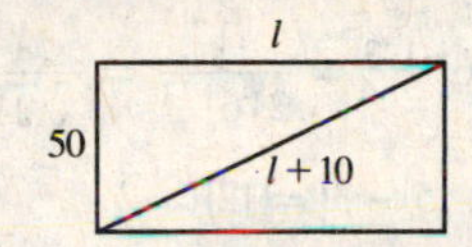

67. Let h be the height in feet of the structure. The structure is composed of a right cylinder with radius 10 and height $\frac{2}{3}h$ and a cone with base radius 10 and height $\frac{1}{3}h$. Using the formulas for the volume of a cylinder and that of a cone, we obtain the equation $1400\pi=\pi\,(10)^2\left(\frac{2}{3}h\right)+\frac{1}{3}\pi\,(10)^2\left(\frac{1}{3}h\right)\Leftrightarrow 1400\pi=\frac{200\pi}{3}h+\frac{100\pi}{9}h\Leftrightarrow 126=6h+h$ (multiply both sides by $\dfrac{9}{100\pi}$) $\Leftrightarrow 126=7h\Leftrightarrow h=18$. Thus the height of the structure is 18 feet.

69. Pythagoras was born about 569 BC in Samos, Ionia and died about 475 BC.
Euclid was born about 325 BC and died about 265 BC in Alexandria, Egypt.
Archimedes was born in 287 BC in Syracuse, Sicily and died in 212 BC in Syracuse.

CHAPTER P REVIEW

1. (a) Since there are initially 250 tablets and she takes 2 tablets per day, the number of tablets T that are left in the bottle after she has been taking the tablets for x days is $T=250-2x$.

(b) After 30 days, there are $250-2\,(30)=190$ tablets left.

(c) We set $T=0$ and solve: $T=250-2x=0\Leftrightarrow 250=2x\Leftrightarrow x=125$. She will run out after 125 days.

3. (a) 16 is rational. It is an integer, and more precisely, a natural number.

(b) -16 is rational. It is an integer, but because it is negative, it is not a natural number.

(c) $\sqrt{16} = 4$ is rational. It is an integer, and more precisely, a natural number.

(d) $\sqrt{2}$ is irrational.

(e) $\frac{8}{3}$ is rational, but is neither a natural number nor an integer.

(f) $-\frac{8}{2} = -4$ is rational. It is an integer, but because it is negative, it is not a natural number.

5. Commutative Property of addition.

7. Distributive Property.

9. (a) $\frac{5}{6} + \frac{2}{3} = \frac{5}{6} + \frac{4}{6} = \frac{9}{6} = \frac{3}{2}$

(b) $\frac{5}{6} - \frac{2}{3} = \frac{5}{6} - \frac{4}{6} = \frac{1}{6}$

11. (a) $\frac{15}{8} \cdot \frac{12}{5} = \frac{15 \cdot 12}{8 \cdot 5} = \frac{3 \cdot 3}{2 \cdot 1} = \frac{9}{2}$

(b) $\frac{15}{8} \div \frac{12}{5} = \frac{15 \cdot 5}{8 \cdot 12} = \frac{5 \cdot 5}{8 \cdot 4} = \frac{25}{32}$

13. $x \in [-2, 6) \Leftrightarrow -2 \le x < 6$

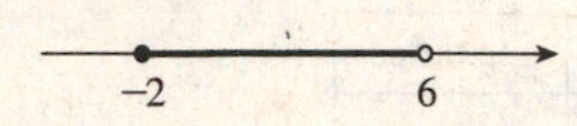

15. $x \in (-\infty, 4] \Leftrightarrow x \le 4$

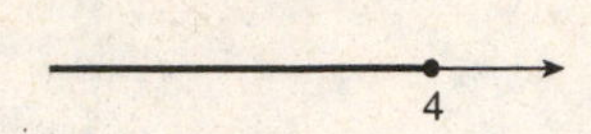

17. $x \ge 5 \Leftrightarrow x \in [5, \infty)$

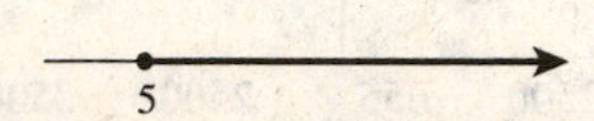

19. $-1 < x \le 5 \Leftrightarrow x \in (-1, 5]$

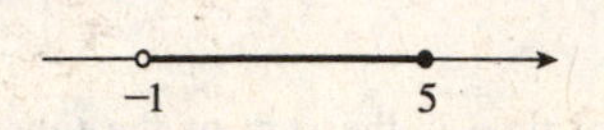

21. (a) $A \cup B = \left\{-1, 0, \frac{1}{2}, 1, 2, 3, 4\right\}$

(b) $A \cap B = \{1\}$

23. (a) $A \cap C = \{1, 2\}$

(b) $B \cap D = \left\{\frac{1}{2}, 1\right\}$

25. $|7 - 10| = |-3| = 3$

27. $2^{1/2}8^{1/2} = \sqrt{2} \cdot \sqrt{8} = \sqrt{16} = 4$

29. $216^{-1/3} = \frac{1}{216^{1/3}} = \frac{1}{\sqrt[3]{216}} = \frac{1}{6}$

31. $\frac{\sqrt{242}}{\sqrt{2}} = \sqrt{\frac{242}{2}} = \sqrt{121} = 11$

33. (a) $|5 - 3| = |2| = 2$

(b) $|-5 - 3| = |-8| = 8$

35. (a) $\sqrt[3]{7} = 7^{1/3}$

(b) $\sqrt[5]{7^4} = 7^{4/5}$

37. (a) $\sqrt[6]{x^5} = x^{5/6}$

(b) $\left(\sqrt{x}\right)^9 = \left(x^{1/2}\right)^9 = x^{9/2}$

39. $\left(2x^3y\right)^2\left(3x^{-1}y^2\right) = 4x^6y^2 \cdot 3x^{-1}y^2 = 4 \cdot 3x^{6-1}y^{2+2}$
$= 12x^5y^4$

41. $\frac{x^4(3x)^2}{x^3} = \frac{x^4 \cdot 9x^2}{x^3} = 9x^{4+2-3} = 9x^3$

43. $\sqrt[3]{\left(x^3y\right)^2 y^4} = \sqrt[3]{x^6y^4y^2} = \sqrt[3]{x^6y^6} = x^2y^2$

45. $\frac{8r^{1/2}s^{-3}}{2r^{-2}s^4} = 4r^{(1/2)-(-2)}s^{-3-4} = 4r^{5/2}s^{-7} = \frac{4r^{5/2}}{s^7}$

47. $78{,}250{,}000{,}000 = 7.825 \times 10^{10}$

49. $\frac{ab}{c} \approx \frac{(0.00000293)\left(1.582 \times 10^{-14}\right)}{2.8064 \times 10^{12}} = \frac{\left(2.93 \times 10^{-6}\right)\left(1.582 \times 10^{-14}\right)}{2.8064 \times 10^{12}} = \frac{2.93 \cdot 1.582}{2.8064} \times 10^{-6-14-12}$
$\approx 1.65 \times 10^{-32}$

51. $2x^2y - 6xy^2 = 2xy(x - 3y)$

53. $x^2 + 5x - 14 = (x + 7)(x - 2)$

55. $3x^2 - 2x - 1 = (3x + 1)(x - 1)$

57. $4t^2 - 13t - 12 = (4t + 3)(t - 4)$

59. $16 - 4t^2 = 4\left(4 - t^2\right) = -4(t + 2)(t - 2)$

61. $x^6 - 1 = \left(x^3 - 1\right)\left(x^3 + 1\right) = (x-1)\left(x^2 + x + 1\right)(x+1)\left(x^2 - x + 1\right)$

63. $x^3 - 27 = (x-3)\left(x^2 + 3x + 9\right)$

65. $4x^3 - 8x^2 + 3x - 6 = 4x^2(x-2) + 3(x-2) = \left(4x^2 + 3\right)(x-2)$

67. $(x+y)^2 - 7(x+y) + 6 = [(x+y) - 6][(x+y) - 1] = (x + y - 6)(x + y + 1)$

69. $(2y - 7)(2y + 7) = 4y^2 - 49$

71. $x^2(x-2) + x(x-2)^2 = x^3 - 2x^2 + x\left(x^2 - 4x + 4\right) = x^3 - 2x^2 + x^3 - 4x^2 + 4x = 2x^3 - 6x^2 + 4x$

73. $\sqrt{x}\left(\sqrt{x} + 1\right)\left(2\sqrt{x} - 1\right) = \left(x + \sqrt{x}\right)\left(2\sqrt{x} - 1\right) = 2x\sqrt{x} - x + 2x - \sqrt{x} = 2x\sqrt{x} + x - \sqrt{x}$

75. $\dfrac{x^2 - 2x - 3}{2x^2 + 5x + 3} = \dfrac{(x-3)(x+1)}{(2x+3)(x+1)} = \dfrac{x-3}{2x+3}$

77. $\dfrac{x^2 + 2x - 3}{x^2 + 8x + 16} \cdot \dfrac{3x + 12}{x - 1} = \dfrac{(x+3)(x-1)}{(x+4)(x+4)} \cdot \dfrac{3(x+4)}{(x-1)} = \dfrac{3(x+3)}{x+4}$

79. $x - \dfrac{1}{x+1} = \dfrac{x(x+1)}{x+1} - \dfrac{1}{x+1} = \dfrac{x^2 + x - 1}{x+1}$

81. $\dfrac{2}{x} + \dfrac{1}{x-2} + \dfrac{3}{(x-2)^2} = \dfrac{2(x-2)^2}{x(x-2)^2} + \dfrac{x(x-2)}{x(x-2)^2} + \dfrac{3x}{x(x-2)^2}$

$$= \frac{2\left(x^2 - 4x + 4\right) + x^2 - 2x + 3x}{x(x-2)^2} = \frac{2x^2 - 8x + 8 + x^2 - 2x + 3x}{x(x-2)^2} = \frac{3x^2 - 7x + 8}{x(x-2)^2}$$

83. $\dfrac{\frac{1}{x} - \frac{1}{2}}{x - 2} = \dfrac{\frac{2}{2x} - \frac{x}{2x}}{x - 2} = \dfrac{2 - x}{2x} \cdot \dfrac{1}{x-2} = \dfrac{-1(x-2)}{2x} \cdot \dfrac{1}{x-2} = \dfrac{-1}{2x}$

85. $\dfrac{3(x+h)^2 - 5(x+h) - \left(3x^2 - 5x\right)}{h} = \dfrac{3x^2 + 6xh + 3h^2 - 5x - 5h - 3x^2 + 5x}{h} = \dfrac{6xh + 3h^2 - 5h}{h}$

$$= \frac{h(6x + 3h - 5)}{h} = 6x + 3h - 5$$

87. $\dfrac{1}{\sqrt{11}} = \dfrac{1}{\sqrt{11}} \cdot \dfrac{\sqrt{11}}{\sqrt{11}} = \dfrac{\sqrt{11}}{11}$

89. $\dfrac{10}{\sqrt{2} - 1} = \dfrac{10}{\sqrt{2} - 1} \cdot \dfrac{\sqrt{2} + 1}{\sqrt{2} + 1} = \dfrac{10 + 10\sqrt{2}}{2 - 1} = 10 + 10\sqrt{2}$

91. $\dfrac{x}{2 + \sqrt{x}} = \dfrac{x}{2 + \sqrt{x}} \cdot \dfrac{2 - \sqrt{x}}{2 - \sqrt{x}} = \dfrac{2x - x\sqrt{x}}{2^2 - \left(\sqrt{x}\right)^2} = \dfrac{x\left(\sqrt{x} - 2\right)}{x - 4}$

93. $\dfrac{x+5}{x+10}$ is defined whenever $x + 10 \neq 0 \Leftrightarrow x \neq -10$, so its domain is $\{x \mid x \neq -10\}$.

95. $\dfrac{\sqrt{x}}{x^2 - 3x - 4}$ is defined whenever $x \geq 0$ (so that $\sqrt{x}$ is defined) and $x^2 - 3x - 4 = (x+1)(x-4) \neq 0 \Leftrightarrow x \neq -1$ and $x \neq 4$. Thus, its domain is $\{x \mid x \geq 0 \text{ and } x \neq 4\}$.

97. This statement is false. For example, take $x = 1$ and $y = 1$. Then LHS $= (x+y)^3 = (1+1)^3 = 2^3 = 8$, while RHS $= x^3 + y^3 = 1^3 + 1^3 = 1 + 1 = 2$, and $8 \neq 2$.

99. This statement is true: $\dfrac{12 + y}{y} = \dfrac{12}{y} + \dfrac{y}{y} = \dfrac{12}{y} + 1$.

101. This statement is false. For example, take $a = -1$. Then LHS $= \sqrt{a^2} = \sqrt{(-1)^2} = \sqrt{1} = 1$, which does not equal $a = -1$. The true statement is $\sqrt{a^2} = |a|$.

103. $3x+12=24 \Leftrightarrow 3x=12 \Leftrightarrow x=4$

105. $7x-6=4x+9 \Leftrightarrow 3x=15 \Leftrightarrow x=5$

107. $\frac{1}{3}x-\frac{1}{2}=2 \Leftrightarrow 2x-3=12 \Leftrightarrow 2x=15 \Leftrightarrow x=\frac{15}{2}$

109. $2(x+3)-4(x-5)=8-5x \Leftrightarrow 2x+6-4x+20=8-5x \Leftrightarrow -2x+26=8-5x \Leftrightarrow 3x=-18 \Leftrightarrow x=-6$

111. $\dfrac{x+1}{x-1}=\dfrac{2x-1}{2x+1} \Leftrightarrow (x+1)(2x+1)=(2x-1)(x-1) \Leftrightarrow 2x^2+3x+1=2x^2-3x+1 \Leftrightarrow 6x=0 \Leftrightarrow x=0$

113. $\dfrac{x+1}{x-1}=\dfrac{3x}{3x-6}=\dfrac{3x}{3(x-2)}=\dfrac{x}{x-2} \Leftrightarrow (x+1)(x-2)=x(x-1) \Leftrightarrow x^2-x-2=x^2-x \Leftrightarrow -2=0$. Since this last equation is never true, there is no real solution to the original equation.

115. $x^2=144 \Rightarrow x=\pm 12$

117. $x^3-27=0 \Leftrightarrow x^3=27 \Rightarrow x=3$.

119. $(x+1)^3=-64 \Leftrightarrow x+1=-4 \Leftrightarrow x=-1-4=-5$.

121. $\sqrt[3]{x}=-3 \Leftrightarrow x=(-3)^3=-27$.

123. $4x^{3/4}-500=0 \Leftrightarrow 4x^{3/4}=500 \Leftrightarrow x^{3/4}=125 \Leftrightarrow x=125^{4/3}=5^4=625$.

125. $A=\dfrac{x+y}{2} \Leftrightarrow 2A=x+y \Leftrightarrow x=2A-y$.

127. Multiply through by t: $J=\dfrac{1}{t}+\dfrac{1}{2t}+\dfrac{1}{3t} \Leftrightarrow tJ=1+\dfrac{1}{2}+\dfrac{1}{3}=\dfrac{11}{6} \Leftrightarrow t=\dfrac{11}{6J}$, $J \neq 0$.

129. Let x be the number of pounds of raisins. Then the number of pounds of nuts is $50-x$.

	Raisins	Nuts	Mixture
Pounds	x	$50-x$	50
Rate (cost per pound)	3.20	2.40	2.72

So $3.20x+2.40(50-x)=2.72(50) \Leftrightarrow 3.20x+120-2.40x=136 \Leftrightarrow 0.8x=16 \Leftrightarrow x=20$. Thus the mixture uses 20 pounds of raisins and $50-20=30$ pounds of nuts.

131. Let x be the amount invested in the account earning 1.5% interest. Then the amount invested in the account earning 2.5% is $7000-x$.

	1.5% Account	2.5% Account	Total
Amount invested	x	$7000-x$	7000
Interest earned	$0.015x$	$0.025(7000-x)$	120.25

From the table, we see that $0.015x+0.025(7000-x)=120.25 \Leftrightarrow 0.015x+175-0.025x=120.25 \Leftrightarrow 54.75=0.01x \Leftrightarrow x=5475$. Thus, Luc invested \$5475 in the account earning 1.5% interest and \$1525 in the account earning 2.5% interest.

133. Let t be the time it would take Abbie to paint a living room if she works alone. It would take Beth $2t$ hours to paint the living room alone, and it would take $3t$ hours for Cathie to paint the living room. Thus Abbie does $\dfrac{1}{t}$ of the job per hour, Beth does $\dfrac{1}{2t}$ of the job per hour, and Cathie does $\dfrac{1}{3t}$ of the job per hour. So $\dfrac{1}{t}+\dfrac{1}{2t}+\dfrac{1}{3t}=1 \Leftrightarrow 6+3+2=6t \Leftrightarrow 6t=11 \Leftrightarrow t=\frac{11}{6}$. So it would Abbie 1 hour 50 minutes to paint the living room alone.

CHAPTER P TEST

1. (a) The cost is $C=9+1.5x$.

(b) There are four extra toppings, so $x=4$ and $C=9+1.5(4)=\$15$.

3. (a) $A \cap B=\{0,1,5\}$

(b) $A \cup B=\left\{-2,0,\frac{1}{2},1,3,5,7\right\}$

5. **(a)** $-2^6 = -64$ **(b)** $(-2)^6 = 64$ **(c)** $2^{-6} = \frac{1}{2^6} = \frac{1}{64}$ **(d)** $\frac{7^{10}}{7^{12}} = 7^{-2} = \frac{1}{49}$

(e) $\left(\frac{3}{2}\right)^{-2} = \left(\frac{2}{3}\right)^2 = \frac{4}{9}$ **(f)** $\frac{\sqrt[5]{32}}{\sqrt{16}} = \frac{2}{4} = \frac{1}{2}$ **(g)** $\sqrt[4]{\frac{3^8}{2^{16}}} = \frac{3^2}{2^4} = \frac{9}{16}$ **(h)** $81^{-3/4} = \left(3^4\right)^{-3/4} = 3^{-3} = \frac{1}{27}$

7. **(a)** $\frac{a^3b^2}{ab^3} = \frac{a^2}{b}$

(b) $\left(2x^3y^{-2}\right)^{-2} = \frac{y^4}{4x^6}$

(c) $\left(2x^{1/2}y^2\right)\left(3x^{1/4}y^{-1}\right)^2 = 2 \cdot 3^2 x^{(1/2)+2(1/4)} y^{2+2(-1)} = 18x$

(d) $\sqrt{20} - \sqrt{125} = \sqrt{4 \cdot 5} - \sqrt{25 \cdot 5} = 2\sqrt{5} - 5\sqrt{5} = -3\sqrt{5}$

(e) $\sqrt{18x^3y^4} = \sqrt{9 \cdot 2 \cdot x^2 \cdot x \cdot (y^2)^2} = 3xy^2\sqrt{2x}$

(f) $\left(\frac{2x^2y}{x^{-3}y^{1/2}}\right)^{-2} = 2^{-2}x^{2(-2)-(-3)(-2)}y^{-2-(1/2)(-2)} = \frac{1}{4x^{10}y}$

9. **(a)** $4x^2 - 25 = (2x - 5)(2x + 5)$

(b) $2x^2 + 5x - 12 = (2x - 3)(x + 4)$

(c) $x^3 - 3x^2 - 4x + 12 = x^2(x - 3) - 4(x - 3) = (x - 3)\left(x^2 - 4\right) = (x - 3)(x - 2)(x + 2)$

(d) $x^4 + 27x = x\left(x^3 + 27\right) = x(x + 3)\left(x^2 - 3x + 9\right)$

(e) $(2x - y)^2 - 10(2x - y) + 25 = (2x - y)^2 - 2(5)(2x - y) + 5^2 = (2x - y - 5)^2$

(f) $x^3y - 4xy = xy\left(x^2 - 4\right) = xy(x - 2)(x + 2)$

11. **(a)** $\frac{6}{\sqrt[3]{4}} = \frac{6}{\sqrt[3]{2^2}} = \frac{6}{\sqrt[3]{2^2}} \cdot \frac{\sqrt[3]{2}}{\sqrt[3]{2}} = \frac{6\sqrt[3]{2}}{2} = 3\sqrt[3]{2}$

(b) $\frac{\sqrt{10}}{\sqrt{5} - 2} = \frac{\sqrt{10}}{\sqrt{5} - 2} \cdot \frac{\sqrt{5} + 2}{\sqrt{5} + 2} = \frac{\sqrt{10}\left(\sqrt{5} + 2\right)}{5 - 4} = \sqrt{50} + 2\sqrt{10} = 5\sqrt{2} + 2\sqrt{10}$

(c) $\frac{1}{1 + \sqrt{x}} = \frac{1}{1 + \sqrt{x}} \cdot \frac{1 - \sqrt{x}}{1 - \sqrt{x}} = \frac{1 - \sqrt{x}}{1 - x}$

13. $E = mc^2 \Leftrightarrow \frac{E}{m} = c^2 \Leftrightarrow c = \sqrt{\frac{E}{m}}$. (We take the positive root because c represents the speed of light, which is positive.)

FOCUS ON MODELING Making the Best Decisions

1. **(a)** The total cost is $\left(\begin{array}{c}\text{cost of}\\\text{copier}\end{array}\right) + \left(\begin{array}{c}\text{maintenance}\\\text{cost}\end{array}\right)\left(\begin{array}{c}\text{number}\\\text{of months}\end{array}\right) + \left(\begin{array}{c}\text{copy}\\\text{cost}\end{array}\right)\left(\begin{array}{c}\text{number}\\\text{of months}\end{array}\right)$. Each month the copy cost is $8000 \cdot 0.03 = 240$. Thus we get $C_1 = 5800 + 25n + 240n = 5800 + 265n$.

(b) In this case the cost is $\left(\begin{array}{c}\text{rental}\\\text{cost}\end{array}\right)\left(\begin{array}{c}\text{number}\\\text{of months}\end{array}\right) + \left(\begin{array}{c}\text{copy}\\\text{cost}\end{array}\right)\left(\begin{array}{c}\text{number}\\\text{of months}\end{array}\right)$. Each month the copy cost is $8000 \cdot 0.06 = 480$. Thus we get $C_2 = 95n + 480n = 575n$.

(c)

Years	n	Purchase	Rental
1	12	8,980	6,900
2	24	12,160	13,800
3	36	15,340	20,700
4	48	18,520	27,600
5	60	21,700	34,500
6	72	24,880	41,400

(d) The cost is the same when $C_1 = C_2$ are equal. So $5800 + 265n = 575n \Leftrightarrow 5800 = 310n \Leftrightarrow n \approx 18.71$ months.

3. (a) The total cost is $\begin{pmatrix}\text{setup}\\\text{cost}\end{pmatrix} + \begin{pmatrix}\text{cost per}\\\text{tire}\end{pmatrix}\begin{pmatrix}\text{number}\\\text{of tires}\end{pmatrix}$. So $C = 8000 + 22x$.

(b) The revenue is $\begin{pmatrix}\text{price per}\\\text{tire}\end{pmatrix}\begin{pmatrix}\text{number}\\\text{of tires}\end{pmatrix}$. So $R = 49x$.

(c) Profit = Revenue − Cost. So $P = R - C = 49x - (8000 + 22x) = 27x - 8000$.

(d) Break even is when profit is zero. Thus $27x - 8000 = 0 \Leftrightarrow 27x = 8000 \Leftrightarrow x \approx 296.3$. So they need to sell at least 297 tires to break even.

5. (a) Design 1 is a square and the perimeter of a square is four times the length of a side. $24 = 4x$, so each side is $x = 6$ feet long. Thus the area is $6^2 = 36\text{ ft}^2$.

Design 2 is a circle with perimeter $2\pi r$ and area πr^2. Thus we must solve $2\pi r = 24 \Leftrightarrow r = \frac{12}{\pi}$. Thus, the area is $\pi\left(\frac{12}{\pi}\right)^2 = \frac{144}{\pi} \approx 45.8\text{ ft}^2$. Design 2 gives the largest area.

(b) In Design 1, the cost is \$3 times the perimeter p, so $120 = 3p$ and the perimeter is 40 feet. By part (a), each side is then $\frac{40}{4} = 10$ feet long. So the area is $10^2 = 100\text{ ft}^2$.

In Design 2, the cost is \$4 times the perimeter p. Because the perimeter is $2\pi r$, we get $120 = 4(2\pi r)$ so $r = \frac{120}{8\pi} = \frac{15}{\pi}$. The area is $\pi r^2 = \pi\left(\frac{15}{\pi}\right)^2 = \frac{225}{\pi} \approx 71.6\text{ ft}^2$. Design 1 gives the largest area.

7. (a)

Data (GB)	Plan A	Plan B	Plan C
1	\$25	\$40	\$60
1.5	$25 + 5(2.00) = \$35$	$40 + 5(1.50) = \$47.50$	$60 + 5(1.00) = \$65$
2	$25 + 10(2.00) = \$45$	$40 + 10(1.50) = \$55$	$60 + 10(1.00) = \$70$
2.5	$25 + 15(2.00) = \$55$	$40 + 15(1.50) = \$62.50$	$60 + 15(1.00) = \$75$
3	$25 + 20(2.00) = \$65$	$40 + 20(1.50) = \$70$	$60 + 20(1.00) = \$80$
3.5	$25 + 25(2.00) = \$75$	$40 + 25(1.50) = \$77.50$	$60 + 25(1.00) = \$85$
4	$25 + 30(2.00) = \$85$	$40 + 30(1.50) = \$85$	$60 + 30(1.00) = \$90$

(b) For Plan A: $C_A = 25 + 2(10x - 10) = 20x + 5$. For Plan B: $C_B = 40 + 1.5(10x - 10) = 15x + 25$. For Plan C: $C_C = 60 + 1(10x - 10) = 10x + 50$. Note that these equations are valid only for $x \geq 1$.

(c) If Gwendolyn uses 2.2 GB, Plan A costs $25 + 12\,(2) = \$49$, Plan B costs $40 + 12\,(1.5) = \$58$, and Plan C costs $60 + 12\,(1) = \$72$.
If she uses 3.7 GB, Plan A costs $25 + 27\,(2) = \$79$, Plan B costs $40 + 27\,(1.5) = \$80.50$, and Plan C costs $60 + 27\,(1) = \$87$.
If she uses 4.9 GB, Plan A costs $25 + 39\,(2) = \$103$, Plan B costs $40 + 39\,(1.5) = \$98.50$, and Plan C costs $60 + 39\,(1) = \$99$.

(d) **(i)** We set $C_A = C_B \Leftrightarrow 20x + 5 = 15x + 25 \Leftrightarrow 5x = 20 \Leftrightarrow x = 4$. Plans A and B cost the same when 4 GB are used.

(ii) We set $C_A = C_C \Leftrightarrow 20x + 5 = 10x + 50 \Leftrightarrow 10x = 45 \Leftrightarrow x = 4.5$. Plans A and C cost the same when 4.5 GB are used.

(iii) We set $C_B = C_C \Leftrightarrow 15x + 25 = 10x + 50 \Leftrightarrow 5x = 25 \Leftrightarrow x = 5$. Plans B and C cost the same when 5 GB are used.

1 EQUATIONS AND GRAPHS

1.1 THE COORDINATE PLANE

1. The point that is 2 units to the left of the y-axis and 4 units above the x-axis has coordinates $(-2, 4)$.

3. The distance between the points (a, b) and (c, d) is $\sqrt{(c-a)^2 + (d-b)^2}$. So the distance between $(1, 2)$ and $(7, 10)$ is $\sqrt{(7-1)^2 + (10-2)^2} = \sqrt{6^2 + 8^2} = \sqrt{36 + 64} = \sqrt{100} = 10$.

5. $A(5, 1)$, $B(1, 2)$, $C(-2, 6)$, $D(-6, 2)$, $E(-4, -1)$, $F(-2, 0)$, $G(-1, -3)$, $H(2, -2)$

7. $(0, 5)$, $(-1, 0)$, $(-1, -2)$, and $\left(\frac{1}{2}, \frac{2}{3}\right)$

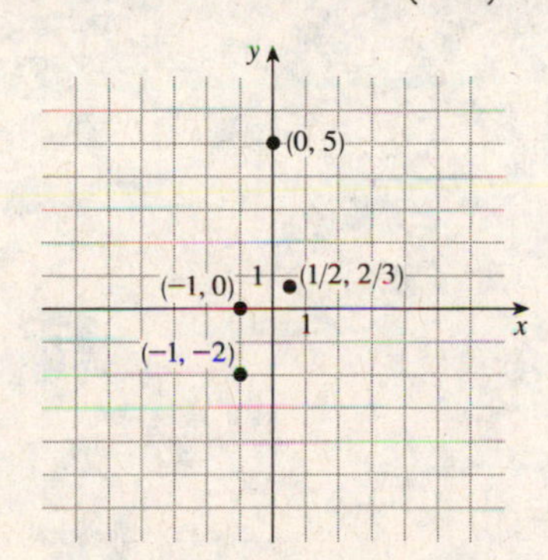

9. $\{(x, y) \mid x \geq 2\}$

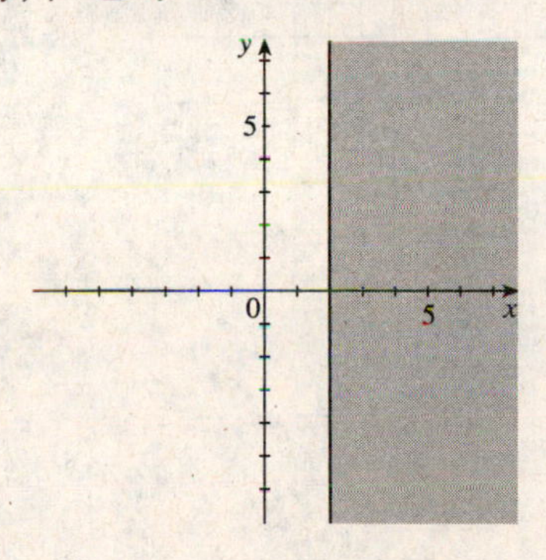

11. $\{(x, y) \mid x = -4\}$

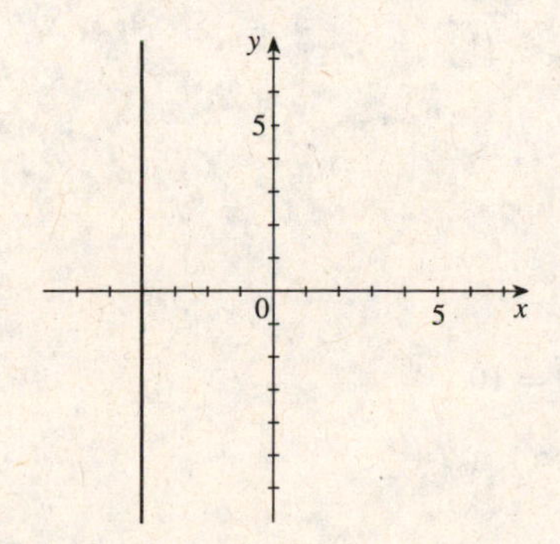

13. $\{(x, y) \mid -3 < x < 3\}$

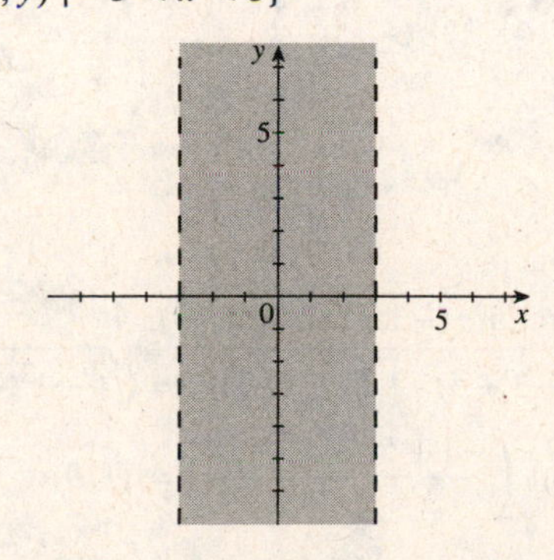

15. $\{(x, y) \mid xy < 0\}$

$= \{(x, y) \mid x < 0 \text{ and } y > 0 \text{ or } x > 0 \text{ and } y < 0\}$

17. $\{(x, y) \mid x \geq 1 \text{ and } y < 3\}$

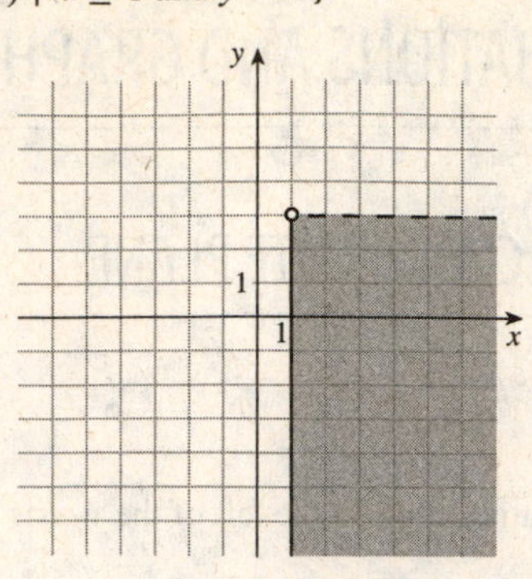

19. $\{(x, y) \mid -1 < x < 1 \text{ and } -2 < y < 2\}$

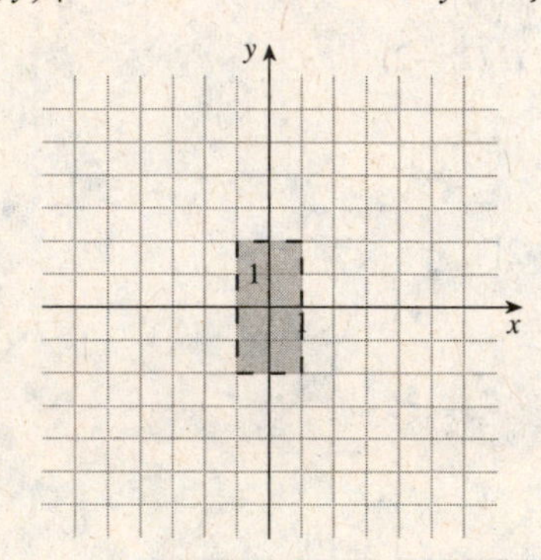

21. The two points are $(0, 2)$ and $(3, 0)$.

(a) $d = \sqrt{(3-0)^2 + (0-(-2))^2} = \sqrt{3^2 + 2^2} = \sqrt{9+4} = \sqrt{13}$

(b) midpoint: $\left(\frac{3+0}{2}, \frac{0+2}{2}\right) = \left(\frac{3}{2}, 1\right)$

23. The two points are $(-3, 3)$ and $(5, -3)$.

(a) $d = \sqrt{(-3-5)^2 + (3-(-3))^2} = \sqrt{(-8)^2 + 6^2} = \sqrt{64+36} = \sqrt{100} = 10$

(b) midpoint: $\left(\frac{-3+5}{2}, \frac{3+(-3)}{2}\right) = (1, 0)$

25. (a)

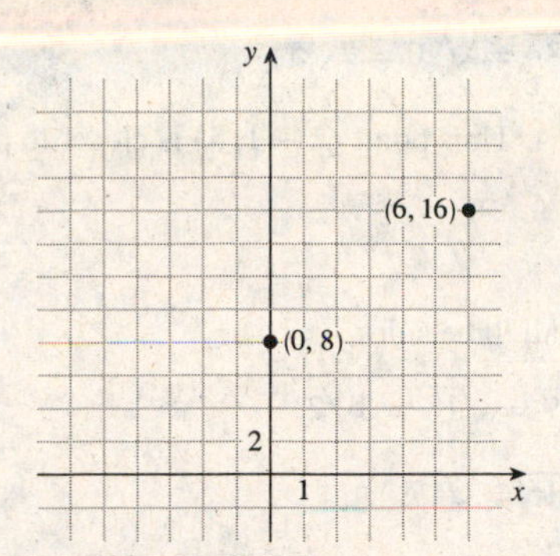

(b) $d = \sqrt{(0-6)^2 + (8-16)^2} = \sqrt{(-6)^2 + (-8)^2} = \sqrt{100} = 10$

(c) Midpoint: $\left(\frac{0+6}{2}, \frac{8+16}{2}\right) = (3, 12)$

27. (a)

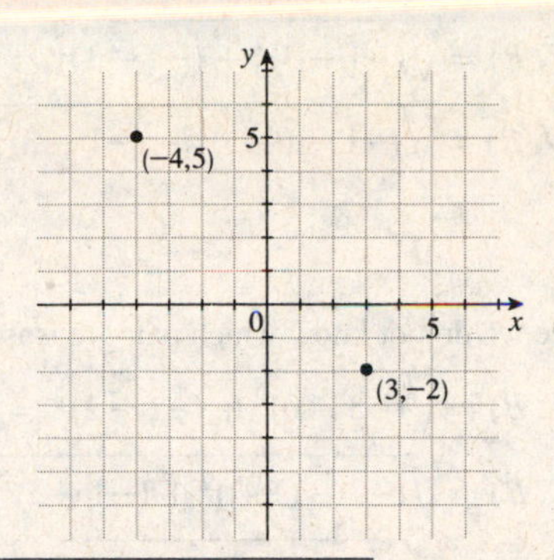

(b) $d = \sqrt{(3-(-4))^2 + (-2-5)^2} = \sqrt{7^2 + (-7)^2} = \sqrt{49+49} = \sqrt{98} = 7\sqrt{2}$

(c) Midpoint: $\left(\frac{-4+3}{2}, \frac{5-2}{2}\right) = \left(-\frac{1}{2}, \frac{3}{2}\right)$

29. (a)

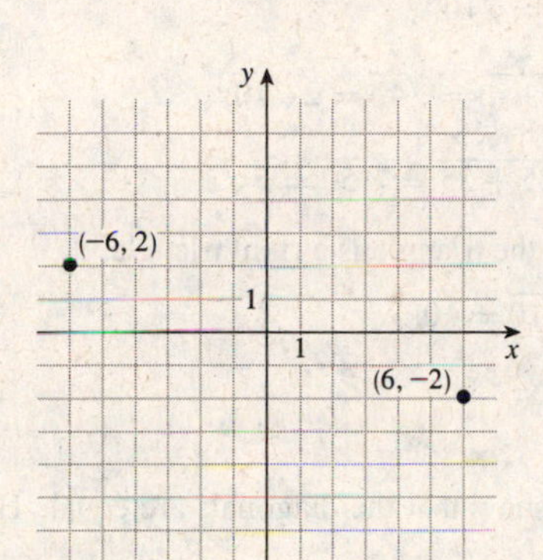

(b) $d = \sqrt{(6-(-6))^2 + (-2-2)^2} = \sqrt{12^2 + (-4)^2} = \sqrt{144+16} = \sqrt{160} = 4\sqrt{10}$

(c) Midpoint: $\left(\frac{6-6}{2}, \frac{-2+2}{2}\right) = (0, 0)$

31. $d(A, B) = \sqrt{(1-5)^2 + (3-3)^2} = \sqrt{(-4)^2} = 4$.

$d(A, C) = \sqrt{(1-1)^2 + (3-(-3))^2} = \sqrt{(6)^2} = 6$. So the area is $4 \cdot 6 = 24$.

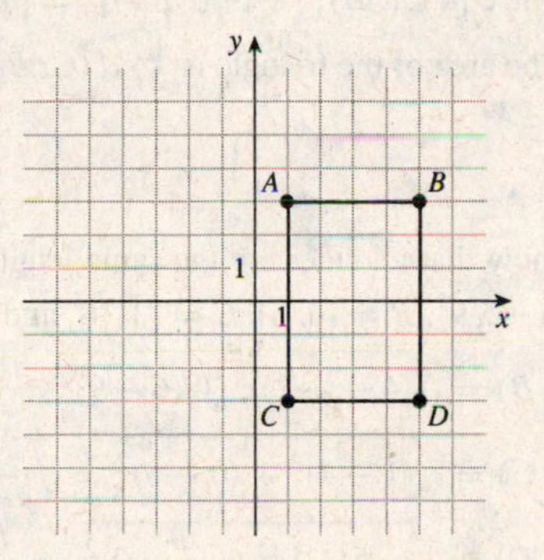

33. From the graph, the quadrilateral $ABCD$ has a pair of parallel sides, so $ABCD$ is a trapezoid. The area is $\left(\frac{b_1+b_2}{2}\right)h$. From the graph we see that

$b_1 = d(A, B) = \sqrt{(1-5)^2 + (0-0)^2} = \sqrt{4^2} = 4$;

$b_2 = d(C, D) = \sqrt{(4-2)^2 + (3-3)^2} = \sqrt{2^2} = 2$; and h is the difference in y-coordinates is $|3-0| = 3$. Thus the area of the trapezoid is $\left(\frac{4+2}{2}\right)3 = 9$.

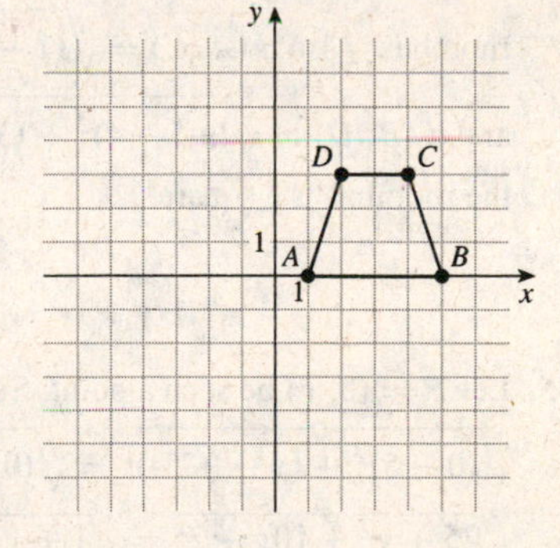

35. $d(0, A) = \sqrt{(6-0)^2 + (7-0)^2} = \sqrt{6^2 + 7^2} = \sqrt{36+49} = \sqrt{85}$.

$d(0, B) = \sqrt{(-5-0)^2 + (8-0)^2} = \sqrt{(-5)^2 + 8^2} = \sqrt{25+64} = \sqrt{89}$.

Thus point $A(6, 7)$ is closer to the origin.

37. $d(P,R) = \sqrt{(-1-3)^2 + (-1-1)^2} = \sqrt{(-4)^2 + (-2)^2} = \sqrt{16+4} = \sqrt{20} = 2\sqrt{5}$.

$d(Q,R) = \sqrt{(-1-(-1))^2 + (-1-3)^2} = \sqrt{0 + (-4)^2} = \sqrt{16} = 4$. Thus point $Q(-1,3)$ is closer to point R.

39. Since we do not know which pair are isosceles, we find the length of all three sides.

$d(A,B) = \sqrt{(-3-0)^2 + (-1-2)^2} = \sqrt{(-3)^2 + (-3)^2} = \sqrt{9+9} = \sqrt{18} = 3\sqrt{2}$.

$d(C,B) = \sqrt{(-3-(-4))^2 + (-1-3)^2} = \sqrt{1^2 + (-4)^2} = \sqrt{1+16} = \sqrt{17}$.

$d(A,C) = \sqrt{(0-(-4))^2 + (2-3)^2} = \sqrt{4^2 + (-1)^2} = \sqrt{16+1} = \sqrt{17}$. So sides AC and CB have the same length.

41. (a) Here we have $A = (2,2)$, $B = (3,-1)$, and $C = (-3,-3)$. So

$d(A,B) = \sqrt{(3-2)^2 + (-1-2)^2} = \sqrt{1^2 + (-3)^2} = \sqrt{1+9} = \sqrt{10}$;

$d(C,B) = \sqrt{(3-(-3))^2 + (-1-(-3))^2} = \sqrt{6^2 + 2^2} = \sqrt{36+4} = \sqrt{40} = 2\sqrt{10}$;

$d(A,C) = \sqrt{(-3-2)^2 + (-3-2)^2} = \sqrt{(-5)^2 + (-5)^2} = \sqrt{25+25} = \sqrt{50} = 5\sqrt{2}$.

Since $[d(A,B)]^2 + [d(C,B)]^2 = [d(A,C)]^2$, we conclude that the triangle is a right triangle.

(b) The area of the triangle is $\frac{1}{2} \cdot d(C,B) \cdot d(A,B) = \frac{1}{2} \cdot \sqrt{10} \cdot 2\sqrt{10} = 10$.

43. We show that all sides are the same length (its a rhombus) and then show that the diagonals are equal. Here we have $A = (-2,9)$, $B = (4,6)$, $C = (1,0)$, and $D = (-5,3)$. So

$d(A,B) = \sqrt{(4-(-2))^2 + (6-9)^2} = \sqrt{6^2 + (-3)^2} = \sqrt{36+9} = \sqrt{45}$;

$d(B,C) = \sqrt{(1-4)^2 + (0-6)^2} = \sqrt{(-3)^2 + (-6)^2} = \sqrt{9+36} = \sqrt{45}$;

$d(C,D) = \sqrt{(-5-1)^2 + (3-0)^2} = \sqrt{(-6)^2 + (-3)^2} = \sqrt{36+9} = \sqrt{45}$;

$d(D,A) = \sqrt{(-2-(-5))^2 + (9-3)^2} = \sqrt{3^2 + 6^2} = \sqrt{9+36} = \sqrt{45}$. So the points form a rhombus. Also $d(A,C) = \sqrt{(1-(-2))^2 + (0-9)^2} = \sqrt{3^2 + (-9)^2} = \sqrt{9+81} = \sqrt{90} = 3\sqrt{10}$,

and $d(B,D) = \sqrt{(-5-4)^2 + (3-6)^2} = \sqrt{(-9)^2 + (-3)^2} = \sqrt{81+9} = \sqrt{90} = 3\sqrt{10}$. Since the diagonals are equal, the rhombus is a square.

45. Let $P = (0,y)$ be such a point. Setting the distances equal we get

$\sqrt{(0-5)^2 + (y-(-5))^2} = \sqrt{(0-1)^2 + (y-1)^2} \Leftrightarrow$

$\sqrt{25 + y^2 + 10y + 25} = \sqrt{1 + y^2 - 2y + 1} \Rightarrow y^2 + 10y + 50 = y^2 - 2y + 2 \Leftrightarrow 12y = -48 \Leftrightarrow y = -4$. Thus, the point is $P = (0,-4)$. Check:

$\sqrt{(0-5)^2 + (-4-(-5))^2} = \sqrt{(-5)^2 + 1^2} = \sqrt{25+1} = \sqrt{26}$;

$\sqrt{(0-1)^2 + (-4-1)^2} = \sqrt{(-1)^2 + (-5)^2} = \sqrt{25+1} = \sqrt{26}$.

47. As indicated by Example 3, we must find a point $S(x_1, y_1)$ such that the midpoints of PR and of QS are the same. Thus $\left(\frac{4+(-1)}{2}, \frac{2+(-4)}{2}\right) = \left(\frac{x_1+1}{2}, \frac{y_1+1}{2}\right)$. Setting the x-coordinates equal, we get $\frac{4+(-1)}{2} = \frac{x_1+1}{2} \Leftrightarrow 4-1 = x_1+1 \Leftrightarrow x_1 = 2$. Setting the y-coordinates equal, we get $\frac{2+(-4)}{2} = \frac{y_1+1}{2} \Leftrightarrow 2-4 = y_1+1 \Leftrightarrow y_1 = -3$. Thus $S = (2, -3)$.

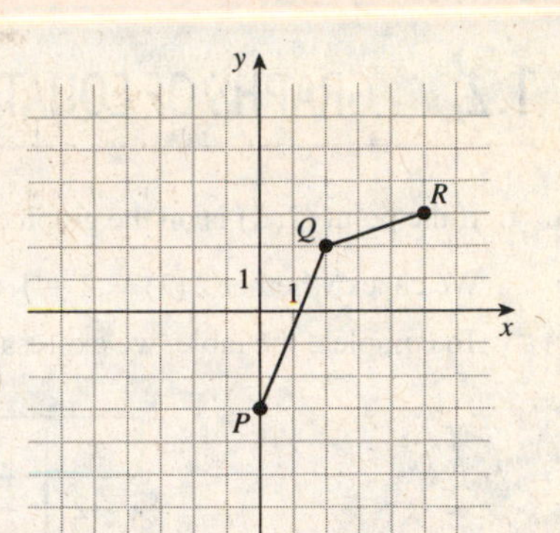

49. **(a)**

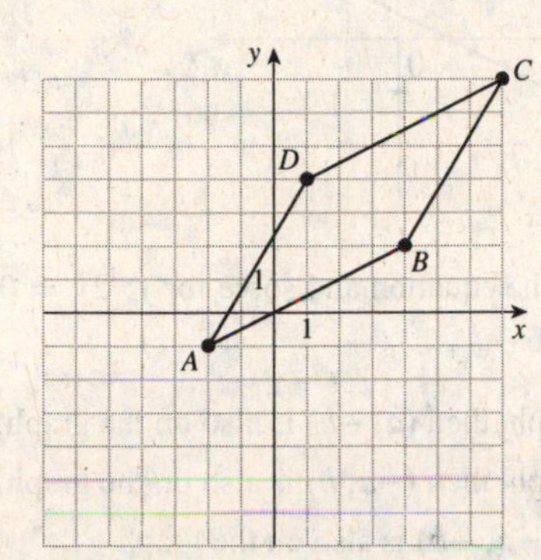

(b) The midpoint of AC is $\left(\frac{-2+7}{2}, \frac{-1+7}{2}\right) = \left(\frac{5}{2}, 3\right)$, the midpoint of BD is $\left(\frac{4+1}{2}, \frac{2+4}{2}\right) = \left(\frac{5}{2}, 3\right)$.

(c) Since the they have the same midpoint, we conclude that the diagonals bisect each other.

51. **(a)** The point $(5, 3)$ is shifted to $(5+3, 3+2) = (8, 5)$.

(b) The point (a, b) is shifted to $(a+3, b+2)$.

(c) Let (x, y) be the point that is shifted to $(3, 4)$. Then $(x+3, y+2) = (3, 4)$. Setting the x-coordinates equal, we get $x+3 = 3 \Leftrightarrow x = 0$. Setting the y-coordinates equal, we get $y+2 = 4 \Leftrightarrow y = 2$. So the point is $(0, 2)$.

(d) $A = (-5, -1)$, so $A' = (-5+3, -1+2) = (-2, 1)$; $B = (-3, 2)$, so $B' = (-3+3, 2+2) = (0, 4)$; and $C = (2, 1)$, so $C' = (2+3, 1+2) = (5, 3)$.

53. **(a)** $d(A, B) = \sqrt{3^2+4^2} = \sqrt{25} = 5$.

(b) We want the distances from $C = (4, 2)$ to $D = (11, 26)$. The walking distance is $|4-11| + |2-26| = 7+24 = 31$ blocks. Straight-line distance is $\sqrt{(4-11)^2+(2-26)^2} = \sqrt{7^2+24^2} = \sqrt{625} = 25$ blocks.

(c) The two points are on the same avenue or the same street.

55. The midpoint of the line segment is $(66, 45)$. The pressure experienced by an ocean diver at a depth of 66 feet is 45 lb/in^2.

57. We need to find a point $S(x_1, y_1)$ such that $PQRS$ is a parallelogram. As indicated by Example 3, this will be the case if the diagonals PR and QS bisect each other. So the midpoints of PR and QS are the same. Thus $\left(\frac{0+5}{2}, \frac{-3+3}{2}\right) = \left(\frac{x_1+2}{2}, \frac{y_1+2}{2}\right)$. Setting the x-coordinates equal, we get $\frac{0+5}{2} = \frac{x_1+2}{2} \Leftrightarrow 0+5 = x_1+2 \Leftrightarrow x_1 = 3$.

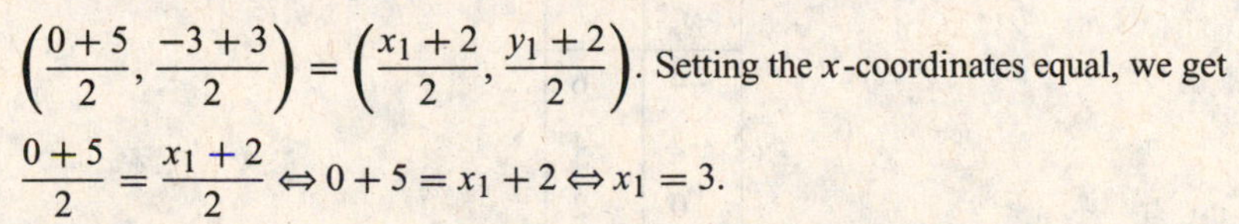

Setting the y-coordinates equal, we get $\frac{-3+3}{2} = \frac{y_1+2}{2} \Leftrightarrow -3+3 = y_1+2 \Leftrightarrow y_1 = -2$. Thus $S = (3, -2)$.

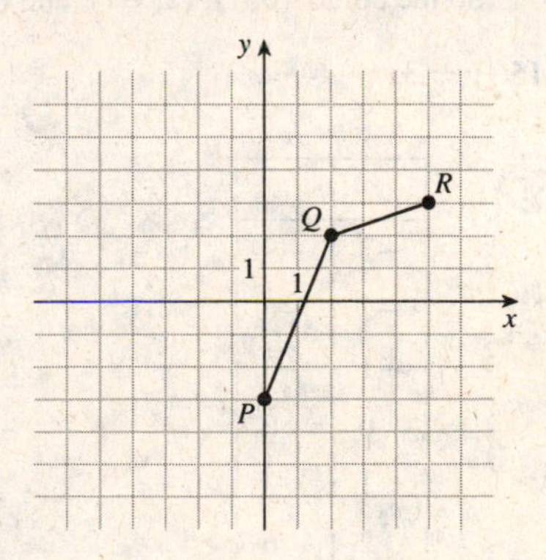

1.2 GRAPHS OF EQUATIONS IN TWO VARIABLES: CIRCLES

1. If the point $(2,3)$ is on the graph of an equation in x and y, then the equation is satisfied when we replace x by 2 and y by 3. We check whether $2(3) \stackrel{?}{=} 2+1 \Leftrightarrow 6 \stackrel{?}{=} 3$. This is false, so the point $(2,3)$ is not on the graph of the equation $2y = x+1$. To complete the table, we express y in terms of x: $2y = x+1 \Leftrightarrow y = \frac{1}{2}(x+1) = \frac{1}{2}x + \frac{1}{2}$.

x	y	(x,y)
-2	$-\frac{1}{2}$	$\left(-2,-\frac{1}{2}\right)$
-1	0	$(-1,0)$
0	$\frac{1}{2}$	$\left(0,\frac{1}{2}\right)$
1	1	$(1,1)$
2	$\frac{3}{2}$	$\left(2,\frac{3}{2}\right)$

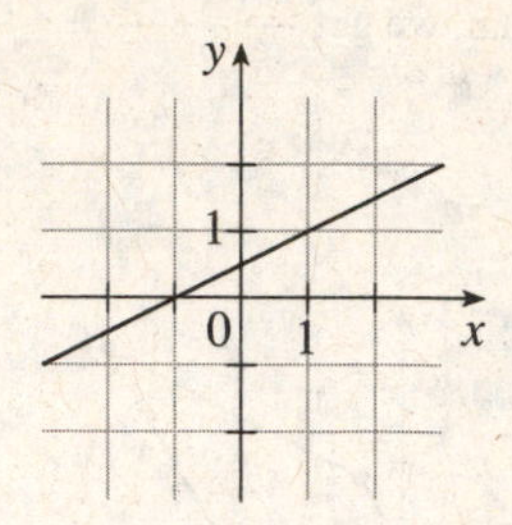

3. To find the y-intercept(s) of the graph of an equation we set x equal to 0 in the equation and solve for y: $2y = 0+1 \Leftrightarrow y = \frac{1}{2}$, so the y-intercept of $2y = x+1$ is $\frac{1}{2}$.

5. (a) If a graph is symmetric with respect to the x-axis and (a,b) is on the graph, then $(a,-b)$ is also on the graph.

(b) If a graph is symmetric with respect to the y-axis and (a,b) is on the graph, then $(-a,b)$ is also on the graph.

(c) If a graph is symmetric about the origin and (a,b) is on the graph, then $(-a,-b)$ is also on the graph.

7. Yes, this is true. If for every point (x,y) on the graph, $(-x,y)$ and $(x,-y)$ are also on the graph, then $(-x,-y)$ must be on the graph as well, and so it is symmetric about the origin.

9. $y = 3-4x$. For the point $(0,3)$: $3 \stackrel{?}{=} 3-4(0) \Leftrightarrow 3 = 3$. Yes. For $(4,0)$: $0 \stackrel{?}{=} 3-4(4) \Leftrightarrow 0 \stackrel{?}{=} -13$. No. For $(1,-1)$: $-1 \stackrel{?}{=} 3-4(1) \Leftrightarrow -1 \stackrel{?}{=} -1$. Yes.
So the points $(0,3)$ and $(1,-1)$ are on the graph of this equation.

11. $x-2y-1=0$. For the point $(0,0)$: $0-2(0)-1 \stackrel{?}{=} 0 \Leftrightarrow -1 \stackrel{?}{=} 0$. No. For $(1,0)$: $1-2(0)-1 \stackrel{?}{=} 0 \Leftrightarrow -1+1 \stackrel{?}{=} 0$. Yes. For $(-1,-1)$: $(-1)-2(-1)-1 \stackrel{?}{=} 0 \Leftrightarrow -1+2-1 \stackrel{?}{=} 0$. Yes.
So the points $(1,0)$ and $(-1,-1)$ are on the graph of this equation.

13. $x^2+2xy+y^2=1$. For the point $(0,1)$: $0^2+2(0)(1)+1^2 \stackrel{?}{=} 1 \Leftrightarrow 1 \stackrel{?}{=} 1$. Yes. For $(2,-1)$: $2^2+2(2)(-1)+(-1)^2 \stackrel{?}{=} 1 \Leftrightarrow 4-4+1 \stackrel{?}{=} 1 \Leftrightarrow 1 = 1$. Yes. For $(-2,3)$: $(-2)^2+2(-2)(3)+3^2 \stackrel{?}{=} 1 \Leftrightarrow 4-12+9 \stackrel{?}{=} 1 \Leftrightarrow 1 \stackrel{?}{=} 1$. Yes.
So the points $(0,1)$, $(2,-1)$, and $(-2,3)$ are on the graph of this equation.

15. $y = 3x$

x	y
-3	-9
-2	-6
-1	-3
0	0
1	3
2	6
3	9

17. $y = 2-x$

x	y
-4	6
-2	4
0	2
2	0
4	-2

19. Solve for y: $2x - y = 6 \Leftrightarrow y = 2x - 6$.

x	y
-2	-10
0	-6
2	-2
4	2
6	6

21. $y = 1 - x^2$

x	y
-3	-8
-2	-3
-1	0
0	1
1	0
2	-3
3	-8

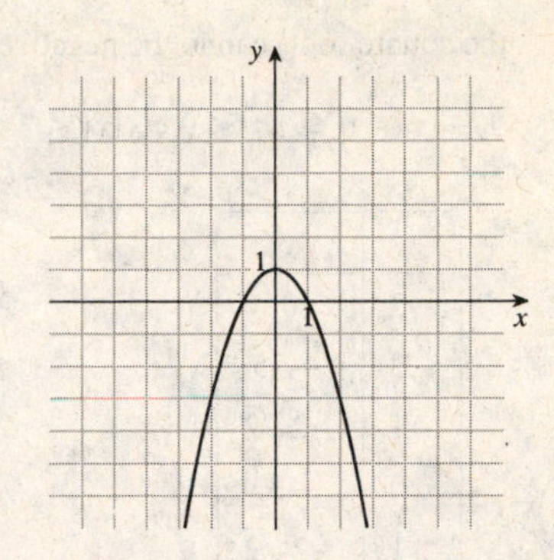

23. $y = x^2 - 2$

x	y
-3	7
-2	2
-1	-1
0	-2
1	-1
2	2
3	7

25. $9y = x^2$. To make a table, we rewrite the equation as

$y = \frac{1}{9}x^2$.

x	y
-9	9
-3	1
0	0
3	1
9	9

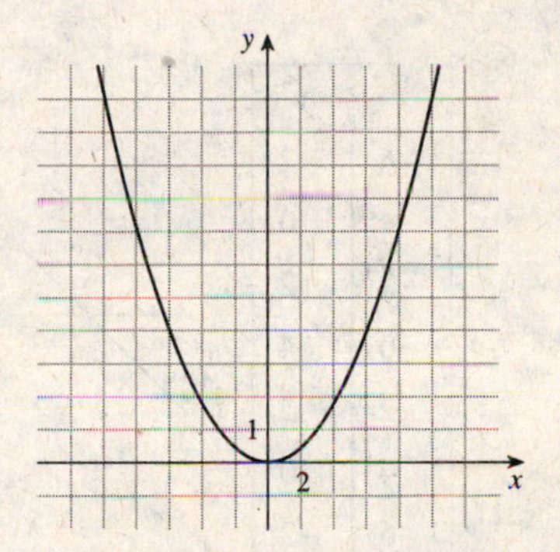

27. $x + y^2 = 4$.

x	y
-12	-4
-5	-3
0	-2
3	-1
4	0
3	1
0	2
-5	3
-12	4

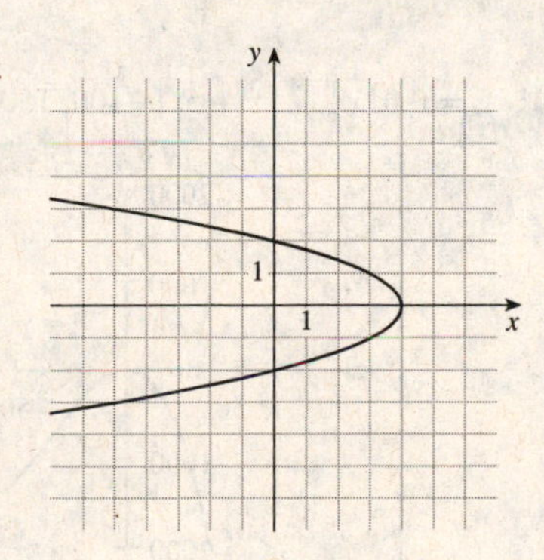

29. $y = \sqrt{x}$.

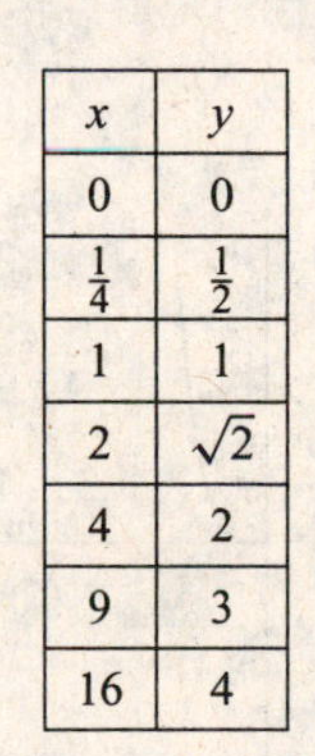

x	y
0	0
$\frac{1}{4}$	$\frac{1}{2}$
1	1
2	$\sqrt{2}$
4	2
9	3
16	4

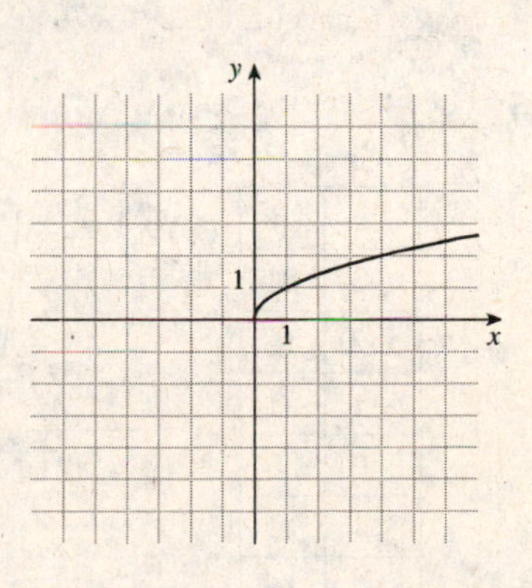

31. $y = -\sqrt{9 - x^2}$. Since the radicand (the expression inside the square root) cannot be negative, we must have $9 - x^2 \geq 0 \Leftrightarrow x^2 \leq 9 \Leftrightarrow |x| \leq 3$.

x	y
-3	0
-2	$-\sqrt{5}$
-1	$-2\sqrt{2}$
0	-3
1	$-2\sqrt{2}$
2	$-\sqrt{5}$
3	0

33. $y = -|x|$.

x	y
-6	-6
-4	-4
-2	-2
0	0
2	-2
4	-4
6	-6

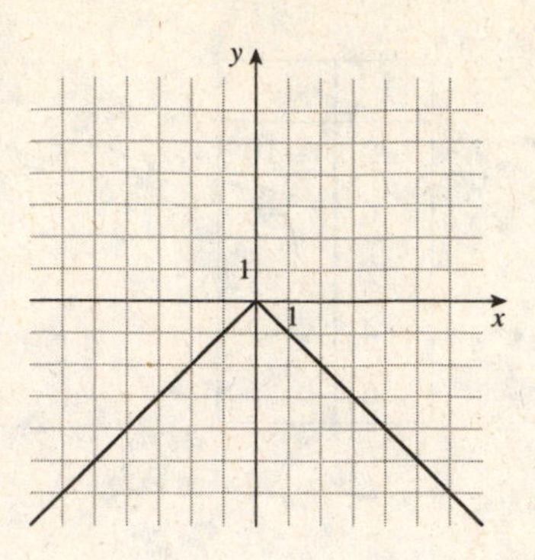

35. $y = 4 - |x|$.

x	y
-6	-2
-4	0
-2	2
0	4
2	2
4	0
6	-2

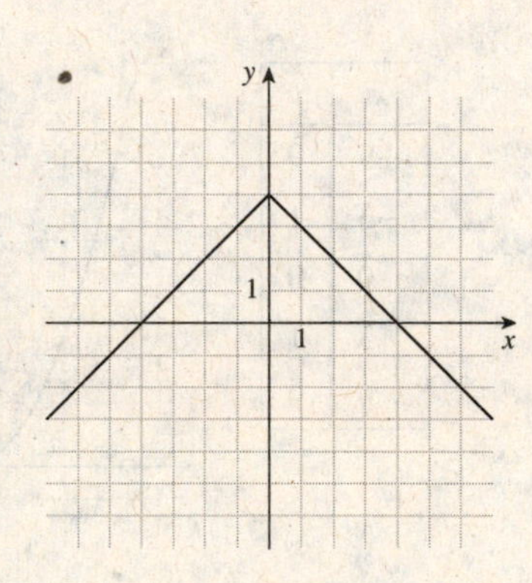

37. $x = y^3$. Since $x = y^3$ is solved for x in terms of y, we insert values for y and find the corresponding values of x in the table below.

x	y
-27	-3
-8	-2
-1	-1
0	0
1	1
8	2
27	3

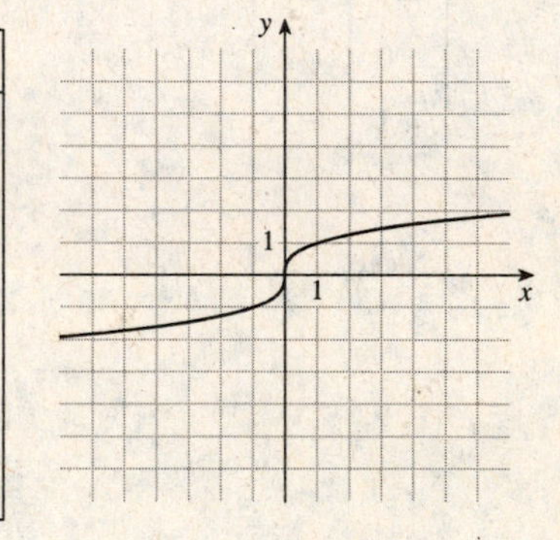

39. $y = x^4$.

x	y
-3	81
-2	16
-1	1
0	0
1	1
2	16
3	81

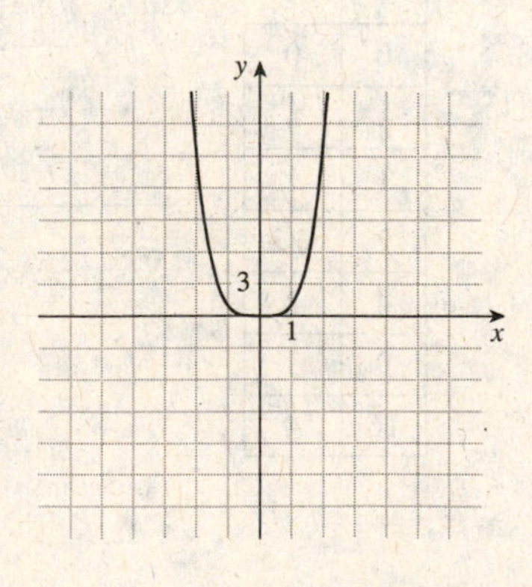

41. $y = 0.01x^3 - x^2 + 5$; $[-100, 150]$ by $[-2000, 2000]$

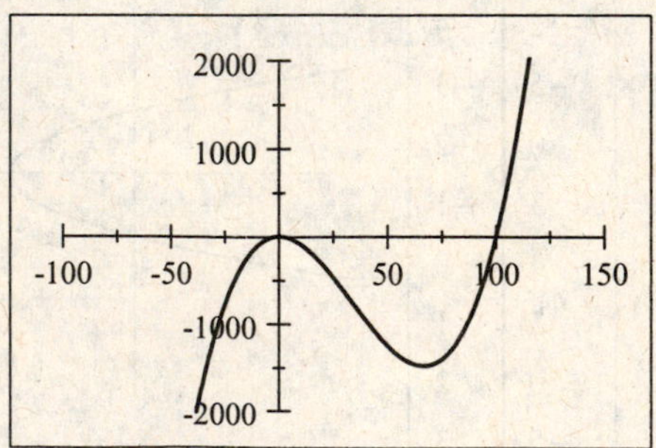

43. $y = \sqrt{12x - 17}$; $[-1, 10]$ by $[-1, 20]$

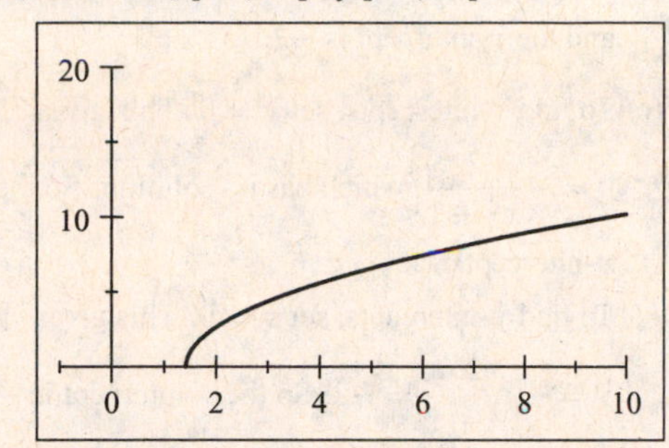

45. $y = \dfrac{x}{x^2 + 25}$; $[-50, 50]$ by $[-0.2, 0.2]$

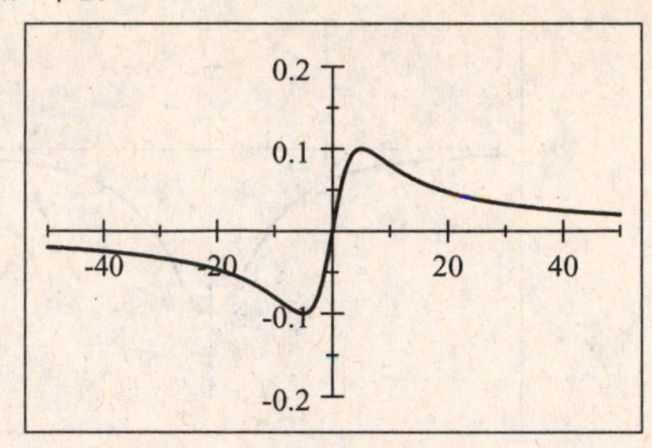

47. $y = x + 6$. To find x-intercepts, set $y = 0$. This gives $0 = x + 6 \Leftrightarrow x = -6$, so the x-intercept is -6.
To find y-intercepts, set $x = 0$. This gives $y = 0 + 6 \Leftrightarrow y = 6$, so the y-intercept is 6.

49. $y = x^2 - 5$. To find x-intercepts, set $y = 0$. This gives $0 = x^2 - 5 \Leftrightarrow x^2 = 5 \Rightarrow x = \pm\sqrt{5}$, so the x-intercepts are $\pm\sqrt{5}$.
To find y-intercepts, set $x = 0$. This gives $y = 0^2 - 5 = -5$, so the y-intercept is -5.

51. $y - 2xy + 2x = 1$. To find x-intercepts, set $y = 0$. This gives $0 - 2x\,(0) + 2x = 1 \Leftrightarrow 2x = 1 \Leftrightarrow x = \dfrac{1}{2}$, so the x-intercept is $\frac{1}{2}$.
To find y-intercepts, set $x = 0$. This gives $y - 2\,(0)\,y + 2\,(0) = 1 \Leftrightarrow y = 1$, so the y-intercept is 1.

53. $y = \sqrt{x + 1}$. To find x-intercepts, set $y = 0$. This gives $0 = \sqrt{x + 1} \Leftrightarrow 0 = x + 1 \Leftrightarrow x = -1$, so the x-intercept is -1.
To find y-intercepts, set $x = 0$. This gives $y = \sqrt{0 + 1} \Leftrightarrow y = 1$, so the y-intercept is 1.

55. $4x^2 + 25y^2 = 100$. To find x-intercepts, set $y = 0$. This gives $4x^2 + 25\,(0)^2 = 100 \Leftrightarrow x^2 = 25 \Leftrightarrow x = \pm 5$, so the x-intercepts are -5 and 5.
To find y-intercepts, set $x = 0$. This gives $4\,(0)^2 + 25y^2 = 100 \Leftrightarrow y^2 = 4 \Leftrightarrow y = \pm 2$, so the y-intercepts are -2 and 2.

57. $y = 4x - x^2$. To find x-intercepts, set $y = 0$. This gives $0 = 4x - x^2 \Leftrightarrow 0 = x\,(4 - x) \Leftrightarrow 0 = x$ or $x = 4$, so the x-intercepts are 0 and 4.
To find y-intercepts, set $x = 0$. This gives $y = 4\,(0) - 0^2 \Leftrightarrow y = 0$, so the y-intercept is 0.

59. $x^4 + y^2 - xy = 16$. To find x-intercepts, set $y = 0$. This gives $x^4 + 0^2 - x\,(0) = 16 \Leftrightarrow x^4 = 16 \Leftrightarrow x = \pm 2$. So the x-intercepts are -2 and 2.
To find y-intercepts, set $x = 0$. This gives $0^4 + y^2 - (0)\,y = 16 \Leftrightarrow y^2 = 16 \Leftrightarrow y = \pm 4$. So the y-intercepts are -4 and 4.

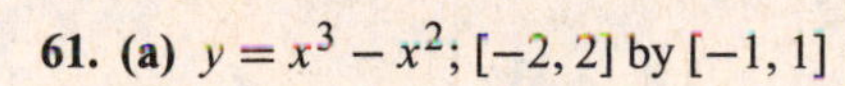

61. (a) $y = x^3 - x^2$; $[-2, 2]$ by $[-1, 1]$

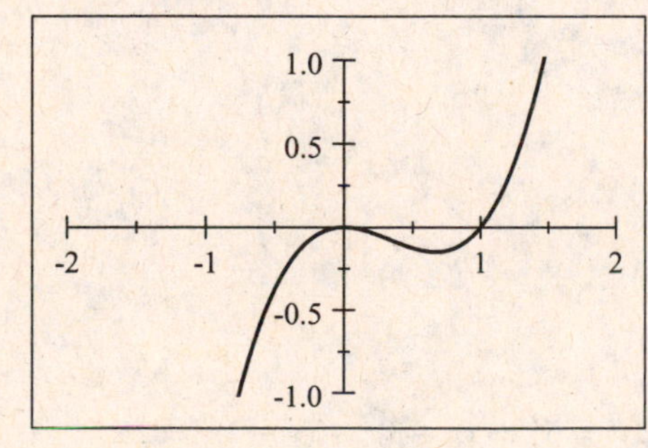

(b) From the graph, it appears that the x-intercepts are 0 and 1 and the y-intercept is 0.

(c) To find x-intercepts, set $y = 0$. This gives $0 = x^3 - x^2 \Leftrightarrow x^2\,(x - 1) = 0 \Leftrightarrow x = 0$ or 1. So the x-intercepts are 0 and 1.
To find y-intercepts, set $x = 0$. This gives $y = 0^3 - 0^2 = 0$. So the y-intercept is 0.

63. **(a)** $y = -\dfrac{2}{x^2+1}$; $[-5, 5]$ by $[-3, 1]$

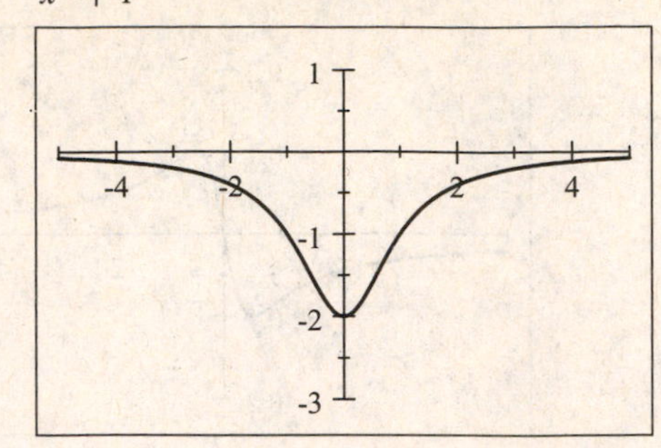

(b) From the graph, it appears that there is no x-intercept and the y-intercept is -2.

(c) To find x-intercepts, set $y = 0$. This gives $0 = -\dfrac{2}{x^2+1}$, which has no solution. So there is no x-intercept.
To find y-intercepts, set $x = 0$. This gives $y = -\dfrac{2}{0^2+1} = -2$. So the y-intercept is -2.

65. **(a)** $y = \sqrt[3]{x}$; $[-5, 5]$ by $[-2, 2]$

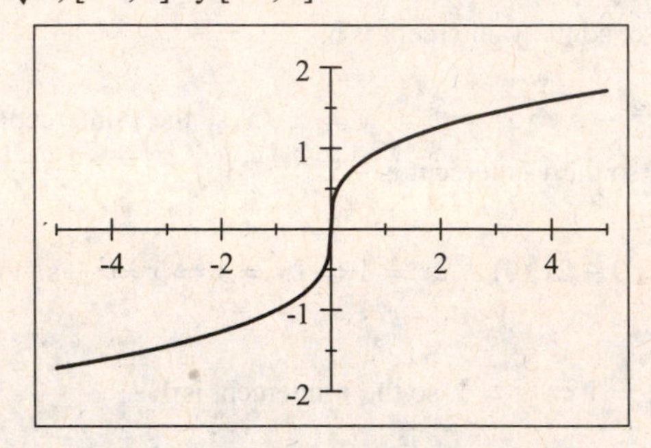

(b) From the graph, it appears that and the x- and y-intercepts are 0.

(c) To find x-intercepts, set $y = 0$. This gives $0 = \sqrt[3]{x} \Leftrightarrow x = 0$. So the x-intercept is 0.
To find y-intercepts, set $x = 0$. This gives $y = \sqrt[3]{0} = 0$. So the y-intercept is 0.

67. $x^2 + y^2 = 9$ has center $(0, 0)$ and radius 3.

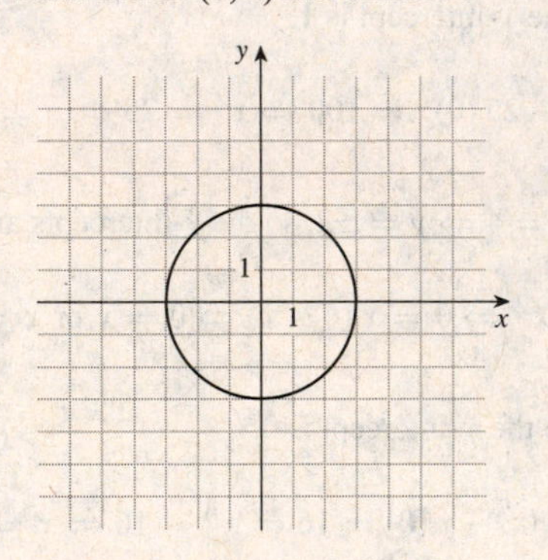

69. $(x-3)^2 + y^2 = 16$ has center $(3, 0)$ and radius 4.

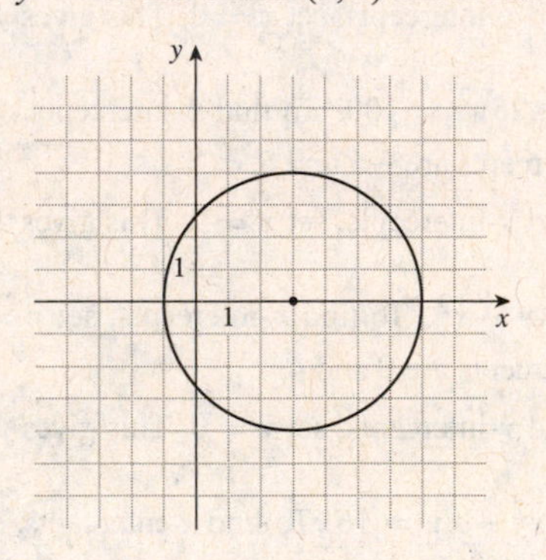

71. $(x+3)^2 + (y-4)^2 = 25$ has center $(-3, 4)$ and radius 5.

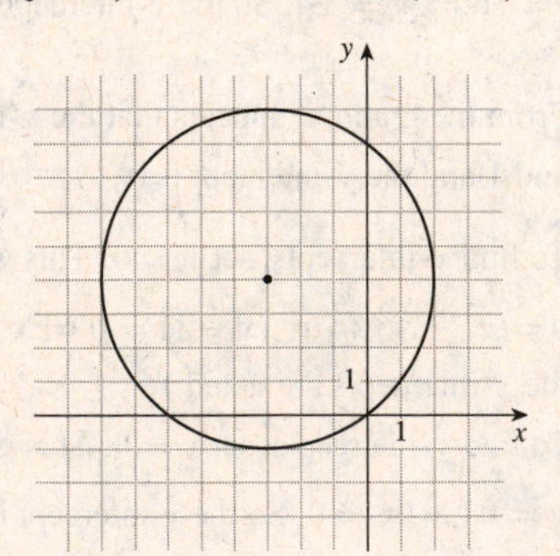

73. Using $h = -3$, $k = 2$, and $r = 5$, we get $(x-(-3))^2 + (y-2)^2 = 5^2 \Leftrightarrow (x+3)^2 + (y-2)^2 = 25$.

75. The equation of a circle centered at the origin is $x^2 + y^2 = r^2$. Using the point $(4, 7)$ we solve for r^2. This gives $(4)^2 + (7)^2 = r^2 \Leftrightarrow 16 + 49 = 65 = r^2$. Thus, the equation of the circle is $x^2 + y^2 = 65$.

77. The center is at the midpoint of the line segment, which is $\left(\frac{-1+5}{2}, \frac{1+9}{2}\right) = (2, 5)$. The radius is one half the diameter, so $r = \frac{1}{2}\sqrt{(-1-5)^2 + (1-9)^2} = \frac{1}{2}\sqrt{36+64} = \frac{1}{2}\sqrt{100} = 5$. Thus, an equation of the circle is $(x-2)^2 + (y-5)^2 = 5^2$ $\Leftrightarrow (x-2)^2 + (y-5)^2 = 25$.

79. Since the circle is tangent to the x-axis, it must contain the point $(7, 0)$, so the radius is the change in the y-coordinates. That is, $r = |-3 - 0| = 3$. So the equation of the circle is $(x-7)^2 + (y-(-3))^2 = 3^2$, which is $(x-7)^2 + (y+3)^2 = 9$.

81. From the figure, the center of the circle is at $(-2, 2)$. The radius is the change in the y-coordinates, so $r = |2 - 0| = 2$. Thus the equation of the circle is $(x-(-2))^2 + (y-2)^2 = 2^2$, which is $(x+2)^2 + (y-2)^2 = 4$.

83. Completing the square gives $x^2 + y^2 - 2x + 4y + 1 = 0 \Leftrightarrow x^2 - 2x + \left(\frac{-2}{2}\right)^2 + y^2 + 4y + \left(\frac{4}{2}\right)^2 = -1 + \left(\frac{-2}{2}\right)^2 + \left(\frac{4}{2}\right)^2$ $\Leftrightarrow x^2 - 2x + 1 + y^2 + 4y + 4 = -1 + 1 + 4 \Leftrightarrow (x-1)^2 + (y+2)^2 = 4$. Thus, the center is $(1, -2)$, and the radius is 2.

85. Completing the square gives $x^2 + y^2 - 4x + 10y + 13 = 0 \Leftrightarrow x^2 - 4x + \left(\frac{-4}{2}\right)^2 + y^2 + 10y + \left(\frac{10}{2}\right)^2 = -13 + \left(\frac{4}{2}\right)^2 + \left(\frac{10}{2}\right)^2$ $\Leftrightarrow x^2 - 4x + 4 + y^2 + 10y + 25 = -13 + 4 + 25 \Leftrightarrow (x-2)^2 + (y+5)^2 = 16$.
Thus, the center is $(2, -5)$, and the radius is 4.

87. Completing the square gives $x^2 + y^2 + x = 0 \Leftrightarrow x^2 + x + \left(\frac{1}{2}\right)^2 + y^2 = \left(\frac{1}{2}\right)^2 \Leftrightarrow x^2 + x + \frac{1}{4} + y^2 = \frac{1}{4} \Leftrightarrow$ $\left(x + \frac{1}{2}\right)^2 + y^2 = \frac{1}{4}$. Thus, the circle has center $\left(-\frac{1}{2}, 0\right)$ and radius $\frac{1}{2}$.

89. Completing the square gives $x^2 + y^2 - \frac{1}{2}x + \frac{1}{2}y = \frac{1}{8} \Leftrightarrow x^2 - \frac{1}{2}x + \left(\frac{-1/2}{2}\right)^2 + y^2 + \frac{1}{2}y + \left(\frac{1/2}{2}\right)^2 = \frac{1}{8} + \left(\frac{-1/2}{2}\right)^2 + \left(\frac{1/2}{2}\right)^2$ $\Leftrightarrow x^2 - \frac{1}{2}x + \frac{1}{16} + y^2 + \frac{1}{2}y + \frac{1}{16} = \frac{1}{8} + \frac{1}{16} + \frac{1}{16} = \frac{2}{8} = \frac{1}{4} \Leftrightarrow \left(x - \frac{1}{4}\right)^2 + \left(y + \frac{1}{4}\right)^2 = \frac{1}{4}$. Thus, the circle has center $\left(\frac{1}{4}, -\frac{1}{4}\right)$ and radius $\frac{1}{2}$.

91. Completing the square gives $x^2 + y^2 + 4x - 10y = 21 \Leftrightarrow$ $x^2 + 4x + \left(\frac{4}{2}\right)^2 + y^2 - 10y + \left(\frac{-10}{2}\right)^2 = 21 + \left(\frac{4}{2}\right)^2 + \left(\frac{-10}{2}\right)^2 \Leftrightarrow (x+2)^2 + (y-5)^2 = 21 + 4 + 25 = 50$.
Thus, the circle has center $(-2, 5)$ and radius $\sqrt{50} = 5\sqrt{2}$.

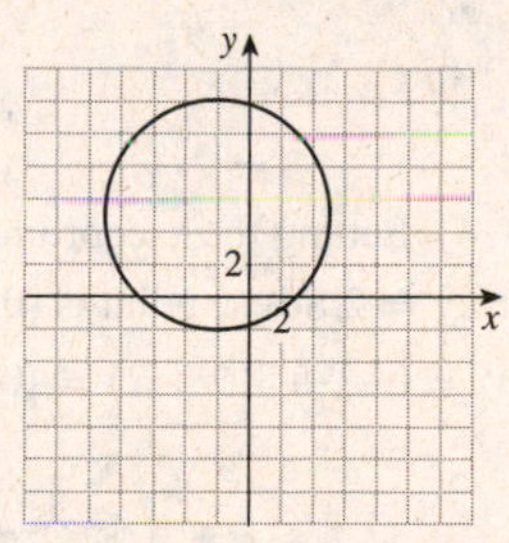

93. Completing the square gives $x^2 + y^2 + 6x - 12y + 45 = 0$ $\Leftrightarrow (x+3)^2 + (y-6)^2 = -45 + 9 + 36 = 0$. Thus, the center is $(-3, 6)$, and the radius is 0. This is a degenerate circle whose graph consists only of the point $(-3, 6)$.

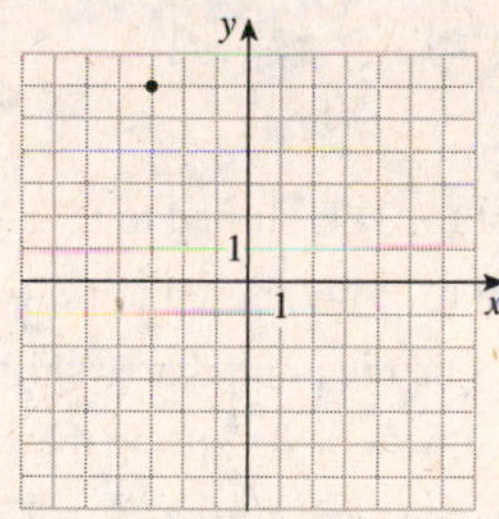

95. x-axis symmetry: $(-y) = x^4 + x^2 \Leftrightarrow y = -x^4 - x^2$, which is not the same as $y = x^4 + x^2$, so the graph is not symmetric with respect to the x-axis.
y-axis symmetry: $y = (-x)^4 + (-x)^2 = x^4 + x^2$, so the graph is symmetric with respect to the y-axis.
Origin symmetry: $(-y) = (-x)^4 + (-x)^2 \Leftrightarrow -y = x^4 + x^2$, which is not the same as $y = x^4 + x^2$, so the graph is not symmetric with respect to the origin.

97. x-axis symmetry: $(-y) = x^3 + 10x \Leftrightarrow y = -x^3 - 10x$, which is not the same as $y = x^3 + 10x$, so the graph is not symmetric with respect to the x-axis.
y-axis symmetry: $y = (-x)^3 + 10(-x) \Leftrightarrow y = -x^3 - 10x$, which is not the same as $y = x^3 + 10x$, so the graph is not symmetric with respect to the y-axis.
Origin symmetry: $(-y) = (-x)^3 + 10(-x) \Leftrightarrow -y = -x^3 - 10x \Leftrightarrow y = x^3 + 10x$, so the graph is symmetric with respect to the origin.

99. x-axis symmetry: $x^4(-y)^4 + x^2(-y)^2 = 1 \Leftrightarrow x^4y^4 + x^2y^2 = 1$, so the graph is symmetric with respect to the x-axis.
y-axis symmetry: $(-x)^4 y^4 + (-x)^2 y^2 = 1 \Leftrightarrow x^4y^4 + x^2y^2 = 1$, so the graph is symmetric with respect to the y-axis.
Origin symmetry: $(-x)^4(-y)^4 + (-x)^2(-y)^2 = 1 \Leftrightarrow x^4y^4 + x^2y^2 = 1$, so the graph is symmetric with respect to the origin.

101. Symmetric with respect to the y-axis.

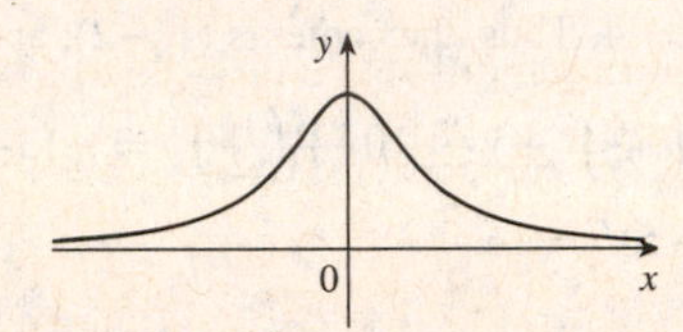

103. Symmetric with respect to the origin.

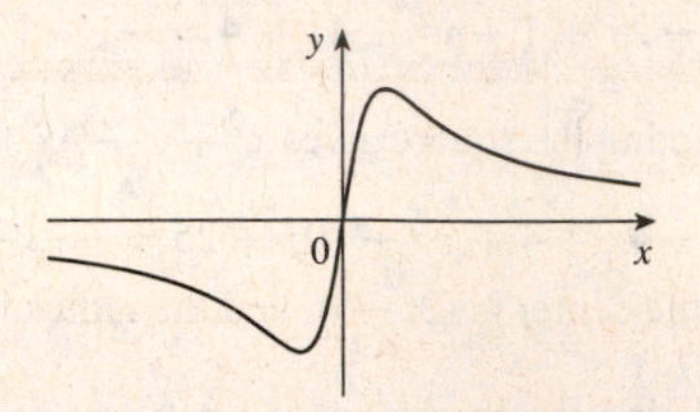

105. $\{(x, y) \mid x^2 + y^2 \leq 1\}$. This is the set of points inside (and on) the circle $x^2 + y^2 = 1$.

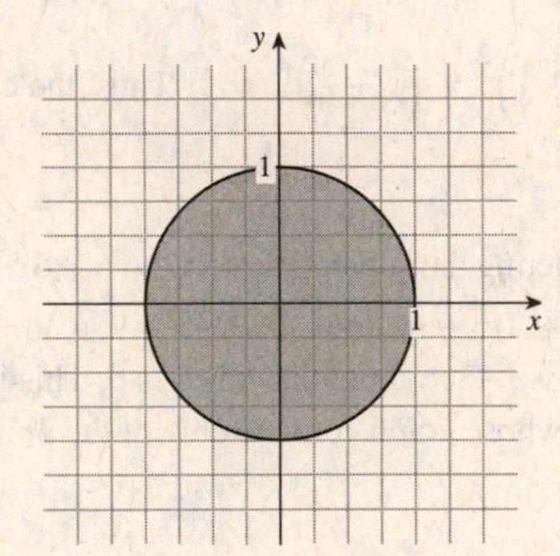

107. Completing the square gives $x^2 + y^2 - 4y - 12 = 0$
$\Leftrightarrow \quad x^2 + y^2 - 4y + \left(\frac{-4}{2}\right)^2 = 12 + \left(\frac{-4}{2}\right)^2 \Leftrightarrow$
$x^2 + (y - 2)^2 = 16$. Thus, the center is $(0, 2)$, and the radius is 4. So the circle $x^2 + y^2 = 4$, with center $(0, 0)$ and radius 2, sits completely inside the larger circle. Thus, the area is $\pi 4^2 - \pi 2^2 = 16\pi - 4\pi = 12\pi$.

109. **(a)** The point $(5, 3)$ is shifted to $(5 + 3, 3 + 2) = (8, 5)$.
(b) The point (a, b) is shifted to $(a + 3, b + 2)$.
(c) Let (x, y) be the point that is shifted to $(3, 4)$. Then $(x + 3, y + 2) = (3, 4)$. Setting the x-coordinates equal, we get $x + 3 = 3 \Leftrightarrow x = 0$. Setting the y-coordinates equal, we get $y + 2 = 4 \Leftrightarrow y = 2$.So the point is $(0, 2)$.
(d) $A = (-5, -1)$, so $A' = (-5 + 3, -1 + 2) = (-2, 1)$; $B = (-3, 2)$, so $B' = (-3 + 3, 2 + 2) = (0, 4)$; and $C = (2, 1)$, so $C' = (2 + 3, 1 + 2) = (5, 3)$.

111. **(a)** In 1980 inflation was 14%; in 1990, it was 6%; in 1999, it was 2%.
(b) Inflation exceeded 6% from 1975 to 1976 and from 1978 to 1982.
(c) Between 1980 and 1985 the inflation rate generally decreased. Between 1987 and 1992 the inflation rate generally increased.
(d) The highest rate was about 14% in 1980. The lowest was about 1% in 2002.

113. Completing the square gives $x^2+y^2+ax+by+c=0 \Leftrightarrow x^2+ax+\left(\frac{a}{2}\right)^2+y^2+by+\left(\frac{b}{2}\right)^2=-c+\left(\frac{a}{2}\right)^2+\left(\frac{b}{2}\right)^2$ $\Leftrightarrow \left(x+\frac{a}{2}\right)^2+\left(y+\frac{b}{2}\right)^2=-c+\frac{a^2+b^2}{4}$. This equation represents a circle only when $-c+\frac{a^2+b^2}{4}>0$. This equation represents a point when $-c+\frac{a^2+b^2}{4}=0$, and this equation represents the empty set when $-c+\frac{a^2+b^2}{4}<0$. When the equation represents a circle, the center is $\left(-\frac{a}{2},-\frac{b}{2}\right)$, and the radius is $\sqrt{-c+\frac{a^2+b^2}{4}}=\frac{1}{2}\sqrt{a^2+b^2-4ac}$.

1.3 LINES

1. We find the "steepness" or slope of a line passing through two points by dividing the difference in the y-coordinates of these points by the difference in the x-coordinates. So the line passing through the points $(0,1)$ and $(2,5)$ has slope $\frac{5-1}{2-0}=2$.

3. The point-slope form of the equation of the line with slope 3 passing through the point $(1,2)$ is $y-2=3(x-1)$.

5. The slope of a horizontal line is 0. The equation of the horizontal line passing through $(2,3)$ is $y=3$.

7. (a) Yes, the graph of $y=-3$ is a horizontal line 3 units below the x-axis.

(b) Yes, the graph of $x=-3$ is a vertical line 3 units to the left of the y-axis.

(c) No, a line perpendicular to a horizontal line is vertical and has undefined slope.

(d) Yes, a line perpendicular to a vertical line is horizontal and has slope 0.

9. $m=\frac{y_2-y_1}{x_2-x_1}=\frac{0-2}{0-(-1)}=\frac{-2}{1}=-2$

11. $m=\frac{y_2-y_1}{x_2-x_1}=\frac{-1-(-2)}{7-2}=\frac{1}{5}$

13. $m=\frac{y_2-y_1}{x_2-x_1}=\frac{4-4}{0-5}=0$

15. $m=\frac{y_2-y_1}{x_2-x_1}=\frac{-5-(-2)}{6-10}=\frac{-3}{-4}=\frac{3}{4}$

17. For ℓ_1, we find two points, $(-1,2)$ and $(0,0)$ that lie on the line. Thus the slope of ℓ_1 is $m=\frac{y_2-y_1}{x_2-x_1}=\frac{2-0}{-1-0}=-2$. For ℓ_2, we find two points $(0,2)$ and $(2,3)$. Thus, the slope of ℓ_2 is $m=\frac{y_2-y_1}{x_2-x_1}=\frac{3-2}{2-0}=\frac{1}{2}$. For ℓ_3 we find the points $(2,-2)$ and $(3,1)$. Thus, the slope of ℓ_3 is $m=\frac{y_2-y_1}{x_2-x_1}=\frac{1-(-2)}{3-2}=3$. For ℓ_4, we find the points $(-2,-1)$ and $(2,-2)$. Thus, the slope of ℓ_4 is $m=\frac{y_2-y_1}{x_2-x_1}=\frac{-2-(-1)}{2-(-2)}=\frac{-1}{4}=-\frac{1}{4}$.

19. First we find two points $(0,4)$ and $(4,0)$ that lie on the line. So the slope is $m=\frac{0-4}{4-0}=-1$. Since the y-intercept is 4, the equation of the line is $y=mx+b=-1x+4$. So $y=-x+4$, or $x+y-4=0$.

21. We choose the two intercepts as points, $(0,-3)$ and $(2,0)$. So the slope is $m=\frac{0-(-3)}{2-0}=\frac{3}{2}$. Since the y-intercept is -3, the equation of the line is $y=mx+b=\frac{3}{2}x-3$, or $3x-2y-6=0$.

23. Using $y=mx+b$, we have $y=3x+(-2)$ or $3x-y-2=0$.

25. Using the equation $y-y_1=m(x-x_1)$, we get $y-3=5(x-2)\Leftrightarrow -5x+y=-7\Leftrightarrow 5x-y-7=0$.

27. Using the equation $y-y_1=m(x-x_1)$, we get $y-7=\frac{2}{3}(x-1)\Leftrightarrow 3y-21=2x-2\Leftrightarrow -2x+3y=19\Leftrightarrow 2x-3y+19=0$.

29. First we find the slope, which is $m=\frac{y_2-y_1}{x_2-x_1}=\frac{6-1}{1-2}=\frac{5}{-1}=-5$. Substituting into $y-y_1=m(x-x_1)$, we get $y-6=-5(x-1)\Leftrightarrow y-6=-5x+5\Leftrightarrow 5x+y-11=0$.

31. We are given two points, $(-2, 5)$ and $(-1, -3)$. Thus, the slope is $m = \dfrac{y_2 - y_1}{x_2 - x_1} = \dfrac{-3 - 5}{-1 - (-2)} = \dfrac{-8}{1} = -8$. Substituting into $y - y_1 = m(x - x_1)$, we get $y - 5 = -8[x - (-2)] \Leftrightarrow y = -8x - 11$ or $8x + y + 11 = 0$.

33. We are given two points, $(1, 0)$ and $(0, -3)$. Thus, the slope is $m = \dfrac{y_2 - y_1}{x_2 - x_1} = \dfrac{-3 - 0}{0 - 1} = \dfrac{-3}{-1} = 3$. Using the y-intercept, we have $y = 3x + (-3)$ or $y = 3x - 3$ or $3x - y - 3 = 0$.

35. Since the equation of a line with slope 0 passing through (a, b) is $y = b$, the equation of this line is $y = 3$.

37. Since the equation of a line with undefined slope passing through (a, b) is $x = a$, the equation of this line is $x = 2$.

39. Any line parallel to $y = 3x - 5$ has slope 3. The desired line passes through $(1, 2)$, so substituting into $y - y_1 = m(x - x_1)$, we get $y - 2 = 3(x - 1) \Leftrightarrow y = 3x - 1$ or $3x - y - 1 = 0$.

41. Since the equation of a horizontal line passing through (a, b) is $y = b$, the equation of the horizontal line passing through $(4, 5)$ is $y = 5$.

43. Since $x + 2y = 6 \Leftrightarrow 2y = -x + 6 \Leftrightarrow y = -\frac{1}{2}x + 3$, the slope of this line is $-\frac{1}{2}$. Thus, the line we seek is given by $y - (-6) = -\frac{1}{2}(x - 1) \Leftrightarrow 2y + 12 = -x + 1 \Leftrightarrow x + 2y + 11 = 0$.

45. Any line parallel to $x = 5$ has undefined slope and an equation of the form $x = a$. Thus, an equation of the line is $x = -1$.

47. First find the slope of $2x + 5y + 8 = 0$. This gives $2x + 5y + 8 = 0 \Leftrightarrow 5y = -2x - 8 \Leftrightarrow y = -\frac{2}{5}x - \frac{8}{5}$. So the slope of the line that is perpendicular to $2x + 5y + 8 = 0$ is $m = -\dfrac{1}{-2/5} = \frac{5}{2}$. The equation of the line we seek is $y - (-2) = \frac{5}{2}(x - (-1)) \Leftrightarrow 2y + 4 = 5x + 5 \Leftrightarrow 5x - 2y + 1 = 0$.

49. First find the slope of the line passing through $(2, 5)$ and $(-2, 1)$. This gives $m = \dfrac{1 - 5}{-2 - 2} = \dfrac{-4}{-4} = 1$, and so the equation of the line we seek is $y - 7 = 1(x - 1) \Leftrightarrow x - y + 6 = 0$.

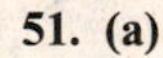

51. (a)

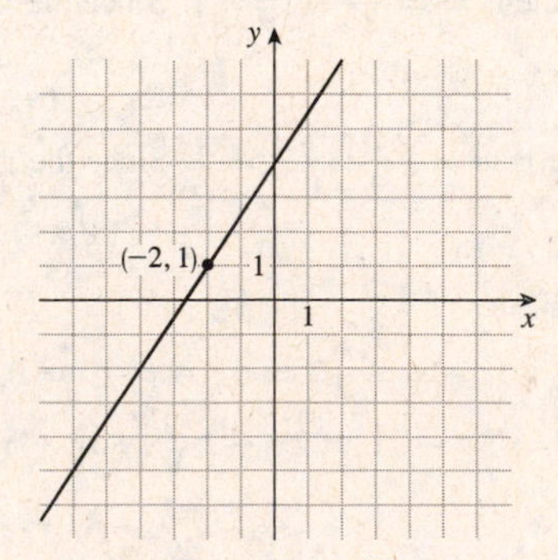

(b) $y - 1 = \frac{3}{2}(x - (-2)) \Leftrightarrow 2y - 2 = 3(x + 2) \Leftrightarrow 2y - 2 = 3x + 6 \Leftrightarrow 3x - 2y + 8 = 0$.

53.

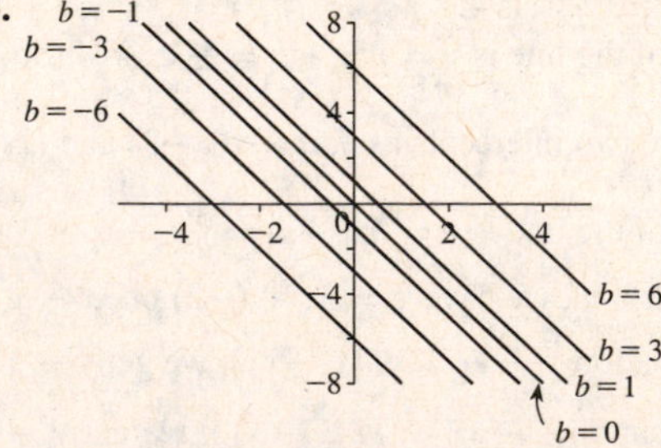

$y = -2x + b$, $b = 0, \pm 1, \pm 3, \pm 6$. They have the same slope, so they are parallel.

55.

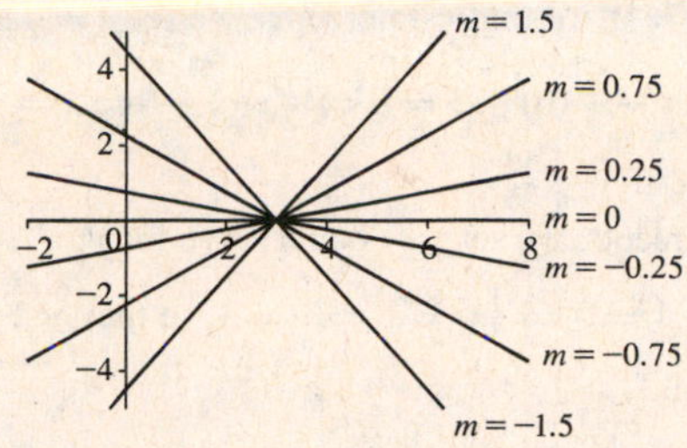

$y = m(x-3)$, $m = 0, \pm 0.25, \pm 0.75, \pm 1.5$. Each of the lines contains the point $(3, 0)$ because the point $(3, 0)$ satisfies each equation $y = m(x-3)$. Since $(3, 0)$ is on the x-axis, we could also say that they all have the same x-intercept.

57. $y = 3 - x = -x + 3$. So the slope is -1 and the y-intercept is 3.

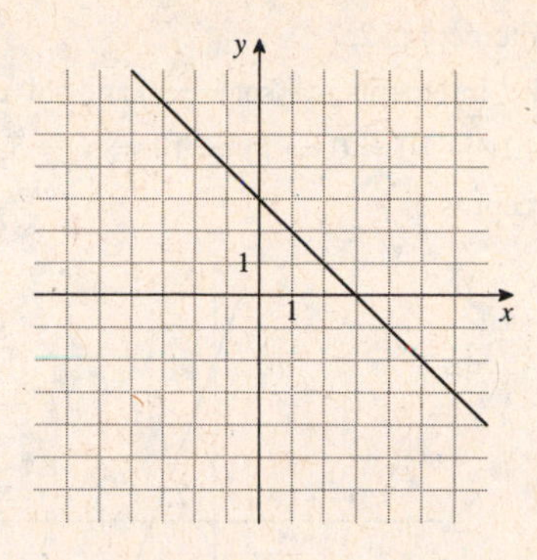

59. $-2x + y = 7 \Leftrightarrow y = 2x + 7$. So the slope is 2 and the y-intercept is 7.

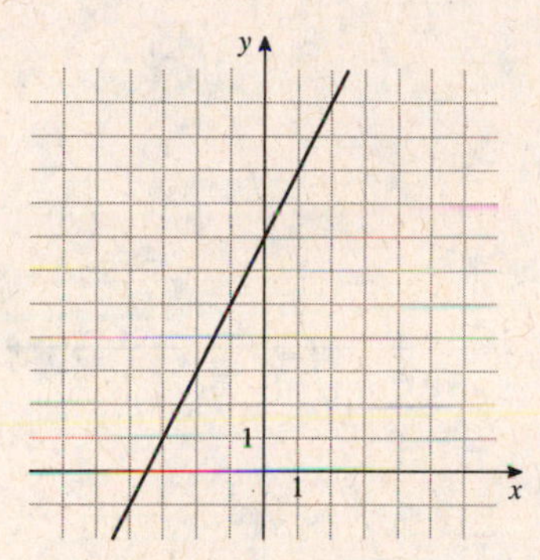

61. $4x + 5y = 10 \Leftrightarrow 5y = -4x + 10 \Leftrightarrow y = -\frac{4}{5}x + 2$. So the slope is $-\frac{4}{5}$ and the y-intercept is 2.

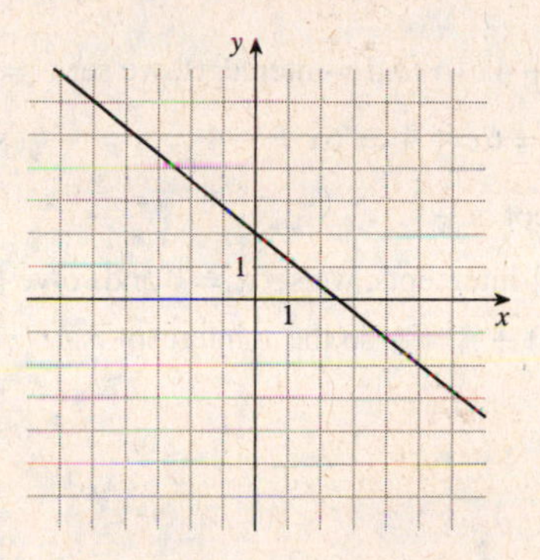

63. $y = 4$ can also be expressed as $y = 0x + 4$. So the slope is 0 and the y-intercept is 4.

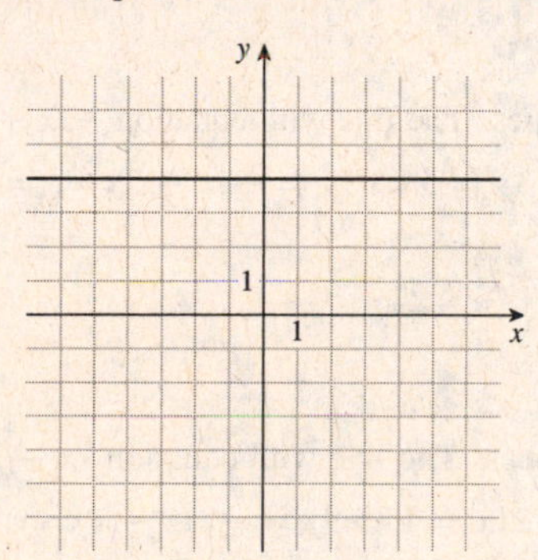

65. $x = 3$ cannot be expressed in the form $y = mx + b$. So the slope is undefined, and there is no y-intercept. This is a vertical line.

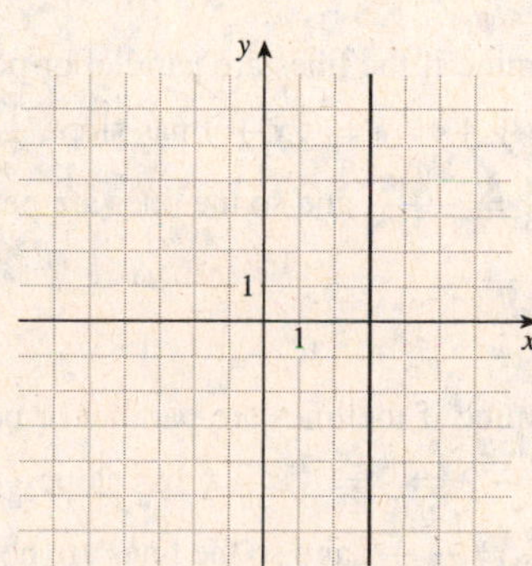

67. $5x + 2y - 10 = 0$. To find x-intercepts, we set $y = 0$ and solve for x: $5x + 2(0) - 10 = 0 \Leftrightarrow 5x = 10 \Leftrightarrow x = 2$, so the x-intercept is 2.

To find y-intercepts, we set $x = 0$ and solve for y: $5(0) + 2y - 10 = 0 \Leftrightarrow 2y = 10 \Leftrightarrow y = 5$, so the y-intercept is 5.

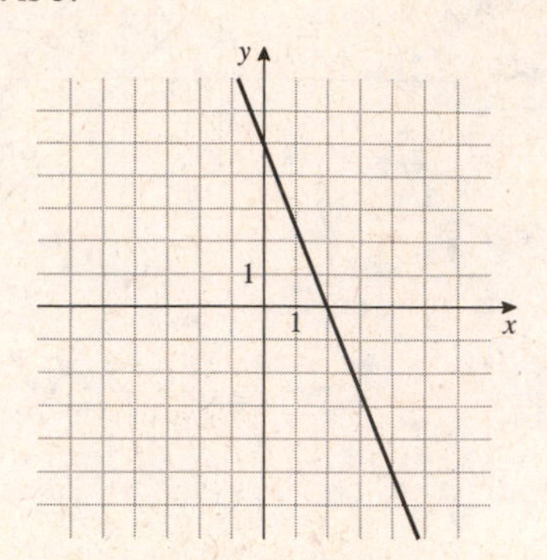

69. $\frac{1}{2}x - \frac{1}{3}y + 1 = 0$. To find x-intercepts, we set $y = 0$ and solve for x: $\frac{1}{2}x - \frac{1}{3}(0) + 1 = 0 \Leftrightarrow \frac{1}{2}x = -1 \Leftrightarrow x = -2$, so the x-intercept is -2.

To find y-intercepts, we set $x = 0$ and solve for y: $\frac{1}{2}(0) - \frac{1}{3}y + 1 = 0 \Leftrightarrow \frac{1}{3}y = 1 \Leftrightarrow y = 3$, so the y-intercept is 3.

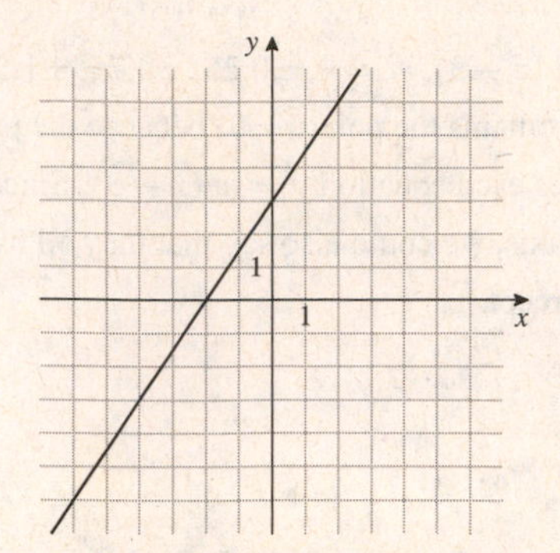

71. $y = 6x + 4$. To find x-intercepts, we set $y = 0$ and solve for x: $0 = 6x + 4 \Leftrightarrow 6x = -4 \Leftrightarrow x = -\frac{2}{3}$, so the x-intercept is $-\frac{2}{3}$.

To find y-intercepts, we set $x = 0$ and solve for y: $y = 6(0) + 4 = 4$, so the y-intercept is 4.

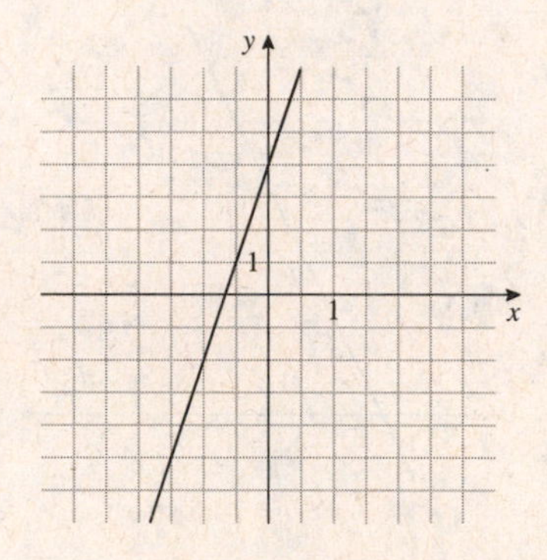

73. To determine if the lines are parallel or perpendicular, we find their slopes. The line with equation $y = 2x + 3$ has slope 2. The line with equation $2y - 4x - 5 = 0 \Leftrightarrow 2y = 4x + 5 \Leftrightarrow y = 2x + \frac{5}{2}$ also has slope 2, and so the lines are parallel.

75. To determine if the lines are parallel or perpendicular, we find their slopes. The line with equation $-3x + 4y = 4 \Leftrightarrow 4y = 3x + 4 \Leftrightarrow y = \frac{3}{4}x + 1$ has slope $\frac{3}{4}$. The line with equation $4x + 3y = 5 \Leftrightarrow 3y = -4x + 5 \Leftrightarrow y = -\frac{4}{3}x + \frac{5}{3}$ has slope $-\frac{4}{3} = -\frac{1}{3/4}$, and so the lines are perpendicular.

77. To determine if the lines are parallel or perpendicular, we find their slopes. The line with equation $7x - 3y = 2 \Leftrightarrow 3y = 7x - 2 \Leftrightarrow y = \frac{7}{3}x - \frac{2}{3}$ has slope $\frac{7}{3}$. The line with equation $9y + 21x = 1 \Leftrightarrow 9y = -21x - 1 \Leftrightarrow y = -\frac{7}{3} - \frac{1}{9}$ has slope $-\frac{7}{3} \neq -\frac{1}{7/3}$, and so the lines are neither parallel nor perpendicular.

79. We first plot the points to find the pairs of points that determine each side. Next we find the slopes of opposite sides. The slope of AB is $\frac{4-1}{7-1} = \frac{3}{6} = \frac{1}{2}$, and the slope of DC is $\frac{10-7}{5-(-1)} = \frac{3}{6} = \frac{1}{2}$. Since these slope are equal, these two sides are parallel. The slope of AD is $\frac{7-1}{-1-1} = \frac{6}{-2} = -3$, and the slope of BC is $\frac{10-4}{5-7} = \frac{6}{-2} = -3$. Since these slope are equal, these two sides are parallel. Hence $ABCD$ is a parallelogram.

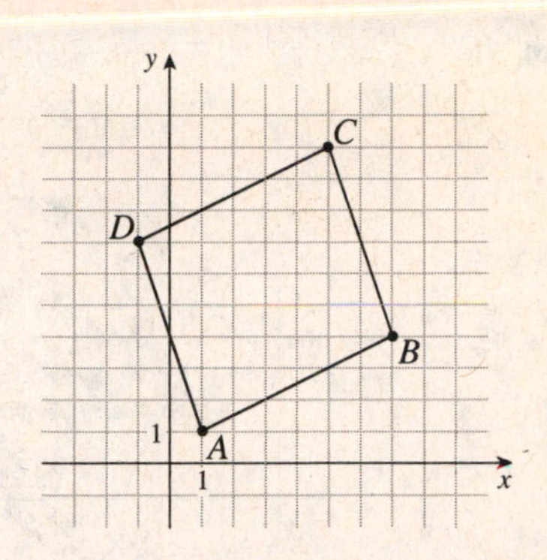

81. We first plot the points to find the pairs of points that determine each side. Next we find the slopes of opposite sides. The slope of AB is $\frac{3-1}{11-1} = \frac{2}{10} = \frac{1}{5}$ and the slope of DC is $\frac{6-8}{0-10} = \frac{-2}{-10} = \frac{1}{5}$. Since these slope are equal, these two sides are parallel. Slope of AD is $\frac{6-1}{0-1} = \frac{5}{-1} = -5$, and the slope of BC is $\frac{3-8}{11-10} = \frac{-5}{1} = -5$. Since these slope are equal, these two sides are parallel. Since (slope of AB) $\times$ (slope of AD) $= \frac{1}{5} \times (-5) = -1$, the first two sides are each perpendicular to the second two sides. So the sides form a rectangle.

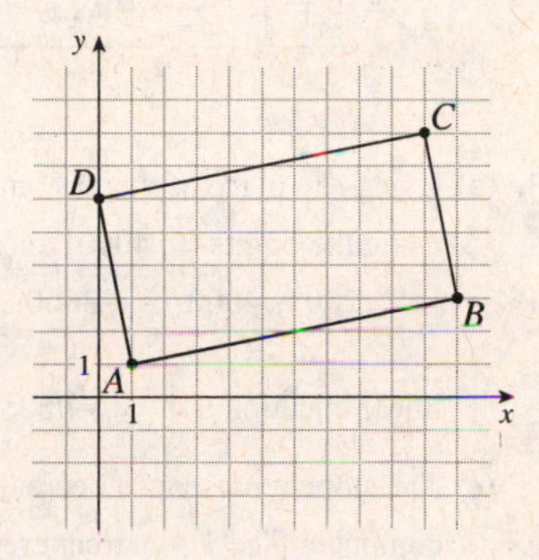

83. We need the slope and the midpoint of the line AB. The midpoint of AB is $\left(\frac{1+7}{2}, \frac{4-2}{2}\right) = (4, 1)$, and the slope of AB is $m = \frac{-2-4}{7-1} = \frac{-6}{6} = -1$. The slope of the perpendicular bisector will have slope $\frac{-1}{m} = \frac{-1}{-1} = 1$. Using the point-slope form, the equation of the perpendicular bisector is $y - 1 = 1(x - 4)$ or $x - y - 3 = 0$.

85. **(a)** We start with the two points $(a, 0)$ and $(0, b)$. The slope of the line that contains them is $\frac{b-0}{0-a} = -\frac{b}{a}$. So the equation of the line containing them is $y = -\frac{b}{a}x + b$ (using the slope-intercept form). Dividing by b (since $b \neq 0$) gives $\frac{y}{b} = -\frac{x}{a} + 1 \Leftrightarrow \frac{x}{a} + \frac{y}{b} = 1$.

(b) Setting $a = 6$ and $b = -8$, we get $\frac{x}{6} + \frac{y}{-8} = 1 \Leftrightarrow 4x - 3y = 24 \Leftrightarrow 4x - 3y - 24 = 0$.

87. **(a)** The slope represents an increase of 0.02° C every year. The T -intercept is the average surface temperature in 1950, or 15° C.

(b) In 2050, $t = 2050 - 1950 = 100$, so $T = 0.02(100) + 15 = 17$ degrees Celsius.

89. (a)

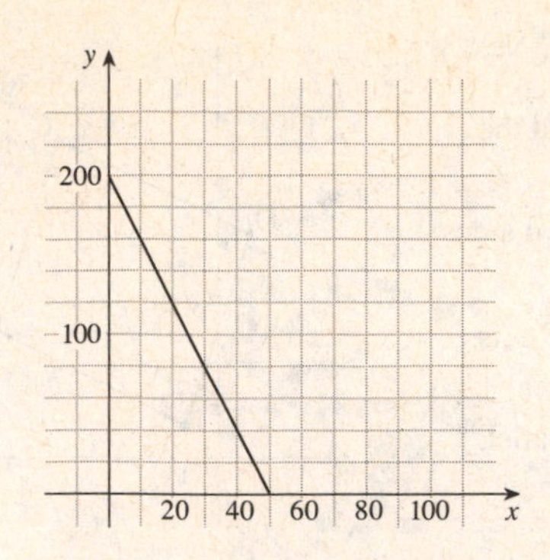

(b) The slope, -4, represents the decline in number of spaces sold for each \$1 increase in rent. The y-intercept is the number of spaces at the flea market, 200, and the x-intercept is the cost per space when the manager rents no spaces, \$50.

91. (a)

C	$-30°$	$-20°$	$-10°$	$0°$	$10°$	$20°$	$30°$
F	$-22°$	$-4°$	$14°$	$32°$	$50°$	$68°$	$86°$

(b) Substituting a for both F and C, we have $a = \frac{9}{5}a + 32 \Leftrightarrow -\frac{4}{5}a = 32 \Leftrightarrow a = -40°$. Thus both scales agree at $-40°$.

93. (a) Using t in place of x and V in place of y, we find the slope of the line using the points $(0, 4000)$ and $(4, 200)$. Thus, the slope is $m = \dfrac{200 - 4000}{4 - 0} = \dfrac{-3800}{4} = -950$. Using the V-intercept, the linear equation is $V = -950t + 4000$.

(b)

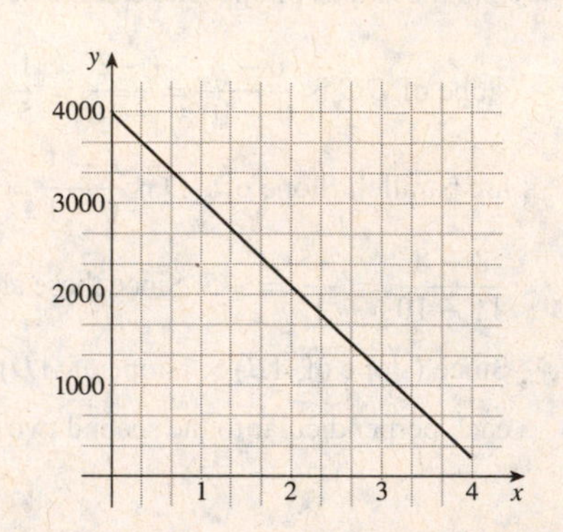

(c) The slope represents a decrease of \$950 each year in the value of the computer. The V -intercept represents the cost of the computer.

(d) When $t = 3$, the value of the computer is given by $V = -950\,(3) + 4000 = 1150$.

95. The temperature is increasing at a constant rate when the slope is positive, decreasing at a constant rate when the slope is negative, and constant when the slope is 0.

1.4 SOLVING QUADRATIC EQUATIONS

1. (a) The Quadratic Formula states that $x = \dfrac{-b \pm \sqrt{b^2 - 4ac}}{2a}$.

(b) In the equation $\frac{1}{2}x^2 - x - 4 = 0$, $a = \frac{1}{2}$, $b = -1$, and $c = -4$. So, the solution of the equation is

$$x = \frac{-(-1) \pm \sqrt{(-1)^2 - 4\left(\frac{1}{2}\right)(-4)}}{2\left(\frac{1}{2}\right)} = \frac{1 \pm 3}{1} = -2 \text{ or } 4.$$

3. For the quadratic equation $ax^2 + bx + c = 0$ the discriminant is $D = b^2 - 4ac$. If $D > 0$, the equation has two real solutions; if $D = 0$, the equation has one real solution; and if $D < 0$, the equation has no real solution.

5. $x^2 - 8x + 15 = 0 \Leftrightarrow (x - 3)(x - 5) = 0 \Leftrightarrow x - 3 = 0$ or $x - 5 = 0$. Thus, $x = 3$ or $x = 5$.

7. $x^2 - x = 6 \Leftrightarrow x^2 - x - 6 = 0 \Leftrightarrow (x + 2)(x - 3) = 0 \Leftrightarrow x + 2 = 0$ or $x - 3 = 0$. Thus, $x = -2$ or $x = 3$.

9. $5x^2 - 9x - 2 = 0 \Leftrightarrow (5x + 1)(x - 2) = 0 \Leftrightarrow 5x + 1 = 0$ or $x - 2 = 0$. Thus, $x = -\frac{1}{5}$ or $x = 2$.

11. $2s^2 = 5s + 3 \Leftrightarrow 2s^2 - 5s - 3 = 0 \Leftrightarrow (2s + 1)(s - 3) = 0 \Leftrightarrow 2s + 1 = 0$ or $s - 3 = 0$. Thus, $s = -\frac{1}{2}$ or $s = 3$.

13. $12z^2 - 44z = 45 \Leftrightarrow 12z^2 - 44z - 45 = 0 \Leftrightarrow (6z + 5)(2z - 9) = 0 \Leftrightarrow 6z + 5 = 0$ or $2z - 9 = 0$. Thus, $z = -\frac{5}{6}$ or $z = \frac{9}{2}$.

15. $x^2 = 5(x+100) \Leftrightarrow x^2 = 5x + 500 \Leftrightarrow x^2 - 5x - 500 = 0 \Leftrightarrow (x-25)(x+20) = 0 \Leftrightarrow x - 25 = 0$ or $x + 20 = 0$. Thus, $x = 25$ or $x = -20$.

17. $x^2 - 8x + 1 = 0 \Leftrightarrow x^2 - 8x = -1 \Leftrightarrow x^2 - 8x + 16 = -1 + 16 \Leftrightarrow (x-4)^2 = 15 \Leftrightarrow x - 4 = \pm\sqrt{15} \Leftrightarrow x = 4 \pm \sqrt{15}$.

19. $x^2 - 6x - 11 = 0 \Leftrightarrow x^2 - 6x = 11 \Leftrightarrow x^2 - 6x + 9 = 11 + 9 \Leftrightarrow (x-3)^2 = 20 \Rightarrow x - 3 = \pm 2\sqrt{5} \Leftrightarrow x = 3 \pm 2\sqrt{5}$.

21. $x^2 + x - \frac{3}{4} = 0 \Leftrightarrow x^2 + x = \frac{3}{4} \Leftrightarrow x^2 + x + \frac{1}{4} = \frac{3}{4} + \frac{1}{4} \Leftrightarrow \left(x + \frac{1}{2}\right)^2 = 1 \Rightarrow x + \frac{1}{2} = \pm 1 \Leftrightarrow x = -\frac{1}{2} \pm 1$. So $x = -\frac{1}{2} - 1 = -\frac{3}{2}$ or $x = -\frac{1}{2} + 1 = \frac{1}{2}$.

23. $x^2 + 22x + 21 = 0 \Leftrightarrow x^2 + 22x = -21 \Leftrightarrow x^2 + 22x + 11^2 = -21 + 11^2 = -21 + 121 \Leftrightarrow (x+11)^2 = 100 \Rightarrow x + 11 = \pm 10 \Leftrightarrow x = -11 \pm 10$. Thus, $x = -1$ or $x = -21$.

25. $5x^2 + 10x - 7 = 0 \Leftrightarrow x^2 + 2x - \frac{7}{5} = 0 \Leftrightarrow x^2 + 2x = \frac{7}{5} \Leftrightarrow x^2 + 2x + 1 = \frac{7}{5} + 1 \Leftrightarrow (x+1)^2 = \frac{12}{5} \Leftrightarrow x + 1 = \pm\sqrt{\frac{12}{5}}$ $\Leftrightarrow x = -1 \pm \frac{2\sqrt{15}}{5}$

27. $2x^2 + 7x + 4 = 0 \Leftrightarrow x^2 + \frac{7}{2}x + 2 = 0 \Leftrightarrow x^2 + \frac{7}{2}x = -2 \Leftrightarrow x^2 + \frac{7}{2}x + \frac{49}{16} = -2 + \frac{49}{16} \Leftrightarrow \left(x + \frac{7}{4}\right)^2 = \frac{17}{16} \Leftrightarrow x + \frac{7}{4} = \pm\sqrt{\frac{17}{16}} \Leftrightarrow x = -\frac{7}{4} \pm \frac{\sqrt{17}}{4}$.

29. $x^2 - 8x + 12 = 0 \Leftrightarrow (x-2)(x-6) = 0 \Leftrightarrow x = 2$ or $x = 6$.

31. $x^2 + 8x - 20 = 0 \Leftrightarrow (x+10)(x-2) = 0 \Leftrightarrow x = -10$ or $x = 2$.

33. $2x^2 + x - 3 = 0 \Leftrightarrow (x-1)(2x+3) = 0 \Leftrightarrow x - 1 = 0$ or $2x + 3 = 0$. If $x - 1 = 0$, then $x = 1$; if $2x + 3 = 0$, then $x = -\frac{3}{2}$.

35. $3x^2 + 6x - 5 = 0 \Leftrightarrow x^2 + 2x - \frac{5}{3} = 0 \Leftrightarrow x^2 + 2x = \frac{5}{3} \Leftrightarrow x^2 + 2x + 1 = \frac{5}{3} + 1 \Leftrightarrow (x+1)^2 = \frac{8}{3} \Rightarrow x + 1 = \pm\sqrt{\frac{8}{3}} \Leftrightarrow x = -1 \pm \frac{2\sqrt{6}}{3}$.

37. $x^2 - \frac{4}{3}x + \frac{4}{9} = 0 \Leftrightarrow 9x^2 - 12x + 4 = 0 \Leftrightarrow (3x-2)^2 = 0 \Leftrightarrow x = \frac{2}{3}$.

39. $4x^2 + 16x - 9 = 0 \Leftrightarrow (2x-1)(2x+9) = 0 \Leftrightarrow 2x - 1 = 0$ or $2x + 9 = 0$. If $2x - 1 = 0$, then $x = \frac{1}{2}$; if $2x + 9 = 0$, then $x = -\frac{9}{2}$.

41. $w^2 = 3(w-1) \Leftrightarrow w^2 - 3w + 3 = 0 \Rightarrow w = \dfrac{-(-3) \pm \sqrt{(-3)^2 - 4(1)(3)}}{2(1)} = \dfrac{3 \pm \sqrt{9-12}}{2} = \dfrac{3 \pm \sqrt{-3}}{2}$. Since the discriminant is less than 0, the equation has no real solution.

43. $10y^2 - 16y + 5 = 0 \Rightarrow$

$$x = \frac{-b \pm \sqrt{b^2 - 4ac}}{2a} = \frac{-(-16) \pm \sqrt{(-16)^2 - 4(10)(5)}}{2(10)} = \frac{16 \pm \sqrt{256 - 200}}{20} = \frac{16 \pm \sqrt{56}}{20} = \frac{8 \pm \sqrt{14}}{10}.$$

45. $3x^2 + 2x + 2 = 0 \Rightarrow x = \dfrac{-b \pm \sqrt{b^2 - 4ac}}{2a} = \dfrac{-(2) \pm \sqrt{(2)^2 - 4(3)(2)}}{2(3)} = \dfrac{-2 \pm \sqrt{4-24}}{6} = \dfrac{-2 \pm \sqrt{-20}}{6}$. Since the discriminant is less than 0, the equation has no real solution.

47. $x^2 - 0.011x - 0.064 = 0 \Rightarrow$

$$x = \frac{-(-0.011) \pm \sqrt{(-0.011)^2 - 4(1)(-0.064)}}{2(1)} = \frac{0.011 \pm \sqrt{0.000121 + 0.256}}{2} \approx \frac{0.011 \pm 0.506}{2}.$$

Thus, $x \approx \dfrac{0.011 + 0.506}{2} = 0.259$ or $x \approx \dfrac{0.011 - 0.506}{2} = -0.248$.

49. $x^2 - 2.450x + 1.501 = 0 \Rightarrow$

$$x = \frac{-(-2.450) \pm \sqrt{(-2.450)^2 - 4(1)(1.501)}}{2(1)} = \frac{2.450 \pm \sqrt{6.0025 - 6.004}}{2} = \frac{2.450 \pm \sqrt{-0.0015}}{2}.$$

Thus, there is no real solution.

51. $h = \frac{1}{2}gt^2 + v_0 t \Leftrightarrow \frac{1}{2}gt^2 + v_0 t - h = 0$. Using the Quadratic Formula,

$$t = \frac{-(v_0) \pm \sqrt{(v_0)^2 - 4\left(\frac{1}{2}g\right)(-h)}}{2\left(\frac{1}{2}g\right)} = \frac{-v_0 \pm \sqrt{v_0^2 + 2gh}}{g}.$$

53. $A = 2x^2 + 4xh \Leftrightarrow 2x^2 + 4xh - A = 0$. Using the Quadratic Formula,

$$x = \frac{-(4h) \pm \sqrt{(4h)^2 - 4(2)(-A)}}{2(2)} = \frac{-4h \pm \sqrt{16h^2 + 8A}}{4} = \frac{-4h \pm \sqrt{4(4h^2 + 2A)}}{4} = \frac{-4h \pm 2\sqrt{4h^2 + 2A}}{4}$$

$$= \frac{2\left(-2h \pm \sqrt{4h^2 + 2A}\right)}{4} = \frac{-2h \pm \sqrt{4h^2 + 2A}}{2}$$

55. $\dfrac{1}{s+a} + \dfrac{1}{s+b} = \dfrac{1}{c} \Leftrightarrow c(s+b) + c(s+a) = (s+a)(s+b) \Leftrightarrow cs + bc + cs + ac = s^2 + as + bs + ab \Leftrightarrow$ $s^2 + (a+b-2c)s + (ab - ac - bc) = 0$. Using the Quadratic Formula,

$$s = \frac{-(a+b-2c) \pm \sqrt{(a+b-2c)^2 - 4(1)(ab - ac - bc)}}{2(1)}$$

$$= \frac{-(a+b-2c) \pm \sqrt{a^2 + b^2 + 4c^2 + 2ab - 4ac - 4bc - 4ab + 4ac + 4bc}}{2}$$

$$= \frac{-(a+b-2c) \pm \sqrt{a^2 + b^2 + 4c^2 - 2ab}}{2}$$

57. $D = b^2 - 4ac = (-6)^2 - 4(1)(1) = 32$. Since D is positive, this equation has two real solutions.

59. $D = b^2 - 4ac = (2.20)^2 - 4(1)(1.21) = 4.84 - 4.84 = 0$. Since $D = 0$, this equation has one real solution.

61. $D = b^2 - 4ac = (5)^2 - 4(4)\left(\frac{13}{8}\right) = 25 - 26 = -1$. Since D is negative, this equation has no real solution.

63. $a^2x^2 + 2ax + 1 = 0 \Leftrightarrow (ax+1)^2 = 0 \Leftrightarrow ax + 1 = 0$. So $ax + 1 = 0$ then $ax = -1 \Leftrightarrow x = -\dfrac{1}{a}$.

65. We want to find the values of k that make the discriminant 0. Thus $k^2 - 4(4)(25) = 0 \Leftrightarrow k^2 = 400 \Leftrightarrow k = \pm 20$.

67. Let n be one number. Then the other number must be $55 - n$, since $n + (55 - n) = 55$. Because the product is 684, we have $(n)(55 - n) = 684 \Leftrightarrow 55n - n^2 = 684 \Leftrightarrow n^2 - 55n + 684 = 0 \Rightarrow$ $n = \frac{-(-55) \pm \sqrt{(-55)^2 - 4(1)(684)}}{2(1)} = \frac{55 \pm \sqrt{3025 - 2736}}{2} = \frac{55 \pm \sqrt{289}}{2} = \frac{55 \pm 17}{2}$. So $n = \frac{55+17}{2} = \frac{72}{2} = 36$ or $n = \frac{55-17}{2} = \frac{38}{2} = 19$. In either case, the two numbers are 19 and 36.

69. Let w be the width of the garden in feet. Then the length is $w + 10$. Thus $875 = w(w+10) \Leftrightarrow w^2 + 10w - 875 = 0 \Leftrightarrow$ $(w+35)(w-25) = 0$. So $w + 35 = 0$ in which case $w = -35$, which is not possible, or $w - 25 = 0$ and so $w = 25$. Thus the width is 25 feet and the length is 35 feet.

71. Let w be the width of the garden in feet. We use the perimeter to express the length l of the garden in terms of width. Since the perimeter is twice the width plus twice the length, we have $200 = 2w + 2l \Leftrightarrow 2l = 200 - 2w \Leftrightarrow l = 100 - w$. Using the formula for area, we have $2400 = w(100 - w) = 100w - w^2 \Leftrightarrow w^2 - 100w + 2400 = 0 \Leftrightarrow (w - 40)(w - 60) = 0$. So $w - 40 = 0 \Leftrightarrow w = 40$, or $w - 60 = 0 \Leftrightarrow w = 60$. If $w = 40$, then $l = 100 - 40 = 60$. And if $w = 60$, then $l = 100 - 60 = 40$. So the length is 60 feet and the width is 40 feet.

73. The shaded area is the sum of the area of a rectangle and the area of a triangle. So $A = y(1) + \frac{1}{2}(y)(y) = \frac{1}{2}y^2 + y$. We are given that the area is 1200 cm^2, so $1200 = \frac{1}{2}y^2 + y \Leftrightarrow y^2 + 2y - 2400 = 0 \Leftrightarrow (y+50)(y-48) = 0$. y is positive, so $y = 48$ cm.

75. Let x be the length of one side of the cardboard, so we start with a piece of cardboard x by x. When 4 inches are removed from each side, the base of the box is $x - 8$ by $x - 8$. Since the volume is 100 in^3, we get $4(x-8)^2 = 100 \Leftrightarrow x^2 - 16x + 64 = 25 \Leftrightarrow x^2 - 16x + 39 = 0 \Leftrightarrow (x-3)(x-13) = 0$. So $x = 3$ or $x = 13$. But $x = 3$ is not possible, since then the length of the base would be $3 - 8 = -5$, and all lengths must be positive. Thus $x = 13$, and the piece of cardboard is 13 inches by 13 inches.

77. Let w be the width of the lot in feet. Then the length is $w + 6$. Using the Pythagorean Theorem, we have $w^2 + (w+6)^2 = (174)^2 \Leftrightarrow w^2 + w^2 + 12w + 36 = 30{,}276 \Leftrightarrow 2w^2 + 12w - 30240 = 0 \Leftrightarrow w^2 + 6w - 15120 = 0 \Leftrightarrow (w+126)(w-120) = 0$. So either $w + 126 = 0$ in which case $w = -126$, which is not possible, or $w - 120 = 0$ in which case $w = 120$. Thus the width is 120 feet and the length is 126 feet.

79. Let x be the rate, in mi/h, at which the salesman drove between Ajax and Barrington.

Direction	Distance	Rate	Time
Ajax $\to$ Barrington	120	x	$\frac{120}{x}$
Barrington $\to$ Collins	150	$x + 10$	$\frac{150}{x+10}$

We have used the equation time $= \dfrac{\text{distance}}{\text{rate}}$ to fill in the "Time" column of the table. Since the second part of the trip took 6 minutes (or $\frac{1}{10}$ hour) more than the first, we can use the time column to get the equation $\dfrac{120}{x} + \dfrac{1}{10} = \dfrac{150}{x+10} \Rightarrow$ $120(10)(x+10) + x(x+10) = 150(10x) \Leftrightarrow 1200x + 12{,}000 + x^2 + 10x = 1500x \Leftrightarrow x^2 - 290x + 12{,}000 = 0 \Leftrightarrow$ $x = \frac{-(-290)\pm\sqrt{(-290)^2-4(1)(12{,}000)}}{2} = \frac{290\pm\sqrt{84{,}100-48{,}000}}{2} = \frac{290\pm\sqrt{36{,}100}}{2} = \frac{290\pm190}{2} = 145 \pm 95$. Hence, the salesman drove either 50 mi/h or 240 mi/h between Ajax and Barrington. (The first choice seems more likely!)

81. Let r be the rowing rate in km/h of the crew in still water. Then their rate upstream was $r - 3$ km/h, and their rate downstream was $r + 3$ km/h.

Direction	Distance	Rate	Time
Upstream	6	$r - 3$	$\frac{6}{r-3}$
Downstream	6	$r + 3$	$\frac{6}{r+3}$

Since the time to row upstream plus the time to row downstream was 2 hours 40 minutes $= \frac{8}{3}$ hour, we get the equation $\dfrac{6}{r-3} + \dfrac{6}{r+3} = \dfrac{8}{3} \Leftrightarrow 6(3)(r+3) + 6(3)(r-3) = 8(r-3)(r+3) \Leftrightarrow 18r + 54 + 18r - 54 = 8r^2 - 72 \Leftrightarrow$ $0 = 8r^2 - 36r - 72 = 4\left(2r^2 - 9r - 18\right) = 4(2r+3)(r-6) \Leftrightarrow 2r + 3 = 0$ or $r - 6 = 0$. If $2r + 3 = 0$, then $r = -\frac{3}{2}$, which is impossible because the rowing rate is positive. If $r - 6 = 0$, then $r = 6$. So the rate of the rowing crew in still water is 6 km/h.

83. Using $h_0 = 288$, we solve $0 = -16t^2 + 288$, for $t \geq 0$. So $0 = -16t^2 + 288 \Leftrightarrow 16t^2 = 288 \Leftrightarrow t^2 = 18 \Rightarrow t = \pm\sqrt{18} = \pm 3\sqrt{2}$. Thus it takes $3\sqrt{2} \approx 4.24$ seconds for the ball the hit the ground.

85. We are given $v_o = 40$ ft/s.

(a) Setting $h = 24$, we have $24 = -16t^2 + 40t \Leftrightarrow 16t^2 - 40t + 24 = 0 \Leftrightarrow 8\left(2t^2 - 5t + 3\right) = 0 \Leftrightarrow 8(2t-3)(t-1) = 0$ $\Leftrightarrow t = 1$ or $t = 1\frac{1}{2}$. Therefore, the ball reaches 24 feet in 1 second (ascending) and again after $1\frac{1}{2}$ seconds (descending).

(b) Setting $h = 48$, we have $48 = -16t^2 + 40t \Leftrightarrow 16t^2 - 40t + 48 = 0 \Leftrightarrow 2t^2 - 5t + 6 = 0 \Leftrightarrow$ $t = \dfrac{5 \pm \sqrt{25 - 48}}{4} = \dfrac{5 \pm \sqrt{-23}}{4}$. However, since the discriminant $D < 0$, there is no real solution, and hence the ball never reaches a height of 48 feet.

(c) The greatest height h is reached only once. So $h = -16t^2 + 40t \Leftrightarrow 16t^2 - 40t + h = 0$ has only one solution. Thus $D = (-40)^2 - 4(16)(h) = 0 \Leftrightarrow 1600 - 64h = 0 \Leftrightarrow h = 25$. So the greatest height reached by the ball is 25 feet.

(d) Setting $h = 25$, we have $25 = -16t^2 + 40t \Leftrightarrow 16t^2 - 40t + 25 = 0 \Leftrightarrow (4t-5)^2 = 0 \Leftrightarrow t = 1\frac{1}{4}$. Thus the ball reaches the highest point of its path after $1\frac{1}{4}$ seconds.

(e) Setting $h = 0$ (ground level), we have $0 = -16t^2 + 40t \Leftrightarrow 2t^2 - 5t = 0 \Leftrightarrow t(2t-5) = 0 \Leftrightarrow t = 0$ (start) or $t = 2\frac{1}{2}$. So the ball hits the ground in $2\frac{1}{2}$ s.

87. (a) The fish population on January 1, 2002 corresponds to $t = 0$, so $F = 1000\left(30 + 17(0) - (0)^2\right) = 30,000$. To find when the population will again reach this value, we set $F = 30,000$, giving $30000 = 1000\left(30 + 17t - t^2\right) = 30000 + 17000t - 1000t^2 \Leftrightarrow 0 = 17000t - 1000t^2 = 1000t(17 - t) \Leftrightarrow t = 0$ or $t = 17$. Thus the fish population will again be the same 17 years later, that is, on January 1, 2019.

(b) Setting $F = 0$, we have $0 = 1000\left(30 + 17t - t^2\right) \Leftrightarrow t^2 - 17t - 30 = 0 \Leftrightarrow$ $t = \dfrac{17 \pm \sqrt{289 + 120}}{-2} = \dfrac{17 \pm \sqrt{409}}{-2} = \dfrac{17 \pm 20.22}{2}$. Thus $t \approx -1.612$ or $t \approx 18.612$. Since $t < 0$ is inadmissible, it follows that the fish in the lake will have died out 18.612 years after January 1, 2002, that is on August 12, 2020.

89. Let w be the uniform width of the lawn. With w cut off each end, the area of the factory is $(240 - 2w)(180 - 2w)$. Since the lawn and the factory are equal in size this area, is $\frac{1}{2} \cdot 240 \cdot 180$. So $21{,}600 = 43{,}200 - 480w - 360w + 4w^2 \Leftrightarrow$ $0 = 4w^2 - 840w + 21{,}600 = 4\left(w^2 - 210w + 5400\right) = 4(w - 30)(w - 180) \Rightarrow w = 30$ or $w = 180$. Since 180 ft is too wide, the width of the lawn is 30 ft, and the factory is 120 ft by 180 ft.

91. Let t be the time, in hours it takes Irene to wash all the windows. Then it takes Henry $t + \frac{3}{2}$ hours to wash all the windows, and the sum of the fraction of the job per hour they can do individually equals the fraction of the job they can do together. Since 1 hour 48 minutes $= 1 + \frac{48}{60} = 1 + \frac{4}{5} = \frac{9}{5}$, we have $\dfrac{1}{t} + \dfrac{1}{t + \frac{3}{2}} = \dfrac{1}{\frac{9}{5}} \Leftrightarrow$ $\dfrac{1}{t} + \dfrac{2}{2t + 3} = \frac{5}{9} \Rightarrow 9(2t + 3) + 2(9t) = 5t(2t + 3) \Leftrightarrow 18t + 27 + 18t = 10t^2 + 15t \Leftrightarrow 10t^2 - 21t - 27 = 0$ $\Leftrightarrow t = \dfrac{-(-21) \pm \sqrt{(-21)^2 - 4(10)(-27)}}{2(10)} = \dfrac{21 \pm \sqrt{441 + 1080}}{20} = \dfrac{21 \pm 39}{20}$. So $t = \dfrac{21 - 39}{20} = -\dfrac{9}{10}$ or $t = \dfrac{21 + 39}{20} = 3$. Since $t < 0$ is impossible, all the windows are washed by Irene alone in 3 hours and by Henry alone in $3 + \frac{3}{2} = 4\frac{1}{2}$ hours.

93. Let x be the distance from the center of the earth to the dead spot (in thousands of miles). Now setting $F = 0$, we have $0 = -\frac{K}{x^2} + \frac{0.012K}{(239-x)^2} \Leftrightarrow \frac{K}{x^2} = \frac{0.012K}{(239-x)^2} \Leftrightarrow K(239-x)^2 = 0.012Kx^2 \Leftrightarrow$ $57121 - 478x + x^2 = 0.012x^2 \Leftrightarrow 0.988x^2 - 478x + 57121 = 0$. Using the Quadratic Formula, we obtain $x = \frac{-(-478)\pm\sqrt{(-478)^2-4(0.988)(57121)}}{2(0.988)} = \frac{478\pm\sqrt{228484-225742.192}}{1.976} = \frac{478\pm\sqrt{2741.808}}{1.976} \approx \frac{478\pm 52.362}{1.976} \approx 241.903 \pm 26.499$. So either $x \approx 241.903 + 26.499 \approx 268$ or $x \approx 241.903 - 26.499 \approx 215$. Since 268 is greater than the distance from the earth to the moon, we reject it; thus $x \approx 215{,}000$ miles.

95. Let x equal the original length of the reed in cubits. Then $x - 1$ is the piece that fits 60 times along the length of the field, that is, the length is $60(x-1)$. The width is $30x$. Then converting cubits to ninda, we have $375 = 60(x-1)\cdot 30x \cdot \frac{1}{12^2} = \frac{25}{2}x(x-1) \Leftrightarrow 30 = x^2 - x \Leftrightarrow x^2 - x - 30 = 0 \Leftrightarrow (x-6)(x+5) = 0$. So $x = 6$ or $x = -5$. Since x must be positive, the original length of the reed is 6 cubits.

1.5 COMPLEX NUMBERS

1. The imaginary number i has the property that $i^2 = -1$.

3. **(a)** The complex conjugate of $3 + 4i$ is $\overline{3+4i} = 3 - 4i$.

(b) $(3+4i)\left(\overline{3+4i}\right) = 3^2 + 4^2 = 25$

5. Yes, every real number a is a complex number of the form $a + 0i$.

7. $5 - 7i$: real part 5, imaginary part -7.

9. $\frac{-2-5i}{3} = -\frac{2}{3} - \frac{5}{3}i$: real part $-\frac{2}{3}$, imaginary part $-\frac{5}{3}$.

11. 3: real part 3, imaginary part 0.

13. $-\frac{2}{3}i$: real part 0, imaginary part $-\frac{2}{3}$.

15. $\sqrt{3} + \sqrt{-4} = \sqrt{3} + 2i$: real part $\sqrt{3}$, imaginary part 2.

17. $(3+2i) + 5i = 3 + (2+5)i = 3 + 7i$

19. $(5-3i) + (-4-7i) = (5-4) + (-3-7)i = 1 - 10i$

21. $(-6+6i) + (9-i) = (-6+9) + (6-1)i = 3 + 5i$

23. $\left(7 - \frac{1}{2}i\right) - \left(5 + \frac{3}{2}i\right) = (7-5) + \left(-\frac{1}{2} - \frac{3}{2}\right)i = 2 - 2i$

25. $(-12+8i) - (7+4i) = -12 + 8i - 7 - 4i = (-12-7) + (8-4)i = -19 + 4i$

27. $4(-1+2i) = -4 + 8i$

29. $(7-i)(4+2i) = 28 + 14i - 4i - 2i^2 = (28+2) + (14-4)i = 30 + 10i$

31. $(6+5i)(2-3i) = 12 - 18i + 10i - 15i^2 = (12+15) + (-18+10)i = 27 - 8i$

33. $(2+5i)(2-5i) = 2^2 - (5i)^2 = 4 - 25(-1) = 29$

35. $(2+5i)^2 = 2^2 + (5i)^2 + 2(2)(5i) = 4 - 25 + 20i = -21 + 20i$

37. $\frac{1}{i} = \frac{1}{i}\cdot\frac{i}{i} = \frac{i}{i^2} = \frac{i}{-1} = -i$

39. $\frac{2-3i}{1-2i} = \frac{2-3i}{1-2i}\cdot\frac{1+2i}{1+2i} = \frac{2+4i-3i-6i^2}{1-4i^2} = \frac{(2+6)+(4-3)i}{1+4} = \frac{8+i}{5}$ or $\frac{8}{5} + \frac{1}{5}i$

41. $\frac{10i}{1-2i} = \frac{10i}{1-2i}\cdot\frac{1+2i}{1+2i} = \frac{10i+20i^2}{1-4i^2} = \frac{-20+10i}{1+4} = \frac{5(-4+2i)}{5} = -4 + 2i$

43. $\frac{4+6i}{3i} = \frac{4+6i}{3i}\cdot\frac{3i}{3i} = \frac{12i+18i^2}{9i^2} = \frac{-18+12i}{-9} = \frac{-18}{-9} + \frac{12}{-9}i = 2 - \frac{4}{3}i$

45. $\frac{1}{1+i} - \frac{1}{1-i} = \frac{1}{1+i}\cdot\frac{1-i}{1-i} - \frac{1}{1-i}\cdot\frac{1+i}{1+i} = \frac{1-i}{1-i^2} - \frac{1+i}{1-i^2} = \frac{1-i}{2} + \frac{-1-i}{2} = -i$

47. $i^3 = i^2 i = -i$

49. $(3i)^5 = 3^5 \left(i^2\right)^2 i = 243(-1)^2 i = 243i$

51. $i^{1000} = \left(i^4\right)^{250} = 1^{250} = 1$

53. $\sqrt{-49} = \sqrt{49}\sqrt{-1} = 7i$

55. $\sqrt{-3}\sqrt{-12} = i\sqrt{3} \cdot 2i\sqrt{3} = 6i^2 = -6$

57. $(3 - \sqrt{-5})(1 + \sqrt{-1}) = \left(3 - i\sqrt{5}\right)(1 + i) = 3 + 3i - i\sqrt{5} - i^2\sqrt{5} = \left(3 + \sqrt{5}\right) + \left(3 - \sqrt{5}\right)i$

59. $\dfrac{2 + \sqrt{-8}}{1 + \sqrt{-2}} = \dfrac{2 + 2i\sqrt{2}}{1 + i\sqrt{2}} = \dfrac{2\left(1 + i\sqrt{2}\right)}{1 + i\sqrt{2}} = 2$

61. $x^2 + 49 = 0 \Leftrightarrow x^2 = -49 \Rightarrow x = \pm 7i$

63. $x^2 - x + 2 = 0 \Leftrightarrow x = \dfrac{-b \pm \sqrt{b^2 - 4ac}}{2a} = \dfrac{-(-1) \pm \sqrt{(-1)^2 - 4(1)(2)}}{2(1)} = \dfrac{1 \pm \sqrt{-7}}{2} = \frac{1}{2} \pm \frac{\sqrt{7}}{2}i$

65. $x^2 + 3x + 7 = 0 \Leftrightarrow x = \dfrac{-3 \pm \sqrt{3^2 - 4(1)(7)}}{2(1)} = \dfrac{-3 \pm \sqrt{-19}}{2} = -\frac{3}{2} \pm \frac{\sqrt{19}}{2}i$

67. $x^2 + x + 1 = 0 \Rightarrow x = \dfrac{-(1) \pm \sqrt{(1)^2 - 4(1)(1)}}{2(1)} = \dfrac{-1 \pm \sqrt{1 - 4}}{2} = \dfrac{-1 \pm \sqrt{-3}}{2} = \dfrac{-1 \pm i\sqrt{3}}{2} = -\frac{1}{2} \pm \frac{\sqrt{3}}{2}i$

69. $2x^2 - 2x + 1 = 0 \Rightarrow x = \dfrac{-(-2) \pm \sqrt{(-2)^2 - 4(2)(1)}}{2(2)} = \dfrac{2 \pm \sqrt{4 - 8}}{4} = \dfrac{2 \pm \sqrt{-4}}{4} = \dfrac{2 \pm 2i}{4} = \frac{1}{2} \pm \frac{1}{2}i$

71. $6x^2 + 12x + 7 = 0 \Rightarrow$

$x = \frac{-(12) \pm \sqrt{(12)^2 - 4(6)(7)}}{2(6)} = \frac{-12 \pm \sqrt{144 - 168}}{12} = \frac{-12 \pm \sqrt{-24}}{12} = \frac{-12 \pm 2i\sqrt{6}}{12} = \frac{-12}{12} \pm \frac{2i\sqrt{6}}{12} = -1 \pm \frac{\sqrt{6}}{6}i$

73. $\overline{z} + \overline{w} = \overline{3 - 4i} + \overline{5 + 2i} = 3 + 4i + 5 - 2i = 8 + 2i$

75. $z \cdot \overline{z} = (3 - 4i)(3 + 4i) = 3^2 + 4^2 = 25$

77. $\text{LHS} = \overline{z} + \overline{w} = \overline{(a + bi)} + \overline{(c + di)} = a - bi + c - di = (a + c) + (-b - d)i = (a + c) - (b + d)i.$

$\text{RHS} = \overline{z + w} = \overline{(a + bi) + (c + di)} = \overline{(a + c) + (b + d)i} = (a + c) - (b + d)i.$

Since LHS = RHS, this proves the statement.

79. $\text{LHS} = (\overline{z})^2 = \left(\overline{(a + bi)}\right)^2 = (a - bi)^2 = a^2 - 2abi + b^2 i^2 = \left(a^2 - b^2\right) - 2abi.$

$\text{RHS} = \overline{z^2} = \overline{(a + bi)^2} = \overline{a^2 + 2abi + b^2 i^2} = \overline{(a^2 - b^2) + 2abi} = \left(a^2 - b^2\right) - 2abi.$

Since LHS = RHS, this proves the statement.

81. $z + \overline{z} = (a + bi) + \overline{(a + bi)} = a + bi + a - bi = 2a$, which is a real number.

83. $z \cdot \overline{z} = (a + bi) \cdot \overline{(a + bi)} = (a + bi) \cdot (a - bi) = a^2 - b^2 i^2 = a^2 + b^2$, which is a real number.

85. Using the Quadratic Formula, the solutions to the equation are $x = \dfrac{-b \pm \sqrt{b^2 - 4ac}}{2a}$. Since both solutions are imaginary, we have $b^2 - 4ac < 0 \Leftrightarrow 4ac - b^2 > 0$, so the solutions are $x = \dfrac{-b}{2a} \pm \dfrac{\sqrt{4ac - b^2}}{2a} i$, where $\sqrt{4ac - b^2}$ is a real number. Thus the solutions are complex conjugates of each other.

1.6 SOLVING OTHER TYPES OF EQUATIONS

Note: In cases where both sides of an equation are squared, the implication symbol $\Leftrightarrow$ is sometimes used loosely. For example, $\sqrt{x} = x - 1$ "$\Leftrightarrow$" $\left(\sqrt{x}\right)^2 = (x - 1)^2$ is valid only for positive x. In these cases, inadmissible solutions are identified later in the solution.

1. (a) To solve the equation $x^3 - 4x^2 = 0$ we *factor* the left-hand side: $x^2(x-4) = 0$, as above.

(b) The solutions of the equation $x^2(x-4) = 0$ are $x = 0$ and $x = 4$.

3. The equation $(x+1)^2 - 5(x+1) + 6 = 0$ is of *quadratic* type. To solve the equation we set $W = x + 1$. The resulting quadratic equation is $W^2 - 5W + 6 = 0 \Leftrightarrow (W-3)(W-2) = 0 \Leftrightarrow W = 2$ or $W = 3 \Leftrightarrow x + 1 = 2$ or $x + 1 = 3 \Leftrightarrow$ $x = 1$ or $x = 2$. You can verify that these are both solutions to the original equation.

5. $x^2 - x = 0 \Leftrightarrow x(x-1) = 0 \Leftrightarrow x = 0$ or $x - 1 = 0$. Thus, the two real solutions are 0 and 1.

7. $x^3 = 25x \Leftrightarrow x^3 - 25x = 0 \Leftrightarrow x\left(x^2 - 25\right) = 0 \Leftrightarrow x(x+5)(x-5) = 0 \Leftrightarrow x = 0$ or $x + 5 = 0$ or $x - 5 = 0$. The three real solutions are -5, 0, and 5.

9. $x^5 - 3x^2 = 0 \Leftrightarrow x^2\left(x^3 - 3\right) = 0 \Leftrightarrow x = 0$ or $x^3 - 3 = 0$. The solutions are 0 and $\sqrt[3]{3}$.

11. $0 = 4z^5 - 10z^2 = 2z^2\left(2z^3 - 5\right)$. If $2z^2 = 0$, then $z = 0$. If $2z^3 - 5 = 0$, then $2z^3 = 5 \Leftrightarrow z = \sqrt[3]{\frac{5}{2}}$. The solutions are 0 and $\sqrt[3]{\frac{5}{2}}$.

13. $0 = x^5 + 8x^2 = x^2\left(x^3 + 8\right) = x^2(x+2)\left(x^2 - 2x + 4\right) \Leftrightarrow x^2 = 0$, $x + 2 = 0$, or $x^2 - 2x + 4 = 0$. If $x^2 = 0$, then $x = 0$; if $x + 2 = 0$, then $x = -2$, and $x^2 - 2x + 4 = 0$ has no real solution. Thus the solutions are $x = 0$ and $x = -2$.

15. $0 = x^3 - 5x^2 + 6x = x\left(x^2 - 5x + 6\right) = x(x-2)(x-3) \Leftrightarrow x = 0$, $x - 2 = 0$, or $x - 3 = 0$. Thus $x = 0$, or $x = 2$, or $x = 3$. The solutions are $x = 0$, $x = 2$, and $x = 3$.

17. $0 = x^4 + 4x^3 + 2x^2 = x^2\left(x^2 + 4x + 2\right)$. So either $x^2 = 0 \Leftrightarrow x = 0$, or using the Quadratic Formula on $x^2 + 4x + 2 = 0$, we have $x = \frac{-4 \pm \sqrt{4^2 - 4(1)(2)}}{2(1)} = \frac{-4 \pm \sqrt{16-8}}{2} = \frac{-4 \pm \sqrt{8}}{2} = \frac{-4 \pm 2\sqrt{2}}{2} = -2 \pm \sqrt{2}$. The solutions are 0, $-2 - \sqrt{2}$, and $-2 + \sqrt{2}$.

19. $(3x+5)^4 - (3x+5)^3 = 0$. Let $y = 3x + 5$. The equation becomes $y^4 - y^3 = 0 \Leftrightarrow$ $y\left(y^3 - 1\right) = y(y-1)\left(y^2 + y + 1\right) = 0$. If $y = 0$, then $3x + 5 = 0 \Leftrightarrow x = -\frac{5}{3}$. If $y - 1 = 0$, then $3x + 5 - 1 = 0$ $\Leftrightarrow x = -\frac{4}{3}$. If $y^2 + y + 1 = 0$, then $(3x+5)^2 + (3x+5) + 1 = 0 \Leftrightarrow 9x^2 + 33x + 31 = 0$. The discriminant is $b^2 - 4ac = 33^2 - 4(9)(31) = -27 < 0$, so this case gives no real solution. The solutions are $x = -\frac{5}{3}$ and $x = -\frac{4}{3}$.

21. $0 = x^3 - 5x^2 - 2x + 10 = x^2(x-5) - 2(x-5) = (x-5)\left(x^2 - 2\right)$. If $x - 5 = 0$, then $x = 5$. If $x^2 - 2 = 0$, then $x^2 = 2 \Leftrightarrow x = \pm\sqrt{2}$. The solutions are 5 and $\pm\sqrt{2}$.

23. $x^3 - x^2 + x - 1 = x^2 + 1 \Leftrightarrow 0 = x^3 - 2x^2 + x - 2 = x^2(x-2) + (x-2) = (x-2)\left(x^2 + 1\right)$. Since $x^2 + 1 = 0$ has no real solution, the only solution comes from $x - 2 = 0 \Leftrightarrow x = 2$.

25. $z + \frac{4}{z+1} = 3 \Leftrightarrow (z+1)\left(z + \frac{4}{z+1}\right) = (z+1)(3) \Leftrightarrow z^2 + z + 4 = 3z + 3 \Leftrightarrow z^2 - 2z + 1 = 0 \Leftrightarrow (z-1)^2 = 0$. The solution is $z = 1$. We must check the original equation to make sure this value of z does not result in a zero denominator.

27. $\frac{1}{x-1} + \frac{1}{x+2} = \frac{5}{4} \Leftrightarrow 4(x-1)(x+2)\left(\frac{1}{x-1} + \frac{1}{x+2}\right) = 4(x-1)(x+2)\left(\frac{5}{4}\right) \Leftrightarrow$ $4(x+2) + 4(x-1) = 5(x-1)(x+2) \Leftrightarrow 4x + 8 + 4x - 4 = 5x^2 + 5x - 10 \Leftrightarrow 5x^2 - 3x - 14 = 0 \Leftrightarrow$ $(5x+7)(x-2) = 0$. If $5x + 7 = 0$, then $x = -\frac{7}{5}$; if $x - 2 = 0$, then $x = 2$. The solutions are $-\frac{7}{5}$ and 2.

29. $\frac{x^2}{x+100} = 50 \Leftrightarrow x^2 = 50(x+100) = 50x + 5000 \Leftrightarrow x^2 - 50x - 5000 = 0 \Leftrightarrow (x-100)(x+50) = 0 \Leftrightarrow x - 100 = 0$ or $x + 50 = 0$. Thus $x = 100$ or $x = -50$. The solutions are 100 and -50.

31. $1+\dfrac{1}{(x+1)(x+2)}=\dfrac{2}{x+1}+\dfrac{1}{x+2} \Leftrightarrow (x+1)(x+2)+1=2(x+2)+(x+1) \Leftrightarrow x^2+3x+2+1=2x+4+x+1$ $\Leftrightarrow x^2-2=0 \Leftrightarrow x=\pm\sqrt{2}$. We verify that these are both solutions to the original equation.

33. $\dfrac{x}{2x+7}-\dfrac{x+1}{x+3}=1 \Leftrightarrow x(x+3)-(x+1)(2x+7)=(2x+7)(x+3) \Leftrightarrow x^2+3x-2x^2-9x-7=2x^2+13x+21$ $\Leftrightarrow 3x^2+19x+28=0 \Leftrightarrow (3x+7)(x+4)=0$. Thus either $3x+7=0$, so $x=-\frac{7}{3}$, or $x=-4$. The solutions are $-\frac{7}{3}$ and -4.

35. $\dfrac{x+\frac{2}{x}}{3+\frac{4}{x}}=5x \Leftrightarrow \left(\dfrac{x+\frac{2}{x}}{3+\frac{4}{x}}\right)\cdot\dfrac{x}{x}=\dfrac{x^2+2}{3x+4}=5x \Leftrightarrow x^2+2=5x(3x+4) \Leftrightarrow x^2+2=15x^2+20x \Leftrightarrow 0=14x^2+20x-2$

$\Leftrightarrow x=\dfrac{-(20)\pm\sqrt{(20)^2-4(14)(-2)}}{2(14)}=\dfrac{-20\pm\sqrt{400+112}}{28}=\dfrac{-20\pm\sqrt{512}}{28}=\dfrac{-20\pm16\sqrt{2}}{28}=\dfrac{-5\pm4\sqrt{2}}{7}$. The solutions are $\dfrac{-5\pm4\sqrt{2}}{7}$.

37. $5=\sqrt{4x-3} \Leftrightarrow 5^2=\left(\sqrt{4x-3}\right)^2 \Leftrightarrow 25=4x-3 \Leftrightarrow 4x=28 \Leftrightarrow x=7$ is a potential solution. Substituting into the original equation, we get $5=\sqrt{4(7)-3} \Leftrightarrow 5=\sqrt{25}$, which is true, so the solution is $x=7$.

39. $\sqrt{2x-1}=\sqrt{3x-5} \Leftrightarrow \left(\sqrt{2x-1}\right)^2=\left(\sqrt{3x-5}\right)^2 \Leftrightarrow 2x-1=3x-5 \Leftrightarrow x=4$. Substituting into the original equation, we get $\sqrt{2(4)-1}=\sqrt{3(4)-5} \Leftrightarrow \sqrt{7}=\sqrt{7}$, which is true, so the solution is $x=4$.

41. $\sqrt{x+2}=x \Leftrightarrow \left(\sqrt{x+2}\right)^2=x^2 \Leftrightarrow x+2=x^2 \Leftrightarrow x^2-x-2=(x+1)(x-2)=0 \Leftrightarrow x=-1$ or $x=2$. Substituting into the original equation, we get $\sqrt{(-1)+2}=-1 \Leftrightarrow \sqrt{1}=-1$, which is false, and $\sqrt{2+2}=2 \Leftrightarrow \sqrt{4}=2$, which is true. So $x=2$ is the only real solution.

43. $\sqrt{2x+1}+1=x \Leftrightarrow \sqrt{2x+1}=x-1 \Leftrightarrow 2x+1=(x-1)^2 \Leftrightarrow 2x+1=x^2-2x+1 \Leftrightarrow 0=x^2-4x=x(x-4)$. Potential solutions are $x=0$ and $x-4 \Leftrightarrow x=4$. These are only potential solutions since squaring is not a reversible operation. We must check each potential solution in the original equation.
Checking $x=0$: $\sqrt{2(0)+1}+1=(0) \Leftrightarrow \sqrt{1}+1=0$ is false.
Checking $x=4$: $\sqrt{2(4)+1}+1=(4) \Leftrightarrow \sqrt{9}+1=4 \Leftrightarrow 3+1=4$ is true. The only solution is $x=4$.

45. $x-\sqrt{x-1}=3 \Leftrightarrow x-3=\sqrt{x-1} \Leftrightarrow (x-3)^2=\left(\sqrt{x-1}\right)^2 \Leftrightarrow x^2-6x+9=x-1 \Leftrightarrow x^2-7x+10=0 \Leftrightarrow$ $(x-2)(x-5)=0$. Potential solutions are $x=2$ and $x=5$. We must check each potential solution in the original equation. Checking $x=2$: $2-\sqrt{2-1}=3$, which is false, so $x=2$ is not a solution. Checking $x=5$: $5-\sqrt{5-1}=3$ $\Leftrightarrow 5-2=3$, which is true, so $x=5$ is the only solution.

47. $\sqrt{3x+1}=2+\sqrt{x+1} \Leftrightarrow \left(\sqrt{3x+1}\right)^2=\left(2+\sqrt{x+1}\right)^2 \Leftrightarrow 3x+1=4+4\sqrt{x+1}+x+1 \Leftrightarrow 2x-4=4\sqrt{x+1} \Leftrightarrow$ $x-2=2\sqrt{x+1} \Leftrightarrow (x-2)^2=\left(2\sqrt{x+1}\right)^2 \Leftrightarrow x^2-4x+4=4(x+1) \Leftrightarrow x^2-8x=0 \Leftrightarrow x(x-8)=0 \Leftrightarrow x=0$ or $x=8$. Substituting each of these solutions into the original equation, we see that $x=0$ is not a solution but $x=8$ is a solution. Thus, $x=8$ is the only solution.

49. $x^4-4x^2+3=0$. Let $y=x^2$. Then the equation becomes $y^2-4y+3=0 \Leftrightarrow (y-1)(y-3)=0$, so $y=1$ or $y=3$. If $y=1$, then $x^2=1 \Leftrightarrow x=\pm1$, and if $y=3$, then $x^2=3 \Leftrightarrow x=\pm\sqrt{3}$.

51. $2x^4+4x^2+1=0$. The LHS is the sum of two nonnegative numbers and a positive number, so $2x^4+4x^2+1\geq 1\neq 0$. This equation has no real solution.

53. $0=x^6-26x^3-27=\left(x^3-27\right)\left(x^3+1\right)$. If $x^3-27=0 \Leftrightarrow x^3=27$, so $x=3$. If $x^3+1=0 \Leftrightarrow x^3=-1$, so $x=-1$. The solutions are 3 and -1.

55. $0=(x+5)^2-3(x+5)-10=[(x+5)-5][(x+5)+2]=x(x+7) \Leftrightarrow x=0$ or $x=-7$. The solutions are 0 and -7.

57. Let $w = \frac{1}{x+1}$. Then $\left(\frac{1}{x+1}\right)^2 - 2\left(\frac{1}{x+1}\right) - 8 = 0$ becomes $w^2 - 2w - 8 = 0 \Leftrightarrow (w-4)(w+2) = 0$. So $w - 4 = 0$ $\Leftrightarrow w = 4$, and $w + 2 = 0 \Leftrightarrow w = -2$. When $w = 4$, we have $\frac{1}{x+1} = 4 \Leftrightarrow 1 = 4x + 4 \Leftrightarrow -3 = 4x \Leftrightarrow x = -\frac{3}{4}$. When $w = -2$, we have $\frac{1}{x+1} = -2 \Leftrightarrow 1 = -2x - 2 \Leftrightarrow 3 = -2x \Leftrightarrow x = -\frac{3}{2}$. Solutions are $-\frac{3}{4}$ and $-\frac{3}{2}$.

59. Let $u = x^{2/3}$. Then $0 = x^{4/3} - 5x^{2/3} + 6$ becomes $u^2 - 5u + 6 = 0 \Leftrightarrow (u-3)(u-2) = 0 \Leftrightarrow u - 3 = 0$ or $u - 2 = 0$. If $u - 3 = 0$, then $x^{2/3} - 3 = 0 \Leftrightarrow x^{2/3} = 3 \Leftrightarrow x = \pm 3^{3/2} = \pm 3\sqrt{3}$. If $u - 2 = 0$, then $x^{2/3} - 2 = 0 \Leftrightarrow x^{2/3} = 2 \Leftrightarrow$ $x = \pm 2^{3/2} = 2\sqrt{2}$. The solutions are $\pm 3\sqrt{3}$ and $\pm 2\sqrt{2}$.

61. $4(x+1)^{1/2} - 5(x+1)^{3/2} + (x+1)^{5/2} = 0 \Leftrightarrow \sqrt{x+1}\left[4 - 5(x+1) + (x+1)^2\right] = 0 \Leftrightarrow$ $\sqrt{x+1}\left(4 - 5x - 5 + x^2 + 2x + 1\right) = 0 \Leftrightarrow \sqrt{x+1}\left(x^2 - 3x\right) = 0 \Leftrightarrow \sqrt{x+1} \cdot x(x-3) = 0 \Leftrightarrow x = -1$ or $x = 0$ or $x = 3$. The solutions are -1, 0, and 3.

63. $x^{3/2} - 10x^{1/2} + 25x^{-1/2} = 0 \Leftrightarrow x^{-1/2}\left(x^2 - 10x + 25\right) = 0 \Leftrightarrow x^{-1/2}(x-5)^2 = 0$. Now $x^{-1/2} \neq 0$, so the only solution is $x = 5$.

65. Let $u = x^{1/6}$. (We choose the exponent $\frac{1}{6}$ because the LCD of 2, 3, and 6 is 6.) Then $x^{1/2} - 3x^{1/3} = 3x^{1/6} - 9 \Leftrightarrow$ $x^{3/6} - 3x^{2/6} = 3x^{1/6} - 9 \Leftrightarrow u^3 - 3u^2 = 3u - 9 \Leftrightarrow 0 = u^3 - 3u^2 - 3u + 9 = u^2(u-3) - 3(u-3) = (u-3)\left(u^2 - 3\right)$. So $u - 3 = 0$ or $u^2 - 3 = 0$. If $u - 3 = 0$, then $x^{1/6} - 3 = 0 \Leftrightarrow x^{1/6} = 3 \Leftrightarrow x = 3^6 = 729$. If $u^2 - 3 = 0$, then $x^{1/3} - 3 = 0 \Leftrightarrow x^{1/3} = 3 \Leftrightarrow x = 3^3 = 27$. The solutions are 729 and 27.

67. $\frac{1}{x^3} + \frac{4}{x^2} + \frac{4}{x} = 0 \Leftrightarrow 1 + 4x + 4x^2 = 0 \Leftrightarrow (1+2x)^2 = 0 \Leftrightarrow 1 + 2x = 0 \Leftrightarrow 2x = -1 \Leftrightarrow x = -\frac{1}{2}$. The solution is $-\frac{1}{2}$.

69. $\sqrt{\sqrt{x+5} + x} = 5$. Squaring both sides, we get $\sqrt{x+5} + x = 25 \Leftrightarrow \sqrt{x+5} = 25 - x$. Squaring both sides again, we get $x + 5 = (25-x)^2 \Leftrightarrow x + 5 = 625 - 50x + x^2 \Leftrightarrow 0 = x^2 - 51x + 620 = (x-20)(x-31)$. Potential solutions are $x = 20$ and $x = 31$. We must check each potential solution in the original equation.

Checking $x = 20$: $\sqrt{\sqrt{20+5} + 20} = 5 \Leftrightarrow \sqrt{\sqrt{25} + 20} = 5 \Leftrightarrow \sqrt{5 + 20} = 5$, which is true, and hence $x = 20$ is a solution.

Checking $x = 31$: $\sqrt{\sqrt{(31)+5} + 31} = 5 \Leftrightarrow \sqrt{\sqrt{36} + 31} = 5 \Leftrightarrow \sqrt{37} = 5$, which is false, and hence $x = 31$ is not a solution. The only real solution is $x = 20$.

71. $x^2\sqrt{x+3} = (x+3)^{3/2} \Leftrightarrow 0 = x^2\sqrt{x+3} - (x+3)^{3/2} \Leftrightarrow 0 = \sqrt{x+3}\left[\left(x^2\right) - (x+3)\right] \Leftrightarrow 0 = \sqrt{x+3}\left(x^2 - x - 3\right)$. If $(x+3)^{1/2} = 0$, then $x + 3 = 0 \Leftrightarrow x = -3$. If $x^2 - x - 3 = 0$, then using the Quadratic Formula $x = \frac{1 \pm \sqrt{13}}{2}$. The solutions are -3 and $\frac{1 \pm \sqrt{13}}{2}$.

73. $\sqrt{x + \sqrt{x+2}} = 2$. Squaring both sides, we get $x + \sqrt{x+2} = 4 \Leftrightarrow \sqrt{x+2} = 4 - x$. Squaring both sides again, we get $x + 2 = (4-x)^2 = 16 - 8x + x^2 \Leftrightarrow 0 = x^2 - 9x + 14 \Leftrightarrow 0 = (x-7)(x-2)$. If $x - 7 = 0$, then $x = 7$. If $x - 2 = 0$, then $x = 2$. So $x = 2$ is a solution but $x = 7$ is not, since it does not satisfy the original equation.

75. $0 = x^4 - 5ax^2 + 4a^2 = \left(a - x^2\right)\left(4a - x^2\right)$. Since a is positive, $a - x^2 = 0 \Leftrightarrow x^2 = a \Leftrightarrow x = \sqrt{a}$. Again, since a is positive, $4a - x^2 = 0 \Leftrightarrow x^2 = 4a \Leftrightarrow x = \pm 2\sqrt{a}$. Thus the four solutions are $\pm\sqrt{a}$ and $\pm 2\sqrt{a}$.

77. $\sqrt{x+a} + \sqrt{x-a} = \sqrt{2}\sqrt{x+6}$. Squaring both sides, we have $x + a + 2\left(\sqrt{x+a}\right)\left(\sqrt{x-a}\right) + x - a = 2(x+6) \Leftrightarrow 2x + 2\left(\sqrt{x+a}\right)\left(\sqrt{x-a}\right) = 2x + 12 \Leftrightarrow 2\left(\sqrt{x+a}\right)\left(\sqrt{x-a}\right) = 12$ $\Leftrightarrow \left(\sqrt{x+a}\right)\left(\sqrt{x-a}\right) = 6$. Squaring both sides again we have $(x+a)(x-a) = 36 \Leftrightarrow x^2 - a^2 = 36 \Leftrightarrow x^2 = a^2 + 36$ $\Leftrightarrow x = \pm\sqrt{a^2 + 36}$. Checking these answers, we see that $x = -\sqrt{a^2 + 36}$ is not a solution (for example, try substituting $a = 8$), but $x = \sqrt{a^2 + 36}$ is a solution.

79. Let x be the number of people originally intended to take the trip. Then originally, the cost of the trip is $\dfrac{900}{x}$. After 5 people cancel, there are now $x-5$ people, each paying $\dfrac{900}{x}+2$. Thus $900=(x-5)\left(\dfrac{900}{x}+2\right)\Leftrightarrow 900=900+2x-\dfrac{4500}{x}-10$ $\Leftrightarrow 0=2x-10-\dfrac{4500}{x}\Leftrightarrow 0=2x^2-10x-4500=(2x-100)(x+45)$. Thus either $2x-100=0$, so $x=50$, or $x+45=0$, $x=-45$. Since the number of people on the trip must be positive, originally 50 people intended to take the trip.

81. We want to solve for t when $P=500$. Letting $u=\sqrt{t}$ and substituting, we have $500=3t+10\sqrt{t}+140\Leftrightarrow$ $500=3u^2+10u+140\Leftrightarrow 0=3u^2+10u-360\Leftrightarrow u=\dfrac{-5\pm\sqrt{1105}}{3}$. Since $u=\sqrt{t}$, we must have $u\geq 0$. So $\sqrt{t}=u=\dfrac{-5+\sqrt{1105}}{3}\approx 9.414\Leftrightarrow t=\approx 88.62$. So it will take 89 days for the fish population to reach 500.

83. Let x be the height of the pile in feet. Then the diameter is $3x$ and the radius is $\frac{3}{2}x$ feet. Since the volume of the cone is 1000 ft^3, we have $\dfrac{\pi}{3}\left(\dfrac{3x}{2}\right)^2 x=1000\Leftrightarrow\dfrac{3\pi x^3}{4}=1000\Leftrightarrow x^3=\dfrac{4000}{3\pi}\Leftrightarrow x=\sqrt[3]{\dfrac{4000}{3\pi}}\approx 7.52$ feet.

85. Let r be the radius of the larger sphere, in mm. Equating the volumes, we have $\frac{4}{3}\pi r^3=\frac{4}{3}\pi\left(2^3+3^3+4^3\right)\Leftrightarrow$ $r^3=2^3+3^3+4^4\Leftrightarrow r^3=99\Leftrightarrow r=\sqrt[3]{99}\approx 4.63$. Therefore, the radius of the larger sphere is about 4.63 mm.

87. Let x be the length, in miles, of the abandoned road to be used. Then the length of the abandoned road not used is $40-x$, and the length of the new road is $\sqrt{10^2+(40-x)^2}$ miles, by the Pythagorean Theorem. Since the cost of the road is cost per mile $\times$ number of miles, we have $100{,}000x+200{,}000\sqrt{x^2-80x+1700}=6{,}800{,}000$ $\Leftrightarrow 2\sqrt{x^2-80x+1700}=68-x$. Squaring both sides, we get $4x^2-320x+6800=4624-136x+x^2\Leftrightarrow$ $3x^2-184x+2176=0\Leftrightarrow x=\frac{184\pm\sqrt{33856-26112}}{6}=\frac{184\pm 88}{6}\Leftrightarrow x=\frac{136}{3}$ or $x=16$. Since $45\frac{1}{3}$ is longer than the existing road, 16 miles of the abandoned road should be used. A completely new road would have length $\sqrt{10^2+40^2}$ (let $x=0$) and would cost $\sqrt{1700}\times 200{,}000\approx 8.3$ million dollars. So no, it would not be cheaper.

89. Let x be the length of the hypotenuse of the triangle, in feet. Then one of the other sides has length $x-7$ feet, and since the perimeter is 392 feet, the remaining side must have length $392-x-(x-7)=399-2x$. From the Pythagorean Theorem, we get $(x-7)^2+(399-2x)^2=x^2\Leftrightarrow 4x^2-1610x+159250=0$. Using the Quadratic Formula, we get

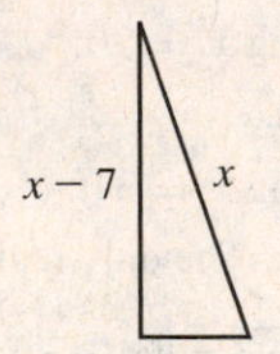

$x=\frac{1610\pm\sqrt{1610^2-4(4)(159250)}}{2(4)}=\frac{1610\pm\sqrt{44100}}{8}=\frac{1610\pm 210}{8}$, and so $x=227.5$ or $x=175$. But if $x=227.5$, then the side of length $x-7$ combined with the hypotenuse already exceeds the perimeter of 392 feet, and so we must have $x=175$. Thus the other sides have length $175-7=168$ and $399-2(175)=49$. The lot has sides of length 49 feet, 168 feet, and 175 feet.

91. Since the total time is 3 s, we have $3=\dfrac{\sqrt{d}}{4}+\dfrac{d}{1090}$. Letting $w=\sqrt{d}$, we have $3=\frac{1}{4}w+\frac{1}{1090}w^2\Leftrightarrow\frac{1}{1090}w^2+\frac{1}{4}w-3=0$ $\Leftrightarrow 2w^2+545w-6540=0\Leftrightarrow w=\dfrac{-545\pm 591.054}{4}$. Since $w\geq 0$, we have $\sqrt{d}=w\approx 11.51$, so $d=132.56$. The well is 132.6 ft deep.

1.7 SOLVING INEQUALITIES

1. (a) If $x<5$, then $x-3<5-3\Rightarrow x-3<2$.

(b) If $x\leq 5$, then $3\cdot x\leq 3\cdot 5\Rightarrow 3x\leq 15$.

(c) If $x \geq 2$, then $-3 \cdot x \leq -3 \cdot 2 \Rightarrow -3x \leq -6$.

(d) If $x < -2$, then $-x > 2$.

3. (a) No. For example, if $x = -2$, then $x(x+1) = -2(-1) = 2 > 0$.

(b) No. For example, if $x = 2$, then $x(x+1) = 2(3) = 6$.

5.

x	$-2+3x \geq \frac{1}{3}$
-5	$-17 \geq \frac{1}{3}$; no
-1	$-5 \geq \frac{1}{3}$; no
0	$-2 \geq 0$; no
$\frac{2}{3}$	$0 \geq \frac{1}{3}$; no
$\frac{5}{6}$	$\frac{1}{2} \geq \frac{1}{3}$; yes
1	$1 \geq \frac{1}{3}$; yes
$\sqrt{5}$	$4.7 \geq \frac{1}{3}$; yes
3	$7 \geq \frac{1}{3}$; yes
5	$13 \geq \frac{1}{3}$; yes

The elements $\frac{5}{6}$, 1, $\sqrt{5}$, 3, and 5 satisfy the inequality.

7.

x	$1 < 2x - 4 \leq 7$
-5	$1 < -14 \leq 7$; no
-1	$1 < -6 \leq 7$; no
0	$1 < -4 \leq 7$; no
$\frac{2}{3}$	$1 < -\frac{8}{3} \leq 7$; no
$\frac{5}{6}$	$1 < -\frac{7}{3} \leq 7$; no
1	$1 < -2 \leq 7$; no
$\sqrt{5}$	$1 < 0.47 \leq 7$; no
3	$1 < 2 \leq 7$; yes
5	$1 < 6 \leq 7$; yes

The elements 3 and 5 satisfy the inequality.

9.

x	$\frac{1}{x} \leq \frac{1}{2}$
-5	$-\frac{1}{5} \leq \frac{1}{2}$; yes
-1	$-1 \leq \frac{1}{2}$; yes
0	$\frac{1}{0}$ is undefined; no
$\frac{2}{3}$	$\frac{3}{2} \leq \frac{1}{2}$; no
$\frac{5}{6}$	$\frac{6}{5} \leq \frac{1}{2}$; no
1	$1 \leq \frac{1}{2}$; no
$\sqrt{5}$	$0.45 \leq \frac{1}{2}$; yes
3	$\frac{1}{3} \leq \frac{1}{2}$; yes
5	$\frac{1}{5} \leq \frac{1}{2}$; yes

The elements -5, -1, $\sqrt{5}$, 3, and 5 satisfy the inequality.

11. $5x \leq 6 \Leftrightarrow x \leq \frac{6}{5}$. Interval: $\left(-\infty, \frac{6}{5}\right]$

Graph: $\frac{6}{5}$

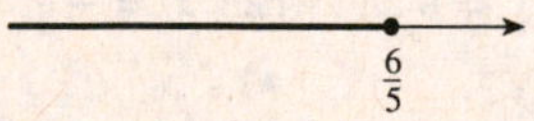

13. $2x - 5 > 3 \Leftrightarrow 2x > 8 \Leftrightarrow x > 4$

Interval: $(4, \infty)$

Graph: 4

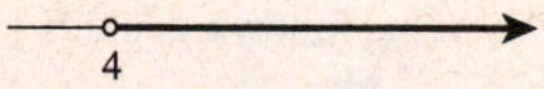

15. $2 - 3x > 8 \Leftrightarrow 3x < 2 - 8 \Leftrightarrow x < -2$

Interval: $(-\infty, -2)$

Graph: -2

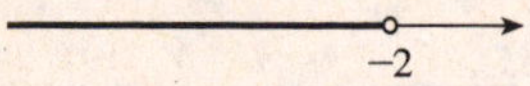

17. $2x + 1 < 0 \Leftrightarrow 2x < -1 \Leftrightarrow x < -\frac{1}{2}$

Interval: $\left(-\infty, -\frac{1}{2}\right)$

Graph: 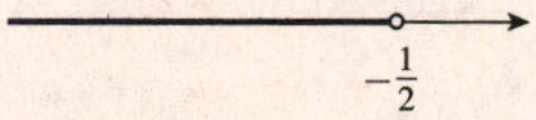 $-\frac{1}{2}$

19. $1 + 4x \le 5 - 2x \Leftrightarrow 6x \le 4 \Leftrightarrow x \le \frac{2}{3}$

Interval: $\left(-\infty, \frac{2}{3}\right]$

Graph: 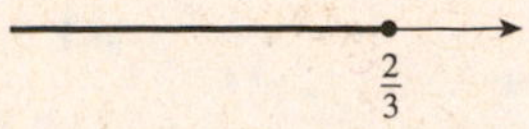

21. $\frac{1}{2}x - \frac{2}{3} > 2 \Leftrightarrow \frac{1}{2}x > \frac{8}{3} \Leftrightarrow x > \frac{16}{3}$

Interval: $\left(\frac{16}{3}, \infty\right)$

Graph: 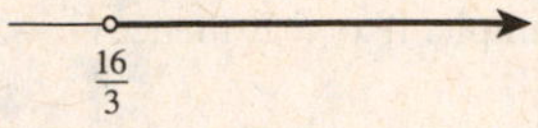

23. $4 - 3x \le -(1 + 8x) \Leftrightarrow 4 - 3x \le -1 - 8x \Leftrightarrow 5x \le -5$
$\Leftrightarrow x \le -1$

Interval: $(-\infty, -1]$

Graph: 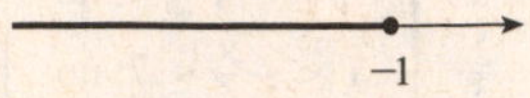

25. $2 \le x + 5 < 4 \Leftrightarrow -3 \le x < -1$

Interval: $[-3, -1)$

Graph: 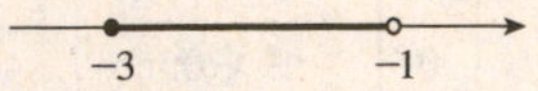

27. $-6 \le 3x - 7 \le 8 \Leftrightarrow 1 \le 3x \le 15 \Leftrightarrow \frac{1}{3} \le x \le 5$

Interval: $\left[\frac{1}{3}, 5\right]$

Graph: 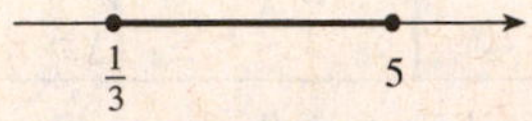

29. $-2 < 8 - 2x \le -1 \Leftrightarrow -10 < -2x \le -9 \Leftrightarrow 5 > x \ge \frac{9}{2}$
$\Leftrightarrow \frac{9}{2} \le x < 5$

Interval: $\left[\frac{9}{2}, 5\right)$

Graph: 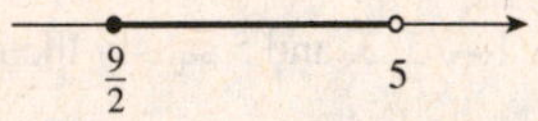

31. $\frac{2}{3} \ge \frac{2x - 3}{12} > \frac{1}{6} \Leftrightarrow 8 \ge 2x - 3 > 2$ (multiply each expression by 12) $\Leftrightarrow 11 \ge 2x > 5 \Leftrightarrow \frac{11}{2} \ge x > \frac{5}{2}$

Interval: $\left(\frac{5}{2}, \frac{11}{2}\right]$

Graph: 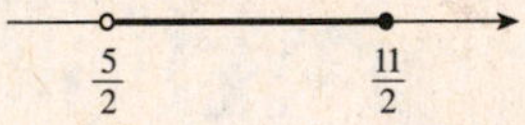

33. $(x + 2)(x - 3) < 0$. The expression on the left of the inequality changes sign where $x = -2$ and where $x = 3$. Thus we must check the intervals in the following table.

Interval	$(-\infty, -2)$	$(-2, 3)$	$(3, \infty)$
Sign of $x + 2$	$-$	$+$	$+$
Sign of $x - 3$	$-$	$-$	$+$
Sign of $(x + 2)(x - 3)$	$+$	$-$	$+$

From the table, the solution set is $\{x \mid -2 < x < 3\}$. Interval: $(-2, 3)$.

Graph: 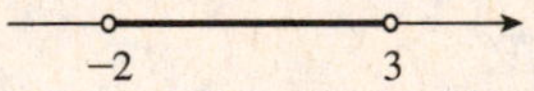

35. $x(2x + 7) \ge 0$. The expression on the left of the inequality changes sign where $x = 0$ and where $x = -\frac{7}{2}$. Thus we must check the intervals in the following table.

Interval	$\left(-\infty, -\frac{7}{2}\right)$	$\left(-\frac{7}{2}, 0\right)$	$(0, \infty)$
Sign of x	$-$	$-$	$+$
Sign of $2x + 7$	$-$	$+$	$+$
Sign of $x(2x + 7)$	$+$	$-$	$+$

From the table, the solution set is $\left\{x \mid x \le -\frac{7}{2} \text{ or } 0 \le x\right\}$.

Interval: $\left(-\infty, -\frac{7}{2}\right] \cup [0, \infty)$.

Graph: 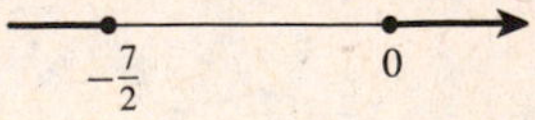

37. $x^2 - 3x - 18 \le 0 \Leftrightarrow (x+3)(x-6) \le 0$. The expression on the left of the inequality changes sign where $x = 6$ and where $x = -3$. Thus we must check the intervals in the following table.

Interval	$(-\infty, -3)$	$(-3, 6)$	$(6, \infty)$
Sign of $x+3$	$-$	$+$	$+$
Sign of $x-6$	$-$	$-$	$+$
Sign of $(x+3)(x-6)$	$+$	$-$	$+$

From the table, the solution set is $\{x \mid -3 \le x \le 6\}$. Interval: $[-3, 6]$.

Graph: −3 6

39. $2x^2 + x \ge 1 \Leftrightarrow 2x^2 + x - 1 \ge 0 \Leftrightarrow (x+1)(2x-1) \ge 0$. The expression on the left of the inequality changes sign where $x = -1$ and where $x = \frac{1}{2}$. Thus we must check the intervals in the following table.

Interval	$(-\infty, -1)$	$\left(-1, \frac{1}{2}\right)$	$\left(\frac{1}{2}, \infty\right)$
Sign of $x+1$	$-$	$+$	$+$
Sign of $2x-1$	$-$	$-$	$+$
Sign of $(x+1)(2x-1)$	$+$	$-$	$+$

From the table, the solution set is $\left\{x \mid x \le -1 \text{ or } \frac{1}{2} \le x\right\}$.

Interval: $(-\infty, -1] \cup \left[\frac{1}{2}, \infty\right)$.

Graph: −1 $\frac{1}{2}$

41. $3x^2 - 3x < 2x^2 + 4 \Leftrightarrow x^2 - 3x - 4 < 0 \Leftrightarrow (x+1)(x-4) < 0$. The expression on the left of the inequality changes sign where $x = -1$ and where $x = 4$. Thus we must check the intervals in the following table.

Interval	$(-\infty, -1)$	$(-1, 4)$	$(4, \infty)$
Sign of $x+1$	$-$	$+$	$+$
Sign of $x-4$	$-$	$-$	$+$
Sign of $(x+1)(x-4)$	$+$	$-$	$+$

From the table, the solution set is $\{x \mid -1 < x < 4\}$. Interval: $(-1, 4)$.

Graph: −1 4

43. $x^2 > 3(x+6) \Leftrightarrow x^2 - 3x - 18 > 0 \Leftrightarrow (x+3)(x-6) > 0$. The expression on the left of the inequality changes sign where $x = 6$ and where $x = -3$. Thus we must check the intervals in the following table.

Interval	$(-\infty, -3)$	$(-3, 6)$	$(6, \infty)$
Sign of $x+3$	$-$	$+$	$+$
Sign of $x-6$	$-$	$-$	$+$
Sign of $(x+3)(x-6)$	$+$	$-$	$+$

From the table, the solution set is $\{x \mid x < -3 \text{ or } 6 < x\}$.

Interval: $(-\infty, -3) \cup (6, \infty)$.

Graph: −3 6

45. $x^2 < 4 \Leftrightarrow x^2 - 4 < 0 \Leftrightarrow (x+2)(x-2) < 0$. The expression on the left of the inequality changes sign where $x = -2$ and where $x = 2$. Thus we must check the intervals in the following table.

Interval	$(-\infty, -2)$	$(-2, 2)$	$(2, \infty)$
Sign of $x+2$	$-$	$+$	$+$
Sign of $x-2$	$-$	$-$	$+$
Sign of $(x+2)(x-2)$	$+$	$-$	$+$

From the table, the solution set is $\{x \mid -2 < x < 2\}$. Interval: $(-2, 2)$.

Graph: −2 2

47. $(x+2)(x-1)(x-3) \le 0$. The expression on the left of the inequality changes sign when $x = -2$, $x = 1$, and $x = 3$. Thus we must check the intervals in the following table.

Interval	$(-\infty, -2)$	$(-2, 1)$	$(1, 3)$	$(3, \infty)$
Sign of $x+2$	$-$	$+$	$+$	$+$
Sign of $x-1$	$-$	$-$	$+$	$+$
Sign of $x-3$	$-$	$-$	$-$	$+$
Sign of $(x+2)(x-1)(x-3)$	$-$	$+$	$-$	$+$

From the table, the solution set is $\{x \mid x \le -2 \text{ or } 1 \le x \le 3\}$. Interval: $(-\infty, -2] \cup [1, 3]$. Graph:

−2 1 3

49. $(x-4)(x+2)^2 < 0$. Note that $(x+2)^2 > 0$ for all $x \neq -2$, so the expression on the left of the original inequality changes sign only when $x = 4$. We check the intervals in the following table.

Interval	$(-\infty, -2)$	$(-2, 4)$	$(4, \infty)$
Sign of $x-4$	$-$	$-$	$+$
Sign of $(x+2)^2$	$+$	$+$	$+$
Sign of $(x-4)(x+2)^2$	$-$	$-$	$+$

From the table, the solution set is $\{x \mid x \neq -2 \text{ and } x < 4\}$. We exclude the endpoint -2 since the original expression cannot be 0. Interval: $(-\infty, -2) \cup (-2, 4)$.

Graph:

−2 4

51. $(x-2)^2(x-3)(x+1) \le 0$. Note that $(x-2)^2 \ge 0$ for all x, so the expression on the left of the original inequality changes sign only when $x = -1$ and $x = 3$. We check the intervals in the following table.

Interval	$(-\infty, -1)$	$(-1, 2)$	$(2, 3)$	$(3, \infty)$
Sign of $(x-2)^2$	$+$	$+$	$+$	$+$
Sign of $x-3$	$-$	$-$	$-$	$+$
Sign of $x+1$	$-$	$+$	$+$	$+$
Sign of $(x-2)^2(x-3)(x+1)$	$+$	$-$	$-$	$+$

From the table, the solution set is $\{x \mid -1 \le x \le 3\}$. Interval: $[-1, 3]$. Graph:

−1 3

53. $x^3 - 4x > 0 \Leftrightarrow x\left(x^2 - 4\right) > 0 \Leftrightarrow x(x+2)(x-2) > 0$. The expression on the left of the inequality changes sign where $x = 0$, $x = -2$ and where $x = 4$. Thus we must check the intervals in the following table.

Interval	$(-\infty, -2)$	$(-2, 0)$	$(0, 2)$	$(2, \infty)$
Sign of x	$-$	$-$	$+$	$+$
Sign of $x+2$	$-$	$+$	$+$	$+$
Sign of $x-2$	$-$	$-$	$-$	$+$
Sign of $x(x+2)(x-2)$	$-$	$+$	$-$	$+$

From the table, the solution set is $\{x \mid -2 < x < 0 \text{ or } x > 2\}$. Interval: $(-2, 0) \cup (2, \infty)$. Graph:

−2 0 2

55. $\dfrac{x+3}{2x-1} \geq 0$. The expression on the left of the inequality changes sign where $x = -3$ and where $x = \frac{1}{2}$. Thus we must check the intervals in the following table.

Interval	$(-\infty, -3)$	$\left(-3, \frac{1}{2}\right)$	$\left(\frac{1}{2}, \infty\right)$
Sign of $x+3$	$-$	$+$	$+$
Sign of $2x-1$	$-$	$-$	$+$
Sign of $\dfrac{x+3}{2x-1}$	$+$	$-$	$+$

From the table, the solution set is $\left\{x \mid x < -3 \text{ or } x > \frac{1}{2}\right\}$. Since the denominator cannot equal 0, $x \neq \frac{1}{2}$.

Interval: $(-\infty, -3] \cup \left(\frac{1}{2}, \infty\right)$.

Graph: −3 1/2

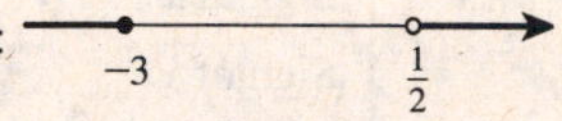

57. $\dfrac{4-x}{x+4} < 0$. The expression on the left of the inequality changes sign where $x = \pm 4$. Thus we must check the intervals in the following table.

Interval	$(-\infty, -4)$	$(-4, 4)$	$(4, \infty)$
Sign of $4-x$	$+$	$+$	$-$
Sign of $x+4$	$-$	$+$	$+$
Sign of $\dfrac{4-x}{x+4}$	$-$	$+$	$-$

From the table, the solution set is $\{x \mid x < -4 \text{ or } x > 4\}$.

Interval: $(-\infty, -4) \cup (4, \infty)$.

Graph: −4 4

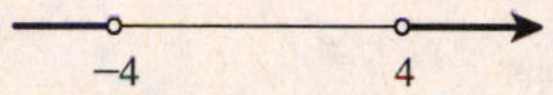

59. $\dfrac{2x+1}{x-5} \leq 3 \Leftrightarrow \dfrac{2x+1}{x-5} - 3 \leq 0 \Leftrightarrow \dfrac{2x+1}{x-5} - \dfrac{3(x-5)}{x-5} \leq 0 \Leftrightarrow \dfrac{-x+16}{x-5} \leq 0$. The expression on the left of the inequality changes sign where $x = 16$ and where $x = 5$. Thus we must check the intervals in the following table.

Interval	$(-\infty, 5)$	$(5, 16)$	$(16, \infty)$
Sign of $-x+16$	$+$	$+$	$-$
Sign of $x-5$	$-$	$+$	$+$
Sign of $\dfrac{-x+16}{x-5}$	$-$	$+$	$-$

From the table, the solution set is $\{x \mid x < 5 \text{ or } x \geq 16\}$. Since the denominator cannot equal 0, we must have $x \neq 5$.

Interval: $(-\infty, 5) \cup [16, \infty)$.

Graph: 5 16

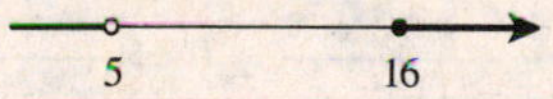

61. $\dfrac{4}{x} < x \Leftrightarrow \dfrac{4}{x} - x < 0 \Leftrightarrow \dfrac{4}{x} - \dfrac{x \cdot x}{x} < 0 \Leftrightarrow \dfrac{4-x^2}{x} < 0 \Leftrightarrow \dfrac{(2-x)(2+x)}{x} < 0$. The expression on the left of the inequality changes sign where $x = 0$, where $x = -2$, and where $x = 2$. Thus we must check the intervals in the following table.

Interval	$(-\infty, -2)$	$(-2, 0)$	$(0, 2)$	$(2, \infty)$
Sign of $2+x$	$-$	$+$	$+$	$+$
Sign of x	$-$	$-$	$+$	$+$
Sign of $2-x$	$+$	$+$	$+$	$-$
Sign of $\dfrac{(2-x)(2+x)}{x}$	$+$	$-$	$+$	$-$

From the table, the solution set is $\{x \mid -2 < x < 0 \text{ or } 2 < x\}$. Interval: $(-2, 0) \cup (2, \infty)$. Graph: −2 0 2

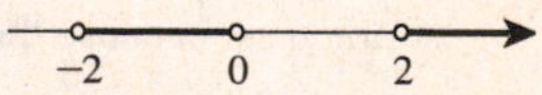

63. $1+\frac{2}{x+1}\leq\frac{2}{x}\Leftrightarrow 1+\frac{2}{x+1}-\frac{2}{x}\leq 0\Leftrightarrow\frac{x(x+1)}{x(x+1)}+\frac{2x}{x(x+1)}-\frac{2(x+1)}{x(x+1)}\leq 0\Leftrightarrow\frac{x^2+x+2x-2x-2}{x(x+1)}\leq 0\Leftrightarrow$

$\frac{x^2+x-2}{x(x+1)}\leq 0\Leftrightarrow\frac{(x+2)(x-1)}{x(x+1)}\leq 0$. The expression on the left of the inequality changes sign where $x=-2$, where $x=-1$, where $x=0$, and where $x=1$. Thus we must check the intervals in the following table.

Interval	$(-\infty,-2)$	$(-2,-1)$	$(-1,0)$	$(0,1)$	$(1,\infty)$
Sign of $x+2$	−	+	+	+	+
Sign of $x-1$	−	−	−	−	+
Sign of x	−	−	−	+	+
Sign of $x+1$	−	−	+	+	+
Sign of $\frac{(x+2)(x-1)}{x(x+1)}$	+	−	+	−	+

Since $x=-1$ and $x=0$ yield undefined expressions, we cannot include them in the solution. From the table, the solution set is $\{x\mid -2\leq x<-1\text{ or }0<x\leq 1\}$. Interval: $[-2,-1)\cup(0,1]$. Graph: −2 −1 0 1

65. $\frac{6}{x-1}-\frac{6}{x}\geq 1\Leftrightarrow\frac{6}{x-1}-\frac{6}{x}-1\geq 0\Leftrightarrow\frac{6x}{x(x-1)}-\frac{6(x-1)}{x(x-1)}-\frac{x(x-1)}{x(x-1)}\geq 0\Leftrightarrow$

$\frac{6x-6x+6-x^2+x}{x(x-1)}\geq 0\Leftrightarrow\frac{-x^2+x+6}{x(x-1)}\geq 0\Leftrightarrow\frac{(-x+3)(x+2)}{x(x-1)}\geq 0$. The expression on the left of the inequality changes sign where $x=3$, where $x=-2$, where $x=0$, and where $x=1$. Thus we must check the intervals in the following table.

Interval	$(-\infty,-2)$	$(-2,0)$	$(0,1)$	$(1,3)$	$(3,\infty)$
Sign of $-x+3$	+	+	+	+	−
Sign of $x+2$	−	+	+	+	+
Sign of x	−	−	+	+	+
Sign of $x-1$	−	−	−	+	+
Sign of $\frac{(-x+3)(x+2)}{x(x-1)}$	−	+	−	+	−

From the table, the solution set is $\{x\mid -2\leq x<0\text{ or }1<x\leq 3\}$. The points $x=0$ and $x=1$ are excluded from the solution set because they make the denominator zero. Interval: $[-2,0)\cup(1,3]$. Graph: −2 0 1 3

67. $\dfrac{x+2}{x+3} < \dfrac{x-1}{x-2} \Leftrightarrow \dfrac{x+2}{x+3} - \dfrac{x-1}{x-2} < 0 \Leftrightarrow \dfrac{(x+2)(x-2)}{(x+3)(x-2)} - \dfrac{(x-1)(x+3)}{(x-2)(x+3)} < 0 \Leftrightarrow$

$\dfrac{x^2-4-x^2-2x+3}{(x+3)(x-2)} < 0 \Leftrightarrow \dfrac{-2x-1}{(x+3)(x-2)} < 0$. The expression on the left of the inequality changes sign where $x = -\frac{1}{2}$, where $x = -3$, and where $x = 2$. Thus we must check the intervals in the following table.

Interval	$(-\infty, -3)$	$\left(-3, -\frac{1}{2}\right)$	$\left(-\frac{1}{2}, 2\right)$	$(2, \infty)$
Sign of $-2x-1$	+	+	−	−
Sign of $x+3$	−	+	+	+
Sign of $x-2$	−	−	−	+
Sign of $\dfrac{-2x-1}{(x+3)(x-2)}$	+	−	+	−

From the table, the solution set is $\left\{x \mid -3 < x < -\frac{1}{2} \text{ or } 2 < x\right\}$. Interval: $\left(-3, -\frac{1}{2}\right) \cup (2, \infty)$.

Graph: −3, $-\frac{1}{2}$, 2

69. $\dfrac{(x-1)(x+2)}{(x-2)^2} \geq 0$. Note that $(x-2)^2 \geq 0$ for all x. The expression on the left of the original inequality changes sign when $x = -2$ and $x = 1$. We check the intervals in the following table.

Interval	$(-\infty, -2)$	$(-2, 1)$	$(1, 2)$	$(2, \infty)$
Sign of $x-1$	−	−	+	+
Sign of $x+2$	−	+	+	+
Sign of $(x-2)^2$	+	+	+	+
Sign of $\dfrac{(x-1)(x+2)}{(x-2)^2}$	+	−	+	+

From the table, and recalling that the point $x = 2$ is excluded from the solution because the expression is undefined at those values, the solution set is $\{x \mid x \leq -2 \text{ or } x \geq 1 \text{ and } x \neq 2\}$. Interval: $(-\infty, -2] \cup [1, 2) \cup (2, \infty)$.

Graph: −2, 1, 2

71. $x^4 > x^2 \Leftrightarrow x^4 - x^2 > 0 \Leftrightarrow x^2\left(x^2-1\right) > 0 \Leftrightarrow x^2(x-1)(x+1) > 0$. The expression on the left of the inequality changes sign where $x = 0$, where $x = 1$, and where $x = -1$. Thus we must check the intervals in the following table.

Interval	$(-\infty, -1)$	$(-1, 0)$	$(0, 1)$	$(1, \infty)$
Sign of x^2	+	+	+	+
Sign of $x-1$	−	−	−	+
Sign of $x+1$	−	+	+	+
Sign of $x^2(x-1)(x+1)$	+	−	−	+

From the table, the solution set is $\{x \mid x < -1 \text{ or } 1 < x\}$. Interval: $(-\infty, -1) \cup (1, \infty)$. Graph: −1, 1

73. For $\sqrt{16-9x^2}$ to be defined as a real number we must have $16-9x^2 \geq 0 \Leftrightarrow (4-3x)(4+3x) \geq 0$. The expression in the inequality changes sign at $x=\frac{4}{3}$ and $x=-\frac{4}{3}$.

Interval	$\left(-\infty,-\frac{4}{3}\right)$	$\left(-\frac{4}{3},\frac{4}{3}\right)$	$\left(\frac{4}{3},\infty\right)$
Sign of $4-3x$	+	+	−
Sign of $4+3x$	−	+	+
Sign of $(4-3x)(4+3x)$	−	+	−

Thus $-\frac{4}{3} \leq x \leq \frac{4}{3}$.

75. For $\left(\dfrac{1}{x^2-5x-14}\right)^{1/2}$ to be defined as a real number we must have $x^2-5x-14>0 \Leftrightarrow (x-7)(x+2)>0$. The expression in the inequality changes sign at $x=7$ and $x=-2$.

Interval	$(-\infty,-2)$	$(-2,7)$	$(7,\infty)$
Sign of $x-7$	−	−	+
Sign of $x+2$	−	+	+
Sign of $(x-7)(x+2)$	+	−	+

Thus $x<-2$ or $7<x$, and the solution set is $(-\infty,-2)\cup(7,\infty)$.

77. $a(bx-c) \geq bc$ (where $a,b,c>0$) $\Leftrightarrow bx-c \geq \dfrac{bc}{a} \Leftrightarrow bx \geq \dfrac{bc}{a}+c \Leftrightarrow x \geq \dfrac{1}{b}\left(\dfrac{bc}{a}+c\right)=\dfrac{c}{a}+\dfrac{c}{b} \Leftrightarrow x \geq \dfrac{c}{a}+\dfrac{c}{b}$.

79. Inserting the relationship $C=\frac{5}{9}(F-32)$, we have $20 \leq C \leq 30 \Leftrightarrow 20 \leq \frac{5}{9}(F-32) \leq 30 \Leftrightarrow 36 \leq F-32 \leq 54 \Leftrightarrow 68 \leq F \leq 86$.

81. Let x be the average number of miles driven per day. Each day the cost of Plan A is $30+0.10x$, and the cost of Plan B is 50. Plan B saves money when $50<30+0.10x \Leftrightarrow 20<0.1x \Leftrightarrow 200<x$. So Plan B saves money when you average more than 200 miles a day.

83. We need to solve $6400 \leq 0.35m+2200 \leq 7100$ for m. So $6400 \leq 0.35m+2200 \leq 7100 \Leftrightarrow 4200 \leq 0.35m \leq 4900 \Leftrightarrow 12{,}000 \leq m \leq 14{,}000$. She plans on driving between 12,000 and 14,000 miles.

85. **(a)** Let x be the number of \$3 increases. Then the number of seats sold is $120-x$. So $P=200+3x \Leftrightarrow 3x=P-200 \Leftrightarrow x=\frac{1}{3}(P-200)$. Substituting for x we have that the number of seats sold is $120-x=120-\frac{1}{3}(P-200)=-\frac{1}{3}P+\frac{560}{3}$.

(b) $90 \leq -\frac{1}{3}P+\frac{560}{3} \leq 115 \Leftrightarrow 270 \leq 360-P+200 \leq 345 \Leftrightarrow 270 \leq -P+560 \leq 345 \Leftrightarrow -290 \leq -P \leq -215 \Leftrightarrow 290 \geq P \geq 215$. Putting this into standard order, we have $215 \leq P \leq 290$. So the ticket prices are between \$215 and \$290.

87. $0.0004 \leq \dfrac{4{,}000{,}000}{d^2} \leq 0.01$. Since $d^2 \geq 0$ and $d \neq 0$, we can multiply each expression by d^2 to obtain $0.0004d^2 \leq 4{,}000{,}000 \leq 0.01d^2$. Solving each pair, we have $0.0004d^2 \leq 4{,}000{,}000 \Leftrightarrow d^2 \leq 10{,}000{,}000{,}000 \Rightarrow d \leq 100{,}000$ (recall that d represents distance, so it is always nonnegative). Solving $4{,}000{,}000 \leq 0.01d^2 \Leftrightarrow 400{,}000{,}000 \leq d^2 \Rightarrow 20{,}000 \leq d$. Putting these together, we have $20{,}000 \leq d \leq 100{,}000$.

89. $128 + 16t - 16t^2 \geq 32 \Leftrightarrow -16t^2 + 16t + 96 \geq 0 \Leftrightarrow -16\left(t^2 - t - 6\right) \geq 0 \Leftrightarrow -16(t-3)(t+2) \geq 0$. The expression on the left of the inequality changes sign at $x = -2$, at $t = 3$, and at $t = -2$. However, $t \geq 0$, so the only endpoint is $t = 3$.

Interval	$(0, 3)$	$(3, \infty)$
Sign of -16	$-$	$-$
Sign of $t - 3$	$-$	$+$
Sign of $t + 2$	$+$	$+$
Sign of $-16(t-3)(t+2)$	$+$	$-$

So $0 \leq t \leq 3$.

91. $240 \geq v + \dfrac{v^2}{20} \Leftrightarrow \frac{1}{20}v^2 + v - 240 \leq 0 \Leftrightarrow \left(\frac{1}{20}v - 3\right)(v + 80) \leq 0$. The expression in the inequality changes sign at $v = 60$ and $v = -80$. However, since v represents the speed, we must have $v \geq 0$.

Interval	$(0, 60)$	$(60, \infty)$
Sign of $\frac{1}{20}v - 3$	$-$	$+$
Sign of $v + 80$	$+$	$+$
Sign of $\left(\frac{1}{20}v - 3\right)(v + 80)$	$-$	$+$

So Kerry must drive between 0 and 60 mi/h.

93. Let x be the length of the garden and w its width. Using the fact that the perimeter is 120 ft, we must have $2x + 2w = 120 \Leftrightarrow w = 60 - x$. Now since the area must be at least 800 ft^2, we have $800 < x(60 - x) \Leftrightarrow 800 < 60x - x^2 \Leftrightarrow x^2 - 60x + 800 < 0 \Leftrightarrow (x - 20)(x - 40) < 0$. The expression in the inequality changes sign at $x = 20$ and $x = 40$. However, since x represents length, we must have $x > 0$.

Interval	$(0, 20)$	$(20, 40)$	$(40, \infty)$
Sign of $x - 20$	$-$	$+$	$+$
Sign of $x - 40$	$-$	$-$	$+$
Sign of $(x - 20)(x - 40)$	$+$	$-$	$+$

The length of the garden should be between 20 and 40 feet.

95. The rule we want to apply here is "$a < b \Rightarrow ac < bc$ if $c > 0$ and $a < b \Rightarrow ac > bc$ if $c < 0$". Thus we cannot simply multiply by x, since we don't yet know if x is positive or negative, so in solving $1 < \dfrac{3}{x}$, we must consider two cases.

Case 1: $x > 0$ Multiplying both sides by x, we have $x < 3$. Together with our initial condition, we have $0 < x < 3$.

Case 2: $x < 0$ Multiplying both sides by x, we have $x > 3$. But $x < 0$ and $x > 3$ have no elements in common, so this gives no additional solution.

Hence, the only solutions are $0 < x < 3$.

97. $\dfrac{a}{b} < \dfrac{c}{d}$, so by Rule 3, $d\dfrac{a}{b} < d\dfrac{c}{d} \Leftrightarrow \dfrac{ad}{b} < c$. Adding a to both sides, we have $\dfrac{ad}{b} + a < c + a$. Rewriting the left-hand side as $\dfrac{ad}{b} + \dfrac{ab}{b} = \dfrac{a(b+d)}{b}$ and dividing both sides by $b + d$ gives $\dfrac{a}{b} < \dfrac{a+c}{b+d}$.

Similarly, $a + c < \dfrac{cb}{d} + c = \dfrac{c(b+d)}{d}$, so $\dfrac{a+c}{b+d} < \dfrac{c}{d}$.

1.8 SOLVING ABSOLUTE VALUE EQUATIONS AND INEQUALITIES

1. The equation $|x| = 3$ has the two solutions -3 and 3.

3. (a) The set of all points on the real line whose distance from zero is less than 3 can be described by the absolute value inequality $|x| < 3$.

(b) The set of all points on the real line whose distance from zero is greater than 3 can be described by the absolute value inequality $|x| > 3$.

5. $|5x| = 20 \Leftrightarrow 5x = \pm 20 \Leftrightarrow x = \pm 4$.

7. $5|x| + 3 = 28 \Leftrightarrow 5|x| = 25 \Leftrightarrow |x| = 5 \Leftrightarrow x = \pm 5$.

9. $|x - 3| = 2$ is equivalent to $x - 3 = \pm 2 \Leftrightarrow x = 3 \pm 2 \Leftrightarrow x = 1$ or $x = 5$.

11. $|x + 4| = 0.5$ is equivalent to $x + 4 = \pm 0.5 \Leftrightarrow x = -4 \pm 0.5 \Leftrightarrow x = -4.5$ or $x = -3.5$.

13. $|2x - 3| = 11$ is equivalent to either $2x - 3 = 11 \Leftrightarrow 2x = 14 \Leftrightarrow x = 7$; or $2x - 3 = -11 \Leftrightarrow 2x = -8 \Leftrightarrow x = -4$. The two solutions are $x = 7$ and $x = -4$.

15. $4 - |3x + 6| = 1 \Leftrightarrow -|3x + 6| = -3 \Leftrightarrow |3x + 6| = 3$, which is equivalent to either $3x + 6 = 3 \Leftrightarrow 3x = -3 \Leftrightarrow x = -1$; or $3x + 6 = -3 \Leftrightarrow 3x = -9 \Leftrightarrow x = -3$. The two solutions are $x = -1$ and $x = -3$.

17. $3|x + 5| + 6 = 15 \Leftrightarrow 3|x + 5| = 9 \Leftrightarrow |x + 5| = 3$, which is equivalent to either $x + 5 = 3 \Leftrightarrow x = -2$; or $x + 5 = -3 \Leftrightarrow x = -8$. The two solutions are $x = -2$ and $x = -8$.

19. $8 + 5\left|\frac{1}{3}x - \frac{5}{6}\right| = 33 \Leftrightarrow 5\left|\frac{1}{3}x - \frac{5}{6}\right| = 25 \Leftrightarrow \left|\frac{1}{3}x - \frac{5}{6}\right| = 5$, which is equivalent to either $\frac{1}{3}x - \frac{5}{6} = 5 \Leftrightarrow \frac{1}{3}x = \frac{35}{6} \Leftrightarrow x = \frac{35}{2}$; or $\frac{1}{3}x - \frac{5}{6} = -5 \Leftrightarrow \frac{1}{3}x = -\frac{25}{6} \Leftrightarrow x = -\frac{25}{2}$. The two solutions are $x = -\frac{25}{2}$ and $x = \frac{35}{2}$.

21. $|x - 1| = |3x + 2|$, which is equivalent to either $x - 1 = 3x + 2 \Leftrightarrow -2x = 3 \Leftrightarrow x = -\frac{3}{2}$; or $x - 1 = -(3x + 2) \Leftrightarrow x - 1 = -3x - 2 \Leftrightarrow 4x = -1 \Leftrightarrow x = -\frac{1}{4}$. The two solutions are $x = -\frac{3}{2}$ and $x = -\frac{1}{4}$.

23. $|x| \le 5 \Leftrightarrow -5 \le x \le 5$. Interval: $[-5, 5]$.

25. $|2x| > 7$ is equivalent to $2x > 7 \Leftrightarrow x > \frac{7}{2}$; or $2x < -7 \Leftrightarrow x < -\frac{7}{2}$. Interval: $\left(-\infty, -\frac{7}{2}\right) \cup \left(\frac{7}{2}, \infty\right)$.

27. $|x - 4| \le 10$ is equivalent to $-10 \le x - 4 \le 10 \Leftrightarrow -6 \le x \le 14$. Interval: $[-6, 14]$.

29. $|x + 1| \ge 1$ is equivalent to $x + 1 \ge 1 \Leftrightarrow x \ge 0$; or $x + 1 \le -1 \Leftrightarrow x \le -2$. Interval: $(-\infty, -2] \cup [0, \infty)$.

31. $|2x + 1| \ge 3$ is equivalent to $2x + 1 \le -3 \Leftrightarrow 2x \le -4 \Leftrightarrow x \le -2$; or $2x + 1 \ge 3 \Leftrightarrow 2x \ge 2 \Leftrightarrow x \ge 1$. Interval: $(-\infty, -2] \cup [1, \infty)$.

33. $|2x - 3| \le 0.4 \Leftrightarrow -0.4 \le 2x - 3 \le 0.4 \Leftrightarrow 2.6 \le 2x \le 3.4 \Leftrightarrow 1.3 \le x \le 1.7$. Interval: $[1.3, 1.7]$.

35. $\left|\dfrac{x-2}{3}\right| < 2 \Leftrightarrow -2 < \dfrac{x-2}{3} < 2 \Leftrightarrow -6 < x - 2 < 6 \Leftrightarrow -4 < x < 8$. Interval: $(-4, 8)$.

37. $|x + 6| < 0.001 \Leftrightarrow -0.001 < x + 6 < 0.001 \Leftrightarrow -6.001 < x < -5.999$. Interval: $(-6.001, -5.999)$.

39. $4|x + 2| - 3 < 13 \Leftrightarrow 4|x + 2| < 16 \Leftrightarrow |x + 2| < 4 \Leftrightarrow -4 < x + 2 < 4 \Leftrightarrow -6 < x < 2$. Interval: $(-6, 2)$.

41. $8 - |2x - 1| \ge 6 \Leftrightarrow -|2x - 1| \ge -2 \Leftrightarrow |2x - 1| \le 2 \Leftrightarrow -2 \le 2x - 1 \le 2 \Leftrightarrow -1 \le 2x \le 3 \Leftrightarrow -\frac{1}{2} \le x \le \frac{3}{2}$. Interval: $\left[-\frac{1}{2}, \frac{3}{2}\right]$.

43. $\frac{1}{2}\left|4x + \frac{1}{3}\right| > \frac{5}{6} \Leftrightarrow \left|4x + \frac{1}{3}\right| > \frac{5}{3}$, which is equivalent to either $4x + \frac{1}{3} > \frac{5}{3} \Leftrightarrow 4x > \frac{4}{3} \Leftrightarrow x > \frac{1}{3}$; or $4x + \dfrac{1}{3} < -\frac{5}{3} \Leftrightarrow 4x < -2 \Leftrightarrow x < -\frac{1}{2}$. Interval: $\left(-\infty, -\frac{1}{2}\right) \cup \left(\frac{1}{3}, \infty\right)$.

45. $1 \le |x| \le 4$. If $x \ge 0$, then this is equivalent to $1 \le x \le 4$. If $x < 0$, then this is equivalent to $1 \le -x \le 4 \Leftrightarrow -1 \ge x \ge -4 \Leftrightarrow -4 \le x \le -1$. Interval: $[-4, -1] \cup [1, 4]$.

47. $\dfrac{1}{|x+7|} > 2 \Leftrightarrow 1 > 2|x + 7|$ $(x \ne -7) \Leftrightarrow |x + 7| < \frac{1}{2} \Leftrightarrow -\frac{1}{2} < x + 7 < \frac{1}{2} \Leftrightarrow -\frac{15}{2} < x < -\frac{13}{2}$ and $x \ne -7$. Interval: $\left(-\frac{15}{2}, -7\right) \cup \left(-7, -\frac{13}{2}\right)$.

49. $|x| < 3$ **51.** $|x - 7| \geq 5$ **53.** $|x| \leq 2$ **55.** $|x| > 3$

57. **(a)** Let x be the thickness of the laminate. Then $|x - 0.020| \leq 0.003$.

(b) $|x - 0.020| \leq 0.003 \Leftrightarrow -0.003 \leq x - 0.020 \leq 0.003 \Leftrightarrow 0.017 \leq x \leq 0.023$.

59. $|x - 1|$ is the distance between x and 1; $|x - 3|$ is the distance between x and 3. So $|x - 1| < |x - 3|$ represents those points closer to 1 than to 3, and the solution is $x < 2$, since 2 is the point halfway between 1 and 3. If $a < b$, then the solution to $|x - a| < |x - b|$ is $x < \dfrac{a+b}{2}$.

1.9 SOLVING EQUATIONS AND INEQUALITIES GRAPHICALLY

1. The solutions of the equation $x^2 - 2x - 3 = 0$ are the x-intercepts of the graph of $y = x^2 - 2x - 3$.

3. **(a)** From the graph, it appears that the graph of $y = x^4 - 3x^3 - x^2 + 3x$ has x-intercepts -1, 0, 1, and 3, so the solutions to the equation $x^4 - 3x^3 - x^2 + 3x = 0$ are $x = -1$, $x = 0$, $x = 1$, and $x = 3$.

(b) From the graph, we see that where $-1 \leq x \leq 0$ or $1 \leq x \leq 3$, the graph lies below the x-axis. Thus, the inequality $x^4 - 3x^3 - x^2 + 3x \leq 0$ is satisfied for $\{x \mid -1 \leq x \leq 0 \text{ or } 1 \leq x \leq 3\} = [-1, 0] \cup [1, 3]$.

5. Algebraically: $x - 4 = 5x + 12 \Leftrightarrow -16 = 4x \Leftrightarrow x = -4$.
Graphically: We graph the two equations $y_1 = x - 4$ and $y_2 = 5x + 12$ in the viewing rectangle $[-6, 4]$ by $[-10, 2]$. Zooming in, we see that the solution is $x = -4$.

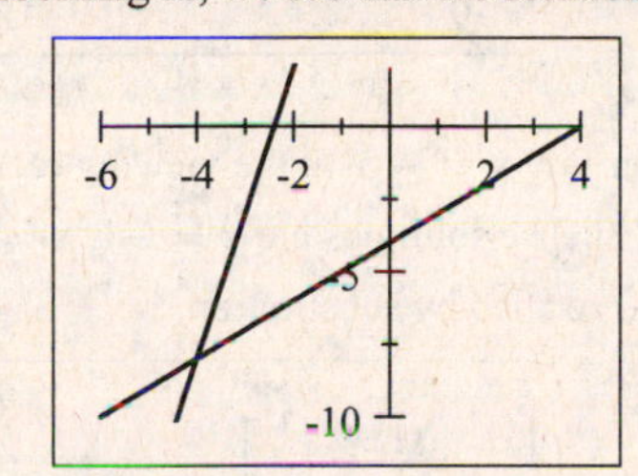

7. Algebraically: $\dfrac{2}{x} + \dfrac{1}{2x} = 7 \Leftrightarrow 2x\left(\dfrac{2}{x} + \dfrac{1}{2x}\right) = 2x\,(7)$
$\Leftrightarrow 4 + 1 = 14x \Leftrightarrow x = \frac{5}{14}$.

Graphically: We graph the two equations $y_1 = \dfrac{2}{x} + \dfrac{1}{2x}$ and $y_2 = 7$ in the viewing rectangle $[-2, 2]$ by $[-2, 8]$. Zooming in, we see that the solution is $x \approx 0.36$.

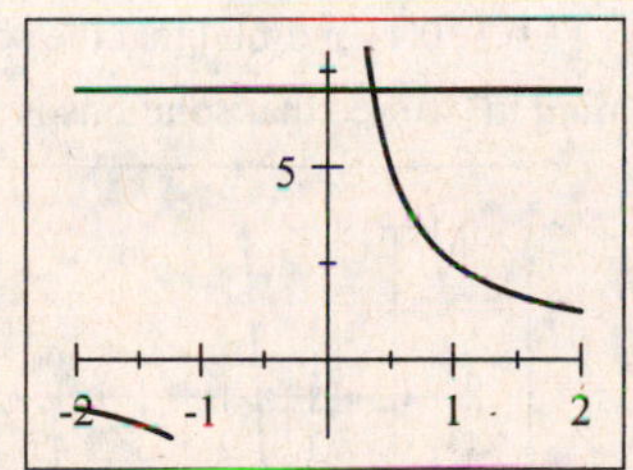

9. Algebraically: $x^2 - 32 = 0 \Leftrightarrow x^2 = 32 \Rightarrow$
$x = \pm\sqrt{32} = \pm 4\sqrt{2}$.
Graphically: We graph the equation $y_1 = x^2 - 32$ and determine where this curve intersects the x-axis. We use the viewing rectangle $[-10, 10]$ by $[-5, 5]$. Zooming in, we see that solutions are $x \approx 5.66$ and $x \approx -5.66$.

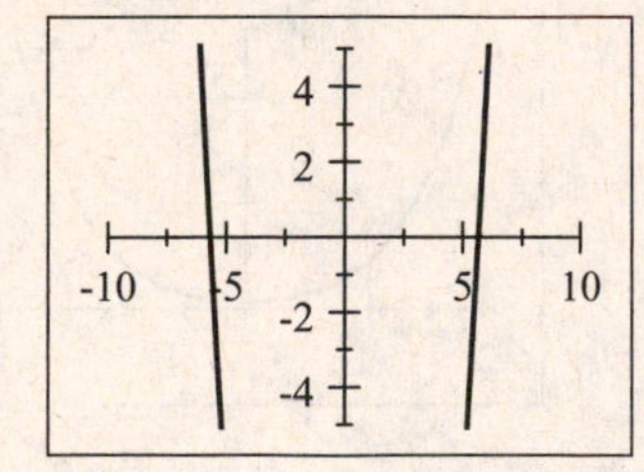

11. Algebraically: $x^2 + 9 = 0 \Leftrightarrow x^2 = -9$, which has no real solution.
Graphically: We graph the equation $y = x^2 + 9$ and see that this curve does not intersect the x-axis. We use the viewing rectangle $[-5, 5]$ by $[-5, 30]$.

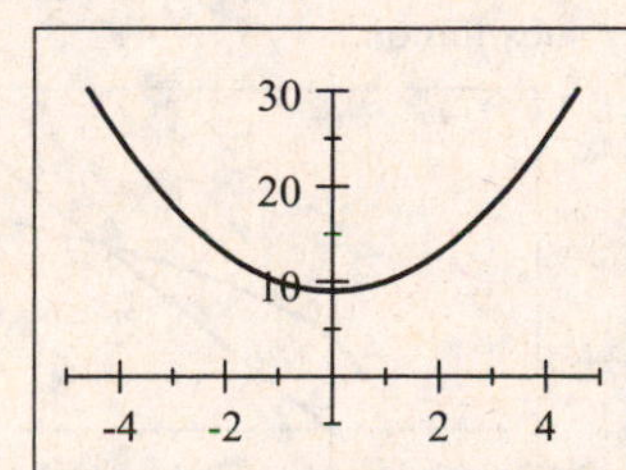

13. Algebraically: $16x^4 = 625 \Leftrightarrow x^4 = \frac{625}{16} \Rightarrow$ $x = \pm\frac{5}{2} = \pm 2.5$.

Graphically: We graph the two equations $y_1 = 16x^4$ and $y_2 = 625$ in the viewing rectangle $[-5, 5]$ by $[610, 640]$. Zooming in, we see that solutions are $x = \pm 2.5$.

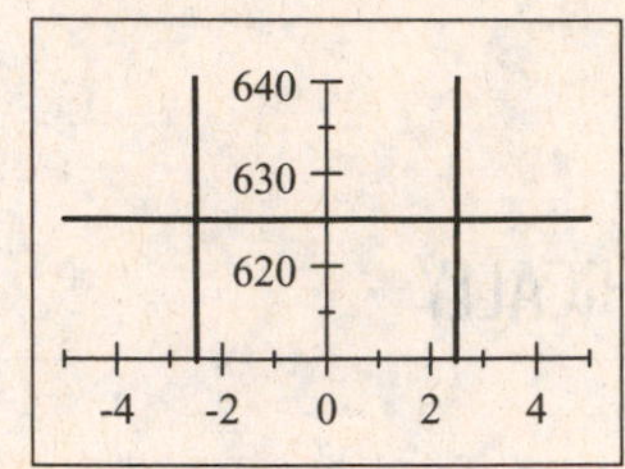

15. Algebraically: $(x - 5)^4 - 80 = 0 \Leftrightarrow (x - 5)^4 = 80 \Rightarrow$ $x - 5 = \pm\sqrt[4]{80} = \pm 2\sqrt[4]{5} \Leftrightarrow x = 5 \pm 2\sqrt[4]{5}$.

Graphically: We graph the equation $y_1 = (x - 5)^4 - 80$ and determine where this curve intersects the x-axis. We use the viewing rectangle $[-1, 9]$ by $[-5, 5]$. Zooming in, we see that solutions are $x \approx 2.01$ and $x \approx 7.99$.

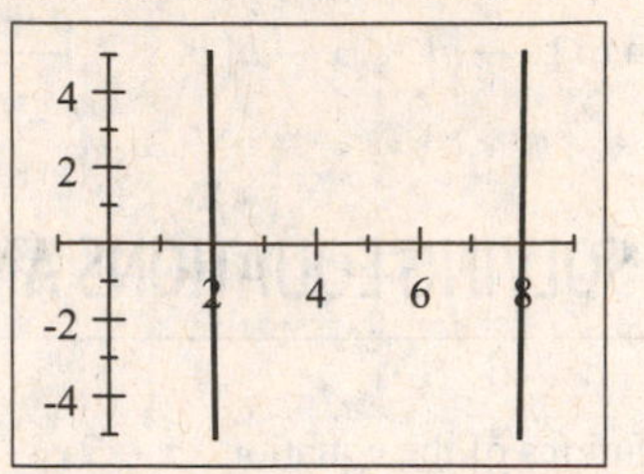

17. We graph $y = x^2 - 7x + 12$ in the viewing rectangle $[0, 6]$ by $[-0.1, 0.1]$. The solutions appear to be exactly $x = 3$ and $x = 4$. [In fact $x^2 - 7x + 12 = (x - 3)(x - 4)$.]

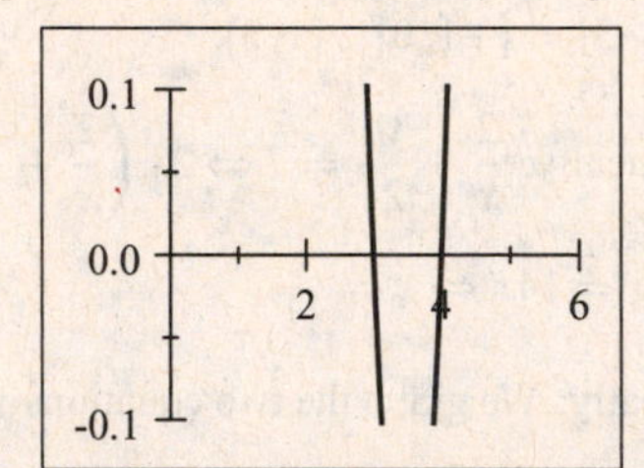

19. We graph $y = x^3 - 6x^2 + 11x - 6$ in the viewing rectangle $[-1, 4]$ by $[-0.1, 0.1]$. The solutions are $x = 1.00$, $x = 2.00$,and $x = 3.00$.

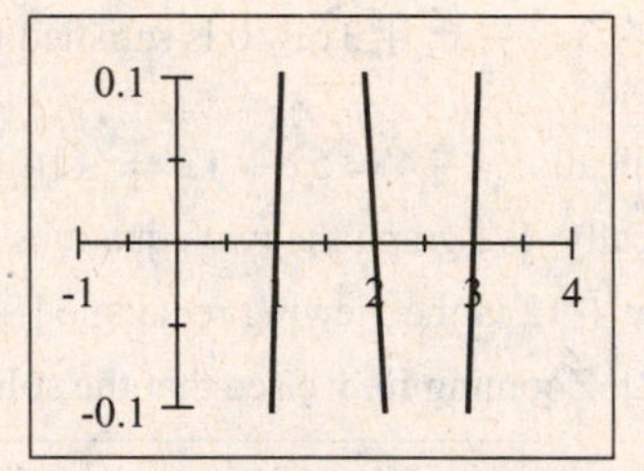

21. We first graph $y = x - \sqrt{x + 1}$ in the viewing rectangle $[-1, 5]$ by $[-0.1, 0.1]$ and find that the solution is near 1.6. Zooming in, we see that solutions is $x \approx 1.62$.

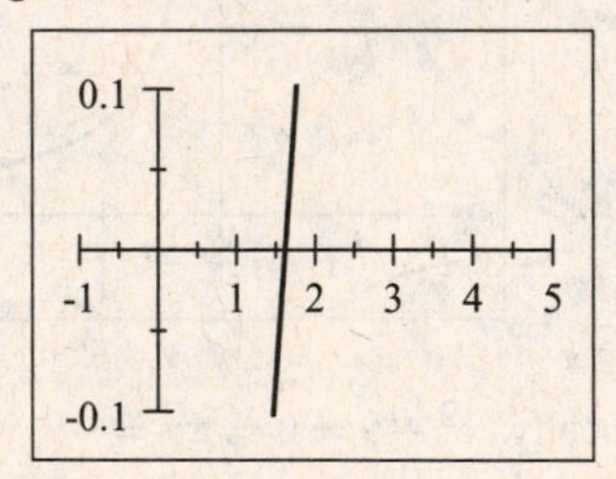

23. We graph $y = x^{1/3} - x$ in the viewing rectangle $[-3, 3]$ by $[-1, 1]$. The solutions are $x = -1$, $x = 0$, and $x = 1$, as can be verified by substitution.

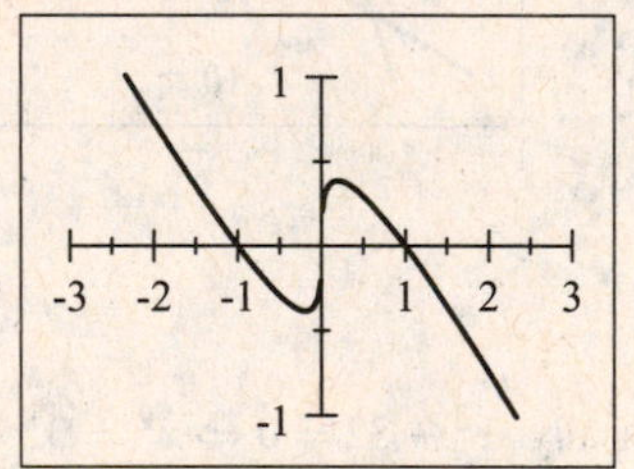

25. We graph $y = \sqrt{2x + 1} + 1$ and $y = x$ in the viewing rectangle $[-3, 6]$ by $[0, 6]$ and see that the only solution to the equation $\sqrt{2x + 1} + 1 = x$ is $x = 4$, which can be verified by substitution.

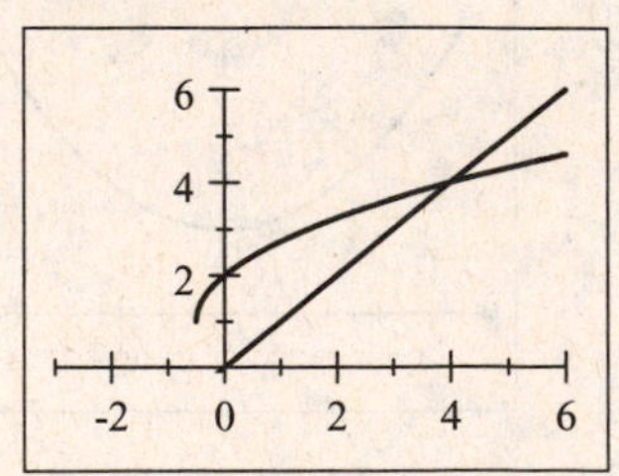

27. We graph $y = 2x^4 + 4x^2 + 1$ in the viewing rectangle $[-2, 2]$ by $[-5, 40]$ and see that the equation $2x^4 + 4x^2 + 1 = 0$ has no solution.

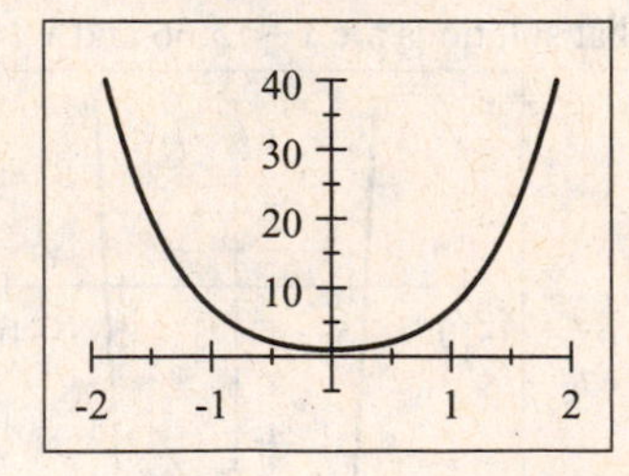

29. $x^3 - 2x^2 - x - 1 = 0$, so we start by graphing the function $y = x^3 - 2x^2 - x - 1$ in the viewing rectangle $[-10, 10]$ by $[-100, 100]$. There appear to be two solutions, one near $x = 0$ and another one between $x = 2$ and $x = 3$. We then use the viewing rectangle $[-1, 5]$ by $[-1, 1]$ and zoom in on the only solution, $x \approx 2.55$.

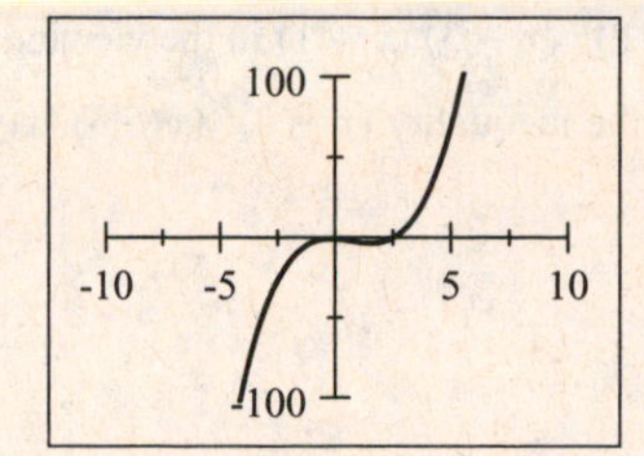

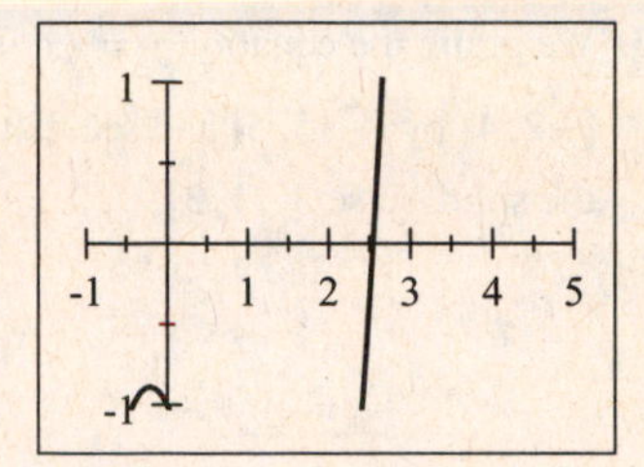

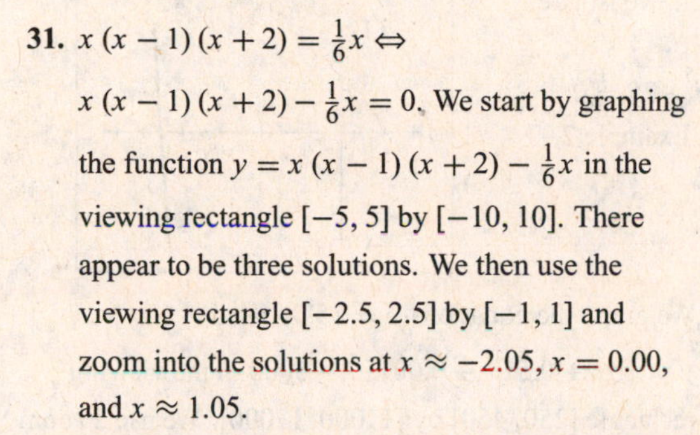

31. $x(x-1)(x+2) = \frac{1}{6}x \Leftrightarrow$
$x(x-1)(x+2) - \frac{1}{6}x = 0$. We start by graphing the function $y = x(x-1)(x+2) - \frac{1}{6}x$ in the viewing rectangle $[-5, 5]$ by $[-10, 10]$. There appear to be three solutions. We then use the viewing rectangle $[-2.5, 2.5]$ by $[-1, 1]$ and zoom into the solutions at $x \approx -2.05$, $x = 0.00$, and $x \approx 1.05$.

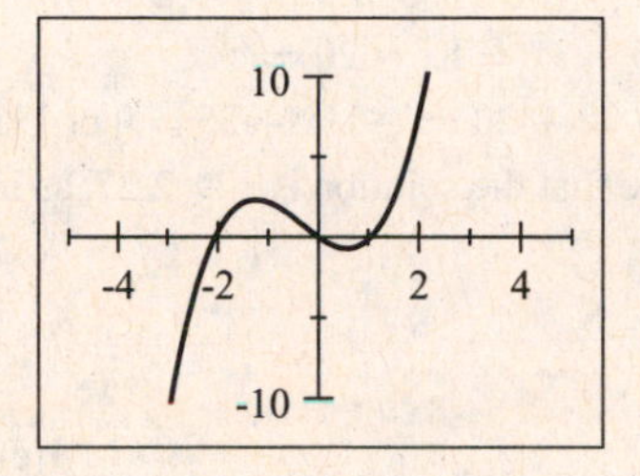

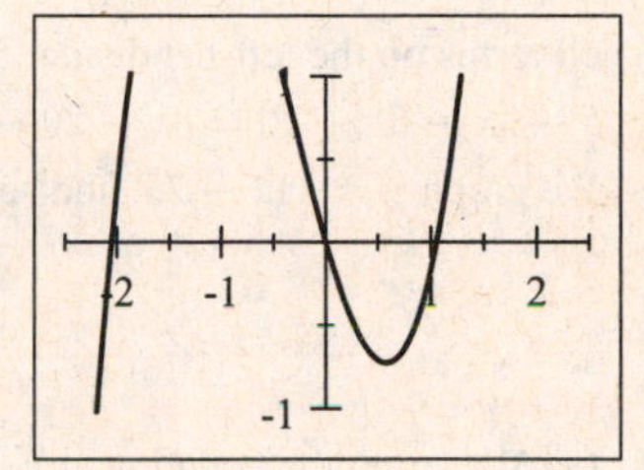

33. We graph $y = x^2$ and $y = 3x + 10$ in the viewing rectangle $[-4, 7]$ by $[-5, 30]$. The solution to the inequality is $[-2, 5]$.

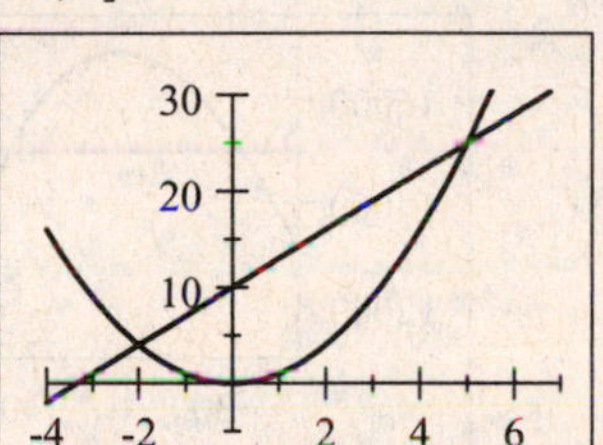

35. Since $x^3 + 11x \le 6x^2 + 6 \Leftrightarrow x^3 - 6x^2 + 11x - 6 \le 0$, we graph $y = x^3 - 6x^2 + 11x - 6$ in the viewing rectangle $[0, 5]$ by $[-5, 5]$. The solution set is $(-\infty, 1.0] \cup [2.0, 3.0]$.

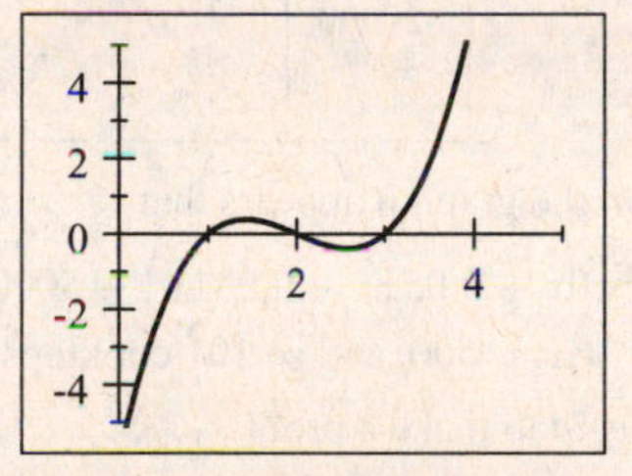

37. Since $x^{1/3} \le x \Leftrightarrow x^{1/3} - x < 0$, we graph $y = x^{1/3} - x$ in the viewing rectangle $[-3, 3]$ by $[-1, 1]$. From this, we find that the solution set is $(-1, 0) \cup (1, \infty)$.

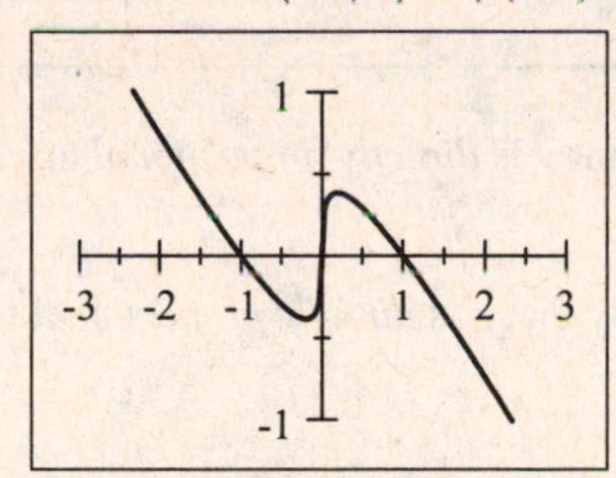

39. Since $(x+1)^2 < (x-1)^2 \Leftrightarrow (x+1)^2 - (x-1)^2 < 0$, we graph $y = (x+1)^2 - (x-1)^2$ in the viewing rectangle $[-2, 2]$ by $[-5, 5]$. The solution set is $(-\infty, 0)$.

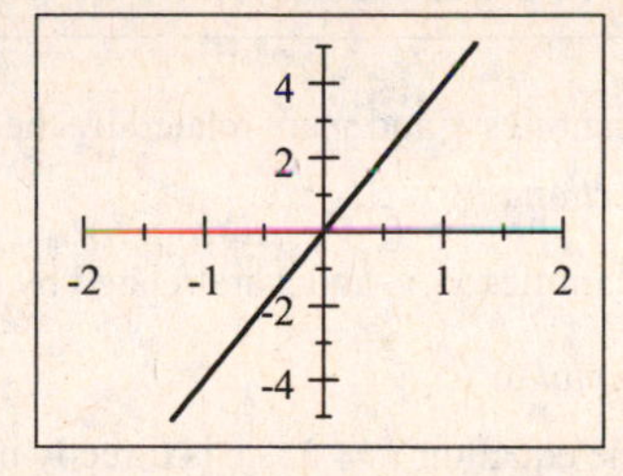

41. We graph the equations $y = 3x^2 - 3x$ and $y = 2x^2 + 4$ in the viewing rectangle $[-2, 6]$ by $[-5, 50]$. We see that the two curves intersect at $x = -1$ and at $x = 4$, and that the first curve is lower than the second for $-1 < x < 4$. Thus, we see that the inequality $3x^2 - 3x < 2x^2 + 4$ has the solution set $(-1, 4)$.

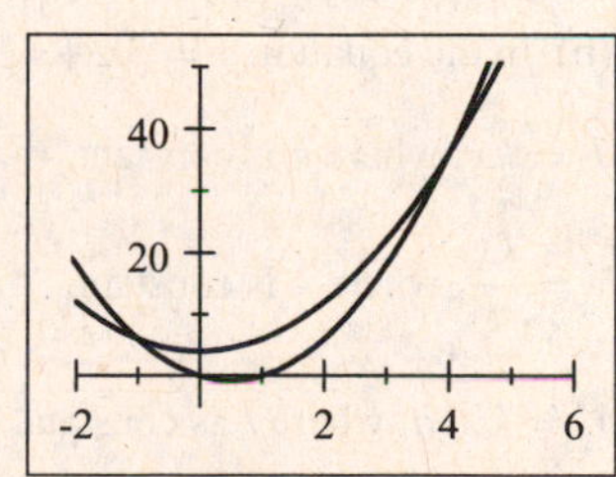

43. We graph the equation $y = (x-2)^2 (x-3)(x+1)$ in the viewing rectangle $[-2, 4]$ by $[-15, 5]$ and see that the inequality $(x-2)^2 (x-3)(x+1) \le 0$ has the solution set $[-1, 3]$.

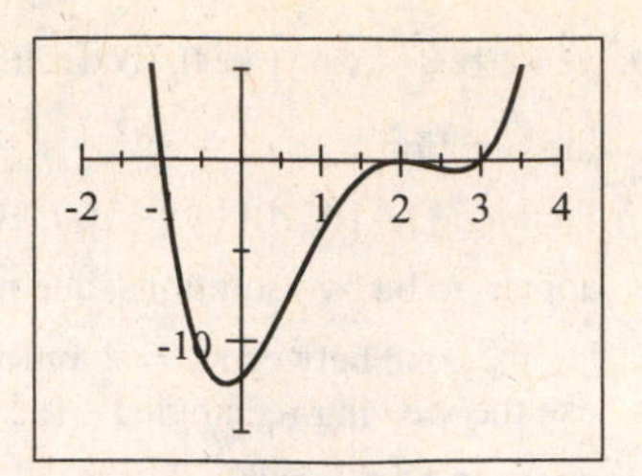

45. To solve $5 - 3x = 8x - 20$ by drawing the graph of a single equation, we isolate all terms on the left-hand side: $5 - 3x = 8x - 20 \Leftrightarrow$
$5 - 3x - 8x + 20 = 8x - 20 - 8x + 20 \Leftrightarrow -11x + 25 = 0$ or $11x - 25 = 0$.
We graph $y = 11x - 25$, and see that the solution is $x \approx 2.27$, as in Example 2.

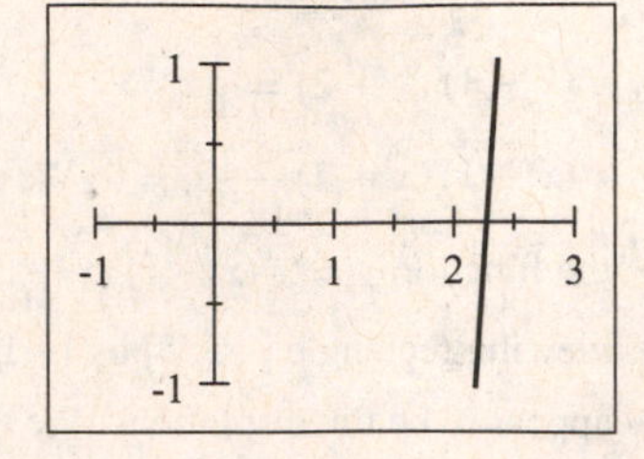

47. **(a)** We graph the equation

$y = 10x + 0.5x^2 - 0.001x^3 - 5000$ in the viewing rectangle $[0, 600]$ by $[-30000, 20000]$.

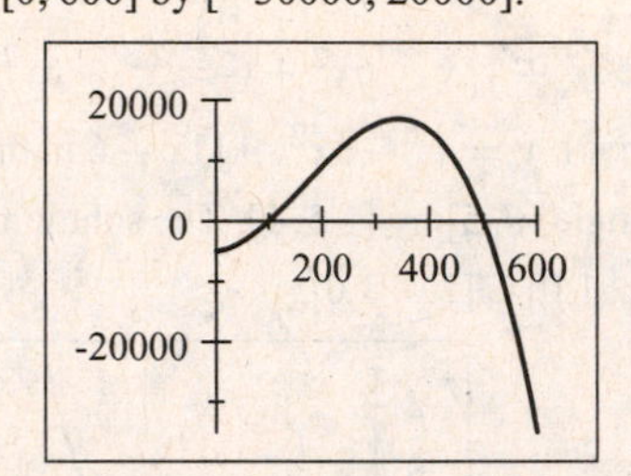

(b) From the graph it appears that

$0 < 10x + 0.05x^2 - 0.001x^3 - 5000$ for $100 < x < 500$, and so 101 cooktops must be produced to *begin* to make a profit.

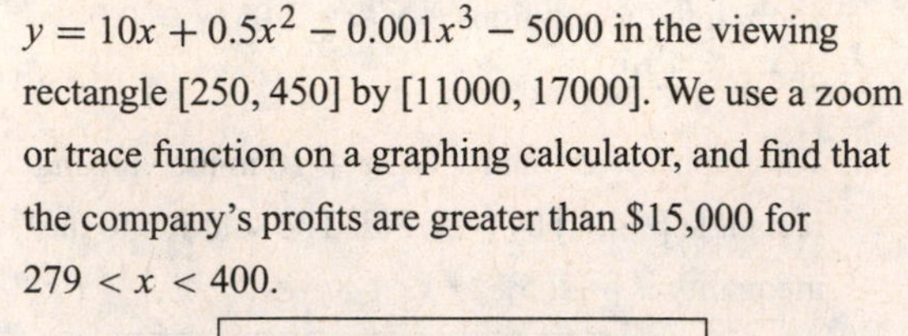
(c) We graph the equations $y = 15{,}000$ and $y = 10x + 0.5x^2 - 0.001x^3 - 5000$ in the viewing rectangle $[250, 450]$ by $[11000, 17000]$. We use a zoom or trace function on a graphing calculator, and find that the company's profits are greater than \$15,000 for $279 < x < 400$.

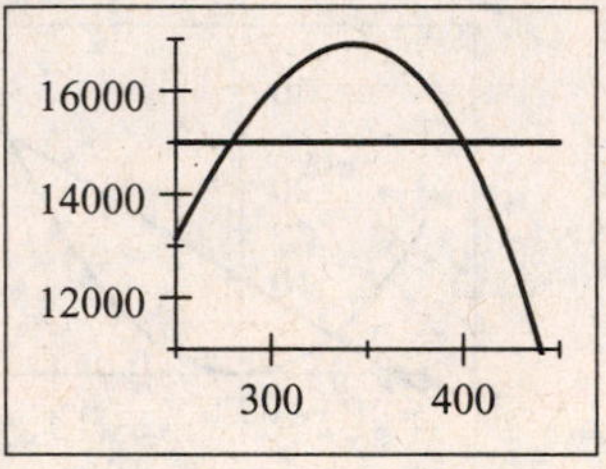

49. Answers will vary.

1.10 MODELING VARIATION

1. If the quantities x and y are related by the equation $y = 3x$ then we say that y is *directly proportional* to x, and the constant of *proportionality* is 3.

3. If the quantities x, y, and z are related by the equation $z = 3\frac{x}{y}$ then we say that z is *directly proportional* to x and *inversely proportional* to y.

5. **(a)** In the equation $y = 3x$, y is directly proportional to x.
(b) In the equation $y = 3x + 1$, y is not proportional to x.

7. $T = kx$, where k is constant.

9. $v = \frac{k}{z}$, where k is constant.

11. $y = \frac{ks}{t}$, where k is constant.

13. $z = k\sqrt{y}$, where k is constant.

15. $V = klwh$, where k is constant.

17. $R = \frac{kP^2t^2}{b^3}$, where k is constant.

19. Since y is directly proportional to x, $y = kx$. Since $y = 42$ when $x = 6$, we have $42 = k(6) \Leftrightarrow k = 7$. So $y = 7x$.

21. A varies inversely as r, so $A = \frac{k}{r}$. Since $A = 7$ when $r = 3$, we have $7 = \frac{k}{3} \Leftrightarrow k = 21$. So $A = \frac{21}{r}$.

23. Since A is directly proportional to x and inversely proportional to t, $A = \frac{kx}{t}$. Since $A = 42$ when $x = 7$ and $t = 3$, we have $42 = \frac{k(7)}{3} \Leftrightarrow k = 18$. Therefore, $A = \frac{18x}{t}$.

25. Since W is inversely proportional to the square of r, $W = \frac{k}{r^2}$. Since $W = 10$ when $r = 6$, we have $10 = \frac{k}{(6)^2} \Leftrightarrow k = 360$. So $W = \frac{360}{r^2}$.

27. Since C is jointly proportional to l, w, and h, we have $C = klwh$. Since $C = 128$ when $l = w = h = 2$, we have $128 = k(2)(2)(2) \Leftrightarrow 128 = 8k \Leftrightarrow k = 16$. Therefore, $C = 16lwh$.

29. $R = \frac{k}{\sqrt{x}}$. Since $R = 2.5$ when $x = 121$, $2.5 = \frac{k}{\sqrt{121}} = \frac{k}{11} \Leftrightarrow k = 27.5$. Thus, $R = \frac{27.5}{\sqrt{x}}$.

31. **(a)** $z = k\frac{x^3}{y^2}$

(b) If we replace x with $3x$ and y with $2y$, then $z = k\frac{(3x)^3}{(2y)^2} = \frac{27}{4}\left(k\frac{x^3}{y^2}\right)$, so z changes by a factor of $\frac{27}{4}$.

33. **(a)** $z = kx^3y^5$

(b) If we replace x with $3x$ and y with $2y$, then $z = k(3x)^3(2y)^5 = 864kx^3y^5$, so z changes by a factor of 864.

35. **(a)** The force F needed is $F = kx$.

(b) Since $F = 30$ N when $x = 9$ cm and the spring's natural length is 5 cm, we have $30 = k(9 - 5) \Leftrightarrow k = 7.5$.

(c) From part (b), we have $F = 7.5x$. Substituting $x = 11 - 5 = 6$ into $F = 7.5x$ gives $F = 7.5(6) = 45$ N.

37. **(a)** $P = ks^3$.

(b) Since $P = 96$ when $s = 20$, we get $96 = k \cdot 20^3 \Leftrightarrow k = 0.012$. So $P = 0.012s^3$.

(c) Substituting $x = 30$, we get $P = 0.012 \cdot 30^3 = 324$ watts.

39. $D = ks^2$. Since $D = 150$ when $s = 40$, we have $150 = k(40)^2$, so $k = 0.09375$. Thus, $D = 0.09375s^2$. If $D = 200$, then $200 = 0.09375s^2 \Leftrightarrow s^2 \approx 2133.3$, so $s \approx 46$ mi/h (for safety reasons we round down).

41. $F = kAs^2$. Since $F = 220$ when $A = 40$ and $s = 5$. Solving for k we have $220 = k(40)(5)^2 \Leftrightarrow 220 = 1000k \Leftrightarrow k = 0.22$. Now when $A = 28$ and $F = 175$ we get $175 = 0.220(28)s^2 \Leftrightarrow 28.4090 = s^2$ so $s = \sqrt{28.4090} = 5.33$ mi/h.

43. **(a)** $P = \frac{kT}{V}$.

(b) Substituting $P = 33.2$, $T = 400$, and $V = 100$, we get $33.2 = \frac{k(400)}{100} \Leftrightarrow k = 8.3$. Thus $k = 8.3$ and the equation is $P = \frac{8.3T}{V}$.

(c) Substituting $T = 500$ and $V = 80$, we have $P = \frac{8.3(500)}{80} = 51.875$ kPa. Hence the pressure of the sample of gas is about 51.9 kPa.

45. **(a)** The loudness L is inversely proportional to the square of the distance d, so $L = \frac{k}{d^2}$.

(b) Substituting $d = 10$ and $L = 70$, we have $70 = \frac{k}{10^2} \Leftrightarrow k = 7000$.

(c) Substituting $2d$ for d, we have $L = \frac{k}{(2d)^2} = \frac{1}{4}\left(\frac{k}{d^2}\right)$, so the loudness is changed by a factor of $\frac{1}{4}$.

(d) Substituting $\frac{1}{2}d$ for d, we have $L = \dfrac{k}{\left(\frac{1}{2}d\right)^2} = 4\left(\dfrac{k}{d^2}\right)$, so the loudness is changed by a factor of 4.

47. (a) $R = \dfrac{kL}{d^2}$

(b) Since $R = 140$ when $L = 1.2$ and $d = 0.005$, we get $140 = \dfrac{k\,(1.2)}{(0.005)^2} \Leftrightarrow k = \frac{7}{2400} = 0.00291\overline{6}$.

(c) Substituting $L = 3$ and $d = 0.008$, we have $R = \dfrac{7}{2400} \cdot \dfrac{3}{(0.008)^2} = \dfrac{4375}{32} \approx 137\ \Omega$.

(d) If we substitute $2d$ for d and $3L$ for L, then $R = \dfrac{k\,(3L)}{(2d)^2} = \dfrac{3}{4}\dfrac{kL}{d^2}$, so the resistance is changed by a factor of $\frac{3}{4}$.

49. (a) For the sun, $E_S = k6000^4$ and for earth $E_E = k300^4$. Thus $\dfrac{E_S}{E_E} = \dfrac{k6000^4}{k300^4} = \left(\frac{6000}{300}\right)^4 = 20^4 = 160{,}000$. So the sun produces 160,000 times the radiation energy per unit area than the Earth.

(b) The surface area of the sun is $4\pi\,(435{,}000)^2$ and the surface area of the Earth is $4\pi\,(3{,}960)^2$. So the sun has $\dfrac{4\pi\,(435{,}000)^2}{4\pi\,(3{,}960)^2} = \left(\dfrac{435{,}000}{3{,}960}\right)^2$ times the surface area of the Earth. Thus the total radiation emitted by the sun is $160{,}000 \times \left(\dfrac{435{,}000}{3{,}960}\right)^2 = 1{,}930{,}670{,}340$ times the total radiation emitted by the Earth.

51. (a) Let T and l be the period and the length of the pendulum, respectively. Then $T = k\sqrt{l}$.

(b) $T = k\sqrt{l} \Rightarrow T^2 = k^2 l \Leftrightarrow l = \dfrac{T^2}{k^2}$. If the period is doubled, the new length is $\dfrac{(2T)^2}{k^2} = 4\dfrac{T^2}{k^2} = 4l$. So we would quadruple the length l to double the period T.

53. (a) Since f is inversely proportional to L, we have $f = \dfrac{k}{L}$, where k is a positive constant.

(b) If we replace L by $2L$ we have $\dfrac{k}{2L} = \frac{1}{2} \cdot \dfrac{k}{L} = \frac{1}{2}f$. So the frequency of the vibration is cut in half.

55. Using $B = k\dfrac{L}{d^2}$ with $k = 0.080$, $L = 2.5 \times 10^{26}$, and $d = 2.4 \times 10^{19}$, we have $B = 0.080\dfrac{2.5 \times 10^{26}}{\left(2.4 \times 10^{19}\right)^2} \approx 3.47 \times 10^{-14}$. The star's apparent brightness is about $3.47 \times 10^{-14}\ \text{W/m}^2$.

57. Examples include radioactive decay and exponential growth in biology.

CHAPTER 1 REVIEW

1. (a)

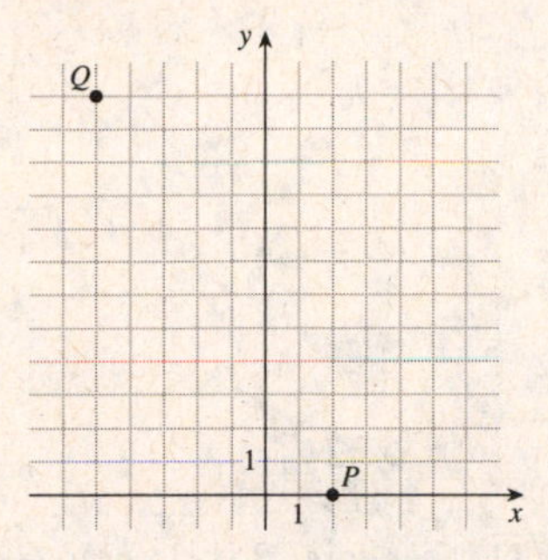

(b) The distance from P to Q is

$$d(P,Q)=\sqrt{(-5-2)^2+(12-0)^2}$$
$$=\sqrt{49+144}=\sqrt{193}$$

(c) The midpoint is $\left(\frac{-5+2}{2},\frac{12+0}{2}\right)=\left(-\frac{3}{2},6\right)$.

(d) The line has slope $m=\frac{12-0}{-5-2}=-\frac{12}{7}$, and has equation $y-0=-\frac{12}{7}(x-2)\Leftrightarrow y=-\frac{12}{7}x+\frac{24}{7}$ $\Leftrightarrow 12x+7y-24=0$.

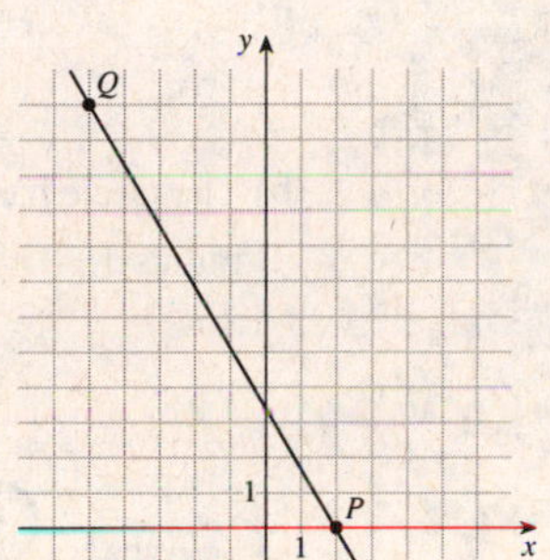

(e) The radius of this circle was found in part (b). It is $r=d(P,Q)=\sqrt{193}$. So an equation is

$(x-2)^2+(y-0)^2=\left(\sqrt{193}\right)^2\Leftrightarrow(x-2)^2+y^2=193.$

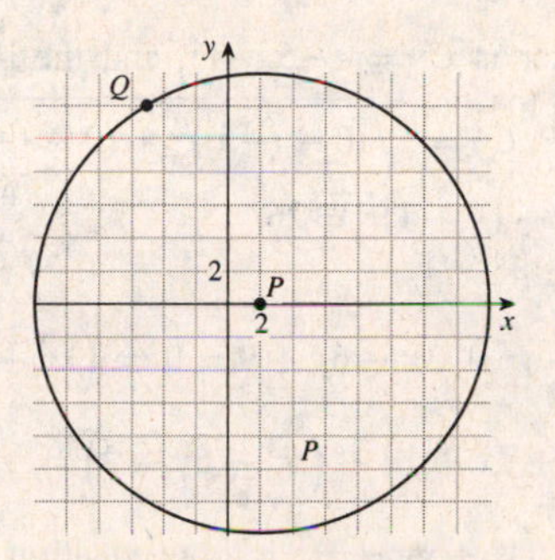

3. (a)

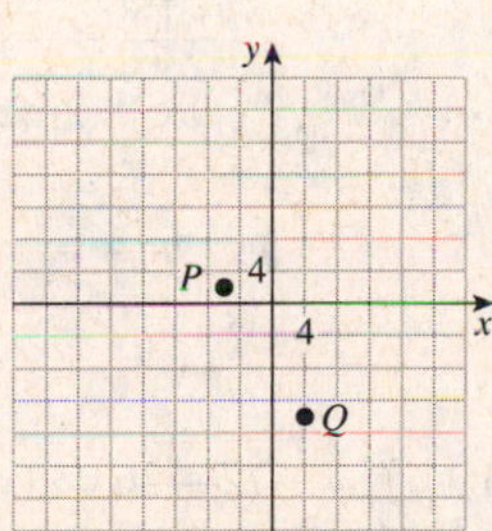

(b) The distance from P to Q is

$$d(P,Q)=\sqrt{(-6-4)^2+[2-(-14)]^2}$$
$$=\sqrt{100+256}=\sqrt{356}=2\sqrt{89}$$

(c) The midpoint is $\left(\frac{-6+4}{2},\frac{2+(-14)}{2}\right)=(-1,-6)$.

(d) The line has slope $m=\frac{2-(-14)}{-6-4}=\frac{16}{-10}=-\frac{8}{5}$ and equation $y-2=-\frac{8}{5}(x+6)\Leftrightarrow$ $y-2=-\frac{8}{5}x-\frac{48}{5}\Leftrightarrow y=-\frac{8}{5}x-\frac{38}{5}$.

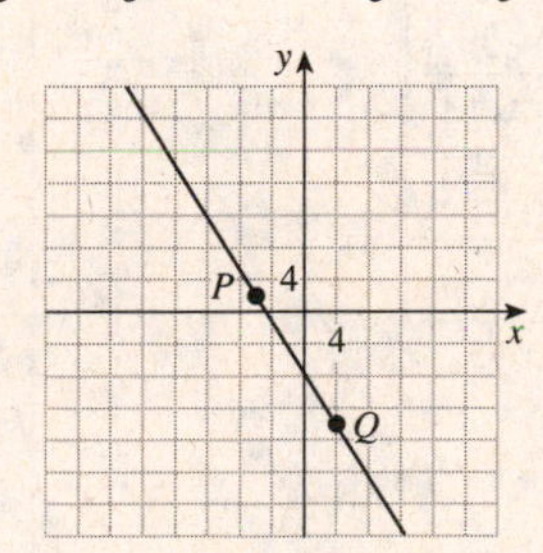

(e) The radius of this circle was found in part (b). It is $r=d(P,Q)=2\sqrt{89}$. So an equation is

$[x-(-6)]^2+(y-2)^2=\left(2\sqrt{89}\right)^2\Leftrightarrow$

$(x+6)^2+(y-2)^2=356.$

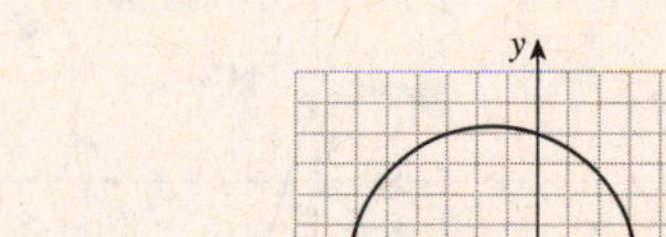

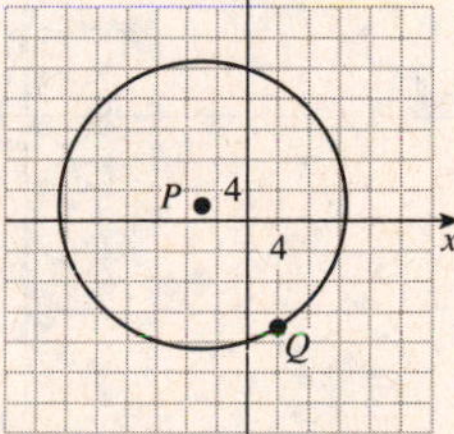

5.

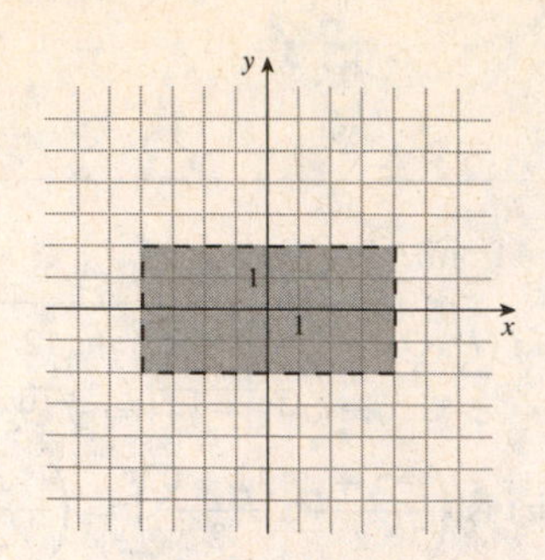

7. $d(A,C) = \sqrt{(4-(-1))^2+(4-(-3))^2} = \sqrt{(4+1)^2+(4+3)^2} = \sqrt{74}$ and
$d(B,C) = \sqrt{(5-(-1))^2+(3-(-3))^2} = \sqrt{(5+1)^2+(3+3)^2} = \sqrt{72}$. Therefore, B is closer to C.

9. The center is $C = (-5,-1)$, and the point $P = (0,0)$ is on the circle. The radius of the circle is
$r = d(P,C) = \sqrt{(0-(-5))^2+(0-(-1))^2} = \sqrt{(0+5)^2+(0+1)^2} = \sqrt{26}$. Thus, the equation of the circle is
$(x+5)^2+(y+1)^2 = 26$.

11. (a) $x^2+y^2+2x-6y+9=0 \Leftrightarrow \left(x^2+2x\right)+\left(y^2-6y\right)=-9 \Leftrightarrow$
$\left(x^2+2x+1\right)+\left(y^2-6y+9\right)=-9+1+9 \Leftrightarrow$
$(x+1)^2+(y-3)^2=1$, an equation of a circle.

(b) The circle has center $(-1,3)$ and radius 1.

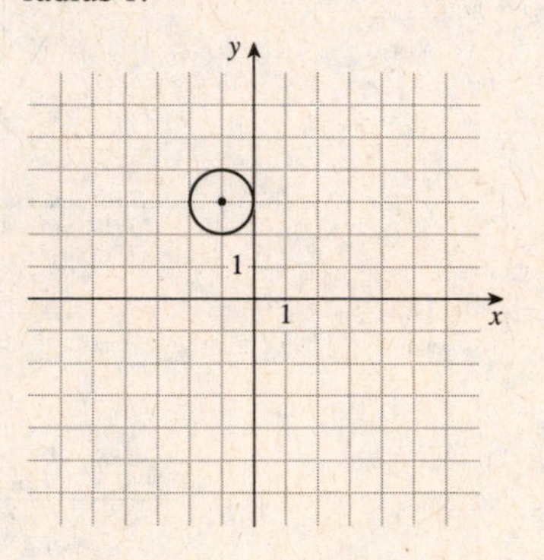

13. (a) $x^2+y^2+72=12x \Leftrightarrow \left(x^2-12x\right)+y^2=-72 \Leftrightarrow \left(x^2-12x+36\right)+y^2=-72+36 \Leftrightarrow (x-6)^2+y^2=-36$.
Since the left side of this equation must be greater than or equal to zero, this equation has no graph.

15. $y = 2-3x$

x	y
-2	8
0	2
$\frac{2}{3}$	0

17. $\frac{x}{2}-\frac{y}{7}=1 \Leftrightarrow y=\frac{7}{2}x-7$

x	y
-2	-14
0	-7
2	0

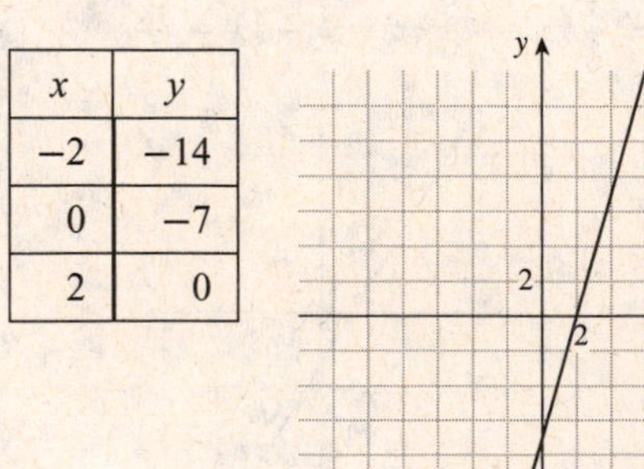

19. $y = 16 - x^2$

x	y
-3	7
-1	15
0	16
1	15
3	7

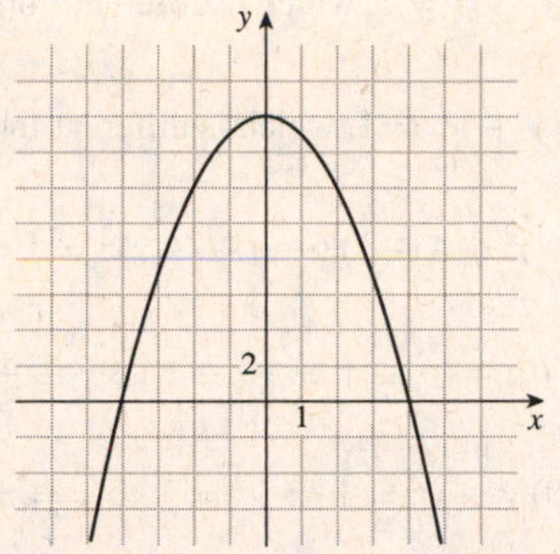

21. $x = \sqrt{y}$

x	y
0	0
1	1
2	4
3	9

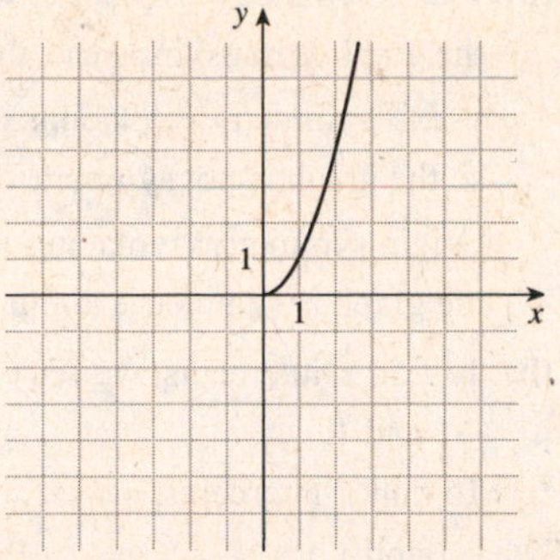

23. $y = 9 - x^2$

(a) x-axis symmetry: replacing y by $-y$ gives $-y = 9 - x^2$, which is not the same as the original equation, so the graph is not symmetric about the x-axis.

y-axis symmetry: replacing x by $-x$ gives $y = 9 - (-x)^2 = 9 - x^2$, which is the same as the original equation, so the graph is symmetric about the y-axis.

Origin symmetry: replacing x by $-x$ and y by $-y$ gives $-y = 9 - (-x)^2 \Leftrightarrow y = -9 + x^2$, which is not the same as the original equation, so the graph is not symmetric about the origin.

(b) To find x-intercepts, we set $y = 0$ and solve for x: $0 = 9 - x^2 \Leftrightarrow x^2 = 9 \Leftrightarrow x = \pm 3$, so the x-intercepts are -3 and 3.

To find y-intercepts, we set $x = 0$ and solve for y: $y = 9 - 0^2 = 9$, so the y-intercept is 9.

25. $x^2 + (y - 1)^2 = 1$

(a) x-axis symmetry: replacing y by $-y$ gives $x^2 + [(-y) - 1]^2 = 1 \Leftrightarrow x^2 + (y + 1)^2 = 1$, so the graph is not symmetric about the x-axis.

y-axis symmetry: replacing x by $-x$ gives $(-x)^2 + (y - 1)^2 = 1 \Leftrightarrow x^2 + (y - 1)^2 = 1$, so the graph is symmetric about the y-axis.

Origin symmetry: replacing x by $-x$ and y by $-y$ gives $(-x)^2 + [(-y) - 1]^2 = 1 \Leftrightarrow x^2 + (y + 1)^2 = 1$, so the graph is not symmetric about the origin.

(b) To find x-intercepts, we set $y = 0$ and solve for x: $x^2 + (0 - 1)^2 = 1 \Leftrightarrow x^2 = 0$, so the x-intercept is 0.

To find y-intercepts, we set $x = 0$ and solve for y: $0^2 + (y - 1)^2 = 1 \Leftrightarrow y - 1 = \pm 1 \Leftrightarrow y = 0$ or 2, so the y-intercepts are 0 and 2.

27. $9x^2 - 16y^2 = 144$

(a) x-axis symmetry: replacing y by $-y$ gives $9x^2 - 16(-y)^2 = 144 \Leftrightarrow 9x^2 - 16y^2 = 144$, so the graph is symmetric about the x-axis.

y-axis symmetry: replacing x by $-x$ gives $9(-x)^2 - 16y^2 = 144 \Leftrightarrow 9x^2 - 16y^2 = 144$, so the graph is symmetric about the y-axis.

Origin symmetry: replacing x by $-x$ and y by $-y$ gives $9(-x)^2 - 16(-y)^2 = 144 \Leftrightarrow 9x^2 - 16y^2 = 144$, so the graph is symmetric about the origin.

(b) To find x-intercepts, we set $y = 0$ and solve for x: $9x^2 - 16(0)^2 = 144 \Leftrightarrow 9x^2 = 144 \Leftrightarrow x = \pm 4$, so the x-intercepts are -4 and 4.

To find y-intercepts, we set $x = 0$ and solve for y: $9(0)^2 - 16y^2 = 144 \Leftrightarrow 16y^2 = -144$, so there is no y-intercept.

29. $x^2 + 4xy + y^2 = 1$

(a) x-axis symmetry: replacing y by $-y$ gives $x^2 + 4x(-y) + (-y)^2 = 1$, which is different from the original equation, so the graph is not symmetric about the x-axis.

y-axis symmetry: replacing x by $-x$ gives $(-x)^2 + 4(-x)y + y^2 = 1$, which is different from the original equation, so the graph is not symmetric about the y-axis.

Origin symmetry: replacing x by $-x$ and y by $-y$ gives $(-x)^2 + 4(-x)(-y) + (-y)^2 = 1 \Leftrightarrow x^2 + 4xy + y^2 = 1$, so the graph is symmetric about the origin.

(b) To find x-intercepts, we set $y = 0$ and solve for x: $x^2 + 4x(0) + 0^2 = 1 \Leftrightarrow x^2 = 1 \Leftrightarrow x = \pm 1$, so the x-intercepts are -1 and 1.

To find y-intercepts, we set $x = 0$ and solve for y: $0^2 + 4(0)y + y^2 = 1 \Leftrightarrow y^2 = 1 \Leftrightarrow y = \pm 1$, so the y-intercepts are -1 and 1.

31. **(a)** We graph $y = x^2 - 6x$ in the viewing rectangle $[-10, 10]$ by $[-10, 10]$.

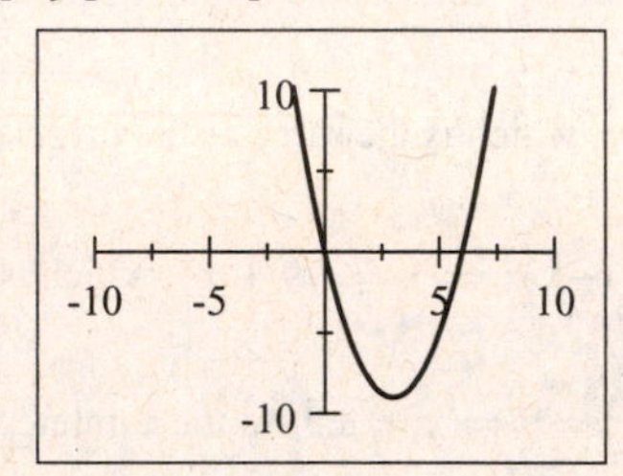

(b) From the graph, we see that the x-intercepts are 0 and 6 and the y-intercept is 0.

33. **(a)** We graph $y = x^3 - 4x^2 - 5x$ in the viewing rectangle $[-4, 8]$ by $[-30, 20]$.

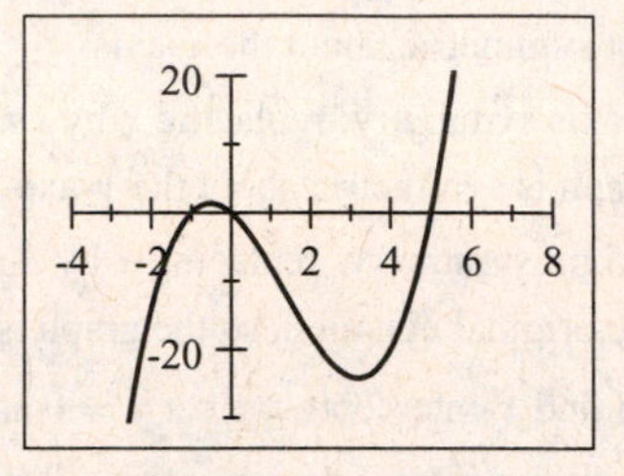

(b) From the graph, we see that the x-intercepts are -1, 0, and 5 and the y-intercept is 0.

35. **(a)** The line that has slope 2 and y-intercept 6 has the slope-intercept equation $y = 2x + 6$.

(b) An equation of the line in general form is $2x - y + 6 = 0$.

(c)

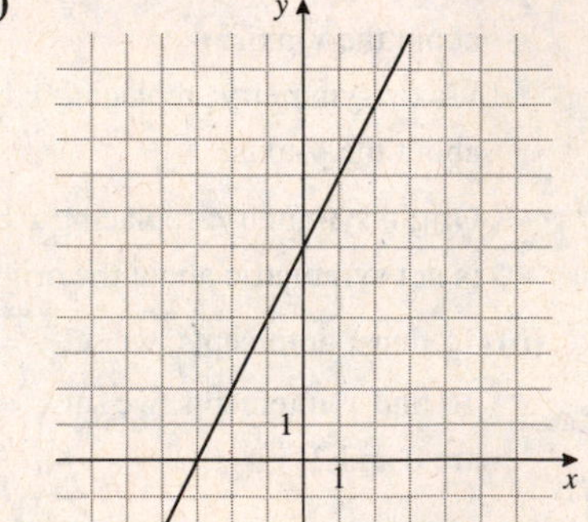

37. **(a)** The line that passes through the points $(-1, -6)$ and $(2, -4)$ has slope $m = \frac{-4 - (-6)}{2 - (-1)} = \frac{2}{3}$, so $y - (-6) = \frac{2}{3}[x - (-1)] \Leftrightarrow y + 6 = \frac{2}{3}x + \frac{2}{3}$ $\Leftrightarrow y = \frac{2}{3}x - \frac{16}{3}$.

(b) $y = \frac{2}{3}x - \frac{16}{3} \Leftrightarrow 3y = 2x - 16 \Leftrightarrow 2x - 3y - 16 = 0$.

(c)

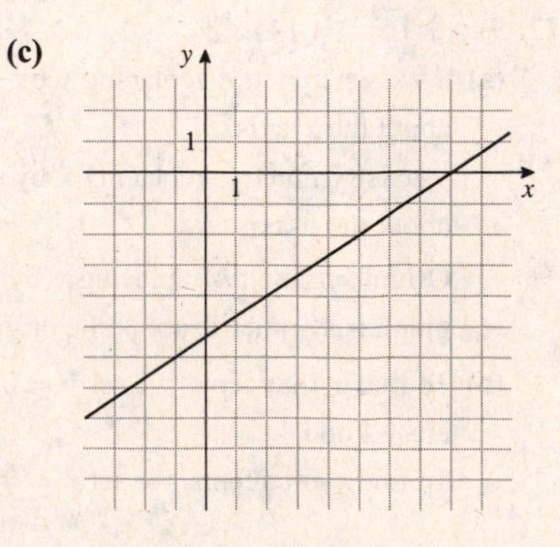

39. (a) The vertical line that passes through the point $(3, -2)$ has equation $x = 3$.

(b) $x = 3 \Leftrightarrow x - 3 = 0$.

(c)

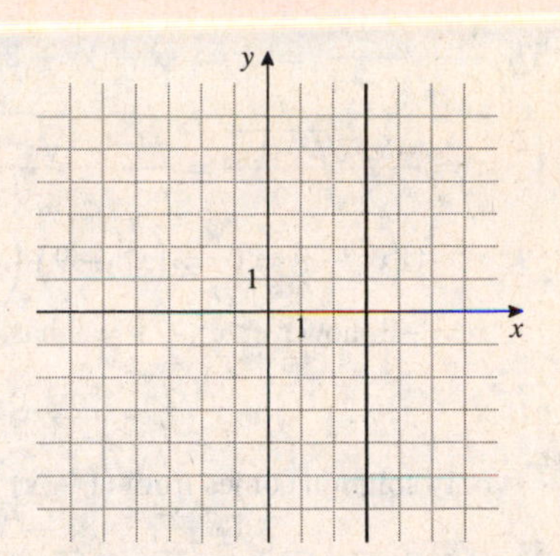

41. (a) $2x - 5y = 10 \Leftrightarrow 5y = 2x - 10 \Leftrightarrow y = \frac{2}{5}x - 2$, so the given line has slope $m = \frac{2}{5}$. Thus, an equation of the line passing through $(1, 1)$ parallel to this line is $y - 1 = \frac{2}{5}(x - 1) \Leftrightarrow y = \frac{2}{5}x + \frac{3}{5}$.

(b) $y = \frac{2}{5}x + \frac{3}{5} \Leftrightarrow 5y = 2x + 3 \Leftrightarrow 2x - 5y + 3 = 0$.

(c)

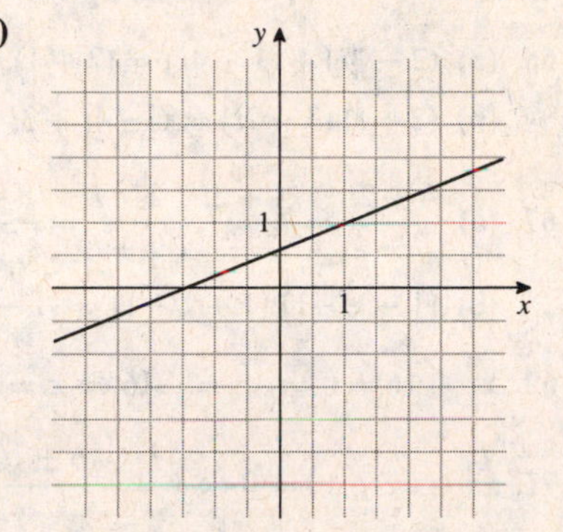

43. (a) The line $y = \frac{1}{2}x - 10$ has slope $\frac{1}{2}$, so a line perpendicular to this one has slope $-\frac{1}{1/2} = -2$. In particular, the line passing through the origin perpendicular to the given line has equation $y = -2x$.

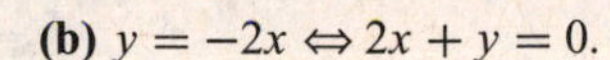
(b) $y = -2x \Leftrightarrow 2x + y = 0$.

(c)

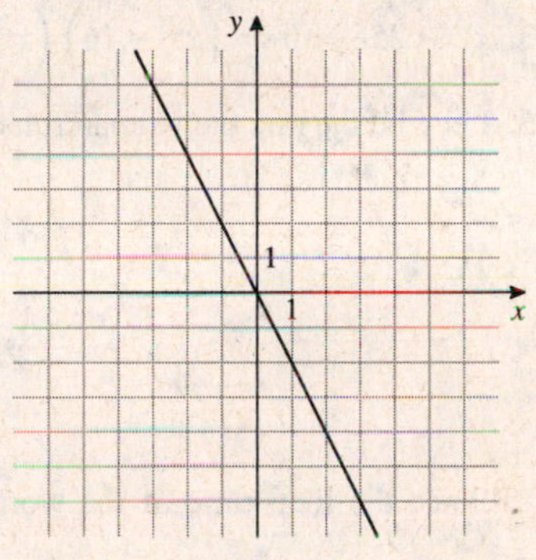

45. The line with equation $y = -\frac{1}{3}x - 1$ has slope $-\frac{1}{3}$. The line with equation $9y + 3x + 3 = 0 \Leftrightarrow 9y = -3x - 3 \Leftrightarrow y = -\frac{1}{3}x - \frac{1}{3}$ also has slope $-\frac{1}{3}$, so the lines are parallel.

47. (a) The slope represents a stretch of 0.3 inches for each one-pound increase in weight. The s-intercept represents the length of the unstretched spring.

(b) When $w = 5$, $s = 0.3(5) + 2.5 = 1.5 + 2.5 = 4.0$ inches.

49. $x^2 - 9x + 14 = 0 \Leftrightarrow (x - 7)(x - 2) = 0 \Leftrightarrow x = 7$ or $x = 2$.

51. $2x^2 + x = 1 \Leftrightarrow 2x^2 + x - 1 = 0 \Leftrightarrow (2x - 1)(x + 1) = 0$. So either $2x - 1 = 0 \Leftrightarrow 2x = 1 \Leftrightarrow x = \frac{1}{2}$; or $x + 1 = 0 \Leftrightarrow x = -1$.

53. $0 = 4x^3 - 25x = x\left(4x^2 - 25\right) = x(2x - 5)(2x + 5) = 0$. So either $x = 0$; or $2x - 5 = 0 \Leftrightarrow 2x = 5 \Leftrightarrow x = \frac{5}{2}$; or $2x + 5 = 0 \Leftrightarrow 2x = -5 \Leftrightarrow x = -\frac{5}{2}$.

55. $3x^2 + 4x - 1 = 0 \Rightarrow$

$$x = \frac{-b \pm \sqrt{b^2 - 4ac}}{2a} = \frac{-(4) \pm \sqrt{(4)^2 - 4(3)(-1)}}{2(-3)} = \frac{-4 \pm \sqrt{16 + 12}}{-6} = \frac{-4 \pm \sqrt{28}}{-6} = \frac{-4 \pm 2\sqrt{7}}{6} = \frac{2\left(-2 \pm \sqrt{7}\right)}{-6} = \frac{-2 \pm \sqrt{7}}{3}.$$

57. $\dfrac{1}{x} + \dfrac{2}{x-1} = 3 \Leftrightarrow (x-1) + 2(x) = 3(x)(x-1) \Leftrightarrow x - 1 + 2x = 3x^2 - 3x \Leftrightarrow 0 = 3x^2 - 6x + 1 \Rightarrow$

$x = \frac{-b \pm \sqrt{b^2-4ac}}{2a} = \frac{-(-6) \pm \sqrt{(-6)^2 - 4(3)(1)}}{2(3)} = \frac{6 \pm \sqrt{36-12}}{6} = \frac{6 \pm \sqrt{24}}{6} = \frac{6 \pm 2\sqrt{6}}{6} = \frac{2\left(3 \pm \sqrt{6}\right)}{6} = \frac{3 \pm \sqrt{6}}{3}$.

59. $x^4 - 8x^2 - 9 = 0 \Leftrightarrow \left(x^2 - 9\right)\left(x^2 + 1\right) = 0 \Leftrightarrow (x-3)(x+3)\left(x^2+1\right) = 0 \Rightarrow x - 3 = 0 \Leftrightarrow x = 3$, or $x + 3 = 0 \Leftrightarrow$ $x = -3$, however $x^2 + 1 = 0$ has no real solution. The solutions are $x = \pm 3$.

61. $x^{-1/2} - 2x^{1/2} + x^{3/2} = 0 \Leftrightarrow x^{-1/2}\left(1 - 2x + x^2\right) = 0 \Leftrightarrow x^{-1/2}(1-x)^2 = 0$. Since $x^{-1/2} - 1/\sqrt{x}$ is never 0, the only solution comes from $(1-x)^2 = 0 \Leftrightarrow 1 - x = 0 \Leftrightarrow x = 1$.

63. $|x - 7| = 4 \Leftrightarrow x - 7 = \pm 4 \Leftrightarrow x = 7 \pm 4$, so $x = 11$ or $x = 3$.

65. **(a)** $(2 - 3i) + (1 + 4i) = (2 + 1) + (-3 + 4)i = 3 + i$

(b) $(2 + i)(3 - 2i) = 6 - 4i + 3i - 2i^2 = 6 - i + 2 = 8 - i$

67. **(a)** $\dfrac{4+2i}{2-i} = \dfrac{4+2i}{2-i} \cdot \dfrac{2+i}{2+i} = \dfrac{8+8i+2i^2}{4-i^2} = \dfrac{8+8i-2}{4+1} = \dfrac{6+8i}{5} = \frac{6}{5} + \frac{8}{5}i$

(b) $\left(1 - \sqrt{-1}\right)\left(1 + \sqrt{-1}\right) = (1 - i)(1 + i) = 1 + i - i - i^2 = 1 + 1 = 2$

69. $x^2 + 16 = 0 \Leftrightarrow x^2 = -16 \Leftrightarrow x = \pm 4i$

71. $x^2 + 6x + 10 = 0 \Leftrightarrow x = \dfrac{-b \pm \sqrt{b^2 - 4ac}}{2a} = \dfrac{-6 \pm \sqrt{6^2 - 4(1)(10)}}{2(1)} = \dfrac{-6 \pm \sqrt{36 - 40}}{2} = -3 \pm i$

73. $x^4 - 256 = 0 \Leftrightarrow \left(x^2 - 16\right)\left(x^2 + 16\right) = 0 \Leftrightarrow x = \pm 4$ or $x = \pm 4i$

75. Let r be the rate the woman runs in mi/h. Then she cycles at $r + 8$ mi/h.

	Rate	Time	Distance
Cycle	$r+8$	$\dfrac{4}{r+8}$	4
Run	r	$\dfrac{2.5}{r}$	2.5

Since the total time of the workout is 1 hour, we have $\dfrac{4}{r+8} + \dfrac{2.5}{r} = 1$. Multiplying by $2r(r+8)$, we get $4(2r) + 2.5(2)(r+8) = 2r(r+8) \Leftrightarrow 8r + 5r + 40 = 2r^2 + 16r \Leftrightarrow 0 = 2r^2 + 3r - 40 \Rightarrow$

$r = \frac{-3 \pm \sqrt{(3)^2 - 4(2)(-40)}}{2(2)} = \frac{-3 \pm \sqrt{9+320}}{4} = \frac{-3 \pm \sqrt{329}}{4}$. Since $r \geq 0$, we reject the negative value. She runs at $r = \frac{-3 + \sqrt{329}}{4} \approx 3.78$ mi/h.

77. Let x be the length of one side in cm. Then $28 - x$ is the length of the other side. Using the Pythagorean Theorem, we have $x^2 + (28 - x)^2 = 20^2 \Leftrightarrow x^2 + 784 - 56x + x^2 = 400 \Leftrightarrow 2x^2 - 56x + 384 = 0 \Leftrightarrow 2\left(x^2 - 28x + 192\right) = 0 \Leftrightarrow$ $2(x - 12)(x - 16) = 0$. So $x = 12$ or $x = 16$. If $x = 12$, then the other side is $28 - 12 = 16$. Similarly, if $x = 16$, then the other side is 12. The sides are 12 cm and 16 cm.

79. $3x - 2 > -11 \Leftrightarrow 3x > -9 \Leftrightarrow x > -3$.
Interval: $(-3, \infty)$.

Graph: -3

81. $3 - x \leq 2x - 7 \Leftrightarrow 10 \leq 3x \Leftrightarrow \frac{10}{3} \leq x$

Interval: $\left[\frac{10}{3}, \infty\right)$

Graph: $\frac{10}{3}$

83. $x^2+4x-12>0 \Leftrightarrow (x-2)(x+6)>0$. The expression on the left of the inequality changes sign where $x=2$ and where $x=-6$. Thus we must check the intervals in the following table.

Interval	$(-\infty,-6)$	$(-6,2)$	$(2,\infty)$
Sign of $x-2$	$-$	$-$	$+$
Sign of $x+6$	$-$	$+$	$+$
Sign of $(x-2)(x+6)$	$+$	$-$	$+$

Interval: $(-\infty,-6)\cup(2,\infty)$.

Graph: −6 2

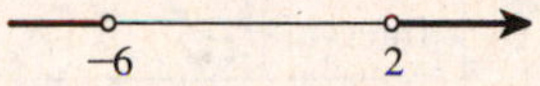

85. $\frac{2x+5}{x+1}\le 1 \Leftrightarrow \frac{2x+5}{x+1}-1\le 0 \Leftrightarrow \frac{2x+5}{x+1}-\frac{x+1}{x+1}\le 0 \Leftrightarrow \frac{x+4}{x+1}\le 0$. The expression on the left of the inequality changes sign where $x=-1$ and where $x=-4$. Thus we must check the intervals in the following table.

Interval	$(-\infty,-4)$	$(-4,-1)$	$(-1,\infty)$
Sign of $x+4$	$-$	$+$	$+$
Sign of $x+1$	$-$	$-$	$+$
Sign of $\frac{x+4}{x+1}$	$+$	$-$	$+$

We exclude $x=-1$, since the expression is not defined at this value. Thus the solution is $[-4,-1)$.

Graph: −4 −1

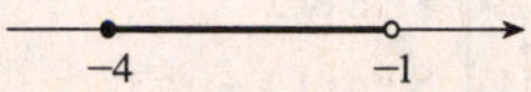

87. $\frac{x-4}{x^2-4}\le 0 \Leftrightarrow \frac{x-4}{(x-2)(x+2)}\le 0$. The expression on the left of the inequality changes sign where $x=-2$, where $x=2$, and where $x=4$. Thus we must check the intervals in the following table.

Interval	$(-\infty,-2)$	$(-2,2)$	$(2,4)$	$(4,\infty)$
Sign of $x-4$	$-$	$-$	$-$	$+$
Sign of $x-2$	$-$	$-$	$+$	$+$
Sign of $x+2$	$-$	$+$	$+$	$+$
Sign of $\frac{x-4}{(x-2)(x+2)}$	$-$	$+$	$-$	$+$

Since the expression is not defined when $x=\pm 2$, we exclude these values and the solution is $(-\infty,-2)\cup(2,4]$.

Graph: −2 2 4

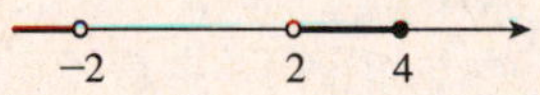

89. $|x-5|\le 3 \Leftrightarrow -3\le x-5\le 3 \Leftrightarrow 2\le x\le 8$.
Interval: $[2,8]$

Graph: 2 8

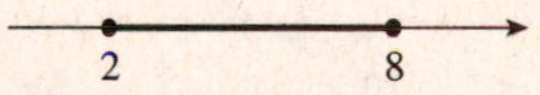

91. $|2x+1|\ge 1$ is equivalent to $2x+1\ge 1$ or $2x+1\le -1$. *Case 1:* $2x+1\ge 1 \Leftrightarrow 2x\ge 0 \Leftrightarrow x\ge 0$. *Case 2:* $2x+1\le -1$ $\Leftrightarrow 2x\le -2 \Leftrightarrow x\le -1$. Interval: $(-\infty,-1]\cup[0,\infty)$. Graph: −1 0

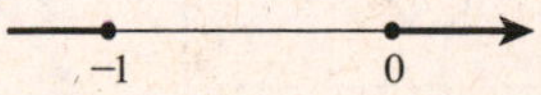

93. **(a)** For $\sqrt{24 - x - 3x^2}$ to define a real number, we must have $24 - x - 3x^2 \geq 0 \Leftrightarrow (8 - 3x)(3 + x) \geq 0$. The expression on the left of the inequality changes sign where $8 - 3x = 0 \Leftrightarrow -3x = -8 \Leftrightarrow x = \frac{8}{3}$; or where $x = -3$. Thus we must check the intervals in the following table.

Interval	$(-\infty, -3)$	$\left(-3, \frac{8}{3}\right)$	$\left(\frac{8}{3}, \infty\right)$
Sign of $8 - 3x$	+	+	−
Sign of $3 + x$	−	+	+
Sign of $(8 - 3x)(3 + x)$	−	+	−

Interval: $\left[-3, \frac{8}{3}\right]$.

Graph:

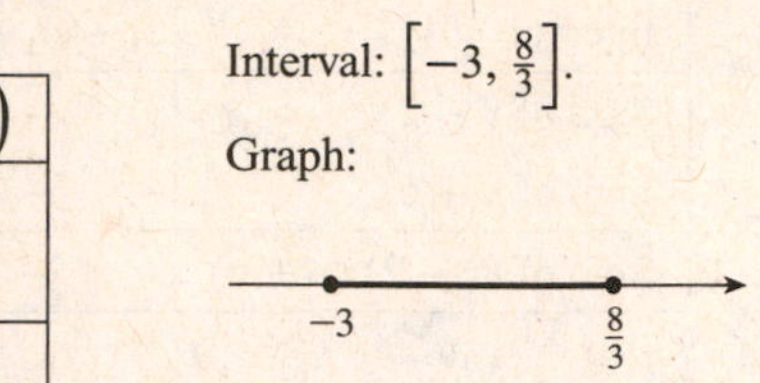

(b) For $\dfrac{1}{\sqrt[4]{x - x^4}}$ to define a real number we must have $x - x^4 > 0 \Leftrightarrow x\left(1 - x^3\right) > 0 \Leftrightarrow x(1 - x)\left(1 + x + x^2\right) > 0$.

The expression on the left of the inequality changes sign where $x = 0$; or where $x = 1$; or where $1 + x + x^2 = 0 \Rightarrow x = \frac{-1 \pm \sqrt{1^2 - 4(1)(1)}}{2(1)} = \frac{1 \pm \sqrt{1-4}}{2}$ which is imaginary. We check the intervals in the following table.

Interval	$(-\infty, 0)$	$(0, 1)$	$(1, \infty)$
Sign of x	−	+	+
Sign of $1 - x$	+	+	−
Sign of $1 + x + x^2$	+	+	+
Sign of $x(1 - x)\left(1 + x + x^2\right)$	−	+	−

Interval: $(0, 1)$.

Graph:

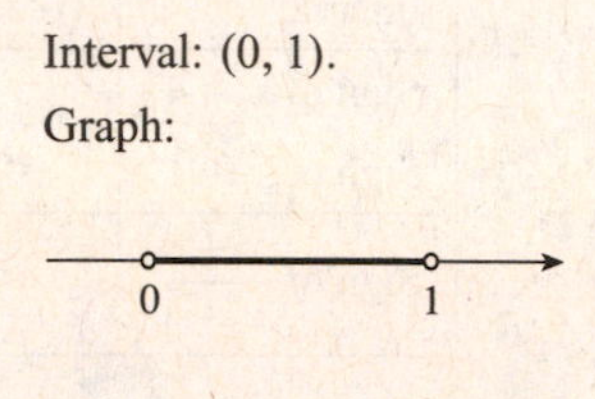

95. From the graph, we see that the graphs of $y = x^2 - 4x$ and $y = x + 6$ intersect at $x = -1$ and $x = 6$, so these are the solutions of the equation $x^2 - 4x = x + 6$.

97. From the graph, we see that the graph of $y = x^2 - 4x$ lies below the graph of $y = x + 6$ for $-1 < x < 6$, so the inequality $x^2 - 4x \leq x + 6$ is satisfied on the interval $[-1, 6]$.

99. From the graph, we see that the graph of $y = x^2 - 4x$ lies above the x-axis for $x < 0$ and for $x > 4$, so the inequality $x^2 - 4x \geq 0$ is satisfied on the intervals $(-\infty, 0]$ and $[4, \infty)$.

101. $x^2 - 4x = 2x + 7$. We graph the equations $y_1 = x^2 - 4x$ and $y_2 = 2x + 7$ in the viewing rectangle $[-10, 10]$ by $[-5, 25]$. Using a zoom or trace function, we get the solutions $x = -1$ and $x = 7$.

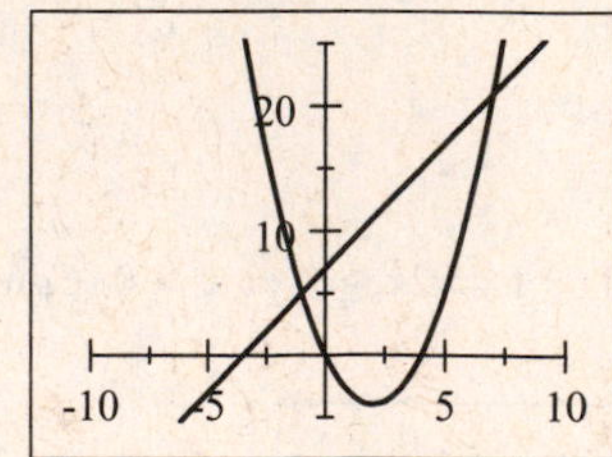

103. $x^4 - 9x^2 = x - 9$. We graph the equations $y_1 = x^4 - 9x^2$ and $y_2 = x - 9$ in the viewing rectangle $[-5, 5]$ by $[-25, 10]$. Using a zoom or trace function, we get the solutions $x \approx -2.72$, $x \approx -1.15$, $x = 1.00$, and $x \approx 2.87$.

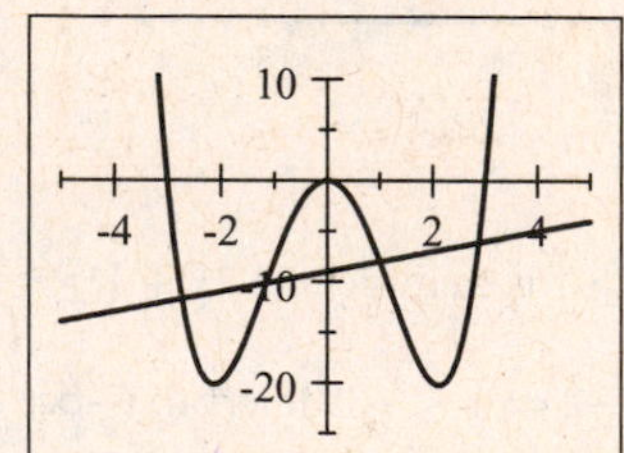

105. $4x - 3 \geq x^2$. We graph the equations $y_1 = 4x - 3$ and $y_2 = x^2$ in the viewing rectangle $[-5, 5]$ by $[0, 15]$. Using a zoom or trace function, we find the points of intersection are at $x = 1$ and $x = 3$. Since we want $4x - 3 \geq x^2$, the solution is the interval $[1, 3]$.

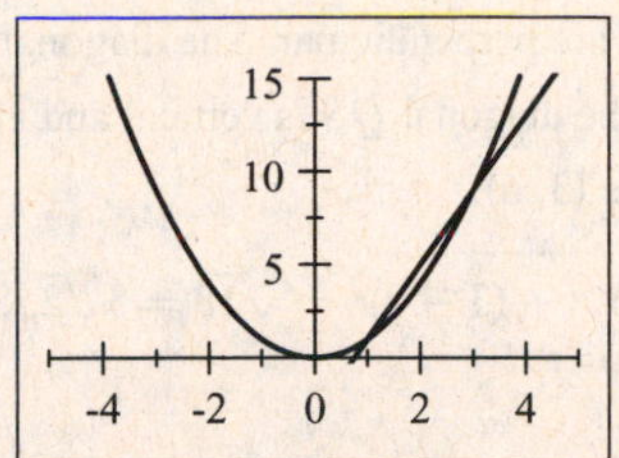

107. $x^4 - 4x^2 < \frac{1}{2}x - 1$. We graph the equations $y_1 = x^4 - 4x^2$ and $y_2 = \frac{1}{2}x - 1$ in the viewing rectangle $[-5, 5]$ by $[-5, 5]$. We find the points of intersection are at $x \approx -1.85$, $x \approx -0.60$, $x \approx 0.45$, and $x = 2.00$. Since we want $x^4 - 4x^2 < \frac{1}{2}x - 1$, the solution is $(-1.85, -0.60) \cup (0.45, 2.00)$.

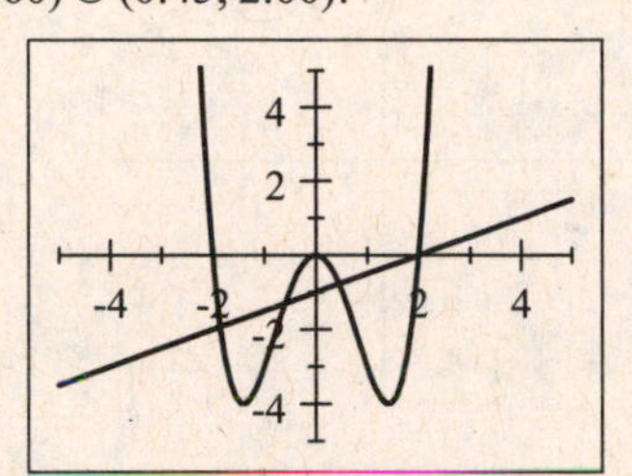

109. Here the center is at $(0, 0)$, and the circle passes through the point $(-5, 12)$, so the radius is $r = \sqrt{(-5 - 0)^2 + (12 - 0)^2} = \sqrt{25 + 144} = \sqrt{169} = 13$. The equation of the circle is $x^2 + y^2 = 13^2 \Leftrightarrow x^2 + y^2 = 169$. The line shown is the tangent that passes through the point $(-5, 12)$, so it is perpendicular to the line through the points $(0, 0)$ and $(-5, 12)$. This line has slope $m_1 = \dfrac{12 - 0}{-5 - 0} = -\dfrac{12}{5}$. The slope of the line we seek is $m_2 = -\dfrac{1}{m_1} = -\dfrac{1}{-12/5} = \dfrac{5}{12}$. Thus, an equation of the tangent line is $y - 12 = \frac{5}{12}(x + 5) \Leftrightarrow y - 12 = \frac{5}{12}x + \frac{25}{12} \Leftrightarrow y = \frac{5}{12}x + \frac{169}{12} \Leftrightarrow 5x - 12y + 169 = 0$.

111. Since M varies directly as z we have $M = kz$. Substituting $M = 120$ when $z = 15$, we find $120 = k(15) \Leftrightarrow k = 8$. Therefore, $M = 8z$.

113. **(a)** The intensity I varies inversely as the square of the distance d, so $I = \dfrac{k}{d^2}$.

(b) Substituting $I = 1000$ when $d = 8$, we get $1000 = \dfrac{k}{(8)^2} \Leftrightarrow k = 64{,}000$.

(c) From parts (a) and (b), we have $I = \dfrac{64{,}000}{d^2}$. Substituting $d = 20$, we get $I = \dfrac{64{,}000}{(20)^2} = 160$ candles.

115. Let v be the terminal velocity of the parachutist in mi/h and w be his weight in pounds. Since the terminal velocity is directly proportional to the square root of the weight, we have $v = k\sqrt{w}$. Substituting $v = 9$ when $w = 160$, we solve for k. This gives $9 = k\sqrt{160} \Leftrightarrow k = \dfrac{9}{\sqrt{160}} \approx 0.712$. Thus $v = 0.712\sqrt{w}$. When $w = 240$, the terminal velocity is $v = 0.712\sqrt{240} \approx 11$ mi/h.

CHAPTER 1 TEST

1. (a)

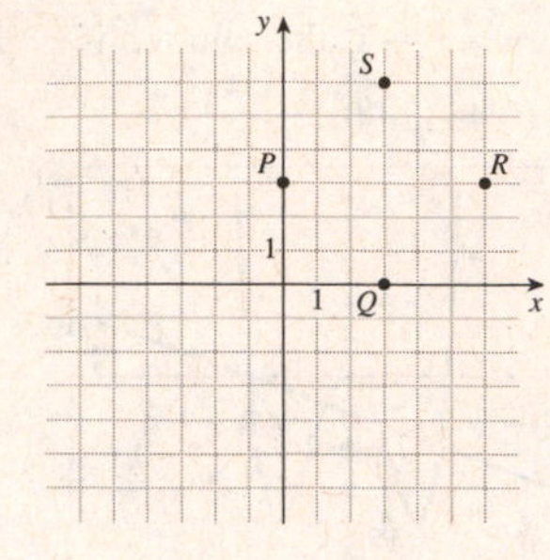

There are several ways to determine the coordinates of S. The diagonals of a square have equal length and are perpendicular. The diagonal PR is horizontal and has length is 6 units, so the diagonal QS is vertical and also has length 6. Thus, the coordinates of S are $(3,6)$.

(b) The length of PQ is $\sqrt{(0-3)^2+(3-0)^2}=\sqrt{18}=3\sqrt{2}$. So the area of $PQRS$ is $\left(3\sqrt{2}\right)^2=18$.

3. (a)

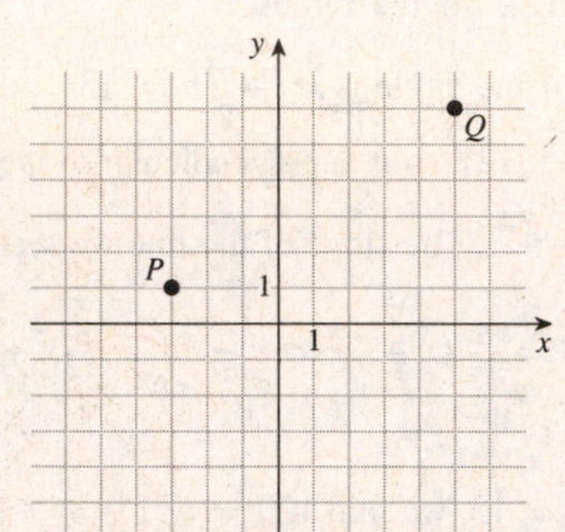

(b) The distance between P and Q is

$$d(P,Q)=\sqrt{(-3-5)^2+(1-6)^2}=\sqrt{64+25}=\sqrt{89}.$$

(c) The midpoint is $\left(\frac{-3+5}{2},\frac{1+6}{2}\right)=\left(1,\frac{7}{2}\right)$.

(d) The slope of the line is $\frac{1-6}{-3-5}=\frac{-5}{-8}=\frac{5}{8}$.

(e) The perpendicular bisector of PQ contains the midpoint, $\left(1,\frac{7}{2}\right)$, and it slope is the negative reciprocal of $\frac{5}{8}$. Thus the slope is $-\frac{1}{5/8}=-\frac{8}{5}$. Hence the equation is $y-\frac{7}{2}=-\frac{8}{5}(x-1)\Leftrightarrow y=-\frac{8}{5}x+\frac{8}{5}+\frac{7}{2}=-\frac{8}{5}x+\frac{51}{10}$. That is, $y=-\frac{8}{5}x+\frac{51}{10}$.

(f) The center of the circle is the midpoint, $\left(1,\frac{7}{2}\right)$, and the length of the radius is $\frac{1}{2}\sqrt{89}$. Thus the equation of the circle whose diameter is PQ is $(x-1)^2+\left(y-\frac{7}{2}\right)^2=\left(\frac{1}{2}\sqrt{89}\right)^2\Leftrightarrow(x-1)^2+\left(y-\frac{7}{2}\right)^2=\frac{89}{4}$.

5. (a) $x=4-y^2$. To test for symmetry about the x-axis, we replace y with $-y$: $x=4-(-y)^2\Leftrightarrow x=4-y^2$, so the graph is symmetric about the x-axis.

To test for symmetry about the y-axis, we replace x with $-x$: $-x=4-y^2$ is different from the original equation, so the graph is not symmetric about the y-axis.

For symmetry about the origin, we replace x with $-x$ and y with $-y$: $-x=4-(-y)^2\Leftrightarrow -x=4-y^2$, which is different from the original equation, so the graph is not symmetric about the origin.

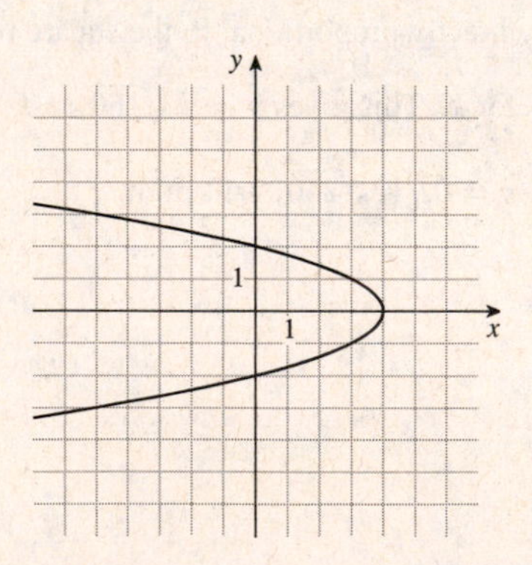

To find x-intercepts, we set $y=0$ and solve for x: $x=4-0^2=4$, so the x-intercept is 4.

To find y-intercepts, we set $x=0$ and solve for y:: $0=4-y^2\Leftrightarrow y^2=4\Leftrightarrow y=\pm2$, so the y-intercepts are -2 and 2.

(b) $y = |x - 2|$. To test for symmetry about the x-axis, we replace y with $-y$: $-y = |x - 2|$ is different from the original equation, so the graph is not symmetric about the x-axis.
To test for symmetry about the y-axis, we replace x with $-x$: $y = |-x - 2| = |x + 2|$ is different from the original equation, so the graph is not symmetric about the y-axis.
To test for symmetry about the origin, we replace x with $-x$ and y with $-y$: $-y = |-x - 2| \Leftrightarrow y = -|x + 2|$, which is different from the original equation, so the graph is not symmetric about the origin.
To find x-intercepts, we set $y = 0$ and solve for x: $0 = |x - 2| \Leftrightarrow x - 2 = 0 \Leftrightarrow x = -2$, so the x-intercept is 2.
To find y-intercepts, we set $x = 0$ and solve for y: $y = |0 - 2| = |-2| = 2$, so the y-intercept is 2.

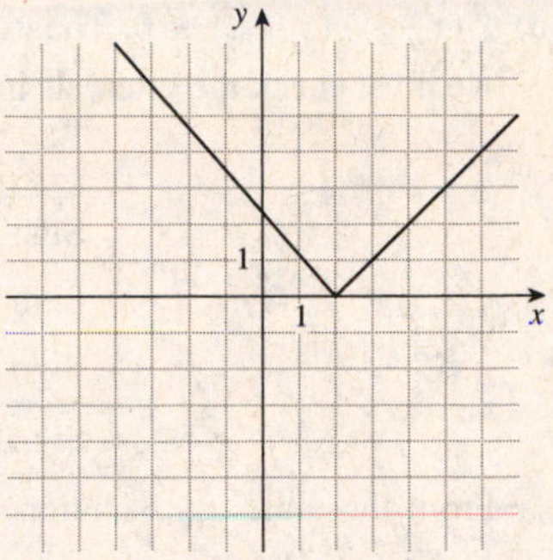

7. (a) $3x + y - 10 = 0 \Leftrightarrow y = -3x + 10$, so the slope of the line we seek is -3. Using the point-slope, $y - (-6) = -3(x - 3) \Leftrightarrow y + 6 = -3x + 9 \Leftrightarrow 3x + y - 3 = 0$.

(b) Using the intercept form we get $\frac{x}{6} + \frac{y}{4} = 1 \Leftrightarrow 2x + 3y = 12 \Leftrightarrow 2x + 3y - 12 = 0$.

9. (a) $x^2 - x - 12 = 0 \Leftrightarrow (x - 4)(x + 3) = 0$. So $x = 4$ or $x = -3$.

(b) $2x^2 + 4x + 1 = 0 \Rightarrow x = \frac{-4 \pm \sqrt{4^2 - 4(2)(1)}}{2(2)} = \frac{-4 \pm \sqrt{16 - 8}}{4} = \frac{-4 \pm \sqrt{8}}{4} = \frac{-4 \pm 2\sqrt{2}}{4} = \frac{-2 \pm \sqrt{2}}{2}$.

(c) $3 - \sqrt{x - 3} = x \Leftrightarrow 3 - x = \sqrt{x - 3} \Leftrightarrow (3 - x)^2 = \left(\sqrt{3 - x}\right)^2 \Leftrightarrow x^2 - 6x + 9 = 3 - x \Leftrightarrow x^2 - 5x + 6 = (x - 2)(x - 3) = 0$. Thus, $x = 2$ and $x = 3$ are potential solutions. Checking in the original equation, we see that only $x = 3$ is valid.

(d) $x^{1/2} - 3x^{1/4} + 2 = 0$. Let $u = x^{1/4}$, then we have $u^2 - 3u + 2 = 0 \Leftrightarrow (u - 2)(u - 1) = 0$. So either $u - 2 = 0$ or $u - 1 = 0$. If $u - 2 = 0$, then $u = 2 \Leftrightarrow x^{1/4} = 2 \Leftrightarrow x = 2^4 = 16$. If $u - 1 = 0$, then $u = 1 \Leftrightarrow x^{1/4} = 1 \Leftrightarrow x = 1$. So $x = 1$ or $x = 16$.

(e) $x^4 - 3x^2 + 2 = 0 \Leftrightarrow \left(x^2 - 1\right)\left(x^2 - 2\right) = 0$. So $x^2 - 1 = 0 \Leftrightarrow x = \pm 1$ or $x^2 - 2 = 0 \Leftrightarrow x = \pm\sqrt{2}$. Thus the solutions are $x = -1$, $x = 1$, $x = -\sqrt{2}$, and $x = \sqrt{2}$.

(f) $3|x - 4| - 10 = 0 \Leftrightarrow 3|x - 4| = 10 \Leftrightarrow |x - 4| = \frac{10}{3} \Leftrightarrow x - 4 = \pm\frac{10}{3} \Leftrightarrow x = 4 \pm \frac{10}{3}$. So $x = 4 - \frac{10}{3} = \frac{2}{3}$ or $x = 4 + \frac{10}{3} = \frac{22}{3}$. Thus the solutions are $x = \frac{2}{3}$ and $x = \frac{22}{3}$.

11. Using the Quadratic Formula, $2x^2 + 4x + 3 = 0 \Leftrightarrow x = \frac{-4 \pm \sqrt{4^2 - 4(2)(3)}}{2(2)} = \frac{-4 \pm \sqrt{-8}}{4} = -1 \pm \frac{\sqrt{2}}{2}i$.

13. (a) $-4 < 5 - 3x \le 17 \Leftrightarrow -9 < -3x \le 12 \Leftrightarrow 3 > x \ge -4$. Expressing in standard form we have: $-4 \le x < 3$.

Interval: $[-4, 3)$. Graph: 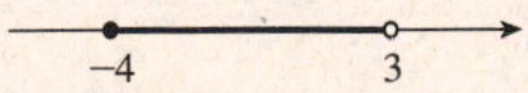

(b) $x(x-1)(x+2) > 0$. The expression on the left of the inequality changes sign when $x = 0$, $x = 1$, and $x = -2$. Thus we must check the intervals in the following table.

Interval	$(-\infty, -2)$	$(-2, 0)$	$(0, 1)$	$(1, \infty)$
Sign of x	$-$	$-$	$+$	$+$
Sign of $x-1$	$-$	$-$	$-$	$+$
Sign of $x+2$	$-$	$+$	$+$	$+$
Sign of $x(x-1)(x-2)$	$-$	$+$	$-$	$+$

From the table, the solution set is $\{x \mid -2 < x < 0 \text{ or } 1 < x\}$. Interval: $(-2, 0) \cup (1, \infty)$.

Graph: −2 0 1

(c) $|x-4| < 3$ is equivalent to $-3 < x - 4 < 3 \Leftrightarrow 1 < x < 7$. Interval: $(1, 7)$. Graph: 1 7

(d) $\dfrac{2x-3}{x+1} \le 1 \Leftrightarrow \dfrac{2x-3}{x+1} - 1 \le 0 \Leftrightarrow \dfrac{2x-3}{x+1} - \dfrac{x+1}{x+1} \le 0 \Leftrightarrow \dfrac{x-4}{x+1} \le 0$. The expression on the left of the inequality changes sign where $x = -4$ and where $x = -1$. Thus we must check the intervals in the following table.

Interval	$(-\infty, -1)$	$(-1, 4)$	$(4, \infty)$
Sign of $x-4$	$-$	$-$	$+$
Sign of $x+1$	$-$	$+$	$+$
Sign of $\dfrac{x-4}{x+1}$	$+$	$-$	$+$

Since $x = -1$ makes the expression in the inequality undefined, we exclude this value. Interval: $(-1, 4]$.

Graph: −1 4

15. For $\sqrt{6x - x^2}$ to be defined as a real number $6x - x^2 \ge 0 \Leftrightarrow x(6-x) \ge 0$. The expression on the left of the inequality changes sign when $x = 0$ and $x = 6$. Thus we must check the intervals in the following table.

Interval	$(-\infty, 0)$	$(0, 6)$	$(6, \infty)$
Sign of x	$-$	$+$	$+$
Sign of $6-x$	$+$	$+$	$-$
Sign of $x(6-x)$	$-$	$+$	$-$

From the table, we see that $\sqrt{6x - x^2}$ is defined when $0 \le x \le 6$.

17. (a) $M = k\dfrac{wh^2}{L}$

(b) Substituting $w = 4$, $h = 6$, $L = 12$, and $M = 4800$, we have $4800 = k\dfrac{(4)\left(6^2\right)}{12} \Leftrightarrow k = 400$. Thus $M = 400\dfrac{wh^2}{L}$.

(c) Now if $L = 10$, $w = 3$, and $h = 10$, then $M = 400\dfrac{(3)\left(10^2\right)}{10} = 12{,}000$. So the beam can support 12,000 pounds.

FOCUS ON MODELING Fitting Lines to Data

1. (a)

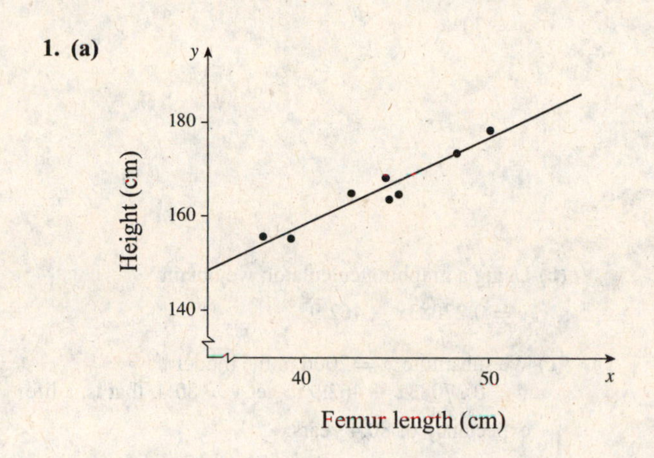

(b) Using a graphing calculator, we obtain the regression line $y = 1.8807x + 82.65$.

(c) Using $x = 58$ in the equation $y = 1.8807x + 82.65$, we get $y = 1.8807\,(58) + 82.65 \approx 191.7$ cm.

3. (a)

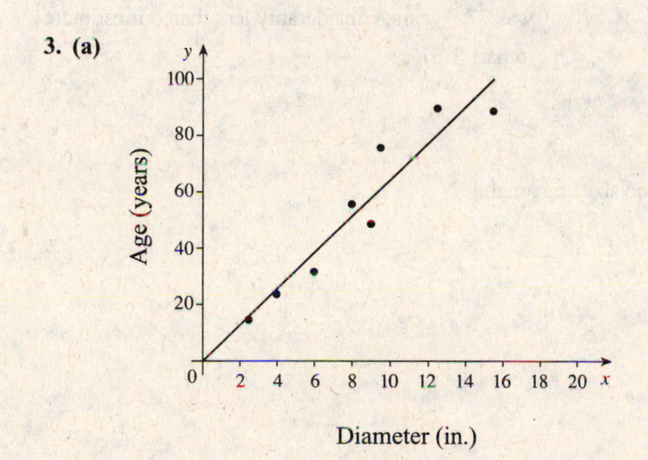

(b) Using a graphing calculator, we obtain the regression line $y = 6.451x - 0.1523$.

(c) Using $x = 18$ in the equation $y = 6.451x - 0.1523$, we get $y = 6.451\,(18) - 0.1523 \approx 116$ years.

5. (a)

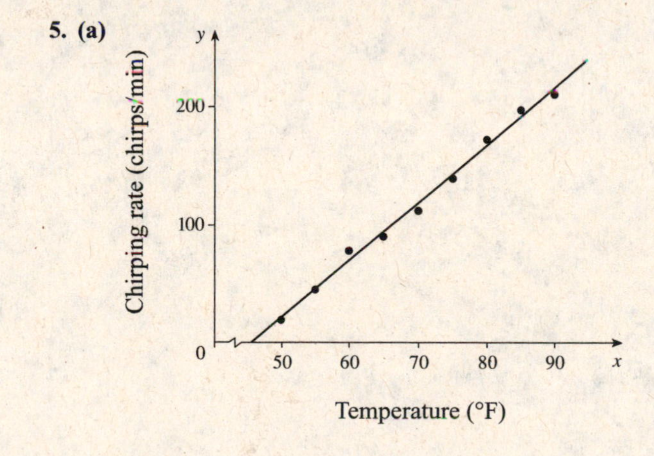

(b) Using a graphing calculator, we obtain the regression line $y = 4.857x - 220.97$.

(c) Using $x = 100°$ F in the equation $y = 4.857x - 220.97$, we get $y \approx 265$ chirps per minute.

7. (a)

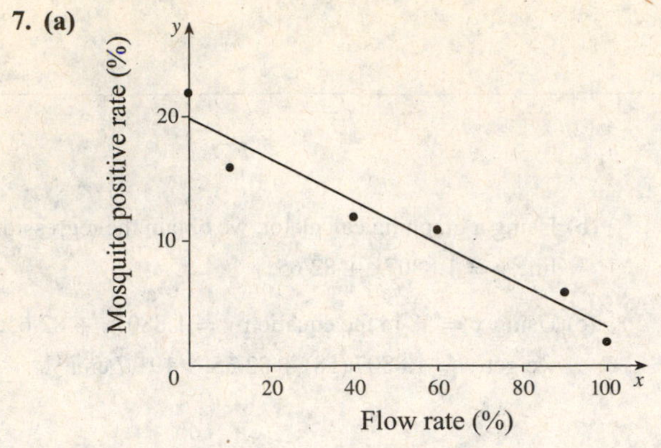

(b) Using a graphing calculator, we obtain the regression line $y = -0.168x + 19.89$.

(c) Using the regression line equation $y = -0.168x + 19.89$, we get $y \approx 8.13\%$ when $x = 70\%$.

9. (a)

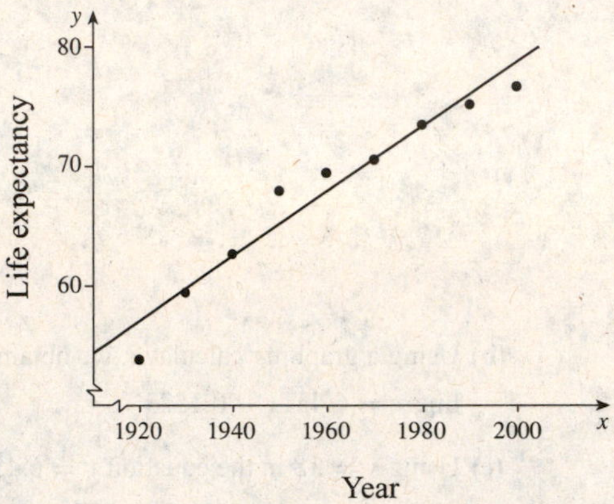

(b) Using a graphing calculator, we obtain $y = 0.27083x - 462.9$.

(c) We substitute $x = 2006$ in the model $y = 0.27083x - 462.9$ to get $y = 80.4$, that is, a life expectancy of 80.4 years.

(d) The life expectancy of a child born in the US in 2006 was 77.7 years, considerably less than our estimate in part (b).

11. Students should find a fairly strong correlation between shoe size and height.

2 FUNCTIONS

2.1 FUNCTIONS

1. If $f(x) = x^3 + 1$, then

(a) the value of f at $x = -1$ is $f(-1) = (-1)^3 + 1 = 0$.

(b) the value of f at $x = 2$ is $f(2) = 2^3 + 1 = 9$.

(c) the net change in the value of f between $x = -1$ and $x = 2$ is $f(2) - f(-1) = 9 - 0 = 9$.

3. (a) $f(x) = x^2 - 3x$ and $g(x) = \dfrac{x-5}{x}$ have 5 in their domain because they are defined when $x = 5$. However, $h(x) = \sqrt{x-10}$ is undefined when $x = 5$ because $\sqrt{5-10} = \sqrt{-5}$, so 5 is not in the domain of h.

(b) $f(5) = 5^2 - 3(5) = 25 - 15 = 10$ and $g(5) = \dfrac{5-5}{5} = \dfrac{0}{5} = 0$.

5. A function f is a rule that assigns to each element x in a set A exactly *one* element called $f(x)$ in a set B. Table (i) defines y as a function of x, but table (ii) does not, because $f(1)$ is not uniquely defined.

7. Multiplying x by 3 gives $3x$, then subtracting 5 gives $f(x) = 3x - 5$.

9. Subtracting 1 gives $x - 1$, then squaring gives $f(x) = (x-1)^2$.

11. $f(x) = 2x + 3$: Multiply by 2, then add 3.

13. $h(x) = 5(x+1)$: Add 1, then multiply by 5.

15. Machine diagram for $f(x) = \sqrt{x-1}$.

1 → subtract 1, then take square root → 0

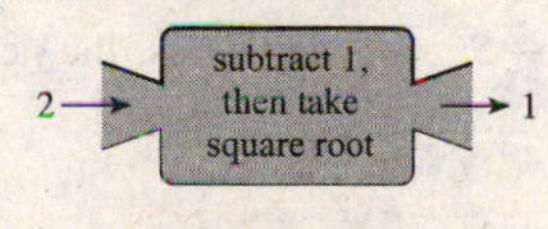

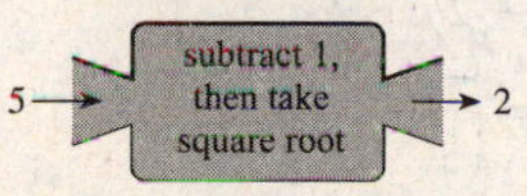

17. $f(x) = 2(x-1)^2$

x	$f(x)$
-1	$2(-1-1)^2 = 8$
0	$2(-1)^2 = 2$
1	$2(1-1)^2 = 0$
2	$2(2-1)^2 = 2$
3	$2(3-1)^2 = 8$

19. $f(x) = x^2 - 6$; $f(-3) = (-3)^2 - 6 = 9 - 6 = 3$; $f(3) = 3^2 - 6 = 9 - 6 = 3$; $f(0) = 0^2 - 6 = -6$; $f\left(\frac{1}{2}\right) = \left(\frac{1}{2}\right)^2 - 6 = \frac{1}{4} - 6 = -\frac{23}{4}$.

21. $f(x) = \dfrac{1-2x}{3}$; $f(2) = \dfrac{1-2(2)}{3} = -1$; $f(-2) = \dfrac{1-2(-2)}{3} = \dfrac{5}{3}$; $f\left(\frac{1}{2}\right) = \dfrac{1-2\left(\frac{1}{2}\right)}{3} = 0$; $f(a) = \dfrac{1-2a}{3}$; $f(-a) = \dfrac{1-2(-a)}{3} = \dfrac{1+2a}{3}$; $f(a-1) = \dfrac{1-2(a-1)}{3} = \dfrac{3-2a}{3}$.

23. $f(x) = x^2 + 2x$; $f(0) = 0^2 + 2(0) = 0$; $f(3) = 3^2 + 2(3) = 9 + 6 = 15$; $f(-3) = (-3)^2 + 2(-3) = 9 - 6 = 3$; $f(a) = a^2 + 2(a) = a^2 + 2a$; $f(-x) = (-x)^2 + 2(-x) = x^2 - 2x$; $f\left(\dfrac{1}{a}\right) = \left(\dfrac{1}{a}\right)^2 + 2\left(\dfrac{1}{a}\right) = \dfrac{1}{a^2} + \dfrac{2}{a}$.

25. $g(x) = \frac{1-x}{1+x}$; $g(2) = \frac{1-(2)}{1+(2)} = \frac{-1}{3} = -\frac{1}{3}$; $g(-1) = \frac{1-(-1)}{1+(-1)}$, which is undefined; $g\left(\frac{1}{2}\right) = \frac{1-\left(\frac{1}{2}\right)}{1+\left(\frac{1}{2}\right)} = \frac{\frac{1}{2}}{\frac{3}{2}} = \frac{1}{3}$;

$g(a) = \frac{1-(a)}{1+(a)} = \frac{1-a}{1+a}$; $g(a-1) = \frac{1-(a-1)}{1+(a-1)} = \frac{1-a+1}{1+a-1} = \frac{2-a}{a}$; $g\left(x^2-1\right) = \frac{1-\left(x^2-1\right)}{1+\left(x^2-1\right)} = \frac{2-x^2}{x^2}$.

27. $k(x) = -x^2 - 2x + 3$; $k(0) = -0^2 - 2(0) + 3 = 3$; $k(2) = -2^2 - 2(2) + 3 = -5$; $k(-2) = -(-2)^2 - 2(-2) + 3 = 3$; $k\left(\sqrt{2}\right) = -\left(\sqrt{2}\right)^2 - 2\left(\sqrt{2}\right) + 3 = 1 - 2\sqrt{2}$; $k(a+2) = -(a+2)^2 - 2(a+2) + 3 = -a^2 - 6a - 5$; $k(-x) = -(-x)^2 - 2(-x) + 3 = -x^2 + 2x + 3$; $k\left(x^2\right) = -\left(x^2\right)^2 - 2\left(x^2\right) + 3 = -x^4 - 2x^2 + 3$.

29. $f(x) = 2|x-1|$; $f(-2) = 2|-2-1| = 2(3) = 6$; $f(0) = 2|0-1| = 2(1) = 2$; $f\left(\frac{1}{2}\right) = 2\left|\frac{1}{2}-1\right| = 2\left(\frac{1}{2}\right) = 1$; $f(2) = 2|2-1| = 2(1) = 2$; $f(x+1) = 2|(x+1)-1| = 2|x|$; $f\left(x^2+2\right) = 2\left|\left(x^2+2\right)-1\right| = 2\left|x^2+1\right| = 2x^2+2$ (since $x^2+1>0$).

31. Since $-2 < 0$, we have $f(-2) = (-2)^2 = 4$. Since $-1 < 0$, we have $f(-1) = (-1)^2 = 1$. Since $0 \geq 0$, we have $f(0) = 0 + 1 = 1$. Since $1 \geq 0$, we have $f(1) = 1 + 1 = 2$. Since $2 \geq 0$, we have $f(2) = 2 + 1 = 3$.

33. Since $-4 \leq -1$, we have $f(-4) = (-4)^2 + 2(-4) = 16 - 8 = 8$. Since $-\frac{3}{2} \leq -1$, we have $f\left(-\frac{3}{2}\right) = \left(-\frac{3}{2}\right)^2 + 2\left(-\frac{3}{2}\right) = \frac{9}{4} - 3 = -\frac{3}{4}$. Since $-1 \leq -1$, we have $f(-1) = (-1)^2 + 2(-1) = 1 - 2 = -1$. Since $-1 < 0 \leq 1$, we have $f(0) = 0$. Since $25 > 1$, we have $f(25) = -1$.

35. $f(x+2) = (x+2)^2 + 1 = x^2 + 4x + 4 + 1 = x^2 + 4x + 5$; $f(x) + f(2) = x^2 + 1 + (2)^2 + 1 = x^2 + 1 + 4 + 1 = x^2 + 6$.

37. $f\left(x^2\right) = x^2 + 4$; $[f(x)]^2 = [x+4]^2 = x^2 + 8x + 16$.

39. $f(x) = 3x - 2$, so $f(1) = 3(1) - 2 = 1$ and $f(5) = 3(5) - 2 = 13$. Thus, the net change is $f(5) - f(1) = 13 - 1 = 12$.

41. $g(t) = 1 - t^2$, so $g(-2) = 1 - (-2)^2 = 1 - 4 = -3$ and $g(5) = 1 - 5^2 = -24$. Thus, the net change is $g(5) - g(-2) = -24 - (-3) = -21$.

43. $f(a) = 5 - 2a$; $f(a+h) = 5 - 2(a+h) = 5 - 2a - 2h$;
$\frac{f(a+h)-f(a)}{h} = \frac{5-2a-2h-(5-2a)}{h} = \frac{5-2a-2h-5+2a}{h} = \frac{-2h}{h} = -2$.

45. $f(a) = 5$; $f(a+h) = 5$; $\frac{f(a+h)-f(a)}{h} = \frac{5-5}{h} = 0$.

47. $f(a) = \frac{a}{a+1}$; $f(a+h) = \frac{a+h}{a+h+1}$;

$$\frac{f(a+h)-f(a)}{h} = \frac{\frac{a+h}{a+h+1} - \frac{a}{a+1}}{h} = \frac{\frac{(a+h)(a+1)}{(a+h+1)(a+1)} - \frac{a(a+h+1)}{(a+h+1)(a+1)}}{h}$$

$$= \frac{\frac{(a+h)(a+1) - a(a+h+1)}{(a+h+1)(a+1)}}{h} = \frac{a^2 + a + ah + h - \left(a^2 + ah + a\right)}{h(a+h+1)(a+1)}$$

$$= \frac{1}{(a+h+1)(a+1)}$$

49. $f(a) = 3 - 5a + 4a^2$;

$$f(a+h) = 3 - 5(a+h) + 4(a+h)^2 = 3 - 5a - 5h + 4\left(a^2 + 2ah + h^2\right)$$
$$= 3 - 5a - 5h + 4a^2 + 8ah + 4h^2;$$

$$\frac{f(a+h) - f(a)}{h} = \frac{\left(3 - 5a - 5h + 4a^2 + 8ah + 4h^2\right) - \left(3 - 5a + 4a^2\right)}{h}$$
$$= \frac{3 - 5a - 5h + 4a^2 + 8ah + 4h^2 - 3 + 5a - 4a^2}{h} = \frac{-5h + 8ah + 4h^2}{h}$$
$$= \frac{h(-5 + 8a + 4h)}{h} = -5 + 8a + 4h.$$

51. $f(x) = 3x$. Since there is no restriction, the domain is all real numbers, $(-\infty, \infty)$. Since every real number y is three times the real number $\frac{1}{3}y$, the range is all real numbers $(-\infty, \infty)$.

53. $f(x) = 3x$, $-2 \le x \le 6$. The domain is $[-2, 6]$, $f(-2) = 3(-2) = -6$, and $f(6) = 3(6) = 18$, so the range is $[-6, 18]$.

55. $f(x) = \dfrac{1}{x-3}$. Since the denominator cannot equal 0 we have $x - 3 \ne 0 \Leftrightarrow x \ne 3$. Thus the domain is $\{x \mid x \ne 3\}$. In interval notation, the domain is $(-\infty, 3) \cup (3, \infty)$.

57. $f(x) = \dfrac{x+2}{x^2-1}$. Since the denominator cannot equal 0 we have $x^2 - 1 \ne 0 \Leftrightarrow x^2 \ne 1 \Rightarrow x \ne \pm 1$. Thus the domain is $\{x \mid x \ne \pm 1\}$. In interval notation, the domain is $(-\infty, -1) \cup (-1, 1) \cup (1, \infty)$.

59. $f(x) = \sqrt{x+1}$. We must have $x + 1 \ge 0 \Leftrightarrow x \ge -1$. Thus, the domain is $[-1, \infty)$.

61. $f(t) = \sqrt[3]{t-1}$. Since the odd root is defined for all real numbers, the domain is the set of real numbers, $(-\infty, \infty)$.

63. $f(x) = \sqrt{1-2x}$. Since the square root is defined as a real number only for nonnegative numbers, we require that $1 - 2x \ge 0 \Leftrightarrow x \le \frac{1}{2}$. So the domain is $\{x \mid x \le \frac{1}{2}\}$. In interval notation, the domain is $\left(-\infty, \frac{1}{2}\right]$.

65. $g(x) = \dfrac{\sqrt{2+x}}{3-x}$. We require $2 + x \ge 0$, and the denominator cannot equal 0. Now $2 + x \ge 0 \Leftrightarrow x \ge -2$, and $3 - x \ne 0$ $\Leftrightarrow x \ne 3$. Thus the domain is $\{x \mid x \ge -2 \text{ and } x \ne 3\}$, which can be expressed in interval notation as $[-2, 3) \cup (3, \infty)$.

67. $g(x) = \sqrt[4]{x^2 - 6x}$. Since the input to an even root must be nonnegative, we have $x^2 - 6x \ge 0 \Leftrightarrow x(x-6) \ge 0$. We make a table:

	$(-\infty, 0)$	$(0, 6)$	$(6, \infty)$
Sign of x	$-$	$+$	$+$
Sign of $x - 6$	$-$	$-$	$+$
Sign of $x(x-6)$	$+$	$-$	$+$

Thus the domain is $(-\infty, 0] \cup [6, \infty)$.

69. $f(x) = \dfrac{3}{\sqrt{x-4}}$. Since the input to an even root must be nonnegative and the denominator cannot equal 0, we have $x - 4 > 0 \Leftrightarrow x > 4$. Thus the domain is $(4, \infty)$.

71. $f(x) = \dfrac{(x+1)^2}{\sqrt{2x-1}}$. Since the input to an even root must be nonnegative and the denominator cannot equal 0, we have $2x - 1 > 0 \Leftrightarrow x > \frac{1}{2}$. Thus the domain is $\left(\frac{1}{2}, \infty\right)$.

73. To evaluate $f(x)$, divide the input by 3 and add $\frac{2}{3}$ to the result.

(a) $f(x) = \dfrac{x}{3} + \dfrac{2}{3}$

(b)

x	$f(x)$
2	$\frac{4}{3}$
4	2
6	$\frac{8}{3}$
8	$\frac{10}{3}$

(c)

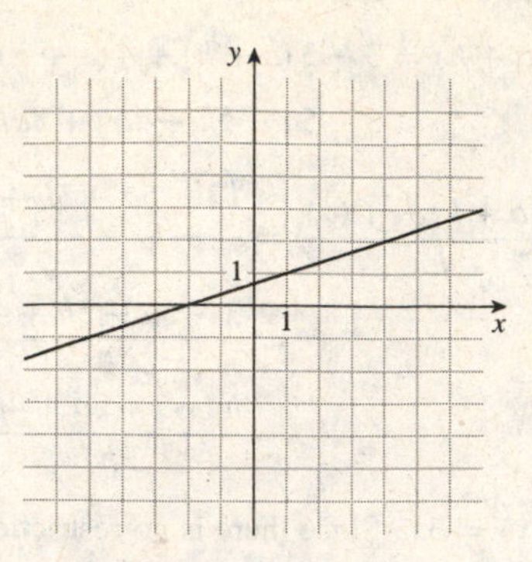

75. Let $T(x)$ be the amount of sales tax charged in Lemon County on a purchase of x dollars. To find the tax, take 8% of the purchase price.

(a) $T(x) = 0.08x$

(b)

x	$T(x)$
2	0.16
4	0.32
6	0.48
8	0.64

(c)

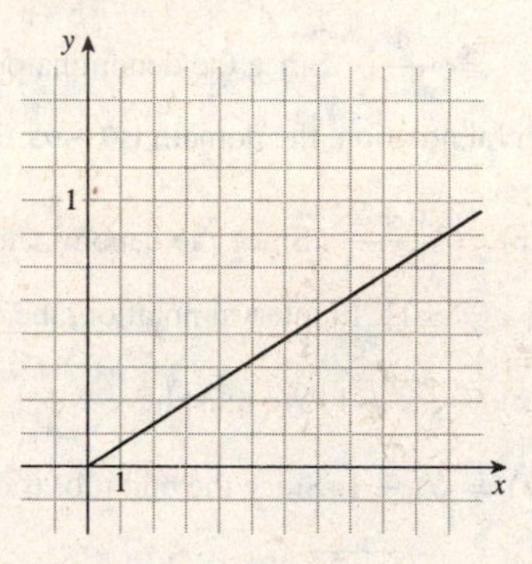

77. $f(x) = \begin{cases} 1 & \text{if } x \text{ is rational} \\ 5 & \text{if } x \text{ is irrational} \end{cases}$ The domain of f is all real numbers, since every real number is either rational or irrational; and the range of f is $\{1, 5\}$.

79. **(a)** $V(0) = 50\left(1 - \frac{0}{20}\right)^2 = 50$ and $V(20) = 50\left(1 - \frac{20}{20}\right)^2 = 0$.

(b) $V(0) = 50$ represents the volume of the full tank at time $t = 0$, and $V(20) = 0$ represents the volume of the empty tank twenty minutes later.

(c)

x	$V(x)$
0	50
5	28.125
10	12.5
15	3.125
20	0

(d) The net change in V as t changes from 0 minutes to 20 minutes is $V(20) - V(0) = 0 - 50 = -50$ gallons.

81. **(a)** $L(0.5c) = 10\sqrt{1 - \dfrac{(0.5c)^2}{c^2}} \approx 8.66$ m, $L(0.75c) = 10\sqrt{1 - \dfrac{(0.75c)^2}{c^2}} \approx 6.61$ m, and $L(0.9c) = 10\sqrt{1 - \dfrac{(0.9c)^2}{c^2}} \approx 4.36$ m.

(b) It will appear to get shorter.

83. **(a)** $v(0.1) = 18500\left(0.25 - 0.1^2\right) = 4440$,

$v(0.4) = 18500\left(0.25 - 0.4^2\right) = 1665$.

(b) They tell us that the blood flows much faster (about 2.75 times faster) 0.1 cm from the center than 0.1 cm from the edge.

(c)

r	$v(r)$
0	4625
0.1	4440
0.2	3885
0.3	2960
0.4	1665
0.5	0

(d) The net change in V as r changes from 0.1 cm to 0.5 cm is $V(0.5) - V(0.1) = 0 - 4440 = -4440$ cm/s.

85. **(a)** Since $0 \le 5{,}000 \le 10{,}000$ we have $T(5{,}000) = 0$. Since $10{,}000 < 12{,}000 \le 20{,}000$ we have $T(12{,}000) = 0.08(12{,}000) = 960$. Since $20{,}000 < 25{,}000$ we have $T(25{,}000) = 1600 + 0.15(25{,}000) = 5350$.

(b) There is no tax on \$5000, a tax of \$960 on \$12,000 income, and a tax of \$5350 on \$25,000.

87. **(a)** $T(x) = \begin{cases} 75x & \text{if } 0 \le x \le 2 \\ 150 + 50(x-2) & \text{if } x > 2 \end{cases}$

(b) $T(2) = 75(2) = 150$; $T(3) = 150 + 50(3-2) = 200$; and $T(5) = 150 + 50(5-2) = 300$.

(c) The total cost of the lodgings.

89. We assume the grass grows linearly.

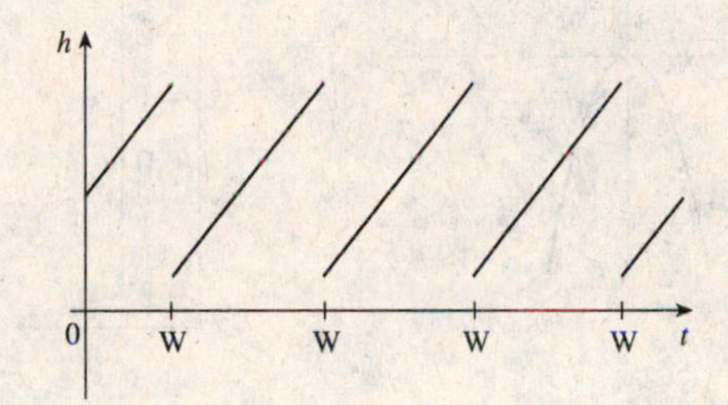

91.

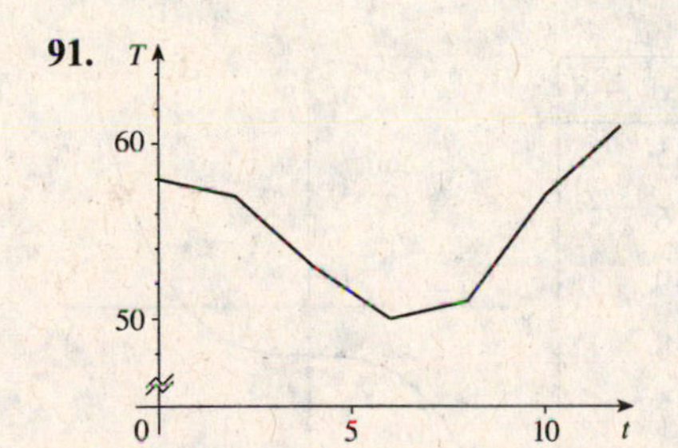

93. Answers will vary.

95. Answers will vary.

2.2 GRAPHS OF FUNCTIONS

1. To graph the function f we plot the points $(x, f(x))$ in a coordinate plane. To graph $f(x) = x^2 - 2$, we plot the points $\left(x, x^2 - 2\right)$. So, the point $\left(3, 3^2 - 2\right) = (3, 7)$ is on the graph of f. The height of the graph of f above the x-axis when $x = 3$ is 7.

x	$f(x)$	(x, y)
-2	2	$(-2, 2)$
-1	-1	$(-1, -1)$
0	-2	$(0, -2)$
1	-1	$(1, -1)$
2	2	$(2, 2)$

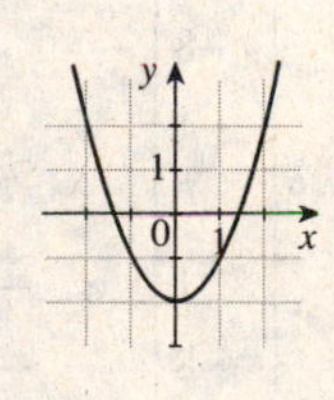

3. If the point $(3, 7)$ is on the graph of f, then $f(3) = 7$.

5.

x	$f(x) = x + 2$
-6	-4
-4	-2
-2	0
0	2
2	4
4	6
6	8

7.

x	$f(x) = -x + 3$, $-3 \le x \le 3$
-3	6
-2	5
0	3
1	2
2	1
3	0

9.

x	$f(x) = -x^2$
± 4	-16
± 3	-9
± 2	-4
± 1	-1
0	0

11.

x	$g(x) = -(x + 1)^2$
-5	-16
-3	-4
-2	-1
-1	0
0	-1
1	-4
3	-16

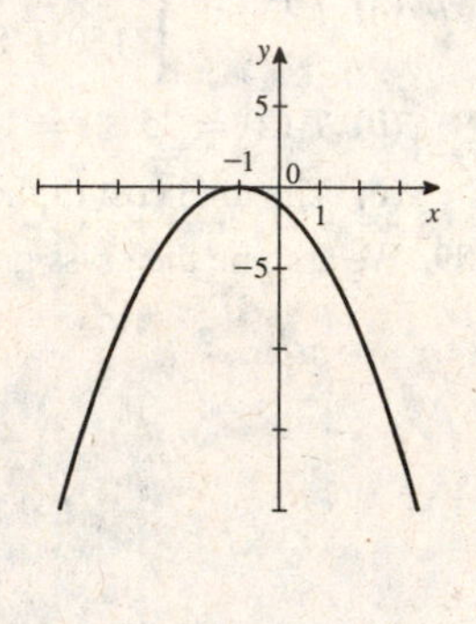

13.

x	$r(x) = 3x^4$
-3	243
-2	48
-1	3
0	0
1	3
2	48
3	243

15.

x	$g(x) = x^3 - 8$
-3	-35
-2	-16
-1	-9
0	-8
1	-7
2	0
3	19

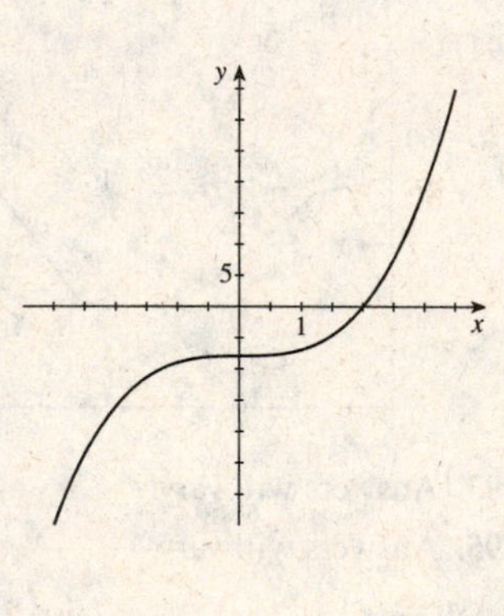

17.

x	$k(x) = \sqrt[3]{-x}$
-27	3
-8	2
-1	1
0	0
1	-1
8	-2
27	-3

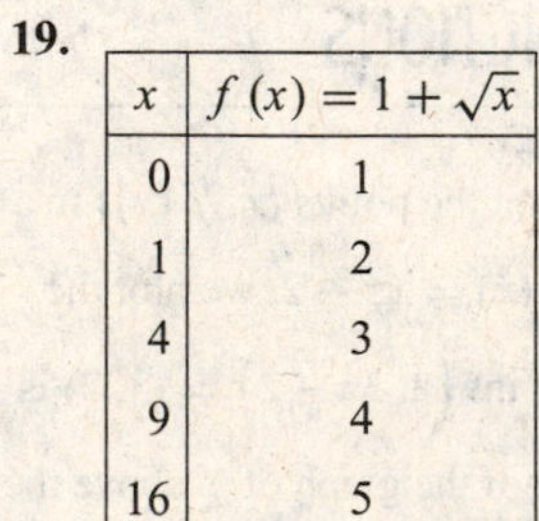

19.

x	$f(x) = 1 + \sqrt{x}$
0	1
1	2
4	3
9	4
16	5
25	6

21.

x	$C(t) = \frac{1}{t^2}$
-2	$\frac{1}{4}$
-1	1
$-\frac{1}{2}$	4
$-\frac{1}{4}$	16
0	—
$\frac{1}{4}$	16
$\frac{1}{2}$	4
1	1
2	$\frac{1}{4}$

23.

x	$H(x) = \lvert 2x \rvert$
± 5	10
± 4	8
± 3	6
± 2	4
± 1	2
0	0

25.

x	$G(x) = \lvert x \rvert + x$
-5	0
-2	0
0	0
1	2
2	4
5	10

27.

x	$f(x) = \lvert 2x - 2 \rvert$
-5	12
-2	8
0	2
1	0
2	2
5	8

29. $f(x) = 8x - x^2$

(a) $[-5, 5]$ by $[-5, 5]$

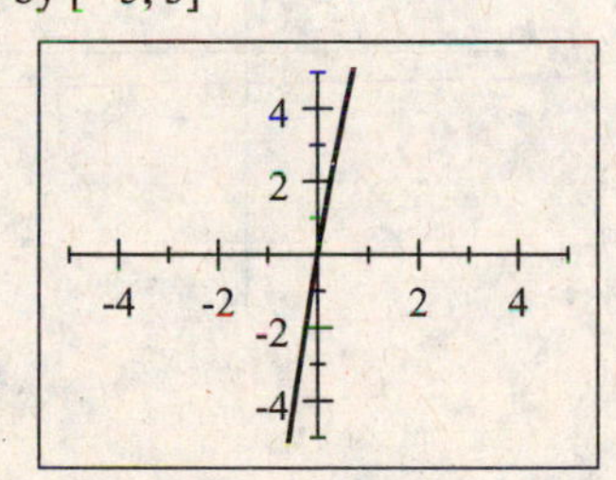

(b) $[-10, 10]$ by $[-10, 10]$

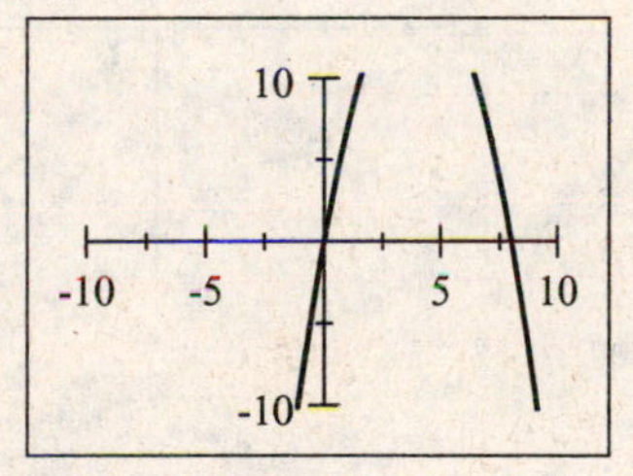

(c) $[-2, 10]$ by $[-5, 20]$

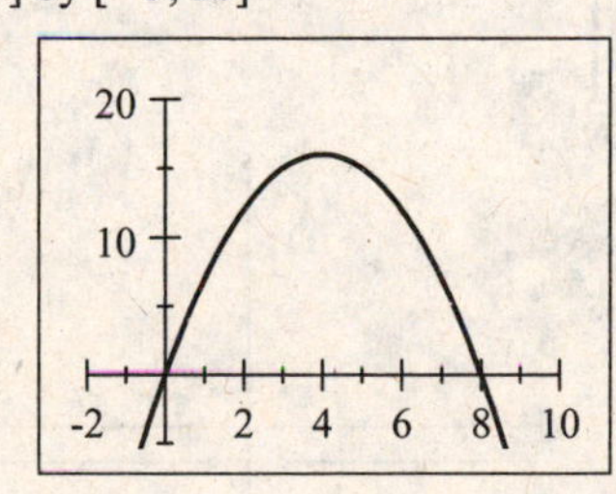

(d) $[-10, 10]$ by $[-100, 100]$

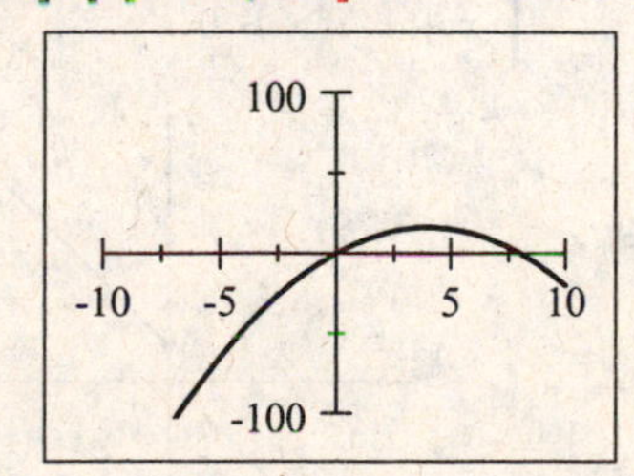

The viewing rectangle in part (c) produces the most appropriate graph of the equation.

31. $h(x) = x^3 - 5x - 4$

(a) $[-2, 2]$ by $[-2, 2]$

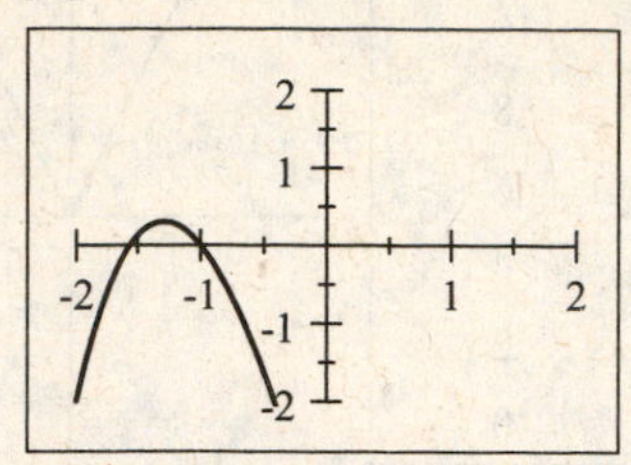

(b) $[-3, 3]$ by $[-10, 10]$

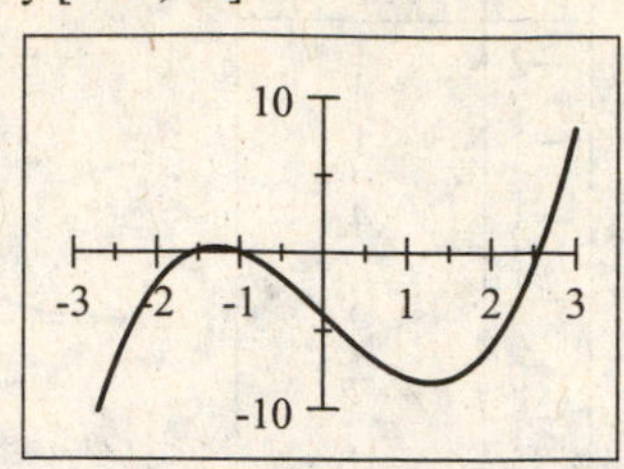

(c) $[-3, 3]$ by $[-10, 5]$

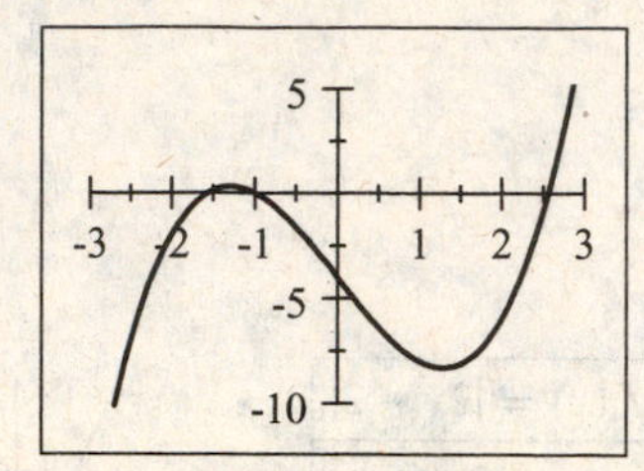

(d) $[-10, 10]$ by $[-10, 10]$

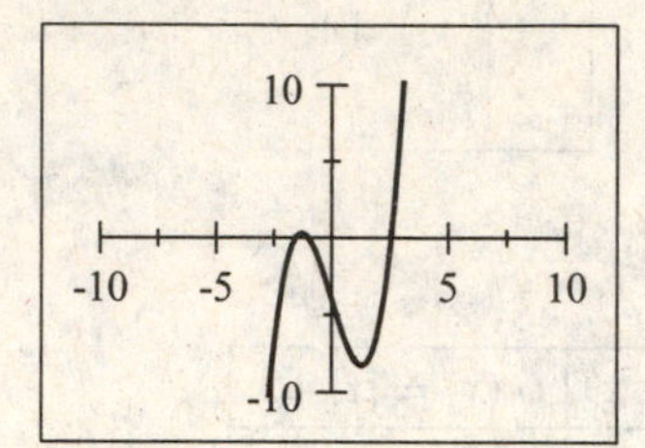

The viewing rectangle in part (c) produces the most appropriate graph of the equation.

33. $f(x) = \begin{cases} 0 & \text{if } x < 2 \\ 1 & \text{if } x \geq 2 \end{cases}$

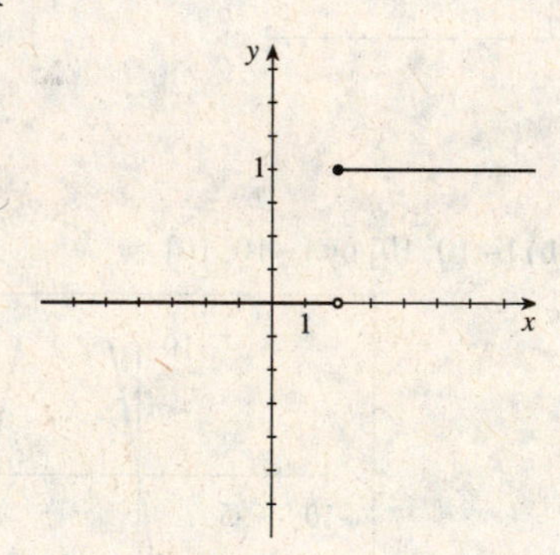

35. $f(x) = \begin{cases} 3 & \text{if } x < 2 \\ x - 1 & \text{if } x \geq 2 \end{cases}$

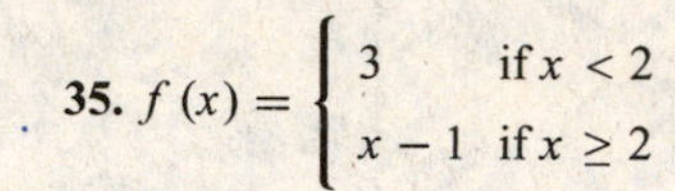
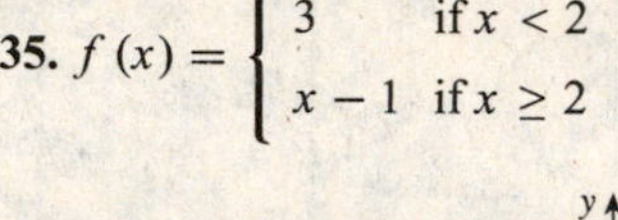

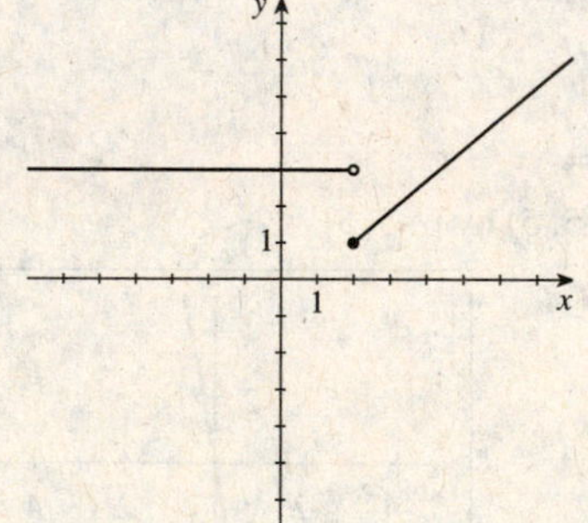

37. $f(x) = \begin{cases} x & \text{if } x \leq 0 \\ x + 1 & \text{if } x > 0 \end{cases}$

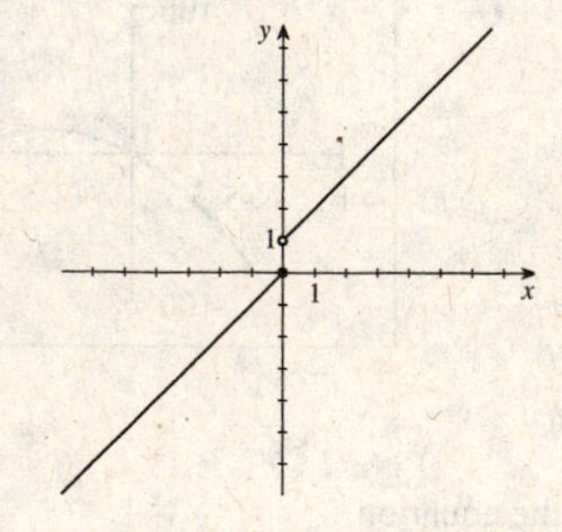

39. $f(x) = \begin{cases} -1 & \text{if } x < -1 \\ 1 & \text{if } -1 \leq x \leq 1 \\ -1 & \text{if } x > 1 \end{cases}$

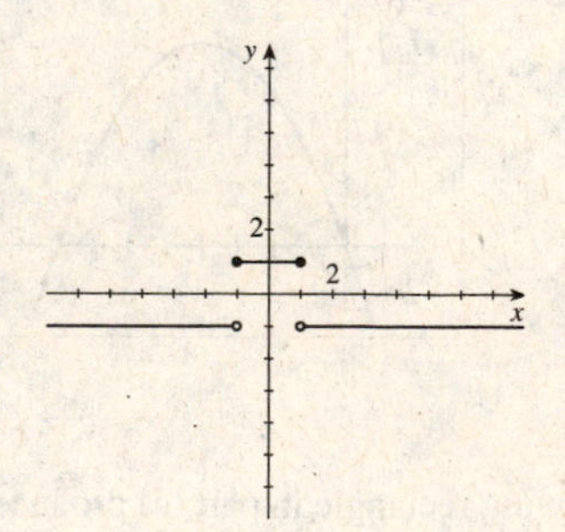

41. $f(x)=\begin{cases}2 & \text{if } x\le -1\\ x^2 & \text{if } x>-1\end{cases}$

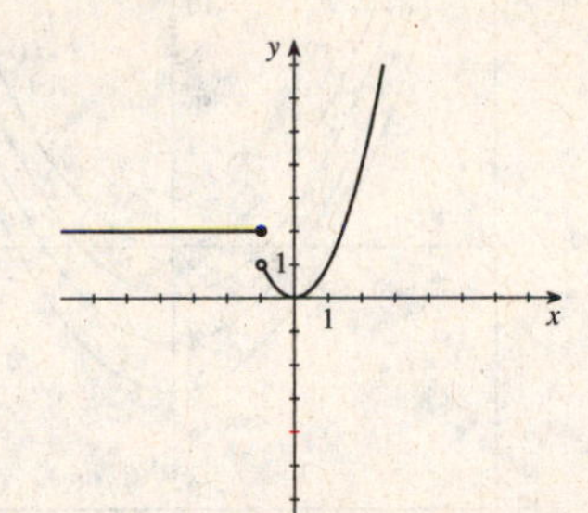

43. 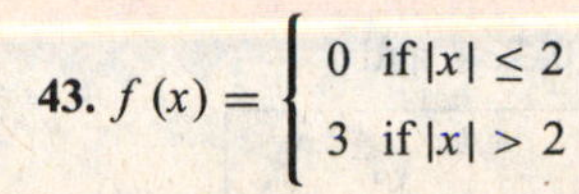
$f(x)=\begin{cases}0 & \text{if } |x|\le 2\\ 3 & \text{if } |x|>2\end{cases}$

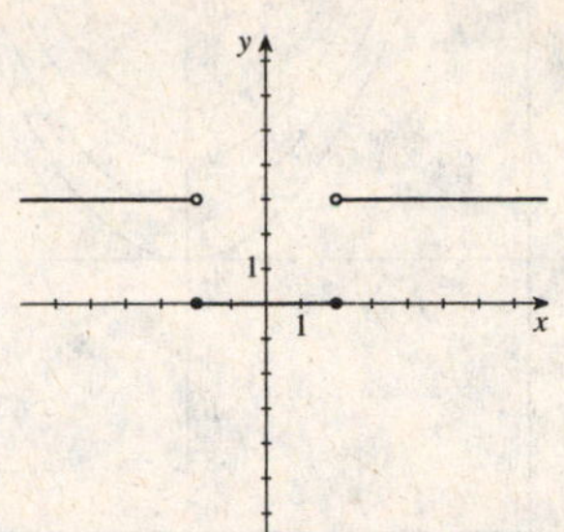

45. $f(x)=\begin{cases}4 & \text{if } x<-2\\ x^2 & \text{if } -2\le x\le 2\\ -x+6 & \text{if } x>2\end{cases}$

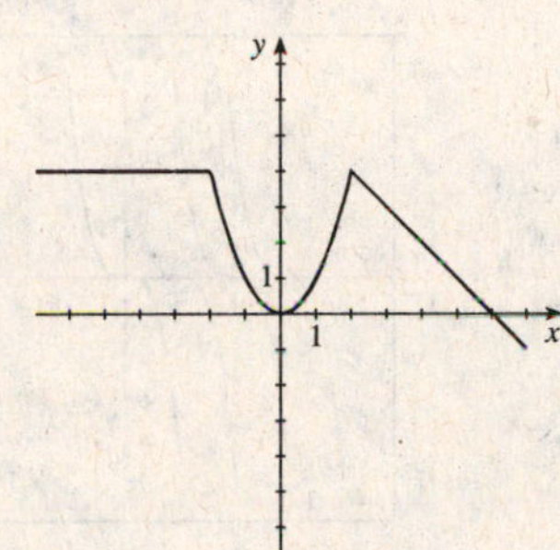

47. 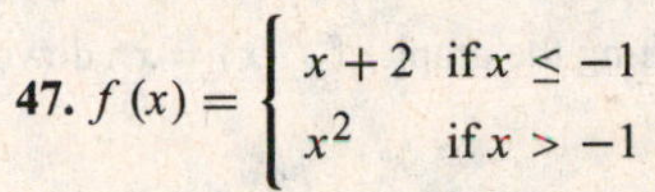
$f(x)=\begin{cases}x+2 & \text{if } x\le -1\\ x^2 & \text{if } x>-1\end{cases}$

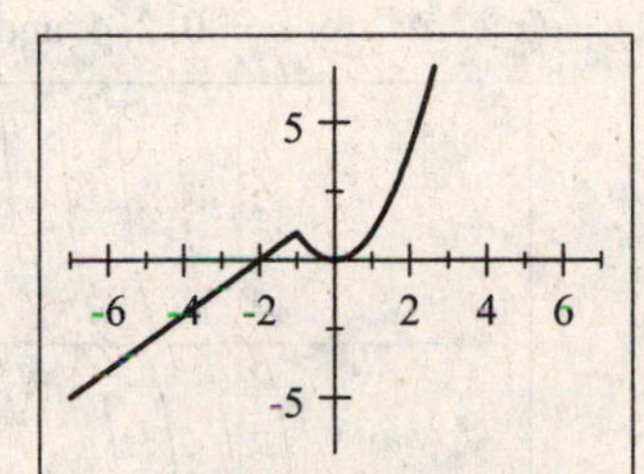

49. $f(x)=\begin{cases}-2 & \text{if } x<-2\\ x & \text{if } -2\le x\le 2\\ 2 & \text{if } x>2\end{cases}$

51. The curves in parts (a) and (c) are graphs of a function of x, by the Vertical Line Test.

53. The given curve is the graph of a function of x, by the Vertical Line Test. Domain: $[-3, 2]$. Range: $[-2, 2]$.

55. No, the given curve is not the graph of a function of x, by the Vertical Line Test.

57. Solving for y in terms of x gives $3x-5y=7 \Leftrightarrow y=\frac{3}{5}x-\frac{7}{5}$. This defines y as a function of x.

59. Solving for y in terms of x gives $x=y^2 \Leftrightarrow y=\pm\sqrt{x}$. The last equation gives two values of y for a given value of x. Thus, this equation does not define y as a function of x.

61. Solving for y in terms of x gives $2x-4y^2=3 \Leftrightarrow 4y^2=2x-3 \Leftrightarrow y=\pm\frac{1}{2}\sqrt{2x-3}$. The last equation gives two values of y for a given value of x. Thus, this equation does not define y as a function of x.

63. Solving for y in terms of x using the Quadratic Formula gives $2xy-5y^2=4 \Leftrightarrow 5y^2-2xy+4=0 \Leftrightarrow$ $y=\dfrac{-(-2x)\pm\sqrt{(-2x)^2-4(5)(4)}}{2(5)}=\dfrac{2x\pm\sqrt{4x^2-80}}{10}=\dfrac{x\pm\sqrt{x^2-20}}{5}$. The last equation gives two values of y for a given value of x. Thus, this equation does not define y as a function of x.

65. Solving for y in terms of x gives $2|x|+y=0 \Leftrightarrow y=-2|x|$. This defines y as a function of x.

67. Solving for y in terms of x gives $x=y^3 \Leftrightarrow y=\sqrt[3]{x}$. This defines y as a function of x.

69. (a) $f(x) = x^2 + c$, for $c = 0, 2, 4$, and 6.

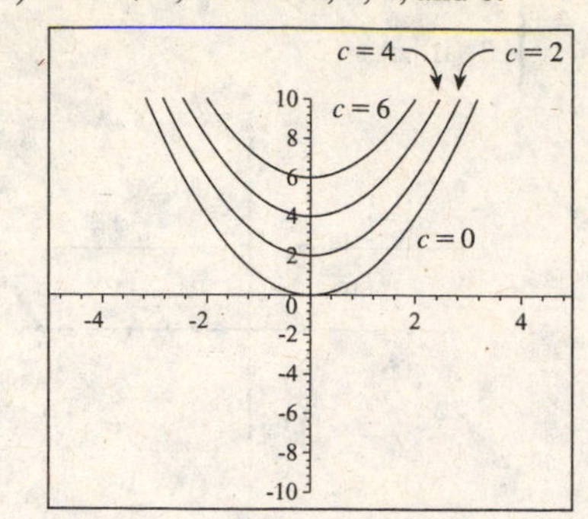

(b) $f(x) = x^2 + c$, for $c = 0, -2, -4$, and -6.

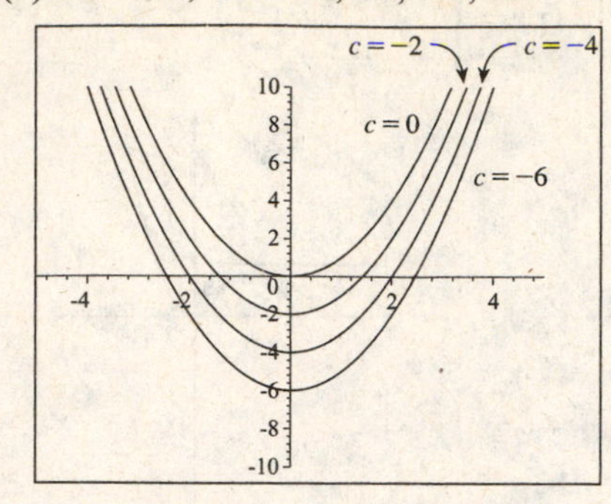

(c) The graphs in part (a) are obtained by shifting the graph of $f(x) = x^2$ upward c units, $c > 0$. The graphs in part (b) are obtained by shifting the graph of $f(x) = x^2$ downward c units.

71. (a) $f(x) = (x - c)^3$, for $c = 0, 2, 4$, and 6.

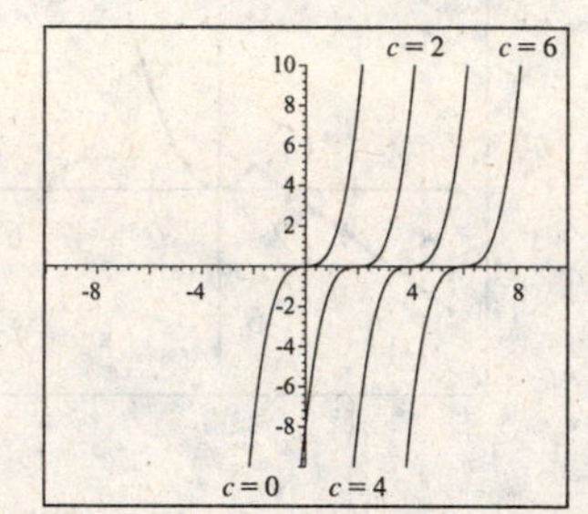

(b) $f(x) = (x - c)^3$, for $c = 0, -2, -4$, and -6.

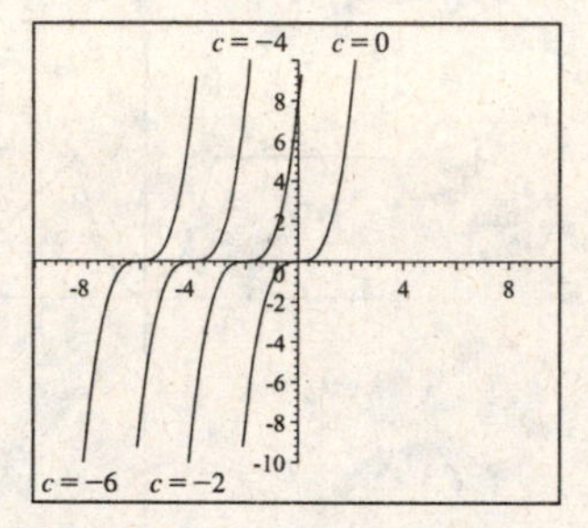

(c) The graphs in part (a) are obtained by shifting the graph of $f(x) = x^3$ to the right c units, $c > 0$. The graphs in part (b) are obtained by shifting the graph of $f(x) = x^3$ to the left $|c|$ units, $c < 0$.

73. (a) $f(x) = x^c$, for $c = \frac{1}{2}, \frac{1}{4}$, and $\frac{1}{6}$.

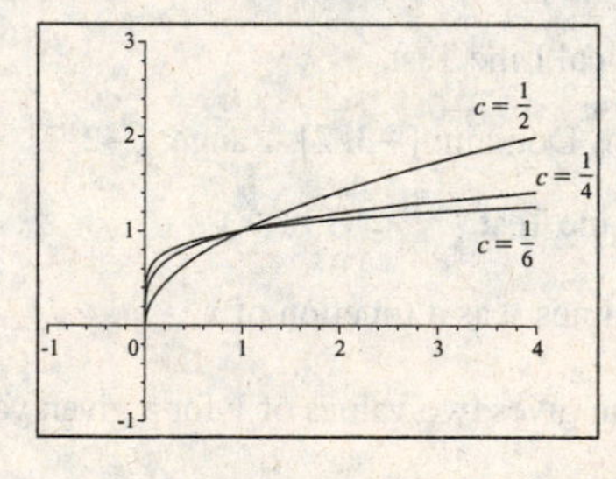

(b) $f(x) = x^c$, for $c = 1, \frac{1}{3}$, and $\frac{1}{5}$.

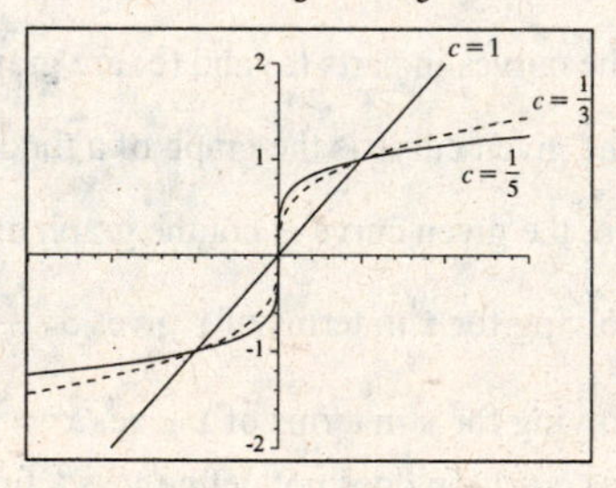

(c) Graphs of even roots are similar to $y = \sqrt{x}$, graphs of odd roots are similar to $y = \sqrt[3]{x}$. As c increases, the graph of $y = \sqrt[c]{x}$ becomes steeper near $x = 0$ and flatter when $x > 1$.

75. The slope of the line segment joining the points $(-2, 1)$ and $(4, -6)$ is $m = \dfrac{-6 - 1}{4 - (-2)} = -\frac{7}{6}$. Using the point-slope form, we have $y - 1 = -\frac{7}{6}(x + 2) \Leftrightarrow y = -\frac{7}{6}x - \frac{7}{3} + 1 \Leftrightarrow y = -\frac{7}{6}x - \frac{4}{3}$. Thus the function is $f(x) = -\frac{7}{6}x - \frac{4}{3}$ for $-2 \le x \le 4$.

77. First solve the circle for y: $x^2 + y^2 = 9 \Leftrightarrow y^2 = 9 - x^2 \Rightarrow y = \pm\sqrt{9 - x^2}$. Since we seek the top half of the circle, we choose $y = \sqrt{9 - x^2}$. So the function is $f(x) = \sqrt{9 - x^2}$, $-3 \le x \le 3$.

79. We graph $T(r) = \frac{0.5}{r^2}$ for $10 \le r \le 100$. As the balloon is inflated, the skin gets thinner, as we would expect.

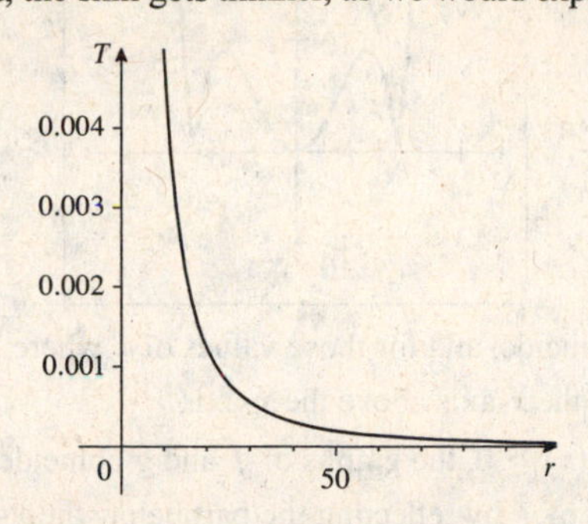

81. (a) $E(x) = \begin{cases} 6.00 + 0.10x & \text{if } 0 \le x \le 300 \\ 36.00 + 0.06(x - 300) & \text{if } 300 < x \end{cases}$

(b)

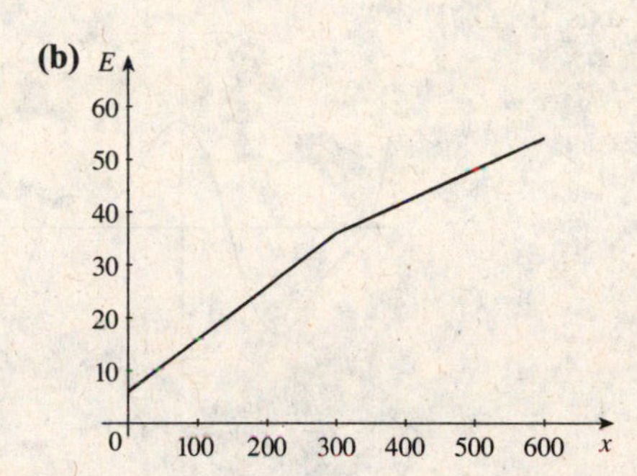

83. $P(x) = \begin{cases} 0.49 & \text{if } 0 < x \le 1 \\ 0.70 & \text{if } 1 < x \le 2 \\ 0.91 & \text{if } 2 < x \le 3 \\ 1.12 & \text{if } 3 < x \le 3.5 \end{cases}$

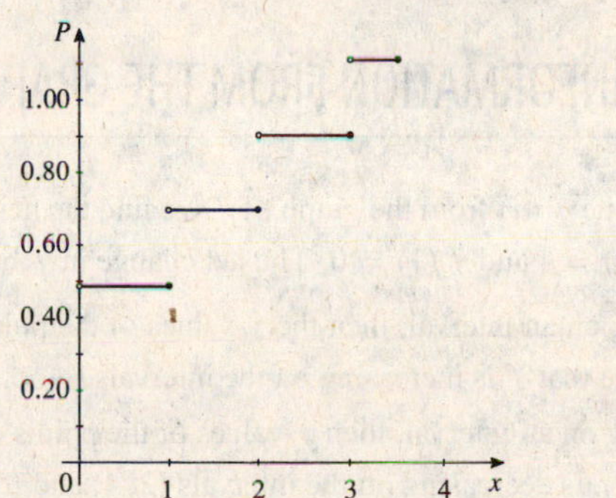

85. Answers will vary. Some examples are almost anything we purchase based on weight, volume, length, or time, for example gasoline. Although the amount delivered by the pump is continuous, the amount we pay is rounded to the penny. An example involving time would be the cost of a telephone call.

87. (a) The graphs of $f(x) = x^2 + x - 6$ and $g(x) = \left|x^2 + x - 6\right|$ are shown in the viewing rectangle $[-10, 10]$ by $[-10, 10]$.

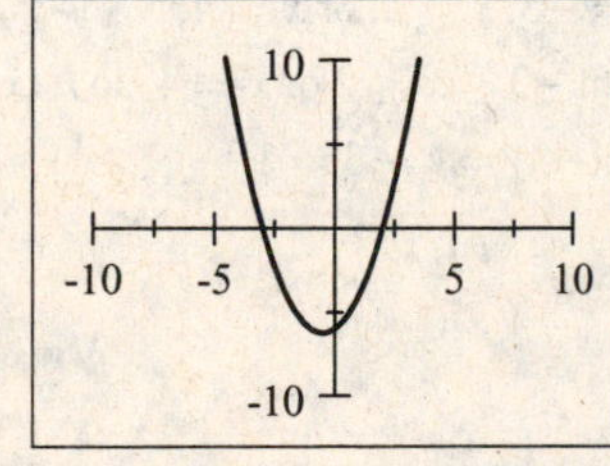

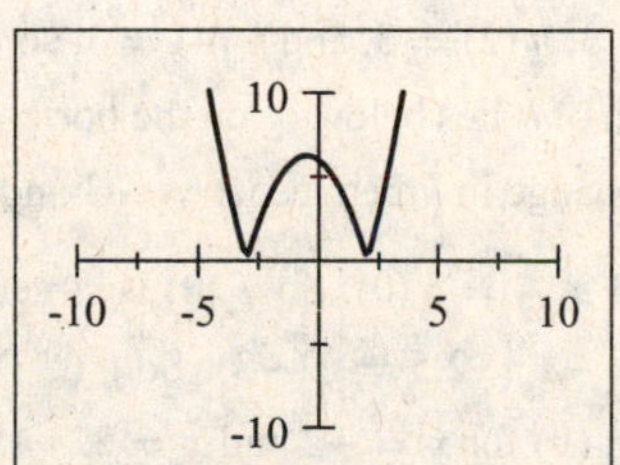

For those values of x where $f(x) \ge 0$, the graphs of f and g coincide, and for those values of x where $f(x) < 0$, the graph of g is obtained from that of f by reflecting the part below the x-axis about the x-axis.

(b) The graphs of $f(x) = x^4 - 6x^2$ and $g(x) = \left|x^4 - 6x^2\right|$ are shown in the viewing rectangle $[-5, 5]$ by $[-10, 15]$.

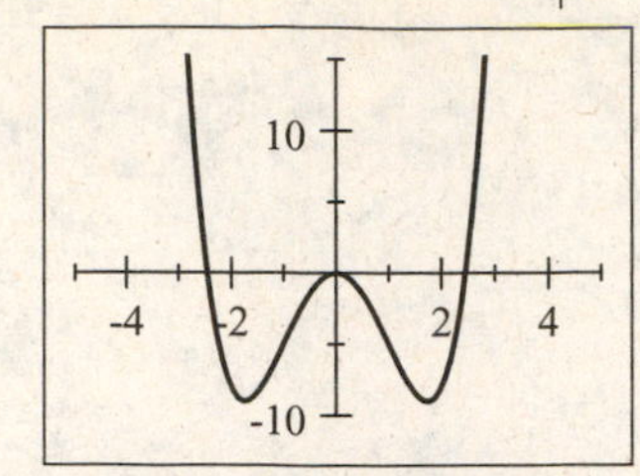

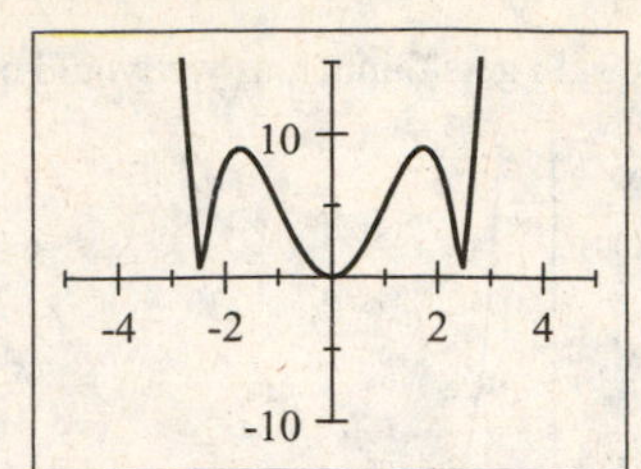

For those values of x where $f(x) \geq 0$, the graphs of f and g coincide, and for those values of x where $f(x) < 0$, the graph of g is obtained from that of f by reflecting the part below the x-axis above the x-axis.

(c) In general, if $g(x) = |f(x)|$, then for those values of x where $f(x) \geq 0$, the graphs of f and g coincide, and for those values of x where $f(x) < 0$, the graph of g is obtained from that of f by reflecting the part below the x-axis above the x-axis.

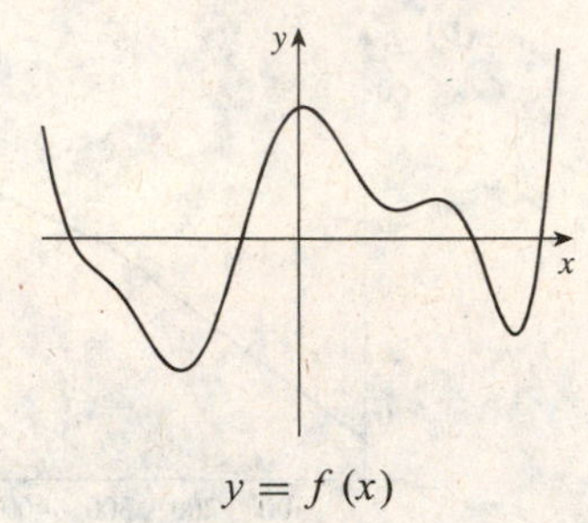

$y = f(x)$

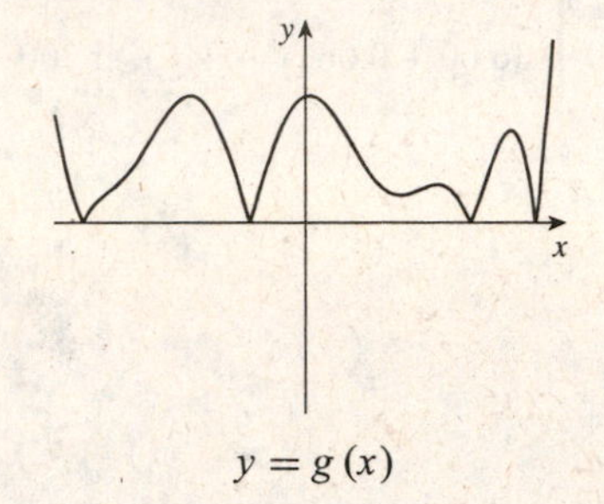

$y = g(x)$

2.3 GETTING INFORMATION FROM THE GRAPH OF A FUNCTION

1. To find a function value $f(a)$ from the graph of f we find the height of the graph above the x-axis at $x = a$. From the graph of f we see that $f(3) = 4$ and $f(1) = 0$. The net change in f between $x = 1$ and $x = 3$ is $f(3) - f(1) = 4 - 0 = 4$.

3. (a) If f is increasing on an interval, then the y-values of the points on the graph *rise* as the x-values increase. From the graph of f we see that f is increasing on the intervals $(-\infty, 2)$ and $(4, 5)$.

(b) If f is decreasing on an interval, then y-values of the points on the graph *fall* as the x-values increase. From the graph of f we see that f is decreasing on the intervals $(2, 4)$ and $(5, \infty)$.

5. The solutions of the equation $f(x) = 0$ are the x-intercepts of the graph of f. The solution of the inequality $f(x) \geq 0$ is the set of x-values at which the graph of f is on or above the x-axis. From the graph of f we find that the solutions of the equation $f(x) = 0$ are $x = 1$ and $x = 7$, and the solution of the inequality $f(x) \geq 0$ is the interval $[1, 7]$.

7. (a) $h(-2) = 1, h(0) = -1, h(2) = 3$, and $h(3) = 4$.

(b) Domain: $[-3, 4]$. Range: $[-1, 4]$.

(c) $h(-3) = 3$, $h(2) = 3$, and $h(4) = 3$, so $h(x) = 3$ when $x = -3, x = 2$, or $x = 4$.

(d) The graph of h lies below or on the horizontal line $y = 3$ when $-3 \leq x \leq 2$ or $x = 4$, so $h(x) \leq 3$ for those values of x.

(e) The net change in h between $x = -3$ and $x = 3$ is $h(3) - h(-3) = 4 - 3 = 1$.

9. (a) $f(0) = 3 > \frac{1}{2} = g(0)$. So $f(0)$ is larger.

(b) $f(-3) \approx -1 < 2.5 = g(-3)$. So $g(-3)$ is larger.

(c) $f(x) = g(x)$ for $x = -2$ and $x = 2$.

(d) $f(x) \leq g(x)$ for $-4 \leq x \leq -2$ and $2 \leq x \leq 3$; that is, on the intervals $[-4, -2]$ and $[2, 3]$.

(e) $f(x) > g(x)$ for $-2 < x < 2$; that is, on the interval $(-2, 2)$.

11. (a)

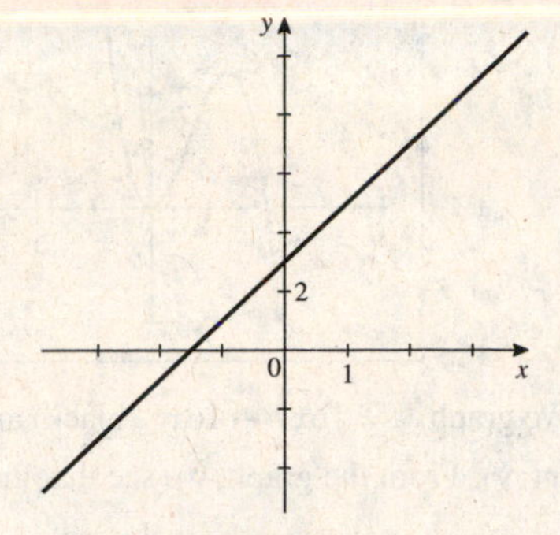

(b) Domain: $(-\infty, \infty)$; Range: $(-\infty, \infty)$

13. (a)

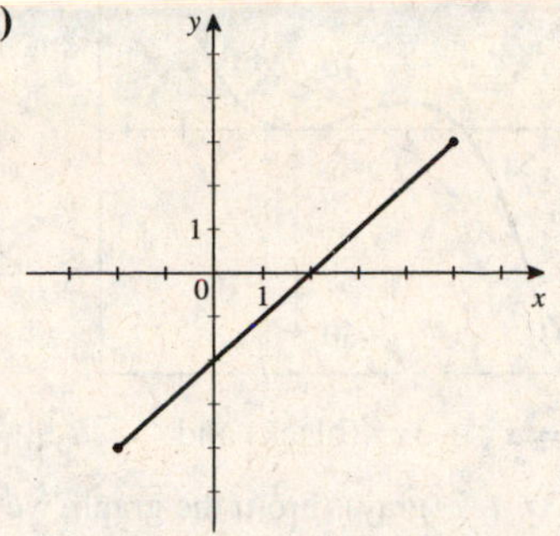

(b) Domain: $[-2, 5]$; Range $[-4, 3]$

15. (a)

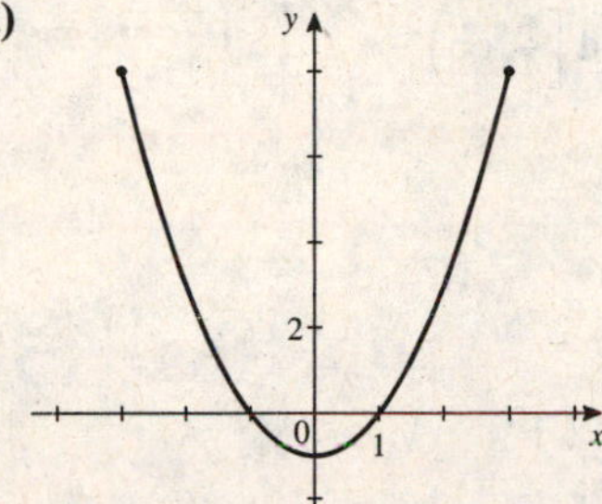

(b) Domain: $[-3, 3]$; Range: $[-1, 8]$

17. (a)

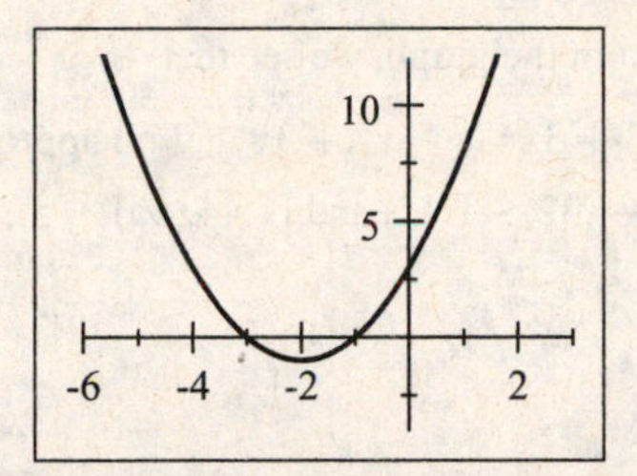

(b) Domain: $(-\infty, \infty)$; Range: $[-1, \infty)$

19. (a)

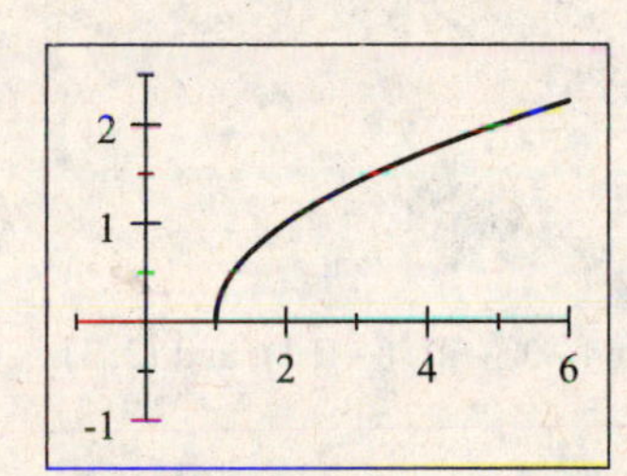

(b) Domain: $[1, \infty)$; Range: $[0, \infty)$

21. (a)

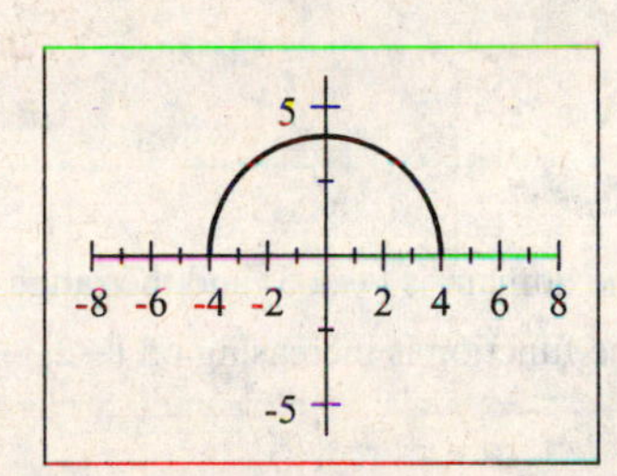

(b) Domain: $[-4, 4]$; Range: $[0, 4]$

23.

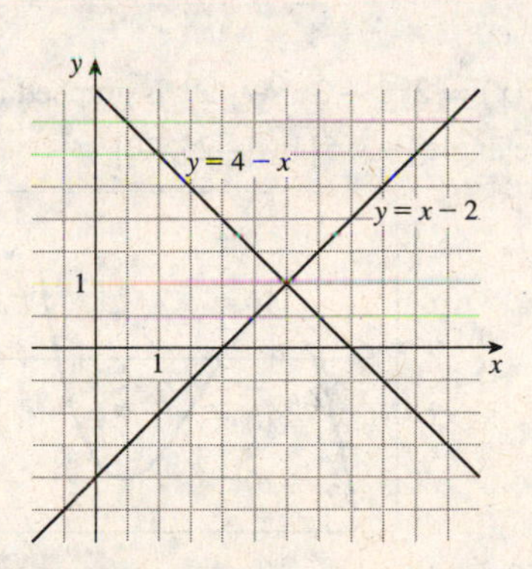

(a) From the graph, we see that $x - 2 = 4 - x$ when $x = 3$.

(b) From the graph, we see $x - 2 > 4 - x$ when $x > 3$.

25.

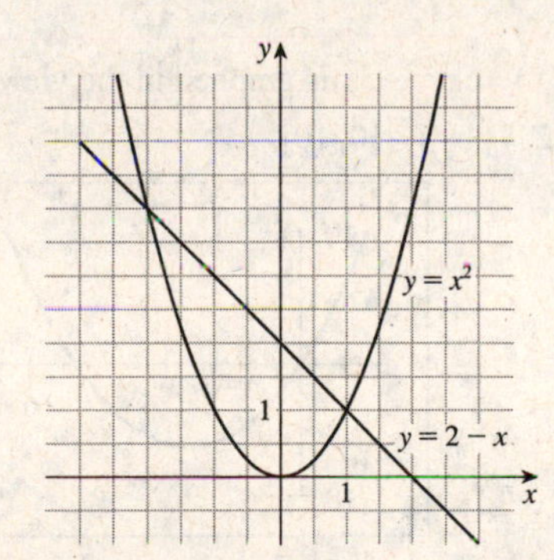

(a) From the graph, we see that $x^2 = 2 - x$ when $x = -2$ or $x = 1$.

(b) From the graph, we see that $x^2 \leq 2 - x$ when $-2 \leq x \leq 1$.

27.

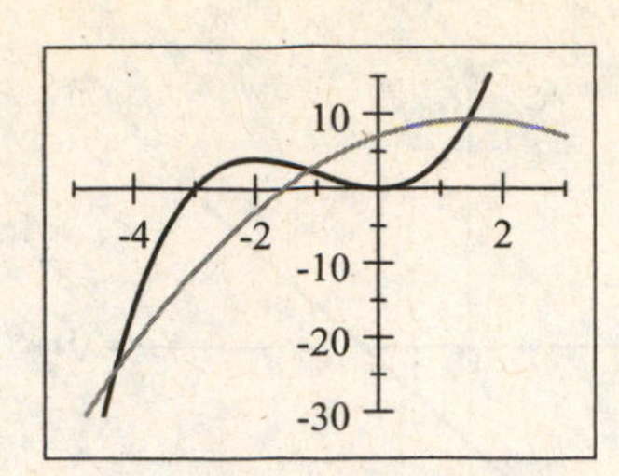

(a) We graph $y = x^3 + 3x^2$ (black) and $y = -x^2 + 3x + 7$ (gray). From the graph, we see that the graphs intersect at $x \approx -4.32$, $x \approx -1.12$, and $x \approx 1.44$.

(b) From the graph, we see that $x^3 + 3x^2 \geq -x^2 + 3x + 7$ on approximately $[-4.32, -1.12]$ and $[1.44, \infty)$.

29.

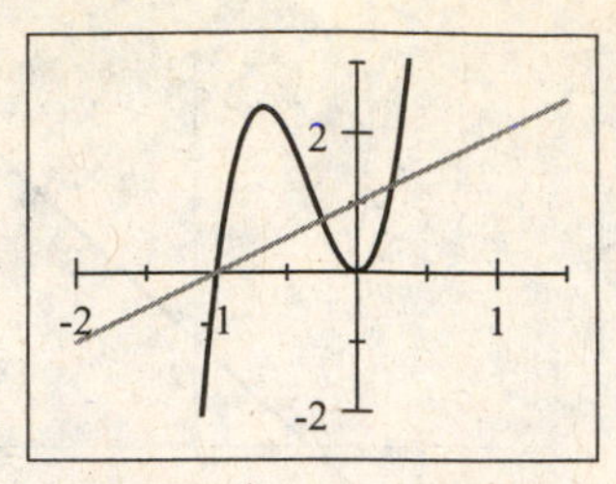

(a) We graph $y = 16x^3 + 16x^2$ (black) and $y = x + 1$ (gray). From the graph, we see that the graphs intersect at $x = -1$, $x = -\frac{1}{4}$, and $x = \frac{1}{4}$.

(b) From the graph, we see that $16x^3 + 16x^2 \geq x + 1$ on $\left[-1, -\frac{1}{4}\right]$ and $\left[\frac{1}{4}, \infty\right)$.

31. **(a)** The domain is $[-1, 4]$ and the range is $[-1, 3]$.

(b) The function is increasing on $(-1, 1)$ and $(2, 4)$ and decreasing on $(1, 2)$.

33. **(a)** The domain is $[-3, 3]$ and the range is $[-2, 2]$.

(b) The function is increasing on $(-2, -1)$ and $(1, 2)$ and decreasing on $(-3, -2)$, $(-1, 1)$, and $(2, 3)$.

35. **(a)** $f(x) = x^2 - 5x$ is graphed in the viewing rectangle $[-2, 7]$ by $[-10, 10]$.

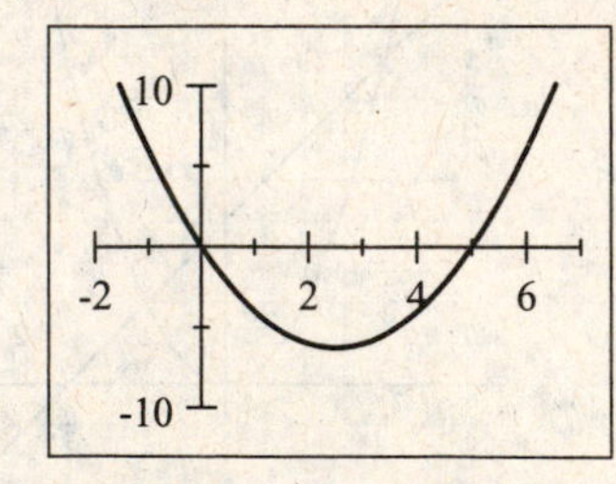

(b) The domain is $(-\infty, \infty)$ and the range is $[-6.25, \infty)$.

(c) The function is increasing on $(2.5, \infty)$. It is decreasing on $(-\infty, 2.5)$.

37. **(a)** $f(x) = 2x^3 - 3x^2 - 12x$ is graphed in the viewing rectangle $[-3, 5]$ by $[-25, 20]$.

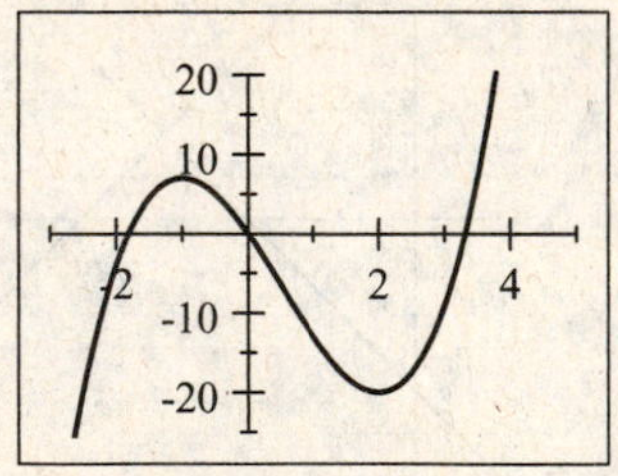

(b) The domain and range are $(-\infty, \infty)$.

(c) The function is increasing on $(-\infty, -1)$ and $(2, \infty)$. It is decreasing on $(-1, 2)$.

39. **(a)** $f(x) = x^3 + 2x^2 - x - 2$ is graphed in the viewing rectangle $[-5, 5]$ by $[-3, 3]$.

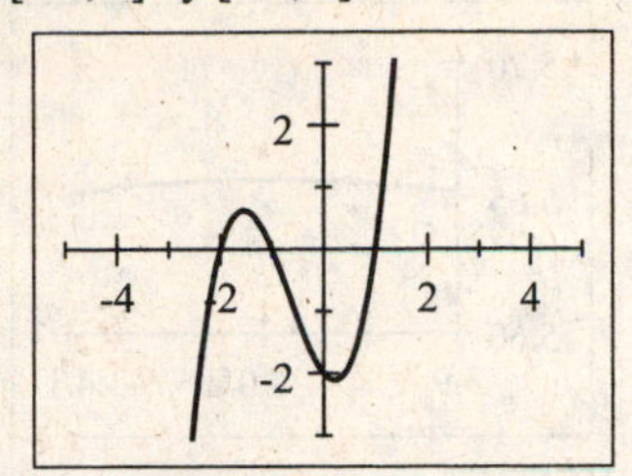

(b) The domain and range are $(-\infty, \infty)$.

(c) The function is increasing on $(-\infty, -1.55)$ and $(0.22, \infty)$. It is decreasing on $(-1.55, 0.22)$.

41. **(a)** $f(x) = x^{2/5}$ is graphed in the viewing rectangle $[-10, 10]$ by $[-5, 5]$.

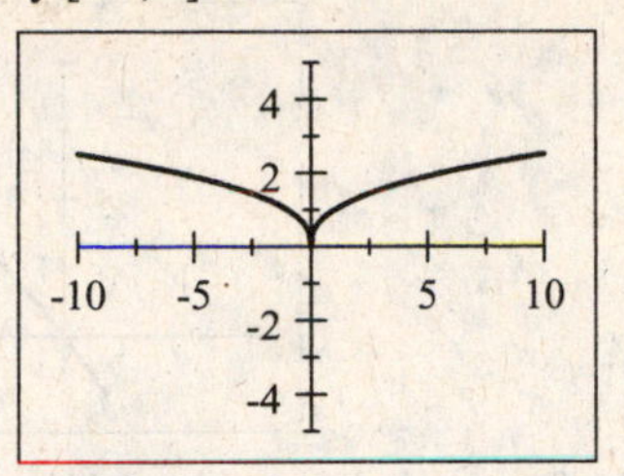

(b) The domain is $(-\infty, \infty)$ and the range is $[0, \infty)$.

(c) The function is increasing on $(0, \infty)$. It is decreasing on $(-\infty, 0)$.

43. **(a)** Local maximum: 2 at $x = 0$. Local minimum: -1 at $x = -2$ and 0 at $x = 2$.

(b) The function is increasing on $(-2, 0)$ and $(2, \infty)$ and decreasing on $(-\infty, -2)$ and $(0, 2)$.

45. **(a)** Local maximum: 0 at $x = 0$ and 1 at $x = 3$. Local minimum: -2 at $x = -2$ and -1 at $x = 1$.

(b) The function is increasing on $(-2, 0)$ and $(1, 3)$ and decreasing on $(-\infty, -2)$, $(0, 1)$, and $(3, \infty)$.

47. **(a)** In the first graph, we see that $f(x) = x^3 - x$ has a local minimum and a local maximum. Smaller x- and y-ranges show that $f(x)$ has a local maximum of about 0.38 when $x \approx -0.58$ and a local minimum of about -0.38 when $x \approx 0.58$.

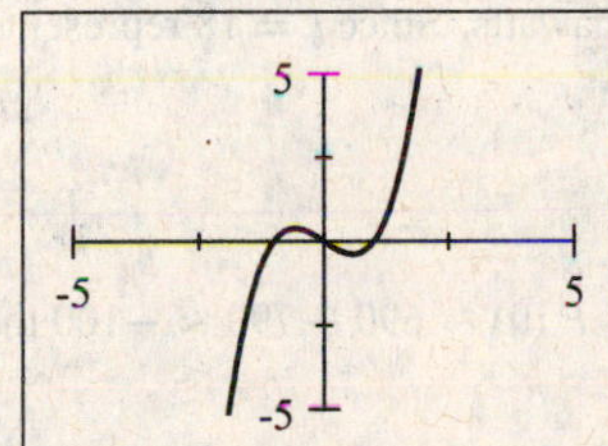

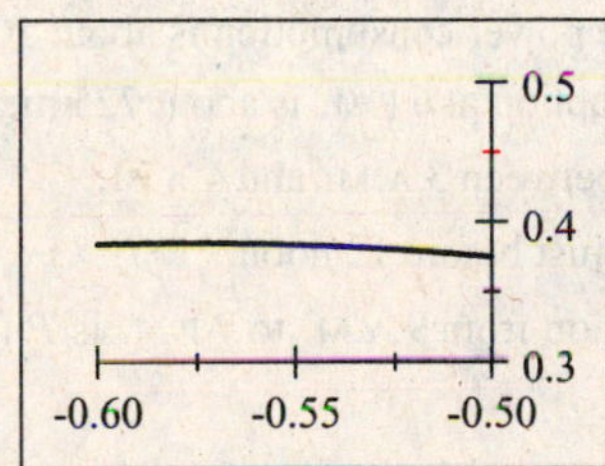

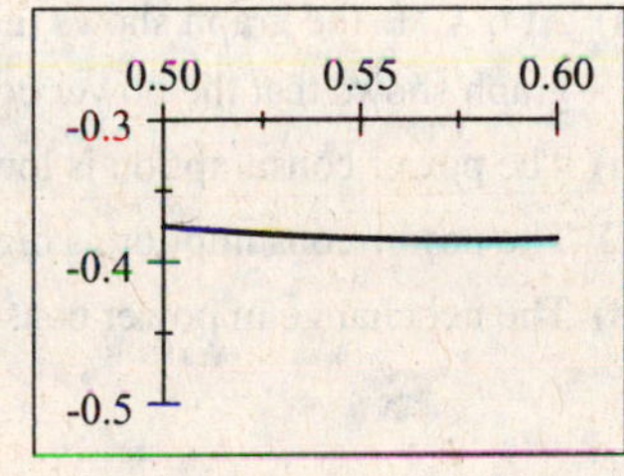

(b) The function is increasing on $(-\infty, -0.58)$ and $(0.58, \infty)$ and decreasing on $(-0.58, 0.58)$.

49. **(a)** In the first graph, we see that $g(x) = x^4 - 2x^3 - 11x^2$ has two local minimums and a local maximum. The local maximum is $g(x) = 0$ when $x = 0$. Smaller x- and y-ranges show that local minima are $g(x) \approx -13.61$ when $x \approx -1.71$ and $g(x) \approx -73.32$ when $x \approx 3.21$.

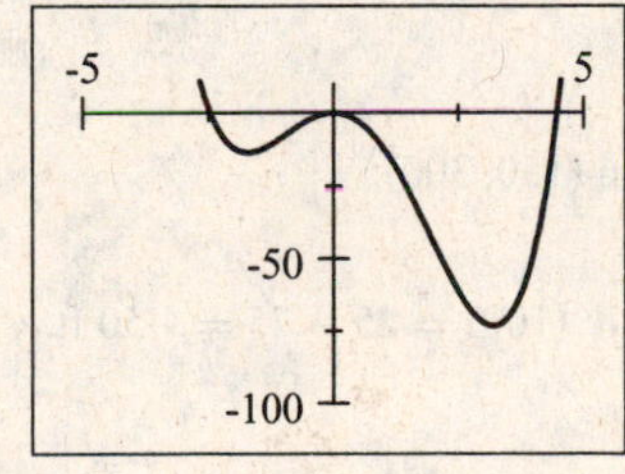

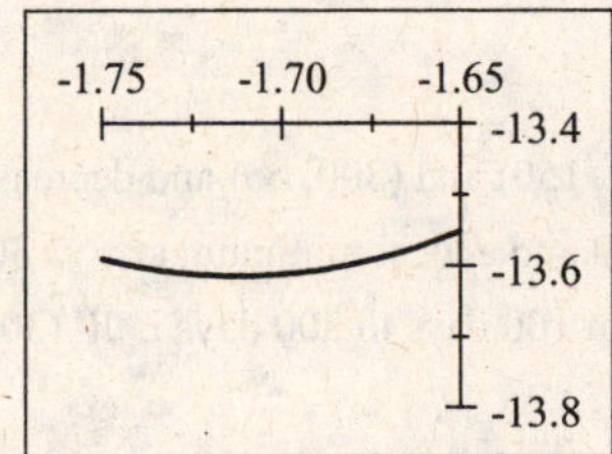

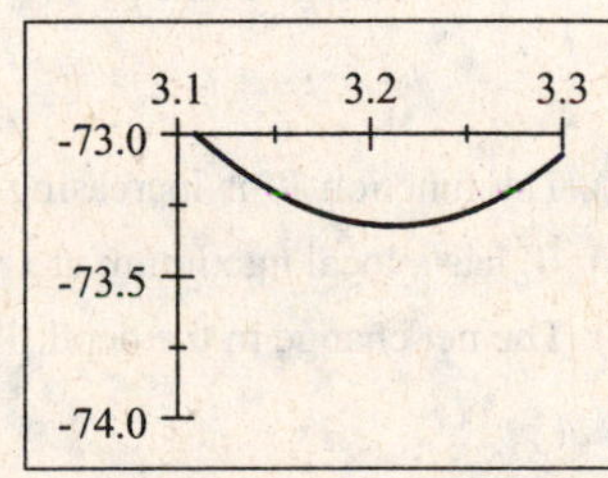

(b) The function is increasing on $(-1.71, 0)$ and $(3.21, \infty)$ and decreasing on $(-\infty, -1.71)$ and $(0, 3.21)$.

51. (a) In the first graph, we see that $U(x) = x\sqrt{6-x}$ has only a local maximum. Smaller x- and y-ranges show that $U(x)$ has a local maximum of about 5.66 when $x \approx 4.00$.

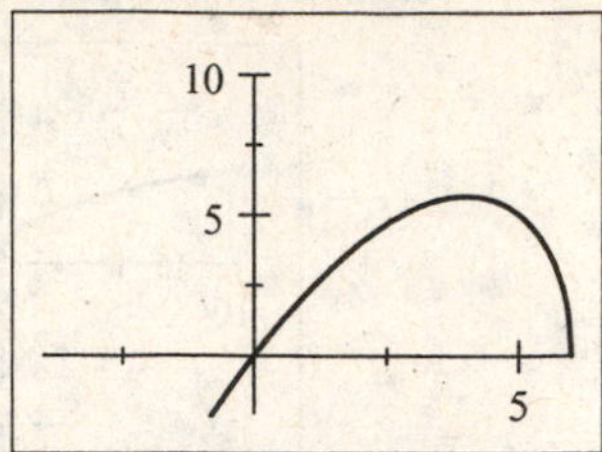

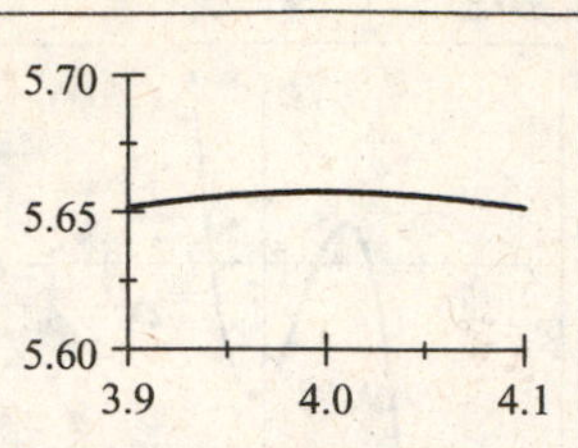

(b) The function is increasing on $(-\infty, 4.00)$ and decreasing on $(4.00, 6)$.

53. (a) In the first graph, we see that $V(x) = \dfrac{1-x^2}{x^3}$ has a local minimum and a local maximum. Smaller x- and y-ranges show that $V(x)$ has a local maximum of about 0.38 when $x \approx -1.73$ and a local minimum of about -0.38 when $x \approx 1.73$.

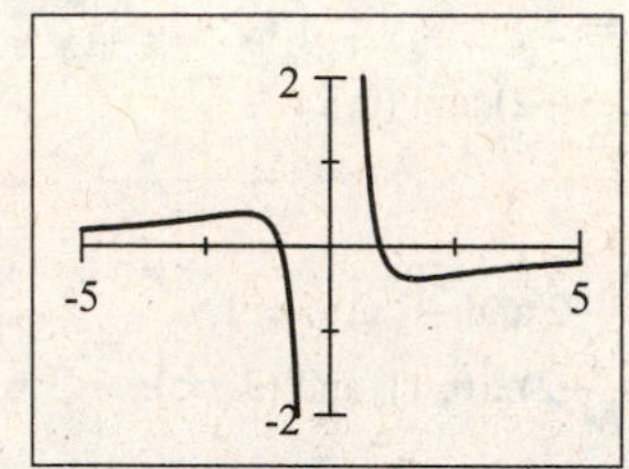

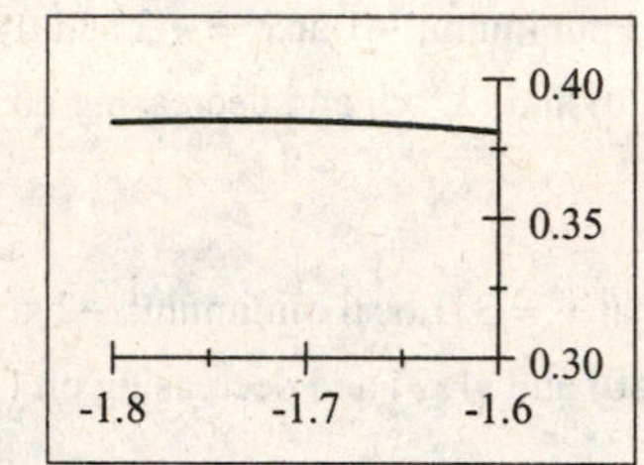

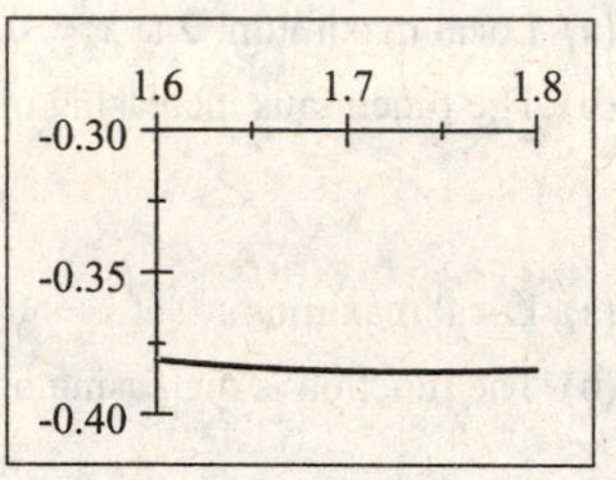

(b) The function is increasing on $(-\infty, -1.73)$ and $(1.73, \infty)$ and decreasing on $(-1.73, 0)$ and $(0, 1.73)$.

55. (a) At 6 A.M. the graph shows that the power consumption is about 500 megawatts. Since $t = 18$ represents 6 P.M., the graph shows that the power consumption at 6 P.M. is about 725 megawatts.

(b) The power consumption is lowest between 3 A.M. and 4 A.M..

(c) The power consumption is highest just before 12 noon.

(d) The net change in power consumption from 9 A.M. to 7 P.M. is $P(19) - P(9) \approx 690 - 790 \approx -100$ megawatts.

57. (a) This person appears to be gaining weight steadily until the age of 21 when this person's weight gain slows down. The person continues to gain weight until the age of 30, at which point this person experiences a sudden weight loss. Weight gain resumes around the age of 32, and the person dies at about age 68. Thus, the person's weight W is increasing on $(0, 30)$ and $(32, 68)$ and decreasing on $(30, 32)$.

(b) The sudden weight loss could be due to a number of reasons, among them major illness, a weight loss program, etc.

(c) The net change in the person's weight from age 10 to age 20 is $W(20) - W(10) = 150 - 50 = 100$ lb.

59. (a) The function W is increasing on $(0, 150)$ and $(300, \infty)$ and decreasing on $(150, 300)$.

(b) W has a local maximum at $x = 150$ and a local minimum at $x = 300$.

(c) The net change in the depth W from 100 days to 300 days is $W(300) - W(100) = 25 - 75 = -50$ ft.

61. Runner A won the race. All runners finished the race. Runner B fell, but got up and finished the race.

63. (a)

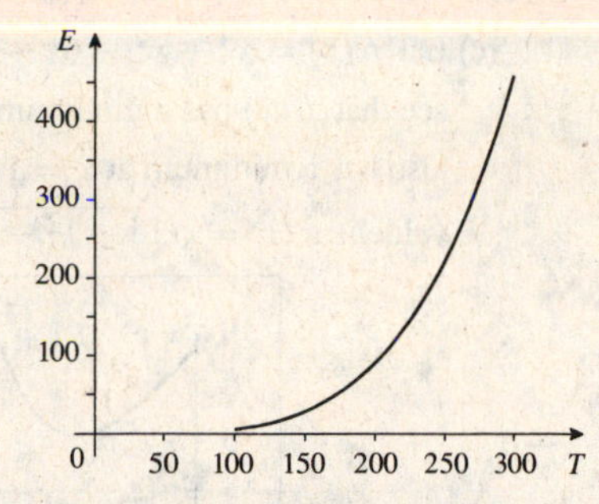

(b) As the temperature T increases, the energy E increases. The rate of increase gets larger as the temperature increases.

65. In the first graph, we see the general location of the minimum of $E(v) = 2.73v^3 \frac{10}{v-5}$. In the second graph, we isolate the minimum, and from this graph, we see that energy is minimized when $v \approx 7.5$ mi/h.

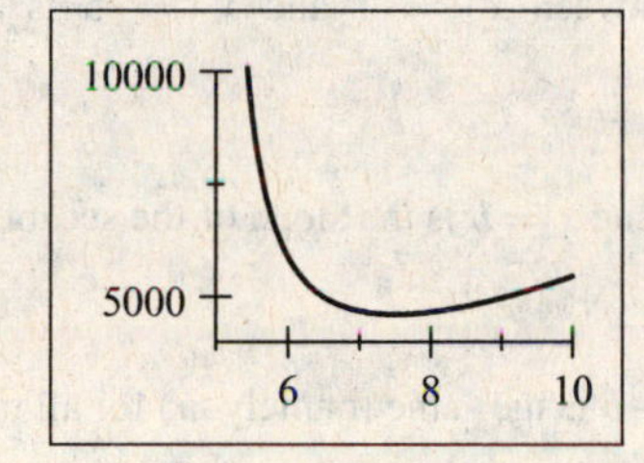

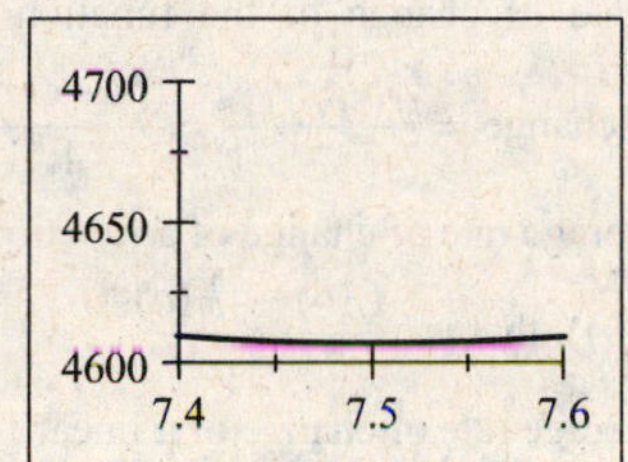

67. (a) $f(x)$ is always increasing, and $f(x) > 0$ for all x.

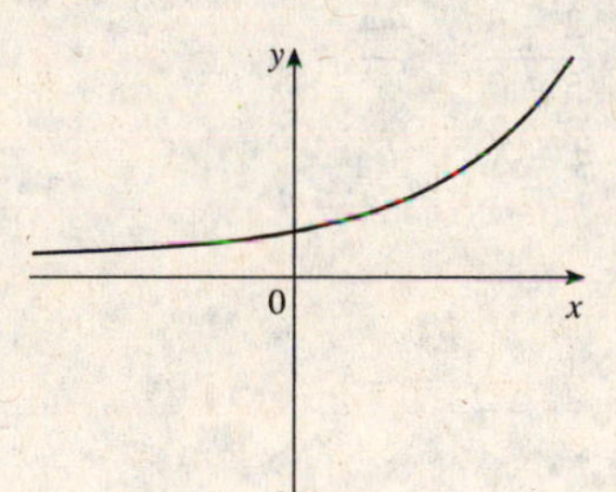

(b) $f(x)$ is always decreasing, and $f(x) > 0$ for all x.

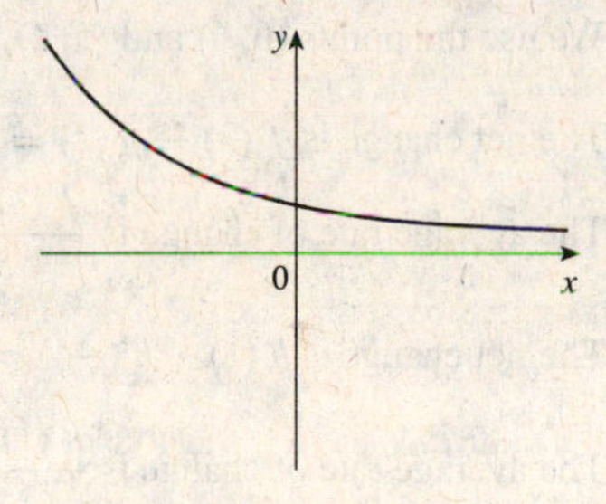

(c) $f(x)$ is always increasing, and $f(x) < 0$ for all x.

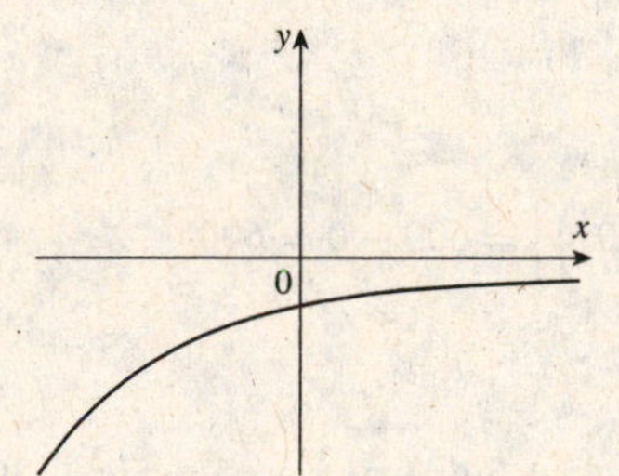

(d) $f(x)$ is always decreasing, and $f(x) < 0$ for all x.

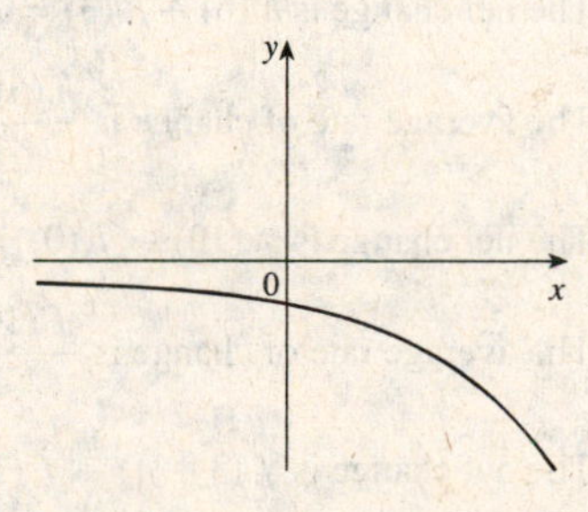

69. **(a)** If $x = a$ is a local maximum of $f(x)$ then $f(a) \geq f(x) \geq 0$ for all x around $x = a$. So $[g(a)]^2 \geq [g(x)]^2$ and thus $g(a) \geq g(x)$. Similarly, if $x = b$ is a local minimum of $f(x)$, then $f(x) \geq f(b) \geq 0$ for all x around $x = b$. So $[g(x)]^2 \geq [g(b)]^2$ and thus $g(x) \geq g(b)$.

(b) Using the distance formula,

$$g(x) = \sqrt{(x-3)^2 + (x^2-0)^2} = \sqrt{x^4 + x^2 - 6x + 9}$$

(c) Let $f(x) = x^4 + x^2 - 6x + 9$. From the graph, we see that $f(x)$ has a minimum at $x = 1$. Thus $g(x)$ also has a minimum at $x = 1$ and this minimum value is $g(1) = \sqrt{1^4 + 1^2 - 6(1) + 9} = \sqrt{5}$.

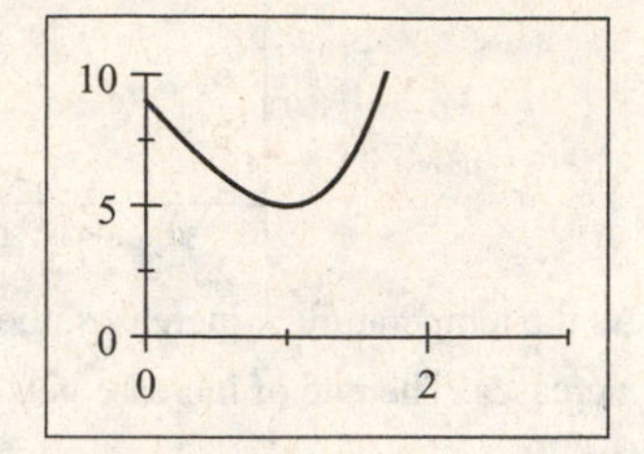

2.4 AVERAGE RATE OF CHANGE OF A FUNCTION

1. If you travel 100 miles in two hours then your average speed for the trip is average speed $= \dfrac{100 \text{ miles}}{2 \text{ hours}} = 50$ mi/h.

3. The average rate of change of the function $f(x) = x^2$ between $x = 1$ and $x = 5$ is average rate of change $= \frac{f(5)-f(1)}{5-1} = \dfrac{5^2 - 1^2}{4} = \dfrac{25-1}{4} = \dfrac{24}{4} = 6$.

5. **(a)** Yes, the average rate of change of a function between $x = a$ and $x = b$ is the slope of the secant line through $(a, f(a))$ and $(b, f(b))$; that is, $\dfrac{f(b) - f(a)}{b - a}$.

(b) Yes, the average rate of change of a linear function $y = mx + b$ is the same (namely m) for all intervals.

7. **(a)** The net change is $f(4) - f(1) = 5 - 3 = 2$.

(b) We use the points $(1, 3)$ and $(4, 5)$, so the average rate of change is $\dfrac{5-3}{4-1} = \dfrac{2}{3}$.

9. **(a)** The net change is $f(5) - f(0) = 2 - 6 = -4$.

(b) We use the points $(0, 6)$ and $(5, 2)$, so the average rate of change is $\dfrac{2-6}{5-0} = \dfrac{-4}{5}$.

11. **(a)** The net change is $f(3) - f(2) = [3(3) - 2] - [3(2) - 2] = 7 - 4 = 3$.

(b) The average rate of change is $\dfrac{f(3) - f(2)}{3-2} = \dfrac{3}{1} = 3$.

13. **(a)** The net change is $h(1) - h(-4) = \left[-1 + \frac{3}{2}\right] - \left[-(-4) + \frac{3}{2}\right] = \frac{1}{2} - \frac{11}{2} = -5$.

(b) The average rate of change is $\dfrac{h(1) - h(-4)}{1 - (-4)} = \dfrac{-5}{5} = -1$.

15. **(a)** The net change is $h(6) - h(3) = \left[2(6)^2 - 6\right] - \left[2(3)^2 - 3\right] = 66 - 15 = 51$.

(b) The average rate of change is $\dfrac{h(6) - h(3)}{6-3} = \dfrac{51}{3} = 17$.

17. **(a)** The net change is $f(10) - f(0) = \left[10^3 - 4\left(10^2\right)\right] - \left[0^3 - 4\left(0^2\right)\right] = 600 - 0 = 600$.

(b) The average rate of change is $\dfrac{f(10) - f(0)}{10 - 0} = \dfrac{600}{10} = 60$.

19. **(a)** The net change is $f(3+h) - f(3) = \left[5(3+h)^2\right] - \left[5(3)^2\right] = 45 + 30h + 5h^2 - 45 = 5h^2 + 30h$.

(b) The average rate of change is $\dfrac{f(3+h)-f(3)}{(3+h)-3}=\dfrac{5h^2+30h}{h}=5h+30$.

21. (a) The net change is $g(a)-g(1)=\dfrac{1}{a}-\dfrac{1}{1}=\dfrac{1-a}{a}$.

(b) The average rate of change is $\dfrac{g(a)-g(1)}{a-1}=\dfrac{\frac{1-a}{a}}{a-1}=\dfrac{1-a}{a(a-1)}=-\dfrac{1}{a}$.

23. (a) The net change is $f(a+h)-f(a)=\dfrac{2}{a+h}-\dfrac{2}{a}=-\dfrac{2h}{a(a+h)}$.

(b) The average rate of change is

$$\frac{f(a+h)-f(a)}{(a+h)-a}=\frac{-\frac{2h}{a(a+h)}}{h}=-\frac{2h}{ah(a+h)}=-\frac{2}{a(a+h)}.$$

25. (a) The average rate of change is

$$\frac{f(a+h)-f(a)}{(a+h)-a}=\frac{\left[\frac{1}{2}(a+h)+3\right]-\left[\frac{1}{2}a+3\right]}{h}=\frac{\frac{1}{2}a+\frac{1}{2}h+3-\frac{1}{2}a-3}{h}=\frac{\frac{1}{2}h}{h}=\frac{1}{2}.$$

(b) The slope of the line $f(x)=\frac{1}{2}x+3$ is $\frac{1}{2}$, which is also the average rate of change.

27. The function f has a greater average rate of change between $x=0$ and $x=1$. The function g has a greater average rate of change between $x=1$ and $x=2$. The functions f and g have the same average rate of change between $x=0$ and $x=1.5$.

29. The average rate of change is $\dfrac{W(200)-W(100)}{200-100}=\dfrac{50-75}{200-100}=\dfrac{-25}{100}=-\dfrac{1}{4}$ ft/day.

31. (a) The average rate of change of population is $\dfrac{1{,}591-856}{2001-1998}=\dfrac{735}{3}=245$ persons/yr.

(b) The average rate of change of population is $\dfrac{826-1{,}483}{2004-2002}=\dfrac{-657}{2}=-328.5$ persons/yr.

(c) The population was increasing from 1997 to 2001.

(d) The population was decreasing from 2001 to 2006.

33. (a) The average rate of change of sales is $\dfrac{635-495}{2013-2003}=\dfrac{140}{10}=14$ players/yr.

(b) The average rate of change of sales is $\dfrac{513-495}{2004-2003}=\dfrac{18}{1}=18$ players/yr.

(c) The average rate of change of sales is $\dfrac{410-513}{2005-2004}=\dfrac{-103}{1}=-103$ players/yr.

(d)

Year	DVD players sold	Change in sales from previous year
2003	495	—
2004	513	18
2005	410	−103
2006	402	−8
2007	520	118
2008	580	60
2009	631	51
2010	719	88
2011	624	−95
2012	582	−42
2013	635	53

Sales increased most quickly between 2006 and 2007, and decreased most quickly between 2004 and 2005.

35. The average rate of change of the temperature of the soup over the first 20 minutes is $\frac{T(20)-T(0)}{20-0}=\frac{119-200}{20-0}=\frac{-81}{20}=-4.05^\circ$ F/min. Over the next 20 minutes, it is $\frac{T(40)-T(20)}{40-20}=\frac{89-119}{40-20}=-\frac{30}{20}=-1.5^\circ$ F/min. The first 20 minutes had a higher average rate of change of temperature (in absolute value).

37. **(a)** For all three skiers, the average rate of change is $\frac{d(10)-d(0)}{10-0}=\frac{100}{10}=10$.

(b) Skier A gets a great start, but slows at the end of the race. Skier B maintains a steady pace. Runner C is slow at the beginning, but accelerates down the hill.

39.

$t=a$	$t=b$	Average Speed $=\frac{f(b)-f(a)}{b-a}$
3	3.5	$\frac{16(3.5)^2-16(3)^2}{3.5-3}=104$
3	3.1	$\frac{16(3.1)^2-16(3)^2}{3.1-3}=97.6$
3	3.01	$\frac{16(3.01)^2-16(3)^2}{3.01-3}=96.16$
3	3.001	$\frac{16(3.001)^2-16(3)^2}{3.001-3}=96.016$
3	3.0001	$\frac{16(3.0001)^2-16(3)^2}{3.0001-3}=96.0016$

From the table it appears that the average speed approaches 96 ft/s as the time intervals get smaller and smaller. It seems reasonable to say that the speed of the object is 96 ft/s at the instant $t=3$.

2.5 LINEAR FUNCTIONS AND MODELS

1. If f is a function with constant rate of change, then
(a) f is a linear function of the form $f(x) = ax + b$.
(b) The graph of f is a line.

3. From the graph, we see that $y(2) = 50$ and $y(0) = 20$, so the slope of the graph is $m = \dfrac{y(2) - y(0)}{2 - 0} = \dfrac{50 - 20}{2} = 15$ gal/min.

5. If a linear function has positive rate of change, its graph slopes upward.

7. $f(x) = 3 + \frac{1}{3}x = \frac{1}{3}x + 3$ is linear with $a = \frac{1}{3}$ and $b = 3$.

9. $f(x) = x(4 - x) = 4x - x^2$ is not of the form $f(x) = ax + b$ for constants a and b, so it is not linear.

11. $f(x) = \dfrac{x + 1}{5} = \frac{1}{5}x + \frac{1}{5}$ is linear with $a = \frac{1}{5}$ and $b = \frac{1}{5}$.

13. $f(x) = (x + 1)^2 = x^2 + 2x + 1$ is not of the form $f(x) = ax + b$ for constants a and b, so it is not linear.

15.

x	$f(x) = 2x - 5$
-1	-7
0	-5
1	-3
2	-1
3	1
4	3

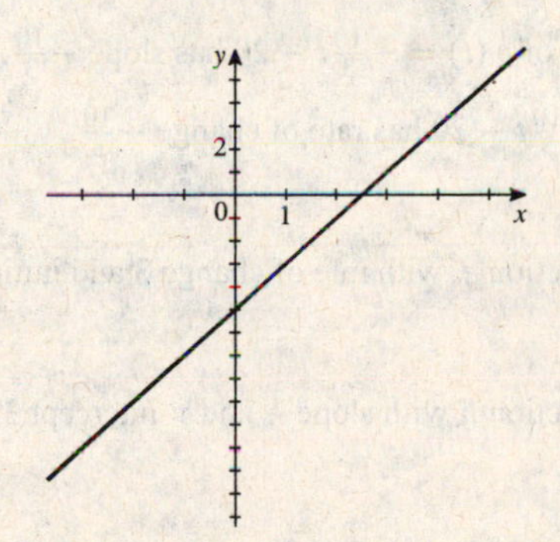

The slope of the graph of $f(x) = 2x - 5$ is 2.

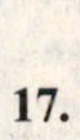

17.

t	$r(t) = -\frac{2}{3}t + 2$
-1	2.67
0	2
1	1.33
2	0.67
3	0
4	-0.67

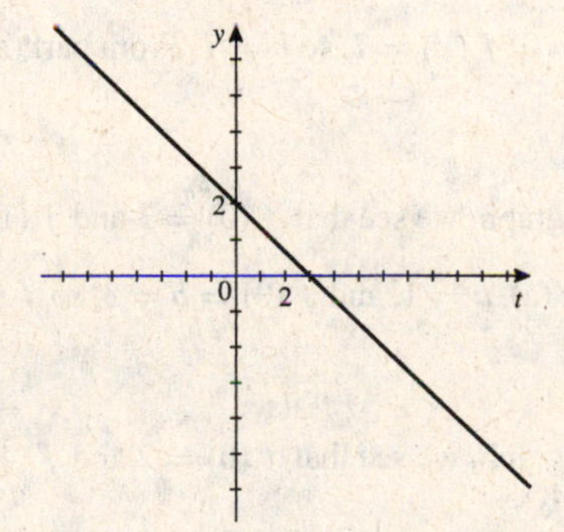

The slope of the graph of $r(t) = -\frac{2}{3}t + 2$ is $-\frac{2}{3}$.

19. (a)

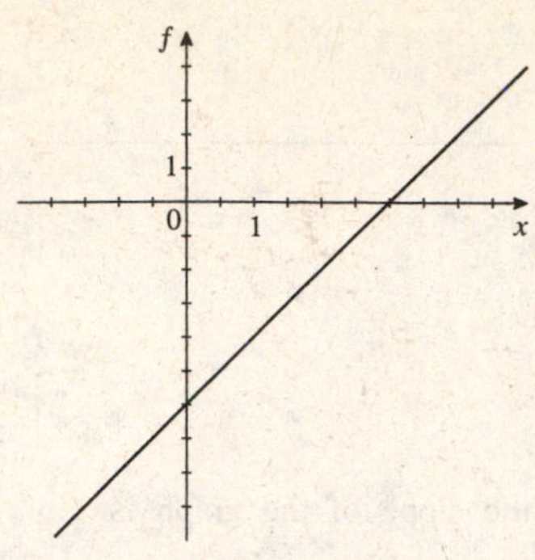

(b) The graph of $f(x) = 2x - 6$ has slope 2.

(c) $f(x) = 2x - 6$ has rate of change 2.

21. (a)

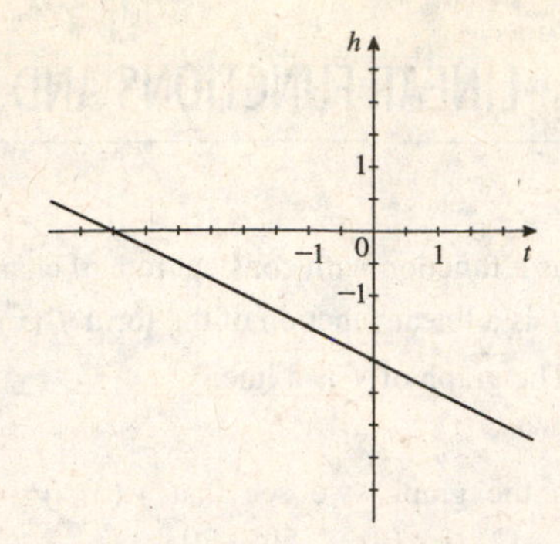

(b) The graph of $h(t) = -0.5t - 2$ has slope -0.5.

(c) $h(t) = -0.5t - 2$ has rate of change -0.5.

23. (a)

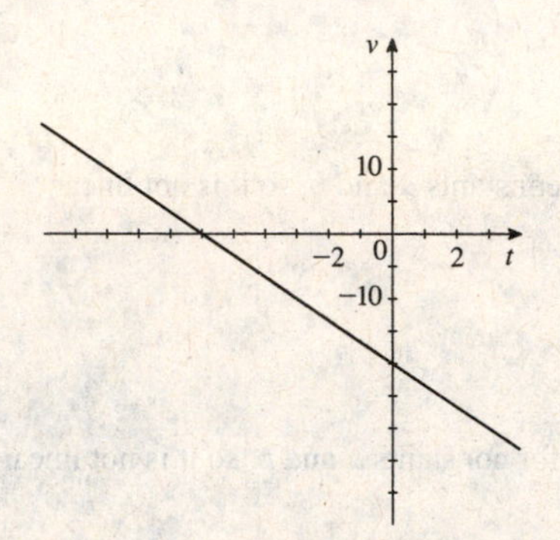

(b) The graph of $v(t) = -\frac{10}{3}t - 20$ has slope $-\frac{10}{3}$.

(c) $v(t) = -\frac{10}{3}t - 20$ has rate of change $-\frac{10}{3}$.

25. (a)

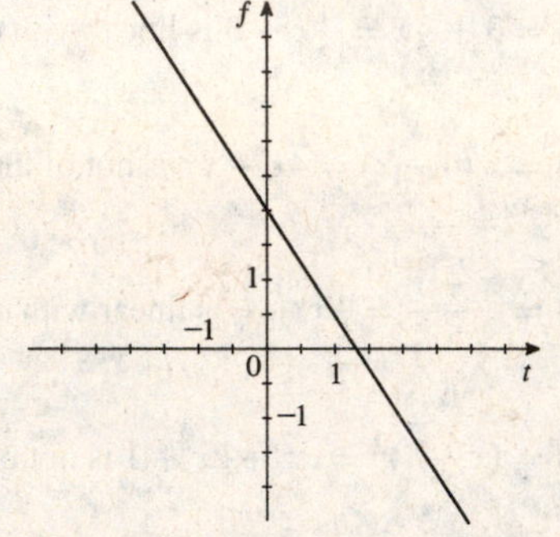

(b) The graph of $f(t) = -\frac{3}{2}t + 2$ has slope $-\frac{3}{2}$.

(c) $f(t) = -\frac{3}{2}t + 2$ has rate of change $-\frac{3}{2}$.

27. The linear function f with rate of change 3 and initial value -1 has equation $f(x) = 3x - 1$.

29. The linear function h with slope $\frac{1}{2}$ and y-intercept 3 has equation $h(x) = \frac{1}{2}x + 3$.

31. (a) From the table, we see that for every increase of 2 in the value of x, $f(x)$ increases by 3. Thus, the rate of change of f is $\frac{3}{2}$.

(b) When $x = 0$, $f(x) = 7$, so $b = 7$. From part (a), $a = \frac{3}{2}$, and so $f(x) = \frac{3}{2}x + 7$.

33. (a) From the graph, we see that $f(0) = 3$ and $f(1) = 4$, so the rate of change of f is $\dfrac{4-3}{1-0} = 1$.

(b) From part (a), $a = 1$, and $f(0) = b = 3$, so $f(x) = x + 3$.

35. (a) From the graph, we see that $f(0) = 2$ and $f(4) = 0$, so the rate of change of f is $\dfrac{0-2}{4-0} = -\dfrac{1}{2}$.

(b) From part (a), $a = -\frac{1}{2}$, and $f(0) = b = 2$, so $f(x) = -\frac{1}{2}x + 2$.

37.

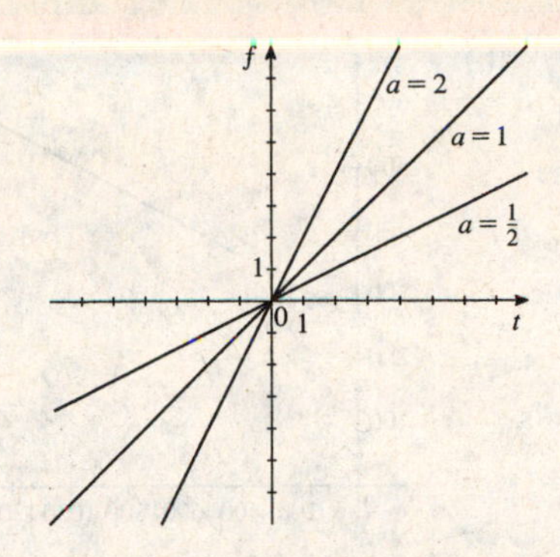

Increasing the value of a makes the graph of f steeper. In other words, it increases the rate of change of f.

39. **(a)**

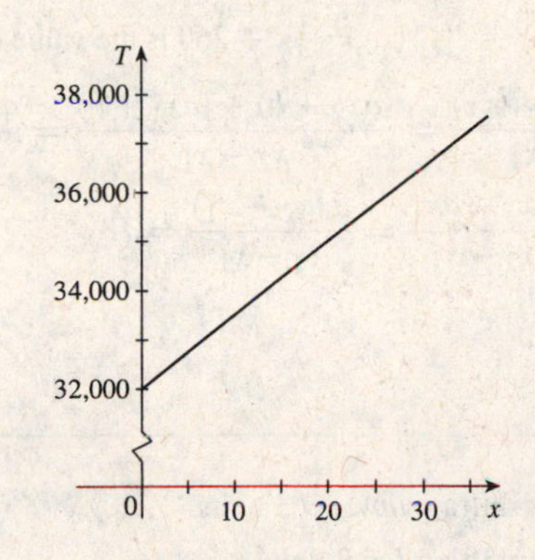

(b) The slope of $T(x) = 150x + 32{,}000$ is the value of a, 150.

(c) The amount of trash is changing at a rate equal to the slope of the graph, 150 thousand tons per year.

41. **(a)** Let $V(t) = at + b$ represent the volume of hydrogen. The balloon is being filled at the rate of 0.5 ft^3/s, so $a = 0.5$, and initially it contains 2 ft^3, so $b = 2$. Thus, $V(t) = 0.5t + 2$.

(b) We solve $V(t) = 15 \Leftrightarrow 0.5t + 2 = 15 \Leftrightarrow 0.5t = 13 \Leftrightarrow t = 26$. Thus, it takes 26 seconds to fill the balloon.

43. **(a)** Let $H(x) = ax + b$ represent the height of the ramp. The maximum rise is 1 inch per 12 inches, so $a = \frac{1}{12}$. The ramp starts on the ground, so $b = 0$. Thus, $H(x) = \frac{1}{12}x$.

(b) We find $H(150) = \frac{1}{12}(150) = 12.5$. Thus, the ramp reaches a height of 12.5 inches.

45. **(a)** From the graph, we see that the slope of Jari's trip is steeper than that of Jade. Thus, Jari is traveling faster.

(b) The points $(0, 0)$ and $(6, 7)$ are on Jari's graph, so her speed is $\dfrac{7-0}{6-0} = \dfrac{7}{6}$ miles per minute or $60\left(\dfrac{7}{6}\right) = 70$ mi/h.

The points $(0, 10)$ and $(6, 16)$ are on Jade's graph, so her speed is $60 \cdot \dfrac{16-10}{6-0} = 60$ mi/h.

(c) t is measured in minutes, so Jade's speed is 60 mi/h $\cdot \frac{1}{60}$ h/min $=$ 1 mi/min and Jari's speed is 70 mi/h $\cdot \frac{1}{60}$ h/min $= \frac{7}{6}$ mi/min. Thus, Jade's distance is modeled by $f(t) = 1(t-0) + 10 = t + 10$ and Jari's distance is modeled by $g(t) = \frac{7}{6}(t-0) + 0 = \frac{7}{6}t$.

47. Let x be the horizontal distance and y the elevation. The slope is $-\frac{6}{100}$, so if we take $(0, 0)$ as the starting point, the elevation is $y = -\frac{6}{100}x$. We have descended 1000 ft, so we substitute $y = -1000$ and solve for x: $-1000 = -\frac{6}{100}x \Leftrightarrow x \approx 16{,}667$ ft. Converting to miles, the horizontal distance is $\frac{1}{5280}(16{,}667) \approx 3.16$ mi.

49. (a) Let $C(x) = ax + b$ be the cost of driving x miles. In May Lynn drove 480 miles at a cost of \$380, and in June she drove 800 miles at a cost of \$460. Thus, the points $(480, 380)$ and $(800, 460)$ are on the graph, so the slope is $a = \dfrac{460 - 380}{800 - 480} = \dfrac{1}{4}$. We use the point $(480, 380)$ to find the value of b: $380 = \frac{1}{4}(480) + b \Leftrightarrow b = 260$. Thus, $C(x) = \frac{1}{4}x + 260$.

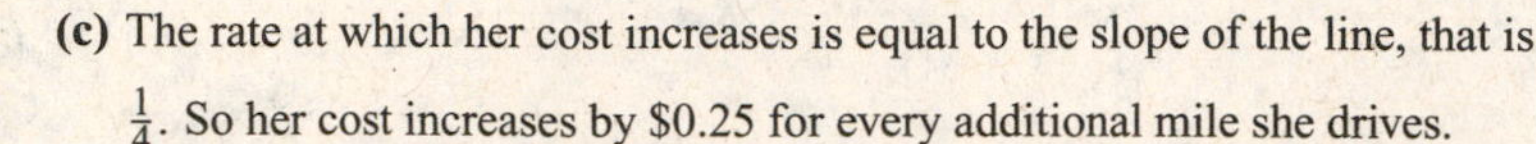

(c) The rate at which her cost increases is equal to the slope of the line, that is $\frac{1}{4}$. So her cost increases by \$0.25 for every additional mile she drives.

(b)

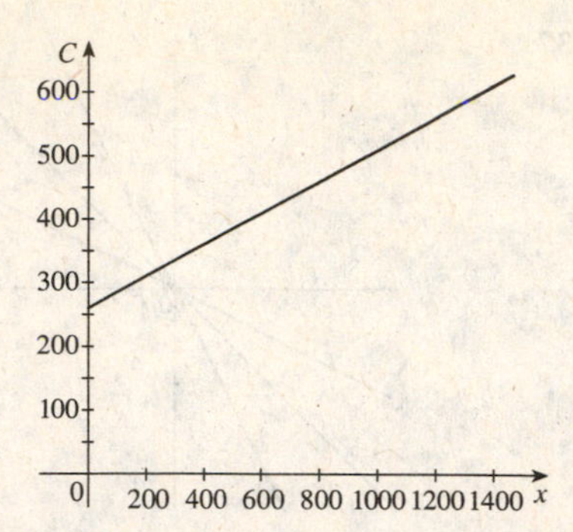

The slope of the graph of $C(x) = \frac{1}{4}x + 260$ is the value of a, $\frac{1}{4}$.

51. (a) By definition, the average rate of change between x_1 and x_2 is $\dfrac{f(x_2) - f(x_1)}{x_2 - x_1} = \dfrac{(ax_2 + b) - (ax_1 + b)}{x_2 - x_1} = \dfrac{ax_2 - ax_1}{x_2 - x_1}$.

(b) Factoring the numerator and cancelling, the average rate of change is $\dfrac{ax_2 - ax_1}{x_2 - x_1} = \dfrac{a(x_2 - x_1)}{x_2 - x_1} = a$.

2.6 TRANSFORMATIONS OF FUNCTIONS

1. (a) The graph of $y = f(x) + 3$ is obtained from the graph of $y = f(x)$ by shifting *upward* 3 units.

(b) The graph of $y = f(x + 3)$ is obtained from the graph of $y = f(x)$ by shifting *left* 3 units.

3. (a) The graph of $y = -f(x)$ is obtained from the graph of $y = f(x)$ by reflecting in the *x-axis*.

(b) The graph of $y = f(-x)$ is obtained from the graph of $y = f(x)$by reflecting in the *y-axis*.

5. If f is an even function, then $f(-x) = f(x)$ and the graph of f is symmetric about the y-axis.

7. (a) The graph of $y = f(x) - 1$ can be obtained by shifting the graph of $y = f(x)$ downward 1 unit.

(b) The graph of $y = f(x - 2)$ can be obtained by shifting the graph of $y = f(x)$ to the right 2 units.

9. (a) The graph of $y = f(-x)$ can be obtained by reflecting the graph of $y = f(x)$ in the y-axis.

(b) The graph of $y = 3f(x)$ can be obtained by stretching the graph of $y = f(x)$ vertically by a factor of 3.

11. (a) The graph of $y = f(x - 5) + 2$ can be obtained by shifting the graph of $y = f(x)$ to the right 5 units and upward 2 units.

(b) The graph of $y = f(x + 1) - 1$ can be obtained by shifting the graph of $y = f(x)$ to the left 1 unit and downward 1 unit.

13. (a) The graph of $y = -f(x) + 5$ can be obtained by reflecting the graph of $y = f(x)$ in the x-axis, then shifting the resulting graph upward 5 units.

(b) The graph of $y = 3f(x) - 5$ can be obtained by stretching the graph of $y = f(x)$ vertically by a factor of 3, then shifting the resulting graph downward 5 units.

15. (a) The graph of $y = 2f(x + 5) - 1$ can be obtained by shifting the graph of $y = f(x)$ to the left 5 units, stretching vertically by a factor of 2, then shifting downward 1 unit.

(b) The graph of $y = \frac{1}{4}f(x - 3) + 5$ can be obtained by shifting the graph of $y = f(x)$ to the right 3 units, shrinking vertically by a factor of $\frac{1}{4}$, then shifting upward 5 units.

17. (a) The graph of $y = f(4x)$ can be obtained by shrinking the graph of $y = f(x)$ horizontally by a factor of $\frac{1}{4}$.

(b) The graph of $y = f\left(\frac{1}{4}x\right)$ can be obtained by stretching the graph of $y = f(x)$ horizontally by a factor of 4.

19. (a) The graph of $g(x) = (x+2)^2$ is obtained by shifting the graph of $f(x)$ to the left 2 units.

(b) The graph of $g(x) = x^2 + 2$ is obtained by shifting the graph of $f(x)$ upward 2 units.

21. (a) The graph of $g(x) = |x+2| - 2$ is obtained by shifting the graph of $f(x)$ to the left 2 units and downward 2 units.

(b) The graph of $g(x) = g(x) = |x-2| + 2$ is obtained from by shifting the graph of $f(x)$ to the right 2 units and upward 2 units.

23. (a)

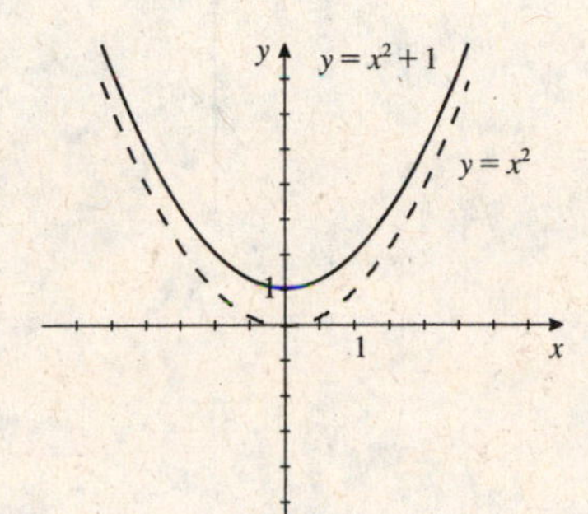

(b)

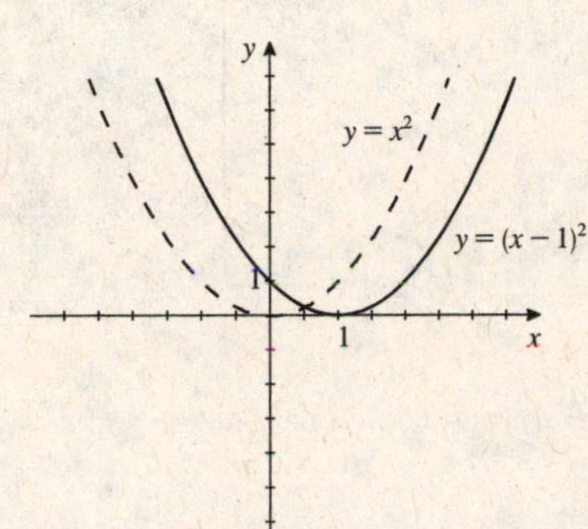

(c)

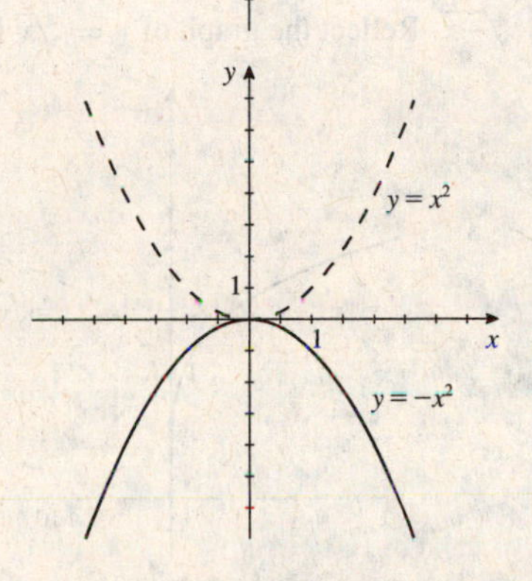

(d)

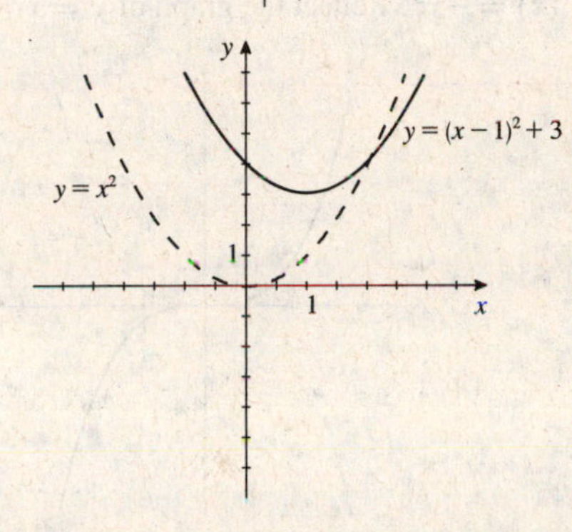

25. The graph of $y = |x+1|$ is obtained from that of $y = |x|$ by shifting to the left 1 unit, so it has graph II.

27. The graph of $y = |x| - 1$ is obtained from that of $y = |x|$ by shifting downward 1 unit, so it has graph I.

29. $f(x) = x^2 + 3$. Shift the graph of $y = x^2$ upward 3 units.

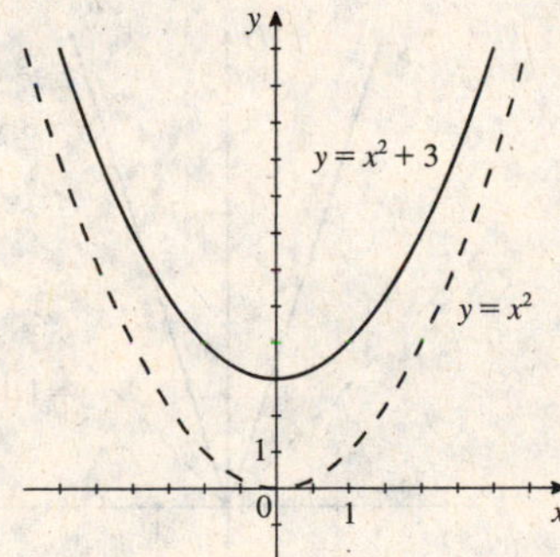

31. $f(x) = |x| - 1$. Shift the graph of $y = |x|$ downward 1 unit.

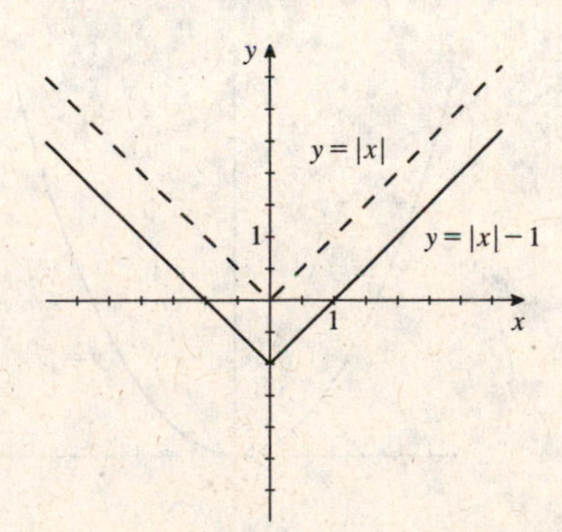

33. $f(x) = (x-5)^2$. Shift the graph of $y = x^2$ to the right 5 units.

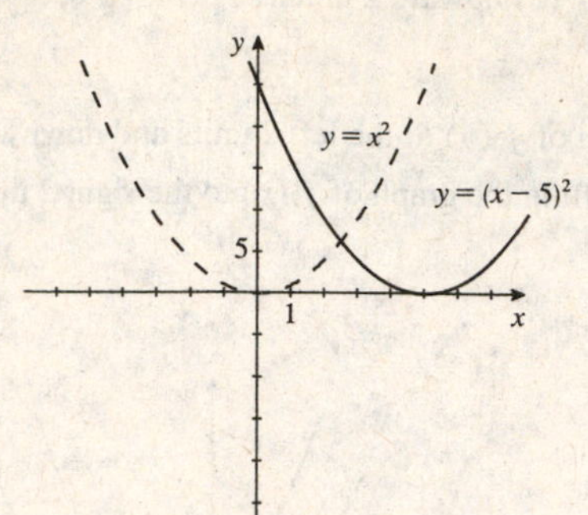

35. $f(x) = |x+2|$. Shift the graph of $y = |x|$ to the left 2 units.

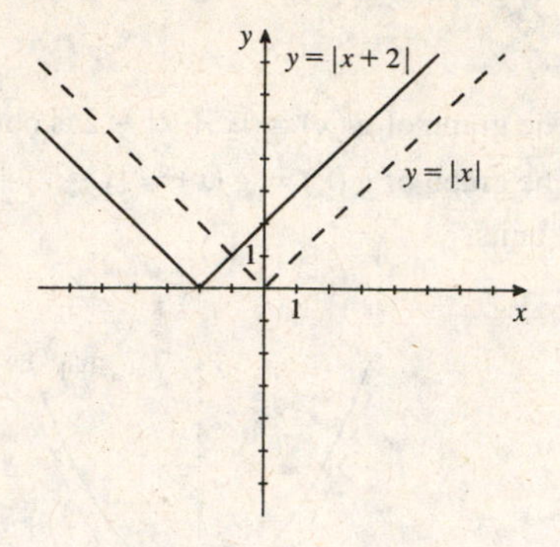

37. $f(x) = -x^3$. Reflect the graph of $y = x^3$ in the x-axis.

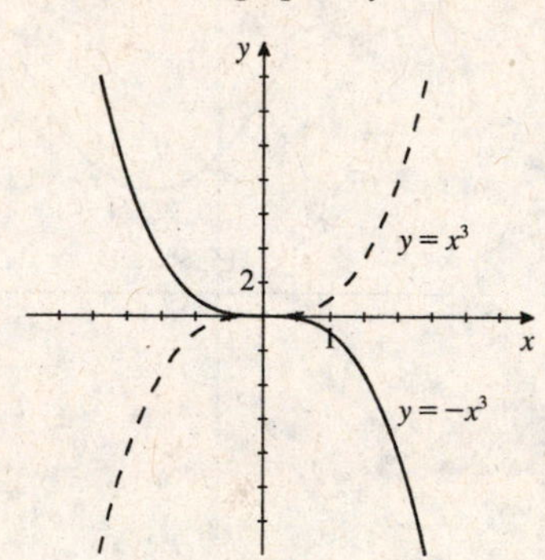

39. $y = \sqrt[4]{-x}$. Reflect the graph of $y = \sqrt[4]{x}$ in the y-axis.

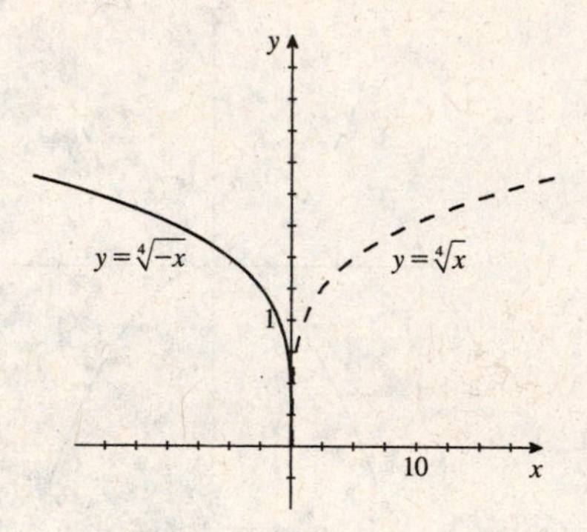

41. $y = \frac{1}{4}x^2$. Shrink the graph of $y = x^2$ vertically by a factor of $\frac{1}{4}$.

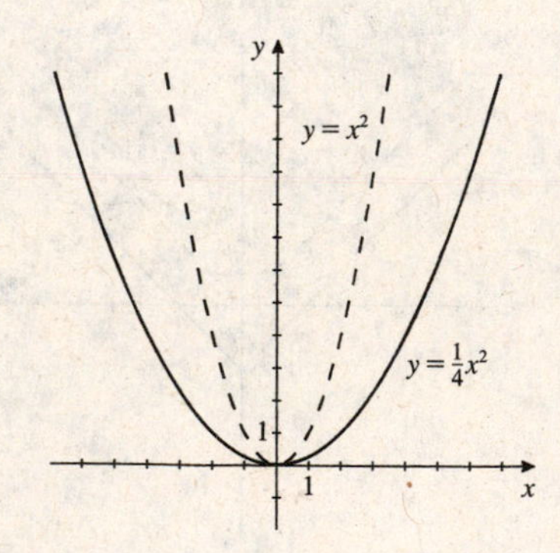

43. $y = 3|x|$. Stretch the graph of $y = |x|$ vertically by a factor of 3.

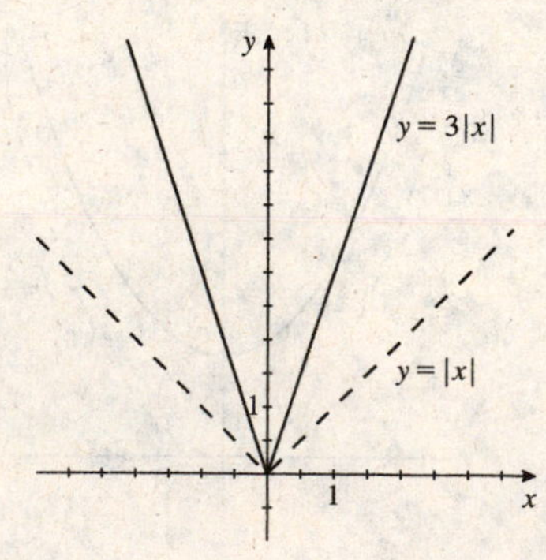

45. $y = (x-3)^2 + 5$. Shift the graph of $y = x^2$ to the right 3 units and upward 5 units.

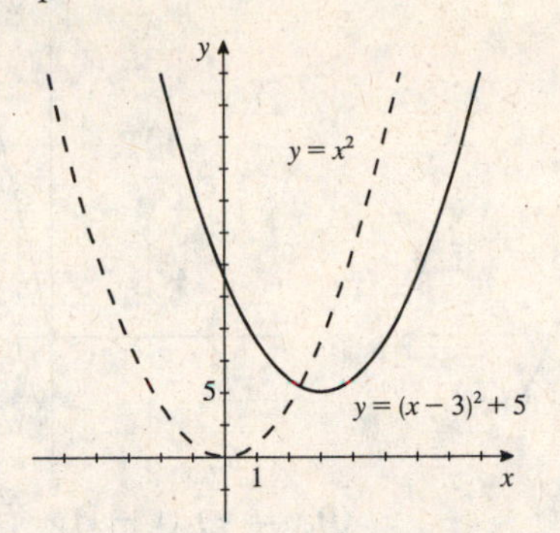

47. $y = 3 - \frac{1}{2}(x-1)^2$. Shift the graph of $y = x^2$ to the right one unit, shrink vertically by a factor of $\frac{1}{2}$, reflect in the x-axis, then shift upward 3 units.

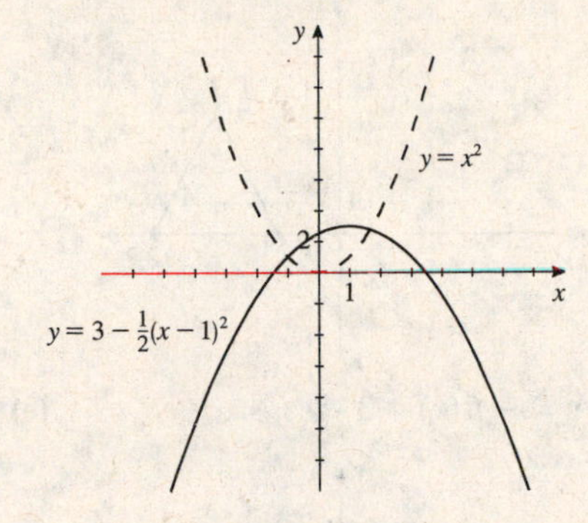

49. $y = |x+2| + 2$. Shift the graph of $y = |x|$ to the left 2 units and upward 2 units.

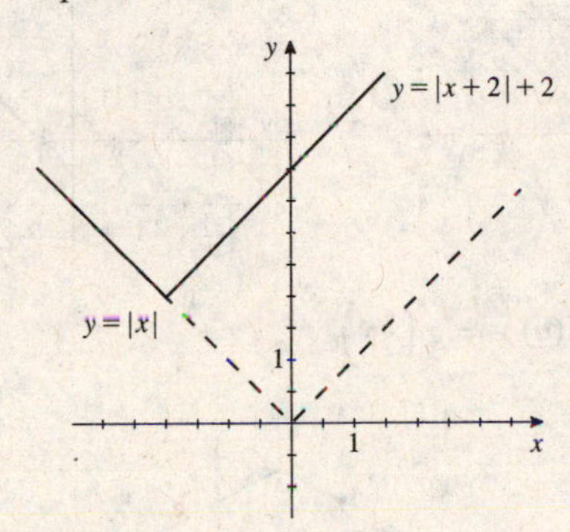

51. $y = \frac{1}{2}\sqrt{x+4} - 3$. Shrink the graph of $y = \sqrt{x}$ vertically by a factor of $\frac{1}{2}$, then shift the result to the left 4 units and downward 3 units.

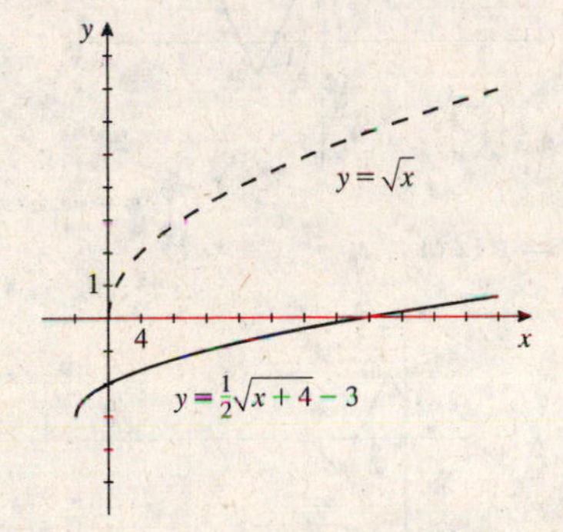

53. $y = f(x) - 3$. When $f(x) = x^2$, $y = x^2 - 3$.

55. $y = f(x+2)$. When $f(x) = \sqrt{x}$, $y = \sqrt{x+2}$.

57. $y = f(x+2) - 5$. When $f(x) = |x|$, $y = |x+2| - 5$.

59. $y = f(-x) + 1$. When $f(x) = \sqrt[4]{x}$, $y = \sqrt[4]{-x} + 1$.

61. $y = 2f(x-3) - 2$. When $f(x) = x^2$, $y = 2(x-3)^2 - 2$.

63. $g(x) = f(x-2) = (x-2)^2 = x^2 - 4x + 4$

65. $g(x) = f(x+1) + 2 = |x+1| + 2$

67. $g(x) = -f(x+2) = -\sqrt{x+2}$

69. **(a)** $y = f(x-4)$ is graph #3.

(b) $y = f(x) + 3$ is graph #1.

(c) $y = 2f(x+6)$ is graph #2.

(d) $y = -f(2x)$ is graph #4.

71. (a) $y = f(x-2)$

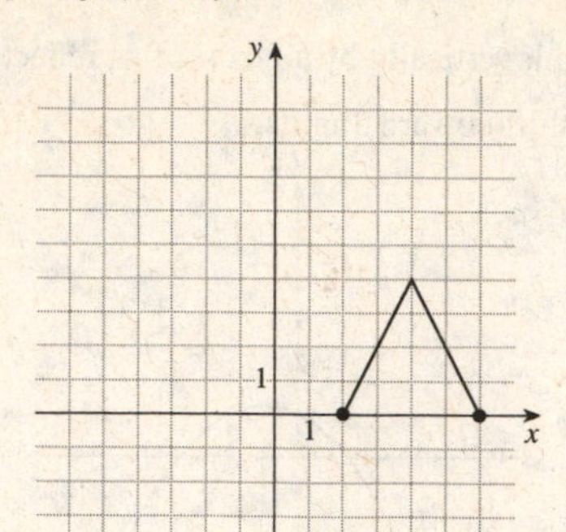

(b) $y = f(x) - 2$

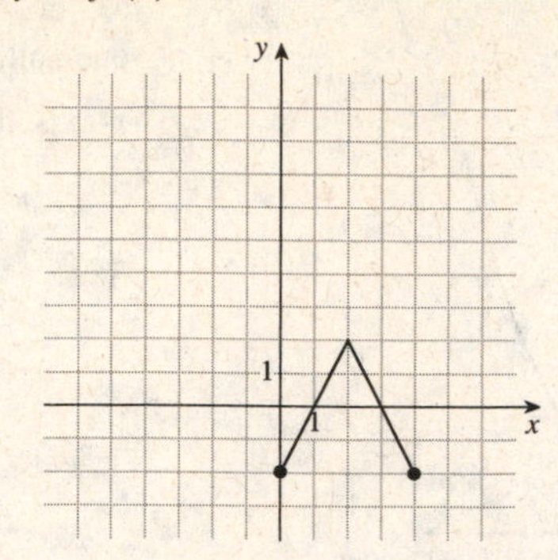

(c) $y = 2f(x)$

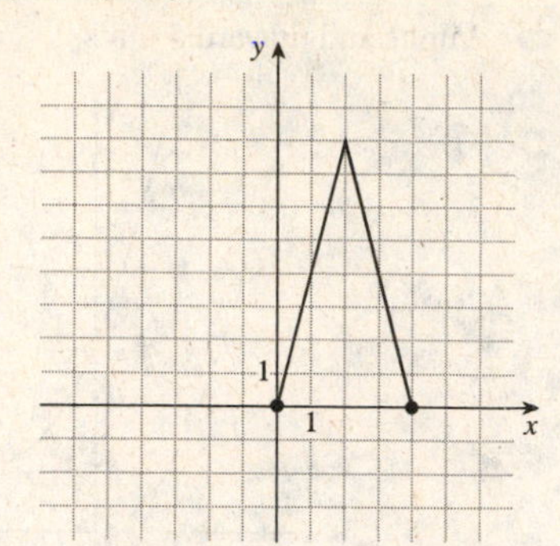

(d) $y = -f(x) + 3$

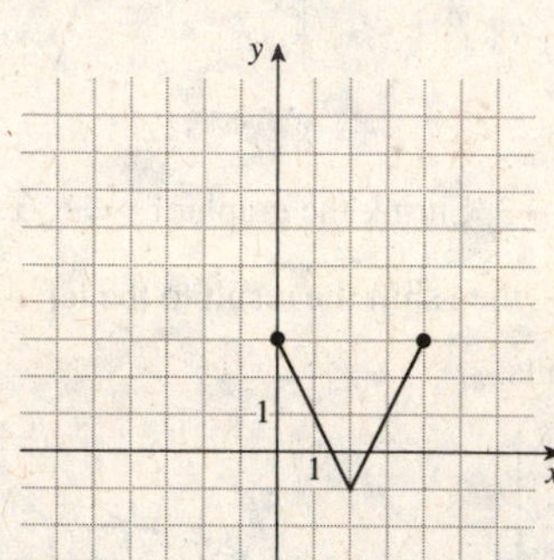

(e) $y = f(-x)$

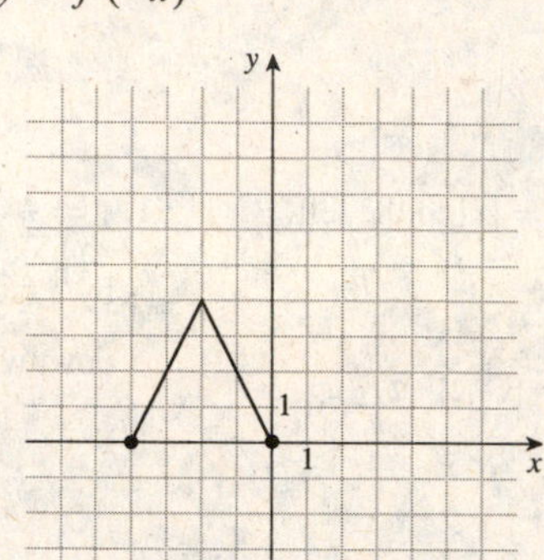

(f) $y = \frac{1}{2}f(x-1)$

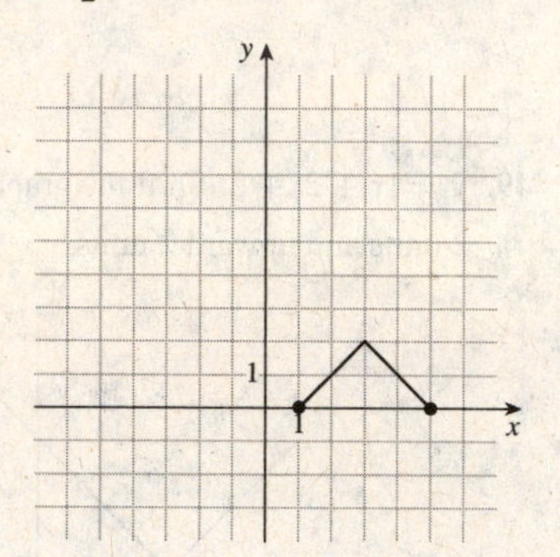

73. (a) $y = g(2x)$

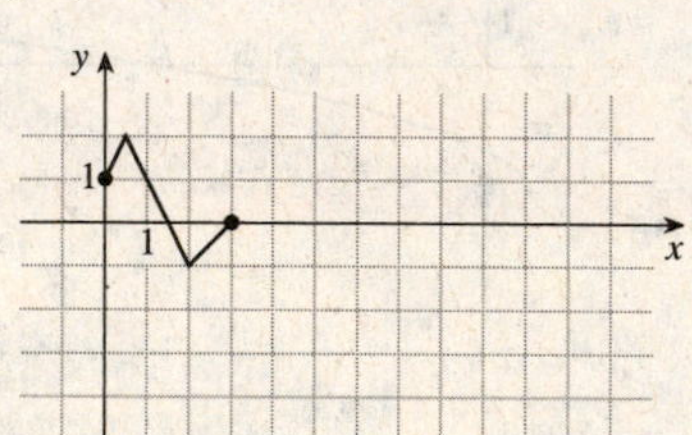

(b) $y = g\left(\frac{1}{2}x\right)$

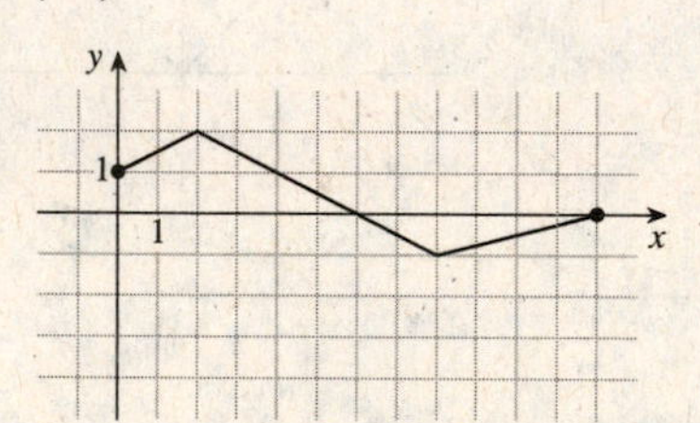

75. $y = [\![2x]\!]$

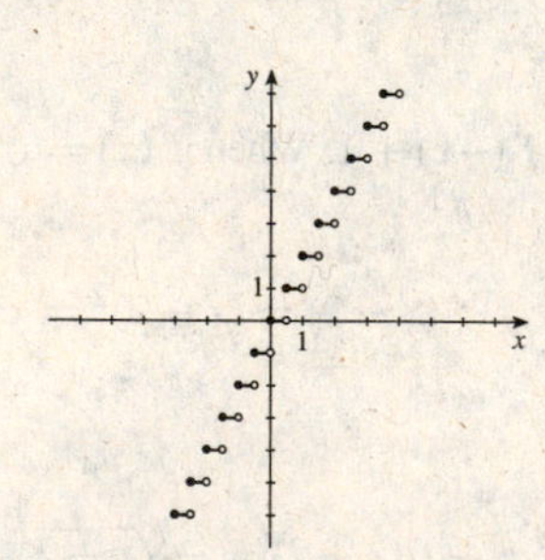

77.

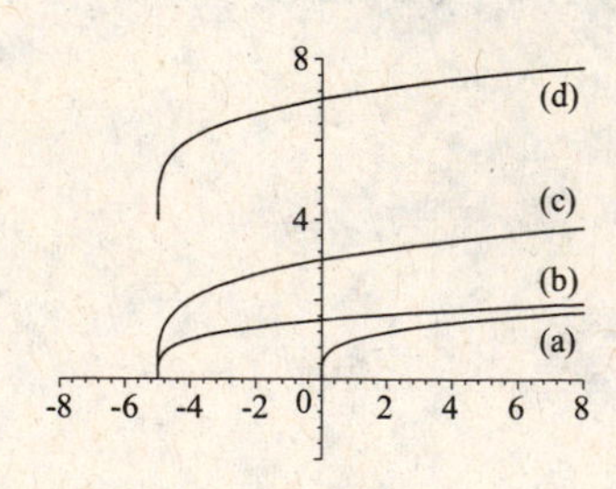

For part (b), shift the graph in (a) to the left 5 units; for part (c), shift the graph in (a) to the left 5 units, and stretch it vertically by a factor of 2; for part (d), shift the graph in (a) to the left 5 units, stretch it vertically by a factor of 2, and then shift it upward 4 units.

79. 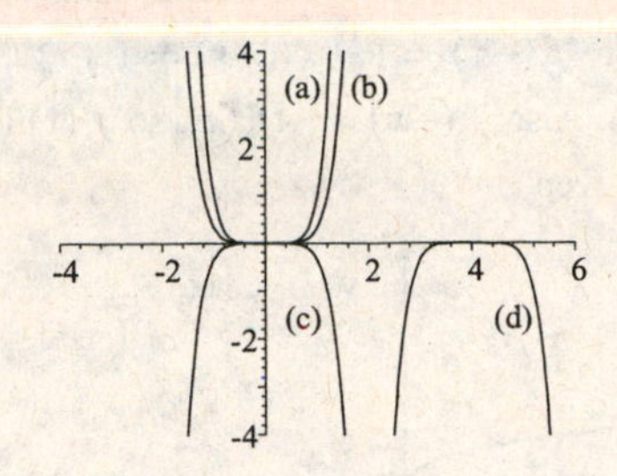

For part (b), shrink the graph in (a) vertically by a factor of $\frac{1}{3}$; for part (c), shrink the graph in (a) vertically by a factor of $\frac{1}{3}$, and reflect it in the x-axis; for part (d), shift the graph in (a) to the right 4 units, shrink vertically by a factor of $\frac{1}{3}$, and then reflect it in the x-axis.

81. **(a)** $y = f(x) = \sqrt{2x - x^2}$

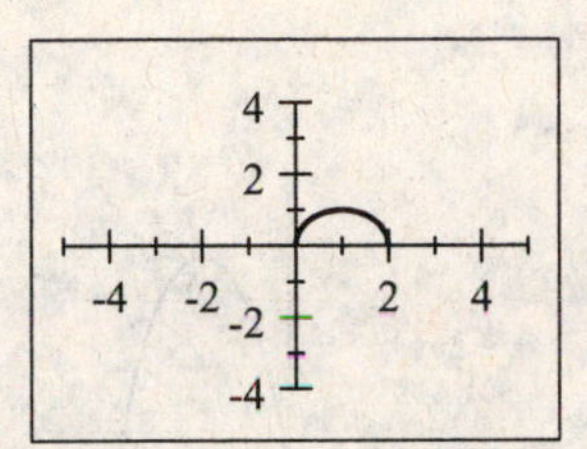

(b) $y = f(2x) = \sqrt{2(2x) - (2x)^2}$
$= \sqrt{4x - 4x^2}$

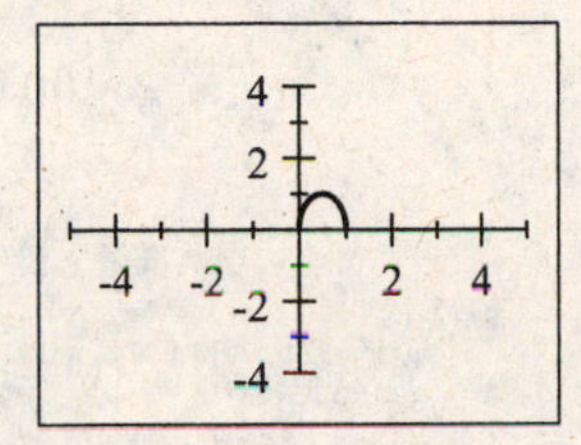

(c) $y = f\left(\frac{1}{2}x\right) = \sqrt{2\left(\frac{1}{2}x\right) - \left(\frac{1}{2}x\right)^2}$
$= \sqrt{x - \frac{1}{4}x^2}$

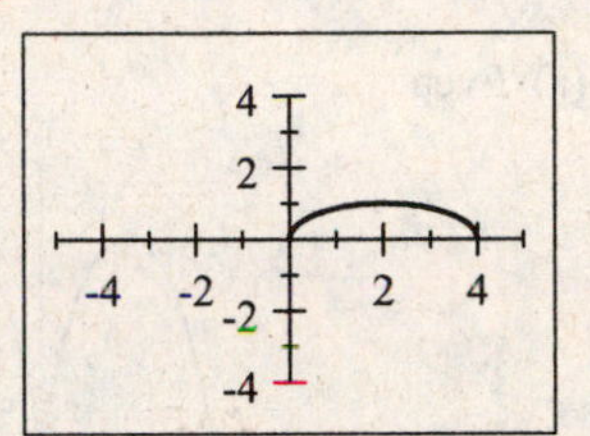

The graph in part (b) is obtained by horizontally shrinking the graph in part (a) by a factor of $\frac{1}{2}$ (so the graph is half as wide). The graph in part (c) is obtained by horizontally stretching the graph in part (a) by a factor of 2 (so the graph is twice as wide).

83. $f(x) = x^4$. $f(-x) = (-x)^4 = x^4 = f(x)$. Thus $f(x)$ is even.

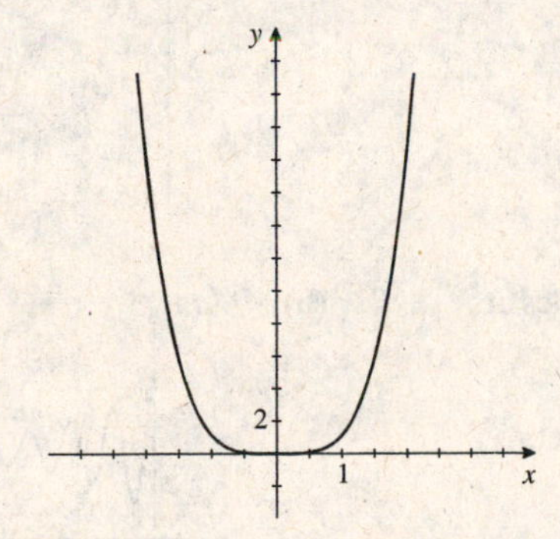

85. $f(x) = x^2 + x$. $f(-x) = (-x)^2 + (-x) = x^2 - x$. Thus $f(-x) \neq f(x)$. Also, $f(-x) \neq -f(x)$, so $f(x)$ is neither odd nor even.

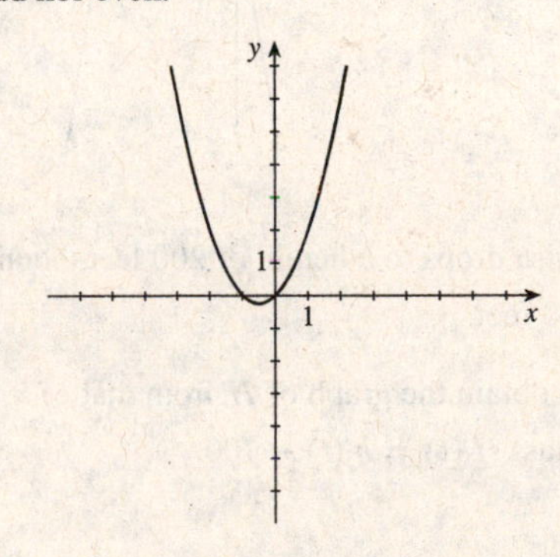

87. $f(x) = x^3 - x$.

$$f(-x) = (-x)^3 - (-x) = -x^3 + x$$
$$= -\left(x^3 - x\right) = -f(x).$$

Thus $f(x)$ is odd.

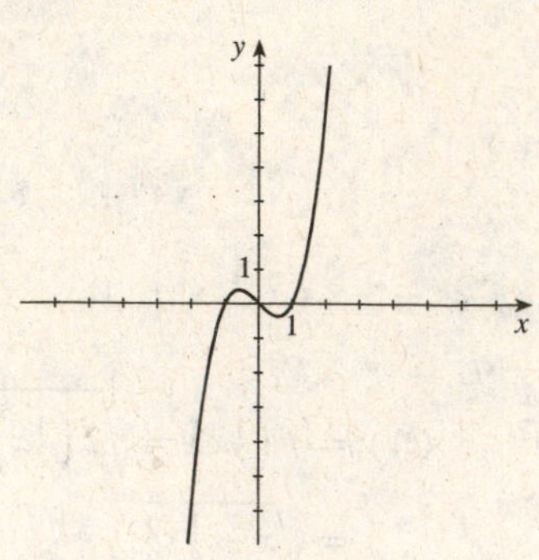

89. $f(x) = 1 - \sqrt[3]{x}$. $f(-x) = 1 - \sqrt[3]{(-x)} = 1 + \sqrt[3]{x}$. Thus $f(-x) \neq f(x)$. Also $f(-x) \neq -f(x)$, so $f(x)$ is neither odd nor even.

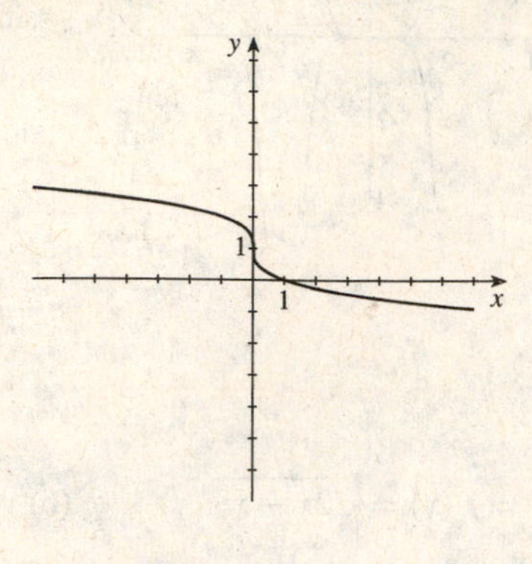

91. **(a)** Even

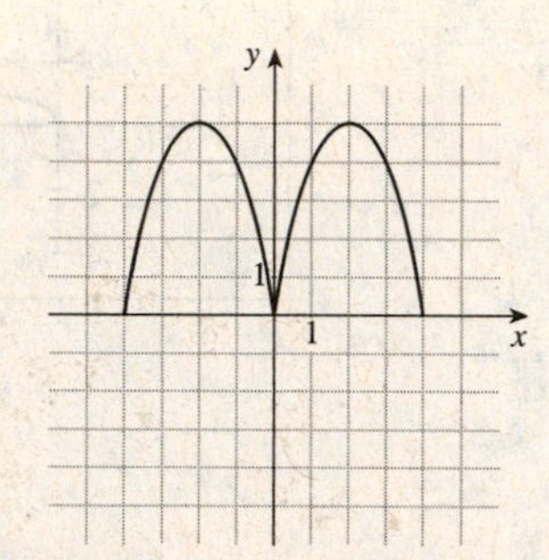

(b) Odd

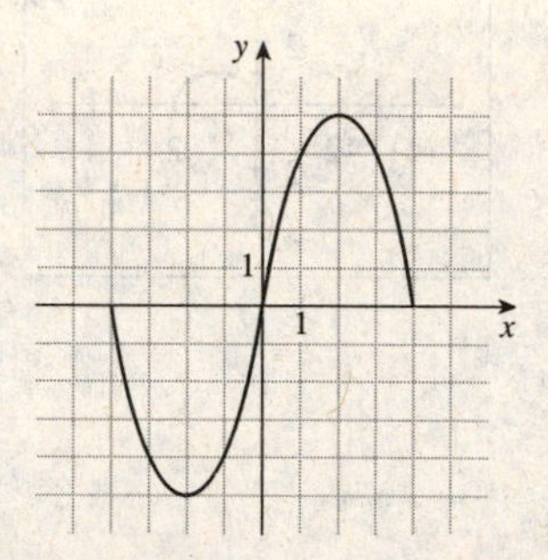

93. Since $f(x) = x^2 - 4 < 0$, for $-2 < x < 2$, the graph of $y = g(x)$ is found by sketching the graph of $y = f(x)$ for $x \leq -2$ and $x \geq 2$, then reflecting in the x-axis the part of the graph of $y = f(x)$ for $-2 < x < 2$.

95. **(a)** $f(x) = 4x - x^2$

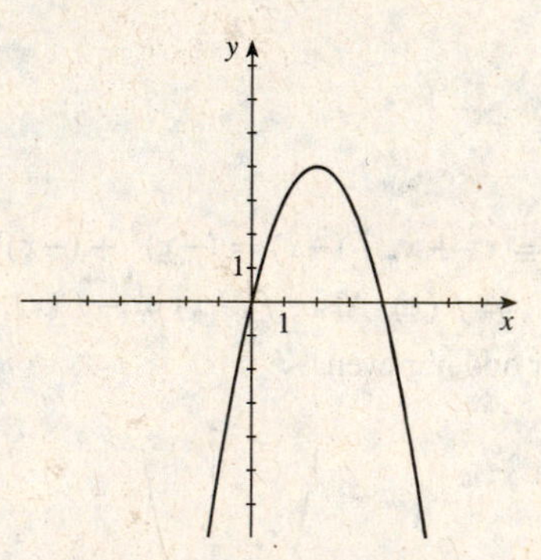

(b) $f(x) = \left|4x - x^2\right|$

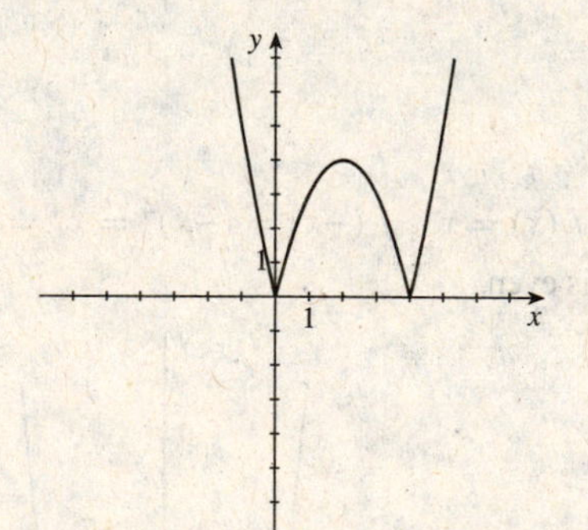

97. **(a)** Luisa drops to a height of 200 feet, bounces up and down, then settles at 350 feet.

(b)

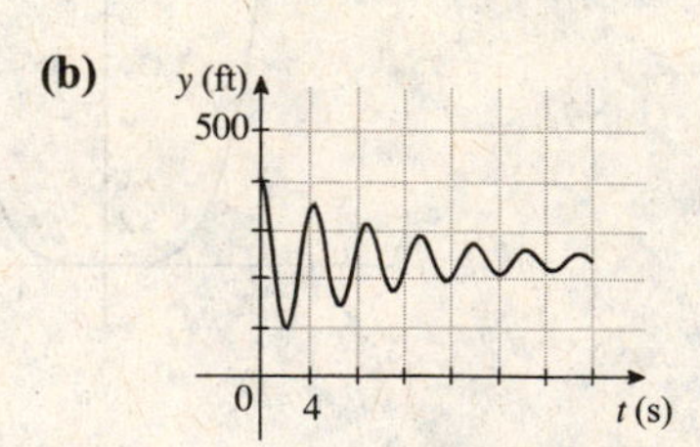

(c) To obtain the graph of H from that of h, we shift downward 100 feet. Thus, $H(t) = h(t) - 100$.

99. (a) The trip to the park corresponds to the first piece of the graph. The class travels 800 feet in 10 minutes, so their average speed is $\frac{800}{10} = 80$ ft/min. The second (horizontal) piece of the graph stretches from $t = 10$ to $t = 30$, so the class spends 20 minutes at the park. The park is 800 feet from the school.

(b)

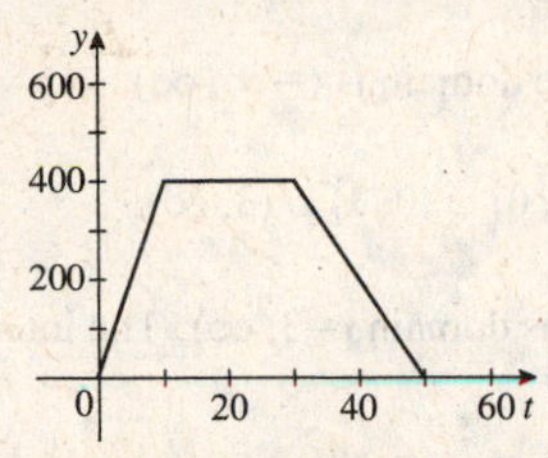

The new graph is obtained by shrinking the original graph vertically by a factor of 0.50. The new average speed is 40 ft/min, and the new park is 400 ft from the school.

(c)

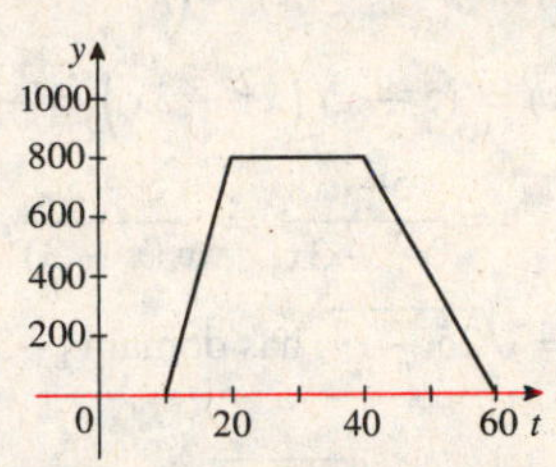

This graph is obtained by shifting the original graph to the right 10 minutes. The class leaves ten minutes later than it did in the original scenario.

101. To obtain the graph of $g(x)$ from that of $f(x)$, we reflect the graph about the y-axis, then reflect about the x-axis, then shift upward 6 units.

103. f even implies $f(-x) = f(x)$; g even implies $g(-x) = g(x)$; f odd implies $f(-x) = -f(x)$; and g odd implies $g(-x) = -g(x)$.

If f and g are both even, then $(fg)(-x) = f(-x) \cdot g(-x) = f(x) \cdot g(x) = (fg)(x)$. Thus fg is even.

If f and g are both odd, then $(fg)(-x) = f(-x) \cdot g(-x) = -f(x) \cdot (-g(x)) = f(x) \cdot g(x) = (fg)(x)$. Thus fg is even

If f if odd and g is even, then $(fg)(-x) = f(-x) \cdot g(-x) = f(x) \cdot (-g(x)) = -f(x) \cdot g(x) = -(fg)(x)$. Thus fg is odd.

2.7 COMBINING FUNCTIONS

1. From the graphs of f and g in the figure, we find $(f+g)(2) = f(2) + g(2) = 3 + 5 = 8$, $(f-g)(2) = f(2) - g(2) = 3 - 5 = -2$, $(fg)(2) = f(2)g(2) = 3 \cdot 5 = 15$, and $\left(\frac{f}{g}\right)(2) = \frac{f(2)}{g(2)} = \frac{3}{5}$.

3. If the rule of the function f is "add one" and the rule of the function g is "multiply by 2" then the rule of $f \circ g$ is "*multiply by* 2, *then add one*" and the rule of $g \circ f$ is "*add one, then multiply by* 2."

5. (a) The function $(f+g)(x)$ is defined for all values of x that are in the domains of both f and g.

(b) The function $(fg)(x)$ is defined for all values of x that are in the domains of both f and g.

(c) The function $(f/g)(x)$ is defined for all values of x that are in the domains of both f and g, and $g(x)$ is not equal to 0.

7. $f(x) = x$ has domain $(-\infty, \infty)$. $g(x) = 2x$ has domain $(-\infty, \infty)$. The intersection of the domains of f and g is $(-\infty, \infty)$.

$(f+g)(x) = x + 2x = 3x$, and the domain is $(-\infty, \infty)$. $(f-g)(x) = x - 2x = -x$, and the domain is $(-\infty, \infty)$.

$(fg)(x) = x(2x) = 2x^2$, and the domain is $(-\infty, \infty)$. $\left(\frac{f}{g}\right)(x) = \frac{x}{2x} = \frac{1}{2}$, and the domain is $(-\infty, 0) \cup (0, \infty)$.

9. $f(x) = x^2 + x$ and $g(x) = x^2$ each have domain $(-\infty, \infty)$. The intersection of the domains of f and g is $(-\infty, \infty)$.

$(f+g)(x) = 2x^2 + x$, and the domain is $(-\infty, \infty)$. $(f-g)(x) = x$, and the domain is $(-\infty, \infty)$.

$(fg)(x) = x^4 + x^3$, and the domain is $(-\infty, \infty)$. $\left(\frac{f}{g}\right)(x) = \frac{x^2+x}{x^2} = 1 + \frac{1}{x}$, and the domain is $(-\infty, 0) \cup (0, \infty)$.

11. $f(x) = 5 - x$ and $g(x) = x^2 - 3x$ each have domain $(-\infty, \infty)$. The intersection of the domains of f and g is $(-\infty, \infty)$.
$(f+g)(x) = (5-x) + \left(x^2 - 3x\right) = x^2 - 4x + 5$, and the domain is $(-\infty, \infty)$.
$(f-g)(x) = (5-x) - \left(x^2 - 3x\right) = -x^2 + 2x + 5$, and the domain is $(-\infty, \infty)$.
$(fg)(x) = (5-x)\left(x^2 - 3x\right) = -x^3 + 8x^2 - 15x$, and the domain is $(-\infty, \infty)$.
$\left(\dfrac{f}{g}\right)(x) = \dfrac{5-x}{x^2-3x} = \dfrac{5-x}{x(x-3)}$, and the domain is $(-\infty, 0) \cup (0, 3) \cup (3, \infty)$.

13. $f(x) = \sqrt{25 - x^2}$, has domain $[-5, 5]$. $g(x) = \sqrt{x+3}$, has domain $[-3, \infty)$. The intersection of the domains of f and g is $[-3, 5]$.
$(f+g)(x) = \sqrt{25-x^2} + \sqrt{x+3}$, and the domain is $[-3, 5]$.
$(f-g)(x) = \sqrt{25-x^2} - \sqrt{x+3}$, and the domain is $[-3, 5]$.
$(fg)(x) = \sqrt{\left(25-x^2\right)(x+3)}$, and the domain is $[-3, 5]$.
$\left(\dfrac{f}{g}\right)(x) = \sqrt{\dfrac{25-x^2}{x+3}}$, and the domain is $(-3, 5]$.

15. $f(x) = \dfrac{2}{x}$ has domain $x \neq 0$. $g(x) = \dfrac{4}{x+4}$, has domain $x \neq -4$. The intersection of the domains of f and g is $\{x \mid x \neq 0, -4\}$; in interval notation, this is $(-\infty, -4) \cup (-4, 0) \cup (0, \infty)$.
$(f+g)(x) = \dfrac{2}{x} + \dfrac{4}{x+4} = \dfrac{2}{x} + \dfrac{4}{x+4} = \dfrac{2(3x+4)}{x(x+4)}$, and the domain is $(-\infty, -4) \cup (-4, 0) \cup (0, \infty)$.
$(f-g)(x) = \dfrac{2}{x} - \dfrac{4}{x+4} = -\dfrac{2(x-4)}{x(x+4)}$, and the domain is $(-\infty, -4) \cup (-4, 0) \cup (0, \infty)$.
$(fg)(x) = \dfrac{2}{x} \cdot \dfrac{4}{x+4} = \dfrac{8}{x(x+4)}$, and the domain is $(-\infty, -4) \cup (-4, 0) \cup (0, \infty)$.
$\left(\dfrac{f}{g}\right)(x) = \dfrac{\dfrac{2}{x}}{\dfrac{4}{x+4}} = \dfrac{x+4}{2x}$, and the domain is $(-\infty, -4) \cup (-4, 0) \cup (0, \infty)$.

17. $f(x) = \sqrt{x} + \sqrt{3-x}$. The domain of $\sqrt{x}$ is $[0, \infty)$, and the domain of $\sqrt{3-x}$ is $(-\infty, 3]$. Thus, the domain of f is $(-\infty, 3] \cap [0, \infty) = [0, 3]$.

19. $h(x) = (x-3)^{-1/4} = \dfrac{1}{(x-3)^{1/4}}$. Since $1/4$ is an even root and the denominator can not equal 0, $x - 3 > 0 \Leftrightarrow x > 3$. So the domain is $(3, \infty)$.

21.

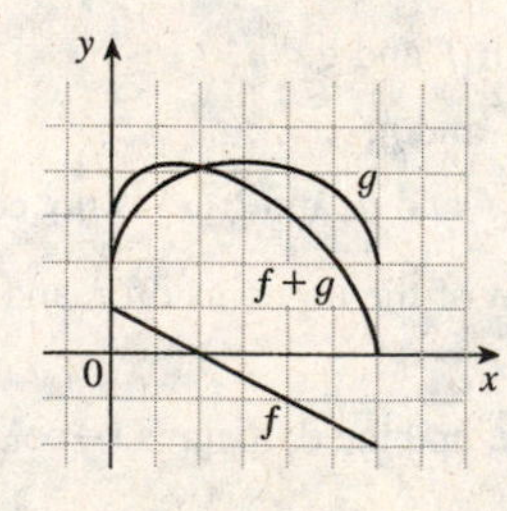

23.

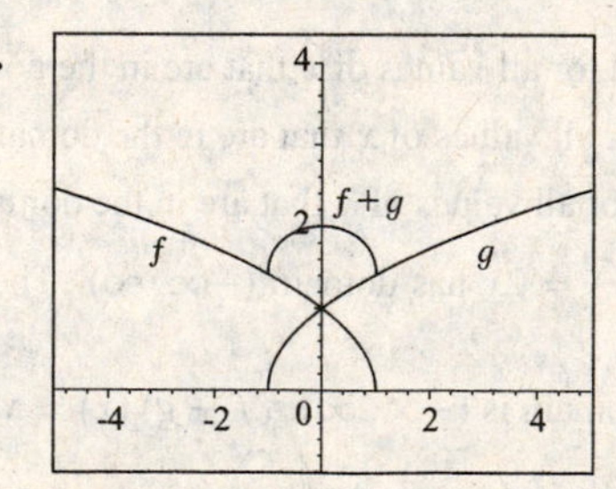

25.

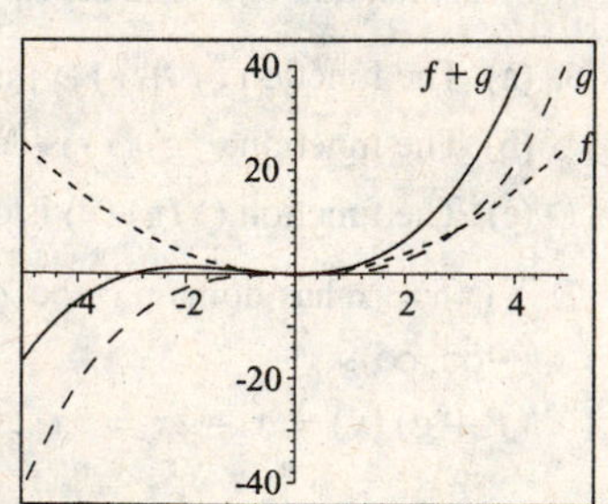

27. $f(x) = 2x - 3$ and $g(x) = 4 - x^2$.

(a) $f(g(0)) = f\left(4 - (0)^2\right) = f(4) = 2(4) - 3 = 5$

(b) $g(f(0)) = g(2(0) - 3) = g(-3) = 4 - (-3)^2 = -5$

29. **(a)** $(f \circ g)(-2) = f(g(-2)) = f\left(4 - (-2)^2\right) = f(0) = 2(0) - 3 = -3$

(b) $(g \circ f)(-2) = g(f(-2)) = g(2(-2) - 3) = g(-7) = 4 - (-7)^2 = -45$

31. (a) $(f \circ g)(x) = f(g(x)) = f\left(4 - x^2\right) = 2\left(4 - x^2\right) - 3 = 8 - 2x^2 - 3 = 5 - 2x^2$

(b) $(g \circ f)(x) = g(f(x)) = g(2x - 3) = 4 - (2x - 3)^2 = 4 - \left(4x^2 - 12x + 9\right) = -4x^2 + 12x - 5$

33. $f(g(2)) = f(5) = 4$

35. $(g \circ f)(4) = g(f(4)) = g(2) = 5$

37. $(g \circ g)(-2) = g(g(-2)) = g(1) = 4$

39. From the table, $g(2) = 5$ and $f(5) = 6$, so $f(g(2)) = 6$.

41. From the table, $f(1) = 2$ and $f(2) = 3$, so $f(f(1)) = 3$.

43. From the table, $g(6) = 4$ and $f(4) = 1$, so $(f \circ g)(6) = 1$.

45. From the table, $f(5) = 6$ and $f(6) = 3$, so $(f \circ f)(5) = 3$.

47. $f(x) = 2x + 3$, has domain $(-\infty, \infty)$; $g(x) = 4x - 1$, has domain $(-\infty, \infty)$.
$(f \circ g)(x) = f(4x - 1) = 2(4x - 1) + 3 = 8x + 1$, and the domain is $(-\infty, \infty)$.
$(g \circ f)(x) = g(2x + 3) = 4(2x + 3) - 1 = 8x + 11$, and the domain is $(-\infty, \infty)$.
$(f \circ f)(x) = f(2x + 3) = 2(2x + 3) + 3 = 4x + 9$, and the domain is $(-\infty, \infty)$.
$(g \circ g)(x) = g(4x - 1) = 4(4x - 1) - 1 = 16x - 5$, and the domain is $(-\infty, \infty)$.

49. $f(x) = x^2$, has domain $(-\infty, \infty)$; $g(x) = x + 1$, has domain $(-\infty, \infty)$.
$(f \circ g)(x) = f(x + 1) = (x + 1)^2 = x^2 + 2x + 1$, and the domain is $(-\infty, \infty)$.
$(g \circ f)(x) = g\left(x^2\right) = \left(x^2\right) + 1 = x^2 + 1$, and the domain is $(-\infty, \infty)$.
$(f \circ f)(x) = f\left(x^2\right) = \left(x^2\right)^2 = x^4$, and the domain is $(-\infty, \infty)$.
$(g \circ g)(x) = g(x + 1) = (x + 1) + 1 = x + 2$, and the domain is $(-\infty, \infty)$.

51. $f(x) = \dfrac{1}{x}$, has domain $\{x \mid x \neq 0\}$; $g(x) = 2x + 4$, has domain $(-\infty, \infty)$.
$(f \circ g)(x) = f(2x + 4) = \dfrac{1}{2x + 4}$. $(f \circ g)(x)$ is defined for $2x + 4 \neq 0 \Leftrightarrow x \neq -2$. So the domain is $\{x \mid x \neq -2\} = (-\infty, -2) \cup (-2, \infty)$.
$(g \circ f)(x) = g\left(\dfrac{1}{x}\right) = 2\left(\dfrac{1}{x}\right) + 4 = \dfrac{2}{x} + 4$, the domain is $\{x \mid x \neq 0\} = (-\infty, 0) \cup (0, \infty)$.
$(f \circ f)(x) = f\left(\dfrac{1}{x}\right) = \dfrac{1}{\left(\dfrac{1}{x}\right)} = x$. $(f \circ f)(x)$ is defined whenever both $f(x)$ and $f(f(x))$ are defined; that is, whenever $\{x \mid x \neq 0\} = (-\infty, 0) \cup (0, \infty)$.
$(g \circ g)(x) = g(2x + 4) = 2(2x + 4) + 4 = 4x + 8 + 4 = 4x + 12$, and the domain is $(-\infty, \infty)$.

53. $f(x) = |x|$, has domain $(-\infty, \infty)$; $g(x) = 2x + 3$, has domain $(-\infty, \infty)$
$(f \circ g)(x) = f(2x + 4) = |2x + 3|$, and the domain is $(-\infty, \infty)$.
$(g \circ f)(x) = g(|x|) = 2|x| + 3$, and the domain is $(-\infty, \infty)$.
$(f \circ f)(x) = f(|x|) = ||x|| = |x|$, and the domain is $(-\infty, \infty)$.
$(g \circ g)(x) = g(2x + 3) = 2(2x + 3) + 3 = 4x + 6 + 3 = 4x + 9$. Domain is $(-\infty, \infty)$.

55. $f(x)=\dfrac{x}{x+1}$, has domain $\{x \mid x \neq -1\}$; $g(x)=2x-1$, has domain $(-\infty,\infty)$

$(f\circ g)(x)=f(2x-1)=\dfrac{2x-1}{(2x-1)+1}=\dfrac{2x-1}{2x}$, and the domain is $\{x \mid x\neq 0\}=(-\infty,0)\cup(0,\infty)$.

$(g\circ f)(x)=g\left(\dfrac{x}{x+1}\right)=2\left(\dfrac{x}{x+1}\right)-1=\dfrac{2x}{x+1}-1$, and the domain is $\{x \mid x\neq -1\}=(-\infty,-1)\cup(-1,\infty)$

$(f\circ f)(x)=f\left(\dfrac{x}{x+1}\right)=\dfrac{\frac{x}{x+1}}{\frac{x}{x+1}+1}\cdot\dfrac{x+1}{x+1}=\dfrac{x}{x+x+1}=\dfrac{x}{2x+1}$. $(f\circ f)(x)$ is defined whenever both $f(x)$ and $f(f(x))$ are defined; that is, whenever $x\neq -1$ and $2x+1\neq 0 \Rightarrow x\neq -\frac{1}{2}$, which is $(-\infty,-1)\cup\left(-1,-\frac{1}{2}\right)\cup\left(-\frac{1}{2},\infty\right)$.

$(g\circ g)(x)=g(2x-1)=2(2x-1)-1=4x-2-1=4x-3$, and the domain is $(-\infty,\infty)$.

57. $f(x)=\dfrac{x}{x+1}$, has domain $\{x \mid x\neq -1\}$; $g(x)=\dfrac{1}{x}$ has domain $\{x \mid x\neq 0\}$.

$(f\circ g)(x)=f\left(\dfrac{1}{x}\right)=\dfrac{\frac{1}{x}}{\frac{1}{x}+1}=\dfrac{1}{x\left(\frac{1}{x}+1\right)}=\dfrac{1}{x+1}$. $(f\circ g)(x)$ is defined whenever both $g(x)$ and $f(g(x))$ are defined, so the domain is $\{x \mid x\neq -1,0\}$.

$(g\circ f)(x)=g\left(\dfrac{x}{x+1}\right)=\dfrac{1}{\frac{x}{x+1}}=\dfrac{x+1}{x}$. $(g\circ f)(x)$ is defined whenever both $f(x)$ and $g(f(x))$ are defined, so the domain is $\{x \mid x\neq -1,0\}$.

$(f\circ f)(x)=f\left(\dfrac{x}{x+1}\right)=\dfrac{\frac{x}{x+1}}{\frac{x}{x+1}+1}=\dfrac{x}{(x+1)\left(\frac{x}{x+1}+1\right)}=\dfrac{x}{2x+1}$. $(f\circ f)(x)$ is defined whenever both $f(x)$ and $f(f(x))$ are defined, so the domain is $\left\{x \mid x\neq -1,-\frac{1}{2}\right\}$.

$(g\circ g)(x)=g\left(\dfrac{1}{x}\right)=\dfrac{1}{\frac{1}{x}}=x$. $(g\circ g)(x)$ is defined whenever both $g(x)$ and $g(g(x))$ are defined, so the domain is $\{x \mid x\neq 0\}$.

59. $(f\circ g\circ h)(x)=f(g(h(x)))=f(g(x-1))=f\left(\sqrt{x-1}\right)=\sqrt{x-1}-1$

61. $(f\circ g\circ h)(x)=f(g(h(x)))=f\left(g\left(\sqrt{x}\right)\right)=f\left(\sqrt{x}-5\right)=\left(\sqrt{x}-5\right)^4+1$

For Exercises 63–72, many answers are possible.

63. $F(x)=(x-9)^5$. Let $f(x)=x^5$ and $g(x)=x-9$, then $F(x)=(f\circ g)(x)$.

65. $G(x)=\dfrac{x^2}{x^2+4}$. Let $f(x)=\dfrac{x}{x+4}$ and $g(x)=x^2$, then $G(x)=(f\circ g)(x)$.

67. $H(x)=\left|1-x^3\right|$. Let $f(x)=|x|$ and $g(x)=1-x^3$, then $H(x)=(f\circ g)(x)$.

69. $F(x)=\dfrac{1}{x^2+1}$. Let $f(x)=\dfrac{1}{x}$, $g(x)=x+1$, and $h(x)=x^2$, then $F(x)=(f\circ g\circ h)(x)$.

71. $G(x)=\left(4+\sqrt[3]{x}\right)^9$. Let $f(x)=x^9$, $g(x)=4+x$, and $h(x)=\sqrt[3]{x}$, then $G(x)=(f\circ g\circ h)(x)$.

73. Yes. If $f(x)=m_1x+b_1$ and $g(x)=m_2x+b_2$, then $(f\circ g)(x)=f(m_2x+b_2)=m_1(m_2x+b_2)+b_1=m_1m_2x+m_1b_2+b_1$, which is a linear function, because it is of the form $y=mx+b$. The slope is m_1m_2.

75. The price per sticker is $0.15-0.000002x$ and the number sold is x, so the revenue is $R(x)=(0.15-0.000002x)x=0.15x-0.000002x^2$.

77. **(a)** Because the ripple travels at a speed of 60 cm/s, the distance traveled in t seconds is the radius, so $g(t)=60t$.

(b) The area of a circle is πr^2, so $f(r) = \pi r^2$.

(c) $f \circ g = \pi (g(t))^2 = \pi (60t)^2 = 3600\pi t^2$ cm^2. This function represents the area of the ripple as a function of time.

79. Let r be the radius of the spherical balloon in centimeters. Since the radius is increasing at a rate of 2 cm/s, the radius is $r = 2t$ after t seconds. Therefore, the surface area of the balloon can be written as $S = 4\pi r^2 = 4\pi (2t)^2 = 4\pi \left(4t^2\right) = 16\pi t^2$.

81. **(a)** $f(x) = 0.90x$

(b) $g(x) = x - 100$

(c) $(f \circ g)(x) = f(x - 100) = 0.90(x - 100) = 0.90x - 90$. $f \circ g$ represents applying the \$100 coupon, then the 10% discount. $(g \circ f)(x) = g(0.90x) = 0.90x - 100$. $g \circ f$ represents applying the 10% discount, then the \$100 coupon. So applying the 10% discount, then the \$100 coupon gives the lower price.

83. $A(x) = 1.05x$. $(A \circ A)(x) = A(A(x)) = A(1.05x) = 1.05(1.05x) = (1.05)^2 x$.
$(A \circ A \circ A)(x) = A(A \circ A(x)) = A\left((1.05)^2 x\right) = 1.05\left[(1.05)^2 x\right] = (1.05)^3 x$.
$(A \circ A \circ A \circ A)(x) = A(A \circ A \circ A(x)) = A\left((1.05)^3 x\right) = 1.05\left[(1.05)^3 x\right] = (1.05)^4 x$. A represents the amount in the account after 1 year; $A \circ A$ represents the amount in the account after 2 years; $A \circ A \circ A$ represents the amount in the account after 3 years; and $A \circ A \circ A \circ A$ represents the amount in the account after 4 years. We can see that if we compose n copies of A, we get $(1.05)^n x$.

2.8 ONE-TO-ONE FUNCTIONS AND THEIR INVERSES

1. A function f is one-to-one if different inputs produce *different* outputs. You can tell from the graph that a function is one-to-one by using the *Horizontal Line* Test.

3. **(a)** Proceeding backward through the description of f, we can describe f^{-1} as follows: "Take the third root, subtract 5, then divide by 3."

(b) $f(x) = (3x + 5)^3$ and $f^{-1}(x) = \dfrac{\sqrt[3]{x} - 5}{3}$.

5. If the point $(3, 4)$ is on the graph of f, then the point $(4, 3)$ is on the graph of f^{-1}. [This is another way of saying that $f(3) = 4 \Leftrightarrow f^{-1}(4) = 3$.]

7. By the Horizontal Line Test, f is not one-to-one.

9. By the Horizontal Line Test, f is one-to-one.

11. By the Horizontal Line Test, f is not one-to-one.

13. $f(x) = -2x + 4$. If $x_1 \neq x_2$, then $-2x_1 \neq -2x_2$ and $-2x_1 + 4 \neq -2x_2 + 4$. So f is a one-to-one function.

15. $g(x) = \sqrt{x}$. If $x_1 \neq x_2$, then $\sqrt{x_1} \neq \sqrt{x_2}$ because two different numbers cannot have the same square root. Therefore, g is a one-to-one function.

17. $h(x) = x^2 - 2x$. Because $h(0) = 0$ and $h(2) = (2) - 2(2) = 0$ we have $h(0) = h(2)$. So f is not a one-to-one function.

19. $f(x) = x^4 + 5$. Every nonzero number and its negative have the same fourth power. For example, $(-1)^4 = 1 = (1)^4$, so $f(-1) = f(1)$. Thus f is not a one-to-one function.

21. $r(t) = t^6 - 3$, $0 \leq t \leq 5$. If $t_1 \neq t_2$, then $t_1^6 \neq t_2^6$ because two different positive numbers cannot have the same sixth power. Thus, $t_1^6 - 3 \neq t_2^6 - 3$. So r is a one-to-one function.

23. $f(x) = \dfrac{1}{x^2}$. Every nonzero number and its negative have the same square. For example, $\dfrac{1}{(-1)^2} = 1 = \dfrac{1}{(1)^2}$, so $f(-1) = f(1)$. Thus f is not a one-to-one function.

25. **(a)** $f(2) = 7$. Since f is one-to-one, $f^{-1}(7) = 2$.

(b) $f^{-1}(3) = -1$. Since f is one-to-one, $f(-1) = 3$.

27. $f(x)=5-2x$. Since f is one-to-one and $f(1)=5-2(1)=3$, then $f^{-1}(3)=1$. (Find 1 by solving the equation $5-2x=3$.)

29. **(a)** Because $f(6)=2$, $f^{-1}(2)=6$. **(b)** Because $f(2)=5$, $f^{-1}(5)=2$. **(c)** Because $f(0)=6$, $f^{-1}(6)=0$.

31. From the table, $f(4)=5$, so $f^{-1}(5)=4$.

33. $f^{-1}(f(1))=1$

35. From the table, $f(6)=1$, so $f^{-1}(1)=6$. Also, $f(2)=6$, so $f^{-1}(6)=1$. Thus, $f^{-1}\left(f^{-1}(1)\right)=f^{-1}(6)=1$.

37. $f(g(x))=f(x+6)=(x+6)-6=x$ for all x.
$g(f(x))=g(x-6)=(x-6)+6=x$ for all x. Thus f and g are inverses of each other.

39. $f(g(x))=f\left(\dfrac{x-4}{3}\right)=3\left(\dfrac{x-4}{3}\right)+4=x-4+4=x$ for all x.
$g(f(x))=g(3x+4)=\dfrac{(3x+4)-4}{3}=x$ for all x. Thus f and g are inverses of each other.

41. $f(g(x))=f\left(\dfrac{1}{x}\right)=\dfrac{1}{1/x}=x$ for all $x\neq 0$. Since $f(x)=g(x)$, we also have $g(f(x))=x$ for all $x\neq 0$. Thus f and g are inverses of each other.

43. $f(g(x))=f\left(\sqrt{x+9}\right)=\left(\sqrt{x+9}\right)^2-9=x+9-9=x$ for all $x\geq -9$.
$g(f(x))=g\left(x^2-9\right)=\sqrt{(x^2-9)+9}=\sqrt{x^2}=x$ for all $x\geq 0$. Thus f and g are inverses of each other.

45. $f(g(x))=f\left(\dfrac{1}{x}+1\right)=\dfrac{1}{\left(\dfrac{1}{x}+1\right)-1}=x$ for all $x\neq 0$.
$g(f(x))=g\left(\dfrac{1}{x-1}\right)=\dfrac{1}{\left(\dfrac{1}{x-1}\right)}+1=(x-1)+1=x$ for all $x\neq 1$. Thus f and g are inverses of each other.

47. $f(g(x))=f\left(\dfrac{2x+2}{x-1}\right)=\dfrac{\frac{2x+2}{x-1}+2}{\frac{2x+2}{x-1}-2}=\dfrac{2x+2+2(x-1)}{2x+2-2(x-1)}=\dfrac{4x}{4}=x$ for all $x\neq 1$.
$g(f(x))=g\left(\dfrac{x+2}{x-2}\right)=\dfrac{2\left(\frac{x+2}{x-2}\right)+2}{\frac{x+2}{x-2}-1}=\dfrac{2(x+2)+2(x-2)}{x+2-1(x-2)}=\dfrac{4x}{4}=x$ for all $x\neq 2$. Thus f and g are inverses of each other.

49. $f(x)=3x+5$. $y=3x+5\Leftrightarrow 3x=y-5\Leftrightarrow x=\frac{1}{3}(y-5)=\frac{1}{3}y-\frac{5}{3}$. So $f^{-1}(x)=\frac{1}{3}x-\frac{5}{3}$.

51. $f(x)=5-4x^3$. $y=5-4x^3\Leftrightarrow 4x^3=5-y\Leftrightarrow x^3=\frac{1}{4}(5-y)\Leftrightarrow x=\sqrt[3]{\frac{1}{4}(5-y)}$. So $f^{-1}(x)=\sqrt[3]{\frac{1}{4}(5-x)}$.

53. $f(x)=\dfrac{1}{x+2}$. $y=\dfrac{1}{x+2}\Leftrightarrow x+2=\dfrac{1}{y}\Leftrightarrow x=\dfrac{1}{y}-2$. So $f^{-1}(x)=\dfrac{1}{x}-2$.

55. $f(x)=\dfrac{x}{x+4}$. $y=\dfrac{x}{x+4}\Leftrightarrow y(x+4)=x\Leftrightarrow xy+4y=x\Leftrightarrow x-xy=4y\Leftrightarrow x(1-y)=4y\Leftrightarrow x=\dfrac{4y}{1-y}$. So $f^{-1}(x)=\dfrac{4x}{1-x}$.

57. $f(x)=\dfrac{2x+5}{x-7}$. $y=\dfrac{2x+5}{x-7}\Leftrightarrow y(x-7)=2x+5\Leftrightarrow xy-7y=2x+5\Leftrightarrow xy-2x=7y+5\Leftrightarrow x(y-2)=7y+5$ $\Leftrightarrow x=\dfrac{7y+5}{y-2}$. So $f^{-1}(x)=\dfrac{7x+5}{x-2}$.

59. $f(x)=\dfrac{2x+3}{1-5x}$. $y=\dfrac{2x+3}{1-5x}\Leftrightarrow y(1-5x)=2x+3\Leftrightarrow y-5xy=2x+3\Leftrightarrow 2x+5xy=y-3\Leftrightarrow x(2+5y)=y-3$ $\Leftrightarrow x=\dfrac{y-3}{5y+2}$. So $f^{-1}(x)=\dfrac{x-3}{5x+2}$.

61. $f(x) = 4 - x^2, x \geq 0$. $y = 4 - x^2 \Leftrightarrow x^2 = 4 - y \Leftrightarrow x = \sqrt{4 - y}$. So $f^{-1}(x) = \sqrt{4 - x}, x \leq 4$. [Note that $x \geq 0 \Rightarrow f(x) \leq 4$.]

63. $f(x) = x^6, x \geq 0$. $y = x^6 \Leftrightarrow x = \sqrt[6]{y}$ for $x \geq 0$. The range of f is $\{y \mid y \geq 0\}$, so $f^{-1}(x) = \sqrt[6]{x}, x \geq 0$.

65. $f(x) = \dfrac{2 - x^3}{5}$. $y = \dfrac{2 - x^3}{5} \Leftrightarrow 5y = 2 - x^3 \Leftrightarrow x^3 = 2 - 5y \Leftrightarrow x = \sqrt[3]{2 - 5y}$. Thus, $f^{-1}(x) = \sqrt[3]{2 - 5x}$.

67. $f(x) = \sqrt{5 + 8x}$. Note that the range of f (and thus the domain of f^{-1}) is $[0, \infty)$. $y = \sqrt{5 + 8x} \Leftrightarrow y^2 = 5 + 8x \Leftrightarrow 8x = y^2 - 5 \Leftrightarrow x = \dfrac{y^2 - 5}{8}$. Thus, $f^{-1}(x) = \dfrac{x^2 - 5}{8}, x \geq 0$.

69. $f(x) = 2 + \sqrt[3]{x}$. $y = 2 + \sqrt[3]{x} \Leftrightarrow y - 2 = \sqrt[3]{x} \Leftrightarrow x = (y - 2)^3$. Thus, $f^{-1}(x) = (x - 2)^3$.

71. (a), (b) $f(x) = 3x - 6$

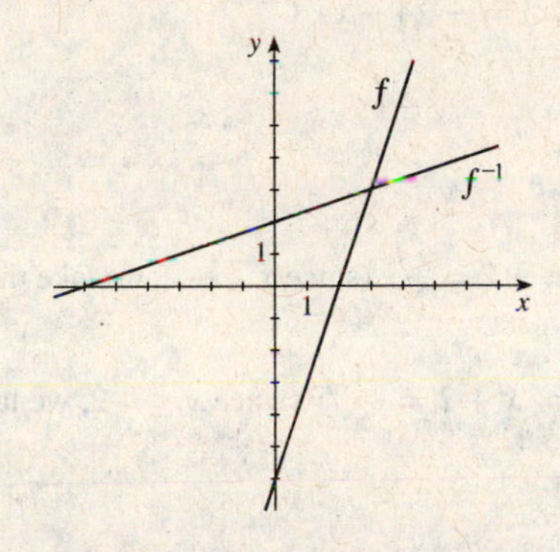

(c) $f(x) = 3x - 6$. $y = 3x - 6 \Leftrightarrow 3x = y + 6 \Leftrightarrow x = \frac{1}{3}(y + 6)$. So $f^{-1}(x) = \frac{1}{3}(x + 6)$.

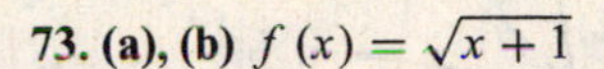

73. (a), (b) $f(x) = \sqrt{x + 1}$

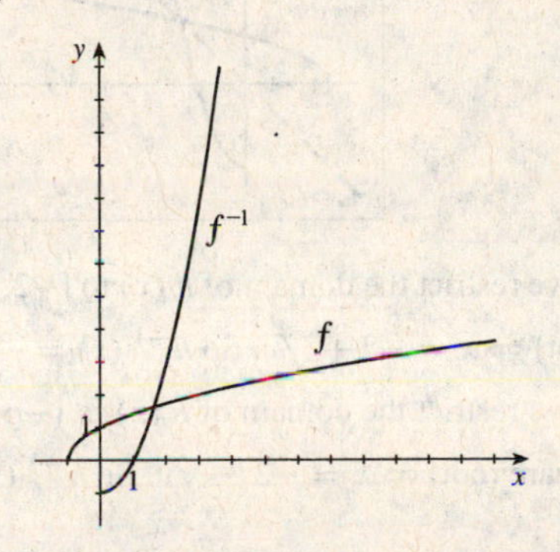

(c) $f(x) = \sqrt{x + 1}, x \geq -1$. $y = \sqrt{x + 1}, y \geq 0 \Leftrightarrow y^2 = x + 1 \Leftrightarrow x = y^2 - 1$ and $y \geq 0$. So $f^{-1}(x) = x^2 - 1, x \geq 0$.

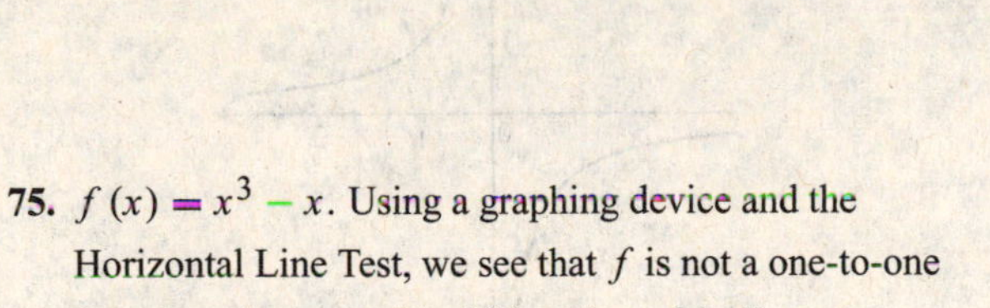

75. $f(x) = x^3 - x$. Using a graphing device and the Horizontal Line Test, we see that f is not a one-to-one function. For example, $f(0) = 0 = f(-1)$.

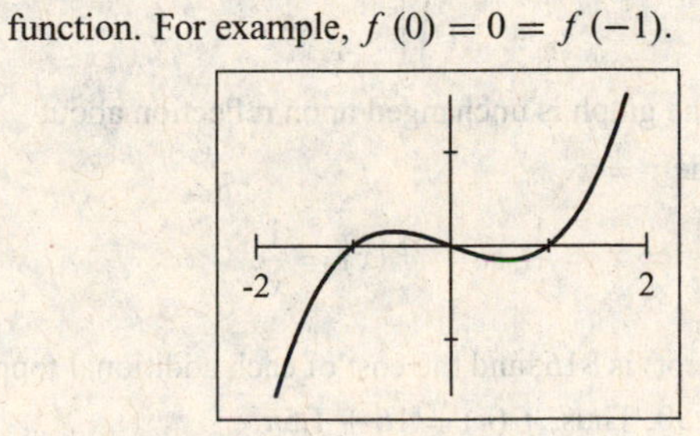

77. $f(x) = \dfrac{x + 12}{x - 6}$. Using a graphing device and the Horizontal Line Test, we see that f is a one-to-one function.

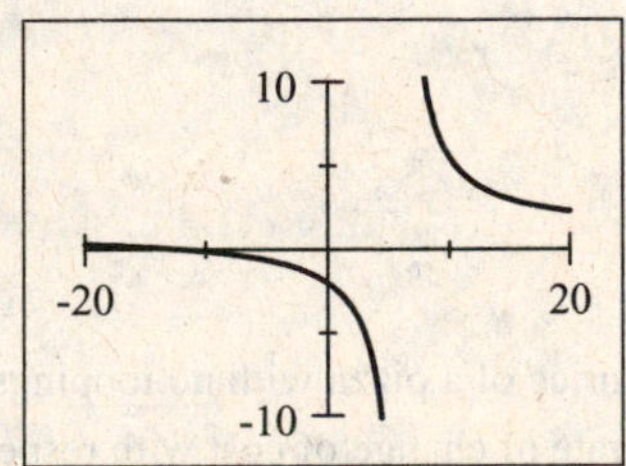

79. $f(x) = |x| - |x - 6|$. Using a graphing device and the Horizontal Line Test, we see that f is not a one-to-one function. For example $f(0) = -6 = f(-2)$.

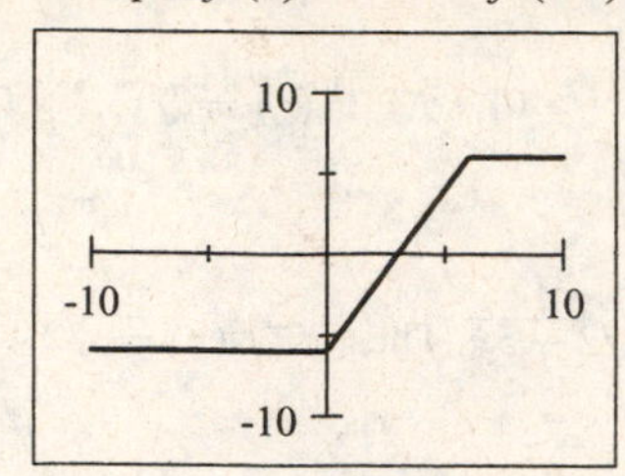

81. (a) $y = f(x) = 2 + x \Leftrightarrow x = y - 2$. So $f^{-1}(x) = x - 2$.

(b)

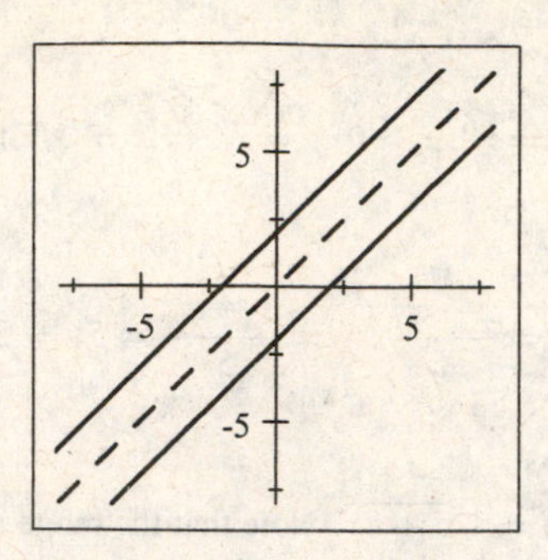

83. (a) $y = g(x) = \sqrt{x+3},\ y \geq 0 \Leftrightarrow x + 3 = y^2,\ y \geq 0$ $\Leftrightarrow x = y^2 - 3,\ y \geq 0$. So $g^{-1}(x) = x^2 - 3,\ x \geq 0$.

(b)

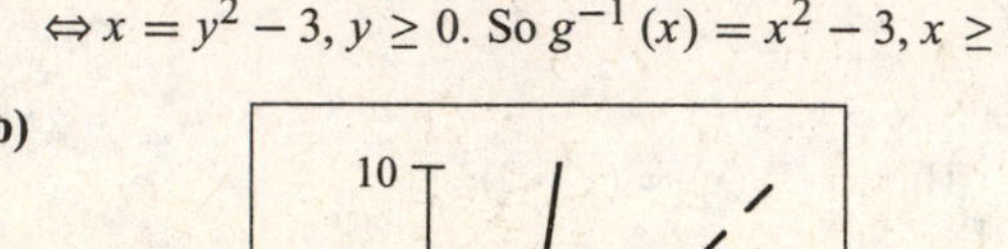

85. If we restrict the domain of $f(x)$ to $[0, \infty)$, then $y = 4 - x^2 \Leftrightarrow x^2 = 4 - y \Rightarrow x = \sqrt{4 - y}$ (since $x \geq 0$, we take the positive square root). So $f^{-1}(x) = \sqrt{4 - x}$.
If we restrict the domain of $f(x)$ to $(-\infty, 0]$, then $y = 4 - x^2 \Leftrightarrow x^2 = 4 - y \Rightarrow x = -\sqrt{4 - y}$ (since $x \leq 0$, we take the negative square root). So $f^{-1}(x) = -\sqrt{4 - x}$.

87. If we restrict the domain of $h(x)$ to $[-2, \infty)$, then $y = (x + 2)^2 \Rightarrow x + 2 = \sqrt{y}$ (since $x \geq -2$, we take the positive square root) $\Leftrightarrow x = -2 + \sqrt{y}$. So $h^{-1}(x) = -2 + \sqrt{x}$.
If we restrict the domain of $h(x)$ to $(-\infty, -2]$, then $y = (x + 2)^2 \Rightarrow x + 2 = -\sqrt{y}$ (since $x \leq -2$, we take the negative square root) $\Leftrightarrow x = -2 - \sqrt{y}$. So $h^{-1}(x) = -2 - \sqrt{x}$.

89.

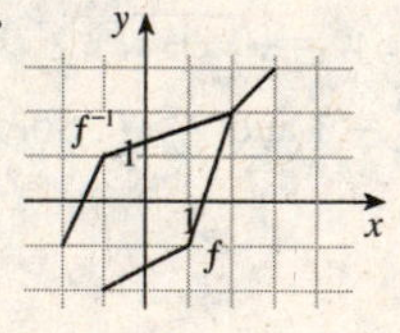

91. (a)

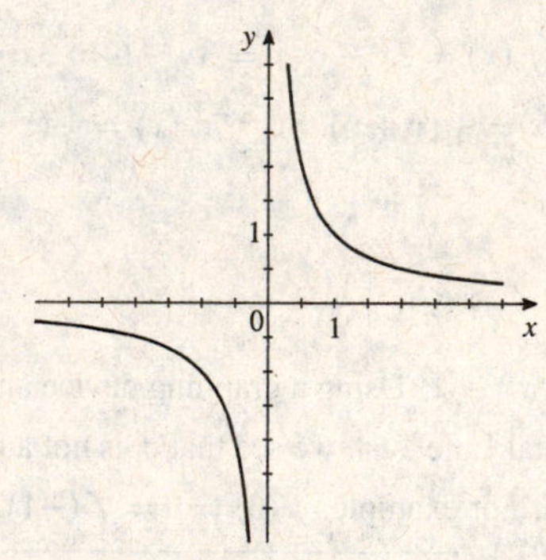

(b) Yes, the graph is unchanged upon reflection about the line $y = x$.

(c) $y = \dfrac{1}{x} \Leftrightarrow x = \dfrac{1}{y}$, so $f^{-1}(x) = \dfrac{1}{x}$.

93. (a) The price of a pizza with no toppings (corresponding to the y-intercept) is \$16, and the cost of each additional topping (the rate of change of cost with respect to number of toppings) is \$1.50. Thus, $f(n) = 16 + 1.5n$.

(b) $p = f(n) = 16 + 1.5n \Leftrightarrow p - 16 = 1.5n \Leftrightarrow n = \frac{2}{3}(p - 16)$. Thus, $n = f^{-1}(p) = \frac{2}{3}(p - 16)$. This function represents the number of toppings on a pizza that costs x dollars.

(c) $f^{-1}(25) = \frac{2}{3}(25 - 16) = \frac{2}{3}(9) = 6$. Thus, a \$25 pizza has 6 toppings.

95. (a) $V = f(t) = 100\left(1 - \frac{t}{40}\right)^2, 0 \le t \le 40$. $V = 100\left(1 - \frac{t}{40}\right)^2 \Leftrightarrow \frac{V}{100} = \left(1 - \frac{t}{40}\right)^2 \Rightarrow 1 - \frac{t}{40} = \pm\sqrt{\frac{V}{100}} \Leftrightarrow$ $\frac{t}{40} = 1 \pm \frac{\sqrt{V}}{10} \Leftrightarrow t = 40 \pm 4\sqrt{V}$. Since $t \le 40$, we must have $t = f^{-1}(V) = 40 - 4\sqrt{V}$. f^{-1} represents time that has elapsed since the tank started to leak.

(b) $f^{-1}(15) = 40 - 4\sqrt{15} \approx 24.5$ minutes. In 24.5 minutes the tank has drained to just 15 gallons of water.

97. (a) $D = f(p) = -3p + 150$. $D = -3p + 150 \Leftrightarrow 3p = 150 - D \Leftrightarrow p = 50 - \frac{1}{3}D$. So $f^{-1}(D) = 50 - \frac{1}{3}D$. $f^{-1}(D)$ represents the price that is associated with demand D.

(b) $f^{-1}(30) = 50 - \frac{1}{3}(30) = 40$. So when the demand is 30 units, the price per unit is \$40.

99. (a) $f^{-1}(U) = 1.02396U$.

(b) $U = f(x) = 0.9766x$. $U = 0.9766x \Leftrightarrow x = 1.0240U$. So $f^{-1}(U) = 1.0240U$. $f^{-1}(U)$ represents the value of U US dollars in Canadian dollars.

(c) $f^{-1}(12{,}250) = 1.0240(12{,}250) = 12{,}543.52$. So \$12,250 in US currency is worth \$12,543.52 in Canadian currency.

101. (a) $f(x) = 0.85x$.

(b) $g(x) = x - 1000$.

(c) $H(x) = (f \circ g)x = f(x - 1000) = 0.85(x - 1000) = 0.85x - 850$.

(d) $P = H(x) = 0.85x - 850$. $P = 0.85x - 850 \Leftrightarrow 0.85x = P + 850 \Leftrightarrow x = 1.176P + 1000$. So $H^{-1}(P) = 1.176P + 1000$. The function H^{-1} represents the original sticker price for a given discounted price P.

(e) $H^{-1}(13{,}000) = 1.176(13{,}000) + 1000 = 16{,}288$. So the original price of the car is \$16,288 when the discounted price (\$1000 rebate, then 15% off) is \$13,000.

103. (a) $f(x) = \frac{2x+1}{5}$ is "multiply by 2, add 1, and then divide by 5 ". So the reverse is "multiply by 5, subtract 1, and then divide by 2 " or $f^{-1}(x) = \frac{5x-1}{2}$. Check: $f \circ f^{-1}(x) = f\left(\frac{5x-1}{2}\right) = \frac{2\left(\frac{5x-1}{2}\right)+1}{5} = \frac{5x-1+1}{5} = \frac{5x}{5} = x$ and $f^{-1} \circ f(x) = f^{-1}\left(\frac{2x+1}{5}\right) = \frac{5\left(\frac{2x+1}{5}\right)-1}{2} = \frac{2x+1-1}{2} = \frac{2x}{2} = x$.

(b) $f(x) = 3 - \frac{1}{x} = \frac{-1}{x} + 3$ is "take the negative reciprocal and add 3 ". Since the reverse of "take the negative reciprocal" is "take the negative reciprocal ", $f^{-1}(x)$ is "subtract 3 and take the negative reciprocal ", that is, $f^{-1}(x) = \frac{-1}{x-3}$. Check: $f \circ f^{-1}(x) = f\left(\frac{-1}{x-3}\right) = 3 - \frac{1}{\frac{-1}{x-3}} = 3 - \left(1 \cdot \frac{x-3}{-1}\right) = 3 + x - 3 = x$ and

$$f^{-1} \circ f(x) = f^{-1}\left(3 - \frac{1}{x}\right) = \frac{-1}{\left(3 - \frac{1}{x}\right) - 3} = \frac{-1}{-\frac{1}{x}} = -1 \cdot \frac{x}{-1} = x.$$

(c) $f(x) = \sqrt{x^3+2}$ is "cube, add 2, and then take the square root". So the reverse is "square, subtract 2, then take the cube root" or $f^{-1}(x) = \sqrt[3]{x^2-2}$. Domain for $f(x)$ is $\left[-\sqrt[3]{2}, \infty\right)$; domain for $f^{-1}(x)$ is $[0, \infty)$. Check:

$f \circ f^{-1}(x) = f\left(\sqrt[3]{x^2-2}\right) = \sqrt{\left(\sqrt[3]{x^2-2}\right)^3 + 2} = \sqrt{x^2-2+2} = \sqrt{x^2} = x$ (on the appropriate domain) and

$f^{-1} \circ f(x) = f^{-1}\left(\sqrt{x^3+2}\right) = \sqrt[3]{\left(\sqrt{x^3+2}\right)^2 - 2} = \sqrt[3]{x^3+2-2} = \sqrt[3]{x^3} = x$ (on the appropriate domain).

(d) $f(x) = (2x-5)^3$ is "double, subtract 5, and then cube". So the reverse is "take the cube root, add 5, and divide by 2" or $f^{-1}(x) = \dfrac{\sqrt[3]{x}+5}{2}$ Domain for both $f(x)$ and $f^{-1}(x)$ is $(-\infty, \infty)$. Check:

$f \circ f^{-1}(x) = f\left(\dfrac{\sqrt[3]{x}+5}{2}\right) = \left[2\left(\dfrac{\sqrt[3]{x}+5}{2}\right) - 5\right]^3 = \left(\sqrt[3]{x}+5-5\right)^3 = \left(\sqrt[3]{x}\right)^3 = \sqrt[3]{x^3} = x$ and

$f^{-1} \circ f(x) = f^{-1}\left((2x-5)^3\right) = \dfrac{\sqrt[3]{(2x-5)^3}+5}{2} = \dfrac{(2x-5)+5}{2} = \dfrac{2x}{2} = x.$

In a function like $f(x) = 3x - 2$, the variable occurs only once and it easy to see how to reverse the operations step by step. But in $f(x) = x^3 + 2x + 6$, you apply two different operations to the variable x (cubing and multiplying by 2) and then add 6, so it is not possible to reverse the operations step by step.

105. (a) We find $g^{-1}(x)$: $y = 2x + 1 \Leftrightarrow 2x = y - 1 \Leftrightarrow x = \frac{1}{2}(y-1)$. So $g^{-1}(x) = \frac{1}{2}(x-1)$. Thus

$f(x) = h \circ g^{-1}(x) = h\left(\frac{1}{2}(x-1)\right) = 4\left[\frac{1}{2}(x-1)\right]^2 + 4\left[\frac{1}{2}(x-1)\right] + 7 = x^2 - 2x + 1 + 2x - 2 + 7 = x^2 + 6.$

(b) $f \circ g = h \Leftrightarrow f^{-1} \circ f \circ g = f^{-1} \circ h \Leftrightarrow I \circ g = f^{-1} \circ h \Leftrightarrow g = f^{-1} \circ h$. Note that we compose with f^{-1} on the left on each side of the equation. We find f^{-1}: $y = 3x + 5 \Leftrightarrow 3x = y - 5 \Leftrightarrow x = \frac{1}{3}(y-5)$. So $f^{-1}(x) = \frac{1}{3}(x-5)$. Thus $g(x) = f^{-1} \circ h(x) = f^{-1}\left(3x^2+3x+2\right) = \frac{1}{3}\left[\left(3x^2+3x+2\right) - 5\right] = \frac{1}{3}\left[3x^2+3x-3\right] = x^2 + x - 1.$

CHAPTER 2 REVIEW

1. "Square, then subtract 5" can be represented by the function $f(x) = x^2 - 5$.

3. $f(x) = 3(x+10)$: "Add 10, then multiply by 3."

5. $g(x) = x^2 - 4x$

x	$g(x)$
−1	5
0	0
1	−3
2	−4
3	−3

7. $C(x) = 5000 + 30x - 0.001x^2$

(a) $C(1000) = 5000 + 30(1000) - 0.001(1000)^2 = \$34{,}000$ and
$C(10{,}000) = 5000 + 30(10{,}000) - 0.001(10{,}000)^2 = \$205{,}000$.

(b) From part (a), we see that the total cost of printing 1000 copies of the book is \$34,000 and the total cost of printing 10,000 copies is \$205,000.

(c) $C(0) = 5000 + 30(0) - 0.001(0)^2 = \$5000$. This represents the fixed costs associated with getting the print run ready.

(d) The net change in C as x changes from 1000 to 10,000 is $C(10{,}000) - C(1000) = 205{,}000 - 34{,}000 = \$171{,}000$, and the average rate of change is $\dfrac{C(10{,}000) - C(1000)}{10{,}000 - 1000} = \dfrac{171{,}000}{9000} = \$19/\text{copy}$.

9. $f(x) = x^2 - 4x + 6$; $f(0) = (0)^2 - 4(0) + 6 = 6$; $f(2) = (2)^2 - 4(2) + 6 = 2$; $f(-2) = (-2)^2 - 4(-2) + 6 = 18$; $f(a) = (a)^2 - 4(a) + 6 = a^2 - 4a + 6$; $f(-a) = (-a)^2 - 4(-a) + 6 = a^2 + 4a + 6$; $f(x+1) = (x+1)^2 - 4(x+1) + 6 = x^2 + 2x + 1 - 4x - 4 + 6 = x^2 - 2x + 3$; $f(2x) = (2x)^2 - 4(2x) + 6 = 4x^2 - 8x + 6$.

11. By the Vertical Line Test, figures (b) and (c) are graphs of functions. By the Horizontal Line Test, figure (c) is the graph of a one-to-one function.

13. Domain: We must have $x + 3 \geq 0 \Leftrightarrow x \geq -3$. In interval notation, the domain is $[-3, \infty)$.
Range: For x in the domain of f, we have $x \geq -3 \Leftrightarrow x + 3 \geq 0 \Leftrightarrow \sqrt{x+3} \geq 0 \Leftrightarrow f(x) \geq 0$. So the range is $[0, \infty)$.

15. $f(x) = 7x + 15$. The domain is all real numbers, $(-\infty, \infty)$.

17. $f(x) = \sqrt{x+4}$. We require $x + 4 \geq 0 \Leftrightarrow x \geq -4$. Thus the domain is $[-4, \infty)$.

19. $f(x) = \dfrac{1}{x} + \dfrac{1}{x+1} + \dfrac{1}{x+2}$. The denominators cannot equal 0, therefore the domain is $\{x \mid x \neq 0, -1, -2\}$.

21. $h(x) = \sqrt{4-x} + \sqrt{x^2 - 1}$. We require the expression inside the radicals be nonnegative. So $4 - x \geq 0 \Leftrightarrow 4 \geq x$; also $x^2 - 1 \geq 0 \Leftrightarrow (x-1)(x+1) \geq 0$. We make a table:

Interval	$(-\infty, -1)$	$(-1, 1)$	$(1, \infty)$
Sign of $x - 1$	$-$	$-$	$+$
Sign of $x + 1$	$-$	$+$	$+$
Sign of $(x-1)(x+1)$	$+$	$-$	$+$

Thus the domain is $(-\infty, 4] \cap \{(-\infty, -1] \cup [1, \infty)\} = (-\infty, -1] \cup [1, 4]$.

23. $f(x) = 1 - 2x$

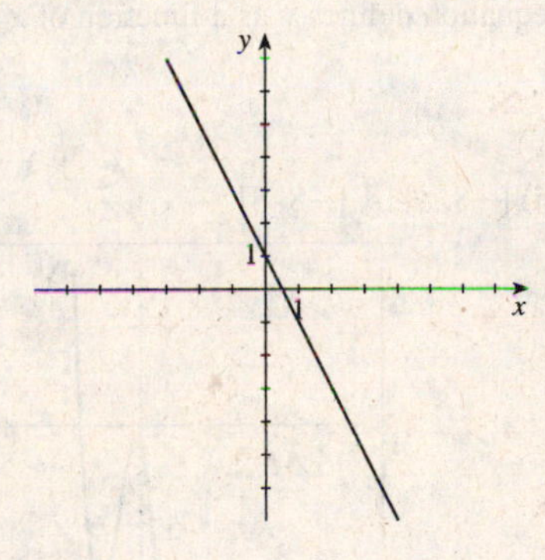

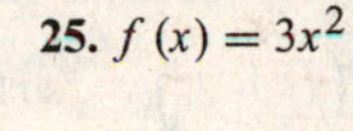

25. $f(x) = 3x^2$

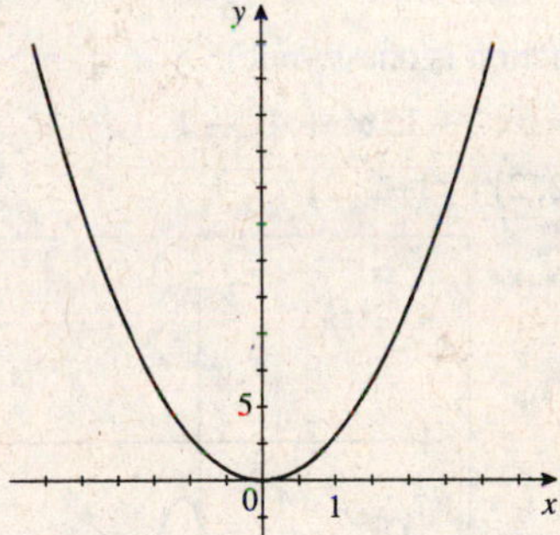

27. $f(x) = 2x^2 - 1$

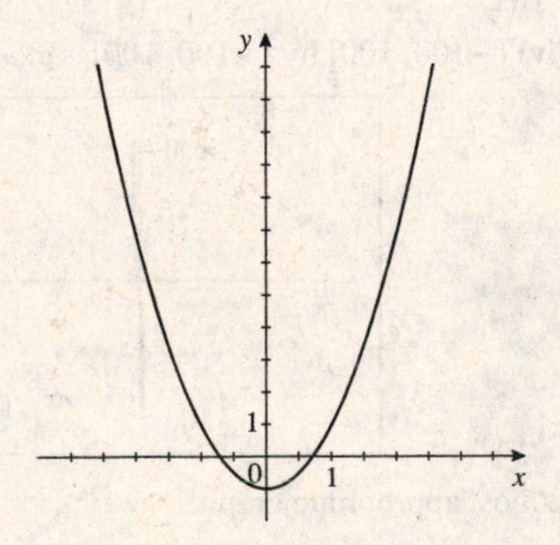

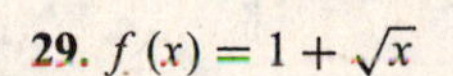

29. $f(x) = 1 + \sqrt{x}$

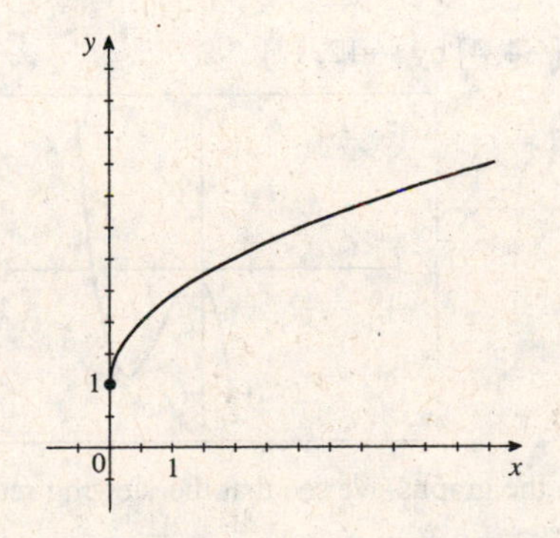

31. $f(x) = \frac{1}{2}x^3$

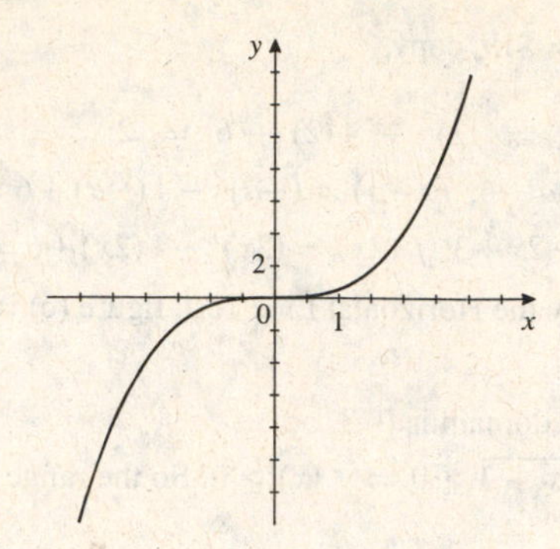

33. $f(x) = -|x|$

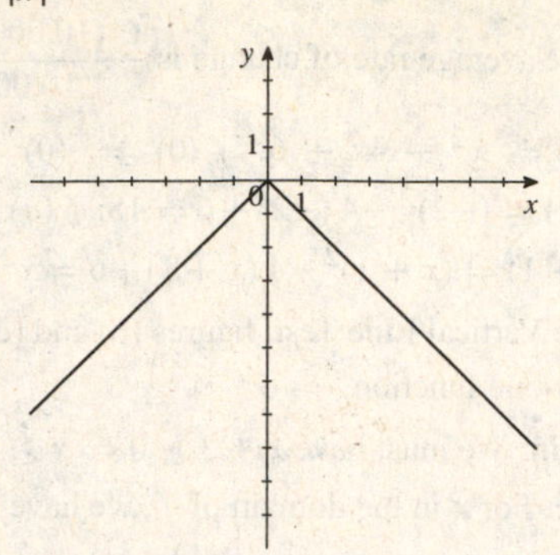

35. $f(x) = -\dfrac{1}{x^2}$

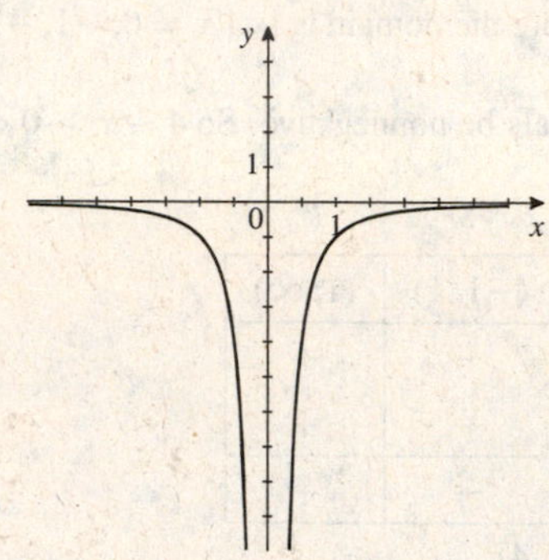

37. $f(x) = \begin{cases} 1 - x & \text{if } x < 0 \\ 1 & \text{if } x \geq 0 \end{cases}$

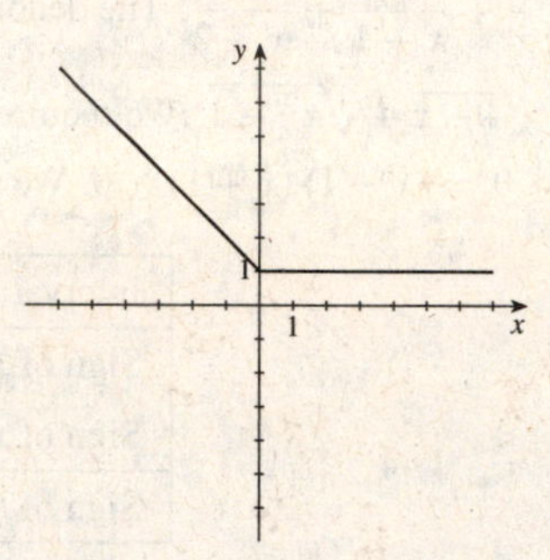

39. $x + y^2 = 14 \Rightarrow y^2 = 14 - x \Rightarrow y = \pm\sqrt{14 - x}$, so the original equation does not define y as a function of x.

41. $x^3 - y^3 = 27 \Leftrightarrow y^3 = x^3 - 27 \Leftrightarrow y = \left(x^3 - 27\right)^{1/3}$, so the original equation defines y as a function of x (since the cube root function is one-to-one).

43. $f(x) = 6x^3 - 15x^2 + 4x - 1$

(i) $[-2, 2]$ by $[-2, 2]$

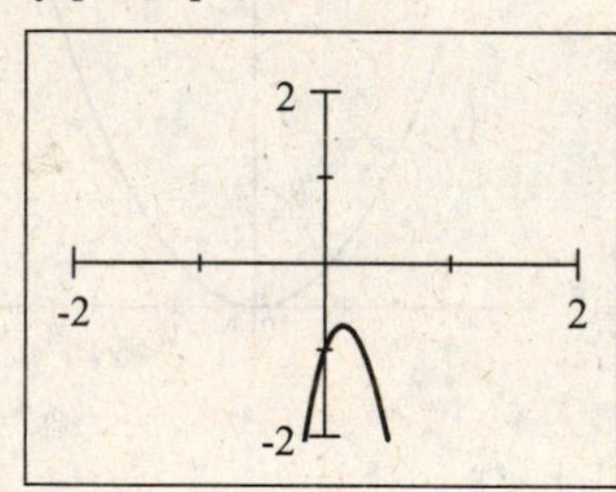

(ii) $[-8, 8]$ by $[-8, 8]$

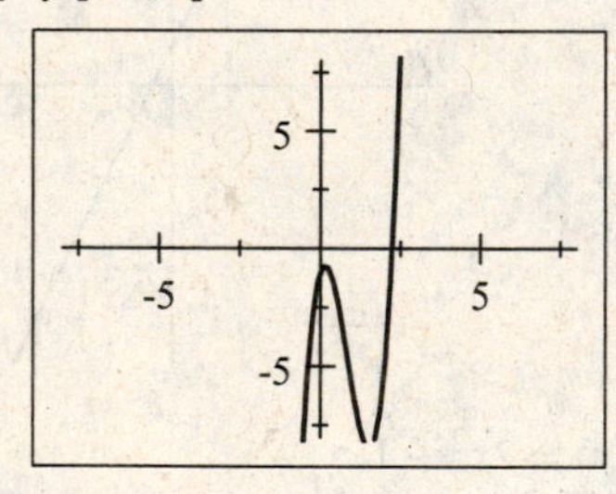

(iii) $[-4, 4]$ by $[-12, 12]$

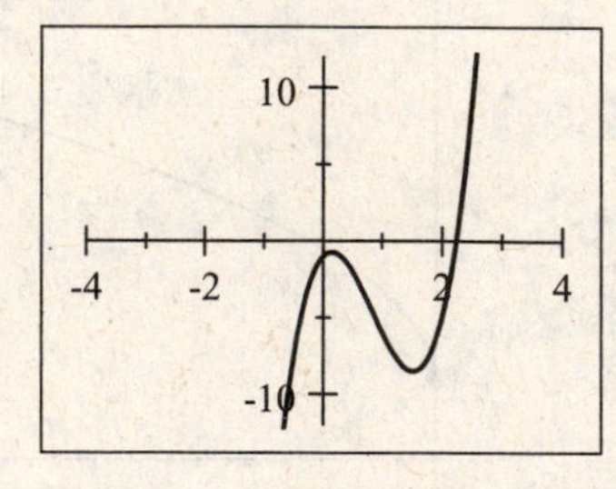

(iv) $[-100, 100]$ by $[-100, 100]$

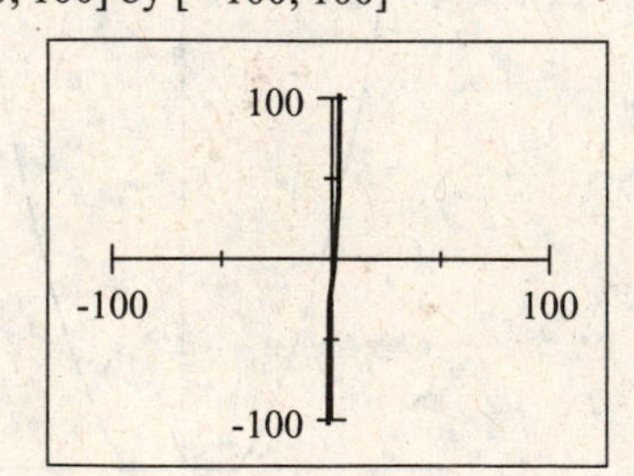

From the graphs, we see that the viewing rectangle in (iii) produces the most appropriate graph.

45. (a) We graph $f(x) = \sqrt{9 - x^2}$ in the viewing rectangle $[-4, 4]$ by $[-1, 4]$.

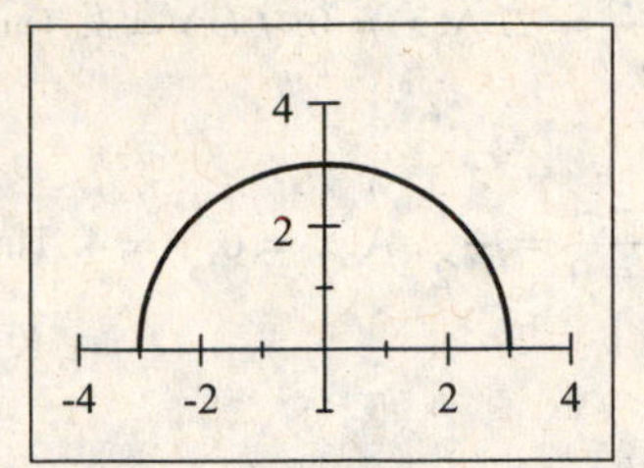

(b) From the graph, the domain of f is $[-3, 3]$ and the range of f is $[0, 3]$.

47. (a) We graph $f(x) = \sqrt{x^3 - 4x + 1}$ in the viewing rectangle $[-5, 5]$ by $[-1, 5]$.

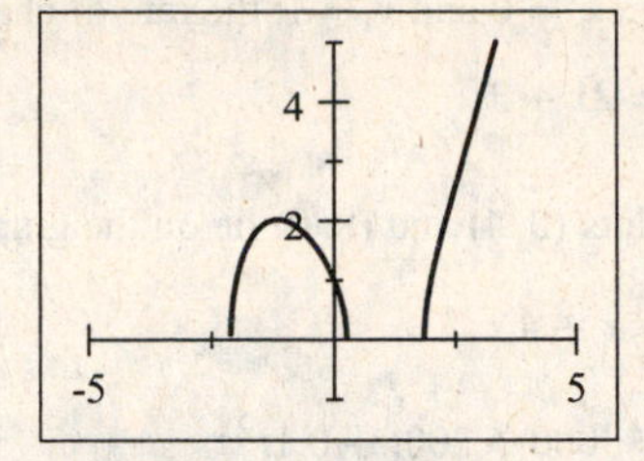

(b) From the graph, the domain of f is approximately $[-2.11, 0.25] \cup [1.86, \infty)$ and the range of f is $[0, \infty)$.

49. $f(x) = x^3 - 4x^2$ is graphed in the viewing rectangle $[-5, 5]$ by $[-20, 10]$. $f(x)$ is increasing on $(-\infty, 0)$ and $(2.67, \infty)$. It is decreasing on $(0, 2.67)$.

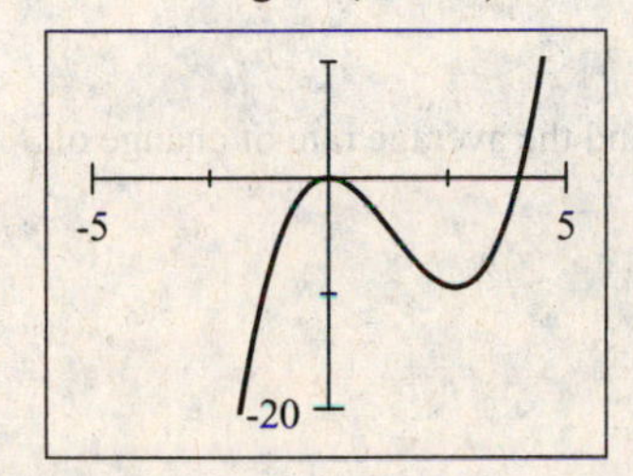

51. The net change is $f(8) - f(4) = 8 - 12 = -4$ and the average rate of change is $\dfrac{f(8) - f(4)}{8 - 4} = \dfrac{-4}{4} = -1$.

53. The net change is $f(2) - f(-1) = 6 - 2 = 4$ and the average rate of change is $\dfrac{f(2) - f(-1)}{2 - (-1)} = \dfrac{4}{3}$.

55. The net change is $f(4) - f(1) = \left[4^2 - 2(4)\right] - \left[1^2 - 2(1)\right] = 8 - (-1) = 9$ and the average rate of change is $\dfrac{f(4) - f(1)}{4 - 1} = \dfrac{9}{3} = 3$.

57. $f(x) = (2 + 3x)^2 = 9x^2 + 12x + 4$ is not linear. It cannot be expressed in the form $f(x) = ax + b$ with constant a and b.

59. (a)

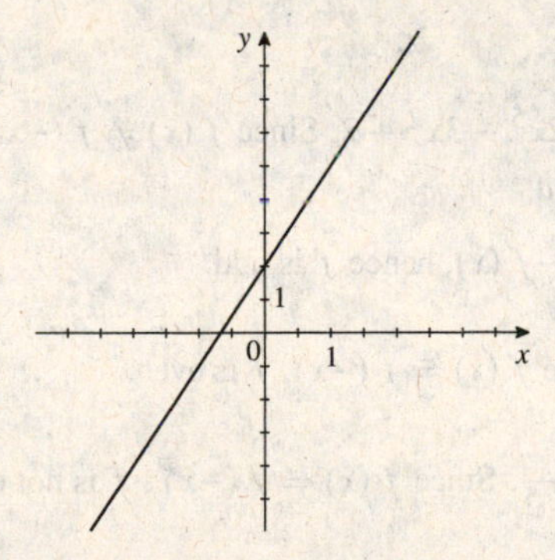

(b) The slope of the graph is the value of a in the equation $f(x) = ax + b = 3x + 2$; that is, 3.

(c) The rate of change is the slope of the graph, 3.

61. The linear function with rate of change -2 and initial value 3 has $a = -2$ and $b = 3$, so $f(x) = -2x + 3$.

63. Between $x = 0$ and $x = 1$, the rate of change is $\dfrac{f(1) - f(0)}{1 - 0} = \dfrac{5 - 3}{1} = 2$. At $x = 0$, $f(x) = 3$. Thus, an equation is $f(x) = 2x + 3$.

65. The points $(0, 4)$ and $(8, 0)$ lie on the graph, so the rate of change is $\dfrac{0 - 4}{8 - 0} = -\dfrac{1}{2}$. At $x = 0$, $y = 4$. Thus, an equation is $y = -\frac{1}{2}x + 4$.

67. $P(t) = 3000 + 200t + 0.1t^2$

(a) $P(10) = 3000 + 200(10) + 0.1(10)^2 = 5010$ represents the population in its 10th year (that is, in 1995), and $P(20) = 3000 + 200(20) + 0.1(20)^2 = 7040$ represents its population in its 20th year (in 2005).

(b) The average rate of change is $\dfrac{P(20) - P(10)}{20 - 10} = \dfrac{7040 - 5010}{10} = \dfrac{2030}{10} = 203$ people/year. This represents the average yearly change in population between 1995 and 2005.

69. $f(x) = \frac{1}{2}x - 6$

(a) The average rate of change of f between $x = 0$ and $x = 2$ is

$\dfrac{f(2) - f(0)}{2 - 0} = \dfrac{\left[\frac{1}{2}(2) - 6\right] - \left[\frac{1}{2}(0) - 6\right]}{2} = \dfrac{-5 - (-6)}{2} = \dfrac{1}{2}$, and the average rate of change of f between $x = 15$ and $x = 50$ is

$\dfrac{f(50) - f(15)}{50 - 15} = \dfrac{\left[\frac{1}{2}(50) - 6\right] - \left[\frac{1}{2}(15) - 6\right]}{35} = \dfrac{19 - \frac{3}{2}}{35} = \dfrac{1}{2}$.

(b) The rates of change are the same.

(c) Yes, f is a linear function with rate of change $\frac{1}{2}$.

71. (a) $y = f(x) + 8$. Shift the graph of $f(x)$ upward 8 units.

(b) $y = f(x + 8)$. Shift the graph of $f(x)$ to the left 8 units.

(c) $y = 1 + 2f(x)$. Stretch the graph of $f(x)$ vertically by a factor of 2, then shift it upward 1 unit.

(d) $y = f(x - 2) - 2$. Shift the graph of $f(x)$ to the right 2 units, then downward 2 units.

(e) $y = f(-x)$. Reflect the graph of $f(x)$ about the y-axis.

(f) $y = -f(-x)$. Reflect the graph of $f(x)$ first about the y-axis, then reflect about the x-axis.

(g) $y = -f(x)$. Reflect the graph of $f(x)$ about the x-axis.

(h) $y = f^{-1}(x)$. Reflect the graph of $f(x)$ about the line $y = x$.

73. (a) $f(x) = 2x^5 - 3x^2 + 2$. $f(-x) = 2(-x)^5 - 3(-x)^2 + 2 = -2x^5 - 3x^2 + 2$. Since $f(x) \neq f(-x)$, f is not even. $-f(x) = -2x^5 + 3x^2 - 2$. Since $-f(x) \neq f(-x)$, f is not odd.

(b) $f(x) = x^3 - x^7$. $f(-x) = (-x)^3 - (-x)^7 = -\left(x^3 - x^7\right) = -f(x)$, hence f is odd.

(c) $f(x) = \dfrac{1 - x^2}{1 + x^2}$. $f(-x) = \dfrac{1 - (-x)^2}{1 + (-x)^2} = \dfrac{1 - x^2}{1 + x^2} = f(x)$. Since $f(x) = f(-x)$, f is even.

(d) $f(x) = \dfrac{1}{x + 2}$. $f(-x) = \dfrac{1}{(-x) + 2} = \dfrac{1}{2 - x}$. $-f(x) = -\dfrac{1}{x + 2}$. Since $f(x) \neq f(-x)$, f is not even, and since $f(-x) \neq -f(x)$, f is not odd.

75. $g(x) = 2x^2 + 4x - 5 = 2\left(x^2 + 2x\right) - 5 = 2\left(x^2 + 2x + 1\right) - 5 - 2 = 2(x + 1)^2 - 7$. So the local minimum value -7 when $x = -1$.

77. $f(x) = 3.3 + 1.6x - 2.5x^3$. In the first viewing rectangle, $[-2, 2]$ by $[-4, 8]$, we see that $f(x)$ has a local maximum and a local minimum. In the next viewing rectangle, $[0.4, 0.5]$ by $[3.78, 3.80]$, we isolate the local maximum value as approximately 3.79 when $x \approx 0.46$. In the last viewing rectangle, $[-0.5, -0.4]$ by $[2.80, 2.82]$, we isolate the local minimum value as 2.81 when $x \approx -0.46$.

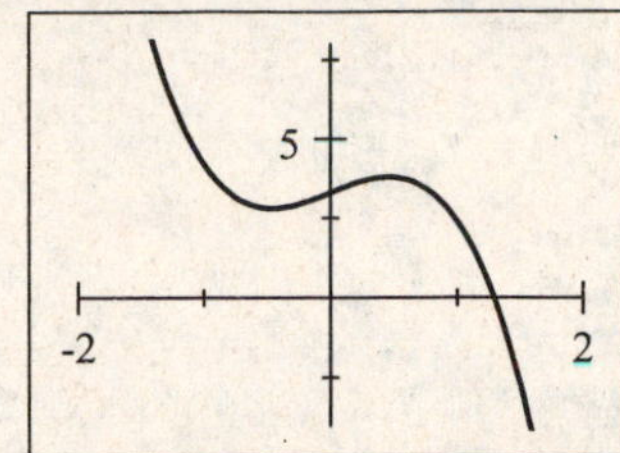

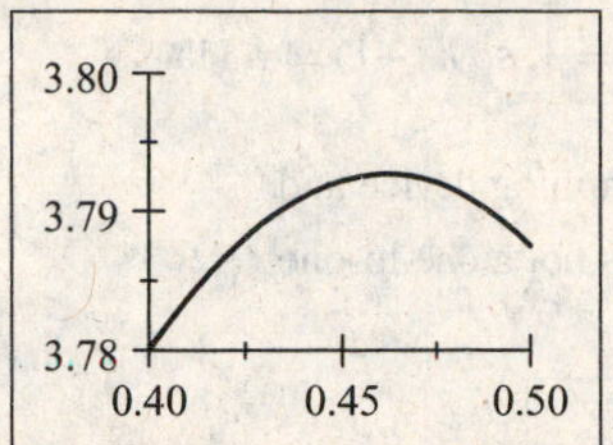

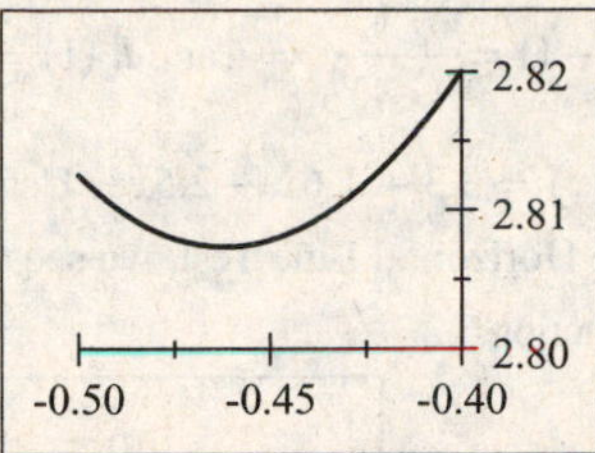

79. $h(t) = -16t^2 + 48t + 32 = -16\left(t^2 - 3t\right) + 32 = -16\left(t^2 - 3t + \frac{9}{4}\right) + 32 + 36$

$= -16\left(t^2 - 3t + \frac{9}{4}\right) + 68 = -16\left(t - \frac{3}{2}\right)^2 + 68$

The stone reaches a maximum height of 68 feet.

81. $f(x) = x + 2$, $g(x) = x^2$

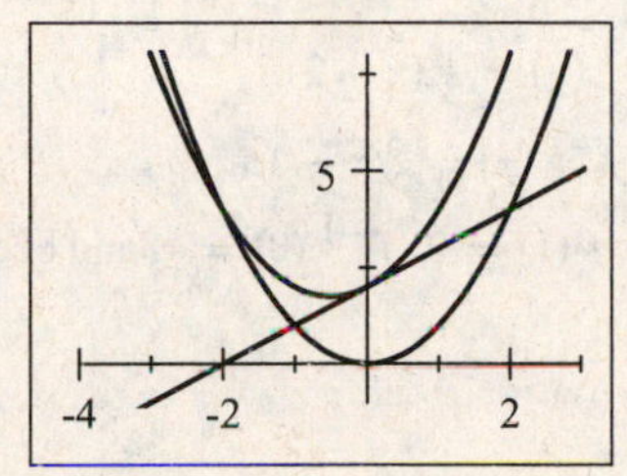

83. $f(x) = x^2 - 3x + 2$ and $g(x) = 4 - 3x$.

(a) $(f + g)(x) = \left(x^2 - 3x + 2\right) + (4 - 3x) = x^2 - 6x + 6$

(b) $(f - g)(x) = \left(x^2 - 3x + 2\right) - (4 - 3x) = x^2 - 2$

(c) $(fg)(x) = \left(x^2 - 3x + 2\right)(4 - 3x) = 4x^2 - 12x + 8 - 3x^3 + 9x^2 - 6x = -3x^3 + 13x^2 - 18x + 8$

(d) $\left(\frac{f}{g}\right)(x) = \frac{x^2 - 3x + 2}{4 - 3x}, x \neq \frac{4}{3}$

(e) $(f \circ g)(x) = f(4 - 3x) = (4 - 3x)^2 - 3(4 - 3x) + 2 = 16 - 24x + 9x^2 - 12 + 9x + 2 = 9x^2 - 15x + 6$

(f) $(g \circ f)(x) = g\left(x^2 - 3x + 2\right) = 4 - 3\left(x^2 - 3x + 2\right) = -3x^2 + 9x - 2$

85. $f(x) = 3x - 1$ and $g(x) = 2x - x^2$.

$(f \circ g)(x) = f\left(2x - x^2\right) = 3\left(2x - x^2\right) - 1 = -3x^2 + 6x - 1$, and the domain is $(-\infty, \infty)$.

$(g \circ f)(x) = g(3x - 1) = 2(3x - 1) - (3x - 1)^2 = 6x - 2 - 9x^2 + 6x - 1 = -9x^2 + 12x - 3$, and the domain is $(-\infty, \infty)$

$(f \circ f)(x) = f(3x - 1) = 3(3x - 1) - 1 = 9x - 4$, and the domain is $(-\infty, \infty)$.

$(g \circ g)(x) = g\left(2x - x^2\right) = 2\left(2x - x^2\right) - \left(2x - x^2\right)^2 = 4x - 2x^2 - 4x^2 + 4x^3 - x^4 = -x^4 + 4x^3 - 6x^2 + 4x$, and domain is $(-\infty, \infty)$.

87. $f(x) = \sqrt{1 - x}$, $g(x) = 1 - x^2$ and $h(x) = 1 + \sqrt{x}$.

$(f \circ g \circ h)(x) = f(g(h(x))) = f\left(g\left(1 + \sqrt{x}\right)\right) = f\left(1 - \left(1 + \sqrt{x}\right)^2\right) = f\left(1 - \left(1 + 2\sqrt{x} + x\right)\right)$

$= f\left(-x - 2\sqrt{x}\right) = \sqrt{1 - \left(-x - 2\sqrt{x}\right)} = \sqrt{1 + 2\sqrt{x} + x} = \sqrt{\left(1 + \sqrt{x}\right)^2} = 1 + \sqrt{x}$

89. $f(x) = 3 + x^3$. If $x_1 \neq x_2$, then $x_1^3 \neq x_2^3$ (unequal numbers have unequal cubes), and therefore $3 + x_1^3 \neq 3 + x_2^3$. Thus f is a one-to-one function.

91. $h(x) = \dfrac{1}{x^4}$. Since the fourth powers of a number and its negative are equal, h is not one-to-one. For example, $h(-1) = \dfrac{1}{(-1)^4} = 1$ and $h(1) = \dfrac{1}{(1)^4} = 1$, so $h(-1) = h(1)$.

93. $p(x) = 3.3 + 1.6x - 2.5x^3$. Using a graphing device and the Horizontal Line Test, we see that p is not a one-to-one function.

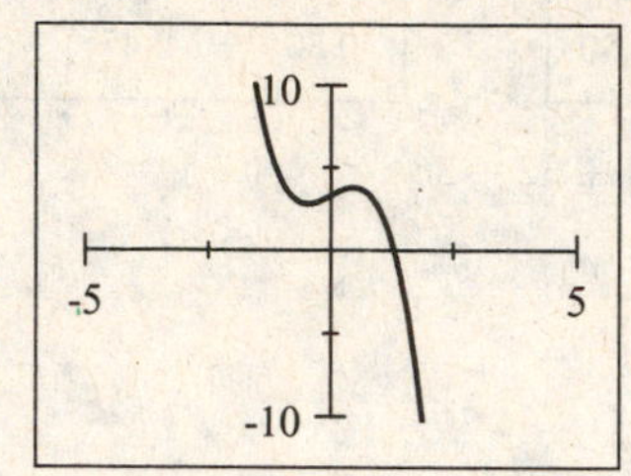

95. $f(x) = 3x - 2 \Leftrightarrow y = 3x - 2 \Leftrightarrow 3x = y + 2 \Leftrightarrow x = \frac{1}{3}(y + 2)$. So $f^{-1}(x) = \frac{1}{3}(x + 2)$.

97. $f(x) = (x + 1)^3 \Leftrightarrow y = (x + 1)^3 \Leftrightarrow x + 1 = \sqrt[3]{y} \Leftrightarrow x = \sqrt[3]{y} - 1$. So $f^{-1}(x) = \sqrt[3]{x} - 1$.

99. The graph passes the Horizontal Line Test, so f has an inverse. Because $f(1) = 0$, $f^{-1}(0) = 1$, and because $f(3) = 4$, $f^{-1}(4) = 3$.

101. (a), (b) $f(x) = x^2 - 4$, $x \geq 0$

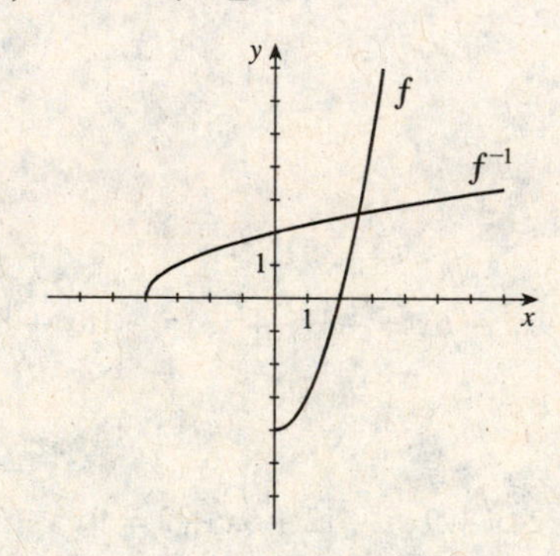

(c) $f(x) = x^2 - 4$, $x \geq 0 \Leftrightarrow y = x^2 - 4$, $y \geq -4$ $\Leftrightarrow x^2 = y + 4 \Leftrightarrow x = \sqrt{y + 4}$. So $f^{-1}(x) = \sqrt{x + 4}$, $x \geq -4$.

CHAPTER 2 TEST

1. By the Vertical Line Test, figures (a) and (b) are graphs of functions. By the Horizontal Line Test, only figure (a) is the graph of a one-to-one function.

3. (a) "Subtract 2, then cube the result" can be expressed algebraically as $f(x) = (x-2)^3$.

(b)

x	$f(x)$
-1	-27
0	-8
1	-1
2	0
3	1
4	8

(c)

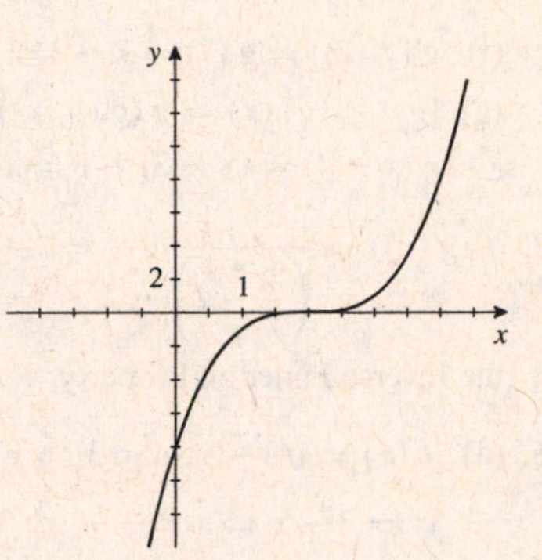

(d) We know that f has an inverse because it passes the Horizontal Line Test. A verbal description for f^{-1} is, "Take the cube root, then add 2."

(e) $y = (x-2)^3 \Leftrightarrow \sqrt[3]{y} = x - 2 \Leftrightarrow x = \sqrt[3]{y} + 2$. Thus, a formula for f^{-1} is $f^{-1}(x) = \sqrt[3]{x} + 2$.

5. $R(x) = -500x^2 + 3000x$

(a) $R(2) = -500(2)^2 + 3000(2) = \$4000$ represents their total sales revenue when their price is \$2 per bar and $R(4) = -500(4)^2 + 3000(4) = \$4000$ represents their total sales revenue when their price is \$4 per bar

(b)

(c) The maximum revenue is \$4500, and it is achieved at a price of $x = \$3$.

7. (a) $f(x) = (x+5)^2 = x^2 + 10x + 25$ is not linear because it cannot be expressed in the form $f(x) = ax + b$ for constants a and b. $g(x) = 1 - 5x$ is linear.

(b)

(c) $g(x)$ has rate of change -5.

9. (a) $y = f(x-3) + 2$. Shift the graph of $f(x)$ to the right 3 units, then shift the graph upward 2 units.

(b) $y = f(-x)$. Reflect the graph of $f(x)$ about the y-axis.

11. $f(x) = x^2 + x + 1$; $g(x) = x - 3$.

(a) $(f+g)(x) = f(x) + g(x) = \left(x^2 + x + 1\right) + (x-3) = x^2 + 2x - 2$

(b) $(f-g)(x) = f(x) - g(x) = \left(x^2+x+1\right) - (x-3) = x^2+4$

(c) $(f \circ g)(x) = f(g(x)) = f(x-3) = (x-3)^2 + (x-3) + 1 = x^2 - 6x + 9 + x - 3 + 1 = x^2 - 5x + 7$

(d) $(g \circ f)(x) = g(f(x)) = g\left(x^2+x+1\right) = \left(x^2+x+1\right) - 3 = x^2 + x - 2$

(e) $f(g(2)) = f(-1) = (-1)^2 + (-1) + 1 = 1$. [We have used the fact that $g(2) = 2 - 3 = -1$.]

(f) $g(f(2)) = g(7) = 7 - 3 = 4$. [We have used the fact that $f(2) = 2^2 + 2 + 1 = 7$.]

(g) $(g \circ g \circ g)(x) = g(g(g(x))) = g(g(x-3)) = g(x-6) = (x-6) - 3 = x - 9$. [We have used the fact that $g(x-3) = (x-3) - 3 = x - 6$.]

13. $f(g(x)) = \dfrac{1}{\left(\dfrac{1}{x}+2\right)-2} = \dfrac{1}{\dfrac{1}{x}} = x$ for all $x \neq 0$, and $g(f(x)) = \dfrac{1}{\dfrac{1}{x-2}} + 2 = x - 2 + 2 = x$ for all $x \neq -2$. Thus, by the Inverse Function Property, f and g are inverse functions.

15. (a) $f(x) = \sqrt{3-x}, x \leq 3 \Leftrightarrow y = \sqrt{3-x} \Leftrightarrow y^2 = 3 - x \Leftrightarrow x = 3 - y^2$. Thus $f^{-1}(x) = 3 - x^2, x \geq 0$.

(b) $f(x) = \sqrt{3-x}, x \leq 3$ and $f^{-1}(x) = 3 - x^2, x \geq 0$

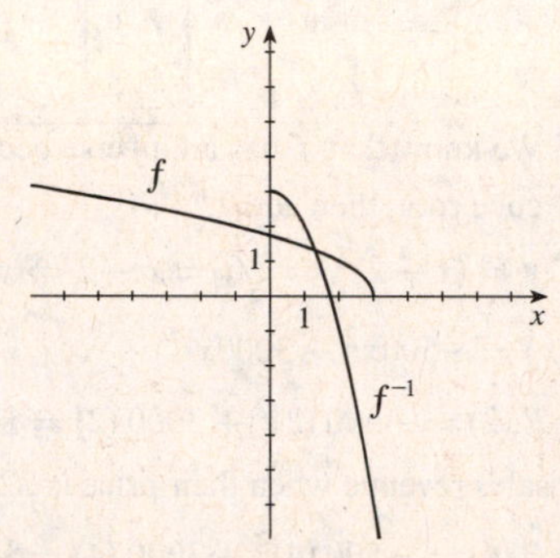

17. The graph passes through the points (0, 1) and (4, 3), so $f(0) = 1$ and $f(4) = 3$.

19. The net change of f between $x = 2$ and $x = 6$ is $f(6) - f(2) = 7 - 2 = 5$ and the average rate of change is $\dfrac{f(6) - f(2)}{6-2} = \dfrac{5}{4}$.

21.

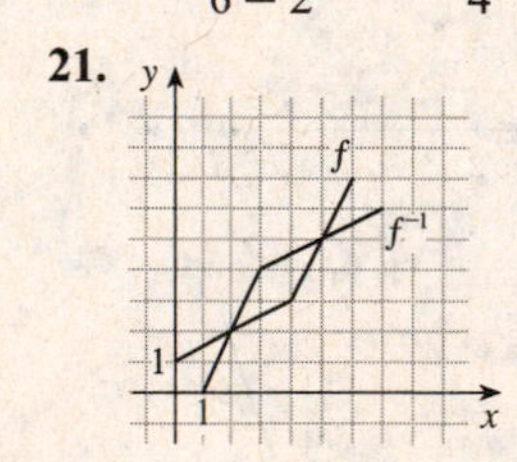

FOCUS ON MODELING Modeling with Functions

1. Let w be the width of the building lot. Then the length of the lot is $3w$. So the area of the building lot is $A(w) = 3w^2$, $w > 0$.

3. Let w be the width of the base of the rectangle. Then the height of the rectangle is $\frac{1}{2}w$. Thus the volume of the box is given by the function $V(w) = \frac{1}{2}w^3$, $w > 0$.

5. Let P be the perimeter of the rectangle and y be the length of the other side. Since $P = 2x + 2y$ and the perimeter is 20, we have $2x + 2y = 20 \Leftrightarrow x + y = 10 \Leftrightarrow y = 10 - x$. Since area is $A = xy$, substituting gives $A(x) = x(10-x) = 10x - x^2$, and since A must be positive, the domain is $0 < x < 10$.

7.

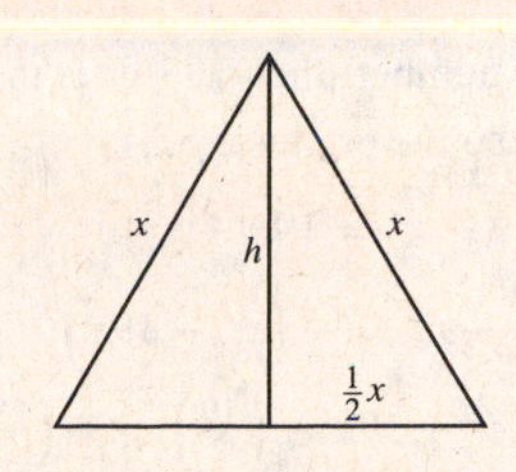

Let h be the height of an altitude of the equilateral triangle whose side has length x, as shown in the diagram. Thus the area is given by $A = \frac{1}{2}xh$. By the Pythagorean Theorem, $h^2 + \left(\frac{1}{2}x\right)^2 = x^2 \Leftrightarrow h^2 + \frac{1}{4}x^2 = x^2 \Leftrightarrow h^2 = \frac{3}{4}x^2 \Leftrightarrow h = \frac{\sqrt{3}}{2}x$.
Substituting into the area of a triangle, we get
$A(x) = \frac{1}{2}xh = \frac{1}{2}x\left(\frac{\sqrt{3}}{2}x\right) = \frac{\sqrt{3}}{4}x^2, x > 0$.

9. We solve for r in the formula for the area of a circle. This gives $A = \pi r^2 \Leftrightarrow r^2 = \frac{A}{\pi} \Rightarrow r = \sqrt{\frac{A}{\pi}}$, so the model is $r(A) = \sqrt{\frac{A}{\pi}}, A > 0$.

11. Let h be the height of the box in feet. The volume of the box is $V = 60$. Then $x^2h = 60 \Leftrightarrow h = \frac{60}{x^2}$. The surface area, S, of the box is the sum of the area of the 4 sides and the area of the base and top. Thus $S = 4xh + 2x^2 = 4x\left(\frac{60}{x^2}\right) + 2x^2 = \frac{240}{x} + 2x^2$, so the model is $S(x) = \frac{240}{x} + 2x^2, x > 0$.

13.

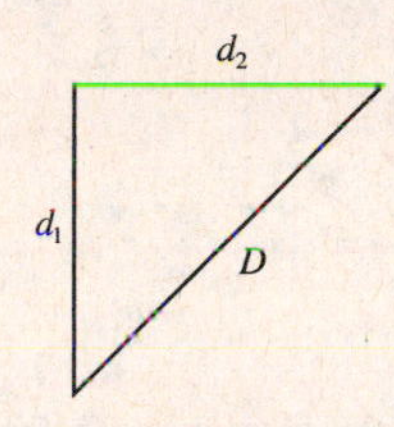

Let d_1 be the distance traveled south by the first ship and d_2 be the distance traveled east by the second ship. The first ship travels south for t hours at 5 mi/h, so $d_1 = 15t$ and, similarly, $d_2 = 20t$. Since the ships are traveling at right angles to each other, we can apply the Pythagorean Theorem to get
$D(t) = \sqrt{d_1^2 + d_2^2} = \sqrt{(15t)^2 + (20t)^2} = \sqrt{225t^2 + 400t^2} = 25t$.

15.

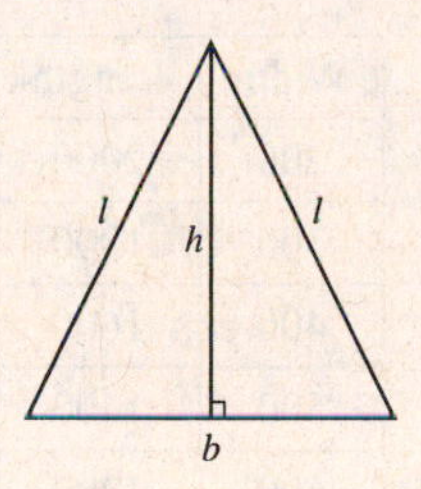

Let b be the length of the base, l be the length of the equal sides, and h be the height in centimeters. Since the perimeter is 8, $2l + b = 8 \Leftrightarrow 2l = 8 - b \Leftrightarrow l = \frac{1}{2}(8 - b)$. By the Pythagorean Theorem, $h^2 + \left(\frac{1}{2}b\right)^2 = l^2 \Leftrightarrow h = \sqrt{l^2 - \frac{1}{4}b^2}$. Therefore the area of the triangle is

$$A = \frac{1}{2}\cdot b\cdot h = \frac{1}{2}\cdot b\sqrt{l^2 - \frac{1}{4}b^2} = \frac{b}{2}\sqrt{\frac{1}{4}(8-b)^2 - \frac{1}{4}b^2}$$
$$= \frac{b}{4}\sqrt{64 - 16b + b^2 - b^2} = \frac{b}{4}\sqrt{64 - 16b} = \frac{b}{4}\cdot 4\sqrt{4-b} = b\sqrt{4-b}$$

so the model is $A(b) = b\sqrt{4-b}, 0 < b < 4$.

17. Let w be the length of the rectangle. By the Pythagorean Theorem, $\left(\frac{1}{2}w\right)^2 + h^2 = 10^2 \Leftrightarrow \frac{w^2}{4} + h^2 = 10^2 \Leftrightarrow w^2 = 4\left(100 - h^2\right) \Leftrightarrow w = 2\sqrt{100 - h^2}$ (since $w > 0$). Therefore, the area of the rectangle is $A = wh = 2h\sqrt{100 - h^2}$, so the model is $A(h) = 2h\sqrt{100 - h^2}, 0 < h < 10$.

19. (a) We complete the table.

First number	Second number	Product
1	18	18
2	17	34
3	16	48
4	15	60
5	14	70
6	13	78
7	12	84
8	11	88
9	10	90
10	9	90
11	8	88

From the table we conclude that the numbers is still increasing, the numbers whose product is a maximum should both be 9.5.

(b) Let x be one number: then $19 - x$ is the other number, and so the product, p, is

$p(x) = x(19 - x) = 19x - x^2$.

(c) $p(x) = 19x - x^2 = -\left(x^2 - 19x\right)$

$= -\left[x^2 - 19x + \left(\frac{19}{2}\right)^2\right] + \left(\frac{19}{2}\right)^2$

$= -(x - 9.5)^2 + 90.25$

So the product is maximized when the numbers are both 9.5.

21. (a) Let x be the width of the field (in feet) and l be the length of the field (in feet). Since the farmer has 2400 ft of fencing we must have $2x + l = 2400$.

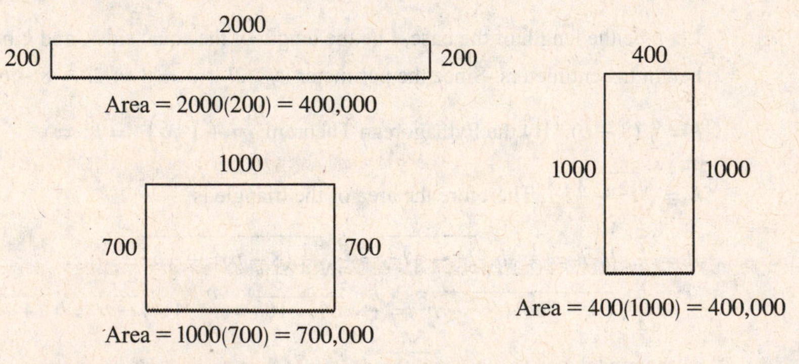

Width	Length	Area
200	2000	400,000
300	1800	540,000
400	1600	640,000
500	1400	700,000
600	1200	720,000
700	1000	700,000
800	800	640,000

It appears that the field of largest area is about 600 ft × 1200 ft.

(b) Let x be the width of the field (in feet) and l be the length of the field (in feet). Since the farmer has 2400 ft of fencing we must have $2x + l = 2400 \Leftrightarrow l = 2400 - 2x$. The area of the fenced-in field is given by $A(x) = l \cdot x = (2400 - 2x)x = -2x^2 + 2400x = -2\left(x^2 - 1200x\right)$.

(c) The area is $A(x) = -2\left(x^2 - 1200x + 600^2\right) + 2\left(600^2\right) = -2(x - 600)^2 + 720{,}000$. So the maximum area occurs when $x = 600$ feet and $l = 2400 - 2(600) = 1200$ feet.

23. **(a)** Let x be the length of the fence along the road. If the area is 1200, we have $1200 = x \cdot$ width, so the width of the garden is $\frac{1200}{x}$. Then the cost of the fence is given by the function $C(x) = 5(x) + 3\left[x + 2 \cdot \frac{1200}{x}\right] = 8x + \frac{7200}{x}$.

(b) We graph the function $y = C(x)$ in the viewing rectangle $[0, 75] \times [0, 800]$. From this we get the cost is minimized when $x = 30$ ft. Then the width is $\frac{1200}{30} = 40$ ft. So the length is 30 ft and the width is 40 ft.

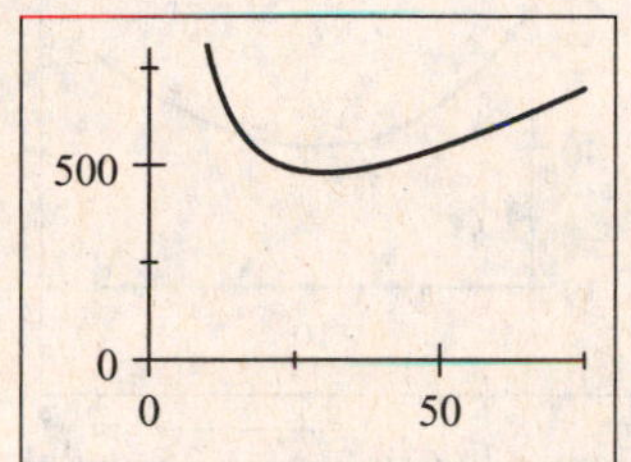

(c) We graph the function $y = C(x)$ and $y = 600$ in the viewing rectangle $[10, 65] \times [450, 650]$. From this we get that the cost is at most \$600 when $15 \le x \le 60$. So the range of lengths he can fence along the road is 15 feet to 60 feet.

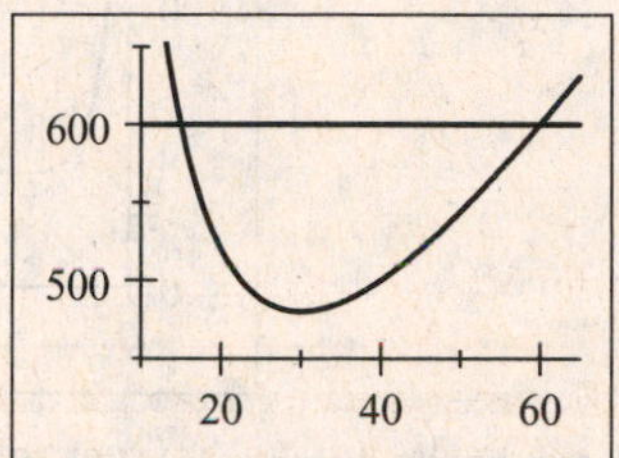

25. **(a)** Let h be the height in feet of the straight portion of the window. The circumference of the semicircle is $C = \frac{1}{2}\pi x$. Since the perimeter of the window is 30 feet, we have $x + 2h + \frac{1}{2}\pi x = 30$. Solving for h, we get $2h = 30 - x - \frac{1}{2}\pi x \Leftrightarrow h = 15 - \frac{1}{2}x - \frac{1}{4}\pi x$. The area of the window is $A(x) = xh + \frac{1}{2}\pi\left(\frac{1}{2}x\right)^2 = x\left(15 - \frac{1}{2}x - \frac{1}{4}\pi x\right) + \frac{1}{8}\pi x^2 = 15x - \frac{1}{2}x^2 - \frac{1}{8}\pi x^2$.

(b) $A(x) = 15x - \frac{1}{8}(\pi + 4)x^2 = -\frac{1}{8}(\pi + 4)\left[x^2 - \frac{120}{\pi + 4}x\right]$

$$= -\tfrac{1}{8}(\pi + 4)\left[x^2 - \frac{120}{\pi + 4}x + \left(\frac{60}{\pi + 4}\right)^2\right] + \frac{450}{\pi + 4} = -\tfrac{1}{8}(\pi + 4)\left(x - \frac{60}{\pi + 4}\right)^2 + \frac{450}{\pi + 4}$$

The area is maximized when $x = \frac{60}{\pi + 4} \approx 8.40$, and hence $h \approx 15 - \frac{1}{2}(8.40) - \frac{1}{4}\pi(8.40) \approx 4.20$.

27. **(a)** Let x be the length of one side of the base and let h be the height of the box in feet. Since the volume of the box is $V = x^2h = 12$, we have $x^2h = 12 \Leftrightarrow h = \frac{12}{x^2}$. The surface area, A, of the box is sum of the area of the four sides and the area of the base. Thus the surface area of the box is given by the formula $A(x) = 4xh + x^2 = 4x\left(\frac{12}{x^2}\right) + x^2 = \frac{48}{x} + x^2, x > 0$.

(b) The function $y = A(x)$ is shown in the first viewing rectangle below. In the second viewing rectangle, we isolate the minimum, and we see that the amount of material is minimized when x (the length and width) is 2.88 ft. Then the height is $h = \frac{12}{x^2} \approx 1.44$ ft.

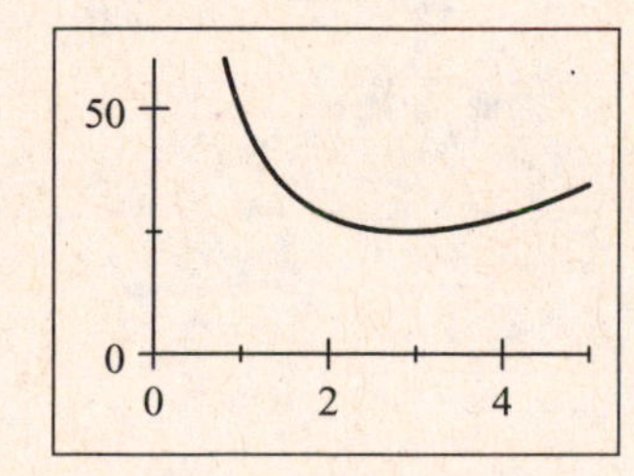

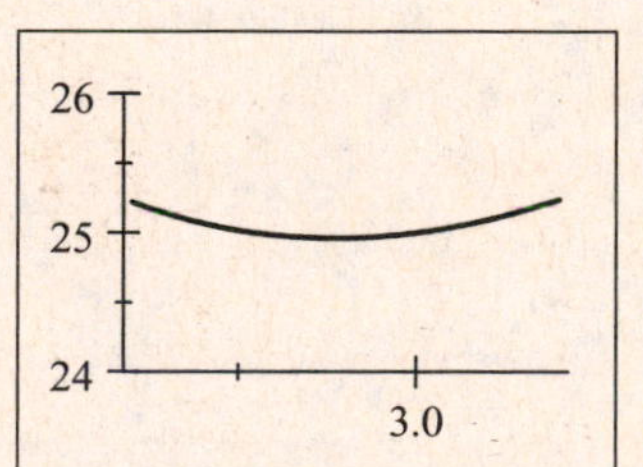

29. (a) Let w be the width of the pen and l be the length in meters. We use the area to establish a relationship between w and l. Since the area is 100 m^2, we have $l \cdot w = 100 \Leftrightarrow l = \dfrac{100}{w}$. So the amount of fencing used is $F = 2l + 2w = 2\left(\dfrac{100}{w}\right) + 2w = \dfrac{200 + 2w^2}{w}$.

(b) Using a graphing device, we first graph F in the viewing rectangle $[0, 40]$ by $[0, 100]$, and locate the approximate location of the minimum value. In the second viewing rectangle, $[8, 12]$ by $[39, 41]$, we see that the minimum value of F occurs when $w = 10$. Therefore the pen should be a square with side 10 m.

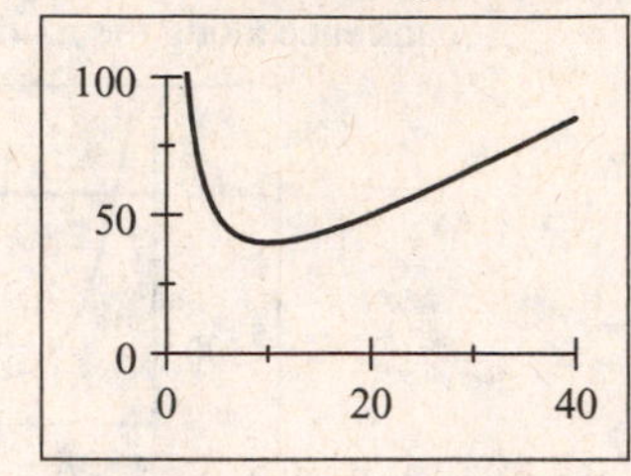

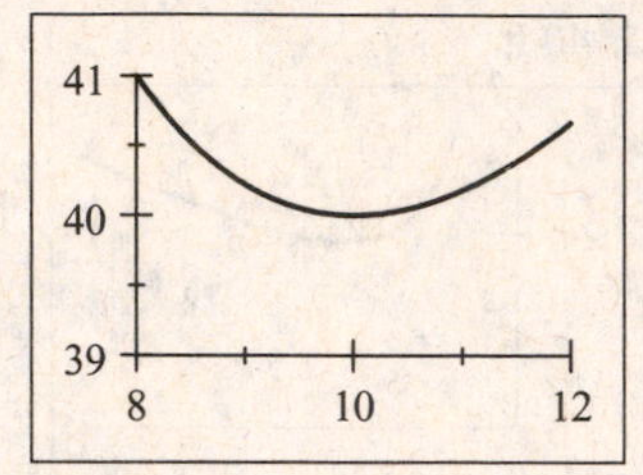

31. (a) Let x be the distance from point B to C, in miles. Then the distance from A to C is $\sqrt{x^2 + 25}$, and the energy used in flying from A to C then C to D is $f(x) = 14\sqrt{x^2 + 25} + 10(12 - x)$.

(b) By using a graphing device, the energy expenditure is minimized when the distance from B to C is about 5.1 miles.

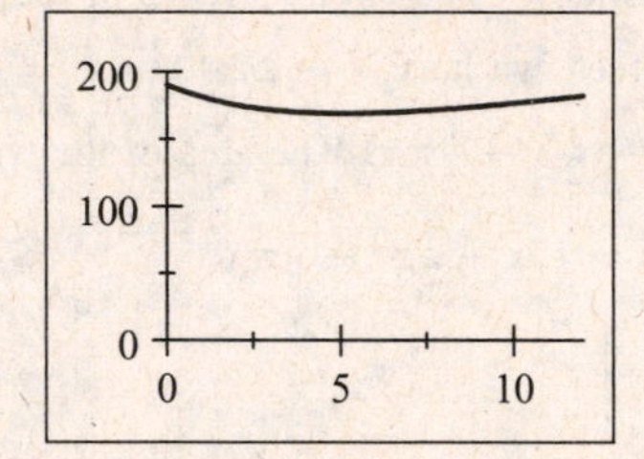

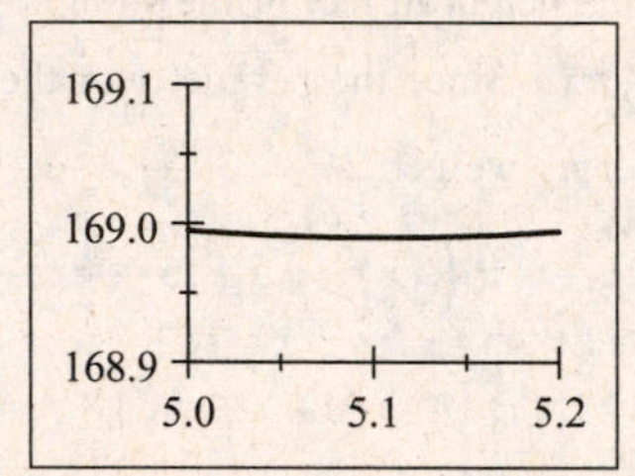

3 POLYNOMIAL AND RATIONAL FUNCTIONS

3.1 QUADRATIC FUNCTIONS AND MODELS

1. To put the quadratic function $f(x) = ax^2 + bx + c$ in standard form we complete the *square*.

3. The graph of $f(x) = 3(x-2)^2 - 6$ is a parabola that opens *upward*, with its vertex at $(2, -6)$, and $f(2) = -6$ is the *minimum* value of f.

5. (a) The vertex is $(3, 4)$, the x-intercepts are 1 and 5, and it appears that the y-intercept is approximately -5.

(b) Maximum value of f: 4

(c) Domain $(-\infty, \infty)$, range: $(-\infty, 4]$

7. (a) The vertex is $(1, -3)$, the x-intercepts are approximately -0.2 and 2.2, and the y-intercept is -1.

(b) Minimum value of f: -3

(c) Domain: $(-\infty, \infty)$, range: $[-3, \infty)$

9. (a) $f(x) = x^2 - 2x + 3 = (x-1)^2 - 1 + 3 = (x-1)^2 + 2$

(b) The vertex is at $(1, 2)$.

x-intercepts: $y = 0 \Rightarrow 0 = (x-1)^2 + 2 \Rightarrow (x-1)^2 = -2$. This has no real solution, so there is no x-intercept.

y-intercept: $x = 0 \Rightarrow y = (0-1)^2 + 2 = 3$. The y-intercept is 3.

(c)

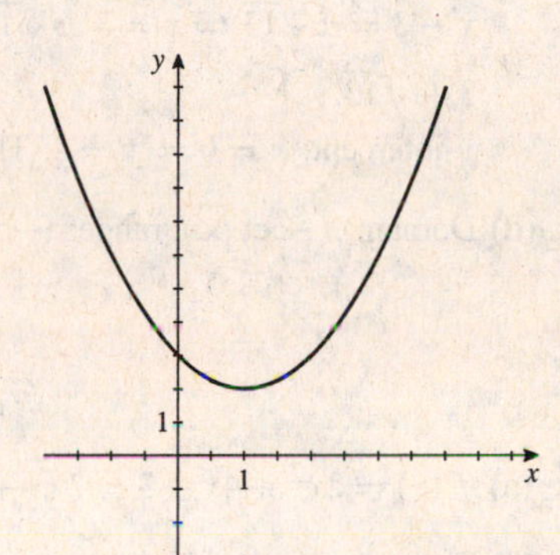

(d) Domain: $(-\infty, \infty)$, range: $[2, \infty)$

11. (a) $f(x) = x^2 - 6x = x^2 - 6x = x^2 - 6x + 9 - 9 = (x-3)^2 - 9$

(b) The vertex is at $(3, -9)$.

x-intercepts: $y = 0 \Rightarrow 0 = x^2 - 6x = x(x-6)$. So $x = 0$ or $x = 6$. The x-intercepts are 0 and 6.

y-intercept: $x = 0 \Rightarrow y = 0$. The y-intercept is 0.

(c)

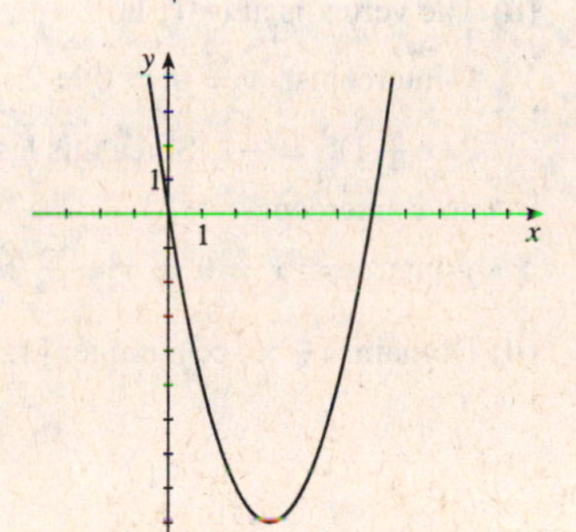

(d) Domain: $(-\infty, \infty)$, range: $[-9, \infty)$

13. (a) $f(x) = 3x^2 + 6x = 3\left(x^2 + 2x\right) = 3(x+1)^2 - 3$

(b) The vertex is at $(-1, -3)$.

x-intercepts: $y = 0 \Rightarrow 0 = 3(x+1)^2 - 3 \Rightarrow (x+1)^2 = 1 \Rightarrow x = -2$ or 0. The x-intercepts are -2 and 0.

y-intercept: $x = 0 \Rightarrow y = 3(0)^2 + 6(0) = 0$. The y-intercept is 0.

(c)

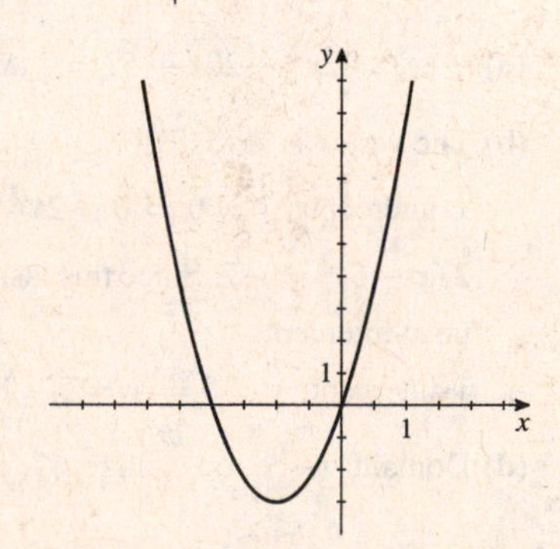

(d) Domain: $(-\infty, \infty)$, range: $[-3, \infty)$

15. **(a)** $f(x) = x^2 + 4x + 3 = (x + 2)^2 - 1$

(b) The vertex is at $(-2, -1)$.

x-intercepts: $y = 0 \Rightarrow 0 = x^2 + 4x + 3 = (x + 1)(x + 3)$. So $x = -1$ or $x = -3$. The x-intercepts are -1 and -3.

y-intercept: $x = 0 \Rightarrow y = 3$. The y-intercept is 3.

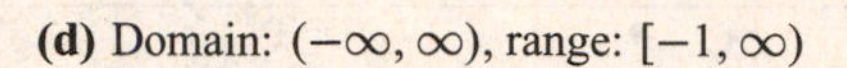

(d) Domain: $(-\infty, \infty)$, range: $[-1, \infty)$

(c)

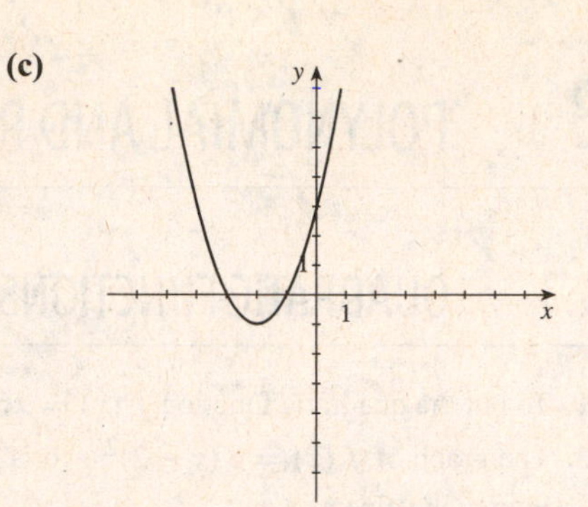

17. **(a)** $f(x) = -x^2 + 6x + 4 = -(x - 3)^2 + 13$

(b) The vertex is at $(3, 13)$.

x-intercepts: $y = 0 \Rightarrow 0 = -(x - 3)^2 + 13 \Leftrightarrow (x - 3)^2 = 13 \Rightarrow$ $x - 3 = \pm\sqrt{13} \Leftrightarrow x = 3 \pm \sqrt{13}$. The x-intercepts are $3 - \sqrt{13}$ and $3 + \sqrt{13}$.

y-intercept: $x = 0 \Rightarrow y = 4$. The y-intercept is 4.

(d) Domain: $(-\infty, \infty)$, range: $(-\infty, 13]$

(c)

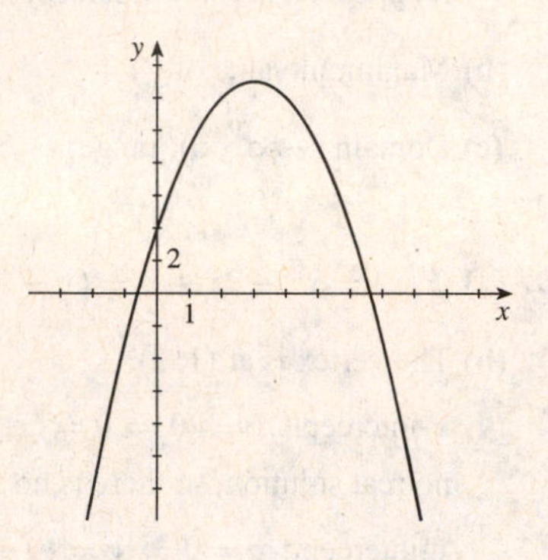

19. **(a)** $f(x) = 2x^2 + 4x + 3 = 2(x + 1)^2 + 1$

(b) The vertex is at $(-1, 1)$.

x-intercepts: $y = 0 \Rightarrow 0 = 2x^2 + 4x + 3 = 2(x + 1)^2 + 1 \Leftrightarrow$ $2(x + 1)^2 = -1$. Since this last equation has no real solution, there is no x-intercept.

y-intercept: $x = 0 \Rightarrow y = 3$. The y-intercept is 3.

(d) Domain: $(-\infty, \infty)$, range: $[1, \infty)$

(c)

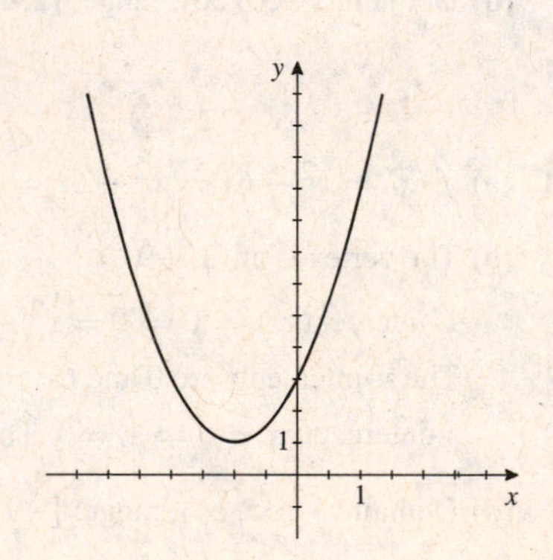

21. **(a)** $f(x) = 2x^2 - 20x + 57 = 2(x - 5)^2 + 7$

(b) The vertex is at $(5, 7)$.

x-intercepts: $y = 0 \Rightarrow 0 = 2x^2 - 20x + 57 = 2(x - 5)^2 + 7 \Leftrightarrow$ $2(x - 5)^2 = -7$. Since this last equation has no real solution, there is no x-intercept.

y-intercept: $x = 0 \Rightarrow y = 57$. The y-intercept is 57.

(d) Domain: $(-\infty, \infty)$, range: $[7, \infty)$

(c)

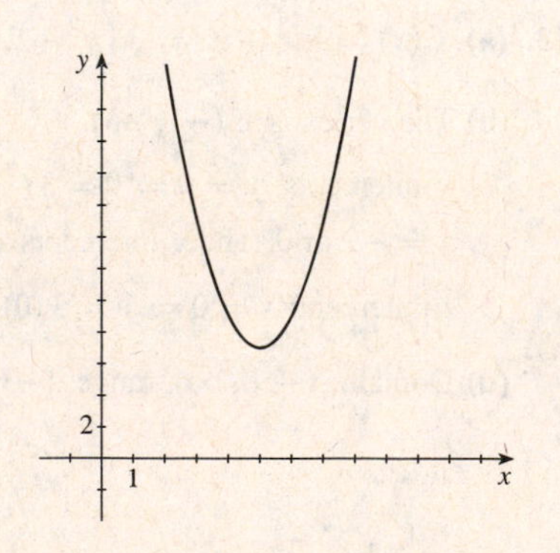

23. **(a)** $f(x) = -4x^2 - 12x + 1 = -4\left(x^2 + 3\right) + 1$
$= -4\left(x + \frac{3}{2}\right)^2 + 9 + 1 = -4\left(x + \frac{3}{2}\right)^2 + 10$

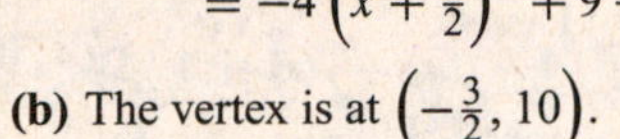

(b) The vertex is at $\left(-\frac{3}{2}, 10\right)$.

x-intercepts: $y = 0 \Rightarrow 0 = -4\left(x + \frac{3}{2}\right)^2 + 10 \Rightarrow \left(x + \frac{3}{2}\right)^2 = \frac{5}{2} \Rightarrow$
$x + \frac{3}{2} = \pm\sqrt{\frac{5}{2}} \Rightarrow x = -\frac{3}{2} \pm \sqrt{\frac{5}{2}} = -\frac{3}{2} \pm \frac{\sqrt{10}}{2}$. The x-intercepts are $-\frac{3}{2} - \frac{\sqrt{10}}{2}$ and $-\frac{3}{2} + \frac{\sqrt{10}}{2}$.

y-intercept: $x = 0 \Rightarrow y = -4(0)^2 - 12(0) + 1 = 1$. The y-intercept is 1.

(c)

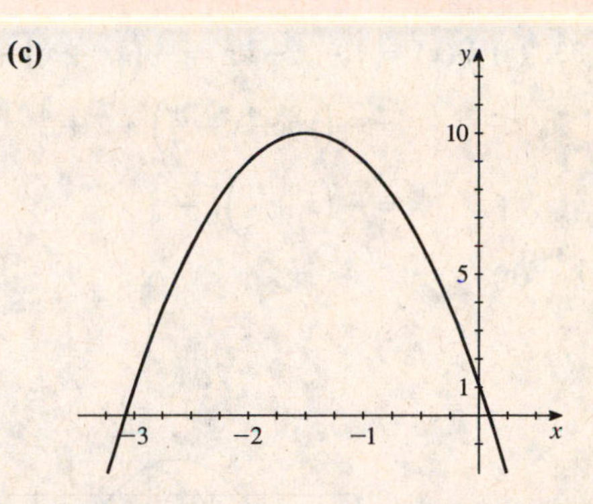

(d) Domain: $(-\infty, \infty)$, range: $(-\infty, 10]$

25. **(a)** $f(x) = x^2 + 2x - 1 = \left(x^2 + 2x\right) - 1$
$= \left(x^2 + 2x + 1\right) - 1 - 1 = (x + 1)^2 - 2$

(b)

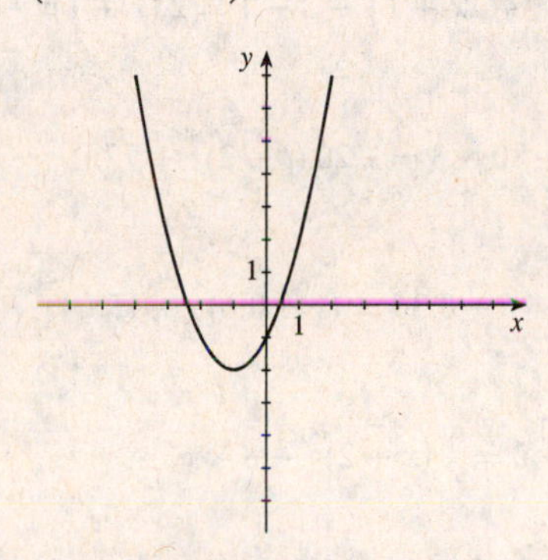

(c) The minimum value is $f(-1) = -2$.

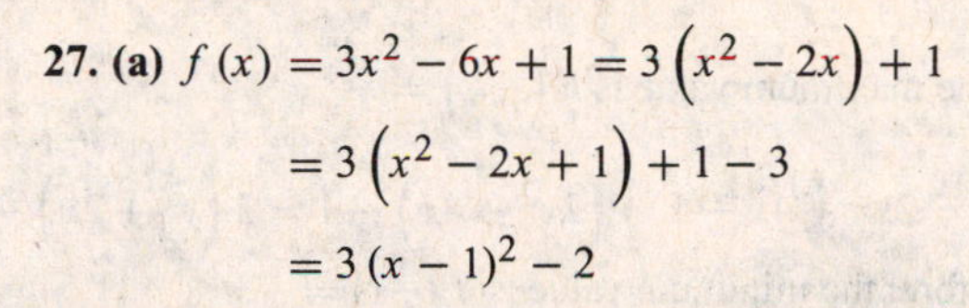

27. **(a)** $f(x) = 3x^2 - 6x + 1 = 3\left(x^2 - 2x\right) + 1$
$= 3\left(x^2 - 2x + 1\right) + 1 - 3$
$= 3(x - 1)^2 - 2$

(b)

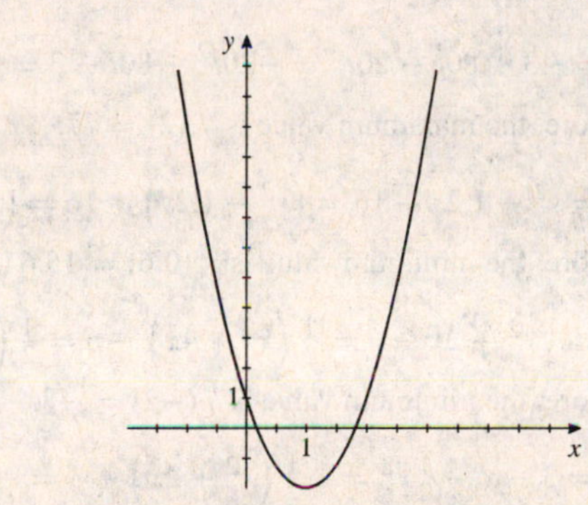

(c) The minimum value is $f(1) = -2$.

29. **(a)** $f(x) = -x^2 - 3x + 3 = -\left(x^2 + 3x\right) + 3$
$= -\left(x^2 + 3x + \frac{9}{4}\right) + 3 + \frac{9}{4}$
$= -\left(x + \frac{3}{2}\right)^2 + \frac{21}{4}$

(b)

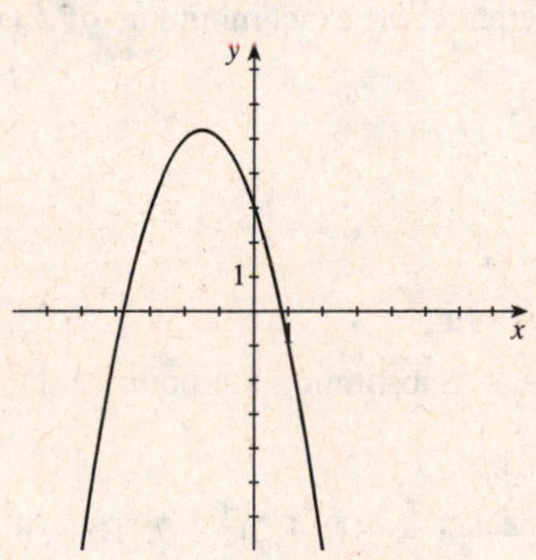

(c) The maximum value is $f\left(-\frac{3}{2}\right) = \frac{21}{4}$.

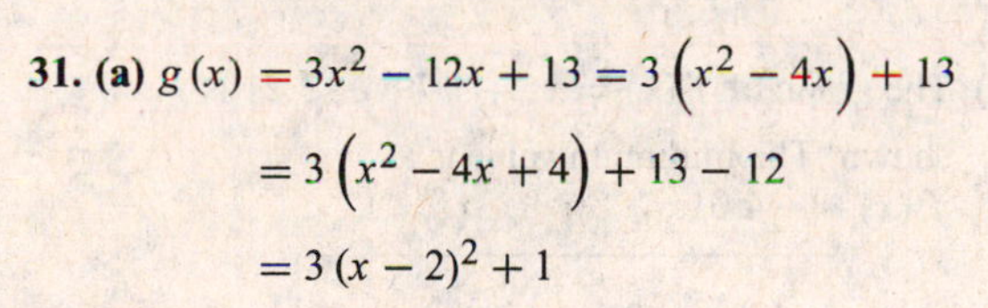

31. **(a)** $g(x) = 3x^2 - 12x + 13 = 3\left(x^2 - 4x\right) + 13$
$= 3\left(x^2 - 4x + 4\right) + 13 - 12$
$= 3(x - 2)^2 + 1$

(b)

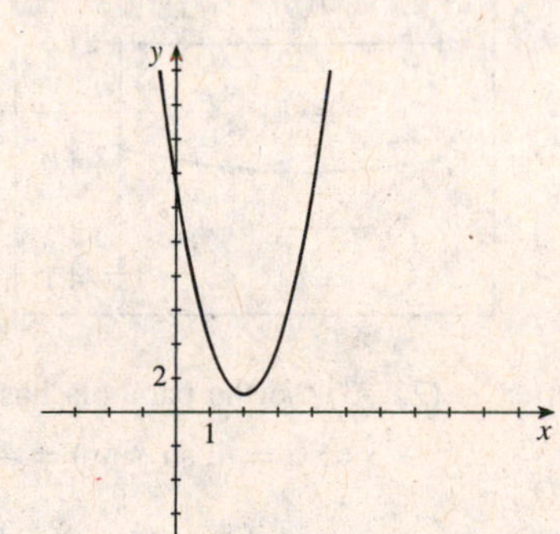

(c) The minimum value is $g(2) = 1$.

33. (a) $h(x) = 1 - x - x^2 = -\left(x^2 + x\right) + 1$
$= -\left(x^2 + x + \frac{1}{4}\right) + 1 + \frac{1}{4}$
$= -\left(x + \frac{1}{2}\right)^2 + \frac{5}{4}$

(b)

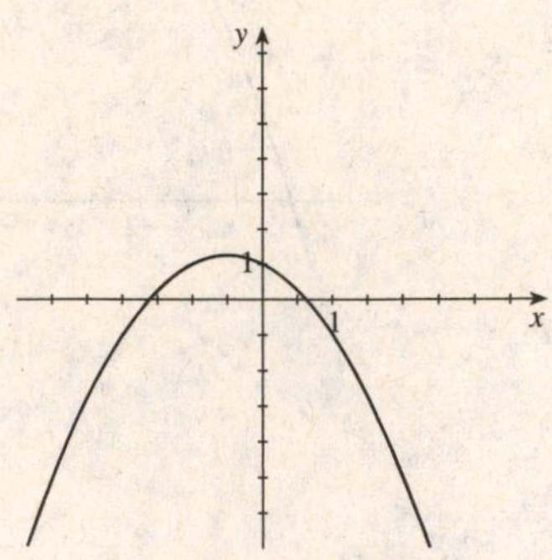

(c) The maximum value is $h\left(-\frac{1}{2}\right) = \frac{5}{4}$.

35. $f(x) = 2x^2 + 4x - 1 = \left(2x^2 + 4x\right) - 1 = 2\left(x^2 + 2x\right) - 1 = 2\left(x^2 + 2x + 1\right) - 2 - 1 = 2(x + 1)^2 - 3$.
Therefore, the minimum value is $f(-1) = -3$.

37. $f(t) = -3 + 80t - 20t^2 = -20t^2 + 80t - 3 = -20\left(t^2 - 4t + 4\right) + 80 - 3 = -20(t - 2)^2 + 77$.
Therefore, the maximum value is $f(2) = 77$.

39. $f(s) = s^2 - 1.2s + 16 = \left(s^2 - 1.2s\right) + 16 = \left(s^2 - 1.2s + 0.36\right) + 16 - 0.36 = (s - 0.6)^2 + 15.64$.
Therefore, the minimum value is $f(0.6) = 15.64$.

41. $h(x) = \frac{1}{2}x^2 + 2x - 6 = \frac{1}{2}\left(x^2 + 4x\right) - 6 = \frac{1}{2}\left(x^2 + 4x + 4\right) - 6 - 2 = \frac{1}{2}(x + 2)^2 - 8$.
Therefore, the minimum value is $h(-2) = -8$.

43. $f(x) = 3 - x - \frac{1}{2}x^2 = -\frac{1}{2}\left(x^2 + 2x\right) + 3 = -\frac{1}{2}\left(x^2 + 2x + 1\right) + 3 + \frac{1}{2} = -\frac{1}{2}(x + 1) + \frac{7}{2}$. Therefore, the maximum value is $f(-1) = \frac{7}{2}$.

45. (a) The graph of $f(x) = x^2 + 1.79x - 3.21$ is shown. The minimum value is $f(x) \approx -4.01$.

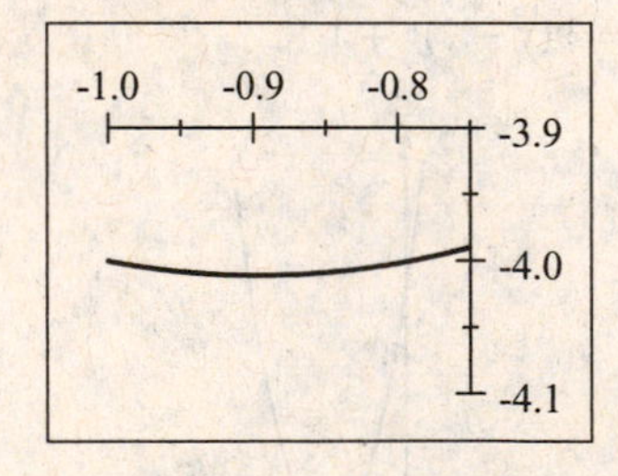

(b) $f(x) = x^2 + 1.79x - 3.21$
$= \left[x^2 + 1.79x + \left(\frac{1.79}{2}\right)^2\right] - 3.21 - \left(\frac{1.79}{2}\right)^2$
$= (x + 0.895)^2 - 4.011025$

Therefore, the exact minimum of $f(x)$ is -4.011025.

47. The vertex is $(2, -3)$, so the parabola has equation $y = a(x - 2)^2 - 3$. Substituting the point $(3, 1)$, we have $1 = a(3 - 2)^2 - 3 \Leftrightarrow a = 4$, so $f(x) = 4(x - 2)^2 - 3$.

49. Substituting $t = x^2$, we have $f(t) = 3 + 4t - t^2 = -\left(t^2 - 4t + 4\right) + 4 + 3 = -(t - 2)^2 + 7$. Thus, the maximum value is 7, when $t = 2$ (or $x = \sqrt{2}$).

51. $y = f(t) = 40t - 16t^2 = -16\left(t^2 - \frac{5}{2}\right) = -16\left[t^2 - \frac{5}{2}t + \left(\frac{5}{4}\right)^2\right] + 16\left(\frac{5}{4}\right)^2 = -16\left(t - \frac{5}{4}\right)^2 + 25$. Thus the maximum height attained by the ball is $f\left(\frac{5}{4}\right) = 25$ feet.

53. $R(x) = 80x - 0.4x^2 = -0.4\left(x^2 - 200x\right) = -0.4\left(x^2 - 200x + 10{,}000\right) + 4{,}000 = -0.4(x - 100)^2 + 4{,}000$. So revenue is maximized at \$4,000 when 100 units are sold.

55. $E(n) = \frac{2}{3}n - \frac{1}{90}n^2 = -\frac{1}{90}\left(n^2 - 60n\right) = -\frac{1}{90}\left(n^2 - 60n + 900\right) + 10 = -\frac{1}{90}(n - 30)^2 + 10$. Since the maximum of the function occurs when $n = 30$, the viewer should watch the commercial 30 times for maximum effectiveness.

57. $A(n) = n(900 - 9n) = -9n^2 + 900n$ is a quadratic function with $a = -9$ and $b = 900$, so by the formula, the maximum or minimum value occurs at $n = -\frac{b}{2a} = -\frac{900}{2(-9)} = 50$ trees, and because $a < 0$, this gives a maximum value.

59. The area of the fenced-in field is given by $A(x) = (2400 - 2x)x = -2x^2 + 2400x$. Thus, by the formula in this section, the maximum or minimum value occurs at $x = -\frac{b}{2a} = -\frac{2400}{2(-2)} = 600$. The maximum area occurs when $x = 600$ feet and $l = 2400 - 2(600) = 1200$ feet.

61. $A(x) = 15x - \frac{1}{8}(\pi + 4)x^2$, so by the formula, the maximum area occurs when $x = -\frac{b}{2a} = -\frac{-15}{2\left[\frac{1}{8}(\pi + 4)\right]} \approx 8.4$ ft and $h \approx 15 - \frac{1}{2}(8.40) - \frac{1}{4}\pi(8.40) \approx 4.2$ ft.

63. **(a)** The area of the corral is $A(x) = x(1200 - x) = 1200x - x^2 = -x^2 + 1200x$.

(b) A is a quadratic function with $a = -1$ and $b = 1200$, so by the formula, it has a maximum or minimum at $x = -\frac{b}{2a} = -\frac{1200}{2(-1)} = 600$, and because $a < 0$, this gives a maximum value. The desired dimensions are 600 ft by 600 ft.

65. **(a)** To model the revenue, we need to find the total attendance. Let x be the ticket price. Then the amount by which the ticket price is lowered is $10 - x$, and we are given that for every dollar it is lowered, the attendance increases by 3000; that is, the increase in attendance is $3000(10 - x)$. Thus, the attendance is $27{,}000 + 3000(10 - x)$, and since each spectator pays \$$x$, the revenue is $R(x) = x[27{,}000 + 3000(10 - x)] = -3000x^2 + 57{,}000x$.

(b) Since R is a quadratic function with $a = -3000$ and $b = 57{,}000$, the maximum occurs at $x = -\frac{b}{2a} = -\frac{57{,}000}{2(-3000)} = 9.5$; that is, when admission is \$9.50.

(c) We solve $R(x) = 0$ for x: $-3000x^2 + 57{,}000x = 0 \Leftrightarrow -3000x(x - 19) = 0 \Leftrightarrow x = 0$ or $x = 19$. Thus, if admission is \$19, nobody will attend and no revenue will be generated.

67. Because $f(x) = (x - m)(x - n) = 0$ when $x = m$ or $x = n$, those are its x-intercepts. By symmetry, we expect that the vertex is halfway between these values; that is, at $x = \frac{m+n}{2}$. We obtain the graph shown at right. Expanding, we see that $f(x) = x^2 - (m + n)x + mn$, a quadratic function with $a = 1$ and $b = -(m + n)$. Because $a > 0$, the minimum value occurs at $x = -\frac{b}{2a} = \frac{m+n}{2}$, the x-value of the vertex, as expected.

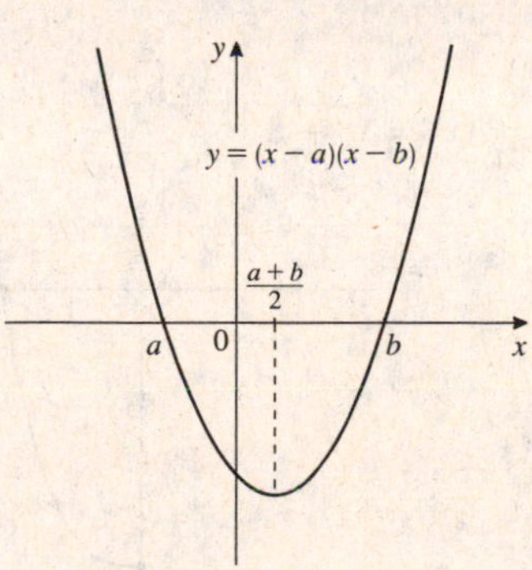

3.2 POLYNOMIAL FUNCTIONS AND THEIR GRAPHS

1. Graph I cannot be that of a polynomial because it is not smooth (it has a cusp.) Graph II could be that of a polynomial function, because it is smooth and continuous. Graph III could not be that of a polynomial function because it has a break. Graph IV could not be that of a polynomial function because it is not smooth.

3. (a) If c is a zero of the polynomial P, then $P(c) = 0$.

(b) If c is a zero of the polynomial P, then $x - c$ is a *factor* of $P(x)$.

(c) If c is a zero of the polynomial P, then c is an x-intercept of the graph of P.

5. (a) $P(x) = x^2 - 4$

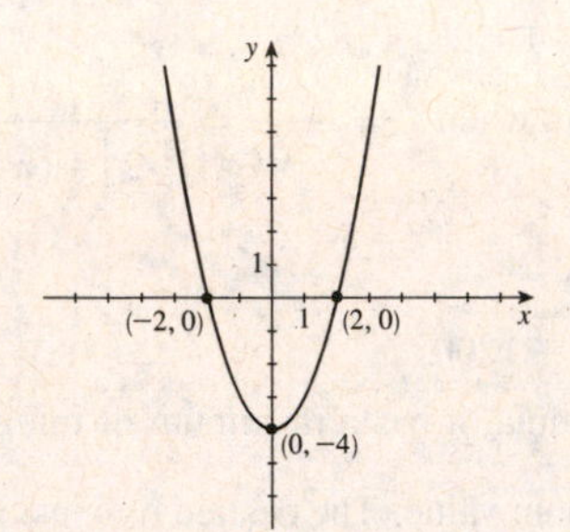

(b) $Q(x) = (x - 4)^2$

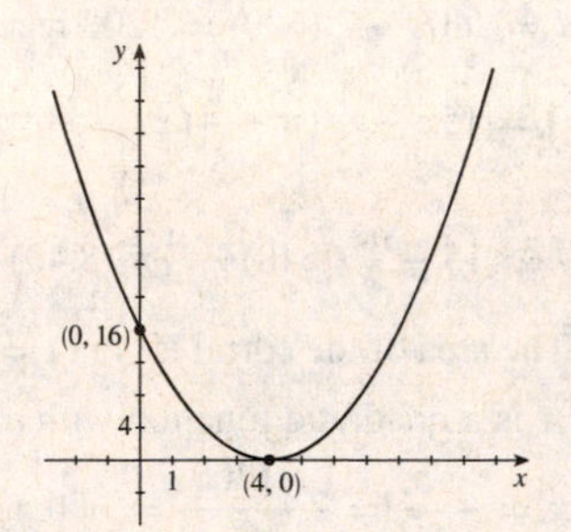

(c) $P(x) = 2x^2 + 3$

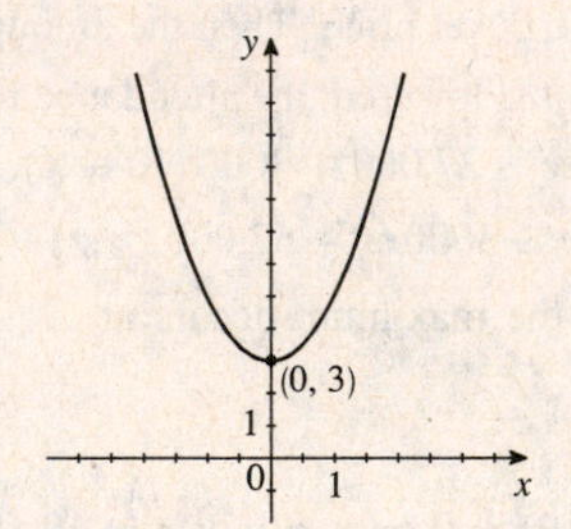

(d) $P(x) = -(x + 2)^2$

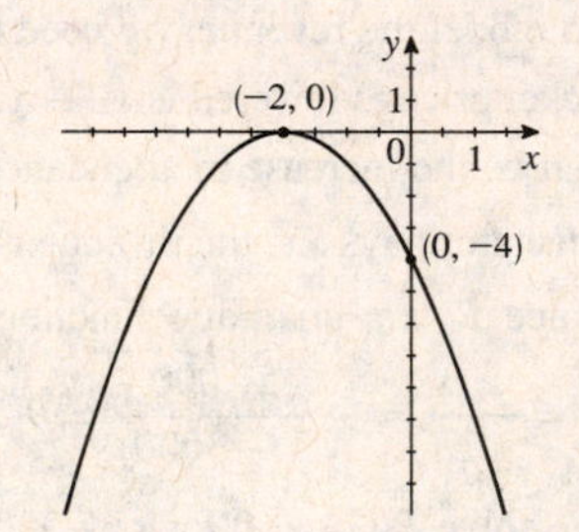

7. (a) $P(x) = x^3 - 8$

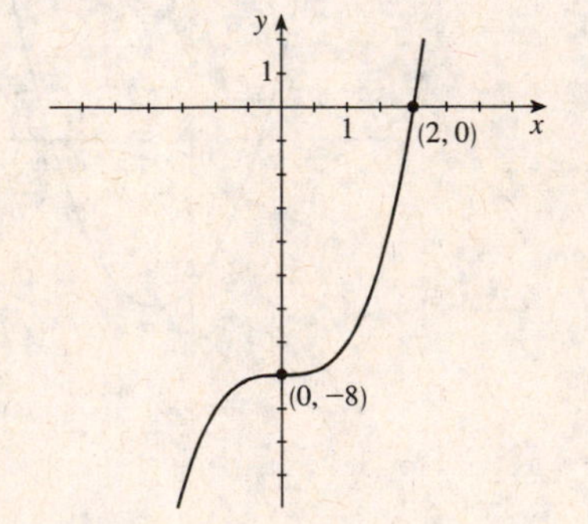

(b) $Q(x) = -x^3 + 27$

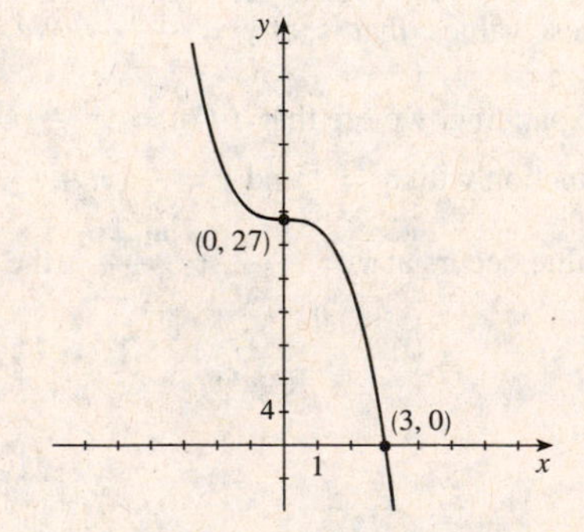

(c) $R(x) = -(x+2)^3$

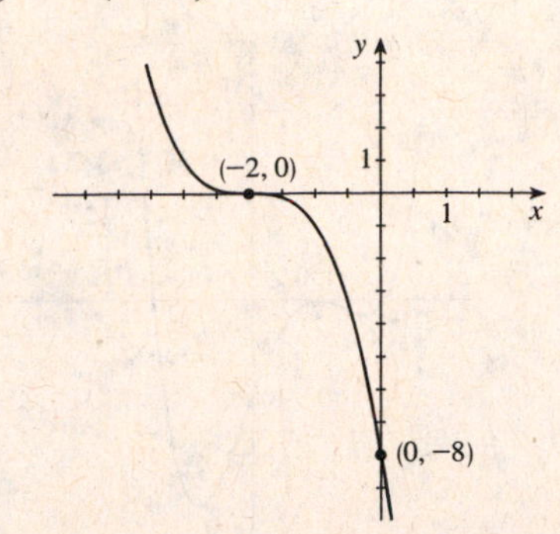

(d) $S(x) = \frac{1}{2}(x-1)^3 + 4$

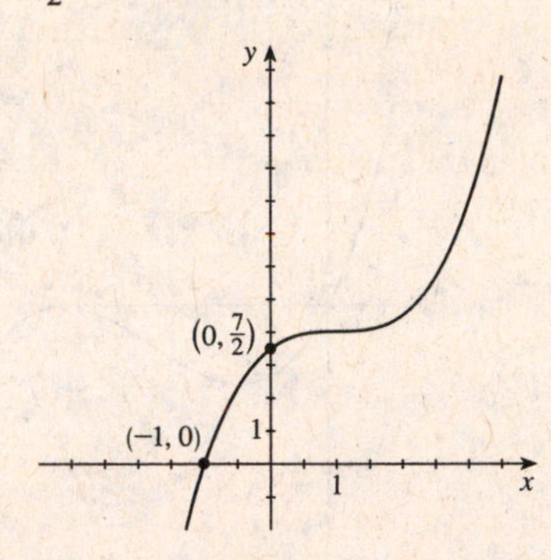

9. (a) $P(x) = x\left(x^2-4\right) = x^3 - 4x$ has odd degree and a positive leading coefficient, so $y \to \infty$ as $x \to \infty$ and $y \to -\infty$ as $x \to -\infty$.

(b) This corresponds to graph III.

11. (a) $R(x) = -x^5 + 5x^3 - 4x$ has odd degree and a negative leading coefficient, so $y \to -\infty$ as $x \to \infty$ and $y \to \infty$ as $x \to -\infty$.

(b) This corresponds to graph V.

13. (a) $T(x) = x^4 + 2x^3$ has even degree and a positive leading coefficient, so $y \to \infty$ as $x \to \infty$ and $y \to \infty$ as $x \to -\infty$.

(b) This corresponds to graph VI.

15. $P(x) = (x-1)(x+2)$

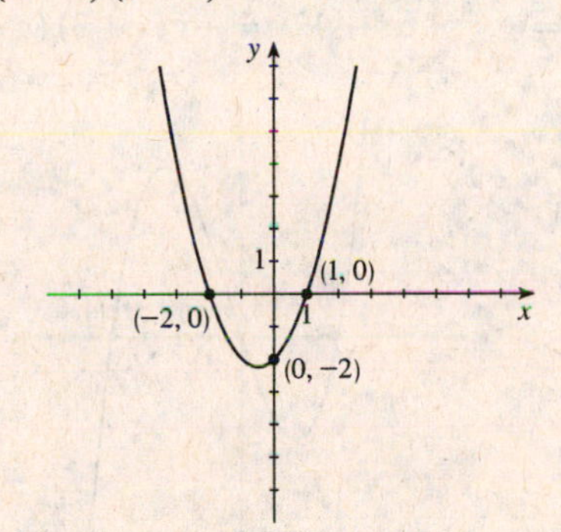

17. $P(x) = -x(x-3)(x+2)$

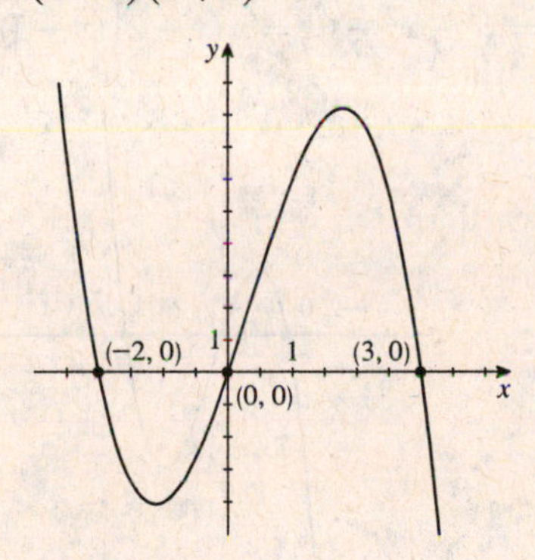

19. $P(x) = -(2x-1)(x+1)(x+3)$

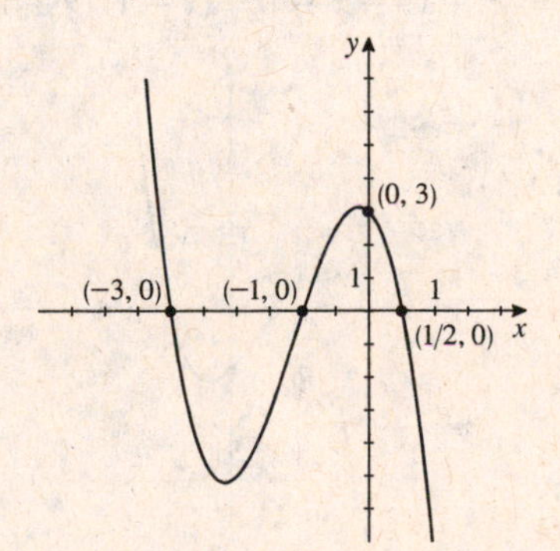

21. $P(x) = (x+2)(x+1)(x-2)(x-3)$

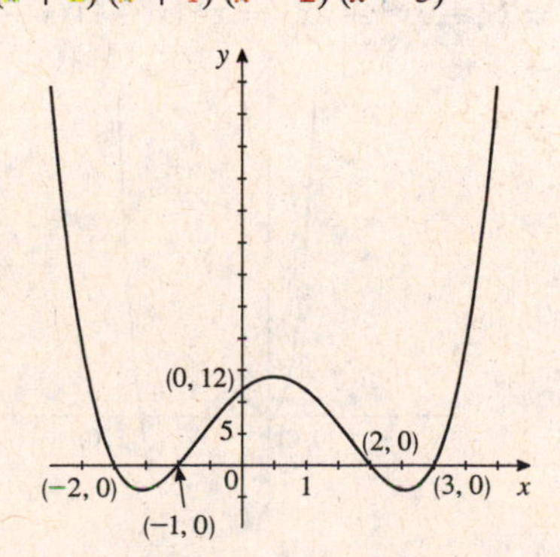

23. $P(x) = -2x(x-2)^2$

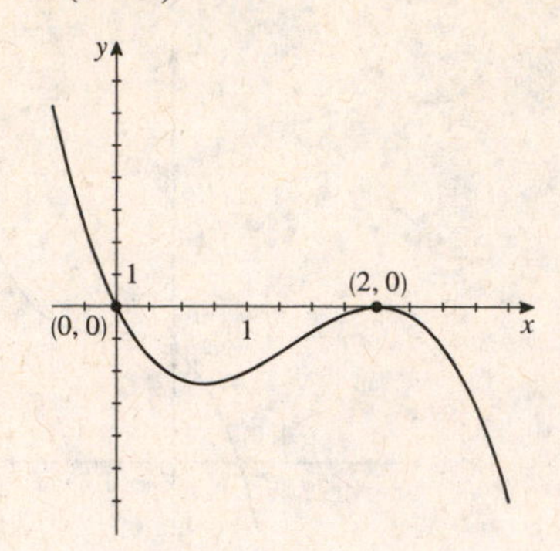

25. $P(x) = (x+1)(x+1)^2(2x-3)$

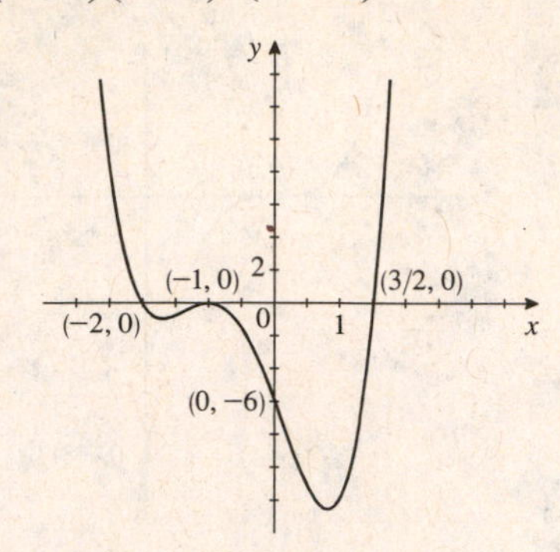

27. $P(x) = \frac{1}{12}(x+2)^2(x-3)^2$

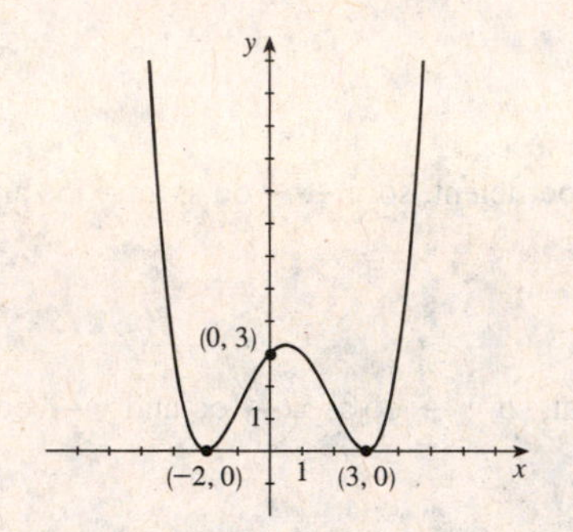

29. $P(x) = x^3(x+2)(x-3)^2$

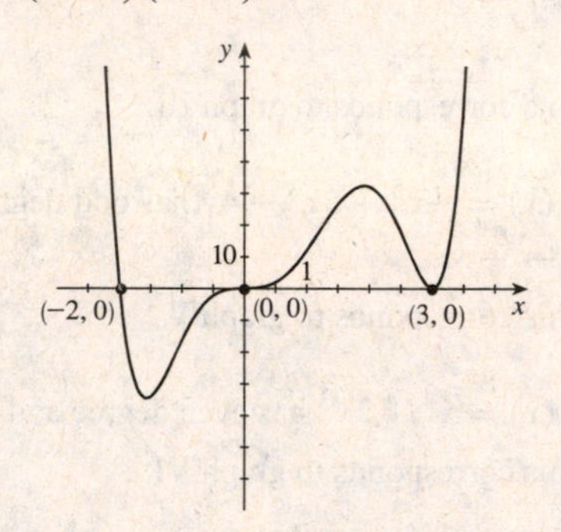

31. $P(x) = x^3 - x^2 - 6x = x(x+2)(x-3)$

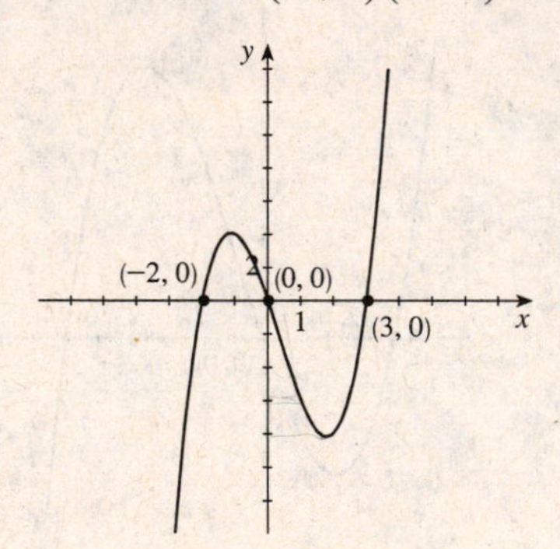

33. $P(x) = -x^3 + x^2 + 12x = -x(x+3)(x-4)$

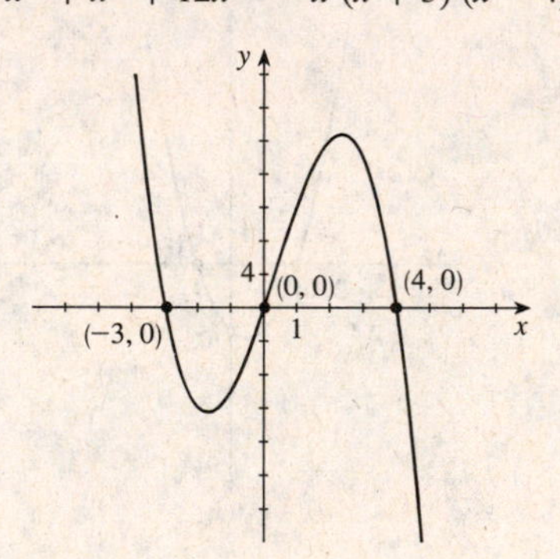

35. $P(x) = x^4 - 3x^3 + 2x^2 = x^2(x-1)(x-2)$

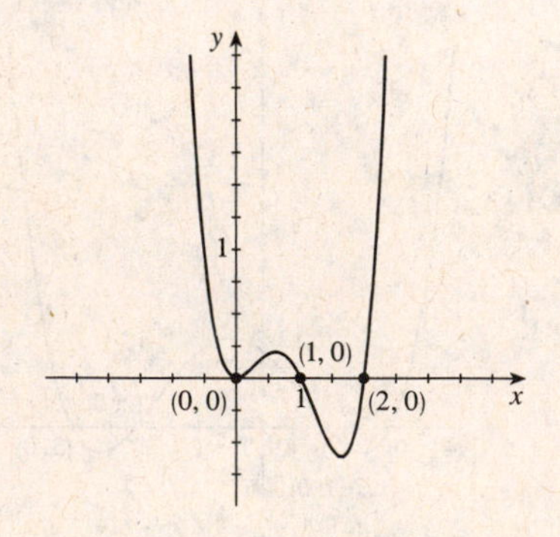

37. $P(x) = x^3 + x^2 - x - 1 = (x-1)(x+1)^2$

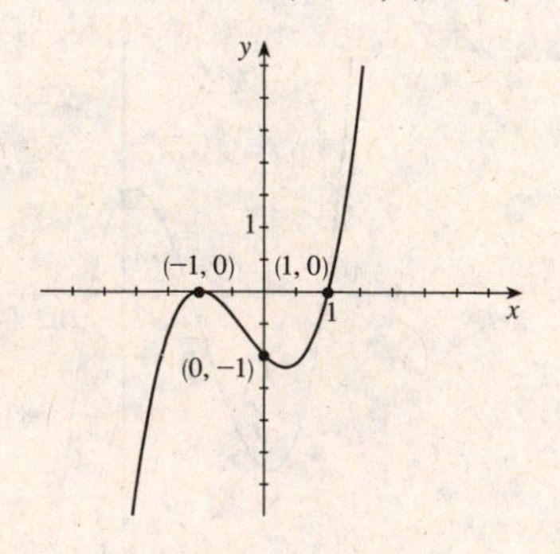

39. $P(x) = 2x^3 - x^2 - 18x + 9$
$= (x-3)(2x-1)(x+3)$

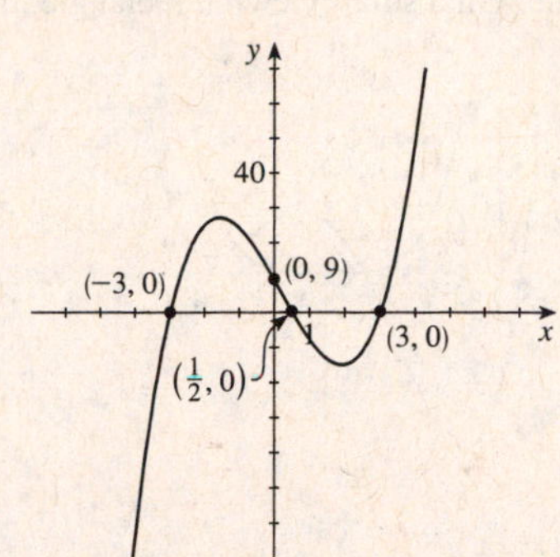

41. $P(x) = x^4 - 2x^3 - 8x + 16$
$= (x-2)^2\left(x^2 + 2x + 4\right)$

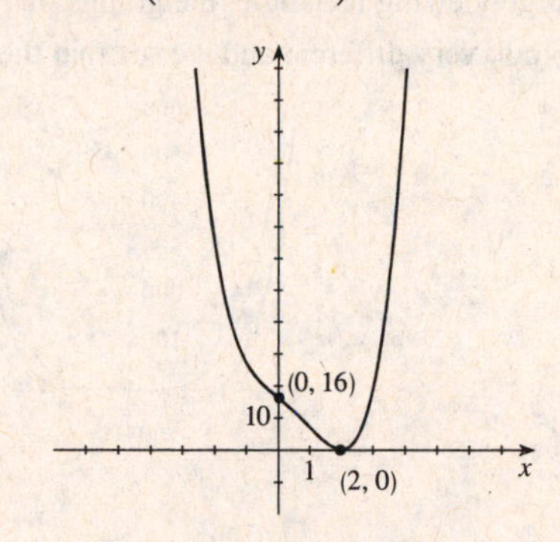

43. $P(x) = x^4 - 3x^2 - 4 = (x-2)(x+2)\left(x^2+1\right)$

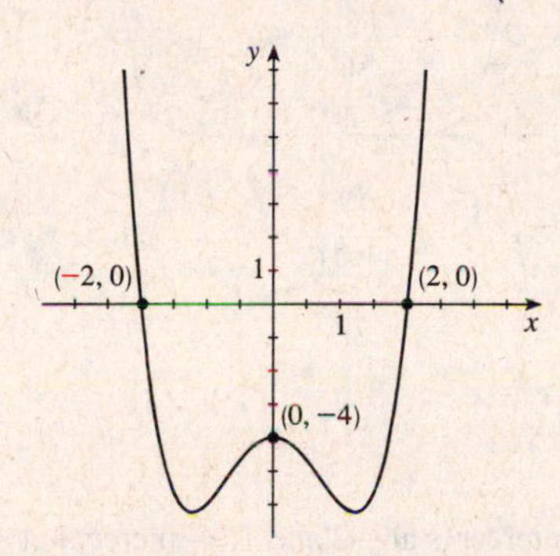

45. $P(x) = 3x^3 - x^2 + 5x + 1$; $Q(x) = 3x^3$. Since P has odd degree and positive leading coefficient, it has the following end behavior: $y \to \infty$ as $x \to \infty$ and $y \to -\infty$ as $x \to -\infty$.
On a large viewing rectangle, the graphs of P and Q look almost the same. On a small viewing rectangle, we see that the graphs of P and Q have different intercepts.

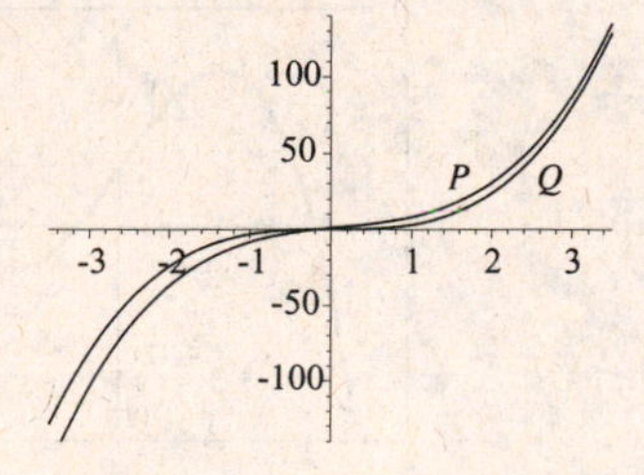

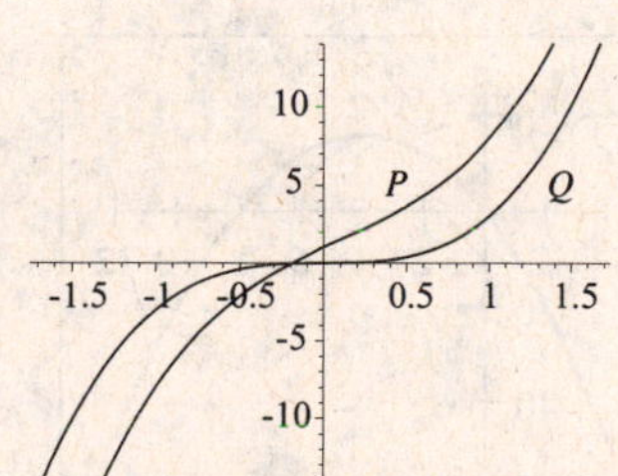

47. $P(x) = x^4 - 7x^2 + 5x + 5$; $Q(x) = x^4$. Since P has even degree and positive leading coefficient, it has the following end behavior: $y \to \infty$ as $x \to \infty$ and $y \to \infty$ as $x \to -\infty$.
On a large viewing rectangle, the graphs of P and Q look almost the same. On a small viewing rectangle, the graphs of P and Q look very different and we see that they have different intercepts.

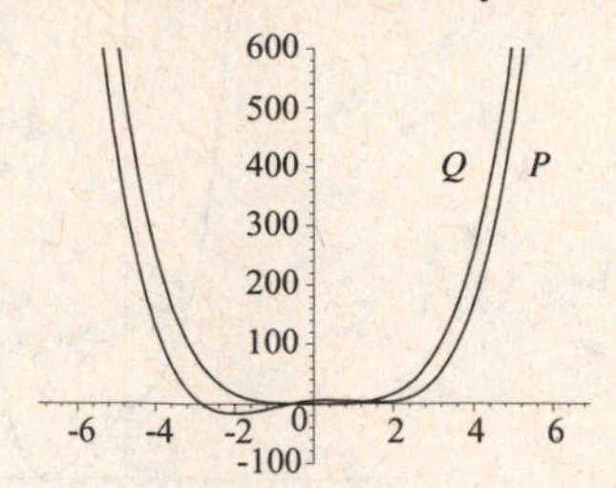

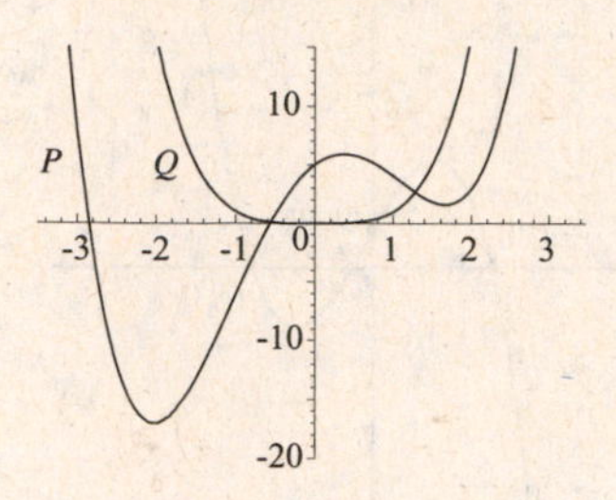

49. $P(x) = x^{11} - 9x^9$; $Q(x) = x^{11}$. Since P has odd degree and positive leading coefficient, it has the following end behavior: $y \to \infty$ as $x \to \infty$ and $y \to -\infty$ as $x \to -\infty$.
On a large viewing rectangle, the graphs of P and Q look like they have the same end behavior. On a small viewing rectangle, the graphs of P and Q look very different and seem (wrongly) to have different end behavior.

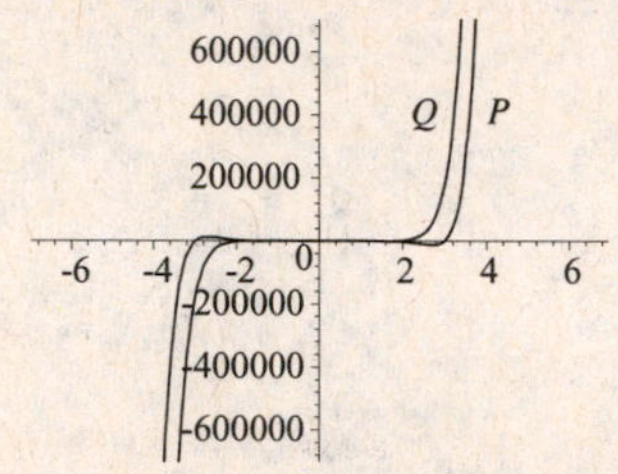

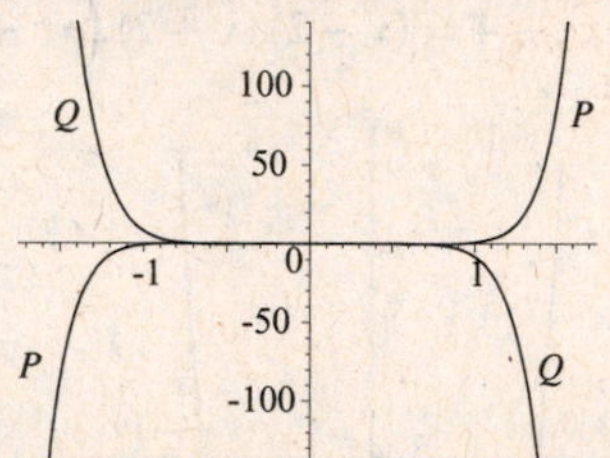

51. **(a)** x-intercepts at 0 and 4, y-intercept at 0.

(b) Local maximum at $(2, 4)$, no local minimum.

53. **(a)** x-intercepts at -2 and 1, y-intercept at -1.

(b) Local maximum at $(1, 0)$, local minimum at $(-1, -2)$.

55. $y = -x^2 + 8x$, $[-4, 12]$ by $[-50, 30]$
No local minimum. Local maximum at $(4, 16)$.
Domain: $(-\infty, \infty)$, range: $(-\infty, 16]$.

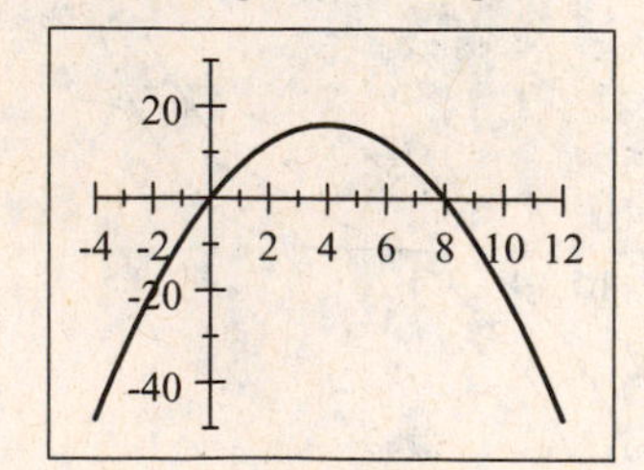

57. $y = x^3 - 12x + 9$, $[-5, 5]$ by $[-30, 30]$
Local maximum at $(-2, 25)$. Local minimum at $(2, -7)$.
Domain: $(-\infty, \infty)$, range: $(-\infty, \infty)$.

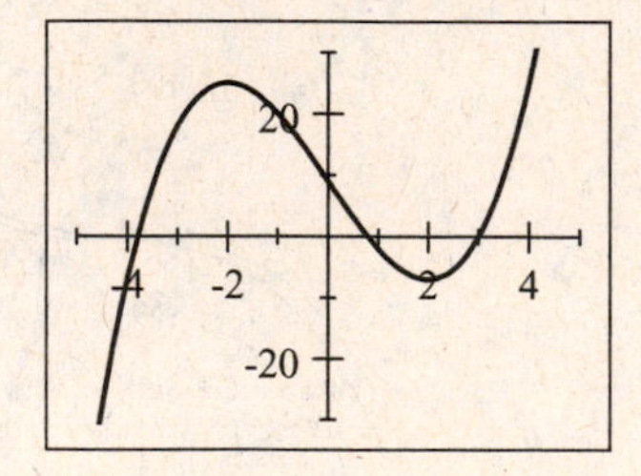

59. $y = x^4 + 4x^3$, $[-5, 5]$ by $[-30, 30]$

Local minimum at $(-3, -27)$. No local maximum.

Domain: $(-\infty, \infty)$, range: $[-27, \infty)$.

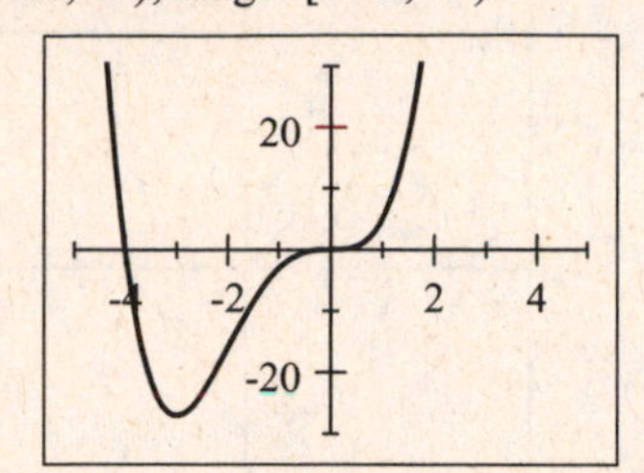

61. $y = 3x^5 - 5x^3 + 3$, $[-3, 3]$ by $[-5, 10]$

Local maximum at $(-1, 5)$. Local minimum at $(1, 1)$.

Domain: $(-\infty, \infty)$, range: $(-\infty, \infty)$.

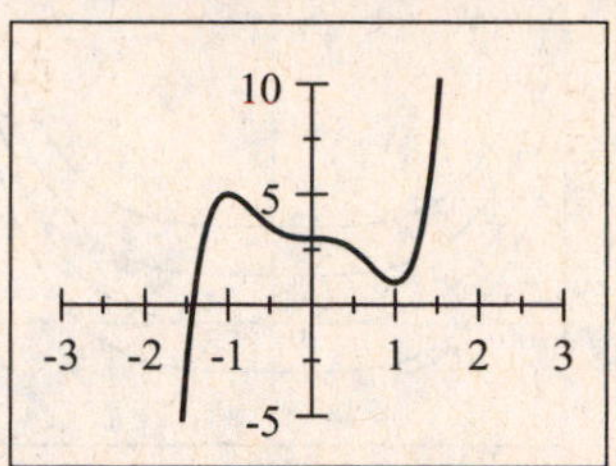

63. $y = -2x^2 + 3x + 5$ has one local maximum at $(0.75, 6.13)$.

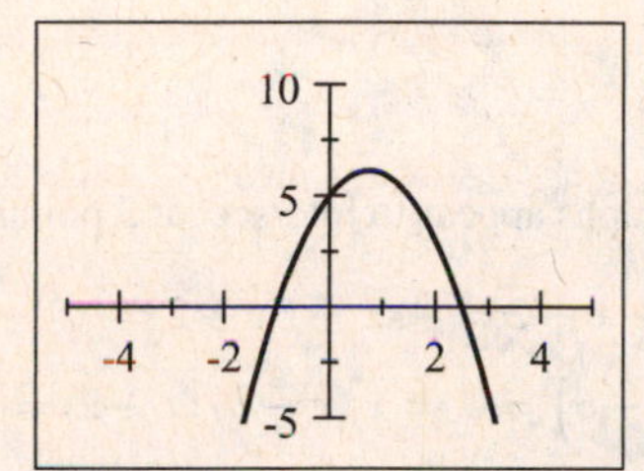

65. $y = x^3 - x^2 - x$ has one local maximum at $(-0.33, 0.19)$ and one local minimum at $(1.00, -1.00)$.

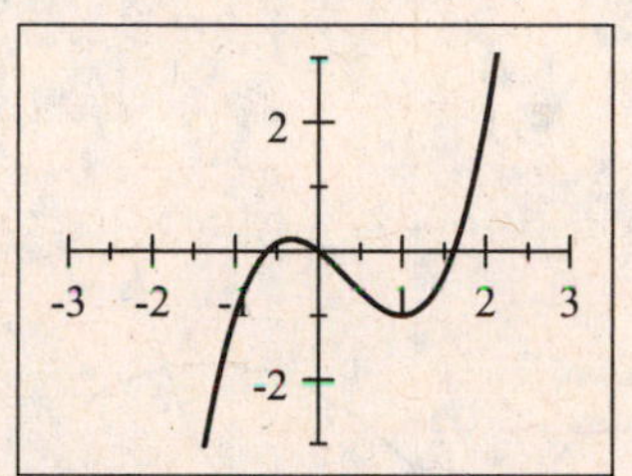

67. $y = x^4 - 5x^2 + 4$ has one local maximum at $(0, 4)$ and two local minima at $(-1.58, -2.25)$ and $(1.58, -2.25)$.

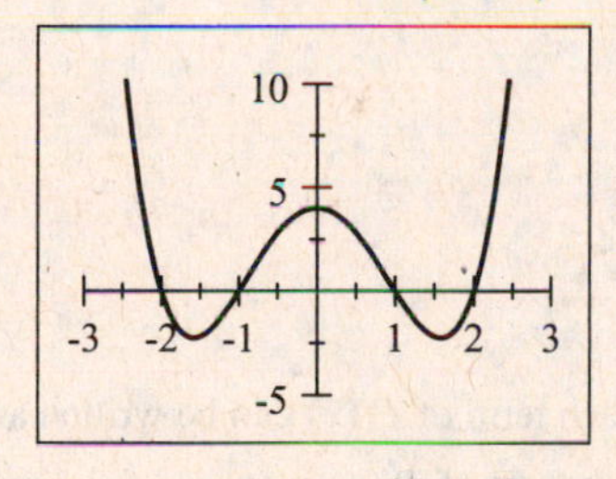

69. $y = (x - 2)^5 + 32$ has no maximum or minimum.

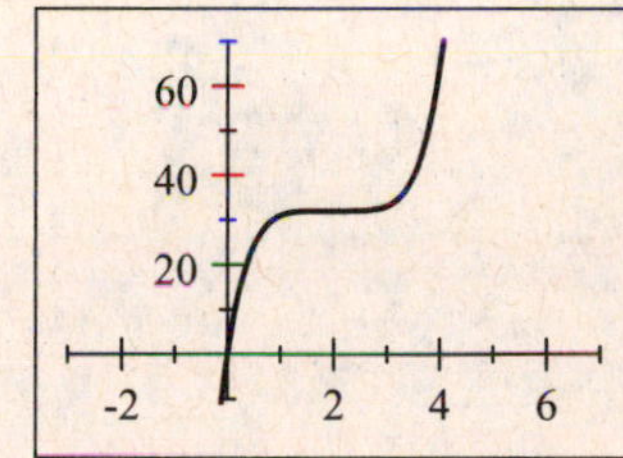

71. $y = x^8 - 3x^4 + x$ has one local maximum at $(0.44, 0.33)$ and two local minima at $(1.09, -1.15)$ and $(-1.12, -3.36)$.

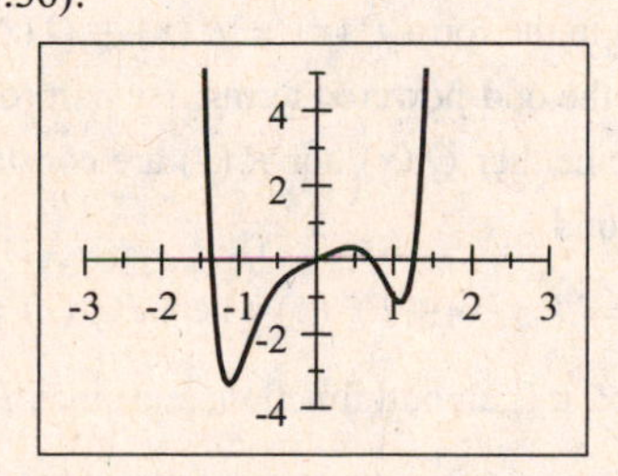

73. $y = cx^3$; $c = 1, 2, 5, \frac{1}{2}$. Increasing the value of c stretches the graph vertically.

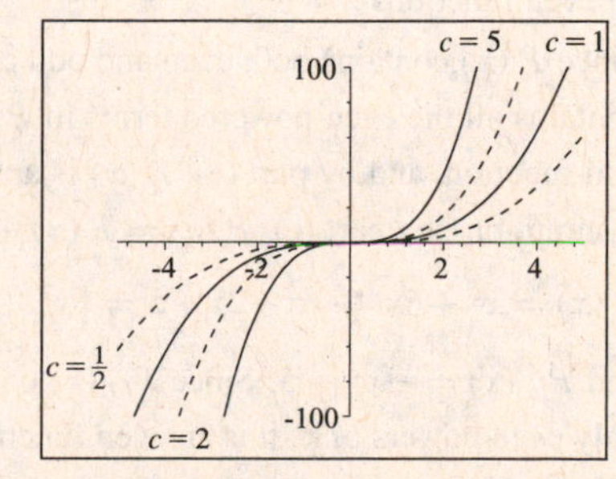

75. $P(x) = x^4 + c$; $c = -1, 0, 1$, and 2. Increasing the value of c moves the graph up.

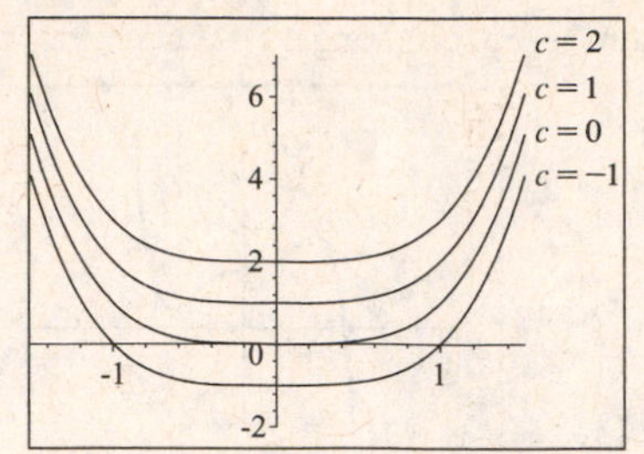

77. $P(x) = x^4 - cx$; $c = 0, 1, 8$, and 27. Increasing the value of c causes a deeper dip in the graph, in the fourth quadrant, and moves the positive x-intercept to the right.

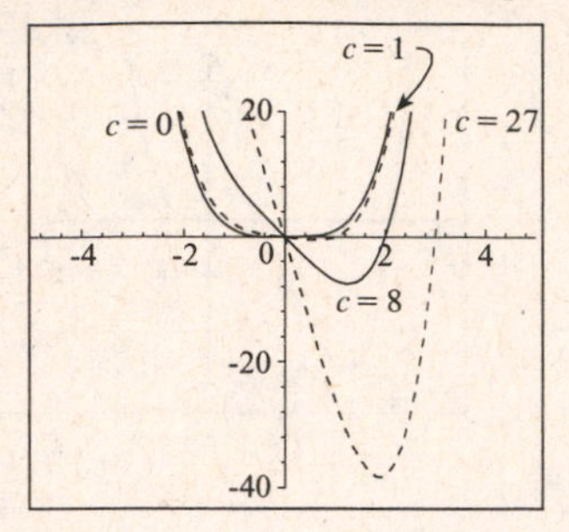

79. **(a)**

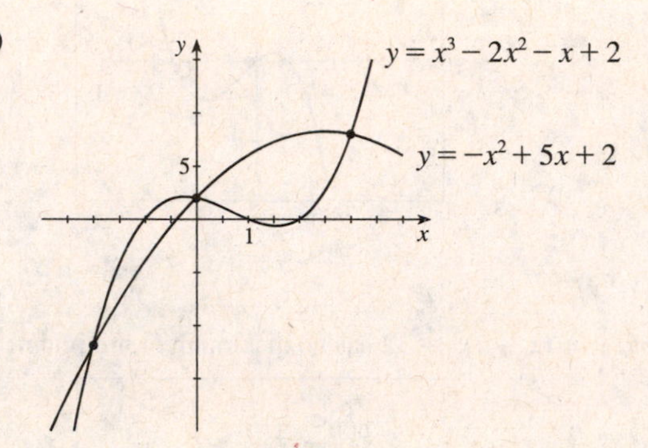

(b) The two graphs appear to intersect at 3 points.

(c) $x^3 - 2x^2 - x + 2 = -x^2 + 5x + 2 \Leftrightarrow x^3 - x^2 - 6x = 0 \Leftrightarrow x\left(x^2 - x - 6\right) = 0 \Leftrightarrow x(x-3)(x+2) = 0$. Then either $x = 0$, $x = 3$, or $x = -2$. If $x = 0$, then $y = 2$; if $x = 3$ then $y = 8$; if $x = -2$, then $y = -12$. Hence the points where the two graphs intersect are $(0, 2)$, $(3, 8)$, and $(-2, -12)$.

81. **(a)** Let $P(x)$ be a polynomial containing only odd powers of x. Then each term of $P(x)$ can be written as Cx^{2n+1}, for some constant C and integer n. Since $C(-x)^{2n+1} = -Cx^{2n+1}$, each term of $P(x)$ is an odd function. Thus by part (a), $P(x)$ is an odd function.

(b) Let $P(x)$ be a polynomial containing only even powers of x. Then each term of $P(x)$ can be written as Cx^{2n}, for some constant C and integer n. Since $C(-x)^{2n} = Cx^{2n}$, each term of $P(x)$ is an even function. Thus by part (b), $P(x)$ is an even function.

(c) Since $P(x)$ contains both even and odd powers of x, we can write it in the form $P(x) = R(x) + Q(x)$, where $R(x)$ contains all the even-powered terms in $P(x)$ and $Q(x)$ contains all the odd-powered terms. By part (d), $Q(x)$ is an odd function, and by part (e), $R(x)$ is an even function. Thus, since neither $Q(x)$ nor $R(x)$ are constantly 0 (by assumption), by part (c), $P(x) = R(x) + Q(x)$ is neither even nor odd.

(d) $P(x) = x^5 + 6x^3 - x^2 - 2x + 5 = \left(x^5 + 6x^3 - 2x\right) + \left(-x^2 + 5\right) = P_O(x) + P_E(x)$ where $P_O(x) = x^5 + 6x^3 - 2x$ and $P_E(x) = -x^2 + 5$. Since $P_O(x)$ contains only odd powers of x, it is an odd function, and since $P_E(x)$ contains only even powers of x, it is an even function.

83. **(a)** $P(x) = (x-1)(x-3)(x-4)$.
Local maximum at $(1.8, 2.1)$.
Local minimum at $(3.6, -0.6)$.

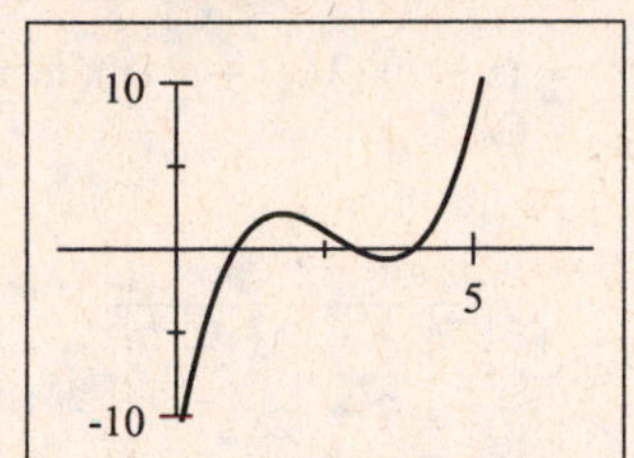

(b) Since $Q(x) = P(x) + 5$, each point on the graph of Q has y-coordinate 5 units more than the corresponding point on the graph of P. Thus Q has a local maximum at $(1.8, 7.1)$ and a local minimum at $(3.5, 4.4)$.

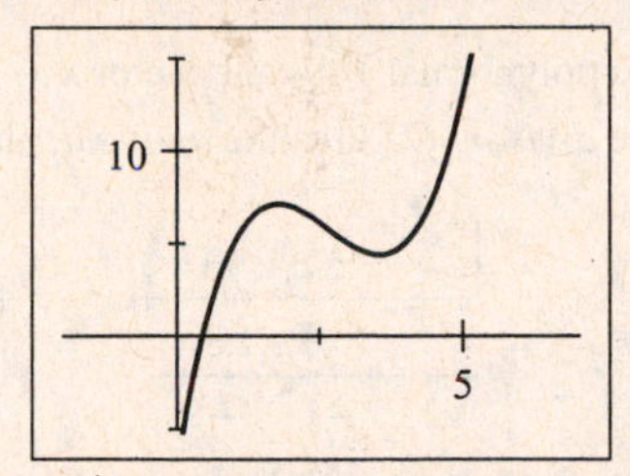

85. Since the polynomial shown has five zeros, it has at least five factors, and so the degree of the polynomial is greater than or equal to 5.

87. $P(x) = 8x + 0.3x^2 - 0.0013x^3 - 372$

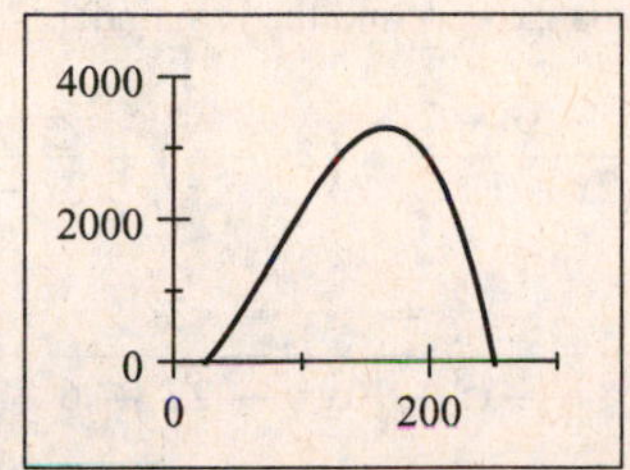

(a) For the firm to break even, $P(x) = 0$. From the graph, we see that $P(x) = 0$ when $x \approx 25.2$. Of course, the firm cannot produce fractions of a blender, so the manufacturer must produce at least 26 blenders a year.

(b) No, the profit does not increase indefinitely. The largest profit is approximately \$ 3276.22, which occurs when the firm produces 166 blenders per year.

89. **(a)** The length of the bottom is $40 - 2x$, the width of the bottom is $20 - 2x$, and the height is x, so the volume of the box is $V = x(20-2x)(40-2x) = 4x^3 - 120x^2 + 800x$.

(b) Since the height and width must be positive, we must have $x > 0$ and $20 - 2x > 0$, and so the domain of V is $0 < x < 10$.

(c) Using the domain from part (b), we graph V in the viewing rectangle $[0, 10]$ by $[0, 1600]$. The maximum volume is $V \approx 1539.6$ when $x = 4.23$.

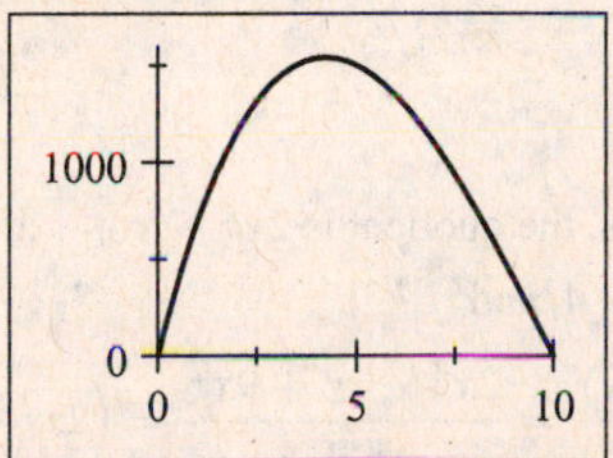

91.

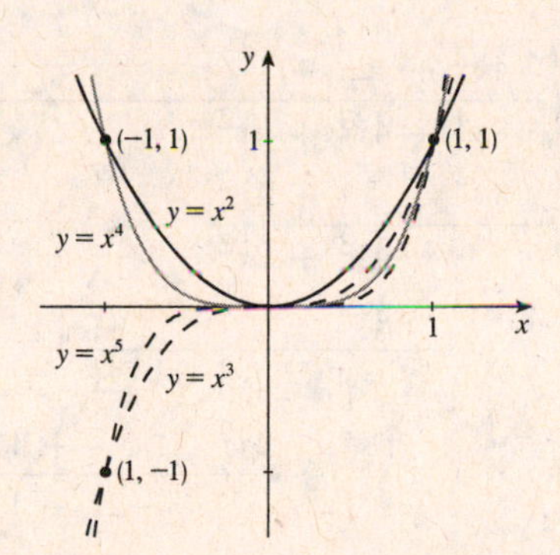

The graph of $y = x^{100}$ is close to the x-axis for $|x| < 1$, but passes through the points $(1, 1)$ and $(-1, 1)$. The graph of $y = x^{101}$ behaves similarly except that the y-values are negative for negative values of x, and it passes through $(-1, -1)$ instead of $(-1, 1)$.

3.3 DIVIDING POLYNOMIALS

1. If we divide the polynomial P by the factor $x - c$, and we obtain the equation $P(x) = (x - c)\,Q(x) + R(x)$, then we say that $x - c$ is the *divisor*, $Q(x)$ is the *quotient*, and $R(x)$ is the *remainder*.

3.

$$\begin{array}{r|rrr} 2 & 2 & -5 & -7 \\ & & 4 & -2 \\ \hline & 2 & -1 & -9 \end{array}$$

Thus, the quotient is $2x - 1$ and the remainder is -9, and $\dfrac{P(x)}{D(x)} = \dfrac{2x^2 - 5x - 7}{x - 2} = (2x - 1) + \dfrac{-9}{x - 2}$.

5.

$$\begin{array}{r|l} & 2x - \tfrac{1}{2} \\ \hline 2x - 1 & 4x^2 - 3x - 7 \\ & 4x^2 - 2x \\ \hline & \quad -x - 7 \\ & \quad -x + \tfrac{1}{2} \\ \hline & \qquad -\tfrac{15}{2} \end{array}$$

Thus, the quotient is $2x - \frac{1}{2}$ and the remainder is $-\frac{15}{2}$, and $\dfrac{P(x)}{D(x)} = \dfrac{4x^2 - 3x - 7}{2x - 1} = \left(2x - \frac{1}{2}\right) + \dfrac{-\frac{15}{2}}{2x - 1}$.

7.

$$\begin{array}{r|l} & 2x^2 - x + 1 \\ \hline x^2 + 4 & 2x^4 - x^3 + 9x^2 \\ & 2x^4 \qquad + 8x^2 \\ \hline & \quad -x^3 + x^2 \\ & \quad -x^3 \qquad - 4x \\ \hline & \qquad x^2 + 4x \\ & \qquad x^2 \qquad + 4 \\ \hline & \qquad\quad 4x - 4 \end{array}$$

Thus, the quotient is $2x^2 - x + 1$ and the remainder is $4x - 4$, and

$$\frac{P(x)}{D(x)} = \frac{2x^4 - x^3 + 9x^2}{x^2 + 4} = \left(2x^2 - x + 1\right) + \frac{4x - 4}{x^2 + 4}.$$

9.

$$\begin{array}{r|l} & -x^2 + x - 3 \\ \hline x + 1 & -x^3 + 0x^2 - 2x + 6 \\ & -x^3 - x^2 \\ \hline & \qquad x^2 - 2x \\ & \qquad x^2 + x \\ \hline & \qquad\quad -3x + 6 \\ & \qquad\quad -3x - 3 \\ \hline & \qquad\qquad\quad 9 \end{array}$$

Thus, the quotient is $-x^2 + x - 3$ and the remainder is 9, so $P(x) = -x^3 - 2x + 6 = (x + 1) \cdot \left(-x^2 + x - 3\right) + 9$.

11.

$$\begin{array}{r|l} & x^2 \qquad - 1 \\ \hline 2x - 3 & 2x^3 - 3x^2 - 2x \\ & 2x^3 - 3x^2 \\ \hline & \qquad\quad -2x \\ & \qquad\quad -2x + 3 \\ \hline & \qquad\qquad\quad -3 \end{array}$$

Thus, the quotient is $x^2 - 1$ and the remainder is -3, and $P(x) = 2x^3 - 3x^2 - 2x = (2x - 3)\left(x^2 - 1\right) - 3$.

13.

$$\begin{array}{r|l} & 4x^2 + 2x + 1 \\ \hline 2x^2 + 1 & 8x^4 + 4x^3 + 6x^2 \\ & 8x^4 \qquad + 4x^2 \\ \hline & \quad 4x^3 + 2x^2 \\ & \quad 4x^3 \qquad + 2x \\ \hline & \qquad 2x^2 - 2x \\ & \qquad 2x^2 \qquad + 1 \\ \hline & \qquad\quad -2x - 1 \end{array}$$

Thus, the quotient is $4x^2 + 2x + 1$ and the remainder is $-2x - 1$, and

$$\begin{aligned} P(x) &= 8x^4 + 4x^3 + 6x^2 \\ &= \left(2x^2 + 1\right) \cdot \left(4x^2 + 2x + 1\right) + (-2x - 1) \end{aligned}$$

15.

$$\begin{array}{rr} & x - 1 \\ x-2 & \overline{)\,x^2 - 3x + 7} \\ & \underline{x^2 - 2x } \\ & -x + 7 \\ & \underline{-x + 2} \\ & 5 \end{array}$$

Thus, the quotient is $x - 1$ and the remainder is 5.

17.

$$\begin{array}{rr} & 2x^2 - 1 \\ 2x+1 & \overline{)\,4x^3 + 2x^2 - 2x - 3} \\ & \underline{4x^3 + 2x^2 } \\ & -2x - 3 \\ & \underline{-2x - 1} \\ & -2 \end{array}$$

Thus, the quotient is $2x^2 - 1$ and the remainder is -2.

19.

$$\begin{array}{rr} & x + 1 \\ x^2-x+3 & \overline{)\,x^3 + 0x^2 + 2x + 1} \\ & \underline{x^3 - x^2 + 3x } \\ & x^2 - x + 1 \\ & \underline{x^2 - x + 3} \\ & -2 \end{array}$$

Thus, the quotient is $x + 1$ and the remainder is -2.

21.

$$\begin{array}{rr} & 3x + 1 \\ 2x^2+0x+5 & \overline{)\,6x^3 + 2x^2 + 22x + 0} \\ & \underline{6x^3 + 15x } \\ & 2x^2 + 7x + 0 \\ & \underline{2x^2 + 5} \\ & 7x - 5 \end{array}$$

Thus, the quotient is $3x + 1$ and the remainder is $7x - 5$.

23.

$$\begin{array}{rr} & x^4 + 1 \\ x^2+1 & \overline{)\,x^6 + 0x^5 + x^4 + 0x^3 + x^2 + 0x + 1} \\ & \underline{x^6 + x^4 } \\ & 0 + x^2 + 1 \\ & \underline{x^2 + 1} \\ & 0 \end{array}$$

Thus, the quotient is $x^4 + 1$ and the remainder is 0.

25. The synthetic division table for this problem takes the following form.

$$\begin{array}{r|rrr} 3 & 2 & -5 & 3 \\ & & 6 & 3 \\ \hline & 2 & 1 & 6 \end{array}$$

Thus, the quotient is $2x + 1$ and the remainder is 6.

27. The synthetic division table for this problem takes the following form.

$$\begin{array}{r|rrr} -1 & 3 & 1 & 0 \\ & & -3 & 2 \\ \hline & 3 & -2 & 2 \end{array}$$

Thus, the quotient is $3x - 2$ and the remainder is 2.

29. Since $x + 2 = x - (-2)$, the synthetic division table for this problem takes the following form.

$$\begin{array}{r|rrrr} -2 & 1 & 2 & 2 & 1 \\ & & -2 & 0 & -4 \\ \hline & 1 & 0 & 2 & -3 \end{array}$$

Thus, the quotient is $x^2 + 2$ and the remainder is -3.

31. Since $x + 3 = x - (-3)$ and $x^3 - 8x + 2 = x^3 + 0x^2 - 8x + 2$, the synthetic division table for this problem takes the following form.

$$\begin{array}{r|rrrr} -3 & 1 & 0 & -8 & 2 \\ & & -3 & 9 & -3 \\ \hline & 1 & -3 & 1 & -1 \end{array}$$

Thus, the quotient is $x^2 - 3x + 1$ and the remainder is -1.

33. Since $x^5 + 3x^3 - 6 = x^5 + 0x^4 + 3x^3 + 0x^2 + 0x - 6$, the synthetic division table for this problem takes the following form.

$$\begin{array}{r|rrrrrr} 1 & 1 & 0 & 3 & 0 & 0 & -6 \\ & & 1 & 1 & 4 & 4 & 4 \\ \hline & 1 & 1 & 4 & 4 & 4 & -2 \end{array}$$

Thus, the quotient is $x^4 + x^3 + 4x^2 + 4x + 4$ and the remainder is -2.

35. The synthetic division table for this problem takes the following form.

$$\begin{array}{r|rrrr} \frac{1}{2} & 2 & 3 & -2 & 1 \\ & & 1 & 2 & 0 \\ \hline & 2 & 4 & 0 & 1 \end{array}$$

Thus, the quotient is $2x^2 + 4x$ and the remainder is 1.

37. Since $x^3 - 27 = x^3 + 0x^2 + 0x - 27$, the synthetic division table for this problem takes the following form.

$$\begin{array}{r|rrrr} 3 & 1 & 0 & 0 & -27 \\ & & 3 & 9 & 27 \\ \hline & 1 & 3 & 9 & 0 \end{array}$$

Thus, the quotient is $x^2 + 3x + 9$ and the remainder is 0.

39. $P(x) = 4x^2 + 12x + 5, c = -1$

$$\begin{array}{r|rrr} -1 & 4 & 12 & 5 \\ & & -4 & -8 \\ \hline & 4 & 8 & -3 \end{array}$$

Therefore, by the Remainder Theorem, $P(-1) = -3$.

41. $P(x) = x^3 + 3x^2 - 7x + 6, c = 2$

$$\begin{array}{r|rrrr} 2 & 1 & 3 & -7 & 6 \\ & & 2 & 10 & 6 \\ \hline & 1 & 5 & 3 & 12 \end{array}$$

Therefore, by the Remainder Theorem, $P(2) = 12$.

43. $P(x) = x^3 + 2x^2 - 7, c = -2$

$$\begin{array}{r|rrrr} -2 & 1 & 2 & 0 & -7 \\ & & -2 & 0 & 0 \\ \hline & 1 & 0 & 0 & -7 \end{array}$$

Therefore, by the Remainder Theorem, $P(-2) = -7$.

45. $P(x) = 5x^4 + 30x^3 - 40x^2 + 36x + 14, c = -7$

$$\begin{array}{r|rrrrr} -7 & 5 & 30 & -40 & 36 & 14 \\ & & -35 & 35 & 35 & -497 \\ \hline & 5 & -5 & -5 & 71 & -483 \end{array}$$

Therefore, by the Remainder Theorem, $P(-7) = -483$.

47. $P(x) = x^7 - 3x^2 - 1$

$= x^7 + 0x^6 + 0x^5 + 0x^4 + 0x^3 - 3x^2 + 0x - 1$

$c = 3$

$$\begin{array}{r|rrrrrrrr} 3 & 1 & 0 & 0 & 0 & 0 & -3 & 0 & -1 \\ & & 3 & 9 & 27 & 81 & 243 & 720 & 2160 \\ \hline & 1 & 3 & 9 & 27 & 81 & 240 & 720 & 2159 \end{array}$$

Therefore by the Remainder Theorem, $P(3) = 2159$.

49. $P(x) = 3x^3 + 4x^2 - 2x + 1, c = \frac{2}{3}$

$$\begin{array}{r|rrrr} \frac{2}{3} & 3 & 4 & -2 & 1 \\ & & 2 & 4 & \frac{4}{3} \\ \hline & 3 & 6 & 2 & \frac{7}{3} \end{array}$$

Therefore, by the Remainder Theorem, $P\left(\frac{2}{3}\right) = \frac{7}{3}$.

51. $P(x) = x^3 + 2x^2 - 3x - 8, c = 0.1$

$$\begin{array}{r|rrrr} 0.1 & 1 & 2 & -3 & -8 \\ & & 0.1 & 0.21 & -0.279 \\ \hline & 1 & 2.1 & -2.79 & -8.279 \end{array}$$

Therefore, by the Remainder Theorem, $P(0.1) = -8.279$.

53. $P(x) = x^3 - 3x^2 + 3x - 1, c = 1$

$$\begin{array}{r|rrrr} 1 & 1 & -3 & 3 & -1 \\ & & 1 & -2 & 1 \\ \hline & 1 & -2 & 1 & 0 \end{array}$$

Since the remainder is 0, $x - 1$ is a factor.

55. $P(x) = 2x^3 + 7x^2 + 6x - 5, c = \frac{1}{2}$

$$\begin{array}{r|rrrr} \frac{1}{2} & 2 & 7 & 6 & -5 \\ & & 1 & 4 & 5 \\ \hline & 2 & 8 & 10 & 0 \end{array}$$

Since the remainder is 0, $x - \frac{1}{2}$ is a factor.

57. $P(x) = x^3 + 2x^2 - 9x - 18, c = -2$

$$\begin{array}{r|rrrr} -2 & 1 & 2 & -9 & -18 \\ & & -2 & 0 & 18 \\ \hline & 1 & 0 & -9 & 0 \end{array}$$

Since the remainder is 0, we know that -2 is a zero and $P(x) = (x+2)\left(x^2 - 9\right) = (x+2)(x+3)(x-3)$.

Hence, the zeros are -3, -2, and 3. .

59. $x^3 - x^2 - 11x + 15, c = 3$

$$\begin{array}{r|rrrr} 3 & 1 & -1 & -11 & 15 \\ & & 3 & 6 & -15 \\ \hline & 1 & 2 & -5 & 0 \end{array}$$

Since the remainder is 0, we know that 3 is a zero and $P(x) = (x-3)\left(x^2 + 2x - 5\right)$. Now $x^2 + 2x - 5 = 0$ when $x = \dfrac{-2 \pm \sqrt{2^2 - 4(1)(-5)}}{2(1)} = -1 \pm \sqrt{6}$. Hence, the zeros are 3 and $-1 \pm \sqrt{6}$.

61. $P(x) = 3x^4 - 8x^3 - 14x^2 + 31x + 6, c = -2, 3$

$$\begin{array}{r|rrrrr} -2 & 3 & -8 & -14 & 31 & 6 \\ & & -6 & 28 & 28 & -6 \\ \hline & 3 & -14 & 14 & 3 & 0 \end{array}$$

Since the remainder is 0, we know that -2 is a zero and $P(x) = (x+2)\left(3x^3 - 14x^2 + 14x + 3\right)$.

$$\begin{array}{r|rrrr} 3 & 3 & -14 & 14 & 3 \\ & & 9 & -15 & -3 \\ \hline & 3 & -5 & -1 & 0 \end{array}$$

Since the remainder is 0, we know that 3 is a zero and $P(x) = (x+2)(x-3)\left(3x^2 - 5x - 1\right)$. Now $3x^2 - 5x - 1 = 0$ when

$$x = \frac{-(-5) \pm \sqrt{(-5)^2 - 4(3)(-1)}}{2(3)} = \frac{5 \pm \sqrt{37}}{6}, \text{ and}$$

hence, the zeros are -2, 3, and $\dfrac{5 \pm \sqrt{37}}{6}$.

63. Since the zeros are $x = -1$, $x = 1$, and $x = 3$, the factors are $x + 1$, $x - 1$, and $x - 3$. Thus $P(x) = (x+1)(x-1)(x-3) = x^3 - 3x^2 - x + 3$.

65. Since the zeros are $x = -1$, $x = 1$, $x = 3$, and $x = 5$, the factors are $x + 1$, $x - 1$, $x - 3$, and $x - 5$. Thus $P(x) = (x+1)(x-1)(x-3)(x-5) = x^4 - 8x^3 + 14x^2 + 8x - 15$.

67. Since the zeros of the polynomial are -2, 0, 1, and 3, it follows that $P(x) = C(x+2)(x)(x-1)(x-3) = Cx^4 - 2Cx^3 - 5Cx^2 + 6Cx$. Since the coefficient of x^3 is to be 4, $-2C = 4$, so $C = -2$. Therefore, $P(x) = -2x^4 + 4x^3 + 10x^2 - 12x$ is the polynomial.

69. Since the polynomial degree 4 and zeros -1, 1, and $\sqrt{2}$ and integer coefficients, the fourth zero must be $-\sqrt{2}$, otherwise the constant term would be irrational. Thus, $P(x) = C(x+1)(x-1)\left(x - \sqrt{2}\right)\left(x + \sqrt{2}\right) = Cx^4 - 3Cx^2 + 2C$. Requiring that the constant term be 6 gives $C = 3$, so $P(x) = 3x^4 - 9x^2 + 6$.

71. The y-intercept is 2 and the zeros of the polynomial are -1, 1, and 2.
It follows that $P(x) = C(x+1)(x-1)(x-2) = C\left(x^3 - 2x^2 - x + 2\right)$. Since $P(0) = 2$ we have $2 = C\left[(0)^3 - 2(0)^2 - (0) + 2\right] \Leftrightarrow 2 = 2C \Leftrightarrow C = 1$ and $P(x) = (x+1)(x-1)(x-2) = x^3 - 2x^2 - x + 2$.

73. The y-intercept is 4 and the zeros of the polynomial are -2 and 1 both being degree two.
It follows that $P(x) = C(x+2)^2(x-1)^2 = C\left(x^4+2x^3-3x^2-4x+4\right)$. Since $P(0) = 4$ we have
$4 = C\left[(0)^4+2(0)^3-3(0)^2-4(0)+4\right] \Leftrightarrow 4 = 4C \Leftrightarrow C = 1$.
Thus $P(x) = (x+2)^2(x-1)^2 = x^4+2x^3-3x^2-4x+4$.

75. **A.** By the Remainder Theorem, the remainder when $P(x) = 6x^{1000} - 17x^{562} + 12x + 26$ is divided by $x+1$ is $P(-1) = 6(-1)^{1000} - 17(-1)^{562} + 12(-1) + 26 = 6 - 17 - 12 + 26 = 3$.
B. If $x-1$ is a factor of $Q(x) = x^{567} - 3x^{400} + x^9 + 2$, then $Q(1)$ must equal 0.
$Q(1) = (1)^{567} - 3(1)^{400} + (1)^9 + 2 = 1 - 3 + 1 + 2 = 1 \neq 0$, so $x-1$ is not a factor.

3.4 REAL ZEROS OF POLYNOMIALS

1. If the polynomial function $P(x) = a_nx^n + a_{n-1}x^{n-1} + \cdots a_1x + a_0$ has integer coefficients, then the only numbers that could possibly be rational zeros of P are all of the form $\frac{p}{q}$, where p is a factor of *the constant coefficient* a_0 and q is a factor of *the leading coefficient* a_n. The possible rational zeros of $P(x) = 6x^3+5x^2-19x-10$ are $\pm1, \pm\frac{1}{2}, \pm\frac{1}{3}, \pm\frac{1}{6}, \pm2, \pm\frac{2}{3}, \pm5, \pm\frac{5}{2}, \pm\frac{5}{3}, \pm\frac{5}{6}, \pm10$, and $\pm\frac{10}{3}$.

3. This is true. If c is a real zero of the polynomial P, then $P(x) = (x-c)Q(x)$, and any other zero of $P(x)$ is also a zero of $Q(x) = P(x)/(x-c)$.

5. $P(x) = x^3 - 4x^2 + 3$ has possible rational zeros ±1 and ±3.

7. $R(x) = 2x^5 + 3x^3 + 4x^2 - 8$ has possible rational zeros $\pm1, \pm2, \pm4, \pm8, \pm\frac{1}{2}$.

9. $T(x) = 4x^4 - 2x^2 - 7$ has possible rational zeros $\pm1, \pm7, \pm\frac{1}{2}, \pm\frac{7}{2}, \pm\frac{1}{4}, \pm\frac{7}{4}$.

11. **(a)** $P(x) = 5x^3 - x^2 - 5x + 1$ has possible rational zeros $\pm1, \pm\frac{1}{5}$.

(b) From the graph, the actual zeros are -1, $\frac{1}{5}$, and 1.

13. **(a)** $P(x) = 2x^4 - 9x^3 + 9x^2 + x - 3$ has possible rational zeros $\pm1, \pm3, \pm\frac{1}{2}, \pm\frac{3}{2}$.

(b) From the graph, the actual zeros are $-\frac{1}{2}$, 1, and 3.

15. $P(x) = x^3 + 2x^2 - 13x + 10$. The possible rational zeros are $\pm1, \pm2, \pm5, \pm10$. $P(x)$ has 2 variations in sign and hence 0 or 2 positive real zeros. $P(-x) = -x^3 + 2x^2 + 13x + 10$ has 1 variation in sign and hence P has 1 negative real zero.

$$\begin{array}{r|rrrr} 1 & 2 & 1 & -13 & 10 \\ & & 2 & 3 & 10 \\ \hline & 2 & 3 & -10 & 0 \end{array} \quad \Rightarrow x = 1 \text{ is a zero.}$$

$P(x) = x^3 + 2x^2 - 13x + 10 = (x-1)\left(x^2+3x-10\right) = (x+5)(x-1)(x-2)$. Therefore, the zeros are -5, 1, and 2.

17. $P(x) = x^3 + 3x^2 - 4$. The possible rational zeros are $\pm1, \pm2, \pm4$. $P(x)$ has 1 variation in sign and hence 1 positive real zero. $P(-x) = -x^3 + 3x^2 - 4$ has 2 variations in sign and hence P has 0 or 2 negative real zeros.

$$\begin{array}{r|rrrr} 1 & 1 & 3 & 0 & -4 \\ & & 1 & 4 & 4 \\ \hline & 1 & 4 & 4 & 0 \end{array} \quad \Rightarrow x = 1 \text{ is a zero.}$$

$P(x) = x^3 + 3x^2 - 4 = (x-1)\left(x^2+4x+4\right) = (x-1)(x+2)^2$. Therefore, the zeros are -2 and 1.

19. $P(x) = x^3 - 6x^2 + 12x - 8$. The possible rational zeros are $\pm1, \pm2, \pm4, \pm8$. $P(x)$ has 3 variations in sign and hence 1 or 3 positive real zeros. $P(-x) = -x^3 - 6x^2 - 12x - 8$ has no variations in sign and hence P has no negative real zero.

$$\begin{array}{r|rrrr} 1 & 1 & -6 & 12 & -8 \\ & & 1 & -5 & 7 \\ \hline & 1 & -5 & 7 & -1 \end{array} \Rightarrow x = 1 \text{ is not a zero.} \qquad \begin{array}{r|rrrr} 2 & 1 & -6 & 12 & -8 \\ & & 2 & -8 & 8 \\ \hline & 1 & -4 & 4 & 0 \end{array} \Rightarrow x = 2 \text{ is a zero.}$$

$P(x) = x^3 - 6x^2 + 12x - 8 = (x - 2)\left(x^2 - 4x + 4\right) = (x - 2)^3$. Therefore, the only zero is $x = 2$.

21. $P(x) = x^3 - 19x - 30$. The possible rational zeros are $\pm1, \pm2, \pm3, \pm5, \pm6, \pm15, \pm30$. $P(x)$ has 1 variations in sign and hence 1 positive real zero. $P(-x) = -x^3 + 19x - 30$ has 2 variations in sign and hence P has 0 or 2 negative real zeros.

$$\begin{array}{r|rrrr} -1 & 1 & 0 & -19 & -30 \\ & & -1 & 1 & 18 \\ \hline & 1 & -1 & -18 & -12 \end{array} \Rightarrow x = -1 \text{ is not a zero.} \qquad \begin{array}{r|rrrr} -2 & 1 & 0 & -19 & -30 \\ & & -2 & 4 & 30 \\ \hline & 1 & -2 & -15 & 0 \end{array} \Rightarrow x = -2 \text{ is a zero.}$$

$$\begin{array}{r|rrrr} -4 & 1 & 12 & 48 & 64 \\ & & -4 & -32 & -64 \\ \hline & 1 & 8 & 16 & 0 \end{array} \Rightarrow x = -4 \text{ is a zero.}$$

$P(x) = x^3 - 19x - 30 = (x + 2)\left(x^2 - 2x - 15\right) = (x + 3)(x + 2)(x - 5)$. Therefore, the zeros are -3, -2, and 5.

23. $P(x) = x^3 + 3x^2 - x - 3$. The possible rational zeros are $\pm1, \pm3$. $P(x)$ has 1 variation in sign and hence 1 positive real zero. $P(-x) = -x^3 + 3x^2 + x - 4$ has 2 variations in sign and hence P has 0 or 2 negative real zeros.

$$\begin{array}{r|rrrr} -1 & 1 & 3 & -1 & -3 \\ & & -1 & -2 & 3 \\ \hline & 1 & 2 & -3 & 0 \end{array} \Rightarrow x = -1 \text{ is a zero.}$$

So $P(x) = x^3 + 3x^2 - x - 3 = (x + 1)\left(x^2 + 2x - 3\right) = (x + 1)(x + 3)(x - 1)$. Therefore, the zeros are -1, -3, and 1.

25. *Method 1:* $P(x) = x^4 - 5x^2 + 4$.The possible rational zeros are $\pm1, \pm2, \pm4$. $P(x)$ has 1 variation in sign and hence 1 positive real zero. $P(-x) = x^4 - 5x^2 + 4$ has 2 variations in sign and hence P has 0 or 2 negative real zeros.

$$\begin{array}{r|rrrrr} 1 & 1 & 0 & -5 & 0 & 4 \\ & & 1 & 1 & -4 & -4 \\ \hline & 1 & 1 & -4 & -4 & 0 \end{array} \Rightarrow x = 1 \text{ is a zero.}$$

Thus $P(x) = x^4 - 5x^2 + 4 = (x - 1)\left(x^3 + x^2 - 4x - 4\right)$. Continuing with the quotient we have:

$$\begin{array}{r|rrrr} -1 & 1 & 1 & -4 & -4 \\ & & -1 & 0 & 4 \\ \hline & 1 & 0 & -4 & 0 \end{array} \Rightarrow x = -1 \text{ is a zero.}$$

$P(x) = x^4 - 5x^2 + 4 = (x - 1)(x + 1)\left(x^2 - 4\right) = (x - 1)(x + 1)(x - 2)(x + 2)$. Therefore, the zeros are $\pm1, \pm2$.

Method 2: Substituting $u = x^2$, the polynomial becomes $P(u) = u^2 - 5u + 4$, which factors: $u^2 - 5u + 4 = (u - 1)(u - 4) = \left(x^2 - 1\right)\left(x^2 - 4\right)$, so either $x^2 = 1$ or $x^2 = 4$. If $x^2 = 1$, then $x = \pm1$; if $x^2 = 4$, then $x = \pm2$. Therefore, the zeros are ±1 and ±2.

27. $P(x) = x^4 + 6x^3 + 7x^2 - 6x - 8$. The possible rational zeros are $\pm1, \pm2, \pm4, \pm8$. $P(x)$ has 1 variation in sign and hence 1 positive real zero. $P(-x) = x^4 - 6x^3 + 7x^2 + 6x - 8$ has 3 variations in sign and hence P has 1 or 3 negative real zeros.

$$\begin{array}{r|rrrrr} 1 & 1 & 6 & 7 & -6 & -8 \\ & & 1 & 7 & 14 & 8 \\ \hline & 1 & 7 & 14 & 8 & 0 \end{array} \quad \Rightarrow x = 1 \text{ is a zero}$$

and there are no other positive zeros. Thus $P(x) = x^4 + 6x^3 + 7x^2 - 6x - 8 = (x-1)\left(x^3 + 7x^2 + 14x + 8\right)$. Continuing by factoring the quotient, we have:

$$\begin{array}{r|rrrr} -1 & 1 & 7 & 14 & 8 \\ & & -1 & -6 & -8 \\ \hline & 1 & 6 & 8 & 0 \end{array} \quad \Rightarrow x = -1 \text{ is a zero.}$$

So $P(x) = x^4 + 6x^3 + 7x^2 - 6x - 8 = (x-1)(x+1)\left(x^2 + 6x + 8\right) = (x-1)(x+1)(x+2)(x+4)$. Therefore, the zeros are -4, -2, and ±1.

29. $P(x) = 4x^4 - 37x^2 + 9$ has possible rational zeros $\pm1, \pm3, \pm9, \pm\frac{1}{2}, \pm\frac{3}{2}, \pm\frac{9}{2}, \pm\frac{1}{4}, \pm\frac{3}{4}, \pm\frac{9}{4}$. Since $P(x)$ has 2 variations in sign, there are 0 or 2 positive real zeros, and since $P(-x) = 4x^4 - 37x^2 + 36$ has 2 variations in sign, P has 0 or 2 negative real zeros.

$$\begin{array}{r|rrrrr} 1 & 4 & 0 & -37 & 0 & 9 \\ & & 4 & 4 & -33 & -33 \\ \hline & 4 & 4 & -33 & -33 & -24 \end{array} \qquad \begin{array}{r|rrrrr} 3 & 4 & 0 & -37 & 0 & 9 \\ & & 12 & 36 & -3 & -9 \\ \hline & 4 & 12 & -1 & -3 & 0 \end{array} \quad \Rightarrow x = 3 \text{ is a zero.}$$

$$\begin{array}{r|rrrr} \frac{1}{2} & 4 & 12 & -1 & -3 \\ & & 2 & 7 & 3 \\ \hline & 4 & 14 & 6 & 0 \end{array} \quad \Rightarrow x = \tfrac{1}{2} \text{ is a zero.}$$

Because P is even, we conclude that the zeros are ±3 and $\pm\frac{1}{2}$ and
$P(x) = 4x^4 - 37x^2 + 9 = (x+3)(2x+1)(2x-1)(x-3)$.
Note: Since $P(x)$ has only even terms, factoring by substitution also works. Let $x^2 = u$; then
$P(u) = 4u^2 - 37u + 9 = (4u-1)(u-9) = \left(4x^2 - 1\right)\left(x^2 - 3\right)$, which gives the same results.

31. $P(x) = 3x^4 - 10x^3 - 9x^2 + 40x - 12$. The possible rational zeros are $\pm1, \pm2, \pm3, \pm4, \pm6, \pm12$. $P(x)$ has 3 variations in sign and hence 1, 3, or 5 positive real zeros. $P(-x) = 3x^4 + 10x^3 - 9x^2 - 40x - 12$ has 1 variation in sign and hence P has 1 negative real zero.

$$\begin{array}{r|rrrrr} 1 & 3 & -10 & -9 & 40 & -12 \\ & & 3 & -7 & -16 & 24 \\ \hline & 3 & -7 & -16 & 24 & 24 \end{array} \quad \Rightarrow x = 1 \text{ is not a zero.} \qquad \begin{array}{r|rrrrr} 2 & 3 & -10 & -9 & 40 & -12 \\ & & 6 & -8 & -34 & 12 \\ \hline & 3 & -4 & -17 & 6 & 0 \end{array} \quad \Rightarrow x = 2 \text{ is a zero.}$$

Thus $P(x) = 3x^4 - 10x^3 - 9x^2 + 40x - 12 = (x-2)\left(3x^3 - 4x^2 - 17x + 6\right)$. Continuing by factoring the quotient, we have

$$\begin{array}{r|rrrr} 3 & 3 & -4 & -17 & 6 \\ & & 9 & 15 & -6 \\ \hline & 3 & 5 & -2 & 0 \end{array} \quad \Rightarrow x = 3 \text{ is a zero.}$$

Thus $P(x) = (x-3)(x-2)\left(3x^2+5x-2\right) = (x-3)(x-2)(3x-1)(x+2)$. Therefore, the zeros are -2, $\frac{1}{3}$, 2, and 3.

33. Factoring by grouping can be applied to this exercise.

$4x^3+4x^2-x-1 = 4x^2(x+1)-(x+1) = (x+1)\left(4x^2-1\right) = (x+1)(2x+1)(2x-1)$.

Therefore, the zeros are -1 and $\pm\frac{1}{2}$.

35. $P(x) = 4x^3-7x+3$. The possible rational zeros are ±1, ±3, $\pm\frac{1}{2}$, $\pm\frac{3}{2}$, $\pm\frac{1}{4}$, $\pm\frac{3}{4}$. Since $P(x)$ has 2 variations in sign, there are 0 or 2 positive zeros. Since $P(-x) = -4x^3+7x+3$ has 1 variation in sign, there is 1 negative zero.

$$\begin{array}{r|rrrr} \frac{1}{2} & 4 & 0 & -7 & 3 \\ & & 2 & 1 & -3 \\ \hline & 4 & 2 & -6 & 0 \end{array} \Rightarrow x = \tfrac{1}{2} \text{ is a zero.}$$

$P(x) = \left(x-\frac{1}{2}\right)\left(4x^2+2x-6\right) = (2x-1)\left(2x^2+x-3\right) = (2x-1)(x-1)(2x+3) = 0$. Thus, the zeros are $-\frac{3}{2}$, $\frac{1}{2}$, and 1.

37. $P(x) = 24x^3+10x^2-13x-6$. The possible rational zeros are ±1, ±2, ±3, ±6, $\pm\frac{1}{2}$, $\pm\frac{3}{2}$, $\pm\frac{1}{3}$, $\pm\frac{2}{3}$, $\pm\frac{1}{4}$, $\pm\frac{3}{4}$, $\pm\frac{1}{6}$, $\pm\frac{1}{8}$, $\pm\frac{3}{8}$, $\pm\frac{1}{12}$, $\pm\frac{1}{24}$. $P(x)$ has 1 variation in sign and hence 1 positive real zero. $P(-x) = -24x^3+10x^2+13x-6$ has 2 variations in sign, so P has 0 or 2 negative real zeros.

$$\begin{array}{r|rrrr} -1 & 24 & 10 & -13 & -6 \\ & & -24 & 14 & -1 \\ \hline & 24 & -14 & 1 & -7 \end{array} \Rightarrow x = -1 \text{ is not a zero.}$$

$$\begin{array}{r|rrrr} -2 & 24 & 10 & -13 & -6 \\ & & -48 & 76 & -126 \\ \hline & 24 & -38 & 63 & -132 \end{array} \Rightarrow x = -2 \text{ is not a zero.}$$

$$\begin{array}{r|rrrr} -3 & 24 & 10 & -13 & -6 \\ & & -72 & 186 & -519 \\ \hline & 24 & -62 & 173 & -525 \end{array} \Rightarrow x = -3 \text{ is not a zero.}$$

$$\begin{array}{r|rrrr} -6 & 24 & 10 & -13 & -6 \\ & & -144 & 804 & -4746 \\ \hline & 24 & -134 & 791 & -4752 \end{array} \Rightarrow x = -6 \text{ is not a zero.}$$

$$\begin{array}{r|rrrr} -\frac{1}{2} & 24 & 10 & -13 & -6 \\ & & -12 & 1 & 6 \\ \hline & 24 & -2 & -12 & 0 \end{array} \Rightarrow x = -\tfrac{1}{2} \text{ is a zero.}$$

Thus $P(x) = 24x^3+10x^2-13x-6 = (2x+1)\left(12x^2-x-6\right) = (3x+2)(2x+1)(4x-3)$ has zeros $-\frac{2}{3}$, $-\frac{1}{2}$, and $\frac{3}{4}$.

39. $P(x) = 2x^4 - 7x^3 + 3x^2 + 8x - 4$. The possible rational zeros are $\pm1, \pm2, \pm4, \pm\frac{1}{2}$. $P(x)$ has 3 variations in sign and hence 1 or 3 positive real zeros. $P(-x) = 2x^4 + 7x^3 + 3x^2 - 8x - 4$ has 1 variation in sign and hence P has 1 negative real zero.

$$\begin{array}{r|rrrrr} 1 & 2 & -7 & 3 & 8 & -4 \\ & & 2 & -5 & -2 & 6 \\ \hline & 2 & -5 & -2 & 6 & 2 \end{array} \Rightarrow x = 1 \text{ is not a zero.} \qquad \begin{array}{r|rrrrr} \frac{1}{2} & 2 & -7 & 3 & 8 & -4 \\ & & 1 & -3 & 0 & 4 \\ \hline & 2 & -6 & 0 & 8 & 0 \end{array} \Rightarrow x = \tfrac{1}{2} \text{ is a zero.}$$

Thus $P(x) = 2x^4 - 7x^3 + 3x^2 + 8x - 4 = \left(x - \frac{1}{2}\right)\left(2x^3 - 6x^2 + 8\right)$. Continuing by factoring the quotient, we have:

$$\begin{array}{r|rrrr} 2 & 2 & -6 & 0 & 8 \\ & & 4 & -4 & -8 \\ \hline & 2 & -2 & -4 & 0 \end{array} \Rightarrow x = 2 \text{ is a zero.}$$

$P(x) = \left(x - \frac{1}{2}\right)(x-2)\left(2x^2 - 2x - 4\right) = 2\left(x - \frac{1}{2}\right)(x-2)\left(x^2 - x - 2\right) = 2\left(x - \frac{1}{2}\right)(x-2)^2(x+1)$. Thus, the zeros are $\frac{1}{2}$, 2, and -1.

41. $P(x) = x^5 + 3x^4 - 9x^3 - 31x^2 + 36$. The possible rational zeros are $\pm1, \pm2, \pm3, \pm4, \pm6, \pm8, \pm9, \pm12, \pm18$. $P(x)$ has 2 variations in sign and hence 0 or 2 positive real zeros. $P(-x) = -x^5 + 3x^4 + 9x^3 - 31x^2 + 36$ has 3 variations in sign and hence P has 1 or 3 negative real zeros.

$$\begin{array}{r|rrrrrr} 1 & 1 & 3 & -9 & -31 & 0 & 36 \\ & & 1 & 4 & -5 & -36 & -36 \\ \hline & 1 & 4 & -5 & -36 & -36 & 0 \end{array} \Rightarrow x = 1 \text{ is a zero.}$$

So $P(x) = x^5 + 3x^4 - 9x^3 - 31x^2 + 36 = (x-1)\left(x^4 + 4x^3 - 5x^2 - 36x - 36\right)$. Continuing by factoring the quotient, we have:

$$\begin{array}{r|rrrrr} 1 & 1 & 4 & -5 & -36 & -36 \\ & & 1 & 5 & 0 & -36 \\ \hline & 1 & 1 & 0 & -36 & -72 \end{array} \qquad \begin{array}{r|rrrrr} 2 & 1 & 4 & -5 & -36 & -36 \\ & & 2 & 12 & 14 & -44 \\ \hline & 1 & 6 & 7 & -22 & -80 \end{array}$$

$$\begin{array}{r|rrrrr} 3 & 1 & 4 & -5 & -36 & -36 \\ & & 3 & 21 & 48 & 36 \\ \hline & 1 & 7 & 16 & 12 & 0 \end{array} \Rightarrow x = 3 \text{ is a zero.}$$

So $P(x) = (x-1)(x-3)\left(x^3 + 7x^2 + 16x + 12\right)$. Since we have 2 positive zeros, there are no more positive zeros, so we continue by factoring the quotient with possible negative zeros.

$$\begin{array}{r|rrrr} -1 & 1 & 7 & 16 & 12 \\ & & -1 & -6 & -10 \\ \hline & 1 & 6 & 10 & 2 \end{array} \qquad \begin{array}{r|rrrr} -2 & 1 & 7 & 16 & 12 \\ & & -2 & -10 & -12 \\ \hline & 1 & 5 & 6 & 0 \end{array} \Rightarrow x = -2 \text{ is a zero.}$$

Then $P(x) = (x-1)(x-3)(x+2)\left(x^2 + 5x + 6\right) = (x-1)(x-3)(x+2)^2(x+3)$. Thus, the zeros are 1, 3, -2, and -3.

43. $P(x) = 3x^5 - 14x^4 - 14x^3 + 36x^2 + 43x + 10$ has possible rational zeros $\pm1, \pm2, \pm5, \pm10, \pm\frac{1}{3}, \pm\frac{2}{3}, \pm\frac{5}{3}, \pm\frac{10}{3}$. Since $P(x)$ has 2 variations in sign, there are 0 or 2 positive real zeros. Since $P(-x) = -3x^5 - 14x^4 + 14x^3 + 36x^2 - 43x + 10$ has 3 variations in sign, P has 1 or 3 negative real zeros.

$$\begin{array}{r|rrrrrr} 1 & 3 & -14 & -14 & 36 & 43 & 10 \\ & & 3 & -11 & -25 & 11 & 54 \\ \hline & 3 & -11 & -25 & 11 & 54 & 64 \end{array} \qquad \begin{array}{r|rrrrrr} 2 & 3 & -14 & -14 & 36 & 43 & 10 \\ & & 6 & -16 & -60 & -48 & -10 \\ \hline & 3 & -8 & -30 & -24 & -5 & 0 \end{array} \Rightarrow x = 2 \text{ is a zero.}$$

$P(x) = (x-2)\left(3x^4 - 8x^3 - 30x^2 - 24x - 5\right)$

$$\begin{array}{r|rrrrr} 2 & 3 & -8 & -30 & -24 & -5 \\ & & 6 & -4 & -68 & -184 \\ \hline & 3 & -2 & -34 & -92 & -189 \end{array} \qquad \begin{array}{r|rrrrr} 5 & 3 & -8 & -30 & -24 & -5 \\ & & 15 & 35 & 25 & 5 \\ \hline & 3 & 7 & 5 & 1 & 0 \end{array} \Rightarrow x = 5 \text{ is a zero.}$$

$P(x) = (x-2)(x-5)\left(3x^3 + 7x^2 + 5x + 1\right)$. Since $3x^3 + 7x^2 + 5x + 1$ has no variation in sign, there are no more positive zeros.

$$\begin{array}{r|rrrr} -1 & 3 & 7 & 5 & 1 \\ & & -3 & -4 & -1 \\ \hline & 3 & 4 & 1 & 0 \end{array} \Rightarrow x = -1 \text{ is a zero.}$$

$P(x) = (x-2)(x-5)(x+1)\left(3x^2 + 4x + 1\right) = (x-2)(x-5)(x+1)(x+1)(3x+1)$. Therefore, the zeros are -1, $-\frac{1}{3}$, 2, and 5.

45. $P(x) = 3x^3 + 5x^2 - 2x - 4$. The possible rational zeros are $\pm1, \pm2, \pm4, \pm\frac{1}{3}, \pm\frac{2}{3}, \pm\frac{4}{3}$. $P(x)$ has 1 variation in sign and hence 1 positive real zero. $P(-x) = -3x^3 + 5x^2 + 2x - 4$ has 2 variations in sign and hence P has 0 or 2 negative real zeros.

$$\begin{array}{r|rrrr} 1 & 3 & 5 & -2 & -4 \\ & & 3 & 2 & 0 \\ \hline & 3 & 2 & 0 & -4 \end{array} \qquad \begin{array}{r|rrrr} -1 & 3 & 5 & -2 & -4 \\ & & -3 & -2 & 4 \\ \hline & 3 & 2 & -4 & 0 \end{array} \Rightarrow x = -1 \text{ is a zero.}$$

So $P(x) = (x+1)\left(3x^2 + 2x - 4\right)$. Using the Quadratic Formula on the second factor, we have $x = \frac{-2 \pm \sqrt{2^2 - 4(3)(-4)}}{2(3)} = \frac{-1 \pm \sqrt{13}}{3}$. Therefore, the zeros of P are -1 and $\frac{-1 \pm \sqrt{13}}{3}$.

47. $P(x) = x^4 - 6x^3 + 4x^2 + 15x + 4$. The possible rational zeros are $\pm1, \pm2, \pm4$. $P(x)$ has 2 variations in sign and hence 0 or 2 positive real zeros. $P(-x) = x^4 + 6x^3 + 4x^2 - 15x + 4$ has 2 variations in sign and hence P has 0 or 2 negative real zeros.

$$\begin{array}{r|rrrrr} 1 & 1 & -6 & 4 & 15 & 4 \\ & & 1 & -5 & -1 & 14 \\ \hline & 1 & -5 & -1 & 14 & 18 \end{array} \qquad \begin{array}{r|rrrrr} 2 & 1 & -6 & 4 & 15 & 4 \\ & & 2 & -8 & -8 & 14 \\ \hline & 1 & -4 & -4 & 7 & 18 \end{array}$$

$$\begin{array}{r|rrrrr} 4 & 1 & -6 & 4 & 15 & 4 \\ & & 4 & -8 & -16 & -4 \\ \hline & 1 & -2 & -4 & -1 & 0 \end{array} \Rightarrow x = 4 \text{ is a zero.}$$

So $P(x) = (x-4)\left(x^3 - 2x^2 - 4x - 1\right)$. Continuing by factoring the quotient, we have:

$$\begin{array}{r|rrrr} 4 & 1 & -2 & -4 & -1 \\ & & 4 & 8 & 16 \\ \hline & 1 & 2 & 4 & 15 \end{array} \Rightarrow x = 4 \text{ is an upper bound.} \qquad \begin{array}{r|rrrr} -1 & 1 & -2 & -4 & -1 \\ & & -1 & 3 & 1 \\ \hline & 1 & -3 & -1 & 0 \end{array} \Rightarrow x = -1 \text{ is a zero.}$$

So $P(x) = (x-4)(x+1)\left(x^2 - 3x - 1\right)$. Using the Quadratic Formula on the third factor, we have:

$x = \frac{-(-3)\pm\sqrt{(-3)^2-4(1)(-1)}}{2(1)} = \frac{3\pm\sqrt{13}}{2}$. Therefore, the zeros are 4, -1, and $\frac{3\pm\sqrt{13}}{2}$.

49. $P(x) = x^4 - 7x^3 + 14x^2 - 3x - 9$. The possible rational zeros are $\pm1, \pm3, \pm9$. $P(x)$ has 3 variations in sign and hence 1 or 3 positive real zeros. $P(-x) = x^4 + 7x^3 + 14x^2 + 3x - 4$ has 1 variation in sign and hence P has 1 negative real zero.

$$\begin{array}{r|rrrrr} 1 & 1 & -7 & 14 & -3 & -9 \\ & & 1 & -6 & 8 & 5 \\ \hline & 1 & -6 & 8 & 5 & 4 \end{array} \qquad \begin{array}{r|rrrrr} 3 & 1 & -7 & 14 & -3 & -9 \\ & & 3 & -12 & 6 & 9 \\ \hline & 1 & -4 & 2 & 3 & 0 \end{array} \Rightarrow x = 3 \text{ is a zero.}$$

So $P(x) = (x-3)\left(x^3 - 4x^2 + 2x + 3\right)$. Since the constant term of the second term is 3, ±9 are no longer possible zeros.

Continuing by factoring the quotient, we have:

$$\begin{array}{r|rrrr} 3 & 1 & -4 & 2 & 3 \\ & & 3 & -3 & -3 \\ \hline & 1 & -1 & -1 & 0 \end{array} \Rightarrow x = 3 \text{ is a zero again.}$$

So $P(x) = (x-3)^2\left(x^2 - x - 1\right)$. Using the Quadratic Formula on the second factor, we have:

$x = \frac{-(-1)\pm\sqrt{(-1)^2-4(1)(-1)}}{2(1)} = \frac{1\pm\sqrt{5}}{2}$. Therefore, the zeros are 3 and $\frac{1\pm\sqrt{5}}{2}$.

51. $P(x) = 4x^3 - 6x^2 + 1$. The possible rational zeros are $\pm1, \pm\frac{1}{2}, \pm\frac{1}{4}$. $P(x)$ has 2 variations in sign and hence 0 or 2 positive real zeros. $P(-x) = -4x^3 - 6x^2 + 1$ has 1 variation in sign and hence P has 1 negative real zero.

$$\begin{array}{r|rrrr} 1 & 4 & -6 & 0 & 1 \\ & & 4 & -2 & -2 \\ \hline & 4 & -2 & -2 & -1 \end{array} \qquad \begin{array}{r|rrrr} \frac{1}{2} & 4 & -6 & 0 & 1 \\ & & 2 & -2 & -1 \\ \hline & 4 & -4 & -2 & 0 \end{array} \Rightarrow x = \tfrac{1}{2} \text{ is a zero.}$$

So $P(x) = \left(x - \frac{1}{2}\right)\left(4x^2 - 4x - 2\right)$. Using the Quadratic Formula on the second factor, we have:

$x = \frac{-(-4)\pm\sqrt{(-4)^2-4(4)(-2)}}{2(4)} = \frac{4\pm\sqrt{48}}{8} = \frac{4\pm4\sqrt{3}}{8} = \frac{1\pm\sqrt{3}}{2}$. Therefore, the zeros are $\frac{1}{2}$ and $\frac{1\pm\sqrt{3}}{2}$.

53. $P(x) = 2x^4 + 15x^3 + 17x^2 + 3x - 1$. The possible rational zeros are $\pm 1, \pm\frac{1}{2}$. $P(x)$ has 1 variation in sign and hence 1 positive real zero. $P(-x) = 2x^4 - 15x^3 + 17x^2 - 3x - 1$ has 3 variations in sign and hence P has 1 or 3 negative real zeros.

$$\begin{array}{r|rrrrr} \frac{1}{2} & 2 & 15 & 17 & 3 & -1 \\ & & 1 & 8 & \frac{25}{2} & \frac{31}{4} \\ \hline & 2 & 16 & 25 & \frac{31}{2} & \frac{27}{4} \end{array} \quad \Rightarrow x = \tfrac{1}{2} \text{ is an upper bound.}$$

$$\begin{array}{r|rrrrr} -\frac{1}{2} & 2 & 15 & 17 & 3 & -1 \\ & & -1 & -7 & -5 & 1 \\ \hline & 2 & 14 & 10 & -2 & 0 \end{array} \quad \Rightarrow x = -\tfrac{1}{2} \text{ is a zero.}$$

So $P(x) = \left(x + \frac{1}{2}\right)\left(2x^3 + 14x^2 + 10x - 2\right) = 2\left(x + \frac{1}{2}\right)\left(x^3 + 7x^2 + 5x - 1\right)$.

$$\begin{array}{r|rrrr} -1 & 1 & 7 & 5 & -1 \\ & & -1 & -6 & 1 \\ \hline & 1 & 6 & -1 & 0 \end{array} \quad \Rightarrow x = -1 \text{ is a zero.}$$

So $P(x) = \left(x + \frac{1}{2}\right)\left(2x^3 + 14x^2 + 10x - 2\right) = 2\left(x + \frac{1}{2}\right)(x + 1)\left(x^2 + 6x - 1\right)$ Using the Quadratic Formula on the third factor, we have $x = \frac{-(6) \pm \sqrt{(6)^2 - 4(1)(-1)}}{2(1)} = \frac{-6 \pm \sqrt{40}}{2} = \frac{-6 \pm 2\sqrt{10}}{2} = -3 \pm \sqrt{10}$. Therefore, the zeros are $-1, -\frac{1}{2}$, and $-3 \pm \sqrt{10}$.

55. (a) $P(x) = x^3 - 3x^2 - 4x + 12$ has possible rational zeros $\pm 1, \pm 2, \pm 3, \pm 4, \pm 6, \pm 12$.

$$\begin{array}{r|rrrr} 1 & 1 & -3 & -4 & 12 \\ & & 1 & -2 & -6 \\ \hline & 1 & -2 & -6 & 6 \end{array}$$

$$\begin{array}{r|rrrr} 2 & 1 & -3 & -4 & 12 \\ & & 2 & -2 & -12 \\ \hline & 1 & -1 & -6 & 0 \end{array} \quad \Rightarrow x = 2 \text{ is a zero.}$$

So $P(x) = (x - 2)\left(x^2 - x - 6\right) = (x - 2)(x + 2)(x - 3)$. The real zeros of P are -2, 2, and 3.

(b)

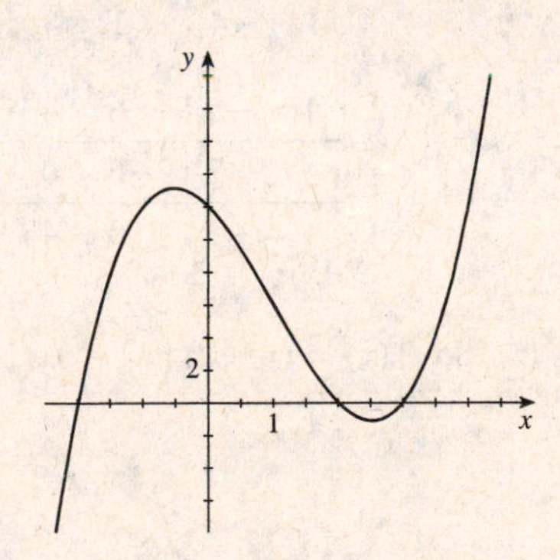

57. (a) $P(x) = 2x^3 - 7x^2 + 4x + 4$ has possible rational zeros $\pm 1, \pm 2, \pm 4, \pm\frac{1}{2}$.

$$\begin{array}{r|rrrr} 1 & 2 & -7 & 4 & 4 \\ & & 2 & -5 & -1 \\ \hline & 2 & -5 & -1 & -3 \end{array} \qquad \begin{array}{r|rrrr} 2 & 2 & -7 & 4 & 4 \\ & & 4 & -6 & -4 \\ \hline & 2 & -3 & -2 & 0 \end{array} \Rightarrow x = 2 \text{ is a zero.}$$

So $P(x) = (x - 2)\left(2x^2 - 3x - 2\right)$. Continuing:

$$\begin{array}{r|rrr} 2 & 2 & -3 & -2 \\ & & 4 & 2 \\ \hline & 2 & 1 & 0 \end{array} \Rightarrow x = 2 \text{ is a zero again.}$$

Thus $P(x) = (x - 2)^2 (2x + 1)$. The real zeros of P are 2 and $-\frac{1}{2}$.

(b)

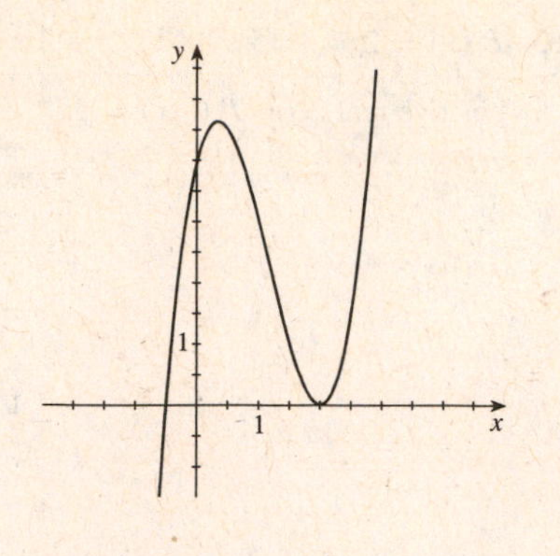

59. (a) $P(x) = x^4 - 5x^3 + 6x^2 + 4x - 8$ has possible rational zeros $\pm 1, \pm 2, \pm 4, \pm 8$.

$$\begin{array}{r|rrrrr} 1 & 1 & -5 & 6 & 4 & -8 \\ & & 1 & -4 & 2 & 6 \\ \hline & 1 & -4 & 2 & 6 & -2 \end{array}$$

$$\begin{array}{r|rrrrr} 2 & 1 & -5 & 6 & 4 & -8 \\ & & 2 & -6 & 0 & 8 \\ \hline & 1 & -3 & 0 & 4 & 0 \end{array} \Rightarrow x = 2 \text{ is a zero.}$$

So $P(x) = (x - 2)\left(x^3 - 3x^2 + 4\right)$ and the possible rational zeros are restricted to $-1, \pm 2, \pm 4$.

$$\begin{array}{r|rrrr} 2 & 1 & -3 & 0 & 4 \\ & & 2 & -2 & -4 \\ \hline & 1 & -1 & -2 & 0 \end{array} \Rightarrow x = 2 \text{ is a zero again.}$$

$P(x) = (x - 2)^2\left(x^2 - x - 2\right) = (x - 2)^2 (x - 2)(x + 1) = (x - 2)^3 (x + 1)$. So the real zeros of P are -1 and 2.

(b)

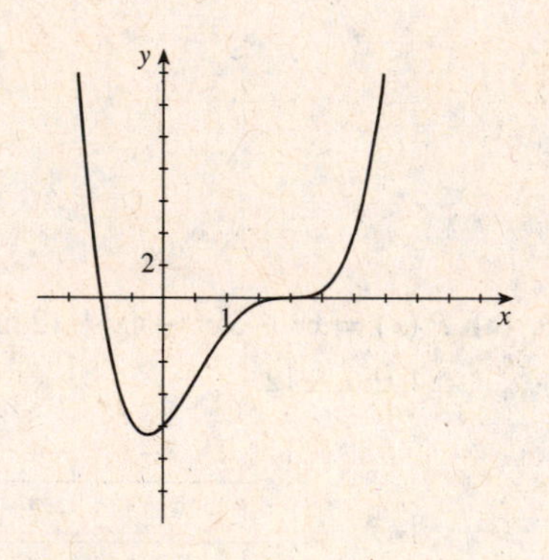

61. (a) $P(x) = x^5 - x^4 - 5x^3 + x^2 + 8x + 4$ has possible rational zeros $\pm1, \pm2, \pm4$.

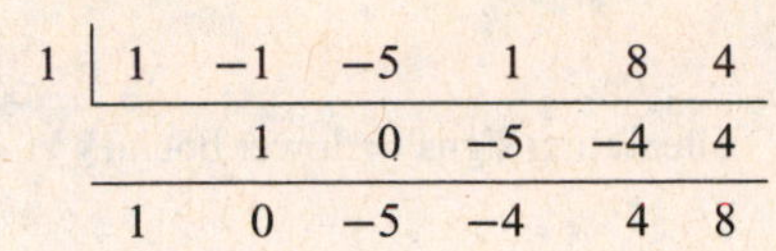

1	1	−1	−5	1	8	4
		1	0	−5	−4	4
	1	0	−5	−4	4	8

2	1	−1	−5	1	8	4	
		2	2	−6	−10	−4	
	1	1	−3	−5	−2	0	$\Rightarrow x = 2$ is a zero.

So $P(x) = (x-2)\left(x^4 + x^3 - 3x^2 - 5x - 2\right)$, and the possible rational zeros are restricted to $-1, \pm2$.

2	1	1	−3	−5	−2	
		2	6	6	2	
	1	3	3	1	0	$\Rightarrow x = 2$ is a zero again.

So $P(x) = (x-2)^2\left(x^3 + 3x^2 + 3x + 1\right)$, and the possible rational zeros are restricted to -1.

−1	1	3	3	1	
		−1	−2	−1	
	1	2	1	0	$\Rightarrow x = -1$ is a zero.

So $P(x) = (x-2)^2(x+1)\left(x^2 + 2x + 1\right) = (x-2)^2(x+1)^3$., and the real zeros of P are -1 and 2.

(b)

63. $P(x) = x^3 - x^2 - x - 3$. Since $P(x)$ has 1 variation in sign, P has 1 positive real zero. Since $P(-x) = -x^3 - x^2 + x - 3$ has 2 variations in sign, P has 2 or 0 negative real zeros. Thus, P has 1 or 3 real zeros.

65. $P(x) = 2x^6 + 5x^4 - x^3 - 5x - 1$. Since $P(x)$ has 1 variation in sign, P has 1 positive real zero. Since $P(-x) = 2x^6 + 5x^4 + x^3 + 5x - 1$ has 1 variation in sign, P has 1 negative real zero. Therefore, P has 2 real zeros.

67. $P(x) = x^5 + 4x^3 - x^2 + 6x$. Since $P(x)$ has 2 variations in sign, P has 2 or 0 positive real zeros. Since $P(-x) = -x^5 - 4x^3 - x^2 - 6x$ has no variation in sign, P has no negative real zero. Therefore, P has a total of 1 or 3 real zeros (since $x = 0$ is a zero, but is neither positive nor negative).

69. $P(x) = 2x^3 + 5x^2 + x - 2$; $a = -3$, $b = 1$

−3	2	5	1	−2	
		−6	3	−12	
	2	−1	4	−14	alternating signs $\Rightarrow$ lower bound.

1	2	5	1	−2	
		2	7	8	
	2	7	8	6	all nonnegative $\Rightarrow$ upper bound.

Therefore $a = -3$ and $b = 1$ are lower and upper bounds.

71. $P(x) = 8x^3 + 10x^2 - 39x + 9$; $a = -3$, $b = 2$

$$\begin{array}{r|rrrr} -3 & 8 & 10 & -39 & 9 \\ & & -24 & 42 & -9 \\ \hline & 8 & -14 & 3 & 0 \end{array} \quad \text{alternating signs} \Rightarrow \text{lower bound.}$$

$$\begin{array}{r|rrrr} 2 & 8 & 10 & -39 & 9 \\ & & 16 & 52 & 26 \\ \hline & 8 & 26 & 13 & 35 \end{array} \quad \text{all nonnegative} \Rightarrow \text{upper bound.}$$

Therefore $a = -3$ and $b = 2$ are lower and upper bounds. Note that $x = -3$ is also a zero.

73. $P(x) = x^4 + 2x^3 + 3x^2 + 5x - 1$; $a = -2$, $b = 1$

$$\begin{array}{r|rrrrr} -2 & 1 & 2 & 3 & 5 & -1 \\ & & -2 & 0 & -6 & 2 \\ \hline & 1 & 0 & 3 & -1 & 1 \end{array} \quad \text{Alternating signs} \Rightarrow \text{lower bound.}$$

$$\begin{array}{r|rrrrr} 1 & 1 & 2 & 3 & 5 & -1 \\ & & 1 & 3 & 6 & 11 \\ \hline & 1 & 3 & 6 & 11 & 10 \end{array} \quad \text{All nonnegative} \Rightarrow \text{upper bound.}$$

Therefore $a = -2$ and $b = 1$ are lower and upper bounds.

75. $P(x) = 2x^4 - 6x^3 + x^2 - 2x + 3$; $a = -1$, $b = 3$

$$\begin{array}{r|rrrrr} -1 & 2 & -6 & 1 & -2 & 3 \\ & & -2 & 8 & -9 & 11 \\ \hline & 2 & -8 & 9 & -11 & 14 \end{array} \quad \text{Alternating signs} \Rightarrow \text{lower bound.}$$

$$\begin{array}{r|rrrrr} 3 & 2 & -6 & 1 & -2 & 3 \\ & & 6 & 0 & 3 & 3 \\ \hline & 2 & 0 & 1 & 1 & 6 \end{array} \quad \text{All nonnegative} \Rightarrow \text{upper bound.}$$

Therefore $a = -1$ and $b = 3$ are lower and upper bounds.

77. $P(x) = x^3 - 3x^2 + 4$ and use the Upper and Lower Bounds Theorem:

$$\begin{array}{r|rrrr} -1 & 1 & -3 & 0 & 4 \\ & & -1 & 4 & -4 \\ \hline & 1 & -4 & 4 & 0 \end{array} \quad \text{alternating signs} \Rightarrow \text{lower bound.}$$

$$\begin{array}{r|rrrr} 3 & 1 & -3 & 0 & 4 \\ & & 3 & 0 & 0 \\ \hline & 1 & 0 & 0 & 4 \end{array} \quad \text{all nonnegative} \Rightarrow \text{upper bound.}$$

Therefore -1 is a lower bound (and a zero) and 3 is an upper bound. (There are many possible solutions.)

79. $P(x) = x^4 - 2x^3 + x^2 - 9x + 2.$

$$\begin{array}{r|rrrrr} 1 & 1 & -2 & 1 & -9 & 2 \\ & & 1 & -1 & 0 & -9 \\ \hline & 1 & -1 & 0 & -9 & -7 \end{array} \qquad \begin{array}{r|rrrrr} 2 & 1 & -2 & 1 & -9 & 2 \\ & & 2 & 0 & 2 & -14 \\ \hline & 1 & 0 & 1 & -7 & -12 \end{array}$$

$$\begin{array}{r|rrrrrl} 3 & 1 & -2 & 1 & -9 & 2 & \\ & & 3 & 3 & 12 & 9 & \\ \hline & 1 & 1 & 4 & 3 & 11 & \text{all positive} \Rightarrow \text{upper bound.} \end{array}$$

$$\begin{array}{r|rrrrrl} -1 & 1 & -2 & 1 & -9 & 2 & \\ & & -1 & 3 & -4 & 13 & \\ \hline & 1 & -3 & 4 & -13 & 15 & \text{alternating signs} \Rightarrow \text{lower bound.} \end{array}$$

Therefore -1 is a lower bound and 3 is an upper bound. (There are many possible solutions.)

81. $P(x) = 2x^4 + 3x^3 - 4x^2 - 3x + 2.$

$$\begin{array}{r|rrrrrl} 1 & 2 & 3 & -4 & -3 & 2 & \\ & & 2 & 5 & 1 & -2 & \\ \hline & 2 & 5 & 1 & -2 & 0 & \Rightarrow x = 1 \text{ is a zero.} \end{array}$$

$P(x) = (x-1)\left(2x^3 + 5x^2 + x - 2\right)$

$$\begin{array}{r|rrrrl} -1 & 2 & 5 & 1 & -2 & \\ & & -2 & -3 & 2 & \\ \hline & 2 & 3 & -2 & 0 & \Rightarrow x = -1 \text{ is a zero.} \end{array}$$

$P(x) = (x-1)(x+1)\left(2x^2 + 3x - 2\right) = (x-1)(x+1)(2x-1)(x+2)$. Therefore, the zeros are $-2, \frac{1}{2}, \pm 1$.

83. *Method 1:* $P(x) = 4x^4 - 21x^2 + 5$ has 2 variations in sign, so by Descartes' rule of signs there are either 2 or 0 positive zeros. If we replace x with $(-x)$, the function does not change, so there are either 2 or 0 negative zeros. Possible rational zeros are $\pm 1, \pm\frac{1}{2}, \pm\frac{1}{4}, \pm 5, \pm\frac{5}{2}, \pm\frac{5}{4}$. By inspection, ± 1 and ± 5 are not zeros, so we must look for non-integer solutions:

$$\begin{array}{r|rrrrrl} \frac{1}{2} & 4 & 0 & -21 & 0 & 5 & \\ & & 2 & 1 & -10 & -5 & \\ \hline & 4 & 2 & -20 & -10 & 0 & \Rightarrow x = \frac{1}{2} \text{ is a zero.} \end{array}$$

$P(x) = \left(x - \frac{1}{2}\right)\left(4x^3 + 2x^2 - 20x - 10\right)$, continuing with the quotient, we have:

$$\begin{array}{r|rrrrl} -\frac{1}{2} & 4 & 2 & -20 & -10 & \\ & & -2 & 0 & 10 & \\ \hline & 4 & 0 & -20 & 0 & \Rightarrow x = -\frac{1}{2} \text{ is a zero.} \end{array}$$

$P(x) = \left(x - \frac{1}{2}\right)\left(x + \frac{1}{2}\right)\left(4x^2 - 20\right) = 0$. If $4x^2 - 20 = 0$, then $x = \pm\sqrt{5}$. Thus the zeros are $\pm\frac{1}{2}, \pm\sqrt{5}$.

Method 2: Substituting $u = x^2$, the equation becomes $4u^2 - 21u + 5 = 0$, which factors: $4u^2 - 21u + 5 = (4u-1)(u-5) = \left(4x^2 - 1\right)\left(x^2 - 5\right)$. Then either we have $x^2 = 5$, so that $x = \pm\sqrt{5}$, or we have $x^2 = \frac{1}{4}$, so that $x = \pm\sqrt{\frac{1}{4}} = \pm\frac{1}{2}$. Thus the zeros are $\pm\frac{1}{2}, \pm\sqrt{5}$.

85. $P(x) = x^5 - 7x^4 + 9x^3 + 23x^2 - 50x + 24$. The possible rational zeros are $\pm1, \pm2, \pm3, \pm4, \pm6, \pm8, \pm12, \pm24$. $P(x)$ has 4 variations in sign and hence 0, 2, or 4 positive real zeros. $P(-x) = -x^5 - 7x^4 - 9x^3 + 23x^2 + 50x + 24$ has 1 variation in sign, and hence P has 1 negative real zero.

$$\begin{array}{r|rrrrrr} 1 & 1 & -7 & 9 & 23 & -50 & 24 \\ & & 1 & -6 & 3 & 26 & -24 \\ \hline & 1 & -6 & 3 & 26 & -24 & 0 \end{array} \Rightarrow x = 1 \text{ is a zero.}$$

$P(x) = (x-1)\left(x^4 - 6x^3 + 3x^2 + 26x - 24\right)$; continuing with the quotient, we try 1 again.

$$\begin{array}{r|rrrrr} 1 & 1 & -6 & 3 & 26 & -24 \\ & & 1 & -5 & -2 & 24 \\ \hline & 1 & -5 & -2 & 24 & 0 \end{array} \Rightarrow x = 1 \text{ is a zero again.}$$

$P(x) = (x-1)^2\left(x^3 - 5x^2 - 2x + 24\right)$; continuing with the quotient, we start by trying 1 again.

$$\begin{array}{r|rrrr} 1 & 1 & -5 & -2 & 24 \\ & & 1 & -4 & -6 \\ \hline & 1 & -4 & -6 & 18 \end{array} \qquad \begin{array}{r|rrrr} 2 & 1 & -5 & -2 & 24 \\ & & 2 & -6 & -16 \\ \hline & 1 & -3 & -8 & 8 \end{array} \qquad \begin{array}{r|rrrr} 3 & 1 & -5 & -2 & 24 \\ & & 3 & -6 & -24 \\ \hline & 1 & -2 & -8 & 0 \end{array} \Rightarrow x = 3 \text{ is a zero.}$$

$P(x) = (x-1)^2(x-3)\left(x^2 - 2x - 8\right) = (x-1)^2(x-3)(x-4)(x+2)$. Therefore, the zeros are $-2, 1, 3, 4$.

87. $P(x) = x^3 - x - 2$. The only possible rational zeros of $P(x)$ are ±1 and ±2.

$$\begin{array}{r|rrrr} 1 & 1 & 0 & -1 & -2 \\ & & 1 & 1 & 0 \\ \hline & 1 & 1 & 0 & -2 \end{array} \qquad \begin{array}{r|rrrr} 2 & 1 & 0 & -1 & -2 \\ & & 2 & 4 & 6 \\ \hline & 1 & 2 & 3 & 4 \end{array} \qquad \begin{array}{r|rrrr} -1 & 1 & 0 & -1 & -2 \\ & & -1 & 1 & 0 \\ \hline & 1 & -1 & 0 & -2 \end{array}$$

Since the row that contains -1 alternates between nonnegative and nonpositive, -1 is a lower bound and there is no need to try -2. Therefore, $P(x)$ does not have any rational zeros.

89. $P(x) = 3x^3 - x^2 - 6x + 12$ has possible rational zeros $\pm1, \pm2, \pm3, \pm4, \pm6, \pm12, \pm\frac{1}{3}, \pm\frac{2}{3}, \pm\frac{4}{3}$.

	3	−1	−6	12	
1	3	2	−4	8	
2	3	5	4	20	all positive $\Rightarrow x = 2$ is an upper bound
−1	3	−4	−2	14	
−2	3	−7	8	−4	alternating signs $\Rightarrow x = -2$ is a lower bound

	3	−1	−6	12
$\frac{1}{3}$	3	0	−6	10
$\frac{2}{3}$	3	1	$-\frac{16}{3}$	$\frac{76}{9}$
$\frac{4}{3}$	3	3	−2	$\frac{28}{3}$
$-\frac{1}{3}$	3	−2	$-\frac{16}{3}$	$\frac{124}{9}$
$-\frac{2}{3}$	3	−3	−4	$\frac{44}{3}$
$-\frac{4}{3}$	3	−5	$\frac{2}{3}$	$\frac{100}{9}$

Therefore, there is no rational zero.

91. $P(x) = x^3 - 3x^2 - 4x + 12$, $[-4, 4]$ by $[-15, 15]$. The possible rational zeros are $\pm 1, \pm 2, \pm 3, \pm 4, \pm 6, \pm 12$. By observing the graph of P, the rational zeros are $x = -2, 2, 3$.

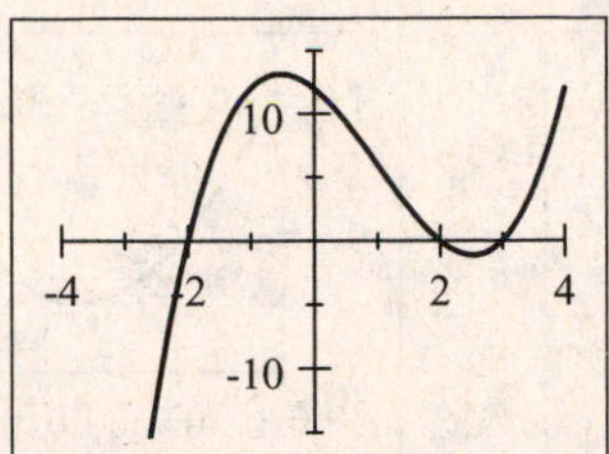

93. $P(x) = 2x^4 - 5x^3 - 14x^2 + 5x + 12$, $[-2, 5]$ by $[-40, 40]$. The possible rational zeros are $\pm 1, \pm 2, \pm 3, \pm 4, \pm 6, \pm 12, \pm\frac{1}{2}, \pm\frac{3}{2}$. By observing the graph of P, the zeros are $-\frac{3}{2}, -1, 1, 4$.

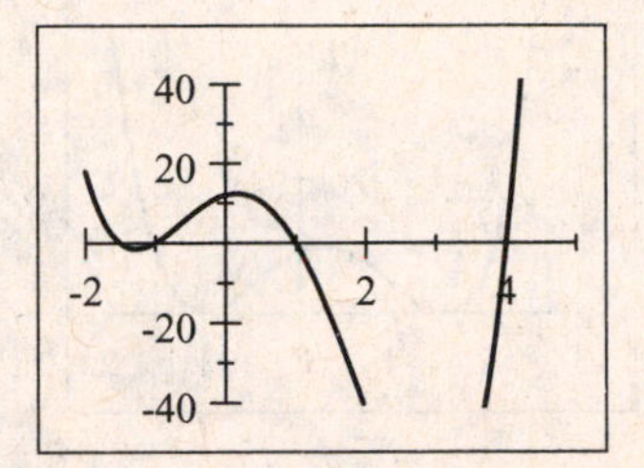

95. $x^4 - x - 4 = 0$. Possible rational solutions are $\pm 1, \pm 2, \pm 4$.

1	1	0	0	-1	-4
		1	1	1	0
	1	1	1	0	-4

2	1	0	0	-1	-4	
		2	4	8	14	
	1	2	4	7	10	$\Rightarrow x = 2$ is an upper bound.

-1	1	0	0	-1	-4
		-1	1	-1	2
	1	-1	1	-2	-2

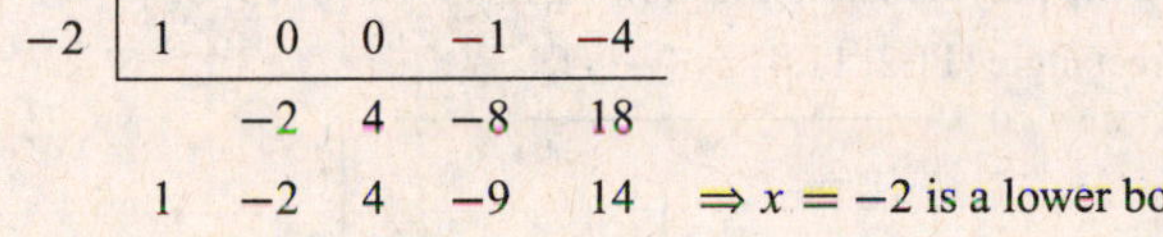

-2	1	0	0	-1	-4	
		-2	4	-8	18	
	1	-2	4	-9	14	$\Rightarrow x = -2$ is a lower bound.

Therefore, we graph the function $P(x) = x^4 - x - 4$ in the viewing rectangle $[-2, 2]$ by $[-5, 20]$ and see there are two solutions. In the viewing rectangle $[-1.3, -1.25]$ by $[-0.1, 0.1]$, we find the solution $x \approx -1.28$. In the viewing rectangle $[1.5.1.6]$ by $[-0.1, 0.1]$, we find the solution $x \approx 1.53$. Thus the solutions are $x \approx -1.28, 1.53$.

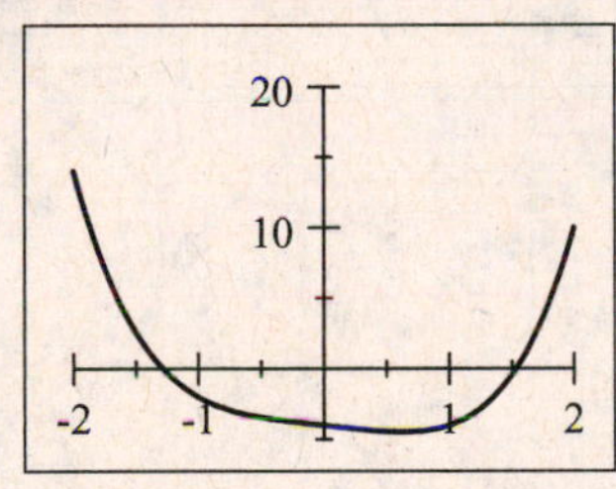

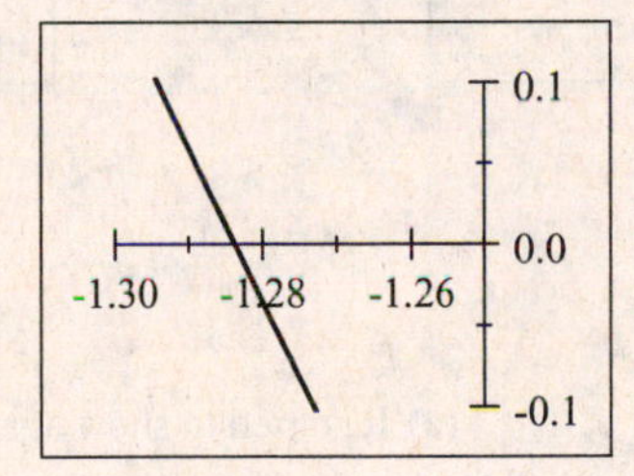

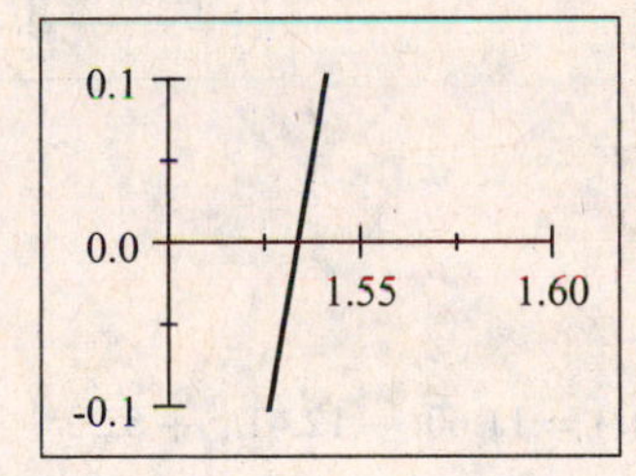

97. $4.00x^4 + 4.00x^3 - 10.96x^2 - 5.88x + 9.09 = 0$.

1	4	4	-10.96	-5.88	9.09
		4	8	-2.96	-8.84
	4	8	-2.96	-8.84	0.25

2	4	4	-10.96	-5.88	9.09	
		8	24	26.08	40.40	
	4	12	13.04	20.2	49.49	$\Rightarrow x = 2$ is an upper bound.

-2	4	4	-10.96	-5.88	9.09
		-8	8	5.92	-0.08
	4	-4	-2.96	0.04	9.01

-3	4	4	-10.96	-5.88	9.09	
		-12	24	-39.12	135	
	4	-8	13.04	-45	144.09	$\Rightarrow x = -3$ is a lower bound.

Therefore, we graph the function $P(x) = 4.00x^4 + 4.00x^3 - 10.96x^2 - 5.88x + 9.09$ in the viewing rectangle $[-3, 2]$ by $[-10, 40]$. There appear to be two solutions. In the viewing rectangle $[-1.6, -1.4]$ by $[-0.1, 0.1]$, we find the solution $x \approx -1.50$. In the viewing rectangle $[0.8, 1.2]$ by $[0, 1]$, we see that the graph comes close but does not go through the x-axis. Thus there is no solution here. Therefore, the only solution is $x \approx -1.50$.

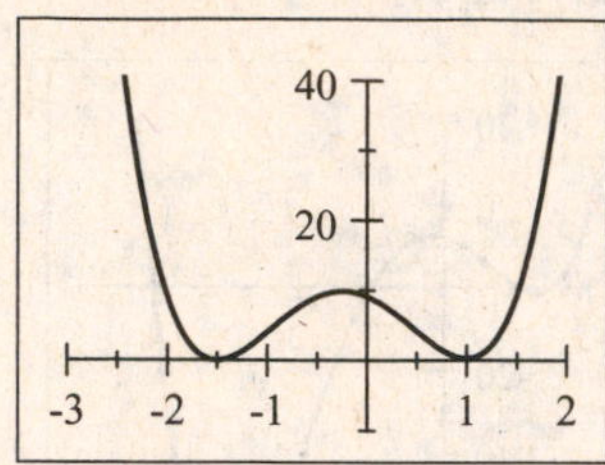

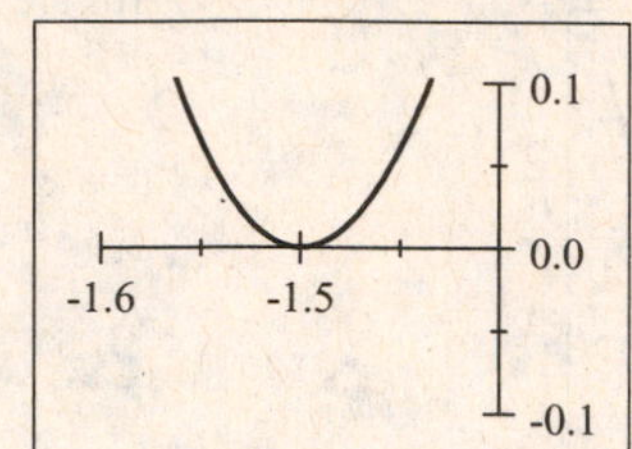

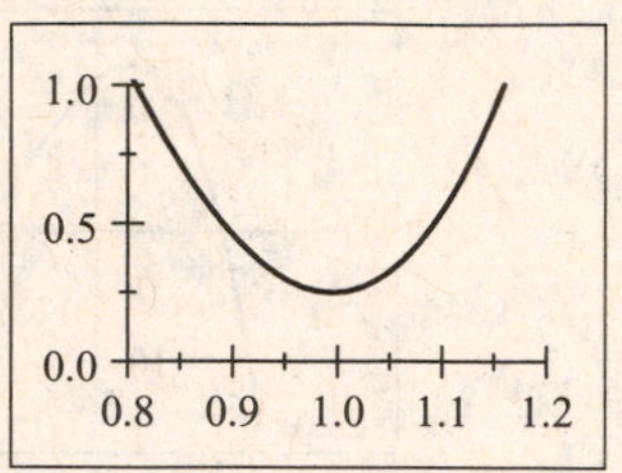

99. Let r be the radius of the silo. The volume of the hemispherical roof is $\frac{1}{2}\left(\frac{4}{3}\pi r^3\right) = \frac{2}{3}\pi r^3$. The volume of the cylindrical section is $\pi\left(r^2\right)(30) = 30\pi r^2$. Because the total volume of the silo is 15,000 ft^3, we get the following equation: $\frac{2}{3}\pi r^3 + 30\pi r^2 = 15000 \Leftrightarrow \frac{2}{3}\pi r^3 + 30\pi r^2 - 15000 = 0 \Leftrightarrow \pi r^3 + 45\pi r^2 - 22500 = 0$. Using a graphing device, we first graph the polynomial in the viewing rectangle $[0, 15]$ by $[-10000, 10000]$. The solution, $r \approx 11.28$ ft., is shown in the viewing rectangle $[11.2, 11.4]$ by $[-1, 1]$.

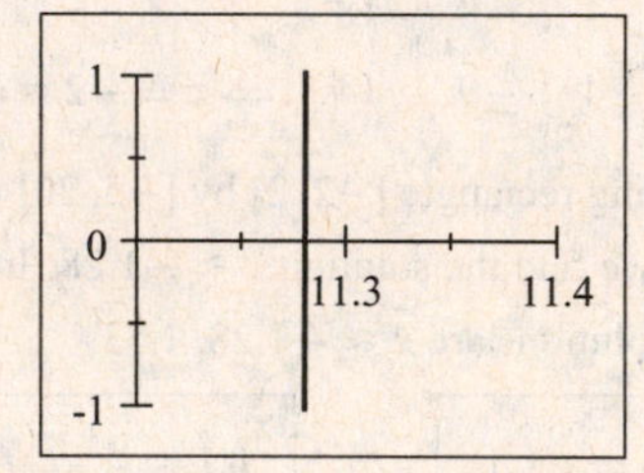

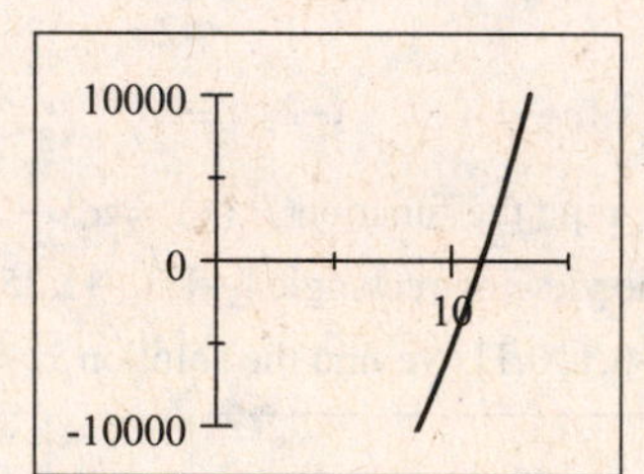

101. $h(t) = 11.60t - 12.41t^2 + 6.20t^3 - 1.58t^4 + 0.20t^5 - 0.01t^6$ is shown in the viewing rectangle $[0, 10]$ by $[0, 6]$.

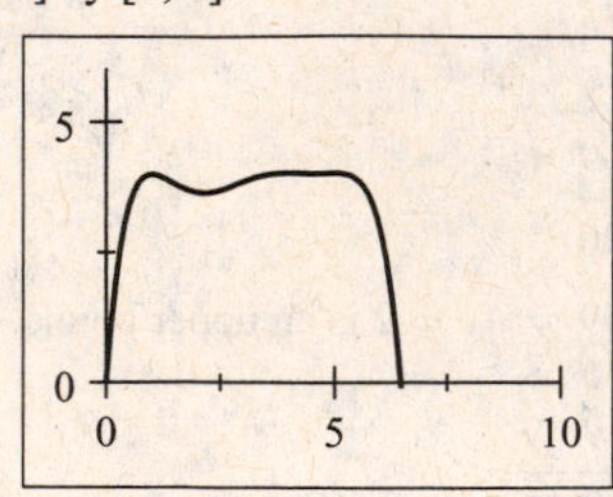

(a) It started to snow again.

(b) No, $h(t) \leq 4$.

(c) The function $h(t)$ is shown in the viewing rectangle $[6, 6.5]$ by $[0, 0.5]$. The x-intercept of the function is a little less than 6.5, which means that the snow melted just before midnight on Saturday night.

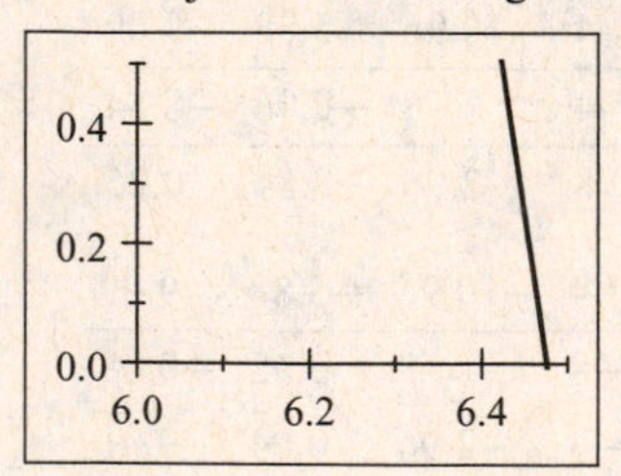

103. Let r be the radius of the cone and cylinder and let h be the height of the cone. Since the height and diameter are equal, we get $h = 2r$. So the volume of the cylinder is $V_1 = \pi r^2 \cdot (\text{cylinder height}) = 20\pi r^2$, and the volume of the cone is $V_2 = \frac{1}{3}\pi r^2 h = \frac{1}{3}\pi r^2 (2r) = \frac{2}{3}\pi r^3$. Since the total volume is $\dfrac{500\pi}{3}$, it follows that $\frac{2}{3}\pi r^3 + 20\pi r^2 = \dfrac{500\pi}{3} \Leftrightarrow r^3 + 30r^2 - 250 = 0$. By Descartes' Rule of Signs, there is 1 positive zero. Since r is between 2.76 and 2.765 (see the table), the radius should be 2.76 m (correct to two decimals).

r	$r^3 + 30r^2 - 250$
1	−219
2	−122
3	47
2.7	−11.62
2.76	−2.33
2.77	1.44
2.765	1.44
2.8	7.15

105. Let b be the width of the base, and let l be the length of the box. Then the length plus girth is $l + 4b = 108$, and the volume is $V = lb^2 = 2200$. Solving the first equation for l and substituting this value into the second equation yields $l = 108 - 4b$ $\Rightarrow V = (108 - 4b)\, b^2 = 2200 \Leftrightarrow 4b^3 - 108b^2 + 2200 = 0 \Leftrightarrow 4\left(b^3 - 27b^2 + 550\right) = 0$. Now $P(b) = b^3 - 27b^2 + 550$ has two variations in sign, so there are 0 or 2 positive real zeros. We also observe that since $l > 0, b < 27$, so $b = 27$ is an upper bound. Thus the possible positive rational real zeros are 1, 2, 3, 10, 11, 22, 25.

$$\begin{array}{r|rrrr} 1 & 1 & -27 & 0 & 550 \\ & & 1 & -26 & -26 \\ \hline & 1 & -26 & -26 & 524 \end{array} \qquad \begin{array}{r|rrrr} 2 & 1 & -27 & 0 & 550 \\ & & 2 & -50 & -100 \\ \hline & 1 & -25 & -50 & 450 \end{array}$$

$$\begin{array}{r|rrrr} 5 & 1 & -27 & 0 & 550 \\ & & 5 & -110 & -550 \\ \hline & 1 & -22 & -110 & 0 \end{array} \quad \Rightarrow b = 5 \text{ is a zero.}$$

$P(b) = (b - 5)\left(b^2 - 22b - 110\right)$. The other zeros are $b = \frac{22 \pm \sqrt{484 - 4(1)(-110)}}{2} = \frac{22 \pm \sqrt{924}}{2} = \frac{22 \pm 30.397}{2}$. The positive answer from this factor is $b \approx 26.20$.Thus we have two possible solutions, $b = 5$ or $b \approx 26.20$. If $b = 5$, then $l = 108 - 4(5) = 88$; if $b \approx 26.20$, then $l = 108 - 4(26.20) = 3.20$. Thus the length of the box is either 88 in. or 3.20 in.

107. **(a)** Substituting $X - \dfrac{a}{3}$ for x we have

$$\begin{aligned} x^3 + ax^2 + bx + c &= \left(X - \frac{a}{3}\right)^3 + a\left(X - \frac{a}{3}\right)^2 + b\left(X - \frac{a}{3}\right) + c \\ &= X^3 - aX^2 + \frac{a^2}{3}X + \frac{a^3}{27} + a\left(X^2 - \frac{2a}{3}X + \frac{a^2}{9}\right) + bX - \frac{ab}{3} + c \\ &= X^3 - aX^2 + \frac{a^2}{3}X + \frac{a^3}{27} + aX^2 - \frac{2a^2}{3}X + \frac{a^3}{9} + bX - \frac{ab}{3} + c \\ &= X^3 + (-a + a)X^2 + \left(-\frac{a^2}{3} - \frac{2a^2}{3} + b\right)X + \left(\frac{a^3}{27} + \frac{a^3}{9} - \frac{ab}{3} + c\right) \\ &= X^3 + \left(b - a^2\right)X + \left(\frac{4a^3}{27} - \frac{ab}{3} + c\right) \end{aligned}$$

(b) $x^3 + 6x^2 + 9x + 4 = 0$. Setting $a = 6$, $b = 9$, and $c = 4$, we have: $X^3 + \left(9 - 6^2\right)X + (32 - 18 + 4) = X^3 - 27X + 18$.

109. (a) Since $z > b$, we have $z - b > 0$. Since all the coefficients of $Q(x)$ are nonnegative, and since $z > 0$, we have $Q(z) > 0$ (being a sum of positive terms). Thus, $P(z) = (z - b) \cdot Q(z) + r > 0$, since the sum of a positive number and a nonnegative number.

(b) In part (a), we showed that if b satisfies the conditions of the first part of the Upper and Lower Bounds Theorem and $z > b$, then $P(z) > 0$. This means that no real zero of P can be larger than b, so b is an upper bound for the real zeros.

(c) Suppose $-b$ is a negative lower bound for the real zeros of $P(x)$. Then clearly b is an upper bound for $P_1(x) = P(-x)$. Thus, as in Part (a), we can write $P_1(x) = (x - b) \cdot Q(x) + r$, where $r > 0$ and the coefficients of Q are all nonnegative, and $P(x) = P_1(-x) = (-x - b) \cdot Q(-x) + r = (x + b) \cdot [-Q(-x)] + r$. Since the coefficients of $Q(x)$ are all nonnegative, the coefficients of $-Q(-x)$ will be alternately nonpositive and nonnegative, which proves the second part of the Upper and Lower Bounds Theorem.

3.5 COMPLEX ZEROS AND THE FUNDAMENTAL THEOREM OF ALGEBRA

1. The polynomial $P(x) = 5x^2 (x - 4)^3 (x + 7)$ has degree 6. It has zeros 0, 4, and -7. The zero 0 has multiplicity 2, and the zero 4 has multiplicity 3.

3. A polynomial of degree $n \geq 1$ has exactly n zeros, if a zero of multiplicity m is counted m times.

5. $P(x) = x^4 + 1$

(a) True, P has degree 4, so by the Zeros Theorem it has four (not necessarily distinct) complex zeros.

(b) True, by the Complete Factorization Theorem, this is true.

(c) False, the fourth power of any real number is nonnegative, so $P(x) \geq 1$ for all real x and P has no real zeros.

7. (a) $x^4 + 4x^2 = 0 \Leftrightarrow x^2 \left(x^2 + 4\right) = 0$. So $x = 0$ or $x^2 + 4 = 0$. If $x^2 + 4 = 0$ then $x^2 = -4 \Leftrightarrow x = \pm 2i$. Therefore, the solutions are $x = 0$ and $\pm 2i$.

(b) To get the complete factorization, we factor the remaining quadratic factor $P(x) = x^2 (x + 4) = x^2 (x - 2i)(x + 2i)$.

9. (a) $x^3 - 2x^2 + 2x = 0 \Leftrightarrow x\left(x^2 - 2x + 2\right) = 0$. So $x = 0$ or $x^2 - 2x + 2 = 0$. If $x^2 - 2x + 2 = 0$ then $x = \frac{-(-2) \pm \sqrt{(-2)^2 - 4(1)(2)}}{2} = \frac{2 \pm \sqrt{-4}}{2} = \frac{2 \pm 2i}{2} = 1 \pm i$. Therefore, the solutions are $x = 0, 1 \pm i$.

(b) Since $1 - i$ and $1 + i$ are zeros, $x - (1 - i) = x - 1 + i$ and $x - (1 + i) = x - 1 - i$ are the factors of $x^2 - 2x + 2$. Thus the complete factorization is $P(x) = x\left(x^2 - 2x + 2\right) = x(x - 1 + i)(x - 1 - i)$.

11. (a) $x^4 + 2x^2 + 1 = 0 \Leftrightarrow \left(x^2 + 1\right)^2 = 0 \Leftrightarrow x^2 + 1 = 0 \Leftrightarrow x^2 = -1 \Leftrightarrow x = \pm i$. Therefore the zeros of P are $x = \pm i$.

(b) Since $-i$ and i are zeros, $x + i$ and $x - i$ are the factors of $x^2 + 1$. Thus the complete factorization is $P(x) = \left(x^2 + 1\right)^2 = [(x + i)(x - i)]^2 = (x + i)^2 (x - i)^2$.

13. (a) $x^4 - 16 = 0 \Leftrightarrow 0 = \left(x^2 - 4\right)\left(x^2 + 4\right) = (x - 2)(x + 2)\left(x^2 + 4\right)$. So $x = \pm 2$ or $x^2 + 4 = 0$. If $x^2 + 4 = 0$ then $x^2 = -4 \Rightarrow x = \pm 2i$. Therefore the zeros of P are $x = \pm 2, \pm 2i$.

(b) Since $-i$ and i are zeros, $x + i$ and $x - i$ are the factors of $x^2 + 1$. Thus the complete factorization is $P(x) = (x - 2)(x + 2)\left(x^2 + 4\right) = (x - 2)(x + 2)(x - 2i)(x + 2i)$.

15. (a) $x^3 + 8 = 0 \Leftrightarrow (x + 2)\left(x^2 - 2x + 4\right) = 0$. So $x = -2$ or $x^2 - 2x + 4 = 0$. If $x^2 - 2x + 4 = 0$ then $x = \frac{-(-2) \pm \sqrt{(-2)^2 - 4(1)(4)}}{2} = \frac{2 \pm \sqrt{-12}}{2} = \frac{2 \pm 2i\sqrt{3}}{2} = 1 \pm i\sqrt{3}$. Therefore, the zeros of P are $x = -2, 1 \pm i\sqrt{3}$.

(b) Since $1-i\sqrt{3}$ and $1+i\sqrt{3}$ are the zeros from the $x^2-2x+4=0$, $x-\left(1-i\sqrt{3}\right)$ and $x-\left(1+i\sqrt{3}\right)$ are the factors of x^2-2x+4. Thus the complete factorization is

$$P(x)=(x+2)\left(x^2-2x+4\right)=(x+2)\left[x-\left(1-i\sqrt{3}\right)\right]\left[x-\left(1+i\sqrt{3}\right)\right]$$
$$=(x+2)\left(x-1+i\sqrt{3}\right)\left(x-1-i\sqrt{3}\right)$$

17. (a) $x^6-1=0\Leftrightarrow 0=\left(x^3-1\right)\left(x^3+1\right)=(x-1)\left(x^2+x+1\right)(x+1)\left(x^2-x+1\right)$. Clearly, $x=\pm1$ are solutions. If $x^2+x+1=0$, then $x=\frac{-1\pm\sqrt{1-4(1)(1)}}{2}=\frac{-1\pm\sqrt{-3}}{2}=-\frac{1}{2}\pm\frac{\sqrt{-3}}{2}$ so $x=-\frac{1}{2}\pm i\frac{\sqrt{3}}{2}$. And if $x^2-x+1=0$, then $x=\frac{1\pm\sqrt{1-4(1)(1)}}{2}=\frac{1\pm\sqrt{-3}}{2}=\frac{1}{2}\pm\frac{\sqrt{-3}}{2}=\frac{1}{2}\pm i\frac{\sqrt{3}}{2}$. Therefore, the zeros of P are $x=\pm1, -\frac{1}{2}\pm i\frac{\sqrt{3}}{2}, \frac{1}{2}\pm i\frac{\sqrt{3}}{2}$.

(b) The zeros of $x^2+x+1=0$ are $-\frac{1}{2}-i\frac{\sqrt{3}}{2}$ and $-\frac{1}{2}+i\frac{\sqrt{3}}{2}$, so x^2+x+1 factors as $\left[x-\left(-\frac{1}{2}-i\frac{\sqrt{3}}{2}\right)\right]\left[x-\left(-\frac{1}{2}+i\frac{\sqrt{3}}{2}\right)\right]=\left(x+\frac{1}{2}+i\frac{\sqrt{3}}{2}\right)\left(x+\frac{1}{2}-i\frac{\sqrt{3}}{2}\right)$. Similarly,since the zeros of $x^2-x+1=0$ are $\frac{1}{2}-i\frac{\sqrt{3}}{2}$ and $\frac{1}{2}+i\frac{\sqrt{3}}{2}$, so x^2-x+1 factors as $\left[x-\left(\frac{1}{2}-i\frac{\sqrt{3}}{2}\right)\right]\left[x-\left(\frac{1}{2}+i\frac{\sqrt{3}}{2}\right)\right]=\left(x-\frac{1}{2}+i\frac{\sqrt{3}}{2}\right)\left(x-\frac{1}{2}-i\frac{\sqrt{3}}{2}\right)$. Thus the complete factorization is

$$P(x)=(x-1)\left(x^2+x+1\right)(x+1)\left(x^2-x+1\right)$$
$$=(x-1)(x+1)\left(x+\frac{1}{2}+i\frac{\sqrt{3}}{2}\right)\left(x+\frac{1}{2}-i\frac{\sqrt{3}}{2}\right)\left(x-\frac{1}{2}+i\frac{\sqrt{3}}{2}\right)\left(x-\frac{1}{2}-i\frac{\sqrt{3}}{2}\right)$$

19. $P(x)=x^2+25=(x-5i)(x+5i)$. The zeros of P are $5i$ and $-5i$, both multiplicity 1.

21. $Q(x)=x^2+2x+2$. Using the Quadratic Formula $x=\frac{-(2)\pm\sqrt{(2)^2-4(1)(2)}}{2(1)}=\frac{-2\pm\sqrt{-4}}{2}=\frac{-2\pm2i}{2}=-1\pm i$. So $Q(x)=(x+1-i)(x+1+i)$. The zeros of Q are $-1-i$ (multiplicity 1) and $-1+i$ (multiplicity 1).

23. $P(x)=x^3+4x=x\left(x^2+4\right)=x(x-2i)(x+2i)$. The zeros of P are 0, $2i$, and $-2i$ (all multiplicity 1).

25. $Q(x)=x^4-1=\left(x^2-1\right)\left(x^2+1\right)=(x-1)(x+1)\left(x^2+1\right)=(x-1)(x+1)(x-i)(x+i)$. The zeros of Q are $1, -1, i$, and $-i$ (all of multiplicity 1).

27. $P(x)=16x^4-81=\left(4x^2-9\right)\left(4x^2+9\right)=(2x-3)(2x+3)(2x-3i)(2x+3i)$. The zeros of P are $\frac{3}{2}, -\frac{3}{2}, \frac{3}{2}i$, and $-\frac{3}{2}i$ (all of multiplicity 1).

29. $P(x)=x^3+x^2+9x+9=x^2(x+1)+9(x+1)=(x+1)\left(x^2+9\right)=(x+1)(x-3i)(x+3i)$. The zeros of P are -1, $3i$, and $-3i$ (all of multiplicity 1).

31. $Q(x)=x^4+2x^2+1=\left(x^2+1\right)^2=(x-i)^2(x+i)^2$. The zeros of Q are i and $-i$ (both of multiplicity 2).

33. $P(x)=x^4+3x^2-4=\left(x^2-1\right)\left(x^2+4\right)=(x-1)(x+1)(x-2i)(x+2i)$. The zeros of P are $1, -1, 2i$, and $-2i$ (all of multiplicity 1).

35. $P(x)=x^5+6x^3+9x=x\left(x^4+6x^2+9\right)=x\left(x^2+3\right)^2=x\left(x-i\sqrt{3}\right)^2\left(x+i\sqrt{3}\right)^2$. The zeros of P are 0 (multiplicity 1), $i\sqrt{3}$ (multiplicity 2), and $-i\sqrt{3}$ (multiplicity 2).

37. Since $1+i$ and $1-i$ are conjugates, the factorization of the polynomial must be $P(x)=a(x-[1+i])(x-[1-i])=a\left(x^2-2x+2\right)$. If we let $a=1$, we get $P(x)=x^2-2x+2$.

39. Since $2i$ and $-2i$ are conjugates, the factorization of the polynomial must be $Q(x) = b(x-3)(x-2i)(x+2i] = b(x-3)\left(x^2+4\right) = b\left(x^3-3x^2+4x-12\right)$. If we let $b = 1$, we get $Q(x) = x^3 - 3x^2 + 4x - 12$.

41. Since i is a zero, by the Conjugate Roots Theorem, $-i$ is also a zero. So the factorization of the polynomial must be $P(x) = a(x-2)(x-i)(x+i) = a\left(x^3 - 2x^2 + x - 2\right)$. If we let $a = 1$, we get $P(x) = x^3 - 2x^2 + x - 2$.

43. Since the zeros are $1-2i$ and 1 (with multiplicity 2), by the Conjugate Roots Theorem, the other zero is $1+2i$. So a factorization is

$$\begin{aligned} R(x) &= c(x-[1-2i])(x-[1+2i])(x-1)^2 = c([x-1]+2i)([x-1]-2i)(x-1)^2 \\ &= c\left([x-1]^2-[2i]^2\right)\left(x^2-2x+1\right) = c\left(x^2-2x+1+4\right)\left(x^2-2x+1\right) = c\left(x^2-2x+5\right)\left(x^2-2x+1\right) \\ &= c\left(x^4-2x^3+x^2-2x^3+4x^2-2x+5x^2-10x+5\right) = c\left(x^4-4x^3+10x^2-12x+5\right) \end{aligned}$$

If we let $c = 1$ we get $R(x) = x^4 - 4x^3 + 10x^2 - 12x + 5$.

45. Since the zeros are i and $1+i$, by the Conjugate Roots Theorem, the other zeros are $-i$ and $1-i$. So a factorization is

$$\begin{aligned} T(x) &= C(x-i)(x+i)(x-[1+i])(x-[1-i]) \\ &= C\left(x^2-i^2\right)([x-1]-i)([x-1]+i) = C\left(x^2+1\right)\left(x^2-2x+1-i^2\right) = C\left(x^2+1\right)\left(x^2-2x+2\right) \\ &= C\left(x^4-2x^3+2x^2+x^2-2x+2\right) = C\left(x^4-2x^3+3x^2-2x+2\right) = Cx^4-2Cx^3+3Cx^2-2Cx+2C \end{aligned}$$

Since the constant coefficient is 12, it follows that $2C = 12 \Leftrightarrow C = 6$, and so $T(x) = 6\left(x^4-2x^3+3x^2-2x+2\right) = 6x^4-12x^3+18x^2-12x+12$.

47. $P(x) = x^3+2x^2+4x+8 = x^2(x+2)+4(x+2) = (x+2)\left(x^2+4\right) = (x+2)(x-2i)(x+2i)$. Thus the zeros are -2 and $\pm 2i$.

49. $P(x) = x^3-2x^2+2x-1$. By inspection, $P(1) = 1-2+2-1 = 0$, and hence $x = 1$ is a zero.

$$\begin{array}{r|rrrr} 1 & 1 & -2 & 2 & -1 \\ & & 1 & -1 & 1 \\ \hline & 1 & -1 & 1 & 0 \end{array}$$

Thus $P(x) = (x-1)\left(x^2-x+1\right)$. So $x = 1$ or $x^2-x+1 = 0$.

Using the Quadratic Formula, we have $x = \frac{1 \pm \sqrt{1-4(1)(1)}}{2} = \frac{1 \pm i\sqrt{3}}{2}$. Hence, the zeros are 1 and $\frac{1 \pm i\sqrt{3}}{2}$.

51. $P(x) = x^3-3x^2+3x-2$.

$$\begin{array}{r|rrrr} 2 & 1 & -3 & 3 & -2 \\ & & 2 & -2 & 2 \\ \hline & 1 & -1 & 1 & 0 \end{array}$$

Thus $P(x) = (x-2)\left(x^2-x+1\right)$. So $x = 2$ or $x^2-x+1 = 0$

Using the Quadratic Formula we have $x = \frac{1 \pm \sqrt{1-4(1)(1)}}{2} = \frac{1 \pm i\sqrt{3}}{2}$. Hence, the zeros are 2, and $\frac{1 \pm i\sqrt{3}}{2}$.

53. $P(x) = 2x^3 + 7x^2 + 12x + 9$ has possible rational zeros $\pm1, \pm3, \pm9, \pm\frac{1}{2}, \pm\frac{3}{2}, \pm\frac{9}{2}$. Since all coefficients are positive, there are no positive real zeros.

$$\begin{array}{r|rrrr} -1 & 2 & 7 & 12 & 9 \\ & & -2 & -5 & -7 \\ \hline & 2 & 5 & 7 & 2 \end{array} \qquad \begin{array}{r|rrrr} -2 & 2 & 7 & 12 & 9 \\ & & -4 & -6 & -12 \\ \hline & 2 & 3 & 6 & -3 \end{array}$$

There is a zero between -1 and -2.

$$\begin{array}{r|rrrr} -\frac{3}{2} & 2 & 7 & 12 & 9 \\ & & -3 & -6 & -9 \\ \hline & 2 & 4 & 6 & 0 \end{array} \Rightarrow x = -\tfrac{3}{2} \text{ is a zero.}$$

$P(x) = \left(x + \frac{3}{2}\right)\left(2x^2 + 4x + 6\right) = 2\left(x + \frac{3}{2}\right)\left(x^2 + 2x + 3\right)$. Now $x^2 + 2x + 3$ has zeros $x = \frac{-2\pm\sqrt{4-4(3)(1)}}{2} = \frac{-2\pm2\sqrt{-2}}{2} = -1 \pm i\sqrt{2}$. Hence, the zeros are $-\frac{3}{2}$ and $-1 \pm i\sqrt{2}$.

55. $P(x) = x^4 + x^3 + 7x^2 + 9x - 18$. Since $P(x)$ has one change in sign, we are guaranteed a positive zero, and since $P(-x) = x^4 - x^3 + 7x^2 - 9x - 18$, there are 1 or 3 negative zeros.

$$\begin{array}{r|rrrrr} 1 & 1 & 1 & 7 & 9 & -18 \\ & & 1 & 2 & 9 & 18 \\ \hline & 1 & 2 & 9 & 18 & 0 \end{array}$$

Therefore, $P(x) = (x - 1)\left(x^3 + 2x^2 + 9x + 18\right)$. Continuing with the quotient, we try negative zeros.

$$\begin{array}{r|rrrr} -1 & 1 & 2 & 9 & 18 \\ & & -1 & -1 & -8 \\ \hline & 1 & 1 & 8 & 10 \end{array} \qquad \begin{array}{r|rrrr} -2 & 1 & 2 & 9 & 18 \\ & & -2 & 0 & -18 \\ \hline & 1 & 0 & 9 & 0 \end{array}$$

$P(x) = (x - 1)(x + 2)\left(x^2 + 9\right) = (x - 1)(x + 2)(x - 3i)(x + 3i)$. Therefore,the zeros are $1, -2$, and $\pm3i$.

57. We see a pattern and use it to factor by grouping. This gives

$$\begin{aligned} P(x) &= x^5 - x^4 + 7x^3 - 7x^2 + 12x - 12 = x^4(x - 1) + 7x^2(x - 1) + 12(x - 1) = (x - 1)\left(x^4 + 7x^2 + 12\right) \\ &= (x - 1)\left(x^2 + 3\right)\left(x^2 + 4\right) = (x - 1)\left(x - i\sqrt{3}\right)\left(x + i\sqrt{3}\right)(x - 2i)(x + 2i) \end{aligned}$$

Therefore,the zeros are $1, \pm i\sqrt{3}$, and $\pm2i$.

59. $P(x) = x^4 - 6x^3 + 13x^2 - 24x + 36$ has possible rational zeros $\pm1, \pm2, \pm3, \pm4, \pm6, \pm9, \pm12, \pm18$. $P(x)$ has 4 variations in sign and $P(-x)$ has no variation in sign.

$$\begin{array}{r|rrrrr} 1 & 1 & -6 & 13 & -24 & 36 \\ & & 1 & -5 & 8 & -16 \\ \hline & 1 & -5 & 8 & -16 & 20 \end{array} \quad \begin{array}{r|rrrrr} 2 & 1 & -6 & 13 & -24 & 36 \\ & & 2 & -8 & 10 & -28 \\ \hline & 1 & -4 & 5 & -14 & 8 \end{array} \quad \begin{array}{r|rrrrr} 3 & 1 & -6 & 13 & -24 & 36 \\ & & 3 & -9 & 12 & -36 \\ \hline & 1 & -3 & 4 & -12 & 0 \end{array} \Rightarrow x = 3 \text{ is a zero.}$$

Continuing:

$$\begin{array}{r|rrrr} 3 & 1 & -3 & 4 & -12 \\ & & 3 & 0 & 12 \\ \hline & 1 & 0 & 4 & 0 \end{array} \Rightarrow x = 3 \text{ is a zero.}$$

$P(x) = (x - 3)^2\left(x^2 + 4\right) = (x - 3)^2(x - 2i)(x + 2i)$. Therefore,the zeros are 3 (multiplicity 2) and $\pm2i$.

61. $P(x) = 4x^4 + 4x^3 + 5x^2 + 4x + 1$ has possible rational zeros $\pm 1, \pm\frac{1}{2}, \pm\frac{1}{4}$. Since there is no variation in sign, all real zeros (if there are any) are negative.

$$\begin{array}{r|rrrrr} -1 & 4 & 4 & 5 & 4 & 1 \\ & & -4 & 0 & -5 & 1 \\ \hline & 4 & 0 & 5 & -1 & 2 \end{array} \qquad \begin{array}{r|rrrrr} -\frac{1}{2} & 4 & 4 & 5 & 4 & 1 \\ & & -2 & -1 & -2 & -1 \\ \hline & 4 & 2 & 4 & 2 & 0 \end{array} \Rightarrow x = -\tfrac{1}{2} \text{ is a zero.}$$

$P(x) = \left(x + \frac{1}{2}\right)\left(4x^3 + 2x^2 + 4x + 2\right)$. Continuing:

$$\begin{array}{r|rrrr} -\frac{1}{2} & 4 & 2 & 4 & 2 \\ & & -2 & 0 & -2 \\ \hline & 4 & 0 & 4 & 0 \end{array} \Rightarrow x = -\tfrac{1}{2} \text{ is a zero again.}$$

$P(x) = \left(x + \frac{1}{2}\right)^2\left(4x^2 + 4\right)$. Thus, the zeros of $P(x)$ are $-\frac{1}{2}$ (multiplicity 2) and $\pm i$.

63. $P(x) = x^5 - 3x^4 + 12x^3 - 28x^2 + 27x - 9$ has possible rational zeros $\pm 1, \pm 3, \pm 9$. $P(x)$ has 4 variations in sign and $P(-x)$ has 1 variation in sign.

$$\begin{array}{r|rrrrrr} 1 & 1 & -3 & 12 & -28 & 27 & -9 \\ & & 1 & -2 & 10 & -18 & 9 \\ \hline & 1 & -2 & 10 & -18 & 9 & 0 \end{array} \Rightarrow x = 1 \text{ is a zero.}$$

$$\begin{array}{r|rrrrr} 1 & 1 & -2 & 10 & -18 & 9 \\ & & 1 & -1 & 9 & -9 \\ \hline & 1 & -1 & 9 & -9 & 0 \end{array} \Rightarrow x = 1 \text{ is a zero.} \qquad \begin{array}{r|rrrr} 1 & 1 & -1 & 9 & -9 \\ & & 1 & 0 & 9 \\ \hline & 1 & 0 & 9 & 0 \end{array} \Rightarrow x = 1 \text{ is a zero.}$$

$P(x) = (x-1)^3\left(x^2+9\right) = (x-1)^3(x-3i)(x+3i)$. Therefore,the zeros are 1 (multiplicity 3) and $\pm 3i$.

65. (a) $P(x) = x^3 - 5x^2 + 4x - 20 = x^2(x-5) + 4(x-5) = (x-5)\left(x^2+4\right)$

(b) $P(x) = (x-5)(x-2i)(x+2i)$

67. (a) $P(x) = x^4 + 8x^2 - 9 = \left(x^2-1\right)\left(x^2+9\right) = (x-1)(x+1)\left(x^2+9\right)$

(b) $P(x) = (x-1)(x+1)(x-3i)(x+3i)$

69. (a) $P(x) = x^6 - 64 = \left(x^3-8\right)\left(x^3+8\right) = (x-2)\left(x^2+2x+4\right)(x+2)\left(x^2-2x+4\right)$

(b) $P(x) = (x-2)(x+2)\left(x+1-i\sqrt{3}\right)\left(x+1+i\sqrt{3}\right)\left(x-1-i\sqrt{3}\right)\left(x-1+i\sqrt{3}\right)$

71. (a) $x^4 - 2x^3 - 11x^2 + 12x = x\left(x^3 - 2x^2 - 11x + 12\right) = 0$. We first find the bounds for our viewing rectangle.

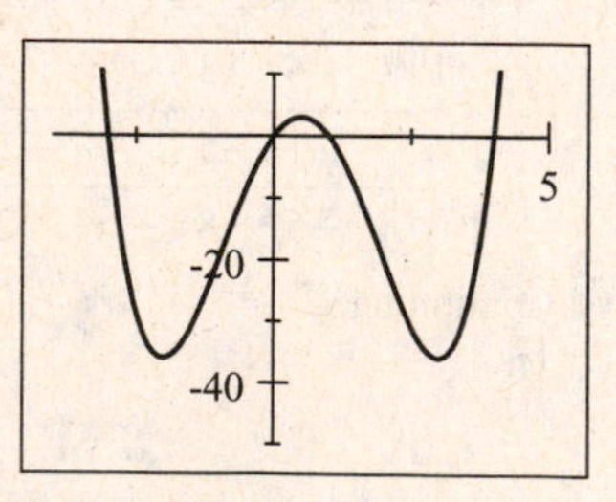

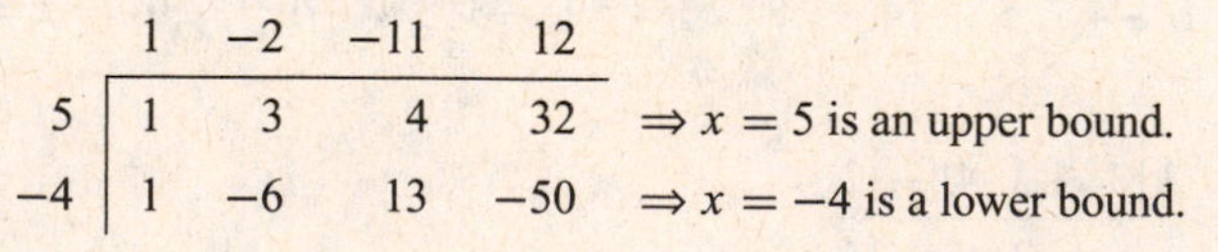

$$\begin{array}{r|rrrrl} & 1 & -2 & -11 & 12 & \\ \hline 5 & 1 & 3 & 4 & 32 & \Rightarrow x = 5 \text{ is an upper bound.} \\ -4 & 1 & -6 & 13 & -50 & \Rightarrow x = -4 \text{ is a lower bound.} \end{array}$$

We graph $P(x) = x^4 - 2x^3 - 11x^2 + 12x$ in the viewing rectangle $[-4, 5]$ by $[-50, 10]$ and see that it has 4 real solutions. Since this matches the degree of $P(x)$, $P(x)$ has no nonreal solution.

(b) $x^4 - 2x^3 - 11x^2 + 12x - 5 = 0$. We use the same bounds for our viewing rectangle, $[-4, 5]$ by $[-50, 10]$, and see that $R(x) = x^4 - 2x^3 - 11x^2 + 12x - 5$ has 2 real solutions. Since the degree of $R(x)$ is 4, $R(x)$ must have 2 nonreal solutions.

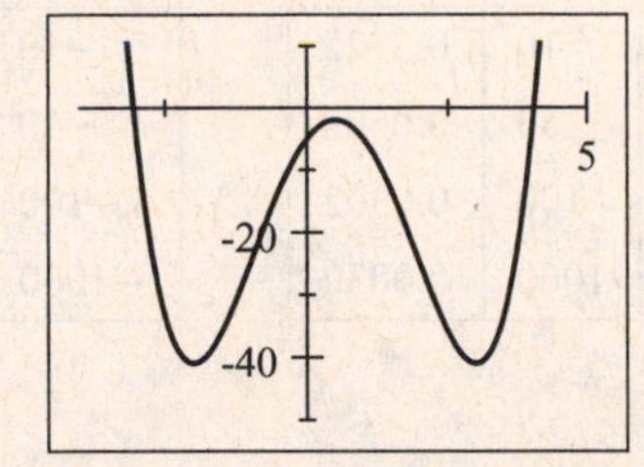

(c) $x^4 - 2x^3 - 11x^2 + 12x + 40 = 0$. We graph $T(x) = x^4 - 2x^3 - 11x^2 + 12x + 40$ in the viewing rectangle $[-4, 5]$ by $[-10, 50]$, and see that T has no real solution. Since the degree of T is 4, T must have 4 nonreal solutions.

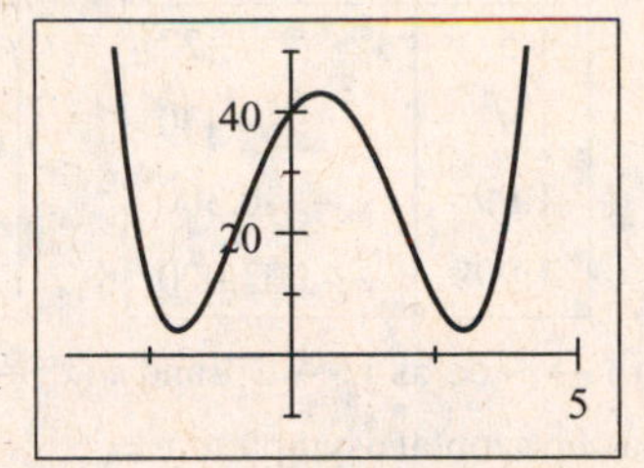

73. (a) $P(x) = x^2 - (1+i)x + (2+2i)$. So $P(2i) = (2i)^2 - (1+i)(2i) + 2 + 2i = -4 - 2i + 2 + 2 + 2i = 0$, and $P(1-i) = (1-i)^2 - (1+i)(1-i) + (2+2i) = 1 - 2i - 1 - 1 - 1 + 2 + 2i = 0$.
Therefore, $2i$ and $1-i$ are solutions of the equation $x^2 - (1+i)x + (2+2i) = 0$. However, $P(-2i) = (-2i)^2 - (1+i)(-2i) + 2 + 2i = -4 + 2i - 2 + 2 + 2i = -4 + 4i$, and $P(1+i) = (1+i)^2 - (1+i)(1+i) + 2 + 2i = 2 + 2i$. Since, $P(-2i) \neq 0$ and $P(1+i) \neq 0$, $-2i$ and $1+i$ are not solutions.

(b) This does not violate the Conjugate Roots Theorem because the coefficients of the polynomial $P(x)$ are not all real.

75. Because P has real coefficients, the imaginary zeros come in pairs: $a \pm bi$ (by the Conjugate Roots Theorem), where $b \neq 0$. Thus there must be an even number of nonreal zeros. Since P is of odd degree, it has an odd number of zeros (counting multiplicity). It follows that P has at least one real zero.

3.6 RATIONAL FUNCTIONS

1. If the rational function $y = r(x)$ has the vertical asymptote $x = 2$, then as $x \to 2^+$, either $y \to \infty$ or $y \to -\infty$.

3. The function $r(x) = \dfrac{(x+1)(x-2)}{(x+2)(x-3)}$ has x-intercepts -1 and 2.

5. The function r has vertical asymptotes $x = -2$ and $x = 3$.

7. $r(x) = \dfrac{x^2 + x}{(x+1)(2x-4)} = \dfrac{x(x+1)}{(x+1)(2x-4)} = \dfrac{x}{2(x-2)}$ for $x \neq -1$.

(a) False. r does not have vertical asymptote $x = -1$. It has a "hole" at $\left(-1, \frac{1}{6}\right)$.

(b) True, r has vertical asymptote $x = 2$.

(c) False, r has a horizontal asymptote $y = \frac{1}{2}$ but not a horizontal asymptote $y = 1$.

(d) True, r has horizontal asymptote $y = \frac{1}{2}$.

9. $r(x) = \dfrac{x}{x-2}$

(a)

x	$r(x)$
1.5	−3
1.9	−19
1.99	−199
1.999	−1999

x	$r(x)$
2.5	5
2.1	21
2.01	201
2.001	2001

x	$r(x)$
10	1.25
50	1.042
100	1.020
1000	1.002

x	$r(x)$
−10	0.833
−50	0.962
−100	0.980
−1000	0.998

(b) $r(x) \to -\infty$ as $x \to 2^-$ and $r(x) \to \infty$ as $x \to 2^+$.

(c) r has horizontal asymptote $y = 1$.

11. $r(x) = \dfrac{3x - 10}{(x-2)^2}$

(a)

x	$r(x)$
1.5	−22
1.9	−430
1.99	−40,300
1.999	−4,003,000

x	$r(x)$
2.5	−10
2.1	−370
2.01	−39,700
2.001	−3,997,000

x	$r(x)$
10	0.3125
50	0.0608
100	0.0302
1000	0.0030

x	$r(x)$
−10	−0.2778
−50	−0.0592
−100	−0.0298
−1000	−0.0030

(b) $r(x) \to -\infty$ as $x \to 2^-$ and $r(x) \to -\infty$ as $x \to 2^+$.

(c) r has horizontal asymptote $y = 0$.

In Exercises 13–20, let $f(x) = \dfrac{1}{x}$.

13. $r(x) = \dfrac{1}{x-1} = f(x-1)$. From this form we see that the graph of r is obtained from the graph of f by shifting 1 unit to the right. Thus r has vertical asymptote $x = 1$ and horizontal asymptote $y = 0$. The domain of r is $(-\infty, 1) \cup (1, \infty)$ and its range is $(-\infty, 0) \cup (0, \infty)$.

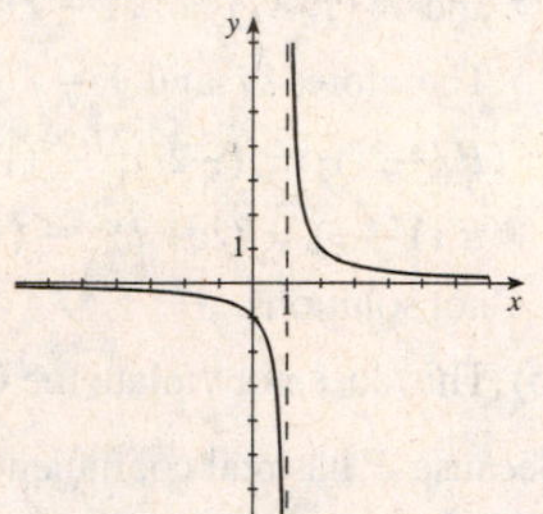

15. $s(x) = \dfrac{3}{x+1} = 3\left(\dfrac{1}{x+1}\right) = 3f(x+1)$. From this form we see that the graph of s is obtained from the graph of f by shifting 1 unit to the left and stretching vertically by a factor of 3. Thus s has vertical asymptote $x = -1$ and horizontal asymptote $y = 0$. The domain of s is $(-\infty, -1) \cup (-1, \infty)$ and its range is $(-\infty, 0) \cup (0, \infty)$.

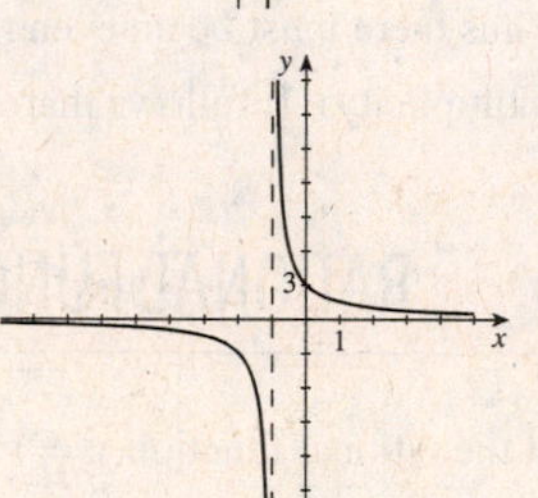

17. $t(x) = \dfrac{2x-3}{x-2} = 2 + \dfrac{1}{x-2} = f(x-2) + 2$ (see the long division at right). From this form we see that the graph of t is obtained from the graph of f by shifting 2 units to the right and 2 units vertically. Thus t has vertical asymptote $x = 2$ and horizontal asymptote $y = 2$. The domain of t is $(-\infty, 2) \cup (2, \infty)$ and its range is $(-\infty, 2) \cup (2, \infty)$.

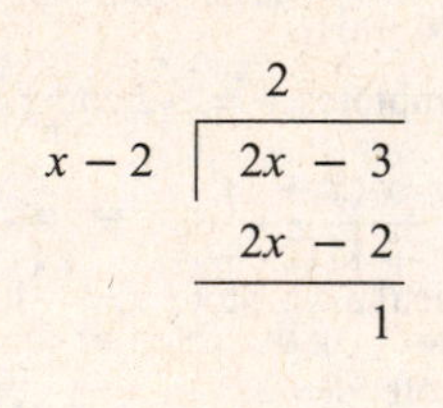

19. $r(x) = \dfrac{x+2}{x+3} = 1 - \dfrac{1}{x+3} = -f(x+3) + 1$ (see the long division at right). From this form we see that the graph of r is obtained from the graph of f by shifting 3 units to the left, reflect about the x-axis, and then shifting vertically 1 unit. Thus r has vertical asymptote $x = -3$ and horizontal asymptote $y = 1$. The domain of r is $(-\infty, -3) \cup (-3, \infty)$ and its range is $(-\infty, 1) \cup (1, \infty)$.

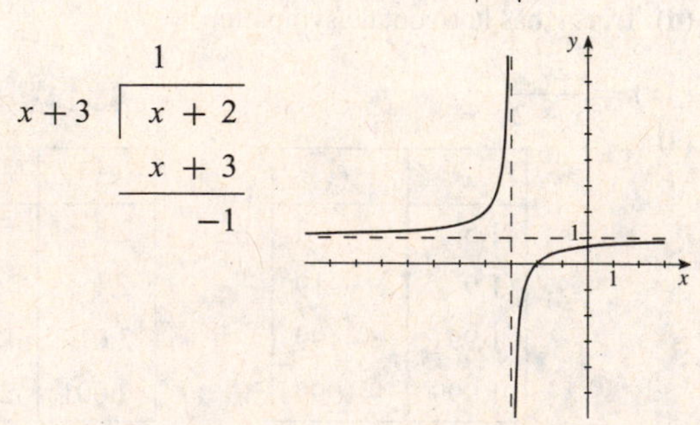

21. $r(x) = \dfrac{x-1}{x+4}$. When $x = 0$, we have $r(0) = -\frac{1}{4}$, so the y-intercept is $-\frac{1}{4}$. The numerator is 0 when $x = 1$, so the x-intercept is 1.

23. $t(x) = \dfrac{x^2 - x - 2}{x - 6}$. When $x = 0$, we have $t(0) = \dfrac{-2}{-6} = \frac{1}{3}$, so the y-intercept is $\frac{1}{3}$. The numerator is 0 when $x^2 - x - 2 = (x-2)(x+1) = 0$ or when $x = 2$ or $x = -1$, so the x-intercepts are 2 and -1.

25. $r(x) = \dfrac{x^2 - 9}{x^2}$. Since 0 is not in the domain of $r(x)$, there is no y-intercept. The numerator is 0 when $x^2 - 9 = (x-3)(x+3) = 0$ or when $x = \pm 3$, so the x-intercepts are ± 3.

27. From the graph, the x-intercept is 3, the y-intercept is 3, the vertical asymptote is $x = 2$, and the horizontal asymptote is $y = 2$.

29. From the graph, the x-intercepts are -1 and 1, the y-intercept is about $\frac{1}{4}$, the vertical asymptotes are $x = -2$ and $x = 2$, and the horizontal asymptote is $y = 1$.

31. $r(x) = \dfrac{5}{x-2}$ has a vertical asymptote where $x - 2 = 0 \Leftrightarrow x = 2$, and $y = 0$ is a horizontal asymptote because the degree of the denominator is greater than that of the numerator.

33. $r(x) = \dfrac{3x+1}{4x^2+1}$ has no vertical asymptote since $4x^2 + 1 > 0$ for all x. $y = 0$ is a horizontal asymptote because the degree of the denominator is greater than that of the numerator.

35. $s(x) = \dfrac{6x^2+1}{2x^2+x-1}$ has vertical asymptotes where $2x^2 + x - 1 = 0 \Leftrightarrow (x+1)(2x-1) = 0 \Leftrightarrow x = -1$ or $x = \frac{1}{2}$, and horizontal asymptote $y = \frac{6}{2} = 3$.

37. $r(x) = \dfrac{(x+1)(2x-3)}{(x-2)(4x+7)}$ has vertical asymptotes where $(x-2)(4x+7) = 0 \Leftrightarrow x = -\frac{7}{4}$ or $x = 2$, and horizontal asymptote $y = \dfrac{1 \cdot 2}{1 \cdot 4} = \dfrac{1}{2}$.

39. $r(x) = \dfrac{6x^3 - 2}{2x^3 + 5x^2 + 6x} = \dfrac{6x^3 - 2}{x(2x^2 + 5x + 6)}$. Because the quadratic in the denominator has no real zero, r has vertical asymptote $x = 0$ and horizontal asymptote $y = \frac{6}{2} = 3$.

41. $y = \dfrac{x^2+2}{x-1}$. A vertical asymptote occurs when $x - 1 = 0 \Leftrightarrow x = 1$. There is no horizontal asymptote because the degree of the numerator is greater than the degree of the denominator.

43. $y = \dfrac{4x-4}{x+2}$. When $x = 0$, $y = -2$, so the y-intercept is -2. When $y = 0$, $4x - 4 = 0 \Leftrightarrow x = 1$, so the x-intercept is 1. Since the degree of the numerator and denominator are the same, the horizontal asymptote is $y = \frac{4}{1} = 4$. A vertical asymptote occurs when $x = -2$. As $x \to -2^+$, $y = \dfrac{4x-4}{x+2} \to -\infty$, and as $x \to -2^-$, $y = \dfrac{4x-4}{x+2} \to \infty$. The domain is $\{x \mid x \neq -2\}$ and the range is $\{y \mid y \neq 4\}$.

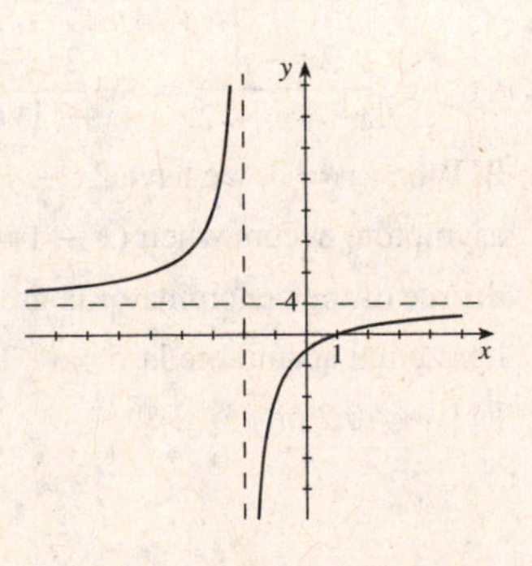

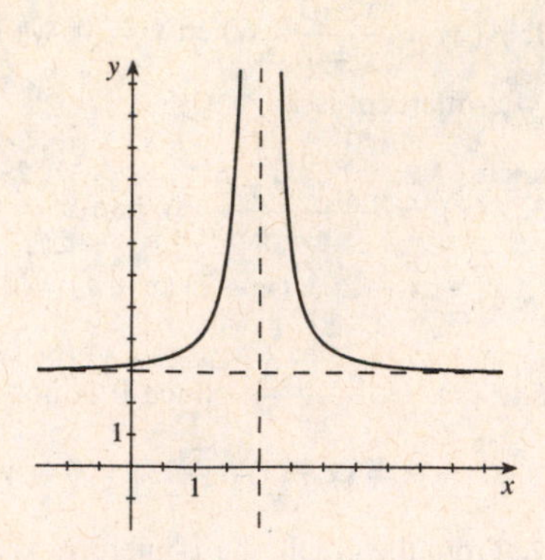

45. $r(x) = \dfrac{3x^2 - 12x + 13}{x^2 - 4x + 4} = \dfrac{3\left(x^2 - 4x + 4\right) + 1}{x^2 - 4x + 4} = 3 + \dfrac{1}{(x-2)^2}$. When $x = 0$, $y = \frac{13}{4}$, so the y-intercept is $\frac{13}{4}$. There is no x-intercept since $\dfrac{1}{(x-2)^2}$ is positive on its domain. There is a vertical asymptote at $x = 2$. The horizontal asymptote is $y = 3$. The domain is $\{x \mid x \neq 2\}$ and the range is $\{y \mid y > 3\}$.

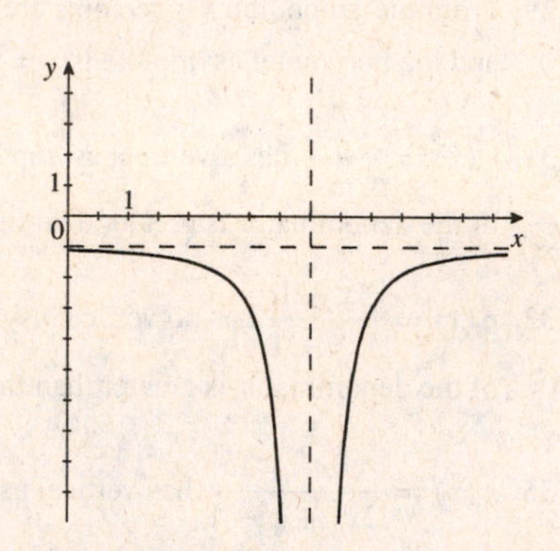

47. $r(x) = \dfrac{-x^2 + 8x - 18}{x^2 - 8x + 16} = \dfrac{-\left(x^2 - 8x + 16\right) - 2}{x^2 - 8x + 16} = -1 - \dfrac{2}{(x-4)^2}$. When $x = 0$, $y = -\frac{9}{8}$, and so the y-intercept is $-\frac{9}{8}$. There is no x-intercept since $\dfrac{2}{(x-4)^2}$ is positive on its domain. There is a vertical asymptote at $x = 4$. The horizontal asymptote is $y = -1$. The domain is $\{x \mid x \neq 4\}$ and the range is $\{y \mid y < -1\}$.

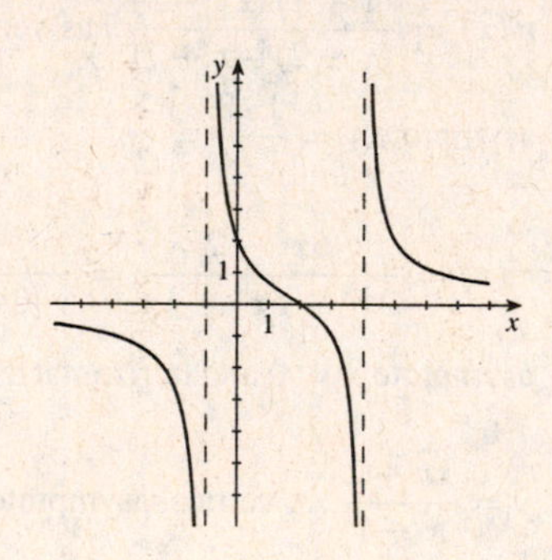

49. $s(x) = \dfrac{4x - 8}{(x-4)(x+1)}$. When $x = 0$, $y = \dfrac{-8}{(-4)(1)} = 2$, so the y-intercept is 2. When $y = 0$, $4x - 8 = 0 \Leftrightarrow x = 2$, so the x-intercept is 2. The vertical asymptotes are $x = -1$ and $x = 4$, and because the degree of the numerator is less than the degree of the denominator, the horizontal asymptote is $y = 0$. The domain is $\{x \mid x \neq -1, 4\}$ and the range is $\mathbb{R}$.

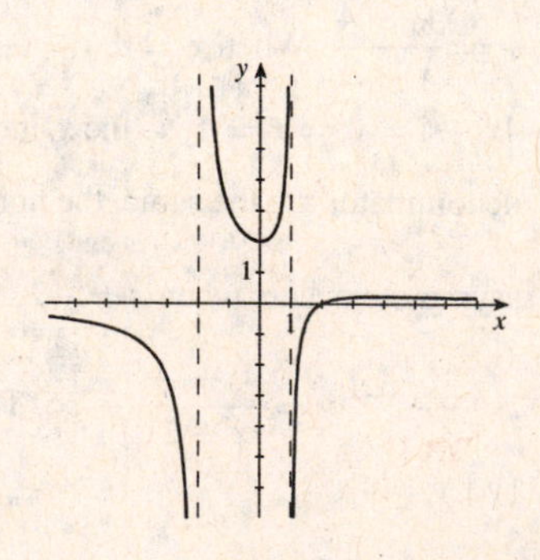

51. $s(x) = \dfrac{2x - 4}{x^2 + x - 2} = \dfrac{2(x-2)}{(x-1)(x+2)}$. When $x = 0$, $y = 2$, so the y-intercept is 2. When $y = 0$, we have $2x - 4 = 0 \Leftrightarrow x = 2$, so the x-intercept is 2. A vertical asymptote occurs when $(x-1)(x+2) = 0 \Leftrightarrow x = 1$ and $x = -2$. Because the degree of the denominator is greater than the degree of the numerator, the horizontal asymptote is $y = 0$. The domain is $\{x \mid x \neq -2, 1\}$ and the range is $\{y \mid y \leq 0.2 \text{ or } y \geq 2\}$.

53. $r(x) = \dfrac{(x-1)(x+2)}{(x+1)(x-3)}$. When $x = 0$, $y = \frac{2}{3}$, so the y-intercept is $\frac{2}{3}$. When $y = 0$, $(x-1)(x+2) = 0 \Rightarrow x = -2, 1$, so, the x-intercepts are -2 and 1. The vertical asymptotes are $x = -1$ and $x = 3$, and because the degree of the numerator and denominator are the same the horizontal asymptote is $y = \frac{1}{1} = 1$. The domain is $\{x \mid x \neq -1, 3\}$ and the range is $\mathbb{R}$.

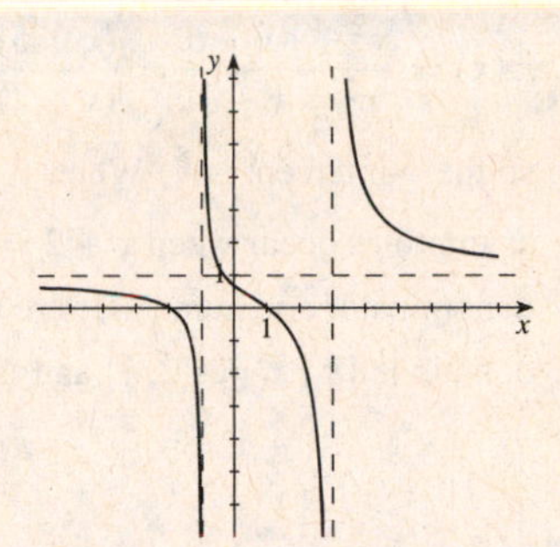

55. $r(x) = \dfrac{2x^2 + 2x - 4}{x^2 + x} = \dfrac{2(x+2)(x-1)}{x(x+1)}$. Vertical asymptotes occur at $x = 0$ and $x = -1$. Since x cannot equal zero, there is no y-intercept. When $y = 0$, we have $x = -2$ or 1, so the x-intercepts are -2 and 1. Because the degree of the denominator and numerator are the same, the horizontal asymptote is $y = \frac{2}{1} = 2$. The domain is $\{x \mid x \neq -1, 0\}$ and the range is $\{y \mid y < 2 \text{ or } y \geq 18.4\}$.

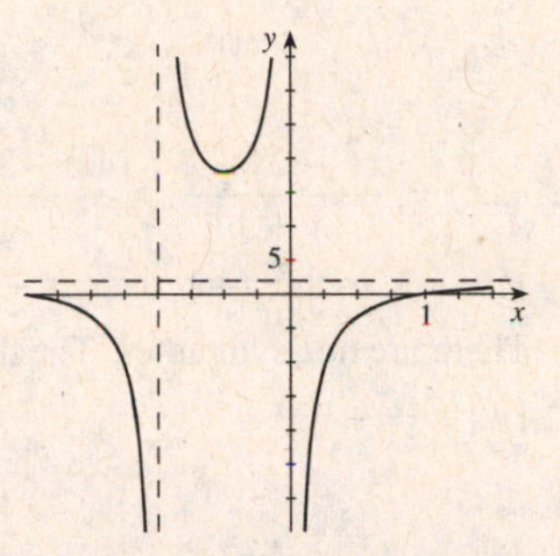

57. $s(x) = \dfrac{x^2 - 2x + 1}{x^3 - 3x^2} = \dfrac{(x-1)^2}{x^2(x-3)}$. Since $x = 0$ is not in the domain of $s(x)$, there is no y-intercept. The x-intercept occurs when $y = 0 \Leftrightarrow x^2 - 2x + 1 = (x-1)^2 = 0 \Rightarrow x = 1$, so the x-intercept is 1. Vertical asymptotes occur when $x = 0, 3$. Since the degree of the numerator is less than the degree of the denominator, the horizontal asymptote is $y = 0$. The domain is $\{x \mid x \neq 0, 3\}$ and the range is $\mathbb{R}$.

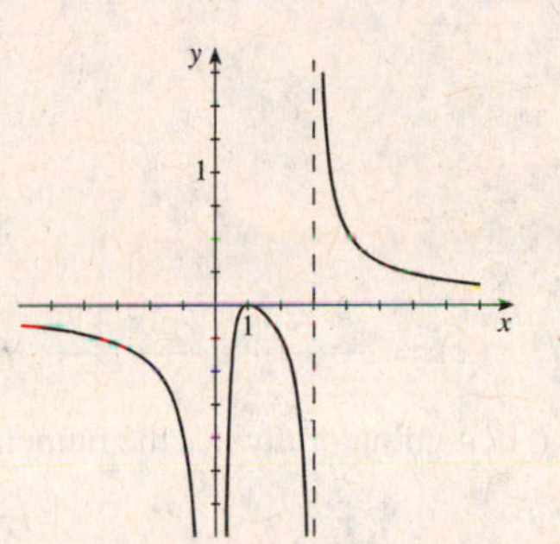

59. $r(x) = \dfrac{x^2 - 2x + 1}{x^2 + 2x + 1} = \dfrac{(x-1)^2}{(x+1)^2} = \left(\dfrac{x-1}{x+1}\right)^2$. When $x = 0$, $y = 1$, so the y-intercept is 1. When $y = 0$, $x = 1$, so the x-intercept is 1. A vertical asymptote occurs at $x + 1 = 0 \Leftrightarrow x = -1$. Because the degree of the numerator and denominator are the same the horizontal asymptote is $y = \frac{1}{1} = 1$. The domain is $\{x \mid x \neq -1\}$ and the range is $\{y \mid y \geq 0\}$.

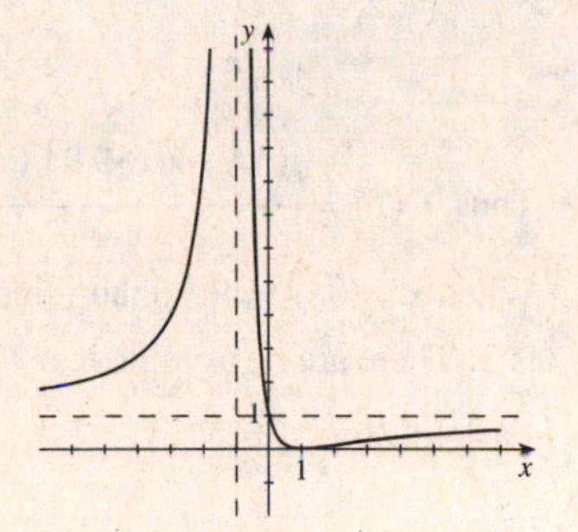

61. $r(x) = \dfrac{5x^2 + 5}{x^2 + 4x + 4} = \dfrac{5\left(x^2 + 1\right)}{(x+2)^2}$. When $x = 0$, we have $y = \dfrac{5}{4}$, so the y-intercept is $\frac{5}{4}$. Since $x^2 + 1 > 0$ for all real x, y never equals zero, and there is no x-intercept. The vertical asymptote is $x = -2$. Because the degree of the denominator and numerator are the same, the horizontal asymptote occurs at $y = \frac{5}{1} = 5$. The domain is $\{x \mid x \neq -2\}$ and the range is $\{y \mid y \geq 1.0\}$.

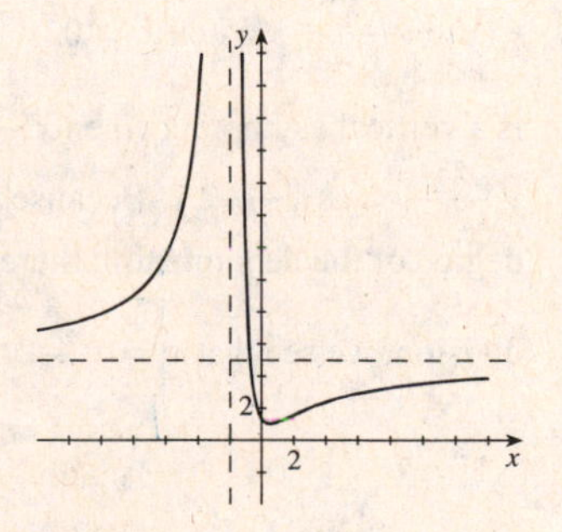

63. $r(x) = \dfrac{x^2 + 4x - 5}{x^2 + x - 2} = \dfrac{(x+5)(x-1)}{(x+2)(x-1)} = \dfrac{x+5}{x+2}$ for $x \neq 1$. When $x = 0$, $y = \frac{5}{2}$, so the y-intercept is $\frac{5}{2}$. When $y = 0$, $x = -5$, so the x-intercept is -5. Vertical asymptotes occur when $x + 2 = 0 \Leftrightarrow x = -2$. Because the degree of the denominator and numerator are the same, the horizontal asymptote is $y = 1$. The domain is $\{x \mid x \neq -2, 1\}$ and the range is $\{y \mid y \neq 1, 2\}$.

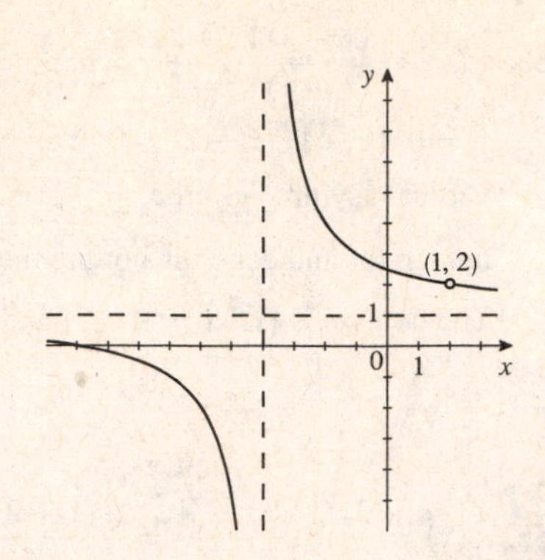

65. $r(x) = \dfrac{x^2 - 2x - 3}{x+1} = \dfrac{(x-3)(x+1)}{x+1} = x - 3$ for $x \neq -1$. When $x = 0$, $y = -3$, so the y-intercept is -3. When $y = 0$, $x = 3$, so the x-intercept is 3. There are no asymptotes. The domain is $\{x \mid x \neq -1\}$ and the range is $\{y \mid y \neq -4\}$.

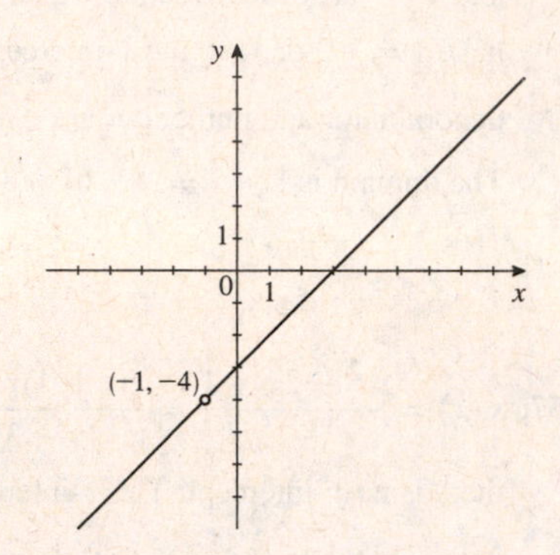

67. $r(x) = \dfrac{x^3 - 5x^2 + 3x + 9}{x+1}$. We use synthetic division to check whether the denominator divides the numerator:

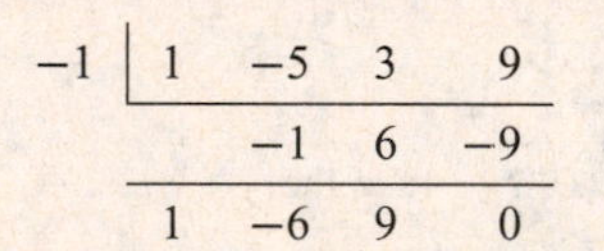

−1	1	−5	3	9
		−1	6	−9
	1	−6	9	0

Thus, $r(x) = \dfrac{\left(x^2 - 6x + 9\right)(x+1)}{x+1} = x^2 - 6x + 9 = (x-3)^2$ for $x \neq -1$. When $x = 0$, $y = 9$, so the y-intercept is 9. When $y = 0$, $x = 3$, so the x-intercept is 3. There are no asymptotes. The domain is $\{x \mid x \neq -1\}$ and the range is $\{y \mid y \geq 0\}$.

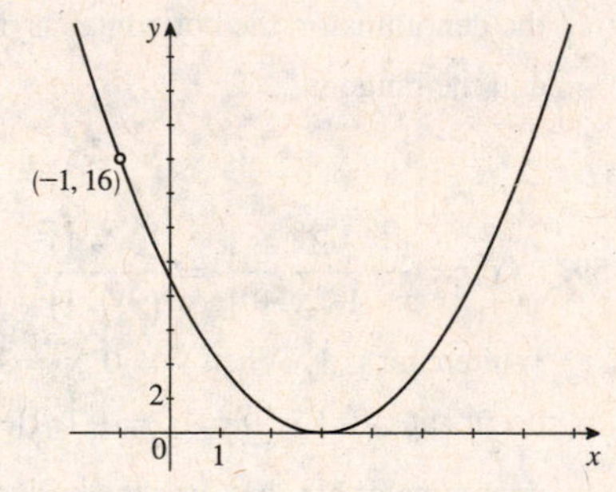

69. $r(x) = \dfrac{x^2}{x-2}$. When $x = 0$, $y = 0$, so the graph passes through the origin. There is a vertical asymptote when $x - 2 = 0 \Leftrightarrow x = 2$, with $y \to \infty$ as $x \to 2^+$, and $y \to -\infty$ as $x \to 2^-$. Because the degree of the numerator is greater than the degree of the denominator, there is no horizontal asymptotes. By using long division, we see that $y = x + 2 + \dfrac{4}{x-2}$, so $y = x + 2$ is a slant asymptote.

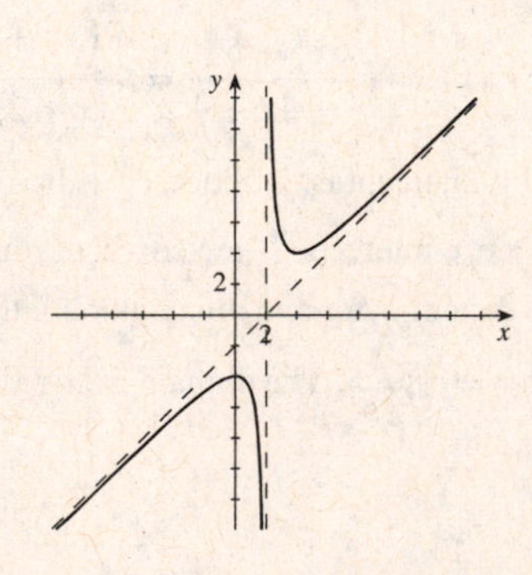

71. $r(x) = \dfrac{x^2 - 2x - 8}{x} = \dfrac{(x-4)(x+2)}{x}$. The vertical asymptote is $x = 0$, thus, there is no y-intercept. If $y = 0$, then $(x-4)(x+2) = 0 \Rightarrow x = -2, 4$, so the x-intercepts are -2 and 4. Because the degree of the numerator is greater than the degree of the denominator, there are no horizontal asymptotes. By using long division, we see that $y = x - 2 - \dfrac{8}{x}$, so $y = x - 2$ is a slant asymptote.

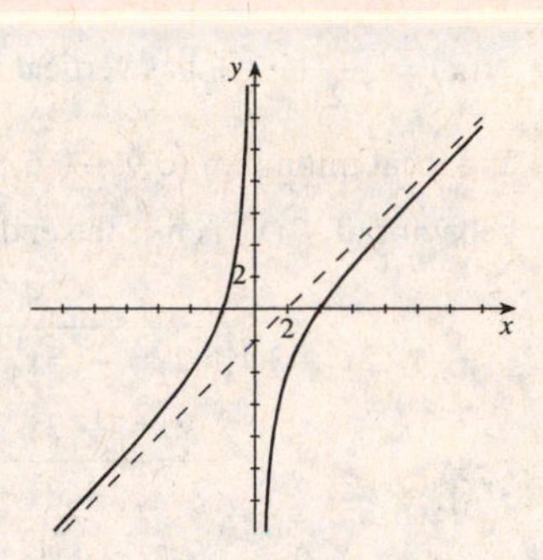

73. $r(x) = \dfrac{x^2 + 5x + 4}{x - 3} = \dfrac{(x+4)(x+1)}{x-3}$. When $x = 0$, $y = -\frac{4}{3}$, so the y-intercept is $-\frac{4}{3}$. When $y = 0$, $(x+4)(x+1) = 0 \Leftrightarrow x = -4, -1$, so the two x-intercepts are -4 and -1. A vertical asymptote occurs when $x = 3$, with $y \to \infty$ as $x \to 3^+$, and $y \to -\infty$ as $x \to 3^-$. Using long division, we see that $y = x + 8 + \dfrac{28}{x-3}$, so $y = x + 8$ is a slant asymptote.

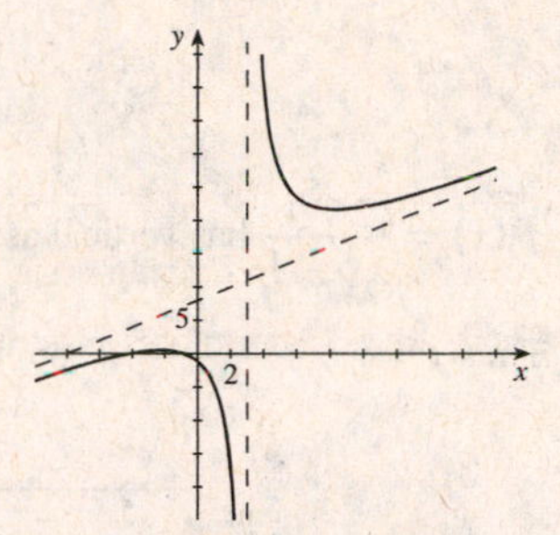

75. $r(x) = \dfrac{x^3 + x^2}{x^2 - 4} = \dfrac{x^2(x+1)}{(x-2)(x+2)}$. When $x = 0$, $y = 0$, so the graph passes through the origin. Moreover, when $y = 0$, we have $x^2(x+1) = 0 \Rightarrow x = 0, -1$, so the x-intercepts are 0 and -1. Vertical asymptotes occur when $x = \pm 2$; as $x \to \pm 2^-$, $y = -\infty$ and as $x \to \pm 2^+$, $y \to \infty$. Because the degree of the numerator is greater than the degree of the denominator, there is no horizontal asymptote. Using long division, we see that $y = x + 1 + \dfrac{4x + 4}{x^2 - 4}$, so $y = x + 1$ is a slant asymptote.

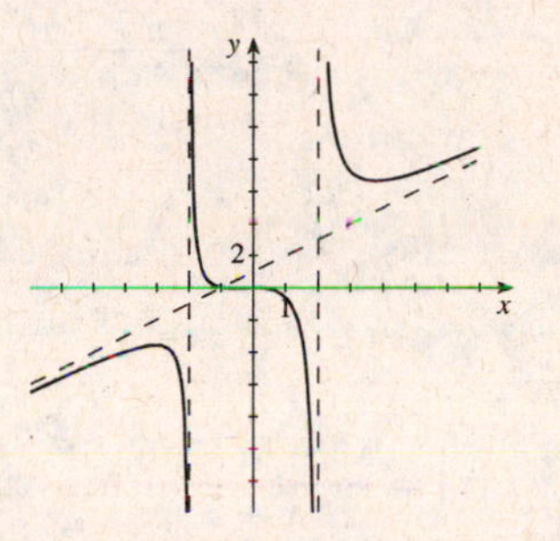

77. $f(x) = \dfrac{2x^2 + 6x + 6}{x + 3}$, $g(x) = 2x$. f has vertical asymptote $x = -3$.

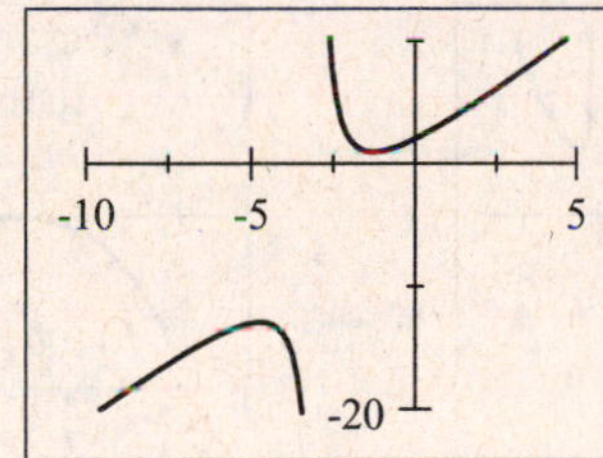

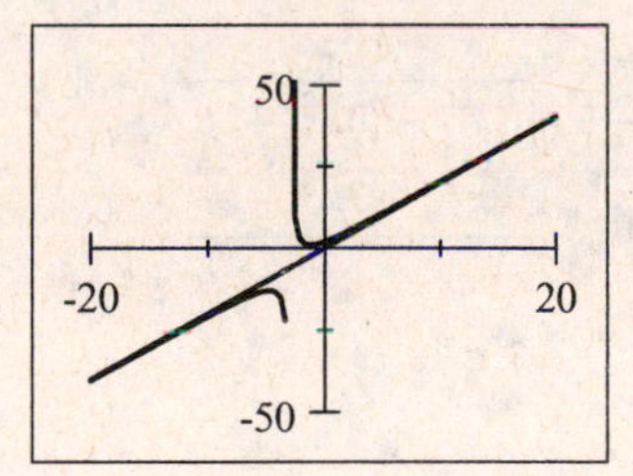

79. $f(x) = \dfrac{x^3 - 2x^2 + 16}{x - 2}$, $g(x) = x^2$. f has vertical asymptote $x = 2$.

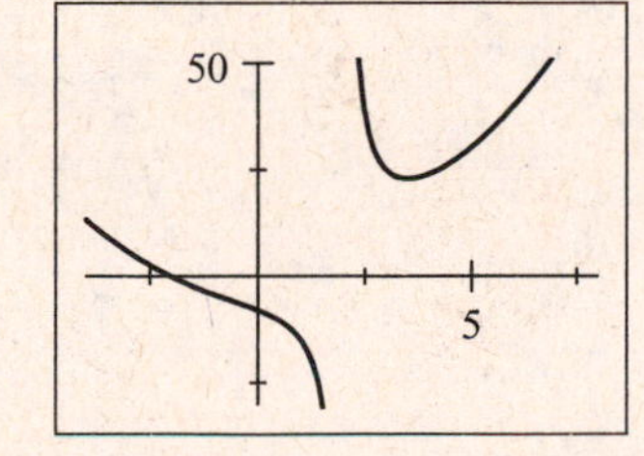

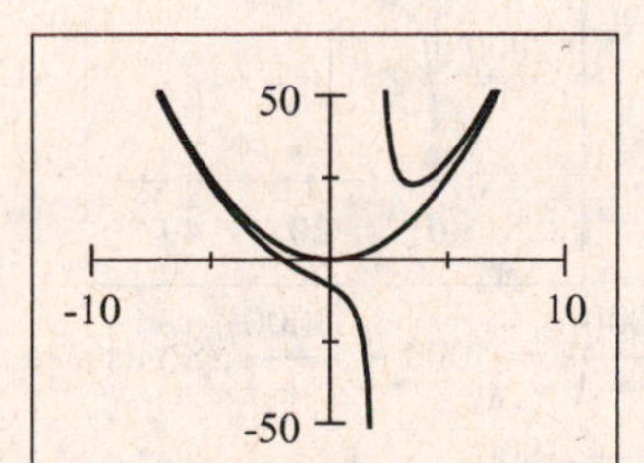

81. $f(x) = \dfrac{2x^2 - 5x}{2x + 3}$ has vertical asymptote $x = -1.5$, x-intercepts 0 and 2.5, y-intercept 0, local maximum $(-3.9, -10.4)$, and local minimum $(0.9, -0.6)$. Using long division, we get $f(x) = x - 4 + \dfrac{12}{2x + 3}$. From the graph, we see that the end behavior of $f(x)$ is like the end behavior of $g(x) = x - 4$.

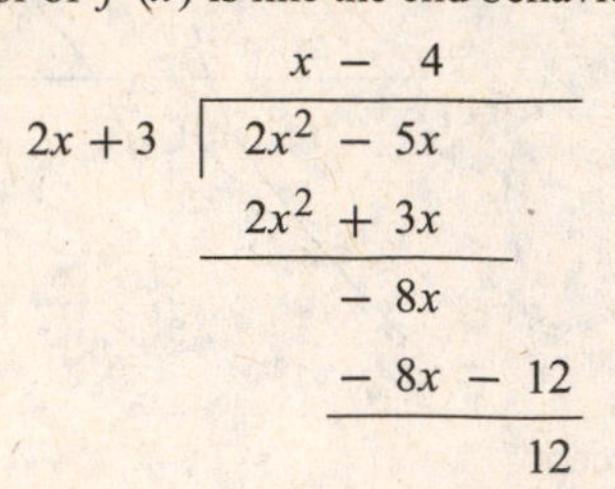

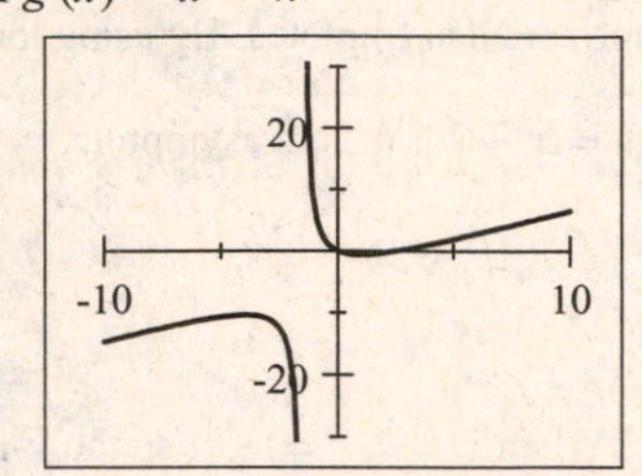

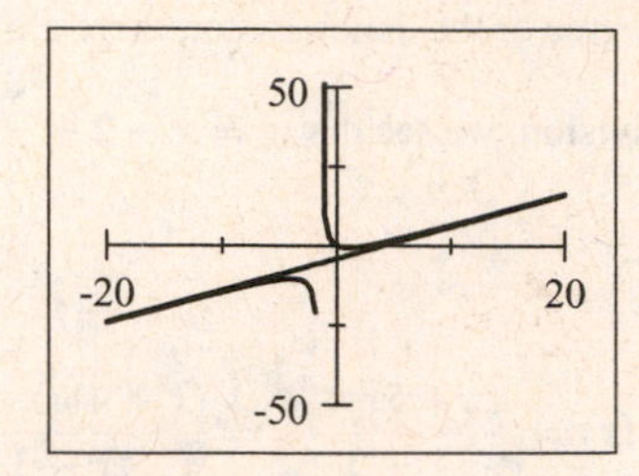

83. $f(x) = \dfrac{x^5}{x^3 - 1}$ has vertical asymptote $x = 1$, x-intercept 0, y-intercept 0, and local minimum $(1.4, 3.1)$. Thus $y = x^2 + \dfrac{x^2}{x^3 - 1}$. From the graph we see that the end behavior of $f(x)$ is like the end behavior of $g(x) = x^2$.

$$\begin{array}{r|l} & x^2 \\ \hline x^3 - 1 & x^5 \\ & x^5 - x^2 \\ \hline & x^2 \end{array}$$

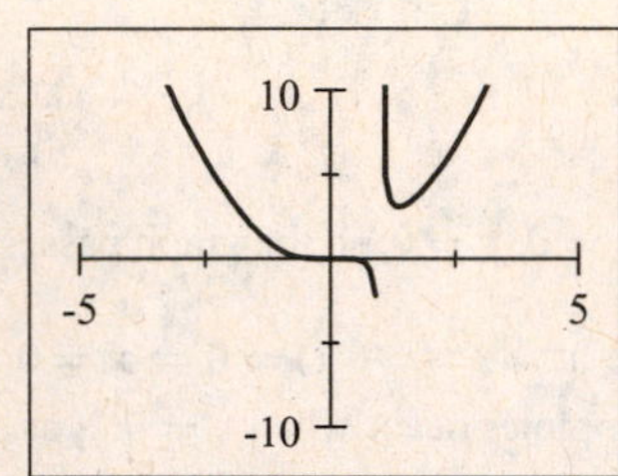

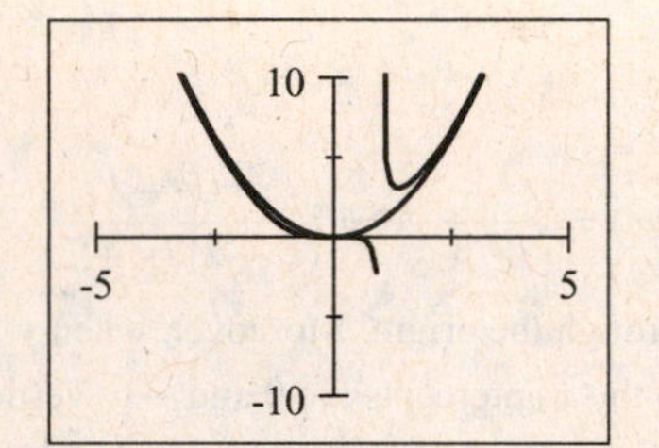

Graph of f Graph of f and g

85. $f(x) = \dfrac{x^4 - 3x^3 + 6}{x - 3}$ has vertical asymptote $x = 3$, x-intercepts 1.6 and 2.7, y-intercept -2, local maxima $(-0.4, -1.8)$ and $(2.4, 3.8)$, and local minima $(0.6, -2.3)$ and $(3.4, 54.3)$. Thus $y = x^3 + \dfrac{6}{x - 3}$. From the graphs, we see that the end behavior of $f(x)$ is like the end behavior of $g(x) = x^3$.

$$\begin{array}{r|l} & x^3 \\ \hline x - 3 & x^4 - 3x^3 + 6 \\ & x^4 - 3x^3 \\ \hline & 6 \end{array}$$

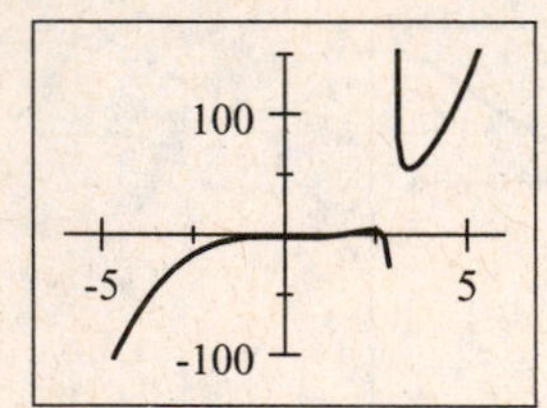

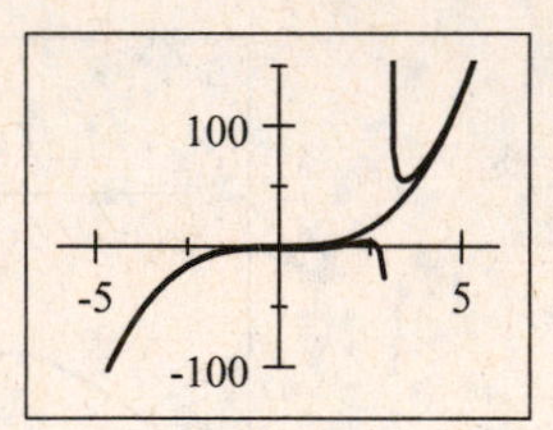

87. (a)

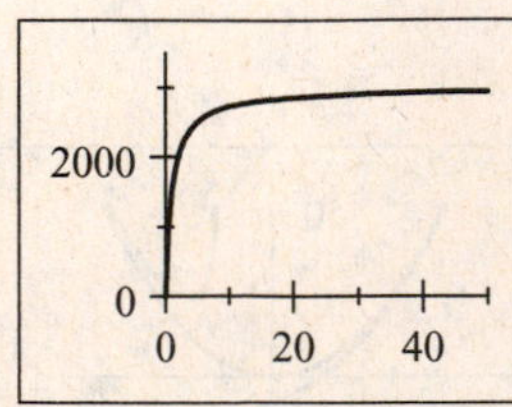

(b) $p(t) = \dfrac{3000t}{t + 1} = 3000 - \dfrac{3000}{t + 1}$. So as $t \to \infty$, we have $p(t) \to 3000$.

89. $c(t) = \dfrac{5t}{t^2 + 1}$

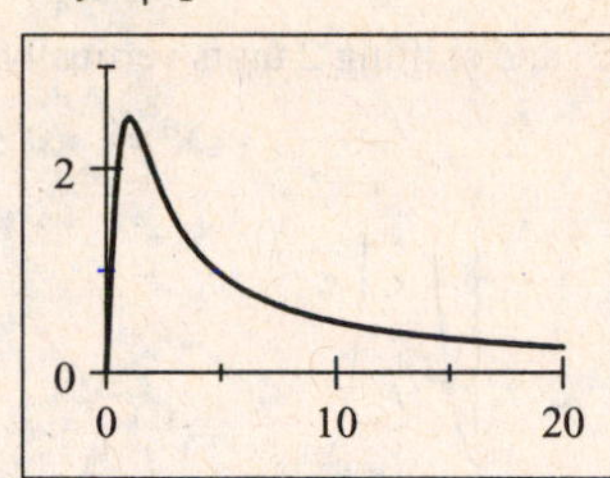

(a) The highest concentration of drug is 2.50 mg/L, and it is reached 1 hour after the drug is administered.

(b) The concentration of the drug in the bloodstream goes to 0.

(c) From the first viewing rectangle, we see that an approximate solution is near $t = 15$. Thus we graph $y = \dfrac{5t}{t^2 + 1}$ and $y = 0.3$ in the viewing rectangle $[14, 18]$ by $[0, 0.5]$. So it takes about 16.61 hours for the concentration to drop below 0.3 mg/L.

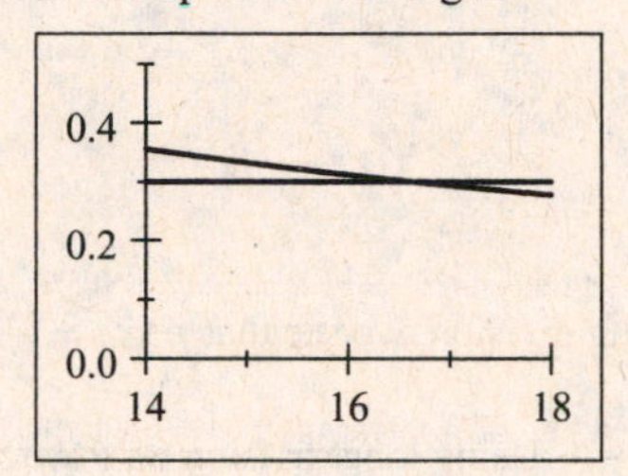

91. $P(v) = P_0\left(\dfrac{s_0}{s_0 - v}\right) \Rightarrow P(v) = 440\left(\dfrac{332}{332 - v}\right)$

If the speed of the train approaches the speed of sound, the pitch of the whistle becomes very loud. This would be experienced as a "sonic boom"— an effect seldom heard with trains.

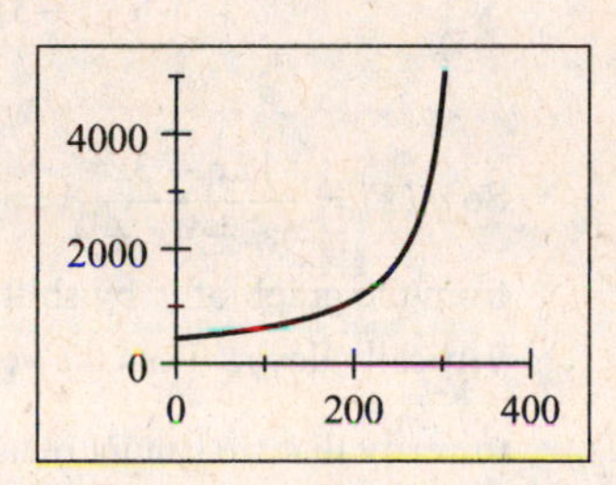

93. Vertical asymptote $x = 3$: $p(x) = \dfrac{1}{x - 3}$. Vertical asymptote $x = 3$ and horizontal asymptote $y = 2$: $r(x) = \dfrac{2x}{x - 3}$.

Vertical asymptotes $x = 1$ and $x = -1$, horizontal asymptote 0, and x-intercept 4: $q(x) = \dfrac{x - 4}{(x - 1)(x + 1)}$. Of course, other answers are possible.

95. **(a)** Let $f(x) = \dfrac{1}{x^2}$. Then $r(x) = \dfrac{1}{(x - 2)^2} = f(x - 2)$. From this form we see that the graph of r is obtained from the graph of f by shifting 2 units to the right. Thus r has vertical asymptote $x = 2$ and horizontal asymptote $y = 0$.

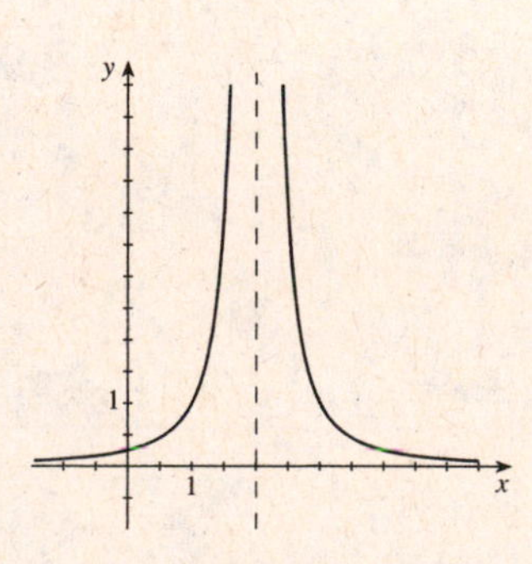

(b) $s(x) = \dfrac{2x^2 + 4x + 5}{x^2 + 2x + 1} = 2 + \dfrac{3}{(x+1)^2} = 3f(x+1) + 2$. From this form we see that the graph of s is obtained from the graph of f by shifting 1 unit to the left, stretching vertically by a factor of 3, and shifting 2 units vertically. Thus r has vertical asymptote $x = -1$ and horizontal asymptote $y = 2$.

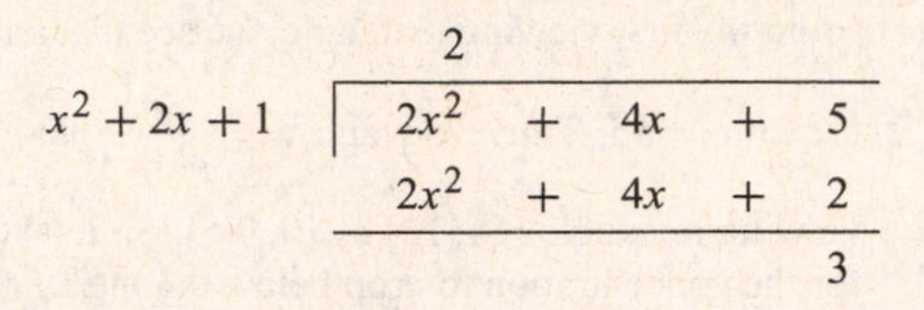

$$\begin{array}{r|rrrrr} & & & & & 2 \\ \hline x^2+2x+1 & 2x^2 & + & 4x & + & 5 \\ & 2x^2 & + & 4x & + & 2 \\ \hline & & & & & 3 \end{array}$$

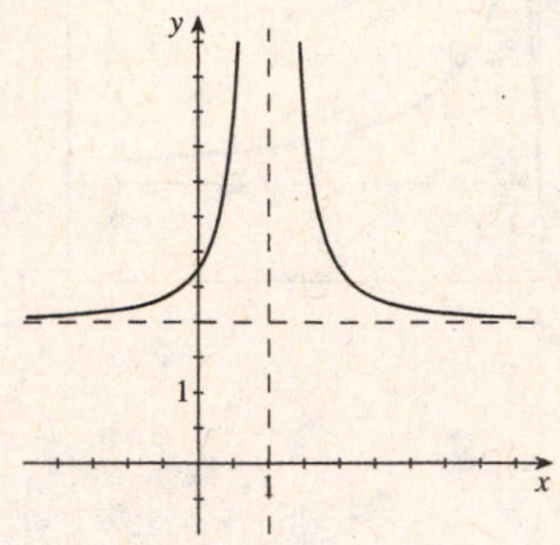

(c) Using long division, we see that $p(x) = \dfrac{2 - 3x^2}{x^2 - 4x + 4} = -3 + \dfrac{-12x + 14}{x^2 - 4x + 4}$ which cannot be graphed by transforming $f(x) = \dfrac{1}{x^2}$. Using long division on q we have:

$$\begin{array}{r|rrrrr} & & & & & -3 \\ \hline x^2-4x+4 & -3x^2 & + & 0x & + & 2 \\ & -3x^2 & + & 12x & - & 12 \\ \hline & & - & 12x & + & 14 \end{array} \qquad \begin{array}{r|rrrrr} & & & & & -3 \\ \hline x^2-4x+4 & -3x^2 & + & 12x & + & 0 \\ & -3x^2 & + & 12x & - & 12 \\ \hline & & & & & 12 \end{array}$$

So $q(x) = \dfrac{12x - 3x^2}{x^2 - 4x + 4} = -3 + \dfrac{12}{(x-2)^2} = 12f(x-2) - 3$. From this form we see that the graph of q is obtained from the graph of f by shifting 2 units to the right, stretching vertically by a factor of 12, and then shifting 3 units vertically down. Thus the vertical asymptote is $x = 2$ and the horizontal asymptote is $y = -3$. We show $y = p(x)$ just to verify that we cannot obtain $p(x)$ from $y = \dfrac{1}{x^2}$.

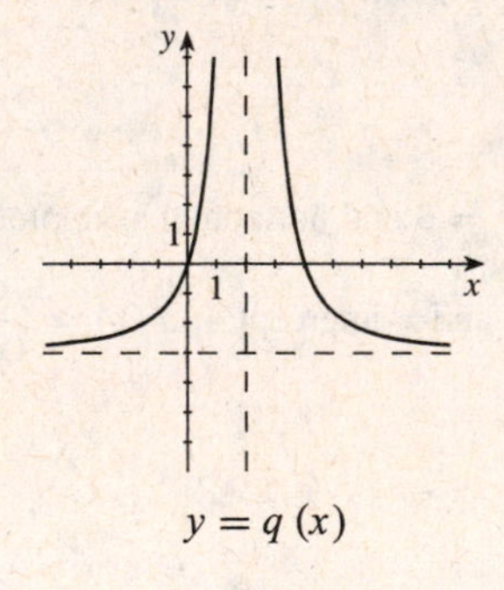

$y = q(x)$

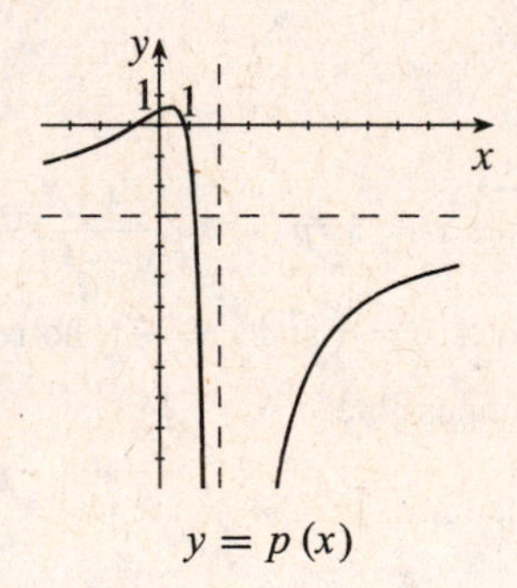

$y = p(x)$

3.7 POLYNOMIAL AND RATIONAL INEQUALITIES

1. To solve a polynomial inequality, we factor the polynomial into irreducible factors and find all the real *zeros* of the polynomial. Then we find the intervals determined by the real *zeros* and use test points in each interval to find the sign of the polynomial on that interval.

Sign of	$(-\infty, -2)$	$(-2, 0)$	$(0, 1)$	$(1, \infty)$
x	$-$	$-$	$+$	$+$
$x+2$	$-$	$+$	$+$	$+$
$x-1$	$-$	$-$	$-$	$+$
$P(x) = x(x+2)(x-1)$	$-$	$+$	$-$	$+$

From the table, we see that $P(x) \geq 0$ on the intervals $[-2, 0]$ and $[1, \infty)$.

3. The inequality $(x-3)(x+5)(2x+5) < 0$ already has all terms on one side and the polynomial is factored. The intervals determined by the zeros $3, -5$, and $-\frac{5}{2}$ are $(-\infty, -5)$, $\left(-5, -\frac{5}{2}\right)$, $\left(-\frac{5}{2}, 3\right)$, and $(3, \infty)$. We make a sign diagram:

Sign of	$(-\infty, -5)$	$\left(-5, -\frac{5}{2}\right)$	$\left(-\frac{5}{2}, 3\right)$	$(3, \infty)$
$x-3$	$-$	$-$	$-$	$+$
$x+5$	$-$	$+$	$+$	$+$
$2x+5$	$-$	$-$	$+$	$+$
$P(x) = (x-3)(x+5)(2x+5)$	$-$	$+$	$-$	$+$

None of the endpoints satisfies the inequality. The solution is $(-\infty, -5) \cup \left(-\frac{5}{2}, 3\right)$.

5. The inequality $(x+5)^2(x+3)(x-1) > 0$ already has all terms on one side and the polynomial is factored. The intervals determined by the zeros $-5, -3$, and 1 are $(-\infty, -5)$, $(-5, -3)$, $(-3, 1)$, and $(1, \infty)$. We make a sign diagram:

Sign of	$(-\infty, -5)$	$(-5, -3)$	$(-3, 1)$	$(1, \infty)$
$(x+5)^2$	$+$	$+$	$+$	$+$
$x+3$	$-$	$-$	$+$	$+$
$x-1$	$-$	$-$	$-$	$+$
$P(x) = (x+5)^2(x+3)(x-1)$	$+$	$+$	$-$	$+$

None of the endpoints satisfies the inequality. The solution is $(-\infty, -5) \cup (-5, -3) \cup (1, \infty)$.

7. We start by moving all terms to one side and factoring: $x^3 + 4x^2 \geq 4x + 16 \Leftrightarrow x^3 + 4x^2 - 4x - 16 = (x+4)(x+2)(x-2) \geq 0$. The intervals determined by the zeros $-4, -2$, and 2 are $(-\infty, -4)$, $(-4, -2)$, $(-2, 2)$, and $(2, \infty)$. We make a sign diagram:

Sign of	$(-\infty, -4)$	$(-4, -2)$	$(-2, 2)$	$(2, \infty)$
$x+4$	$-$	$+$	$+$	$+$
$x+2$	$-$	$-$	$+$	$+$
$x-2$	$-$	$-$	$-$	$+$
$P(x) = (x+4)(x+2)(x-2)$	$-$	$+$	$-$	$+$

All of the endpoints satisfy the inequality. The solution is $[-4, -2] \cup [2, \infty)$.

9. We start by moving all terms to one side and factoring: $2x^3 - x^2 < 9 - 18x \Leftrightarrow 2x^3 - x^2 + 18x - 9 = (2x - 1)\left(x^2 + 9\right) < 0$. Note that $x^2 + 9 > 0$ for all x, so the sign of $P(x) = (2x - 1)\left(x^2 + 9\right)$ is negative where $2x - 1$ is negative and positive where $2x - 1$ is positive. The endpoint $x = \frac{1}{2}$ does not satisfy the inequality, so the solution is $\left(-\infty, \frac{1}{2}\right)$.

11. All the terms are on the left size. We factor, using the substitution $t = x^2$: $x^4 - 7x^2 - 18 < 0 \Leftrightarrow t^2 - 7t - 18 = (t + 2)(t - 9) < 0 \Leftrightarrow \left(x^2 + 2\right)\left(x^2 - 9\right) < 0 \Leftrightarrow \left(x^2 + 2\right)(x + 3)(x - 3) < 0$. The first factor is positive everywhere, so we test $(-\infty, -3)$, $(-3, 3)$, and $(3, \infty)$:

Sign of	$(-\infty, -3)$	$(-3, 3)$	$(3, \infty)$
$x + 3$	$-$	$+$	$+$
$x - 3$	$-$	$-$	$+$
$P(x) = \left(x^2 + 2\right)(x + 3)(x - 3)$	$+$	$-$	$+$

Neither of the endpoints satisfies the inequality. The solution is $(-3, 3)$.

13. All the terms are on the left size. To factor, note that the possible rational zeros of $P(x) = x^3 + x^2 - 17x + 15$ are $\pm 1, \pm 3, \pm 5, \pm 15$.

$$\begin{array}{r|rrrr} 1 & 1 & 1 & -17 & 15 \\ & & 1 & 2 & -15 \\ \hline & 1 & 2 & -15 & 0 \end{array}$$

Thus, $P(x) = (x - 1)\left(x^2 + 2x - 15\right) = (x + 5)(x - 1)(x - 3)$. The zeros are -5, 1, and 3, so we test $(-\infty, -5)$, $(-5, 1)$, $(1, 3)$, and $(3, \infty)$:

Sign of	$(-\infty, -5)$	$(-5, 1)$	$(1, 3)$	$(3, \infty)$
$x + 5$	$-$	$+$	$+$	$+$
$x - 1$	$-$	$-$	$+$	$+$
$x - 3$	$-$	$-$	$-$	$+$
$P(x)$	$-$	$+$	$-$	$+$

All of the endpoints satisfy the inequality $P(x) \geq 0$, so the solution is $[-5, 1] \cup [3, \infty)$.

15. We start by moving all terms to one side and factoring: $x\left(1 - x^2\right)^3 > 7\left(1 - x^2\right)^3 \Leftrightarrow (x - 7)\left(1 - x^2\right)^3 > 0 \Leftrightarrow P(x) = (x - 7)(1 - x)^3(1 + x)^3 > 0$. The intervals determined by the zeros -1, 1, and 7 are $(-\infty, -1)$, $(-1, 1)$, $(1, 7)$, and $(7, \infty)$. We make a sign diagram:

Sign of	$(-\infty, -1)$	$(-1, 1)$	$(1, 7)$	$(7, \infty)$
$x - 7$	$-$	$-$	$-$	$+$
$(1 - x)^3$	$+$	$+$	$-$	$-$
$(1 + x)^3$	$-$	$+$	$+$	$+$
$P(x)$	$+$	$-$	$+$	$-$

None of the endpoints satisfies the inequality. The solution is $(-\infty, -1) \cup (1, 7)$.

17. $r(x) = \dfrac{x-1}{x-10} < 0$. Since all nonzero terms are already on one side of the inequality symbol and there is no factoring needed, we find the intervals determined by the cut points 1 and 10. These are $(-\infty, 1)$, $(1, 10)$, and $(10, \infty)$. We make a sign diagram:

Sign of	$(-\infty, 1)$	$(1, 10)$	$(10, \infty)$
$x-1$	$-$	$+$	$+$
$x-10$	$-$	$-$	$+$
$r(x)$	$+$	$-$	$+$

The cut point 1 does not satisfy the inequality, and the cut point 10 is not in the domain of r. Thus, the solution is $(1, 10)$.

19. $r(x) = \dfrac{2x+5}{x^2+2x-35} \geq 0$. Since all nonzero terms are already on one side, we factor: $r(x) = \dfrac{2x+5}{x^2+2x-35} = \dfrac{2x+5}{(x+7)(x-5)}$. Thus, the cut points are -7, $-\frac{5}{2}$, and 5. The intervals determined by these points are $(-\infty, -7)$, $\left(-7, -\frac{5}{2}\right)$, $\left(-\frac{5}{2}, 5\right)$, and $(5, \infty)$. We make a sign diagram:

Sign of	$(-\infty, -7)$	$\left(-7, -\frac{5}{2}\right)$	$\left(-\frac{5}{2}, 5\right)$	$(5, \infty)$
$2x+5$	$-$	$-$	$+$	$+$
$x+7$	$-$	$+$	$+$	$+$
$x-5$	$-$	$-$	$-$	$+$
$r(x)$	$-$	$+$	$-$	$+$

The cut point $-\frac{5}{2}$ satisfies equality, but the cut points -7 and 5 are not in the domain of r. Thus, the solution is $\left(-7, -\frac{5}{2}\right] \cup (5, \infty)$.

21. $r(x) = \dfrac{x}{x^2+2x-2} \leq 0$. Since all nonzero terms are already on one side but the denominator cannot be factored, we use the Quadratic Formula: $x^2+2x-2=0 \Leftrightarrow x = \dfrac{-2 \pm \sqrt{2^2-4(1)(-2)}}{2} = -1 \pm \sqrt{3}$. Thus, the cut points are $-1-\sqrt{3}$, 0, and $-1+\sqrt{3}$. The intervals determined by these points are $\left(-\infty, -1-\sqrt{3}\right)$, $\left(-1-\sqrt{3}, 0\right)$, $\left(0, \sqrt{3}-1\right)$, and $\left(\sqrt{3}-1, \infty\right)$. We make a sign diagram:

Sign of	$\left(-\infty, -1-\sqrt{3}\right)$	$\left(-1-\sqrt{3}, 0\right)$	$\left(0, \sqrt{3}-1\right)$	$\left(\sqrt{3}-1, \infty\right)$
x	$-$	$-$	$+$	$+$
x^2+2x-2	$+$	$-$	$-$	$+$
$r(x)$	$-$	$+$	$-$	$+$

The cut point 0 satisfies equality, but the cut points $-1 \pm \sqrt{3}$ are not in the domain of r. Thus, the solution is $\left(-\infty, -1-\sqrt{3}\right) \cup \left[0, \sqrt{3}-1\right)$.

23. $r(x)=\dfrac{x^2+2x-3}{3x^2-7x-6}=\dfrac{(x+3)(x-1)}{(3x+2)(x-3)}>0$. The intervals determined by the cut points are $(-\infty,-3)$, $\left(-3,-\frac{2}{3}\right)$, $\left(-\frac{2}{3},1\right)$, $(1,3)$, and $(3,\infty)$.

Sign of	$(-\infty,-3)$	$\left(-3,-\frac{2}{3}\right)$	$\left(-\frac{2}{3},1\right)$	$(1,3)$	$(3,\infty)$
$x+3$	$-$	$+$	$+$	$+$	$+$
$3x+2$	$-$	$-$	$+$	$+$	$+$
$x-1$	$-$	$-$	$-$	$+$	$+$
$x-3$	$-$	$-$	$-$	$-$	$+$
$r(x)$	$+$	$-$	$+$	$-$	$+$

None of the cut points satisfies the strict inequality, so the solution is $(-\infty,-3)\cup\left(-\frac{2}{3},1\right)\cup(3,\infty)$.

25. $r(x)=\dfrac{x^3+3x^2-9x-27}{x+4}=\dfrac{x\left(x^2+6x+9\right)-3\left(x^2+6x+9\right)}{x+4}=\dfrac{(x-3)(x+3)^2}{x+4}\le 0$. The second factor in the numerator is positive for all x, so the intervals determined by the cut points are $(-\infty,-4)$, $(-4,3)$, and $(3,\infty)$.

Sign of	$(-\infty,-4)$	$(-4,3)$	$(3,\infty)$
$x+4$	$-$	$+$	$+$
$x-3$	$-$	$-$	$+$
$r(x)$	$+$	$-$	$+$

The cut point 3 satisfies equality, but -4 is not in the domain of r. Thus, the solution is $(-4,3]$.

27. We start by moving all terms to one side and simplifying: $\dfrac{x-3}{2x+5}\ge 1\Leftrightarrow\dfrac{x-3}{2x+5}-1\ge 0\Leftrightarrow\dfrac{x-3-(2x+5)}{2x+5}\ge 0\Leftrightarrow$ $r(x)=\dfrac{-(x+8)}{2x+5}\ge 0$. The intervals determined by the cut points are $(-\infty,-8)$, $\left(-8,-\frac{5}{2}\right)$, and $\left(-\frac{5}{2},\infty\right)$.

Sign of	$(-\infty,-8)$	$\left(-8,-\frac{5}{2}\right)$	$\left(-\frac{5}{2},\infty\right)$
$-(x+8)$	$+$	$-$	$-$
$2x+5$	$-$	$-$	$+$
$r(x)$	$-$	$+$	$-$

The cut point -8 satisfies equality, but $-\frac{5}{2}$ is not in the domain of r. Thus, the solution is $\left[-8,-\frac{5}{2}\right)$.

29. We start by moving all terms to one side and simplifying: $2+\frac{1}{1-x}\leq\frac{3}{x}\Leftrightarrow 2+\frac{1}{1-x}-\frac{3}{x}\leq 0\Leftrightarrow$

$\frac{2(1-x)(x)+x-3(1-x)}{x(1-x)}\leq 0\Leftrightarrow r(x)=\frac{-2x^2+6x-3}{x(1-x)}=\frac{2x^2-6x+3}{x(x-1)}\leq 0$. The numerator is 0 when

$x=\frac{-(-6)\pm\sqrt{(-6)^2-4(2)(3)}}{2(2)}=\frac{3\pm\sqrt{3}}{2}$, so the intervals determined by the cut points are $(-\infty,0)$, $\left(0,\frac{3-\sqrt{3}}{2}\right)$, $\left(\frac{3-\sqrt{3}}{2},1\right)$, $\left(1,\frac{3+\sqrt{3}}{2}\right)$, and $\left(\frac{3+\sqrt{3}}{2},\infty\right)$.

Sign of	$(-\infty,0)$	$\left(0,\frac{3-\sqrt{3}}{2}\right)$	$\left(\frac{3-\sqrt{3}}{2},1\right)$	$\left(1,\frac{3+\sqrt{3}}{2}\right)$	$\left(\frac{3+\sqrt{3}}{2},\infty\right)$
x	$-$	$+$	$+$	$+$	$+$
$x-1$	$-$	$-$	$-$	$+$	$+$
$2x^2-6x+3$	$+$	$+$	$-$	$-$	$+$
$r(x)$	$+$	$-$	$+$	$-$	$+$

The cut points $\frac{3\pm\sqrt{3}}{2}$ satisfy equality, but 0 and 1 are not in the domain of r, so the solution is $\left(0,\frac{3-\sqrt{3}}{2}\right]\cup\left(1,\frac{3-\sqrt{3}}{2}\right]$.

31. $r(x)=\frac{(x-1)^2}{(x+1)(x+2)}>0$. The numerator is nonnegative for all x, but note that $x=1$ fails to satisfy the strict inequality. The intervals determined by the cut points are $(-\infty,-2)$, $(-2,-1)$, and $(-1,\infty)$.

Sign of	$(-\infty,-2)$	$(-2,-1)$	$(-1,\infty)$
$(x-1)^2$	$+$	$+$	$+$
$x+2$	$-$	$+$	$+$
$x+1$	$-$	$-$	$+$
$r(x)$	$+$	$-$	$+$ (except at $x=1$)

The cut points -2 and -1 are not in the domain of r, so the solution is $(-\infty,-2)\cup(-1,1)\cup(1,\infty)$.

33. We start by moving all terms to one side and factoring: $\frac{6}{x-1}-\frac{6}{x}\geq 1\Leftrightarrow\frac{6}{x-1}-\frac{6}{x}-1\geq 0\Leftrightarrow$

$\frac{6x-6(x-1)-x(x-1)}{x(x-1)}\geq 0\Leftrightarrow\frac{x^2-x-6}{x(x-1)}\geq 0\Leftrightarrow r(x)=-\frac{(x+2)(x-3)}{x(x-1)}\geq 0$. The intervals determined by the cut points are $(-\infty,-2)$, $(-2,0)$, $(0,1)$, $(1,3)$, and $(3,\infty)$.

Sign of	$(-\infty,-2)$	$(-2,0)$	$(0,1)$	$(1,3)$	$(3,\infty)$
$x+2$	$-$	$+$	$+$	$+$	$+$
x	$-$	$-$	$+$	$+$	$+$
$x-1$	$-$	$-$	$-$	$+$	$+$
$x-3$	$-$	$-$	$-$	$-$	$+$
$r(x)$ (note negative sign)	$-$	$+$	$-$	$+$	$-$

The cut points 0 and 1 are inadmissible in the original inequality, so the solution is $[-2,0)\cup(1,3]$.

35. We start by moving all terms to one side and factoring: $\frac{x+2}{x+3} < \frac{x-1}{x-2} \Leftrightarrow \frac{x+2}{x+3} - \frac{x-1}{x-2} < 0 \Leftrightarrow$ $\frac{(x+2)(x-2)-(x-1)(x+3)}{(x+3)(x-2)} < 0 \Leftrightarrow r(x) = \frac{-2\left(x+\frac{1}{2}\right)}{(x+3)(x-2)} < 0$. The intervals determined by the cut points are $(-\infty,-3)$, $\left(-3,-\frac{1}{2}\right)$, $\left(-\frac{1}{2},2\right)$, and $(2,\infty)$.

Sign of	$(-\infty,-3)$	$\left(-3,-\frac{1}{2}\right)$	$\left(-\frac{1}{2},2\right)$	$(2,\infty)$
$x+3$	$-$	$+$	$+$	$+$
$x+\frac{1}{2}$	$-$	$-$	$+$	$+$
$x-2$	$-$	$-$	$-$	$+$
$r(x)$ (note negative sign)	$+$	$-$	$+$	$-$

The cut points fail to satisfy the strict inequality, so the solution is $\left(-3,-\frac{1}{2}\right)\cup(2,\infty)$.

37. The graph of f lies above that of g where $f(x) > g(x)$; that is, where $x^2 > 3x+10 \Leftrightarrow x^2-3x-10>0 \Leftrightarrow$ $(x+2)(x-5)>0$. We make a sign diagram:

Sign of	$(-\infty,-2)$	$(-2,5)$	$(5,\infty)$
$x+2$	$-$	$+$	$+$
$x-5$	$-$	$-$	$+$
$(x+2)(x-5)$	$+$	$-$	$+$

Thus, the graph of f lies above the graph of g on $(-\infty,-2)$ and $(5,\infty)$.

39. The graph of f lies above that of g where $f(x) > g(x)$; that is, where $4x > \frac{1}{x} \Leftrightarrow 4x - \frac{1}{x} > 0 \Leftrightarrow \frac{4x^2-1}{x} > 0 \Leftrightarrow$ $r(x) = \frac{(2x+1)(2x-1)}{x} > 0$. We make a sign diagram:

Sign of	$\left(-\infty,-\frac{1}{2}\right)$	$\left(-\frac{1}{2},0\right)$	$\left(0,\frac{1}{2}\right)$	$\left(\frac{1}{2},\infty\right)$
$2x+1$	$-$	$+$	$+$	$+$
x	$-$	$-$	$+$	$+$
$2x-1$	$-$	$-$	$-$	$+$
$r(x)$	$-$	$+$	$-$	$+$

Thus, the graph of f lies above the graph of g on $\left(-\frac{1}{2},0\right)$ and $\left(\frac{1}{2},\infty\right)$.

41. $f(x) = \sqrt{6+x-x^2}$ is defined where $6+x-x^2 = -(x+2)(x-3) \geq 0$. We make a sign diagram:

Sign of	$(-\infty,-2)$	$(-2,3)$	$(3,\infty)$
$x+2$	$-$	$+$	$+$
$x-3$	$-$	$-$	$+$
$-(x+2)(x-3)$	$-$	$+$	$-$

Thus, the domain of f is $[-2,3]$.

43. $h(x) = \sqrt[4]{x^4 - 1}$ is defined where $x^4 - 1 = (x-1)(x+1)\left(x^2+1\right) \geq 0$. The last factor is positive for all x. We make a sign diagram:

Sign of	$(-\infty, -1)$	$(-1, 1)$	$(1, \infty)$
$x + 1$	$-$	$+$	$+$
$x - 1$	$-$	$-$	$+$
$x^2 + 1$	$+$	$+$	$+$
$(x-1)(x+1)\left(x^2+1\right)$	$+$	$-$	$+$

Thus, the domain of h is $(-\infty, -1] \cup [1, \infty)$.

45.

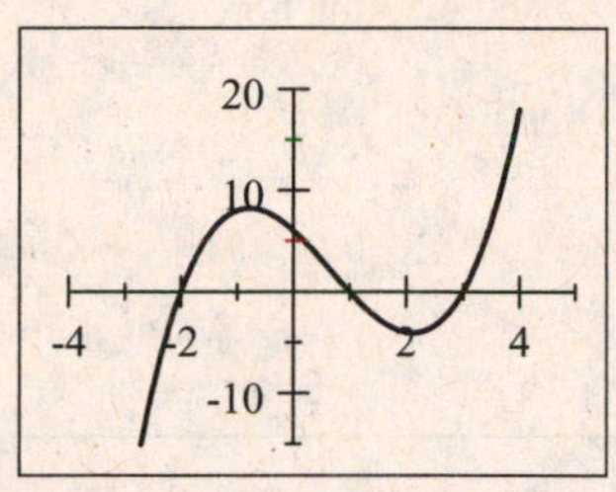

From the graph, we see that $x^3 - 2x^2 - 5x + 6 \geq 0$ on $[-2, 1] \cup [3, \infty)$.

47.

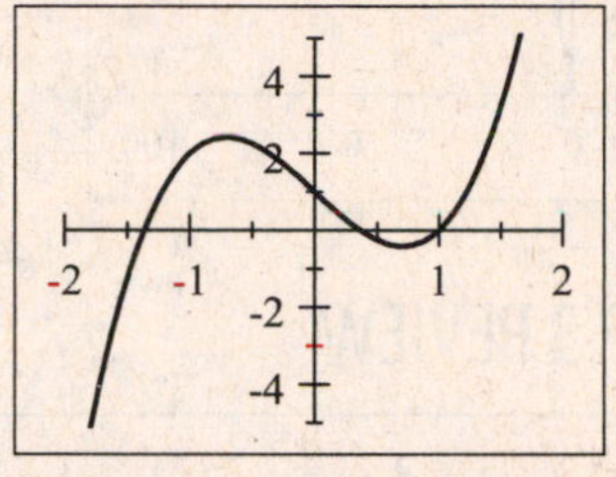

From the graph, we see that $2x^3 - 3x + 1 < 0$ on approximately $(-\infty, -1.37) \cup (0.37, 1)$.

49.

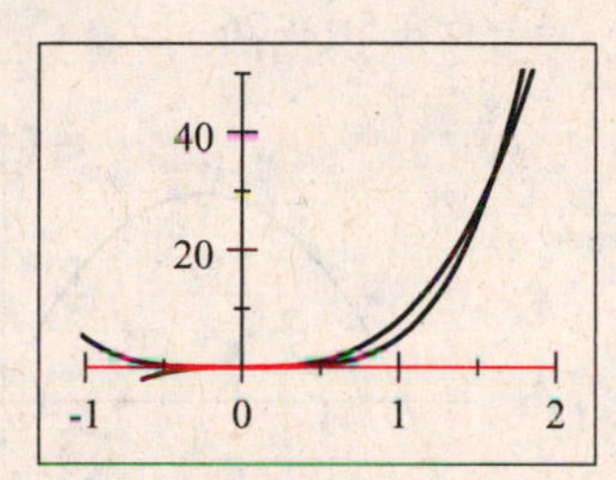

From the graph, we see that $5x^4 < 8x^3$ on $\left(0, \frac{8}{5}\right)$.

51. $\dfrac{(1-x)^2}{\sqrt{x}} \geq 4\sqrt{x}(x-1) \Leftrightarrow \dfrac{(1-x)^2}{\sqrt{x}} - 4\sqrt{x}(x-1) \geq 0 \Leftrightarrow \dfrac{(1-x)^2 - \sqrt{x}\left(4\sqrt{x}\right)(x-1)}{\sqrt{x}} \geq 0 \Leftrightarrow$ $\dfrac{1 - 2x + x^2 - 4x^2 + 4x}{\sqrt{x}} \geq 0 \Leftrightarrow \dfrac{-3x^2 + 2x + 1}{\sqrt{x}} \geq 0 \Leftrightarrow r(x) = \dfrac{(1-x)(3x+1)}{\sqrt{x}} \geq 0$. The domain of r is $(0, \infty)$, and both $3x + 1$ and $\sqrt{x}$ are positive there. $1 - x \geq 0$ for $x \leq 1$, so the solution is $(0, 1]$.

53. We want to solve $P(x) = (x-a)(x-b)(x-c)(x-d) \geq 0$, where $a < b < c < d$. We make a sign diagram with cut points a, b, c, and d:

Sign of	$(-\infty, a)$	(a, b)	(b, c)	(c, d)	(d, ∞)
$x - a$	$-$	$+$	$+$	$+$	$+$
$x - b$	$-$	$-$	$+$	$+$	$+$
$x - c$	$-$	$-$	$-$	$+$	$+$
$x - d$	$-$	$-$	$-$	$-$	$+$
$P(x)$	$+$	$-$	$+$	$-$	$+$

Each cut points satisfies equality, so the solution is $(-\infty, a] \cup [b, c] \cup [d, \infty)$.

55. We want to solve the inequality $T(x) < 300 \Leftrightarrow \dfrac{500{,}000}{x^2+400} < 300 \Leftrightarrow \dfrac{500{,}000}{x^2+400} - 300 < 0 \Leftrightarrow$
$\dfrac{500{,}000 - 300\left(x^2+400\right)}{x^2+400} < 0 \Leftrightarrow \dfrac{-300x^2+380{,}000}{x^2+400} < 0 \Leftrightarrow \dfrac{300\left(\frac{3800}{3} - x^2\right)}{x^2+400}$. Because x represents distance, it is positive, and the denominator is positive for all x. Thus, the inequality holds where $x^2 > \frac{3800}{3} \Leftrightarrow x > \sqrt{\frac{3800}{3}} \approx 35.6$. The temperature is below 300 °C at distances greater than 35.6 meters from the center of the fire.

57.

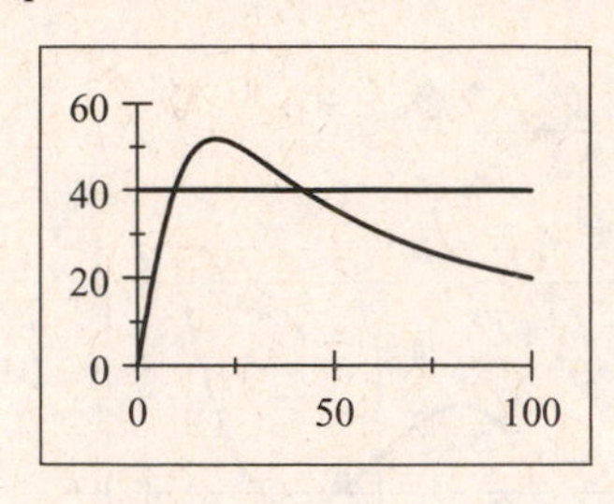

We graph $N(x) = \dfrac{88x}{17 + 17\left(\frac{x}{20}\right)^2}$ and $y = 40$ in the viewing rectangle $[0, 100]$ by $[0, 60]$, and see that $N(x) > 40$ for approximately $9.5 < x < 42.3$. Thus, cars can travel at between 9.5 and 42.3 mi/h.

CHAPTER 3 REVIEW

1. (a) $f(x) = x^2 + 6x + 2 = \left(x^2+6x\right) + 2$
$= \left(x^2+6x+9\right) + 2 - 9$
$= (x+3)^2 - 7$

(b)

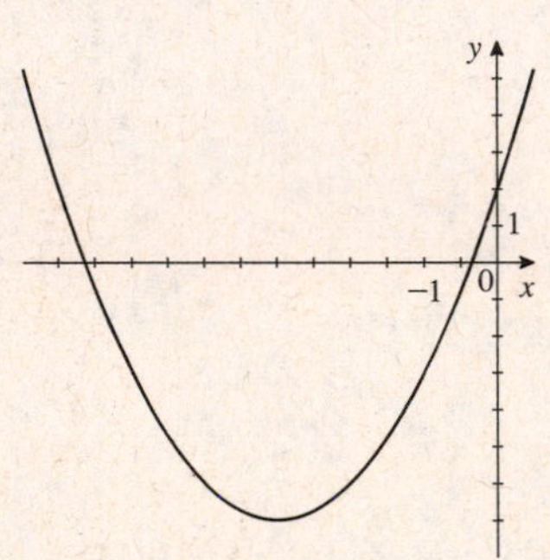

3. (a) $f(x) = 1 - 10x - x^2 = -\left(x^2+10x\right) + 1$
$= -\left(x^2+10x+25\right) + 1 + 25$
$= -(x+5)^2 + 26$

(b)

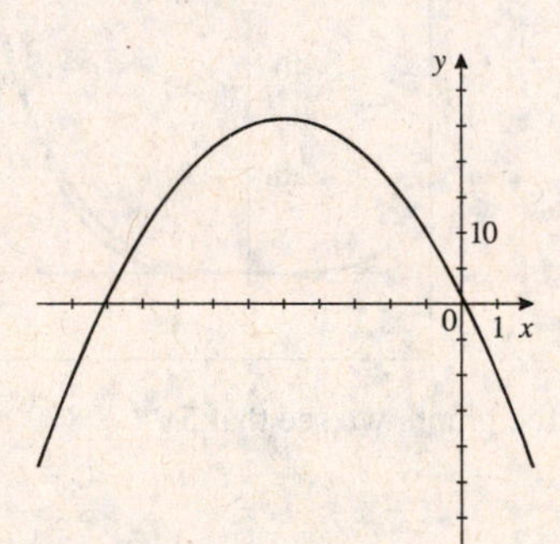

5. $f(x) = -x^2 + 3x - 1 = -\left(x^2-3x\right) - 1 = -\left(x^2-3x+\frac{9}{4}\right) - 1 + \frac{9}{4} = -\left(x-\frac{3}{2}\right)^2 + \frac{5}{4}$ has the maximum value $\frac{5}{4}$ when $x = \frac{3}{2}$.

7. We write the height function in standard form: $h(t) = -16t^2 + 48t + 32 = -16\left(t^2-3t\right) + 32 = -16\left(t^2-3t+\frac{9}{4}\right) + 32 + 36 = -16\left(t-\frac{3}{2}\right)^2 + 68$. The stone reaches a maximum height of 68 ft.

9. $P(x) = -x^3 + 64$

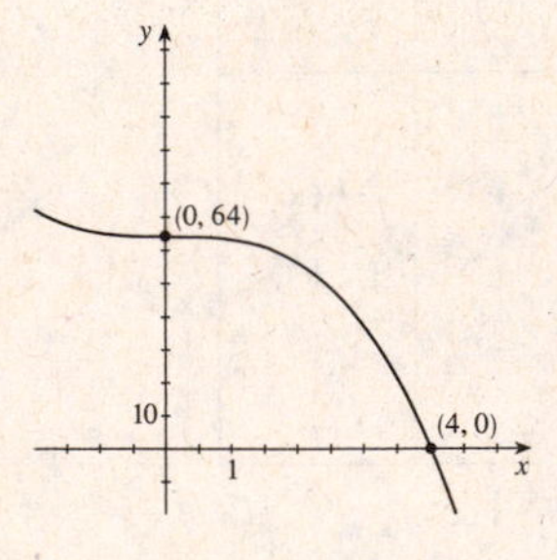

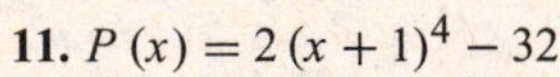

11. $P(x) = 2(x+1)^4 - 32$

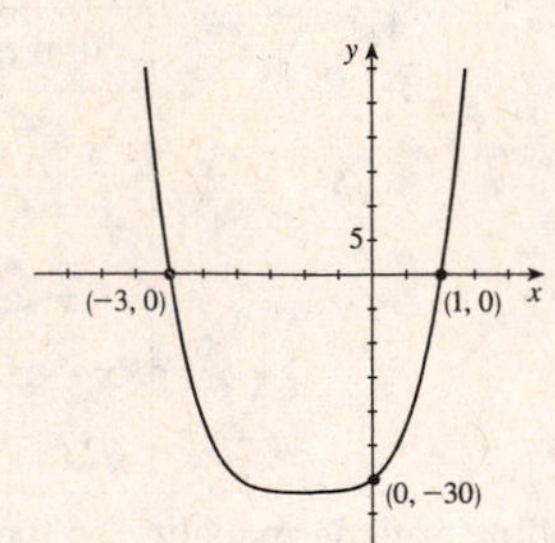

13. $P(x) = 32 + (x-1)^5$

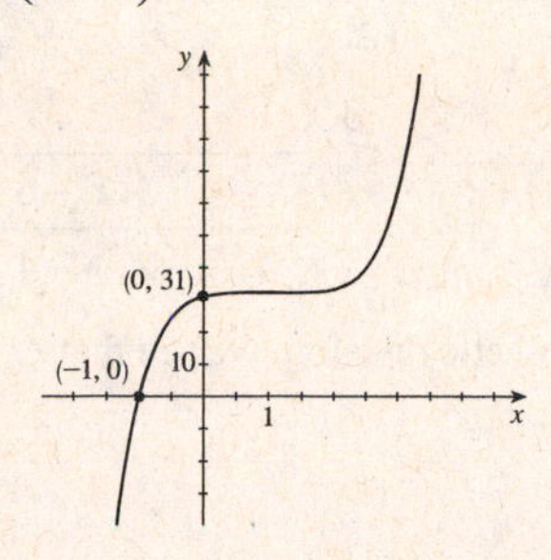

15. (a) $P(x) = (x-3)(x+1)(x-5) = x^3 - 7x^2 + 7x + 15$ has odd degree and a positive leading coefficient, so $y \to \infty$ as $x \to \infty$ and $y \to -\infty$ as $x \to -\infty$.

(b)

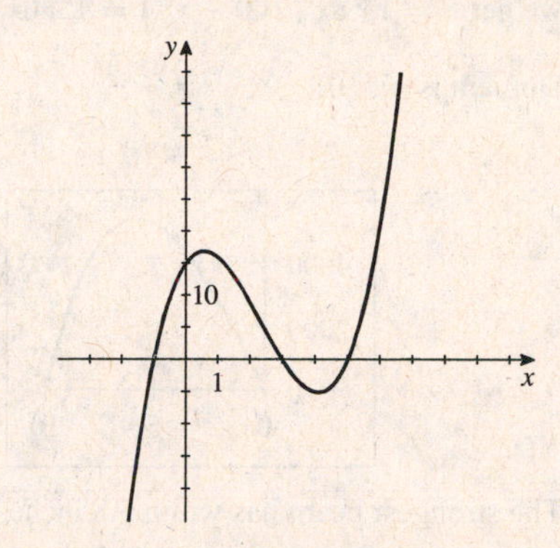

17. (a) $P(x) = -(x-1)^2(x-4)(x+2)^2$

$$= -x^5 + 2x^4 + 11x^3 - 8x^2 - 20x + 16$$

has odd degree and a negative leading coefficient, so $y \to -\infty$ as $x \to \infty$ and $y \to \infty$ as $x \to -\infty$.

(b)

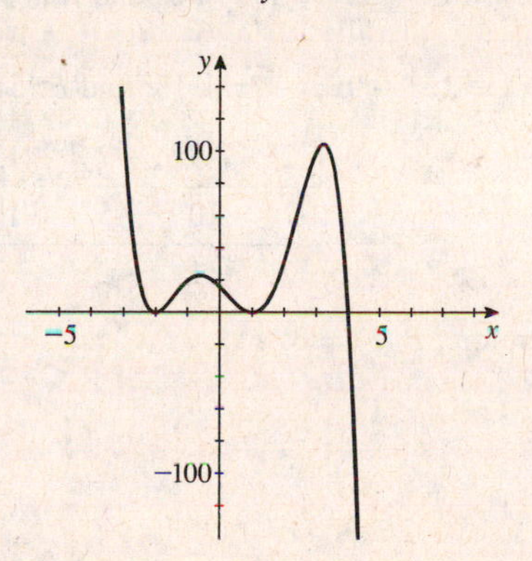

19. (a) $P(x) = x^3(x-2)^2$. The zeros of P are 0 and 2, with multiplicities 3 and 2, respectively.

(b) We sketch the graph using the guidelines on page 295.

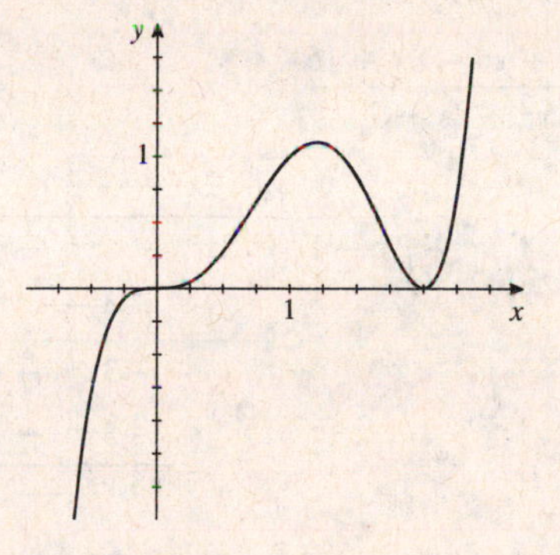

21. $P(x) = x^3 - 4x + 1$. x-intercepts: -2.1, 0.3, and 1.9. y-intercept: 1. Local maximum is $(-1.2, 4.1)$. Local minimum is $(1.2, -2.1)$. $y \to \infty$ as $x \to \infty$; $y \to -\infty$ as $x \to -\infty$.

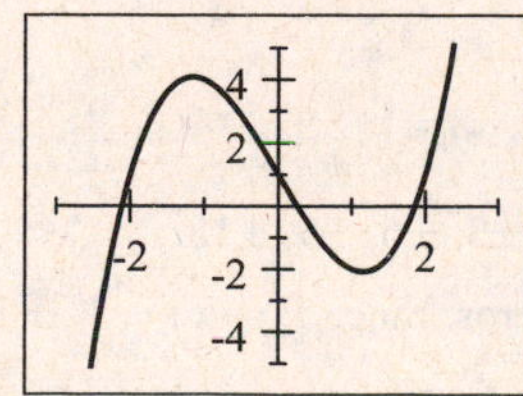

23. $P(x) = 3x^4 - 4x^3 - 10x - 1$. x-intercepts: -0.1 and 2.1. y-intercept: -1. Local maximum is $(1.4, -14.5)$. There is no local maximum. $y \to \infty$ as $x \to \pm\infty$.

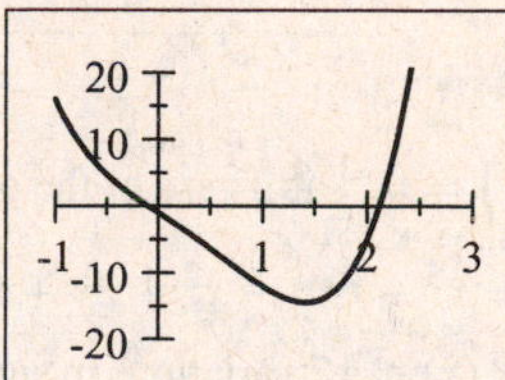

25. **(a)** Use the Pythagorean Theorem and solving for y^2 we have, $x^2 + y^2 = 10^2 \Leftrightarrow y^2 = 100 - x^2$. Substituting we get $S = 13.8x\left(100 - x^2\right) = 1380x - 13.8x^3$.

(b) Domain is [0, 10].

(c)

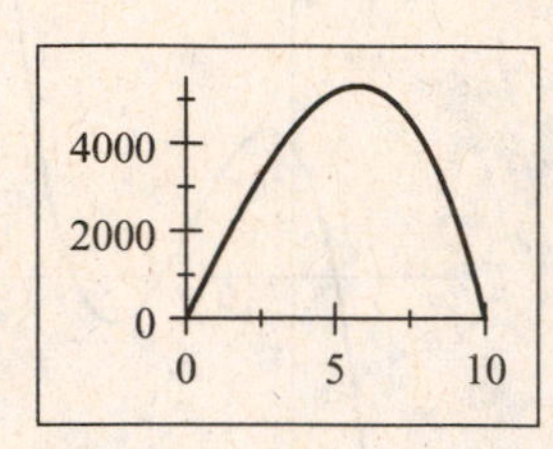

(d) The strongest beam has width 5.8 inches.

27. $\dfrac{x^2 - 5x + 2}{x - 3}$

$$\begin{array}{r|rrr} 3 & 1 & -5 & 2 \\ & & 3 & -6 \\ \hline & 1 & -2 & -4 \end{array}$$

Using synthetic division, we see that $Q(x) = x - 2$ and $R(x) = -4$.

29. $\dfrac{2x^3 - x^2 + 3x - 4}{x + 5}$

$$\begin{array}{r|rrrr} -5 & 2 & -1 & 3 & -4 \\ & & -10 & 55 & -290 \\ \hline & 2 & -11 & 58 & -294 \end{array}$$

Using synthetic division, we see that $Q(x) = 2x^2 - 11x + 58$ and $R(x) = -294$.

31. $\dfrac{x^4 - 8x^2 + 2x + 7}{x + 5}$

$$\begin{array}{r|rrrrr} -5 & 1 & 0 & -8 & 2 & 7 \\ & & -5 & 25 & -85 & 415 \\ \hline & 1 & -5 & 17 & -83 & 422 \end{array}$$

Using synthetic division, we see that $Q(x) = x^3 - 5x^2 + 17x - 83$ and $R(x) = 422$.

33. $\dfrac{2x^3 + x^2 - 8x + 15}{x^2 + 2x - 1}$

$$\begin{array}{r|l} & 2x - 3 \\ \hline x^2 + 2x - 1 & 2x^3 + x^2 - 8x + 15 \\ & 2x^3 + 4x^2 - 2x \\ \hline & \quad -3x^2 - 6x + 15 \\ & \quad -3x^2 - 6x + 3 \\ \hline & \qquad\qquad 12 \end{array}$$

Therefore, $Q(x) = 2x - 3$, and $R(x) = 12$.

35. $P(x) = 2x^3 - 9x^2 - 7x + 13$; find $P(5)$.

$$\begin{array}{r|rrrr} 5 & 2 & -9 & -7 & 13 \\ & & 10 & 5 & -10 \\ \hline & 2 & 1 & -2 & 3 \end{array}$$

Therefore, $P(5) = 3$.

37. The remainder when dividing $P(x) = x^{500} + 6x^{101} - x^2 - 2x + 4$ by $x - 1$ is $P(1) = (1)^{500} + 6(1)^{201} - (1)^2 - 2(1) + 4 = 8$.

39. $\frac{1}{2}$ is a zero of $P(x) = 2x^4 + x^3 - 5x^2 + 10x - 4$ if $P\left(\frac{1}{2}\right) = 0$.

$$\begin{array}{r|rrrrr} \frac{1}{2} & 2 & 1 & -5 & 10 & -4 \\ & & 1 & 1 & -2 & 4 \\ \hline & 2 & 2 & -4 & 8 & 0 \end{array}$$

Since $P\left(\frac{1}{2}\right) = 0$, $\frac{1}{2}$ is a zero of the polynomial.

41. **(a)** $P(x) = x^5 - 6x^3 - x^2 + 2x + 18$ has possible rational zeros $\pm1, \pm2, \pm3, \pm6, \pm9, \pm18$.

(b) Since $P(x)$ has 2 variations in sign, there are either 0 or 2 positive real zeros. Since $P(-x) = -x^5 + 6x^3 - x^2 - 2x + 18$ has 3 variations in sign, there are 1 or 3 negative real zeros.

43. (a) $P(x) = 3x^7 - x^5 + 5x^4 + x^3 + 8$ has possible rational zeros $\pm 1, \pm 2, \pm 4, \pm 8, \pm\frac{1}{3}, \pm\frac{2}{3}, \pm\frac{4}{3}, \pm\frac{8}{3}$.

(b) Since $P(x)$ has 2 variations in sign, there are either 0 or 2 positive real zeros. Since $P(-x) = -3x^7 + x^5 + 5x^4 - x^3 + 8$ has 3 variations in sign, there are 1 or 3 negative real zeros.

45. (a) $P(x) = x^3 - 16x = x\left(x^2 - 16\right)$

$= x(x-4)(x+4)$

has zeros $-4, 0, 4$ (all of multiplicity 1).

(b)

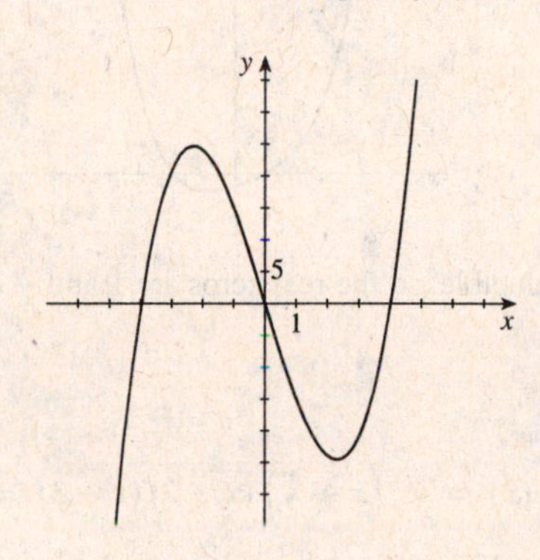

47. (a) $P(x) = x^4 + x^3 - 2x^2 = x^2\left(x^2 + x - 2\right)$

$= x^2(x+2)(x-1)$

The zeros are 0 (multiplicity 2), -2 (multiplicity 1), and 1 (multiplicity 1).

(b)

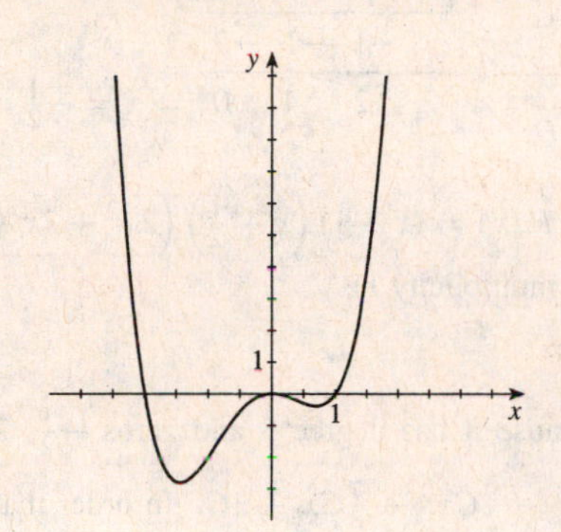

49. (a) $P(x) = x^4 - 2x^3 - 7x^2 + 8x + 12$. The possible rational zeros are $\pm 1, \pm 2, \pm 3, \pm 4, \pm 6, \pm 12$. P has 2 variations in sign, so it has either 2 or 0 positive real zeros.

$$\begin{array}{r|rrrrr} 1 & 1 & -2 & -7 & 8 & 12 \\ & & 1 & -1 & -8 & 0 \\ \hline & 1 & -1 & -8 & 0 & 12 \end{array} \qquad \begin{array}{r|rrrrr} 2 & 1 & -2 & -7 & 8 & 12 \\ & & 2 & 0 & -14 & -12 \\ \hline & 2 & 0 & -7 & -6 & 0 \end{array} \Rightarrow x = 2 \text{ is a root.}$$

$P(x) = x^4 - 2x^3 - 7x^2 + 8x + 12 = (x-2)\left(x^3 - 7x - 6\right)$. Continuing:

$$\begin{array}{r|rrrr} 2 & 1 & 0 & -6 & -6 \\ & & 2 & 4 & -4 \\ \hline & 1 & 2 & -2 & -10 \end{array} \qquad \begin{array}{r|rrrr} 3 & 1 & 0 & -7 & -6 \\ & & 3 & 9 & 6 \\ \hline & 1 & 3 & 2 & 0 \end{array}$$

so $x = 3$ is a root and

$P(x) = (x-2)(x-3)\left(x^2 + 3x + 2\right)$

$= (x-2)(x-3)(x+1)(x+2)$

Therefore the real roots are $-2, -1, 2$, and 3 (all of multiplicity 1).

(b)

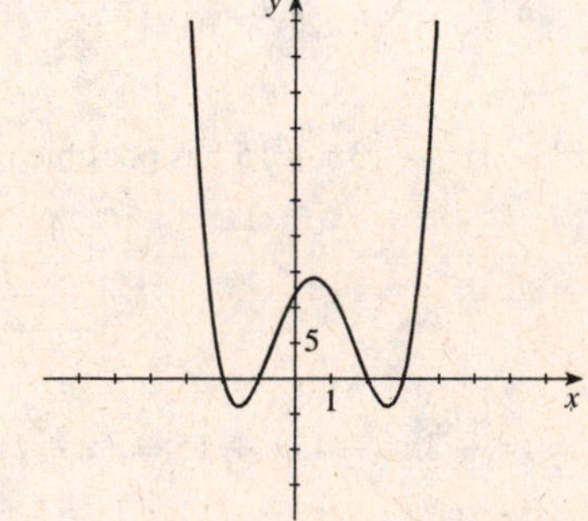

51. (a) $P(x) = 2x^4 + x^3 + 2x^2 - 3x - 2$. The possible rational roots are $\pm 1, \pm 2, \pm \frac{1}{2}$. P has one variation in sign, and hence 1 positive real root. $P(-x)$ has 3 variations in sign and hence either 3 or 1 negative real roots.

1	2	1	2	−3	−2	
		2	3	5	2	
	2	3	5	2	0	$\Rightarrow x = 1$ is a zero.

$P(x) = 2x^4 + x^3 + 2x^2 - 3x - 2 = (x - 1)\left(2x^3 + 3x^2 + 5x + 2\right)$.

Continuing:

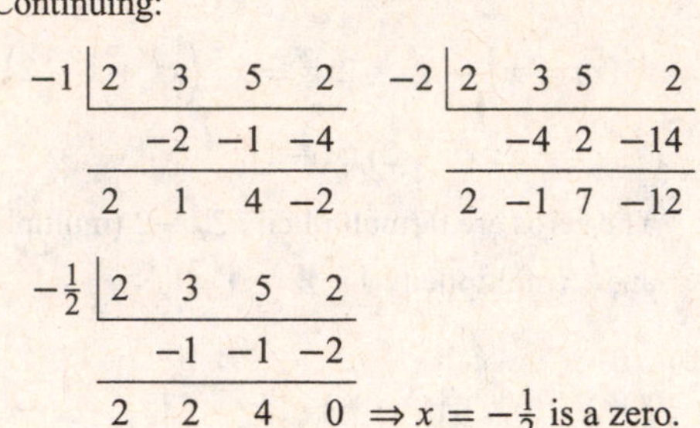

−1	2	3	5	2
		−2	−1	−4
	2	1	4	−2

−2	2	3	5	2
		−4	2	−14
	2	−1	7	−12

$-\frac{1}{2}$	2	3	5	2	
		−1	−1	−2	
	2	2	4	0	$\Rightarrow x = -\frac{1}{2}$ is a zero.

(b)

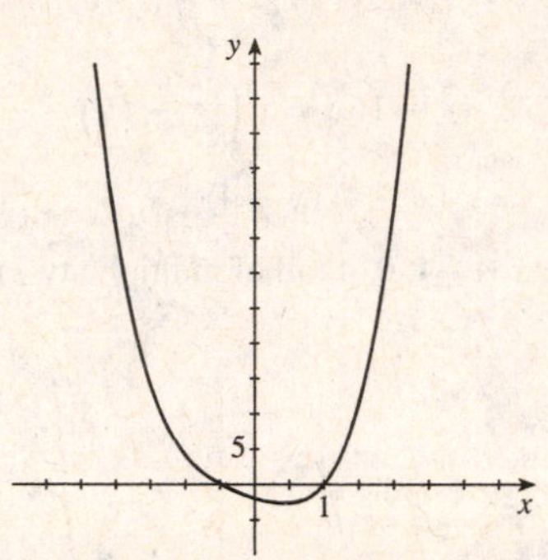

$P(x) = (x - 1)\left(x + \frac{1}{2}\right)\left(2x^2 + 2x + 4\right)$. The quadratic is irreducible, so the real zeros are 1 and $-\frac{1}{2}$ (each of multiplicity 1).

53. Because it has degree 3 and zeros $-\frac{1}{2}$, 2, and 3, we can write $P(x) = C\left(x + \frac{1}{2}\right)(x - 2)(x - 3) = Cx^3 - \frac{9}{2}Cx^2 + \frac{7}{2}Cx + 3C$. In order that the constant coefficient be 12, we must have $3C = 12 \Leftrightarrow C = 4$, so $P(x) = 4x^3 - 18x^2 + 14x + 12$.

55. No, there is no polynomial of degree 4 with integer coefficients that has zeros i, $2i$, $3i$ and $4i$. Since the imaginary zeros of polynomial equations with real coefficients come in complex conjugate pairs, there would have to be 8 zeros, which is impossible for a polynomial of degree 4.

57. $P(x) = x^3 - x^2 + x - 1$ has possible rational zeros ± 1.

1	1	−1	1	−1	
		1	0	1	
	1	0	1	0	$\Rightarrow x = 1$ is a zero.

So $P(x) = (x - 1)\left(x^2 + 1\right)$. Therefore, the zeros are 1 and $\pm i$.

59. $P(x) = x^3 - 3x^2 - 13x + 15$ has possible rational zeros $\pm 1, \pm 3, \pm 5, \pm 15$.

1	1	−3	−13	15	
		1	−2	−15	
	1	−2	−15	0	$\Rightarrow x = 1$ is a zero.

So $P(x) = x^3 - 3x^2 - 13x + 15 = (x - 1)\left(x^2 - 2x - 15\right) = (x - 1)(x - 5)(x + 3)$. Therefore, the zeros are −3, 1, and 5.

61. $P(x) = x^4 + 6x^3 + 17x^2 + 28x + 20$ has possible rational zeros $\pm1, \pm2, \pm4, \pm5, \pm10, \pm20$. Since all of the coefficients are positive, there are no positive real zeros.

$$\begin{array}{r|rrrrr} -1 & 1 & 6 & 17 & 28 & 20 \\ & & -1 & -5 & -12 & -16 \\ \hline & 1 & 5 & 12 & 16 & 4 \end{array} \qquad \begin{array}{r|rrrrr} -2 & 1 & 6 & 17 & 28 & 20 \\ & & -2 & -8 & -18 & -20 \\ \hline & 1 & 4 & 9 & 10 & 0 \end{array} \Rightarrow x = -2 \text{ is a zero.}$$

$P(x) = x^4 + 6x^3 + 17x^2 + 28x + 20 = (x + 2)\left(x^3 + 4x^2 + 9x + 10\right)$. Continuing with the quotient, we have

$$\begin{array}{r|rrrr} -2 & 1 & 4 & 9 & 10 \\ & & -2 & -4 & -10 \\ \hline & 1 & 2 & 5 & 0 \end{array} \Rightarrow x = -2 \text{ is a zero.}$$

Thus $P(x) = x^4 + 6x^3 + 17x^2 + 28x + 20 = (x + 2)^2\left(x^2 + 2x + 5\right)$. Now $x^2 + 2x + 5 = 0$ when $x = \frac{-2\pm\sqrt{4-4(5)(1)}}{2} = \frac{-2\pm4i}{2} = -1 \pm 2i$. Thus, the zeros are -2 (multiplicity 2) and $-1 \pm 2i$.

63. $P(x) = x^5 - 3x^4 - x^3 + 11x^2 - 12x + 4$ has possible rational zeros $\pm1, \pm2, \pm4$.

$$\begin{array}{r|rrrrrr} 1 & 1 & -3 & -1 & 11 & -12 & 4 \\ & & 1 & -2 & -3 & 8 & -4 \\ \hline & 1 & -2 & -3 & 8 & -4 & 0 \end{array} \Rightarrow x = 1 \text{ is a zero.}$$

$P(x) = x^5 - 3x^4 - x^3 + 11x^2 - 12x + 4 = (x - 1)\left(x^4 - 2x^3 - 3x^2 + 8x - 4\right)$. Continuing with the quotient, we have

$$\begin{array}{r|rrrrr} 1 & 1 & -2 & -3 & 8 & -4 \\ & & 1 & -1 & -4 & 4 \\ \hline & 1 & -1 & -4 & 4 & 0 \end{array} \Rightarrow x = 1 \text{ is a zero.}$$

$$\begin{aligned} x^5 - 3x^4 - x^3 + 11x^2 - 12x + 4 &= (x - 1)^2\left(x^3 - x^2 - 4x + 4\right) = (x - 1)^3\left(x^2 - 4\right) \\ &= (x - 1)^3(x - 2)(x + 2) \end{aligned}$$

Therefore, the zeros are 1 (multiplicity 3), -2, and 2.

65. $P(x) = x^6 - 64 = \left(x^3 - 8\right)\left(x^3 + 8\right) = (x - 2)\left(x^2 + 2x + 4\right)(x + 2)\left(x^2 - 2x + 4\right)$. Now using the Quadratic Formula to find the zeros of $x^2 + 2x + 4$, we have $x = \frac{-2\pm\sqrt{4-4(4)(1)}}{2} = \frac{-2\pm2i\sqrt{3}}{2} = -1 \pm i\sqrt{3}$, and using the Quadratic Formula to find the zeros of $x^2 - 2x + 4$, we have $x = \frac{2\pm\sqrt{4-4(4)(1)}}{2} = \frac{2\pm2i\sqrt{3}}{2} = 1 \pm i\sqrt{3}$. Therefore, the zeros are $2, -2, 1 \pm i\sqrt{3}$, and $-1 \pm i\sqrt{3}$.

67. $P(x) = 6x^4 - 18x^3 + 6x^2 - 30x + 36 = 6\left(x^4 - 3x^3 + x^2 - 5x + 6\right)$ has possible rational zeros $\pm1, \pm2, \pm3, \pm6$.

1	6	−18	6	−30	36	
		6	−12	−6	−36	
	6	−12	−6	−36	0	$\Rightarrow x = 1$ is a zero.

So $P(x) = 6x^4 - 18x^3 + 6x^2 - 30x + 36 = (x-1)\left(6x^3 - 12x^2 - 6x - 36\right) = 6(x-1)\left(x^3 - 2x^2 - x - 6\right)$. Continuing with the quotient we have

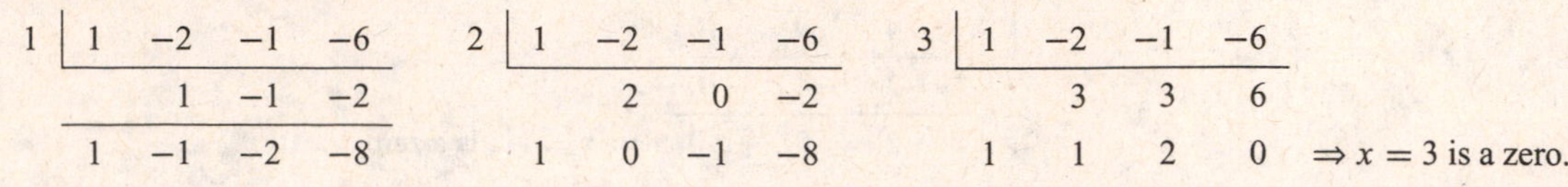

1	1	−2	−1	−6
		1	−1	−2
	1	−1	−2	−8

2	1	−2	−1	−6
		2	0	−2
	1	0	−1	−8

3	1	−2	−1	−6	
		3	3	6	
	1	1	2	0	$\Rightarrow x = 3$ is a zero.

So $P(x) = 6x^4 - 18x^3 + 6x^2 - 30x + 36 = 6(x-1)(x-3)\left(x^2 + x + 2\right)$. Now $x^2 + x + 2 = 0$ when $x = \frac{-1\pm\sqrt{1-4(1)(2)}}{2} = \frac{-1\pm i\sqrt{7}}{2}$, and so the zeros are 1, 3, and $\frac{-1\pm i\sqrt{7}}{2}$.

69. $2x^2 = 5x + 3 \Leftrightarrow 2x^2 - 5x - 3 = 0$. The solutions are $x = -0.5, 3$.

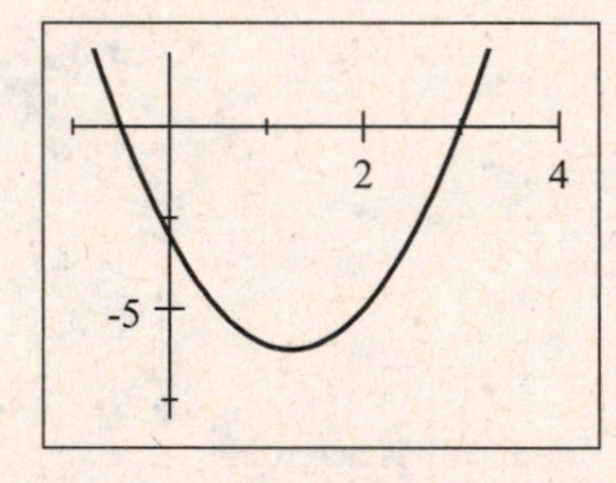

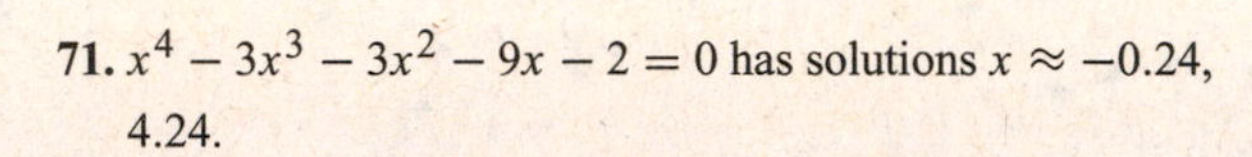

71. $x^4 - 3x^3 - 3x^2 - 9x - 2 = 0$ has solutions $x \approx -0.24$, 4.24.

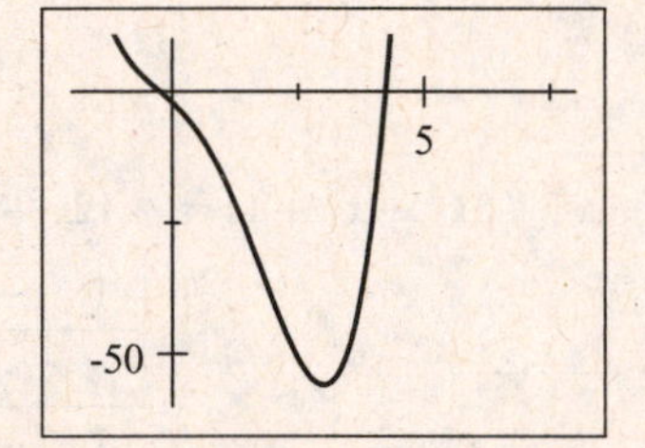

73. $P(x) = x^3 - 2x - 4$

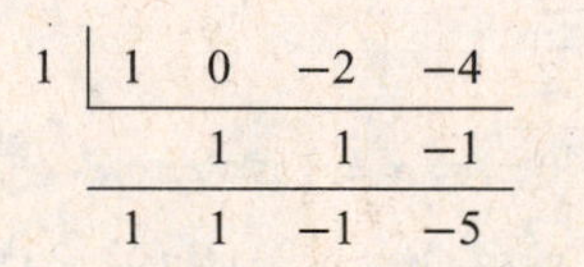

1	1	0	−2	−4
		1	1	−1
	1	1	−1	−5

2	1	0	−2	−4
		2	4	4
	1	2	2	0

$P(x) = x^3 - 2x - 4 = (x-2)\left(x^2 + 2x + 2\right)$. Since $x^2 + 2x + 2 = 0$ has no real solution, the only real zero of P is $x = 2$.

75. (a) $r(x) = \dfrac{3}{x+4}$. The vertical asymptote is $x = -4$. Because the denominator has higher degree than the numerator, the horizontal asymptote is $y = 0$. When $x = 0$, $y = \frac{3}{4}$, so the y-intercept is $\frac{3}{4}$. There is no x-intercept because the numerator is never 0. The domain of r is $(-\infty, -4) \cup (-4, \infty)$ and its range is $(-\infty, 0) \cup (0, \infty)$.

(b) If $f(x) = \dfrac{1}{x}$, then $r(x) = \dfrac{3}{x+4} = 3\left(\dfrac{1}{x+4}\right) = 3f(x+4)$, so we obtain the graph of r by shifting the graph of f to the left 4 units and stretching vertically by a factor of 3.

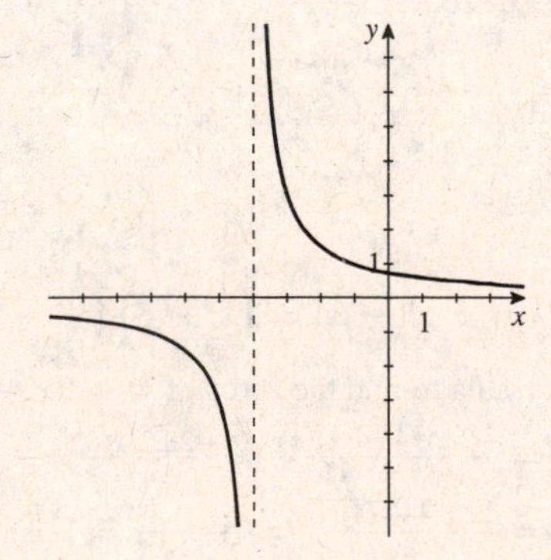

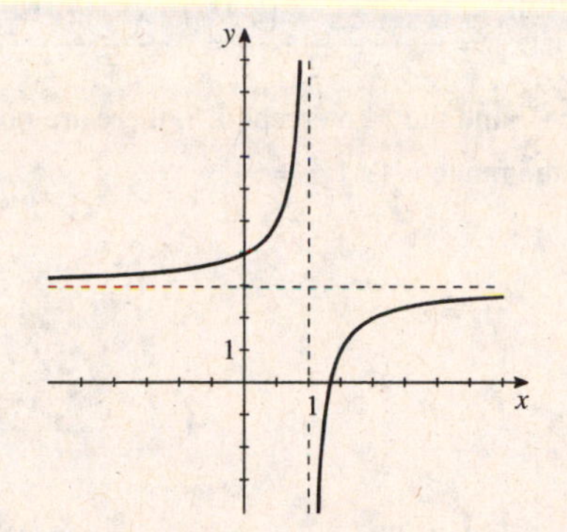

77. **(a)** $r(x) = \dfrac{3x-4}{x-1}$. The vertical asymptote is $x = 1$. Because the denominator has the same degree as the numerator, the horizontal asymptote is $y = \frac{3}{1} = 3$. When $x = 0$, $y = -4$, so the y-intercept is -4. When $y = 0$, $3x - 4 = 0 \Leftrightarrow x = \frac{4}{3}$, so the x-intercept is $\frac{4}{3}$. The domain of r is $(-\infty, 1) \cup (1, \infty)$ and its range is $(-\infty, 3) \cup (3, \infty)$.

(b) If $f(x) = \dfrac{1}{x}$, then

$r(x) = \dfrac{3(x-1)-1}{x-1} = 3 - \dfrac{1}{x-1} = 3 - f(x-1)$. Thus, we obtain the graph of r by shifting the graph of f to the right 1 unit, reflecting in the x-axis, and shifting upward 3 units.

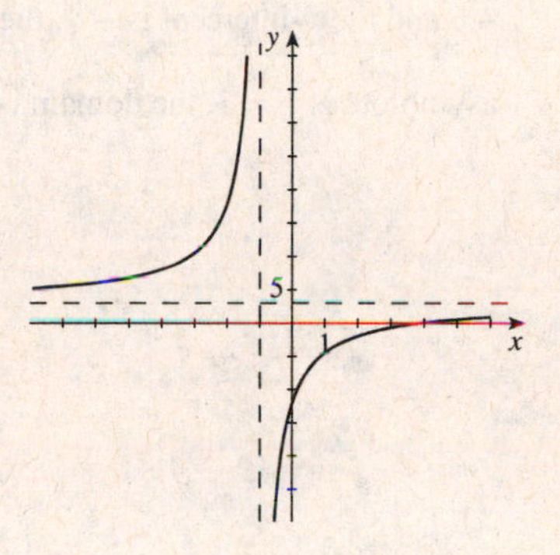

79. $r(x) = \dfrac{3x-12}{x+1}$. When $x = 0$, we have $r(0) = \dfrac{-12}{1} = -12$, so the y-intercept is -12. Since $y = 0$, when $3x - 12 = 0 \Leftrightarrow x = 4$, the x-intercept is 4. The vertical asymptote is $x = -1$. Because the denominator has the same degree as the numerator, the horizontal asymptote is $y = \frac{3}{1} = 3$. The domain of r is $(-\infty, -1) \cup (-1, \infty)$ and its range is $(-\infty, 3) \cup (3, \infty)$.

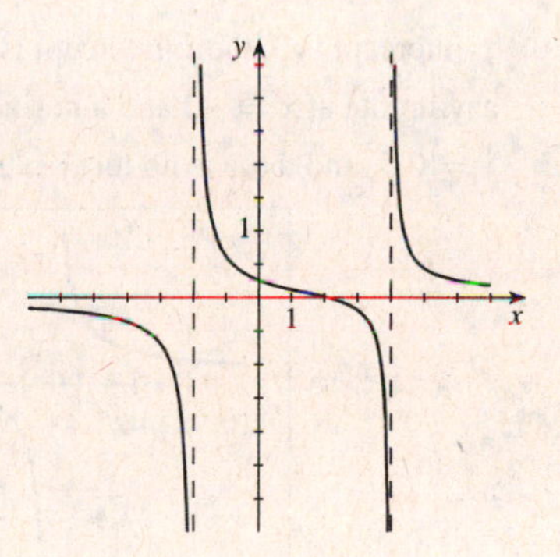

81. $r(x) = \dfrac{x-2}{x^2-2x-8} = \dfrac{x-2}{(x+2)(x-4)}$. When $x = 0$, we have $r(0) = \frac{-2}{-8} = \frac{1}{4}$, so the y-intercept is $\frac{1}{4}$. When $y = 0$, we have $x - 2 = 0 \Leftrightarrow x = 2$, so the x-intercept is 2. There are vertical asymptotes at $x = -2$ and $x = 4$. The domain of r is $(-\infty, -2) \cup (-2, 4) \cup (4, \infty)$ and its range is $(-\infty, \infty)$.

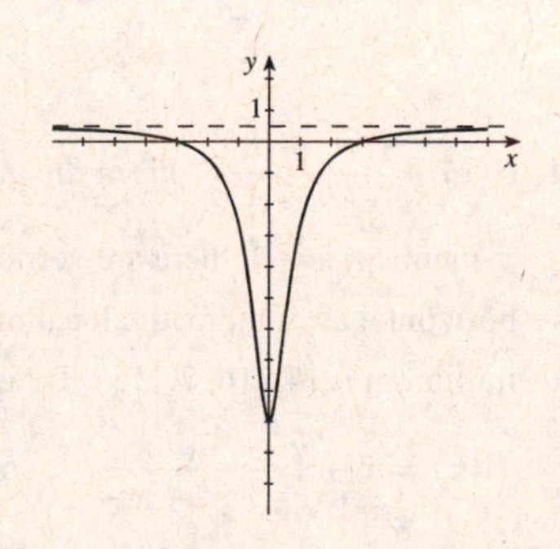

83. $r(x) = \dfrac{x^2-9}{2x^2+1} = \dfrac{(x+3)(x-3)}{2x^2+1}$. When $x = 0$, we have $r(0) = \frac{-9}{1}$, so the y-intercept is -9. When $y = 0$, we have $x^2 - 9 = 0 \Leftrightarrow x = \pm 3$ so the x-intercepts are -3 and 3. Since $2x^2 + 1 > 0$, the denominator is never zero so there are no vertical asymptotes. The horizontal asymptote is at $y = \frac{1}{2}$ because the degree of the denominator and numerator are the same. The domain of r is $(-\infty, \infty)$ and its range is $\left[-9, \frac{1}{2}\right)$.

85. $r(x) = \dfrac{x^2 + 5x - 14}{x - 2} = \dfrac{(x+7)(x-2)}{x-2} = x + 7$ for $x \neq 2$. The x-intercept is -7 and the y-intercept is 7, there are no asymptotes, the domain is $\{x \mid x \neq 2\}$, and the range is $\{y \mid y \neq 9\}$.

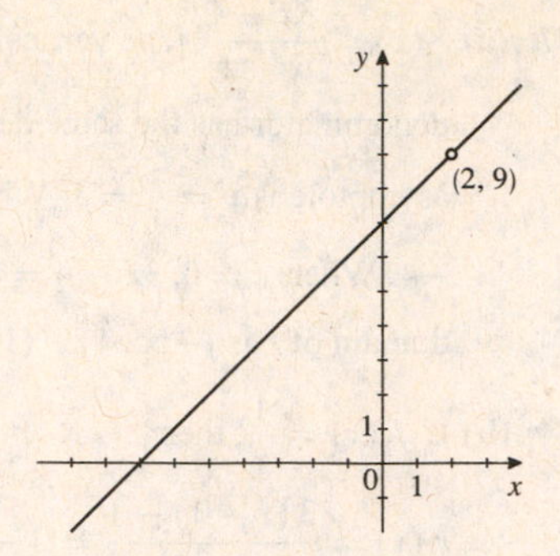

87. $r(x) = \dfrac{x^2 + 3x - 18}{x^2 - 8x + 15} = \dfrac{(x+6)(x-3)}{(x-5)(x-3)} = \dfrac{x+6}{x+5}$ for $x \neq 3$. The x-intercept is -6 and the y-intercept is $-\frac{6}{5}$, the vertical asymptote is $x = 5$, the horizontal asymptote is $y = 1$, the domain is $\{x \mid x \neq 3, 5\}$, and the range is $\left\{y \mid y \neq 1, -\frac{9}{2}\right\}$.

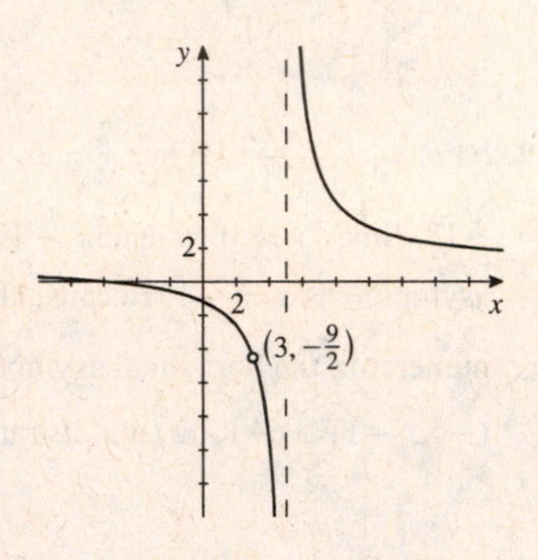

89. $r(x) = \dfrac{x - 3}{2x + 6}$. From the graph we see that the x-intercept is 3, the y-intercept is -0.5, there is a vertical asymptote at $x = -3$ and a horizontal asymptote at $y = 0.5$, and there is no local extremum.

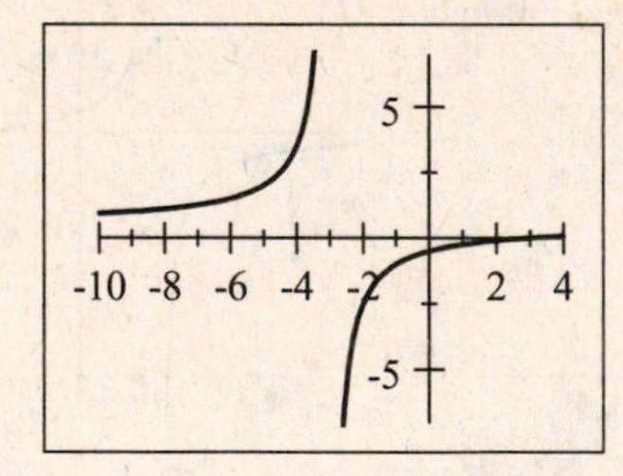

91. $r(x) = \dfrac{x^3 + 8}{x^2 - x - 2}$. From the graph we see that the x-intercept is -2, the y-intercept is -4, there are vertical asymptotes at $x = -1$ and $x = 2$, there is no horizontal asymptote, the local maximum is $(0.425, -3.599)$, and the local minimum is $(4.216, 7.175)$. By using long division, we see that $f(x) = x + 1 + \dfrac{10 - x}{x^2 - x - 2}$, so f has a slant asymptote of $y = x + 1$.

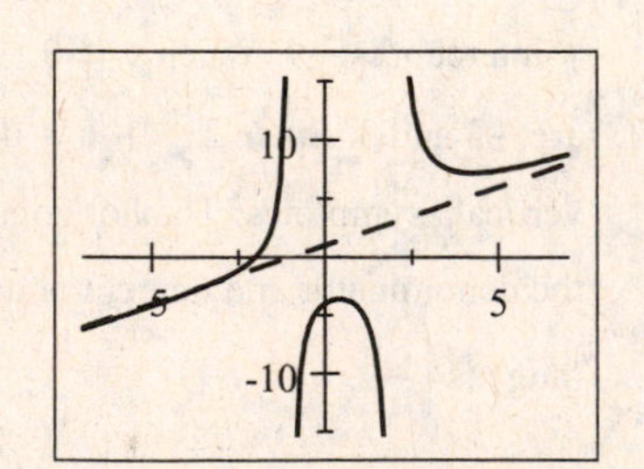

93. $2x^2 \geq x + 3 \Leftrightarrow 2x^2 - x - 3 \geq 0 \Leftrightarrow P(x) = (x+1)(2x-3) \geq 0$. The cut points occur where $x + 1 = 0$ and where $2x - 3 = 0$; that is, at $x = -1$ and $x = \frac{3}{2}$. We make a sign diagram:

Sign of	$(-\infty, -1)$	$\left(-1, \frac{3}{2}\right)$	$\left(\frac{3}{2}, \infty\right)$
$x+1$	$-$	$+$	$+$
$2x-3$	$-$	$-$	$+$
$P(x)$	$+$	$-$	$+$

Both endpoints satisfy the inequality. The solution is $(-\infty, -1] \cup \left[\frac{3}{2}, \infty\right)$.

95. $x^4 - 7x^2 - 18 < 0 \Leftrightarrow \left(x^2 - 9\right)\left(x^2 + 2\right) < 0 \Leftrightarrow P(x) = (x+3)(x-3)\left(x^2+2\right) < 0$. The last factor is positive for all x. We make a sign diagram:

Sign of	$(-\infty, -3)$	$(-3, 3)$	$(3, \infty)$
$x+3$	$-$	$+$	$+$
$x-3$	$-$	$-$	$+$
$P(x)$	$+$	$-$	$+$

Neither endpoint satisfies the strict inequality. The solution is $(-3, 3)$.

97. $\dfrac{5}{x^3 - x^2 - 4x + 4} < 0 \Leftrightarrow \dfrac{5}{x\left(x^2-4\right) - \left(x^2-4\right)} < 0 \Leftrightarrow r(x) = \dfrac{5}{(x-1)(x-2)(x+2)} < 0$. We make a sign diagram:

Sign of	$(-\infty, -2)$	$(-2, 1)$	$(1, 2)$	$(2, \infty)$
$x+2$	$-$	$+$	$+$	$+$
$x-1$	$-$	$-$	$+$	$+$
$x-2$	$-$	$-$	$-$	$+$
$r(x)$	$-$	$+$	$-$	$+$

None of the endpoints is in the domain of r. The solution is $(-\infty, -2) \cup (1, 2)$.

99. $\dfrac{1}{x-2} + \dfrac{2}{x+3} \geq \dfrac{3}{x} \Leftrightarrow \dfrac{1}{x-2} + \dfrac{2}{x+3} - \dfrac{3}{x} \geq 0 \Leftrightarrow \dfrac{x(x+3) + 2x(x-2) - 3(x-2)(x+3)}{x(x-2)(x+3)} \geq 0 \Leftrightarrow$
$\dfrac{-4x+18}{x(x-2)(x+3)} \geq 0 \Leftrightarrow r(x) = \dfrac{2(9-2x)}{x(x-2)(x+3)} \geq 0$. We make a sign diagram:

Sign of	$(-\infty, -3)$	$(-3, 0)$	$(0, 2)$	$\left(2, \frac{9}{2}\right)$	$\left(\frac{9}{2}, \infty\right)$
$x+3$	$-$	$+$	$+$	$+$	$+$
x	$-$	$-$	$+$	$+$	$+$
$x-2$	$-$	$-$	$-$	$+$	$+$
$9-2x$	$+$	$+$	$+$	$+$	$-$
$r(x)$	$-$	$+$	$-$	$+$	$-$

The cut point $\frac{9}{2}$ satisfies equality, but the other cut points are not in the domain of r. The solution is $(-3, 0) \cup \left(2, \frac{9}{2}\right]$.

101. $f(x) = \sqrt{24 - x - 3x^2} = \sqrt{(x+3)(8-3x)}$ is defined where $r(x) = (x+3)(8-3x) \geq 0$. We make a sign diagram:

Sign of	$(-\infty, -3)$	$\left(-3, \frac{8}{3}\right)$	$\left(\frac{8}{3}, \infty\right)$
$x+3$	$-$	$+$	$+$
$8-3x$	$+$	$+$	$-$
$r(x)$	$-$	$+$	$-$

Both cut points satisfy equality, so the domain of f is $\left[-3, \frac{8}{3}\right]$.

103.

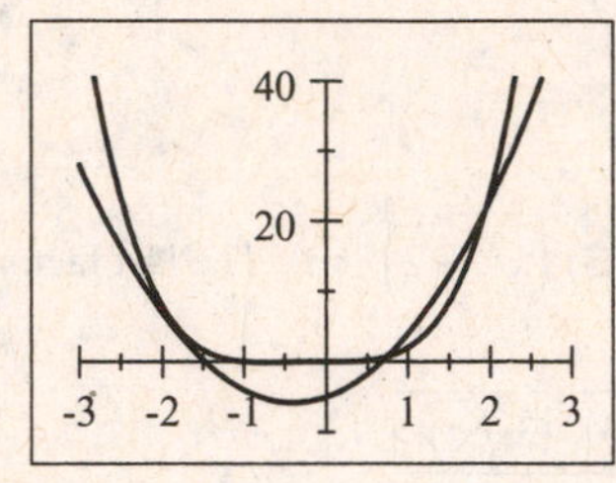

From the graph, we see that $x^4 + x^3 \leq 5x^2 + 4x - 5$ on approximately $[0.74, 1.95]$.

105. (a) We use synthetic division to show that -1 is a zero of $P(x) = 2x^4 + 5x^3 + x + 4$:

$$\begin{array}{r|rrrrr} -1 & 2 & 5 & 0 & 1 & 4 \\ & & -2 & -3 & 3 & -4 \\ \hline & 2 & 3 & -3 & 4 & 0 \end{array}$$

Thus, -1 is a zero and $P(x) = (x+1)\left(2x^3 + 3x^2 - 3x + 4\right)$.

(b) $P(x)$ has no change of sign, and hence no positive real zeros. But $P(x) = (x+1)Q(x)$, so Q cannot have a positive real zero either.

CHAPTER 3 TEST

1.

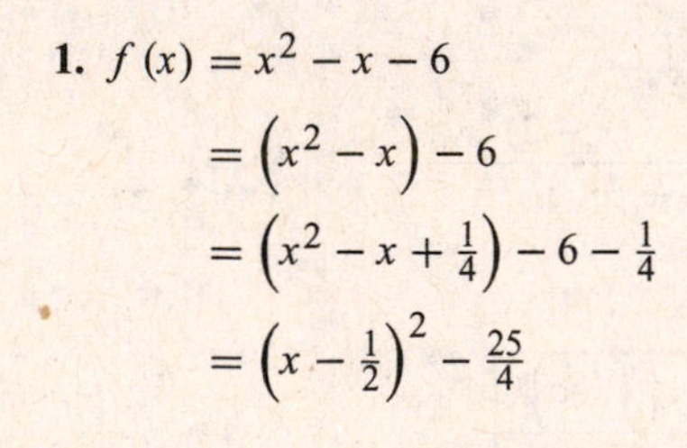

$$\begin{aligned} f(x) &= x^2 - x - 6 \\ &= \left(x^2 - x\right) - 6 \\ &= \left(x^2 - x + \tfrac{1}{4}\right) - 6 - \tfrac{1}{4} \\ &= \left(x - \tfrac{1}{2}\right)^2 - \tfrac{25}{4} \end{aligned}$$

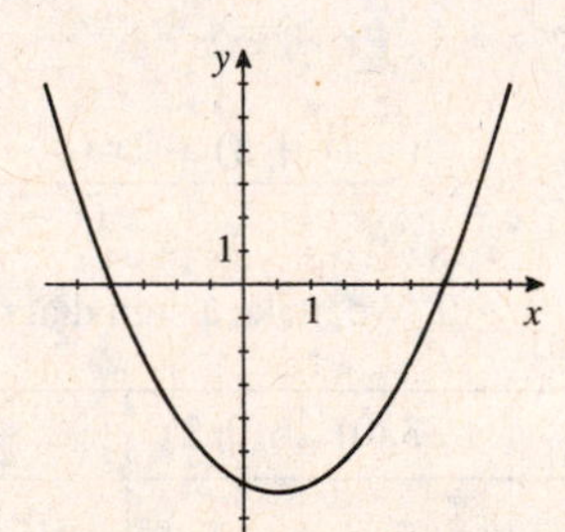

3. (a) We write the function in standard form: $h(x) = 10x - 0.01x^2 = -0.01\left(x^2 - 1000x\right) = -0.01\left(x^2 - 1000x + 500^2\right) + 0.01\left(500^2\right) = -0.01(x - 500)^2 + 2500$. Thus, the maximum height reached by the cannonball is 2500 feet.

(b) By the symmetry of the parabola, we see that the cannonball's height will be 0 again (and thus it will splash into the water) when $x = 1000$ ft.

5. (a)

$$\begin{array}{r|rrrrr} 2 & 1 & 0 & -4 & 2 & 5 \\ & & 2 & 4 & 0 & 4 \\ \hline & 1 & 2 & 0 & 2 & 9 \end{array}$$

Therefore, the quotient is $Q(x) = x^3 + 2x^2 + 2$, and the remainder is $R(x) = 9$.

(b)

$$\begin{array}{r|l} & x^3 + 2x^2 \qquad + \tfrac{1}{2} \\ \hline 2x^2 - 1 & 2x^5 + 4x^4 - x^3 - x^2 + 0x + 7 \\ & 2x^5 \qquad - x^3 \\ \hline & \qquad 4x^4 \qquad - x^2 \\ & \qquad 4x^4 \qquad - 2x^2 \\ \hline & \qquad\qquad x^2 \quad + 7 \\ & \qquad\qquad x^2 \quad - \tfrac{1}{2} \\ \hline & \qquad\qquad\qquad \tfrac{15}{2} \end{array}$$

Therefore, the quotient is $Q(x) = x^3 + 2x^2 + \frac{1}{2}$ and the remainder is $R(x) = \frac{15}{2}$.

7. $P(x) = x^3 - x^2 - 4x - 6$. Possible rational zeros are: $\pm 1, \pm 2, \pm 3, \pm 6$.

$$\begin{array}{r|rrrr} 1 & 1 & -1 & -4 & -6 \\ & & 1 & 0 & -4 \\ \hline & 1 & 0 & -4 & -10 \end{array} \qquad \begin{array}{r|rrrr} 2 & 1 & -1 & -4 & -6 \\ & & 2 & 2 & -4 \\ \hline & 1 & 1 & -2 & -10 \end{array} \qquad \begin{array}{r|rrrr} 3 & 1 & -1 & -4 & -6 \\ & & 3 & 6 & 6 \\ \hline & 1 & 2 & 2 & 0 \end{array} \Rightarrow x = 3 \text{ is a zero.}$$

So $P(x) = (x-3)\left(x^2 + 2x + 2\right)$. Using the Quadratic Formula on the second factor, we have $x = \frac{-2 \pm \sqrt{2^2 - 4(1)(2)}}{2(1)} = \frac{-2 \pm \sqrt{-4}}{2} = \frac{-2 \pm 2\sqrt{-1}}{2} = -1 \pm i$. So zeros of $P(x)$ are 3, $-1 - i$, and $-1 + i$.

9. Since $3i$ is a zero of $P(x)$, $-3i$ is also a zero of $P(x)$. And since -1 is a zero of multiplicity 2, $P(x) = (x+1)^2 (x - 3i)(x + 3i) = \left(x^2 + 2x + 1\right)\left(x^2 + 9\right) = x^4 + 2x^3 + 10x^2 + 18x + 9$.

11. $r(x) = \frac{2x - 1}{x^2 - x - 2}$, $s(x) = \frac{x^3 + 27}{x^2 + 4}$, $t(x) = \frac{x^3 - 9x}{x + 2}$, $u(x) = \frac{x^2 + x - 6}{x^2 - 25}$, and $w(x) = \frac{x^3 + 6x^2 + 9x}{x + 3}$.

(a) $r(x)$ has the horizontal asymptote $y = 0$ because the degree of the denominator is greater than the degree of the numerator. $u(x)$ has the horizontal asymptote $y = \frac{1}{1} = 1$ because the degrees of the numerator and the denominator are the same.

(b) The degree of the numerator of $s(x)$ is one more than the degree of the denominator, so s has a slant asymptote.

(c) The denominator of $s(x)$ is never 0, so s has no vertical asymptote. $w(x) = \frac{x\left(x^2 + 6x + 9\right)}{x + 3} = \frac{x(x+3)^2}{x+3} = x(x+3)$ for $x \neq -3$, so w has no vertical asymptote.

(d) From part (c), w has a "hole" at $(-3, 0)$.

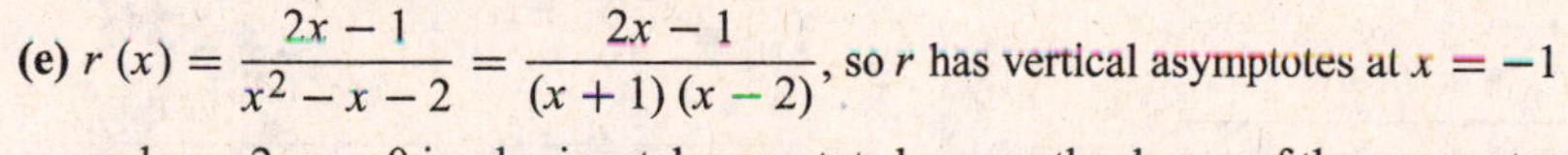

(e) $r(x) = \frac{2x - 1}{x^2 - x - 2} = \frac{2x - 1}{(x+1)(x-2)}$, so r has vertical asymptotes at $x = -1$ and $x = 2$. $y = 0$ is a horizontal asymptote because the degree of the numerator is less than the degree of the denominator.

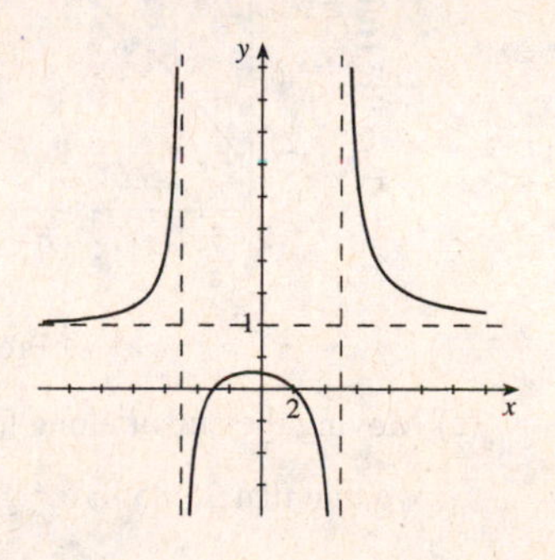

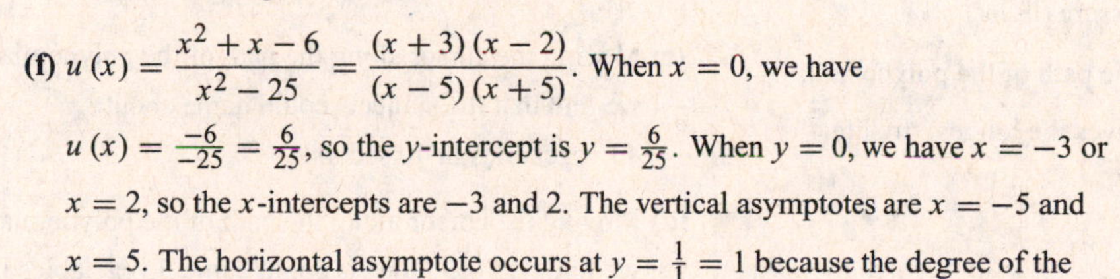

(f) $u(x) = \frac{x^2 + x - 6}{x^2 - 25} = \frac{(x+3)(x-2)}{(x-5)(x+5)}$. When $x = 0$, we have $u(x) = \frac{-6}{-25} = \frac{6}{25}$, so the y-intercept is $y = \frac{6}{25}$. When $y = 0$, we have $x = -3$ or $x = 2$, so the x-intercepts are -3 and 2. The vertical asymptotes are $x = -5$ and $x = 5$. The horizontal asymptote occurs at $y = \frac{1}{1} = 1$ because the degree of the denominator and numerator are the same.

(g)

$$\begin{array}{r|l} & x^2 - 2x - 5 \\ \hline x+2 & x^3 + 0x^2 - 9x + 0 \\ & \underline{x^3 + 2x^2} \\ & \quad -2x^2 - 9x \\ & \quad \underline{-2x^2 - 4x} \\ & \qquad -5x + 0 \\ & \qquad \underline{-5x - 10} \\ & \qquad\quad -10 \end{array}$$

Thus $P(x) = x^2 - 2x - 5$ and $t(x) = \dfrac{x^3 - 9x}{x+2}$ have the same end behavior.

13. $f(x) = \dfrac{1}{\sqrt{4 - 2x - x^2}}$ is defined where $4 - 2x - x^2 > 0$. Using the Quadratic Formula to solve $-x^2 - 2x + 4 = 0$, we have $x = \dfrac{-(-2) \pm \sqrt{(-2)^2 - 4(-1)(4)}}{2(-1)} = -1 \pm \sqrt{5}$. The radicand is positive between these two roots, so the domain of f is $\left(-1 - \sqrt{5}, -1 + \sqrt{5}\right)$.

FOCUS ON MODELING Fitting Polynomial Curves to Data

1. (a) Using a graphing calculator, we obtain the quadratic polynomial $y = -0.275428x^2 + 19.7485x - 273.5523$ (where miles are measured in thousands).

(b)

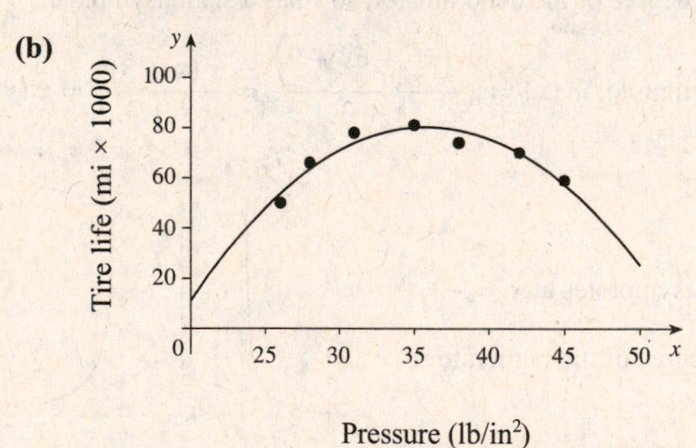

(c) Moving the cursor along the path of the polynomial, we find that 35.85 lb/in^2 gives the longest tire life.

3. (a) Using a graphing calculator, we obtain the cubic polynomial

$$y = 0.00203709x^3 - 0.104522x^2 + 1.966206x + 1.45576.$$

(b)

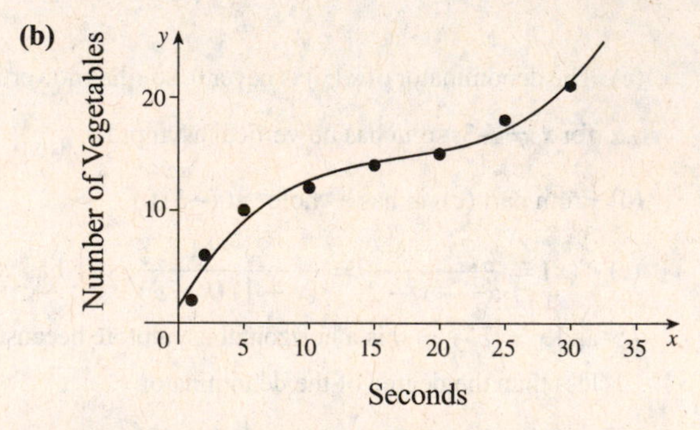

(c) Moving the cursor along the path of the polynomial, we find that the subjects could name about 43 vegetables in 40 seconds.

(d) Moving the cursor along the path of the polynomial, we find that the subjects could name 5 vegetables in about 2.0 seconds.

5. (a) Using a graphing calculator, we obtain the quadratic polynomial

$y = 0.0120536x^2 - 0.490357x + 4.96571.$

(b)

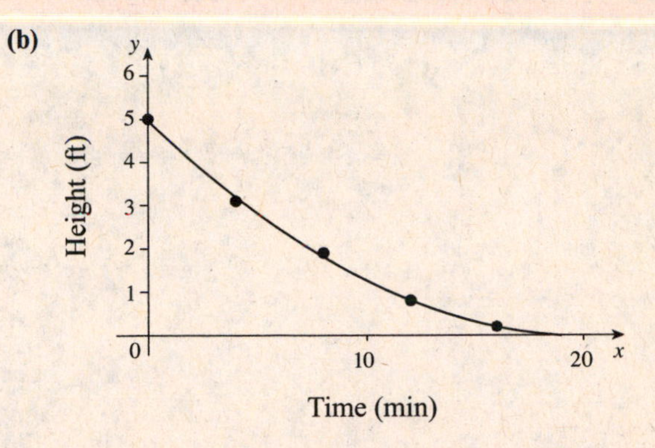

(c) Moving the cursor along the path of the polynomial, we find that the tank should drain in 19.0 minutes.

4 EXPONENTIAL AND LOGARITHMIC FUNCTIONS

4.1 EXPONENTIAL FUNCTIONS

1. The function $f(x) = 5^x$ is an exponential function with base 5; $f(-2) = 5^{-2} = \frac{1}{25}$, $f(0) = 5^0 = 1$, $f(2) = 5^2 = 25$, and $f(6) = 5^6 = 15{,}625$.

3. **(a)** To obtain the graph of $g(x) = 2^x - 1$ we start with the graph of $f(x) = 2^x$ and shift it *downward* 1 unit.

(b) To obtain the graph of $h(x) = 2^{x-1}$ we start with the graph of $f(x) = 2^x$ and shift it to the *right* 1 unit.

5. The exponential function $f(x) = \left(\frac{1}{2}\right)^x$ has the *horizontal* asymptote $y = 0$. This means that as $x \to \infty$, we have $\left(\frac{1}{2}\right)^x \to 0$.

7. $f(x) = 4^x$; $f\left(\frac{1}{2}\right) = 4^{1/2} = \sqrt{4} = 2$, $f\left(\sqrt{5}\right) \approx 22.195$, $f(-2) = 0.063$, $f(0.3) \approx 1.516$.

9. $g(x) = \left(\frac{1}{3}\right)^{x+1}$; $g\left(\frac{1}{2}\right) \approx 0.192$, $g\left(\sqrt{2}\right) \approx 0.070$, $g(-3.5) \approx 15.588$, $g(-1.4) \approx 1.552$.

11. $f(x) = 2^x$

x	y
-4	$\frac{1}{16}$
-2	$\frac{1}{4}$
0	1
2	4
4	16

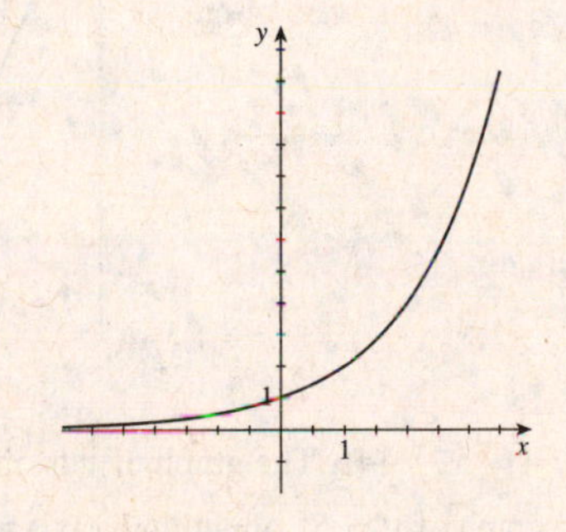

13. $f(x) = \left(\frac{1}{3}\right)^x$

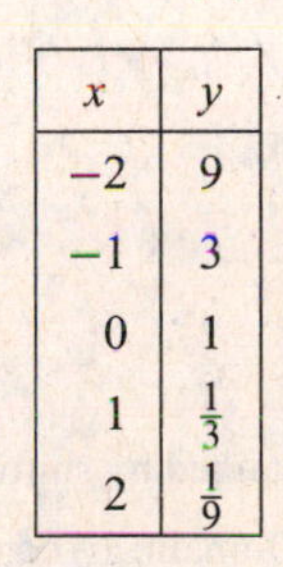

x	y
-2	9
-1	3
0	1
1	$\frac{1}{3}$
2	$\frac{1}{9}$

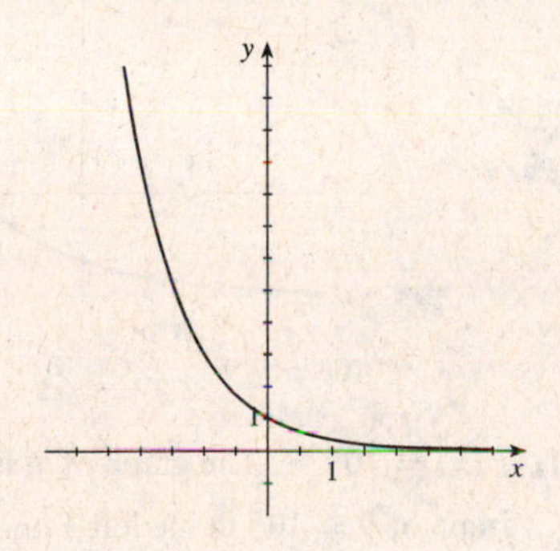

15. $g(x) = 3(1.3)^x$

x	y
-2	1.775
-1	2.308
0	3.0
1	3.9
2	5.07
3	6.591
4	8.568

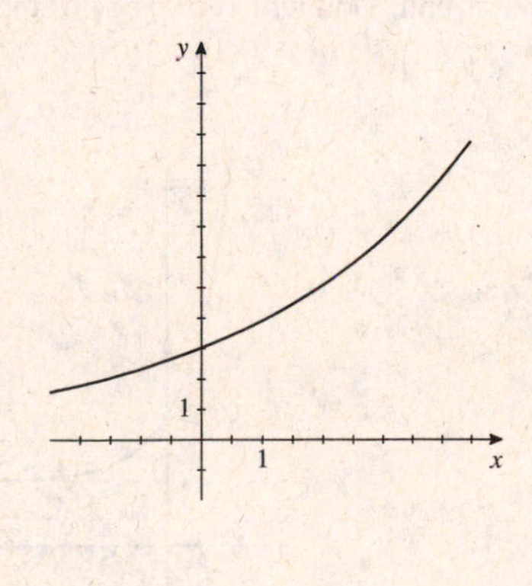

17. $f(x) = 2^x$ and $g(x) = 2^{-x}$

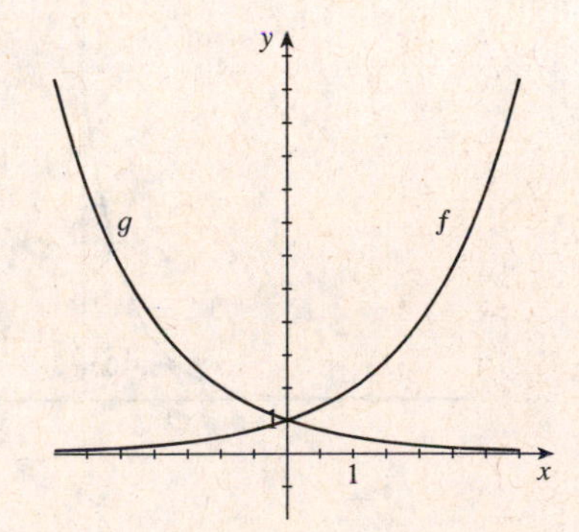

19. $f(x) = 4^x$ and $g(x) = 7^x$.

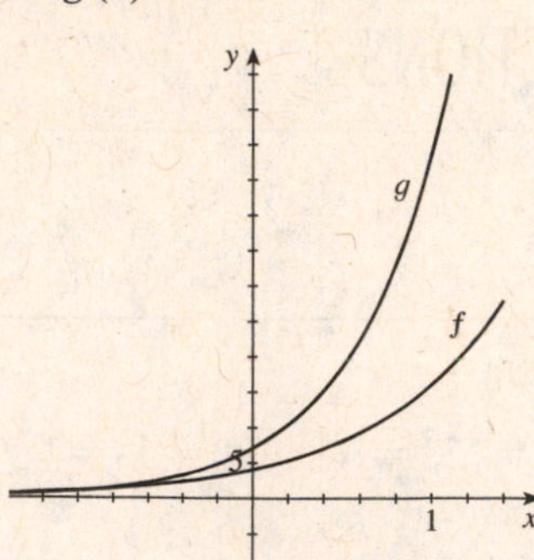

21. From the graph, $f(2) = a^2 = 9$, so $a = 3$. Thus $f(x) = 3^x$.

23. From the graph, $f(2) = a^2 = \frac{1}{16}$, so $a = \frac{1}{4}$. Thus $f(x) = \left(\frac{1}{4}\right)^x$.

25. The graph of $f(x) = 5^{x+1}$ is obtained from that of $y = 5^x$ by shifting 1 unit to the left, so it has graph II.

27. $g(x) = 2^x - 3$. The graph of g is obtained by shifting the graph of $y = 2^x$ downward 3 units. Domain: $(-\infty, \infty)$. Range: $(-3, \infty)$. Asymptote: $y = -3$.

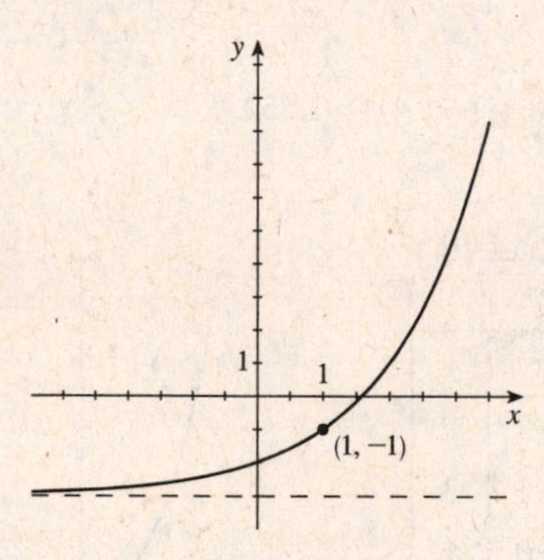

29. The graph of $f(x) = -3^x$ is obtained by reflecting the graph of $y = 3^x$ about the x-axis. Domain: $(-\infty, \infty)$. Range: $(-\infty, 0)$. Asymptote: $y = 0$.

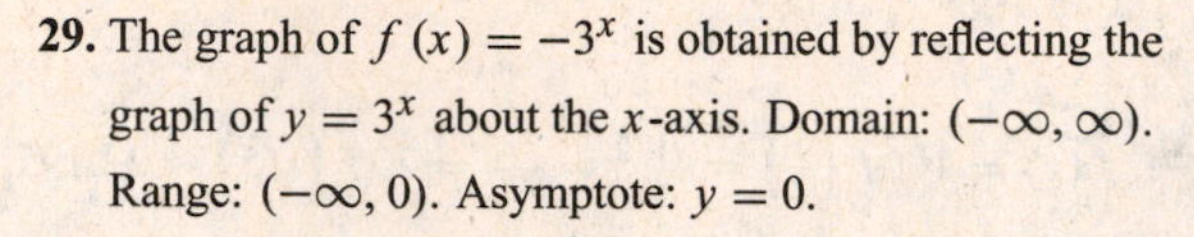

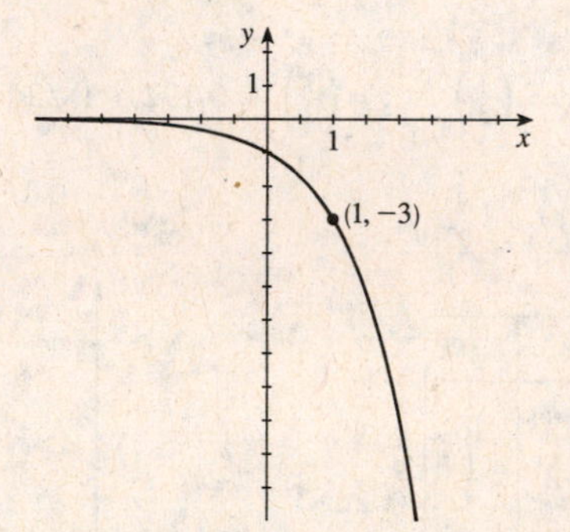

31. $f(x) = 10^{x+3}$. The graph of f is obtained by shifting the graph of $y = 10^x$ to the left 3 units. Domain: $(-\infty, \infty)$. Range: $(0, \infty)$. Asymptote: $y = 0$.

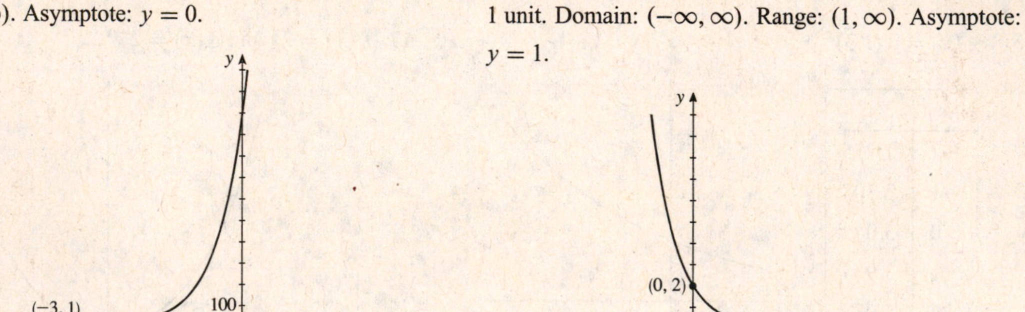

33. $y = 5^{-x} + 1$. The graph of y is obtained by reflecting the graph of $y = 5^x$ about the x-axis and then shifting upward 1 unit. Domain: $(-\infty, \infty)$. Range: $(1, \infty)$. Asymptote: $y = 1$.

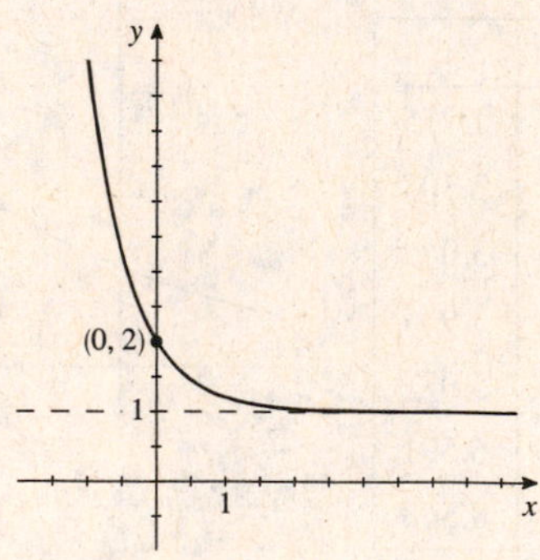

35. $y = 2 - \left(\frac{1}{3}\right)^x$. The graph is obtained by reflecting the graph of $y = \left(\frac{1}{3}\right)^x$ about the x-axis and shifting upward 2 units. Domain: $(-\infty, \infty)$. Range: $(-\infty, 2)$. Asymptote: $y = 2$.

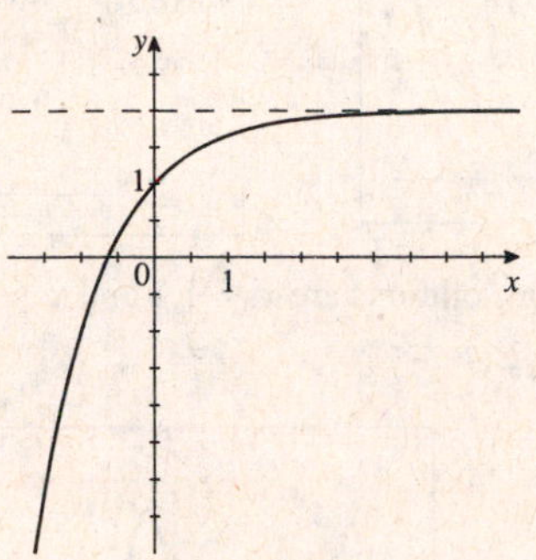

37. $h(x) = 2^{x-4} + 1$. The graph of h is obtained by shifting the graph of $y = 2^x$ to the right 4 units and upward 1 unit. Domain: $(-\infty, \infty)$. Range: $(1, \infty)$. Asymptote: $y = 1$.

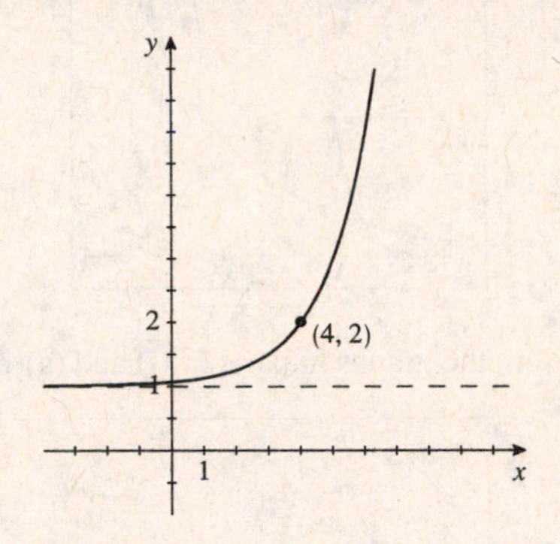

39. $g(x) = 1 - 3^{-x} = -3^{-x} + 1$. The graph of g is obtained by reflecting the graph of $y = 3^x$ about the x- and y-axes and then shifting upward 1 unit. Domain: $(-\infty, \infty)$. Range: $(-\infty, 1)$. Asymptote: $y = 1$.

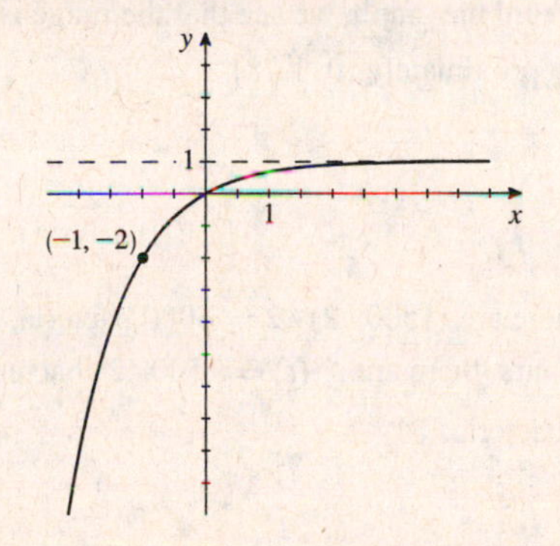

41. (a)

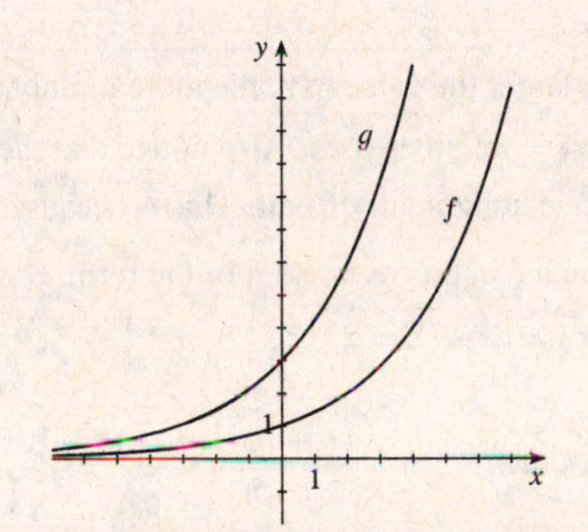

(b) Since $g(x) = 3(2^x) = 3f(x)$ and $f(x) > 0$, the height of the graph of $g(x)$ is always three times the height of the graph of $f(x) = 2^x$, so the graph of g is steeper than the graph of f.

43.

x	$f(x) = x^3$	$g(x) = 3^x$
0	0	1
1	1	3
2	8	9
3	27	27
4	64	81
6	216	729
8	512	6561
10	1000	59,049

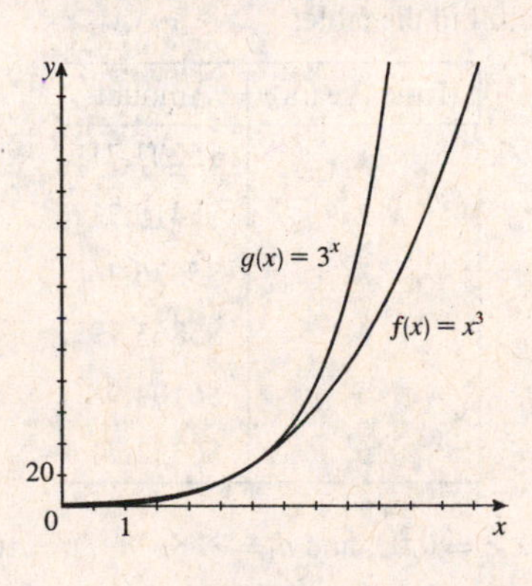

From the graph and the table, we see that the graph of g ultimately increases much more quickly than the graph of f.

45. **(a)** From the graphs below, we see that the graph of f ultimately increases much more quickly than the graph of g.

(i) $[0, 5]$ by $[0, 20]$

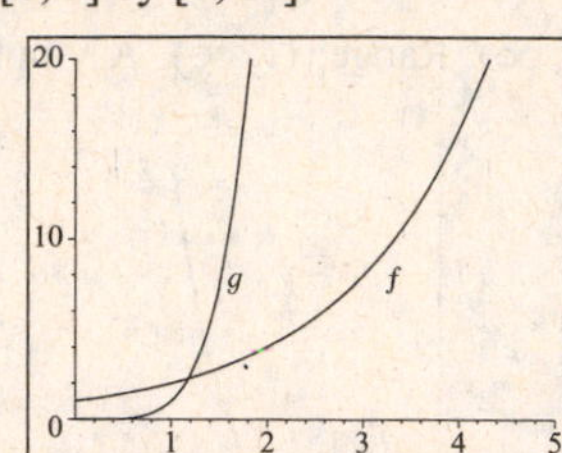

(ii) $[0, 25]$ by $[0, 10^7]$

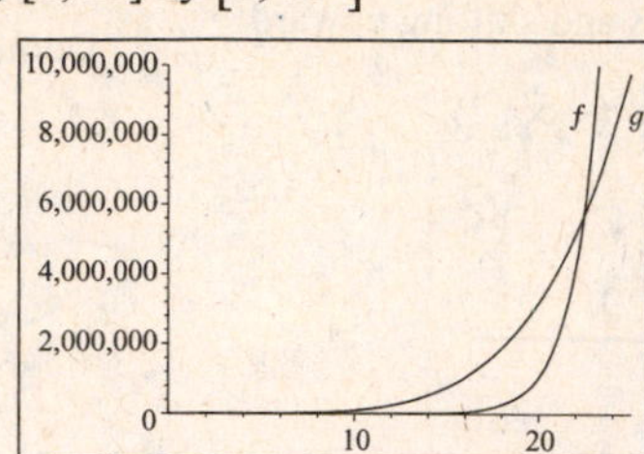

(iii) $[0, 50]$ by $[0, 10^8]$

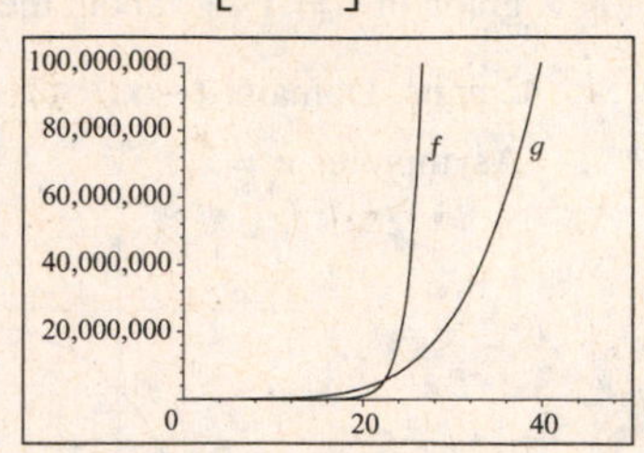

(b) From the graphs in parts (a)(i) and (a)(ii), we see that the approximate solutions are $x \approx 1.2$ and $x \approx 22.4$.

47.

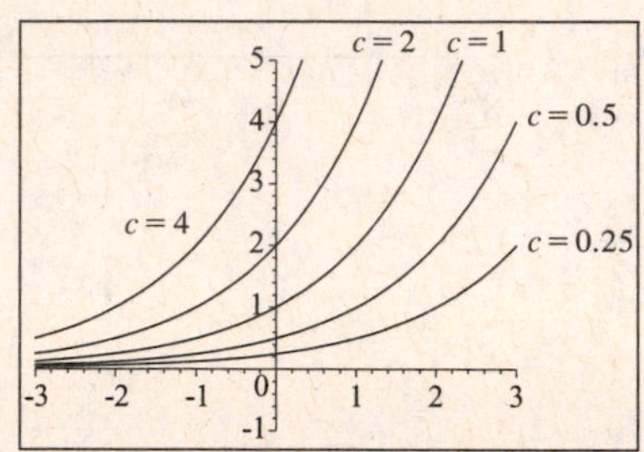

The larger the value of c, the more rapidly the graph of $f(x) = c2^x$ increases. Also notice that the graphs are just shifted horizontally 1 unit. This is because of our choice of c; each c in this exercise is of the form 2^k. So $f(x) = 2^k \cdot 2^x = 2^{x+k}$.

49. $y = 10^{x-x^2}$

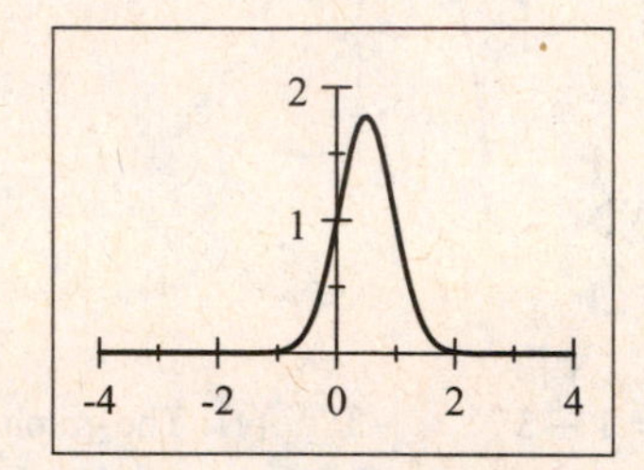

(a) From the graph, we see that the function is increasing on $(-\infty, 0.50)$ and decreasing on $(0.50, \infty)$.

(b) From the graph, we see that the range is approximately $(0, 1.78]$.

51. $f(x) = 10^x$, so $\dfrac{f(x+h) - f(x)}{h} = \dfrac{10^{x+h} - 10^x}{h} = 10^x \left(\dfrac{10^h - 1}{h}\right)$.

53. **(a)** After 1 hour, there are $1500 \cdot 2 = 3000$ bacteria. After 2 hours, there are $(1500 \cdot 2) \cdot 2 = 6000$ bacteria. After 3 hours, there are $(1500 \cdot 2 \cdot 2) \cdot 2 = 12{,}000$ bacteria. We see that after t hours, there are $N(t) = 1500 \cdot 2^t$ bacteria.

(b) After 24 hours, there are $N(24) = 1500 \cdot 2^{24} = 25{,}165{,}824{,}000$ bacteria.

55. Using the formula $A(t) = P(1+i)^k$ with $P = 5000$, $i = 4\%$ per year $= \dfrac{0.04}{12}$ per month, and $k = 12 \cdot$ number of years, we fill in the table:

Time (years)	Amount
1	\$5203.71
2	\$5415.71
3	\$5636.36
4	\$5865.99
5	\$6104.98
6	\$6353.71

57. $P = 10{,}000$, $r = 0.03$, and $n = 2$. So $A(t) = 10{,}000 \left(1 + \frac{0.03}{2}\right)^{2t} = 10{,}000 \cdot 1.015^{2t}$.

(a) $A(5) = 10000 \cdot 1.015^{10} \approx 11{,}605.41$, and so the value of the investment is \$11,605.41.

(b) $A(10) = 10000 \cdot 1.015^{20} \approx 13{,}468.55$, and so the value of the investment is \$13,468.55.

(c) $A(15) = 10000 \cdot 1.015^{30} \approx 15{,}630.80$, and so the value of the investment is \$15,630.80.

59. $P = 500$, $r = 0.0375$, and $n = 4$. So $A(t) = 500\left(1 + \frac{0.0375}{4}\right)^{4t}$.

(a) $A(1) = 500\left(1 + \frac{0.0375}{4}\right)^{4} \approx 519.02$, and so the value of the investment is \$519.02.

(b) $A(2) = 500\left(1 + \frac{0.0375}{4}\right)^{8} \approx 538.75$, and so the value of the investment is \$538.75.

(c) $A(10) = 500\left(1 + \frac{0.0375}{4}\right)^{40} \approx 726.23$, and so the value of the investment is \$726.23.

61. We must solve for P in the equation $10000 = P\left(1 + \frac{0.09}{2}\right)^{2(3)} = P(1.045)^6 \Leftrightarrow 10000 = 1.3023P \Leftrightarrow P = 7678.96$. Thus, the present value is \$7,678.96.

63. $r_{\text{APY}} = \left(1 + \frac{r}{n}\right)^n - 1$. Here $r = 0.08$ and $n = 12$, so $r_{\text{APY}} = \left(1 + \frac{0.08}{12}\right)^{12} - 1 \approx (1.0066667)^{12} - 1 \approx 0.083000$. Thus, the annual percentage yield is about 8.3%.

65. **(a)** In this case the payment is \$1 million.

(b) In this case the total pay is $2 + 2^2 + 2^3 + \cdots + 2^{30} > 2^{30}$ cents $= \$10{,}737{,}418.24$. Since this is much more than method (a), method (b) is more profitable.

4.2 THE NATURAL EXPONENTIAL FUNCTION

1. The function $f(x) = e^x$ is called the *natural* exponential function. The number e is approximately equal to 2.71828.

3. $h(x) = e^x$; $h(1) \approx 2.718$, $h(\pi) \approx 23.141$, $h(-3) \approx 0.050$, $h\left(\sqrt{2}\right) \approx 4.113$

5. $f(x) = 1.5e^x$

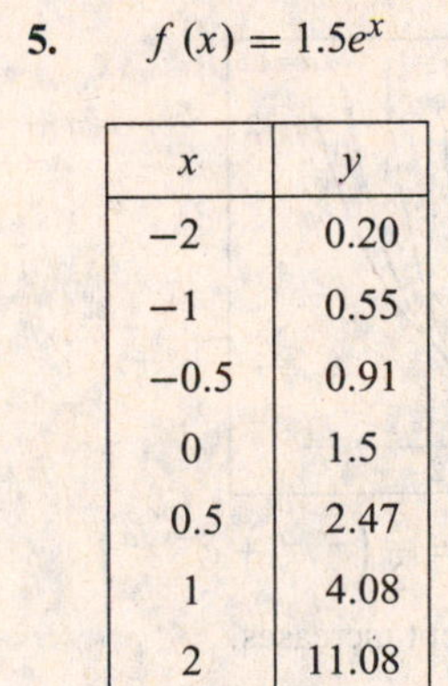

x	y
-2	0.20
-1	0.55
-0.5	0.91
0	1.5
0.5	2.47
1	4.08
2	11.08

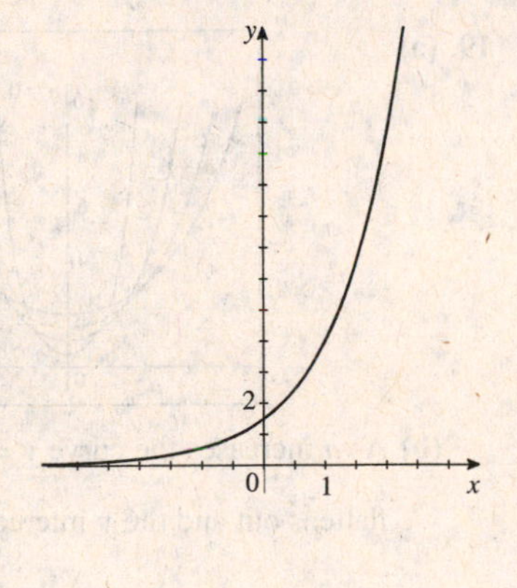

7. $g(x) = 2 + e^x$. The graph of g is obtained from the graph of $y = e^x$ by shifting it upward 2 units. Domain: $(-\infty, \infty)$. Range: $(2, \infty)$. Asymptote: $y = 2$.

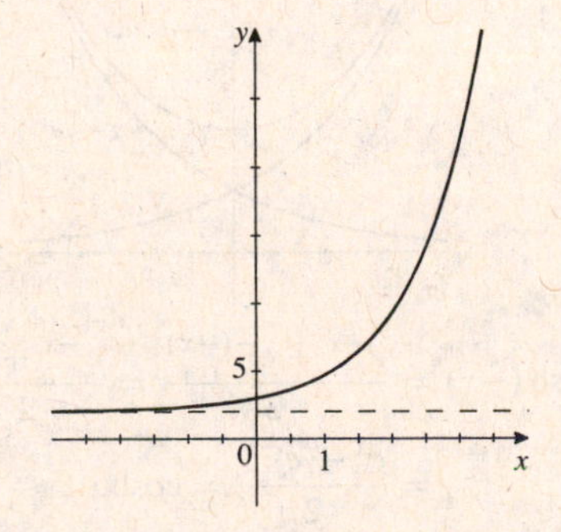

9. $y = -e^x$. The graph of $y = -e^x$ is obtained from the graph of $y = e^x$ by reflecting it about the x-axis. Domain: $(-\infty, \infty)$. Range: $(-\infty, 0)$. Asymptote: $y = 0$.

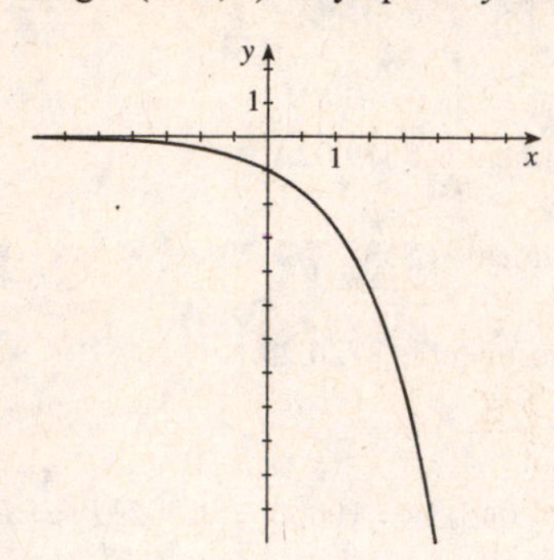

11. $y = e^{-x} - 1$. The graph of $y = e^{-x} - 1$ is obtained from the graph of $y = e^x$ by reflecting it about the y-axis then shifting downward 1 unit. Domain: $(-\infty, \infty)$. Range: $(-1, \infty)$. Asymptote: $y = -1$.

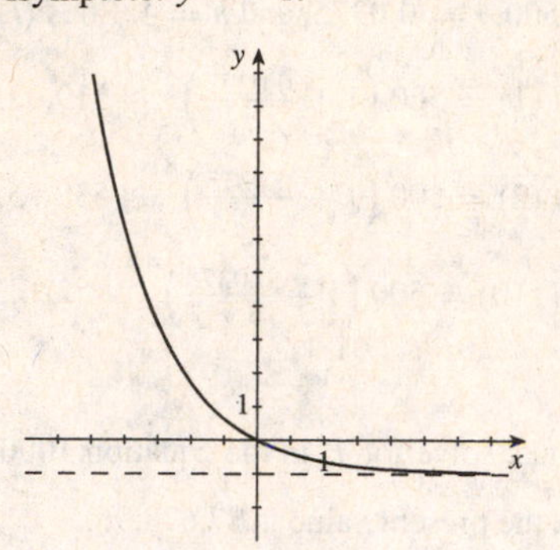

13. $y = e^{x-2}$. The graph of $y = e^{x-2}$ is obtained from the graph of $y = e^x$ by shifting it to the right 2 units. Domain: $(-\infty, \infty)$. Range: $(0, \infty)$. Asymptote: $y = 0$.

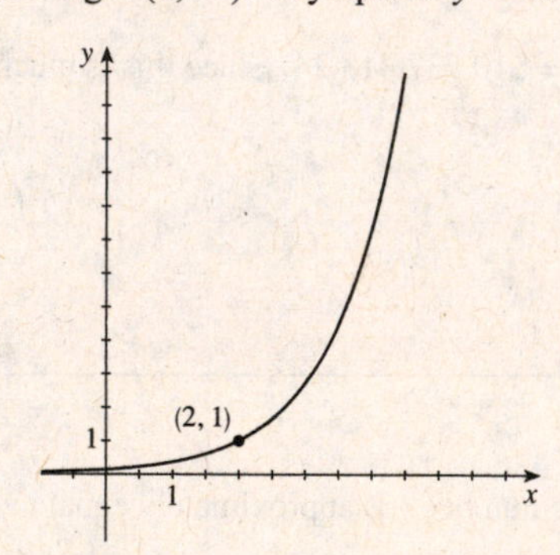

15. $h(x) = e^{x+1} - 3$. The graph of h is obtained from the graph of $y = e^x$ by shifting it to the left 1 unit and downward 3 units. Domain: $(-\infty, \infty)$. Range: $(-3, \infty)$. Asymptote: $y = -3$.

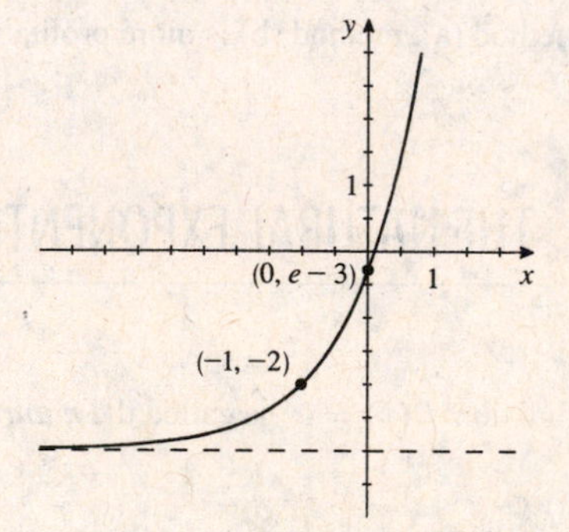

17. (a)

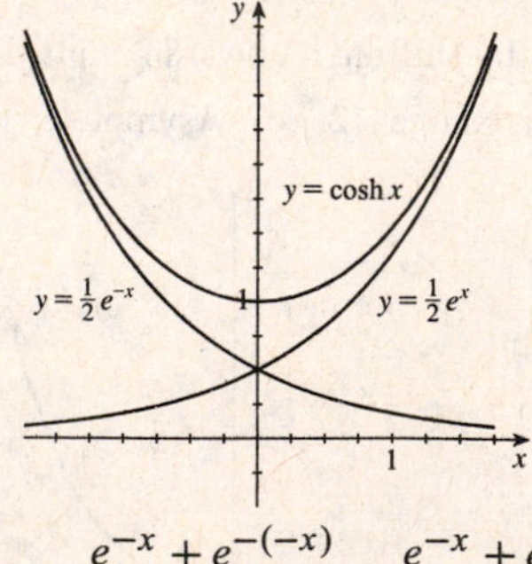

(b) $\cosh(-x) = \dfrac{e^{-x} + e^{-(-x)}}{2} = \dfrac{e^{-x} + e^{x}}{2}$

$= \dfrac{e^{x} + e^{-x}}{2} = \cosh x$

19. (a)

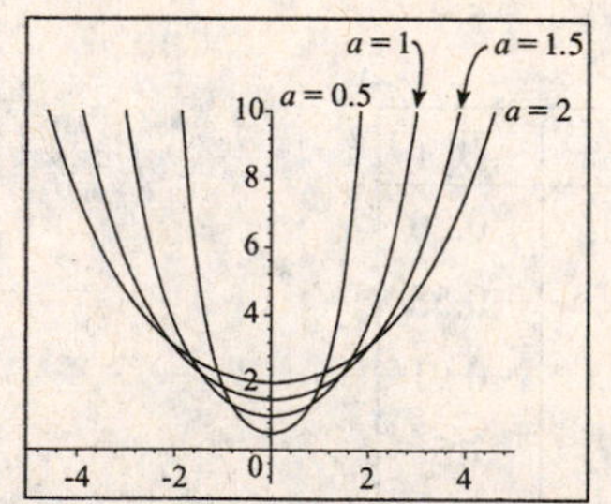

(b) As a increases the curve $y = \frac{a}{2}\left(e^{x/a} + e^{-x/a}\right)$ flattens out and the y intercept increases.

21. $g(x) = x^x$. Notice that $g(x)$ is only defined for $x \geq 0$. The graph of $g(x)$ is shown in the viewing rectangle $[0, 1.5]$ by $[0, 1.5]$. From the graph, we see that there is a local minimum of about 0.69 when $x \approx 0.37$.

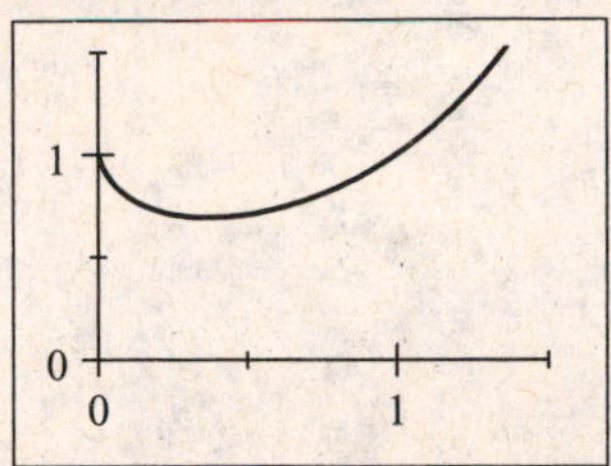

23. $D(t) = 50e^{-0.2t}$. So when $t = 3$ we have $D(3) = 50e^{-0.2(3)} \approx 27.4$ milligrams.

25. $v(t) = 180\left(1 - e^{-0.2t}\right)$

(a) $v(0) = 180\left(1 - e^{0}\right) = 180(1 - 1) = 0$.

(b) $v(5) = 180\left(1 - e^{-0.2(5)}\right) \approx 180(0.632) = 113.76$ ft/s. So the velocity after 5 s is about 113.8 ft/s.

$v(10) = 180\left(1 - e^{-0.2(10)}\right) \approx 180(0.865) = 155.7$ ft/s. So the velocity after 10 s is about 155.7 ft/s.

(c)

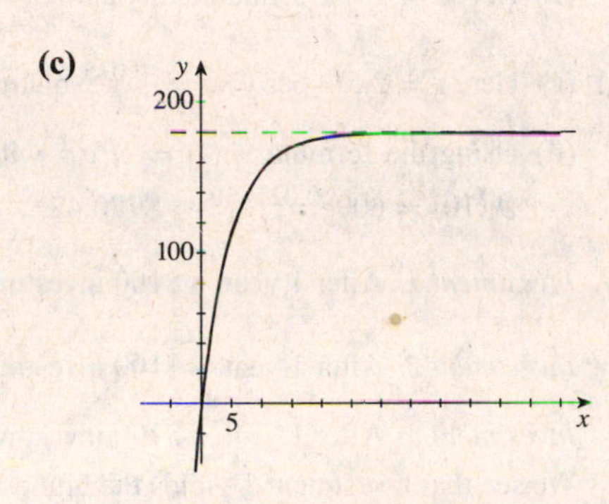

(d) The terminal velocity is 180 ft/s.

27. $P(t) = \dfrac{1200}{1 + 11e^{-0.2t}}$

(a) $P(0) = \dfrac{1200}{1 + 11e^{-0.2(0)}} = \dfrac{1200}{1 + 11} = 100$.

(b) $P(10) = \dfrac{1200}{1 + 11e^{-0.2(10)}} \approx 482$. $P(20) = \dfrac{1200}{1 + 11e^{-0.2(20)}} \approx 999$. $P(30) = \dfrac{1200}{1 + 11e^{-0.2(30)}} \approx 1168$.

(c) As $t \to \infty$ we have $e^{-0.2t} \to 0$, so $P(t) \to \dfrac{1200}{1 + 0} = 1200$. The graph shown confirms this.

29. $P(t) = \dfrac{73.2}{6.1 + 5.9e^{-0.02t}}$

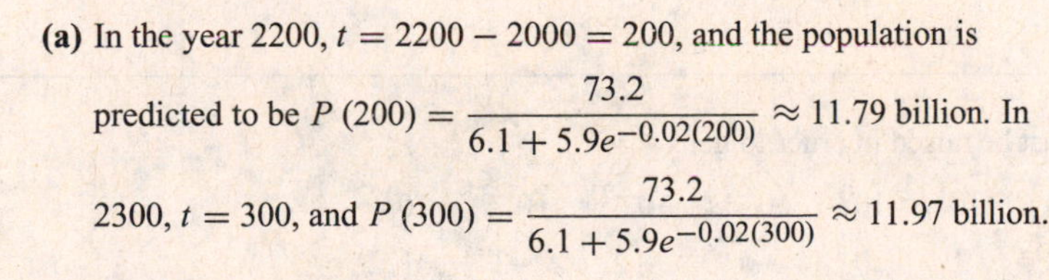

(a) In the year 2200, $t = 2200 - 2000 = 200$, and the population is predicted to be $P(200) = \dfrac{73.2}{6.1 + 5.9e^{-0.02(200)}} \approx 11.79$ billion. In 2300, $t = 300$, and $P(300) = \dfrac{73.2}{6.1 + 5.9e^{-0.02(300)}} \approx 11.97$ billion.

(b)

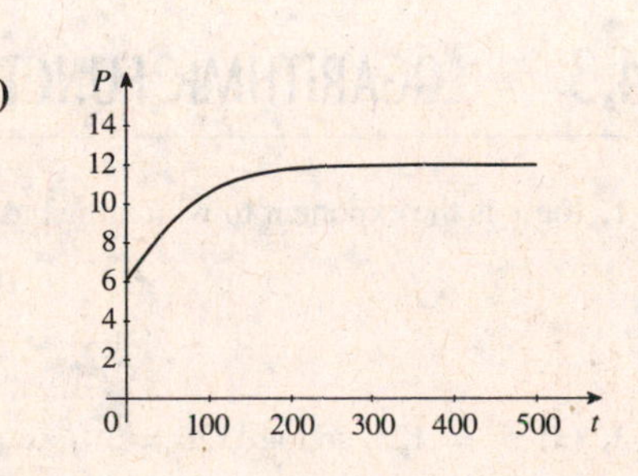

(c) As t increases, the denominator approaches 6.1, so according to this model, the world population approaches $\frac{73.2}{6.1} = 12$ billion people.

31. Using the formula $A(t) = Pe^{rt}$ with $P = 7000$ and $r = 3\% = 0.03$, we fill in the table:

Time (years)	Amount
1	\$7213.18
2	\$7432.86
3	\$7659.22
4	\$7892.48
5	\$8132.84
6	\$8380.52

33. We use the formula $A(t) = Pe^{rt}$ with $P = 2000$ and $r = 3.5\% = 0.035$.

(a) $A(2) = 2000e^{0.035 \cdot 2} \approx \$2145.02$ **(b)** $A(42) = 2000e^{0.035 \cdot 4} \approx \$2300.55$ **(c)** $A(12) = 2000e^{0.035 \cdot 12} \approx \$3043.92$

35. (a) Using the formula $A(t) = P(1+i)^k$ with $P = 600$, $i = 2.5\%$ per year $= 0.025$, and $k = 10$, we calculate $A(10) = 600(1.025)^{10} \approx \$768.05$.

(b) Here $i = \frac{0.025}{2}$ semiannually and $k = 10 \cdot 2 = 20$, so $A(10) = 600\left(1 + \frac{0.025}{2}\right)^{20} \approx \$769.22$.

(c) Here $i = 2.5\%$ per year $= \frac{0.025}{4}$ quarterly and $k = 10 \cdot 4 = 40$, so $A(10) = 600\left(1 + \frac{0.025}{4}\right)^{40} \approx \$769.82$.

(d) Using the formula $A(t) = Pe^{rt}$ with $P = 600$, $r = 2.5\% = 0.025$, and $t = 10$, we have $A(10) = 600e^{0.025 \cdot 10} \approx \$770.42$.

37. *Investment 1:* After 1 year, a \$100 investment grows to $A(1) = 100\left(1 + \frac{0.025}{2}\right)^2 \approx 102.52$.

Investment 2: After 1 year, a \$100 investment grows to $A(1) = 100\left(1 + \frac{0.0225}{4}\right)^4 = 102.27$.

Investment 3: After 1 year, a \$100 investment grows to $A(1) = 100e^{0.02} \approx 102.02$.

We see that Investment 1 yields the highest return.

39. (a) $A(t) = Pe^{rt} = 5000e^{0.09t}$

(b)

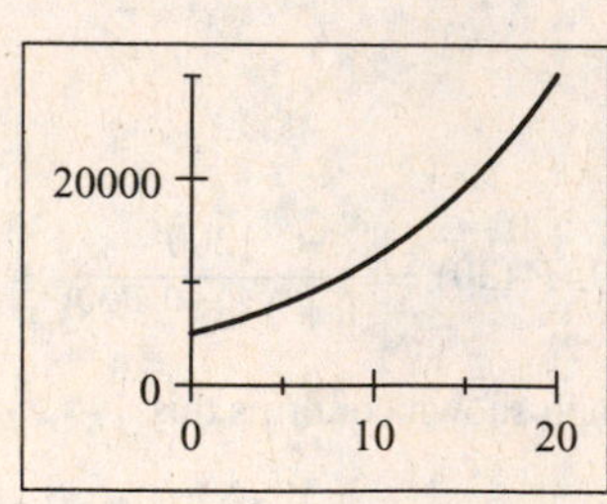

(c) $A(t) = 25{,}000$ when $t \approx 17.88$ years.

4.3 LOGARITHMIC FUNCTIONS

1. $\log x$ is the exponent to which the base 10 must be raised in order to get x.

x	10^3	10^2	10^1	10^0	10^{-1}	10^{-2}	10^{-3}	$10^{1/2}$
$\log x$	3	2	1	0	-1	-2	-3	$1/2$

3. (a) $5^3 = 125$, so $\log_5 125 = 3$. **(b)** $\log_5 25 = 2$, so $5^2 = 25$.

5. The natural logarithmic function $f(x) = \ln x$ has the *vertical* asymptote $x = 0$.

7.

Logarithmic form	Exponential form
$\log_8 8 = 1$	$8^1 = 8$
$\log_8 64 = 2$	$8^2 = 64$
$\log_8 4 = \frac{2}{3}$	$8^{2/3} = 4$
$\log_8 512 = 3$	$8^3 = 512$
$\log_8 \frac{1}{8} = -1$	$8^{-1} = \frac{1}{8}$
$\log_8 \frac{1}{64} = -2$	$8^{-2} = \frac{1}{64}$

9. **(a)** $3^4 = 81$

(b) $3^0 = 1$

11. **(a)** $8^{1/3} = 2$

(b) $10^{-2} = 0.01$

13. **(a)** $3^x = 5$

(b) $7^2 = 3y$

15. **(a)** $e^{3y} = 5$

(b) $e^{-1} = t + 1$

17. **(a)** $\log_{10} 10{,}000 = 4$

(b) $\log_5 \left(\frac{1}{25}\right) = -2$

19. **(a)** $\log_8 \frac{1}{8} = -1$

(b) $\log_2 \left(\frac{1}{8}\right) = -3$

21. **(a)** $\log_4 70 = x$

(b) $\log_3 w = 5$

23. **(a)** $\ln 2 = x$

(b) $\ln y = 3$

25. **(a)** $\log_2 2 = 1$

(b) $\log_5 1 = \log_5 5^0 = 0$

(c) $\log_6 6^5 = 5$

27. **(a)** $\log_6 36 = \log_6 6^2 = 2$

(b) $\log_9 81 = \log_9 9^2 = 2$

(c) $\log_7 7^{10} = 10$

29. **(a)** $\log_3 \left(\frac{1}{27}\right) = \log_3 3^{-3} = -3$

(b) $\log_{10} \sqrt{10} = \log_{10} 10^{1/2} = \frac{1}{2}$

(c) $\log_5 0.2 = \log_5 \left(\frac{1}{5}\right) = \log_5 5^{-1} = -1$

31. **(a)** $3^{\log_3 5} = 5$

(b) $5^{\log_5 27} = 27$

(c) $e^{\ln 10} = 10$

33. **(a)** $\log_8 0.25 = \log_8 8^{-2/3} = -\frac{2}{3}$

(b) $\ln e^4 = 4$

(c) $\ln \left(\frac{1}{e}\right) = \ln e^{-1} = -1$

35. **(a)** $\log_4 x = 3 \Leftrightarrow x = 4^3 = 64$

(b) $\log_{10} 0.01 = x \Leftrightarrow 10^x = 0.01 \Leftrightarrow x = -2$

37. **(a)** $\ln x = 3 \Leftrightarrow x = e^3$

(b) $\ln e^2 = x \Leftrightarrow x = 2 \ln e = 2$

39. **(a)** $\log_7 \left(\frac{1}{49}\right) = x \Leftrightarrow 7^x = \frac{1}{49} \Leftrightarrow x = -2$

(b) $\log_2 x = 5 \Leftrightarrow 2^5 = x \Leftrightarrow x = 32$

41. **(a)** $\log_2 \left(\frac{1}{2}\right) = x \Leftrightarrow 2^x = \frac{1}{2} \Leftrightarrow x = -1$

(b) $\log_{10} x = -3 \Leftrightarrow 10^{-3} = x \Leftrightarrow x = \frac{1}{1000}$

43. **(a)** $\log_x 16 = 4 \Leftrightarrow x^4 = 16 \Leftrightarrow x = 2$

(b) $\log_x 8 = \frac{3}{2} \Leftrightarrow x^{3/2} = 8 \Leftrightarrow x = 8^{2/3} = 4$

45. **(a)** $\log 2 \approx 0.3010$

(b) $\log 35.2 \approx 1.5465$

(c) $\log \left(\frac{2}{3}\right) \approx -0.1761$

47. **(a)** $\ln 5 \approx 1.6094$

(b) $\ln 25.3 \approx 3.2308$

(c) $\ln \left(1 + \sqrt{3}\right) \approx 1.0051$

49.

x	$f(x)$
$\frac{1}{3^3}$	-3
$\frac{1}{3^2}$	-2
$\frac{1}{3}$	-1
1	0
3	1
3^2	2

$f(x) = \log_3 x$

51.

x	$f(x)$
$\frac{1}{10^3}$	-6
$\frac{1}{10^2}$	-4
$\frac{1}{10}$	-2
1	0
10	2
10^2	4

$f(x) = 2\log x$

53. Since the point (5, 1) is on the graph, we have $1 = \log_a 5 \Leftrightarrow a^1 = 5$. Thus the function is $y = \log_5 x$.

55. Since the point $\left(3, \frac{1}{2}\right)$ is on the graph, we have $\frac{1}{2} = \log_a 3 \Leftrightarrow a^{1/2} = 3 \Leftrightarrow a = 9$. Thus the function is $y = \log_9 x$.

57. The graph of $f(x) = 2 + \ln x$ is obtained from that of $y = \ln x$ by shifting it upward 2 units, as in graph I.

59. The graph of $y = \log_4 x$ is obtained from that of $y = 4^x$ by reflecting it in the line $y = x$.

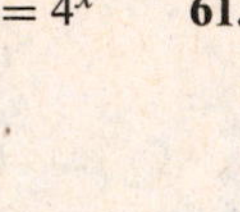

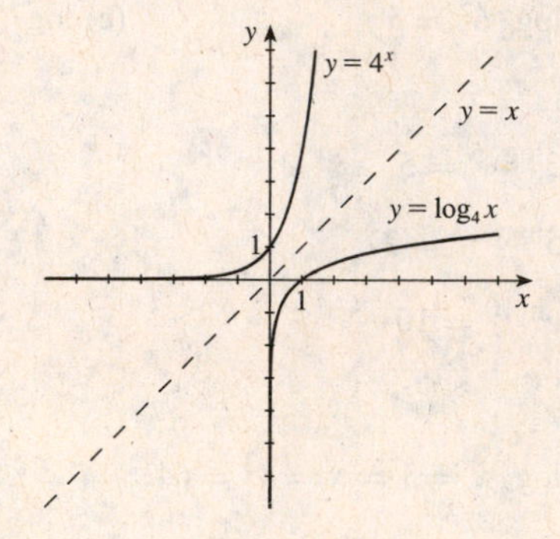

61. The graph of $g(x) = \log_5(-x)$ is obtained from that of $y = \log_5 x$ by reflecting it about the y-axis.
Domain: $(-\infty, 0)$. Range: $(-\infty, \infty)$.
Vertical asymptote: $x = 0$.

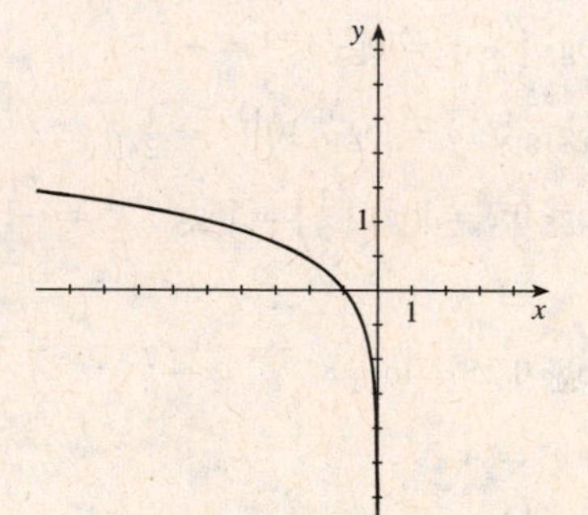

63. The graph of $f(x) = \log_2(x - 4)$ is obtained from that of $y = \log_2 x$ by shifting it to the right 4 units.
Domain: $(4, \infty)$. Range: $(-\infty, \infty)$.
Vertical asymptote: $x = 4$.

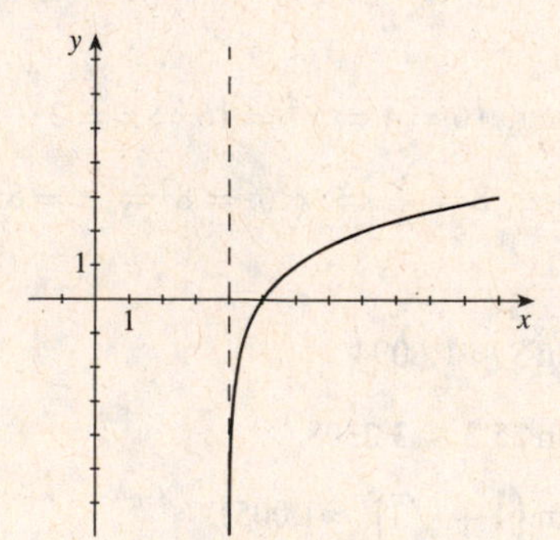

65. The graph of $h(x) = \ln(x + 5)$ is obtained from that of $y = \ln x$ by shifting to the left 5 units. Domain: $(-5, \infty)$.
Range: $(-\infty, \infty)$. Vertical asymptote: $x = -5$.

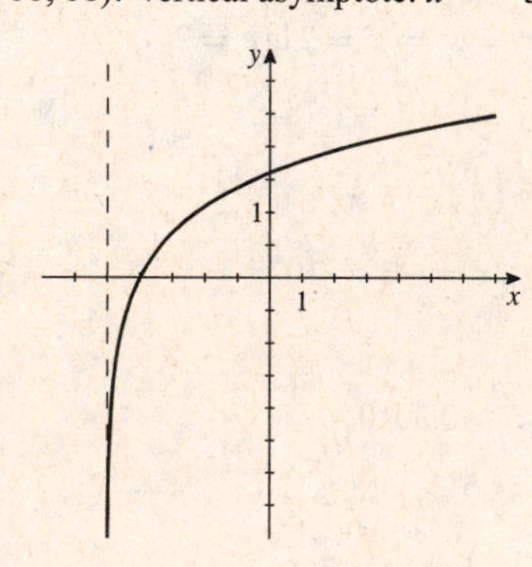

67. The graph of $y = 2 + \log_3 x$ is obtained from that of $y = \log_3 x$ by shifting upward 2 units. Domain: $(0, \infty)$. Range: $(-\infty, \infty)$. Vertical asymptote: $x = 0$.

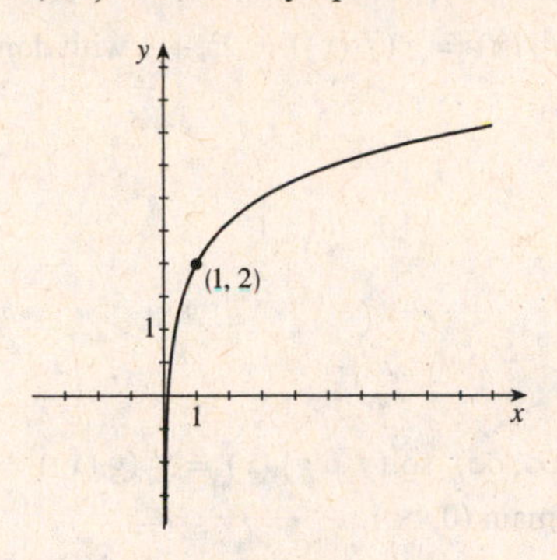

69. The graph of $y = \log_3 (x - 1) - 2$ is obtained from that of $y = \log_3 x$ by shifting to the right 1 unit and then downward 2 units. Domain: $(1, \infty)$. Range: $(-\infty, \infty)$. Vertical asymptote: $x = 1$.

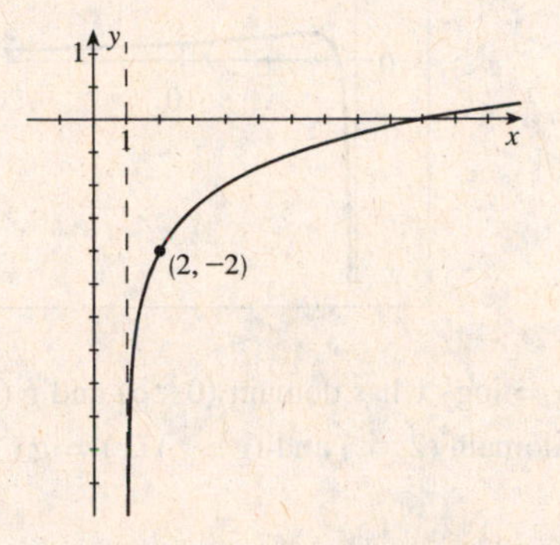

71. The graph of $y = |\ln x|$ is obtained from that of $y = \ln x$ by reflecting the part of the graph for $0 < x < 1$ about the x-axis. Domain: $(0, \infty)$. Range: $[0, \infty)$. Vertical asymptote: $x = 0$.

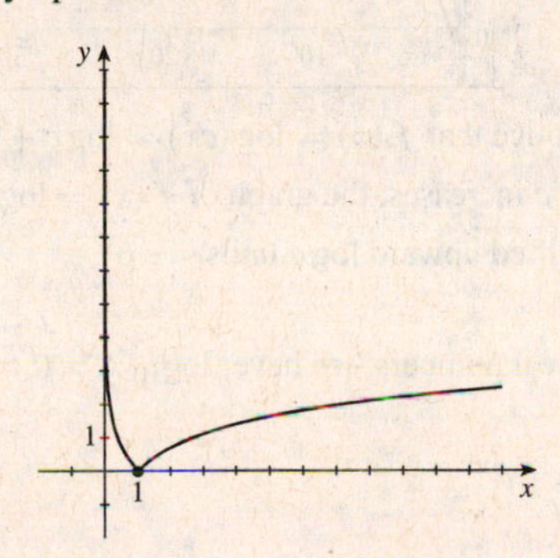

73. $f(x) = \log_{10} (x + 3)$. We require that $x + 3 > 0 \Leftrightarrow x > -3$, so the domain is $(-3, \infty)$.

75. $g(x) = \log_3 \left(x^2 - 1\right)$. We require that $x^2 - 1 > 0 \Leftrightarrow x^2 > 1 \Rightarrow x < -1$ or $x > 1$, so the domain is $(-\infty, -1) \cup (1, \infty)$.

77. $h(x) = \ln x + \ln (2 - x)$. We require that $x > 0$ and $2 - x > 0 \Leftrightarrow x > 0$ and $x < 2 \Leftrightarrow 0 < x < 2$, so the domain is $(0, 2)$.

79. $y = \log_{10} \left(1 - x^2\right)$ has domain $(-1, 1)$, vertical asymptotes $x = -1$ and $x = 1$, and local maximum $y = 0$ at $x = 0$.

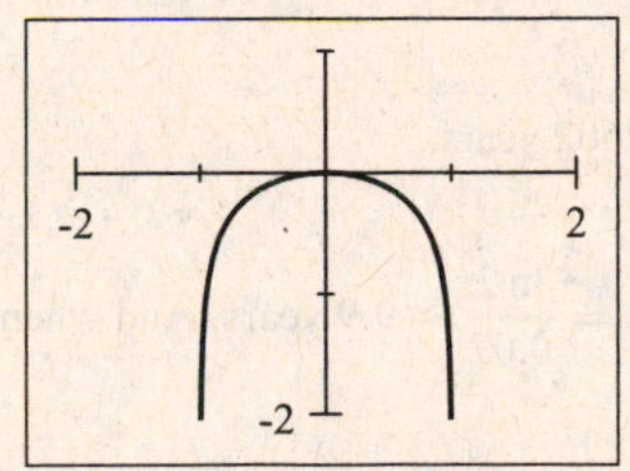

81. $y = x + \ln x$ has domain $(0, \infty)$, vertical asymptote $x = 0$, and no local maximum or minimum.

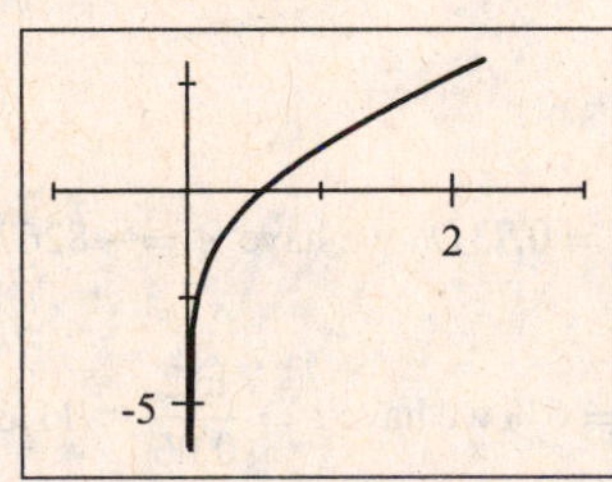

83. $y = \dfrac{\ln x}{x}$ has domain $(0, \infty)$, vertical asymptote $x = 0$, horizontal asymptote $y = 0$, and local maximum $y \approx 0.37$ at $x \approx 2.72$.

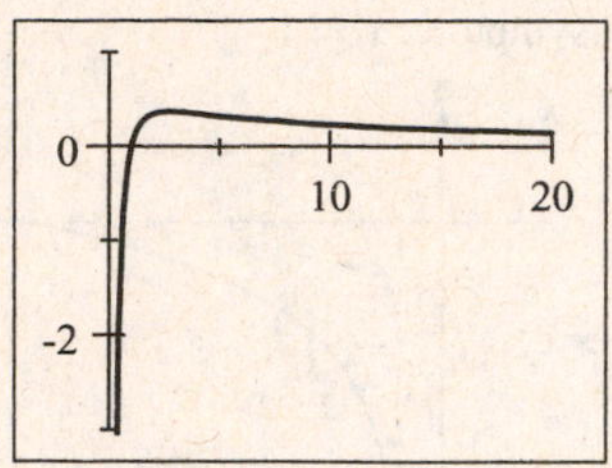

85. $f(x) = 2^x$ and $g(x) = x + 1$ both have domain $(-\infty, \infty)$, so $(f \circ g)(x) = f(g(x)) = 2^{g(x)} = 2^{x+1}$ with domain $(-\infty, \infty)$ and $(g \circ f)(x) = g(f(x)) = 2^x + 1$ with domain $(-\infty, \infty)$.

87. $f(x) = \log_2 x$ has domain $(0, \infty)$ and $g(x) = x - 2$ has domain $(-\infty, \infty)$, so $(f \circ g)(x) = f(g(x)) = \log_2(x - 2)$ with domain $(2, \infty)$ and $(g \circ f)(x) = g(f(x)) = \log_2 x - 2$ with domain $(0, \infty)$.

89. The graph of $g(x) = \sqrt{x}$ grows faster than the graph of $f(x) = \ln x$.

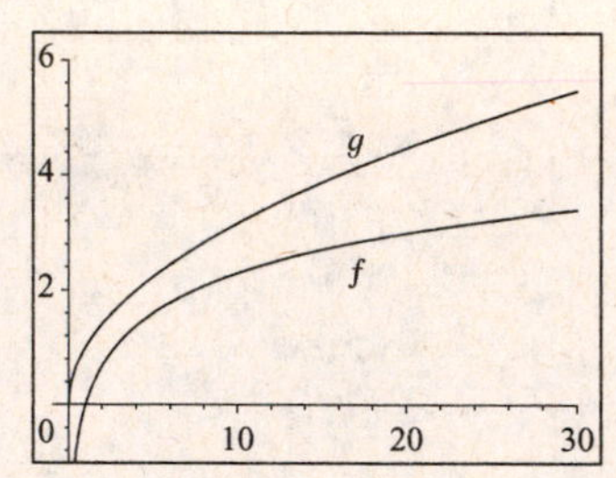

91. (a)

(b) Notice that $f(x) = \log(cx) = \log c + \log x$, so as c increases, the graph of $f(x) = \log(cx)$ is shifted upward $\log c$ units.

93. (a) $f(x) = \log_2(\log_{10} x)$. Since the domain of $\log_2 x$ is the positive real numbers, we have: $\log_{10} x > 0 \Leftrightarrow x > 10^0 = 1$. Thus the domain of $f(x)$ is $(1, \infty)$.

(b) $y = \log_2(\log_{10} x) \Leftrightarrow 2^y = \log_{10} x \Leftrightarrow 10^{2^y} = x$. Thus $f^{-1}(x) = 10^{2^x}$.

95. (a) $f(x) = \dfrac{2^x}{1 + 2^x}$. $y = \dfrac{2^x}{1 + 2^x} \Leftrightarrow y + y2^x = 2^x$ $\Leftrightarrow y = 2^x - y2^x = 2^x(1 - y) \Leftrightarrow 2^x = \dfrac{y}{1 - y}$ $\Leftrightarrow x = \log_2\left(\dfrac{y}{1 - y}\right)$. Thus $f^{-1}(x) = \log_2\left(\dfrac{x}{1 - x}\right)$.

(b) $\dfrac{x}{1 - x} > 0$. Solving this using the methods from Chapter 1, we start with the endpoints, 0 and 1.

Interval	$(-\infty, 0)$	$(0, 1)$	$(1, \infty)$
Sign of x	$-$	$+$	$+$
Sign of $1 - x$	$+$	$+$	$-$
Sign of $\dfrac{x}{1 - x}$	$-$	$+$	$-$

Thus the domain of $f^{-1}(x)$ is $(0, 1)$.

97. Using $D = 0.73D_0$ we have $A = -8267 \ln\left(\dfrac{D}{D_0}\right) = -8267 \ln 0.73 \approx 2602$ years.

99. When $r = 6\%$ we have $t = \dfrac{\ln 2}{0.06} \approx 11.6$ years. When $r = 7\%$ we have $t = \dfrac{\ln 2}{0.07} \approx 9.9$ years. And when $r = 8\%$ we have $t = \dfrac{\ln 2}{0.08} \approx 8.7$ years.

101. Using $A = 100$ and $W = 5$ we find the ID to be $\dfrac{\log(2A/W)}{\log 2} = \dfrac{\log(2 \cdot 100/5)}{\log 2} = \dfrac{\log 40}{\log 2} \approx 5.32$. Using $A = 100$ and $W = 10$ we find the ID to be $\dfrac{\log(2A/W)}{\log 2} = \dfrac{\log(2 \cdot 100/10)}{\log 2} = \dfrac{\log 20}{\log 2} \approx 4.32$. So the smaller icon is $\dfrac{5.23}{4.32} \approx 1.23$ times harder.

103. $\log\left(\log 10^{100}\right) = \log 100 = 2$

$\log\left(\log\left(\log 10^{\text{googol}}\right)\right) = \log(\log(\text{googol})) = \log\left(\log 10^{100}\right) = \log(100) = 2$

105. The numbers between 1000 and 9999 (inclusive) each have 4 digits, while $\log 1000 = 3$ and $\log 10{,}000 = 4$. Since $[\![\log x]\!] = 3$ for all integers x where $1000 \le x < 10{,}000$, the number of digits is $[\![\log x]\!] + 1$. Likewise, if x is an integer where $10^{n-1} \le x < 10^n$, then x has n digits and $[\![\log x]\!] = n - 1$. Since $[\![\log x]\!] = n - 1 \Leftrightarrow n = [\![\log x]\!] + 1$, the number of digits in x is $[\![\log x]\!] + 1$.

4.4 LAWS OF LOGARITHMS

1. The logarithm of a product of two numbers is the same as the *sum* of the logarithms of these numbers. So $\log_5(25 \cdot 125) = \log_5 25 + \log_5 125 = 2 + 3 = 5$.

3. The logarithm of a number raised to a power is the same as the *power* times the logarithm of the number. So $\log_5\left(25^{10}\right) = 10 \cdot \log_5 25 = 10 \cdot 2 = 20$.

5. $2\log x + \log y - \log z = \log x^2 + \log y - \log z = \log\left(x^2 y\right) - \log z = \log\left(\dfrac{x^2 y}{z}\right)$.

7. (a) False. $\log(A + B) \ne \log A + \log B$.

(b) True. $\log AB = \log A + \log B$.

9. $\log 50 + \log 200 = \log(50 \cdot 200) = 4$

11. $\log_2 60 - \log_2 15 = \log_2 \frac{60}{15} = 2$

13. $\frac{1}{4}\log_3 81 = \frac{1}{4}(4) = 1$

15. $\log_5 \sqrt{5} = \log_5\left(5^{1/2}\right) = \frac{1}{2}$

17. $\log_2 6 - \log_2 15 + \log_2 20 = \log_2 \frac{6}{15} + \log_2 20 = \log_2\left(\frac{2}{5} \cdot 20\right) = \log_2 8 = \log_2 2^3 = 3$

19. $\log_4 16^{100} = \log_4\left(4^2\right)^{100} = \log_4 4^{200} = 200$

21. $\log\left(\log 10^{10{,}000}\right) = \log(10{,}000\log 10) = \log(10{,}000 \cdot 1) = \log(10{,}000) = \log 10^4 = 4\log 10 = 4$

23. $\log_3 8x = \log_3 8 + \log_3 x$

25. $\log_3 2xy = \log_3 2 + \log_3 x + \log_3 y$

27. $\ln a^3 = 3\ln a$

29. $\log_2(xy)^{10} = 10\log_2(xy) = 10\left(\log_2 x + \log_2 y\right)$

31. $\log_2\left(AB^2\right) = \log_2 A + \log_2 B^2 = \log_2 A + 2\log_2 B$

33. $\log_3 \dfrac{2x}{y} = \log_3 2 + \log_3 x - \log_3 y$

35. $\log_5\left(\dfrac{3x^2}{y^3}\right) = \log_5 3 + 2\log_5 x - 3\log_5 y$

37. $\log_3 \dfrac{\sqrt{3x^5}}{y} = \frac{1}{2} + \frac{5}{2}\log_3 x - \log_3 y$

39. $\log\left(\dfrac{x^3 y^4}{z^6}\right) = \log\left(x^3 y^4\right) - \log z^6 = 3\log x + 4\log y - 6\log z$

41. $\ln\sqrt{x^4 + 2} = \frac{1}{2}\ln\left(x^4 + 2\right)$

43. $\ln\left(x\sqrt{\dfrac{y}{z}}\right) = \ln x + \frac{1}{2}\ln\left(\dfrac{y}{z}\right) = \ln x + \frac{1}{2}(\ln y - \ln z)$

45. $\log \sqrt[4]{x^2+y^2} = \frac{1}{4}\log\left(x^2+y^2\right)$

47. $\log \sqrt{\dfrac{x^2+4}{\left(x^2+1\right)\left(x^3-7\right)^2}} = \frac{1}{2}\log \dfrac{x^2+4}{\left(x^2+1\right)\left(x^3-7\right)^2} = \frac{1}{2}\left[\log\left(x^2+4\right) - \log\left(x^2+1\right)\left(x^3-7\right)^2\right]$

$= \frac{1}{2}\left[\log\left(x^2+4\right) - \log\left(x^2+1\right) - 2\log\left(x^3-7\right)\right]$

49. $\log_4 6 + 2\log_4 7 = \log_4 6 + \log_4 7^2 = \log_4\left(6\cdot 7^2\right) = \log_4 294$

51. $2\log x - 3\log(x+1) = \log x^2 - \log(x+1)^3 = \log \dfrac{x^2}{(x+1)^3}$

53. $4\log x - \frac{1}{3}\log\left(x^2+1\right) + 2\log(x-1) = \log x^4 - \log\sqrt[3]{x^2+1} + \log(x-1)^2$

$= \log\left(\dfrac{x^4}{\sqrt[3]{x^2+1}}\right) + \log(x-1)^2 = \log\left(\dfrac{x^4(x-1)^2}{\sqrt[3]{x^2+1}}\right)$

55. $\ln(a+b) + \ln(a-b) - 2\ln c = \ln((a+b)(a-b)) - \ln\left(c^2\right) = \ln \dfrac{a^2-b^2}{c^2}$

57. $\frac{1}{3}\log(x+2)^3 + \frac{1}{2}\left[\log x^4 - \log\left(x^2-x-6\right)^2\right] = 3\cdot\frac{1}{3}\log(x+2) + \frac{1}{2}\log\dfrac{x^4}{\left(x^2-x-6\right)^2}$

$= \log(x+2) + \log\left(\dfrac{x^4}{[(x-3)(x+2)]^2}\right)^{1/2} = \log(x+2) + \log\dfrac{x^2}{(x-3)(x+2)} = \log\dfrac{x^2(x+2)}{(x-3)(x+2)} = \log\dfrac{x^2}{x-3}$

59. $\log_2 5 = \dfrac{\log 5}{\log 2} \approx 2.321928$

61. $\log_3 16 = \dfrac{\log 16}{\log 3} \approx 2.523719$

63. $\log_7 2.61 = \dfrac{\log 2.61}{\log 7} \approx 0.493008$

65. $\log_4 125 = \dfrac{\log 125}{\log 4} \approx 3.482892$

67. $\log_3 x = \dfrac{\log_e x}{\log_e 3} = \dfrac{\ln x}{\ln 3} = \dfrac{1}{\ln 3}\ln x$. The graph of $y = \dfrac{1}{\ln 3}\ln x$ is shown in the viewing rectangle $[-1, 4]$ by $[-3, 2]$.

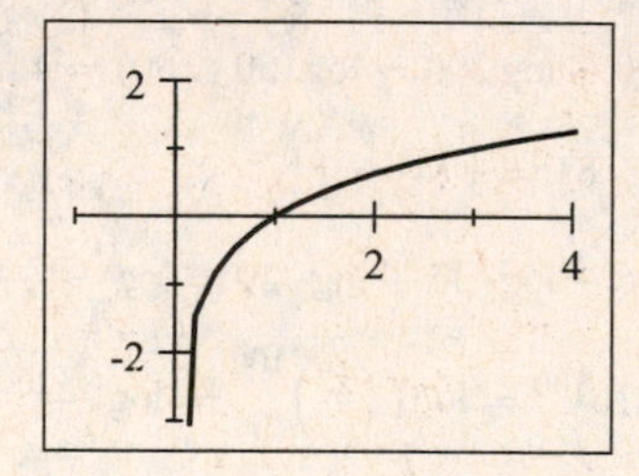

69. $\log e = \dfrac{\ln e}{\ln 10} = \dfrac{1}{\ln 10}$

71. $-\ln\left(x-\sqrt{x^2-1}\right) = \ln\left(\dfrac{1}{x-\sqrt{x^2-1}}\right) = \ln\left(\dfrac{1}{x-\sqrt{x^2-1}}\cdot\dfrac{x+\sqrt{x^2-1}}{x+\sqrt{x^2-1}}\right) = \ln\left(\dfrac{x+\sqrt{x^2-1}}{x^2-\left(x^2-1\right)}\right)$

$= \ln\left(x+\sqrt{x^2-1}\right)$

73. **(a)** $\log P = \log c - k\log W \Leftrightarrow \log P = \log c - \log W^k \Leftrightarrow \log P = \log\left(\dfrac{c}{W^k}\right) \Leftrightarrow P = \dfrac{c}{W^k}$.

(b) Using $k = 2.1$ and $c = 8000$, when $W = 2$ we have $P = \dfrac{8000}{2^{2.1}} \approx 1866$ and when $W = 10$ we have $P = \dfrac{8000}{10^{2.1}} \approx 64$.

75. **(a)** $M = -2.5\log(B/B_0) = -2.5\log B + 2.5\log B_0$.

(b) Suppose B_1 and B_2 are the brightness of two stars such that $B_1 < B_2$ and let M_1 and M_2 be their respective magnitudes. Since log is an increasing function, we have $\log B_1 < \log B_2$. Then $\log B_1 < \log B_2 \Leftrightarrow \log B_1 - \log B_0 < \log B_2 - \log B_0 \Leftrightarrow \log(B_1/B_0) < \log(B_2/B_0) \Leftrightarrow -2.5\log(B_1/B_0) > -2.5\log(B_2/B_0) \Leftrightarrow M_1 > M_2$. Thus the brighter star has less magnitudes.

(c) Let B_1 be the brightness of the star Albiero. Then $100B_1$ is the brightness of Betelgeuse, and its magnitude is
$M = -2.5 \log(100B_1/B_0) = -2.5\left[\log 100 + \log(B_1/B_0)\right] = -2.5\left[2 + \log(B_1/B_0)\right] = -5 - 2.5\log(B_1/B_0)$
$= -5 +$ magnitude of Albiero

77. The error is on the first line: $\log 0.1 < 0$, so $2\log 0.1 < \log 0.1$.

4.5 EXPONENTIAL AND LOGARITHMIC EQUATIONS

1. (a) First we isolate e^x to get the equivalent equation $e^x = 25$.
(b) Next, we take the natural logarithm of each side to get the equivalent equation $x = \ln 25$.
(c) Now we use a calculator to find $x \approx 3.219$.

3. Because the function 5^x is one-to-one, $5^{x-1} = 125 \Leftrightarrow 5^{x-1} = 5^3 \Leftrightarrow x - 1 = 3 \Leftrightarrow x = 4$.

5. Because the function 5^x is one-to-one, $5^{2x-3} = 1 \Leftrightarrow 5^{2x-3} = 5^0 \Leftrightarrow 2x - 3 = 0 \Leftrightarrow x = \frac{3}{2}$.

7. Because the function 7^x is one-to-one, $7^{2x-3} = 7^{6+5x} \Leftrightarrow 2x - 3 = 6 + 5x \Leftrightarrow 3x = -9 \Leftrightarrow x = -3$.

9. Because the function 6^x is one-to-one, $6^{x^2-1} = 6^{1-x^2} \Leftrightarrow x^2 - 1 = 1 - x^2 \Leftrightarrow 2x^2 = 2 \Leftrightarrow x = \pm 1$.

11. (a) $10^x = 25 \Leftrightarrow \log 10^x = \log 25 \Leftrightarrow x \log 10 = \log 25 \Leftrightarrow x = \log 25 = 2\log 5$
(b) $x \approx 1.397940$

13. (a) $e^{-5x} = 10 \Leftrightarrow \ln e^{-5x} = \ln 10 \Leftrightarrow -5x \ln e = \ln 10 \Leftrightarrow -5x = \ln 10 \Leftrightarrow x = -\frac{1}{5}\ln 10$
(b) $x \approx -0.460517$

15. (a) $2^{1-x} = 3 \Leftrightarrow \log 2^{1-x} = \log 3 \Leftrightarrow (1-x)\log 2 = \log 3 \Leftrightarrow 1 - x = \frac{\log 3}{\log 2} \Leftrightarrow x = 1 - \frac{\log 3}{\log 2}$
(b) $x \approx -0.584963$

17. (a) $3e^x = 10 \Leftrightarrow e^x = \frac{10}{3} \Leftrightarrow x = \ln\left(\frac{10}{3}\right)$
(b) $x \approx 1.203973$

19. (a) $300(1.025)^{12t} = 1000 \Leftrightarrow \left(\frac{41}{40}\right)^{12t} = \frac{10}{3} \Leftrightarrow 12t \ln\frac{41}{40} = \ln\frac{10}{3} \Leftrightarrow 12t = \dfrac{\ln\frac{10}{3}}{\ln\frac{41}{40}} \Leftrightarrow t = \dfrac{\ln\frac{10}{3}}{12\ln\frac{41}{40}}$
(b) $x \approx 4.063202$

21. (a) $e^{1-4x} = 2 \Leftrightarrow 1 - 4x = \ln 2 \Leftrightarrow -4x = -1 + \ln 2 \Leftrightarrow x = \frac{1}{4}(1 - \ln 2)$
(b) $x \approx 0.076713$

23. (a) $2^{5-7x} = 15 \Leftrightarrow (5 - 7x)\ln 2 = \ln 15 \Leftrightarrow 5 - 7x = \dfrac{\ln 15}{\ln 2} \Leftrightarrow 7x = 5 - \dfrac{\ln 15}{\ln 2} \Leftrightarrow x = \dfrac{5}{7} - \dfrac{\ln 15}{7\ln 2}$
(b) $x \approx 0.156158$

25. (a) $3^{x/14} = 0.1 \Leftrightarrow \log 3^{x/14} = \log 0.1 \Leftrightarrow \left(\frac{x}{14}\right)\log 3 = \log 0.1 \Leftrightarrow x = \dfrac{14\log 0.1}{\log 3}$
(b) $x \approx -29.342646$

27. (a) $4\left(1 + 10^{5x}\right) = 9 \Leftrightarrow 1 + 10^{5x} = \frac{9}{4} \Leftrightarrow 10^{5x} = \frac{5}{4} \Leftrightarrow 5x = \log\frac{5}{4} \Leftrightarrow x = \frac{1}{5}\log\frac{5}{4}$
(b) $x \approx 0.019382$

29. (a) $8 + e^{1-4x} = 20 \Leftrightarrow e^{1-4x} = 12 \Leftrightarrow 1 - 4x = \ln 12 \Leftrightarrow x = \dfrac{1 - \ln 12}{4}$
(b) $x \approx -0.371227$

31. (a) $4^x + 2^{1+2x} = 50 \Leftrightarrow 2^{2x} + 2^{1+2x} = 50 \Leftrightarrow 2^{2x}(1 + 2) = 50 \Leftrightarrow 2^{2x} = \frac{50}{3} \Leftrightarrow 2x\ln 2 = \ln 50 - \ln 3 \Leftrightarrow x = \dfrac{\ln\frac{50}{3}}{\ln 4}$
(b) $x \approx 2.029447$

33. (a) $5^x = 4^{x+1} \Leftrightarrow \log 5^x = \log 4^{x+1} \Leftrightarrow x \log 5 = (x+1) \log 4 = x \log 4 + \log 4 \Leftrightarrow x \log 5 - x \log 4 = \log 4 \Leftrightarrow$ $x (\log 5 - \log 4) = \log 4 \Leftrightarrow x = \dfrac{\log 4}{\log \frac{5}{4}}$

(b) $x \approx 6.212567$

35. (a) $2^{3x+1} = 3^{x-2} \Leftrightarrow \log 2^{3x+1} = \log 3^{x-2} \Leftrightarrow (3x+1) \log 2 = (x-2) \log 3 \Leftrightarrow 3x \log 2 + \log 2 = x \log 3 - 2 \log 3$ $\Leftrightarrow 3x \log 2 - x \log 3 = -\log 2 - 2 \log 3 \Leftrightarrow x (3 \log 2 - \log 3) = -(\log 2 + 2 \log 3) \Leftrightarrow$ $x = -\frac{\log 2 + 2 \log 3}{3 \log 2 - \log 3} = -\dfrac{\log \left(2 \cdot 3^2\right)}{\log \left(2^3/3\right)} = -\dfrac{\log 18}{\log \frac{8}{3}}$

(b) $x \approx -2.946865$

37. (a) $\dfrac{50}{1+e^{-x}} = 4 \Leftrightarrow 50 = 4 + 4e^{-x} \Leftrightarrow 46 = 4e^{-x} \Leftrightarrow 11.5 = e^{-x} \Leftrightarrow \ln 11.5 = -x \Leftrightarrow x = -\ln 11.5$

(b) $x \approx -2.442347$

39. $e^{2x} - 3e^x + 2 = 0 \Leftrightarrow (e^x - 1)(e^x - 2) = 0 \Rightarrow e^x - 1 = 0$ or $e^x - 2 = 0$. If $e^x - 1 = 0$, then $e^x = 1 \Leftrightarrow x = \ln 1 = 0$. If $e^x - 2 = 0$, then $e^x = 2 \Leftrightarrow x = \ln 2 \approx 0.6931$. So the solutions are $x = 0$ and $x \approx 0.6931$.

41. $e^{4x} + 4e^{2x} - 21 = 0 \Leftrightarrow \left(e^{2x} + 7\right)\left(e^{2x} - 3\right) = 0 \Rightarrow e^{2x} = -7$ or $e^{2x} = 3$. Now $e^{2x} = -7$ has no solution, since $e^{2x} > 0$ for all x. But we can solve $e^{2x} = 3 \Leftrightarrow 2x = \ln 3 \Leftrightarrow x = \frac{1}{2} \ln 3 \approx 0.5493$. So the only solution is $x \approx 0.5493$.

43. $2^x - 10\left(2^{-x}\right) + 3 = 0 \Leftrightarrow 2^{-x}\left(2^{2x} - 10 + 3 \cdot 2^x\right) = 0 \Leftrightarrow 2^{-x}\left(2^x + 5\right)\left(2^x - 2\right) = 0$. The first two factors are positive everywhere, so we solve $2^x - 2 = 0 \Leftrightarrow x = 1$.

45. $x^2 2^x - 2^x = 0 \Leftrightarrow 2^x\left(x^2 - 1\right) = 0 \Rightarrow 2^x = 0$ (never) or $x^2 - 1 = 0$. If $x^2 - 1 = 0$, then $x^2 = 1 \Leftrightarrow x = \pm 1$. So the only solutions are $x = \pm 1$.

47. $4x^3 e^{-3x} - 3x^4 e^{-3x} = 0 \Leftrightarrow x^3 e^{-3x}(4 - 3x) = 0 \Rightarrow x = 0$ or $e^{-3x} = 0$ (never) or $4 - 3x = 0$. If $4 - 3x = 0$, then $3x = 4 \Leftrightarrow x = \frac{4}{3}$. So the solutions are $x = 0$ and $x = \frac{4}{3}$.

49. $\log x + \log(x - 1) = \log(4x) \Leftrightarrow \log[x(x-1)] = \log(4x) \Leftrightarrow x^2 - x = 4x \Leftrightarrow x^2 - 5x = 0 \Leftrightarrow x(x-5) = 0 \Rightarrow x = 0$ or $x = 5$. So the possible solutions are $x = 0$ and $x = 5$. However, when $x = 0$, $\log x$ is undefined. Thus the only solution is $x = 5$.

51. $2 \log x = \log 2 + \log(3x - 4) \Leftrightarrow \log\left(x^2\right) = \log(6x - 8) \Leftrightarrow x^2 = 6x - 8 \Leftrightarrow x^2 - 6x + 8 = 0 \Leftrightarrow (x-4)(x-2) = 0 \Leftrightarrow$ $x = 4$ or $x = 2$. Thus the solutions are $x = 4$ and $x = 2$.

53. $\log_2 3 + \log_2 x = \log_2 5 + \log_2 (x - 2) \Leftrightarrow \log_2 (3x) = \log_2 (5x - 10) \Leftrightarrow 3x = 5x - 10 \Leftrightarrow 2x = 10 \Leftrightarrow x = 5$

55. $\ln x = 10 \Leftrightarrow x = e^{10} \approx 22{,}026$

57. $\log x = -2 \Leftrightarrow x = 10^{-2} = 0.01$

59. $\log(3x + 5) = 2 \Leftrightarrow 3x + 5 = 10^2 = 100 \Leftrightarrow 3x = 95 \Leftrightarrow x = \frac{95}{3} \approx 31.6667$

61. $4 - \log(3 - x) = 3 \Leftrightarrow \log(3 - x) = 1 \Leftrightarrow 3 - x = 10 \Leftrightarrow x = -7$

63. $\log_2 x + \log_2 (x - 3) = 2 \Leftrightarrow \log_2 [x(x-3)] = 2 \Leftrightarrow x^2 - 3x = 2^2 \Leftrightarrow x^2 - 3x - 4 = 0 \Leftrightarrow (x-4)(x+1) \Leftrightarrow x = -1$ or $x = 4$. Since $\log(-1 - 3) = \log(-4)$ is undefined, the only solution is $x = 4$.

65. $\log_9 (x - 5) + \log_9 (x + 3) = 1 \Leftrightarrow \log_9 [(x-5)(x+3)] = 1 \Leftrightarrow (x-5)(x+3) = 9^1 \Leftrightarrow x^2 - 2x - 24 = 0 \Leftrightarrow$ $(x-6)(x+4) = 0 \Rightarrow x = 6$ or -4. However, $x = -4$ is inadmissible, so $x = 6$ is the only solution.

67. $\log_5 (x + 1) - \log_5 (x - 1) = 2 \Leftrightarrow \log_5 \left(\dfrac{x+1}{x-1}\right) = 2 \Leftrightarrow \dfrac{x+1}{x-1} = 5^2 \Leftrightarrow x + 1 = 25x - 25 \Leftrightarrow 24x = 26 \Leftrightarrow x = \frac{13}{12}$

69. $\ln x = 3 - x \Leftrightarrow \ln x + x - 3 = 0$. Let $f(x) = \ln x + x - 3$. We need to solve the equation $f(x) = 0$. From the graph of f, we get $x \approx 2.21$.

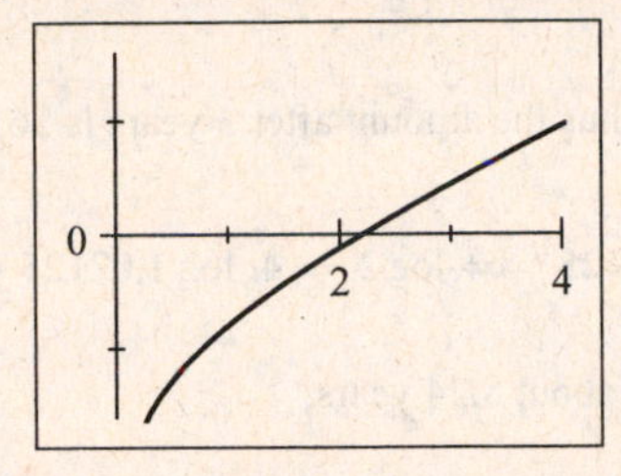

71. $x^3 - x = \log_{10}(x+1) \Leftrightarrow x^3 - x - \log_{10}(x+1) = 0$. Let $f(x) = x^3 - x - \log_{10}(x+1)$. We need to solve the equation $f(x) = 0$. From the graph of f, we get $x = 0$ or $x \approx 1.14$.

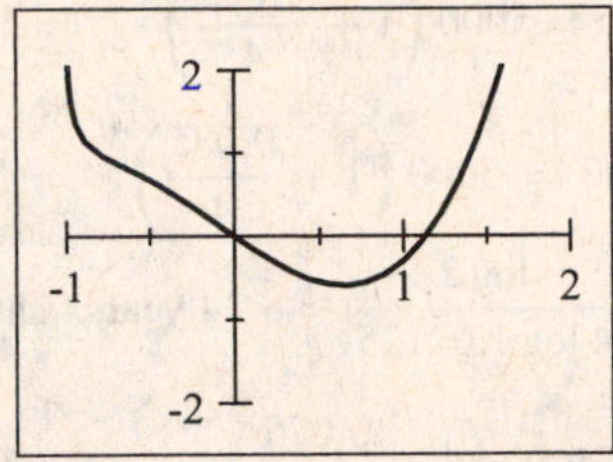

73. $e^x = -x \Leftrightarrow e^x + x = 0$. Let $f(x) = e^x + x$. We need to solve the equation $f(x) = 0$. From the graph of f, we get $x \approx -0.57$.

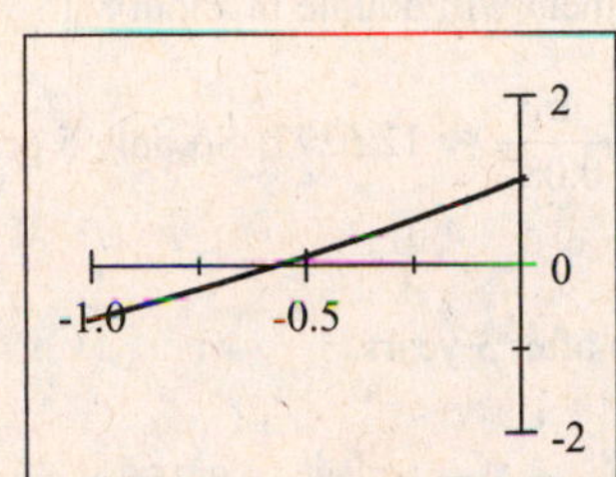

75. $4^{-x} = \sqrt{x} \Leftrightarrow 4^{-x} - \sqrt{x} = 0$. Let $f(x) = 4^{-x} - \sqrt{x}$. We need to solve the equation $f(x) = 0$. From the graph of f, we get $x \approx 0.36$.

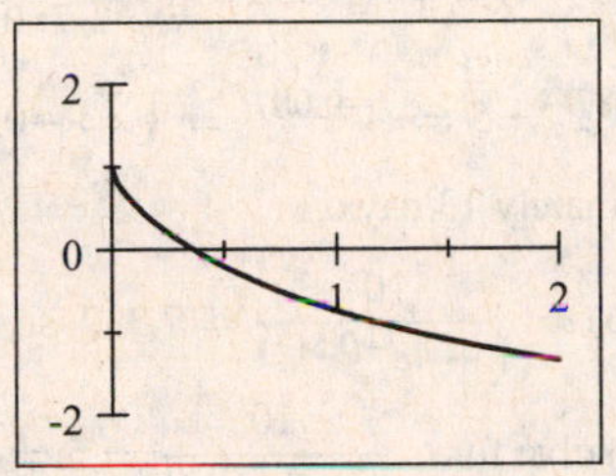

77. $2^{2/\log_5 x} = \frac{1}{16} \Leftrightarrow \log_2 2^{2/\log_5 x} = \log_2\left(\frac{1}{16}\right) \Leftrightarrow \dfrac{2}{\log_5 x} = -4 \Leftrightarrow \log_5 x = -\frac{1}{2} \Leftrightarrow x = 5^{-1/2} = \frac{1}{\sqrt{5}} \approx 0.4472$

79. $\log(x-2) + \log(9-x) < 1 \Leftrightarrow \log[(x-2)(9-x)] < 1 \Leftrightarrow \log\left(-x^2 + 11x - 18\right) < 1 \Rightarrow -x^2 + 11x - 18 < 10^1$ $\Leftrightarrow 0 < x^2 - 11x + 28 \Leftrightarrow 0 < (x-7)(x-4)$. Also, since the domain of a logarithm is positive we must have $0 < -x^2 + 11x - 18 \Leftrightarrow 0 < (x-2)(9-x)$. Using the methods from Chapter 1 with the endpoints 2, 4, 7, 9 for the intervals, we make the following table:

Interval	$(-\infty, 2)$	$(2, 4)$	$(4, 7)$	$(7, 9)$	$(9, \infty)$
Sign of $x - 7$	−	−	−	+	+
Sign of $x - 4$	−	−	+	+	+
Sign of $x - 2$	−	+	+	+	+
Sign of $9 - x$	+	+	+	+	−
Sign of $(x-7)(x-4)$	+	+	−	+	+
Sign of $(x-2)(9-x)$	−	+	+	+	−

Thus the solution is $(2, 4) \cup (7, 9)$.

81. $2 < 10^x < 5 \Leftrightarrow \log 2 < x < \log 5 \Leftrightarrow 0.3010 < x < 0.6990$. Hence the solution to the inequality is approximately the interval $(0.3010, 0.6990)$.

83. To find the inverse of $f(x) = 2^{2x}$, we set $y = f(x)$ and solve for x. $y = 2^{2x} \Leftrightarrow \ln y = \ln\left(2^{2x}\right) = 2x \ln 2 \Leftrightarrow x = \dfrac{\ln y}{2\ln 2}$.

Interchange x and y: $y = \dfrac{\ln x}{2\ln 2}$. Thus, $f^{-1}(x) = \dfrac{\ln x}{2\ln 2}$.

85. To find the inverse of $f(x) = \log_2(x-1)$, we set $y = f(x)$ and solve for x. $y = \log_2(x-1) \Leftrightarrow 2^y = 2^{\log_2(x-1)} = x-1$ $\Leftrightarrow x = 2^y + 1$. Interchange x and y: $y = 2^x + 1$. Thus, $f^{-1}(x) = 2^x + 1$.

87. $\log(x+3) = \log x + \log 3 \Leftrightarrow \log(x+3) = \log(3x) \Leftrightarrow x + 3 = 3x \Leftrightarrow 2x = 3 \Leftrightarrow x = \frac{3}{2}$

89. (a) $A(3) = 5000\left(1 + \dfrac{0.085}{4}\right)^{4(3)} = 5000\left(1.02125^{12}\right) = 6435.09$. Thus the amount after 3 years is \$6,435.09.

(b) $10000 = 5000\left(1 + \dfrac{0.085}{4}\right)^{4t} = 5000\left(1.02125^{4t}\right) \Leftrightarrow 2 = 1.02125^{4t} \Leftrightarrow \log 2 = 4t \log 1.02125 \Leftrightarrow$

$t = \dfrac{\log 2}{4 \log 1.02125} \approx 8.24$ years. Thus the investment will double in about 8.24 years.

91. $8000 = 5000\left(1 + \dfrac{0.075}{4}\right)^{4t} = 5000\left(1.01875^{4t}\right) \Leftrightarrow 1.6 = 1.01875^{4t} \Leftrightarrow \log 1.6 = 4t \log 1.01875 \Leftrightarrow$

$t = \dfrac{\log 1.6}{4 \log 1.01875} \approx 6.33$ years. The investment will increase to \$8000 in approximately 6 years and 4 months.

93. $2 = e^{0.085t} \Leftrightarrow \ln 2 = 0.085t \Leftrightarrow t = \dfrac{\ln 2}{0.085} \approx 8.15$ years. Thus the investment will double in about 8.15 years.

95. $15e^{-0.087t} = 5 \Leftrightarrow e^{-0.087t} = \frac{1}{3} \Leftrightarrow -0.087t = \ln\left(\frac{1}{3}\right) = -\ln 3 \Leftrightarrow t = \dfrac{\ln 3}{0.087} \approx 12.6277$. So only 5 grams remain after approximately 13 days.

97. (a) $P(3) = \dfrac{10}{1 + 4e^{-0.8(3)}} = 7.337$, so there are approximately 7337 fish after 3 years.

(b) We solve for t. $\dfrac{10}{1 + 4e^{-0.8t}} = 5 \Leftrightarrow 1 + 4e^{-0.8t} = \frac{10}{5} = 2 \Leftrightarrow 4e^{-0.8t} = 1 \Leftrightarrow e^{-0.8t} = 0.25 \Leftrightarrow -0.8t = \ln 0.25 \Leftrightarrow$

$t = \dfrac{\ln 0.25}{-0.8} = 1.73$. So the population will reach 5000 fish in about 1 year and 9 months.

99. (a) $\ln\left(\dfrac{P}{P_0}\right) = -\dfrac{h}{k} \Leftrightarrow \dfrac{P}{P_0} = e^{-h/k} \Leftrightarrow P = P_0 e^{-h/k}$. Substituting $k = 7$ and $P_0 = 100$ we get $P = 100e^{-h/7}$.

(b) When $h = 4$ we have $P = 100e^{-4/7} \approx 56.47$ kPa.

101. (a) $I = \frac{60}{13}\left(1 - e^{-13t/5}\right) \Leftrightarrow \frac{13}{60}I = 1 - e^{-13t/5} \Leftrightarrow e^{-13t/5} = 1 - \frac{13}{60}I \Leftrightarrow -\frac{13}{5}t = \ln\left(1 - \frac{13}{60}I\right) \Leftrightarrow$

$t = -\frac{5}{13}\ln\left(1 - \frac{13}{60}I\right)$.

(b) Substituting $I = 2$, we have $t = -\frac{5}{13}\ln\left[1 - \frac{13}{60}(2)\right] \approx 0.218$ seconds.

103. Since $9^1 = 9$, $9^2 = 81$, and $9^3 = 729$, the solution of $9^x = 20$ must be between 1 and 2 (because 20 is between 9 and 81), whereas the solution to $9^x = 100$ must be between 2 and 3 (because 100 is between 81 and 729).

105. (a) $(x-1)^{\log(x-1)} = 100(x-1) \Leftrightarrow \log\left((x-1)^{\log(x-1)}\right) = \log(100(x-1)) \Leftrightarrow$

$[\log(x-1)]\log(x-1) = \log 100 + \log(x-1) \Leftrightarrow [\log(x-1)]^2 - \log(x-1) - 2 = 0 \Leftrightarrow$

$[\log(x-1) - 2][\log(x-1) + 1] = 0$. Thus either $\log(x-1) = 2 \Leftrightarrow x = 101$ or $\log(x-1) = -1 \Leftrightarrow x = \frac{11}{10}$.

(b) $\log_2 x + \log_4 x + \log_8 x = 11 \Leftrightarrow \log_2 x + \log_2 \sqrt{x} + \log_2 \sqrt[3]{x} = 11 \Leftrightarrow \log_2\left(x\sqrt{x}\sqrt[3]{x}\right) = 11 \Leftrightarrow \log_2\left(x^{11/6}\right) = 11$ $\Leftrightarrow \frac{11}{6}\log_2 x = 11 \Leftrightarrow \log_2 x = 6 \Leftrightarrow x = 2^6 = 64$

(c) $4^x - 2^{x+1} = 3 \Leftrightarrow (2^x)^2 - 2(2^x) - 3 = 0 \Leftrightarrow (2^x - 3)(2^x + 1) = 0 \Leftrightarrow$ either $2^x = 3 \Leftrightarrow x = \dfrac{\ln 3}{\ln 2}$ or $2^x = -1$, which has no real solution. So $x = \dfrac{\ln 3}{\ln 2}$ is the only real solution.

4.6 MODELING WITH EXPONENTIAL FUNCTIONS

1. (a) Here $n_0 = 10$ and $a = 1.5$ hours, so $n(t) = 10 \cdot 2^{t/1.5} = 10 \cdot 2^{2t/3}$.

(b) After 35 hours, there will be $n(35) = 10 \cdot 2^{2(35)/3} \approx 1.06 \times 10^8$ bacteria.

(c) $n(t) = 10 \cdot 2^{2t/3} = 10{,}000 \Leftrightarrow 2^{2t/3} = 1000 \Leftrightarrow \ln\left(2^{2t/3}\right) = \ln 1000 \Leftrightarrow \frac{2t}{3}\ln 2 = \ln 1000 \Leftrightarrow t = \frac{3}{2}\frac{\ln 1000}{\ln 2} \approx 14.9$, so the bacteria count will reach 10,000 in about 14.9 hours.

3. (a) A model for the squirrel population is $n(t) = n_0 \cdot 2^{t/6}$. We are given that $n(30) = 100{,}000$, so $n_0 \cdot 2^{30/6} = 100{,}000 \Leftrightarrow$ $n_0 = \frac{100{,}000}{2^5} = 3125$. Initially, there were approximately 3125 squirrels.

(b) In 10 years, we will have $t = 40$, so the population will be $n(40) = 3125 \cdot 2^{40/6} \approx 317{,}480$ squirrels.

(c)

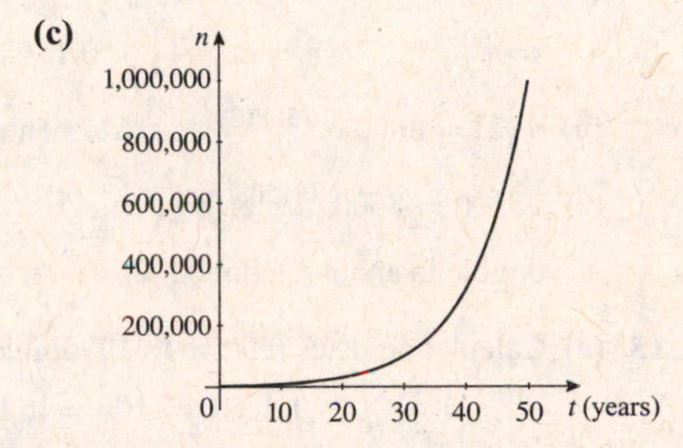

5. (a) $r = 0.08$ and $n(0) = 18{,}000$. Thus the population is given by the formula $n(t) = 18{,}000e^{0.08t}$.

(b) $t = 2021 - 2013 = 8$. Then we have $n(8) = 18{,}000e^{0.08(8)} = 18{,}000e^{0.64} \approx 34{,}100$. Thus there should be about 34,100 foxes in the region by the year 2021.

(c) Solving $n(t) = 25{,}000$, we get $18{,}000e^{0.08t} = 25{,}000 \Leftrightarrow$ $18e^{0.08t} = 25 \Leftrightarrow \ln\left(18e^{0.08t}\right) = \ln 25 \Leftrightarrow \ln 18 + 0.08t = \ln 25 \Leftrightarrow t = \frac{1}{0.08}(\ln 25 - \ln 18) \approx 4.1$, so the fox population will reach 25,000 after about 4.1 years.

(d)

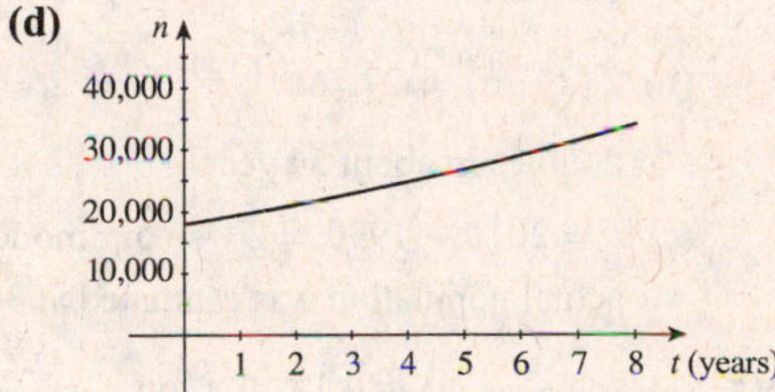

7. $n(t) = n_0e^{rt}$; $n_0 = 110$ million, $t = 2036 - 2011 = 25$.

(a) $r = 0.03$; $n(25) = 110{,}000{,}000e^{0.03(25)} = 110{,}000{,}000e^{0.75} \approx 232{,}870{,}000$. Thus at a 3% growth rate, the projected population will be approximately 233 million people by the year 2036.

(b) $r = 0.02$; $n(25) = 110{,}000{,}000e^{0.02(25)} = 110{,}000{,}000e^{0.50} \approx 181{,}359{,}340$. Thus at a 2% growth rate, the projected population will be approximately 181 million people by the year 2036.

9. (a) The doubling time is 18 years and the initial population is 112,000, so a model is $n(t) = 112{,}000 \cdot 2^{t/18}$.

(b) We need to find the relative growth rate r. Since the population is $2 \cdot 112{,}000 = 224{,}000$ when $t = 18$, we have $224{,}000 = 112{,}000e^{18r}$ $\Leftrightarrow 2 = e^{18r} \Leftrightarrow \ln 2 = 18r \Leftrightarrow r = \frac{\ln 2}{18} \approx 0.0385$. Thus, a model is $n(t) = 112{,}000e^{0.0385t}$.

(c)

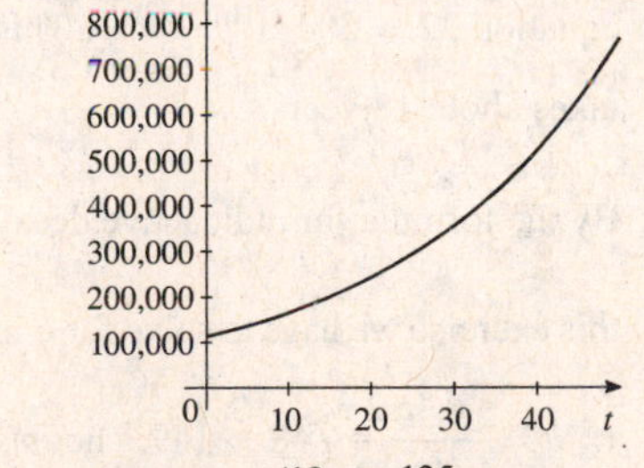

(d) Using the model in part (a), we solve the equation $n(t) = 112{,}000 \cdot 2^{t/18} = 500{,}000 \Leftrightarrow 2^{t/18} = \frac{125}{28} \Leftrightarrow$ $\ln 2^{t/18} = \ln \frac{125}{28} \Leftrightarrow \frac{t}{18}\ln 2 = \ln\frac{125}{28} \Leftrightarrow t = \frac{18\ln\frac{125}{28}}{\ln 2} \approx 38.85$. Therefore, it takes about 38.85 years for the population to reach 500,000.

11. (a) The deer population in 2010 was 20,000.

(b) Using the model $n(t) = 20{,}000e^{rt}$ and the point $(4, 31000)$, we have $31{,}000 = 20{,}000e^{4r} \Leftrightarrow 1.55 = e^{4r} \Leftrightarrow 4r = \ln 1.55 \Leftrightarrow r = \frac{1}{4}\ln 1.55 \approx 0.1096$. Thus $n(t) = 20{,}000e^{0.1096t}$

(c) $n(8) = 20{,}000e^{0.1096(8)} \approx 48{,}218$, so the projected deer population in 2018 is about 48,000.

(d) $100{,}000 = 20{,}000e^{0.1096t} \Leftrightarrow 5 = e^{0.1096t} \Leftrightarrow 0.1096t = \ln 5 \Leftrightarrow t = \dfrac{\ln 5}{0.1096} \approx 14.68$. Thus, it takes about 14.7 years for the deer population to reach 100,000.

13. (a) Using the formula $n(t) = n_0e^{rt}$ with $n_0 = 8600$ and $n(1) = 10000$, we solve for r, giving $10000 = n(1) = 8600e^{r} \Leftrightarrow \frac{50}{43} = e^{r} \Leftrightarrow r = \ln\left(\frac{50}{43}\right) \approx 0.1508$. Thus $n(t) = 8600e^{0.1508t}$.

(b) $n(2) = 8600e^{0.1508(2)} \approx 11627$. Thus the number of bacteria after two hours is about 11,600.

(c) $17200 = 8600e^{0.1508t} \Leftrightarrow 2 = e^{0.1508t} \Leftrightarrow 0.1508t = \ln 2 \Leftrightarrow t = \dfrac{\ln 2}{0.1508} \approx 4.596$. Thus the number of bacteria will double in about 4.6 hours.

15. (a) Calculating dates relative to 1990 gives $n_0 = 29.76$ and $n(10) = 33.87$. Then $n(10) = 29.76e^{10r} = 33.87 \Leftrightarrow e^{10r} = \frac{33.87}{29.76} \approx 1.1381 \Leftrightarrow 10r = \ln 1.1381 \Leftrightarrow r = \frac{1}{10}\ln 1.1381 \approx 0.012936$. Thus $n(t) = 29.76e^{0.012936t}$ million people.

(b) $2(29.76) = 29.76e^{0.012936t} \Leftrightarrow 2 = e^{0.012936t} \Leftrightarrow \ln 2 = 0.012936t \Leftrightarrow t = \dfrac{\ln 2}{0.012936} \approx 53.58$, so the population doubles in about 54 years.

(c) $t = 2010 - 1990 = 20$, so our model gives the 2010 population as $n(20) \approx 29.76e^{0.012936(20)} \approx 38.55$ million. The actual population was estimated at 36.96 million in 2009.

17. (a) Because the half-life is 1600 years and the sample weighs 22 mg initially, a suitable model is $m(t) = 22 \cdot 2^{-t/1600}$.

(b) From the formula for radioactive decay, we have $m(t) = m_0e^{-rt}$, where $m_0 = 22$ and $r = \dfrac{\ln 2}{h} = \dfrac{\ln 2}{1600} \approx 0.000433$. Thus, the amount after t years is given by $m(t) = 22e^{-0.000433t}$.

(c) $m(4000) = 22e^{-0.000433(4000)} \approx 3.89$, so the amount after 4000 years is about 4 mg.

(d) We have to solve for t in the equation $18 = 22\,e^{-0.000433t}$. This gives $18 = 22e^{-0.000433t} \Leftrightarrow \frac{9}{11} = e^{-0.000433t} \Leftrightarrow -0.000433t = \ln\left(\frac{9}{11}\right) \Leftrightarrow t = \dfrac{\ln\left(\frac{9}{11}\right)}{-0.000433} \approx 463.4$, so it takes about 463 years.

19. By the formula in the text, $m(t) = m_0e^{-rt}$ where $r = \dfrac{\ln 2}{h}$, so $m(t) = 50e^{-[(\ln 2)/28]t}$. We need to solve for t in the equation $32 = 50e^{-[(\ln 2)/28]t}$. This gives $e^{-[(\ln 2)/28]t} = \frac{32}{50} \Leftrightarrow -\dfrac{\ln 2}{28}t = \ln\left(\frac{32}{50}\right) \Leftrightarrow t = -\dfrac{28}{\ln 2} \cdot \ln\left(\frac{32}{50}\right) \approx 18.03$, so it takes about 18 years.

21. By the formula for radioactive decay, we have $m(t) = m_0e^{-rt}$, where $r = \dfrac{\ln 2}{h}$, in other words $m(t) = m_0e^{-[(\ln 2)/h]t}$. In this exercise we have to solve for h in the equation $200 = 250e^{-[(\ln 2)/h]\cdot 48} \Leftrightarrow 0.8 = e^{-[(\ln 2)/h]\cdot 48} \Leftrightarrow \ln(0.8) = -\dfrac{\ln 2}{h}\cdot 48 \Leftrightarrow h = -\dfrac{\ln 2}{\ln 0.8} \cdot 48 \approx 149.1$ hours. So the half-life is approximately 149 hours.

23. By the formula in the text, $m(t) = m_0e^{-[(\ln 2)/h]\cdot t}$, so we have $0.65 = 1 \cdot e^{-[(\ln 2)/5730]\cdot t} \Leftrightarrow \ln(0.65) = -\dfrac{\ln 2}{5730}t \Leftrightarrow t = -\dfrac{5730\ln 0.65}{\ln 2} \approx 3561$. Thus the artifact is about 3560 years old.

25. **(a)** $T(0) = 65 + 145e^{-0.05(0)} = 65 + 145 = 210°$ F.

(b) $T(10) = 65 + 145e^{-0.05(10)} \approx 152.9$. Thus the temperature after 10 minutes is about 153° F.

(c) $100 = 65 + 145e^{-0.05t} \Leftrightarrow 35 = 145e^{-0.05t} \Leftrightarrow 0.2414 = e^{-0.05t} \Leftrightarrow \ln 0.2414 = -0.05t \Leftrightarrow t = -\dfrac{\ln 0.2414}{0.05} \approx 28.4$.
Thus the temperature will be 100° F in about 28 minutes.

27. Using Newton's Law of Cooling, $T(t) = T_s + D_0e^{-kt}$ with $T_s = 75$ and $D_0 = 185 - 75 = 110$. So $T(t) = 75 + 110e^{-kt}$.

(a) Since $T(30) = 150$, we have $T(30) = 75 + 110e^{-30k} = 150 \Leftrightarrow 110e^{-30k} = 75 \Leftrightarrow e^{-30k} = \frac{15}{22} \Leftrightarrow -30k = \ln\left(\frac{15}{22}\right)$
$\Leftrightarrow k = -\frac{1}{30}\ln\left(\frac{15}{22}\right)$. Thus we have $T(45) = 75 + 110e^{(45/30)\ln(15/22)} \approx 136.9$, and so the temperature of the turkey after 45 minutes is about 137° F.

(b) The temperature will be 100°F when $75 + 110e^{(t/30)\ln(15/22)} = 100 \Leftrightarrow e^{(t/30)\ln(15/22)} = \dfrac{25}{110} = \frac{5}{22} \Leftrightarrow$
$\left(\dfrac{t}{30}\right)\ln\left(\frac{15}{22}\right) = \ln\left(\frac{5}{22}\right) \Leftrightarrow t = 30\dfrac{\ln\left(\frac{5}{22}\right)}{\ln\left(\frac{15}{22}\right)} \approx 116.1$. So the temperature will be 100° F after about 2 hours.

4.7 LOGARITHMIC SCALES

1. **(a)** $\text{pH} = -\log[\text{H}^+] = -\log\left(5.0 \times 10^{-3}\right) \approx 2.3$

(b) $\text{pH} = -\log[\text{H}^+] = -\log\left(3.2 \times 10^{-4}\right) \approx 3.5$

(c) $\text{pH} = -\log[\text{H}^+] = -\log\left(5.0 \times 10^{-9}\right) \approx 8.3$

3. **(a)** $\text{pH} = -\log[\text{H}^+] = 3.0 \Leftrightarrow [\text{H}^+] = 10^{-3}$ M

(b) $\text{pH} = -\log[\text{H}^+] = 6.5 \Leftrightarrow [\text{H}^+] = 10^{-6.5} \approx 3.2 \times 10^{-7}$ M

5. $4.0 \times 10^{-7} \le [\text{H}^+] \le 1.6 \times 10^{-5} \Leftrightarrow \log\left(4.0 \times 10^{-7}\right) \le \log[\text{H}^+] \le \log\left(1.6 \times 10^{-5}\right) \Leftrightarrow$
$-\log\left(4.0 \times 10^{-7}\right) \ge \text{pH} \ge -\log\left(1.6 \times 10^{-5}\right) \Leftrightarrow 6.4 \ge \text{pH} \ge 4.8$. Therefore the range of pH readings for cheese is approximately 4.8 to 6.4.

7. **(a)** For the California wine, we have $\text{pH} = -\log[\text{H}^+] \Leftrightarrow \log[\text{H}^+] = -3.2 \Leftrightarrow [\text{H}^+] = 10^{-3.2} \approx 6.3 \times 10^{-4}$ M. For the Italian wine, $\text{pH} = -\log[\text{H}^+] \Leftrightarrow \log[\text{H}^+] = -2.9 \Leftrightarrow [\text{H}^+] = 10^{-2.9} \approx 1.3 \times 10^{-3}$ M.

(b) The California wine has lower hydrogen ion concentration.

9. **(a)** $M = \log\dfrac{I}{S}$ with $S = 10^{-4}$ and $I = 31.25$, so $M = \log\dfrac{31.25}{10^{-4}} \approx 5.5$.

(b) $M = \log\dfrac{I}{S} \Leftrightarrow 10^M = \dfrac{I}{S} \Leftrightarrow I = S \cdot 10^M$. We have $M = 4.8$ and $S = 10^{-4}$, so $I = 10^{-4} \cdot 10^{4.8} \approx 6.3$.

11. Let I_0 be the intensity of the smaller earthquake and I_1 the intensity of the larger earthquake. Then $I_1 = 20I_0$. Notice that $M_0 = \log\dfrac{I_0}{S} = \log I_0 - \log S$ and $M_1 = \log\dfrac{I_1}{S} = \log\dfrac{20I_0}{S} = \log 20 + \log I_0 - \log S$. Then $M_1 - M_0 = \log 20 + \log I_0 - \log S - \log I_0 + \log S = \log 20 \approx 1.3$. Therefore the magnitude is 1.3 times larger.

13. Let the subscript J represent the Japan earthquake and S represent the San Francisco earthquake. Then $M_J = \log\dfrac{I_J}{S} = 9.1$
$\Leftrightarrow I_J = S \cdot 10^{9.1}$ and $M_S = \log\dfrac{I_S}{S} = 8.3 \Leftrightarrow I_S = S \cdot 10^{8.3}$. So $\dfrac{I_J}{I_S} = \dfrac{S \cdot 10^{9.1}}{S \cdot 10^{8.3}} = 10^{0.8} \approx 6.3$, and hence the Japan earthquake was about six times more intense than the San Francisco earthquake.

15. $\beta = 10\log\frac{I}{I_0} = 10\log\frac{2.0\times10^{-5}}{1.0\times10^{-12}} = 10\log\left(2\times10^{7}\right) = 10\left(\log 2+\log 10^{7}\right) = 10\left(\log 2+7\right)\approx 73$. Therefore the intensity level was 73 dB.

17. $\beta = 10\log\frac{I}{I_0} \Leftrightarrow 70 = 10\log\frac{I}{1.0\times10^{-12}} \Leftrightarrow \log I + 12 = 7 \Leftrightarrow \log I = -5$, so the intensity was 10^{-5} watts/m^2.

19. **(a)** The intensity is 3.1×10^{-5} W/m^2, so $\beta = 10\log\frac{I}{I_0} = 10\log\frac{3.1\times10^{-5}}{1.0\times10^{-12}} = 10\log\left(3.1\times10^{7}\right)\approx 75$ dB.

(b) Here $\beta = 90$ dB, so $90 = 10\log\frac{I}{10^{-12}} \Leftrightarrow \log I + 12 = 9 \Leftrightarrow I = 10^{-3}$ W/m^2.

(c) The ratio of the intensities is $\frac{I_e}{I_s} = \frac{10^{-3}}{3.1\times10^{-5}}\approx 32.3$.

21. **(a)** $\beta_1 = 10\log\frac{I_1}{I_0}$ and $I_1 = \frac{k}{d_1^2} \Leftrightarrow \beta_1 = 10\log\frac{k}{d_1^2 I_0} = 10\left[\log\left(\frac{k}{I_0}\right) - 2\log d_1\right] = 10\log\frac{k}{I_0} - 20\log d_1$. Similarly,

$\beta_2 = 10\log\frac{k}{I_0} - 20\log d_2$. Substituting the expression for β_1 gives

$\beta_2 = 10\log\frac{k}{I_0} - 20\log d_1 + 20\log d_1 - 20\log d_2 = \beta_1 + 20\log d_1 - 20\log d_2 = \beta_1 + 20\log\left(\frac{d_1}{d_2}\right)$.

(b) $\beta_1 = 120$, $d_1 = 2$, and $d_2 = 10$. Then $\beta_2 = \beta_1 + 20\log\frac{d_1}{d_2} = 120 + 20\log\frac{2}{10} = 120 + 20\log 0.2 \approx 106$, and so the intensity level at 10 m is approximately 106 dB.

CHAPTER 4 REVIEW

1. $f(x) = 5^x$; $f(-1.5)\approx 0.089$, $f\left(\sqrt{2}\right)\approx 9.739$, $f(2.5)\approx 55.902$

3. $g(x) = 4e^{x-2}$; $g(-0.7)\approx 0.269$, $g(1)\approx 1.472$, $g(\pi)\approx 12.527$

5. $f(x) = 3^{x-2}$. domain $(-\infty,\infty)$, range $(0,\infty)$, asymptote $y = 0$.

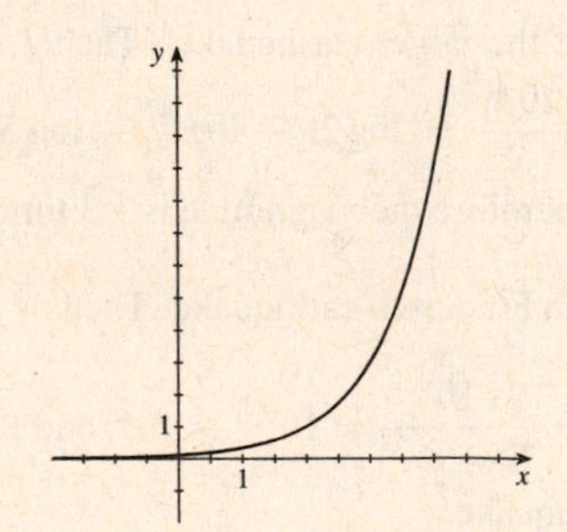

7. $g(x) = 3 + 2^x$. Domain $(-\infty,\infty)$, range $(3,\infty)$, asymptote $y = 3$.

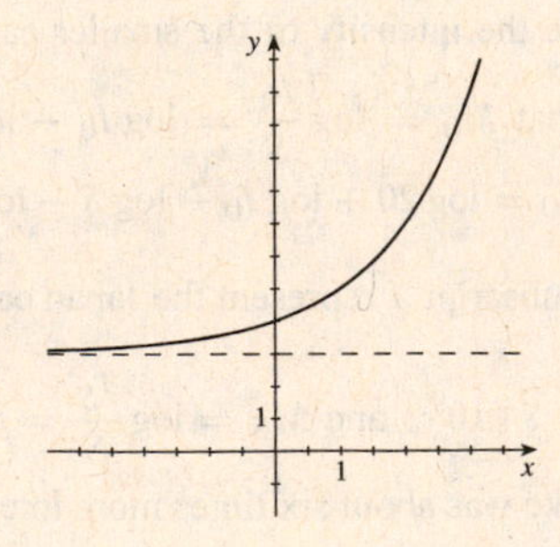

9. $F(x) = e^{x-1} + 1$. Domain $(-\infty, \infty)$, range $(1, \infty)$, asymptote $y = 1$.

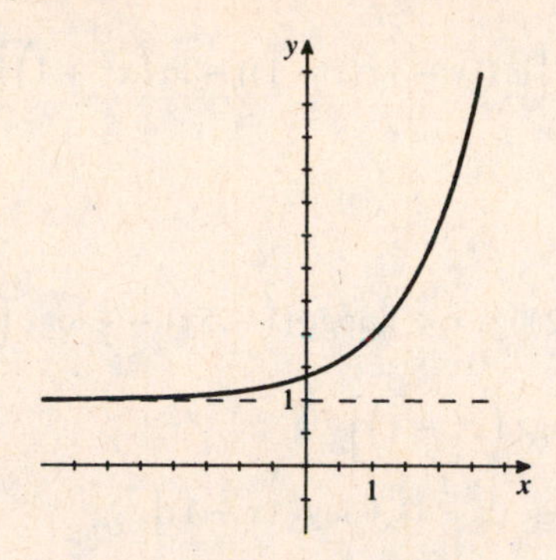

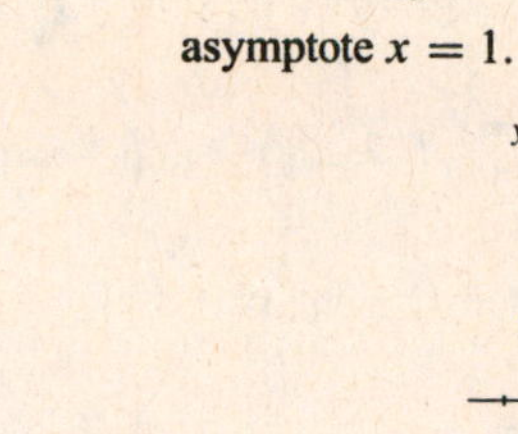

11. $f(x) = \log_3(x - 1)$. Domain $(1, \infty)$, range $(-\infty, \infty)$, asymptote $x = 1$.

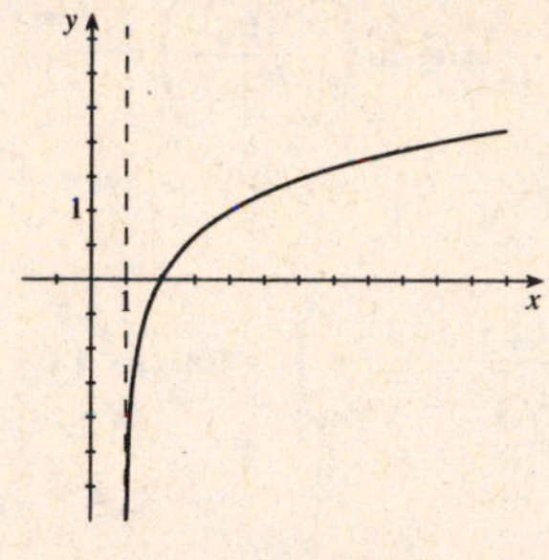

13. $f(x) = 2 - \log_2 x$. Domain $(0, \infty)$, range $(-\infty, \infty)$, asymptote $x = 0$.

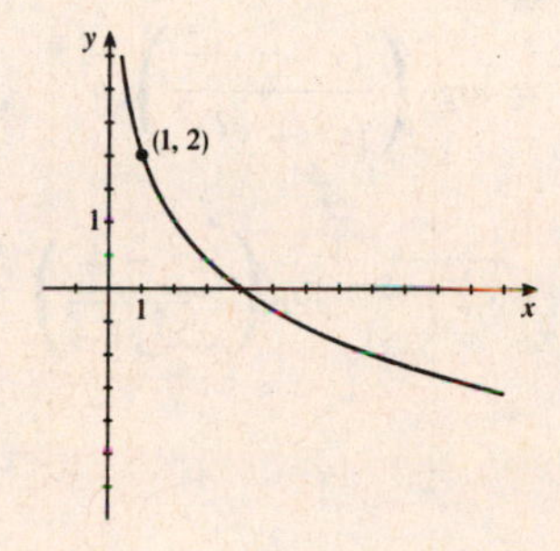

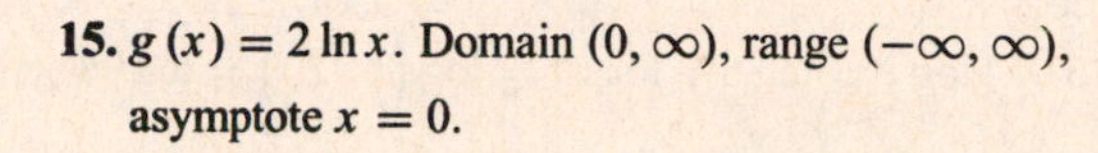

15. $g(x) = 2\ln x$. Domain $(0, \infty)$, range $(-\infty, \infty)$, asymptote $x = 0$.

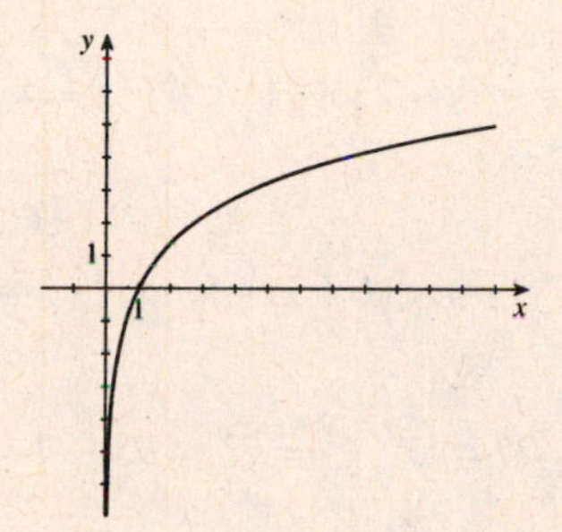

17. $f(x) = 10^{x^2} + \log(1 - 2x)$. Since $\log u$ is defined only for $u > 0$, we require $1 - 2x > 0 \Leftrightarrow -2x > -1 \Leftrightarrow x < \frac{1}{2}$, and so the domain is $\left(-\infty, \frac{1}{2}\right)$.

19. $h(x) = \ln\left(x^2 - 4\right)$. We must have $x^2 - 4 > 0$ (since $\ln y$ is defined only for $y > 0$) $\Leftrightarrow x^2 - 4 > 0 \Leftrightarrow (x - 2)(x + 2) > 0$. The endpoints of the intervals are -2 and 2.

Interval	$(-\infty, -2)$	$(-2, 2)$	$(2, \infty)$
Sign of $x - 2$	$-$	$-$	$+$
Sign of $x + 2$	$-$	$+$	$+$
Sign of $(x - 2)(x + 2)$	$+$	$-$	$+$

Thus the domain is $(-\infty, -2) \cup (2, \infty)$.

21. $\log_2 1024 = 10 \Leftrightarrow 2^{10} = 1024$

23. $\log x = y \Leftrightarrow 10^y = x$

25. $2^6 = 64 \Leftrightarrow \log_2 64 = 6$

27. $10^x = 74 \Leftrightarrow \log_{10} 74 = x \Leftrightarrow \log 74 = x$

29. $\log_2 128 = \log_2\left(2^7\right) = 7$

31. $10^{\log 45} = 45$

33. $\ln\left(e^6\right) = 6$

35. $\log_3 \frac{1}{27} = \log_3 3^{-3} = -3$

37. $\log_5 \sqrt{5} = \log_5 5^{1/2} = \frac{1}{2}$

39. $\log 25 + \log 4 = \log(25 \cdot 4) = \log 10^2 = 2$

41. $\log_2\left(16^{23}\right) = \log_2\left(2^4\right)^{23} = \log_2 2^{92} = 92$

43. $\log_8 6 - \log_8 3 + \log_8 2 = \log_8\left(\frac{6}{3} \cdot 2\right) = \log_8 4 = \log_8 8^{2/3} = \frac{2}{3}$

45. $\log\left(AB^2C^3\right) = \log A + 2\log B + 3\log C$

47. $\ln\sqrt{\dfrac{x^2-1}{x^2+1}} = \dfrac{1}{2}\ln\left(\dfrac{x^2-1}{x^2+1}\right) = \frac{1}{2}\left[\ln\left(x^2-1\right) - \ln\left(x^2+1\right)\right] = \frac{1}{2}\left[\ln\left((x-1)(x+1)\right) - \ln\left(x^2+1\right)\right]$

$= \frac{1}{2}\left[\ln(x-1) + \ln(x+1) - \ln\left(x^2+1\right)\right]$

49. $\log_5\left(\dfrac{x^2(1-5x)^{3/2}}{\sqrt{x^3-x}}\right) = \log_5 x^2(1-5x)^{3/2} - \log_5\sqrt{x\left(x^2-1\right)} = 2\log_5 x + \frac{3}{2}\log_5(1-5x) - \frac{1}{2}\log_5\left(x^3-x\right)$

$= 2\log_5 x + \frac{3}{2}\log_5(1-5x) - \frac{1}{2}\left[\log_5 x + \log_5\left(x^2-1\right)\right]$

$= 2\log_5 x + \frac{3}{2}\log_5(1-5x) - \frac{1}{2}\left[\log_5 x + \log_5(x-1) + \log_5(x+1)\right]$

51. $\log 6 + 4\log 2 = \log 6 + \log 2^4 = \log\left(6\cdot 2^4\right) = \log 96$

53. $\frac{3}{2}\log_2(x-y) - 2\log_2\left(x^2+y^2\right) = \log_2(x-y)^{3/2} - \log_2\left(x^2+y^2\right)^2 = \log_2\left(\dfrac{(x-y)^{3/2}}{\left(x^2+y^2\right)^2}\right)$

55. $\log(x-2) + \log(x+2) - \frac{1}{2}\log\left(x^2+4\right) = \log[(x-2)(x+2)] - \log\sqrt{x^2+4} = \log\left(\dfrac{x^2-4}{\sqrt{x^2+4}}\right)$

57. $3^{2x-7} = 27 \Leftrightarrow 3^{2x-7} = 3^3 \Leftrightarrow 2x-7 = 3 \Leftrightarrow 2x = 10 \Leftrightarrow x = 5$

59. $2^{3x-5} = 7 \Leftrightarrow \log_2\left(2^{3x-5}\right) = \log_2 7 \Leftrightarrow 3x-5 = \log_2 7 \Leftrightarrow x = \frac{1}{3}\left(\log_2 7 + 5\right)$. Using the Change of Base Formula, we have $\log_2 7 = \frac{\log 7}{\log 2} \approx 2.807$, so $x \approx \frac{1}{3}(2.807+5) \approx 2.60$.

61. $4^{1-x} = 3^{2x+5} \Leftrightarrow \log 4^{1-x} = \log 3^{2x+5} \Leftrightarrow (1-x)\log 4 = (2x+5)\log 3 \Leftrightarrow \log 4 - 5\log 3 = 2x\log 3 + x\log 4 \Leftrightarrow$ $x(\log 3 + \log 4) = \log 4 - 5\log 3 \Leftrightarrow x = \dfrac{\log 4 - 5\log 3}{2\log 3 + \log 4} \approx -1.15$

63. $x^2e^{2x} + 2xe^{2x} = 8e^{2x} \Leftrightarrow e^{2x}\left(x^2+2x-8\right) = 0 \Leftrightarrow x^2+2x-8 = 0$ (since $e^{2x} \neq 0$) $\Leftrightarrow (x+4)(x-2) = 0 \Leftrightarrow x = -4$ or $x = 2$

65. $\log x + \log(x+1) = \log 12 \Leftrightarrow \log(x(x+1)) = \log 12 \Leftrightarrow x(x+1) = 12 \Leftrightarrow x^2+x-12 = 0 \Leftrightarrow (x-3)(x+4) = 0 \Leftrightarrow$ $x = -4$ or 3. Since $\log(-4)$ is undefined, the only solution is $x = 3$.

67. $\log_2(1-x) = 4 \Leftrightarrow 1-x = 2^4 \Leftrightarrow x = 1-16 = -15$

69. $\log_3(x-8) + \log_3 x = 2 \Leftrightarrow \log_3(x(x-8)) = 2 \Leftrightarrow x(x-8) = 9 \Leftrightarrow x^2-8x-9 = 0 \Leftrightarrow (x-9)(x+1) = 0 \Leftrightarrow$ $x = -1$ or 9. We reject -1 because it does not satisfy the original equation, so the only solution is $x = 9$.

71. $5^{-2x/3} = 0.63 \Leftrightarrow \dfrac{-2x}{3}\log 5 = \log 0.63 \Leftrightarrow x = -\dfrac{3\log 0.63}{2\log 5} \approx 0.430618$

73. $5^{2x+1} = 3^{4x-1} \Leftrightarrow (2x+1)\log 5 = (4x-1)\log 3 \Leftrightarrow 2x\log 5 + \log 5 = 4x\log 3 - \log 3 \Leftrightarrow$ $x(2\log 5 - 4\log 3) = -\log 3 - \log 5 \Leftrightarrow x = \dfrac{\log 3 + \log 5}{4\log 3 - 2\log 5} \approx 2.303600$

75. $y = e^{x/(x+2)}$. Vertical asymptote $x = -2$, horizontal asymptote $y \approx 2.72$, no maximum or minimum.

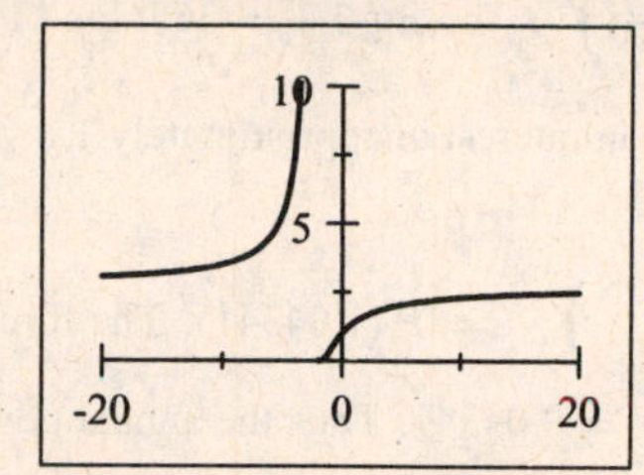

77. $y = \log\left(x^3 - x\right)$. Vertical asymptotes $x = -1$, $x = 0$, $x = 1$, no horizontal asymptote, local maximum of about -0.41 when $x \approx -0.58$.

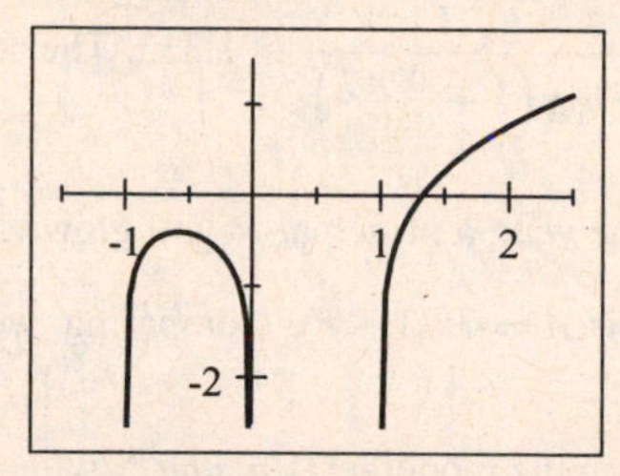

79. $3 \log x = 6 - 2x$. We graph $y = 3 \log x$ and $y = 6 - 2x$ in the same viewing rectangle. The solution occurs where the two graphs intersect. From the graphs, we see that the solution is $x \approx 2.42$.

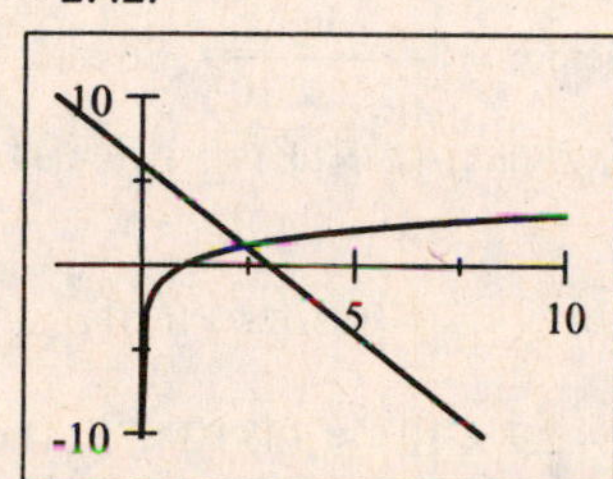

81. $\ln x > x - 2$.We graph the function $f(x) = \ln x - x + 2$, and we see that the graph lies above the x-axis for $0.16 < x < 3.15$. So the approximate solution of the given inequality is $0.16 < x < 3.15$.

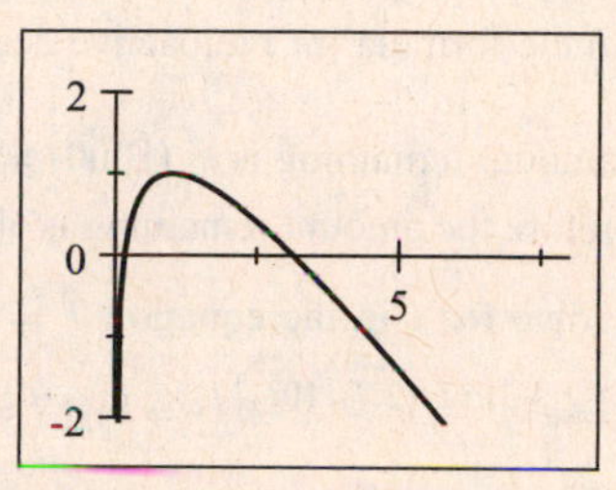

83. $f(x) = e^x - 3e^{-x} - 4x$. We graph the function $f(x)$, and we see that the function is increasing on $(-\infty, 0)$ and $(1.10, \infty)$ and that it is decreasing on $(0, 1.10)$.

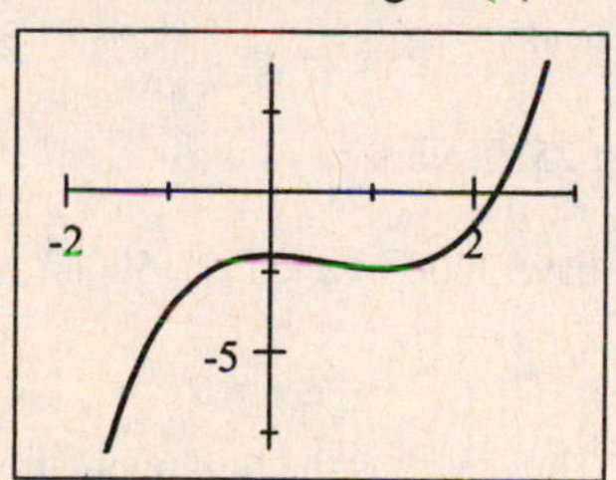

85. $\log_4 15 = \dfrac{\log 15}{\log 4} = 1.953445$

87. $\log_9 0.28 = \dfrac{\log 0.28}{\log 9} \approx -0.579352$

89. Notice that $\log_4 258 > \log_4 256 = \log_4 4^4 = 4$ and so $\log_4 258 > 4$. Also $\log_5 620 < \log_5 625 = \log_5 5^4 = 4$ and so $\log_5 620 < 4$. Then $\log_4 258 > 4 > \log_5 620$ and so $\log_4 258$ is larger.

91. $P = 12{,}000$, $r = 0.10$, and $t = 3$. Then $A = P\left(1 + \frac{r}{n}\right)^{nt}$.

(a) For $n = 2$, $A = 12{,}000\left(1 + \frac{0.10}{2}\right)^{2(3)} = 12{,}000\left(1.05^6\right) \approx \$16{,}081.15$.

(b) For $n = 12$, $A = 12{,}000\left(1 + \frac{0.10}{12}\right)^{12(3)} \approx \$16{,}178.18$.

(c) For $n = 365$, $A = 12{,}000\left(1 + \frac{0.10}{365}\right)^{365(3)} \approx \$16{,}197.64$.

(d) For $n = \infty$, $A = Pe^{rt} = 12{,}000e^{0.10(3)} \approx \$16{,}198.31$.

93. We use the formula $A = P\left(1 + \frac{r}{n}\right)^{nt}$ with $P = 100{,}000$, $r = 0.052$, $n = 365$, and $A = 100{,}000 + 10{,}000 = 110{,}000$, and solve for t: $110{,}000 = 100{,}000\left(1 + \frac{0.052}{365}\right)^{365t} \Leftrightarrow 1.1 = \left(1 + \frac{0.052}{365}\right)^{365t} \Leftrightarrow \log 1.1 = 365t \log\left(1 + \frac{0.052}{365}\right) \Leftrightarrow$ $t = \dfrac{\log 1.1}{365 \log\left(1 + \frac{0.052}{365}\right)} \approx 1.833$. The account will accumulate \$10,000 in interest in approximately 1.8 years.

95. After one year, a principal P will grow to the amount $A = P\left(1 + \dfrac{0.0425}{365}\right)^{365} = P(1.04341)$. The formula for simple interest is $A = P(1 + r)$. Comparing, we see that $1 + r = 1.04341$, so $r = 0.04341$. Thus the annual percentage yield is 4.341%.

97. **(a)** Using the model $n(t) = n_0 e^{rt}$, with $n_0 = 30$ and $r = 0.15$, we have the formula $n(t) = 30e^{0.15t}$.

(b) $n(4) = 30e^{0.15(4)} \approx 55$.

(c) $500 = 30e^{0.15t} \Leftrightarrow \frac{50}{3} = e^{0.15t} \Leftrightarrow 0.15t = \ln\left(\frac{50}{3}\right) \Leftrightarrow t = \dfrac{1}{0.15}\ln\left(\frac{50}{3}\right) \approx 18.76$. So the stray cat population will reach 500 in about 19 years.

99. **(a)** From the formula for radioactive decay, we have $m(t) = 10e^{-rt}$, where $r = -\dfrac{\ln 2}{2.7 \times 10^5}$. So after 1000 years the amount remaining is $m(1000) = 10 \cdot e^{[-\ln 2/(2.7\times 10^5)]\cdot 1000} = 10e^{-(\ln 2)/(2.7\times 10^2)} = 10e^{-(\ln 2)/270} \approx 9.97$. Therefore the amount remaining is about 9.97 mg.

(b) We solve for t in the equation $7 = 10e^{-[\ln 2/(2.7\times 10^5)]\cdot t}$. We have $7 = 10e^{-[\ln 2/(2.7\times 10^5)]\cdot t} \Leftrightarrow$ $0.7 = e^{-[\ln 2/(2.7\times 10^5)]\cdot t} \Leftrightarrow \ln 0.7 = -\dfrac{\ln 2}{2.7 \times 10^5} \cdot t \Leftrightarrow t = -\dfrac{\ln 0.7}{\ln 2} \cdot 2.7 \times 10^5 \approx 138{,}934.75$. Thus it takes about 139,000 years.

101. **(a)** From the formula for radioactive decay, $r = \dfrac{\ln 2}{1590} \approx 0.0004359$ and $n(t) = 150 \cdot e^{-0.0004359t}$.

(b) $n(1000) = 150 \cdot e^{-0.0004359\cdot 1000} \approx 97.00$, and so the amount remaining is about 97.00 mg.

(c) Find t so that $50 = 150 \cdot e^{-0.0004359t}$. We have $50 = 150 \cdot e^{-0.0004359t} \Leftrightarrow \frac{1}{3} = e^{-0.0004359t} \Leftrightarrow$ $t = -\dfrac{1}{0.0004359}\ln\left(\frac{1}{3}\right) \approx 2520$. Thus only 50 mg remain after about 2520 years.

103. **(a)** Using $n_0 = 1500$ and $n(5) = 3200$ in the formula $n(t) = n_0 e^{rt}$, we have $3200 = n(5) = 1500e^{5r} \Leftrightarrow e^{5r} = \frac{32}{15} \Leftrightarrow$ $5r = \ln\left(\frac{32}{15}\right) \Leftrightarrow r = \frac{1}{5}\ln\left(\frac{32}{15}\right) \approx 0.1515$. Thus $n(t) = 1500 \cdot e^{0.1515t}$.

(b) We have $t = 2020 - 2009 = 11$ so $n(11) = 1500e^{0.1515\cdot 11} \approx 7940$. Thus in 2009 the bird population should be about 7940.

105. $[\mathrm{H}^+] = 1.3 \times 10^{-8}$ M. Then $\mathrm{pH} = -\log[\mathrm{H}^+] = -\log\left(1.3 \times 10^{-8}\right) \approx 7.9$, and so fresh egg whites are basic.

107. Let I_0 be the intensity of the smaller earthquake and I_1 be the intensity of the larger earthquake. Then $I_1 = 35I_0$. Since $M = \log\left(\dfrac{I}{S}\right)$, we have $M_0 = \log\left(\dfrac{I_0}{S}\right) = 6.5$ and

$M_1 = \log\left(\dfrac{I_1}{S}\right) = \log\left(\dfrac{35I_0}{S}\right) = \log 35 + \log\left(\dfrac{I_0}{S}\right) = \log 35 + M_0 = \log 35 + 6.5 \approx 8.04$. So the magnitude on the Richter scale of the larger earthquake is approximately 8.0.

CHAPTER 4 TEST

1. (a)

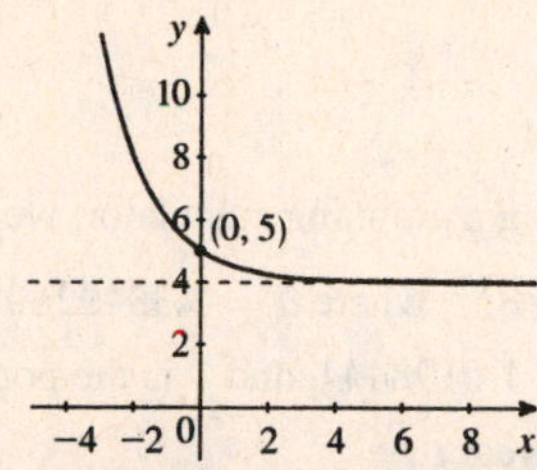

$f(x) = 2^{-x} + 4$ has domain $(-\infty, \infty)$, range $(4, \infty)$, and horizontal asymptote $y = 4$.

(b)

$g(x) = \log_3(x+3)$ has domain $(-3, \infty)$, range $(-\infty, \infty)$, and vertical asymptote $x = -3$.

3. (a) $6^{2x} = 25 \Leftrightarrow \log_6 6^{2x} = \log_6 25 \Leftrightarrow 2x = \log_6 25$

(b) $\ln A = 3 \Leftrightarrow e^{\ln A} = e^3 \Leftrightarrow A = e^3$

5. (a) $\log\left(\dfrac{xy^3}{z^2}\right) = \log x + \log y^3 - \log z^2 = \log x + 3\log y - 2\log z$

(b) $\ln\sqrt{\dfrac{x}{y}} = \ln\left(\left(\dfrac{x}{y}\right)^{1/2}\right) = \dfrac{1}{2}\ln\left(\dfrac{x}{y}\right) = \frac{1}{2}\ln x - \frac{1}{2}\ln y$

(c) $\log\sqrt[3]{\dfrac{x+2}{x^4(x^2+4)}} = \dfrac{1}{3}\log\left(\dfrac{x+2}{x^4(x^2+4)}\right) = \frac{1}{3}\left[\log(x+2) - \left(4\log x + \log\left(x^2+4\right)\right)\right]$

$= \frac{1}{3}\log(x+2) - \frac{4}{3}\log x - \frac{1}{3}\log\left(x^2+4\right)$

7. (a) $3^{4x} = 3^{100} \Leftrightarrow 4x = 100 \Leftrightarrow x = 25$

(b) $e^{3x-2} = e^{x^2} \Leftrightarrow 3x - 2 = x^2 \Leftrightarrow x^2 - 3x + 2 = 0 \Leftrightarrow (x-1)(x-2) = 0 \Leftrightarrow x = 1$ or $x = 2$

(c) $5^{x/10} + 1 = 7 \Leftrightarrow 5^{x/10} = 6 \Leftrightarrow \log_5 5^{x/10} = \log_5 6 \Leftrightarrow x/10 = \log_5 6 \Leftrightarrow x = 10\log_5 6 \approx 11.13$

(d) $10^{x+3} = 6^{2x} \Leftrightarrow \log 10^{x+3} = \log 6^{2x} \Leftrightarrow x + 3 = \dfrac{\log_6 6^{2x}}{\log_6 10} \Leftrightarrow (\log_6 10)(x+3) = 2x \Leftrightarrow (2 - \log_6 10)\,x = 3\log_6 10$

$\Leftrightarrow x = \dfrac{3\log_6 10}{2 - \log_6 10} \approx 5.39$

9. Using the Change of Base Formula, we have $\log_{12} 27 = \dfrac{\log 27}{\log 12} \approx 1.326$.

11. (a) $A(t) = 12{,}000\left(1 + \frac{0.056}{12}\right)^{12t}$, where t is in years.

(b) $A(t) = 12{,}000\left(1 + \frac{0.056}{365}\right)^{365t}$. So $A(3) = 12{,}000\left(1 + \dfrac{0.056}{365}\right)^{365(3)} = \$14{,}195.06$.

(c) $A(t) = 12{,}000e^{0.056t}$. So $20{,}000 = 12{,}000e^{0.056t} \Leftrightarrow 5 = 3e^{0.056t} \Leftrightarrow \ln 5 = \ln\left(3e^{0.056t}\right) \Leftrightarrow \ln 5 = \ln 3 + 0.056t \Leftrightarrow$

$t = \frac{1}{0.056}(\ln 5 - \ln 3) \approx 9.12$. Thus, the amount will grow to \$20,000 in approximately 9.12 years.

13. Let the subscripts J and P represent the two earthquakes. Then we have $M_J = \log\left(\dfrac{I_J}{S}\right) = 6.4 \Leftrightarrow 10^{6.4} = \dfrac{I_J}{S} \Leftrightarrow$

$10^{6.4}S = I_J$. Similarly, $M_P = \log\left(\dfrac{I_P}{S}\right) = 3.1 \Leftrightarrow 10^{3.1} = \dfrac{I_P}{S} \Leftrightarrow 10^{3.1}S = I_P$. So $\dfrac{I_J}{I_P} = \dfrac{10^{6.4}S}{10^{3.1}S} = 10^{3.3} \approx 1995.3$,

and so the Japan earthquake was about 1995 times more intense than the Pennsylvania earthquake.

FOCUS ON MODELING Fitting Exponential and Power Curves to Data

1. (a)

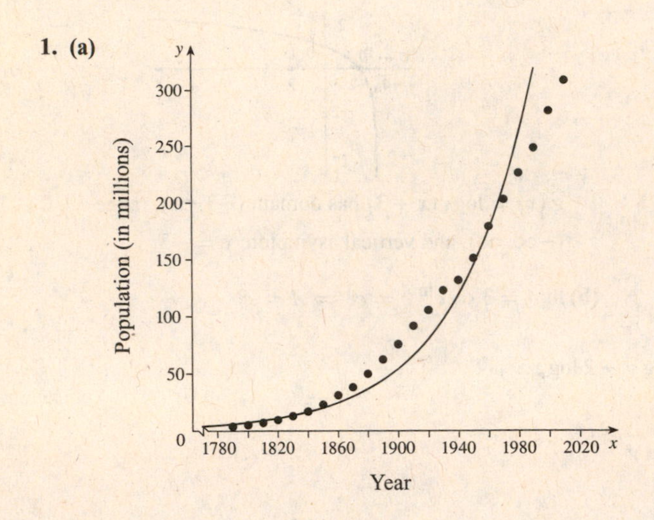

(b) Using a graphing calculator, we obtain the model $y = ab^t$, where $a = 3.3349260 \times 10^{-15}$ and $b = 1.0198444$, and y is the population (in millions) in the year t.

(c) Substituting $t = 2020$ into the model of part (b), we get $y = ab^{2020} \approx 577.6$ million.

(d) According to the model, the population in 1965 was $y = ab^{1965} \approx 196.0$ million.

3. (a) Yes.

(b)

Year t	Health Expenditures E (\$bn)	$\ln E$
1970	74.3	4.30811
1980	255.8	5.54440
1985	444.6	6.09718
1987	519.1	6.25210
1990	724.3	6.58521
1992	857.9	6.75449
1994	972.7	6.88008
1996	1081.8	6.98638
1998	1208.9	7.09747
2000	1377.2	7.22781
2002	1638.0	7.40123
2005	2035.4	7.61845
2008	2411.7	7.78809
2010	2599.0	7.86288
2012	2793.4	7.93501

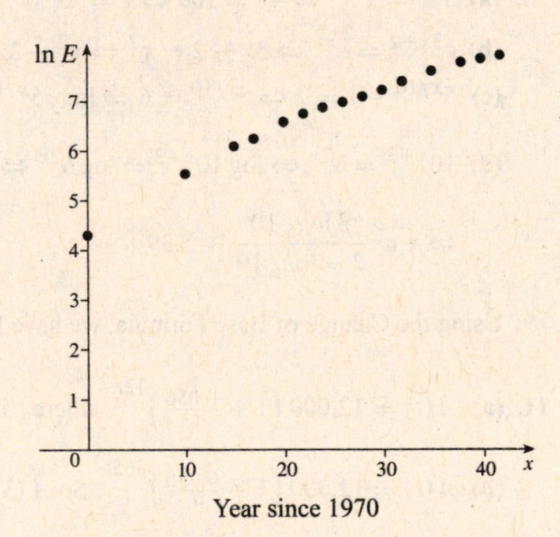

Yes, the scatter plot appears to be roughly linear.

(c) Let t be the number of years elapsed since 1970 . Then $\ln E = 4.7473926 + 0.08213193t$, where E is expenditure in billions of dollars.

(d) $E \approx e^{4.7473926+0.08213193t} \approx 115.28330e^{0.08213193t}$

(e) In 2020 we have $t = 2020 - 1970 = 50$, so the estimated 2020 health-care expenditures are $115.28330e^{0.08213193(50)} \approx 7002.3$ billion dollars.

5. (a) Using a graphing calculator, we find that $I_0 = 22.7586444$ and $k = 0.1062398$.

(b)

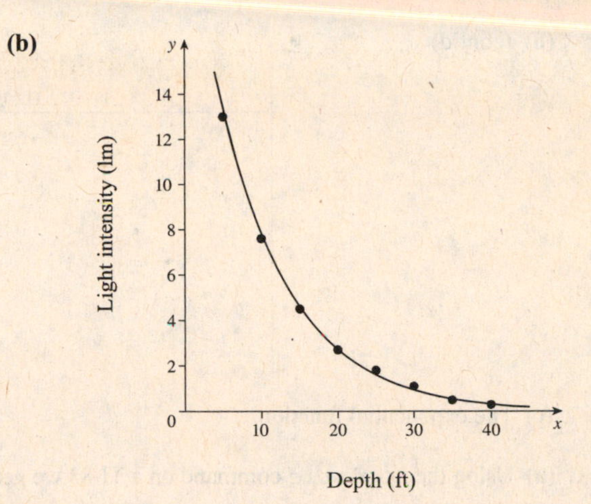

(c) We solve $0.15 = 22.7586444e^{-0.1062398x}$ for x:
$0.15 = 22.7586444e^{-0.1062398x} \Leftrightarrow$
$0.006590902 = e^{-0.1062398x} \Leftrightarrow$
$-5.022065 = -0.1062398x \Leftrightarrow x \approx 47.27$. So light intensity drops below 0.15 lumens below around 47.27 feet.

7. (a) Let A be the area of the cave and S the number of species of bat. Using a graphing calculator, we obtain the power function model $S = 0.14A^{0.64}$.

(b)

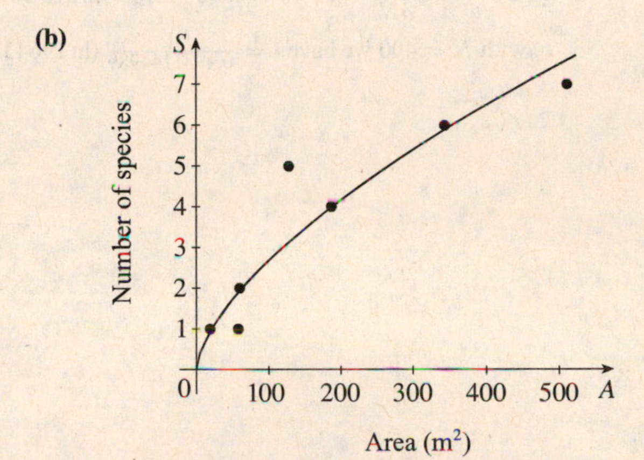

The model fits the data reasonably well.

(c) According to the model, there are $S = 0.14\,(205)^{0.64} \approx 4$ species of bat living in the El Sapo cave.

9. (a)

(b)

x	y	$\ln x$	$\ln y$
2	0.08	0.69315	−2.52573
4	0.12	1.38629	−2.12026
6	0.18	1.79176	−1.71480
8	0.25	2.07944	−1.38629
10	0.36	2.30259	−1.02165
12	0.52	2.48491	−0.65393
14	0.73	2.63906	−0.31471
16	1.06	2.77259	0.05827

(b) (cont'd)

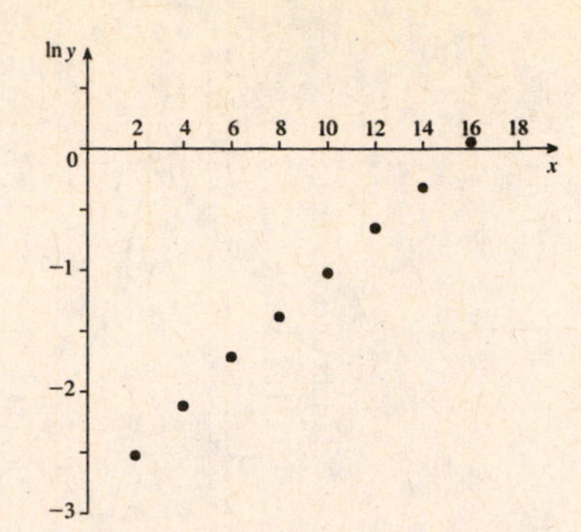

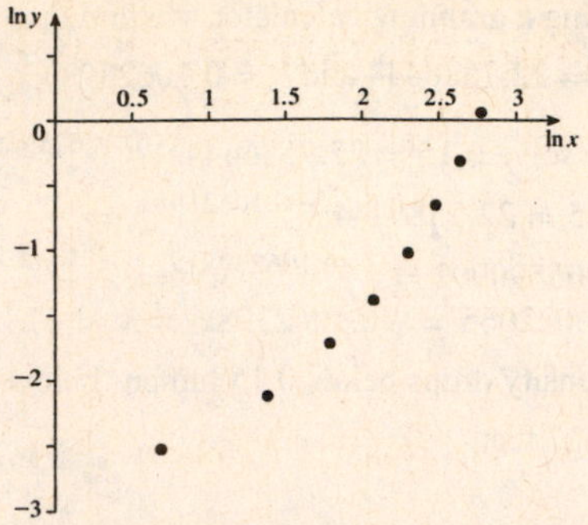

(c) The exponential function.

(d) $y = a \cdot b^x$ where $a = 0.057697$ and $b = 1.200236$.

11. **(a)** Using the `Logistic` command on a TI-83 we get $y = \dfrac{c}{1 + ae^{-bx}}$ where $a = 49.10976596$, $b = 0.4981144989$, and $c = 500.855793$.

(b) Using the model $N = \dfrac{c}{1 + ae^{-bt}}$ we solve for t. So $N = \dfrac{c}{1 + ae^{-bt}} \Leftrightarrow 1 + ae^{-bt} = \dfrac{c}{N} \Leftrightarrow ae^{-bt} = \left(\dfrac{c}{N}\right) - 1 = \dfrac{c - N}{N}$ $\Leftrightarrow e^{-bt} = \dfrac{c - N}{aN} \Leftrightarrow -bt = \ln(c - N) - \ln aN \Leftrightarrow t = \dfrac{1}{b}[\ln aN - \ln(c - N)]$. Substituting the values for a, b, and c, with $N = 400$ we have $t = \frac{1}{0.4981144989}(\ln 19643.90638 - \ln 100.855793) \approx 10.58$ days.

5 TRIGONOMETRIC FUNCTIONS: RIGHT TRIANGLE APPROACH

5.1 ANGLE MEASURE

1. (a) The radian measure of an angle θ is the length of the *arc* that subtends the angle in a circle of radius 1.

(b) To convert degrees to radians we multiply by $\frac{\pi}{180}$.

(c) To convert radians to degrees we multiply by $\frac{180}{\pi}$.

3. (a) The angular speed of the point is $\omega = \dfrac{\theta}{t}$.

(b) The linear speed of the point is $v = \dfrac{s}{t}$.

(c) The linear speed v and the angular speed ome are related by the equation $v = r\omega$.

5. $15° = 15° \cdot \frac{\pi}{180°}\text{ rad} = \frac{\pi}{12}\text{ rad} \approx 0.262\text{ rad}$

7. $54° = 54° \cdot \frac{\pi}{180°}\text{ rad} = \frac{3\pi}{10}\text{ rad} \approx 0.942\text{ rad}$

9. $-45° = -45° \cdot \frac{\pi}{180°}\text{ rad} = -\frac{\pi}{4}\text{ rad} \approx -0.785\text{ rad}$

11. $100° = 100° \cdot \frac{\pi}{180°}\text{ rad} = \frac{5\pi}{9}\text{ rad} \approx 1.745\text{ rad}$

13. $1000° = 1000° \cdot \frac{\pi}{180°}\text{ rad} = \frac{50\pi}{9}\text{ rad} \approx 17.453\text{ rad}$

15. $-70° = -70° \cdot \frac{\pi}{180°}\text{ rad} = -\frac{7\pi}{18}\text{ rad} \approx -1.222\text{ rad}$

17. $\frac{5\pi}{3} = \frac{5\pi}{3} \cdot \frac{180°}{\pi} = 300°$

19. $\frac{5\pi}{6} = \frac{5\pi}{6} \cdot \frac{180°}{\pi} = 150°$

21. $3 = 3 \cdot \frac{180°}{\pi} = \frac{540°}{\pi} \approx 171.9°$

23. $-1.2 = -1.2 \cdot \frac{180°}{\pi} = -\frac{216°}{\pi} = -68.8°$

25. $\frac{\pi}{10} = \frac{\pi}{10} \cdot \frac{180°}{\pi} = 18°$

27. $-\frac{2\pi}{15} = -\frac{2\pi}{15} \cdot \frac{180°}{\pi} = -24°$

29. $50°$ is coterminal with $50° + 360° = 410°$, $50° + 720° = 770°$, $50° - 360° = -310°$, and $50° - 720° = -670°$. (Other answers are possible.)

31. $\frac{3\pi}{4}$ is coterminal with $\frac{3\pi}{4} + 2\pi = \frac{11\pi}{4}$, $\frac{3\pi}{4} + 4\pi = \frac{19\pi}{4}$, $\frac{3\pi}{4} - 2\pi = -\frac{5\pi}{4}$, and $\frac{3\pi}{4} - 4\pi = -\frac{13\pi}{4}$. (Other answers are possible.)

33. $-\frac{\pi}{4}$ is coterminal with $-\frac{\pi}{4} + 2\pi = \frac{7\pi}{4}$, $-\frac{\pi}{4} + 4\pi = \frac{15\pi}{4}$, $-\frac{\pi}{4} - 2\pi = -\frac{9\pi}{4}$, and $-\frac{\pi}{4} - 4\pi = -\frac{17\pi}{4}$. (Other answers are possible.)

35. Since $430° - 70° = 360°$, the angles are coterminal.

37. Since $\frac{17\pi}{6} - \frac{5\pi}{6} = \frac{12\pi}{6} = 2\pi$; the angles are coterminal.

39. Since $875° - 155° = 720° = 2 \cdot 360°$, the angles are coterminal.

41. Since $400° - 360° = 40°$, the angles $400°$ and $40°$ are coterminal.

43. Since $780° - 2 \cdot 360° = 60°$, the angles $780°$ and $60°$ are coterminal.

45. Since $-800° + 3 \cdot 360° = 280°$, the angles $-800°$ and $280°$ are coterminal.

47. Since $\frac{19\pi}{6} - 2\pi = \frac{7\pi}{6}$, the angles $\frac{19\pi}{6}$ and $\frac{7\pi}{6}$ are coterminal.

49. Since $25\pi - 12 \cdot 2\pi = \pi$, the angles 25π and π are coterminal.

51. Since $\frac{17\pi}{4} - 2 \cdot 2\pi = \frac{\pi}{4}$, the angles $\frac{17\pi}{4}$ and $\frac{\pi}{4}$ are coterminal.

53. Using the formula $s = \theta r$, $s = \frac{5\pi}{6} \cdot 9 = \frac{15\pi}{2}$.

55. $\theta = \dfrac{s}{r} = \dfrac{10}{5} = 2\text{ rad} = 2 \cdot \dfrac{180°}{\pi} \approx 114.6°$

57. Solving for s, we have $s = r\theta$, so the length of the arc is $5 \cdot 3 = 15$ cm.

59. Solving for θ, we have $\theta = \dfrac{s}{r} = \dfrac{14}{9}\text{ rad} = \dfrac{14}{9} \cdot \dfrac{180°}{\pi} \approx 89.1°$.

61. $r = \dfrac{s}{\theta} = \dfrac{15}{5\pi/6} = \dfrac{18}{\pi} \approx 5.73$ m

63. **(a)** $A = \frac{1}{2}r^2\theta = \frac{1}{2} \cdot 8^2 \cdot 80° \cdot \frac{\pi}{180°} = 32 \cdot \frac{4\pi}{9} = \frac{128\pi}{9} \approx 44.68$

(b) $A = \frac{1}{2}r^2\theta = \frac{1}{2} \cdot 10^2 \cdot 0.5 = 25$

65. $\theta = \frac{2\pi}{3}$ rad and $r = 10$ m, so $A = \frac{1}{2}r^2\theta = \frac{1}{2}(10)^2\left(\frac{2\pi}{3}\right) = \frac{100\pi}{3} \approx 104.7\text{ m}^2$.

67. $A = 70\text{ m}^2$ and $\theta = 140° = 140° \cdot \dfrac{\pi}{180°} = \dfrac{7\pi}{9}$ rad. Thus, $A = \frac{1}{2}r^2\theta \Leftrightarrow 70 = \dfrac{1}{2} \cdot \dfrac{7\pi}{9}r^2 \Rightarrow$
$r = \sqrt{2 \cdot 70 \cdot \dfrac{9}{7\pi}} = \dfrac{6\sqrt{5\pi}}{\pi} \approx 7.6$ m.

69. $r = 80$ mi and $A = 1600\text{ mi}^2$, so $A = \frac{1}{2}r^2\theta \Leftrightarrow 1600 = \frac{1}{2}(80)^2\theta \Leftrightarrow \theta = \frac{1}{2}$ rad.

71. Referring to the figure, we have $AC = 3 + 1 = 4$, $BC = 1 + 2 = 3$, and $AB = 2 + 3 = 5$. Since $AB^2 = AC^2 + BC^2$, then by the Pythagorean Theorem, the triangle is a right triangle. Therefore, $\theta = \frac{\pi}{2}$ and $A = \frac{1}{2}r^2\theta = \frac{1}{2} \cdot 1^2 \cdot \frac{\pi}{2} = \frac{\pi}{4}\text{ ft}^2$.

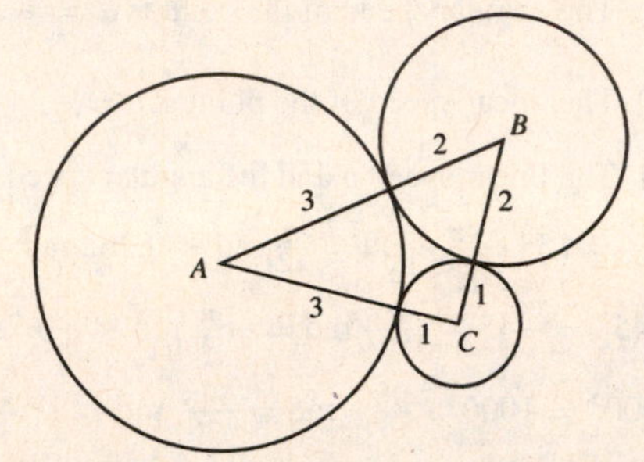

73. Between 1:00 P.M. and 1:00 P.M., the minute hand traverses three-quarters of a complete revolution, or $\frac{3}{4}(2\pi) = \frac{3\pi}{2}$ rad, while the hour hand moves three-quarters of the way from 12 to 1, which is itself one-twelfth of a revolution. So the hour hand traverses $\frac{3}{4}\left(\frac{1}{12}\right)(2\pi) = \frac{\pi}{8}$ rad.

75. The circumference of each wheel is $\pi d = 28\pi$ in. If the wheels revolve 10,000 times, the distance traveled is $10{,}000 \cdot 28\pi\text{ in.} \cdot \dfrac{1\text{ ft}}{12\text{ in.}} \cdot \dfrac{1\text{ mi}}{5280\text{ ft}} \approx 13.88$ mi.

77. We find the measure of the angle in degrees and then convert to radians. $\theta = 40.5° - 25.5° = 15°$ and $15 \cdot \frac{\pi}{180°}$ rad $= \frac{\pi}{12}$ rad. Then using the formula $s = \theta r$, we have $s = \frac{\pi}{12} \cdot 3960 = 330\pi \approx 1036.725$ and so the distance between the two cities is roughly 1037 mi.

79. In one day, the earth travels $\frac{1}{365}$ of its orbit which is $\frac{2\pi}{365}$ rad. Then $s = \theta r = \frac{2\pi}{365} \cdot 93{,}000{,}000 \approx 1{,}600{,}911.3$, so the distance traveled is approximately 1.6 million miles.

81. The central angle is 1 minute $= \left(\frac{1}{60}\right)° = \frac{1}{60} \cdot \frac{\pi}{180°}$ rad $= \frac{\pi}{10{,}800}$ rad. Then $s = \theta r = \frac{\pi}{10{,}800} \cdot 3960 \approx 1.152$, and so a nautical mile is approximately 1.152 mi.

83. The area is equal to the area of the large sector (with radius 34 in.) minus the area of the small sector (with radius 14 in.) Thus, $A = \frac{1}{2}r_1^2\theta - \frac{1}{2}r_2^2\theta = \frac{1}{2}\left(34^2 - 14^2\right)\left(135° \cdot \frac{\pi}{180°}\right) \approx 1131\text{ in.}^2$.

85. **(a)** The angular speed is $\omega = \dfrac{45 \cdot 2\pi\text{ rad}}{1\text{ min}} = 90\pi$ rad/min.

(b) The linear speed is $v = \dfrac{45 \cdot 2\pi \cdot 16}{1} = 1440\pi$ in./min ≈ 4523.9 in./min.

87. $v = \dfrac{8 \cdot 2\pi \cdot 2}{15} = \dfrac{32\pi}{15} \approx 6.702$ ft/s.

89. 23 h 56 min 4 s = 23.9344 hr. So the linear speed is $\dfrac{1 \cdot 2\pi \cdot (3960)}{1\text{ day}} \cdot \dfrac{1\text{ day}}{23.9344\text{ hr}} \approx 1039.57$ mi/h.

91. $v = \dfrac{100 \cdot 2\pi \cdot 0.20\text{ m}}{60\text{ s}} = \dfrac{2\pi}{3} \approx 2.09$ m/s.

93. **(a)** The circumference of the opening is the length of the arc subtended by the angle θ on the flat piece of paper, that is, $C = s = r\theta = 6 \cdot \frac{5\pi}{3} = 10\pi \approx 31.4$ cm.

(b) Solving for r, we find $r = \dfrac{C}{2\pi} = \dfrac{10\pi}{2\pi} = 5$ cm.

(c) By the Pythagorean Theorem, $h^2 = 6^2 - 5^2 = 11$, so $h = \sqrt{11} \approx 3.3$ cm.

(d) The volume of a cone is $V = \frac{1}{3}\pi r^2 h$. In this case $V = \frac{1}{3}\pi \cdot 5^2 \cdot \sqrt{11} \approx 86.8\text{ cm}^3$.

95. Answers will vary, although of course everybody prefers radians.

5.2 TRIGONOMETRY OF RIGHT TRIANGLES

1. **(a)**

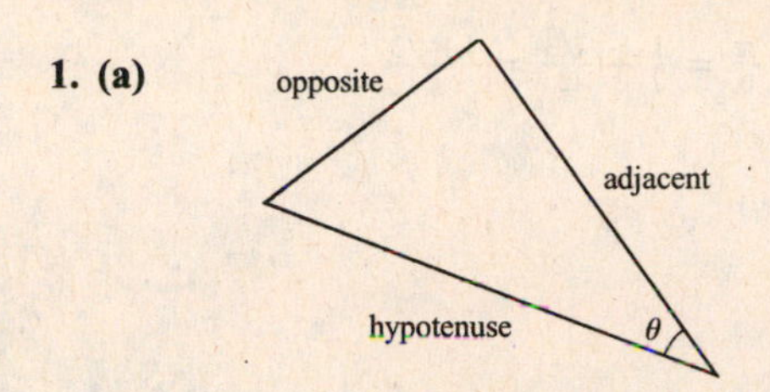

(b) $\sin\theta = \dfrac{\text{opposite}}{\text{hypotenuse}}$, $\cos\theta = \dfrac{\text{adjacent}}{\text{hypotenuse}}$, and $\tan\theta = \dfrac{\text{opposite}}{\text{adjacent}}$.

(c) The trigonometric ratios do not depend on the size of the triangle because all right triangles with angle θ are *similar*.

3. $\sin\theta = \frac{4}{5}$, $\cos\theta = \frac{3}{5}$, $\tan\theta = \frac{4}{3}$, $\csc\theta = \frac{5}{4}$, $\sec\theta = \frac{5}{3}$, $\cot\theta = \frac{3}{4}$

5. The remaining side is obtained by the Pythagorean Theorem: $\sqrt{41^2 - 40^2} = \sqrt{81} = 9$. Then $\sin\theta = \frac{40}{41}$, $\cos\theta = \frac{9}{41}$, $\tan\theta = \frac{40}{9}$, $\csc\theta = \frac{41}{40}$, $\sec\theta = \frac{41}{9}$, $\cot\theta = \frac{9}{40}$

7. The remaining side is obtained by the Pythagorean Theorem: $\sqrt{3^2 + 2^2} = \sqrt{13}$. Then $\sin\theta = \frac{2}{\sqrt{13}} = \frac{2\sqrt{13}}{13}$, $\cos\theta = \frac{3}{\sqrt{13}} = \frac{3\sqrt{13}}{13}$, $\tan\theta = \frac{2}{3}$, $\csc\theta = \frac{\sqrt{13}}{2}$, $\sec\theta = \frac{\sqrt{13}}{3}$, $\cot\theta = \frac{3}{2}$

9. $c = \sqrt{5^2 + 3^2} = \sqrt{34}$

(a) $\sin\alpha = \cos\beta = \frac{3}{\sqrt{34}} = \frac{3\sqrt{34}}{34}$

(b) $\tan\alpha = \cot\beta = \frac{3}{5}$

(c) $\sec\alpha = \csc\beta = \frac{\sqrt{34}}{5}$

11. **(a)** $\sin 22^\circ \approx 0.37461$

(b) $\cot 23^\circ \approx 2.35585$

13. **(a)** $\sec 13^\circ \approx 1.02630$

(b) $\tan 51^\circ \approx 1.23490$

15. Since $\sin 30^\circ = \dfrac{x}{25}$, we have $x = 25\sin 30^\circ = 25 \cdot \frac{1}{2} = \frac{25}{2}$.

17. Since $\sin 60^\circ = \dfrac{x}{13}$, we have $x = 13\sin 60^\circ = 13 \cdot \frac{\sqrt{3}}{2} = \frac{13\sqrt{3}}{2}$.

19. Since $\tan 36^\circ = \dfrac{12}{x}$, we have $x = \dfrac{12}{\tan 36^\circ} \approx 16.51658$.

21. $\dfrac{x}{28} = \cos\theta \Leftrightarrow x = 28\cos\theta$, and $\dfrac{y}{28} = \sin\theta \Leftrightarrow y = 28\sin\theta$.

23. $\tan\theta = \frac{5}{6}$. Then the third side is $x = \sqrt{5^2+6^2} = \sqrt{61}$. The other five ratios are $\sin\theta = \frac{5\sqrt{61}}{61}$, $\cos\theta = \frac{6\sqrt{61}}{61}$, $\csc\theta = \frac{\sqrt{61}}{5}$, $\sec\theta = \frac{\sqrt{61}}{6}$, and $\cot\theta = \frac{6}{5}$.

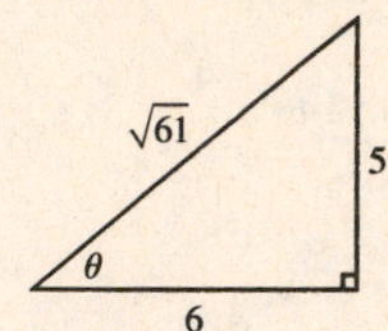

25. $\cot\theta = 1$. Then the third side is $r = \sqrt{1^2+1^2} = \sqrt{2}$. The other five ratios are $\sin\theta = \frac{1}{\sqrt{2}} = \frac{\sqrt{2}}{2}$, $\cos\theta = \frac{1}{\sqrt{2}} = \frac{\sqrt{2}}{2}$, $\tan\theta = 1$, $\csc\theta = \sqrt{2}$, and $\sec\theta = \sqrt{2}$.

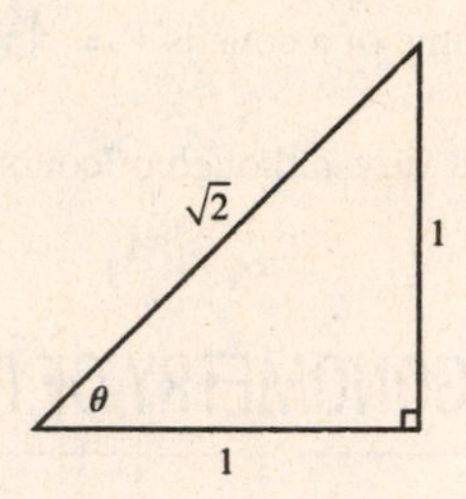

27. $\csc\theta = \frac{11}{6}$. The third side is $x = \sqrt{11^2-6^2} = \sqrt{85}$. The other five ratios are $\sin\theta = \frac{6}{11}$, $\cos\theta = \frac{\sqrt{85}}{11}$, $\tan\theta = \frac{6\sqrt{85}}{85}$, $\sec\theta = \frac{11\sqrt{85}}{85}$, and $\cot\theta = \frac{\sqrt{85}}{6}$.

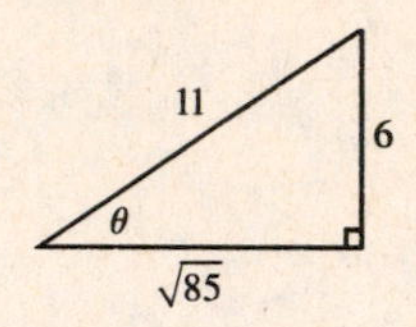

29. $\sin\frac{\pi}{6} + \cos\frac{\pi}{6} = \frac{1}{2} + \frac{\sqrt{3}}{2} = \frac{1+\sqrt{3}}{2}$

31. $\sin 30^\circ \cos 60^\circ + \sin 60^\circ \cos 30^\circ = \frac{1}{2}\cdot\frac{1}{2} + \frac{\sqrt{3}}{2}\cdot\frac{\sqrt{3}}{2} = \frac{1}{4} + \frac{3}{4} = 1$

33. $(\cos 30^\circ)^2 - (\sin 30^\circ)^2 = \left(\frac{\sqrt{3}}{2}\right)^2 - \left(\frac{1}{2}\right)^2 = \frac{3}{4} - \frac{1}{4} = \frac{1}{2}$

35. $\left(\cos\frac{\pi}{4} + \sin\frac{\pi}{6}\right)^2 = \left(\frac{\sqrt{2}}{2} + \frac{1}{2}\right)^2 = \frac{1}{2} + \frac{1}{4} + 2\left(\frac{1}{2}\right)\left(\frac{\sqrt{2}}{2}\right) = \frac{3}{4} + \frac{\sqrt{2}}{2}$

37. This is an isosceles right triangle, so the other leg has length $16\tan 45^\circ = 16$, the hypotenuse has length $\dfrac{16}{\sin 45^\circ} = 16\sqrt{2} \approx 22.63$, and the other angle is $90^\circ - 45^\circ = 45^\circ$.

39. The other leg has length $35\tan 52^\circ \approx 44.80$, the hypotenuse has length $\dfrac{35}{\cos 52^\circ} \approx 56.85$, and the other angle is $90^\circ - 52^\circ = 38^\circ$.

41. The adjacent leg has length $33.5\cos\frac{\pi}{8} \approx 30.95$, the opposite leg has length $33.5\sin\frac{\pi}{8} \approx 12.82$, and the other angle is $\frac{\pi}{2} - \frac{\pi}{8} = \frac{3\pi}{8}$.

43. The adjacent leg has length $\dfrac{106}{\tan\frac{\pi}{5}} \approx 145.90$, the hypotenuse has length $\dfrac{106}{\sin\frac{\pi}{5}} \approx 180.34$, and the other angle is $\frac{\pi}{2} - \frac{\pi}{5} = \frac{3\pi}{10}$.

45. $\sin\theta \approx \frac{1}{2.24} \approx 0.45$. $\cos\theta \approx \frac{2}{2.24} \approx 0.89$, $\tan\theta = \frac{1}{2}$, $\csc\theta \approx 2.24$, $\sec\theta \approx \frac{2.24}{2} \approx 1.12$, $\cot\theta \approx 2.00$.

47. $x = \dfrac{100}{\tan 60^\circ} + \dfrac{100}{\tan 30^\circ} \approx 230.9$

49. Let h be the length of the shared side. Then $\sin 60^\circ = \dfrac{50}{h} \Leftrightarrow h = \dfrac{50}{\sin 60^\circ} \approx 57.735 \Leftrightarrow \sin 65^\circ = \dfrac{h}{x} \Leftrightarrow x = \dfrac{h}{\sin 65^\circ} \approx 63.7$

51. 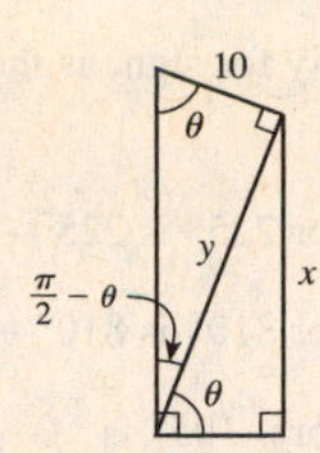

From the diagram, $\sin\theta = \frac{x}{y}$ and $\tan\theta = \frac{y}{10}$, so $x = y\sin\theta = 10\sin\theta\tan\theta$.

53. Let h be the height, in feet, of the Empire State Building. Then $\tan 11° = \frac{h}{5280} \Leftrightarrow h = 5280 \cdot \tan 11° \approx 1026$ ft.

55. **(a)** Let h be the distance, in miles, that the beam has diverged. Then $\tan 0.5° = \frac{h}{240{,}000} \Leftrightarrow$ $h = 240{,}000 \cdot \tan 0.5° \approx 2100$ mi.

(b) Since the deflection is about 2100 mi whereas the radius of the moon is about 1000 mi, the beam will not strike the moon.

57. Let h represent the height, in feet, that the ladder reaches on the building. Then $\sin 72° = \frac{h}{20} \Leftrightarrow h = 20\sin 72° \approx 19$ ft.

59. Let h be the height, in feet, of the kite above the ground. Then $\sin 50° = \frac{h}{450} \Leftrightarrow h = 450\sin 50° \approx 345$ ft.

61. Let h_1 be the height of the window in feet and h_2 be the height from the window to the top of the tower. Then $\tan 25° = \frac{h_1}{325}$ $\Leftrightarrow h_1 = 325 \cdot \tan 25° \approx 152$ ft. Also, $\tan 39° = \frac{h_2}{325} \Leftrightarrow h_2 = 325 \cdot \tan 39° \approx 263$ ft. Therefore, the height of the window is approximately 152 ft and the height of the tower is approximately $152 + 263 = 415$ ft.

63. Let d_1 be the distance, in feet, between a point directly below the plane and one car, and d_2 be the distance, in feet, between the same point and the other car. Then $\tan 52° = \frac{d_1}{5150} \Leftrightarrow d_1 = 5150 \cdot \tan 52° \approx 6591.7$ ft. Also, $\tan 38° = \frac{d_2}{5150} \Leftrightarrow$ $d_2 = 5150 \cdot \tan 38° \approx 4023.6$ ft. So in this case, the distance between the two cars is about 2570 ft.

65. Let x be the horizontal distance, in feet, between a point on the ground directly below the top of the mountain and the point on the plain closest to the mountain.. Let h be the height, in feet, of the mountain. Then $\tan 35° = \frac{h}{x}$ and $\tan 32° = \frac{h}{x+1000}$. So $h = x\tan 35° = (x+1000)\tan 32° \Leftrightarrow x = \frac{1000 \cdot \tan 32°}{\tan 35° - \tan 32°} \approx 8294.2$. Thus $h \approx 8294.2 \cdot \tan 35° \approx 5808$ ft.

67. Let d be the distance, in miles, from the earth to the sun. Then $\sec 89.85° = \frac{d}{240{,}000} \Leftrightarrow$ $d = 240{,}000 \cdot \sec 89.85° \approx 91.7$ million miles.

69. Let r represent the radius, in miles, of the earth. Then $\sin 60.276° = \frac{r}{r+600} \Leftrightarrow (r+600)\sin 60.276° = r \Leftrightarrow$ $600\sin 60.276° = r(1 - \sin 60.276°) \Leftrightarrow r = \frac{600\sin 60.276°}{1-\sin 60.276°} \approx 3960.099$. So the earth's radius is about 3960 mi.

71. Let d be the distance, in AU, between Venus and the sun. Then $\sin 46.3° = \frac{d}{1} = d$, so $d = \sin 46.3° \approx 0.723$ AU.

5.3 TRIGONOMETRIC FUNCTIONS OF ANGLES

1. If the angle θ is in standard position and $P(x, y)$ is a point on the terminal side of θ, and r is the distance from the origin to P, then $\sin\theta = \frac{y}{r}$, $\cos\theta = \frac{x}{r}$, and $\tan\theta = \frac{y}{x}$.

3. **(a)** If θ is in standard position, then the reference angle $\overline{\theta}$ is the acute angle formed by the terminal side of θ and the *x-axis*. So the reference angle for $\theta = 100°$ is $\overline{\theta} = 80°$ and that for $\theta = 190°$ is $\overline{\theta} = 10°$.

(b) If θ is any angle, the value of a trigonometric function of θ is the same, except possibly for sign, as the value of the trigonometric function of $\overline{\theta}$. So $\sin 100^\circ = \sin 80^\circ$ and $\sin 190^\circ = -\sin 10^\circ$.

5. (a) The reference angle for 120° is $180^\circ - 120^\circ = 60^\circ$.

(b) The reference angle for 200° is $200^\circ - 180^\circ = 20^\circ$.

(c) The reference angle for 285° is $360^\circ - 285^\circ = 75^\circ$.

7. (a) The reference angle for 225° is $225^\circ - 180^\circ = 45^\circ$.

(b) The reference angle for 810° is $810^\circ - 720^\circ = 90^\circ$.

(c) The reference angle for -105° is $180^\circ - 105^\circ = 75^\circ$.

9. (a) The reference angle for $\frac{7\pi}{10}$ is $\pi - \frac{7\pi}{10} = \frac{3\pi}{10}$.

(b) The reference angle for $\frac{9\pi}{8}$ is $\frac{9\pi}{8} - \pi = \frac{\pi}{8}$.

(c) The reference angle for $\frac{10\pi}{3}$ is $\frac{10\pi}{3} - 3\pi = \frac{\pi}{3}$.

11. (a) The reference angle for $\frac{5\pi}{7}$ is $\pi - \frac{5\pi}{7} = \frac{2\pi}{7}$.

(b) The reference angle for -1.4π is $1.4\pi - \pi = 0.4\pi$.

(c) The reference angle for 1.4 is 1.4 because $1.4 < \frac{\pi}{2}$.

13. $\cos 150^\circ = -\cos 30^\circ = -\frac{\sqrt{3}}{2}$

15. $\tan 330^\circ = -\tan 30^\circ = -\frac{\sqrt{3}}{3}$

17. $\cot(-120^\circ) = \cot 60^\circ = \frac{\sqrt{3}}{3}$

19. $\csc(-630^\circ) = \csc 90^\circ = \frac{1}{\sin 90^\circ} = 1$

21. $\cos 570^\circ = -\cos 30^\circ = -\frac{\sqrt{3}}{2}$

23. $\tan 750^\circ = \tan 30^\circ = \frac{1}{\sqrt{3}} = \frac{\sqrt{3}}{3}$

25. $\sin \frac{3\pi}{2} = -\sin \frac{\pi}{2} = -1$

27. $\tan\left(-\frac{4\pi}{3}\right) = -\tan \frac{\pi}{3} = -\sqrt{3}$

29. $\csc\left(-\frac{5\pi}{6}\right) = -\csc\left(\frac{\pi}{6}\right) = -2$

31. $\sec \frac{17\pi}{3} = \sec \frac{\pi}{3} = \dfrac{1}{\cos \frac{\pi}{3}} = 2$

33. $\cot\left(-\frac{\pi}{4}\right) = -\cot \frac{\pi}{4} = \dfrac{-1}{\tan \frac{\pi}{4}} = -1$

35. $\tan \frac{5\pi}{2} = \tan \frac{\pi}{2}$ which is undefined.

37. Since $\sin\theta < 0$ and $\cos\theta < 0$, θ is in quadrant III.

39. $\sec\theta > 0 \Rightarrow \cos\theta > 0$. Also $\tan\theta < 0 \Rightarrow \dfrac{\sin\theta}{\cos\theta} < 0 \Leftrightarrow \sin\theta < 0$ (since $\cos\theta > 0$). Since $\sin\theta < 0$ and $\cos\theta > 0$, θ is in quadrant IV.

41. Since $\sin\theta$ is negative in quadrant III, $\sin\theta = -\sqrt{1 - \cos^2\theta}$ and we have $\tan\theta = \dfrac{\sin\theta}{\cos\theta} = -\dfrac{\sqrt{1-\cos^2\theta}}{\cos\theta}$.

43. $\cos^2\theta + \sin^2\theta = 1 \Leftrightarrow \cos\theta = \sqrt{1 - \sin^2\theta}$ because $\cos\theta > 0$ in quadrant IV.

45. $\sec^2\theta = 1 + \tan^2\theta \Leftrightarrow \sec\theta = -\sqrt{1 + \tan^2\theta}$ because $\sec\theta < 0$ in quadrant II.

47. $\sin\theta = -\frac{4}{5}$. Then $x = -\sqrt{5^2 - 4^2} = -\sqrt{9} = -3$, since θ is in quadrant IV. Thus, $\cos\theta = -\frac{3}{5}$, $\tan\theta = -\frac{4}{3}$, $\csc\theta = -\frac{5}{4}$, $\sec\theta = \frac{5}{3}$, and $\cot\theta = -\frac{3}{4}$.

49. $\cos\theta = \frac{7}{12}$, so $\sin\theta = -\sqrt{1 - \left(\frac{7}{12}\right)^2} = -\frac{\sqrt{95}}{12}$, $\tan\theta = -\frac{\sqrt{95}}{7}$, $\csc\theta = -\frac{12\sqrt{95}}{95}$, $\sec\theta = \frac{12}{7}$, and $\cot\theta = -\frac{7\sqrt{95}}{95}$

51. $\csc\theta = 2$. Then $\sin\theta = \frac{1}{2}$ and $x = \sqrt{2^2 - 1^2} = \sqrt{3}$. So $\sin\theta = \frac{1}{2}$, $\cos\theta = \frac{\sqrt{3}}{2}$, $\tan\theta = \frac{1}{\sqrt{3}} = \frac{\sqrt{3}}{3}$, $\sec\theta = \frac{2}{\sqrt{3}} = \frac{2\sqrt{3}}{3}$, and $\cot\theta = \sqrt{3}$.

53. $\cos\theta = -\frac{2}{7}$. Then $y = \sqrt{7^2 - 2^2} = \sqrt{45} = 3\sqrt{5}$, and so $\sin\theta = \frac{3\sqrt{5}}{7}$, $\tan\theta = -\frac{3\sqrt{5}}{2}$, $\csc\theta = \frac{7}{3\sqrt{5}} = \frac{7\sqrt{5}}{15}$, $\sec\theta = -\frac{7}{2}$, and $\cot\theta = -\frac{2}{3\sqrt{5}} = -\frac{2\sqrt{5}}{15}$.

55. If $\theta = \frac{\pi}{3}$, then $\sin 2\theta = \sin \frac{2\pi}{3} = \frac{\sqrt{3}}{2}$ and $2\sin\theta = 2\sin\frac{\pi}{3} = \sqrt{3}$.

57. The area is $\frac{1}{2}(7)(9)\sin 72^\circ \approx 30.0$.

59. The area is $\frac{1}{2}(10)^2 \sin 60^\circ = 25\sqrt{3} \approx 43.3$.

61. Let the angle be θ. Then $A = \frac{1}{2}xy\sin\theta \Rightarrow 16 = \frac{1}{2}(5)(7)\sin\theta \Leftrightarrow \sin\theta = \frac{32}{35} \Leftrightarrow \theta = \sin^{-1}\left(\frac{32}{35}\right) = 1.154$, or approximately 66.1°.

63. For the sector defined by the two sides, $A_1 = \frac{1}{2}r^2\theta = \frac{1}{2}\cdot 2^2 \cdot 120^\circ \cdot \frac{\pi}{180^\circ} = \frac{4\pi}{3}$. For the triangle defined by the two sides, $A_2 = \frac{1}{2}ab\sin\theta = \frac{1}{2}\cdot 2\cdot 2\cdot \sin 120^\circ = 2\sin 60^\circ = \sqrt{3}$. Thus the area of the region is $A_1 - A_2 = \frac{4\pi}{3} - \sqrt{3} \approx 2.46$.

65. **(a)** $\tan\theta = \dfrac{h}{1\text{ mile}}$, so $h = \tan\theta \cdot 1\text{ mile} \cdot \dfrac{5280\text{ ft}}{1\text{ mile}} = 5280\tan\theta$ ft.

(b)

θ	20°	60°	80°	85°
h	1922	9145	29,944	60,351

67. **(a)** From the figure in the text, we express depth and width in terms of θ.

Since $\sin\theta = \dfrac{\text{depth}}{20}$ and $\cos\theta = \dfrac{\text{width}}{20}$, we have depth $= 20\sin\theta$ and width $= 20\cos\theta$. Thus, the cross-section area of the beam is $A(\theta) = (\text{depth})(\text{width}) = (20\cos\theta)(20\sin\theta) = 400\cos\theta\sin\theta$.

(b)

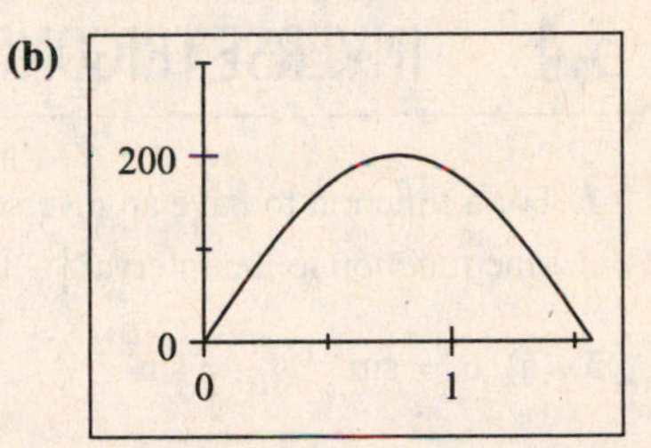

(c) The beam with the largest cross-sectional area is the square beam, $10\sqrt{2}$ by $10\sqrt{2}$ (about 14.14 by 14.14).

69. **(a)** On Earth, the range is $R = \dfrac{v_0^2\sin(2\theta)}{g} = \dfrac{12^2\sin\frac{\pi}{3}}{32} = \dfrac{9\sqrt{3}}{4} \approx 3.897$ ft and the height is

$$H = \frac{v_0^2\sin^2\theta}{2g} = \frac{12^2\sin^2\frac{\pi}{6}}{2\cdot 32} = \frac{9}{16} = 0.5625 \text{ ft.}$$

(b) On the moon, $R = \dfrac{12^2\sin\frac{\pi}{3}}{5.2} \approx 23.982$ ft and $H = \dfrac{12^2\sin^2\frac{\pi}{6}}{2\cdot 5.2} \approx 3.462$ ft

71. **(a)** $W = 3.02 - 0.38\cot\theta + 0.65\csc\theta$

(b) From the graph, it appears that W has its minimum value at about $\theta = 0.946 \approx 54.2^\circ$.

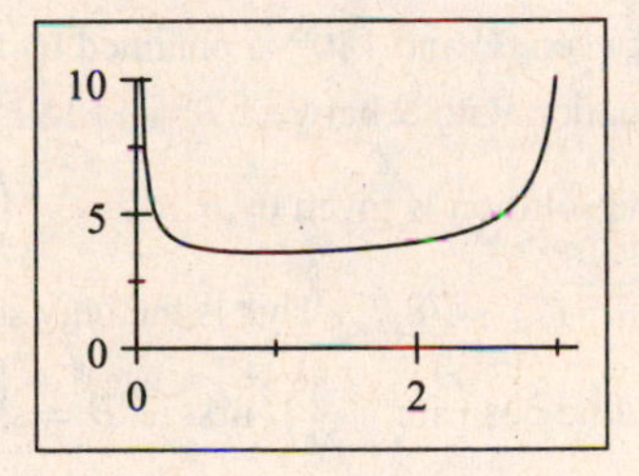

73. We have $\sin\alpha = k\sin\beta$, where $\alpha = 59.4^\circ$ and $k = 1.33$. Substituting, $\sin 59.4^\circ = 1.33\sin\beta$ $\Rightarrow \sin\beta = \dfrac{\sin 59.4^\circ}{1.33} \approx 0.6472$. Using a calculator, we find that $\beta \approx \sin^{-1} 0.6472 \approx 40.3^\circ$, so $\theta = 4\beta - 2\alpha \approx 4(40.3^\circ) - 2(59.4^\circ) = 42.4^\circ$.

75. $\cos\theta = \dfrac{\text{adj}}{\text{hyp}} = \dfrac{|OP|}{|OR|} = \dfrac{|OP|}{1} = |OP|$. Since QS is tangent to the circle at R, $\triangle ORQ$ is a right triangle. Then $\tan\theta = \dfrac{\text{opp}}{\text{adj}} = \dfrac{|RQ|}{|OR|} = |RQ|$ and $\sec\theta = \dfrac{\text{hyp}}{\text{adj}} = \dfrac{|OQ|}{|OR|} = |OQ|$. Since $\angle SOQ$ is a right angle $\triangle SOQ$ is a right triangle and $\angle OSR = \theta$. Then $\csc\theta = \dfrac{\text{hyp}}{\text{opp}} = \dfrac{|OS|}{|OR|} = |OS|$ and $\cot\theta = \dfrac{\text{adj}}{\text{opp}} = \dfrac{|SR|}{|OR|} = |SR|$. Summarizing, we have $\sin\theta = |PR|$, $\cos\theta = |OP|$, $\tan\theta = |RQ|$, $\sec\theta = |OQ|$, $\csc\theta = |OS|$, and $\cot\theta = |SR|$.

77. Let x represent the desired real number, in radians. We wish to find the smallest nonzero solution of the equation $\sin x = \sin \frac{180x}{\pi}$. (Since x is in radians, $\frac{180x}{\pi}$ is in degrees.) Now if a is in degrees with $0 \le a \le 180$ and $\sin a = b$, then either $a = b$ or $180 - a = b$. The graph shows that the first positive solution occurs when $0 < \dfrac{180x}{\pi} \le 180$, so we have $180 - \dfrac{180x}{\pi} = x$.

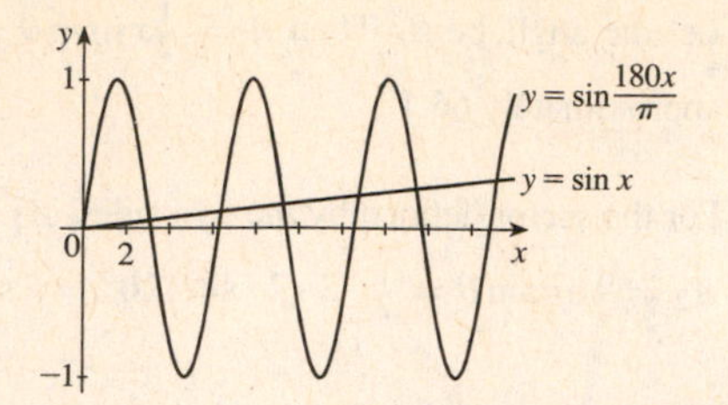

Solving this equation, we get $x = \dfrac{180\pi}{180 + \pi} \approx 3.088$. You can verify that the value of $\sin \dfrac{180\pi}{180 + \pi}$ is the same whether your calculator is set to radian mode or degree mode.

5.4 INVERSE TRIGONOMETRIC FUNCTIONS AND TRIANGLES

1. For a function to have an inverse, it must be *one-to-one*. To define the inverse sine function we restrict the *domain* of the sine function to the interval $\left[-\frac{\pi}{2}, \frac{\pi}{2}\right]$.

3. (a) $\theta = \sin^{-1} \frac{8}{10} = \sin^{-1} \frac{4}{5}$ **(b)** $\theta = \cos^{-1} \frac{6}{10} = \cos^{-1} \frac{3}{5}$ **(c)** $\theta = \tan^{-1} \frac{8}{6} = \tan^{-1} \frac{4}{3}$

5. (a) $\sin^{-1} 1 = \frac{\pi}{2}$ **(b)** $\cos^{-1} 0 = \frac{\pi}{2}$ **(c)** $\tan^{-1} \sqrt{3} = \frac{\pi}{3}$

7. (a) $\sin^{-1}\left(-\frac{\sqrt{2}}{2}\right) = -\frac{\pi}{4}$ **(b)** $\cos^{-1}\left(-\frac{\sqrt{2}}{2}\right) = \frac{3\pi}{4}$ **(c)** $\tan^{-1}(-1) = -\frac{\pi}{4}$

7. $\sin^{-1}(0.30) = 0.30469$ **11.** $\cos^{-1}\left(\frac{1}{3}\right) \approx 1.23096$ **13.** $\tan^{-1} 3 \approx 1.24905$ **15.** $\cos^{-1} 3$ is undefined.

17. $\sin\theta = \frac{6}{10} = \frac{3}{5}$, so $\theta = \sin^{-1} \frac{3}{5} \approx 36.9°$. **19.** $\tan\theta = \frac{9}{13}$, so $\theta = \tan^{-1} \frac{9}{13} \approx 34.7°$.

21. $\sin\theta = \frac{4}{7}$, so $\theta = \sin^{-1} \frac{4}{7} \approx 34.8°$.

23. We use $\sin^{-1}$ to find one solution in the interval $[-90°, 90°]$. $\sin\theta = \frac{2}{3} \Rightarrow \theta = \sin^{-1} \frac{2}{3} \approx 41.8°$. Another solution with θ between $0°$ and $180°$ is obtained by taking the supplement of the angle: $180° - 41.8° = 138.2°$. So the solutions of the equation with θ between $0°$ and $180°$ are approximately $\theta = 41.8°$ and $\theta = 138.2°$.

25. One solution is given by $\theta = \cos^{-1}\left(-\frac{2}{5}\right) \approx 113.6°$. This is the only solution, because $\cos x$ is one-to-one on $[0, 180°]$.

27. $\tan^{-1}(5) \approx 78.7°$. This is the only solution on $[0, 180°]$.

29. To find $\cos\left(\sin^{-1} \frac{4}{5}\right)$, first let $\theta = \sin^{-1} \frac{4}{5}$. Then θ is the number in the interval $\left[-\frac{\pi}{2}, \frac{\pi}{2}\right]$ whose sine is $\frac{4}{5}$. We draw a right triangle with θ as one of its acute angles, with opposite side 4 and hypotenuse 5. The remaining leg of the triangle is found by the Pythagorean Theorem to be 3. From the figure we get $\cos\left(\sin^{-1} \frac{4}{5}\right) = \sin\theta = \frac{3}{5}$.

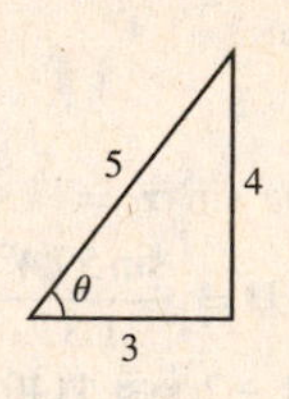

Another method: By the cancellation properties of inverse functions, $\sin\left(\sin^{-1} \frac{4}{5}\right)$ is exactly $\frac{4}{5}$. To find $\cos\left(\sin^{-1} \frac{4}{5}\right)$, we first write the cosine function in terms of the sine function. Let $u = \sin^{-1} \frac{4}{5}$. Since $0 \le u \le \frac{\pi}{2}$, $\cos u$ is positive, and since $\cos^2 u + \sin^2 u = 1$, we can write $\cos u = \sqrt{1 - \sin^2 u} = \sqrt{1 - \sin^2\left(\sin^{-1} \frac{4}{5}\right)} = \sqrt{1 - \left(\frac{4}{5}\right)^2} = \sqrt{1 - \frac{16}{25}} = \sqrt{\frac{9}{25}} = \frac{3}{5}$. Therefore, $\cos\left(\sin^{-1} \frac{4}{5}\right) = \frac{3}{5}$.

31. To find $\sec\left(\sin^{-1}\frac{12}{13}\right)$, we draw a right triangle with angle θ, opposite side 12, and hypotenuse 13. From the figure we see that $\sec\left(\sin^{-1}\frac{12}{13}\right) = \sec\theta = \frac{13}{5}$.

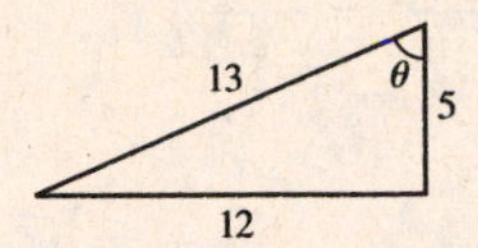

33. To find $\tan\left(\sin^{-1}\frac{12}{13}\right)$, we draw a right triangle with angle θ, opposite side 12, and hypotenuse 13. From the figure we see that $\tan\left(\sin^{-1}\frac{12}{13}\right) = \tan\theta = \frac{12}{5}$.

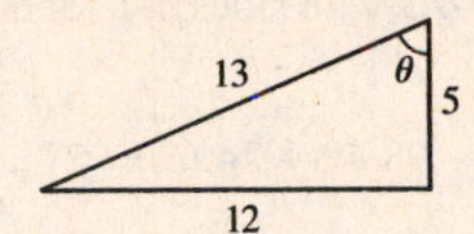

35. We want to find $\cos\left(\sin^{-1}x\right)$. Let $\theta = \sin^{-1}x$, so $\sin\theta = x$. We sketch a right triangle with an acute angle θ, opposite side x, and hypotenuse 1. By the Pythagorean Theorem, the remaining leg is $\sqrt{1-x^2}$. From the figure we have $\cos\left(\sin^{-1}x\right) = \cos\theta = \sqrt{1-x^2}$.

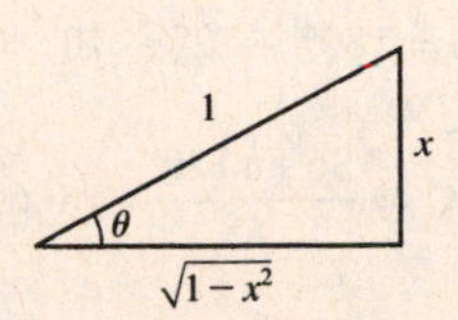

Another method: Let $u = \sin^{-1}x$. We need to find $\cos u$ in terms of x. To do so, we write cosine in terms of sine. Note that $-\frac{\pi}{2} \le u \le \frac{\pi}{2}$ because $u = \sin^{-1}x$. Now $\cos u = \sqrt{1-\sin^2 u}$ is positive because u lies in the interval $\left[-\frac{\pi}{2}, \frac{\pi}{2}\right]$. Substituting $u = \sin^{-1}x$ and using the cancellation property $\sin(\sin^{-1}x) = x$ gives $\cos(\sin^{-1}x) = \sqrt{1-x^2}$.

37. We want to find $\tan\left(\sin^{-1}x\right)$. Let $\theta = \sin^{-1}x$, so $\sin\theta = x$. We sketch a right triangle with an acute angle θ, opposite side x, and hypotenuse 1. By the Pythagorean Theorem, the remaining leg is $\sqrt{1-x^2}$. From the figure we have $\tan\left(\sin^{-1}x\right) = \tan\theta = \dfrac{x}{\sqrt{1-x^2}}$.

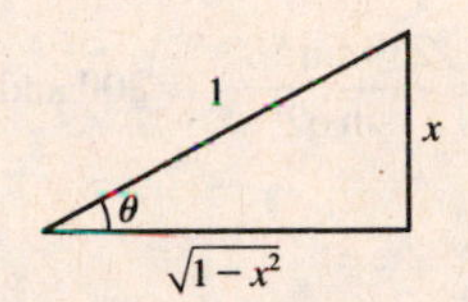

39. Let θ represent the angle of elevation of the ladder. Let h represent the height, in feet, that the ladder reaches on the building. Then $\cos\theta = \frac{6}{20} = 0.3 \Leftrightarrow \theta = \cos^{-1}0.3 \approx 1.266$ rad $\approx 72.5^\circ$. By the Pythagorean Theorem, $h^2 + 6^2 = 20^2 \Leftrightarrow h = \sqrt{400-36} = \sqrt{364} \approx 19$ ft.

41. **(a)** Solving $\tan\theta = h/2$ for h, we have $h = 2\tan\theta$.

(b) Solving $\tan\theta = h/2$ for θ we have $\theta = \tan^{-1}(h/2)$.

43. **(a)** Solving $\sin\theta = h/680$ for θ we have $\theta = \sin^{-1}(h/680)$.

(b) Set $h = 500$ to get $\theta = \sin^{-1}\left(\frac{500}{680}\right) \approx 0.826$ rad $\approx 47.3^\circ$.

45. **(a)** $\theta = \sin^{-1}\left(\dfrac{1}{(2\cdot 3+1)\tan 10^\circ}\right) = \sin^{-1}\left(\dfrac{1}{7\tan 10^\circ}\right) \approx \sin^{-1}0.8102 \approx 54.1^\circ$

(b) For $n = 2$, $\theta = \sin^{-1}\left(\dfrac{1}{5\tan 15^\circ}\right) \approx 48.3^\circ$. For $n = 3$, $\theta = \sin^{-1}\left(\dfrac{1}{7\tan 15^\circ}\right) \approx 32.2^\circ$. For $n = 4$, $\theta = \sin^{-1}\left(\dfrac{1}{9\tan 15^\circ}\right) \approx 24.5^\circ$. $n = 0$ and $n = 1$ are outside of the domain for $\beta = 15^\circ$, because $\dfrac{1}{\tan 15^\circ} \approx 3.732$ and $\dfrac{1}{3\tan 15^\circ} \approx 1.244$, neither of which is in the domain of $\sin^{-1}$.

5.5 THE LAW OF SINES

1. In triangle ABC with sides a, b, and c the Law of Sines states that $\dfrac{\sin A}{a} = \dfrac{\sin B}{b} = \dfrac{\sin C}{c}$.

3. $\angle C = 180° - 98.4° - 24.6° = 57°$. $x = \dfrac{376 \sin 57°}{\sin 98.4°} \approx 318.75$.

5. $\angle C = 180° - 52° - 70° = 58°$. $x = \dfrac{26.7 \sin 52°}{\sin 58°} \approx 24.8$.

7. $\sin C = \dfrac{36 \sin 120°}{45} \approx 0.693 \Leftrightarrow \angle C \approx \sin^{-1} 0.693 \approx 44°$.

9. $\angle C = 180° - 46° - 20° = 114°$. Then $a = \dfrac{65 \sin 46°}{\sin 114°} \approx 51$ and $b = \dfrac{65 \sin 20°}{\sin 114°} \approx 24$.

11. $\angle B = 68°$, so $\angle A = 180° - 68° - 68° = 44°$ and $a = \dfrac{12 \sin 44°}{\sin 68°} \approx 8.99$.

13. $\angle C = 180° - 50° - 68° = 62°$. Then

$a = \dfrac{230 \sin 50°}{\sin 62°} \approx 200$ and $b = \dfrac{230 \sin 68°}{\sin 62°} \approx 242$.

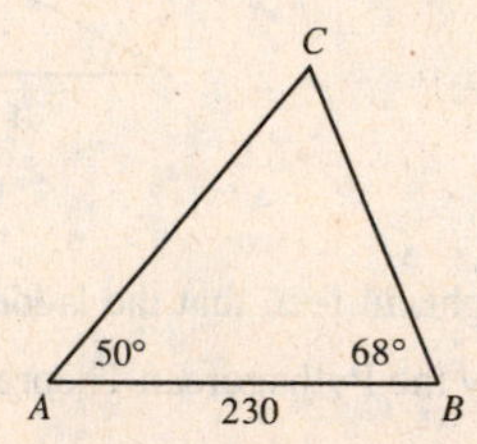

15. $\angle B = 180° - 30° - 65° = 85°$. Then

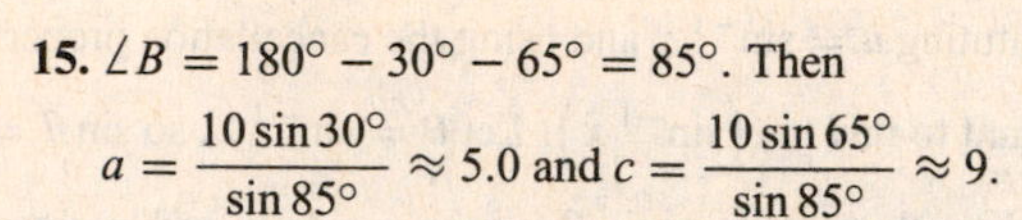

$a = \dfrac{10 \sin 30°}{\sin 85°} \approx 5.0$ and $c = \dfrac{10 \sin 65°}{\sin 85°} \approx 9$.

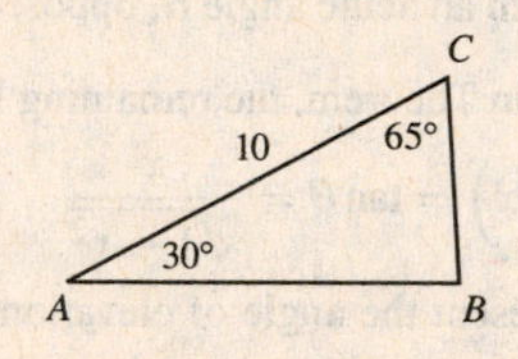

17. $\angle A = 180° - 51° - 29° = 100°$. Then

$a = \dfrac{44 \sin 100°}{\sin 29°} \approx 89$ and $c = \dfrac{44 \sin 51°}{\sin 29°} \approx 71$.

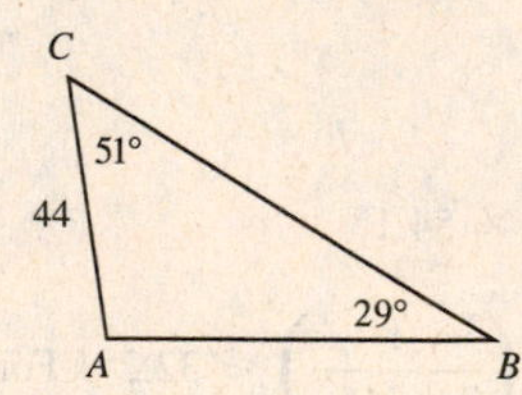

19. Since $\angle A > 90°$ there is only one triangle.

$\sin B = \dfrac{15 \sin 110°}{28} \approx 0.503 \Leftrightarrow$

$\angle B \approx \sin^{-1} 0.503 \approx 30°$. Then

$\angle C \approx 180° - 110° - 30° = 40°$, and so

$c = \dfrac{28 \sin 40°}{\sin 110°} \approx 19$. Thus $\angle B \approx 30°$, $\angle C \approx 40°$, and

$c \approx 19$.

21. $\angle A = 125°$ is the largest angle, but since side a is not the longest side, there can be no such triangle.

23. $\sin C = \dfrac{30 \sin 25°}{25} \approx 0.507 \Leftrightarrow \angle C_1 \approx \sin^{-1} 0.507 \approx 30.47°$ or $\angle C_2 \approx 180° - 39.47° = 149.53°$.

If $\angle C_1 = 30.47°$, then $\angle A_1 \approx 180° - 25° - 30.47° = 124.53°$ and $a_1 = \dfrac{25 \sin 124.53°}{\sin 25°} \approx 48.73$.

If $\angle C_2 = 149.53°$, then $\angle A_2 \approx 180° - 25° - 149.53° = 5.47°$ and $a_2 = \dfrac{25 \sin 5.47°}{\sin 25°} \approx 5.64$.

Thus, one triangle has $\angle A_1 \approx 125°$, $\angle C_1 \approx 30°$, and $a_1 \approx 49$; the other has $\angle A_2 \approx 5°$, $\angle C_2 \approx 150°$, and $a_2 \approx 5.6$.

25. $\sin B = \dfrac{100 \sin 50°}{50} \approx 1.532$. Since $|\sin \theta| \le 1$ for all θ, there can be no such angle B, and thus no such triangle.

27. $\sin A = \dfrac{26 \sin 29^\circ}{15} \approx 0.840 \Leftrightarrow \angle A_1 \approx \sin^{-1} 0.840 \approx 57.2^\circ$ or $\angle A_2 \approx 180^\circ - 57.2^\circ = 122.8^\circ$.

If $\angle A_1 \approx 57.2^\circ$, then $\angle B_1 = 180^\circ - 29^\circ - 57.2^\circ = 93.8^\circ$ and $b_1 \approx \dfrac{15 \sin 93.8^\circ}{\sin 29^\circ} \approx 30.9$.

If $\angle A_2 \approx 122.8^\circ$, then $\angle B_2 = 180 - 29^\circ - 122.8^\circ = 28.2^\circ$ and $b_2 \approx \dfrac{15 \sin 28.1^\circ}{\sin 29^\circ} \approx 14.6$.

Thus, one triangle has $\angle A_1 \approx 57.2^\circ$, $\angle B_1 \approx 93.8^\circ$, and $b_1 \approx 30.9$; the other has $\angle A_2 \approx 122.8^\circ$, $\angle B_2 \approx 28.2^\circ$, and $b_2 \approx 14.6$.

29. **(a)** From $\triangle ABC$ and the Law of Sines we get $\dfrac{\sin 30^\circ}{20} = \dfrac{\sin B}{28} \Leftrightarrow \sin B = \dfrac{28 \sin 30^\circ}{20} = 0.7$, so $\angle B \approx \sin^{-1} 0.7 \approx 44.427^\circ$. Since $\triangle BCD$ is isosceles, $\angle B = \angle BDC \approx 44.427^\circ$. Thus, $\angle BCD = 180^\circ - 2\angle B \approx 91.146^\circ \approx 91.1^\circ$.

(b) From $\triangle ABC$ we get $\angle BCA = 180^\circ - \angle A - \angle B \approx 180^\circ - 30^\circ - 44.427^\circ = 105.573^\circ$. Hence $\angle DCA = \angle BCA - \angle BCD \approx 105.573^\circ - 91.146^\circ = 14.4^\circ$.

31. **(a)** Let a be the distance from satellite to the tracking station A in miles. Then the subtended angle at the satellite is $\angle C = 180^\circ - 93^\circ - 84.2^\circ = 2.8^\circ$, and so $a = \dfrac{50 \sin 84.2^\circ}{\sin 2.8^\circ} \approx 1018$ mi.

(b) Let d be the distance above the ground in miles. Then $d = 1018.3 \sin 87^\circ \approx 1017$ mi.

33. $\angle C = 180^\circ - 82^\circ - 52^\circ = 46^\circ$, so by the Law of Sines, $\dfrac{|AC|}{\sin 52^\circ} = \dfrac{|AB|}{\sin 46^\circ} \Leftrightarrow |AC| = \dfrac{|AB| \sin 52^\circ}{\sin 46^\circ}$, so substituting we have $|AC| = \dfrac{200 \sin 52^\circ}{\sin 46^\circ} \approx 219$ ft.

35.

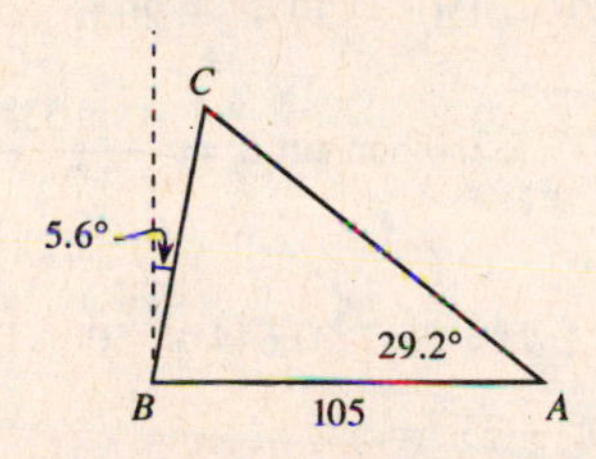

We draw a diagram. A is the position of the tourist and C is the top of the tower. $\angle B = 90^\circ - 5.6^\circ = 84.4^\circ$ and so $\angle C = 180^\circ - 29.2^\circ - 84.4^\circ = 66.4^\circ$. Thus, by the Law of Sines, the length of the tower is $|BC| = \dfrac{105 \sin 29.2^\circ}{\sin 66.4^\circ} \approx 55.9$ m.

37. The angle subtended by the top of the tree and the sun's rays is $\angle A = 180^\circ - 90^\circ - 52^\circ = 38^\circ$. Thus the height of the tree is $h = \dfrac{215 \sin 30^\circ}{\sin 38^\circ} \approx 175$ ft.

39. Call the balloon's position R. Then in $\triangle PQR$, we see that $\angle P = 62^\circ - 32^\circ = 30^\circ$, and $\angle Q = 180^\circ - 71^\circ + 32^\circ = 141^\circ$. Therefore, $\angle R = 180^\circ - 30^\circ - 141^\circ = 9^\circ$. So by the Law of Sines, $\dfrac{|QR|}{\sin 30^\circ} = \dfrac{|PQ|}{\sin 9^\circ} \Leftrightarrow |QR| = 60 \cdot \dfrac{\sin 30^\circ}{\sin 9^\circ} \approx 192$ m.

41. Let d be the distance from the earth to Venus, and let β be the angle formed by sun, Venus, and earth. By the Law of Sines, $\dfrac{\sin \beta}{1} = \dfrac{\sin 39.4^\circ}{0.723} \approx 0.878$, so either $\beta \approx \sin^{-1} 0.878 \approx 61.4^\circ$ or $\beta \approx 180^\circ - \sin^{-1} 0.878 \approx 118.6^\circ$.

In the first case, $\dfrac{d}{\sin(180^\circ - 39.4^\circ - 61.4^\circ)} = \dfrac{0.723}{\sin 39.4^\circ} \Leftrightarrow d \approx 1.119$ AU; in the second case, $\dfrac{d}{\sin(180^\circ - 39.4^\circ - 118.6^\circ)} = \dfrac{0.723}{\sin 39.4^\circ} \Leftrightarrow d \approx 0.427$ AU.

43. By the area formula from Section 5.3, the area of $\triangle ABC$ is $A = \frac{1}{2}ab \sin C$. Because we are given a and the three angles, we need to find b in terms of these quantities. By the Law of Sines, $\dfrac{\sin B}{b} = \dfrac{\sin A}{a} \Leftrightarrow b = \dfrac{a \sin B}{\sin A}$. Thus,

$$A = \tfrac{1}{2}ab \sin C = \tfrac{1}{2}a\left(\frac{a \sin B}{\sin A}\right)\sin C = \frac{a^2 \sin B \sin C}{2 \sin A}.$$

45.

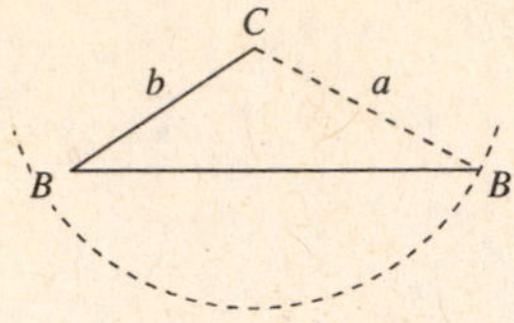

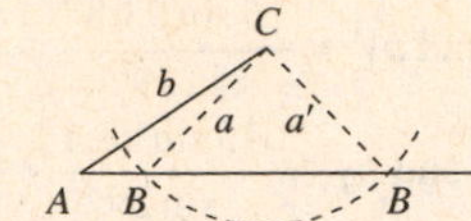

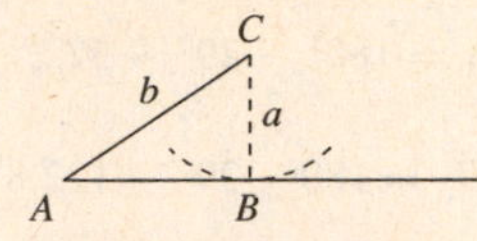

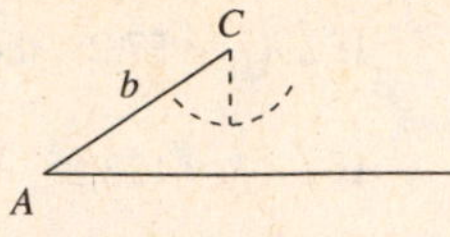

$a \geq b$: One solution $\quad b > a > b \sin A$: Two solutions $\quad a = b \sin A$: One solution $\quad a < b \sin A$: No solution

$\angle A = 30°$, $b = 100$, $\sin A = \frac{1}{2}$. If $a \geq b = 100$ then there is one triangle. If $100 > a > 100 \sin 30° = 50$, then there are two possible triangles. If $a = 50$, then there is one (right) triangle. And if $a < 50$, then no triangle is possible.

5.6 THE LAW OF COSINES

1. For triangle ABC with sides a, b, and c the Law of Cosines states $c^2 = a^2 + b^2 - 2ab \cos C$.

3. $x^2 = 21^2 + 42^2 - 2 \cdot 21 \cdot 42 \cdot \cos 39° = 441 + 1764 - 1764 \cos 39° \approx 834.115$ and so $x \approx \sqrt{834.115} \approx 28.9$.

5. $x^2 = 25^2 + 25^2 - 2 \cdot 25 \cdot 25 \cdot \cos 140° = 625 + 625 - 1250 \cos 140° \approx 2207.556$ and so $x \approx \sqrt{2207.556} \approx 47$.

7. $37.83^2 = 68.01^2 + 42.15^2 - 2 \cdot 68.01 \cdot 42.15 \cdot \cos \theta$. Then $\cos \theta = \dfrac{37.83^2 - 68.01^2 - 42.15^2}{-2 \cdot 68.01 \cdot 42.15} \approx 0.867 \Leftrightarrow$ $\theta \approx \cos^{-1} 0.867 \approx 29.89°$.

9. $x^2 = 24^2 + 30^2 - 2 \cdot 24 \cdot 30 \cdot \cos 30° = 576 + 900 - 1440 \cos 30° \approx 228.923$ and so $x \approx \sqrt{228.923} \approx 15$.

11. $c^2 = 10^2 + 18^2 - 2 \cdot 10 \cdot 18 \cdot \cos 120° = 100 + 324 - 360 \cos 120° = 604$ and so $c \approx \sqrt{604} \approx 24.576$. Then $\sin A \approx \dfrac{18 \sin 120°}{24.576} \approx 0.634295 \Leftrightarrow \angle A \approx \sin^{-1} 0.634295 \approx 39.4°$, and $\angle B \approx 180° - 120° - 39.4° = 20.6°$.

13. $c^2 = 3^2 + 4^2 - 2 \cdot 3 \cdot 4 \cdot \cos 53° = 9 + 16 - 24 \cos 53° \approx 10.556 \Leftrightarrow c \approx \sqrt{10.556} \approx 3.2$. Then $\sin B = \dfrac{4 \sin 53°}{3.25} \approx 0.983$ $\Leftrightarrow \angle B \approx \sin^{-1} 0.983 \approx 79°$ and $\angle A \approx 180° - 53° - 79° = 48°$.

15. $20^2 = 25^2 + 22^2 - 2 \cdot 25 \cdot 22 \cdot \cos A \Leftrightarrow \cos A = \frac{20^2 - 25^2 - 22^2}{-2 \cdot 25 \cdot 22} \approx 0.644 \Leftrightarrow \angle A \approx \cos^{-1} 0.644 \approx 50°$. Then $\sin B \approx \frac{25 \sin 49.9°}{20} \approx 0.956 \Leftrightarrow \angle B \approx \sin^{-1} 0.956 \approx 73°$, and so $\angle C \approx 180° - 50° - 73° = 57°$.

17. $\sin C = \frac{162 \sin 40°}{125} \approx 0.833 \Leftrightarrow \angle C_1 \approx \sin^{-1} 0.833 \approx 56.4°$ or $\angle C_2 \approx 180° - 56.4° \approx 123.6°$.

If $\angle C_1 \approx 56.4°$, then $\angle A_1 \approx 180° - 40° - 56.4° = 83.6°$ and $a_1 = \frac{125 \sin 83.6°}{\sin 40°} \approx 193$.

If $\angle C_2 \approx 123.6°$, then $\angle A_2 \approx 180° - 40° - 123.6° = 16.4°$ and $a_2 = \frac{125 \sin 16.4°}{\sin 40°} \approx 54.9$.

Thus, one triangle has $\angle A \approx 83.6°$, $\angle C \approx 56.4°$, and $a \approx 193$; the other has $\angle A \approx 16.4°$, $\angle C \approx 123.6°$, and $a \approx 54.9$.

19. $\sin B = \frac{65 \sin 55°}{50} \approx 1.065$. Since $|\sin \theta| \leq 1$ for all θ, there is no such $\angle B$, and hence there is no such triangle.

21. $\angle B = 180° - 35° - 85° = 60°$. Then $x = \frac{3 \sin 35°}{\sin 60°} \approx 2$.

23. $x = \frac{50 \sin 30°}{\sin 100°} \approx 25.4$

25. $b^2 = 110^2 + 138^2 - 2(110)(138) \cdot \cos 38° = 12{,}100 + 19{,}044 - 30{,}360 \cos 38° \approx 7220.0$ and so $b \approx 85.0$. Therefore, using the Law of Cosines again, we have $\cos \theta = \frac{110^2 + 85^2 - 128^2}{2(110)(138)} \Leftrightarrow \theta \approx 89.15°$.

27. $x^2 = 38^2 + 48^2 - 2 \cdot 38 \cdot 48 \cdot \cos 30° = 1444 + 2304 - 3648 \cos 30° \approx 588.739$ and so $x \approx 24.3$.

29. The semiperimeter is $s = \frac{9+12+15}{2} = 18$, so by Heron's Formula the area is $A = \sqrt{18(18-9)(18-12)(18-15)} = \sqrt{2916} = 54$.

31. The semiperimeter is $s = \frac{7+8+9}{2} = 12$, so by Heron's Formula the area is $A = \sqrt{12(12-7)(12-8)(12-9)} = \sqrt{720} = 12\sqrt{5} \approx 26.8$.

33. The semiperimeter is $s = \frac{3+4+6}{2} = \frac{13}{2}$, so by Heron's Formula the area is $A = \sqrt{\frac{13}{2}\left(\frac{13}{2}-3\right)\left(\frac{13}{2}-4\right)\left(\frac{13}{2}-6\right)} = \sqrt{\frac{455}{16}} = \frac{\sqrt{455}}{4} \approx 5.33$.

35. We draw a diagonal connecting the vertices adjacent to the 100° angle. This forms two triangles. Consider the triangle with sides of length 5 and 6 containing the 100° angle. The area of this triangle is $A_1 = \frac{1}{2}(5)(6)\sin 100° \approx 14.77$. To use Heron's Formula to find the area of the second triangle, we need to find the length of the diagonal using the Law of Cosines: $c^2 = a^2 + b^2 - 2ab\cos C = 5^2 + 6^2 - 2\cdot 5\cdot 6\cos 100° \approx 71.419 \Rightarrow c \approx 8.45$. Thus the second triangle has semiperimeter $s = \dfrac{8+7+8.45}{2} \approx 11.7255$ and area $A_2 = \sqrt{11.7255\,(11.7255-8)\,(11.7255-7)\,(11.7255-8.45)} \approx 26.00$. The area of the quadrilateral is the sum of the areas of the two triangles: $A = A_1 + A_2 \approx 14.77 + 26.00 = 40.77$.

37.

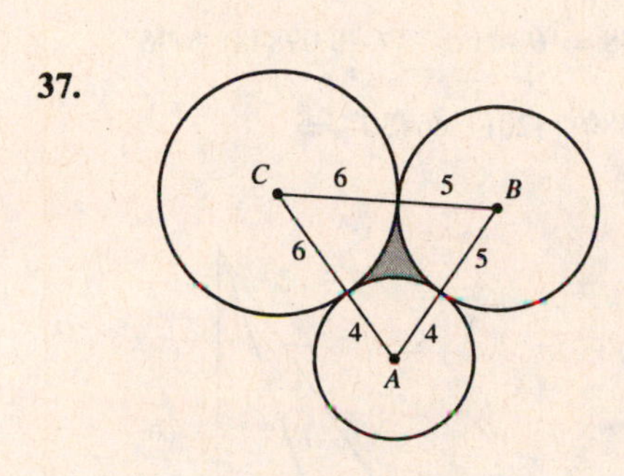

Label the centers of the circles A, B, and C, as in the figure. By the Law of Cosines, $\cos A = \dfrac{AB^2 + AC^2 - BC^2}{2(AB)(AC)} = \dfrac{9^2 + 10^2 - 11^2}{2(9)(10)} = \frac{1}{3} \Rightarrow \angle A \approx 70.53°$.

Now, by the Law of Sines, $\dfrac{\sin 70.53°}{11} = \dfrac{\sin B}{AC} = \dfrac{\sin C}{AB}$. So

$\sin B = \frac{10}{11}\sin 70.53° \approx 0.85710 \Rightarrow B \approx \sin^{-1} 0.85710 \approx 58.99°$ and

$\sin C = \frac{9}{11}\sin 70.53° \approx 0.77139 \Rightarrow C \approx \sin^{-1} 0.77139 \approx 50.48°$. The area of ΔABC is $\frac{1}{2}(AB)(AC)\sin A = \frac{1}{2}(9)(10)(\sin 70.53°) \approx 42.426$.

The area of sector A is given by $S_A = \pi R^2 \cdot \dfrac{\theta}{360°} = \pi(4)^2 \cdot \dfrac{70.53°}{360°} \approx 9.848$. Similarly, the areas of sectors B and C are $S_B \approx 12.870$ and $S_C \approx 15.859$. Thus, the area enclosed between the circles is $A = \Delta ABC - S_A - S_B - S_C \Rightarrow A \approx 42.426 - 9.848 - 12.870 - 15.859 \approx 3.85\text{ cm}^2$.

39. Let c be the distance across the lake, in miles. Then $c^2 = 2.82^2 + 3.56^2 - 2(2.82)(3.56)\cdot\cos 40.3° \approx 5.313 \Leftrightarrow c \approx 2.30$ mi.

41. In half an hour, the faster car travels 25 miles while the slower car travels 15 miles. The distance between them is given by the Law of Cosines: $d^2 = 25^2 + 15^2 - 2(25)(15)\cdot\cos 65° \Rightarrow d = \sqrt{25^2 + 15^2 - 2(25)(15)\cdot\cos 65°} = 5\sqrt{25 + 9 - 30\cdot\cos 65°} \approx 23.1$ mi.

43. The pilot travels a distance of $625\cdot 1.5 = 937.5$ miles in her original direction and $625\cdot 2 = 1250$ miles in the new direction. Since she makes a course correction of 10° to the right, the included angle is $180° - 10° = 170°$. From the figure, we use the Law of Cosines to get the expression $d^2 = 937.5^2 + 1250^2 - 2(937.5)(1250)\cdot\cos 170° \approx 4{,}749{,}549.42$, so $d \approx 2179$ miles. Thus, the pilot's distance from her original position is approximately 2179 miles.

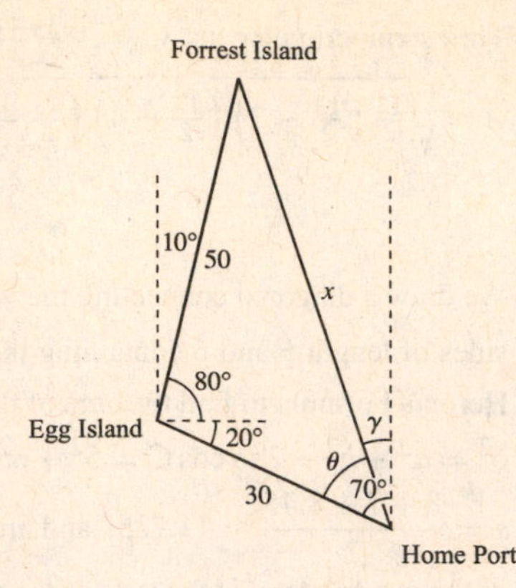

45. **(a)** The angle subtended at Egg Island is 100°. Thus using the Law of Cosines, the distance from Forrest Island to the fisherman's home port is

$$\begin{aligned} x^2 &= 30^2 + 50^2 - 2 \cdot 30 \cdot 50 \cdot \cos 100^\circ \\ &= 900 + 2500 - 3000 \cos 100^\circ \approx 3920.945 \end{aligned}$$

and so $x \approx \sqrt{3920.945} \approx 62.62$ miles.

(b) Let θ be the angle shown in the figure. Using the Law of Sines, $\sin\theta = \dfrac{50 \sin 100^\circ}{62.62} \approx 0.7863 \Leftrightarrow \theta \approx \sin^{-1} 0.7863 \approx 51.8^\circ$. Then $\gamma = 90^\circ - 20^\circ - 51.8^\circ = 18.2^\circ$. Thus the bearing to his home port is S 18.2° E.

47. The largest angle is the one opposite the longest side; call this angle θ. Then by the Law of Cosines, $44^2 = 36^2 + 22^2 - 2(36)(22) \cdot \cos\theta \Leftrightarrow \cos\theta = \dfrac{36^2 + 22^2 - 44^2}{2(36)(22)} = -0.09848 \Rightarrow \theta \approx \cos^{-1}(-0.09848) \approx 96^\circ$.

49. Let d be the distance between the kites. Then $d^2 \approx 380^2 + 420^2 - 2(380)(420) \cdot \cos 30^\circ \Rightarrow d \approx \sqrt{380^2 + 420^2 - 2(380)(420) \cdot \cos 30^\circ} \approx 211$ ft.

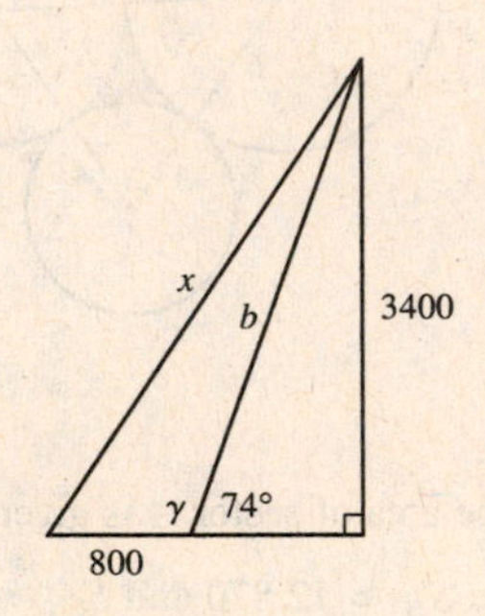

51. *Solution 1:* From the figure, we see that $\gamma = 106^\circ$ and $\sin 74^\circ = \dfrac{3400}{b} \Leftrightarrow$ $b = \dfrac{3400}{\sin 74^\circ} \approx 3537$. Thus, $x^2 = 800^2 + 3537^2 - 2(800)(3537)\cos 106^\circ \Rightarrow$ $x = \sqrt{800^2 + 3537^2 - 2(800)(3537)\cos 106^\circ} \Rightarrow x \approx 3835$ ft.

Solution 2: Notice that $\tan 74^\circ = \dfrac{3400}{a} \Leftrightarrow a = \dfrac{3400}{\tan 74^\circ} \approx 974.9$. By the Pythagorean Theorem, $x^2 = (a + 800)^2 + 3400^2$. So $x = \sqrt{(974.9 + 800)^2 + 3400^2} \approx 3835$ ft.

53. By Heron's formula, $A = \sqrt{s(s-a)(s-b)(s-c)}$, where $s = \dfrac{a+b+c}{2} = \dfrac{112 + 148 + 190}{2} = 225$. Thus, $A = \sqrt{225(225-112)(225-148)(225-190)} \approx 8277.7 \text{ ft}^2$. Since the land value is \$20 per square foot, the value of the lot is approximately $8277.7 \cdot 20 = \$165{,}554$.

55. In any ΔABC, the Law of Cosines gives $a^2 = b^2 + c^2 - 2bc \cdot \cos A$, $b^2 = a^2 + c^2 - 2ac \cdot \cos B$, and $c^2 = a^2 + b^2 - 2ab \cdot \cos C$. Adding the second and third equations gives

$$\begin{aligned} b^2 &= a^2 + c^2 - 2ac \cdot \cos B \\ c^2 &= a^2 + b^2 - 2ab \cdot \cos C \\ b^2 + c^2 &= 2a^2 + b^2 + c^2 - 2a(c\cos B + b\cos C) \end{aligned}$$

Thus $2a^2 - 2a(c\cos B + b\cos C) = 0$, and so $2a(a - c\cos B + b\cos C) = 0$.
Since $a \neq 0$ we must have $a - (c\cos B + b\cos C) = 0 \Leftrightarrow a = b\cos C + c\cos B$. The other laws follow from the symmetry of a, b, and c.

CHAPTER 5 REVIEW

1. **(a)** $30° = 30 \cdot \frac{\pi}{180} = \frac{\pi}{6} \approx 0.52$ rad

(b) $150° = 150 \cdot \frac{\pi}{180} = \frac{5\pi}{6} \approx 2.62$ rad

(c) $-20° = -20 \cdot \frac{\pi}{180} = -\frac{\pi}{9} \approx -0.35$ rad

(d) $-225° = -225 \cdot \frac{\pi}{180} = -\frac{5\pi}{4} \approx -3.93$ rad

3. **(a)** $\frac{5\pi}{6}$ rad $= \frac{5\pi}{6} \cdot \frac{180}{\pi} = 150°$

(b) $-\frac{\pi}{9}$ rad $= -\frac{\pi}{9} \cdot \frac{180}{\pi} = -20°$

(c) $-\frac{4\pi}{3}$ rad $= -\frac{4\pi}{3} \cdot \frac{180}{\pi} = -240°$

(d) 4 rad $= 4 \cdot \frac{180}{\pi} = \frac{720}{\pi} \approx 229.2°$

5. $r = 10$ m, $\theta = \frac{2\pi}{5}$ rad. Then $s = r\theta = 10 \cdot \frac{2\pi}{5} = 4\pi \approx 12.6$ m.

7. $s = 25$ ft, $\theta = 50° = 50 \cdot \frac{\pi}{180} = \frac{5\pi}{18}$ rad. Then $r = \frac{s}{\theta} = 25 \cdot \frac{18}{5\pi} = \frac{90}{\pi} \approx 28.6$ ft.

9. Since the diameter is 28 in, $r = 14$ in. In one revolution, the arc length (distance traveled) is $s = \theta r = 2\pi \cdot 14 = 28\pi$ in. The total distance traveled is 60 mi/h $\cdot$ 0.5 h $=$ 30 mi $=$ 30 mi $\cdot$ 5280 ft/mi $\cdot$ 12 in./ft $=$ 1,900,800 in. The number of revolution is $1{,}900{,}800 \text{ in} \cdot \dfrac{1 \text{ rev}}{28\pi \text{ in.}} \approx 21608.7$ rev. Therefore the car wheel will make approximately 21,609 revolutions.

11. $r = 5$ m, $\theta = 2$ rad. Then $A = \frac{1}{2}r^2\theta = \frac{1}{2} \cdot 5^2 \cdot 2 = 25 \text{ m}^2$.

13. $A = 125 \text{ ft}^2$, $r = 25$ ft. Then $\theta = \dfrac{2A}{r^2} = \dfrac{2 \cdot 125}{25^2} = \frac{250}{625} = 0.4 \text{ rad} \approx 22.9°$

15. The angular speed is $\omega = \dfrac{150 \cdot 2\pi \text{ rad}}{1 \text{ min}} = 300\pi \text{ rad/min} \approx 942.5 \text{ rad/min}$. The linear speed is $v = \dfrac{150 \cdot 2\pi \cdot 8}{1} = 2400\pi \text{ in./min} \approx 7539.8 \text{ in./min} \approx 628.3 \text{ ft/min}$.

17. $r = \sqrt{5^2 + 7^2} = \sqrt{74}$. Then $\sin\theta = \frac{5}{\sqrt{74}}$, $\cos\theta = \frac{7}{\sqrt{74}}$, $\tan\theta = \frac{5}{7}$, $\csc\theta = \frac{\sqrt{74}}{5}$, $\sec\theta = \frac{\sqrt{74}}{7}$, and $\cot\theta = \frac{7}{5}$.

19. $\dfrac{x}{5} = \cos 40° \quad \Leftrightarrow \quad x = 5\cos 40° \approx 3.83$, and $\frac{y}{5} = \sin 40° \quad \Leftrightarrow \quad y = 5\sin 40° \approx 3.21$.

21. $\dfrac{1}{x} = \sin 20° \quad \Leftrightarrow \quad x = \frac{1}{\sin 20°} \approx 2.92$, and $\frac{x}{y} = \cos 20° \quad \Leftrightarrow \quad y = \frac{x}{\cos 20°} \approx \frac{2.924}{0.9397} \approx 3.11$.

23. $A = 90° - 20° = 70°$, $a = 3\cos 20° \approx 2.819$, and $b = 3\sin 20° \approx 1.026$.

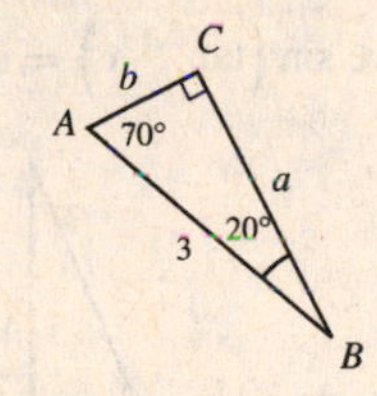

25. $c = \sqrt{25^2 - 7^2} = 24$, $A = \sin^{-1}\frac{7}{24} \approx 0.2960 \approx 17.0°$, and $C = \sin^{-1}\frac{24}{25} \approx 1.2870 \approx 73.7°$.

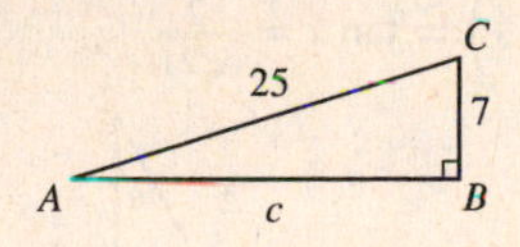

27. $\tan\theta = \dfrac{1}{a} \Leftrightarrow a = \dfrac{1}{\tan\theta} = \cot\theta$, $\sin\theta = \dfrac{1}{b} \Leftrightarrow b = \dfrac{1}{\sin\theta} = \csc\theta$

29.

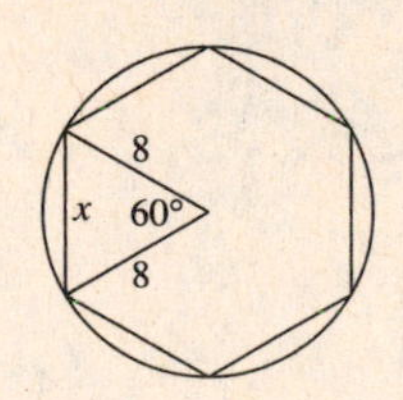

One side of the hexagon together with radial line segments through its endpoints forms a triangle with two sides of length 8 m and subtended angle 60°. Let x be the length of one such side (in meters). By the Law of Cosines, $x^2 = 8^2 + 8^2 - 2 \cdot 8 \cdot 8 \cdot \cos 60° = 64 \quad \Leftrightarrow \quad x = 8$. Thus the perimeter of the hexagon is $6x = 6 \cdot 8 = 48$ m.

31. Let r represent the radius, in miles, of the moon. Then $\tan\frac{\theta}{2}=\frac{r}{r+|AB|}$, $\theta=0.518°$ $\Leftrightarrow$ $r=(r+236{,}900)\cdot\tan 0.259°$ $\Leftrightarrow$ $r(1-\tan 0.259°)=236{,}900\cdot\tan 0.259°$ $\Leftrightarrow r=\frac{236{,}900\cdot\tan 0.259°}{1-\tan 0.259°}\approx 1076$ and so the radius of the moon is roughly 1076 miles.

33. $\sin 315°=-\sin 45°=-\frac{1}{\sqrt{2}}=-\frac{\sqrt{2}}{2}$

35. $\tan(-135°)=\tan 45°=1$

37. $\cot\left(-\frac{22\pi}{3}\right)=\cot\frac{2\pi}{3}=\cot\frac{\pi}{3}=-\frac{1}{\sqrt{3}}=-\frac{\sqrt{3}}{3}$

39. $\cos 585°=\cos 225°=-\cos 45°=-\frac{1}{\sqrt{2}}=-\frac{\sqrt{2}}{2}$

41. $\csc\frac{8\pi}{3}=\csc\frac{2\pi}{3}=\csc\frac{\pi}{3}=\frac{2}{\sqrt{3}}=\frac{2\sqrt{3}}{3}$

43. $\cot(-390°)=\cot(-30°)=-\cot 30°=-\sqrt{3}$

45. $r=\sqrt{(-5)^2+12^2}=\sqrt{169}=13$. Then $\sin\theta=\frac{12}{13}$, $\cos\theta=-\frac{5}{13}$, $\tan\theta=-\frac{12}{5}$, $\csc\theta=\frac{13}{12}$, $\sec\theta=-\frac{13}{5}$, and $\cot\theta=-\frac{5}{12}$.

47. $y-\sqrt{3}x+1=0\Leftrightarrow y=\sqrt{3}x-1$, so the slope of the line is $m=\sqrt{3}$. Then $\tan\theta=m=\sqrt{3}\Leftrightarrow\theta=60°$.

49. Since $\sin\theta$ is positive in quadrant II, $\sin\theta=\sqrt{1-\cos^2\theta}$ and we have $\tan\theta=\frac{\sin\theta}{\cos\theta}=\frac{\sqrt{1-\cos^2\theta}}{\cos\theta}$.

51. $\tan^2\theta=\frac{\sin^2\theta}{\cos^2\theta}=\frac{\sin^2\theta}{1-\sin^2\theta}$

53. $\tan\theta=\frac{\sqrt{7}}{3}$, $\sec\theta=\frac{4}{3}$. Then $\cos\theta=\frac{3}{4}$ and $\sin\theta=\tan\theta\cdot\cos\theta=\frac{\sqrt{7}}{4}$, $\csc\theta=\frac{4}{\sqrt{7}}=\frac{4\sqrt{7}}{7}$, and $\cot\theta=\frac{3}{\sqrt{7}}=\frac{3\sqrt{7}}{7}$.

55. $\sin\theta=\frac{3}{5}$. Since $\cos\theta<0$, θ is in quadrant II. Thus, $x=-\sqrt{5^2-3^2}=-\sqrt{16}=-4$ and so $\cos\theta=-\frac{4}{5}$, $\tan\theta=-\frac{3}{4}$, $\csc\theta=\frac{5}{3}$, $\sec\theta=-\frac{5}{4}$, $\cot\theta=-\frac{4}{3}$.

57. $\tan\theta=-\frac{1}{2}$. $\sec^2\theta=1+\tan^2\theta=1+\frac{1}{4}=\frac{5}{4}\Leftrightarrow\cos^2\theta=\frac{4}{5}\Rightarrow\cos\theta=-\sqrt{\frac{4}{5}}=-\frac{2}{\sqrt{5}}$ since $\cos\theta<0$ in quadrant II. But $\tan\theta=\frac{\sin\theta}{\cos\theta}=-\frac{1}{2}\Leftrightarrow\sin\theta=-\frac{1}{2}\cos\theta=-\frac{1}{2}\left(-\frac{2}{\sqrt{5}}\right)=\frac{1}{\sqrt{5}}$. Therefore, $\sin\theta+\cos\theta=\frac{1}{\sqrt{5}}+\left(-\frac{2}{\sqrt{5}}\right)=-\frac{1}{\sqrt{5}}=-\frac{\sqrt{5}}{5}$.

59. By the Pythagorean Theorem, $\sin^2\theta+\cos^2\theta=1$ for any angle θ.

61. $\sin^{-1}\frac{\sqrt{3}}{2}=\frac{\pi}{3}$

63. Let $u=\sin^{-1}\frac{2}{5}$ and so $\sin u=\frac{2}{5}$. Then from the triangle, $\tan\left(\sin^{-1}\frac{2}{5}\right)=\tan u=\frac{2}{\sqrt{21}}$.

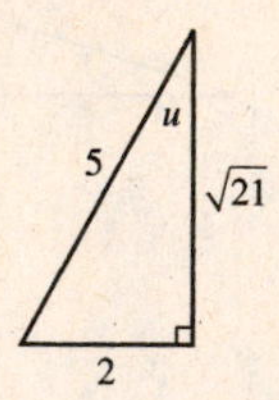

65. Let $\theta=\tan^{-1}x$ $\Leftrightarrow$ $\tan\theta=x$. Then from the triangle, we have $\sin\left(\tan^{-1}x\right)=\sin\theta=\frac{x}{\sqrt{1+x^2}}$.

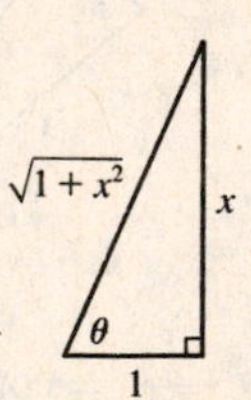

67. $\cos\theta=\frac{x}{3}\Rightarrow\theta=\cos^{-1}\left(\frac{x}{3}\right)$

69. $\angle B=180°-30°-80°=70°$, and so by the Law of Sines, $x=\frac{10\sin 30°}{\sin 70°}\approx 5.32$.

71. $x^2=100^2+210^2-2\cdot 100\cdot 210\cdot\cos 40°\approx 21{,}926.133\Leftrightarrow x\approx 148.07$

73. $x^2=2^2+8^2-2(2)(8)\cos 120°=84\Leftrightarrow x\approx\sqrt{84}\approx 9.17$

75. By the Law of Sines, $\dfrac{\sin\theta}{23} = \dfrac{\sin 25^\circ}{12} \Rightarrow \sin\theta = \dfrac{23\sin 25^\circ}{12} \Rightarrow \theta = \sin^{-1}\left(\dfrac{23\sin 25^\circ}{12}\right) \approx 54.1^\circ$ or $\theta \approx 180^\circ - 54.1^\circ = 125.9^\circ$.

77. By the Law of Cosines, $120^2 = 100^2 + 85^2 - 2(100)(85)\cos\theta$, so $\cos\theta = \dfrac{120^2 - 100^2 - 85^2}{-2(100)(85)} \approx 0.16618$. Thus, $\theta \approx \cos^{-1}(0.16618) \approx 80.4^\circ$.

79. After 2 hours the ships have traveled distances $d_1 = 40$ mi and $d_2 = 56$ mi. The subtended angle is $180^\circ - 32^\circ - 42^\circ = 106^\circ$. Let d be the distance between the two ships in miles. Then by the Law of Cosines, $d^2 = 40^2 + 56^2 - 2(40)(56)\cos 106^\circ \approx 5970.855 \Leftrightarrow d \approx 77.3$ miles.

81. Let d be the distance, in miles, between the points A and B. Then by the Law of Cosines, $d^2 = 3.2^2 + 5.6^2 - 2(3.2)(5.6)\cos 42^\circ \approx 14.966 \Leftrightarrow d \approx 3.9$ mi.

83. $A = \frac{1}{2}ab\sin\theta = \frac{1}{2}(8)(14)\sin 35^\circ \approx 32.12$

CHAPTER 5 TEST

1. $330^\circ = 330 \cdot \frac{\pi}{180} = \frac{11\pi}{6}$ rad. $-135^\circ = -135 \cdot \frac{\pi}{180} = -\frac{3\pi}{4}$ rad.

3. (a) The angular speed is

$$\omega = \frac{120 \cdot 2\pi \text{ rad}}{1 \text{ min}} = 240\pi \text{ rad/min} \approx 753.98 \text{ rad/min}.$$

(b) The linear speed is

$$v = \frac{120 \cdot 2\pi \cdot 16}{1} = 3840\pi \text{ ft/min}$$
$$\approx 12{,}063.7 \text{ ft/min}$$
$$\approx 137 \text{ mi/h}$$

5. $r = \sqrt{3^2 + 2^2} = \sqrt{13}$. Then

$$\tan\theta + \sin\theta = \frac{2}{3} + \frac{2}{\sqrt{13}} = \frac{2\left(\sqrt{13}+3\right)}{3\sqrt{13}} = \frac{26 + 6\sqrt{13}}{39}.$$

7. $\cos\theta = -\frac{1}{3}$ and θ is in quadrant III, so $r = 3$, $x = -1$, and $y = -\sqrt{3^2 - 1^2} = -2\sqrt{2}$. Then $\tan\theta\cot\theta + \csc\theta = \tan\theta \cdot \dfrac{1}{\tan\theta} + \csc\theta = 1 - \frac{3}{2\sqrt{2}} = \frac{2\sqrt{2}-3}{2\sqrt{2}} = \frac{4-3\sqrt{2}}{4}$.

9. $\sec^2\theta = 1 + \tan^2\theta \Leftrightarrow \tan\theta = \pm\sqrt{\sec^2\theta - 1}$. Thus, $\tan\theta = -\sqrt{\sec^2\theta - 1}$ since $\tan\theta < 0$ in quadrant II.

11. (a) $\tan\theta = \dfrac{x}{4} \Rightarrow \theta = \tan^{-1}\left(\dfrac{x}{4}\right)$

(b) $\cos\theta = \dfrac{3}{x} \Rightarrow \theta = \cos^{-1}\left(\dfrac{3}{x}\right)$

13. By the Law of Cosines, $x^2 = 10^2 + 12^2 - 2(10)(12) \cdot \cos 48^\circ \approx 8.409 \Leftrightarrow x \approx 9.1$.

15. Let h be the height of the shorter altitude. Then $\tan 20^\circ = \dfrac{h}{50} \Leftrightarrow h = 50\tan 20^\circ$ and $\tan 28^\circ = \dfrac{x+h}{50} \Leftrightarrow x + h = 50\tan 28^\circ \Leftrightarrow x = 50\tan 28^\circ - h = 50\tan 28^\circ - 50\tan 20^\circ \approx 8.4$.

17. By the Law of Cosines, $9^2 = 8^2 + 6^2 - 2(8)(6)\cos\theta \Rightarrow \cos\theta = \dfrac{8^2 + 6^2 - 9^2}{2(8)(6)} \approx 0.1979$, so $\theta \approx \cos^{-1}(0.1979) \approx 78.6^\circ$.

19. (a) $A(\text{sector}) = \frac{1}{2}r^2\theta = \frac{1}{2} \cdot 10^2 \cdot 72 \cdot \frac{\pi}{180} = 50 \cdot \frac{72\pi}{180}$. $A(\text{triangle}) = \frac{1}{2}r \cdot r\sin\theta = \frac{1}{2} \cdot 10^2 \sin 72^\circ$. Thus, the area of the shaded region is $A(\text{shaded}) = A(\text{sector}) - A(\text{triangle}) = 50\left(\frac{72\pi}{180} - \sin 72^\circ\right) \approx 15.3 \text{ m}^2$.

(b) The shaded region is bounded by two pieces: one piece is part of the triangle, the other is part of the circle. The first part has length $l = \sqrt{10^2 + 10^2 - 2(10)(10) \cdot \cos 72^\circ} = 10\sqrt{2 - 2 \cdot \cos 72^\circ}$. The second has length $s = 10 \cdot 72 \cdot \frac{\pi}{180} = 4\pi$. Thus, the perimeter of the shaded region is $p = l + s = 10\sqrt{2 - 2\cos 72^\circ} + 4\pi \approx 24.3$ m.

21. Label the figure as shown. Now $\angle\beta = 85° - 75° = 10°$, so by the Law of Sines,

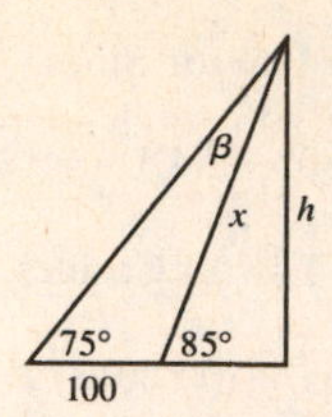

$\dfrac{x}{\sin 75°} = \dfrac{100}{\sin 10°} \quad\Leftrightarrow\quad x = 100 \cdot \dfrac{\sin 75°}{\sin 10°}$. Now $\sin 85° = \dfrac{h}{x} \quad\Leftrightarrow$

$h = x \sin 85° = 100 \cdot \dfrac{\sin 75°}{\sin 10°} \sin 85° \approx 554.$

FOCUS ON MODELING Surveying

1. Let x be the distance between the church and City Hall. To apply the Law of Sines to the triangle with vertices at City Hall, the church, and the first bridge, we first need the measure of the angle at the first bridge, which is $180° - 25° - 30° = 125°$. Then $\dfrac{x}{\sin 125°} = \dfrac{0.86}{\sin 30°} \Leftrightarrow x = \dfrac{0.86 \sin 125°}{\sin 30°} \approx 1.4089$. So the distance between the church and City Hall is about 1.41 miles.

3. First notice that $\angle DBC = 180° - 20° - 95° - 65°$ and $\angle DAC = 180° - 60° - 45° = 75°$. From $\triangle ACD$ we get $\dfrac{|AC|}{\sin 45°} = \dfrac{20}{\sin 75°} \Leftrightarrow |AC| = \dfrac{20 \sin 45°}{\sin 75°} \approx 14.6°$. From $\triangle BCD$ we get $\dfrac{|BC|}{\sin 95°} = \dfrac{20}{\sin 65°} \Leftrightarrow |BC| = \dfrac{20 \sin 95°}{\sin 65°} \approx 22.0$. By applying the Law of Cosines to $\triangle ABC$ we get $|AB|^2 = |AC|^2 + |BC|^2 - 2\,|AC|\,|BC| \cos 40° \approx 14.6^2 + 22.0^2 - 2 \cdot 14.6 \cdot 22.0 \cdot \cos 40° \approx 205$, so $|AB| \approx \sqrt{205} \approx 14.3$ m. Therefore, the distance between A and B is approximately 14.3 m.

5. (a) In $\triangle ABC$, $\angle B = 180° - \beta$, so $\angle C = 180° - \alpha - (180° - \beta) = \beta - \alpha$. By the Law of Sines, $\dfrac{|BC|}{\sin \alpha} = \dfrac{|AB|}{\sin(\beta - \alpha)}$

$\Rightarrow |BC| = |AB| \dfrac{\sin \alpha}{\sin(\beta - \alpha)} = \dfrac{d \sin \alpha}{\sin(\beta - \alpha)}$.

(b) From part (a) we know that $|BC| = \dfrac{d \sin \alpha}{\sin(\beta - \alpha)}$. But $\sin \beta = \dfrac{h}{|BC|} \Leftrightarrow |BC| = \dfrac{h}{\sin \beta}$. Therefore,

$|BC| = \dfrac{d \sin \alpha}{\sin(\beta - \alpha)} = \dfrac{h}{\sin \beta} \Rightarrow h = \dfrac{d \sin \alpha \sin \beta}{\sin(\beta - \alpha)}$.

(c) $h = \dfrac{d \sin \alpha \sin \beta}{\sin(\beta - \alpha)} = \dfrac{800 \sin 25° \sin 29°}{\sin 4°} \approx 2350$ ft

7. We start by labeling the edges and calculating the remaining angles, as shown in the first figure. Using the Law of Sines, we find the following: $\dfrac{a}{\sin 29^\circ} = \dfrac{150}{\sin 60^\circ} \Leftrightarrow a = \dfrac{150 \sin 29^\circ}{\sin 60^\circ} \approx 83.97$, $\dfrac{b}{\sin 91^\circ} = \dfrac{150}{\sin 60^\circ} \Leftrightarrow b = \dfrac{150 \sin 91^\circ}{\sin 60^\circ} \approx 173.18$, $\dfrac{c}{\sin 32^\circ} = \dfrac{173.18}{\sin 87^\circ} \Leftrightarrow c = \dfrac{173.18 \sin 32^\circ}{\sin 87^\circ} \approx 91.90$, $\dfrac{d}{\sin 61^\circ} = \dfrac{173.18}{\sin 87^\circ} \Leftrightarrow e = \dfrac{173.18 \sin 61^\circ}{\sin 87^\circ} \approx 151.67$, $\dfrac{e}{\sin 41^\circ} = \dfrac{151.67}{\sin 51^\circ} \Leftrightarrow e = \dfrac{151.67 \sin 41^\circ}{\sin 51^\circ} \approx 128.04$, $\dfrac{f}{\sin 88^\circ} = \dfrac{151.67}{\sin 51^\circ} \Leftrightarrow f = \dfrac{151.67 \sin 88^\circ}{\sin 51^\circ} \approx 195.04$, $\dfrac{g}{\sin 50^\circ} = \dfrac{195.04}{\sin 92^\circ} \Leftrightarrow g = \dfrac{195.04 \sin 50^\circ}{\sin 92^\circ} \approx 149.50$, and $\dfrac{h}{\sin 38^\circ} = \dfrac{195.04}{\sin 92^\circ} \Leftrightarrow h = \dfrac{195.04 \sin 38^\circ}{\sin 92^\circ} \approx 120.15$. Note that we used two decimal places throughout our calculations. Our results are shown (to one decimal place) in the second figure.

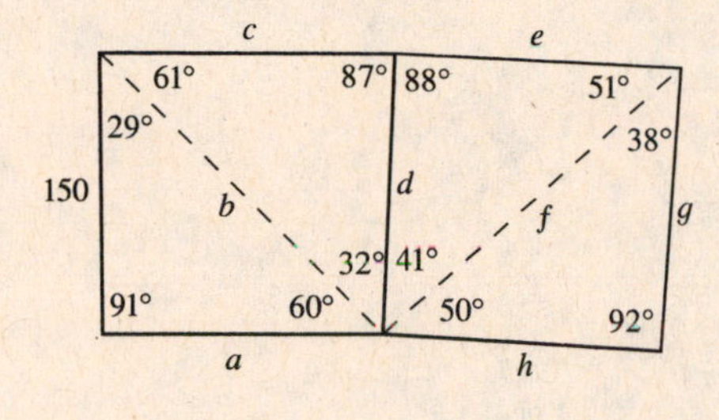

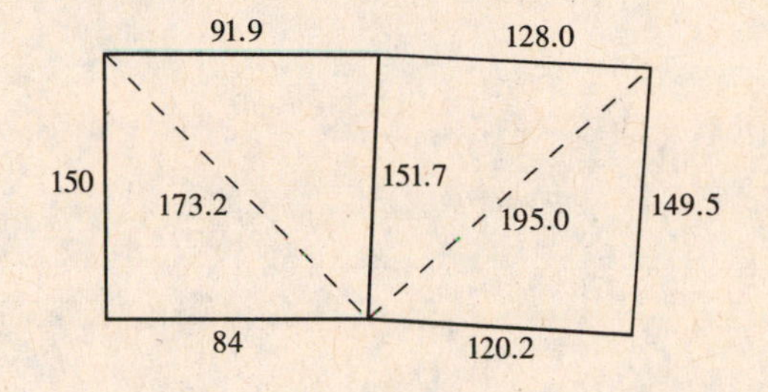

6 TRIGONOMETRIC FUNCTIONS: UNIT CIRCLE APPROACH

6.1 THE UNIT CIRCLE

1. (a) The unit circle is the circle centered at (0, 0) with radius 1.

(b) The equation of the unit circle is $x^2 + y^2 = 1$.

(c) (i) Since $1^2 + 0^2 = 1$, the point is $P(1, 0)$. **(ii)** $P(0, 1)$ **(iii)** $P(-1, 0)$ **(iv)** $P(0, -1)$

3. Since $\left(\frac{3}{5}\right)^2 + \left(-\frac{4}{5}\right)^2 = \frac{9}{25} + \frac{16}{25} = 1$, $P\left(\frac{3}{5}, -\frac{4}{5}\right)$ lies on the unit circle.

5. Since $\left(\frac{3}{4}\right)^2 + \left(-\frac{\sqrt{7}}{4}\right)^2 = \frac{9}{16} + \frac{7}{16} = 1$, $P\left(\frac{3}{4}, -\frac{\sqrt{7}}{4}\right)$ lies on the unit circle.

7. Since $\left(-\frac{\sqrt{5}}{3}\right)^2 + \left(\frac{2}{3}\right)^2 = \frac{5}{9} + \frac{4}{9} = 1$, $P\left(-\frac{\sqrt{5}}{3}, \frac{2}{3}\right)$ lies on the unit circle.

9. $\left(-\frac{3}{5}\right)^2 + y^2 = 1 \Leftrightarrow y^2 = 1 - \frac{9}{25} \Leftrightarrow y^2 = \frac{16}{25} \Leftrightarrow y = \pm\frac{4}{5}$. Since $P(x, y)$ is in quadrant III, y is negative, so the point is $P\left(-\frac{3}{5}, -\frac{4}{5}\right)$.

11. $x^2 + \left(\frac{1}{3}\right)^2 = 1 \Leftrightarrow x^2 = 1 - \frac{1}{9} \Leftrightarrow x^2 = \frac{8}{9} \Leftrightarrow x = \pm\frac{2\sqrt{2}}{3}$. Since P is in quadrant II, x is negative, so the point is $P\left(-\frac{2\sqrt{2}}{3}, \frac{1}{3}\right)$.

13. $x^2 + \left(-\frac{2}{7}\right)^2 = 1 \Leftrightarrow x^2 = 1 - \frac{4}{49} \Leftrightarrow x^2 = \frac{45}{49} \Leftrightarrow x = \pm\frac{3\sqrt{5}}{7}$. Since $P(x, y)$ is in quadrant IV, x is positive, so the point is $P\left(\frac{3\sqrt{5}}{7}, -\frac{2}{7}\right)$.

15. $\left(\frac{5}{13}\right)^2 + y^2 = 1 \Leftrightarrow y^2 = 1 - \frac{25}{169} \Leftrightarrow y^2 = \frac{144}{169} \Leftrightarrow y = \pm\frac{12}{13}$. Since its y-coordinate is negative, the point is $P\left(\frac{5}{13}, -\frac{12}{13}\right)$.

17. $x^2 + \left(\frac{2}{3}\right)^2 = 1 \Leftrightarrow x^2 = 1 - \frac{4}{9} \Leftrightarrow x^2 = \frac{5}{9} \Leftrightarrow x = \pm\frac{\sqrt{5}}{3}$. Since its x-coordinate is negative, the point is $P\left(-\frac{\sqrt{5}}{3}, \frac{2}{3}\right)$.

19. $\left(-\frac{\sqrt{2}}{3}\right)^2 + y^2 = 1 \Leftrightarrow y^2 = 1 - \frac{2}{9} \Leftrightarrow y^2 = \frac{7}{9} \Leftrightarrow y = \pm\frac{\sqrt{7}}{3}$. Since P lies below the x-axis, its y-coordinate is negative, so the point is $P\left(-\frac{\sqrt{2}}{3}, -\frac{\sqrt{7}}{3}\right)$.

21.

t	Terminal Point
0	$(1, 0)$
$\frac{\pi}{4}$	$\left(\frac{\sqrt{2}}{2}, \frac{\sqrt{2}}{2}\right)$
$\frac{\pi}{2}$	$(0, 1)$
$\frac{3\pi}{4}$	$\left(-\frac{\sqrt{2}}{2}, \frac{\sqrt{2}}{2}\right)$
π	$(-1, 0)$

t	Terminal Point
π	$(-1, 0)$
$\frac{5\pi}{4}$	$\left(-\frac{\sqrt{2}}{2}, -\frac{\sqrt{2}}{2}\right)$
$\frac{3\pi}{2}$	$(0, -1)$
$\frac{7\pi}{4}$	$\left(\frac{\sqrt{2}}{2}, -\frac{\sqrt{2}}{2}\right)$
2π	$(1, 0)$

23. $\bar{t} = 0$, so $t = 4\pi$ corresponds to $P(x, y) = (1, 0)$.

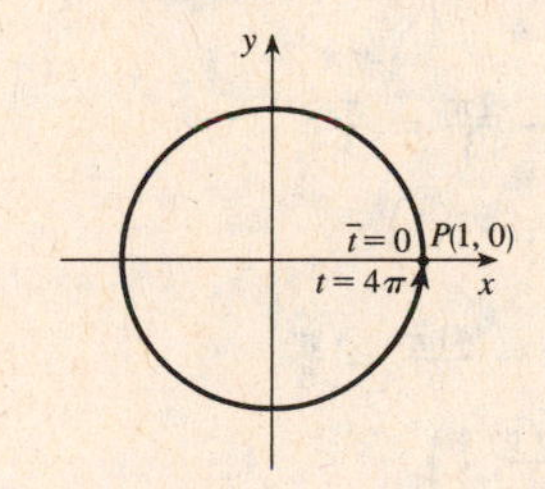

25. $\bar{t} = \frac{\pi}{2}$, so $t = \frac{3\pi}{2}$ corresponds to $P(x, y) = (0, -1)$.

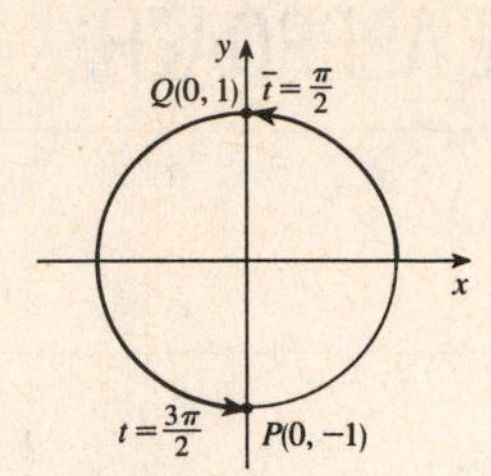

27. $t = -\frac{\pi}{6}$ corresponds to $P(x, y) = \left(\frac{\sqrt{3}}{2}, -\frac{1}{2}\right)$.

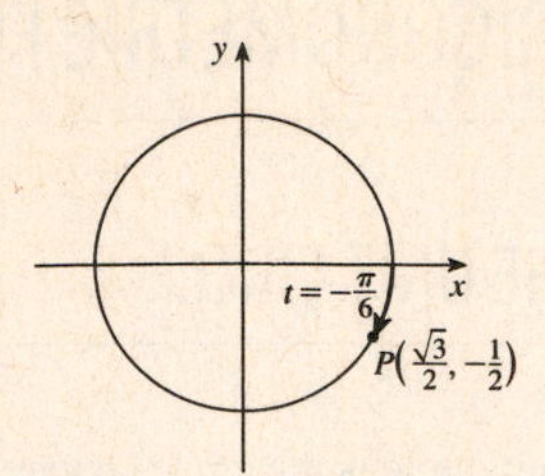

29. $\bar{t} = \frac{\pi}{4}$ and $t = \frac{5\pi}{4}$ corresponds to

$P(x, y) = \left(-\frac{\sqrt{2}}{2}, -\frac{\sqrt{2}}{2}\right)$.

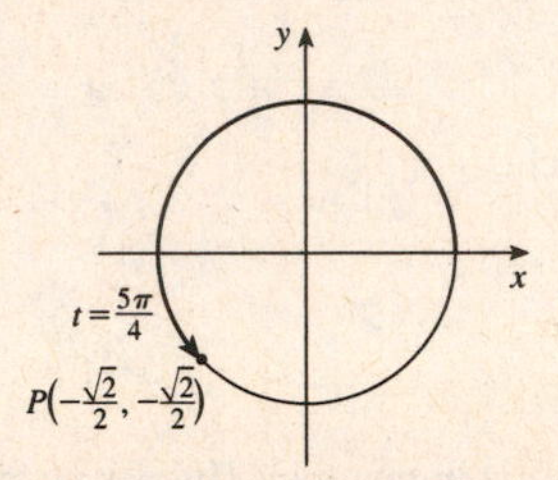

31. $\bar{t} = \frac{\pi}{6}$ and $t = -\frac{7\pi}{6}$ corresponds to

$P(x, y) = \left(-\frac{\sqrt{3}}{2}, \frac{1}{2}\right)$.

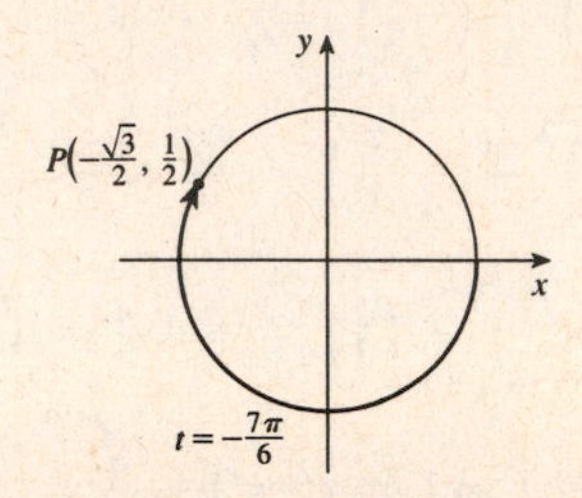

33. $\bar{t} = \frac{\pi}{4}$ and $t = -\frac{7\pi}{4}$ corresponds to

$P(x, y) = \left(\frac{\sqrt{2}}{2}, \frac{\sqrt{2}}{2}\right)$.

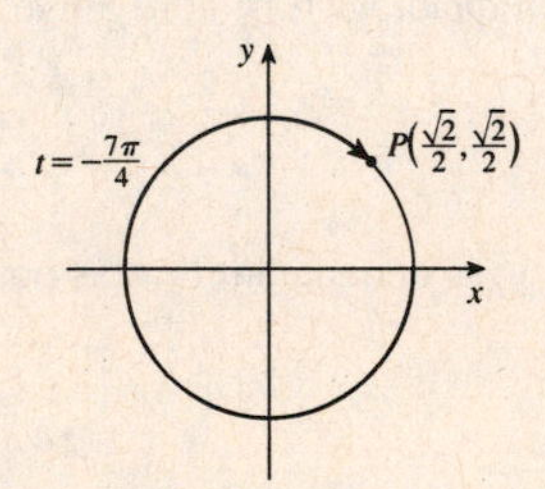

35. $\bar{t} = \frac{\pi}{4}$ and $t = -\frac{3\pi}{4}$ corresponds to

$P(x, y) = \left(-\frac{\sqrt{2}}{2}, -\frac{\sqrt{2}}{2}\right)$.

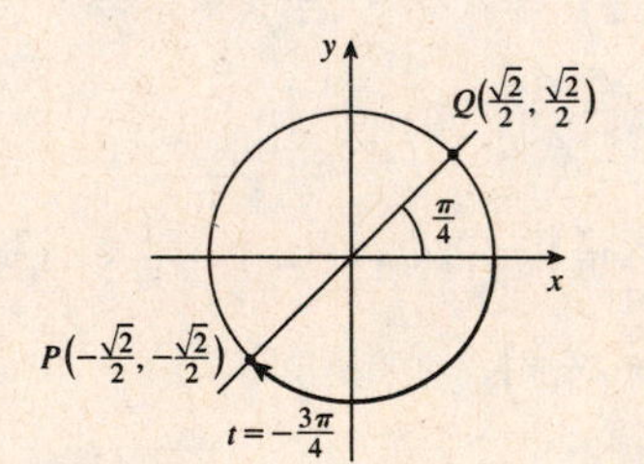

37. **(a)** $\bar{t} = \frac{4\pi}{3} - \pi = \frac{\pi}{3}$

(b) $\bar{t} = 2\pi - \frac{5\pi}{3} = \frac{\pi}{3}$

(c) $\bar{t} = -\pi - \left(-\frac{7\pi}{6}\right) = \frac{\pi}{6}$

(d) $\bar{t} = 3.5 - \pi \approx 0.36$

39. **(a)** $\bar{t} = \pi - \frac{5\pi}{7} = \frac{2\pi}{7}$

(b) $\bar{t} = \pi - \frac{7\pi}{9} = \frac{2\pi}{9}$

(c) $\bar{t} = \pi - 3 \approx 0.142$

(d) $\bar{t} = 2\pi - 5 \approx 1.283$

41. **(a)** $\bar{t} = 2\pi - \frac{11\pi}{6} = \frac{\pi}{6}$

(b) $P\left(\frac{\sqrt{3}}{2}, -\frac{1}{2}\right)$

43. **(a)** $\bar{t} = -\pi - \left(-\frac{4\pi}{3}\right) = \frac{\pi}{3}$

(b) $P\left(-\frac{1}{2}, \frac{\sqrt{3}}{2}\right)$

45. **(a)** $\bar{t} = \pi - \frac{2\pi}{3} = \frac{\pi}{3}$

(b) $P\left(-\frac{1}{2}, -\frac{\sqrt{3}}{2}\right)$

47. **(a)** $\bar{t} = \frac{13\pi}{4} - 3\pi = \frac{\pi}{4}$

(b) $P\left(-\frac{\sqrt{2}}{2}, -\frac{\sqrt{2}}{2}\right)$

49. **(a)** $\bar{t} = 7\pi - \frac{41\pi}{6} = \frac{\pi}{6}$

(b) $P\left(-\frac{\sqrt{3}}{2}, \frac{1}{2}\right)$

51. **(a)** $\bar{t} = 4\pi - \frac{11\pi}{3} = \frac{\pi}{3}$

(b) $P\left(\frac{1}{2}, \frac{\sqrt{3}}{2}\right)$

53. **(a)** $\bar{t} = \frac{16\pi}{3} - 5\pi = \frac{\pi}{3}$

(b) $P\left(-\frac{1}{2}, -\frac{\sqrt{3}}{2}\right)$.

55. $t = 1 \Rightarrow (0.5, 0.8)$

57. $t = -1.1 \Rightarrow (0.5, -0.9)$

59. Let $Q(x, y) = \left(\frac{3}{5}, \frac{4}{5}\right)$ be the terminal point determined by t.

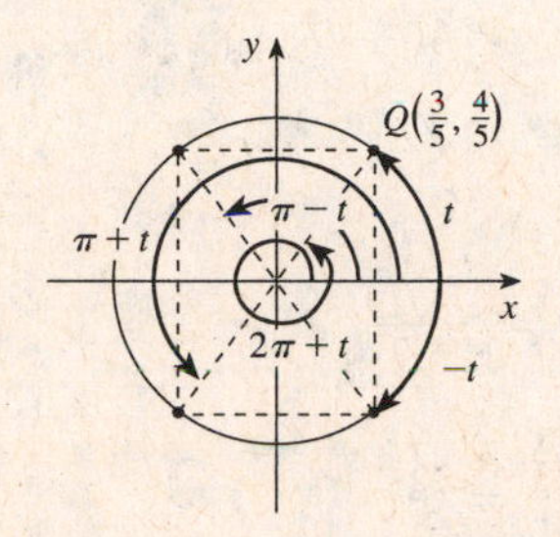

(a) $\pi - t$ determines the point $P(-x, y) = \left(-\frac{3}{5}, \frac{4}{5}\right)$.

(b) $-t$ determines the point $P(x, -y) = \left(\frac{3}{5}, -\frac{4}{5}\right)$.

(c) $\pi + t$ determines the point $P(-x, -y) = \left(-\frac{3}{5}, -\frac{4}{5}\right)$.

(d) $2\pi + t$ determines the point $P(x, y) = \left(\frac{3}{5}, \frac{4}{5}\right)$.

61. The distances PQ and PR are equal because they both subtend arcs of length $\frac{\pi}{3}$. Since $P(x, y)$ is a point on the unit circle, $x^2 + y^2 = 1$. Now $d(P, Q) = \sqrt{(x - x)^2 + (y - (-y))^2} = 2y$ and $d(R, S) = \sqrt{(x - 0)^2 + (y - 1)^2} = \sqrt{x^2 + y^2 - 2y + 1} = \sqrt{2 - 2y}$ (using the fact that $x^2 + y^2 = 1$). Setting these equal gives $2y = \sqrt{2 - 2y} \Rightarrow 4y^2 = 2 - 2y \Leftrightarrow 4y^2 + 2y - 2 = 0 \Leftrightarrow 2(2y - 1)(y + 1) = 0$. So $y = -1$ or $y = \frac{1}{2}$. Since P is in quadrant I, $y = \frac{1}{2}$ is the only viable solution. Again using $x^2 + y^2 = 1$ we have $x^2 + \left(\frac{1}{2}\right)^2 = 1 \Leftrightarrow x^2 = \frac{3}{4} \Rightarrow x = \pm\frac{\sqrt{3}}{2}$. Again, since P is in quadrant I the coordinates must be $\left(\frac{\sqrt{3}}{2}, \frac{1}{2}\right)$.

6.2 TRIGONOMETRIC FUNCTIONS OF REAL NUMBERS

1. If $P(x, y)$ is the terminal point on the unit circle determined by t, then $\sin t = y$, $\cos t = x$, and $\tan t = y/x$.

3.

t	$\sin t$	$\cos t$
0	0	1
$\frac{\pi}{4}$	$\frac{\sqrt{2}}{2}$	$\frac{\sqrt{2}}{2}$
$\frac{\pi}{2}$	1	0
$\frac{3\pi}{4}$	$\frac{\sqrt{2}}{2}$	$-\frac{\sqrt{2}}{2}$
π	0	-1
$\frac{5\pi}{4}$	$-\frac{\sqrt{2}}{2}$	$-\frac{\sqrt{2}}{2}$
$\frac{3\pi}{2}$	-1	0
$\frac{7\pi}{4}$	$-\frac{\sqrt{2}}{2}$	$\frac{\sqrt{2}}{2}$
2π	0	1

5. **(a)** $\sin\frac{7\pi}{6} = -\frac{1}{2}$

(b) $\cos\frac{17\pi}{6} = -\frac{\sqrt{3}}{2}$

(c) $\tan\frac{7\pi}{6} = \frac{\sqrt{3}}{3}$

7. **(a)** $\sin\frac{11\pi}{4} = \frac{\sqrt{2}}{2}$

(b) $\sin\left(-\frac{\pi}{4}\right) = -\frac{\sqrt{2}}{2}$

(c) $\sin\frac{5\pi}{4} = -\frac{\sqrt{2}}{2}$

9. **(a)** $\cos\frac{3\pi}{4} = -\frac{\sqrt{2}}{2}$

(b) $\cos\frac{5\pi}{4} = -\frac{\sqrt{2}}{2}$

(c) $\cos\frac{7\pi}{4} = \frac{\sqrt{2}}{2}$

11. **(a)** $\sin\frac{7\pi}{3} = \frac{\sqrt{3}}{2}$

(b) $\csc\frac{7\pi}{3} = \frac{2\sqrt{3}}{3}$

(c) $\cot\frac{7\pi}{3} = \frac{\sqrt{3}}{3}$

13. **(a)** $\cos\left(-\frac{\pi}{3}\right) = \frac{1}{2}$

(b) $\sec\left(-\frac{\pi}{3}\right) = 2$

(c) $\sin\left(-\frac{\pi}{3}\right) = -\frac{\sqrt{3}}{2}$

15. **(a)** $\cos\left(-\frac{\pi}{6}\right) = \frac{\sqrt{3}}{2}$

(b) $\csc\left(-\frac{\pi}{3}\right) = -\frac{2\sqrt{3}}{3}$

(c) $\tan\left(-\frac{\pi}{6}\right) = -\frac{\sqrt{3}}{3}$

17. **(a)** $\csc\frac{7\pi}{6}=-2$
(b) $\sec\left(-\frac{\pi}{6}\right)=\frac{2\sqrt{3}}{3}$
(c) $\cot\left(-\frac{5\pi}{6}\right)=\sqrt{3}$

19. **(a)** $\sin\frac{4\pi}{3}=-\frac{\sqrt{3}}{2}$
(b) $\sec\frac{11\pi}{6}=\frac{2\sqrt{3}}{3}$
(c) $\cot\left(-\frac{\pi}{3}\right)=-\frac{\sqrt{3}}{3}$

21. **(a)** $\sin 13\pi=0$
(b) $\cos 14\pi=1$
(c) $\tan 15\pi=0$

23. $t=0 \Rightarrow \sin t=0, \cos t=1, \tan t=0, \sec t=1$, $\csc t$ and $\cot t$ are undefined.

25. $t=\pi \Rightarrow \sin t=0, \cos t=-1, \tan t=0, \sec t=-1$, $\csc t$ and $\cot t$ are undefined.

27. $\left(-\frac{3}{5}\right)^2+\left(-\frac{4}{5}\right)^2=\frac{9}{25}+\frac{16}{25}=1$. So $\sin t=-\frac{4}{5}$, $\cos t=-\frac{3}{5}$, and $\tan t=\dfrac{-\frac{4}{5}}{-\frac{3}{5}}=\frac{4}{3}$.

29. $\left(-\frac{1}{3}\right)^2+\left(\frac{2\sqrt{2}}{3}\right)^2=\frac{1}{9}+\frac{8}{9}=1$. So $\sin t=\frac{2\sqrt{2}}{3}$, $\cos t=-\frac{1}{3}$, and $\tan t=\dfrac{\frac{2\sqrt{2}}{3}}{-\frac{1}{3}}=-2\sqrt{2}$.

31. $\left(-\frac{6}{7}\right)^2+\left(\frac{\sqrt{13}}{7}\right)^2=\frac{36}{49}+\frac{13}{49}=1$. So $\sin t=\frac{\sqrt{13}}{7}$, $\cos t=-\frac{6}{7}$, and $\tan t=\dfrac{\frac{\sqrt{13}}{7}}{-\frac{6}{7}}=-\frac{\sqrt{13}}{6}$.

33. $\left(-\frac{5}{13}\right)^2+\left(-\frac{12}{13}\right)^2=\frac{25}{169}+\frac{144}{169}=1$. So $\sin t=-\frac{12}{13}$, $\cos t=-\frac{5}{13}$, and $\tan t\dfrac{-\frac{12}{13}}{-\frac{5}{13}}=\frac{12}{5}$.

35. $\left(-\frac{20}{29}\right)^2+\left(\frac{21}{29}\right)^2=\frac{400}{841}+\frac{441}{841}=1$. So $\sin t=\frac{21}{29}$, $\cos t=-\frac{20}{29}$, and $\tan t=\dfrac{\frac{21}{29}}{-\frac{20}{29}}=-\frac{21}{20}$.

37. **(a)** 0.8 **(b)** 0.84147

39. **(a)** 0.9 **(b)** 0.93204

41. **(a)** 1.0 **(b)** 1.02964

43. **(a)** −0.6 **(b)** −0.57482

45. $\sin t\cdot\cos t$. Since $\sin t$ is positive in quadrant II and $\cos t$ is negative in quadrant II, their product is negative.

47. $\dfrac{\tan t\cdot\sin t}{\cot t}=\tan t\cdot\dfrac{1}{\cot t}\cdot\sin t=\tan t\cdot\tan t\cdot\sin t=\tan^2 t\cdot\sin t$. Since $\tan^2 t$ is always positive and $\sin t$ is negative in quadrant III, the expression is negative in quadrant III.

49. Quadrant II

51. Quadrant II

53. $\sin t=\sqrt{1-\cos^2 t}$

55. $\tan t=\dfrac{\sin t}{\cos t}=\dfrac{\sin t}{\sqrt{1-\sin^2 t}}$

57. $\sec t=-\sqrt{1+\tan^2 t}$

59. $\tan t=\sqrt{\sec^2 t-1}$

61. $\tan^2 t=\dfrac{\sin^2 t}{\cos^2 t}=\dfrac{\sin^2 t}{1-\sin^2 t}$

63. $\sin t=-\frac{4}{5}$ and the terminal point of t is in quadrant IV, so the terminal point determined by t is $P\left(x,-\frac{4}{5}\right)$. Since P is on the unit circle, $x^2+\left(-\frac{4}{5}\right)^2=1$. Solving for x gives $x=\pm\sqrt{1-\frac{16}{25}}=\pm\sqrt{\frac{9}{25}}=\pm\frac{3}{5}$. Since the terminal point is in quadrant IV, $x=\frac{3}{5}$. Thus the terminal point is $P\left(\frac{3}{5},-\frac{4}{5}\right)$. Thus, $\cos t=\frac{3}{5}$, $\tan t=-\frac{4}{3}$, $\csc t=-\frac{5}{4}$, $\sec t=\frac{5}{3}$, $\cot t=-\frac{3}{4}$.

65. $\sec t=3$ and the terminal point of t lies in quadrant IV. Thus, $\cos t=\frac{1}{3}$ and the terminal point determined by t is $P\left(\frac{1}{3},y\right)$. Since P is on the unit circle, $\left(\frac{1}{3}\right)^2+y^2=1$. Solving for y gives $y=\pm\sqrt{1-\frac{1}{9}}=\pm\sqrt{\frac{8}{9}}=\pm\frac{2\sqrt{2}}{3}$. Since the terminal point is in quadrant IV, $y=-\frac{2\sqrt{2}}{3}$. Thus the terminal point is $P\left(\frac{1}{3},-\frac{2\sqrt{2}}{3}\right)$. Therefore, $\sin t=-\frac{2\sqrt{2}}{3}$, $\cos t=\frac{1}{3}$, $\tan t=-2\sqrt{2}$, $\csc t=-\frac{3}{2\sqrt{2}}=-\frac{3\sqrt{2}}{4}$, $\cot t=-\frac{1}{2\sqrt{2}}=-\frac{\sqrt{2}}{4}$.

67. $\tan t = -\frac{12}{5}$ and $\sin t > 0$, so t is in quadrant II. Since $\sec^2 t = \tan^2 t + 1$ we have $\sec^2 t = \left(-\frac{12}{5}\right)^2 + 1 = \frac{169}{25}$, so because secant is negative in quadrant II, $\sec t = -\sqrt{\frac{169}{25}} = -\frac{13}{5}$. Thus, $\cos t = \dfrac{1}{\sec t} = -\frac{5}{13}$ and we have $P\left(-\frac{5}{13}, y\right)$. Since $\tan t \cdot \cos t = \sin t$ we have $\sin t = \left(-\frac{12}{5}\right)\left(-\frac{5}{13}\right) = \frac{12}{13}$. Thus, the terminal point determined by t is $P\left(-\frac{5}{13}, \frac{12}{13}\right)$, and so $\sin t = \frac{12}{13}$, $\cos t = -\frac{5}{13}$, $\csc t = \frac{13}{12}$, $\sec t = -\frac{13}{5}$, $\cot t = -\frac{5}{12}$.

69. $\sin t = -\frac{1}{4}$, $\sec t < 0$, so t is in quadrant III. So the terminal point determined by t is $P\left(x, -\frac{1}{4}\right)$. Since P is on the unit circle, $x^2 + \left(-\frac{1}{4}\right)^2 = 1$. Solving for x gives $x = \pm\sqrt{1 - \frac{1}{16}} = \pm\sqrt{\frac{15}{16}} = \pm\frac{\sqrt{15}}{4}$. Since the terminal point is in quadrant III, $x = -\dfrac{\sqrt{15}}{4}$. Thus, the terminal point determined by t is $P\left(-\frac{\sqrt{15}}{4}, -\frac{1}{4}\right)$, and so $\cos t = -\frac{\sqrt{15}}{4}$, $\tan t = \frac{1}{\sqrt{15}} = \frac{\sqrt{15}}{15}$, $\csc t = -4$, $\sec t = -\frac{4}{\sqrt{15}} = -\frac{4\sqrt{15}}{15}$, $\cot t = \sqrt{15}$.

71. $f(-x) = (-x)^2 \sin(-x) = -x^2 \sin x = -f(x)$, so f is odd.

73. $f(-x) = \sin(-x)\cos(-x) = -\sin x \cos x = -f(x)$, so f is odd.

75. $f(-x) = |-x|\cos(-x) = |x|\cos x = f(x)$, so f is even.

77. $f(-x) = (-x)^3 + \cos(-x) = -x^3 + \cos x$ which is neither $f(x)$ nor $-f(x)$, so f is neither even nor odd.

79.

t	0	0.25	0.50	0.75	1.00	1.25
$y(t)$	4	−2.83	0	2.83	−4	2.83

81. **(a)** $I(0.1) = 0.8e^{-0.3}\sin 1 \approx 0.499$ A

(b) $I(0.5) = 0.8e^{-1.5}\sin 5 \approx -0.171$ A

83. Notice that if $P(t) = (x, y)$, then $P(t + \pi) = (-x, -y)$. Thus,

(a) $\sin(t + \pi) = -y$ and $\sin t = y$. Therefore, $\sin(t + \pi) = -\sin t$.

(b) $\cos(t + \pi) = -x$ and $\cos t = x$. Therefore, $\cos(t + \pi) = -\cos t$.

(c) $\tan(t + \pi) = \dfrac{\sin(t + \pi)}{\cos(t + \pi)} = \dfrac{-y}{-x} = \dfrac{y}{x} = \dfrac{\sin t}{\cos t} = \tan t$.

6.3 TRIGONOMETRIC GRAPHS

1. If a function f is periodic with period p, then $f(t + p) = f(t)$ for every t. The trigonometric functions $y = \sin x$ and $\cos x$ are periodic, with period 2π and amplitude 1.

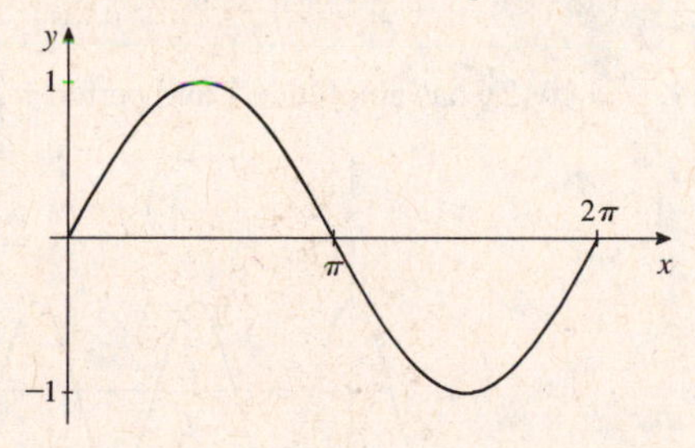

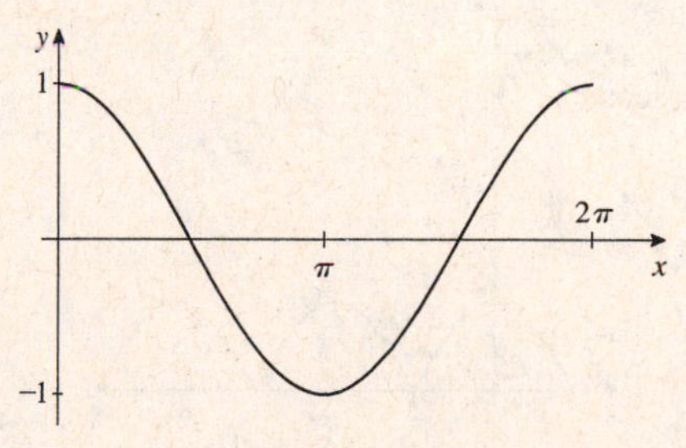

3. The sine and cosine curves $y = a\sin kx$ and $y = a\cos kx$, $k > 0$, have amplitude $|a|$ and period $2\pi/k$. The sine curve $y = 3\sin 2x$ has amplitude $|3| = 3$ and period $2\pi/2 = \pi$.

5. $f(x) = 2 + \sin x$

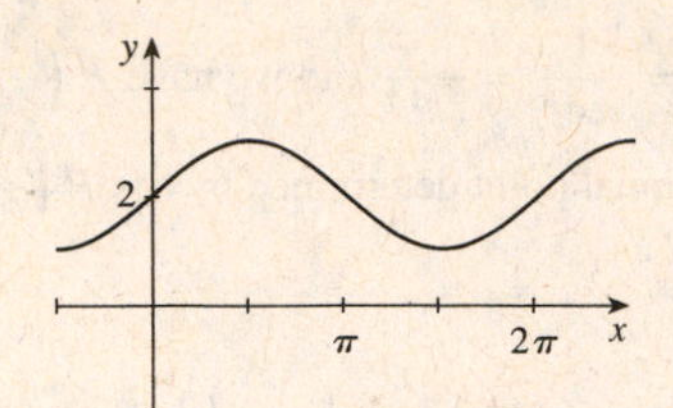

7. $f(x) = -\sin x$

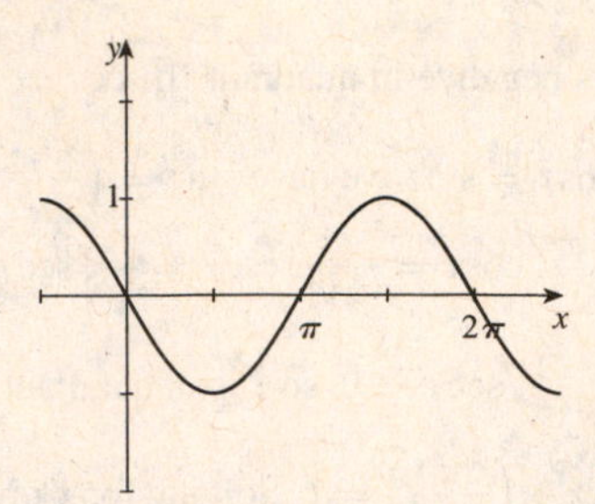

9. $f(x) = -2 + \sin x$

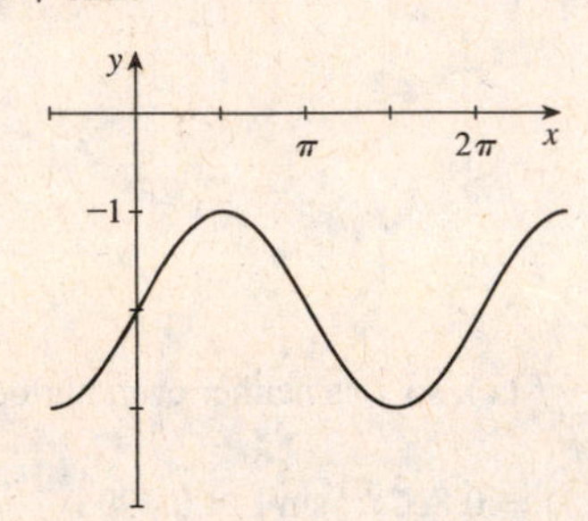

11. $g(x) = 3\cos x$

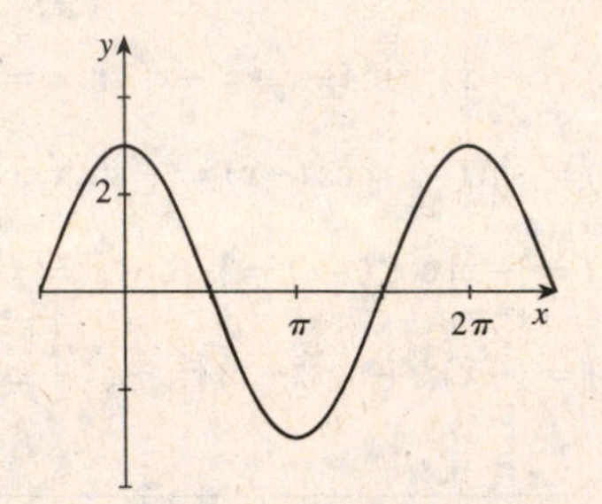

13. $g(x) = -\frac{1}{2}\sin x$

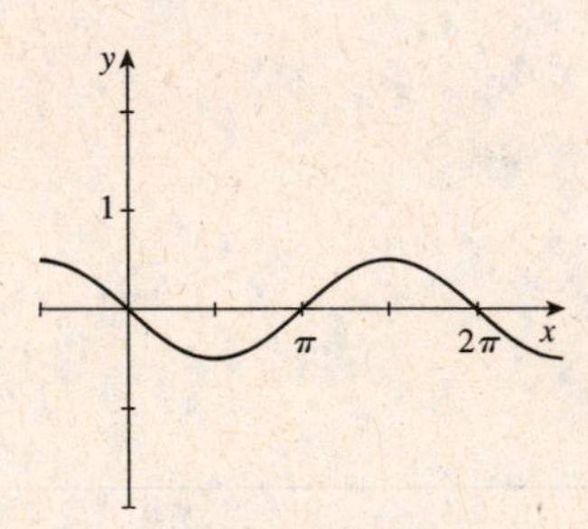

15. $g(x) = 3 + 3\cos x$

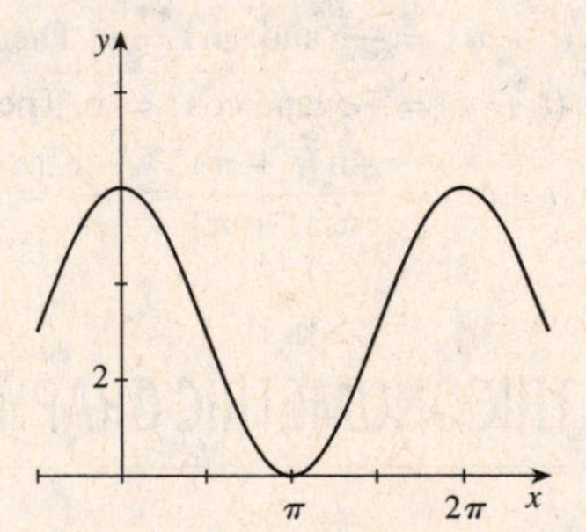

17. $h(x) = |\cos x|$

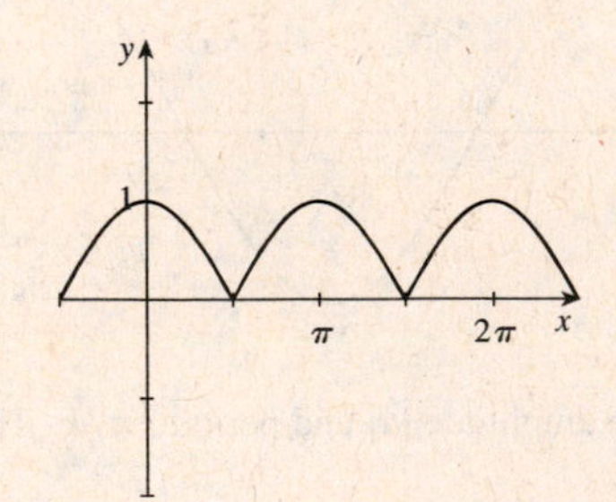

19. $y = \cos 2x$ **has amplitude 1 and period** π**.**

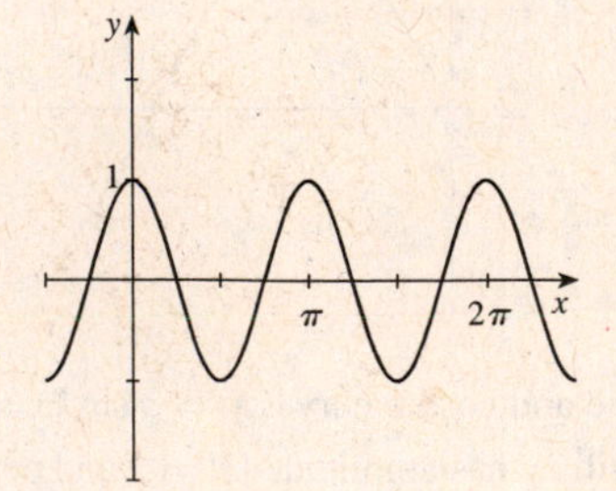

21. $y = -\sin 3x$ has amplitude 1 and period $\frac{2\pi}{3}$.

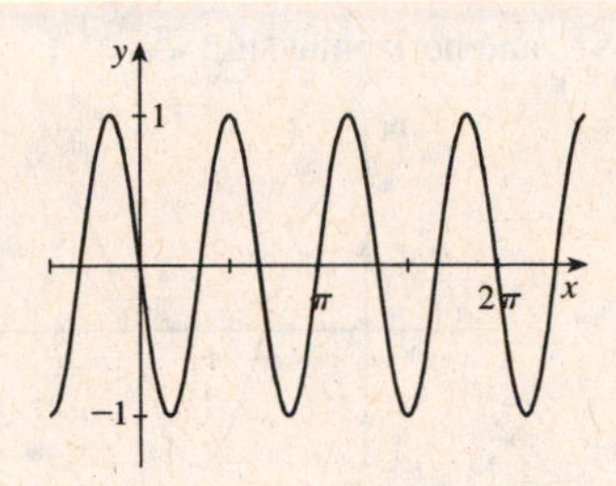

23. $y = -2\cos 3\pi x$ has amplitude 2 and period $\frac{2}{3}$.

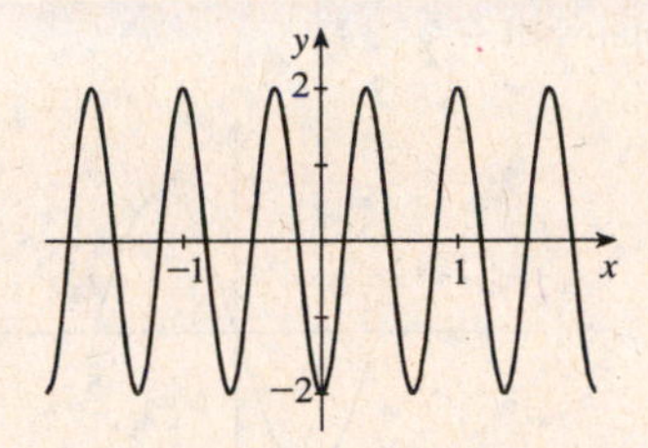

25. $y = 10\sin \frac{1}{2}x$ has amplitude 10 and period 4π.

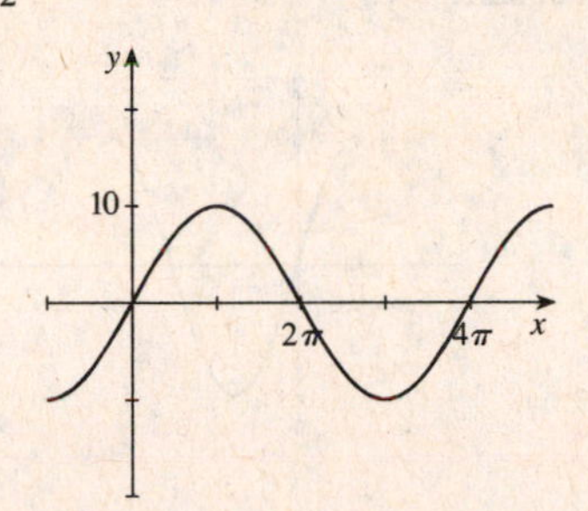

27. $y = -\frac{1}{3}\cos \frac{1}{3}x$ has amplitude $\frac{1}{3}$ and period 6π.

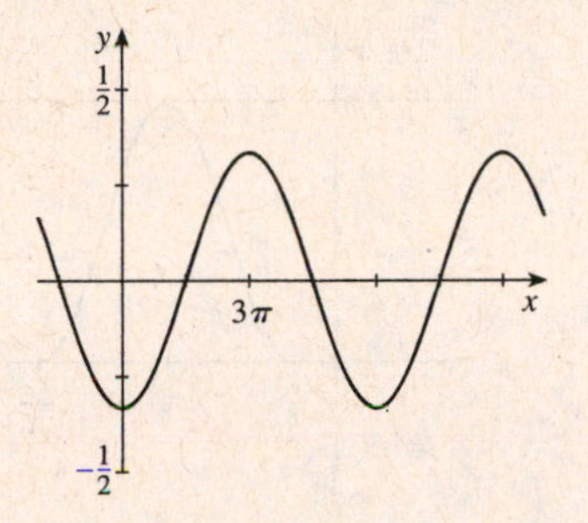

29. $y = -2\sin 2\pi x$ has amplitude 2 and period 1.

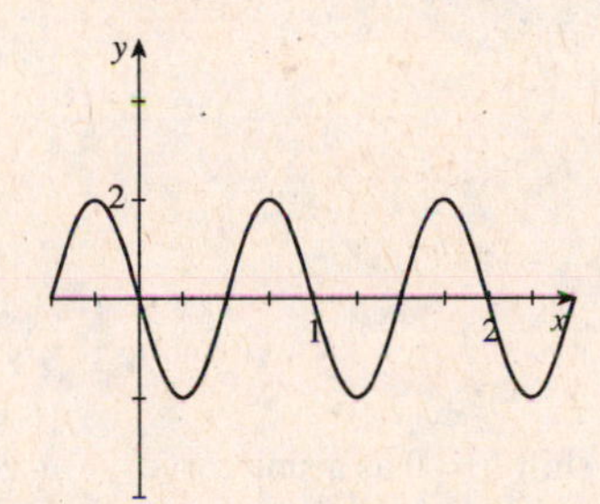

31. $y = 1 + \frac{1}{2}\cos \pi x$ has amplitude $\frac{1}{2}$ and period 2.

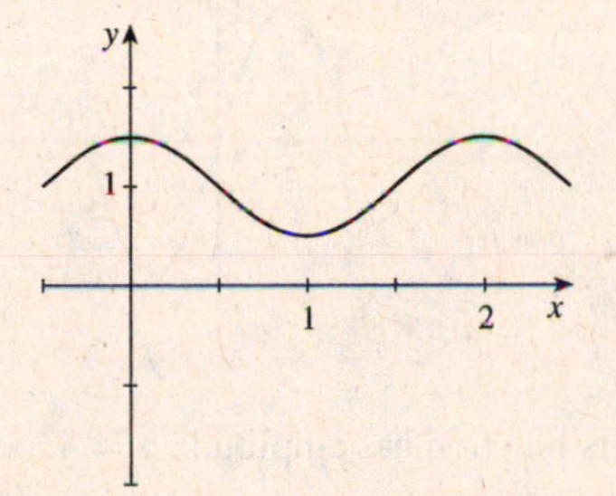

33. $y = \cos\left(x - \frac{\pi}{2}\right)$ has amplitude 1, period 2π, and horizontal shift $\frac{\pi}{2}$.

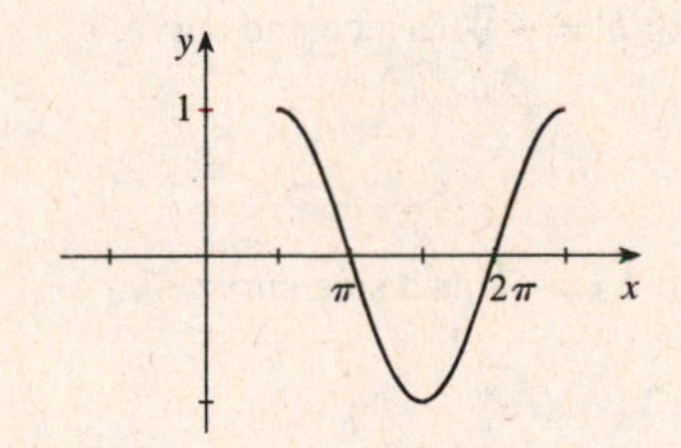

35. $y = -2\sin\left(x - \frac{\pi}{6}\right)$ has amplitude 2, period 2π, and horizontal shift $\frac{\pi}{6}$.

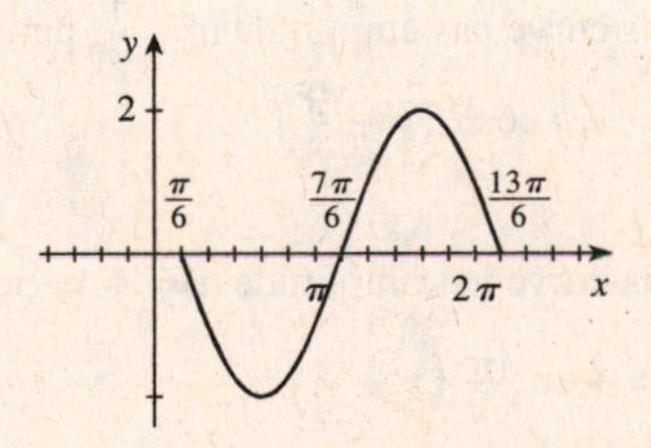

37. $y = -4\sin 2\left(x + \frac{\pi}{2}\right)$ has amplitude 4, period π, and horizontal shift $-\frac{\pi}{2}$.

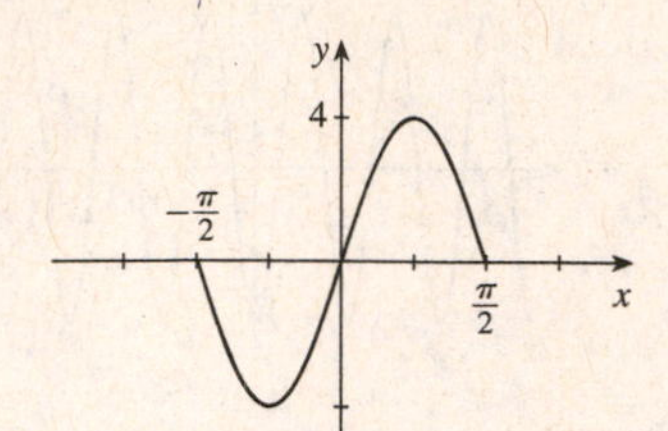

39. $y = 5\cos\left(3x - \frac{\pi}{4}\right) = 5\cos 3\left(x - \frac{\pi}{12}\right)$ has amplitude 5, period $\frac{2\pi}{3}$, and horizontal shift $\frac{\pi}{12}$.

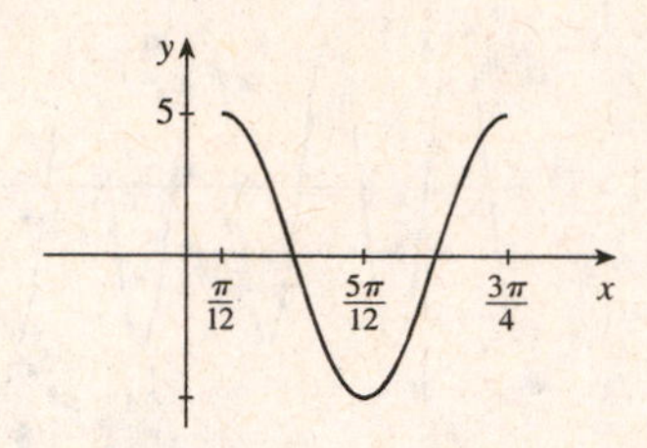

41. $y = \frac{1}{2} - \frac{1}{2}\cos\left(2x - \frac{\pi}{3}\right) = \frac{1}{2} - \frac{1}{2}\cos 2\left(x - \frac{\pi}{6}\right)$ has amplitude $\frac{1}{2}$, period π, and horizontal shift $\frac{\pi}{6}$.

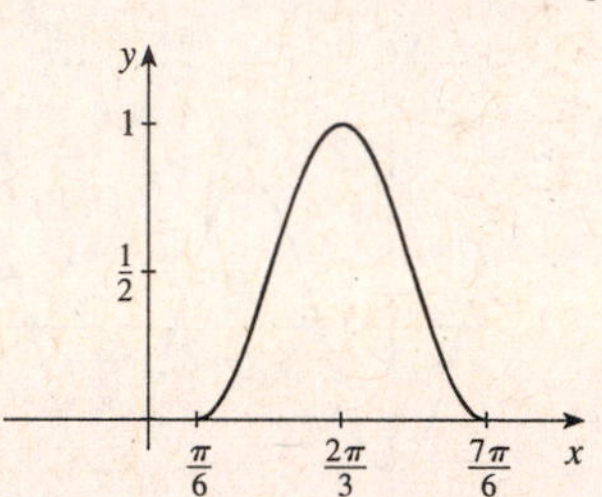

43. $y = 3\cos\pi\left(x + \frac{1}{2}\right)$ has amplitude 3, period 2, and horizontal shift $-\frac{1}{2}$.

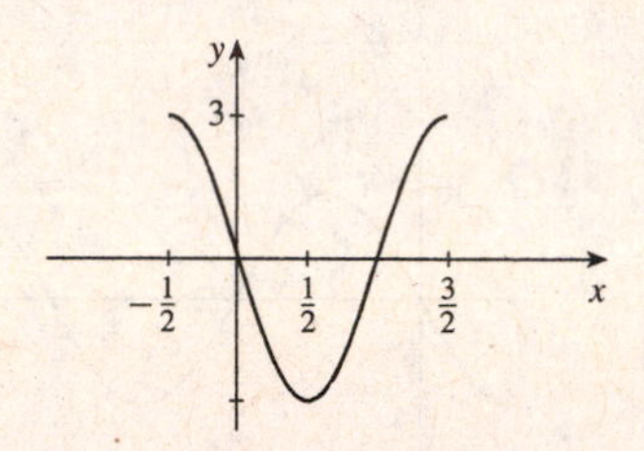

45. $y = \sin(3x + \pi) = \sin 3\left(x + \frac{\pi}{3}\right)$ has amplitude 1, period $\frac{2\pi}{3}$, and horizontal shift $-\frac{\pi}{3}$

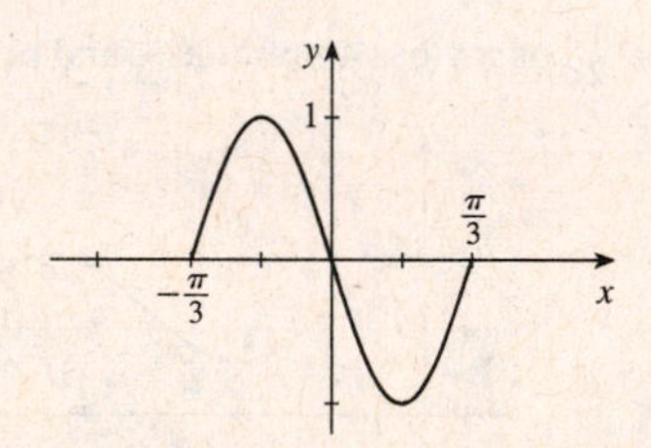

47. **(a)** This function has amplitude $a = 4$, period $\frac{2\pi}{k} = 2\pi$, and horizontal shift $b = 0$ as a sine curve.

(b) $y = a\sin k(x - b) = 4\sin x$

49. **(a)** This curve has amplitude $a = \frac{3}{2}$, period $\frac{2\pi}{k} = \frac{2\pi}{3}$, and horizontal shift $b = 0$ as a cosine curve.

(b) $y = a\cos k(x - b) = \frac{3}{2}\cos 3x$

51. **(a)** This curve has amplitude $a = \frac{1}{2}$, period $\frac{2\pi}{k} = \pi$, and horizontal shift $b = -\frac{\pi}{3}$ as a cosine curve.

(b) $y = -\frac{1}{2}\cos 2\left(x + \frac{\pi}{3}\right)$

53. **(a)** This curve has amplitude $a = 4$, period $\frac{2\pi}{k} = \frac{3}{2}$, and horizontal shift $b = -\frac{1}{2}$ as a sine curve.

(b) $y = 4\sin\frac{4\pi}{3}\left(x + \frac{1}{2}\right)$

55. $f(x) = \cos 100x$, $[-0.1, 0.1]$ by $[-1.5, 1.5]$

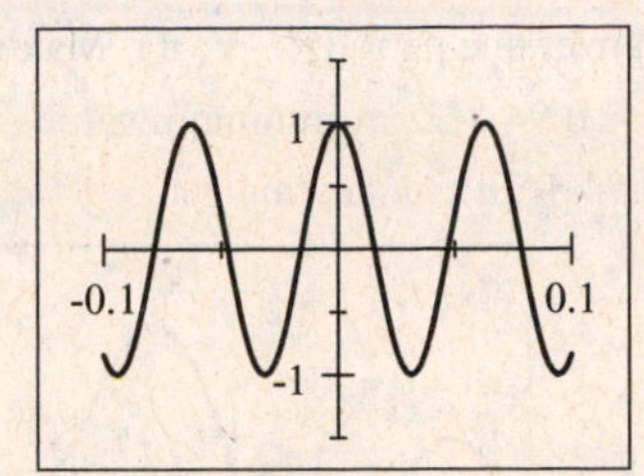

57. $f(x) = \sin \dfrac{x}{40}$, $[-250, 250]$ by $[-1.5, 1.5]$

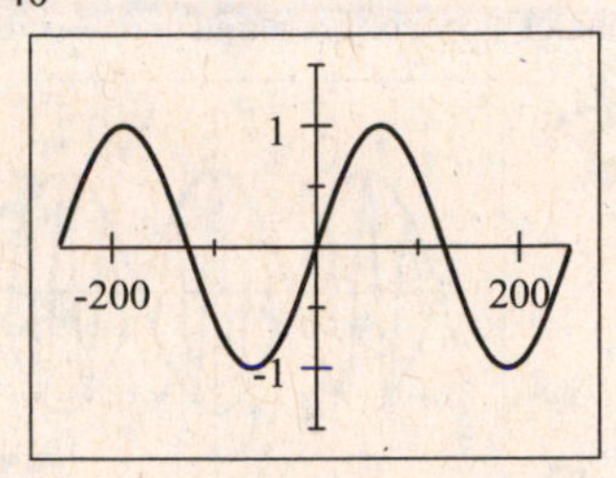

59. $y = \tan 25x$, $[-0.2, 0.2]$ by $[-3, 3]$

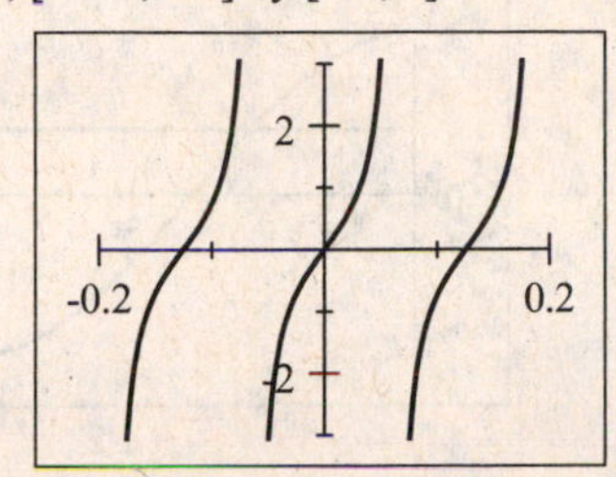

61. $y = \sin^2 20x$, $[-0.5, 0.5]$ by $[-0.2, 1.2]$

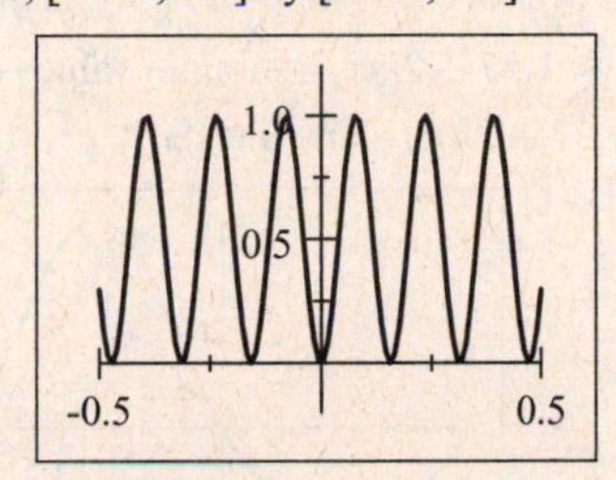

63. $f(x) = x$, $g(x) = \sin x$

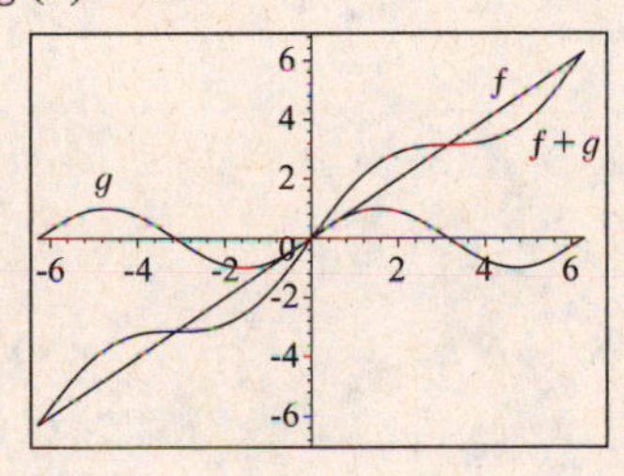

65. $f(x) = \sin 3x$, $g(x) = \cos \frac{1}{2}x$

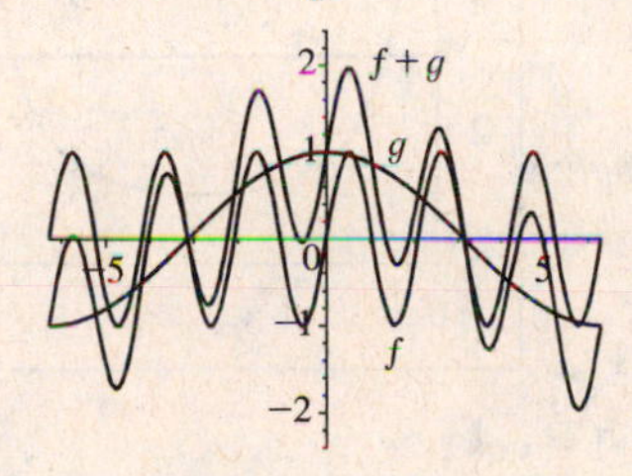

67. $y = x^2 \sin x$ is a sine curve that lies between the graphs of $y = x^2$ and $y = -x^2$.

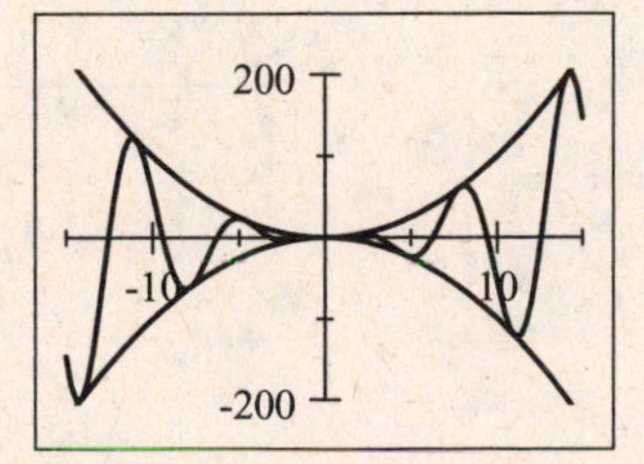

69. $y = \sqrt{x} \sin 5\pi x$ is a sine curve that lies between the graphs of $y = \sqrt{x}$ and $y = -\sqrt{x}$.

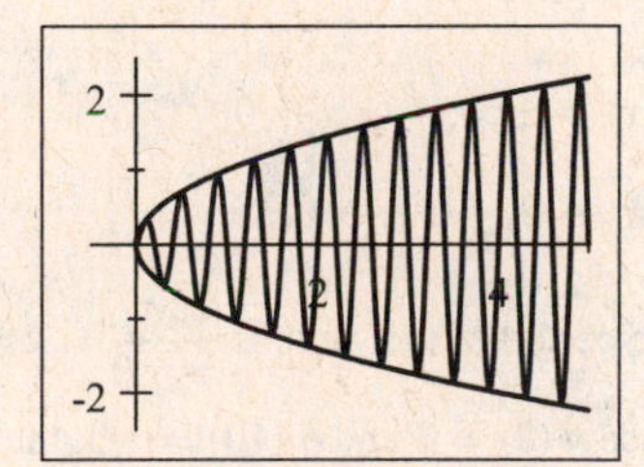

71. $y = \cos 3\pi x \cos 21\pi x$ is a cosine curve that lies between the graphs of $y = \cos 3\pi x$ and $y = -\cos 3\pi x$.

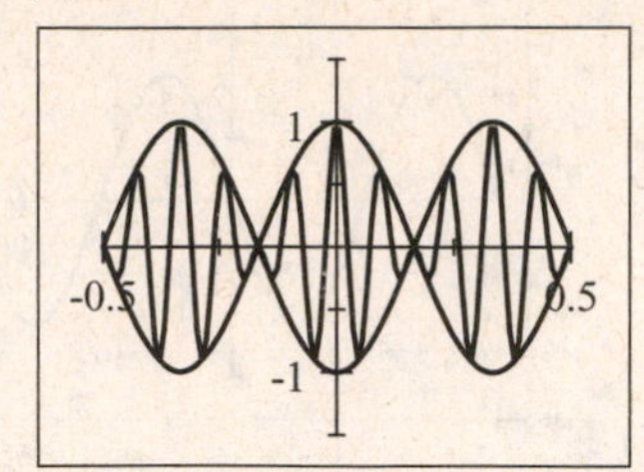

73. $y = \sin x + \sin 2x$. The period is 2π, so we graph the function over one period, $(-\pi, \pi)$. Maximum value 1.76 when $x \approx 0.94 + 2n\pi$, minimum value -1.76 when $x \approx -0.94 + 2n\pi$, n any integer.

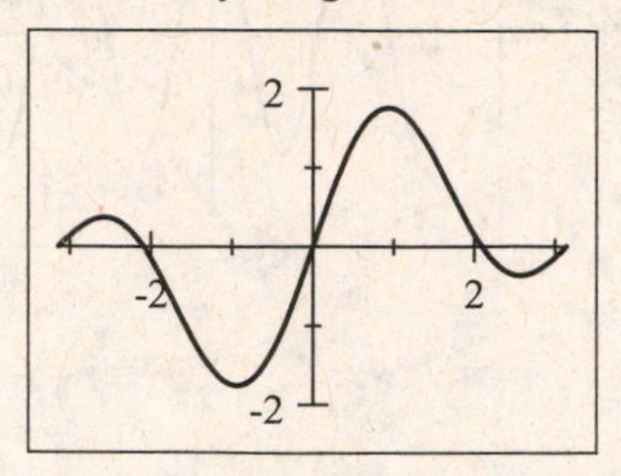

75. $y = 2\sin x + \sin^2 x$. The period is 2π, so we graph the function over one period, $(-\pi, \pi)$. Maximum value 3.00 when $x \approx 1.57 + 2n\pi$, minimum value -1.00 when $x \approx -1.57 + 2n\pi$, n any integer.

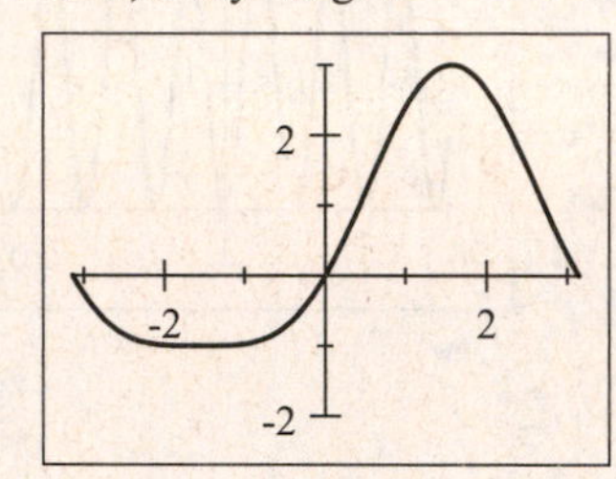

77. $\cos x = 0.4$, $x \in [0, \pi]$. The solution is $x \approx 1.16$.

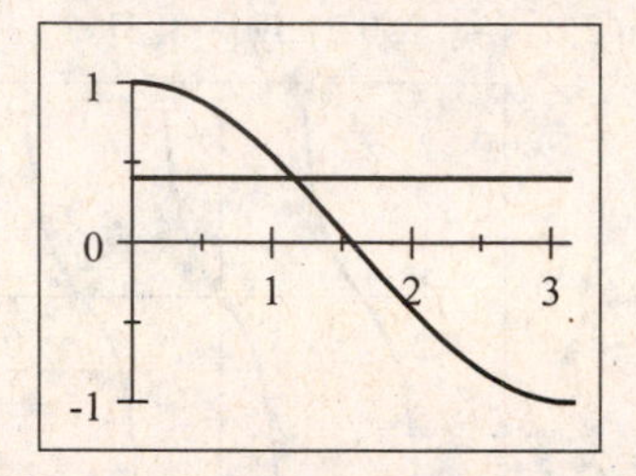

79. $\csc x = 3$, $x \in [0, \pi]$. The solutions are $x \approx 0.34, 2.80$.

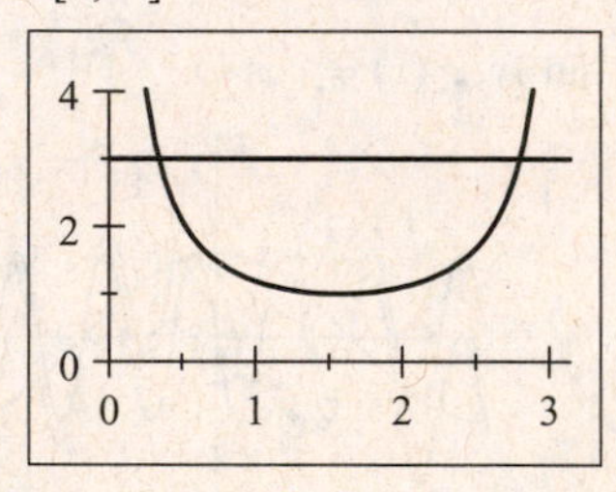

81. $f(x) = \dfrac{1 - \cos x}{x}$

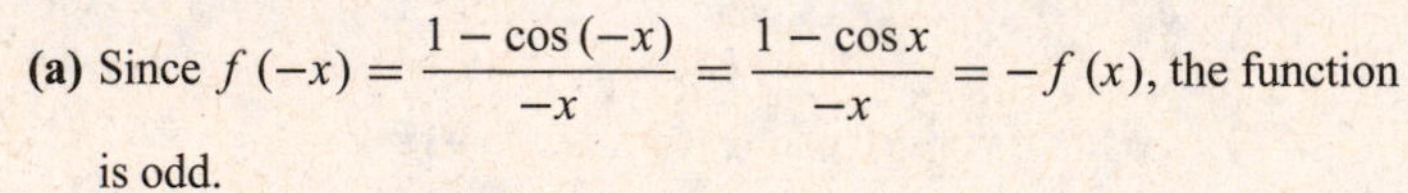

(a) Since $f(-x) = \dfrac{1 - \cos(-x)}{-x} = \dfrac{1 - \cos x}{-x} = -f(x)$, the function is odd.

(b) The function is undefined at $x = 0$, so the x-intercepts occur when $1 - \cos x = 0, x \neq 0 \Leftrightarrow \cos x = 1, x \neq 0 \Leftrightarrow x = \pm 2\pi, \pm 4\pi, \pm 6\pi, \ldots$

(c)

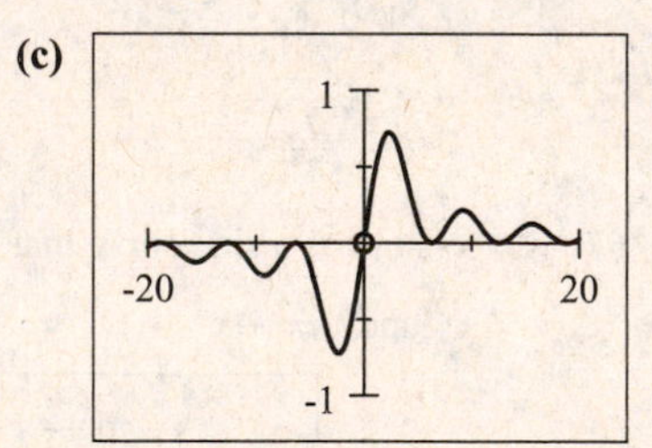

(d) As $x \to \pm\infty$, $f(x) \to 0$.

(e) As $x \to 0$, $f(x) \to 0$.

83. **(a)** The period of the wave is $\dfrac{2\pi}{\pi/10} = 20$ seconds.

(b) Since $h(0) = 3$ and $h(10) = -3$, the wave height is $3 - (-3) = 6$ feet.

85. **(a)** The period of p is $\frac{2\pi}{160\pi} = \frac{1}{80}$ minute.

(b) Since each period represents a heart beat, there are 80 heart beats per minute.

(c)

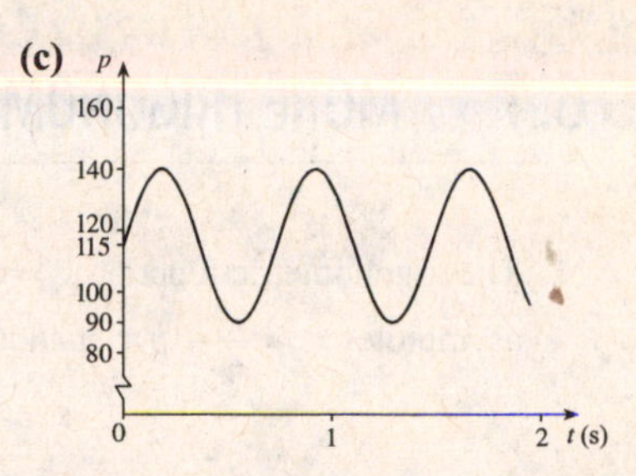

(d) The maximum (or systolic) is $115 + 25 = 140$ and the minimum (or diastolic) is $115 - 25 = 90$. The read would be $140/90$ which higher than normal.

87. **(a)** $y = \sin\left(\sqrt{x}\right)$. This graph looks like a sine function which has been stretched horizontally (stretched more for larger values of x). It is defined only for $x \geq 0$, so it is neither even nor odd.

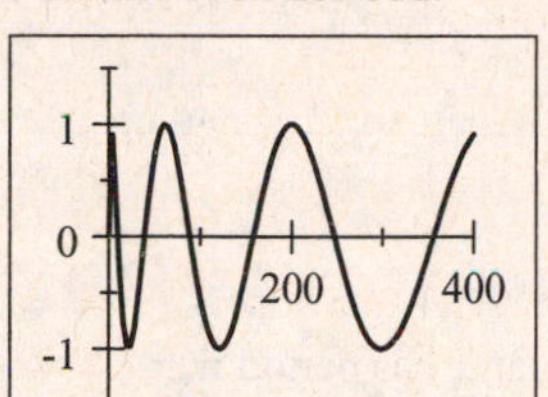

(b) $y = \sin\left(x^2\right)$. This graph looks like a graph of $\sin|x|$ which has been shrunk for $|x| > 1$ (shrunk more for larger values of x) and stretched for $|x| < 1$. It is an even function, whereas $\sin x$ is odd.

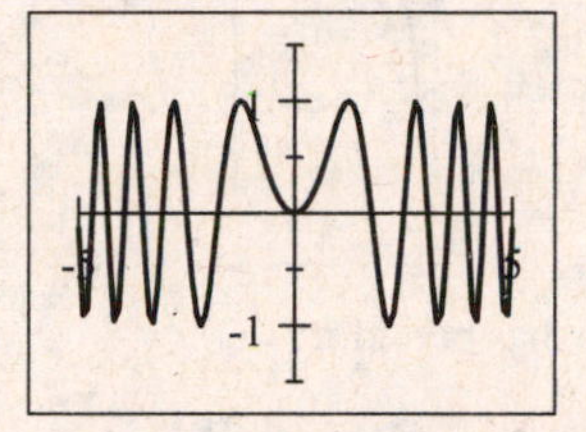

89. **(a)** The graph of $y = |\sin x|$ is shown in the viewing rectangle $[-6.28, 6.28]$ by $[-0.5, 1.5]$. This function is periodic with periodπ.

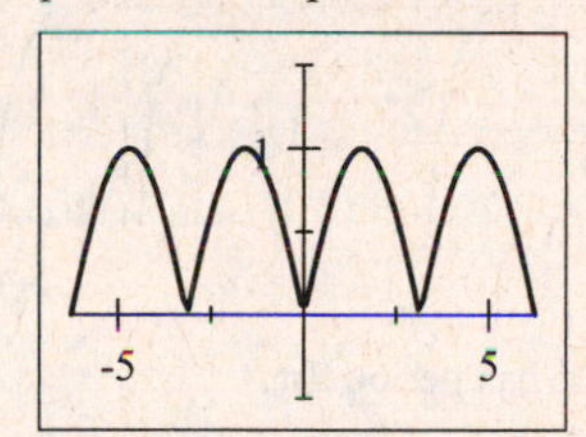

(b) The graph of $y = \sin|x|$ is shown in the viewing rectangle $[-10, 10]$ by $[-1.5, 1.5]$. The function is not periodic. Note that while $\sin|x + 2\pi| = \sin|x|$ for many values of x, it is false for $x \in (-2\pi, 0)$. For example $\sin\left|-\frac{\pi}{2}\right| = \sin\frac{\pi}{2} = 1$ while $\sin\left|-\frac{\pi}{2} + 2\pi\right| = \sin\frac{3\pi}{2} = -1$.

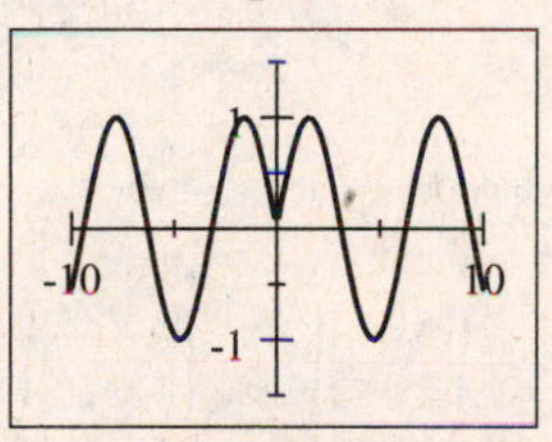

(c) The graph of $y = 2^{\cos x}$ is shown in the viewing rectangle $[-10, 10]$ by $[-1, 3]$. This function is periodic with period $= 2\pi$.

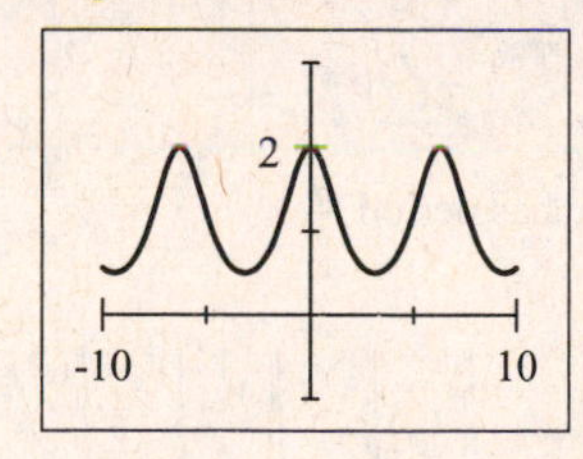

(d) The graph of $y = x - [\![x]\!]$ is shown in the viewing rectangle $[-7.5, 7.5]$ by $[-0.5, 1.5]$. This function is periodic with period1. Be sure to turn off "connected" mode when graphing functions with gaps in their graph.

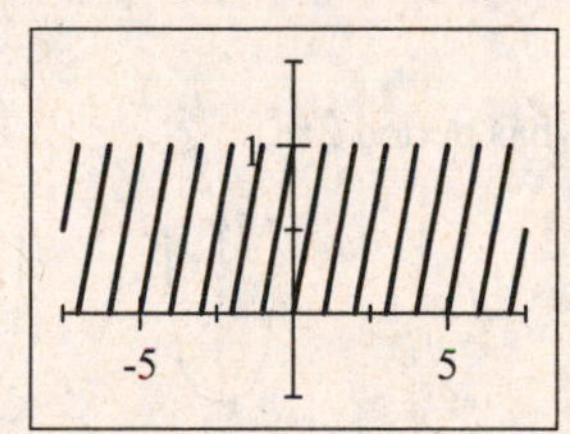

6.4 MORE TRIGONOMETRIC GRAPHS

1. The trigonometric function $y = \tan x$ has period π and asymptotes $x = \frac{\pi}{2} + n\pi$, n an integer.

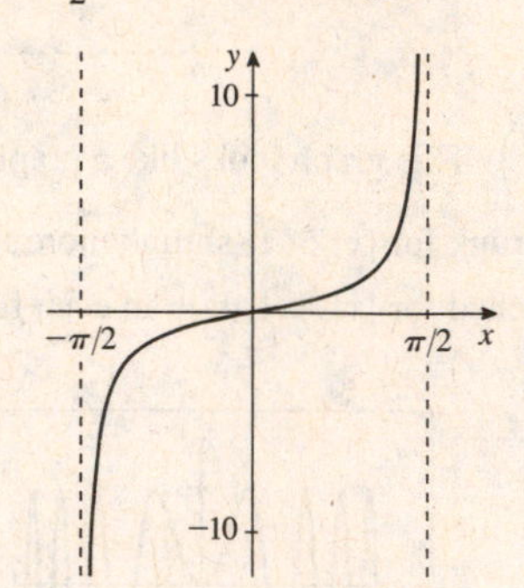

3. $f(x) = \tan\left(x + \frac{\pi}{4}\right)$ corresponds to Graph II. f is undefined at $x = \frac{\pi}{4}$ and $x = \frac{3\pi}{4}$, and Graph II has the shape of a graph of a tangent function.

5. $f(x) = \cot 4x$ corresponds to Graph VI.

7. $f(x) = 2\sec x$ corresponds to Graph IV.

9. $y = 4\tan x$ has period π.

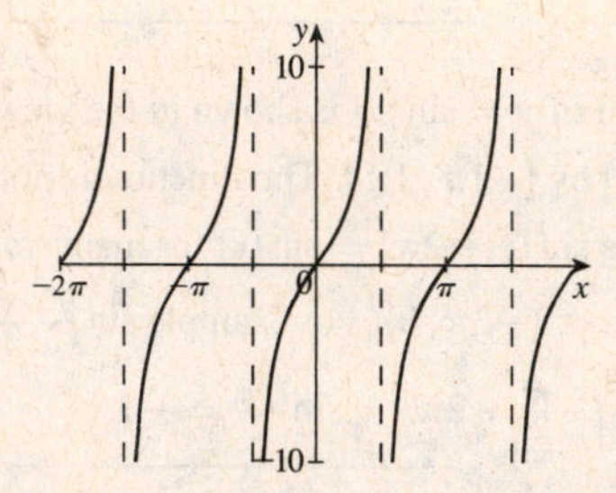

11. $y = -\frac{3}{2}\tan x$ has period π.

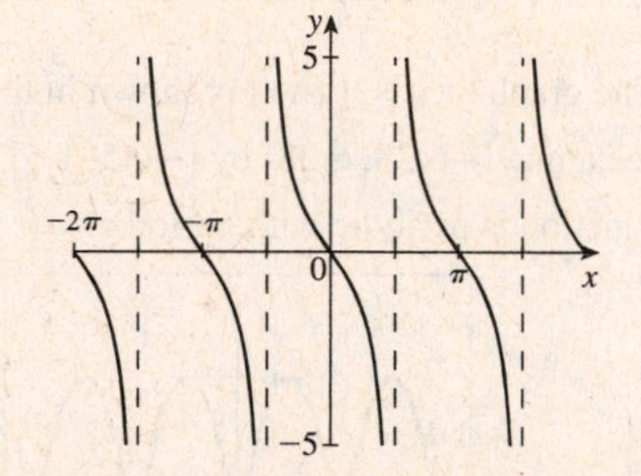

13. $y = 2\cot x$ has period π.

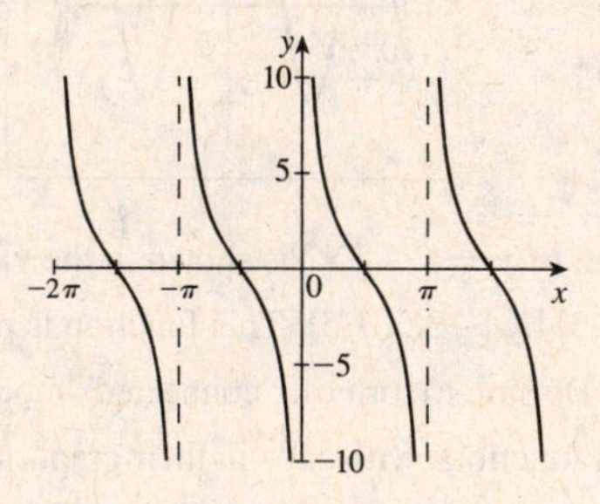

15. $y = 2\csc x$ has period 2π.

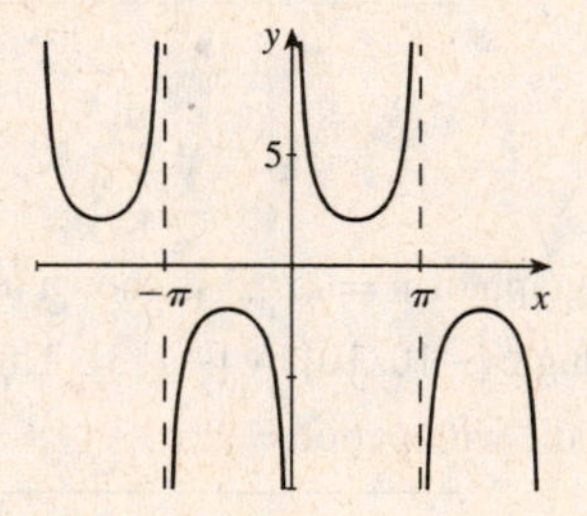

17. $y = 3\sec x$ has period 2π.

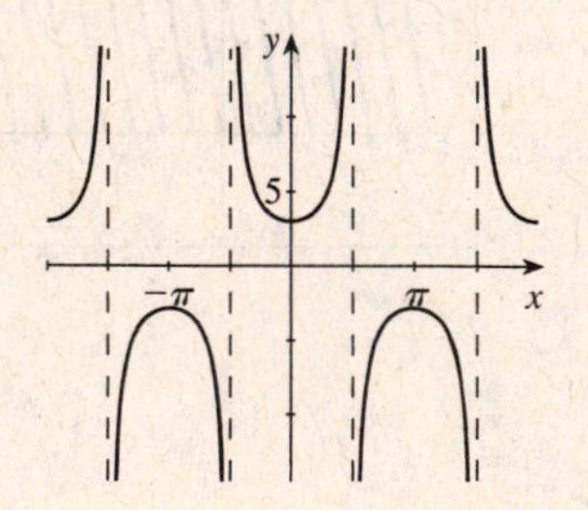

19. $y = \tan 3x$ has period $\frac{\pi}{3}$.

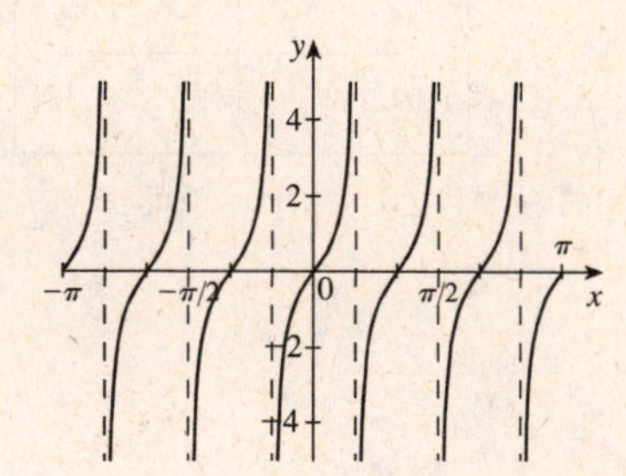

21. $y = -5\tan \pi x$ has period 1.

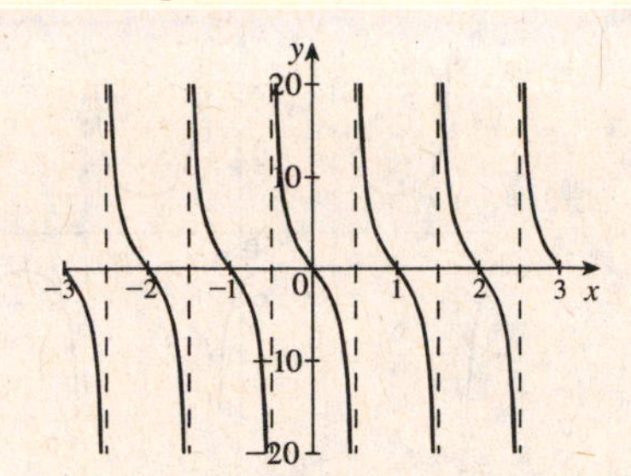

23. $y = 2\cot 3\pi x$ has period $\frac{1}{3}$.

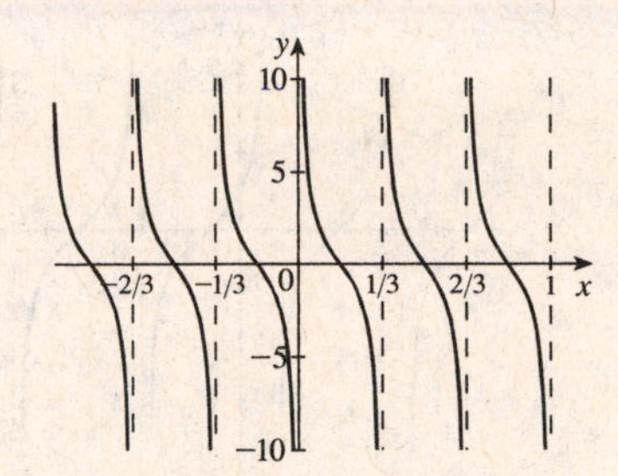

25. $y = \tan\left(\frac{\pi}{4}x\right)$ has period $\dfrac{\pi}{\frac{\pi}{4}} = 4$.

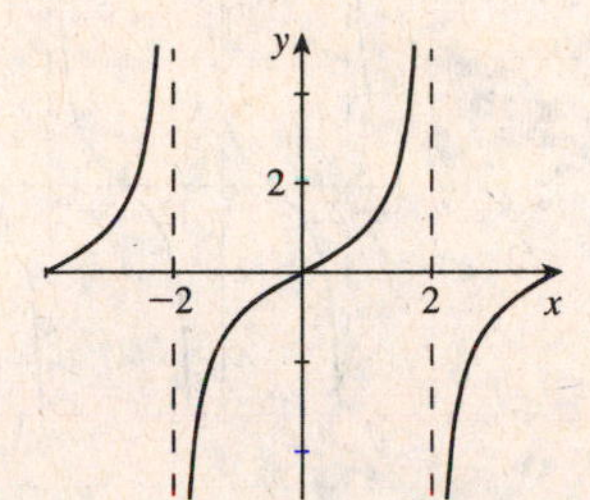

27. $y = 2\tan 3\pi x$ has period $\frac{1}{3}$.

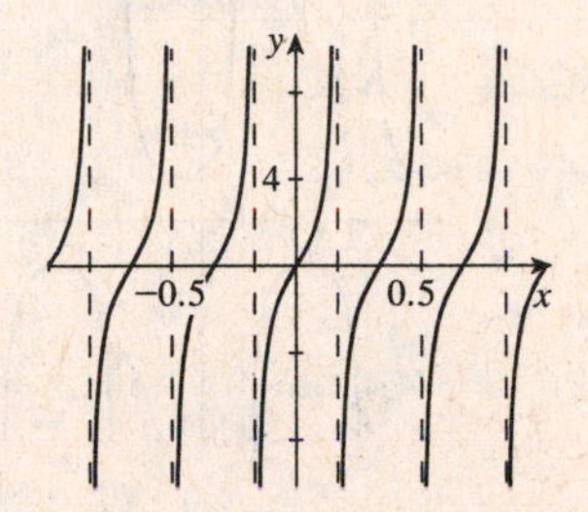

29. $y = \csc 4x$ has period $\frac{2\pi}{4} = \frac{\pi}{2}$.

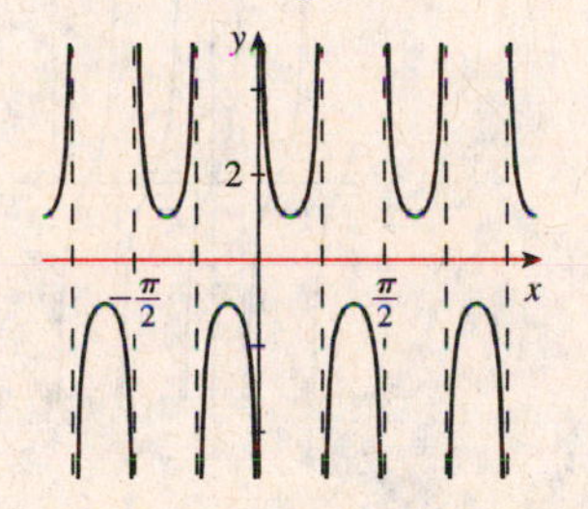

31. $y = \sec 2x$ has period $\dfrac{2\pi}{2} = \pi$.

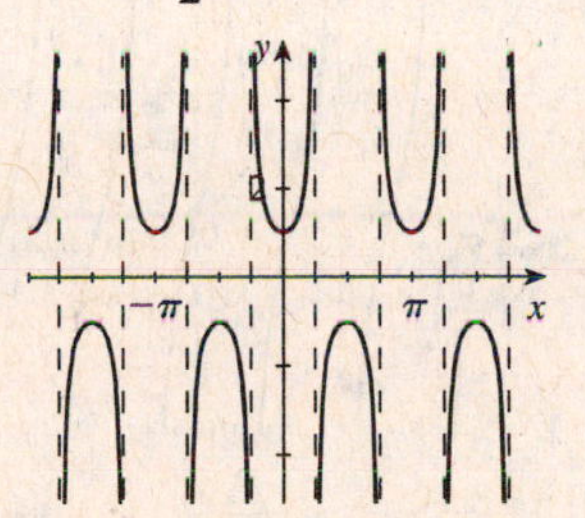

33. $y = 5\csc \frac{3\pi}{2}x$ has period $\dfrac{2\pi}{\frac{3\pi}{2}} = \dfrac{4}{3}$.

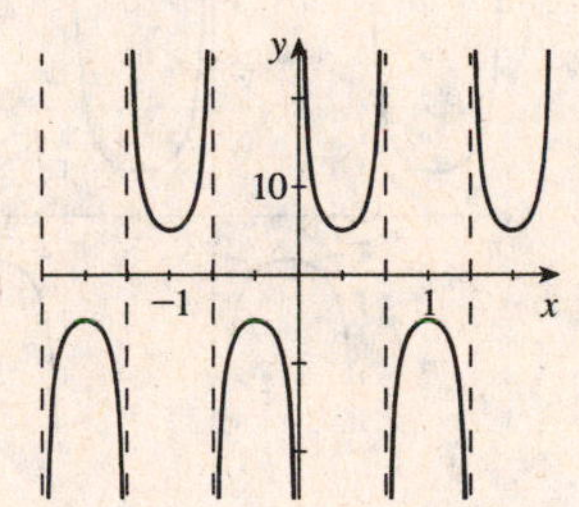

35. $y = \tan\left(x + \frac{\pi}{4}\right)$ has period π.

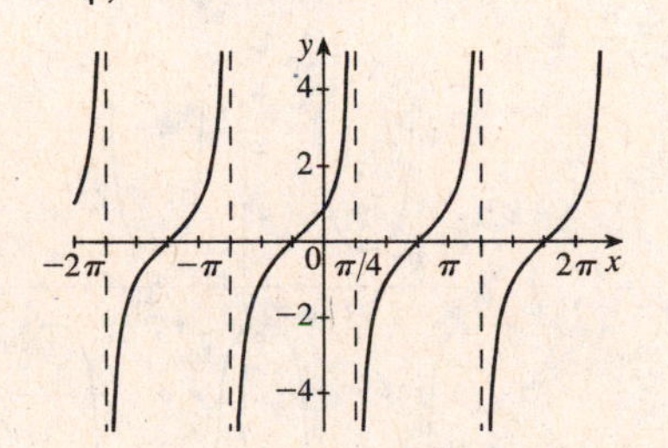

37. $y = \cot\left(x + \frac{\pi}{4}\right)$ has period π.

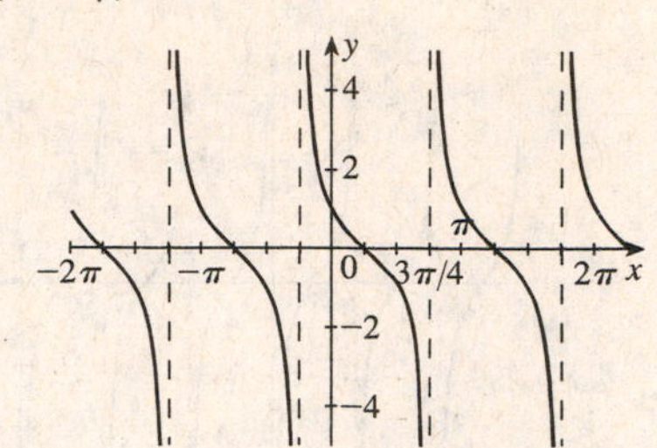

39. $y = \csc\left(x - \frac{\pi}{4}\right)$ has period 2π.

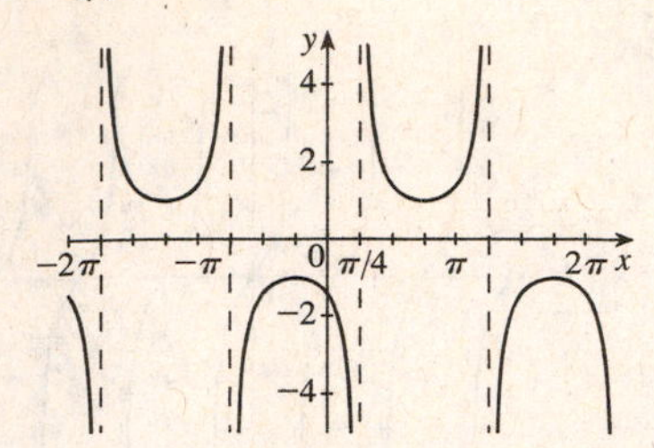

41. $y = \frac{1}{2}\sec\left(x - \frac{\pi}{6}\right)$ has period 2π.

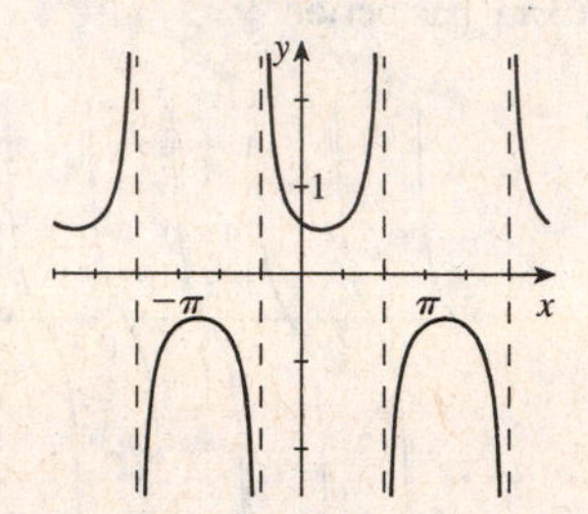

43. $y = \tan 2\left(x - \frac{\pi}{3}\right)$ has period $\frac{\pi}{2}$.

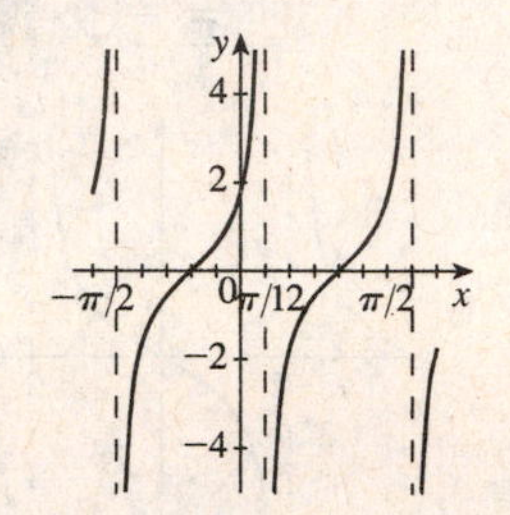

45. $y = 5\cot\left(3x + \frac{\pi}{2}\right)$ has period $\frac{\pi}{3}$.

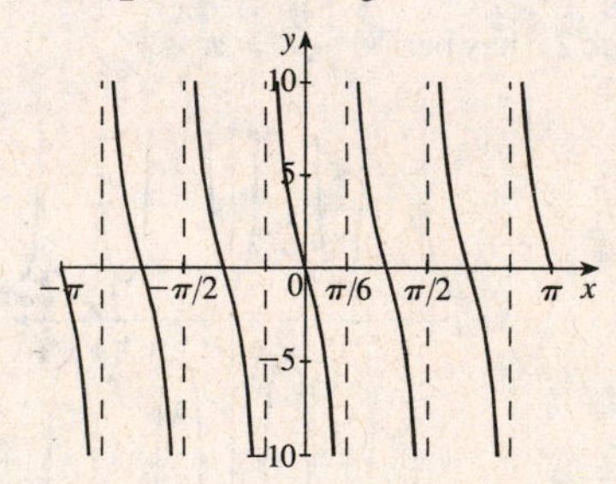

47. $y = \cot\left(2x - \frac{\pi}{2}\right) = \cot 2\left(x - \frac{\pi}{4}\right)$ has period $\frac{\pi}{2}$.

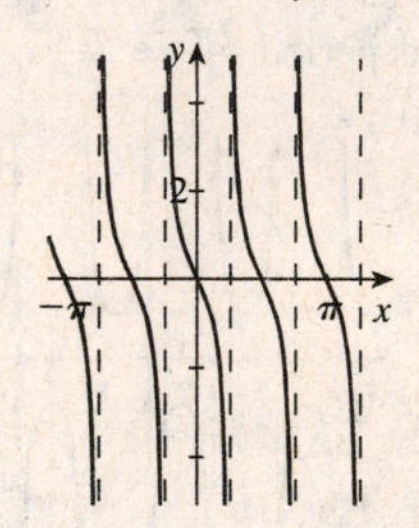

49. $y = 2\csc\left(\pi x - \frac{\pi}{3}\right) = 2\csc\pi\left(x - \frac{1}{3}\right)$ has period $\frac{2\pi}{\pi} = 2$.

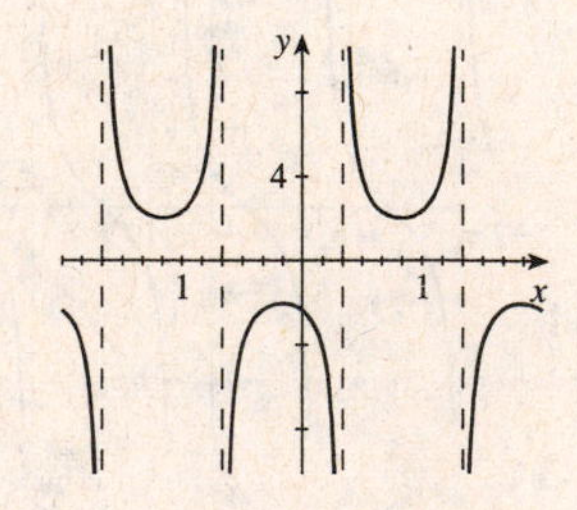

51. $y = \sec 2\left(x - \frac{\pi}{4}\right)$ has period π.

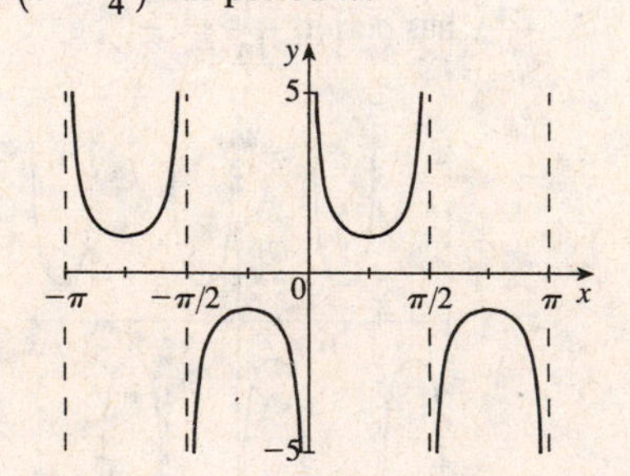

53. $y = 5\sec\left(3x - \frac{\pi}{2}\right) = 5\sec 3\left(x - \frac{\pi}{6}\right)$ has period $\frac{2\pi}{3}$.

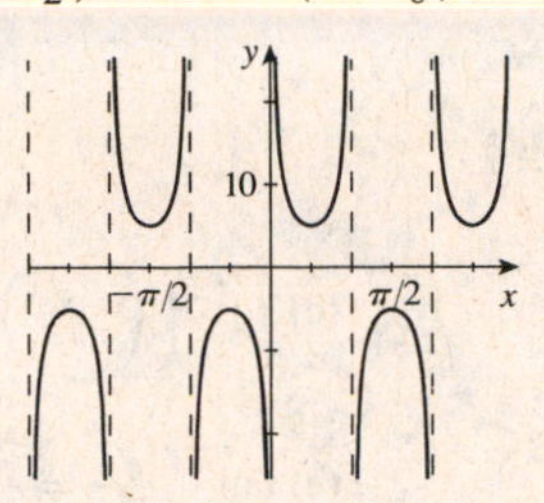

55. $y = \tan\left(\frac{2}{3}x - \frac{\pi}{6}\right) = \tan\frac{2}{3}\left(x - \frac{\pi}{4}\right)$ has period $\pi/\left(\frac{2}{3}\right) = \frac{3\pi}{2}$.

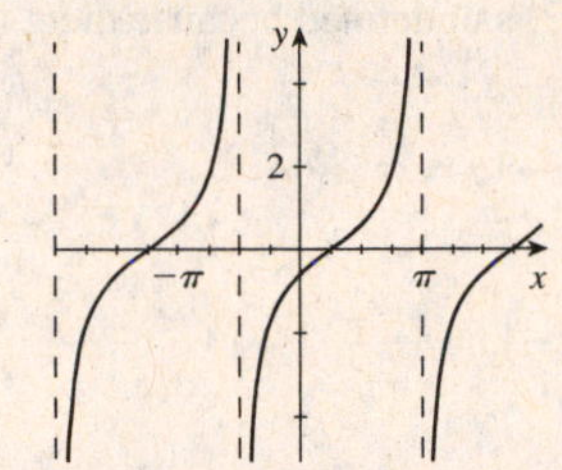

57. $y = 3\sec\pi\left(x + \frac{1}{2}\right)$ has period $\frac{2\pi}{\pi} = 2$.

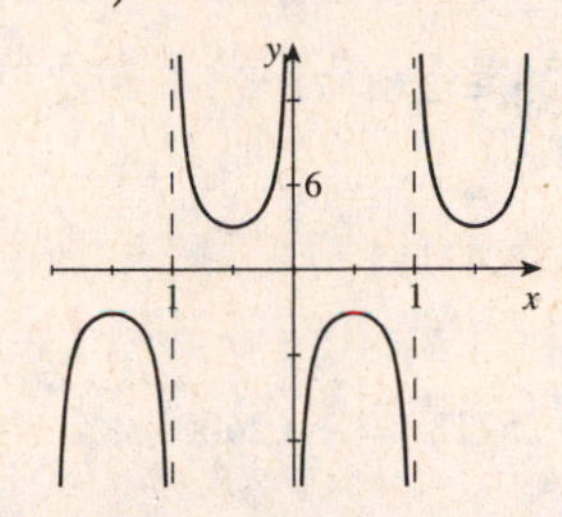

59. $y = -2\tan\left(2x - \frac{\pi}{3}\right) = -2\tan 2\left(x - \frac{\pi}{6}\right)$ has period $\frac{\pi}{2}$.

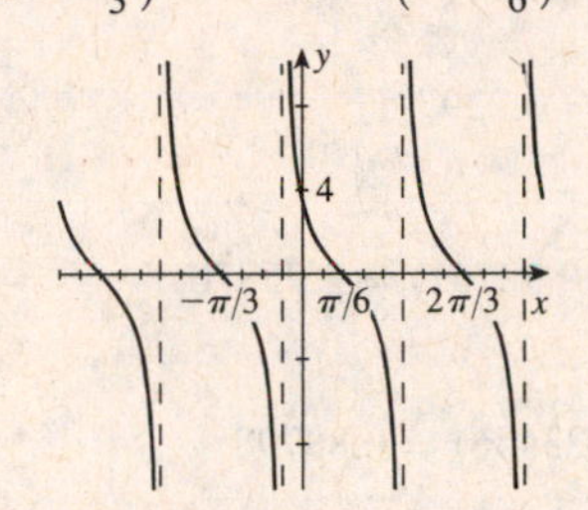

61. **(a)** $d(t) = 3\tan\pi t$, so $d(0.15) \approx 1.53$, $d(0.25) \approx 3.00$, and $d(0.45) \approx 18.94$.

(c) $d \to \infty$ as $t \to \frac{1}{2}$.

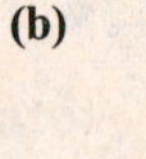

(b)

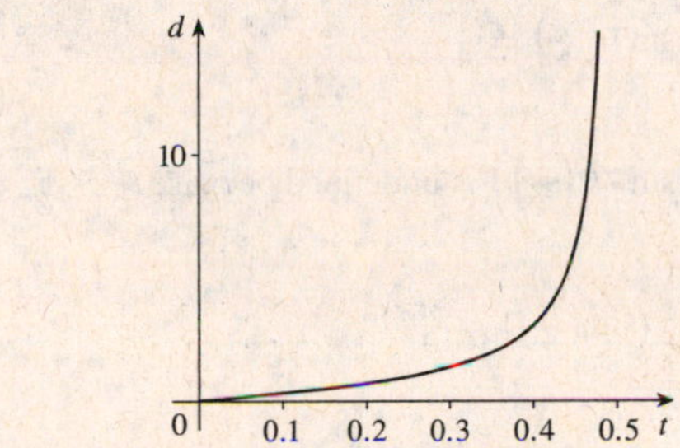

63. **(a)** If f is periodic with period p, then by the definition of a period, $f(x+p) = f(x)$ for all x in the domain of f. Therefore, $\dfrac{1}{f(x+p)} = \dfrac{1}{f(x)}$ for all $f(x) \neq 0$. Thus, $\dfrac{1}{f}$ is also periodic with period p.

(b) Since $\sin x$ has period 2π, it follows from part (a) that $\csc x = \dfrac{1}{\sin x}$ also has period 2π. Similarly, since $\cos x$ has period 2π, we conclude $\sec x = \dfrac{1}{\cos x}$ also has period 2π.

65. The graph of $y = -\cot x$ is the same as the graph of $y = \tan x$ shifted $\frac{\pi}{2}$ units to the right, and the graph of $y = \csc x$ is the same as the graph of $y = \sec x$ shifted $\frac{\pi}{2}$ units to the right.

6.5 INVERSE TRIGONOMETRIC FUNCTIONS AND THEIR GRAPHS

1. **(a)** To define the inverse sine function we restrict the domain of sine to the interval $\left[-\frac{\pi}{2}, \frac{\pi}{2}\right]$. On this interval the sine function is one-to-one and its inverse function $\sin^{-1}$ is defined by $\sin^{-1}x = y \Leftrightarrow \sin y = x$. For example, $\sin^{-1}\frac{1}{2} = \frac{\pi}{6}$ because $\sin\frac{\pi}{6} = \frac{1}{2}$.

(b) To define the inverse cosine function we restrict the domain of cosine to the interval $[0, \pi]$. On this interval the cosine function is one-to-one and its inverse function $\cos^{-1}$ is defined by $\cos^{-1}x = y \Leftrightarrow \cos y = x$. For example, $\cos^{-1}\frac{1}{2} = \frac{\pi}{3}$ because $\cos\frac{\pi}{3} = \frac{1}{2}$.

3. **(a)** $\sin^{-1} 1 = \frac{\pi}{2}$ because $\sin \frac{\pi}{2} = 1$ and $\frac{\pi}{2}$ lies in $\left[-\frac{\pi}{2}, \frac{\pi}{2}\right]$.

(b) $\sin^{-1} \frac{\sqrt{3}}{2} = \frac{\pi}{3}$ because $\sin \frac{\pi}{3} = \frac{\sqrt{3}}{2}$ and $\frac{\pi}{3}$ lies in $\left[-\frac{\pi}{2}, \frac{\pi}{2}\right]$.

(c) $\sin^{-1} 2$ is undefined because there is no real number x such that $\sin x = 2$.

5. **(a)** $\cos^{-1}(-1) = \pi$ **(b)** $\cos^{-1} \frac{1}{2} = \frac{\pi}{3}$ **(c)** $\cos^{-1}\left(-\frac{\sqrt{3}}{2}\right) = \frac{5\pi}{6}$

7. **(a)** $\tan^{-1}(-1) = -\frac{\pi}{4}$ **(b)** $\tan^{-1} \sqrt{3} = \frac{\pi}{3}$ **(c)** $\tan^{-1} \frac{\sqrt{3}}{3} = \frac{\pi}{6}$

9. **(a)** $\cos^{-1}\left(-\frac{1}{2}\right) = \frac{2\pi}{3}$ **(b)** $\sin^{-1}\left(-\frac{\sqrt{2}}{2}\right) = -\frac{\pi}{4}$ **(c)** $\tan^{-1} 1 = \frac{\pi}{4}$

11. $\sin^{-1} \frac{2}{3} = 0.72973$

13. $\cos^{-1}\left(-\frac{3}{7}\right) = 2.01371$

15. $\cos^{-1}(-0.92761) = 2.75876$

17. $\tan^{-1} 10 = 1.47113$

19. $\tan^{-1}(1.23456) = 0.88998$

21. $\sin^{-1}(-0.25713) = -0.26005$

23. $\sin\left(\sin^{-1} \frac{1}{4}\right) = \frac{1}{4}$

25. $\tan\left(\tan^{-1} 5\right) = 5$

27. $\sin\left(\sin^{-1}\left(\frac{3}{2}\right)\right)$ is undefined because $\frac{3}{2} > 1$.

29. $\cos\left(\cos^{-1}\left(-\frac{1}{5}\right)\right) = -\frac{1}{5}$

31. $\sin^{-1}\left(\sin\left(\frac{\pi}{4}\right)\right) = \frac{\pi}{4}$

33. $\sin^{-1}\left(\sin\left(\frac{3\pi}{4}\right)\right) = \frac{\pi}{4}$

35. $\cos^{-1}\left(\cos\left(\frac{5\pi}{6}\right)\right) = \frac{5\pi}{6}$

37. $\cos^{-1}\left(\cos\left(\frac{7\pi}{6}\right)\right) = \frac{5\pi}{6}$

39. $\tan^{-1}\left(\tan\left(\frac{\pi}{4}\right)\right) = \frac{\pi}{4}$

41. $\tan^{-1}\left(\tan \frac{2\pi}{3}\right) = -\frac{\pi}{3}$ because $\tan\left(-\frac{\pi}{3}\right) = \tan \frac{2\pi}{3}$ and $-\frac{\pi}{3}$ lies in $\left(-\frac{\pi}{2}, \frac{\pi}{2}\right)$.

43. $\tan\left(\sin^{-1} \frac{1}{2}\right) = \tan \frac{\pi}{6} = \frac{\sqrt{3}}{3}$ **45.** $\cos\left(\sin^{-1} \frac{\sqrt{3}}{2}\right) = \cos \frac{\pi}{3} = \frac{1}{2}$ **47.** $\sin\left(\tan^{-1}(-1)\right) = \sin\left(-\frac{\pi}{4}\right) = -\frac{\sqrt{2}}{2}$

49. **(a)**

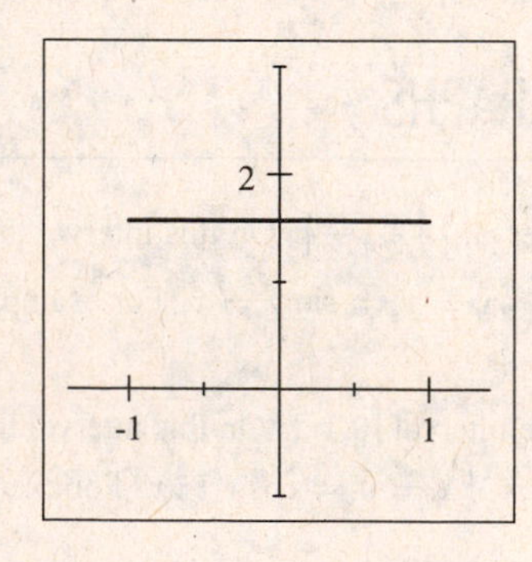

From the graph of $y = \sin^{-1} x + \cos^{-1} x$, it appears that $y \approx 1.57$. We suspect that the actual value is $\frac{\pi}{2}$.

(b) To show that $\sin^{-1} x + \cos^{-1} x = \frac{\pi}{2}$, start with the identity $\sin\left(a - \frac{\pi}{2}\right) = -\cos a$ and take arcsin of both sides to obtain $a - \frac{\pi}{2} = \sin^{-1}(-\cos a)$. Now let $a = \cos^{-1} x$. Then $\cos^{-1} x - \frac{\pi}{2} = \sin^{-1}\left(-\cos\left(\cos^{-1} x\right)\right) = \sin^{-1}(-x) = -\sin^{-1} x$, so $\sin^{-1} x + \cos^{-1} x = \frac{\pi}{2}$.

51. The domain of $f(x) = \sin\left(\sin^{-1} x\right)$ is the same as that of $\sin^{-1} x$, $[-1, 1]$, and the graph of f is the same as that of $y = x$ on $[1, 1]$.

The domain of $g(x) = \sin^{-1}(\sin x)$ is the same as that of $\sin x$, $(-\infty, \infty)$, because for all x, the value of $\sin x$ lies within the domain of $\sin^{-1} x$. $g(x) = \sin^{-1}(\sin x) = x$ for $-\frac{\pi}{2} \le x \le \frac{\pi}{2}$. Because the graph of $y = \sin x$ is symmetric about the line $x = \frac{\pi}{2}$, we can obtain the part of the graph of g for $\frac{\pi}{2} \le x \le \frac{3\pi}{2}$ by reflecting the graph of $y = x$ about this vertical line. The graph of g is periodic with period 2π.

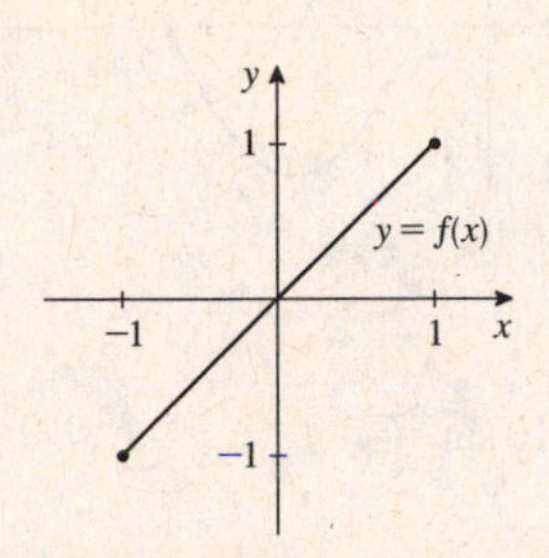

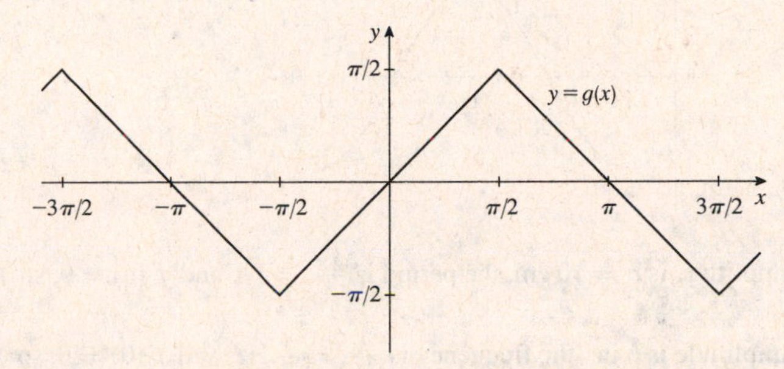

6.6 MODELING HARMONIC MOTION

1. (a) Because $y = 0$ at time $t = 0$, $y = a \sin \omega t$ is an appropriate model.

(b) Because $y = a$ at time $t = 0$, $y = a \cos \omega t$ is an appropriate model.

3. (a) For an object in harmonic motion modeled by $y = A \sin(kt - b)$ the amplitude is $|A|$, the period is $\frac{2\pi}{k}$, and the phase is b. To find the horizontal shift, we factor k to get $y = A \sin k\left(t - \frac{b}{k}\right)$. From this form of the equation we see that the horizontal shift is $\frac{b}{k}$.

(b) For an object in harmonic motion modeled by $y = 5 \sin(4t - \pi)$ the amplitude is 5, the period is $\frac{\pi}{2}$, the phase is π, and the horizontal shift is $\frac{\pi}{4}$.

5. $y = 2 \sin 3t$

(a) Amplitude 2, period $\frac{2\pi}{3}$, frequency $\frac{1}{\text{period}} = \frac{3}{2\pi}$.

(b)

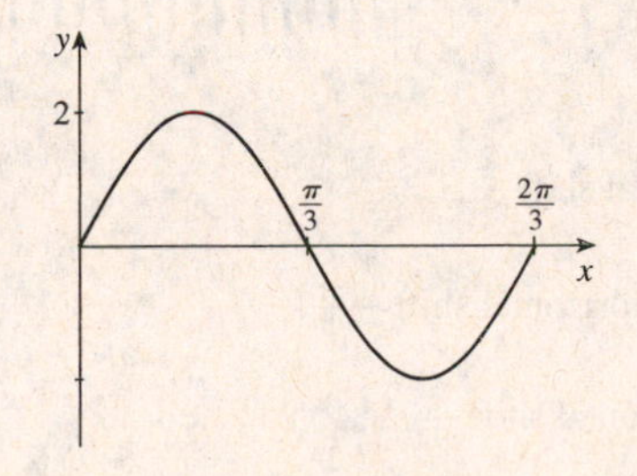

7. $y = -\cos 0.3t$

(a) Amplitude 1, period $\frac{2\pi}{0.3} = \frac{20\pi}{3}$, frequency $\frac{3}{20\pi}$.

(b)

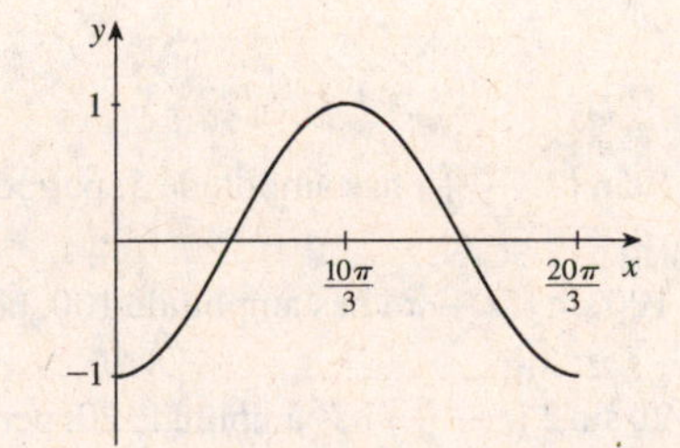

9. $y = -0.25\cos\left(1.5t - \frac{\pi}{3}\right) = -0.25\cos\left(\frac{3}{2}t - \frac{\pi}{3}\right)$
$= -0.25\cos\frac{3}{2}\left(t - \frac{2\pi}{9}\right)$

(a) Amplitude 0.25, period $\frac{2\pi}{3/2} = \frac{4\pi}{3}$, frequency $\frac{3}{4\pi}$.

(b)

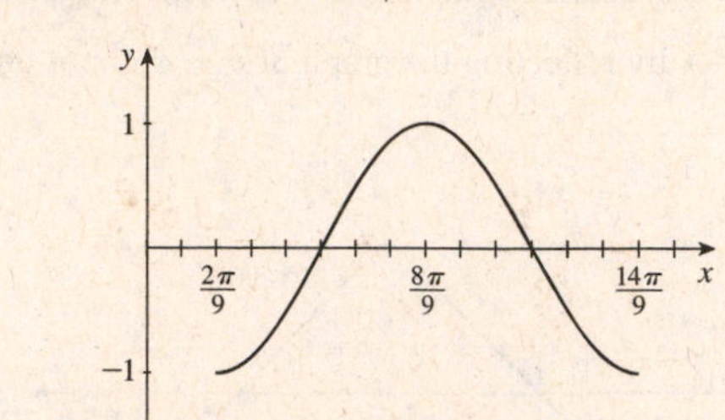

11. $y = 5\cos\left(\frac{2}{3}t + \frac{3}{4}\right) = 5\cos\frac{2}{3}\left(t + \frac{9}{8}\right)$

(a) Amplitude 5, period $\frac{2\pi}{2/3} = 3\pi$, frequency $\frac{1}{3\pi}$.

(b)

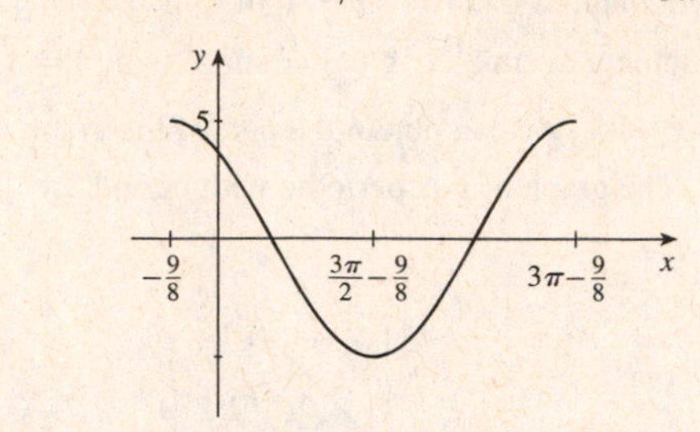

13. The amplitude is $a = 10$ cm, the period is $\frac{2\pi}{k} = 3$ s, and $f(0) = 0$, so $f(t) = 10\sin\frac{2\pi}{3}t$.

15. The amplitude is 6 in., the frequency is $\frac{k}{2\pi} = \frac{5}{\pi}$ Hz, and $f(0) = 0$, so $f(t) = 6\sin 10t$.

17. The amplitude is 60 ft, the period is $\frac{2\pi}{k} = 0.5$ min, and $f(0) = 60$, so $f(t) = 60\cos 4\pi t$.

19. The amplitude is 2.4 m, the frequency is $\frac{k}{2\pi} = 750$ Hz, and $f(0) = 2.4$, so $f(t) = 2.4\cos 1500\pi t$.

21. (a) $k = 2$, $c = 1.5$, and $f = 3 \Rightarrow \omega = 6\pi$, so we have $y = 2e^{-1.5t}\cos 6\pi t$.

(b)

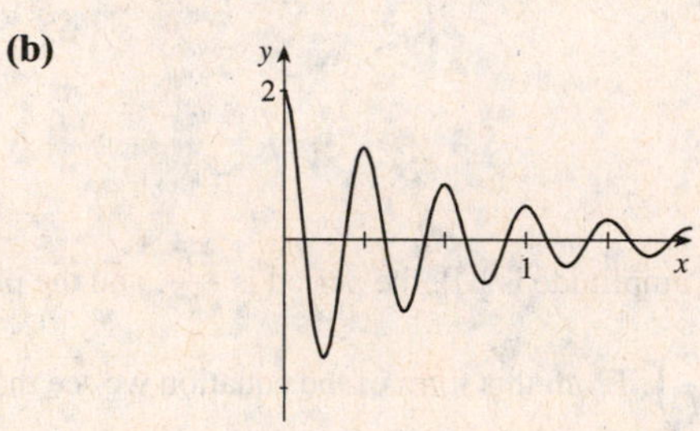

23. (a) $k = 100$, $c = 0.05$, and $p = 4 \Rightarrow \omega = \frac{\pi}{2}$, so we have $y = 100e^{-0.05t}\cos\frac{\pi}{2}t$.

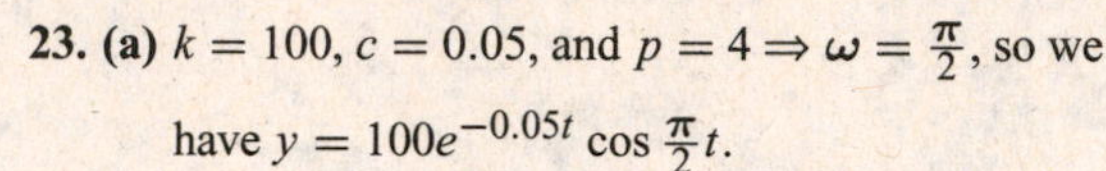

(b)

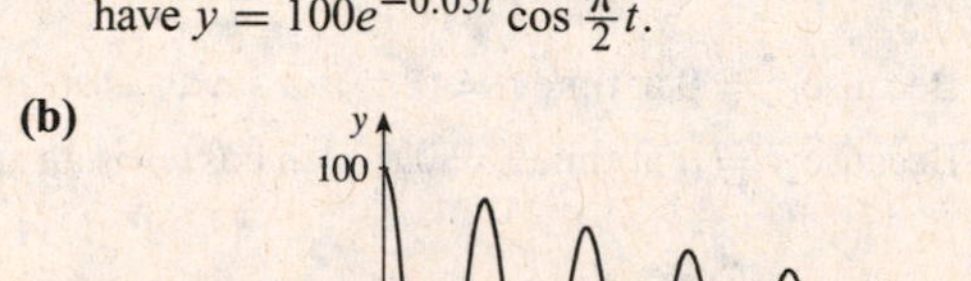

25. (a) $k = 7$, $c = 10$, and $p = \frac{\pi}{6} \Rightarrow \omega = 12$, so we have $y = 7e^{-10t}\sin 12t$.

(b)

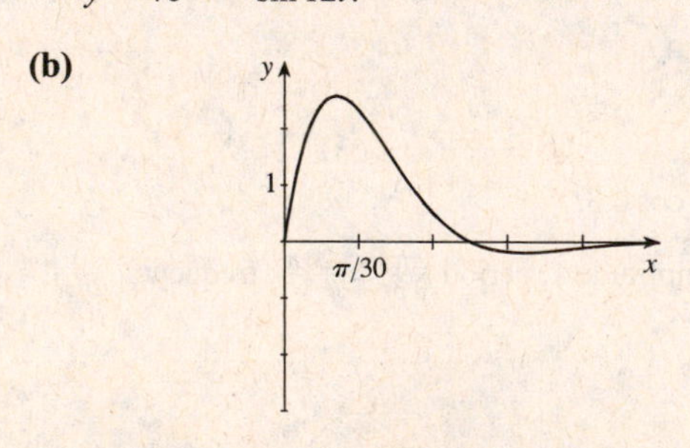

27. (a) $k = 0.3$, $c = 0.2$, and $f = 20 \Rightarrow \omega = 40\pi$, so we have $y = 0.3e^{-0.2t}\sin 40\pi t$.

(b)

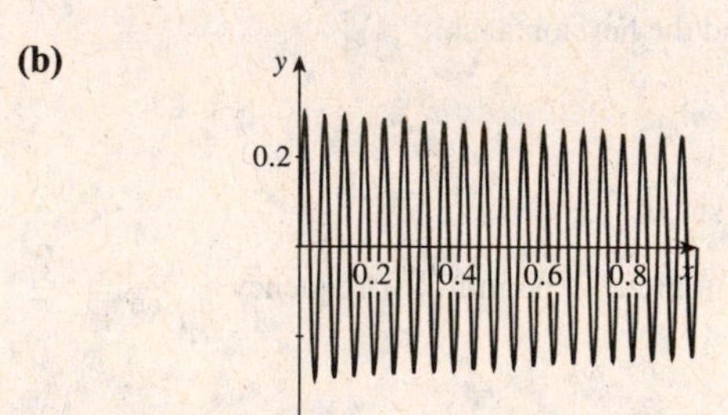

29. $y = 5\sin\left(2t - \frac{\pi}{2}\right)$ has amplitude 5, period π, phase $\frac{\pi}{2}$, and horizontal shift $\frac{\pi}{4}$.

31. $y = 100\sin(5t + \pi)$ has amplitude 100, period $\frac{2\pi}{5}$, phase $-\pi$, and horizontal shift $-\frac{\pi}{5}$.

33. $y = 20\sin 2\left(t - \frac{\pi}{4}\right)$ has amplitude 20, period π, phase $\frac{\pi}{2}$, and horizontal shift $\frac{\pi}{4}$.

35. $y_1 = 10\sin\left(3t - \frac{\pi}{2}\right)$; $y_2 = 10\sin\left(3t - \frac{5\pi}{2}\right)$

(a) y_1 has phase $\frac{\pi}{2}$ and y_2 has phase $\frac{5\pi}{2}$.

(b) The phase difference is $\frac{\pi}{2} - \frac{5\pi}{2} = -2\pi$.

(c) Because the phase difference is a multiple of 2π, the curves are in phase.

(d)

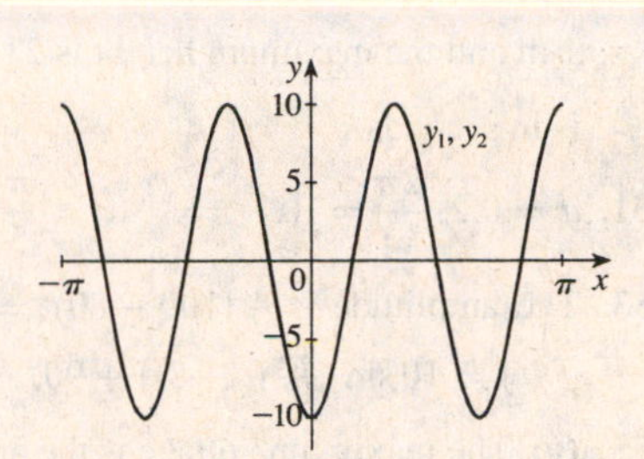

37. $y_1 = 80\sin 5\left(t - \frac{\pi}{10}\right) = 80\sin\left(5t - \frac{\pi}{2}\right)$; $y_2 = 80\sin\left(5t - \frac{\pi}{3}\right)$

(a) y_1 has phase $\frac{\pi}{2}$ and y_2 has phase $\frac{\pi}{3}$.

(b) The phase difference is $\frac{\pi}{2} - \frac{\pi}{3} = \frac{\pi}{6}$.

(c) Because the phase difference is not a multiple of 2π, the curves are out of phase.

(d)

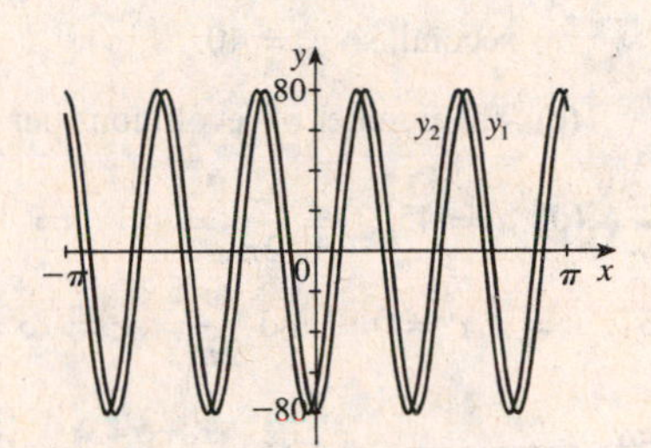

39. $y = 0.2\cos 20\pi t + 8$

(a) The frequency is $\frac{20\pi}{2\pi} = 10$ cycles/min.

(b)

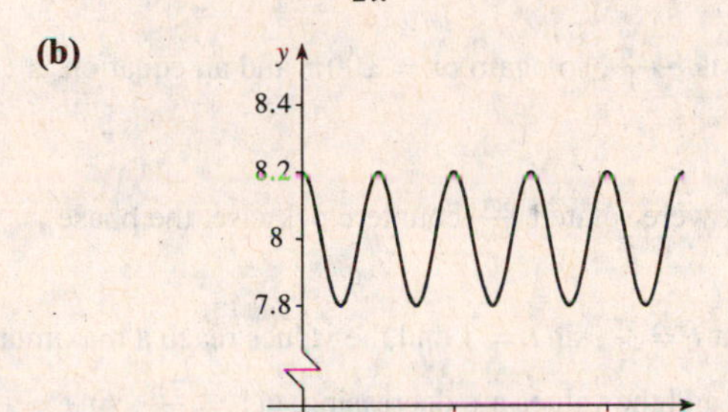

(c) Since $y = 0.2\cos 20\pi t + 8 \leq 0.2(1) + 8 = 8.2$ and when $t = 0$, $y = 8.2$, the maximum displacement is 8.2 m.

41. $p(t) = 115 + 25\sin(160\pi t)$

(a) Amplitude 25, period $\frac{2\pi}{160\pi} = \frac{1}{80} = 0.0125$, frequency $\dfrac{1}{\text{period}} = 80$.

(b)

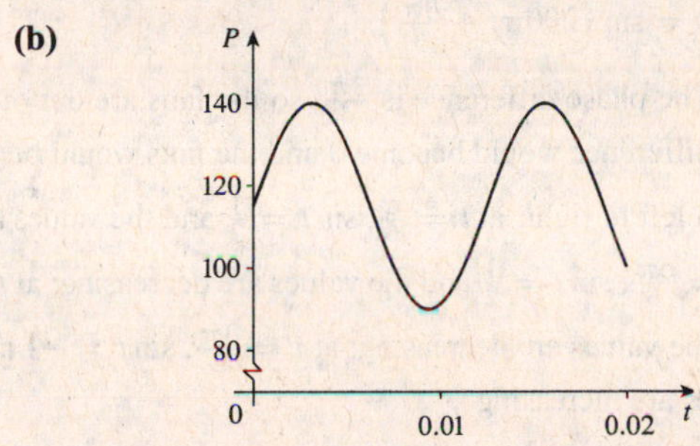

(c) The period decreases and the frequency increases.

43. The graph resembles a sine wave with an amplitude of 5, a period of $\frac{2}{5}$, and no phase shift. Therefore, $a = 5$, $\dfrac{2\pi}{\omega} = \frac{2}{5} \Leftrightarrow \omega = 5\pi$, and a formula is $d(t) = 5\sin 5\pi t$.

45. $a = 21$, $f = \frac{1}{12}$ cycle/hour $\Rightarrow \dfrac{\omega}{2\pi} = \frac{1}{12} \Leftrightarrow \omega = \frac{\pi}{6}$. So, $y = 21\sin\left(\frac{\pi}{6}t\right)$ (assuming the tide is at mean level and rising when $t = 0$).

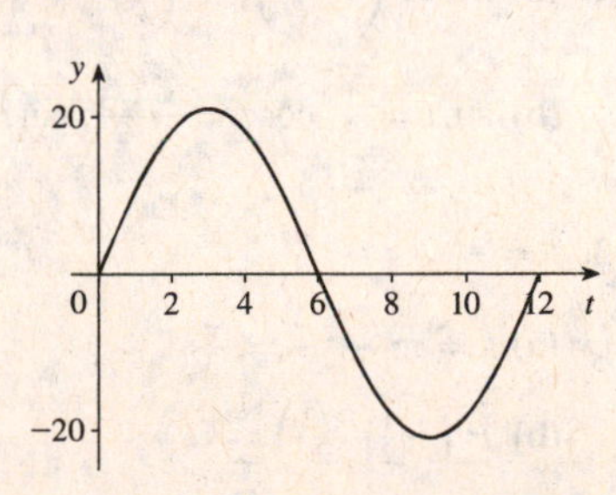

47. Since the mass travels from its highest point (compressed spring) to its lowest point in $\frac{1}{2}$ s, it completes half a period in $\frac{1}{2}$ s. So, $\frac{1}{2}$ (one period) $= \frac{1}{2}$ s $\Rightarrow \frac{1}{2} \cdot \dfrac{2\pi}{\omega} = \frac{1}{2} \Leftrightarrow \omega = 2\pi$. Also, $a = 5$. So $y = 5\cos 2\pi t$.

49. Since the Ferris wheel has a radius of 10 m and the bottom of the wheel is 1 m above the ground, the minimum height is 1 m and the maximum height is 21 m. Then $a = 10$ and $\dfrac{2\pi}{\omega} = 20$ s $\Leftrightarrow$ $\omega = \frac{\pi}{10}$, and so $y = 11 + 10\sin\left(\frac{\pi}{10}t\right)$, where t is in seconds.

51. $a = 0.2$, $\dfrac{2\pi}{\omega} = 10$ $\Leftrightarrow$ $\omega = \dfrac{\pi}{5}$. Then $y = 3.8 + 0.2\sin\left(\frac{\pi}{5}t\right)$.

53. The amplitude is $\frac{1}{2}(100 - 80) = 10$ mmHG, the period is 24 hours, and the phase shift is 8 hours, so $f(t) = 10\sin\left(\frac{\pi}{12}(t-8)\right) + 90$.

55. **(a)** The maximum voltage is the amplitude, that is, $V_{\text{max}} = a = 45$ V.

(b) From the graph we see that 4 cycles are completed every 0.1 seconds, or equivalently, 40 cycles are completed every second, so $f = 40$.

(c) The number of revolutions per second of the armature is the frequency, that is, $\dfrac{\omega}{2\pi} = f = 40$.

(d) $a = 45$, $f = \dfrac{\omega}{2\pi} = 40$ $\Leftrightarrow$ $\omega = 80\pi$. Then $V(t) = 45\cos 80\pi t$.

57. $k = 1$, $c = 0.9$, and $\dfrac{\omega}{2\pi} = \frac{1}{2} \Leftrightarrow \omega = \pi$. Since $f(0) = 0$, $f(t) = e^{-0.9t}\sin\pi t$.

59. $\dfrac{ke^{-ct}}{ke^{-c(t+3)}} = 4 \Leftrightarrow e^{-ct+c(t+3)} = 4$ $\Leftrightarrow$ $e^{3c} = 4$ $\Leftrightarrow$ $3c = \ln 4$ $\Leftrightarrow$ $c = \frac{1}{3}\ln 4 \approx 0.46$.

61. **(a)** For fan A, the amplitude is 1, the frequency is 100, and the phase is 0, so $\dfrac{\omega}{2\pi} = 100 \Leftrightarrow \omega = 200\pi$ and an equation is $y = \sin 200\pi t$.

For fan B, the amplitude is 1, the frequency is 100, and the phase is $-\frac{3\pi}{4}$, so again $\omega = 200\pi$ and an equation is $y = \sin\left(200\pi t + \frac{3\pi}{4}\right)$.

(b) The phase difference is $\frac{3\pi}{4}$, so the fans are out of phase. If fan A were rotated $\frac{3\pi}{4}$ counterclockwise, the phase difference would become 0 and the fans would be in phase.

63. From left to right: at $t = \frac{\pi}{6}$, $\sin t = \frac{1}{2}$ and the values are increasing; at $t = \frac{\pi}{2}$, $\sin t = 1$ and the values reach a maximum; at $t = \frac{5\pi}{6}$, $\sin t = \frac{1}{2}$ and the values are decreasing; at $t = \pi$, $\sin t = 0$ and the values are decreasing; at $t = \frac{7\pi}{6}$, $\sin t = -\frac{1}{2}$ and the values are decreasing; at $t = \frac{3\pi}{2}$, $\sin t = -1$ and the values reach a minimum, and at $t = \frac{11\pi}{6}$, $\sin t = -\frac{1}{2}$ and the values are increasing.

CHAPTER 6 REVIEW

1. **(a)** Since $\left(-\frac{\sqrt{3}}{2}\right)^2 + \left(\frac{1}{2}\right)^2 = \frac{3}{4} + \frac{1}{4} = 1$, the point $P\left(-\frac{\sqrt{3}}{2}, \frac{1}{2}\right)$ lies on the unit circle.

(b) $\sin t = \frac{1}{2}$, $\cos t = -\frac{\sqrt{3}}{2}$, $\tan t = \dfrac{\frac{1}{2}}{-\frac{\sqrt{3}}{2}} = -\frac{\sqrt{3}}{3}$.

3. $t = \frac{2\pi}{3}$

(a) $\bar{t} = \pi - \frac{2\pi}{3} = \frac{\pi}{3}$

(b) $P\left(-\frac{1}{2}, \frac{\sqrt{3}}{2}\right)$

(c) $\sin t = \frac{\sqrt{3}}{2}$, $\cos t = -\frac{1}{2}$, $\tan t = -\sqrt{3}$, $\csc t = \frac{2\sqrt{3}}{3}$, $\sec t = -2$, and $\cot t = -\frac{\sqrt{3}}{3}$.

5. $t = -\frac{11\pi}{4}$

(a) $\bar{t} = 3\pi + \left(-\frac{11\pi}{4}\right) = \frac{\pi}{4}$

(b) $P\left(-\frac{\sqrt{2}}{2}, -\frac{\sqrt{2}}{2}\right)$

(c) $\sin t = -\frac{\sqrt{2}}{2}$, $\cos t = -\frac{\sqrt{2}}{2}$, $\tan t = 1$, $\csc t = -\sqrt{2}$, $\sec t = -\sqrt{2}$, and $\cot t = 1$.

7. (a) $\sin \frac{3\pi}{4} = \sin \frac{\pi}{4} = \frac{\sqrt{2}}{2}$

(b) $\cos \frac{3\pi}{4} = -\cos \frac{\pi}{4} = -\frac{\sqrt{2}}{2}$

9. (a) $\sin 1.1 \approx 0.89121$

(b) $\cos 1.1 \approx 0.45360$

11. (a) $\cos \frac{9\pi}{2} = \cos \frac{\pi}{2} = 0$

(b) $\sec \frac{9\pi}{2}$ is undefined

13. (a) $\tan \frac{5\pi}{2}$ is undefined

(b) $\cot \frac{5\pi}{2} = \cot \frac{\pi}{2} = 0$

15. (a) $\tan \frac{5\pi}{6} = -\frac{\sqrt{3}}{3}$

(b) $\cot \frac{5\pi}{6} = -\sqrt{3}$

17. $\dfrac{\tan t}{\cos t} = \dfrac{\frac{\sin t}{\cos t}}{\cos t} = \dfrac{\sin t}{\cos^2 t} = \dfrac{\sin t}{1 - \sin^2 t}$

19. $\tan t = \dfrac{\sin t}{\cos t} = \dfrac{\sin t}{\pm\sqrt{1 - \sin^2 t}} = \dfrac{\sin t}{\sqrt{1 - \sin^2 t}}$ (because t is in quadrant IV, $\cos t$ is positive).

21. $\sin t = \frac{5}{13}$, $\cos t = -\frac{12}{13}$. Then $\tan t = \dfrac{\frac{5}{13}}{-\frac{12}{13}} = -\frac{5}{12}$, $\csc t = \frac{13}{5}$, $\sec t = -\frac{13}{12}$, and $\cot t = -\frac{12}{5}$.

23. $\cot t = -\frac{1}{2}$, $\csc t = \frac{\sqrt{5}}{2}$. Since $\csc t = \dfrac{1}{\sin t}$, we know $\sin t = \frac{2}{\sqrt{5}} = \frac{2\sqrt{5}}{5}$. Now $\cot t = \dfrac{\cos t}{\sin t}$, so $\cos t = \sin t \cdot \cot t = \frac{2\sqrt{5}}{5} \cdot \left(-\frac{1}{2}\right) = -\frac{\sqrt{5}}{5}$, and $\tan t = \dfrac{1}{\left(-\frac{1}{2}\right)} = -2$ while $\sec t = \dfrac{1}{\cos t} = \dfrac{1}{\left(-\frac{\sqrt{5}}{5}\right)} = -\frac{5}{\sqrt{5}} = -\sqrt{5}$.

25. $\tan t = \frac{1}{4}$, so $\cot t = 4$ and $\dfrac{\sin t}{\cos t} = \dfrac{1}{4} \Leftrightarrow \sin t = \frac{1}{4}\cos t$. Thus, $\cos^2 t + \sin^2 t = 1 \Leftrightarrow \cos^2 t + \left(\frac{1}{4}\cos t\right)^2 = 1 \Leftrightarrow \cos^2 t = \frac{16}{17} \Leftrightarrow \sec^2 t = \frac{17}{16}$. Because cosine and secant are negative in Quadrant III, we have $\sec t = -\frac{\sqrt{17}}{4}$, and thus, $\sec t + \cot t = 4 - \frac{\sqrt{17}}{4}$.

27. $\cos t = \frac{3}{5}$, so because the terminal point is in Quadrant I, $\sin t = \frac{4}{5}$. Thus, $\tan t = \frac{4}{3}$ and $\sec t = \frac{5}{3}$, so $\tan t + \sec t = \frac{4}{3} + \frac{5}{3} = 3$.

29. $y = 10\cos \frac{1}{2}x$

(a) This function has amplitude 10, period $\dfrac{2\pi}{\frac{1}{2}} = 4\pi$, and phase shift 0.

(b)

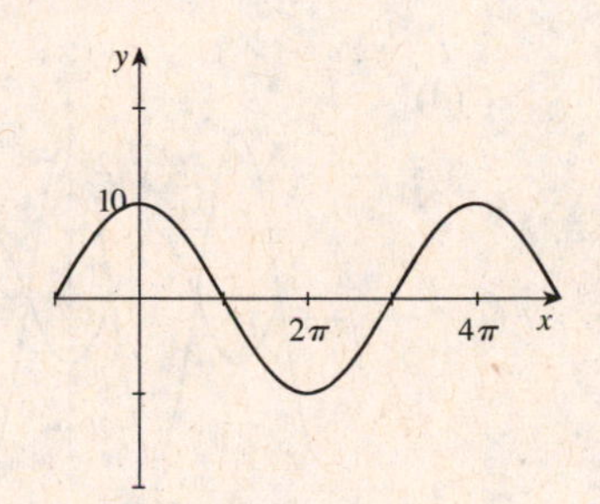

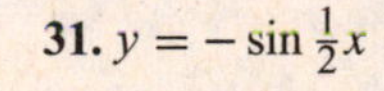
31. $y = -\sin \frac{1}{2}x$

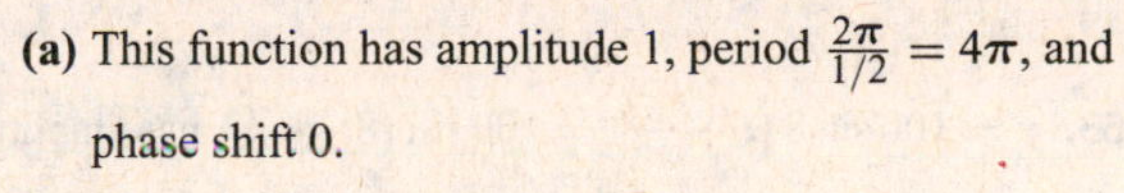
(a) This function has amplitude 1, period $\frac{2\pi}{1/2} = 4\pi$, and phase shift 0.

(b)

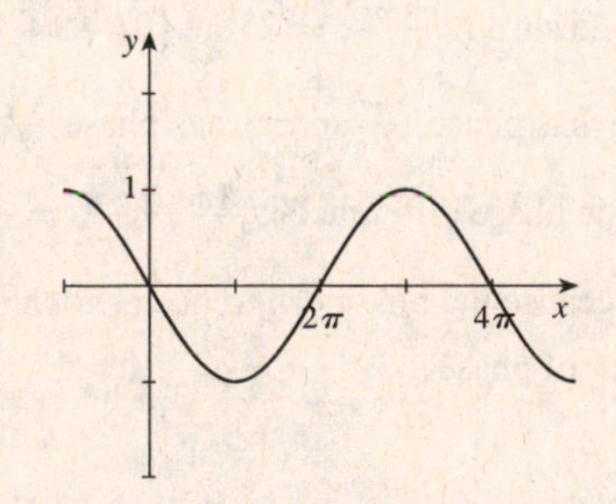

33. $y = 3\sin(2x - 2) = 3\sin 2(x - 1)$

(a) This function has amplitude 3, period $\frac{2\pi}{2} = \pi$, and phase shift 1.

(b)

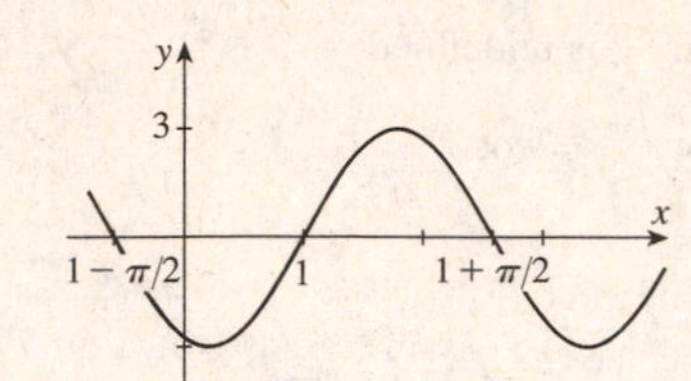

35. $y = -\cos\left(\frac{\pi}{2}x + \frac{\pi}{6}\right) = -\cos\frac{\pi}{2}\left(x + \frac{1}{3}\right)$

(a) This function has amplitude 1, period $\frac{2\pi}{\pi/2} = 4$, and phase shift $-\frac{1}{3}$.

(b)

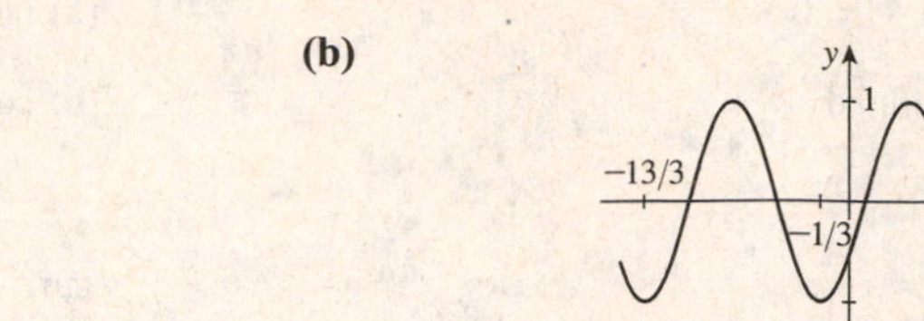

37. From the graph we see that the amplitude is 5, the period is $\frac{\pi}{2}$, and there is no phase shift. Therefore, the function is $y = 5\sin 4x$.

39. From the graph we see that the amplitude is $\frac{1}{2}$, the period is 1, and there is a phase shift of $-\frac{1}{3}$. Therefore, the function is $y = \frac{1}{2}\sin 2\pi\left(x + \frac{1}{3}\right)$.

41. $y = 3\tan x$ has period π.

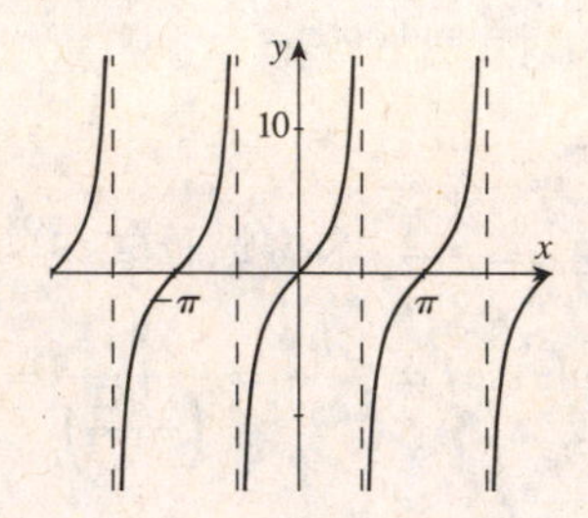

43. $y = 2\cot\left(x - \frac{\pi}{2}\right)$ has period π.

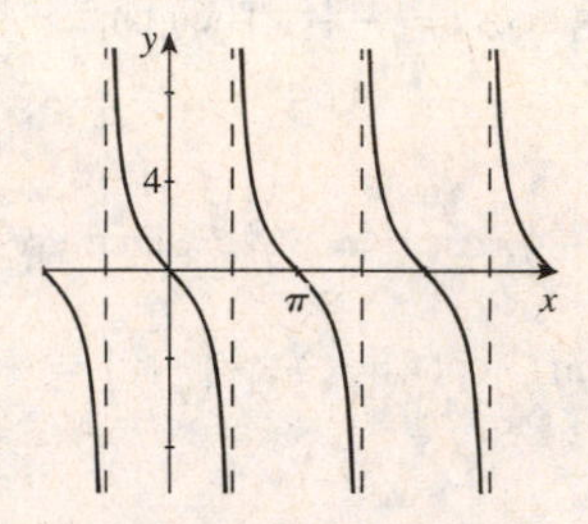

45. $y = 4\csc(2x + \pi) = 4\csc 2\left(x + \frac{\pi}{2}\right)$ has period $\frac{2\pi}{2} = \pi$.

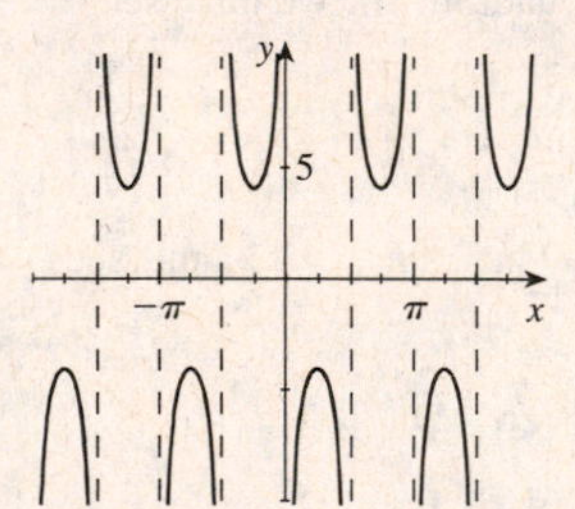

47. $y = \tan\left(\frac{1}{2}x - \frac{\pi}{8}\right) = \tan\frac{1}{2}\left(x - \frac{\pi}{4}\right)$ has period $\frac{\pi}{\frac{1}{2}} = 2\pi$.

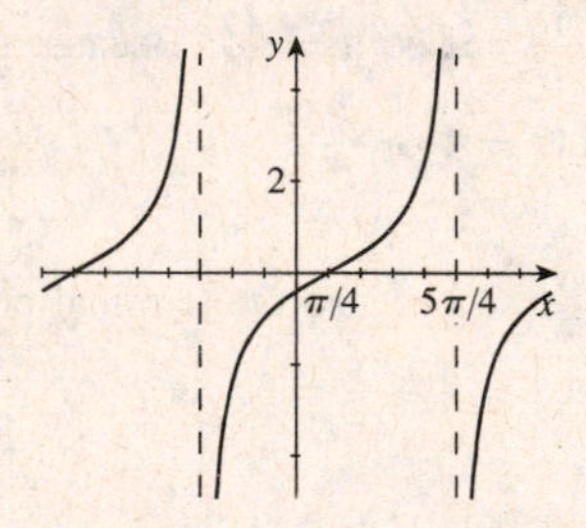

49. $\sin^{-1} 1 = \frac{\pi}{2}$

51. $\sin^{-1}\left(\sin\frac{13\pi}{6}\right) = \frac{\pi}{6}$

53. $y = 100\sin 8\left(t + \frac{\pi}{16}\right) = 100\sin\left(8t + \frac{\pi}{2}\right)$ has amplitude 100, period $\frac{\pi}{4}$, phase $-\frac{\pi}{2}$, and horizontal shift $-\frac{\pi}{16}$.

55. $y_1 = 25\sin 3\left(t - \frac{\pi}{2}\right) = 25\sin\left(3t - \frac{3\pi}{2}\right)$; $y_2 = 10\sin\left(3t - \frac{5\pi}{2}\right)$

(a) y_1 has phase $\frac{3\pi}{2}$ and y_2 has phase $\frac{5\pi}{2}$.

(b) The phase difference is $\frac{3\pi}{2} - \frac{5\pi}{2} = -\pi$.

(c) Because the phase difference is not a multiple of 2π, the curves are out of phase.

(d)

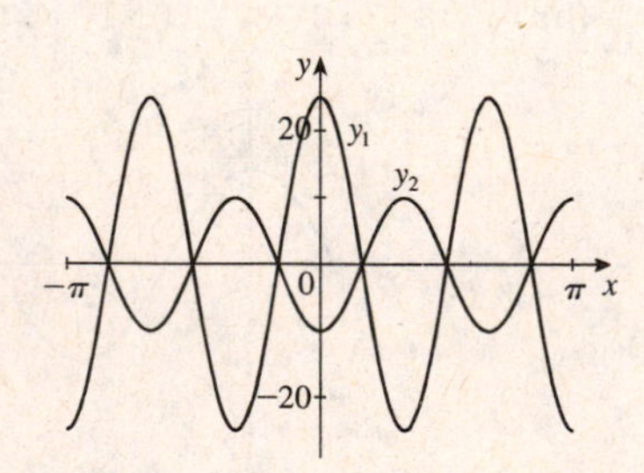

57. (a) $y = |\cos x|$

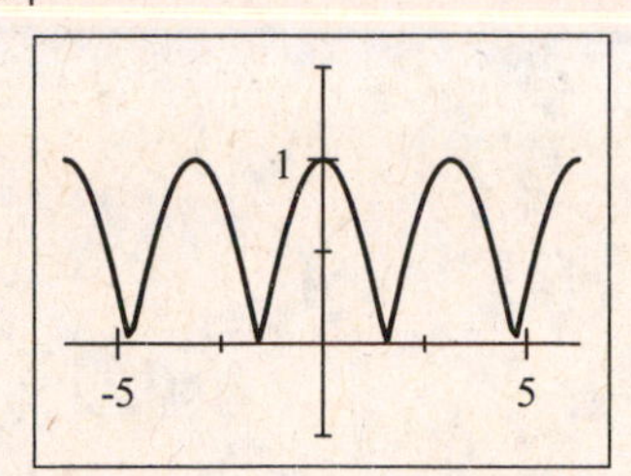

(b) This function has period π.

(c) This function is even.

59. (a) $y = \cos\left(2^{0.1x}\right)$

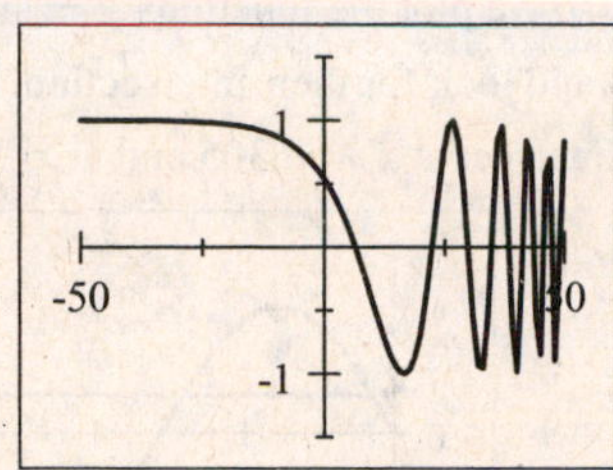

(b) This function is not periodic.

(c) This function is neither even nor odd.

61. (a) $y = |x| \cos 3x$

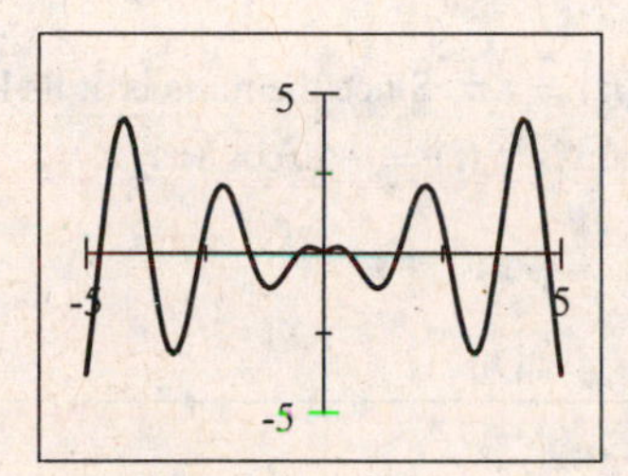

(b) This function is not periodic.

(c) This function is even.

63. $y = x \sin x$ is a sine function whose graph lies between those of $y = x$ and $y = -x$.

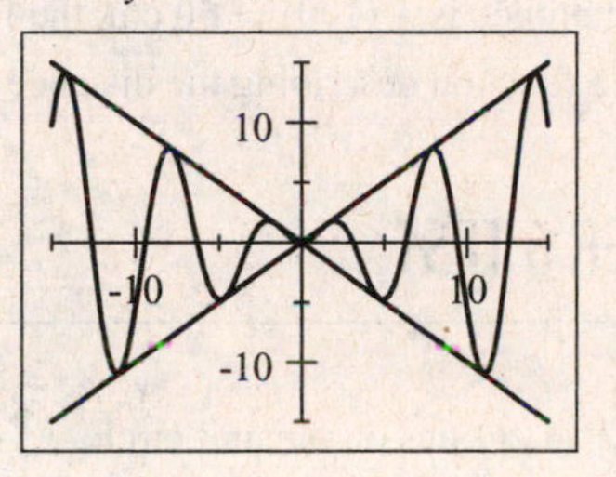

65. $y = x + \sin 4x$ is the sum of the two functions $y = x$ and $y = \sin 4x$.

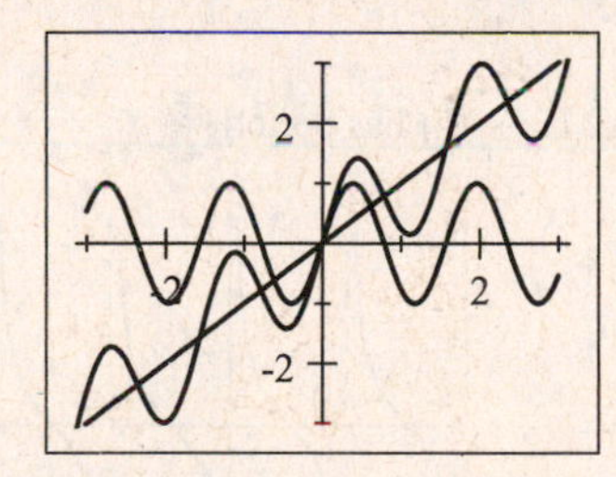

67. We graph $y = f(x) = \cos x + \sin 2x$ in the viewing rectangle $[0, 2\pi] \times [-2, 2]$ and see that the function has local maxima of approximately $f(0.63) \approx 1.76$ and $f(4.14) \approx 0.37$ and local minima of approximately $f(2.51) \approx -1.76$ and $f(5.28) \approx -0.37$. The function is periodic with period 2π.

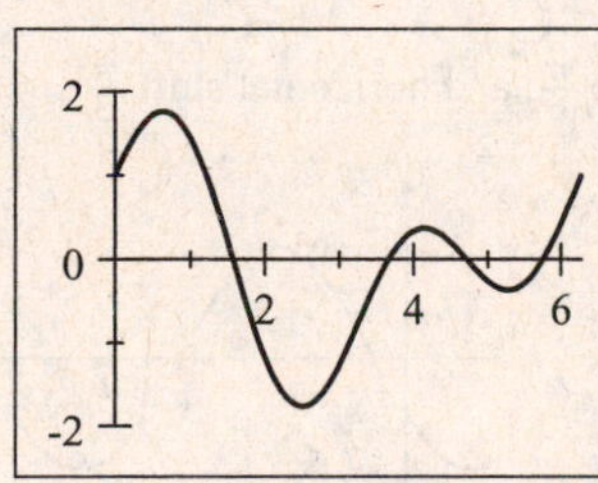

69. We want to find solutions to $\sin x = 0.3$ in the interval $[0, 2\pi]$, so we plot the functions $y = \sin x$ and $y = 0.3$ and look for their intersection. We see that the graphs intersect at $x \approx 0.305$ and at $x \approx 2.837$.

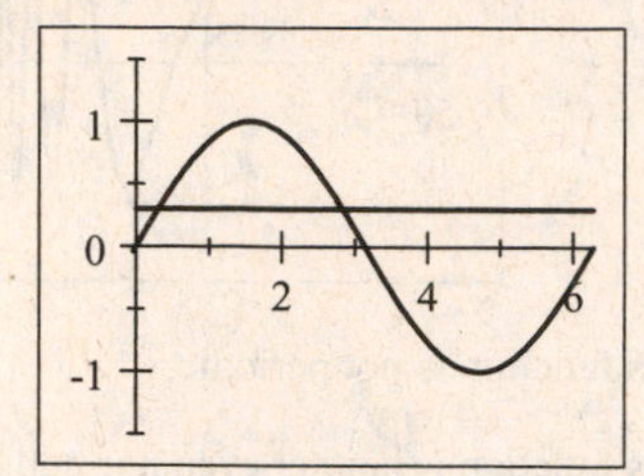

71. $y_1 = \cos(\sin x)$, $y_2 = \sin(\cos x)$

(a)

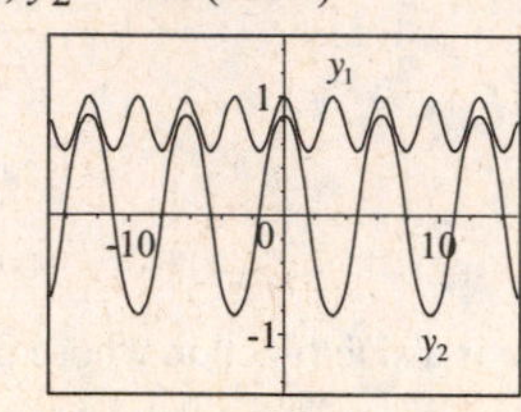

(b) y_1 has period π, while y_2 has period 2π.

(c) $\sin(\cos x) < \cos(\sin x)$ for all x.

73. The amplitude is $\frac{1}{2}(100) = 50$ cm, the frequency is 4 Hz, so $\omega = 4(2\pi) = 8\pi$. Since the mass is at its lowest point when $t = 0$, a function describing the distance of the mass from its rest position is $f(t) = -50\cos 8\pi t$.

CHAPTER 6 TEST

1. Since $P(x, y)$ lies on the unit circle, $x^2 + y^2 = 1 \Rightarrow y = \pm\sqrt{1 - \left(\frac{\sqrt{11}}{6}\right)^2} = \pm\sqrt{\frac{25}{36}} = \pm\frac{5}{6}$. But $P(x, y)$ lies in the fourth quadrant. Therefore y is negative $\Rightarrow$ $y = -\frac{5}{6}$.

3. (a) $\sin\frac{7\pi}{6} = -0.5$

(b) $\cos\frac{13\pi}{4} = -\frac{\sqrt{2}}{2}$

(c) $\tan\left(-\frac{5\pi}{3}\right) = \sqrt{3}$

(d) $\csc\left(\frac{3\pi}{2}\right) = -1$

5. $\cos t = -\frac{8}{17}$, t in quadrant III $\Rightarrow$ $\tan t \cdot \cot t + \csc t = 1 + \dfrac{1}{-\sqrt{1 - \cos^2 t}}$ (since t is in quadrant III)$= 1 - \dfrac{1}{-\sqrt{1 - \frac{64}{289}}} = 1 - \dfrac{1}{\frac{15}{17}} = -\frac{2}{15}$.

7. $y = 2\sin\left(\frac{1}{2}x - \frac{\pi}{6}\right) = \sin\frac{1}{2}\left(x - \frac{\pi}{3}\right)$

(a) This function has amplitude2, period $\dfrac{2\pi}{\frac{1}{2}} = 4\pi$, phase $\frac{\pi}{6}$, and horizontal shift $\frac{\pi}{3}$.

(b)

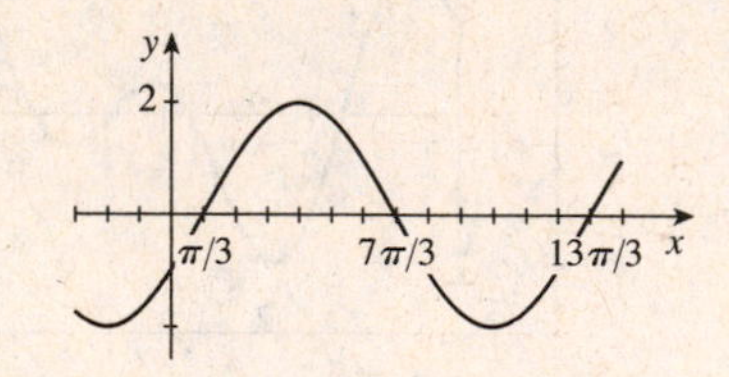

9. $y = \tan 2\left(x - \frac{\pi}{4}\right)$ has period $\frac{\pi}{2}$.

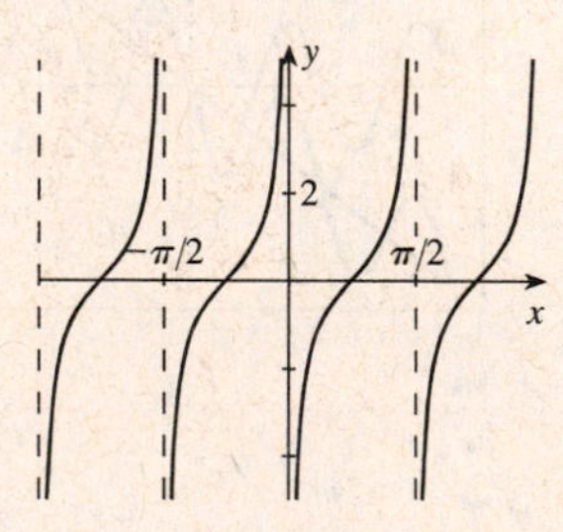

11. From the graph, we see that the amplitude is 2 and the phase shift is $-\frac{\pi}{3}$. Also, the period is π, so $\frac{2\pi}{k} = \pi \Rightarrow k = \frac{2\pi}{\pi} = 2$. Thus, the function is $y = 2\sin 2\left(x + \frac{\pi}{3}\right)$.

13. $y = \dfrac{\cos x}{1 + x^2}$

(a)

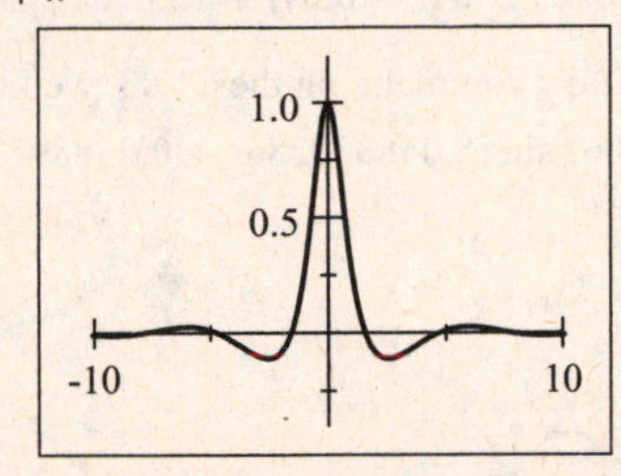

(b) The function is even.

(c) The function has a minimum value of approximately -0.11 when $x \approx \pm 2.54$ and a maximum value of 1 when $x = 0$.

15. (a) The initial amplitude is 16 in. and the frequency is 12 Hz, so a function describing the motion is $y = 16e^{-0.1t}\cos 24\pi t$.

(b)

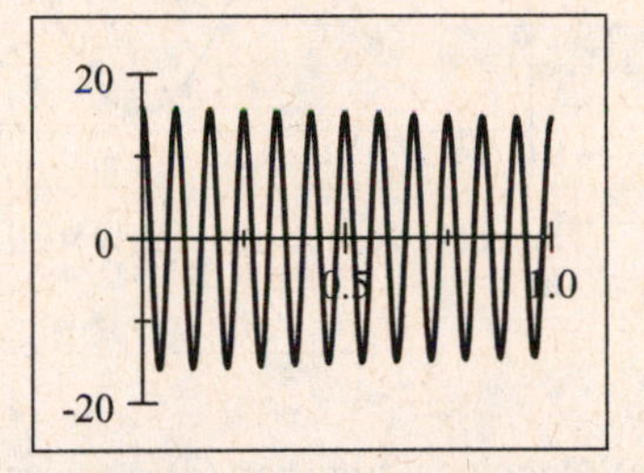

FOCUS ON MODELING Fitting Sinusoidal Curves to Data

1. (a) See the graph in part (c).

(b) Using the method of Example 1, we find the vertical shift $b = \frac{1}{2}(\text{maximum value} + \text{minimum value}) = \frac{1}{2}(2.1 - 2.1) = 0$, the amplitude $a = \frac{1}{2}(\text{maximum value} - \text{minimum value}) = \frac{1}{2}(2.1 - (-2.1)) = 2.1$, the period $\dfrac{2\pi}{\omega} = 2(6 - 0) = 12$ (so $\omega \approx 0.5236$), and the phase shift $c = 0$. Thus, our model is $y = 2.1\cos\frac{\pi}{6}t$.

(c)

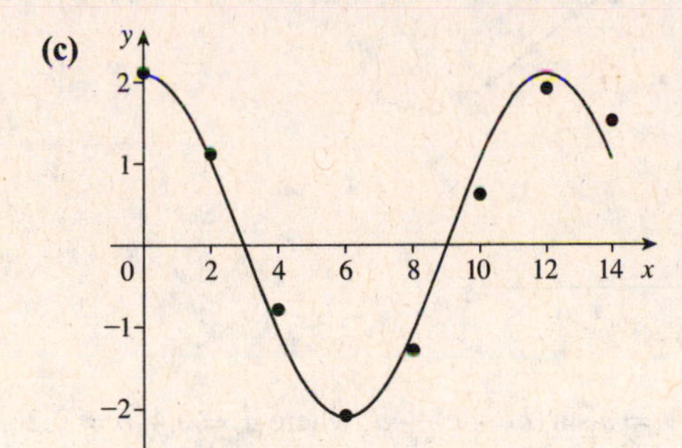

The curve fits the data quite well.

(d) Using the `SinReg` command on the TI-83, we find

$$y = 2.048714222\sin(0.5030795477t + 1.551856108) - 0.0089616507.$$

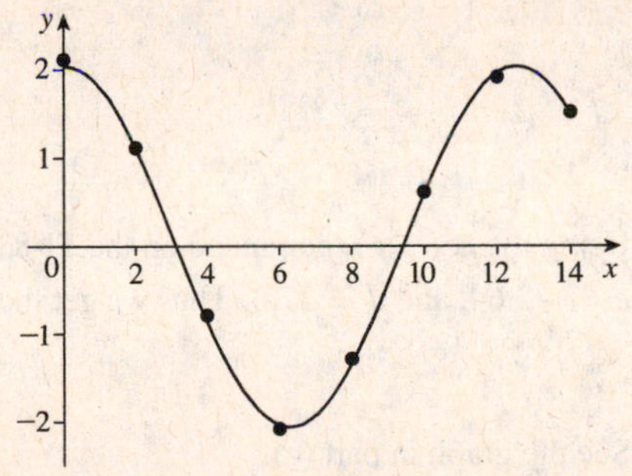

(e) Our model from part (d) is equivalent to $y = 2.05\cos\left(0.50t + 1.55 - \frac{\pi}{2}\right) - 0.01 \approx 2.05\cos(0.50t - 0.02) - 0.01$. This is the same as the function in part (b), correct to one decimal place.

3. (a) See the graph in part (c).

(b) Using the method of Example 1, we find the vertical shift $b = \frac{1}{2}$ (maximum value + minimum value) $= \frac{1}{2}(25.1 + 1.0) = 13.05$, the amplitude $a = \frac{1}{2}$ (maximum value − minimum value) $= \frac{1}{2}(25.1 - 1.0) = 12.05$, the period $\frac{2\pi}{\omega} = 2(1.5 - 0.9) = 1.2$ (so $\omega \approx 5.236$), and the phase shift $c = 0.3$. Thus, our model is $y = 12.05\cos(5.236(t - 0.3)) + 13.05$.

(c)

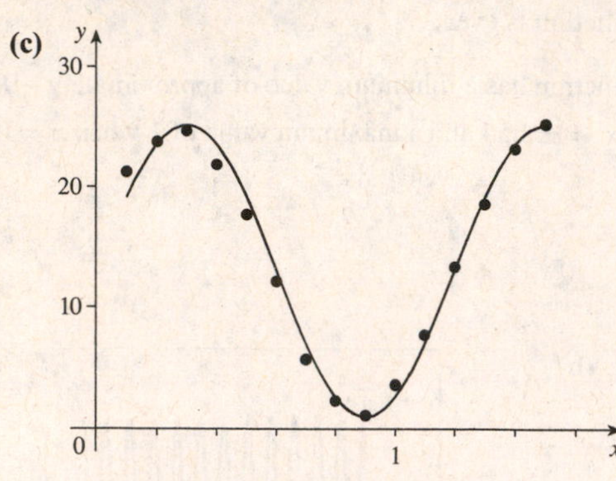

The curve fits the data fairly well.

(d) Using the `SinReg` command on the TI-83, we find $y = 11.71905062\sin(5.048853286t + 0.2388957877) + 12.96070536$.

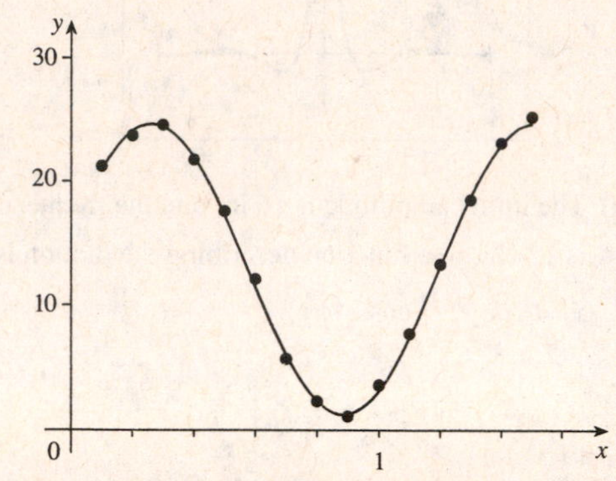

(e) Our model from part (d) is equivalent to $y = 11.72\cos\left(5.05t + 0.24 - \frac{\pi}{2}\right) + 12.96 \approx 11.72\cos(5.05t - 1.33) + 12.96$. This is close but not identical to the function in part (b).

5. (a) See the graph in part (c).

(b) Let t be the time since midnight. We find a function of the form $y = a\sin\omega(t - c) + b$. $a = \frac{1}{2}(37.4 - 36.6) = 0.4$. The period is 24 and so $\omega = \frac{2\pi}{24} \approx 0.26$. $b = \frac{1}{2}(37.4 + 36.6) = 37$. Because the maximum value occurs at $t = 16$, we get $c = 16$. Thus the function is $y = 0.4\cos(0.26(t - 16)) + 37$.

(c)

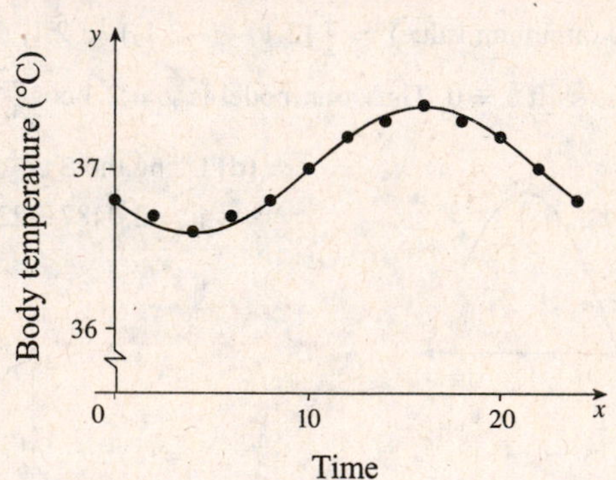

(d) Using the `SinReg` command on the TI-83 we obtain the function $y = a\sin(bt + c) + d$, where $a = 0.4$, $b = 0.26$, $c = -2.64$, and $d = 37.0$. Thus we get the model $y = 0.4\sin(0.26t - 2.64) + 37.0$.

7. (a) See the graph in part (c).

(b) Let t be the time since 1985. We find a function of the form $y = a\sin\omega(t - c) + b$. $a = \frac{1}{2}(63 - 22) = 20.5$. The periodis $2(15 - 9) = 12$ and so $\omega = \frac{2\pi}{20} \approx 0.52$. $b = \frac{1}{2}(63 + 22) = 42.5$. The average value occurs in the ninth year, so $c = 6$. Thus, our model is $y = 20.5\sin(0.52(t - 6)) + 42.5$.

(c)

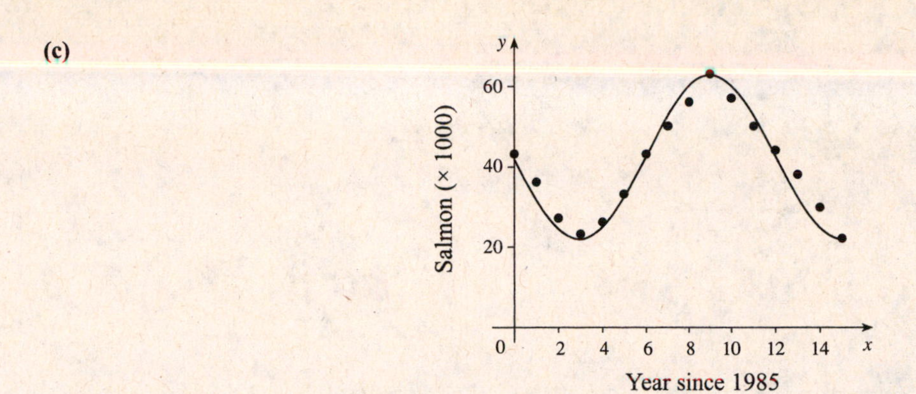

(d) Using the `SinReg` command on the TI-83, we find that for the function $y = a\sin(bt + c) + d$, where $a = 17.8$, $b = 0.52$, $c = 3.11$, and $d = 42.4$. Thus we get the model $y = 17.8\sin(0.52t + 3.11) + 42.4$.

7 ANALYTIC TRIGONOMETRY

7.1 TRIGONOMETRIC IDENTITIES

1. An equation is called an identity if it is valid for *all* values of the variable. The equation $2x = x + x$ is an algebraic identity and the equation $\sin^2 x + \cos^2 x = 1$ is a trigonometric identity.

3. $\cos t \tan t = \cos t \cdot \dfrac{\sin t}{\cos t} = \sin t$

5. $\sin\theta \sec\theta = \sin\theta \cdot \dfrac{1}{\cos\theta} = \tan\theta$

7. $\tan^2 x - \sec^2 x = \dfrac{\sin^2 x}{\cos^2 x} - \dfrac{1}{\cos^2 x} = \dfrac{\sin^2 x - 1}{\cos^2 x} = \dfrac{-\cos^2 x}{\cos^2 x} = -1$

9. $\sin u + \cot u \cos u = \sin u + \dfrac{\cos u}{\sin u} \cdot \cos u = \dfrac{\sin^2 u + \cos^2 u}{\sin u} = \dfrac{1}{\sin u} = \csc u$

11. $\dfrac{\sec\theta - \cos\theta}{\sin\theta} = \dfrac{\dfrac{1}{\cos\theta} - \cos\theta}{\sin\theta} = \dfrac{1 - \cos^2\theta}{\sin\theta\cos\theta} = \dfrac{\sin^2\theta}{\sin\theta\cos\theta} = \dfrac{\sin\theta}{\cos\theta} = \tan\theta$

13. $\dfrac{\sin x \sec x}{\tan x} = \dfrac{\sin x \cdot \dfrac{1}{\cos x}}{\dfrac{\cos x}{\sin x}} = 1$

15. $\dfrac{\sin t + \tan t}{\tan t} = \dfrac{\sin t}{\dfrac{\sin t}{\cos t}} + \dfrac{\tan t}{\tan t} = \cos t + 1$

17. $\cos^3 x + \sin^2 x \cos x = \cos x \left(\cos^2 x + \sin^2 x\right) = \cos x$

19. $\dfrac{\sec^2 x - 1}{\sec^2 x} = \dfrac{\tan^2 x}{\sec^2 x} = \dfrac{\sin^2 x}{\cos^2 x} \cdot \cos^2 x = \sin^2 x$.

Another method: $\dfrac{\sec^2 x - 1}{\sec^2 x} = 1 - \dfrac{1}{\sec^2 x} = 1 - \cos^2 x = \sin^2 x$

21. $\dfrac{1 + \cos y}{1 + \sec y} = \dfrac{1 + \cos y}{1 + \dfrac{1}{\cos y}} = \dfrac{1 + \cos y}{\dfrac{\cos y + 1}{\cos y}} = \dfrac{1 + \cos y}{1} \cdot \dfrac{\cos y}{\cos y + 1} = \cos y$

23. $\dfrac{1 + \sin u}{\cos u} + \dfrac{\cos u}{1 + \sin u} = \dfrac{(1 + \sin u)^2 + \cos^2 u}{\cos u\,(1 + \sin u)} = \dfrac{1 + 2\sin u + \sin^2 u + \cos^2 u}{\cos u\,(1 + \sin u)} = \dfrac{1 + 2\sin u + 1}{\cos u\,(1 + \sin u)}$

$= \dfrac{2 + 2\sin u}{\cos u\,(1 + \sin u)} = \dfrac{2\,(1 + \sin u)}{\cos u\,(1 + \sin u)} = \dfrac{2}{\cos u} = 2\sec u$

25. $\dfrac{\cos x}{\sec x + \tan x} = \dfrac{\cos x}{\dfrac{1}{\cos x} + \dfrac{\sin x}{\cos x}} = \dfrac{\cos^2 x}{1 + \sin x} = \dfrac{1 - \sin^2 x}{1 + \sin x} = \dfrac{(1 + \sin x)(1 - \sin x)}{1 + \sin x} = 1 - \sin x$

27. $\dfrac{1}{1 - \sin\alpha} + \dfrac{1}{1 + \sin\alpha} = \dfrac{1 + \sin\alpha + 1 - \sin\alpha}{1 - \sin^2\alpha} = \dfrac{2}{\cos^2\alpha} = 2\sec^2\alpha$

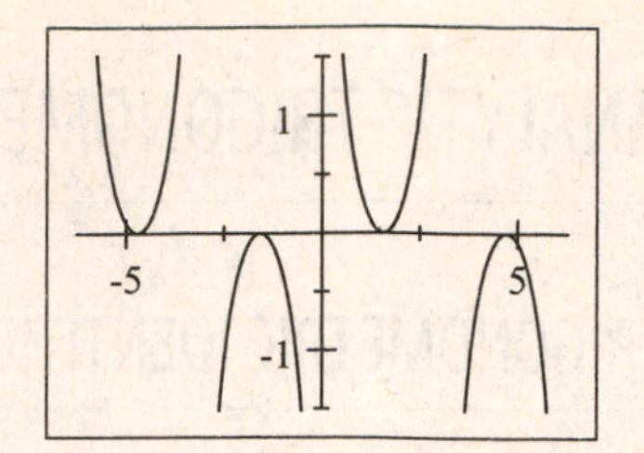

29. (a) $\dfrac{\cos x}{\sec x \sin x} = \dfrac{\cos x}{\dfrac{1}{\cos x} \cdot \sin x} = \dfrac{\cos^2 x}{\sin x} = \dfrac{1 - \sin^2 x}{\sin x}$

$= \dfrac{1}{\sin x} - \dfrac{\sin^2 x}{\sin x} = \csc x - \sin x$

(b) We graph each side of the equation and see that the graphs of $y = \dfrac{\cos x}{\sec x \sin x}$ and $y = \csc x - \sin x$ are identical, confirming that the equation is an identity.

31. $\dfrac{\sin\theta}{\tan\theta} = \dfrac{\sin\theta}{\dfrac{\sin\theta}{\cos\theta}} = \sin\theta \cdot \dfrac{\cos\theta}{\sin\theta} = \cos\theta$

33. $\dfrac{\cos u \sec u}{\tan u} = \cos u \dfrac{1}{\cos u} \cot u = \cot u$

35. $\dfrac{\tan y}{\csc y} = \dfrac{\dfrac{\sin y}{\cos y}}{\dfrac{1}{\sin y}} = \dfrac{\sin^2 y}{\cos y} = \dfrac{1 - \cos^2 y}{\cos y} = \dfrac{1}{\cos y} - \cos y = \dfrac{1}{\cos y} - \dfrac{1}{\sec y}$

37. $\cos(-x) - \sin(-x) = \cos x - (-\sin x) = \cos x + \sin x$

39. $\tan\theta + \cot\theta = \dfrac{\sin\theta}{\cos\theta} + \dfrac{\cos\theta}{\sin\theta} = \dfrac{\sin^2\theta + \cos^2\theta}{\cos\theta\sin\theta} = \dfrac{1}{\cos\theta\sin\theta} = \sec\theta\csc\theta$

41. $(1 - \cos\beta)(1 + \cos\beta) = 1 - \cos^2\beta = \sin^2\beta = \dfrac{1}{\csc^2\beta}$

43. $\dfrac{1}{1 - \sin^2 y} = \dfrac{1}{\cos^2 y} = \sec^2 y = 1 + \tan^2 y$

45. $(\tan x + \cot x)^2 = \tan^2 x + 2\tan x\cot x + \cot^2 x = \tan^2 x + 2 + \cot^2 x = \left(\tan^2 x + 1\right) + \left(\cot^2 x + 1\right) = \sec^2 x + \csc^2 x$

47. $\left(1 - \sin^2 t + \cos^2 t\right)^2 + 4\sin^2 t\cos^2 t = \left(2\cos^2 t\right)^2 + 4\sin^2 t\cos^2 t = 4\cos^2 t\left(\cos^2 t + \sin^2 t\right) = 4\cos^2 t$

49. $\csc x\cos^2 x + \sin x = \dfrac{\cos^2 x}{\sin x} + \dfrac{\sin^2 x}{\sin x} = \dfrac{1}{\sin x} = \csc x$

51. $\dfrac{(\sin x + \cos x)^2}{\sin^2 x - \cos^2 x} = \dfrac{(\sin x + \cos x)^2}{(\sin x + \cos x)(\sin x - \cos x)} = \dfrac{\sin x + \cos x}{\sin x - \cos x} = \dfrac{(\sin x + \cos x)(\sin x - \cos x)}{(\sin x - \cos x)(\sin x - \cos x)} = \dfrac{\sin^2 x - \cos^2 x}{(\sin x - \cos x)^2}$

53. $\dfrac{\sec t - \cos t}{\sec t} = \dfrac{\dfrac{1}{\cos t} - \cos t}{\dfrac{1}{\cos t}} = \dfrac{\dfrac{1}{\cos t} - \cos t}{\dfrac{1}{\cos t}} \cdot \dfrac{\cos t}{\cos t} = \dfrac{1 - \cos^2 t}{1} = \sin^2 t$

55. $\cos^2 x - \sin^2 x = \cos^2 x - \left(1 - \cos^2 x\right) = 2\cos^2 x - 1$

57. $\sin^4\theta - \cos^4\theta = \left(\sin^2\theta\right)^2 - \left(\cos^2\theta\right)^2 = \left(\sin^2\theta - \cos^2\theta\right)\left(\sin^2\theta + \cos^2\theta\right) = \sin^2\theta - \cos^2\theta$

59. $\dfrac{(\sin t + \cos t)^2}{\sin t\cos t} = \dfrac{\sin^2 t + 2\sin t\cos t + \cos^2 t}{\sin t\cos t} = \dfrac{\sin^2 t + \cos^2 t}{\sin t\cos t} + \dfrac{2\sin t\cos t}{\sin t\cos t} = \dfrac{1}{\sin t\cos t} + 2 = 2 + \sec t\csc t$

61. $\dfrac{1 + \tan^2 u}{1 - \tan^2 u} = \dfrac{1 + \dfrac{\sin^2 u}{\cos^2 u}}{1 - \dfrac{\sin^2 u}{\cos^2 u}} = \dfrac{1 + \dfrac{\sin^2 u}{\cos^2 u}}{1 - \dfrac{\sin^2 u}{\cos^2 u}} \cdot \dfrac{\cos^2 u}{\cos^2 u} = \dfrac{\cos^2 u + \sin^2 u}{\cos^2 u - \sin^2 u} = \dfrac{1}{\cos^2 u - \sin^2 u}$

63. $\dfrac{\sec x + \csc x}{\tan x + \cot x} = \dfrac{\dfrac{1}{\cos x} + \dfrac{1}{\sin x}}{\dfrac{\sin x}{\cos x} + \dfrac{\cos x}{\sin x}} = \dfrac{\dfrac{1}{\cos x} + \dfrac{1}{\sin x}}{\dfrac{\sin x}{\cos x} + \dfrac{\cos x}{\sin x}} \cdot \dfrac{\sin x\cos x}{\sin x\cos x} = \dfrac{\sin x + \cos x}{\sin^2 x + \cos^2 x} = \sin x + \cos x$

65. $\dfrac{1-\cos x}{\sin x}+\dfrac{\sin x}{1-\cos x}=\dfrac{1-\cos x}{\sin x}\cdot\dfrac{1-\cos x}{1-\cos x}+\dfrac{\sin x}{1-\cos x}\cdot\dfrac{\sin x}{\sin x}=\dfrac{1-2\cos x+\cos^2 x+\sin^2 x}{\sin x\,(1-\cos x)}$

$=\dfrac{2-2\cos x}{\sin x\,(1-\cos x)}=\dfrac{2\,(1-\cos x)}{\sin x\,(1-\cos x)}=2\csc x$

67. $\tan^2 u-\sin^2 u=\dfrac{\sin^2 u}{\cos^2 u}-\dfrac{\sin^2 u\cos^2 u}{\cos^2 u}=\dfrac{\sin^2 u}{\cos^2 u}\left(1-\cos^2 u\right)=\tan^2 u\sin^2 u$

69. $\dfrac{1+\tan x}{1-\tan x}=\dfrac{1+\dfrac{\sin x}{\cos x}}{1-\dfrac{\sin x}{\cos x}}\cdot\dfrac{\cos x}{\cos x}=\dfrac{\cos x+\sin x}{\cos x-\sin x}$

71. $\dfrac{1}{\sec x+\tan x}+\dfrac{1}{\sec x-\tan x}=\dfrac{\sec x-\tan x+\sec x+\tan x}{(\sec x+\tan x)(\sec x-\tan x)}=\dfrac{2\sec x}{\sec^2 x-\tan^2 x}=\dfrac{2\sec x}{1}=2\sec x$

73. $\dfrac{1+\sin x}{1-\sin x}-\dfrac{1-\sin x}{1+\sin x}=\dfrac{(1+\sin x)^2-(1-\sin x)^2}{(1-\sin x)(1+\sin x)}=\dfrac{1+2\sin x+\sin^2 x-1+2\sin x-\sin^2 x}{1-\sin^2 x}$

$=\dfrac{4\sin x}{\cos^2 x}=4\dfrac{\sin x}{\cos x}\cdot\dfrac{1}{\cos x}=4\tan x\sec x$

75. $\dfrac{\sin^3 x+\cos^3 x}{\sin x+\cos x}=\dfrac{(\sin x+\cos x)\left(\sin^2 x-\sin x\cos x+\cos^2 x\right)}{\sin x+\cos x}=\sin^2-\sin x\cos x+\cos^2 x=1-\sin x\cos x$

77. $\dfrac{1-\cos\alpha}{\sin\alpha}=\dfrac{1-\cos\alpha}{\sin\alpha}\cdot\dfrac{1+\cos\alpha}{1+\cos\alpha}=\dfrac{1-\cos^2\alpha}{\sin\alpha\,(1+\cos\alpha)}=\dfrac{\sin^2\alpha}{\sin\alpha\,(1+\cos\alpha)}=\dfrac{\sin\alpha}{1+\cos\alpha}$

79. $\dfrac{\sin w}{\sin w+\cos w}=\dfrac{\sin w}{\sin w+\cos w}\cdot\dfrac{\dfrac{1}{\cos w}}{\dfrac{1}{\cos w}}=\dfrac{\tan w}{1+\tan w}$

81. $\dfrac{\sec x}{\sec x-\tan x}=\dfrac{\sec x}{\sec x-\tan x}\cdot\dfrac{\sec x+\tan x}{\sec x+\tan x}=\dfrac{\sec x\,(\sec x+\tan x)}{\sec^2 x-\tan^2 x}=\dfrac{\sec x\,(\sec x+\tan x)}{1}=\sec x\,(\sec x+\tan x)$

83. $\dfrac{\cos\theta}{1-\sin\theta}=\dfrac{\cos\theta}{1-\sin\theta}\cdot\dfrac{1+\sin\theta}{1+\sin\theta}=\dfrac{\cos\theta\,(1+\sin\theta)}{1-\sin^2\theta}=\dfrac{\cos\theta\,(1+\sin\theta)}{\cos^2\theta}=\dfrac{1}{\cos\theta}+\dfrac{\sin\theta}{\cos\theta}=\sec\theta+\tan\theta$

85. $\dfrac{1-\sin x}{1+\sin x}=\dfrac{1-\sin x}{1+\sin x}\cdot\dfrac{1-\sin x}{1-\sin x}=\dfrac{1-2\sin x+\sin^2 x}{1-\sin^2 x}=\dfrac{1-2\sin x+\sin^2 x}{\cos^2 x}=\dfrac{1}{\cos^2 x}-\dfrac{2\sin x}{\cos^2 x}+\dfrac{\sin^2 x}{\cos^2 x}$

$=\sec^2 x-2\sec x\tan x+\tan^2 x=(\sec x-\tan x)^2$

87. $\csc x-\cot x=\dfrac{1}{\sin x}-\dfrac{\cos x}{\sin x}=\dfrac{1-\cos x}{\sin x}=\dfrac{1-\cos x}{\sin x}\cdot\dfrac{1+\cos x}{1+\cos x}=\dfrac{1-\cos^2 x}{\sin x\,(1+\cos x)}=\dfrac{\sin^2 x}{\sin x\,(1+\cos x)}$

$=\dfrac{1}{\dfrac{1}{\sin x}(1+\cos x)}=\dfrac{1}{\dfrac{1}{\sin x}+\dfrac{\cos x}{\sin x}}=\dfrac{1}{\csc x+\cot x}$

89. $x=\sin\theta$; then $\dfrac{x}{\sqrt{1-x^2}}=\dfrac{\sin\theta}{\sqrt{1-\sin^2\theta}}=\dfrac{\sin\theta}{\sqrt{\cos^2\theta}}=\dfrac{\sin\theta}{\cos\theta}=\tan\theta$ (since $\cos\theta\geq 0$ for $0\leq\theta\leq\frac{\pi}{2}$).

91. $x=\sec\theta$; then $\sqrt{x^2-1}=\sqrt{\sec^2\theta-1}=\sqrt{\left(\tan^2\theta+1\right)-1}=\sqrt{\tan^2\theta}=\tan\theta$ (since $\tan\theta\geq 0$ for $0\leq\theta<\frac{\pi}{2}$)

93. $x=3\sin\theta$; then $\sqrt{9-x^2}=\sqrt{9-(3\sin\theta)^2}=\sqrt{9-9\sin^2\theta}=\sqrt{9\left(1-\sin^2\theta\right)}=3\sqrt{\cos^2\theta}=3\cos\theta$ (since $\cos\theta\geq 0$ for $0\leq\theta<\frac{\pi}{2}$).

95. $f(x) = \cos^2 x - \sin^2 x$, $g(x) = 1 - 2\sin^2 x$. From the graph, $f(x) = g(x)$ this appears to be an identity. *Proof:*
$f(x) = \cos^2 x - \sin^2 x = \cos^2 x + \sin^2 x - 2\sin^2 x = 1 - 2\sin^2 x = g(x)$. Since $f(x) = g(x)$ for all x, this is an identity.

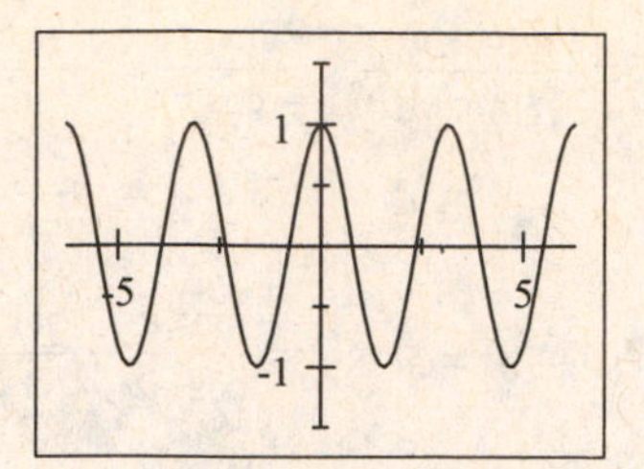

97. $f(x) = (\sin x + \cos x)^2$, $g(x) = 1$. From the graph, $f(x) = g(x)$ does not appear to be an identity. In order to show this, we can set $x = \frac{\pi}{4}$. Then we have
$f\left(\frac{\pi}{4}\right) = \left(\frac{1}{\sqrt{2}} + \frac{1}{\sqrt{2}}\right)^2 = \left(\frac{2}{\sqrt{2}}\right)^2 = \left(\sqrt{2}\right)^2 = 2 \neq 1 = g\left(\frac{\pi}{4}\right)$. Since $f\left(\frac{\pi}{4}\right) \neq g\left(\frac{\pi}{4}\right)$, this is not an identity.

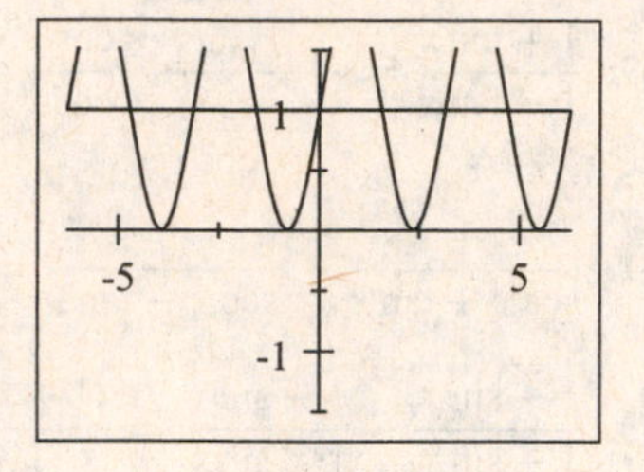

99. $(\sin x \sin y - \cos x \cos y)(\sin x \sin y + \cos x \cos y) = (\sin x \sin y)^2 - (\cos x \cos y)^2$
$= \left(1 - \cos^2 x\right)\sin^2 y - \cos^2 x\left(1 - \sin^2 y\right) = \sin^2 y - \cos^2 x$

101. $(\tan x + \cot x)^4 = \left(\frac{\sin x}{\cos x} + \frac{\cos x}{\sin x}\right)^4 = \left(\frac{\sin^2 x + \cos^2 x}{\sin x \cos x}\right)^4 = \left(\frac{1}{\sin x \cos x}\right)^4 = \sec^4 x \csc^4 x$

103. $\frac{\sin^3 y - \csc^3 y}{\sin y - \csc y} = \frac{(\sin y - \csc y)\left(\sin^2 y + \sin y \csc y + \csc^2 y\right)}{\sin y - \csc y} = \sin^2 y + \csc^2 y + 1$

105. $\ln|\tan x \sin x| = \ln|\tan x| + \ln|\sin x| = \ln\left|\frac{\sin x}{\cos x}\right| + \ln|\sin x| = \ln|\sin x| - \ln|\cos x| + \ln|\sin x|$
$= 2\ln|\sin x| + \ln\left|\frac{1}{\cos x}\right| = 2\ln|\sin x| + \ln|\sec x|$

107. $e^{\sin^2 x} e^{\tan^2 x} = e^{1-\cos^2 x} e^{\sec^2 x - 1} = e^{1-\cos^2 x + \sec^2 x - 1} = e^{\sec^2 x} e^{-\cos^2 x}$

109. This is an identity: LHS $= e^{\sin^2 x + \cos^2 x} = e^1 =$ RHS.

111. Squaring both sides, we have $\sin^2 x + 1 = \sin^2 x + 2\sqrt{\sin^2 x} + 1 \Leftrightarrow \sin x = 0$, which has solutions $x = k\pi$, k an integer.

113. LHS $= (R\cos\theta\sin\phi)^2 + (R\sin\theta\sin\phi)^2 + (R\cos\phi)^2 = (R\sin\phi)^2\left(\cos^2\theta + \sin^2\theta\right) + (R\cos\phi)^2$
$= (R\sin\phi)^2 + (R\cos\phi)^2 =$ RHS

115. **(a)** Choose $x = \frac{\pi}{2}$. Then $\sin 2x = \sin\pi = 0$, whereas $2\sin x = 2\sin\frac{\pi}{2} = 2$.

(b) Choose $x = \frac{\pi}{4}$ and $y = \frac{\pi}{4}$. Then $\sin(x + y) = \sin\frac{\pi}{2} = 1$, whereas $\sin x + \sin y = \sin\frac{\pi}{4} + \sin\frac{\pi}{4} = \frac{1}{\sqrt{2}} + \frac{1}{\sqrt{2}} = \frac{2}{\sqrt{2}}$. Since these are not equal, the equation is not an identity.

(c) Choose $\theta = \frac{\pi}{4}$. Then $\sec^2\theta + \csc^2\theta = \left(\sqrt{2}\right)^2 + \left(\sqrt{2}\right)^2 = 4 \neq 1$.

(d) Choose $x = \frac{\pi}{4}$. Then $\frac{1}{\sin x + \cos x} = \frac{1}{\sin\frac{\pi}{4} + \cos\frac{\pi}{4}} = \frac{1}{\frac{1}{\sqrt{2}} + \frac{1}{\sqrt{2}}} = \frac{1}{\sqrt{2}}$ whereas $\csc x + \sec x = \csc\frac{\pi}{4} + \sec\frac{\pi}{4} = \sqrt{2} + \sqrt{2}$. Since these are not equal, the equation is not an identity.

117. Answers will vary.

7.2 ADDITION AND SUBTRACTION FORMULAS

1. If we know the values of the sine and cosine of x and y we can find the value of $\sin(x+y)$ using the *addition* formula for sine, $\sin(x+y) = \sin x\cos y + \cos x\sin y$.

3. $\sin 75^\circ = \sin(45^\circ + 30^\circ) = \sin 45^\circ \cos 30^\circ + \cos 45^\circ \sin 30^\circ = \frac{\sqrt{2}}{2}\cdot\frac{\sqrt{3}}{2} + \frac{\sqrt{2}}{2}\cdot\frac{1}{2} = \frac{\sqrt{6}+\sqrt{2}}{4}$

5. $\cos 105^\circ = \cos(60^\circ + 45^\circ) = \cos 60^\circ \cos 45^\circ - \sin 60^\circ \sin 45^\circ = \frac{1}{2}\cdot\frac{\sqrt{2}}{2} - \frac{\sqrt{3}}{2}\cdot\frac{\sqrt{2}}{2} = \frac{\sqrt{2}-\sqrt{6}}{4}$

7. $\tan 15^\circ = \tan(45^\circ - 30^\circ) = \dfrac{\tan 45^\circ - \tan 30^\circ}{1 + \tan 45^\circ \tan 30^\circ} = \dfrac{1 - \frac{\sqrt{3}}{3}}{1 + 1\cdot\frac{\sqrt{3}}{3}} = \dfrac{3-\sqrt{3}}{3+\sqrt{3}} = 2 - \sqrt{3}$

9. $\sin\frac{19\pi}{12} = -\sin\frac{7\pi}{12} = -\sin\left(\frac{\pi}{4} + \frac{\pi}{3}\right) = -\sin\frac{\pi}{4}\cos\frac{\pi}{3} - \cos\frac{\pi}{4}\sin\frac{\pi}{3} = -\frac{\sqrt{2}}{2}\cdot\frac{1}{2} - \frac{\sqrt{2}}{2}\cdot\frac{\sqrt{3}}{2} = -\frac{\sqrt{6}+\sqrt{2}}{4}$

11. $\tan\left(-\frac{\pi}{12}\right) = -\tan\frac{\pi}{12} = -\tan\left(\frac{\pi}{3} - \frac{\pi}{4}\right) = -\dfrac{\tan\frac{\pi}{3} - \tan\frac{\pi}{4}}{1 + \tan\frac{\pi}{3}\tan\frac{\pi}{4}} = \dfrac{1-\sqrt{3}}{1+\sqrt{3}} = \sqrt{3} - 2$

13. $\cos\frac{11\pi}{12} = -\cos\frac{\pi}{12} = -\cos\left(\frac{\pi}{3} - \frac{\pi}{4}\right) = -\cos\frac{\pi}{3}\cos\frac{\pi}{4} - \sin\frac{\pi}{3}\sin\frac{\pi}{4} = -\frac{\sqrt{3}}{2}\cdot\frac{\sqrt{2}}{2} - \frac{1}{2}\cdot\frac{\sqrt{2}}{2} = -\frac{\sqrt{6}+\sqrt{2}}{4}$

15. $\sin 18^\circ \cos 27^\circ + \cos 18^\circ \sin 27^\circ = \sin(18^\circ + 27^\circ) = \sin 45^\circ = \frac{1}{\sqrt{2}} = \frac{\sqrt{2}}{2}$

17. $\cos\frac{3\pi}{7}\cos\frac{2\pi}{21} + \sin\frac{3\pi}{7}\sin\frac{2\pi}{21} = \cos\left(\frac{3\pi}{7} - \frac{2\pi}{21}\right) = \cos\frac{7\pi}{21} = \cos\frac{\pi}{3} = \frac{1}{2}$

19. $\dfrac{\tan 73^\circ - \tan 13^\circ}{1 + \tan 73^\circ \tan 13^\circ} = \tan(73^\circ - 13^\circ) = \tan 60^\circ = \sqrt{3}$

21. $\tan\left(\frac{\pi}{2} - u\right) = \dfrac{\sin\left(\frac{\pi}{2} - u\right)}{\cos\left(\frac{\pi}{2} - u\right)} = \dfrac{\sin\frac{\pi}{2}\cos u - \cos\frac{\pi}{2}\sin u}{\cos\frac{\pi}{2}\cos u + \sin\frac{\pi}{2}\sin u} = \dfrac{1\cdot\cos u - 0\cdot\sin u}{0\cdot\cos u + 1\cdot\sin u} = \dfrac{\cos u}{\sin u} = \cot u$

23. $\sec\left(\frac{\pi}{2} - u\right) = \dfrac{1}{\cos\left(\frac{\pi}{2} - u\right)} = \dfrac{1}{\cos\frac{\pi}{2}\cos u + \sin\frac{\pi}{2}\sin u} = \dfrac{1}{0\cdot\cos u + 1\cdot\sin u} = \dfrac{1}{\sin u} = \csc u$

25. $\sin\left(x - \frac{\pi}{2}\right) = \sin x\cos\frac{\pi}{2} - \cos x\sin\frac{\pi}{2} = 0\cdot\sin x - 1\cdot\cos x = -\cos x$

27. $\sin(x - \pi) = \sin x\cos\pi - \cos x\sin\pi = -1\cdot\sin x - 0\cdot\cos x = -\sin x$

29. $\tan(x - \pi) = \dfrac{\tan x - \tan\pi}{1 + \tan x\tan\pi} = \dfrac{\tan x - 0}{1 + \tan x\cdot 0} = \tan x$

31. $\text{LHS} = \sin\left(\frac{\pi}{2} - x\right) = \sin\frac{\pi}{2}\cos x - \cos\frac{\pi}{2}\sin x = 1\cdot\cos x - 0\cdot\sin x = \cos x$ and
$\text{RHS} = \sin\left(\frac{\pi}{2} + x\right) = \sin\frac{\pi}{2}\cos x + \cos\frac{\pi}{2}\sin x = 1\cdot\cos x + 0\cdot\sin x = \cos x$. Therefore, LHS = RHS.

33. $\tan\left(x + \frac{\pi}{3}\right) = \dfrac{\tan x + \tan\frac{\pi}{3}}{1 - \tan x\tan\frac{\pi}{3}} = \dfrac{\sqrt{3} + \tan x}{1 - \sqrt{3}\tan x}$

35. $\sin(x+y) - \sin(x-y) = \sin x\cos y + \cos x\sin y - (\sin x\cos y - \cos x\sin y) = 2\cos x\sin y$

37. $\cot(x - y) = \dfrac{1}{\tan(x-y)} = \dfrac{1 + \tan x\tan y}{\tan x - \tan y} = \dfrac{1 + \dfrac{1}{\cot x}\dfrac{1}{\cot y}}{\dfrac{1}{\cot x} - \dfrac{1}{\cot y}}\cdot\dfrac{\cot x\cot y}{\cot x\cot y} = \dfrac{\cot x\cot y + 1}{\cot y - \cot x}$

39. $\tan x - \tan y = \dfrac{\sin x}{\cos x} - \dfrac{\sin y}{\cos y} = \dfrac{\sin x\cos y - \cos x\sin y}{\cos x\cos y} = \dfrac{\sin(x-y)}{\cos x\cos y}$

41. $\dfrac{\tan x - \tan y}{1 - \tan x\tan y} = \dfrac{(\tan x - \tan y)(\cos x\cos y)}{(1 - \tan x\tan y)(\cos x\cos y)} = \dfrac{\sin x\cos y - \cos x\sin y}{\cos x\cos y - \sin x\sin y} = \dfrac{\sin(x-y)}{\cos(x+y)}$

43.
$$\begin{aligned}\cos(x+y)\cos(x-y) &= (\cos x\cos y - \sin x\sin y)(\cos x\cos y + \sin x\sin y) = \cos^2 x\cos^2 y - \sin^2 x\sin^2 y\\ &= \cos^2 x\left(1 - \sin^2 y\right) - \left(1 - \cos^2 x\right)\sin^2 y = \cos^2 x - \sin^2 y\cos^2 x + \sin^2 y\cos^2 x - \sin^2 y\\ &= \cos^2 x - \sin^2 y\end{aligned}$$

45. $\sin(x+y+z) = \sin((x+y)+z) = \sin(x+y)\cos z + \cos(x+y)\sin z$

$= \cos z(\sin x\cos y + \cos x\sin y) + \sin z(\cos x\cos y - \sin x\sin y)$

$= \sin x\cos y\cos z + \cos x\sin y\cos z + \cos x\cos y\sin z - \sin x\sin y\sin z$

47. We want to write $\cos(\sin^{-1}x - \tan^{-1}y)$ in terms of x and y only. We let $\theta = \sin^{-1}x$ and $\phi = \tan^{-1}y$ and sketch triangles with angles θ and ϕ such that $\sin\theta = x$ and $\tan\phi = y$. From the triangles, we have $\cos\theta = \sqrt{1-x^2}$, $\cos\phi = \dfrac{1}{\sqrt{1+y^2}}$, and $\sin\phi = \dfrac{y}{\sqrt{1+y^2}}$.

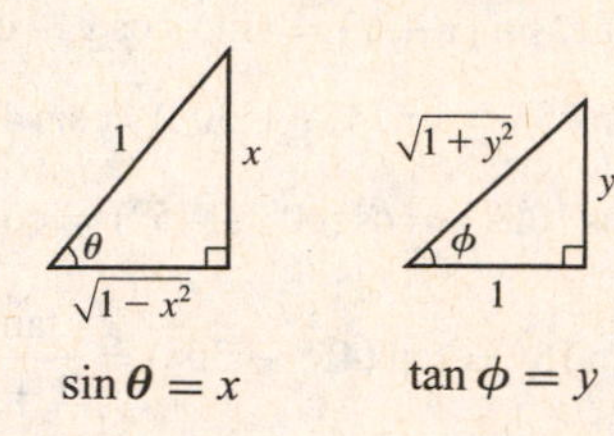

From the subtraction formula for cosine we have

$$\cos\left(\sin^{-1}x - \tan^{-1}y\right) = \cos(\theta-\phi) = \cos\theta\cos\phi + \sin\theta\sin\phi = \sqrt{1-x^2}\cdot\frac{1}{\sqrt{1+y^2}} + x\cdot\frac{y}{\sqrt{1+y^2}} = \frac{\sqrt{1-x^2}+xy}{\sqrt{1+y^2}}$$

49. Let $\theta = \tan^{-1}x$ and $\phi = \tan^{-1}y$. From the triangles, $\cos\theta = \dfrac{1}{\sqrt{1+x^2}}$, $\sin\theta = \dfrac{x}{\sqrt{1+x^2}}$, $\cos\phi = \dfrac{1}{\sqrt{1+y^2}}$, and $\sin\phi = \dfrac{y}{\sqrt{1+y^2}}$, so using the subtraction formula for sine, we have

$\tan\theta = x$ $\quad$ $\tan\phi = y$

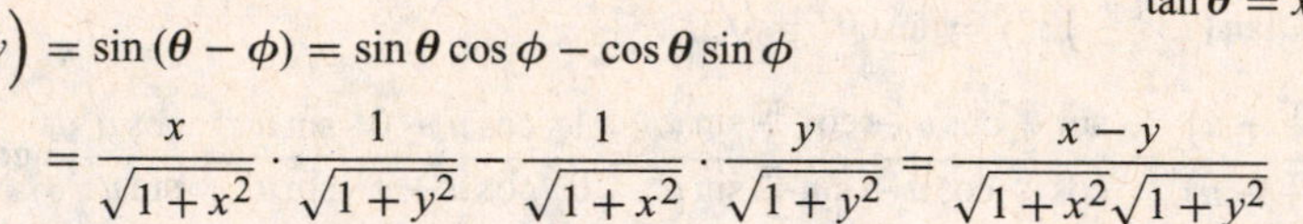

$$\sin\left(\tan^{-1}x - \tan^{-1}y\right) = \sin(\theta-\phi) = \sin\theta\cos\phi - \cos\theta\sin\phi$$

$$= \frac{x}{\sqrt{1+x^2}}\cdot\frac{1}{\sqrt{1+y^2}} - \frac{1}{\sqrt{1+x^2}}\cdot\frac{y}{\sqrt{1+y^2}} = \frac{x-y}{\sqrt{1+x^2}\sqrt{1+y^2}}$$

51. We know that $\cos^{-1}\frac{1}{2} = \frac{\pi}{3}$ and $\tan^{-1}1 = \frac{\pi}{4}$, so the addition formula for sine gives $\sin\left(\cos^{-1}\frac{1}{2} + \tan^{-1}1\right) = \sin\left(\frac{\pi}{3}+\frac{\pi}{4}\right) = \sin\frac{\pi}{3}\cos\frac{\pi}{4} + \cos\frac{\pi}{3}\sin\frac{\pi}{4} = \frac{\sqrt{3}}{2}\cdot\frac{\sqrt{2}}{2} + \frac{1}{2}\cdot\frac{\sqrt{2}}{2} = \frac{\sqrt{6}+\sqrt{2}}{4}$.

53. We sketch triangles such that $\theta = \sin^{-1}\frac{3}{4}$ and $\phi = \cos^{-1}\frac{1}{3}$. From the triangles, we have $\tan\theta = \frac{3}{\sqrt{7}}$ and $\tan\phi = 2\sqrt{2}$, so the subtraction formula for tangent gives

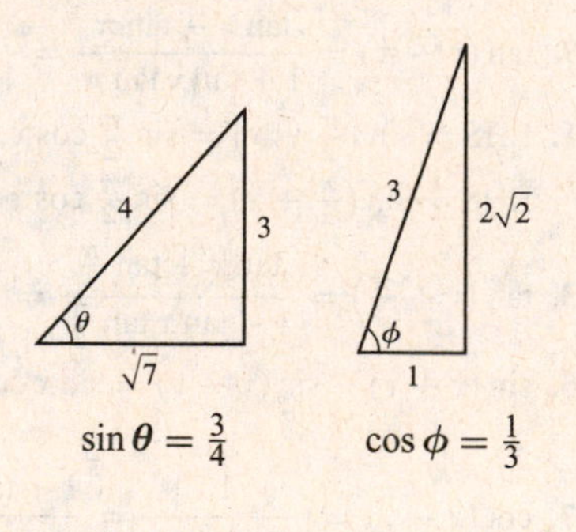

$$\tan\left(\sin^{-1}\tfrac{3}{4} - \cos^{-1}\tfrac{1}{3}\right) = \frac{\tan\theta - \tan\phi}{1+\tan\theta\tan\phi} = \frac{\frac{3}{\sqrt{7}} - 2\sqrt{2}}{1+\frac{3}{\sqrt{7}}\cdot 2\sqrt{2}}$$

$$= \frac{3-2\sqrt{14}}{\sqrt{7}+6\sqrt{2}}$$

55. As in Example 7, we sketch the angles θ and ϕ in standard position with terminal sides in the appropriate quadrants and find the remaining sides using the Pythagorean Theorem. To find $\cos(\theta-\phi)$, we use the addition formula for sine and the triangles we have sketched:

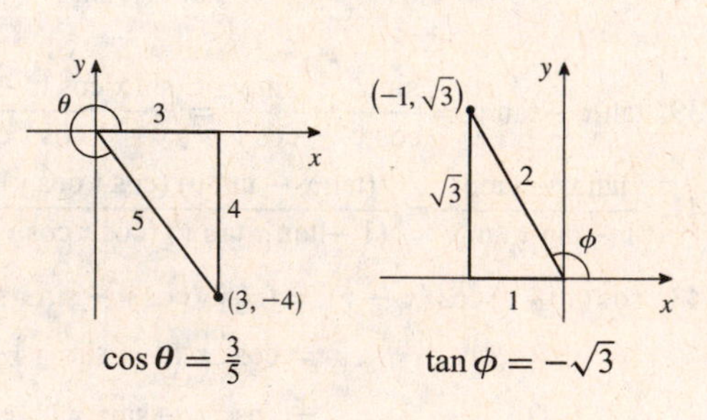

$$\cos(\theta-\phi) = \cos\theta\cos\phi + \sin\theta\sin\phi$$

$$= \tfrac{3}{5}\left(-\tfrac{1}{2}\right) + \left(-\tfrac{4}{5}\right)\tfrac{\sqrt{3}}{2} = -\tfrac{3+4\sqrt{3}}{10}$$

57. Using the addition formula for sine and the triangles shown, we have

$\sin(\theta + \phi) = \sin\theta\cos\phi + \cos\theta\sin\phi$

$= \frac{5}{13}\left(-\frac{2\sqrt{5}}{5}\right) + \frac{12}{13}\left(\frac{\sqrt{5}}{5}\right) = \frac{2\sqrt{5}}{65}$

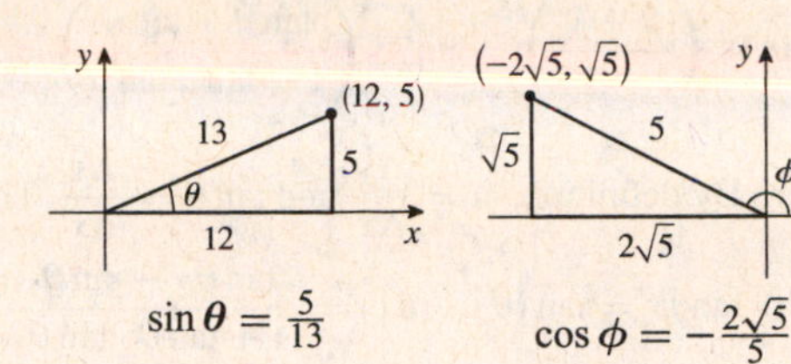

59. $k = \sqrt{A^2 + B^2} = \sqrt{\left(-\sqrt{3}\right)^2 + 1^2} = \sqrt{4} = 2$. Thus, $\sin\phi = \frac{1}{2}$ and $\cos\phi = \frac{-\sqrt{3}}{2} \Rightarrow \phi = \frac{5\pi}{6}$, so $-\sqrt{3}\sin x + \cos x = k\sin(x + \phi) = 2\sin\left(x + \frac{5\pi}{6}\right)$.

61. $k = \sqrt{A^2 + B^2} = \sqrt{5^2 + (-5)^2} = \sqrt{50} = 5\sqrt{2}$. Thus, $\sin\phi = -\frac{5}{5\sqrt{2}} = -\frac{1}{\sqrt{2}}$ and $\cos\phi = \frac{5}{5\sqrt{2}} = \frac{1}{\sqrt{2}} \Rightarrow \phi = \frac{7\pi}{4}$, so $5(\sin 2x - \cos 2x) = k\sin(2x + \phi) = 5\sqrt{2}\sin\left(2x + \frac{7\pi}{4}\right)$.

63. (a) $g(x) = \cos 2x + \sqrt{3}\sin 2x \Rightarrow$

$k = \sqrt{1^2 + \left(\sqrt{3}\right)^2} = \sqrt{4} = 2$, and ϕ satisfies

$\sin\phi = \frac{1}{2}$, $\cos\phi = \frac{\sqrt{3}}{2} \Rightarrow \phi = \frac{\pi}{6}$. Thus, we can write

$g(x) = k\sin(2x + \phi) = 2\sin\left(2x + \frac{\pi}{6}\right) = 2\sin 2\left(x + \frac{\pi}{12}\right)$.

(b) This is a sine curve with amplitude 2, period π, and phase shift $-\frac{\pi}{12}$.

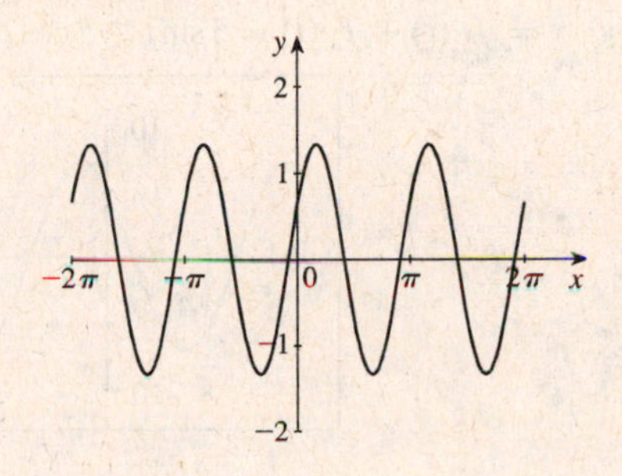

65. $f(x) = \cos x$. Now

$$\frac{f(x+h) - f(x)}{h} = \frac{\cos(x+h) - \cos x}{h} = \frac{\cos x\cos h - \sin x\sin h - \cos x}{h}$$

$$= \frac{-\cos x(1 - \cos h) - \sin h(\sin x)}{h} = -\cos x\left(\frac{1 - \cos h}{h}\right) - \left(\frac{\sin h}{h}\right)\sin x$$

67. (a) $y = \sin^2\left(x + \frac{\pi}{4}\right) + \sin^2\left(x - \frac{\pi}{4}\right)$. From the graph we see that the value of y seems to always be equal to 1.

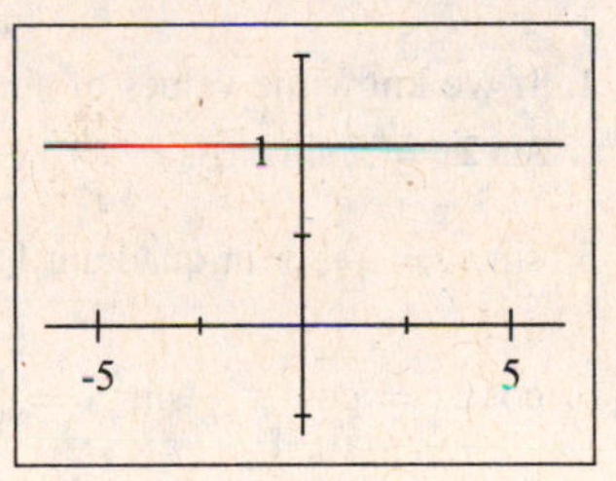

(b) $y = \sin^2\left(x + \frac{\pi}{4}\right) + \sin^2\left(x - \frac{\pi}{4}\right) = \left(\sin x\cos\frac{\pi}{4} + \cos x\sin\frac{\pi}{4}\right)^2 + \left(\sin x\cos\frac{\pi}{4} - \cos x\sin\frac{\pi}{4}\right)^2$

$= \left[\frac{1}{\sqrt{2}}(\sin x + \cos x)\right]^2 + \left[\frac{1}{\sqrt{2}}(\sin x - \cos x)\right]^2 = \frac{1}{2}\left[(\sin x + \cos x)^2 + (\sin x - \cos x)^2\right]$

$= \frac{1}{2}\left[\left(\sin^2 x + 2\sin x\cos x + \cos^2 x\right) + \left(\sin^2 x - 2\sin x\cos x + \cos^2 x\right)\right]$

$= \frac{1}{2}\left[(1 + 2\sin x\cos x) + (1 - 2\sin x\cos x)\right] = \frac{1}{2}\cdot 2 = 1$

69. If $\beta - \alpha = \frac{\pi}{2}$, then $\beta = \alpha + \frac{\pi}{2}$. Now if we let $y = x + \alpha$, then $\sin(x + \alpha) + \cos\left(x + \alpha + \frac{\pi}{2}\right) = \sin y + \cos\left(y + \frac{\pi}{2}\right) = \sin y + (-\sin y) = 0$. Therefore, $\sin(x + \alpha) + \cos(x + \beta) = 0$.

71. $\tan^{-1}\left(\dfrac{x+y}{1-xy}\right) = \tan^{-1}\left(\dfrac{\tan u + \tan v}{1 - \tan u \tan v}\right) = \tan^{-1}(\tan(u+v)) = u + v = \tan^{-1} x + \tan^{-1} y$

73. (a) By definition, $m = \dfrac{\Delta y}{\Delta x}$ and $\tan\theta = \dfrac{\Delta y}{\Delta x}$. Thus, $m = \tan\theta$.

(b) $\tan\psi = \tan(\theta_2 - \theta_1) = \dfrac{\tan\theta_2 - \tan\theta_1}{1 + \tan\theta_2 \tan\theta_1}$. From part (a), we have $m_1 = \tan\theta_1$ and $m_2 = \tan\theta_2$. Then by substitution, $\tan\psi = \dfrac{m_2 - m_1}{1 + m_1 m_2}$.

(c) Let ψ be the unknown angle as in part (b). Since $m_1 = \frac{1}{3}$ and $m_2 = \frac{1}{2}$,

$$\tan\psi = \frac{m_2 - m_1}{1 + m_1 m_2} = \frac{\frac{1}{2} - \frac{1}{3}}{1 + \frac{1}{3}\left(\frac{1}{2}\right)} = \frac{\frac{1}{6}}{\frac{7}{6}} = \frac{1}{7} \Leftrightarrow \psi = \tan^{-1}\frac{1}{7} \approx 0.142 \text{ rad} \approx 8.1°.$$

(d) From part (b), we have $\cot\psi = \dfrac{1 + m_1 m_2}{m_2 - m_1}$. If the two lines are perpendicular then $\psi = 90°$ and so $\cot\psi = 0$. Thus we have $0 = \dfrac{1 + m_1 m_2}{m_2 - m_1} \Leftrightarrow 0 = 1 + m_1 m_2 \Leftrightarrow m_1 m_2 = -1 \Leftrightarrow m_2 = -1/m_1$. Thus m_2 is the negative reciprocal of m_1.

75. (a) $y = f_1(t) + f_2(t) = 5\sin t + 5\cos t = 5(\sin t + \cos t)$

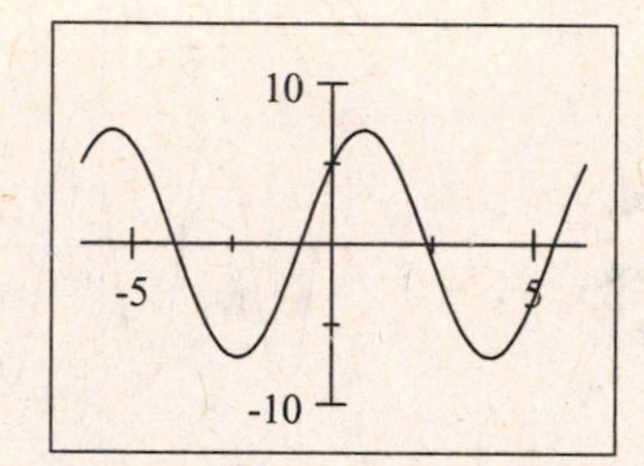

(b) $k = \sqrt{A^2 + B^2} = \sqrt{5^2 + 5^2} = 5\sqrt{2}$. Therefore,

$\sin\phi = \dfrac{B}{\sqrt{A^2 + B^2}} = \dfrac{5}{5\sqrt{2}} = \dfrac{1}{\sqrt{2}}$ and

$\cos\phi = \dfrac{A}{\sqrt{A^2 + B^2}} = \dfrac{1}{\sqrt{2}}$. Thus, $\phi = \frac{\pi}{4}$.

77. $\sin(s+t) = \cos\left[\frac{\pi}{2} - (s+t)\right] = \cos\left[\left(\frac{\pi}{2} - s\right) - t\right] = \cos\left(\frac{\pi}{2} - s\right)\cos t + \sin\left(\frac{\pi}{2} - s\right)\sin t = \sin s \cos t + \cos s \sin t$. The last equality comes from again applying the cofunction identities.

7.3 DOUBLE-ANGLE, HALF-ANGLE, AND PRODUCT-SUM FORMULAS

1. If we know the values of $\sin x$ and $\cos x$, we can find the value of $\sin 2x$ using the *Double-Angle* Formula for Sine: $\sin 2x = 2\sin x \cos x$.

3. $\sin x = \frac{5}{13}$, x in quadrant I $\Rightarrow \cos x = \frac{12}{13}$ and $\tan x = \frac{5}{12}$. Thus, $\sin 2x = 2\sin x \cos x = 2\left(\frac{5}{13}\right)\left(\frac{12}{13}\right) = \frac{120}{169}$,

$\cos 2x = \cos^2 x - \sin^2 x = \left(\frac{12}{13}\right)^2 - \left(\frac{5}{13}\right)^2 = \frac{144-25}{169} = \frac{119}{169}$, and $\tan 2x = \dfrac{\sin 2x}{\cos 2x} = \dfrac{\frac{120}{169}}{\frac{119}{169}} = \frac{120}{169} \cdot \frac{169}{119} = \frac{120}{119}$.

5. $\cos x = \frac{4}{5}$. Then $\sin x = -\frac{3}{5}$ ($\csc x < 0$) and $\tan x = -\frac{3}{4}$. Thus, $\sin 2x = 2\sin x \cos x = 2\left(-\frac{3}{5}\right) \cdot \frac{4}{5} = -\frac{24}{25}$,

$\cos 2x = \cos^2 x - \sin^2 x = \left(\frac{4}{5}\right)^2 - \left(-\frac{3}{5}\right)^2 = \frac{16-9}{25} = \frac{7}{25}$, and $\tan 2x = \dfrac{\sin 2x}{\cos 2x} = \dfrac{-\frac{24}{25}}{\frac{7}{25}} = -\frac{24}{25} \cdot \frac{25}{7} = -\frac{24}{7}$.

7. $\sin x = -\frac{3}{5}$. Then, $\cos x = -\frac{4}{5}$ and $\tan x = \frac{3}{4}$ (x is in quadrant III). Thus, $\sin 2x = 2\sin x \cos x = 2\left(-\frac{3}{5}\right)\left(-\frac{4}{5}\right) = \frac{24}{25}$,

$\cos 2x = \cos^2 x - \sin^2 x = \left(-\frac{4}{5}\right)^2 - \left(-\frac{3}{5}\right)^2 = \frac{16-9}{25} = \frac{7}{25}$, and $\tan 2x = \dfrac{\sin 2x}{\cos 2x} = \dfrac{\frac{24}{25}}{\frac{7}{25}} = \frac{24}{25} \cdot \frac{25}{7} = \frac{24}{7}$.

9. $\tan x = -\frac{1}{3}$ and $\cos x > 0$, so $\sin x < 0$. Thus, $\sin x = -\frac{1}{\sqrt{10}}$ and $\cos x = \frac{3}{\sqrt{10}}$. Thus, $\sin 2x = 2\sin x\cos x = 2\left(-\frac{1}{\sqrt{10}}\right)\left(\frac{3}{\sqrt{10}}\right) = -\frac{6}{10} = -\frac{3}{5}$, $\cos 2x = \cos^2 x - \sin^2 x = \left(\frac{3}{\sqrt{10}}\right)^2 - \left(-\frac{1}{\sqrt{10}}\right)^2 = \frac{8}{10} = \frac{4}{5}$, and $\tan 2x = \dfrac{\sin 2x}{\cos 2x} = \dfrac{-\frac{3}{5}}{\frac{4}{5}} = -\frac{3}{5}\cdot\frac{5}{4} = -\frac{3}{4}$.

11.
$$\begin{aligned}\sin^4 x &= \left(\sin^2 x\right)^2 = \left(\frac{1-\cos 2x}{2}\right)^2 = \tfrac{1}{4} - \tfrac{1}{2}\cos 2x + \tfrac{1}{4}\cos^2 2x \\ &= \tfrac{1}{4} - \tfrac{1}{2}\cos 2x + \tfrac{1}{4}\cdot\frac{1+\cos 4x}{2} = \tfrac{1}{4} - \tfrac{1}{2}\cos 2x + \tfrac{1}{8} + \tfrac{1}{8}\cos 4x = \tfrac{3}{8} - \tfrac{1}{2}\cos 2x + \tfrac{1}{8}\cos 4x \\ &= \tfrac{1}{2}\left(\tfrac{3}{4} - \cos 2x + \tfrac{1}{4}\cos 4x\right)\end{aligned}$$

13. We use the result of Example 4 to get
$\cos^2 x\sin^4 x = \left(\sin^2 x\cos^2 x\right)\sin^2 x = \left(\frac{1}{8} - \frac{1}{8}\cos 4x\right)\cdot\left(\frac{1}{2} - \frac{1}{2}\cos 2x\right) = \frac{1}{16}\left(1 - \cos 2x - \cos 4x + \cos 2x\cos 4x\right)$.

15. Since $\sin^4 x\cos^4 x = \left(\sin^2 x\cos^2 x\right)^2$ we can use the result of Example 4 to get

$$\begin{aligned}\sin^4 x\cos^4 x &= \left(\tfrac{1}{8} - \tfrac{1}{8}\cos 4x\right)^2 = \tfrac{1}{64} - \tfrac{1}{32}\cos 4x + \tfrac{1}{64}\cos^2 4x \\ &= \tfrac{1}{64} - \tfrac{1}{32}\cos 4x + \tfrac{1}{64}\cdot\tfrac{1}{2}\left(1+\cos 8x\right) = \tfrac{1}{64} - \tfrac{1}{32}\cos 4x + \tfrac{1}{128} + \tfrac{1}{128}\cos 8x \\ &= \tfrac{3}{128} - \tfrac{1}{32}\cos 4x + \tfrac{1}{128}\cos 8x = \tfrac{1}{32}\left(\tfrac{3}{4} - \cos 4x + \tfrac{1}{4}\cos 8x\right)\end{aligned}$$

17. $\sin 15^\circ = \sqrt{\frac{1}{2}\left(1-\cos 30^\circ\right)} = \sqrt{\frac{1}{2}\left(1-\frac{\sqrt{3}}{2}\right)} = \sqrt{\frac{1}{4}\left(2-\sqrt{3}\right)} = \frac{1}{2}\sqrt{2-\sqrt{3}}$

19. $\tan 22.5^\circ = \dfrac{1-\cos 45^\circ}{\sin 45^\circ} = \dfrac{1-\frac{\sqrt{2}}{2}}{\frac{\sqrt{2}}{2}} = \sqrt{2}-1$

21. $\cos 165^\circ = -\sqrt{\frac{1}{2}\left(1+\cos 330^\circ\right)} = -\sqrt{\frac{1}{2}\left(1+\cos 30^\circ\right)} = -\sqrt{\frac{1}{2}\left(1+\frac{\sqrt{3}}{2}\right)} = -\frac{1}{2}\sqrt{2+\sqrt{3}}$

23. $\tan\frac{\pi}{8} = \dfrac{1-\cos\frac{\pi}{4}}{\sin\frac{\pi}{4}} = \dfrac{1-\frac{\sqrt{2}}{2}}{\frac{\sqrt{2}}{2}} = \sqrt{2}-1$

25. $\cos\frac{\pi}{12} = \sqrt{\frac{1}{2}\left(1+\cos\frac{\pi}{6}\right)} = \sqrt{\frac{1}{2}\left(1+\frac{\sqrt{3}}{2}\right)} = \frac{1}{2}\sqrt{2+\sqrt{3}}$

27. $\sin\frac{9\pi}{8} = -\sqrt{\frac{1}{2}\left(1-\cos\frac{9\pi}{4}\right)} = -\sqrt{\frac{1}{2}\left(1-\frac{\sqrt{2}}{2}\right)} = -\frac{1}{2}\sqrt{2-\sqrt{2}}$. We have chosen the negative root because $\frac{9\pi}{8}$ is in quadrant III, so $\sin\frac{9\pi}{8} < 0$.

29. (a) $2\sin 18^\circ\cos 18^\circ = \sin 36^\circ$

(b) $2\sin 3\theta\cos 3\theta = \sin 6\theta$

31. (a) $\cos^2 34^\circ - \sin^2 34^\circ = \cos 68^\circ$

(b) $\cos^2 5\theta - \sin^2 5\theta = \cos 10\theta$

33. (a) $\dfrac{\sin 8^\circ}{1+\cos 8^\circ} = \tan\dfrac{8^\circ}{2} = \tan 4^\circ$

(b) $\dfrac{1-\cos 4\theta}{\sin 4\theta} = \tan\dfrac{4\theta}{2} = \tan 2\theta$

35. $\sin(x+x) = \sin x\cos x + \cos x\sin x = 2\sin x\cos x$

37. $\sin x = \frac{3}{5}$. Since x is in quadrant I, $\cos x = \frac{4}{5}$ and $\frac{x}{2}$ is also in quadrant I. Thus,

$\sin\frac{x}{2} = \sqrt{\frac{1}{2}(1-\cos x)} = \sqrt{\frac{1}{2}\left(1-\frac{4}{5}\right)} = \frac{1}{\sqrt{10}} = \frac{\sqrt{10}}{10}$, $\cos\frac{x}{2} = \sqrt{\frac{1}{2}(1+\cos x)} = \sqrt{\frac{1}{2}\left(1+\frac{4}{5}\right)} = \frac{3}{\sqrt{10}} = \frac{3\sqrt{10}}{10}$, and

$\tan\frac{x}{2} = \dfrac{\sin\frac{x}{2}}{\cos\frac{x}{2}} = \frac{1}{\sqrt{10}} \cdot \frac{\sqrt{10}}{3} = \frac{1}{3}$.

39. $\csc x = 3$. Then, $\sin x = \frac{1}{3}$ and since x is in quadrant II, $\cos x = -\dfrac{2\sqrt{2}}{3}$. Since $90^\circ \le x \le 180^\circ$, we have

$45^\circ \le \frac{x}{2} \le 90^\circ$ and so $\frac{x}{2}$ is in quadrant I. Thus, $\sin\frac{x}{2} = \sqrt{\frac{1}{2}(1-\cos x)} = \sqrt{\frac{1}{2}\left(1+\frac{2\sqrt{2}}{3}\right)} = \sqrt{\frac{1}{6}\left(3+2\sqrt{2}\right)}$,

$\cos\frac{x}{2} = \sqrt{\frac{1}{2}(1+\cos x)} = \sqrt{\frac{1}{2}\left(1-\frac{2\sqrt{2}}{3}\right)} = \sqrt{\frac{1}{6}\left(3-2\sqrt{2}\right)}$, and $\tan\frac{x}{2} = \dfrac{\sin\frac{x}{2}}{\cos\frac{x}{2}} = \sqrt{\frac{3+2\sqrt{2}}{3-2\sqrt{2}}} = 3+2\sqrt{2}$.

41. $\sec x = \frac{3}{2}$. Then $\cos x = \frac{2}{3}$ and since x is in quadrant IV, $\sin x = -\frac{\sqrt{5}}{3}$. Since $270^\circ \le x \le 360^\circ$, we have

$135^\circ \le \frac{x}{2} \le 180^\circ$ and so $\frac{x}{2}$ is in quadrant II. Thus, $\sin\frac{x}{2} = \sqrt{\frac{1}{2}(1-\cos x)} = \sqrt{\frac{1}{2}\left(1-\frac{2}{3}\right)} = \frac{1}{\sqrt{6}} = \frac{\sqrt{6}}{6}$,

$\cos\frac{x}{2} = -\sqrt{\frac{1}{2}(1+\cos x)} = -\sqrt{\frac{1}{2}\left(1+\frac{2}{3}\right)} = -\frac{\sqrt{5}}{\sqrt{6}} = -\dfrac{\sqrt{30}}{6}$, and $\tan\frac{x}{2} = \dfrac{\sin\frac{x}{2}}{\cos\frac{x}{2}} = \frac{1}{\sqrt{6}} \cdot \frac{\sqrt{6}}{-\sqrt{5}} = -\frac{1}{\sqrt{5}} = -\frac{\sqrt{5}}{5}$.

43. To write $\sin\left(2\tan^{-1}x\right)$ as an algebraic expression in x, we let $\theta = \tan^{-1}x$ and sketch a suitable triangle. We see that $\sin\theta = \dfrac{x}{\sqrt{1+x^2}}$ and

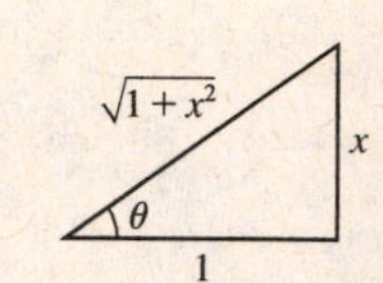

$\cos\theta = \dfrac{1}{\sqrt{1+x^2}}$, so using the double-angle formula for sine, we have

$\sin\left(2\tan^{-1}x\right) = \sin 2\theta = 2\sin\theta\cos\theta = 2\cdot\dfrac{x}{\sqrt{1+x^2}}\cdot\dfrac{1}{\sqrt{1+x^2}} = \dfrac{2x}{1+x^2}$.

45. Using the half-angle formula for sine, we have $\sin\left(\frac{1}{2}\cos^{-1}x\right) = \pm\sqrt{\dfrac{1-\cos\left(\cos^{-1}x\right)}{2}} = \pm\sqrt{\dfrac{1-x}{2}}$. Because $\cos^{-1}$ has range $[0,\pi]$, $\frac{1}{2}\cos^{-1}x$ lies in $\left[0,\frac{\pi}{2}\right]$ and so $\sin\left(\frac{1}{2}\cos^{-1}x\right)$ is positive. Thus, $\sin\left(\frac{1}{2}\cos^{-1}x\right) = \sqrt{\dfrac{1-x}{2}}$.

47. We sketch a triangle with $\theta = \cos^{-1}\frac{7}{25}$ and find that $\sin\theta = \frac{24}{25}$. Thus, using the double-angle formula for sine,

$\sin\left(2\cos^{-1}\frac{7}{25}\right) = \sin 2\theta = 2\sin\theta\cos\theta = 2\cdot\frac{24}{25}\cdot\frac{7}{25} = \frac{336}{625}$.

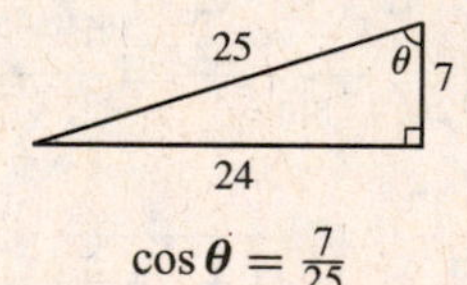

49. Rewriting the given expression and using a double-angle formula for cosine, we have

$\sec\left(2\sin^{-1}\frac{1}{4}\right) = \dfrac{1}{\cos\left(2\sin^{-1}\frac{1}{4}\right)} = \dfrac{1}{1-2\sin^2\left(\sin^{-1}\frac{1}{4}\right)} = \dfrac{1}{1-2\left(\frac{1}{4}\right)^2} = \dfrac{8}{7}$.

51. Using a double-angle formula for cosine, we have $\cos 2\theta = 1-2\sin^2\theta = 1-2\left(-\frac{3}{5}\right)^2 = \frac{7}{25}$.

53. To evaluate $\sin 2\theta$, we first sketch the angle θ in standard position with terminal side in quadrant II and find the remaining side using the Pythagorean Theorem. Using the double-angle formula for sine, we have

$\sin 2\theta = 2\sin\theta\cos\theta = 2\left(\frac{1}{7}\right)\left(-\frac{4\sqrt{3}}{7}\right) = -\frac{8\sqrt{3}}{49}$.

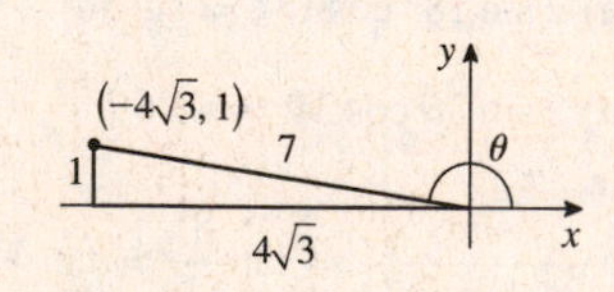

55. $\sin 2x\cos 3x = \frac{1}{2}[\sin(2x+3x)+\sin(2x-3x)] = \frac{1}{2}(\sin 5x - \sin x)$

57. $\cos x \sin 4x = \frac{1}{2}[\sin(4x+x) + \sin(4x-x)] = \frac{1}{2}(\sin 5x + \sin 3x)$

59. $3\cos 4x \cos 7x = 3 \cdot \frac{1}{2}[\cos(4x+7x) + \cos(4x-7x)] = \frac{3}{2}(\cos 11x + \cos 3x)$

61. $\sin 5x + \sin 3x = 2\sin\left(\frac{5x+3x}{2}\right)\cos\left(\frac{5x-3x}{2}\right) = 2\sin 4x \cos x$

63. $\cos 4x - \cos 6x = -2\sin\left(\frac{4x+6x}{2}\right)\sin\left(\frac{4x-6x}{2}\right) = -2\sin 5x \sin(-x) = 2\sin 5x \sin x$

65. $\sin 2x - \sin 7x = 2\cos\left(\frac{2x+7x}{2}\right)\sin\left(\frac{2x-7x}{2}\right) = 2\cos\frac{9x}{2}\sin\left(-\frac{5x}{2}\right) = -2\cos\frac{9x}{2}\sin\frac{5x}{2}$

67. $2\sin 52.5^\circ \sin 97.5^\circ = 2 \cdot \frac{1}{2}\left[\cos(52.5^\circ - 97.5^\circ) - \cos(52.5^\circ + 97.5^\circ)\right] = \cos(-45^\circ) - \cos 150^\circ$
$= \cos 45^\circ - \cos 150^\circ = \frac{\sqrt{2}}{2} + \frac{\sqrt{3}}{2} = \frac{1}{2}\left(\sqrt{2} + \sqrt{3}\right)$

69. $\cos 37.5^\circ \sin 7.5^\circ = \frac{1}{2}(\sin 45^\circ - \sin 30^\circ) = \frac{1}{2}\left(\frac{\sqrt{2}}{2} - \frac{1}{2}\right) = \frac{1}{4}\left(\sqrt{2} - 1\right)$

71. $\cos 255^\circ - \cos 195^\circ = -2\sin\left(\frac{255^\circ + 195^\circ}{2}\right)\sin\left(\frac{255^\circ - 195^\circ}{2}\right) = -2\sin 225^\circ \sin 30^\circ = -2\left(-\frac{\sqrt{2}}{2}\right)\frac{1}{2} = \frac{\sqrt{2}}{2}$

73. $\cos^2 5x - \sin^2 5x = \cos(2 \cdot 5x) = \cos 10x$

75. $(\sin x + \cos x)^2 = \sin^2 x + 2\sin x \cos x + \cos^2 x = 1 + 2\sin x \cos x = 1 + \sin 2x$

77. $\dfrac{2\tan x}{1 + \tan^2 x} = \dfrac{2\tan x}{\sec^2 x} = 2 \cdot \dfrac{\sin x}{\cos x}\cos^2 x = 2\sin x \cos x = \sin 2x$

79. $\tan\left(\frac{x}{2}\right) + \cos x \tan\left(\frac{x}{2}\right) = \dfrac{1 - \cos x}{\sin x} + \cos x\left(\dfrac{1 - \cos x}{\sin x}\right) = \dfrac{1 - \cos x + \cos x - \cos^2 x}{\sin x} = \dfrac{\sin^2 x}{\sin x} = \sin x$

81. $\dfrac{\sin 4x}{\sin x} = \dfrac{2\sin 2x \cos 2x}{\sin x} = \dfrac{2(2\sin x \cos x)(\cos 2x)}{\sin x} = 4\cos x \cos 2x$

83. $\dfrac{2(\tan x - \cot x)}{\tan^2 x - \cot^2 x} = \dfrac{2(\tan x - \cot x)}{(\tan x + \cot x)(\tan x - \cot x)} = \dfrac{2}{\tan x + \cot x} = \dfrac{2}{\dfrac{\sin x}{\cos x} + \dfrac{\cos x}{\sin x}}$

$= \dfrac{2}{\dfrac{\sin x}{\cos x} + \dfrac{\cos x}{\sin x}} \cdot \dfrac{\sin x \cos x}{\sin x \cos x} = \dfrac{2\sin x \cos x}{\sin^2 x + \cos^2 x} = 2\sin x \cos x = \sin 2x$

85. $\cot 2x = \dfrac{1}{\tan 2x} = \dfrac{1}{\dfrac{2\tan x}{1 - \tan^2 x}} = \dfrac{1 - \tan^2 x}{2\tan x}$

87. $\tan 3x = \tan(2x + x) = \dfrac{\tan 2x + \tan x}{1 - \tan 2x \tan x} = \dfrac{\dfrac{2\tan x}{1 - \tan^2 x} + \tan x}{1 - \dfrac{2\tan x}{1 - \tan^2 x}\tan x} = \dfrac{2\tan x + \tan x\left(1 - \tan^2 x\right)}{1 - \tan^2 x - 2\tan x \tan x}$

$= \dfrac{3\tan x - \tan^3 x}{1 - 3\tan^2 x}$

89. $\dfrac{\sin x + \sin 5x}{\cos x + \cos 5x} = \dfrac{2\sin 3x \cos 2x}{2\cos 3x \cos 2x} = \dfrac{\sin 3x}{\cos 3x} = \tan 3x$

91. $\dfrac{\sin 10x}{\sin 9x + \sin x} = \dfrac{2\sin 5x \cos 5x}{2\sin 5x \cos 4x} = \dfrac{\cos 5x}{\cos 4x}$

93. $\dfrac{\sin x + \sin y}{\cos x + \cos y} = \dfrac{2\sin\left(\frac{x+y}{2}\right)\cos\left(\frac{x-y}{2}\right)}{2\cos\left(\frac{x+y}{2}\right)\cos\left(\frac{x-y}{2}\right)} = \dfrac{\sin\left(\frac{x+y}{2}\right)}{\cos\left(\frac{x+y}{2}\right)} = \tan\left(\frac{x+y}{2}\right)$

95. Let $y = \frac{x}{2} + \frac{\pi}{4} \Leftrightarrow 2y = x + \frac{\pi}{2}$. Then

$$\tan^2\left(\frac{x}{2} + \frac{\pi}{4}\right) = \tan^2 y = \frac{1 - \cos 2y}{1 + \cos 2y} = \frac{1 - \cos\left(x + \frac{\pi}{2}\right)}{1 + \cos\left(x + \frac{\pi}{2}\right)} = \frac{1 - (-\sin x)}{1 + (-\sin x)} = \frac{1 + \sin x}{1 - \sin x}$$

97. $\sin 130^\circ - \sin 110^\circ = 2\cos\frac{130^\circ + 110^\circ}{2}\sin\frac{130^\circ - 110^\circ}{2} = 2\cos 120^\circ \sin 10^\circ = 2\left(-\frac{1}{2}\right)\sin 10^\circ = -\sin 10^\circ$

99. $\sin 45^\circ + \sin 15^\circ = 2\sin\left(\frac{45^\circ + 15^\circ}{2}\right)\cos\left(\frac{45^\circ - 15^\circ}{2}\right) = 2\sin 30^\circ \cos 15^\circ = 2 \cdot \frac{1}{2} \cdot \cos 15^\circ$

$= \cos 15^\circ = \sin(90^\circ - 15^\circ) = \sin 75^\circ$ (applying the cofunction identity)

101.

$$\frac{\sin x + \sin 2x + \sin 3x + \sin 4x + \sin 5x}{\cos x + \cos 2x + \cos 3x + \cos 4x + \cos 5x} = \frac{(\sin x + \sin 5x) + (\sin 2x + \sin 4x) + \sin 3x}{(\cos x + \cos 5x) + (\cos 2x + \cos 4x) + \cos 3x}$$

$$= \frac{2\sin 3x \cos 2x + 2\sin 3x \cos x + \sin 3x}{2\cos 3x \cos 2x + 2\cos 3x \cos x + \cos 3x} = \frac{\sin 3x\,(2\cos 2x + 2\cos x + 1)}{\cos 3x\,(2\cos 2x + 2\cos x + 1)} = \tan 3x$$

103. With $u = \sin^{-1} x$ for $0 \le x \le 1$, we can write

$\cos^{-1}\left(1 - 2x^2\right) = \cos^{-1}\left(1 - 2\sin^2 u\right) = \cos^{-1}(\cos 2u) = 2u = 2\sin^{-1} x$.

105. (a) $f(x) = \frac{\sin 3x}{\sin x} - \frac{\cos 3x}{\cos x}$

The function appears to have a constant value of 2 wherever it is defined.

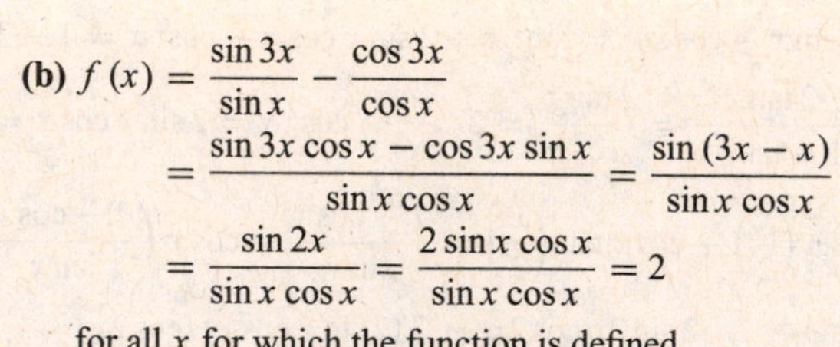

(b) $f(x) = \frac{\sin 3x}{\sin x} - \frac{\cos 3x}{\cos x}$

$= \frac{\sin 3x \cos x - \cos 3x \sin x}{\sin x \cos x} = \frac{\sin(3x - x)}{\sin x \cos x}$

$= \frac{\sin 2x}{\sin x \cos x} = \frac{2\sin x \cos x}{\sin x \cos x} = 2$

for all x for which the function is defined.

107. (a) $y = \sin 6x + \sin 7x$

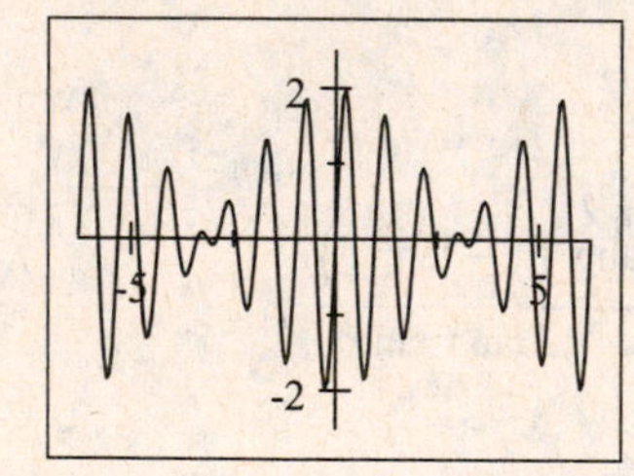

(b) By a sum-to-product formula,

$y = \sin 6x + \sin 7x$

$= 2\sin\left(\frac{6x + 7x}{2}\right)\cos\left(\frac{6x - 7x}{2}\right)$

$= 2\sin\left(\frac{13}{2}x\right)\cos\left(-\frac{1}{2}x\right)$

$= 2\sin\frac{13}{2}x\cos\frac{1}{2}x$

(c) We graph $y = \sin 6x + \sin 7x$, $y = 2\cos\left(\frac{1}{2}x\right)$, and $y = -2\cos\left(\frac{1}{2}x\right)$.

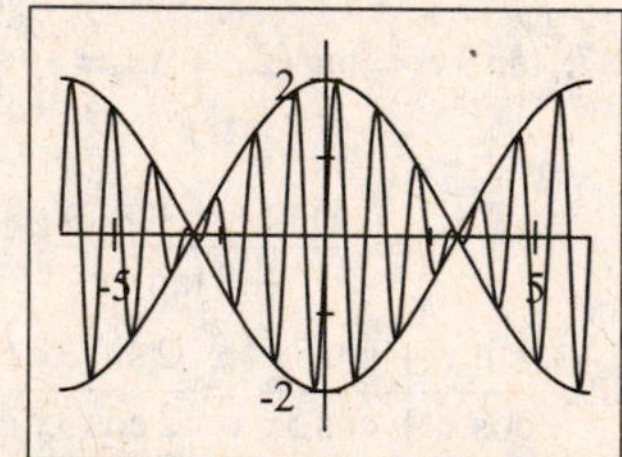

The graph of $y = f(x)$ lies between the other two graphs.

109. (a) $\cos 4x = \cos(2x + 2x) = 2\cos^2 2x - 1 = 2\left(2\cos^2 x - 1\right)^2 - 1 = 8\cos^4 x - 8\cos^2 x + 1$. Thus the desired polynomial is $P(t) = 8t^4 - 8t^2 + 1$.

(b) $\cos 5x = \cos(4x + x) = \cos 4x \cos x - \sin 4x \sin x = \cos x \left(8\cos^4 x - 8\cos^2 x + 1\right) - 2\sin 2x \cos 2x \sin x$

$= 8\cos^5 x - 8\cos^3 x + \cos x - 4\sin x \cos x \left(2\cos^2 x - 1\right)\sin x$ [from part (a)]

$= 8\cos^5 x - 8\cos^3 x + \cos x - 4\cos x \left(2\cos^2 x - 1\right)\sin^2 x$

$= 8\cos^5 x - 8\cos^3 x + \cos x - 4\cos x \left(2\cos^2 x - 1\right)\left(1 - \cos^2 x\right)$

$= 8\cos^5 x - 8\cos^3 x + \cos x + 8\cos^5 x - 12\cos^3 x + 4\cos x = 16\cos^5 x - 20\cos^3 x + 5\cos x$

Thus, the desired polynomial is $P(t) = 16t^5 - 20t^3 + 5t$.

111. Using a product-to-sum formula,

$\text{RHS} = 4\sin A \sin B \sin C = 4\sin A\left\{\frac{1}{2}\left[\cos(B-C) - \cos(B+C)\right]\right\} = 2\sin A\cos(B-C) - 2\sin A\cos(B+C)$.

Using another product-to-sum formula, this is equal to

$2\left\{\frac{1}{2}\left[\sin(A+B-C) + \sin(A-B+C)\right]\right\} - 2\left\{\frac{1}{2}\left[\sin(A+B+C) + \sin(A-B-C)\right]\right\}$

$= \sin(A+B-C) + \sin(A-B+C) - \sin(A+B+C) - \sin(A-B-C)$

Now $A + B + C = \pi$, so $A + B - C = \pi - 2C$, $A - B + C = \pi - 2B$, and $A - B - C = 2A - \pi$.

Thus our expression simplifies to

$\sin(A+B-C) + \sin(A-B+C) - \sin(A+B+C) - \sin(A-B-C)$

$= \sin(\pi - 2C) + \sin(\pi - 2B) + 0 - \sin(2A - \pi) = \sin 2C + \sin 2B + \sin 2A = \text{LHS}$

113. (a) In both logs the length of the adjacent side is $20\cos\theta$ and the length of the opposite side is $20\sin\theta$. Thus the cross-sectional area of the beam is modeled by

$A(\theta) = (20\cos\theta)(20\sin\theta) = 400\sin\theta\cos\theta = 200(2\sin\theta\cos\theta) = 200\sin 2\theta$.

(b) The function $y = \sin u$ is maximized when $u = \frac{\pi}{2}$. So $2\theta = \frac{\pi}{2} \Leftrightarrow \theta = \frac{\pi}{4}$. Thus the maximum cross-sectional area is $A\left(\frac{\pi}{4}\right) = 200\sin 2\left(\frac{\pi}{4}\right) = 200$.

115. (a) $y = f_1(t) + f_2(t) = \cos 11t + \cos 13t$

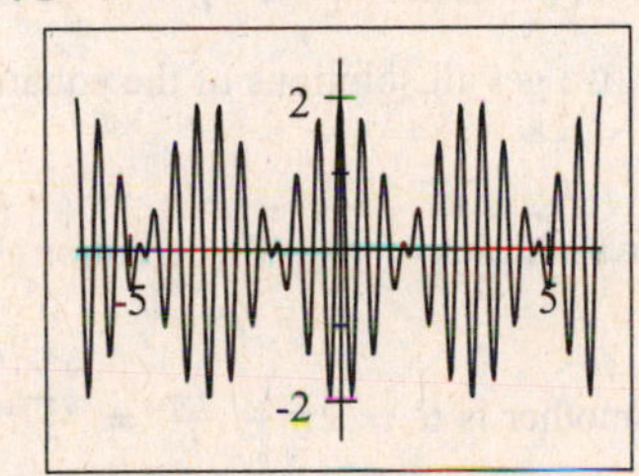

(b) Using the identity

$\cos\alpha + \cos y = 2\cdot\cos\left(\frac{\alpha + y}{2}\right)\cos\left(\frac{\alpha - y}{2}\right)$, we have

$f(t) = \cos 11t + \cos 13t = 2\cdot\cos\left(\frac{11t + 13t}{2}\right)\cos\left(\frac{11t - 13t}{2}\right)$

$= 2\cdot\cos 12t\cdot\cos(-t) = 2\cos 12t\cos t$

(c) We graph $y = \cos 11t + \cos 13t$, $y = 2\cos t$, and $y = -2\cos t$.

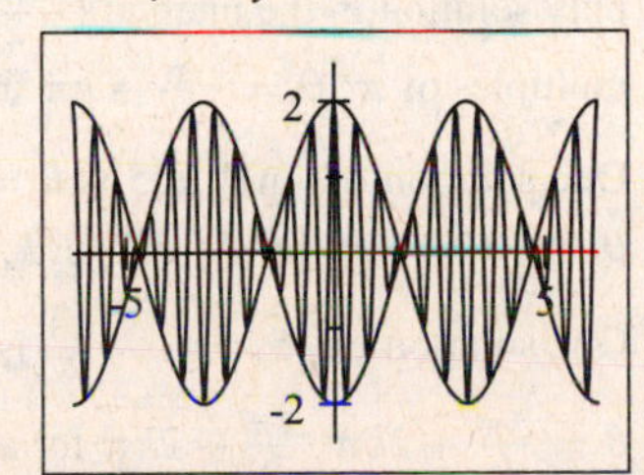

The graph of f lies between the graphs of $y = 2\cos t$ and $y = -2\cos t$. Thus, the loudness of the sound varies between $y = \pm 2\cos t$.

117. We find the area of $\triangle ABC$ in two different ways. First, let AB be the base and CD be the height. Since $\angle BOC = 2\theta$ we see that $CD = \sin 2\theta$. So the area is $\frac{1}{2}(\text{base})(\text{height}) = \frac{1}{2}\cdot 2\cdot\sin 2\theta = \sin 2\theta$. On the other hand, in $\triangle ABC$ we see that $\angle C$ is a right angle. So $BC = 2\sin\theta$ and $AC = 2\cos\theta$, and the area is $\frac{1}{2}(\text{base})(\text{height}) = \frac{1}{2}\cdot(2\sin\theta)(2\cos\theta) = 2\sin\theta\cos\theta$. Equating the two expressions for the area of $\triangle ABC$, we get $\sin 2\theta = 2\sin\theta\cos\theta$.

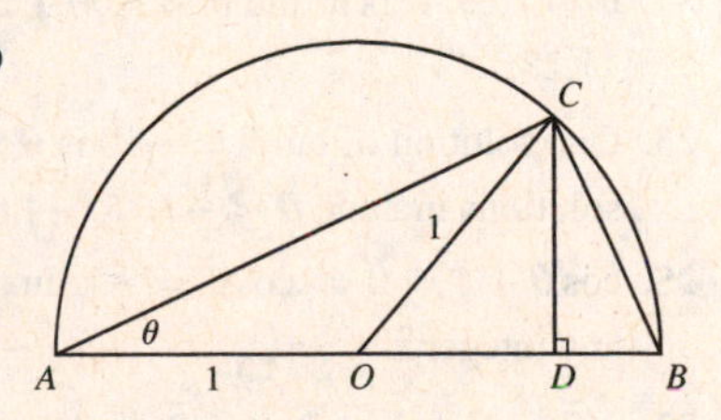

7.4 BASIC TRIGONOMETRIC EQUATIONS

1. Because the trigonometric functions are periodic, if a basic trigonometric equation has one solution, it has *infinitely many* solutions.

3. We can find some of the solutions of $\sin x = 0.3$ graphically by graphing $y = \sin x$ and $y = 0.3$. The solutions shown are $x \approx -9.7, x \approx -6.0, x \approx -3.4, x \approx 0.3, x \approx 2.8, x \approx 6.6$, and $x \approx 9.1$.

5. Because sine has period 2π, we first find the solutions in the interval $[0, 2\pi)$. From the unit circle shown, we see that $\sin\theta = \frac{\sqrt{3}}{2}$ in quadrants I and II, so the solutions are $\theta = \frac{\pi}{3}$ and $\theta = \frac{2\pi}{3}$. We get all solutions of the equation by adding integer multiples of 2π to these solutions: $\theta = \frac{\pi}{3} + 2k\pi$ and $\theta = \frac{2\pi}{3} + 2k\pi$ for any integer k.

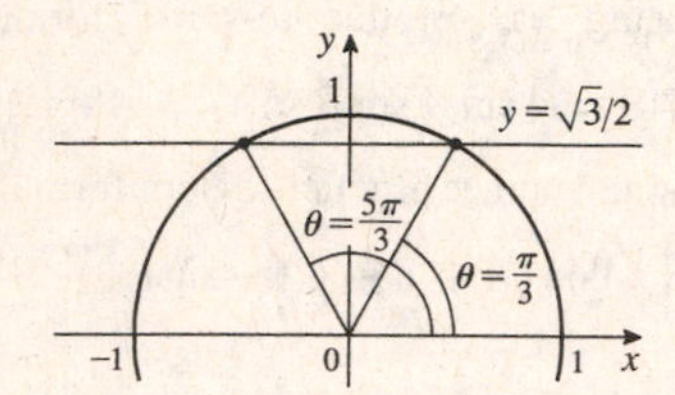

7. The cosine function is negative in quadrants II and III, so the solution of $\cos\theta = -1$ on the interval $[0, 2\pi)$ is $\theta = \pi$. Adding integer multiples of 2π to this solution gives all solutions: $\theta = \pi + 2k\pi = (2k+1)\pi$ for any integer k.

9. The cosine function is positive in quadrants I and IV, so the solutions of $\cos\theta = \frac{1}{4}$ on the interval $[0, 2\pi)$ are $\theta = \cos^{-1}\frac{1}{4} \approx 1.32$ and $\theta = 2\pi - \cos^{-1}\frac{1}{4} \approx 4.97$. Adding integer multiples of 2π to these solutions gives all solutions: $\theta \approx 1.32 + 2k\pi, 4.97 + 2k\pi$ for any integer k.

11. The sine function is negative in quadrants III and IV, so the solutions of $\sin\theta = -0.45$ on the interval $[0, 2\pi)$ are $\theta = \pi + \sin^{-1}(0.45) \approx 3.61$ and $\theta = 2\pi - \sin^{-1}(0.45) \approx 5.82$. Adding integer multiples of 2π to these solutions gives all solutions, $\theta \approx 3.61 + 2k\pi, 5.82 + 2k\pi$ for any integer k.

13. We first find one solution by taking $\tan^{-1}$ of each side of the equation: $\theta = \tan^{-1}\left(-\sqrt{3}\right) = -\frac{\pi}{3}$. By definition, this is the only solution in the interval $\left(-\frac{\pi}{2}, \frac{\pi}{2}\right)$. Since tangent has period π, we get all solutions of the equation by adding integer multiples of π: $\theta = -\frac{\pi}{3} + k\pi$ for any integer k.

15. One solution of $\tan\theta = 5$ is $\theta = \tan^{-1} 5 \approx 1.37$. Adding integer multiples of π to this solution gives all solutions: $\theta \approx 1.37 + k\pi$ for any integer k.

17. One solution of $\cos\theta = -\frac{\sqrt{3}}{2}$ is $\theta = \cos^{-1}\left(-\frac{\sqrt{3}}{2}\right) = \frac{5\pi}{6}$ and another is $\theta = 2\pi - \frac{5\pi}{6} = \frac{7\pi}{6}$. All solutions are $\theta = \frac{5\pi}{6} + 2k\pi, \frac{7\pi}{6} + 2k\pi$ for any integer k. Specific solutions include $\theta = \frac{5\pi}{6} - 2\pi = -\frac{7\pi}{6}, \theta = \frac{7\pi}{6} - 2\pi = -\frac{5\pi}{6}$, $\theta = \frac{5\pi}{6}, \theta = \frac{7\pi}{6}, \theta = \frac{5\pi}{6} + 2\pi = \frac{17\pi}{6}$, and $\theta = \frac{7\pi}{6} + 2\pi = \frac{19\pi}{6}$.

19. One solution of $\sin\theta = \frac{\sqrt{2}}{2}$ is $\theta = \sin^{-1}\frac{\sqrt{2}}{2} = \frac{\pi}{4}$ and another is $\theta = \pi - \frac{\pi}{4} = \frac{3\pi}{4}$. All solutions are $\theta = \frac{\pi}{4} + 2k\pi$ and $\theta = \frac{3\pi}{4} + 2k\pi$ for any integer k. Specific solutions include $\theta = -\frac{7\pi}{4}, -\frac{5\pi}{4}, \frac{\pi}{4}, \frac{3\pi}{4}, \frac{9\pi}{4}$, and $\frac{11\pi}{4}$.

21. One solution of $\cos\theta = 0.28$ is $\theta = \cos^{-1} 0.28 \approx 1.29$ and another is $\theta = 2\pi - \cos^{-1} 0.28 \approx 5.00$. All solutions are $\theta \approx 1.29 + 2k\pi$ and $\theta \approx 5.00 + 2k\pi$ for any integer k. Specific solutions include $\theta \approx -5.00, -1.29, 1.29, 5.00, 7.57$, and 11.28.

23. One solution of $\tan\theta = -10$ is $\theta = \tan^{-1}(-10) \approx -1.47$. All solutions are $\theta \approx -1.47 + k\pi$ for any integer k. Specific solutions include $\theta \approx -7.75, -4.61, -1.47, 1.67, 4.81$, and 7.95.

25. $\cos\theta + 1 = 0 \Leftrightarrow \cos\theta = -1$. In the interval $[0, 2\pi)$ the only solution is $\theta = \pi$.Thus the solutions are $\theta = (2k+1)\pi$ for any integer k.

27. $\sqrt{2}\sin\theta + 1 = 0 \Rightarrow \sqrt{2}\sin\theta = -1 \Leftrightarrow \sin\theta = -\frac{1}{\sqrt{2}}$The solutions in the interval $[0, 2\pi)$ are $\theta = \frac{5\pi}{4}, \frac{7\pi}{4}$. Thus the solutions are $\theta = \frac{5\pi}{4} + 2k\pi, \frac{7\pi}{4} + 2k\pi$ for any integer k.

29. $5\sin\theta - 1 = 0 \Leftrightarrow \sin\theta = \frac{1}{5}$. The solutions in the interval $[0, 2\pi)$ are $\theta = \sin^{-1}\frac{1}{5} \approx 0.20$ and $\theta = \pi - \sin^{-1}\frac{1}{5} \approx 2.94$. Thus the solutions are $\theta \approx 0.20 + 2k\pi, 2.94 + 2k\pi$ for any integer k.

31. $3\tan^2\theta - 1 = 0 \Leftrightarrow \tan^2\theta = \frac{1}{3} \Leftrightarrow \tan\theta = \pm\frac{\sqrt{3}}{3}$. The solutions in the interval $\left(-\frac{\pi}{2}, \frac{\pi}{2}\right)$ are $\theta = \pm\frac{\pi}{6}$, so all solutions are $\theta = -\frac{\pi}{6} + k\pi, \frac{\pi}{6} + k\pi$ for any integer k.

33. $2\cos^2\theta - 1 = 0 \Leftrightarrow \cos^2\theta = \frac{1}{2} \Leftrightarrow \cos\theta = \pm\frac{1}{\sqrt{2}} \Leftrightarrow \theta = \frac{\pi}{4}, \frac{3\pi}{4}, \frac{5\pi}{4}, \frac{7\pi}{4}$ in $[0, 2\pi)$. Thus, the solutions are $\theta = \frac{\pi}{4} + k\pi$, $\frac{3\pi}{4} + k\pi$ for any integer k.

35. $\tan^2\theta - 4 = 0 \Leftrightarrow \tan^2\theta = 4 \Leftrightarrow \tan\theta = \pm 2 \Leftrightarrow \theta = \tan^{-1}(-2) \approx -1.11$ or $\theta = \tan^{-1} 2 \approx 1.11$ in $\left(-\frac{\pi}{2}, \frac{\pi}{2}\right)$. Thus, the solutions are $\theta \approx -1.11 + k\pi, 1.11 + k\pi$ for any integer k.

37. $\sec^2\theta - 2 = 0 \Leftrightarrow \sec^2\theta = 2 \Leftrightarrow \sec\theta = \pm\sqrt{2}$. In the interval $[0, 2\pi)$ the solutions are $\theta = \frac{\pi}{4}, \frac{3\pi}{4}, \frac{5\pi}{4}, \frac{7\pi}{4}$. Thus, the solutions are $\theta = (2k+1)\frac{\pi}{4}$ for any integer k.

39. $\left(\tan^2\theta - 4\right)(2\cos\theta + 1) = 0 \Leftrightarrow \tan^2\theta = 4$ or $2\cos\theta = -1$. From Exercise 35, we know that the first equation has solutions $\theta \approx -1.11 + k\pi, 1.11 + k\pi$ for any integer k. $2\cos\theta = -1 \Leftrightarrow \cos\theta = -\frac{1}{2}$ has solutions $\cos^{-1}\left(-\frac{1}{2}\right) = \frac{2\pi}{3}$ and $\frac{4\pi}{3}$ on $[0, 2\pi)$, so all solutions are $\theta \approx \frac{2\pi}{3} + 2k\pi$ and $\frac{4\pi}{3} + 2k\pi$ for any integer k. Thus, the original equation has solutions $\theta \approx -1.11 + k\pi, 1.11 + k\pi, \frac{2\pi}{3} + 2k\pi$, and $\frac{4\pi}{3} + 2k\pi$ for any integer k.

41. $4\cos^2\theta - 4\cos\theta + 1 = 0 \Leftrightarrow (2\cos\theta - 1)^2 = 0 \Leftrightarrow 2\cos\theta - 1 = 0 \Leftrightarrow \cos\theta = \frac{1}{2} \Leftrightarrow \theta = \frac{\pi}{3} + 2k\pi, \frac{5\pi}{3} + 2k\pi$ for any integer k.

43. $3\sin^2\theta - 7\sin\theta + 2 = 0 \Rightarrow (3\sin\theta - 1)(\sin\theta - 2) = 0 \Rightarrow 3\sin\theta - 1 = 0$ or $\sin\theta - 2 = 0$. Since $|\sin\theta| \leq 1$, $\sin\theta - 2 = 0$ has no solution. Thus $3\sin\theta - 1 = 0 \Rightarrow \sin\theta = \frac{1}{3} \Rightarrow \theta \approx 0.33984$ and $\theta \approx \pi - 0.33984 \approx 2.80176$ are the solutions in $[0, 2\pi)$, and all solutions are $\theta \approx 0.33984 + 2k\pi, 2.80176 + 2k\pi$ for any integer k.

45. $2\cos^2\theta - 7\cos\theta + 3 = 0 \Leftrightarrow (2\cos\theta - 1)(\cos\theta - 3) = 0 \Leftrightarrow \cos\theta = \frac{1}{2}$ or $\cos\theta = 3$ (which is inadmissible) $\Leftrightarrow \theta = \frac{\pi}{3}$, $\frac{5\pi}{3}$. Therefore, the solutions are $\theta = \frac{\pi}{3} + 2k\pi, \frac{5\pi}{3} + 2k\pi$ for any integer k.

47. $\cos^2\theta - \cos\theta - 6 = 0 \Leftrightarrow (\cos\theta + 2)(\cos x - 3) = 0 \Leftrightarrow \cos x = -2$ or $\cos x = 3$, neither of which has a solution. Thus, the original equation has no solution.

49. $\sin^2\theta = 2\sin\theta + 3 \Leftrightarrow \sin^2\theta - 2\sin\theta - 3 = 0 \Leftrightarrow (\sin\theta - 3)(\sin\theta + 1) = 0 \Leftrightarrow \sin\theta - 3 = 0$ or $\sin\theta + 1 = 0$. Since $|\sin\theta| \leq 1$ for all θ, there is no solution for $\sin\theta - 3 = 0$. Hence $\sin\theta + 1 = 0 \Leftrightarrow \sin\theta = -1 \Leftrightarrow \theta = \frac{3\pi}{2} + 2k\pi$ for any integer k.

51. $\cos\theta(2\sin\theta + 1) = 0 \Leftrightarrow \cos\theta = 0$ or $\sin\theta = -\frac{1}{2} \Leftrightarrow \theta = \frac{\pi}{2} + k\pi, \frac{7\pi}{6} + 2k\pi, \frac{11\pi}{6} + 2k\pi$ for any integer k.

53. $\cos\theta\sin\theta - 2\cos\theta = 0 \Leftrightarrow \cos\theta(\sin\theta - 2) = 0 \Leftrightarrow \cos\theta = 0$ or $\sin\theta - 2 = 0$. Since $|\sin\theta| \leq 1$ for all θ, there is no solution for $\sin\theta - 2 = 0$. Hence, $\cos\theta = 0 \Leftrightarrow \theta = \frac{\pi}{2} + 2k\pi, \frac{3\pi}{2} + 2k\pi \Leftrightarrow \theta = \frac{\pi}{2} + k\pi$ for any integer k.

55. $3\tan\theta\sin\theta - 2\tan\theta = 0 \Leftrightarrow \tan\theta(3\sin\theta - 2) = 0 \Leftrightarrow \tan\theta = 0$ or $\sin\theta = \frac{2}{3}$. $\tan\theta = 0$ has solution $\theta = 0$ on $\left(-\frac{\pi}{2}, \frac{\pi}{2}\right)$ and $\sin\theta = \frac{2}{3}$ has solutions $\theta = \sin^{-1}\frac{2}{3} \approx 0.73$ and $\theta = \pi - \sin^{-1}\frac{2}{3} \approx 2.41$ on $[0, 2\pi)$, so the original equation has solutions $\theta = k\pi$, $\theta \approx 0.73 + 2k\pi, 2.41 + 2k\pi$ for any integer k.

57. We substitute $\theta_1 = 70°$ and $\dfrac{v_1}{v_2} = 1.33$ into Snell's Law to get $\dfrac{\sin 70°}{\sin\theta_2} = 1.33 \Leftrightarrow \sin\theta_2 = \dfrac{\sin 70°}{1.33} = 0.7065 \Rightarrow$ $\theta_2 \approx 44.95°$.

59. **(a)** $F = \frac{1}{2}(1 - \cos\theta) = 0 \Rightarrow \cos\theta = 1 \Rightarrow \theta = 0$

(b) $F = \frac{1}{2}(1 - \cos\theta) = 0.25 \Rightarrow 1 - \cos\theta = 0.5 \Rightarrow \cos\theta = 0.5 \Rightarrow \theta = 60°$ or $120°$

(c) $F = \frac{1}{2}(1 - \cos\theta) = 0.5 \Rightarrow 1 - \cos\theta = 1 \Rightarrow \cos\theta = 0 \Rightarrow \theta = 90°$ or $270°$

(d) $F = \frac{1}{2}(1 - \cos\theta) = 1 \Rightarrow 1 - \cos\theta = 2 \Rightarrow \cos\theta = -1 \Rightarrow \theta = 180°$

7.5 MORE TRIGONOMETRIC EQUATIONS

1. Using a Pythagorean identity, we calculate $\sin x + \sin^2 x + \cos^2 x = 1 \Leftrightarrow \sin x + 1 = 1 \Leftrightarrow \sin x = 0$, whose solutions are $x = k\pi$ for any integer k.

3. $2\cos^2\theta + \sin\theta = 1 \Leftrightarrow 2\left(1 - \sin^2\theta\right) + \sin\theta - 1 = 0 \Leftrightarrow -2\sin^2\theta + \sin\theta + 1 = 0 \Leftrightarrow 2\sin^2\theta - \sin\theta - 1 = 0$. From Exercise 7.4.42, the solutions are $\theta = \frac{7\pi}{6} + 2k\pi, \frac{11\pi}{6} + 2k\pi, \frac{\pi}{2} + 2k\pi$ for any integer k.

5. $\tan^2\theta - 2\sec\theta = 2 \Leftrightarrow \sec^2\theta - 1 - 2\sec\theta = 2 \Leftrightarrow \sec^2\theta - 2\sec\theta - 3 = 0 \Leftrightarrow (\sec\theta - 3)(\sec\theta + 1) = 0 \Leftrightarrow \sec\theta = 3$ or $\sec\theta = -1$. If $\sec\theta = 3$, then $\cos\theta = \frac{1}{3}$, which has solutions $\theta = \cos^{-1}\frac{1}{3} \approx 1.23$ and $\theta = 2\pi - \cos^{-1}\frac{1}{3} \approx 5.05$ on $[0, 2\pi)$. If $\sec\theta = -1$, then $\cos\theta = -1$, which has solution $\theta = \pi$ on $[0, 2\pi)$. Thus, solutions are $\theta = (2k+1)\pi$, $\theta \approx 1.23 + 2k\pi, 5.05 + 2k\pi$ for any integer k.

7. $2\sin 2\theta - 3\sin\theta = 0 \Leftrightarrow 2(2\sin\theta\cos\theta) - 3\sin\theta = 0 \Leftrightarrow \sin\theta(4\cos\theta - 3) = 0 \Leftrightarrow \sin\theta = 0$ or $\cos\theta = \frac{3}{4}$. The first equation has solutions $\theta = 0, \pi$ on $[0, 2\pi)$, and the second has solutions $\theta = \cos^{-1}\frac{3}{4} \approx 0.72$ and $\theta = 2\pi - \cos^{-1}\frac{3}{4} \approx 5.56$ on $[0, 2\pi)$. Thus, solutions are $\theta = k\pi, \theta \approx 0.72 + 2k\pi, 5.56 + 2k\pi$ for any integer k.

9. $\cos 2\theta = 3\sin\theta - 1 \Leftrightarrow 1 - 2\sin^2\theta = 3\sin\theta - 1 \Leftrightarrow 2\sin^2\theta + 3\sin\theta - 2 = 0 \Leftrightarrow (\sin\theta + 2)(2\sin\theta - 1) = 0$ $\Leftrightarrow \sin\theta = -2$ or $\sin\theta = \frac{1}{2}$. The first equation has no solution and the second has solutions $\theta = \sin^{-1}\frac{1}{2} = \frac{\pi}{6}$ and $\theta = \pi - \sin^{-1}\frac{1}{2} = \frac{5\pi}{6}$ on $[0, 2\pi)$, so the original equation has solutions $\theta = \frac{\pi}{6} + 2k\pi, \frac{5\pi}{6} + 2k\pi$ for any integer k.

11. $2\sin^2\theta - \cos\theta = 1 \Leftrightarrow 2\left(1 - \cos^2\theta\right) - \cos\theta - 1 = 0 \Leftrightarrow -2\cos^2\theta - \cos\theta + 1 = 0 \Leftrightarrow (2\cos\theta - 1)(\cos\theta + 1) = 0 \Leftrightarrow$ $2\cos\theta - 1 = 0$ or $\cos\theta + 1 = 0 \Leftrightarrow \cos\theta = \frac{1}{2}$ or $\cos\theta = -1 \Leftrightarrow \theta = \frac{\pi}{3} + 2k\pi, \frac{5\pi}{3} + 2k\pi, (2k+1)\pi$ for any integer k.

13. $\sin\theta - 1 = \cos\theta \Leftrightarrow \sin\theta + \cos\theta = 1$. Squaring both sides, we have $\sin^2\theta + \cos^2\theta + 2\sin\theta\cos\theta = 1 \Leftrightarrow \sin 2\theta = 0$, which has solutions $\theta = 0, \frac{\pi}{2}, \pi, \frac{3\pi}{2}$ in $[0, 2\pi)$. Checking in the original equation, we see that only $\theta = \frac{\pi}{2}$ and $\theta = \pi$ are valid. (The extraneous solutions were introduced by squaring both sides.) Thus, the solutions are $\theta = (2k+1)\pi, \frac{\pi}{2} + 2k\pi$ for any integer k.

15. $\tan\theta + 1 = \sec\theta \Leftrightarrow \dfrac{\sin\theta}{\cos\theta} + 1 = \dfrac{1}{\cos\theta} \Leftrightarrow \sin\theta + \cos\theta = 1$. Squaring both sides, we have $\sin^2\theta + \cos^2\theta + 2\sin\theta\cos\theta = 1$ $\Leftrightarrow \sin 2\theta = 0$, which has solutions $\theta = 0, \frac{\pi}{2}, \pi, \frac{3\pi}{2}$ on $[0, 2\pi)$. Checking in the original equation, we see that only $\theta = 0$ is valid. Thus, the solutions are $\theta = 2k\pi$ for any integer k.

17. **(a)** $2\cos 3\theta = 1 \Leftrightarrow \cos 3\theta = \frac{1}{2} \Rightarrow 3\theta = \frac{\pi}{3}, \frac{5\pi}{3}$ for 3θ in $[0, 2\pi)$. Thus, solutions are $\frac{\pi}{9} + \frac{2}{3}k\pi, \frac{5\pi}{9} + \frac{2}{3}k\pi$ for any integer k.

(b) We take $k = 0, 1, 2$ in the expressions in part (a) to obtain the solutions $\theta = \frac{\pi}{9}, \frac{5\pi}{9}, \frac{7\pi}{9}, \frac{11\pi}{9}, \frac{13\pi}{9}, \frac{17\pi}{9}$ in $[0, 2\pi)$.

19. **(a)** $2\cos 2\theta + 1 = 0 \Leftrightarrow \cos 2\theta = -\frac{1}{2} \Leftrightarrow 2\theta = \frac{2\pi}{3} + 2k\pi, \frac{4\pi}{3} + 2k\pi \Leftrightarrow \theta = \frac{\pi}{3} + k\pi, \frac{2\pi}{3} + k\pi$ for any integer k.

(b) The solutions in $[0, 2\pi)$ are $\frac{\pi}{3}, \frac{2\pi}{3}, \frac{4\pi}{3}, \frac{5\pi}{3}$.

21. **(a)** $\sqrt{3}\tan 3\theta + 1 = 0 \Leftrightarrow \tan 3\theta = -\frac{1}{\sqrt{3}} \Leftrightarrow 3\theta = \frac{5\pi}{6} + k\pi \Leftrightarrow \theta = \frac{5\pi}{18} + \frac{1}{3}k\pi$ for any integer k.

(b) The solutions in $[0, 2\pi)$ are $\frac{5\pi}{18}, \frac{11\pi}{18}, \frac{17\pi}{18}, \frac{23\pi}{18}, \frac{29\pi}{18}, \frac{35\pi}{18}$.

23. **(a)** $\cos\frac{\theta}{2} - 1 = 0 \Leftrightarrow \cos\frac{\theta}{2} = 1 \Leftrightarrow \frac{\theta}{2} = 2k\pi \Leftrightarrow \theta = 4k\pi$ for any integer k.

(b) The only solution in $[0, 2\pi)$ is $\theta = 0$.

25. **(a)** $2\sin\frac{\theta}{3} + \sqrt{3} = 0 \Leftrightarrow 2\sin\frac{\theta}{3} = -\sqrt{3} \Leftrightarrow \sin\frac{\theta}{3} = -\frac{\sqrt{3}}{2} \Leftrightarrow \frac{\theta}{3} = \frac{4\pi}{3} + 2k\pi, \frac{5\pi}{3} + 2k\pi \Leftrightarrow \theta = 4\pi + 6k\pi, 5\pi + 6k\pi$ for any integer k.

(b) There is no solution in $[0, 2\pi)$.

27. **(a)** $\sin 2\theta = 3\cos 2\theta \Leftrightarrow \tan 2\theta = 3 \Leftrightarrow \theta = \frac{1}{2}\tan^{-1} 3 \approx 0.62$ on $\left(-\frac{\pi}{4}, \frac{\pi}{4}\right)$. Thus, solutions are $\theta \approx 0.62 + \frac{1}{2}k\pi$ for any integer k.

(b) The solutions in $[0, 2\pi)$ are $\theta \approx 0.62, 2.19, 3.76, 5.33$.

29. **(a)** $1 - 2\sin\theta = \cos 2\theta \Leftrightarrow 1 - 2\sin\theta = 1 - 2\sin^2\theta \Leftrightarrow 2\sin^2\theta - 2\sin\theta = 0 \Leftrightarrow 2\sin\theta(\sin\theta - 1) = 0 \Leftrightarrow \sin\theta = 0$ or $\sin\theta = 1 \Leftrightarrow \theta = 0, \pi$, or $\frac{\pi}{2}$ in $[0, 2\pi)$. Thus, the solutions are $\theta = k\pi$, $\theta = \frac{\pi}{2} + 2\pi k$ for any integer k.

(b) The solutions in $[0, 2\pi)$ are $\theta = 0, \frac{\pi}{2}, \pi$.

31. **(a)** $3\tan^3\theta - 3\tan^2\theta - \tan\theta + 1 = 0 \Leftrightarrow (\tan\theta - 1)\left(3\tan^2\theta - 1\right) = 0 \Leftrightarrow \tan\theta = 1$ or $3\tan^2\theta = 1 \Leftrightarrow \tan\theta = 1$ or $\tan\theta = \pm\frac{1}{\sqrt{3}} \Leftrightarrow \theta = \frac{\pi}{6} + k\pi, \frac{\pi}{4} + k\pi, \frac{5\pi}{6} + k\pi$ for any integer k.

(b) The solutions in $[0, 2\pi)$ are $\theta = \frac{\pi}{6}, \frac{\pi}{4}, \frac{5\pi}{6}, \frac{7\pi}{6}, \frac{5\pi}{4}, \frac{11\pi}{6}$.

33. **(a)** $2\sin\theta\tan\theta - \tan\theta = 1 - 2\sin\theta \Leftrightarrow 2\sin\theta\tan\theta - \tan\theta + 2\sin\theta - 1 = 0 \Leftrightarrow (2\sin\theta - 1)(\tan\theta + 1) = 0 \Leftrightarrow$ $2\sin\theta - 1 = 0$ or $\tan\theta + 1 = 0 \Leftrightarrow \sin\theta = \frac{1}{2}$ or $\tan\theta = -1 \Leftrightarrow \theta = \frac{\pi}{6} + 2k\pi, \frac{5\pi}{6} + 2k\pi, \frac{3\pi}{4} + k\pi$ for any integer k.

(b) The solutions in $[0, 2\pi)$ are $\frac{\pi}{6}, \frac{3\pi}{4}, \frac{5\pi}{6}, \frac{7\pi}{4}$.

35. **(a)**

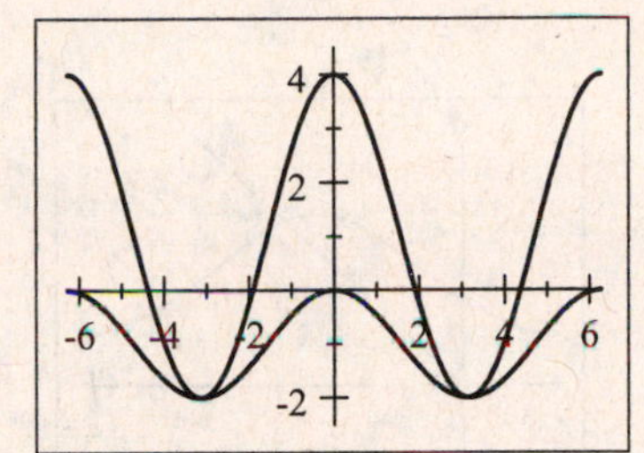

The points of intersection are approximately $(\pm 3.14, -2)$.

(b) $f(x) = 3\cos x + 1$; $g(x) = \cos x - 1$. $f(x) = g(x)$ when $3\cos x + 1 = \cos x - 1 \Leftrightarrow 2\cos x = -2 \Leftrightarrow \cos x = -1 \Leftrightarrow$ $x = \pi + 2k\pi = (2k + 1)\pi$. The points of intersection are $((2k + 1)\pi, -2)$ for any integer k.

37. **(a)**

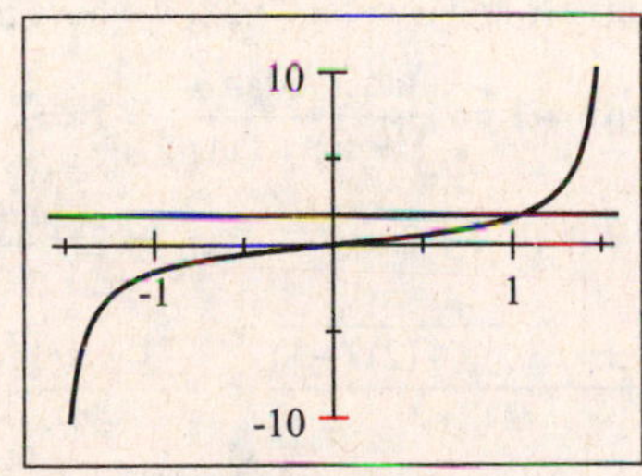

The point of intersection is approximately $(1.04, 1.73)$.

(b) $f(x) = \tan x$; $g(x) = \sqrt{3}$. $f(x) = g(x)$ when $\tan x = \sqrt{3} \Leftrightarrow$ $x = \frac{\pi}{3} + k\pi$. The intersection points are $\left(\frac{\pi}{3} + k\pi, \sqrt{3}\right)$ for any integer k.

39. $\cos\theta\cos 3\theta - \sin\theta\sin 3\theta = 0 \Leftrightarrow \cos(\theta + 3\theta) = 0 \Leftrightarrow \cos 4\theta = 0 \Leftrightarrow 4\theta = \frac{\pi}{2}, \frac{3\pi}{2}, \frac{5\pi}{2}, \frac{7\pi}{2}, \frac{9\pi}{2}, \frac{11\pi}{2}, \frac{13\pi}{2}, \frac{15\pi}{2}$ in $[0, 8\pi) \Leftrightarrow \theta = \frac{\pi}{8}, \frac{3\pi}{8}, \frac{5\pi}{8}, \frac{7\pi}{8}, \frac{9\pi}{8}, \frac{11\pi}{8}, \frac{13\pi}{8}, \frac{15\pi}{8}$ in $[0, 2\pi)$.

41. $\sin 2\theta\cos\theta - \cos 2\theta\sin\theta = \frac{\sqrt{3}}{2} \Leftrightarrow \sin(2\theta - \theta) = \frac{\sqrt{3}}{2} \Leftrightarrow \sin\theta = \frac{\sqrt{3}}{2} \Leftrightarrow \theta = \frac{\pi}{3}, \frac{2\pi}{3}$ in $[0, 2\pi)$.

43. $\sin 2\theta + \cos\theta = 0 \Leftrightarrow 2\sin\theta\cos\theta + \cos\theta = 0 \Leftrightarrow \cos\theta(2\sin\theta + 1) = 0 \Leftrightarrow \cos\theta = 0$ or $\sin\theta = -\frac{1}{2} \Leftrightarrow \theta = \frac{\pi}{2}, \frac{7\pi}{6}, \frac{3\pi}{2}, \frac{11\pi}{6}$ in $[0, 2\pi)$.

45. $\cos 2\theta + \cos\theta = 2 \Leftrightarrow 2\cos^2\theta - 1 + \cos\theta - 2 = 0 \Leftrightarrow 2\cos^2\theta + \cos\theta - 3 = 0 \Leftrightarrow (2\cos\theta + 3)(\cos\theta - 1) = 0 \Leftrightarrow$ $2\cos\theta + 3 = 0$ or $\cos\theta - 1 = 0 \Leftrightarrow \cos\theta = -\frac{3}{2}$ (which is impossible) or $\cos\theta = 1 \Leftrightarrow \theta = 0$ in $[0, 2\pi)$.

47. $\cos 2\theta - \cos^2\theta = 0 \Leftrightarrow 2\cos^2\theta - 1 - \cos^2\theta = 0 \Leftrightarrow \cos^2\theta = 1 \Leftrightarrow \theta = k\pi$ for any integer k. On $[0, 2\pi)$, the solutions are $\theta = 0, \pi$.

49. $\cos 2\theta - \cos 4\theta = 0 \Leftrightarrow \cos 2\theta - \left(2\cos^2 2\theta - 1\right) = 0 \Leftrightarrow (\cos 2\theta - 1)(2\cos 2\theta + 1) = 0$. The first factor has zeros at $\theta = 0, \pi$ and the second has zeros at $\theta = \frac{\pi}{3}, \frac{2\pi}{3}, \frac{4\pi}{3}, \frac{5\pi}{3}$. Thus, solutions of the original equation are are $\theta = 0, \frac{\pi}{3}, \frac{2\pi}{3}, \pi, \frac{4\pi}{3}, \frac{5\pi}{3}$ in $[0, 2\pi)$.

51. $\cos\theta - \sin\theta = \sqrt{2}\sin\frac{\theta}{2} \Leftrightarrow \cos\theta - \sin\theta = \sqrt{2}\left(\pm\sqrt{\frac{1-\cos\theta}{2}}\right)$. Squaring both sides, we have $\cos^2\theta + \sin^2\theta - 2\sin\theta\cos\theta = 1 - \cos\theta \Leftrightarrow 1 - 2\sin\theta\cos\theta = 1 - \cos\theta \Leftrightarrow$ either $2\sin\theta = 1$ or $\cos\theta = 0 \Leftrightarrow \theta = \frac{\pi}{6}, \frac{\pi}{2}, \frac{5\pi}{6}, \frac{3\pi}{2}$ in $[0, 2\pi)$. Of these, only $\frac{\pi}{6}$ and $\frac{3\pi}{2}$ satisfy the original equation.

53. $\sin\theta + \sin 3\theta = 0 \Leftrightarrow 2\sin 2\theta\cos(-\theta) = 0 \Leftrightarrow 2\sin 2\theta\cos\theta = 0 \Leftrightarrow \sin 2\theta = 0$ or $\cos\theta = 0 \Leftrightarrow 2\theta = k\pi$ or $\theta = k\frac{\pi}{2} \Leftrightarrow \theta = \frac{1}{2}k\pi$ for any integer k.

55. $\cos 4\theta + \cos 2\theta = \cos\theta \Leftrightarrow 2\cos 3\theta\cos\theta = \cos\theta \Leftrightarrow \cos\theta(2\cos 3\theta - 1) = 0 \Leftrightarrow \cos\theta = 0$ or $\cos 3\theta = \frac{1}{2} \Leftrightarrow \theta = \frac{\pi}{2}$ or $3\theta = \frac{\pi}{3} + 2k\pi, \frac{5\pi}{3} + 2k\pi, \frac{7\pi}{3} + 2k\pi, \frac{11\pi}{3} + 2k\pi, \frac{13\pi}{3} + 2k\pi, \frac{17\pi}{3} + 2k\pi \Leftrightarrow \theta = \frac{\pi}{2} + k\pi, \frac{\pi}{9} + \frac{2}{3}k\pi, \frac{5\pi}{9} + \frac{2}{3}k\pi$ for any integer k.

57. $\sin 2x = x$

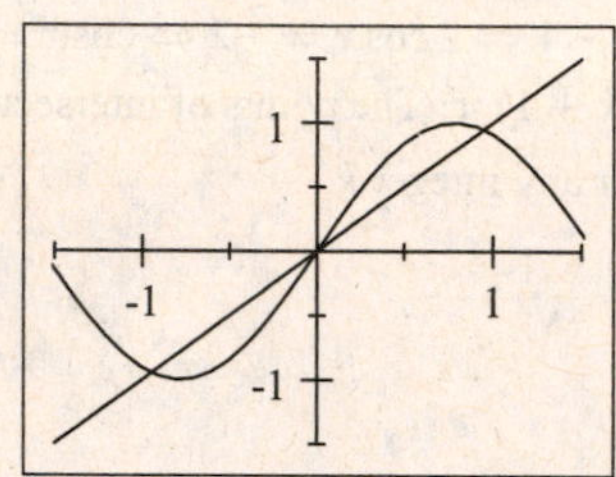

The three solutions are $x = 0$ and $x \approx \pm 0.95$.

59. $2^{\sin x} = x$

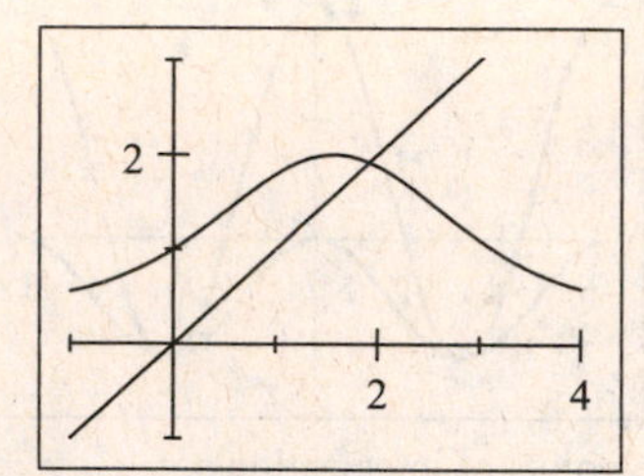

The only solution is $x \approx 1.92$.

61. $\frac{\cos x}{1 + x^2} = x^2$

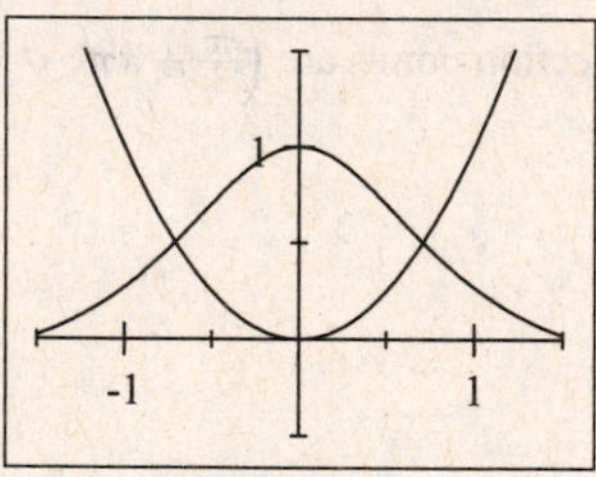

The two solutions are $x \approx \pm 0.71$.

63. With $u = \tan^{-1} x$ and $v = \tan^{-1} 2x$, we have $u + v = \frac{\pi}{4}$ $\Leftrightarrow \tan(u + v) = 1 \Leftrightarrow \frac{\tan u + \tan v}{1 - \tan u \tan v} = 1 \Leftrightarrow$ $\frac{x + 2x}{1 - x(2x)} = 1 \Leftrightarrow x + 2x = 1 - 2x^2 \Leftrightarrow 2x^2 + 3x - 1 = 0$ $\Leftrightarrow x = \frac{-3 \pm \sqrt{3^2 - 4(2)(-1)}}{2(2)} = \frac{-3 \pm \sqrt{17}}{4}$. Because $\tan x$ is not one-to-one, we must check both roots, and find that only $\frac{\sqrt{17}-3}{4}$ is a solution to the original equation.

65. We substitute $v_0 = 2200$ and $R(\theta) = 5000$ and solve for θ. So $5000 = \frac{(2200)^2\sin 2\theta}{32} \Leftrightarrow 5000 = 151250\sin 2\theta \Leftrightarrow \sin 2\theta = 0.03308 \Rightarrow 2\theta = 1.89442^\circ$ or $2\theta = 180^\circ - 1.89442^\circ = 178.10558^\circ$. If $2\theta = 1.89442^\circ$, then $\theta = 0.94721^\circ$, and if $2\theta = 178.10558^\circ$, then $\theta = 89.05279^\circ$.

67. **(a)** $10 = 12 + 2.83\sin\left(\frac{2\pi}{3}(t - 80)\right) \Leftrightarrow 2.83\sin\left(\frac{2\pi}{3}(t - 80)\right) = -2 \Leftrightarrow \sin\left(\frac{2\pi}{3}(t - 80)\right) = -0.70671$. Now $\sin\theta = -0.70671$ and $\theta = -0.78484$. If $\frac{2\pi}{3}(t - 80) = -0.78484 \Leftrightarrow t - 80 = 45.6 \Leftrightarrow t = 34.4$. Now in the interval $[0, 2\pi)$, we have $\theta = \pi + 0.78484 \approx 3.92644$ and $\theta = 2\pi - 0.78484 \approx 5.49834$. If $\frac{2\pi}{3}(t - 80) = 3.92644 \Leftrightarrow t - 80 = 228.1 \Leftrightarrow t = 308.1$. And if $\frac{2\pi}{3}(t - 80) = 5.49834 \Leftrightarrow t - 80 = 319.4 \Leftrightarrow t = 399.4$ $(399.4 - 365 = 34.4)$. So according to this model, there should be 10 hours of sunshine on the 34th day (February 3) and on the 308th day (November 4).

(b) Since $L(t) = 12 + 2.83 \sin\left(\frac{2\pi}{3}(t - 80)\right) \geq 10$ for $t \in [34, 308]$, the number of days with more than 10 hours of daylight is $308 - 34 + 1 = 275$ days.

69. $\sin(\cos x)$ is a function of a function, that is, a composition of trigonometric functions (see Section 2.6). Most of the other equations involve sums, products, differences, or quotients of trigonometric functions.
$\sin(\cos x) = 0 \Leftrightarrow \cos x = 0$ or $\cos x = \pi$. However, since $|\cos x| \leq 1$, the only solution is $\cos x = 0 \Rightarrow x = \frac{\pi}{2} + k\pi$. The graph of $f(x) = \sin(\cos x)$ is shown.

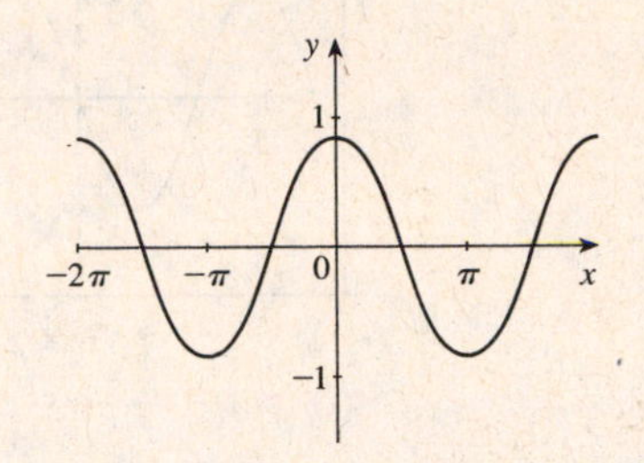

CHAPTER 7 REVIEW

1. $\sin\theta(\cot\theta + \tan\theta) = \sin\theta\left(\frac{\cos\theta}{\sin\theta} + \frac{\sin\theta}{\cos\theta}\right) = \cos\theta + \frac{\sin^2\theta}{\cos\theta} = \frac{\cos^2\theta + \sin^2\theta}{\cos\theta} = \frac{1}{\cos\theta} = \sec\theta$

3. $\cos^2 x \csc x - \csc x = \left(1 - \sin^2 x\right)\csc x - \csc x = \csc x - \sin^2 x \csc x - \csc x = -\sin^2 x \cdot \frac{1}{\sin x} = -\sin x$

5. $\frac{\cos^2 x - \tan^2 x}{\sin^2 x} = \frac{\cos^2 x}{\sin^2 x} - \frac{\tan^2 x}{\sin^2 x} = \cot^2 x - \frac{1}{\cos^2 x} = \cot^2 x - \sec^2 x$

7. $\frac{\cos^2 x}{1 - \sin x} = \frac{\cos x}{\frac{1}{\cos x}(1 - \sin x)} = \frac{\cos x}{\frac{1}{\cos x} - \frac{\sin x}{\cos x}} = \frac{\cos x}{\sec x - \tan x}$

9. $\sin^2 x \cot^2 x + \cos^2 x \tan^2 x = \sin^2 x \cdot \frac{\cos^2 x}{\sin^2 x} + \cos^2 x \cdot \frac{\sin^2 x}{\cos^2 x} = \cos^2 x + \sin^2 x = 1$

11. $\frac{\sin 2x}{1 + \cos 2x} = \frac{2\sin x \cos x}{1 + 2\cos^2 x - 1} = \frac{2\sin x \cos x}{2\cos^2 x} = \frac{2\sin x}{2\cos x} = \tan x$

13. $\csc x - \tan\frac{x}{2} = \csc x - \frac{1 - \cos x}{\sin x} = \csc x - (\csc x - \cot x) = \cot x$

15. $\frac{\sin 2x}{\sin x} - \frac{\cos 2x}{\cos x} = \frac{2\sin x \cos x}{\sin x} - \frac{2\cos^2 x - 1}{\cos x} = 2\cos x - 2\cos x + \frac{1}{\cos x} = \sec x$

17. $\frac{\sec x - 1}{\sin x \sec x} = \frac{\frac{1}{\cos x} - 1}{\sin x \frac{1}{\cos x}} = \frac{\frac{1}{\cos x} - 1}{\sin x \frac{1}{\cos x}} \cdot \frac{\cos x}{\cos x} = \frac{1 - \cos x}{\sin x} = \tan\frac{x}{2}$

19. $\left(\cos\frac{x}{2} - \sin\frac{x}{2}\right)^2 = \cos^2\frac{x}{2} - 2\sin\frac{x}{2}\cos\frac{x}{2} + \sin^2\frac{x}{2} = \sin^2\frac{x}{2} + \cos^2\frac{x}{2} - 2\sin\frac{x}{2}\cos\frac{x}{2} = 1 - \sin\left(2 \cdot \frac{x}{2}\right) = 1 - \sin x$

21. $\frac{\sin(x + y) + \sin(x - y)}{\cos(x + y) + \cos(x - y)} = \frac{2\sin\left(\frac{(x + y) + (x - y)}{2}\right)\cos\left(\frac{(x + y) - (x - y)}{2}\right)}{2\cos\left(\frac{(x + y) + (x - y)}{2}\right)\cos\left(\frac{(x + y) - (x - y)}{2}\right)} = \frac{2\sin x \cos y}{2\cos x \cos y} = \frac{\sin x}{\cos x} = \tan x$

23. **(a)** 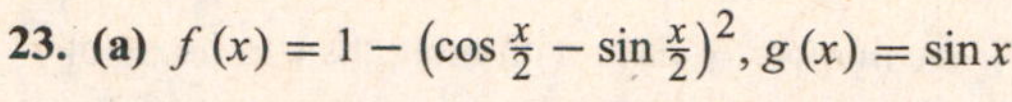
$f(x) = 1 - \left(\cos\frac{x}{2} - \sin\frac{x}{2}\right)^2, g(x) = \sin x$

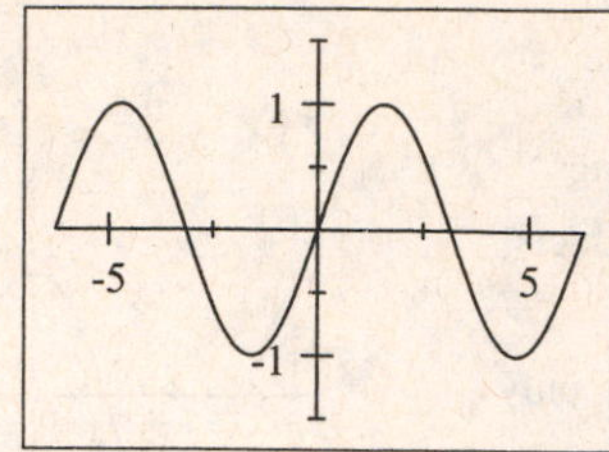

(b) The graphs suggest that $f(x) = g(x)$ is an identity. To prove this, expand $f(x)$ and simplify, using the double-angle formula for sine:

$$\begin{aligned} f(x) &= 1 - \left(\cos\frac{x}{2} - \sin\frac{x}{2}\right)^2 \\ &= 1 - \left(\cos^2\frac{x}{2} - 2\cos\frac{x}{2}\sin\frac{x}{2} + \sin^2\frac{x}{2}\right) \\ &= 1 + 2\cos\frac{x}{2}\sin\frac{x}{2} - \left(\cos^2\frac{x}{2} + \sin^2\frac{x}{2}\right) \\ &= 1 + \sin x - (1) = \sin x = g(x) \end{aligned}$$

25. **(a)** $f(x) = \tan x \tan\frac{x}{2}, g(x) = \dfrac{1}{\cos x}$

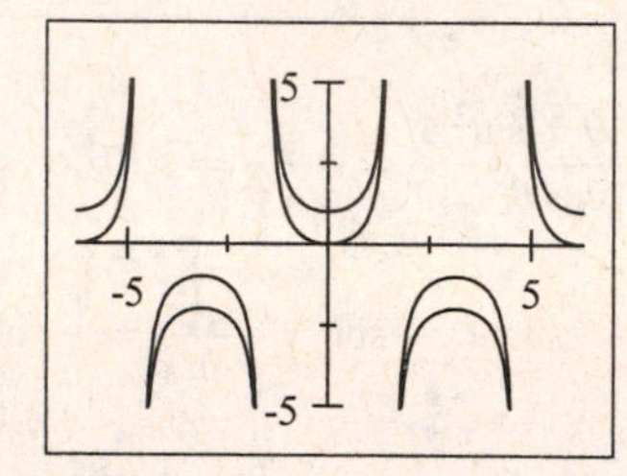

(b) The graphs suggest that $f(x) \neq g(x)$ in general. For example, choose $x = \frac{\pi}{3}$ and evaluate: $f\left(\frac{\pi}{3}\right) = \tan\frac{\pi}{3}\tan\frac{\pi}{6} = \sqrt{3} \cdot \frac{1}{\sqrt{3}} = 1$, whereas $g\left(\frac{\pi}{3}\right) = \dfrac{1}{\frac{1}{2}} = 2$, so $f(x) \neq g(x)$.

27. **(a)** $f(x) = 2\sin^2 3x + \cos 6x$

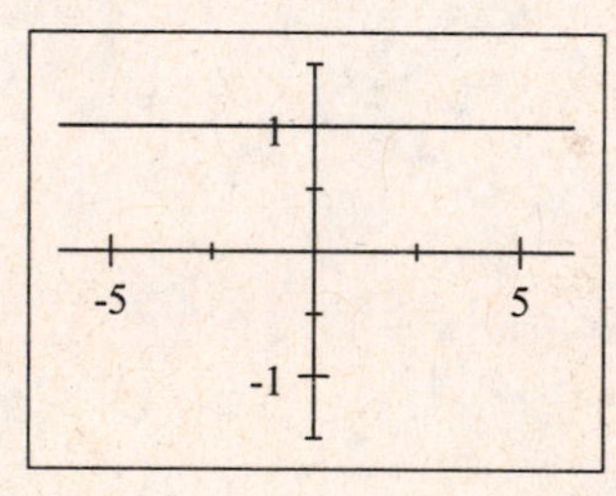

(b) The graph suggests that $f(x) = 1$ for all x. To prove this, we use the double angle formula to note that $\cos 6x = \cos(2(3x)) = 1 - 2\sin^2 3x$, so $f(x) = 2\sin^2 3x + \left(1 - 2\sin^2 3x\right) = 1$.

29. $4\sin\theta - 3 = 0 \Leftrightarrow 4\sin\theta = 3 \Leftrightarrow \sin\theta = \frac{3}{4} \Leftrightarrow \theta = \sin^{-1}\frac{3}{4} \approx 0.8481$ or $\theta = \pi - \sin^{-1}\frac{3}{4} \approx 2.2935$.

31. $\cos x \sin x - \sin x = 0 \Leftrightarrow \sin x(\cos x - 1) = 0 \Leftrightarrow \sin x = 0$ or $\cos x = 1 \Leftrightarrow x = 0, \pi$ or $x = 0$. Therefore, the solutions are $x = 0$ and π.

33. $2\sin^2 x - 5\sin x + 2 = 0 \Leftrightarrow (2\sin x - 1)(\sin x - 2) = 0 \Leftrightarrow \sin x = \frac{1}{2}$ or $\sin x = 2$ (which is inadmissible) $\Leftrightarrow x = \frac{\pi}{6}, \frac{5\pi}{6}$. Thus, the solutions in $[0, 2\pi)$ are $x = \frac{\pi}{6}$ and $\frac{5\pi}{6}$.

35. $2\cos^2 x - 7\cos x + 3 = 0 \Leftrightarrow (2\cos x - 1)(\cos x - 3) = 0 \Leftrightarrow \cos x = \frac{1}{2}$ or $\cos x = 3$ (which is inadmissible) $\Leftrightarrow x = \frac{\pi}{3}, \frac{5\pi}{3}$. Therefore, the solutions in $[0, 2\pi)$ are $x = \frac{\pi}{3}, \frac{5\pi}{3}$.

37. Note that $x = \pi$ is not a solution because the denominator is zero. $\dfrac{1 - \cos x}{1 + \cos x} = 3 \Leftrightarrow 1 - \cos x = 3 + 3\cos x \Leftrightarrow -4\cos x = 2 \Leftrightarrow \cos x = -\frac{1}{2} \Leftrightarrow x = \frac{2\pi}{3}, \frac{4\pi}{3}$ in $[0, 2\pi)$.

39. Factor by grouping: $\tan^3 x + \tan^2 x - 3\tan x - 3 = 0 \Leftrightarrow (\tan x + 1)\left(\tan^2 x - 3\right) = 0 \Leftrightarrow \tan x = -1$ or $\tan x = \pm\sqrt{3} \Leftrightarrow x = \frac{3\pi}{4}, \frac{7\pi}{4}$ or $x = \frac{\pi}{3}, \frac{2\pi}{3}, \frac{4\pi}{3}, \frac{5\pi}{3}$. Therefore, the solutions in $[0, 2\pi)$ are $x = \frac{\pi}{3}, \frac{2\pi}{3}, \frac{3\pi}{4}, \frac{4\pi}{3}, \frac{5\pi}{3}, \frac{7\pi}{4}$.

41. $\tan\frac{1}{2}x+2\sin 2x=\csc x \Leftrightarrow \dfrac{1-\cos x}{\sin x}+4\sin x\cos x=\dfrac{1}{\sin x} \Leftrightarrow 1-\cos x+4\sin^2 x\cos x=1 \Leftrightarrow 4\sin^2 x\cos x-\cos x=0$ $\Leftrightarrow \cos x\left(4\sin^2 x-1\right)=0 \Leftrightarrow \cos x=0$ or $\sin x=\pm\frac{1}{2} \Leftrightarrow x=\frac{\pi}{2},\frac{3\pi}{2}$ or $x=\frac{\pi}{6},\frac{5\pi}{6},\frac{7\pi}{6},\frac{11\pi}{6}$. Thus, the solutions in $[0,2\pi)$ are $x=\frac{\pi}{6},\frac{\pi}{2},\frac{5\pi}{6},\frac{7\pi}{6},\frac{3\pi}{2},\frac{11\pi}{6}$.

43. $\tan x+\sec x=\sqrt{3} \Leftrightarrow \dfrac{\sin x}{\cos x}+\dfrac{1}{\cos x}=\sqrt{3} \Leftrightarrow \sin x+1=\sqrt{3}\cos x \Leftrightarrow \sqrt{3}\cos x-\sin x=1 \Leftrightarrow \frac{\sqrt{3}}{2}\cos x-\frac{1}{2}\sin x=\frac{1}{2}$ $\Leftrightarrow \cos\frac{\pi}{6}\cos x-\sin\frac{\pi}{6}\sin x=\frac{1}{2} \Leftrightarrow \cos\left(x+\frac{\pi}{6}\right)=\frac{1}{2} \Leftrightarrow x+\frac{\pi}{6}=\frac{\pi}{3},\frac{5\pi}{3} \Leftrightarrow x=\frac{\pi}{6},\frac{3\pi}{2}$. However, $x=\frac{3\pi}{2}$ is inadmissible because $\sec\frac{3\pi}{2}$ is undefined. Thus, the only solution in $[0,2\pi)$ is $x=\frac{\pi}{6}$.

45. We graph $f(x)=\cos x$ and $g(x)=x^2-1$ in the viewing rectangle $[0,6.5]$ by $[-2,2]$. The two functions intersect at only one point, $x\approx 1.18$.

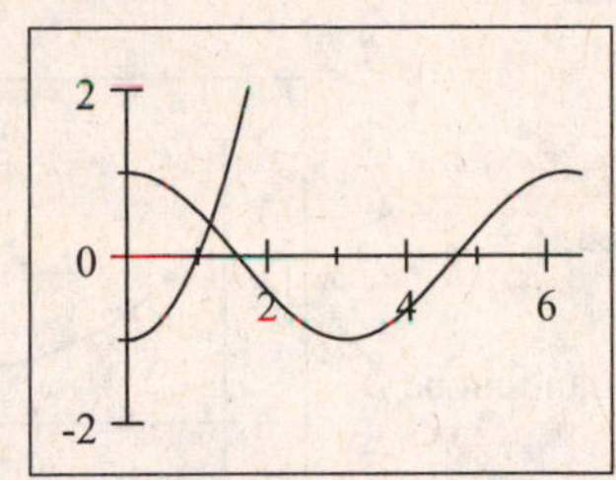

47. **(a)** $2000=\dfrac{(400)^2\sin^2\theta}{64} \Leftrightarrow \sin^2\theta=0.8 \Leftrightarrow \sin\theta\approx 0.8944 \Leftrightarrow \theta\approx 63.4^\circ$

(b) $\dfrac{(400)^2\sin^2\theta}{64}=2500\sin^2\theta\le 2500$. Therefore it is impossible for the projectile to reach a height of 3000 ft.

(c) The function $M(\theta)=2500\sin^2\theta$ is maximized when $\sin^2\theta=1$, so $\theta=90^\circ$. The projectile will travel the highest when it is shot straight up.

49. Since 15° is in quadrant I, $\cos 15^\circ=\sqrt{\dfrac{1+\cos 30^\circ}{2}}=\sqrt{\dfrac{2+\sqrt{3}}{4}}=\dfrac{\sqrt{2+\sqrt{3}}}{2}$.

Another method: $\cos 15^\circ=\cos(45^\circ-30^\circ)=\cos 45^\circ\cos 30^\circ+\sin 45^\circ\sin 30^\circ=\frac{\sqrt{2}}{2}\frac{\sqrt{3}}{2}+\frac{\sqrt{2}}{2}\frac{1}{2}=\frac{\sqrt{6}+\sqrt{2}}{4}$, which is equal to $\frac{1}{2}\sqrt{2+\sqrt{3}}$.

51. $\tan\frac{\pi}{8}=\dfrac{1-\cos\frac{\pi}{4}}{\sin\frac{\pi}{4}}=\dfrac{1-\frac{1}{\sqrt{2}}}{\frac{1}{\sqrt{2}}}=\left(1-\frac{1}{\sqrt{2}}\right)\sqrt{2}=\sqrt{2}-1$

53. $\sin 5^\circ\cos 40^\circ+\cos 5^\circ\sin 40^\circ=\sin(5^\circ+40^\circ)=\sin 45^\circ=\frac{1}{\sqrt{2}}=\frac{\sqrt{2}}{2}$

55. $\cos^2\frac{\pi}{8}-\sin^2\frac{\pi}{8}=\cos\left(2\left(\frac{\pi}{8}\right)\right)=\cos\frac{\pi}{4}=\frac{1}{\sqrt{2}}=\frac{\sqrt{2}}{2}$

57. We use a product-to-sum formula: $\cos 37.5^\circ\cos 7.5^\circ=\frac{1}{2}(\cos 45^\circ+\cos 30^\circ)=\frac{1}{2}\left(\frac{\sqrt{2}}{2}+\frac{\sqrt{3}}{2}\right)=\frac{1}{4}\left(\sqrt{2}+\sqrt{3}\right)$.

In Exercises 59–63, x and y are in quadrant I, so we know that $\sec x=\frac{3}{2}\Rightarrow\cos x=\frac{2}{3}$, so $\sin x=\frac{\sqrt{5}}{3}$ and $\tan x=\frac{\sqrt{5}}{2}$. Also, $\csc y=3\Rightarrow\sin y=\frac{1}{3}$, and so $\cos y=\frac{2\sqrt{2}}{3}$, and $\tan y=\frac{1}{2\sqrt{2}}=\frac{\sqrt{2}}{4}$.

59. $\sin(x+y)=\sin x\cos y+\cos x\sin y=\frac{\sqrt{5}}{3}\cdot\frac{2\sqrt{2}}{3}+\frac{2}{3}\cdot\frac{1}{3}=\frac{2}{9}\left(1+\sqrt{10}\right)$.

61. $\tan(x+y)=\dfrac{\tan x+\tan y}{1-\tan x\tan y}=\dfrac{\frac{\sqrt{5}}{2}+\frac{\sqrt{2}}{4}}{1-\left(\frac{\sqrt{5}}{2}\right)\left(\frac{\sqrt{2}}{4}\right)}=\dfrac{\frac{\sqrt{5}}{2}+\frac{\sqrt{2}}{4}}{1-\left(\frac{\sqrt{5}}{2}\right)\left(\frac{\sqrt{2}}{4}\right)}\cdot\dfrac{8}{8}=\dfrac{2\left(2\sqrt{5}+\sqrt{2}\right)}{8-\sqrt{10}}\cdot\dfrac{8+\sqrt{10}}{8+\sqrt{10}}=\frac{2}{3}\left(\sqrt{2}+\sqrt{5}\right)$

63. $\cos\frac{y}{2} = \sqrt{\frac{1+\cos y}{2}} = \sqrt{\frac{1+\left(\frac{2\sqrt{2}}{3}\right)}{2}} = \sqrt{\frac{3+2\sqrt{2}}{6}}$ (since cosine is positive in quadrant I)

65. We sketch a triangle such that $\theta = \cos^{-1}\frac{3}{7}$. We see that $\tan\theta = \frac{2\sqrt{10}}{3}$, and the double-angle formula for tangent gives

$$\tan 2\theta = \frac{2\tan\theta}{1-\tan^2\theta} = \frac{2\cdot\frac{2\sqrt{10}}{3}}{1-\left(\frac{2\sqrt{10}}{3}\right)^2} = \frac{\frac{4\sqrt{10}}{3}}{1-\frac{40}{9}} = -\frac{12\sqrt{10}}{31}.$$

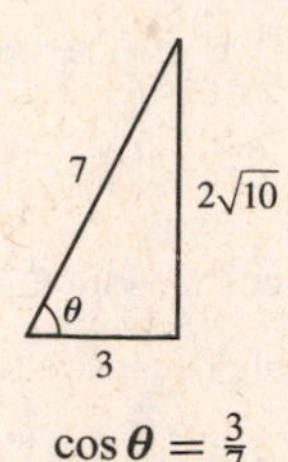

$\cos\theta = \frac{3}{7}$

67. The double-angle formula for tangent gives $\tan\left(2\tan^{-1}x\right) = \frac{2\tan\left(\tan^{-1}x\right)}{1-\tan^2\left(\tan^{-1}x\right)} = \frac{2x}{1-x^2}$.

69. **(a)** $\tan\theta = \frac{10}{x} \Leftrightarrow \theta = \tan^{-1}\left(\frac{10}{x}\right)$

(b) $\theta = \tan^{-1}\left(\frac{10}{x}\right)$, for $x > 0$. Since the road sign can first be seen when $\theta = 2°$, we have $2° = \tan^{-1}\left(\frac{10}{x}\right) \Leftrightarrow x = \frac{10}{\tan 2°} \approx 286.4$ ft. Thus, the sign can first be seen at a height of 286.4 ft.

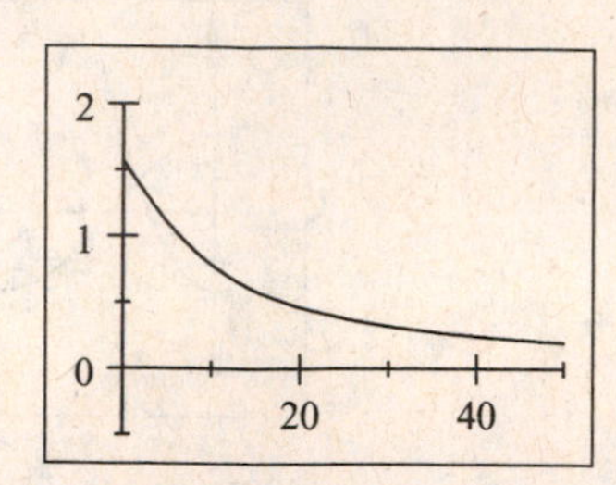

CHAPTER 7 TEST

1. $\tan\theta\sin\theta + \cos\theta = \frac{\sin\theta}{\cos\theta}\sin\theta + \cos\theta = \frac{\sin^2\theta}{\cos\theta} + \frac{\cos^2\theta}{\cos\theta} = \frac{1}{\cos\theta} = \sec\theta$

3. $\frac{2\tan x}{1+\tan^2 x} = \frac{2\tan x}{\sec^2 x} = \frac{2\sin x}{\cos x}\cdot\cos^2 x = 2\sin x\cos x = \sin 2x$

5. $2\sin^2(3x) = 1-\cos(2(3x)) = 1-\cos(6x)$

7. $\left(\sin\left(\frac{x}{2}\right)+\cos\left(\frac{x}{2}\right)\right)^2 = \left(\sqrt{\frac{1-\cos x}{2}}+\sqrt{\frac{1+\cos x}{2}}\right)^2 = \frac{1-\cos x}{2}+\frac{1+\cos x}{2}+2\sqrt{\frac{1-\cos^2 x}{4}} = 1+\sin x$

Another method: $\left(\sin\left(\frac{x}{2}\right)+\cos\left(\frac{x}{2}\right)\right)^2 = \sin^2\left(\frac{x}{2}\right)+2\sin\left(\frac{x}{2}\right)\cos\left(\frac{x}{2}\right)+\cos^2\left(\frac{x}{2}\right) = 1+2\sin\left(\frac{x}{2}\right) = 1+\sin x$

9. **(a)** $\sin 8°\cos 22° + \cos 8°\sin 22° = \sin(8°+22°) = \sin 30° = \frac{1}{2}$

(b) $\sin 75° = \sin(45°+30°) = \sin 45°\cos 30° + \cos 45°\sin 30° = \frac{\sqrt{2}}{2}\cdot\frac{\sqrt{3}}{2}+\frac{\sqrt{2}}{2}\cdot\frac{1}{2} = \frac{1}{4}\left(\sqrt{6}+\sqrt{2}\right)$

Another method: Since 75° is in quadrant I, $\sin 75° = \sqrt{\frac{1-\cos 150°}{2}} = \sqrt{\frac{1-\left(-\frac{\sqrt{3}}{2}\right)}{2}} = \sqrt{\frac{2+\sqrt{3}}{4}} = \frac{\sqrt{2+\sqrt{3}}}{2}$, which is equal to $\frac{1}{4}\left(\sqrt{6}+\sqrt{2}\right)$.

(c) $\sin\frac{\pi}{12} = \sqrt{\frac{1-\cos\frac{\pi}{6}}{2}} = \sqrt{\frac{1-\frac{\sqrt{3}}{2}}{2}} = \sqrt{\frac{2-\sqrt{3}}{4}} = \frac{1}{2}\sqrt{2-\sqrt{3}}$

Another method: Since $\frac{\pi}{12}$ is in quadrant I,

$\sin\frac{\pi}{12} = \sin\left(\frac{\pi}{3}-\frac{\pi}{4}\right) = \sin\frac{\pi}{3}\cos\frac{\pi}{4} - \cos\frac{\pi}{3}\sin\frac{\pi}{4} = \frac{\sqrt{3}}{2}\frac{\sqrt{2}}{2} - \frac{1}{2}\frac{\sqrt{2}}{2} = \frac{1}{4}\left(\sqrt{6}-\sqrt{2}\right)$, which is equal to $\frac{1}{2}\sqrt{2-\sqrt{3}}$.

11. $\sin 3x \cos 5x = \frac{1}{2}[\sin(3x+5x)+\sin(3x-5x)] = \frac{1}{2}(\sin 8x - \sin 2x)$

13. $\sin\theta = -\frac{4}{5}$. Since θ is in quadrant III, $\cos\theta = -\frac{3}{5}$. Then $\tan\frac{\theta}{2} = \frac{1-\cos\theta}{\sin\theta} = \frac{1-\left(-\frac{3}{5}\right)}{-\frac{4}{5}} = -\frac{1+\frac{3}{5}}{\frac{4}{5}} = -\frac{5+3}{4} = -2.$

15. $(2\cos\theta - 1)(\sin\theta - 1) = 0 \Leftrightarrow \cos\theta = \frac{1}{2}$ or $\sin\theta = 1$. The first equation has solutions $\theta = \frac{\pi}{3} \approx 1.05$ and $\theta = \frac{5\pi}{3} \approx 5.24$ on $[0, 2\pi)$, while the second has the solution $\theta = \frac{\pi}{2} \approx 1.57$.

17. $\sin 2\theta - \cos\theta = 0 \Leftrightarrow 2\sin\theta\cos\theta - \cos\theta = 0 \Leftrightarrow \cos\theta(2\sin\theta - 1) = 0 \Leftrightarrow \cos\theta = 0$ or $\sin\theta = \frac{1}{2} \Leftrightarrow \theta = \frac{\pi}{2}, \frac{3\pi}{2}$ or $\theta = \frac{\pi}{6}, \frac{5\pi}{6}$. Therefore, the solutions in $[0, 2\pi)$ are $\theta = \frac{\pi}{6} \approx 0.52$, $\frac{\pi}{2} \approx 1.57$, $\frac{5\pi}{6} \approx 2.62$, $\frac{3\pi}{2} \approx 4.71$.

19. $2\cos^2 x + \cos 2x = 0 \Leftrightarrow 2\cos^2 x + \left(2\cos^2 x - 1\right) = 0 \Leftrightarrow \cos x = \pm\frac{1}{2}$. The solutions in $[0, 4\pi)$ are $x = \frac{\pi}{3} \approx 1.05$, $x = \frac{2\pi}{3} \approx 2.09$, $x = \frac{4\pi}{3} \approx 4.19$, and $x = \frac{5\pi}{3} \approx 5.24$.

21. Let $u = \tan^{-1}\frac{9}{40}$ so $\tan u = \frac{9}{40}$. From the triangle, $\cos u = \frac{40}{41}$, so using a double-angle formula for cosine, $\cos 2u = 2\cos^2 u - 1 = 2\left(\frac{40}{41}\right)^2 - 1 = \frac{1519}{1681}$.

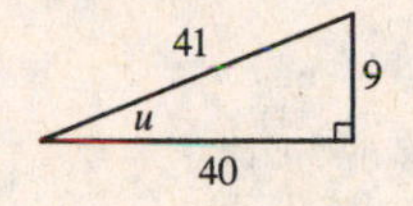

FOCUS ON MODELING Traveling and Standing Waves

1. (a) Substituting $x = 0$, we get $y(0,t) = 5\sin\left(2\cdot 0 - \frac{\pi}{2}t\right) = 5\sin\left(-\frac{\pi}{2}t\right) = -5\sin\frac{\pi}{2}t$.

(b)

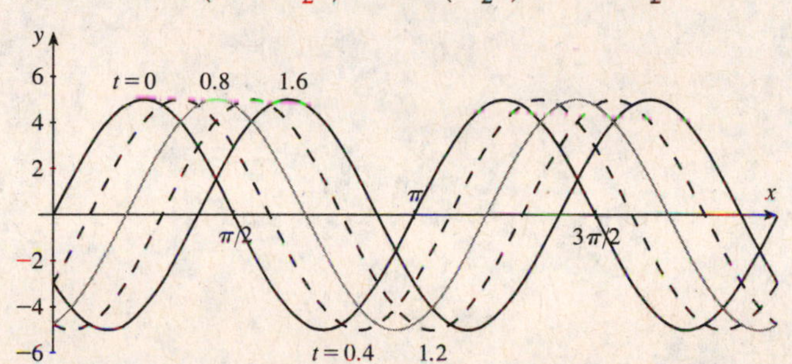

(c) We express the function in the standard form $y(x,t) = A\sin k(x - vt)$: $y(x,t) = 5\sin\left(2x - \frac{\pi}{2}t\right) = 5\sin 2\left(x - \frac{\pi}{4}t\right)$. Comparing this to the standard form, we see that the velocity of the wave is $v = \frac{\pi}{4}$.

3. From the graph, we see that the amplitude is $A = 2.7$ and the period is 9.2, so $k = \frac{2\pi}{9.2} \approx 0.68$. Since $v = 6$, we have $kv = \frac{2\pi}{9.2}\cdot 6 \approx 4.10$, so the equation we seek is $y(x,t) = 2.7\sin(0.68x - 4.10t)$.

5. From the graphs, we see that the amplitude is $A = 0.6$. The nodes occur at $x = 0, 1, 2, 3$. Since $\sin\alpha x = 0$ when $\alpha x = k\pi$ (k any integer), we have $\alpha = \pi$. Then since the frequency is $\beta/2\pi$, we get $20 = \beta/2\pi \Leftrightarrow \beta = 40\pi$. Thus, an equation for this model is $f(x,t) = 0.6\sin\pi x\cos 40\pi t$.

7. (a) The first standing wave has $\alpha = 1$, the second has $\alpha = 2$, the third has $\alpha = 3$, and the fourth has $\alpha = 4$.

(b) α is equal to the number of nodes minus 1. The first string has two nodes and $\alpha = 1$; the second string has three nodes and $\alpha = 2$, and so forth. Thus, the next two values of α would be $\alpha = 5$ and $\alpha = 6$, as sketched below.

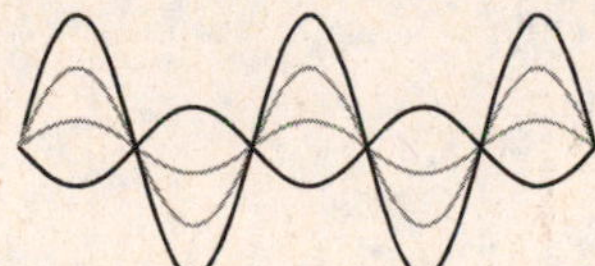

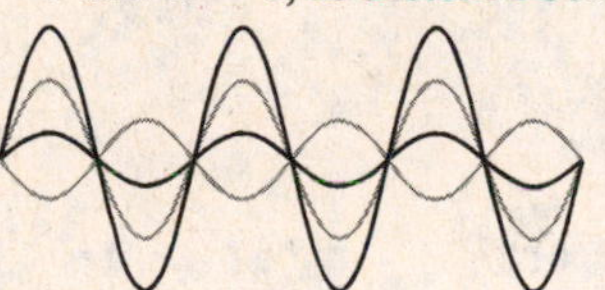

(c) Since the frequency is $\beta/2\pi$, we have $440 = \beta/2\pi \Leftrightarrow \beta = 880\pi$.

(d) The first standing wave has equation $y = \sin x \cos 880\pi t$, the second has equation $y = \sin 2x \cos 880\pi t$, the third has equation $y = \sin 3x \cos 880\pi t$, and the fourth has equation $y = \sin 4x \cos 880\pi t$.

8 POLAR COORDINATES AND PARAMETRIC EQUATIONS

8.1 POLAR COORDINATES

1. We can describe the location of a point in the plane using different *coordinate* systems. The point P shown in the figure has rectangular coordinates $(1, 1)$ and polar coordinates $\left(\sqrt{2}, \frac{\pi}{4}\right)$.

3. Yes; both $\left(2, \frac{\pi}{6}\right)$ and $\left(-2, \frac{7\pi}{6}\right)$ correspond to the point $\left(\sqrt{3}, 1\right)$ in Cartesian coordinates.

5.

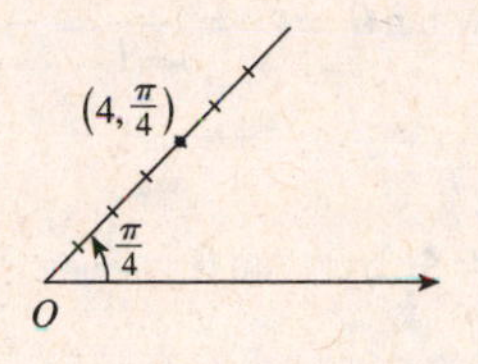

7.

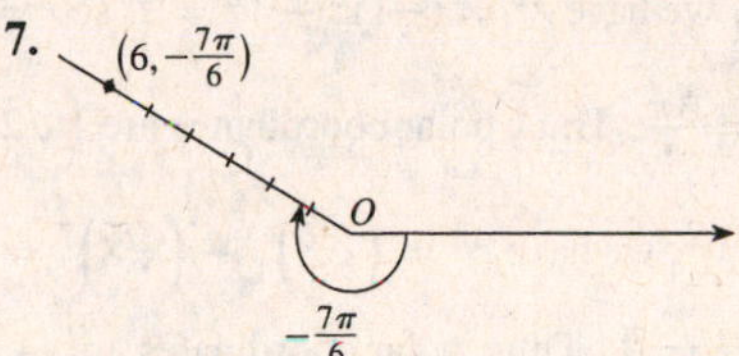

9.

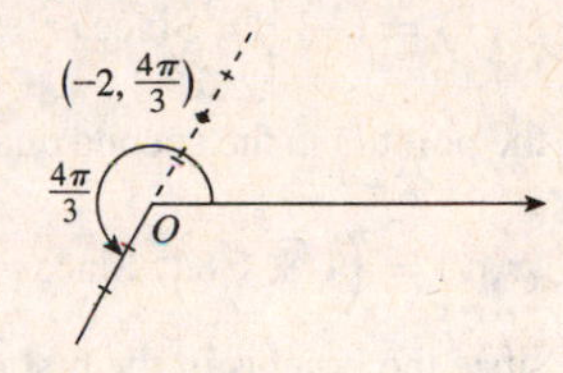

Answers to Exercises 11–16 will vary.

11. $\left(3, \frac{\pi}{2}\right)$ has polar coordinates $\left(3, \frac{5\pi}{2}\right)$ or $\left(-3, \frac{3\pi}{2}\right)$

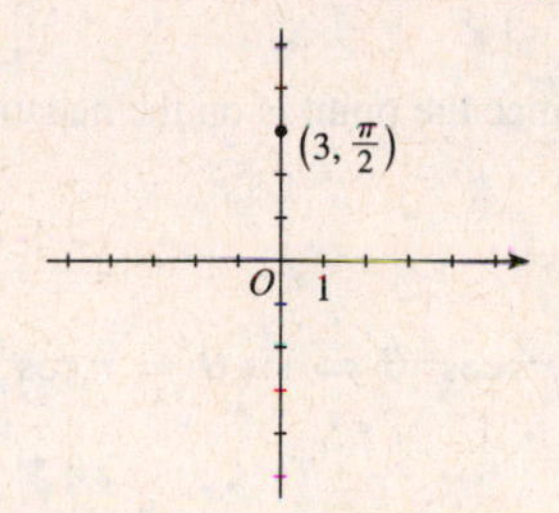

13. $\left(-1, \frac{7\pi}{6}\right)$ has polar coordinates $\left(1, \frac{\pi}{6}\right)$ or $\left(-1, -\frac{5\pi}{6}\right)$.

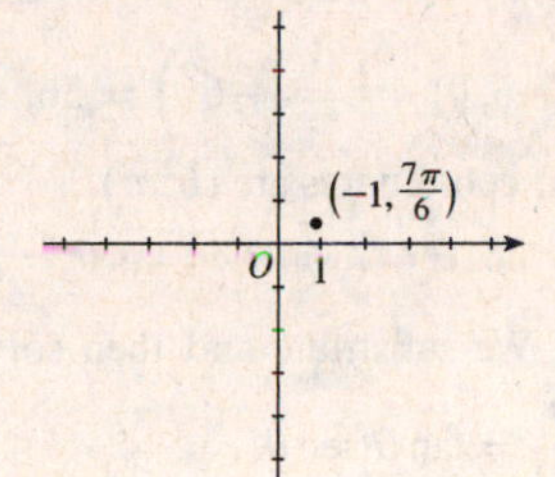

15. $(-5, 0)$ has polar coordinates $(5, \pi)$ or $(-5, 2\pi)$.

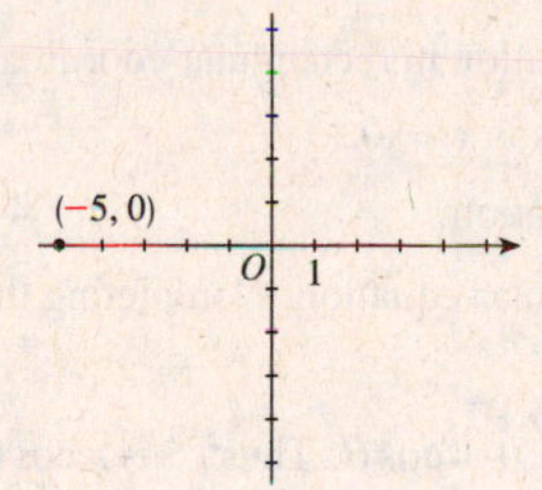

17. Q has coordinates $\left(4, \frac{3\pi}{4}\right)$.

19. Q has coordinates $\left(-4, -\frac{\pi}{4}\right) = \left(4, \frac{3\pi}{4}\right)$.

21. P has coordinates $\left(4, -\frac{23\pi}{4}\right) = \left(4, \frac{\pi}{4}\right)$.

23. P has coordinates $\left(-4, \frac{101\pi}{4}\right) = \left(-4, \frac{5\pi}{4}\right) = \left(4, \frac{\pi}{4}\right)$.

25. $P = (-3, 3)$ in rectangular coordinates, so $r^2 = x^2 + y^2 = (-3)^2 + 3^2 = 18$ and we can take $r = 3\sqrt{2}$. $\tan\theta = \frac{y}{x} = \frac{3}{-3} = -1$, so since P is in quadrant 2 we take $\theta = \frac{3\pi}{4}$. Thus, polar coordinates for P are $\left(3\sqrt{2}, \frac{3\pi}{4}\right)$.

27. Here $r = 5$ and $\theta = -\frac{2\pi}{3}$, so $x = r\cos\theta = 5\cos\left(-\frac{2\pi}{3}\right) = -\frac{5}{2}$ and $y = r\sin\theta = 5\sin\left(-\frac{2\pi}{3}\right) = -\frac{5\sqrt{3}}{2}$. R has rectangular coordinates $\left(-\frac{5}{2}, -\frac{5\sqrt{3}}{2}\right)$.

29. $(r,\theta)=\left(4,\frac{\pi}{6}\right)$. So $x=r\cos\theta=4\cos\frac{\pi}{6}=4\cdot\frac{\sqrt{3}}{2}=2\sqrt{3}$ and $y=r\sin\theta=4\sin\frac{\pi}{6}=4\cdot\frac{1}{2}=2$. Thus, the rectangular coordinates are $\left(2\sqrt{3},2\right)$.

31. $(r,\theta)=\left(\sqrt{2},-\frac{\pi}{4}\right)$. So $x=r\cos\theta=\sqrt{2}\cos\left(-\frac{\pi}{4}\right)=\sqrt{2}\cdot\frac{1}{\sqrt{2}}=1$, and $y=r\sin\theta=\sqrt{2}\sin\left(-\frac{\pi}{4}\right)=\sqrt{2}\left(-\frac{1}{\sqrt{2}}\right)=-1$. Thus, the rectangular coordinates are $(1,-1)$.

33. $(r,\theta)=(5,5\pi)$. So $x=r\cos\theta=5\cos 5\pi=-5$, and $y=r\sin\theta=5\sin 5\pi=0$. Thus, the rectangular coordinates are $(-5,0)$.

35. $(r,\theta)=\left(6\sqrt{2},\frac{11\pi}{6}\right)$. So $x=r\cos\theta=6\sqrt{2}\cos\frac{11\pi}{6}=3\sqrt{6}$ and $y=r\sin\theta=6\sqrt{2}\sin\frac{11\pi}{6}=-3\sqrt{2}$. Thus, the rectangular coordinates are $\left(3\sqrt{6},-3\sqrt{2}\right)$.

37. $(x,y)=(-1,1)$. Since $r^2=x^2+y^2$, we have $r^2=(-1)^2+1^2=2$, so $r=\sqrt{2}$. Now $\tan\theta=\frac{y}{x}=\frac{1}{-1}=-1$, so, since the point is in the second quadrant, $\theta=\frac{3\pi}{4}$. Thus, polar coordinates are $\left(\sqrt{2},\frac{3\pi}{4}\right)$.

39. $(x,y)=\left(\sqrt{8},\sqrt{8}\right)$. Since $r^2=x^2+y^2$, we have $r^2=\left(\sqrt{8}\right)^2+\left(\sqrt{8}\right)^2=16$, so $r=4$. Now $\tan\theta=\frac{y}{x}=\frac{\sqrt{8}}{\sqrt{8}}=1$, so, since the point is in the first quadrant, $\theta=\frac{\pi}{4}$. Thus, polar coordinates are $\left(4,\frac{\pi}{4}\right)$.

41. $(x,y)=(3,4)$. Since $r^2=x^2+y^2$, we have $r^2=3^2+4^2=25$, so $r=5$. Now $\tan\theta=\frac{y}{x}=\frac{4}{3}$, so, since the point is in the first quadrant, $\theta=\tan^{-1}\frac{4}{3}$. Thus, polar coordinates are $\left(5,\tan^{-1}\frac{4}{3}\right)$.

43. $(x,y)=(-6,0)$. $r^2=\left(-6^2\right)=36$, so $r=6$. Now $\tan\theta=\frac{y}{x}=0$, so since the point is on the negative x-axis, $\theta=\pi$. Thus, polar coordinates are $(6,\pi)$.

45. $x=y\Leftrightarrow r\cos\theta=r\sin\theta\Leftrightarrow\tan\theta=1$, and so $\theta=\frac{\pi}{4}$.

47. $y=x^2$. We substitute and then solve for r: $r\sin\theta=(r\cos\theta)^2=r^2\cos^2\theta\Leftrightarrow\sin\theta=r\cos^2\theta\Leftrightarrow$ $r=\dfrac{\sin\theta}{\cos^2\theta}=\tan\theta\sec\theta$.

49. $x=4$. We substitute and then solve for r: $r\cos\theta=4\Leftrightarrow r=\dfrac{4}{\cos\theta}=4\sec\theta$.

51. $r=7$. But $r^2=x^2+y^2$, so $x^2+y^2=r^2=49$. Hence, the equivalent equation in rectangular coordinates is $x^2+y^2=49$.

53. $\theta=-\frac{\pi}{2}\Rightarrow\cos\theta=0$, so an equivalent equation in rectangular coordinates is $x=0$.

55. $r\cos\theta=6$. But $x=r\cos\theta$, and so $x=6$ is an equivalent rectangular equation.

57. $r=4\sin\theta\Leftrightarrow r^2=4r\sin\theta$. Thus, $x^2+y^2=4y$ is an equivalent rectangular equation. Completing the square, it can be written as $x^2+(y-2)^2=4$.

59. $r=1+\cos\theta$. If we multiply both sides of this equation by r we get $r^2=r+r\cos\theta$. Thus $r^2-r\cos\theta=r$, and squaring both sides gives $\left(r^2-r\cos\theta\right)^2=r^2$, or $\left(x^2+y^2-x\right)^2=x^2+y^2$ in rectangular coordinates.

61. $r=1+2\sin\theta$. If we multiply both sides of this equation by r we get $r^2=r+2r\sin\theta$. Thus $r^2-2r\sin\theta=r$, and squaring both sides gives $\left(r^2-2r\sin\theta\right)^2=r^2$, or $\left(x^2+y^2-2y\right)^2=x^2+y^2$ in rectangular coordinates.

63. $r=\dfrac{1}{\sin\theta-\cos\theta}\Rightarrow r(\sin\theta-\cos\theta)=1\Leftrightarrow r\sin\theta-r\cos\theta=1$, and since $r\cos\theta=x$ and $r\sin\theta=y$, we get $y-x=1$.

65. $r=\dfrac{4}{1+2\sin\theta}\Leftrightarrow r(1+2\sin\theta)=4\Leftrightarrow r+2r\sin\theta=4$. Thus $r=4-2r\sin\theta$. Squaring both sides, we get $r^2=(4-2r\sin\theta)^2$. Substituting, $x^2+y^2=(4-2y)^2\Leftrightarrow x^2+y^2=16-16y+4y^2\Leftrightarrow x^2-3y^2+16y-16=0$.

67. $r^2 = \tan\theta$. Substituting $r^2 = x^2 + y^2$ and $\tan\theta = \dfrac{y}{x}$, we get $x^2 + y^2 = \dfrac{y}{x}$.

69. $\sec\theta = 2 \Leftrightarrow \cos\theta = \frac{1}{2} \Leftrightarrow \theta = \pm\frac{\pi}{3} \Leftrightarrow \tan\theta = \pm\sqrt{3} \Leftrightarrow \dfrac{y}{x} = \pm\sqrt{3} \Leftrightarrow y = \pm\sqrt{3}x \Leftrightarrow y^2 = 3x^2 \Leftrightarrow y^2 - 3x^2 = 0$.

71. **(a)** In rectangular coordinates, the points (r_1, θ_1) and (r_2, θ_2) are $(x_1, y_1) = (r_1\cos\theta_1, r_1\sin\theta_1)$ and $(x_2, y_2) = (r_2\cos\theta_2, r_2\sin\theta_2)$. Then, the distance between the points is

$$\begin{aligned} D &= \sqrt{(x_1 - x_2)^2 + (y_1 - y_2)^2} = \sqrt{(r_1\cos\theta_1 - r_2\cos\theta_2)^2 + (r_1\sin\theta_1 - r_2\sin\theta_2)^2} \\ &= \sqrt{r_1^2\left(\cos^2\theta_1 + \sin^2\theta_1\right) + r_2^2\left(\cos^2\theta_2 + \sin^2\theta_2\right) - 2r_1r_2\left(\cos\theta_1\cos\theta_2 + \sin\theta_1\sin\theta_2\right)} \\ &= \sqrt{r_1^2 + r_2^2 - 2r_1r_2\cos(\theta_2 - \theta_1)} \end{aligned}$$

(b) The distance between the points $\left(3, \frac{3\pi}{4}\right)$ and $\left(-1, \frac{7\pi}{6}\right)$ is

$$D = \sqrt{3^2 + (-1)^2 - 2(3)(-1)\cos\left(\tfrac{7\pi}{6} - \tfrac{3\pi}{4}\right)} = \sqrt{9 + 1 + 6\cos\tfrac{5\pi}{12}} \approx 3.40$$

8.2 GRAPHS OF POLAR EQUATIONS

1. To plot points in polar coordinates we use a grid consisting of *circles* centered at the pole and *rays* emanating from the pole.

3. VI

5. II

7. I

9. Polar axis: $2 - \sin(-\theta) = 2 + \sin\theta \neq r$, so the graph is not symmetric about the polar axis.

Pole: $2 - \sin(\theta + \pi) = 2 - (\sin\pi\cos\theta + \cos\pi\sin\theta) = 2 - (-\sin\theta) = 2 + \sin\theta \neq r$, so the graph is not symmetric about the pole.

Line $\theta = \frac{\pi}{2}$: $2 - \sin(\pi - \theta) = 2 - (\sin\pi\cos\theta - \cos\pi\sin\theta) = 2 - \sin\theta = r$, so the graph is symmetric about $\theta = \frac{\pi}{2}$.

11. Polar axis: $3\sec(-\theta) = 3\sec\theta = r$, so the graph is symmetric about the polar axis.

Pole: $3\sec(\theta + \pi) = \dfrac{3}{\cos(\theta + \pi)} = \dfrac{1}{\cos\pi\cos\theta - \sin\pi\sin\theta} = \dfrac{3}{-\cos\theta} = -3\sec\theta \neq r$, so the graph is not symmetric about the pole.

Line $\theta = \frac{\pi}{2}$: $3\sec(\pi - \theta) = \dfrac{3}{\cos(\pi - \theta)} = \dfrac{1}{\cos\pi\cos\theta + \sin\pi\sin\theta} = \dfrac{3}{-\cos\theta} = -3\sec\theta \neq r$, so the graph is not symmetric about $\theta = \frac{\pi}{2}$.

13. Polar axis: $\dfrac{4}{3 - 2\sin(-\theta)} = \dfrac{4}{3 + 2\sin\theta} \neq r$, so the graph is not symmetric about the polar axis.

Pole: $\dfrac{4}{3 - 2\sin(\theta + \pi)} = \dfrac{4}{3 - 2(\sin\pi\cos\theta + \cos\pi\sin\theta)} = \dfrac{4}{3 - 2(-\sin\theta)} = \dfrac{4}{3 + 2\sin\theta} \neq r$, so the graph is not symmetric about the pole.

Line $\theta = \frac{\pi}{2}$: $\dfrac{4}{3 - 2\sin(\pi - \theta)} = \dfrac{4}{3 - 2(\sin\pi\cos\theta - \cos\pi\sin\theta)} = \dfrac{4}{3 - 2\sin\theta} = r$, so the graph is symmetric about $\theta = \frac{\pi}{2}$.

15. Polar axis: $4\cos 2(-\theta) = 4\cos 2\theta = r^2$, so the graph is symmetric about the polar axis.

Pole: $(-r)^2 = r^2$, so the graph is symmetric about the pole.

Line $\theta = \frac{\pi}{2}$: $4\cos 2(\pi - \theta) = 4\cos(2\pi - 2\theta) = 4\cos(-2\theta) = 4\cos 2\theta = r^2$, so the graph is symmetric about $\theta = \frac{\pi}{2}$.

17. $r = 2 \Rightarrow r^2 = 4 \Rightarrow x^2 + y^2 = 4$ is an equation of a circle with radius 2 centered at the origin.

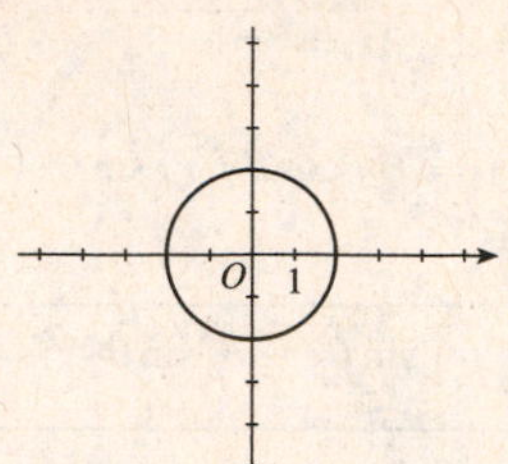

19. $\theta = -\frac{\pi}{2} \Rightarrow \cos\theta = 0 \Rightarrow x = 0$ is an equation of a vertical line.

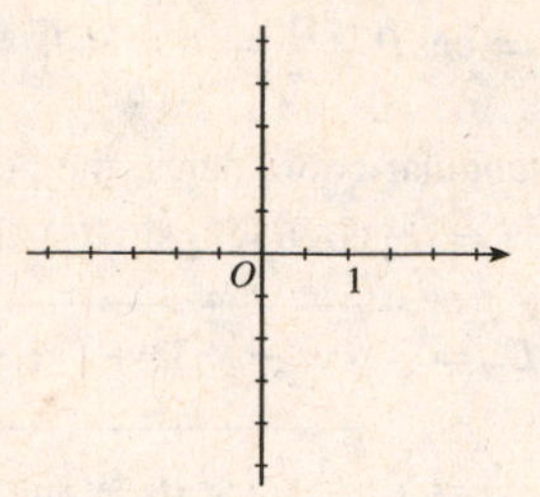

21. $r = 6\sin\theta \Rightarrow r^2 = 6r\sin\theta \Rightarrow x^2 + y^2 = 6y \Rightarrow$ $x^2 + (y-3)^2 = 9$, a circle of radius 3 centered at $(0, 3)$.

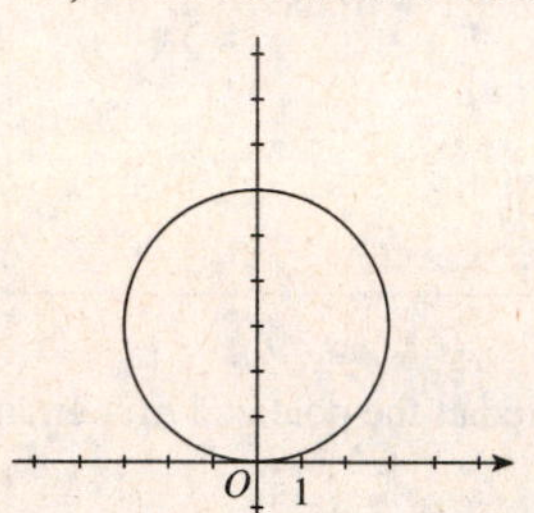

23. $r = -2\cos\theta$. Circle.

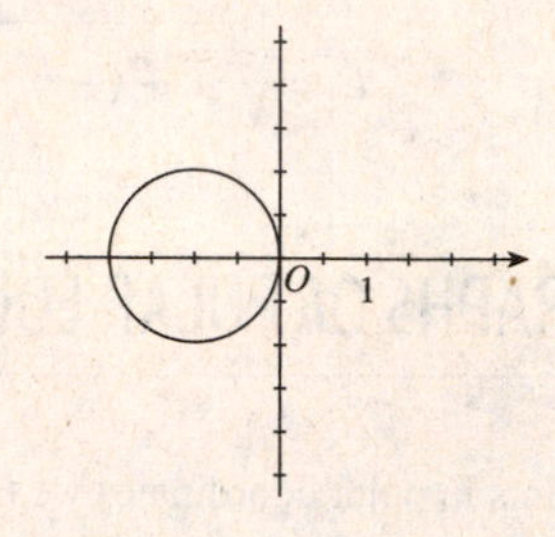

25. $r = 2 - 2\cos\theta$. Cardioid.

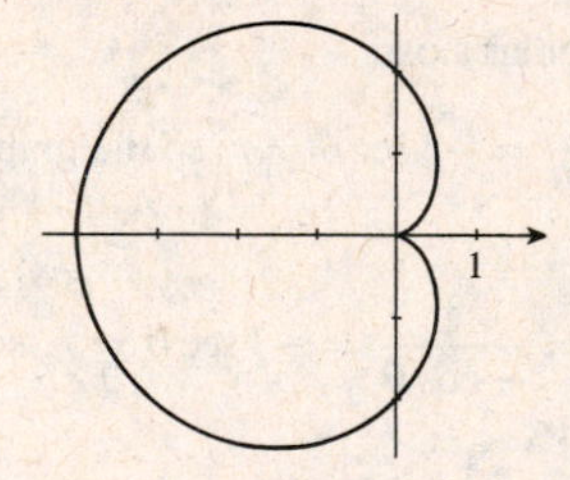

27. $r = -3(1 + \sin\theta)$. Cardioid.

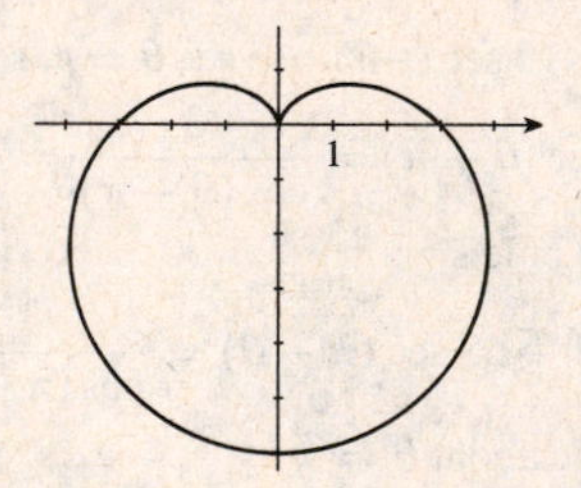

29. $r = \sin 2\theta$

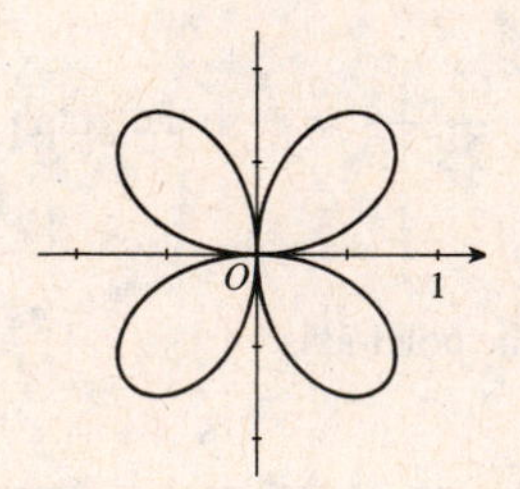

31. $r = -\cos 5\theta$

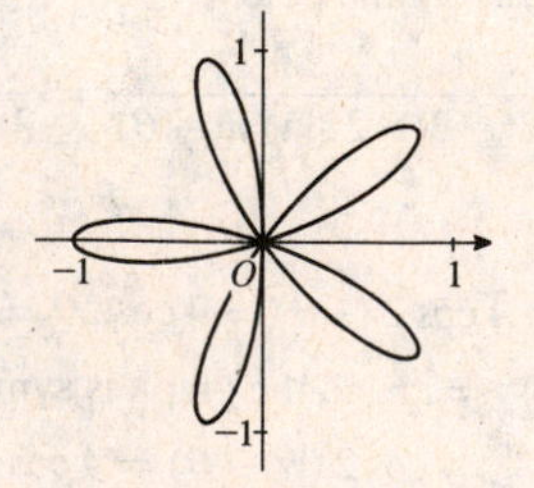

33. $r = 2\sin 5\theta$

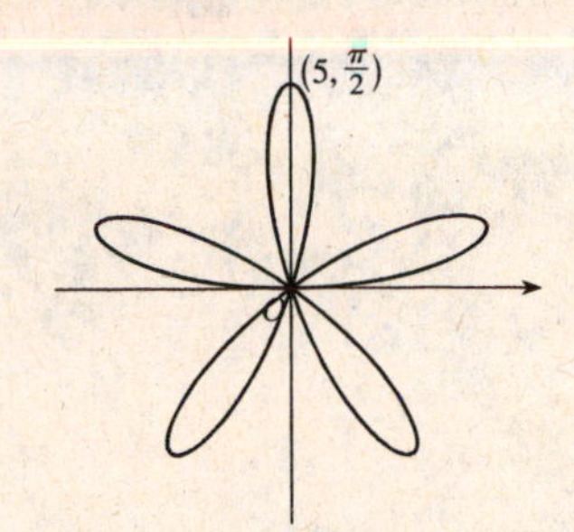

35. $r = \sqrt{3} - 2\sin\theta$

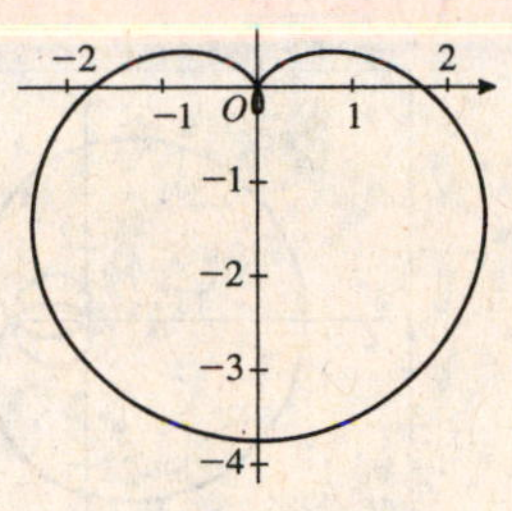

37. $r = \sqrt{3} + \cos\theta$

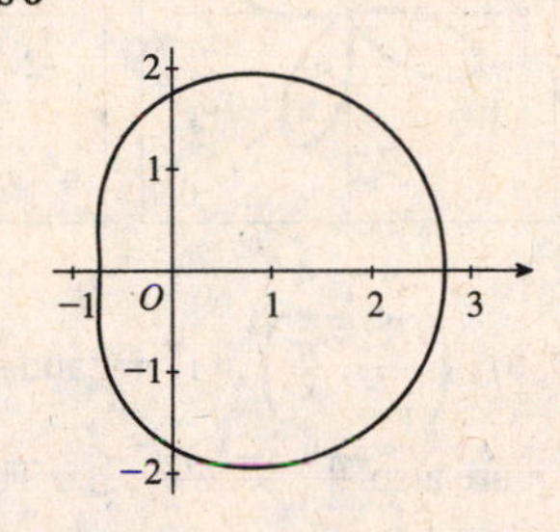

39. $r = 2 - 2\sqrt{2}\cos\theta$

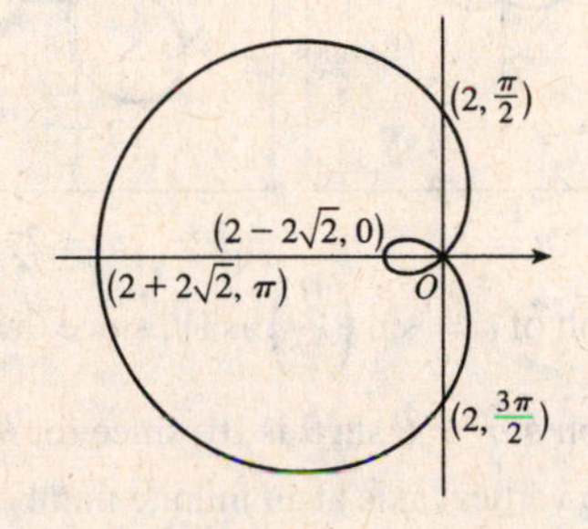

41. $r^2 = \cos 2\theta$

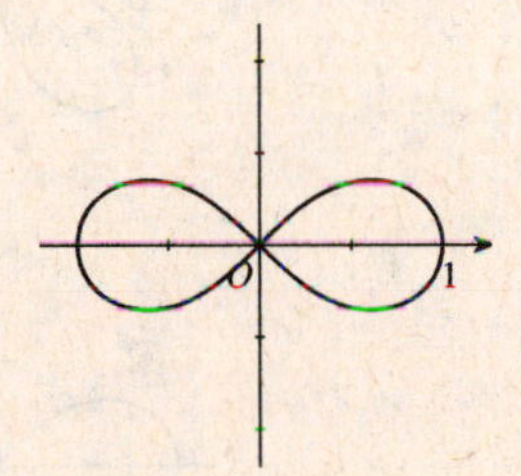

43. $r = \theta, \theta \geq 0$

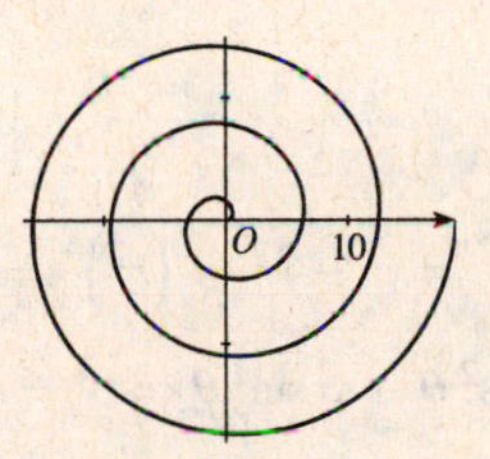

45. $r = 2 + \sec\theta$

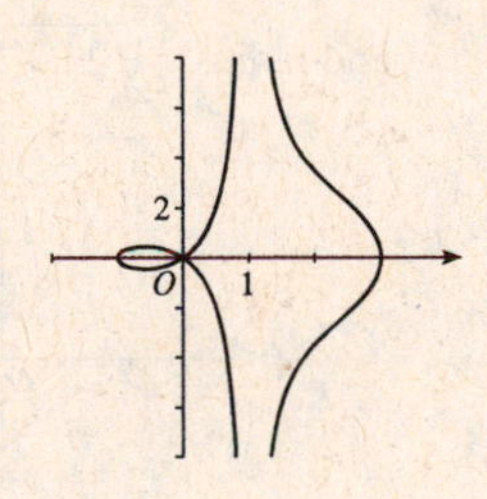

47. $r = \cos\left(\frac{\theta}{2}\right), \theta \in [0, 4\pi]$

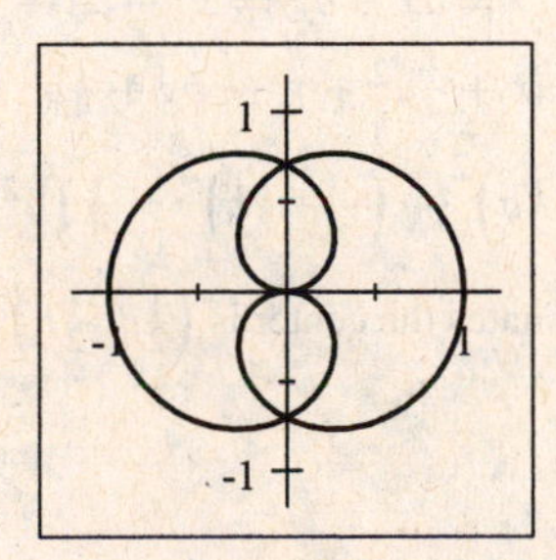

49. $r = 1 + 2\sin\left(\dfrac{\theta}{2}\right)$, $\theta \in [0, 4\pi]$

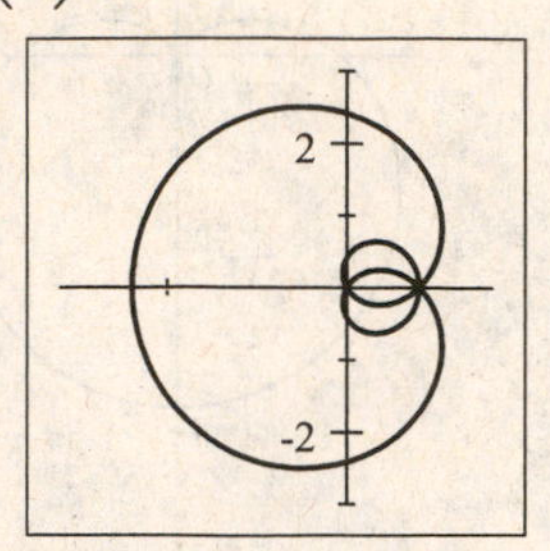

51. $r = 1 + \sin n\theta$. There are n loops.

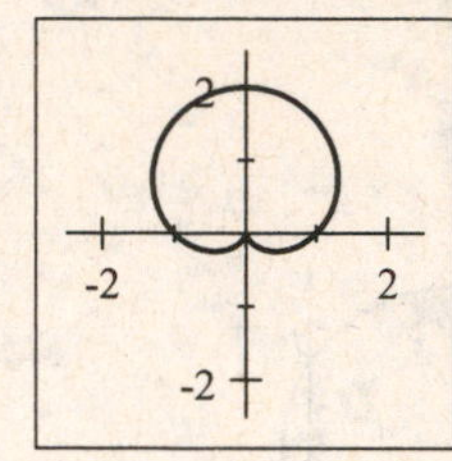

$n = 1$

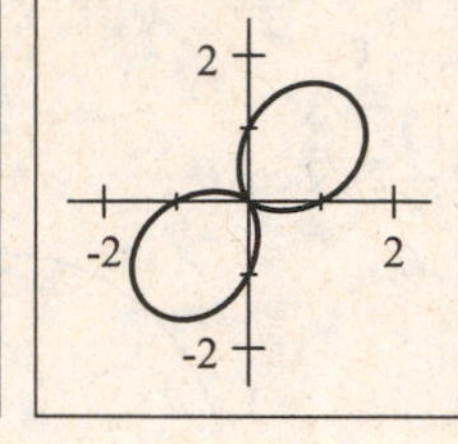

$n = 2$

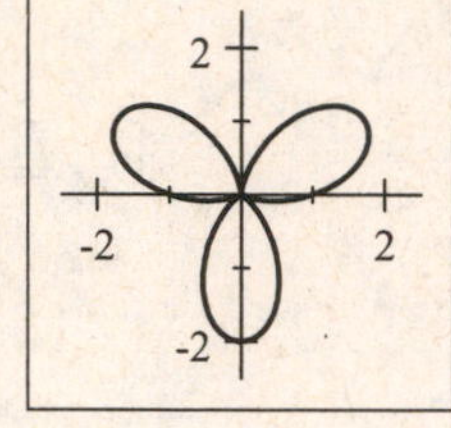

$n = 3$

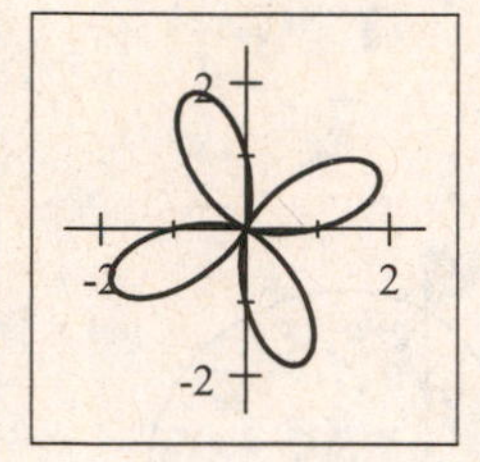

$n = 4$

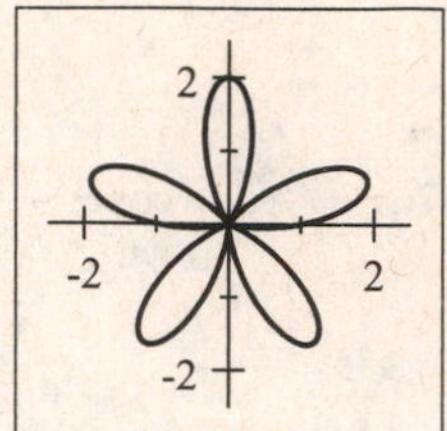

$n = 5$

53. The graph of $r = \sin\left(\dfrac{\theta}{2}\right)$ is IV, since the graph must contain the points $(0, 0)$, $\left(\dfrac{1}{\sqrt{2}}, \dfrac{\pi}{2}\right)$, $(1, \pi)$, and so on.

55. The graph of $r = \theta\sin\theta$ is III, since for $\theta = \frac{\pi}{2}, \frac{5\pi}{2}, \frac{7\pi}{2}, \ldots$ the values of r are also $\frac{\pi}{2}, \frac{5\pi}{2}, \frac{7\pi}{2}, \ldots$. Thus the graph must cross the vertical axis at an infinite number of points.

57. $\left(x^2 + y^2\right)^3 = 4x^2y^2 \Leftrightarrow \left(r^2\right)^3 = 4\,(r\cos\theta)^2\,(r\sin\theta)^2 \Leftrightarrow$

$r^6 = 4r^4\cos^2\theta\sin^2\theta \Leftrightarrow r^2 = 4\cos^2\theta\sin^2\theta \Leftrightarrow r = 2\cos\theta\sin\theta = \sin 2\theta$. The equation is $r = \sin 2\theta$, a rose.

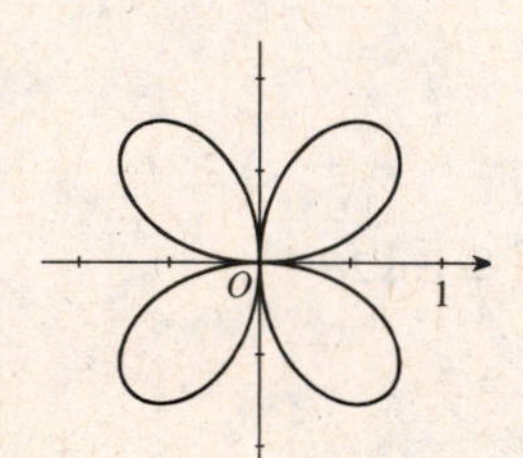

59. $\left(x^2 + y^2\right)^2 = x^2 - y^2 \Leftrightarrow \left(r^2\right)^2 = (r\cos\theta)^2 - (r\sin\theta)^2 \Leftrightarrow$

$r^4 = r^2\cos^2\theta - r^2\sin^2\theta \Leftrightarrow r^4 = r^2\left(\cos^2\theta - \sin^2\theta\right) \Leftrightarrow$

$r^2 = \cos^2\theta - \sin^2\theta = \cos 2\theta$. The graph is $r^2 = \cos 2\theta$, a leminiscate.

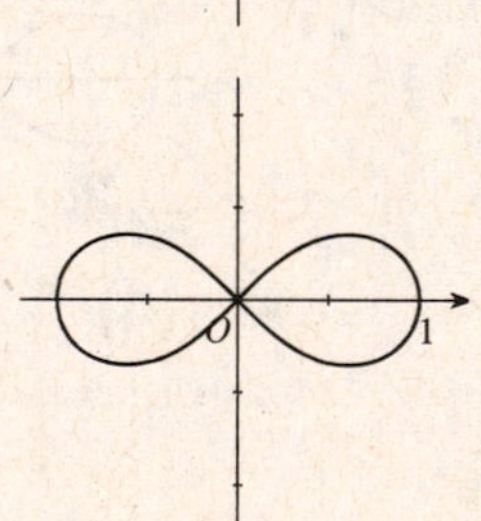

61. (a) $r = a\cos\theta + b\sin\theta \Leftrightarrow r^2 = ar\cos\theta + br\sin\theta \Leftrightarrow$

$x^2 + y^2 = ax + by \Leftrightarrow x^2 - ax + y^2 - by = 0 \Leftrightarrow$

$x^2 - ax + \frac{1}{4}a^2 + y^2 - by + \frac{1}{4}b^2 = \frac{1}{4}a^2 + \frac{1}{4}b^2 \quad \Leftrightarrow$

$\left(x - \frac{1}{2}a\right)^2 + \left(y - \frac{1}{2}b\right)^2 = \frac{1}{4}\left(a^2 + b^2\right)$. Thus, in rectangular coordinates the center is $\left(\frac{1}{2}a, \frac{1}{2}b\right)$ and the radius is $\frac{1}{2}\sqrt{a^2 + b^2}$.

(b) $r = 2\sin\theta + 2\cos\theta$ has center $(1, 1)$ and radius $\frac{1}{2}\sqrt{2^2 + 2^2} = \sqrt{2}$.

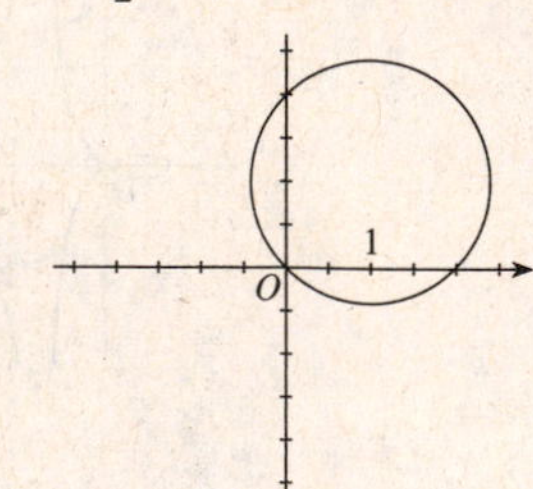

63. (a)

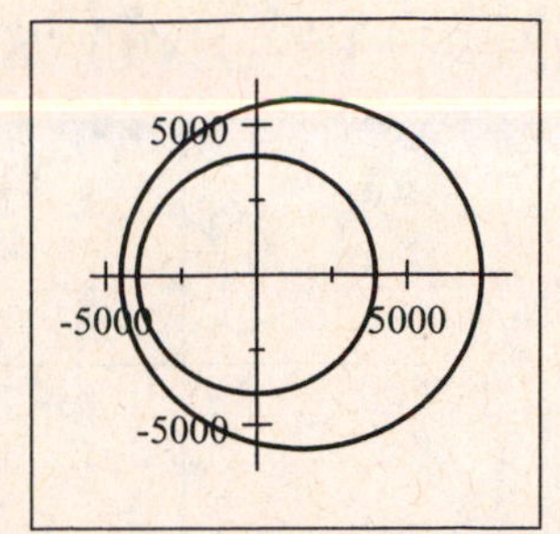

At $\theta = 0$, the satellite is at the "rightmost" point in its orbit, $(5625, 0)$. As θ increases, it travels counterclockwise. Note that it is moving fastest when $\theta = \pi$.

(b) The satellite is closest to earth when $\theta = \pi$. Its height above the earth's surface at this point is $22{,}500/(4 - \cos\pi) - 3960 = 4500 - 3960 = 540$ mi.

65. The graphs of $r = 1 + \sin\left(\theta - \frac{\pi}{6}\right)$ and $r = 1 + \sin\left(\theta - \frac{\pi}{3}\right)$ have the same shape as $r = 1 + \sin\theta$, rotated through angles of $\frac{\pi}{6}$ and $\frac{\pi}{3}$, respectively. Similarly, the graph of $r = f(\theta - \alpha)$ is the graph of $r = f(\theta)$ rotated by the angle α.

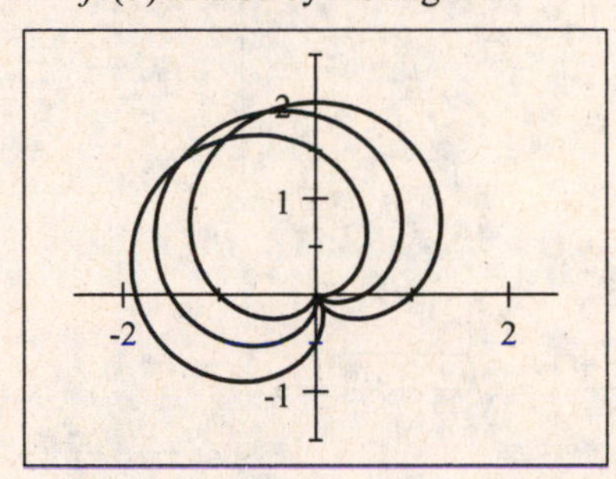

67. $y = 2 \Leftrightarrow r\sin\theta = 2 \Leftrightarrow r = 2\csc\theta$. The rectangular coordinate system gives the simpler equation here. It is easier to study lines in rectangular coordinates.

8.3 POLAR FORM OF COMPLEX NUMBERS; DE MOIVRE'S THEOREM

1. A complex number $z = a + bi$ has two parts: a is the *real* part and b is the *imaginary* part. To graph $a + bi$ we graph the ordered pair (a, b) in the complex plane.

3. (a) The complex number $z = -1 + i$ in polar form is $z = \sqrt{2}\left(\cos\frac{3\pi}{4} + i\sin\frac{3\pi}{4}\right)$.

(b) The complex number $z = 2\left(\cos\frac{\pi}{6} + i\sin\frac{\pi}{6}\right)$ in rectangular form is $z = \sqrt{3} + i$.

(c) The complex number z can be expressed in rectangular form as $1 + i$ or in polar form as $\sqrt{2}\left(\cos\frac{\pi}{4} + i\sin\frac{\pi}{4}\right)$.

5. $|4i| = \sqrt{0^2 + 4^2} = 4$

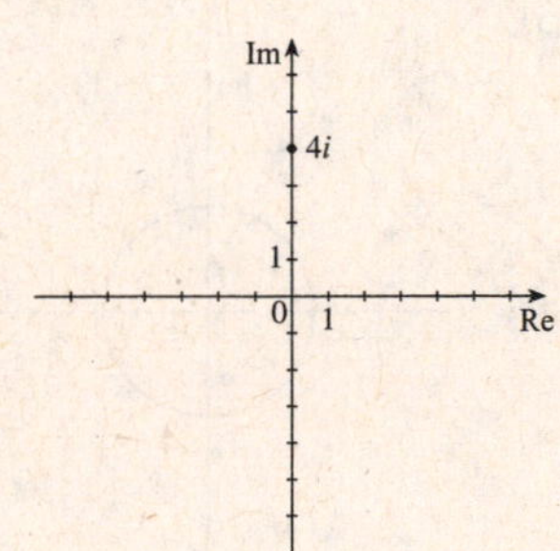

7. $|-2| = \sqrt{4 + 0} = 2$

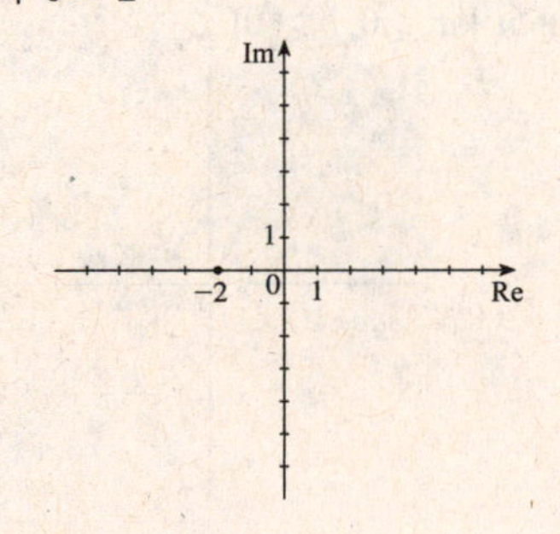

9. $|5 + 2i| = \sqrt{5^2 + 2^2} = \sqrt{29}$

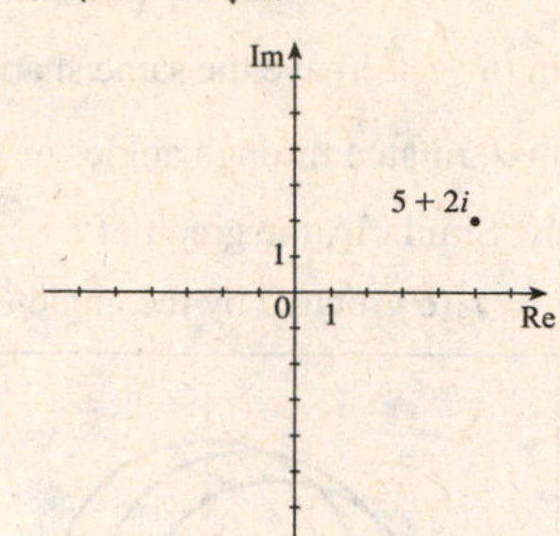

11. $\left|\sqrt{3} + i\right| = \sqrt{3 + 1} = 2$

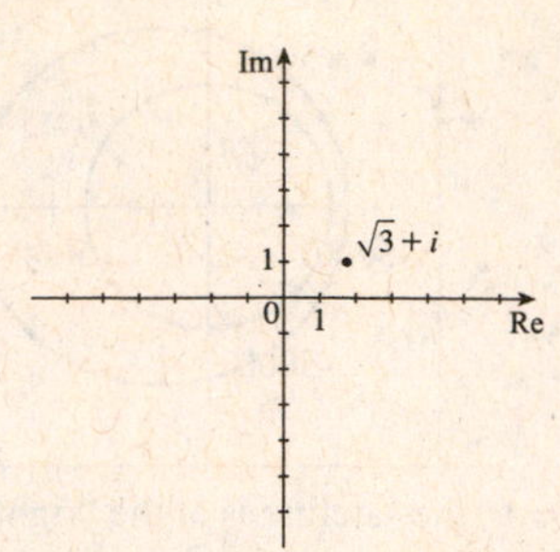

13. $\left|\dfrac{3 + 4i}{5}\right| = \sqrt{\frac{9}{25} + \frac{16}{25}} = 1$

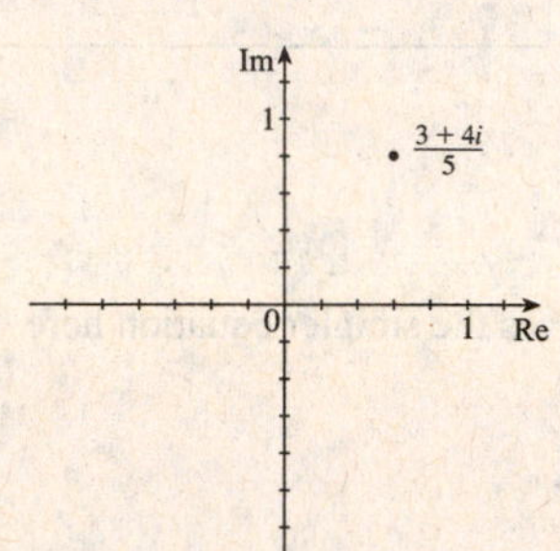

15. $z = 1 + i, 2z = 2 + 2i, -z = -1 - i, \frac{1}{2}z = \frac{1}{2} + \frac{1}{2}i$

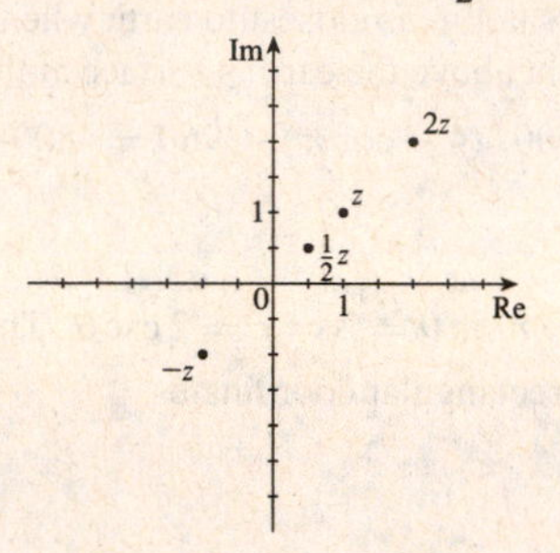

17. $z = 8 + 2i, \overline{z} = 8 - 2i$

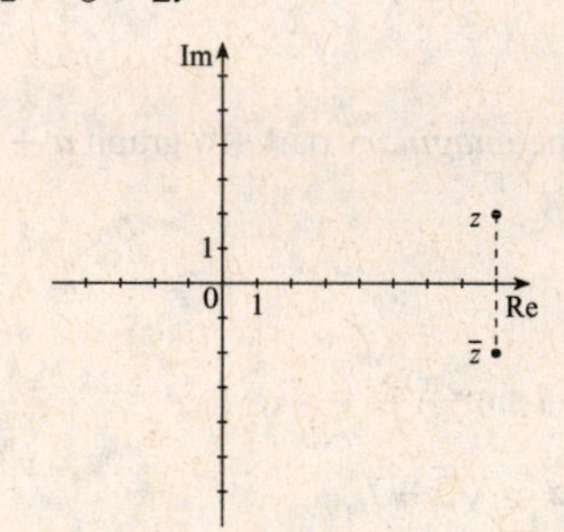

19. $z_1 = 2 - i, z_2 = 2 + i, z_1 + z_2 = 2 - i + 2 + i = 4,$

$z_1 z_2 = (2 - i)(2 + i) = 4 - i^2 = 5$

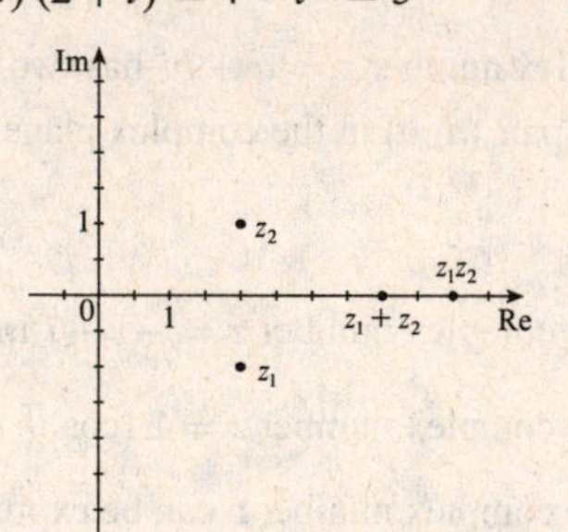

21. $\{z = a + bi \mid a \leq 0, b \geq 0\}$

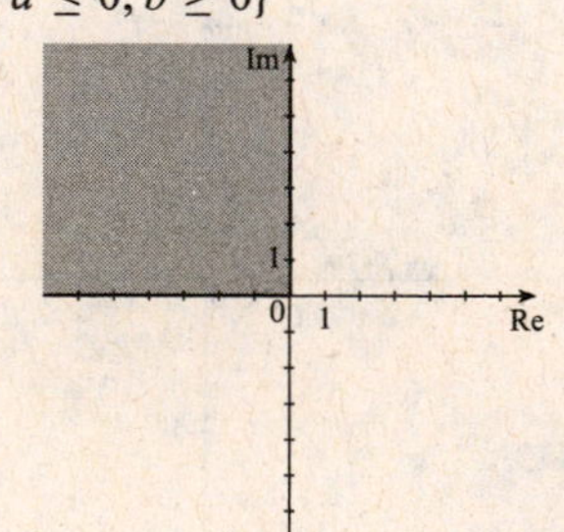

23. $\{z \mid |z| = 3\}$

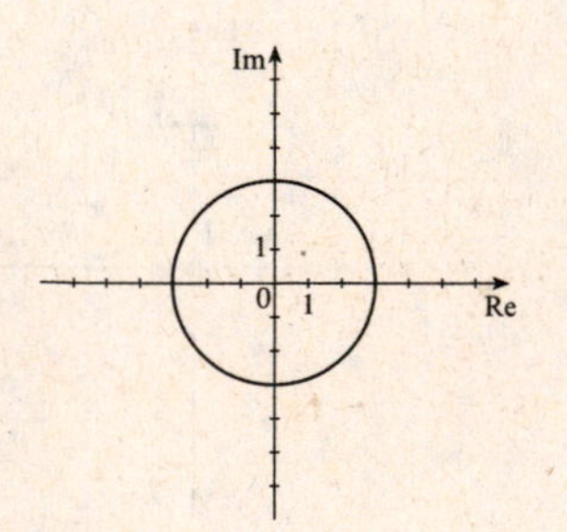

25. $\{z \mid |z| < 2\}$

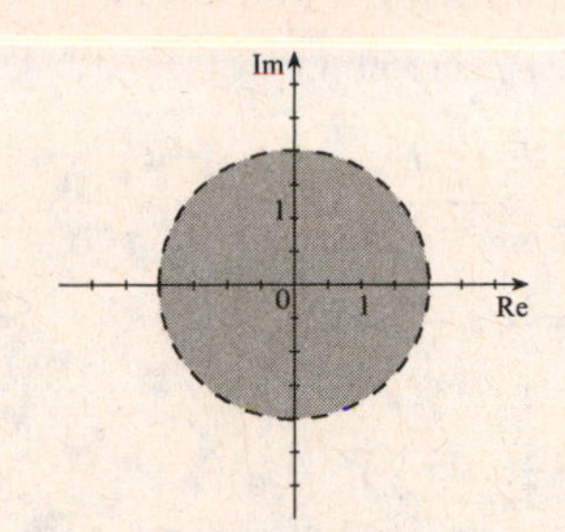

27. $\{z = a + bi \mid a + b < 2\}$

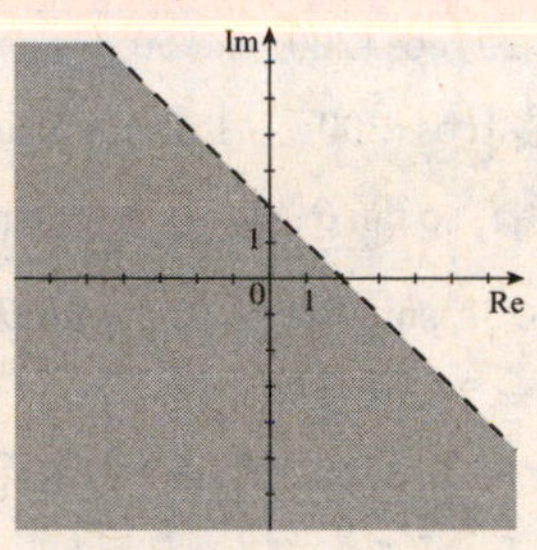

29. $1+i$. Then $\tan\theta = \frac{1}{1} = 1$ with θ in quadrant I $\Rightarrow \theta = \frac{\pi}{4}$, and $r = \sqrt{1^2 + 1^2} = \sqrt{2}$. Hence, $1+i = \sqrt{2}\left(\cos\frac{\pi}{4} + i\sin\frac{\pi}{4}\right)$.

31. $-2 + 2i$. Then $\tan\theta = \frac{2}{-2} = -1$ with θ in quadrant II $\Rightarrow \theta = \frac{3\pi}{4}$, and $r = \sqrt{2^2 + 2^2} = 2\sqrt{2}$. Hence, $-2 + 2i = 2\sqrt{2}\left(\cos\frac{3\pi}{4} + i\sin\frac{3\pi}{4}\right)$.

33. $-\sqrt{3} - i$. Then $\tan\theta = \frac{-1}{-\sqrt{3}} = \frac{\sqrt{3}}{3}$ with θ in quadrant III $\Rightarrow \theta = \frac{7\pi}{6}$, and $r = \sqrt{\left(-\sqrt{3}\right)^2 + (-1)^2} = 2$. Hence, $-\sqrt{3} - i = 2\left(\cos\frac{7\pi}{6} + i\sin\frac{7\pi}{6}\right)$.

35. $2\sqrt{3} - 2i$. Then $\tan\theta = \frac{-2}{2\sqrt{3}} = -\frac{\sqrt{3}}{3}$ with θ in quadrant IV $\Rightarrow \theta = \frac{11\pi}{6}$, and $r = \sqrt{\left(2\sqrt{3}\right)^2 + (-2)^2} = 4$. Hence, $2\sqrt{3} - 2i = 4\left(\cos\frac{11\pi}{6} + i\sin\frac{11\pi}{6}\right)$.

37. $2i$. Then $\tan\theta$ is undefined, $\theta = \frac{\pi}{2}$, and $r = 2$. Hence, $2i = 2\left(\cos\frac{\pi}{2} + i\sin\frac{\pi}{2}\right)$.

39. -3. Then $\tan\theta = 0$, $\theta = \pi$, and $r = 3$. Hence, $-3 = 3(\cos\pi + i\sin\pi)$.

41. $-\sqrt{6} + \sqrt{2}i$. Then $\tan\theta = \frac{\sqrt{2}}{-\sqrt{6}} = -\frac{\sqrt{3}}{3}$ with θ in quadrant II $\Rightarrow \theta = \frac{5\pi}{6}$, and $r = \sqrt{\left(-\sqrt{6}\right)^2 + \left(\sqrt{2}\right)^2} = 2\sqrt{2}$. Hence, $-\sqrt{6} + \sqrt{2}i = 2\sqrt{2}\left(\cos\frac{5\pi}{6} + i\sin\frac{5\pi}{6}\right)$.

43. $4 + 3i$. Then $\tan\theta = \frac{3}{4}$ with θ in quadrant I $\Rightarrow \theta = \tan^{-1}\frac{3}{4} \approx 0.6435$, and $r = \sqrt{4^2 + 3^2} = 5$. Hence, $4 + 3i = 5\left[\cos\left(\tan^{-1}\frac{3}{4}\right) + i\sin\left(\tan^{-1}\frac{3}{4}\right)\right]$.

45. $4\left(\sqrt{3} - i\right) = 4\sqrt{3} - 4i$. Then $\tan\theta = \frac{-4}{4\sqrt{3}} = -\frac{\sqrt{3}}{3}$ with θ in quadrant IV $\Rightarrow \theta = \frac{11\pi}{6}$, and $r = \sqrt{\left(4\sqrt{3}\right)^2 + (-4)^2} = 8$. Hence, $4\left(\sqrt{3} - i\right) = 8\left(\cos\frac{11\pi}{6} + i\sin\frac{11\pi}{6}\right)$.

47. $-3(1 - i) = -3 + 3i$. Then $\tan\theta = \frac{3}{-3} = -1$ with θ in quadrant II $\Rightarrow \theta = \frac{3\pi}{4}$, and $r = \sqrt{(-3)^2 + 3^2} = 3\sqrt{2}$. Hence, $-3(1 - i) = 3\sqrt{2}\left(\cos\frac{3\pi}{4} + i\sin\frac{3\pi}{4}\right)$.

49. $z_1 = 3\left(\cos\frac{\pi}{3} + i\sin\frac{\pi}{3}\right)$, $z_2 = 2\left(\cos\frac{\pi}{6} + i\sin\frac{\pi}{6}\right)$, $z_1z_2 = 3\cdot 2\left[\cos\left(\frac{\pi}{3} + \frac{\pi}{6}\right) + i\sin\left(\frac{\pi}{3} + \frac{\pi}{6}\right)\right] = 6\left(\cos\frac{\pi}{2} + i\sin\frac{\pi}{2}\right)$
$z_1/z_2 = \frac{3}{2}\left[\cos\left(\frac{\pi}{3} - \frac{\pi}{6}\right) + i\sin\left(\frac{\pi}{3} - \frac{\pi}{6}\right)\right] = \frac{3}{2}\left(\cos\frac{\pi}{6} + i\sin\frac{\pi}{6}\right)$

51. $z_1 = \sqrt{2}\left(\cos\frac{5\pi}{3} + i\sin\frac{5\pi}{3}\right)$, $z_2 = 2\sqrt{2}\left(\cos\frac{3\pi}{2} + i\sin\frac{3\pi}{2}\right)$,
$z_1z_2 = \sqrt{2}\cdot 2\sqrt{2}\left[\cos\left(\frac{5\pi}{3} + \frac{3\pi}{2}\right) + i\sin\left(\frac{5\pi}{3} + \frac{3\pi}{2}\right)\right] = 4\left(\cos\frac{7\pi}{6} + i\sin\frac{7\pi}{6}\right)$,
$z_1/z_2 = \frac{\sqrt{2}}{2\sqrt{2}}\left[\cos\left(\frac{5\pi}{3} - \frac{3\pi}{2}\right) + i\sin\left(\frac{5\pi}{3} - \frac{3\pi}{2}\right)\right] = \frac{1}{2}\left(\cos\frac{\pi}{6} + i\sin\frac{\pi}{6}\right)$

53. $z_1 = 4(\cos 120^\circ + i\sin 120^\circ)$, $z_2 = 2(\cos 30^\circ + i\sin 30^\circ)$,
$z_1z_2 = 4\cdot 2\left[\cos(120^\circ + 30^\circ) + i\sin(120^\circ + 30^\circ)\right] = 8(\cos 150^\circ + i\sin 150^\circ)$,
$z_1/z_2 = \frac{4}{2}\left[\cos(120^\circ - 30^\circ) + i\sin(120^\circ - 30^\circ)\right] = 2(\cos 90^\circ + i\sin 90^\circ)$

55. $z_1 = 4(\cos 200° + i\sin 200°)$, $z_2 = 25(\cos 150° + i\sin 150°)$,
$z_1 z_2 = 4 \cdot 25\left[\cos(200° + 150°) + i\sin(200° + 150°)\right] = 100(\cos 350° + i\sin 350°)$,
$z_1/z_2 = \frac{4}{25}\left[\cos(200° - 150°) + i\sin(200° - 150°)\right] = \frac{4}{25}(\cos 50° + i\sin 50°)$

57. $z_1 = \sqrt{3} + i$, so $\tan\theta_1 = \frac{1}{\sqrt{3}}$ with θ_1 in quadrant I $\Rightarrow \theta_1 = \frac{\pi}{6}$, and $r_1 = \sqrt{3+1} = 2$.
$z_2 = 1 + \sqrt{3}i$, so $\tan\theta_2 = \sqrt{3}$ with θ_2 in quadrant I $\Rightarrow \theta_2 = \frac{\pi}{3}$, and $r_1 = \sqrt{1+3} = 2$.
Hence, $z_1 = 2\left(\cos\frac{\pi}{6} + i\sin\frac{\pi}{6}\right)$ and $z_2 = 2\left(\cos\frac{\pi}{3} + i\sin\frac{\pi}{3}\right)$.
Thus, $z_1 z_2 = 2 \cdot 2\left[\cos\left(\frac{\pi}{6} + \frac{\pi}{3}\right) + i\sin\left(\frac{\pi}{6} + \frac{\pi}{3}\right)\right] = 4\left(\cos\frac{\pi}{2} + i\sin\frac{\pi}{2}\right)$,
$z_1/z_2 = \frac{2}{2}\left[\cos\left(\frac{\pi}{6} - \frac{\pi}{3}\right) + i\sin\left(\frac{\pi}{6} - \frac{\pi}{3}\right)\right] = \cos\left(-\frac{\pi}{6}\right) + i\sin\left(-\frac{\pi}{6}\right)$, and $1/z_1 = \frac{1}{2}\left[\cos\left(-\frac{\pi}{6}\right) + i\sin\left(-\frac{\pi}{6}\right)\right]$.

59. $z_1 = 2\sqrt{3} - 2i$, so $\tan\theta_1 = \frac{-2}{2\sqrt{3}} = -\frac{1}{\sqrt{3}}$ with θ_1 in quadrant IV $\Rightarrow \theta_1 = \frac{11\pi}{6}$, and $r_1 = \sqrt{12+4} = 4$.
$z_2 = -1 + i$, so $\tan\theta_2 = -1$ with θ_2 in quadrant II $\Rightarrow \theta_2 = \frac{3\pi}{4}$, and $r_2 = \sqrt{1+1} = \sqrt{2}$.
Hence, $z_1 = 4\left(\cos\frac{11\pi}{6} + i\sin\frac{11\pi}{6}\right)$ and $z_2 = \sqrt{2}\left(\cos\frac{3\pi}{4} + i\sin\frac{3\pi}{4}\right)$.
Thus, $z_1 z_2 = 4 \cdot \sqrt{2}\left[\cos\left(\frac{11\pi}{6} + \frac{3\pi}{4}\right) + i\sin\left(\frac{11\pi}{6} + \frac{3\pi}{4}\right)\right] = 4\sqrt{2}\left(\cos\frac{7\pi}{12} + i\sin\frac{7\pi}{12}\right)$,
$z_1/z_2 = \frac{4}{\sqrt{2}}\left[\cos\left(\frac{11\pi}{6} - \frac{3\pi}{4}\right) + i\sin\left(\frac{11\pi}{6} - \frac{3\pi}{4}\right)\right] = 2\sqrt{2}\left(\cos\frac{13\pi}{12} + i\sin\frac{13\pi}{12}\right)$, and
$1/z_1 = \frac{1}{4}\left(\cos\left(-\frac{11\pi}{6}\right) + i\sin\left(-\frac{11\pi}{6}\right)\right) = \frac{1}{4}\left(\cos\frac{\pi}{6} + i\sin\frac{\pi}{6}\right)$.

61. $z_1 = 5 + 5i$, so $\tan\theta_1 = \frac{5}{5} = 1$with θ_1 in quadrant I $\Rightarrow \theta_1 = \frac{\pi}{4}$, and $r_1 = \sqrt{25+25} = 5\sqrt{2}$.
$z_2 = 4$, so $\theta_2 = 0$, and $r_2 = 4$.
Hence, $z_1 = 5\sqrt{2}\left(\cos\frac{\pi}{4} + i\sin\frac{\pi}{4}\right)$ and $z_2 = 4(\cos 0 + i\sin 0)$.
Thus, $z_1 z_2 = 5\sqrt{2} \cdot 4\left[\cos\left(\frac{\pi}{4} + 0\right) + i\sin\left(\frac{\pi}{4} + 0\right)\right] = 20\sqrt{2}\left(\cos\frac{\pi}{4} + i\sin\frac{\pi}{4}\right)$, $z_1/z_2 = \frac{5\sqrt{2}}{4}\left(\cos\frac{\pi}{4} + i\sin\frac{\pi}{4}\right)$, and
$1/z_1 = \frac{1}{5\sqrt{2}}\left(\cos\left(-\frac{\pi}{4}\right) + i\sin\left(-\frac{\pi}{4}\right)\right) = \frac{\sqrt{2}}{10}\left(\cos\left(-\frac{\pi}{4}\right) + i\sin\left(-\frac{\pi}{4}\right)\right)$.

63. $z_1 = -20$, so $\theta_1 = \pi$, and $r_1 = 20$.
$z_2 = \sqrt{3} + i$, so $\tan\theta_2 = \frac{1}{\sqrt{3}}$ with θ_2 in quadrant I $\Rightarrow \theta_2 = \frac{\pi}{6}$, and $r_2 = \sqrt{3+1} = 2$.
Hence, $z_1 = 20(\cos\pi + i\sin\pi)$ and $z_2 = 2\left(\cos\frac{\pi}{6} + i\sin\frac{\pi}{6}\right)$.
Thus, $z_1 z_2 = 20 \cdot 2\left[\cos\left(\pi + \frac{\pi}{6}\right) + i\sin\left(\pi + \frac{\pi}{6}\right)\right] = 40\left(\cos\frac{7\pi}{6} + i\sin\frac{7\pi}{6}\right)$,
$z_1/z_2 = \frac{20}{2}\left[\cos\left(\pi - \frac{\pi}{6}\right) + i\sin\left(\pi - \frac{\pi}{6}\right)\right] = 10\left(\cos\frac{5\pi}{6} + i\sin\frac{5\pi}{6}\right)$, and
$1/z_1 = \frac{1}{20}\left[\cos(-\pi) + i\sin(-\pi)\right] = \frac{1}{20}(\cos\pi + i\sin\pi)$.

65. $-\sqrt{3} + i = 2\left(\cos\frac{5\pi}{6} + i\sin\frac{5\pi}{6}\right)$, so $\left(-\sqrt{3} + i\right)^6 = 2^6\left(\cos\frac{30\pi}{6} + i\sin\frac{30\pi}{6}\right) = 64(\cos 5\pi + i\sin 5\pi) = -64$.

67. $-\sqrt{2} - \sqrt{2}i = 2\left(\cos\frac{5\pi}{4} + i\sin\frac{5\pi}{4}\right)$, so $\left(-\sqrt{2} - \sqrt{2}i\right)^5 = 2^5\left(\cos\frac{25\pi}{4} + i\sin\frac{25\pi}{4}\right) = 16\sqrt{2} + 16\sqrt{2}i$.

69. $r = \sqrt{\frac{1}{2} + \frac{1}{2}} = 1$ and $\tan\theta = 1 \Rightarrow \theta = \frac{\pi}{4}$. Thus $\frac{\sqrt{2}}{2} + \frac{\sqrt{2}}{2}i = \cos\frac{\pi}{4} + i\sin\frac{\pi}{4}$. Therefore,
$\left(\frac{\sqrt{2}}{2} + \frac{\sqrt{2}}{2}i\right)^{12} = \cos 12\left(\frac{\pi}{4}\right) + i\sin 12\left(\frac{\pi}{4}\right) = \cos 3\pi + i\sin 3\pi = -1$.

71. $r = \sqrt{4+4} = 4\sqrt{2}$ and $\tan\theta = -1$ with θ in quadrant IV $\Rightarrow \theta = \frac{7\pi}{4}$. Thus $2 - 2i = 2\sqrt{2}\left(\cos\frac{7\pi}{4} + i\sin\frac{7\pi}{4}\right)$, so
$(2 - 2i)^8 = \left(2\sqrt{2}\right)^8(\cos 14\pi + i\sin 14\pi) = 4096(1 - 0i) = 4096$.

73. $r = \sqrt{1+1} = \sqrt{2}$ and $\tan\theta = 1$ with θ in quadrant III $\Rightarrow \theta = \frac{5\pi}{4}$. Thus $-1 - i = \sqrt{2}\left(\cos\frac{5\pi}{4} + i\sin\frac{5\pi}{4}\right)$, so
$(-1 - i)^7 = \left(\sqrt{2}\right)^7\left(\cos\frac{35\pi}{4} + i\sin\frac{35\pi}{4}\right) = 8\sqrt{2}\left(\cos\frac{3\pi}{4} + i\sin\frac{3\pi}{4}\right) = 8\sqrt{2}\left(\frac{1}{\sqrt{2}} - i\frac{1}{\sqrt{2}}\right) = 8(-1 + i)$.

75. $r = \sqrt{12+4} = 4$ and $\tan\theta = \frac{2}{2\sqrt{3}} = \frac{1}{\sqrt{3}} \Rightarrow \theta = \frac{\pi}{6}$. Thus $2\sqrt{3} + 2i = 4\left(\cos\frac{\pi}{6} + i\sin\frac{\pi}{6}\right)$, so

$$\left(2\sqrt{3}+2i\right)^{-5} = \left(\tfrac{1}{4}\right)^5\left(\cos\tfrac{-5\pi}{6} + i\sin\tfrac{-5\pi}{6}\right) = \tfrac{1}{1024}\left(-\tfrac{\sqrt{3}}{2} - \tfrac{1}{2}i\right) = \tfrac{1}{2048}\left(-\sqrt{3} - i\right)$$

77. $r = \sqrt{48+16} = 8$ and $\tan\theta = \frac{4}{4\sqrt{3}} = \frac{1}{\sqrt{3}} \Rightarrow \theta = \frac{\pi}{6}$. Thus

$4\sqrt{3}+4i = 8\left(\cos\frac{\pi}{6} + i\sin\frac{\pi}{6}\right)$. So,

$$\left(4\sqrt{3}+4i\right)^{1/2} = \sqrt{8}\left[\cos\left(\frac{\pi/6+2k\pi}{2}\right) + i\sin\left(\frac{\pi/6+2k\pi}{2}\right)\right] \text{ for } k = 0, 1.$$

Thus the two roots are $w_0 = 2\sqrt{2}\left(\cos\frac{\pi}{12} + i\sin\frac{\pi}{12}\right)$ and

$w_1 = 2\sqrt{2}\left(\cos\frac{13\pi}{12} + i\sin\frac{13\pi}{12}\right)$.

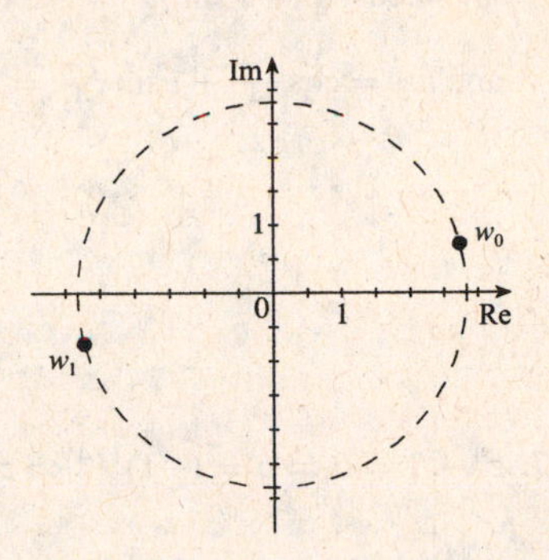

79. $-81i = 81\left(\cos\frac{3\pi}{2} + i\sin\frac{3\pi}{2}\right)$. Thus,

$$(-81i)^{1/4} = 81^{1/4}\left[\cos\left(\frac{3\pi/2+2k\pi}{4}\right) + i\sin\left(\frac{3\pi/2+2k\pi}{4}\right)\right] \text{ for } k = 0, 1,$$

2, 3. The four roots are $w_0 = 3\left(\cos\frac{3\pi}{8} + i\sin\frac{3\pi}{8}\right)$,

$w_1 = 3\left(\cos\frac{7\pi}{8} + i\sin\frac{7\pi}{8}\right)$, $w_2 = 3\left(\cos\frac{11\pi}{8} + i\sin\frac{11\pi}{8}\right)$, and

$w_3 = 3\left(\cos\frac{15\pi}{8} + i\sin\frac{15\pi}{8}\right)$.

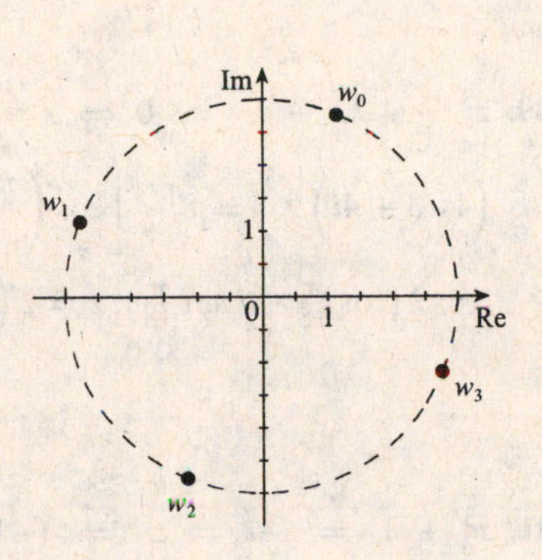

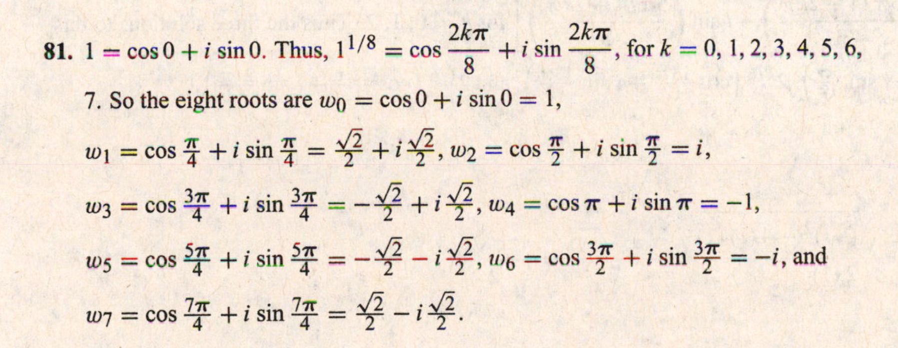

81. $1 = \cos 0 + i\sin 0$. Thus, $1^{1/8} = \cos\dfrac{2k\pi}{8} + i\sin\dfrac{2k\pi}{8}$, for $k = 0, 1, 2, 3, 4, 5, 6,$

7. So the eight roots are $w_0 = \cos 0 + i\sin 0 = 1$,

$w_1 = \cos\frac{\pi}{4} + i\sin\frac{\pi}{4} = \frac{\sqrt{2}}{2} + i\frac{\sqrt{2}}{2}$, $w_2 = \cos\frac{\pi}{2} + i\sin\frac{\pi}{2} = i$,

$w_3 = \cos\frac{3\pi}{4} + i\sin\frac{3\pi}{4} = -\frac{\sqrt{2}}{2} + i\frac{\sqrt{2}}{2}$, $w_4 = \cos\pi + i\sin\pi = -1$,

$w_5 = \cos\frac{5\pi}{4} + i\sin\frac{5\pi}{4} = -\frac{\sqrt{2}}{2} - i\frac{\sqrt{2}}{2}$, $w_6 = \cos\frac{3\pi}{2} + i\sin\frac{3\pi}{2} = -i$, and

$w_7 = \cos\frac{7\pi}{4} + i\sin\frac{7\pi}{4} = \frac{\sqrt{2}}{2} - i\frac{\sqrt{2}}{2}$.

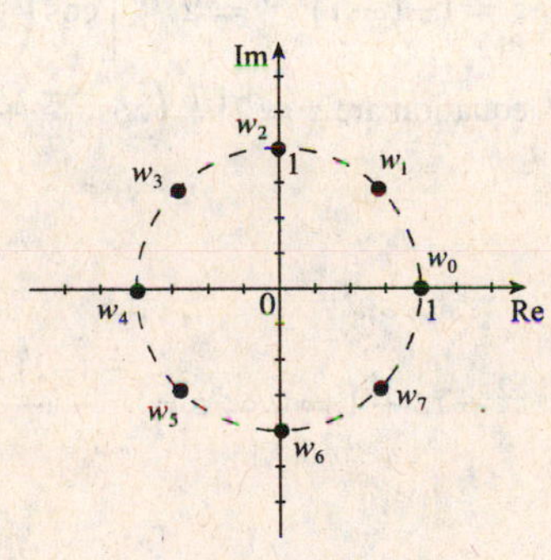

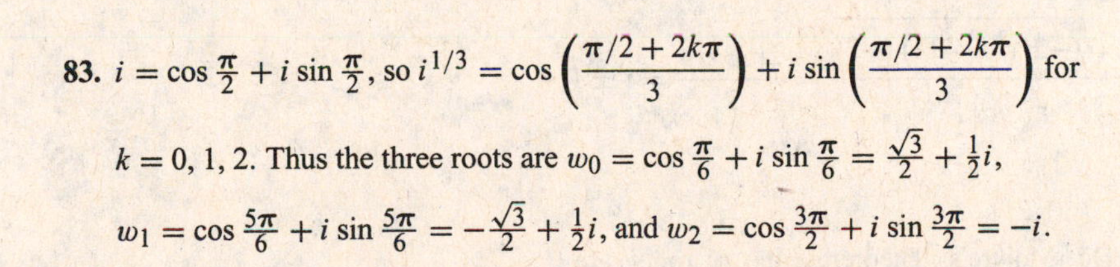

83. $i = \cos\frac{\pi}{2} + i\sin\frac{\pi}{2}$, so $i^{1/3} = \cos\left(\dfrac{\pi/2+2k\pi}{3}\right) + i\sin\left(\dfrac{\pi/2+2k\pi}{3}\right)$ for

$k = 0, 1, 2$. Thus the three roots are $w_0 = \cos\frac{\pi}{6} + i\sin\frac{\pi}{6} = \frac{\sqrt{3}}{2} + \frac{1}{2}i$,

$w_1 = \cos\frac{5\pi}{6} + i\sin\frac{5\pi}{6} = -\frac{\sqrt{3}}{2} + \frac{1}{2}i$, and $w_2 = \cos\frac{3\pi}{2} + i\sin\frac{3\pi}{2} = -i$.

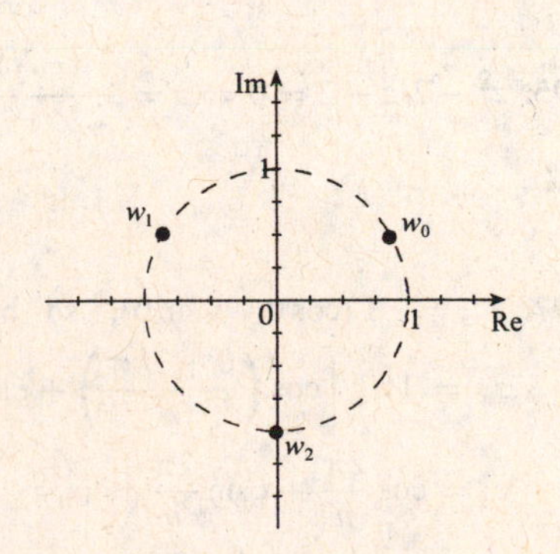

85. $-1 = \cos\pi + i\sin\pi$. Then $(-1)^{1/4} = \cos\left(\frac{\pi + 2k\pi}{4}\right) + i\sin\left(\frac{\pi + 2k\pi}{4}\right)$ for $k = 0, 1, 2, 3$. So the four roots are $w_0 = \cos\frac{\pi}{4} + i\sin\frac{\pi}{4} = \frac{\sqrt{2}}{2} + i\frac{\sqrt{2}}{2}$, $w_1 = \cos\frac{3\pi}{4} + i\sin\frac{3\pi}{4} = -\frac{\sqrt{2}}{2} + i\frac{\sqrt{2}}{2}$, $w_2 = \cos\frac{5\pi}{4} + i\sin\frac{5\pi}{4} = -\frac{\sqrt{2}}{2} - i\frac{\sqrt{2}}{2}$, and $w_3 = \cos\frac{7\pi}{4} + i\sin\frac{7\pi}{4} = \frac{\sqrt{2}}{2} - i\frac{\sqrt{2}}{2}$.

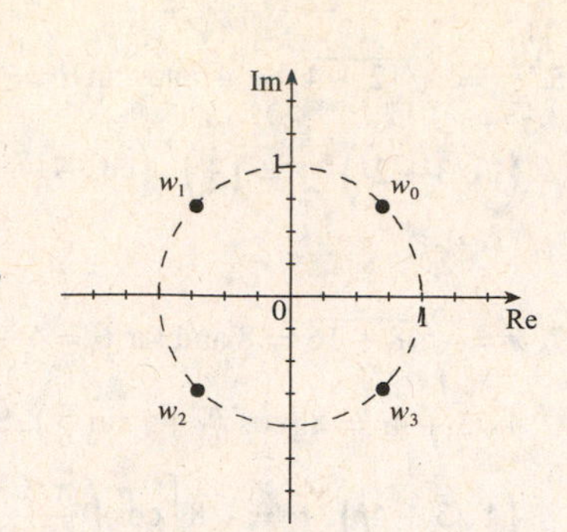

87. $z^4 + 1 = 0 \Leftrightarrow z = (-1)^{1/4} \Leftrightarrow z = \frac{\sqrt{2}}{2} \pm \frac{\sqrt{2}}{2}i, z = -\frac{\sqrt{2}}{2} \pm \frac{\sqrt{2}}{2}i$ (from Exercise 85)

89. $z^3 - 4\sqrt{3} - 4i = 0 \Leftrightarrow z = \left(4\sqrt{3} + 4i\right)^{1/3}$. Since $4\sqrt{3} + 4i = 8\left(\cos\frac{\pi}{6} + i\sin\frac{\pi}{6}\right)$, $\left(4\sqrt{3} + 4i\right)^{1/3} = 8^{1/3}\left[\cos\left(\frac{\pi/6 + 2k\pi}{3}\right) + i\sin\left(\frac{\pi/6 + 2k\pi}{3}\right)\right]$, for $k = 0, 1, 2$. Thus the three roots are $z = 2\left(\cos\frac{\pi}{18} + i\sin\frac{\pi}{18}\right)$, $z = 2\left(\cos\frac{13\pi}{18} + i\sin\frac{13\pi}{8}\right)$, and $z = 2\left(\cos\frac{25\pi}{18} + i\sin\frac{25\pi}{18}\right)$.

91. $z^3 + 1 = -i \Rightarrow z = (-1 - i)^{1/3}$. Since $-1 - i = \sqrt{2}\left(\cos\frac{5\pi}{4} + i\sin\frac{5\pi}{4}\right)$, $z = (-1 - i)^{1/3} = 2^{1/6}\left[\cos\left(\frac{5\pi/4 + 2k\pi}{3}\right) + i\sin\left(\frac{5\pi/4 + 2k\pi}{3}\right)\right]$ for $k = 0, 1, 2$. Thus the three solutions to this equation are $z = 2^{1/6}\left(\cos\frac{5\pi}{12} + i\sin\frac{5\pi}{12}\right)$, $2^{1/6}\left(\cos\frac{13\pi}{12} + i\sin\frac{13\pi}{12}\right)$, and $2^{1/6}\left(\cos\frac{21\pi}{12} + i\sin\frac{21\pi}{12}\right)$.

93. $z^2 - iz + 1 = 0 \Leftrightarrow z = \dfrac{-(-i) \pm \sqrt{(-i)^2 - 4(1)(1)}}{2(1)} = \dfrac{i \pm \sqrt{-5}}{2} = \dfrac{1 \pm \sqrt{5}}{2}i$

95. $z^2 - 2iz - 2 = 0 \Leftrightarrow z = \dfrac{-(-2i) \pm \sqrt{(-2i)^2 - 4(1)(-2)}}{2} = \dfrac{2i \pm \sqrt{4}}{2} = i \pm 1$

97. $1 = 1(\cos 0 + i\sin 0)$, so by De Moivre's Theorem, its n roots are $z_k = 1^{1/n}\left[\cos\left(\frac{0 + 2k\pi}{n}\right) + i\sin\left(\frac{0 + 2k\pi}{n}\right)\right] = \cos\frac{2k\pi}{n} + i\sin\frac{2k\pi}{n}$ for $k = 0, 1, 2, \ldots n - 1$. So $z_0 = 1$, $z_1 = \cos\frac{2\pi}{n} + i\sin\frac{2\pi}{n} = w$, $z_2 = \cos\frac{4\pi}{n} + i\sin\frac{4\pi}{n} = w^2$, and so on.

99. The cube roots of 1 are $w^0 = 1$, $w^1 = \cos\frac{2\pi}{3} + i\sin\frac{2\pi}{3} = -\frac{1}{2} + \frac{\sqrt{3}}{2}i$, and $w^2 = \cos\frac{4\pi}{3} + i\sin\frac{4\pi}{3} = \frac{1}{2} - \frac{\sqrt{3}}{2}i$, so their sum is $w^0 + w^1 + w^2 = 1 + \left(-\frac{1}{2} + \frac{\sqrt{3}}{2}i\right) + \left(\frac{1}{2} - \frac{\sqrt{3}}{2}i\right) = 0$.

The fourth roots of 1 are $w^0 = 1$, $w^1 = i$, $w^2 = -1$, and $w^3 = -i$, so their sum is $w^0 + w^1 + w^2 + w^3 = 1 + i - 1 - i = 0$.

The fifth roots of 1 are $w^0 = 1$, $w^1 = \cos\frac{2\pi}{5} + i\sin\frac{2\pi}{5}$, $w^2 = \cos\frac{4\pi}{5} + i\sin\frac{4\pi}{5}$, $w^3 = \cos\frac{6\pi}{5} + i\sin\frac{6\pi}{5}$, and $w^4 = \cos\frac{8\pi}{5} + i\sin\frac{8\pi}{5}$, so their sum is

$$1 + \left(\cos\frac{2\pi}{5} + i\sin\frac{2\pi}{5}\right) + \left(\cos\frac{4\pi}{5} + i\sin\frac{4\pi}{5}\right) + \left(\cos\frac{6\pi}{5} + i\sin\frac{6\pi}{5}\right) + \left(\cos\frac{8\pi}{5} + i\sin\frac{8\pi}{5}\right)$$

$$= 1 + 2\cos\frac{2\pi}{5} + 2\cos\frac{6\pi}{5} \quad \text{(most terms cancel)} \quad = 1 + 2\left[\left(\frac{\sqrt{5}}{4} - \frac{1}{4}\right) + \left(-\frac{\sqrt{5}}{4} - \frac{1}{4}\right)\right] = 0$$

The sixth roots of 1 are $w^0 = 1$, $w^1 = \cos\frac{\pi}{3} + i\sin\frac{\pi}{3} = \frac{1}{2} + \frac{\sqrt{3}}{2}i$, $w^2 = \cos\frac{2\pi}{3} + i\sin\frac{2\pi}{3} = -\frac{1}{2} + \frac{\sqrt{3}}{2}i$, $w^3 = -1$, $w^4 = \cos\frac{4\pi}{3} + i\sin\frac{4\pi}{3} = -\frac{1}{2} - \frac{\sqrt{3}}{2}i$, and $w^5 = \cos\frac{5\pi}{3} + i\sin\frac{5\pi}{3} = \frac{1}{2} - \frac{\sqrt{3}}{2}i$, so their sum is $1 + \left(\frac{1}{2} + \frac{\sqrt{3}}{2}i\right) + \left(-\frac{1}{2} + \frac{\sqrt{3}}{2}i\right) - 1 + \left(-\frac{1}{2} - \frac{\sqrt{3}}{2}i\right) + \left(\frac{1}{2} - \frac{\sqrt{3}}{2}i\right) = 0$.

The eight roots of 1 are $w^0 = 1$, $w^1 = \cos\frac{\pi}{4} + i\sin\frac{\pi}{4} = \frac{\sqrt{2}}{2} + \frac{\sqrt{2}}{2}i$, $w^2 = i$, $w^3 = \cos\frac{3\pi}{4} + i\sin\frac{3\pi}{4} = -\frac{\sqrt{2}}{2} + \frac{\sqrt{2}}{2}i$, $w^4 = -1$, $w^5 = \cos\frac{5\pi}{4} + i\sin\frac{5\pi}{4} = -\frac{\sqrt{2}}{2} - \frac{\sqrt{2}}{2}i$, $w^6 = -i$, and $w^7 = \cos\frac{7\pi}{4} + i\sin\frac{7\pi}{4} = \frac{\sqrt{2}}{2} - \frac{\sqrt{2}}{2}i$, so their sum is $1 + \left(\frac{\sqrt{2}}{2} + \frac{\sqrt{2}}{2}i\right) + i + \left(-\frac{\sqrt{2}}{2} + \frac{\sqrt{2}}{2}\right) - 1 + \left(-\frac{\sqrt{2}}{2} - \frac{\sqrt{2}}{2}i\right) - i + \left(\frac{\sqrt{2}}{2} - \frac{\sqrt{2}}{2}i\right) = 0$.

It seems that the sum of any set of nth roots is 0.

To prove this, factor $w^n - 1 = \left(w^1 - 1\right)\left(1 + w^1 + w^2 + w^3 + \cdots + w^{n-1}\right)$. Since this is 0 and $w^1 \neq 0$, we must have $1 + w^1 + w^2 + w^3 + \cdots + w^{n-1} = 0$.

101. $\frac{z_1}{z_2} = \frac{r_1(\cos\theta_1 + i\sin\theta_1)}{r_2(\cos\theta_2 + i\sin\theta_2)} \cdot \frac{\cos\theta_2 - i\sin\theta_2}{\cos\theta_2 - i\sin\theta_2} = \frac{r_1}{r_2} \frac{\cos\theta_1\cos\theta_2 - i^2\sin\theta_1\cos\theta_2 + i\sin\theta_1\cos\theta_2 - i\sin\theta_2\cos\theta_1}{\cos^2\theta_2 - i^2\sin^2\theta_2}$

$= \frac{r_1}{r_2}\left[\cos(\theta_1 - \theta_2) + i\sin(\theta_1 - \theta_2)\right]$

8.4 PLANE CURVES AND PARAMETRIC EQUATIONS

1. (a) The parametric equations $x = f(t)$ and $y = g(t)$ give the coordinates of a point $(x, y) = (f(t), g(t))$ for appropriate values of t. The variable t is called a *parameter*.

(b) When $t = 0$ the object is at $\left(0, 0^2\right) = (0, 0)$ and when $t = 1$ the object is at $\left(1, 1^2\right) = (1, 1)$.

(c) If we eliminate the parameter in part (b) we get the equation $y = x^2$. We see from this equation that the path of the moving object is a *parabola*.

3. (a) $x = 2t$, $y = t + 6$

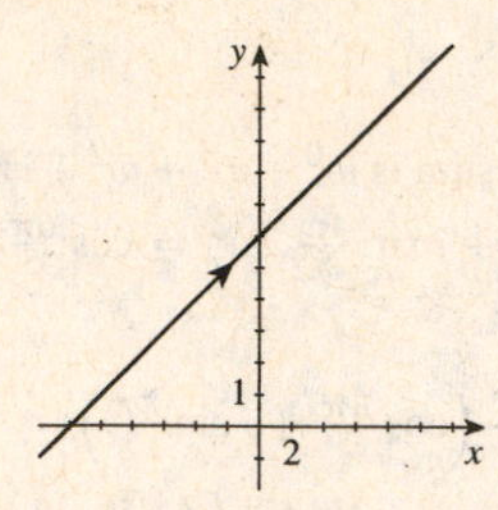

(b) Since $x = 2t$, $t = \frac{x}{2}$ and so $y = \frac{x}{2} + 6 \Leftrightarrow$ $x - 2y + 12 = 0$.

5. (a) $x = t^2$, $y = t - 2$, $2 \le t \le 4$

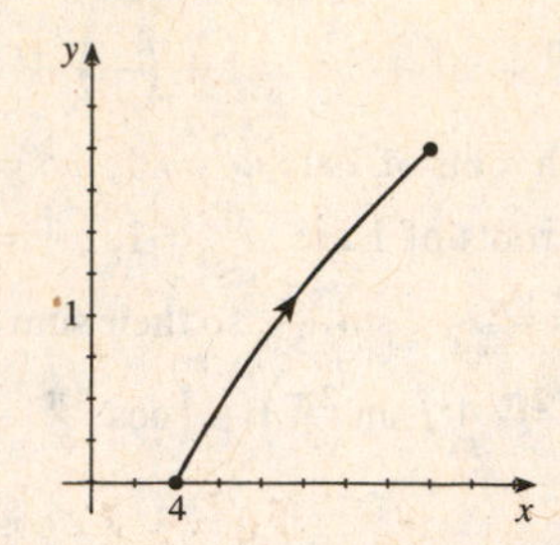

(b) Since $y = t - 2 \Leftrightarrow t = y + 2$, we have $x = t^2$ $\Leftrightarrow x = (y + 2)^2$, and since $2 \le t \le 4$, we have $4 \le x \le 16$.

7. (a) $x = \sqrt{t}$, $y = 1 - t \Rightarrow t \ge 0$

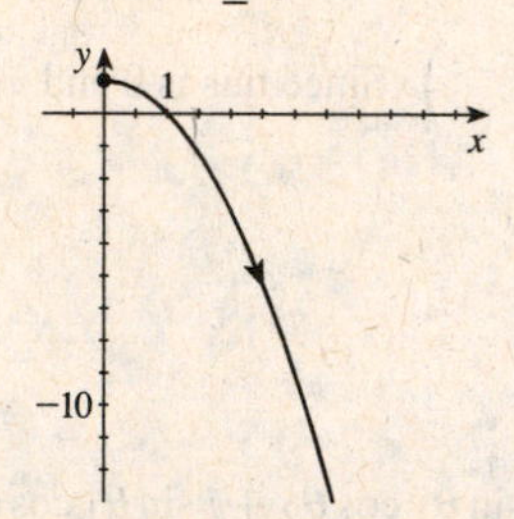

(b) Since $x = \sqrt{t}$, we have $x^2 = t$, and so $y = 1 - x^2$ with $x \ge 0$.

9. (a) $x = \frac{1}{t}$, $y = t + 1$

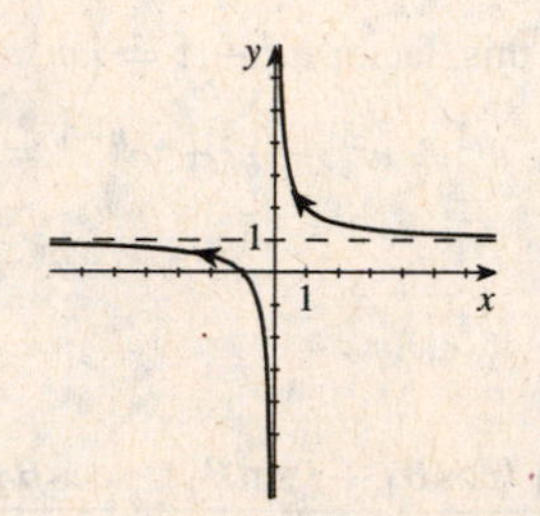

(b) Since $x = \frac{1}{t}$ we have $t = \frac{1}{x}$ and so $y = \frac{1}{x} + 1$.

11. (a) $x = 4t^2$, $y = 8t^3$

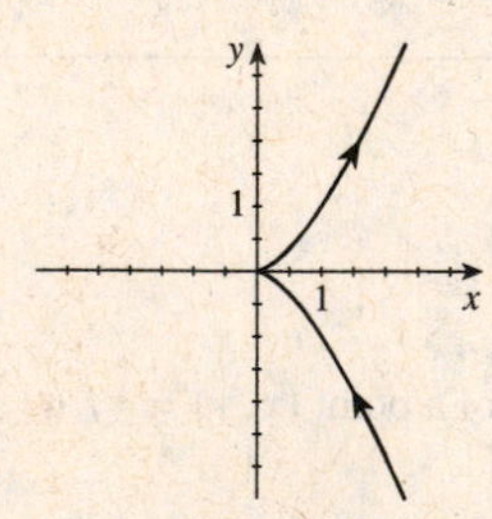

(b) Since $y = 8t^3 \Leftrightarrow y^2 = 64t^6 = \left(4t^2\right)^3 = x^3$, we have $y^2 = x^3$.

13. (a) $x = 2\sin t$, $y = 2\cos t$, $0 \le t \le \pi$

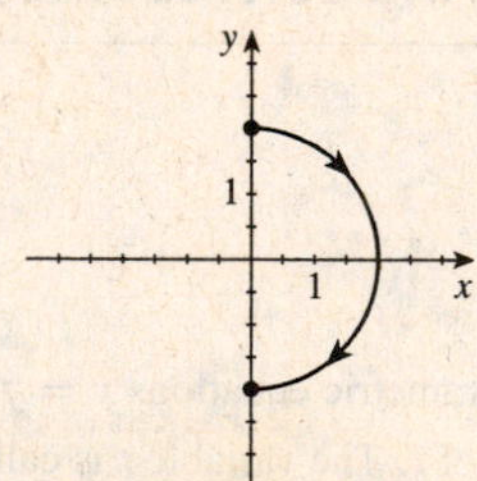

(b) $x^2 = (2\sin t)^2 = 4\sin^2 t$ and $y^2 = 4\cos^2 t$. Hence, $x^2 + y^2 = 4\sin^2 t + 4\cos^2 t = 4 \Leftrightarrow x^2 + y^2 = 4$, where $x \ge 0$.

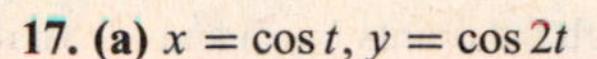

15. (a) $x = \sin^2 t$, $y = \sin^4 t$

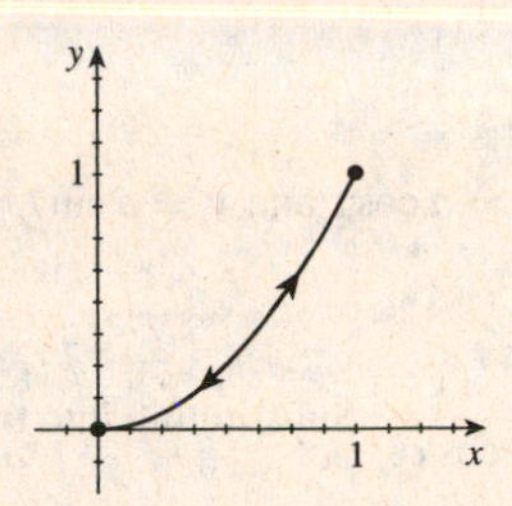

(b) Since $x = \sin^2 t$ we have $x^2 = \sin^4 t$ and so $y = x^2$. But since $0 \le \sin^2 t \le 1$ we only get the part of this parabola for which $0 \le x \le 1$.

17. (a) $x = \cos t$, $y = \cos 2t$

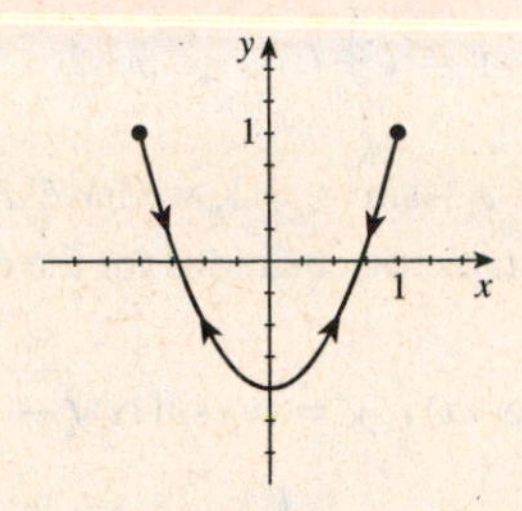

(b) Since $x = \cos t$ we have $x^2 = \cos^2 t$, so $2x^2 - 1 = 2\cos^2 t - 1 = \cos 2t = y$. Hence, the rectangular equation is $y = 2x^2 - 1$, $-1 \le x \le 1$.

19. (a) $x = \sec t$, $y = \tan t$, $0 \le t < \frac{\pi}{2} \Rightarrow x \ge 1$ and $y \ge 0$.

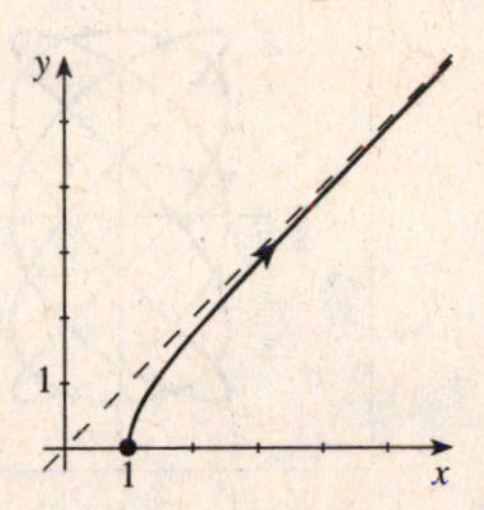

(b) $x^2 = \sec^2 t$, $y^2 = \tan^2 t$, and $y^2 + 1 = \tan^2 t + 1 = \sec^2 t = x^2$. Therefore, $y^2 + 1 = x^2 \Leftrightarrow x^2 - y^2 = 1, x \ge 1, y \ge 0$.

21. (a) $x = \tan t$, $y = \cot t$, $0 < t < \frac{\pi}{2} \Rightarrow x > 0$.

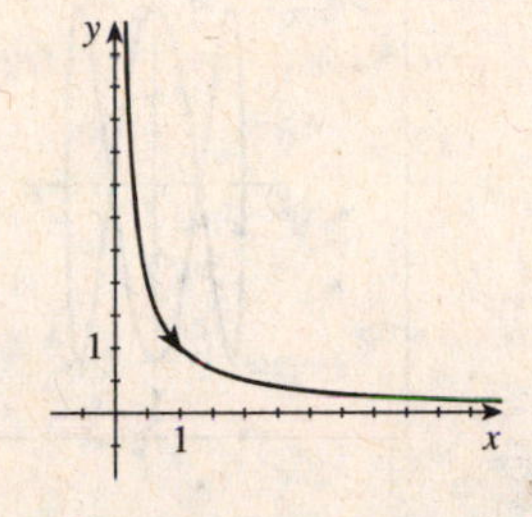

(b) $xy = \tan t \cdot \cot t = 1$, so $y = 1/x$ for $x > 0$.

23. (a) $x = e^{2t}$, $y = e^t$, so $y > 0$.

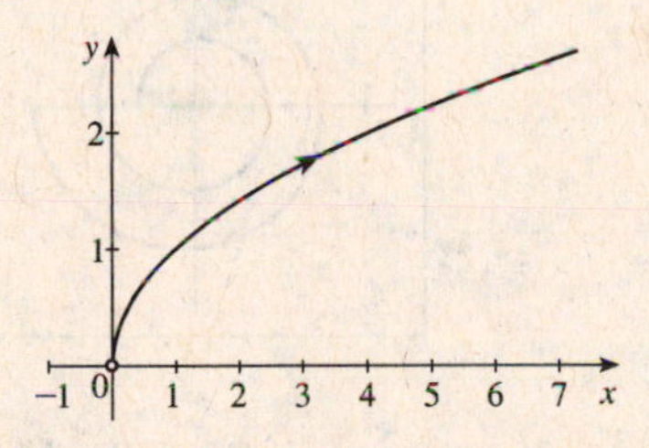

(b) $x = e^{2t} = (e^t)^2 = y^2$, $y > 0$.

25. (a) $x = \cos^2 t$, $y = \sin^2 t$, so $0 \le x \le 1$ and $0 \le y \le 1$.

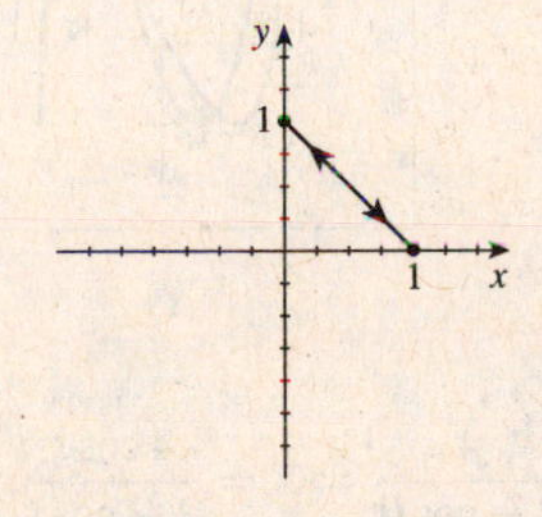

(b) $x + y = \cos^2 t + \sin^2 t = 1$. Hence, the equation is $x + y = 1$ for $0 \le x \le 1$ and $0 \le y \le 1$.

27. $x = 3\cos t$, $y = 3\sin t$. The radius of the circle is 3, the position at time 0 is $(x(0), y(0)) = (3\cos 0, 3\sin 0) = (3, 0)$ and the orientation is counterclockwise (because x is decreasing and y is increasing initially). $(x, y) = (3, 0)$ again when $t = 2\pi$, so it takes 2π units of time to complete one revolution.

29. $x = \sin 2t$, $y = \cos 2t$. The radius of the circle is 1, the position at time 0 is $(x(0), y(0)) = (\sin 0, \cos 0) = (0, 1)$ and the orientation is clockwise (because x is increasing and y is decreasing initially). $(x, y) = (0, 1)$ again when $t = \pi$, so it takes π units of time to complete one revolution.

31. Since the line passes through the point $(4, -1)$ and has slope $\frac{1}{2}$, parametric equations for the line are $x = 4 + t$, $y = -1 + \frac{1}{2}t$.

33. Since the line passes through the points (6, 7) and (7, 8), its slope is $\dfrac{8-7}{7-6} = 1$. Thus, parametric equations for the line are $x = 6 + t$, $y = 7 + t$.

35. Since $\cos^2 t + \sin^2 t = 1$, we have $a^2 \cos^2 t + a^2 \sin^2 t = a^2$. If we let $x = a \cos t$ and $y = a \sin t$, then $x^2 + y^2 = a^2$. Hence, parametric equations for the circle are $x = a \cos t$, $y = a \sin t$.

37. $x = (v_0 \cos \alpha)\, t$, $y = (v_0 \sin \alpha)\, t - 16t^2$. From the equation for x, $t = \dfrac{x}{v_0 \cos \alpha}$. Substituting into the equation for y gives $y = (v_0 \sin \alpha) \dfrac{x}{v_0 \cos \alpha} - 16 \left(\dfrac{x}{v_0 \cos \alpha} \right)^2 = x \tan \alpha - \dfrac{16x^2}{v_0^2 \cos^2 \alpha}$. Thus the equation is of the form $y = c_1 x - c_2 x^2$, where c_1 and c_2 are constants, so its graph is a parabola.

39. $x = \sin t$, $y = 2 \cos 3t$

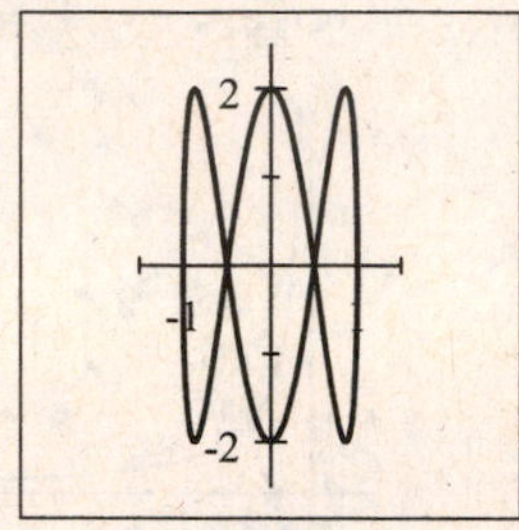

41. $x = 3 \sin 5t$, $y = 5 \cos 3t$

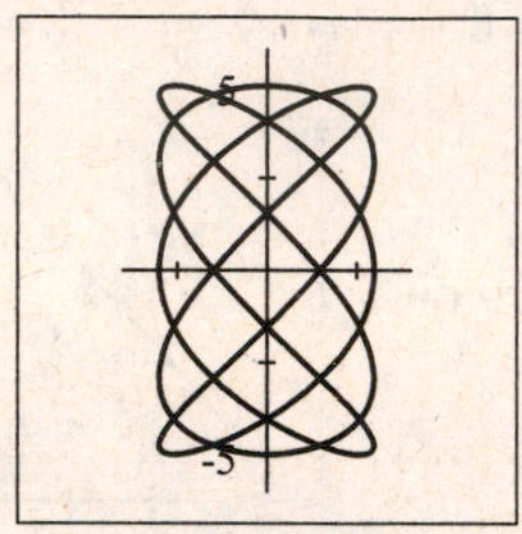

43. $x = \sin(\cos t)$, $y = \cos t^{3/2}$, $0 \le t \le 2\pi$

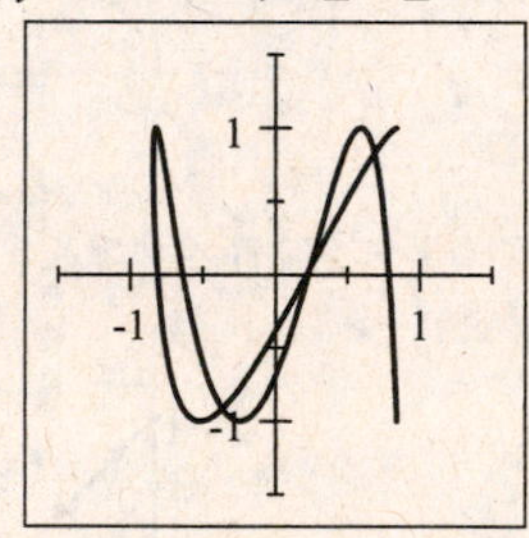

45. **(a)** $r = 2^{\theta/12}$, $0 \le \theta \le 4\pi \Rightarrow x = 2^{t/12} \cos t$, $y = 2^{t/12} \sin t$

(b)

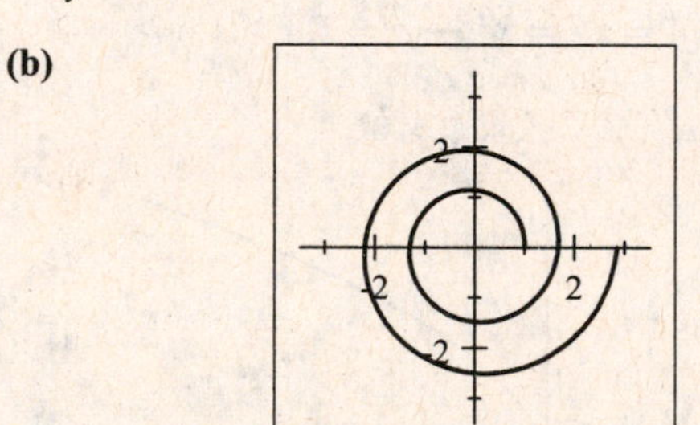

47. **(a)** $r = \dfrac{4}{2 - \cos \theta} \Leftrightarrow x = \dfrac{4 \cos t}{2 - \cos t}$, $y = \dfrac{4 \sin t}{2 - \cos t}$

(b)

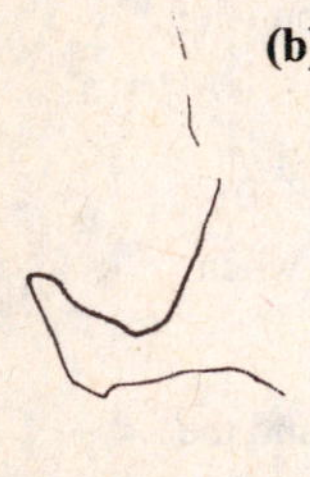

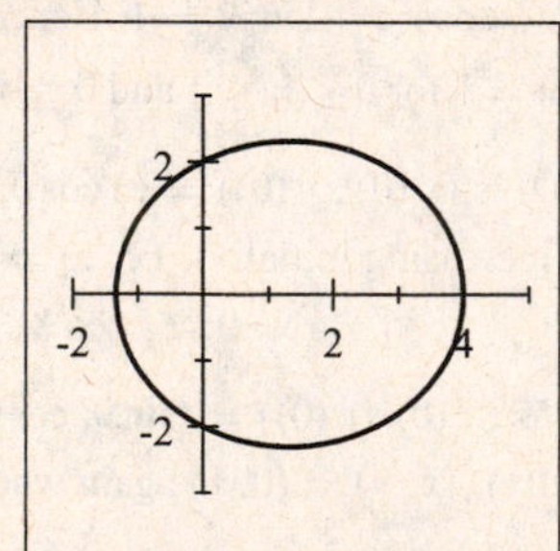

49. $x = t^3 - 2t$, $y = t^2 - t$ is Graph III, since $y = t^2 - t = \left(t^2 - t + \frac{1}{4}\right) - \frac{1}{4} = \left(t - \frac{1}{2}\right)^2 - \frac{1}{4}$, and so $y \ge -\frac{1}{4}$ on this curve, while x is unbounded.

51. $x = t + \sin 2t$, $y = t + \sin 3t$ is Graph II, since the values of x and y oscillate about their values on the line $x = t$, $y = t \Leftrightarrow y = x$.

53. (a) It is apparent that $x = |OQ|$ and $y = |QP| = |ST|$. From the diagram, $x = |OQ| = a\cos\theta$ and $y = |ST| = b\sin\theta$. Thus, parametric equations are $x = a\cos\theta$ and $y = b\sin\theta$.

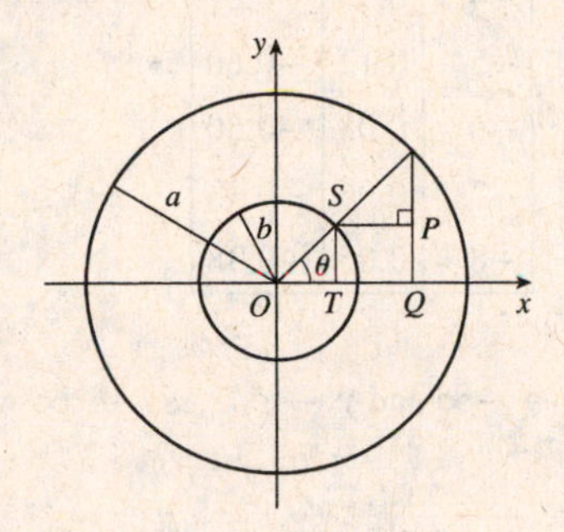

(b) The curve is graphed with $a = 3$ and $b = 2$.

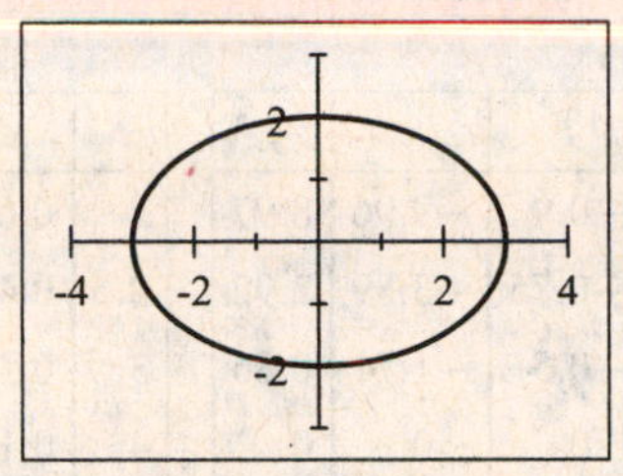

(c) To eliminate θ we rearrange: $\sin\theta = y/b \Rightarrow \sin^2\theta = (y/b)^2$ and $\cos\theta = x/a \Rightarrow \cos^2\theta = (x/a)^2$. Adding the two equations: $\sin^2\theta + \cos^2\theta = 1 = x^2/a^2 + y^2/b^2$. As indicated in part (b), the curve is an ellipse.

55. (a) If we modify Figure 8 so that $|PC| = b$, then by the same reasoning as in Example 6, we see that $x = |OT| - |PQ| = a\theta - b\sin\theta$ and $y = |TC| - |CQ| = a - b\cos\theta$. We graph the case where $a = 3$ and $b = 2$.

(b)

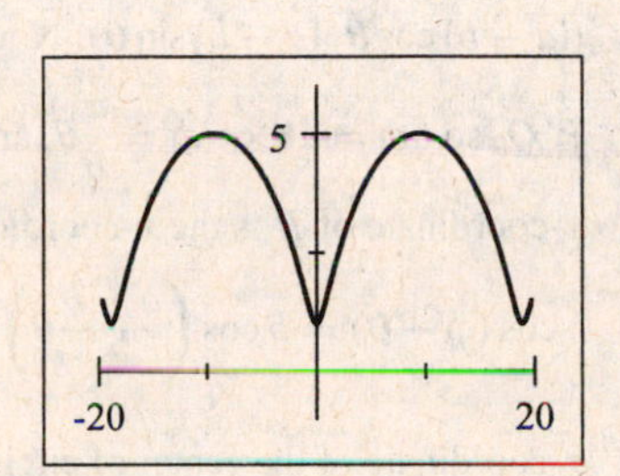

57. $x = a\tan\theta \Leftrightarrow \tan\theta = \dfrac{x}{a} \Rightarrow \tan^2\theta = \dfrac{x^2}{a^2}$. Also, $y = b\sec\theta \Leftrightarrow \sec\theta = \dfrac{y}{b} \Rightarrow \sec^2\theta = \dfrac{y^2}{b^2}$. Since $\tan^2\theta = \sec^2\theta - 1$, we have $\dfrac{x^2}{a^2} = \dfrac{y^2}{b^2} - 1 \Leftrightarrow \dfrac{y^2}{b^2} - \dfrac{x^2}{a^2} = 1$, which is the equation of a hyperbola.

59. $x = t\cos t$, $y = t\sin t$, $t \geq 0$

t	x	y
0	0	0
$\frac{\pi}{4}$	$\frac{\pi\sqrt{2}}{8}$	$\frac{\pi\sqrt{2}}{8}$
$\frac{\pi}{2}$	0	$\frac{\pi}{2}$
$\frac{3\pi}{4}$	$-\frac{3\pi\sqrt{2}}{8}$	$\frac{3\pi\sqrt{2}}{8}$
π	$-\pi$	0

t	x	y
$\frac{5\pi}{4}$	$-\frac{5\pi\sqrt{2}}{8}$	$-\frac{5\pi\sqrt{2}}{8}$
$\frac{3\pi}{2}$	0	$-\frac{3\pi}{2}$
$\frac{7\pi}{4}$	$\frac{7\pi\sqrt{2}}{8}$	$-\frac{7\pi\sqrt{2}}{8}$
2π	2π	0

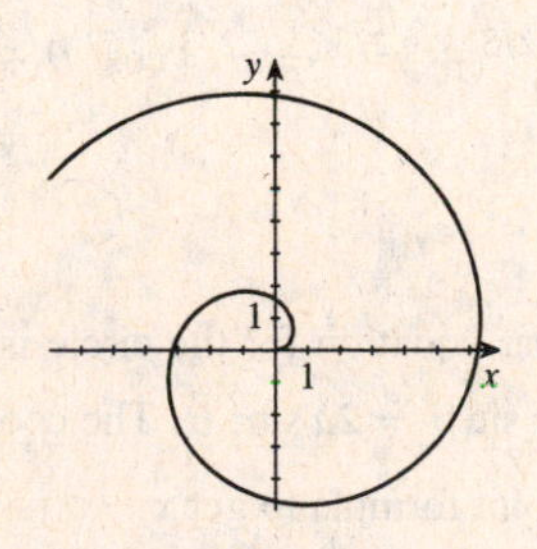

61. $x = \dfrac{3t}{1+t^3}, y = \dfrac{3t^2}{1+t^3}, t \neq -1$

t	x	y
−0.9	−9.96	8.97
−0.75	−3.89	2.92
−0.5	−1.71	0.86
0	0	0
0.5	1.33	0.67
1	1.5	1.5
1.5	1.03	1.54

t	x	y
2	0.67	1.33
2.5	0.45	1.13
3	0.32	0.96
4	0.18	0.74
5	0.12	0.60
6	0.08	0.50

t	x	y
−1.1	9.97	−10.97
−1.25	3.93	−4.92
−1.5	1.89	−2.84
−2	0.86	−1.71
−2.5	0.51	−1.28
−3	0.35	−1.04
−3.5	0.25	−0.87

t	x	y
−4	0.19	−0.76
−4.5	0.15	−0.67
−5	0.12	−0.60
−6	0.08	−0.50
−7	0.06	−0.43
−8	0.05	−0.38

As $t \to -1^-$ we have $x \to \infty$ and $y \to -\infty$. As $t \to -1^+$ we have $x \to -\infty$ and $y \to \infty$. As $t \to \infty$ we have $x \to 0^+$ and $y \to 0^+$. As $t \to -\infty$ we have $x \to 0^+$ and $y \to 0^-$.

63. (a) We first note that the center of circle C (the small circle) has coordinates $([a-b]\cos\theta, [a-b]\sin\theta)$. Now the arc PQ has the same length as the arc $P'Q$, so $b\phi = a\theta \Leftrightarrow \phi = \dfrac{a}{b}\theta$, and so $\phi - \theta = \dfrac{a}{b}\theta - \theta = \dfrac{a-b}{b}\theta$. Thus the x-coordinate of P is the x-coordinate of the center of circle C plus $b\cos(\phi-\theta) = b\cos\left(\dfrac{a-b}{b}\theta\right)$, and the y-coordinate of P is the y-coordinate of the center of circle C minus $b \cdot \sin(\phi - \theta) = b\sin\left(\dfrac{a-b}{b}\theta\right)$.

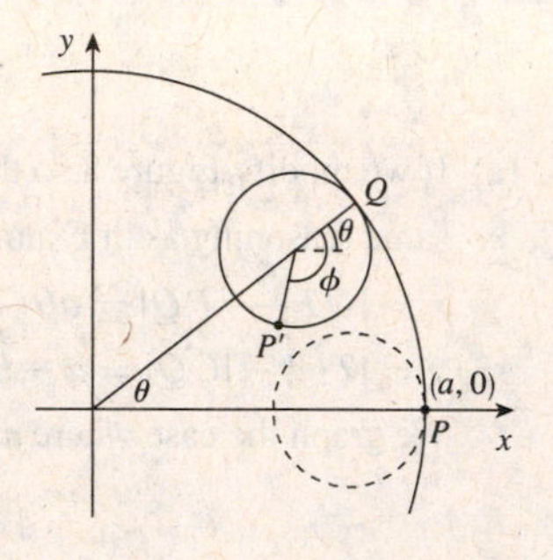

So $x = (a-b)\cos\theta + b\cos\left(\dfrac{a-b}{b}\theta\right)$ and

$y = (a-b)\sin\theta - b\sin\left(\dfrac{a-b}{b}\theta\right)$.

(b) If $a = 4b$, $b = \dfrac{a}{4}$, and $x = \frac{3}{4}a\cos\theta + \frac{1}{4}a\cos 3\theta$, $y = \frac{3}{4}a\sin\theta - \frac{1}{4}a\sin 3\theta$.

From Example 2 in Section 7.3, $\cos 3\theta = 4\cos^3\theta - 3\cos\theta$. Similarly, one can prove that $\sin 3\theta = 3\sin\theta - 4\sin^3\theta$. Substituting, we get

$x = \frac{3}{4}a\cos\theta + \frac{1}{4}a\left(4\cos^3\theta - 3\cos\theta\right) = a\cos^3\theta$

$y = \frac{3}{4}a\sin\theta - \frac{1}{4}a\left(3\sin\theta - 4\sin^3\theta\right) = a\sin^3\theta$. Thus,

$x^{2/3} + y^{2/3} = a^{2/3}\cos^2\theta + a^{2/3}\sin^2\theta = a^{2/3}$, so $x^{2/3} + y^{2/3} = a^{2/3}$.

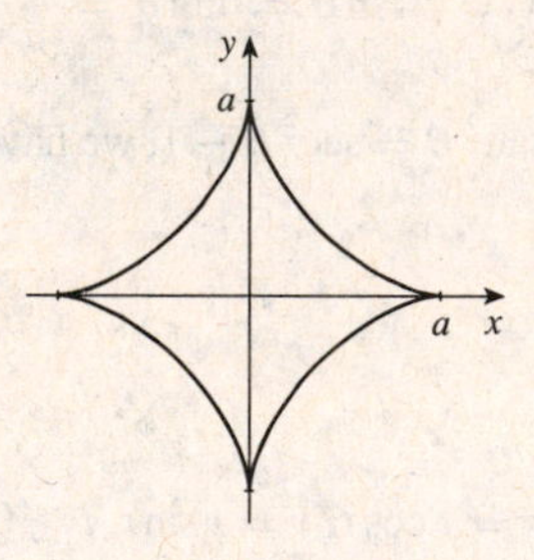

65. A polar equation for the circle is $r = 2a\sin\theta$. Thus the coordinates of Q are $x = r\cos\theta = 2a\sin\theta\cos\theta$ and $y = r\sin\theta = 2a\sin^2\theta$. The coordinates of R are $x = 2a\cot\theta$ and $y = 2a$. Since P is the midpoint of QR, we use the midpoint formula to get $x = a(\sin\theta\cos\theta + \cot\theta)$ and $y = a\left(1 + \sin^2\theta\right)$.

67. We use the equation for y from Example 6 and solve for θ. Thus for $0 \le \theta \le \pi$, $y = a\,(1 - \cos\theta) \Leftrightarrow \dfrac{a-y}{a} = \cos\theta \Leftrightarrow$ $\theta = \cos^{-1}\left(\dfrac{a-y}{a}\right)$. Substituting into the equation for x, we get $x = a\left[\cos^{-1}\left(\dfrac{a-y}{a}\right) - \sin\left(\cos^{-1}\left(\dfrac{a-y}{a}\right)\right)\right]$.

However, $\sin\left(\cos^{-1}\left(\dfrac{a-y}{a}\right)\right) = \sqrt{1 - \left(\dfrac{a-y}{a}\right)^2} = \dfrac{\sqrt{2ay - y^2}}{a}$. Thus, $x = a\left[\cos^{-1}\left(\dfrac{a-y}{a}\right) - \dfrac{\sqrt{2ay - y^2}}{a}\right]$,

and we have $\dfrac{\sqrt{2ay - y^2} + x}{a} = \cos^{-1}\left(\dfrac{a-y}{a}\right) \Rightarrow 1 - \dfrac{y}{a} = \cos\left(\dfrac{\sqrt{2ay - y^2} + x}{a}\right) \Rightarrow$

$y = a\left[1 - \cos\left(\dfrac{\sqrt{2ay - y^2} + x}{a}\right)\right]$.

69. **(a)** In the figure, since OQ and QT are perpendicular and OT and TD are perpendicular, the angles formed by their intersections are equal, that is, $\theta = \angle DTQ$. Now the coordinates of T are $(\cos\theta, \sin\theta)$. Since $|TD|$ is the length of the string that has been unwound from the circle, it must also have arc length θ, so $|TD| = \theta$. Thus the x-displacement from T to D is $\theta \cdot \sin\theta$ while the y-displacement from T to D is $\theta \cdot \cos\theta$. So the coordinates of D are $x = \cos\theta + \theta\sin\theta$ and $y = \sin\theta - \theta\cos\theta$.

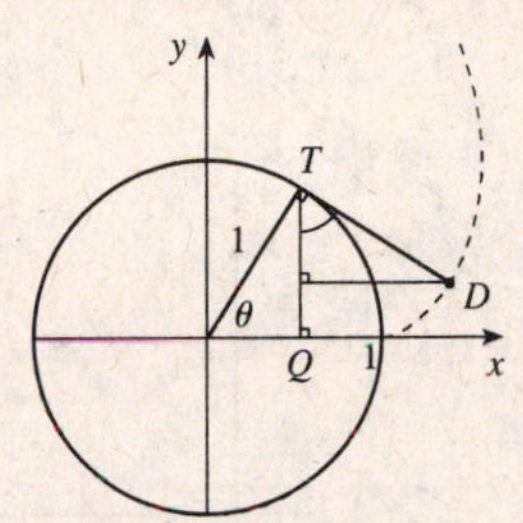

(b)

71. C: $x = t$, $y = t^2$; D: $x = \sqrt{t}$, $y = t$, $t \ge 0$ E: $x = \sin t$, $y = 1 - \cos^2 t$ F: $x = e^t$, $y = e^{2t}$

(a) For C, $x = t$, $y = t^2 \Rightarrow y = x^2$.

For D, $x = \sqrt{t}$, $y = t \Rightarrow y = x^2$.

For E, $x = \sin t \Rightarrow x^2 = \sin^2 t = 1 - \cos^2 t = y$ and so $y = x^2$.

For F, $x = e^t \Rightarrow x^2 = e^{2t} = y$ and so $y = x^2$. Therefore, the points on all four curves satisfy the same rectangular equation.

(b) Curve C is the entire parabola $y = x^2$. Curve D is the right half of the parabola because $t \ge 0$ and so $x \ge 0$. Curve E is the portion of the parabola for $-1 \le x \le 1$. Curve F is the portion of the parabola where $x > 0$, since $e^t > 0$ for all t.

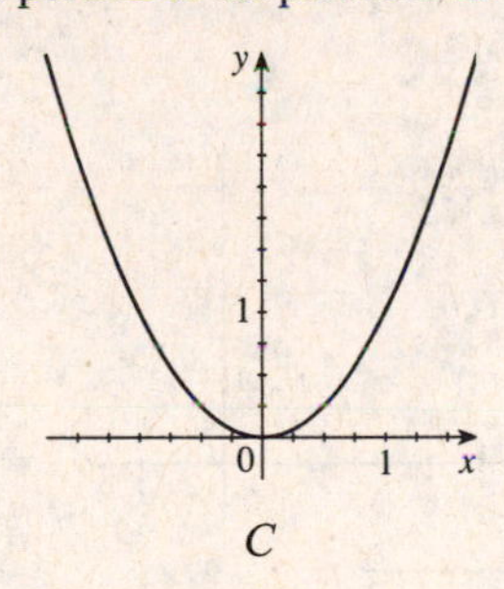

C

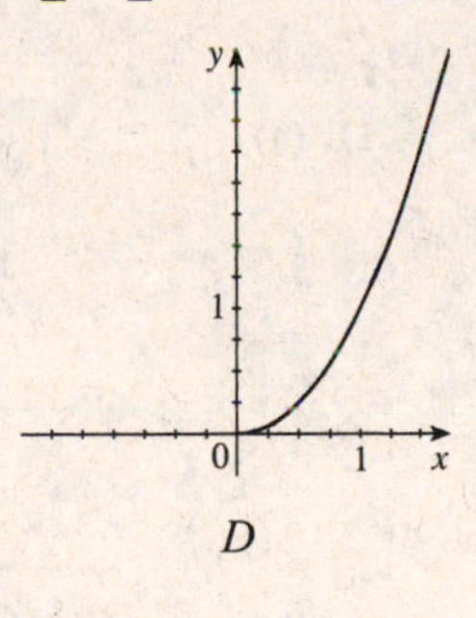

D

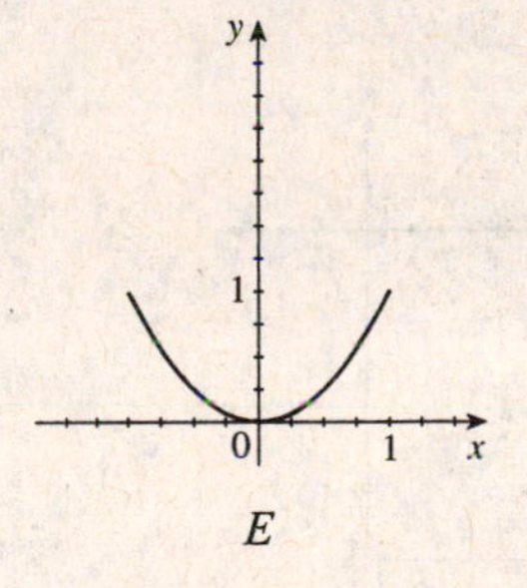

E

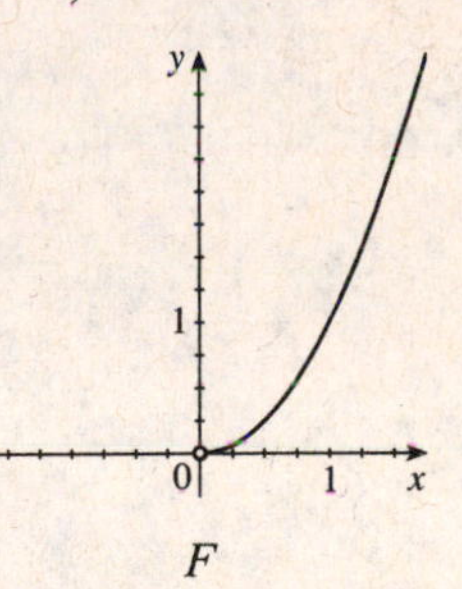

F

CHAPTER 8 REVIEW

1. (a)

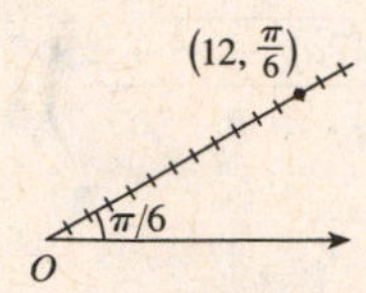

(b) $x = 12\cos\frac{\pi}{6} = 12 \cdot \frac{\sqrt{3}}{2} = 6\sqrt{3}$,

$y = 12\sin\frac{\pi}{6} = 12 \cdot \frac{1}{2} = 6$. Thus, the rectangular coordinates of P are $\left(6\sqrt{3}, 6\right)$.

3. (a)

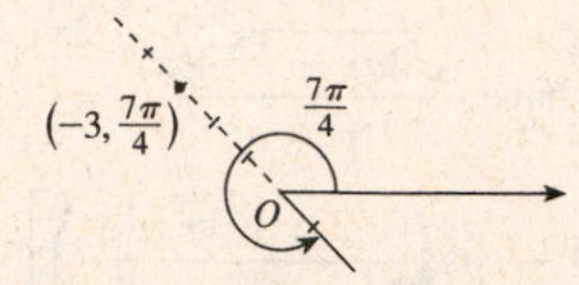

(b) $x = -3\cos\frac{7\pi}{4} = -3\left(\frac{\sqrt{2}}{2}\right) = -\frac{3\sqrt{2}}{2}$,

$y = -3\sin\frac{7\pi}{4} = -3\left(-\frac{\sqrt{2}}{2}\right) = \frac{3\sqrt{2}}{2}$. Thus, the rectangular coordinates of P are $\left(-\frac{3\sqrt{2}}{2}, \frac{3\sqrt{2}}{2}\right)$.

5. (a)

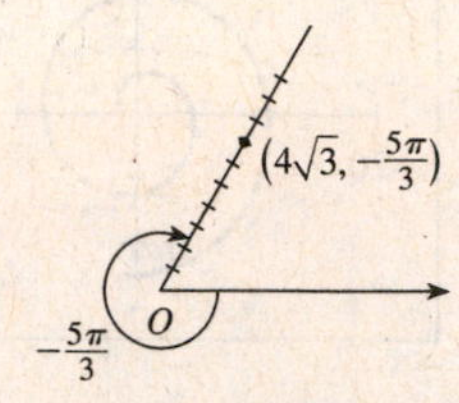

(b) $x = 4\sqrt{3}\cos\left(-\frac{5\pi}{3}\right) = 4\sqrt{3}\left(\frac{1}{2}\right) = 2\sqrt{3}$,

$y = 4\sqrt{3}\sin\left(-\frac{5\pi}{3}\right) = 4\sqrt{3}\left(\frac{\sqrt{3}}{2}\right) = 6$. Thus, the rectangular coordinates of P are $\left(2\sqrt{3}, 6\right)$.

7. (a)

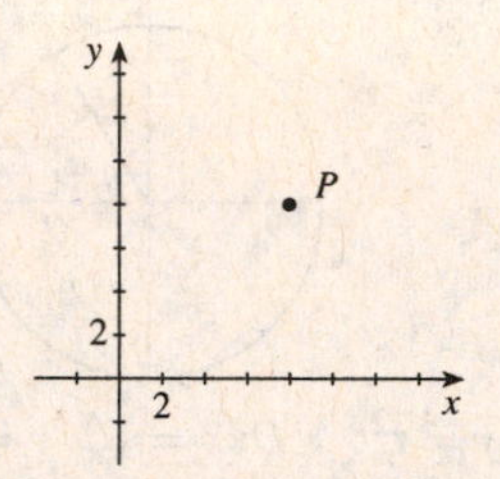

(b) $r = \sqrt{8^2 + 8^2} = \sqrt{128} = 8\sqrt{2}$ and $\overline{\theta} = \tan^{-1}\frac{8}{8}$. Since P is in quadrant I, $\theta = \frac{\pi}{4}$. Polar coordinates for P are $\left(8\sqrt{2}, \frac{\pi}{4}\right)$.

(c) $\left(-8\sqrt{2}, \frac{5\pi}{4}\right)$

9. (a)

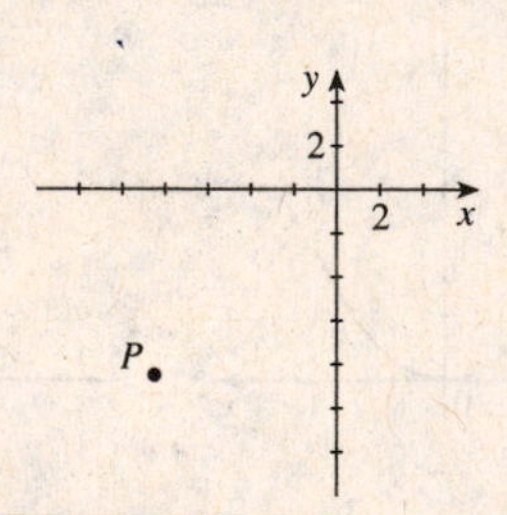

(b) $r = \sqrt{\left(-6\sqrt{2}\right)^2 + \left(-6\sqrt{2}\right)^2} = \sqrt{144} = 12$ and

$\overline{\theta} = \tan^{-1}\frac{-6\sqrt{2}}{-6\sqrt{2}} = \frac{\pi}{4}$. Since P is in quadrant III,

$\theta = \frac{5\pi}{4}$. Polar coordinates for P are $\left(12, \frac{5\pi}{4}\right)$.

(c) $\left(-12, \frac{\pi}{4}\right)$

11. (a)

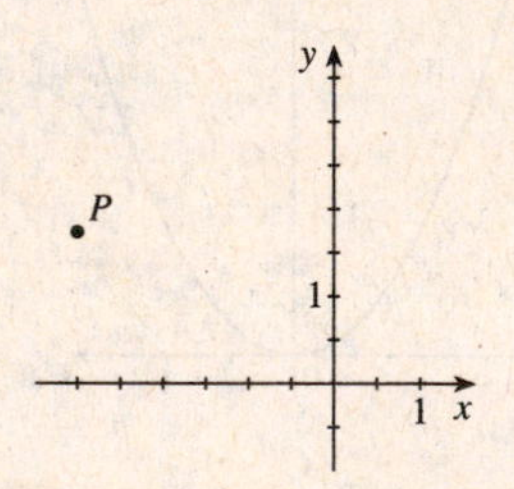

(b) $r = \sqrt{(-3)^2 + \left(\sqrt{3}\right)^2} = \sqrt{12} = 2\sqrt{3}$ and

$\overline{\theta} = \tan^{-1}\frac{\sqrt{3}}{-3}$. Since P is in quadrant II, $\theta = \frac{5\pi}{6}$.

Polar coordinates for P are $\left(2\sqrt{3}, \frac{5\pi}{6}\right)$.

(c) $\left(-2\sqrt{3}, -\frac{\pi}{6}\right)$

13. (a) $x + y = 4 \Leftrightarrow r\cos\theta + r\sin\theta = 4 \Leftrightarrow$

$$r(\cos\theta + \sin\theta) = 4 \Leftrightarrow r = \frac{4}{\cos\theta + \sin\theta}$$

(b) The rectangular equation is easier to graph.

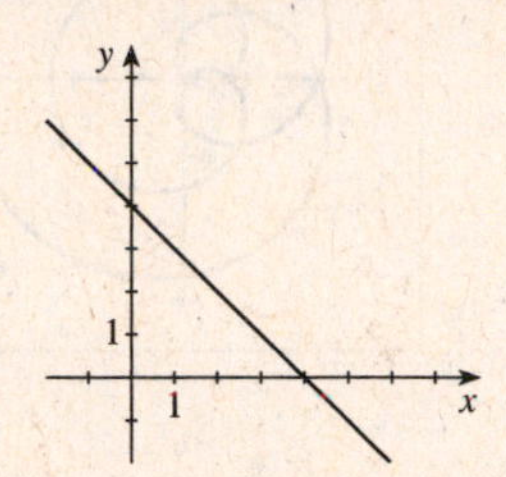

15. (a) $x^2 + y^2 = 4x + 4y \Leftrightarrow r^2 = 4r\cos\theta + 4r\sin\theta \Leftrightarrow$

$$r^2 = r(4\cos\theta + 4\sin\theta) \Leftrightarrow r = 4\cos\theta + 4\sin\theta$$

(b) The polar equation is easier to graph.

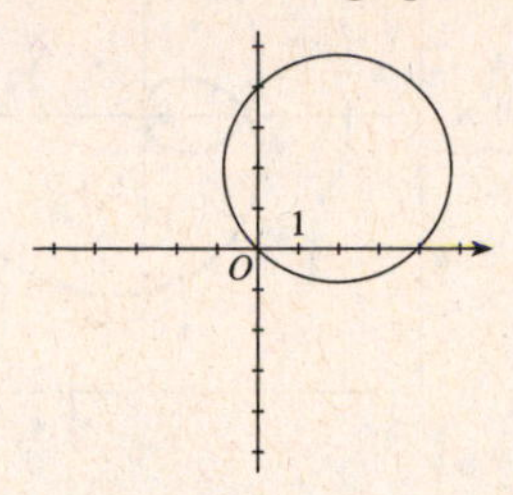

17. (a)

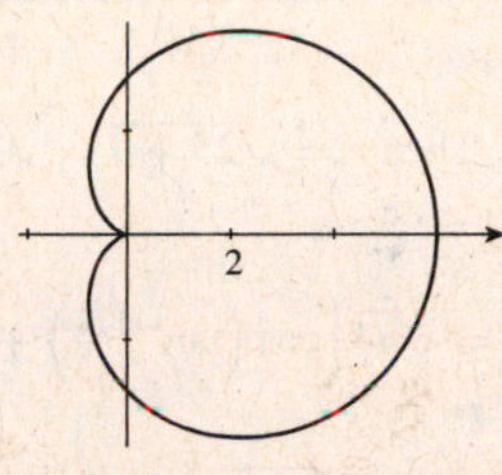

(b) $r = 3 + 3\cos\theta \Leftrightarrow r^2 = 3r + 3r\cos\theta$, which gives

$x^2 + y^2 = 3\sqrt{x^2 + y^2} + 3x \Leftrightarrow$

$x^2 - 3x + y^2 = 3\sqrt{x^2 + y^2}$. Squaring both sides

gives $\left(x^2 - 3x + y^2\right)^2 = 9\left(x^2 + y^2\right)$.

19. (a)

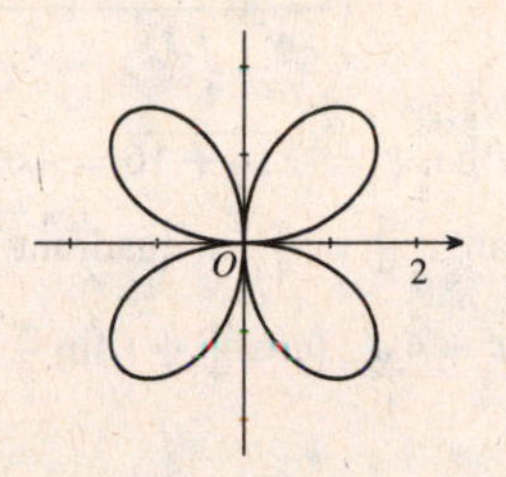

(b) $r = 2\sin 2\theta \Leftrightarrow r = 2 \cdot 2\sin\theta\cos\theta \Leftrightarrow$

$r^3 = 4r^2\sin\theta\cos\theta \Leftrightarrow$

$\left(r^2\right)^{3/2} = 4(r\sin\theta)(r\cos\theta)$ and so, since

$x = r\cos\theta$ and $y = r\sin\theta$, we get

$\left(x^2 + y^2\right)^3 = 16x^2y^2$.

21. (a)

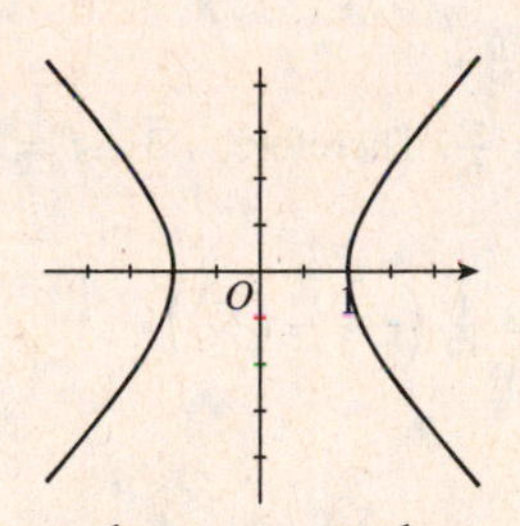

(b) $r^2 = \sec 2\theta = \dfrac{1}{\cos 2\theta} = \dfrac{1}{\cos^2\theta - \sin^2\theta} \Leftrightarrow$

$r^2\left(\cos^2\theta - \sin^2\theta\right) = 1 \Leftrightarrow$

$r^2\cos^2\theta - r^2\sin^2\theta = 1 \Leftrightarrow$

$(r\cos\theta)^2 - (r\sin\theta)^2 = 1 \Leftrightarrow x^2 - y^2 = 1$.

23. (a)

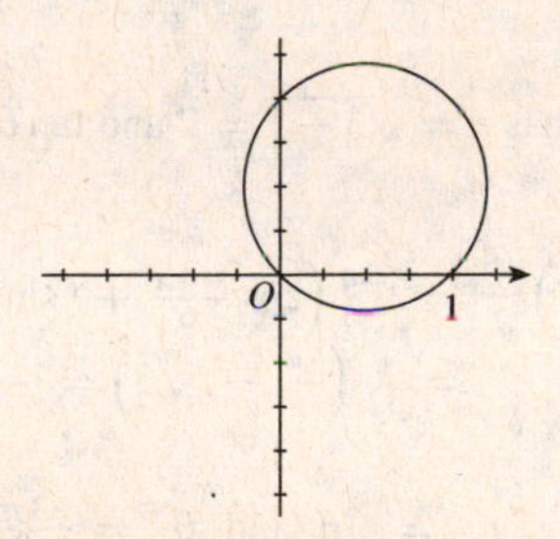

(b) $r = \sin\theta + \cos\theta \Leftrightarrow r^2 = r\sin\theta + r\cos\theta$, so

$x^2 + y^2 = y + x \quad \Leftrightarrow$

$\left(x^2 - x + \frac{1}{4}\right) + \left(y^2 - y + \frac{1}{4}\right) = \frac{1}{2} \Leftrightarrow$

$\left(x - \frac{1}{2}\right)^2 + \left(y - \frac{1}{2}\right)^2 = \frac{1}{2}$.

25. $r = \cos(\theta/3)$, $\theta \in [0, 3\pi]$.

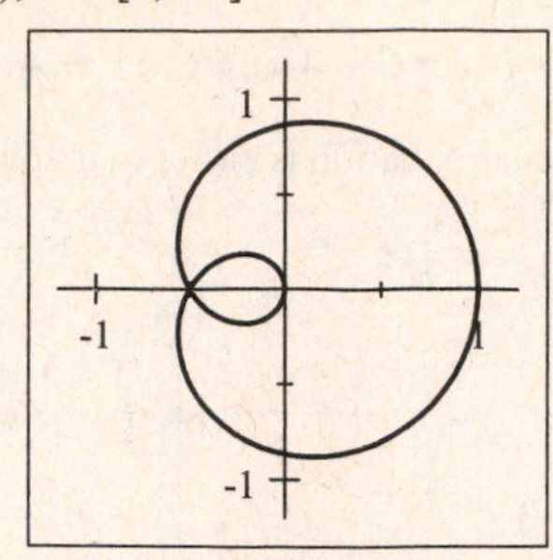

27. $r = 1 + 4\cos(\theta/3)$, $\theta \in [0, 6\pi]$.

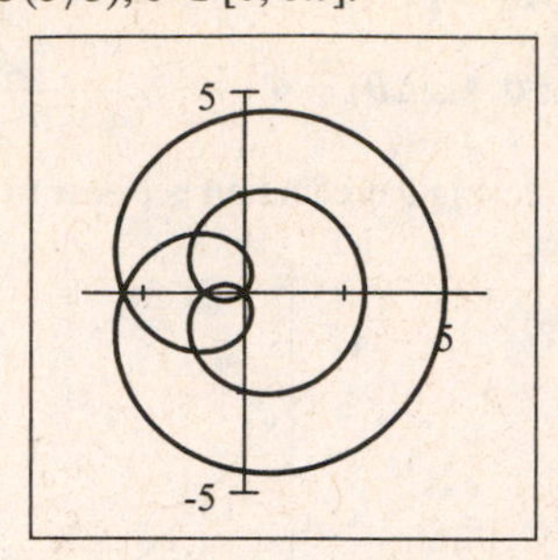

29. (a)

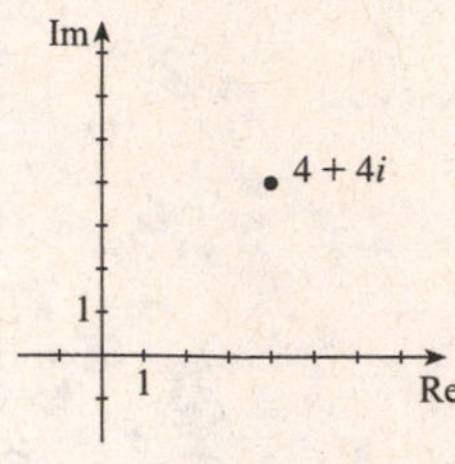

(b) $4 + 4i$ has $r = \sqrt{16 + 16} = 4\sqrt{2}$, and

$\theta = \tan^{-1} \frac{4}{4} = \frac{\pi}{4}$ (in quadrant I).

(c) $4 + 4i = 4\sqrt{2}\left(\cos \frac{\pi}{4} + i \sin \frac{\pi}{4}\right)$

31. (a)

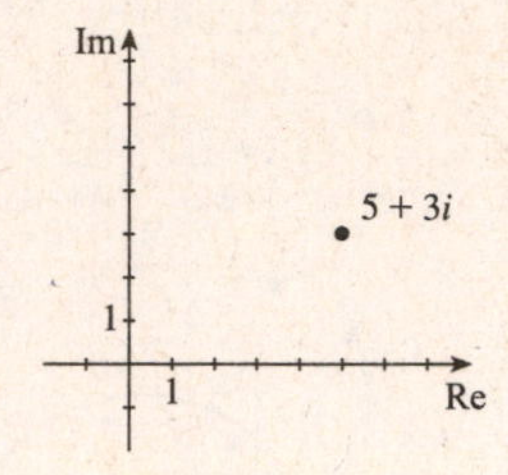

(b) $5 + 3i$. Then $r = \sqrt{25 + 9} = \sqrt{34}$, and

$\theta = \tan^{-1} \frac{3}{5}$.

(c) $5 + 3i = \sqrt{34}\left[\cos\left(\tan^{-1} \frac{3}{5}\right) + i \sin\left(\tan^{-1} \frac{3}{5}\right)\right]$

33. (a)

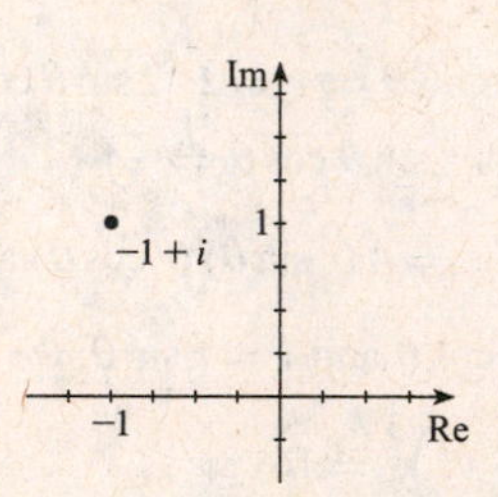

(b) $-1 + i$ has $r = \sqrt{1 + 1} = \sqrt{2}$ and $\tan\theta = \dfrac{1}{-1}$ with

θ in quadrant II $\Leftrightarrow \theta = \frac{3\pi}{4}$.

(c) $-1 + i = \sqrt{2}\left(\cos \frac{3\pi}{4} + i \sin \frac{3\pi}{4}\right)$

35. $1 - \sqrt{3}i$ has $r = \sqrt{1 + 3} = 2$ and $\tan\theta = \frac{-\sqrt{3}}{1} = -\sqrt{3}$ with θ in quadrant III $\Leftrightarrow \theta = \frac{5\pi}{3}$. Therefore,

$1 - \sqrt{3}i = 2\left(\cos \frac{5\pi}{3} + i \sin \frac{5\pi}{3}\right)$, and so

$$\begin{aligned}\left(1 - \sqrt{3}i\right)^4 &= 2^4\left(\cos \tfrac{20\pi}{3} + i \sin \tfrac{20\pi}{3}\right)\\ &= 16\left(\cos \tfrac{2\pi}{3} + i \sin \tfrac{2\pi}{3}\right)\\ &= 16\left(-\tfrac{1}{2} + i\tfrac{\sqrt{3}}{2}\right) = 8\left(-1 + i\sqrt{3}\right).\end{aligned}$$

37. $\sqrt{3} + i$ has $r = \sqrt{3 + 1} = 2$ and $\tan\theta = \frac{1}{\sqrt{3}}$ with θ in quadrant I $\Leftrightarrow \theta = \frac{\pi}{6}$. Therefore, $\sqrt{3} + i = 2\left(\cos \frac{\pi}{6} + i \sin \frac{\pi}{6}\right)$, and so

$$\begin{aligned}\left(\sqrt{3} + i\right)^{-4} &= 2^{-4}\left(\cos \tfrac{-4\pi}{6} + i \sin \tfrac{-4\pi}{6}\right) = \tfrac{1}{16}\left(\cos \tfrac{2\pi}{3} - i \sin \tfrac{2\pi}{3}\right) = \tfrac{1}{16}\left(-\tfrac{1}{2} - i\tfrac{\sqrt{3}}{2}\right)\\ &= \tfrac{1}{32}\left(-1 - i\sqrt{3}\right) = -\tfrac{1}{32}\left(1 + i\sqrt{3}\right)\end{aligned}$$

39. $-16i$ has $r = 16$ and $\theta = \frac{3\pi}{2}$. Thus, $-16i = 16\left(\cos \frac{3\pi}{2} + i \sin \frac{3\pi}{2}\right)$ and so $(-16i)^{1/2} = 16^{1/2}\left[\cos\left(\dfrac{3\pi + 4k\pi}{4}\right) + i \sin\left(\dfrac{3\pi + 4k\pi}{4}\right)\right]$ for $k = 0, 1$. Thus the roots are $w_0 = 4\left(\cos \frac{3\pi}{4} + i \sin \frac{3\pi}{4}\right) = 4\left(-\frac{1}{\sqrt{2}} + i\frac{1}{\sqrt{2}}\right) = 2\sqrt{2}(-1 + i)$ and $w_1 = 4\left(\cos \frac{7\pi}{4} + i \sin \frac{7\pi}{4}\right) = 4\left(\frac{1}{\sqrt{2}} - i\frac{1}{\sqrt{2}}\right) = 2\sqrt{2}(1 - i)$.

41. $1 = \cos 0 + i\sin 0$. Then $1^{1/6} = 1\left(\cos\dfrac{2k\pi}{6} + i\sin\dfrac{2k\pi}{6}\right)$ for $k = 0, 1, 2, 3, 4, 5$. Thus the six roots are $w_0 = 1(\cos 0 + i\sin 0) = 1$, $w_1 = 1\left(\cos\frac{\pi}{3} + i\sin\frac{\pi}{3}\right) = \frac{1}{2} + i\frac{\sqrt{3}}{2}$, $w_2 = 1\left(\cos\frac{2\pi}{3} + i\sin\frac{2\pi}{3}\right) = -\frac{1}{2} + i\frac{\sqrt{3}}{2}$, $w_3 = 1(\cos\pi + i\sin\pi) = -1$, $w_4 = 1\left(\cos\frac{4\pi}{3} + i\sin\frac{4\pi}{3}\right) = -\frac{1}{2} - i\frac{\sqrt{3}}{2}$, and $w_5 = 1\left(\cos\frac{5\pi}{3} + i\sin\frac{5\pi}{3}\right) = \frac{1}{2} - i\frac{\sqrt{3}}{2}$.

43. (a)

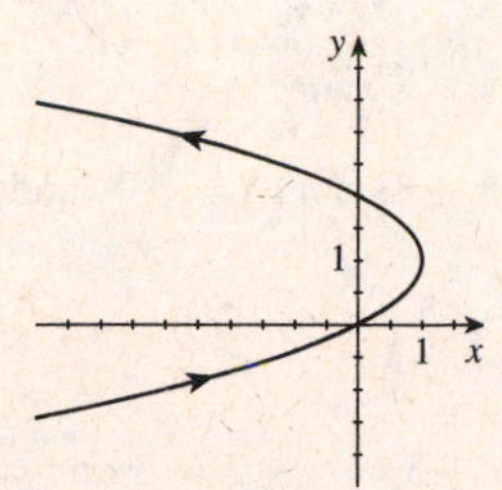

(b) $x = 1 - t^2$, $y = 1 + t \Leftrightarrow t = y - 1$. Substituting for t gives $x = 1 - (y-1)^2 \Leftrightarrow x = 2y - y^2$ in rectangular coordinates.

45. (a)

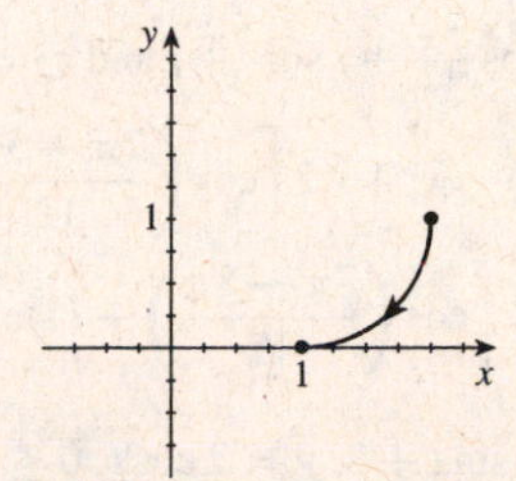

(b) $x = 1 + \cos t \Leftrightarrow \cos t = x - 1$, and $y = 1 - \sin t \Leftrightarrow \sin t = 1 - y$. Since $\cos^2 t + \sin^2 t = 1$, it follows that $(x-1)^2 + (1-y)^2 = 1 \Leftrightarrow (x-1)^2 + (y-1)^2 = 1$. Since t is restricted by $0 \le t \le \frac{\pi}{2}$, $1 + \cos 0 \le x \le 1 + \cos\frac{\pi}{2} \Leftrightarrow 1 \le x \le 2$, and similarly, $0 \le y \le 1$. (This is the lower right quarter of the circle.)

47. $x = \cos 2t$, $y = \sin 3t$

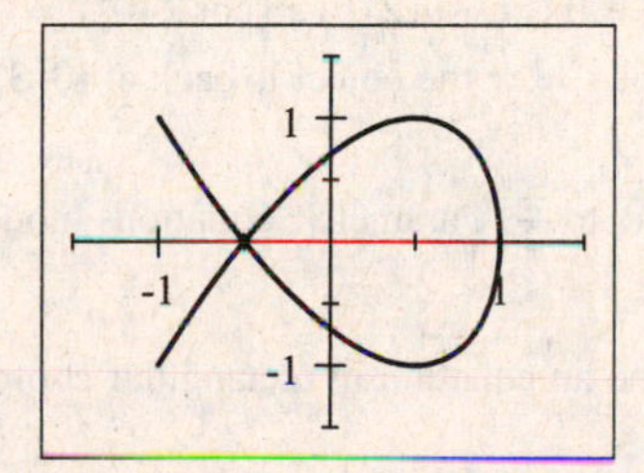

49. The coordinates of Q are $x = \cos\theta$ and $y = \sin\theta$. The coordinates of R are $x = 1$ and $y = \tan\theta$. Hence, the midpoint P is $\left(\dfrac{1+\cos\theta}{2}, \dfrac{\sin\theta + \tan\theta}{2}\right)$, so parametric equations for the curve are $x = \dfrac{1+\cos\theta}{2}$ and $y = \dfrac{\sin\theta + \tan\theta}{2}$.

CHAPTER 8 TEST

1. (a) $x = 8\cos\frac{5\pi}{4} = 8\left(-\frac{\sqrt{2}}{2}\right) = -4\sqrt{2}$, $y = 8\sin\frac{5\pi}{4} = 8\left(-\frac{\sqrt{2}}{2}\right) = -4\sqrt{2}$. So the point has rectangular coordinates $\left(-4\sqrt{2}, -4\sqrt{2}\right)$.

(b) $P = \left(-6, 2\sqrt{3}\right)$ in rectangular coordinates. So $\tan\theta = \frac{2\sqrt{3}}{-6}$ and the reference angle is $\overline{\theta} = \frac{\pi}{6}$. Since P is in quadrant II, we have $\theta = \frac{5\pi}{6}$. Next, $r^2 = (-6)^2 + \left(2\sqrt{3}\right)^2 = 36 + 12 = 48$, so $r = 4\sqrt{3}$. Thus, polar coordinates for the point are $\left(4\sqrt{3}, \frac{5\pi}{6}\right)$ or $\left(-4\sqrt{3}, \frac{11\pi}{6}\right)$.

3. 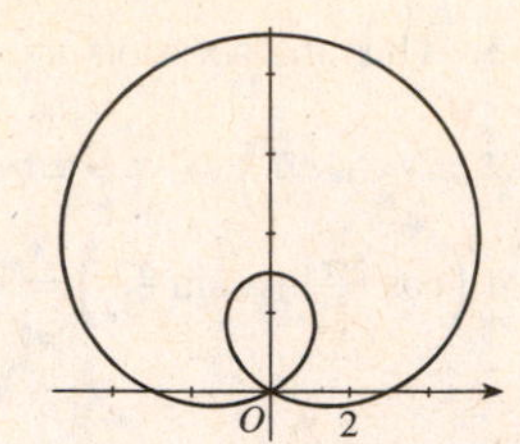

The curve is a limaçon.

5. $z_1 = 4\left(\cos\frac{7\pi}{12} + i\sin\frac{7\pi}{12}\right)$ and $z_2 = 2\left(\cos\frac{5\pi}{12} + i\sin\frac{5\pi}{12}\right)$.

Then $z_1 z_2 = 4 \cdot 2\left[\cos\left(\frac{7\pi + 5\pi}{12}\right) + i\sin\left(\frac{7\pi + 5\pi}{12}\right)\right] = 8(\cos\pi + i\sin\pi) = -8$ and

$z_1/z_2 = \frac{4}{2}\left[\cos\left(\frac{7\pi - 5\pi}{12}\right) + i\sin\left(\frac{7\pi - 5\pi}{12}\right)\right] = 2\left(\cos\frac{\pi}{6} + i\sin\frac{\pi}{6}\right) = 2\left(\frac{\sqrt{3}}{2} + \frac{1}{2}i\right) = \sqrt{3} + i$.

7. **(a)** $x = 3\sin t + 3$, $y = 2\cos t$, $0 \le t \le \pi$. From the work of part (b), we see that this is the half-ellipse shown.

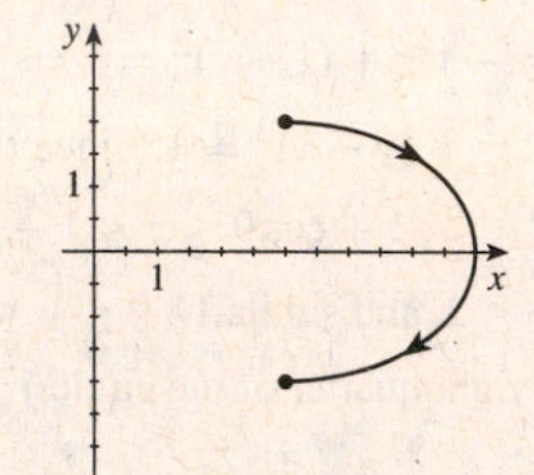

(b) $x = 3\sin t + 3 \Leftrightarrow x - 3 = 3\sin t \Leftrightarrow \frac{x-3}{3} = \sin t$.

Squaring both sides gives $\frac{(x-3)^2}{9} = \sin^2 t$. Similarly, $y = 2\cos t \Leftrightarrow \frac{y}{2} = \cos t$, and squaring both sides gives $\frac{y^2}{4} = \cos^2 t$. Since $\sin^2 t + \cos^2 t = 1$, it follows that $\frac{(x-3)^2}{9} + \frac{y^2}{4} = 1$. Since $0 \le t \le \pi$, $\sin t \ge 0$, so $3\sin t \ge 0 \Rightarrow 3\sin t + 3 \ge 3$, and so $x \ge 3$. Thus the curve consists of only the right half of the ellipse.

9. **(a)** $x = 3\sin 2t$, $y = 3\cos 2t$. The radius is 3 and the position at time $t = 0$ is $(3\sin 2(0), 3\cos 2(0)) = (0, 3)$. Initially x is increasing and y is decreasing, so the motion is clockwise. At time $t = \pi$ the object is back at $(0, 3)$, so it takes π units of time to complete one revolution.

(b) If the speed is doubled, the time to complete one revolution is halved, to $\frac{\pi}{2}$. Parametric equations modeling this motion are $x = 3\sin 4t$, $y = 3\cos 4t$.

(c) $x^2 + y^2 = (3\sin 2t)^2 + (3\cos 2t)^2 = 9\left(\sin^2 2t + \cos^2 2t\right) = 9$, so an equation in rectangular coordinates is $x^2 + y^2 = 9$.

(d) In polar coordinates, an equation is $r = 3$.

FOCUS ON MODELING The Path of a Projectile

1. From $x = (v_0\cos\theta)\,t$, we get $t = \frac{x}{v_0\cos\theta}$. Substituting this value for t into the equation for y, we get

$y = (v_0\sin\theta)\,t - \frac{1}{2}gt^2 \Leftrightarrow y = (v_0\sin\theta)\left(\frac{x}{v_0\cos\theta}\right) - \frac{1}{2}g\left(\frac{x}{v_0\cos\theta}\right)^2 \Leftrightarrow y = (\tan\theta)\,x - \frac{g}{2v_0^2\cos^2\theta}x^2$. This shows that y is a quadratic function of x, so its graph is a parabola as long as $\theta \ne 90°$. When $\theta = 90°$, the path of the projectile is a straight line up (then down).

3. **(a)** We find the time at which the projectile hits the ground using the equation $t = \frac{2v_0\sin\theta}{g}$, where $v_0 = 1000$ and $g = -16$. Since the rocket is fired 5° from the *vertical* axis and θ is measured from the horizontal, we have $\theta = 90° - 5° = 85°$. Then $t = \frac{2000\sin 85°}{32} \approx 62.26$, so the rocket is in the air for 62.26 seconds.

(b) Substituting the given values into $y = (v_0 \sin\theta) t - \frac{1}{2} g t^2$, we get $y = (1000 \sin 85°) t - 16t^2 \approx 996t - 16t^2$. Then $y = f(t) = at^2 + bt + c$, where $a = -16$, $b = 996$, and $c = 0$. So y is a quadratic function whose maximum value is attained at $t = -\frac{b}{2a} = -\frac{996}{2(-16)} = 31.125$, and the maximum value is $f(31.125) = 996(31.125) - 16(31.125)^2 \approx 15{,}500$ (see Section 3.1 for a guide to finding the maximum value of a quadratic function). Thus, the rocket reaches a maximum height of 15,500 feet.

(c) We use the equation $x = (v_0 \cos\theta) t$ to find the horizontal distance traveled after t seconds. Since the rocket hits the ground after 62.26 seconds, substituting into the equation for horizontal distance gives $x = (1000 \cos 85°) 62.26 \approx 5426$. Thus, the rocket travels a horizontal distance of 5426 feet.

(d)

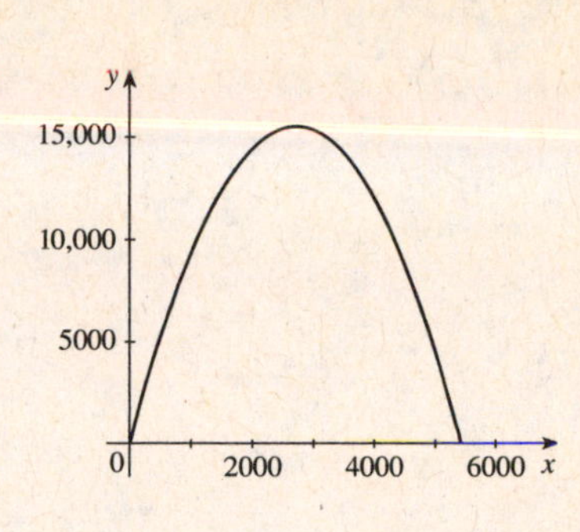

5. We use the equation of the parabola from Exercise 1 and find its vertex:

$$y = (\tan\theta) x - \frac{g}{2v_0^2 \cos^2\theta} x^2 \Leftrightarrow y = -\frac{g}{2v_0^2 \cos^2\theta}\left[x^2 - \frac{2v_0^2 \sin\theta\cos\theta x}{g}\right] \Leftrightarrow$$

$$y = -\frac{g}{2v_0^2 \cos^2\theta}\left[x^2 - \frac{2v_0^2 \sin\theta\cos\theta x}{g} + \left(\frac{v_0^2 \sin\theta\cos\theta}{g}\right)^2\right] + \frac{g}{2v_0^2 \cos^2\theta} \cdot \left(\frac{v_0^2 \sin\theta\cos\theta}{g}\right)^2 \Leftrightarrow$$

$y = -\dfrac{g}{2v_0^2 \cos^2\theta}\left[x - \dfrac{v_0^2 \sin\theta\cos\theta}{g}\right]^2 + \dfrac{v_0^2 \sin^2\theta}{2g}$. Thus the vertex is at $\left(\dfrac{v_0^2 \sin\theta\cos\theta}{g}, \dfrac{v_0^2 \sin^2\theta}{2g}\right)$, so the maximum height is $\dfrac{v_0^2 \sin^2\theta}{2g}$.

7. In Exercise 6 we derived the equations $x = (v_0 \cos\theta - w) t$, $y = (v_0 \sin\theta) t - \frac{1}{2} g t^2$. We plot the graphs for the given values of v_0, w, and θ in the figure to the right. The projectile will be blown backwards if the horizontal component of its velocity is less than the speed of the wind, that is, $32 \cos\theta < 24 \Leftrightarrow \cos\theta < \frac{3}{4} \Rightarrow \theta > 41.4°$.

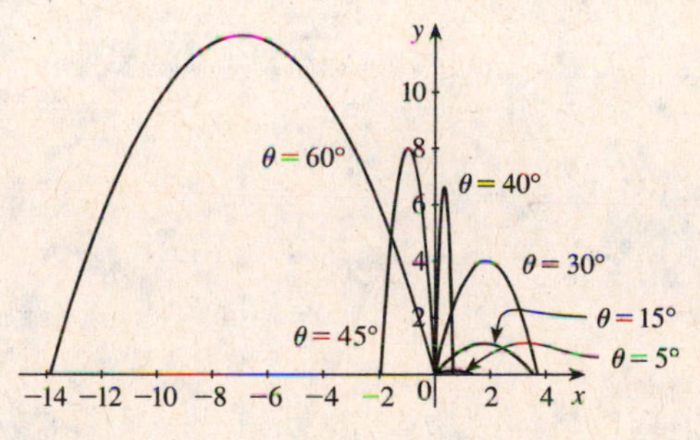

The optimal firing angle appears to be between 15° and 30°. We graph the trajectory for $\theta = 20°$, $\theta = 23°$, and $\theta = 25°$. The solution appears to be close to 23°.

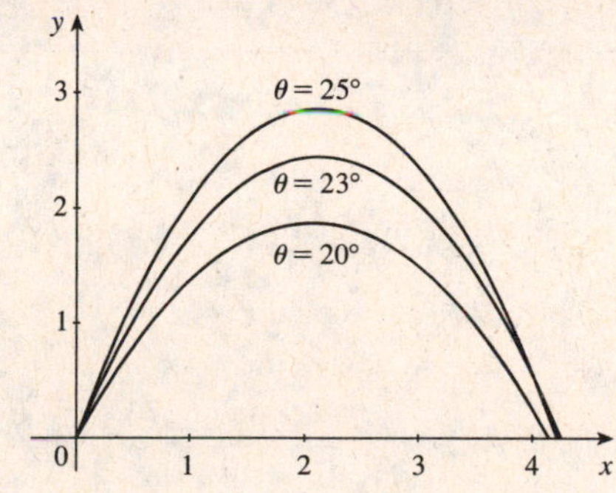

9 VECTORS IN TWO AND THREE DIMENSIONS

9.1 VECTORS IN TWO DIMENSIONS

1. (a) The vector **u** has initial point A and terminal point B.

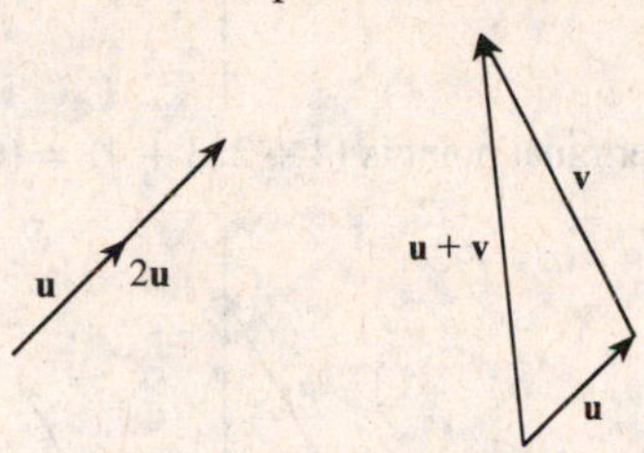

(b) The vector **u** has initial point (2, 1) and terminal point (4, 3). In component form we write $\mathbf{u} = \langle 2, 2\rangle$ and $\mathbf{v} = \langle -3, 6\rangle$. Then $2\mathbf{u} = \langle 4, 4\rangle$ and $\mathbf{u} + \mathbf{v} = \langle -1, 8\rangle$.

3. $2\mathbf{u} = 2\langle -2, 3\rangle = \langle -4, 6\rangle$

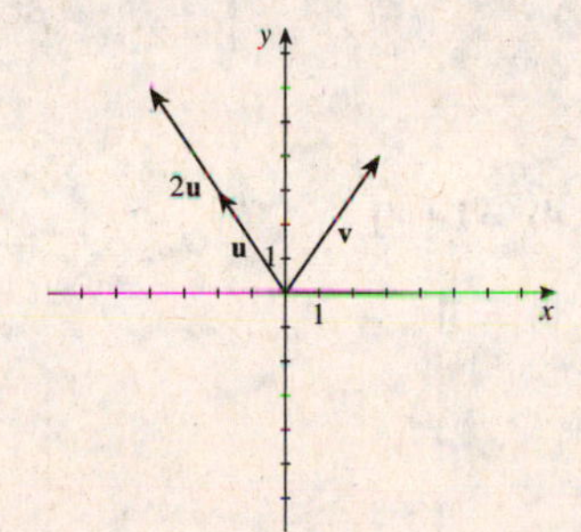

5. $\mathbf{u} + \mathbf{v} = \langle -2, 3\rangle + \langle 3, 4\rangle = \langle -2 + 3, 3 + 4\rangle = \langle 1, 7\rangle$

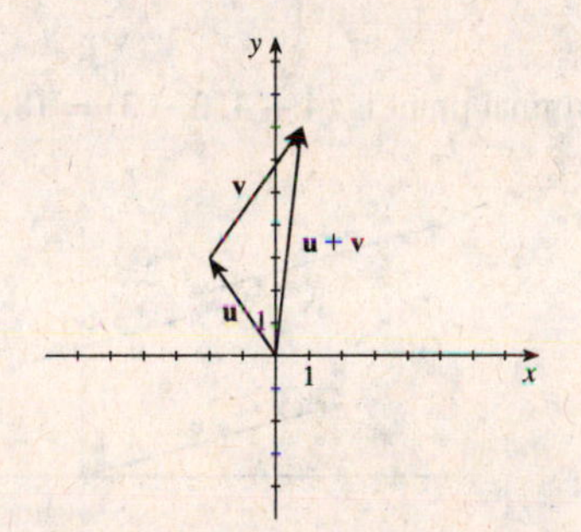

7. $\mathbf{v} - 2\mathbf{u} = \langle 3, 4\rangle - 2\langle -2, 3\rangle = \langle 3 - 2(-2), 4 - 2(3)\rangle$

$= \langle 7, -2\rangle$

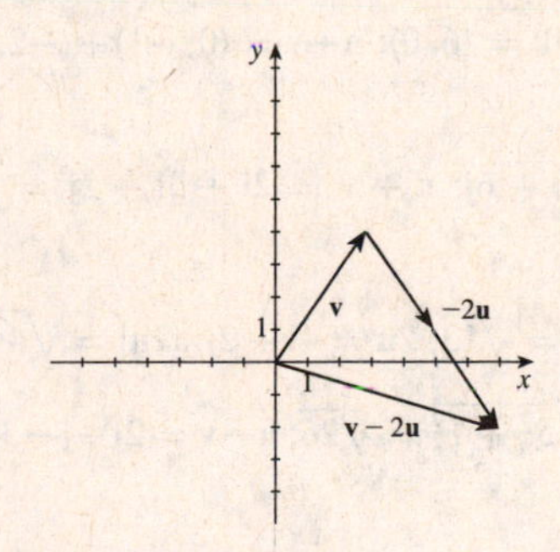

In Solutions 9–17, v represents the vector with initial point P and terminal point Q.

9. $P(2,1)$, $Q(5,4)$. $\mathbf{v} = \langle 5-2, 4-1\rangle = \langle 3,3\rangle$

11. $P(1,2)$, $Q(4,1)$. $\mathbf{v} = \langle 4-1, 1-2\rangle = \langle 3,-1\rangle$

13. $P(3,2)$, $Q(8,9)$. $\mathbf{v} = \langle 8-3, 9-2\rangle = \langle 5,7\rangle$

15. $P(5,3)$, $Q(1,0)$. $\mathbf{v} = \langle 1-5, 0-3\rangle = \langle -4,-3\rangle$

17. $P(-1,-1)$, $Q(-1,1)$.
$\mathbf{v} = \langle -1-(-1), 1-(-1)\rangle = \langle 0,2\rangle$

19.

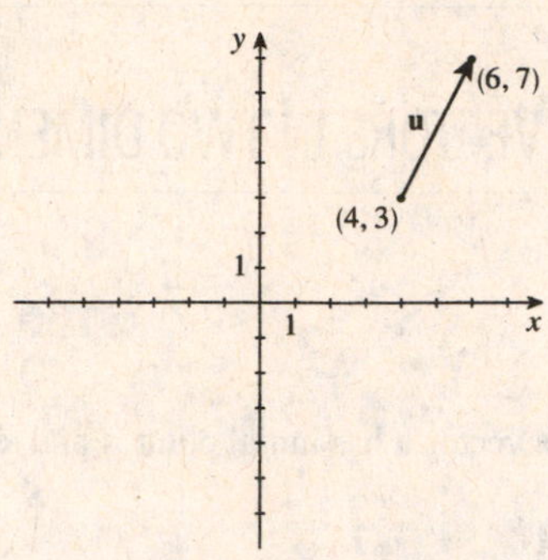

The terminal point is $(4+2, 3+4) = (6,7)$.

21.

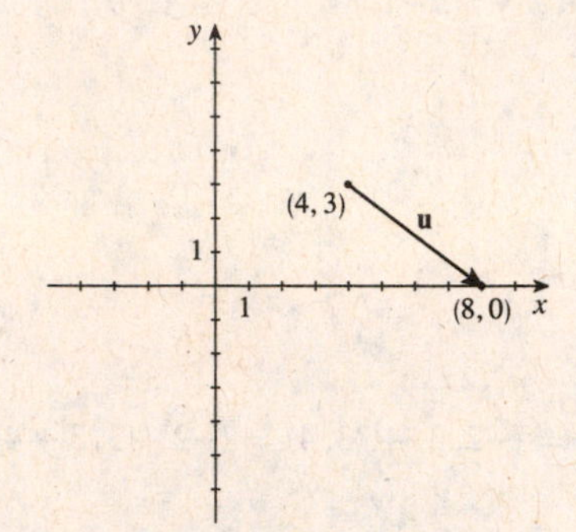

The terminal point is $(4+4, 3-3) = (8,0)$.

23.

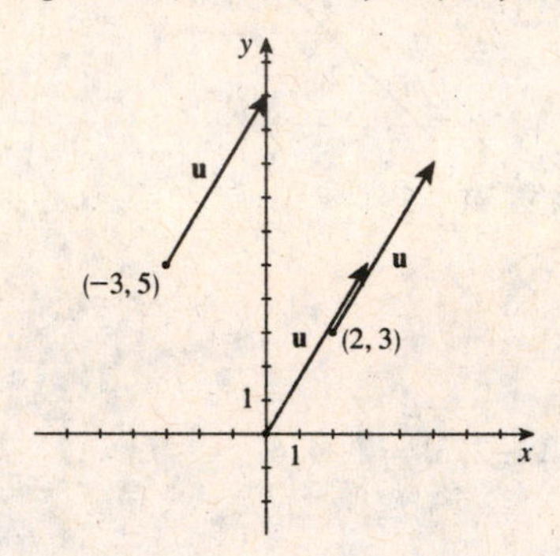

25.

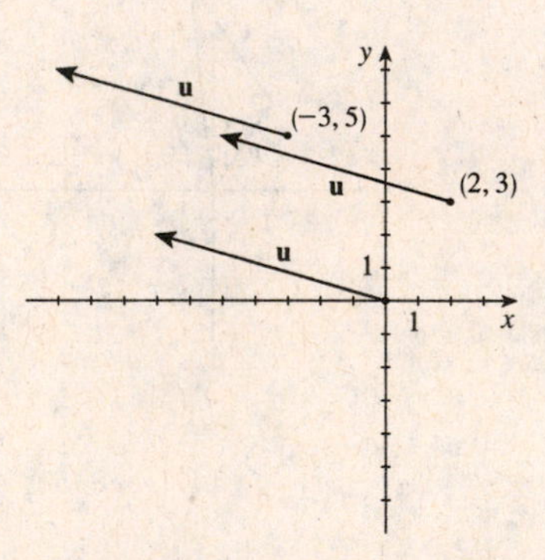

27. $\mathbf{u} = \langle 1,4\rangle = \mathbf{i} + 4\mathbf{j}$

29. $\mathbf{u} = \langle 3,0\rangle = 3\mathbf{i}$

31. $\mathbf{u} = \langle 2,7\rangle$, $\mathbf{v} = \langle 3,1\rangle$. $2\mathbf{u} = 2\cdot\langle 2,7\rangle = \langle 4,14\rangle$; $-3\mathbf{v} = -3\cdot\langle 3,1\rangle = \langle -9,-3\rangle$; $\mathbf{u}+\mathbf{v} = \langle 2,7\rangle + \langle 3,1\rangle = \langle 5,8\rangle$;
$3\mathbf{u} - 4\mathbf{v} = \langle 6,21\rangle - \langle 12,4\rangle = \langle -6,17\rangle$

33. $\mathbf{u} = \langle 0,-1\rangle$, $\mathbf{v} = \langle -2,0\rangle$. $2\mathbf{u} = 2\cdot\langle 0,-1\rangle = \langle 0,-2\rangle$; $-3\mathbf{v} = -3\cdot\langle -2,0\rangle = \langle 6,0\rangle$; $\mathbf{u}+\mathbf{v} = \langle 0,-1\rangle + \langle -2,0\rangle = \langle -2,-1\rangle$;
$3\mathbf{u} - 4\mathbf{v} = \langle 0,-3\rangle - \langle -8,0\rangle = \langle 8,-3\rangle$

35. $\mathbf{u} = 2\mathbf{i}$, $\mathbf{v} = 3\mathbf{i} - 2\mathbf{j}$. $2\mathbf{u} = 2\cdot 2\mathbf{i} = 4\mathbf{i}$; $-3\mathbf{v} = -3(3\mathbf{i} - 2\mathbf{j}) = -9\mathbf{i} + 6\mathbf{j}$; $\mathbf{u}+\mathbf{v} = 2\mathbf{i} + 3\mathbf{i} - 2\mathbf{j} = 5\mathbf{i} - 2\mathbf{j}$;
$3\mathbf{u} - 4\mathbf{v} = 3\cdot 2\mathbf{i} - 4(3\mathbf{i} - 2\mathbf{j}) = -6\mathbf{i} + 8\mathbf{j}$

37. $\mathbf{u} = 2\mathbf{i} + \mathbf{j}$, $\mathbf{v} = 3\mathbf{i} - 2\mathbf{j}$. Then $|\mathbf{u}| = \sqrt{2^2+1^2} = \sqrt{5}$; $|\mathbf{v}| = \sqrt{3^2+2^2} = \sqrt{13}$; $2\mathbf{u} = 4\mathbf{i} + 2\mathbf{j}$; $|2\mathbf{u}| = \sqrt{4^2+2^2} = 2\sqrt{5}$;
$\frac{1}{2}\mathbf{v} = \frac{3}{2}\mathbf{i} - \mathbf{j}$; $\left|\frac{1}{2}\mathbf{v}\right| = \sqrt{\left(\frac{3}{2}\right)^2 + 1^2} = \frac{1}{2}\sqrt{13}$; $\mathbf{u}+\mathbf{v} = 5\mathbf{i} - \mathbf{j}$; $|\mathbf{u}+\mathbf{v}| = \sqrt{5^2+1^2} = \sqrt{26}$; $\mathbf{u}-\mathbf{v} = 2\mathbf{i} + \mathbf{j} - 3\mathbf{i} + 2\mathbf{j} = -\mathbf{i} + 3\mathbf{j}$;
$|\mathbf{u}-\mathbf{v}| = \sqrt{1^2+3^2} = \sqrt{10}$; $|\mathbf{u}| - |\mathbf{v}| = \sqrt{5} - \sqrt{13}$

39. $\mathbf{u} = \langle 10, -1 \rangle$, $\mathbf{v} = \langle -2, -2 \rangle$. Then $|\mathbf{u}| = \sqrt{10^2 + 1^2} = \sqrt{101}$; $|\mathbf{v}| = \sqrt{(-2)^2 + (-2)^2} = 2\sqrt{2}$; $2\mathbf{u} = \langle 20, -2 \rangle$;
$|2\mathbf{u}| = \sqrt{20^2 + 2^2} = \sqrt{404} = 2\sqrt{101}$; $\frac{1}{2}\mathbf{v} = \langle -1, -1 \rangle$; $\left|\frac{1}{2}\mathbf{v}\right| = \sqrt{(-1)^2 + (-1)^2} = \sqrt{2}$; $\mathbf{u} + \mathbf{v} = \langle 8, -3 \rangle$;
$|\mathbf{u} + \mathbf{v}| = \sqrt{8^2 + 3^2} = \sqrt{73}$; $\mathbf{u} - \mathbf{v} = \langle 12, 1 \rangle$; $|\mathbf{u} - \mathbf{v}| = \sqrt{12^2 + 1^2} = \sqrt{145}$; $|\mathbf{u}| - |\mathbf{v}| = \sqrt{101} - 2\sqrt{2}$

In Solutions 41–45, x represents the horizontal component and y the vertical component.

41. $|\mathbf{v}| = 40$, direction $\theta = 30°$. $x = 40 \cos 30° = 20\sqrt{3}$ and $y = 40 \sin 30° = 20$. Thus, $\mathbf{v} = x\mathbf{i} + y\mathbf{j} = 20\sqrt{3}\mathbf{i} + 20\mathbf{j}$.

43. $|\mathbf{v}| = 1$, direction $\theta = 225°$. $x = \cos 225° = -\frac{1}{\sqrt{2}}$ and $y = \sin 225° = -\frac{1}{\sqrt{2}}$. Thus,
$\mathbf{v} = x\mathbf{i} + y\mathbf{j} = -\frac{1}{\sqrt{2}}\mathbf{i} - \frac{1}{\sqrt{2}}\mathbf{j} = -\frac{\sqrt{2}}{2}\mathbf{i} - \frac{\sqrt{2}}{2}\mathbf{j}$.

45. $|\mathbf{v}| = 4$, direction $\theta = 10°$. $x = 4 \cos 10° \approx 3.94$ and $y = 4 \sin 10° \approx 0.69$. Thus,
$\mathbf{v} = x\mathbf{i} + y\mathbf{j} = (4 \cos 10°)\,\mathbf{i} + (4 \sin 10°)\,\mathbf{j} \approx 3.94\mathbf{i} + 0.69\mathbf{j}$.

47. $\mathbf{v} = \langle 3, 4 \rangle$. The magnitude is $|\mathbf{v}| = \sqrt{3^2 + 4^2} = 5$. The direction is θ, where $\tan\theta = \frac{4}{3} \Leftrightarrow \theta = \tan^{-1}\left(\frac{4}{3}\right) \approx 53.13°$.

49. $\mathbf{v} = \langle -12, 5 \rangle$. The magnitude is $|\mathbf{v}| = \sqrt{(-12)^2 + 5^2} = \sqrt{169} = 13$. The direction is θ, where $\tan\theta = -\frac{5}{12}$ with θ in quadrant II $\Leftrightarrow \theta = \pi + \tan^{-1}\left(-\frac{5}{12}\right) \approx 157.38°$.

51. $\mathbf{v} = \mathbf{i} + \sqrt{3}\mathbf{j}$. The magnitude is $|\mathbf{v}| = \sqrt{1^2 + \left(\sqrt{3}\right)^2} = 2$. The direction is θ, where $\tan\theta = \sqrt{3}$ with θ in quadrant I $\Leftrightarrow$
$\theta = \tan^{-1}\sqrt{3} = 60°$.

53. $|\mathbf{v}| = 30$, direction $\theta = 30°$. $x = 30 \cos 30° = 30 \cdot \dfrac{\sqrt{3}}{2} \approx 25.98$, $y = 30 \sin 30° = 15$. So the horizontal component of force is $15\sqrt{3}$ lb and the vertical component is -15 lb.

55. The flow of the river can be represented by the vector $\mathbf{v} = -3\mathbf{j}$ and the swimmer can be represented by the vector $\mathbf{u} = 2\mathbf{i}$. Therefore the true velocity is $\mathbf{u} + \mathbf{v} = 2\mathbf{i} - 3\mathbf{j}$.

57. The speed of the airplane is 300 mi/h, so its velocity relative to the air is $\mathbf{v} = (-300 \cos\theta)\,\mathbf{i} + (-300 \sin\theta)\,\mathbf{j}$. The wind has velocity $\mathbf{w} = 30\mathbf{j}$, so the true course of the airplane is given by

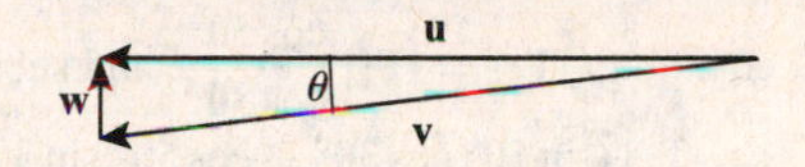

$\mathbf{u} = \mathbf{v}+\mathbf{w} = (-300 \cos\theta)\,\mathbf{i} + (-300 \sin\theta + 30)\,\mathbf{j}$. We want the y-component of the airplane's velocity to be 0, so we solve $-300 \sin\theta + 30 = 0 \Leftrightarrow \sin\theta = \frac{1}{10} \Leftrightarrow \theta \approx 5.74°$. Therefore, the airplane should head in the direction $185.74°$ (or S $84.26°$ W).

59. **(a)** The velocity of the wind is $40\mathbf{j}$.

(b) The velocity of the jet relative to the air is $425\mathbf{i}$.

(c) The true velocity of the jet is $\mathbf{v} = 425\mathbf{i} + 40\mathbf{j} = \langle 425, 40 \rangle$.

(d) The true speed of the jet is $|\mathbf{v}| = \sqrt{425^2 + 40^2} \approx 427$ mi/h, and the true direction is $\theta = \tan^{-1}\left(\frac{40}{425}\right) \approx 5.4° \Rightarrow \theta$ is N $84.6°$ E.

61. If the direction of the plane is N $30°$ W, the airplane's velocity is $\mathbf{u} = \langle \mathbf{u}_x, \mathbf{u}_y \rangle$ where $\mathbf{u}_x = -765 \cos 60° = -382.5$, and $\mathbf{u}_y = 765 \sin 60° \approx 662.51$. If the direction of the wind is N $30°$ E, the wind velocity is $\mathbf{w} = \langle w_x, w_y \rangle$ where $w_x = 55 \cos 60° = 27.5$, and $w_y = 55 \sin 60° \approx 47.63$. Thus, the actual flight path is $\mathbf{v} = \mathbf{u} + \mathbf{w} = \langle -382.5 + 27.5, 662.51 + 47.63 \rangle = \langle -355, 710.14 \rangle$, and so the true speed is $|\mathbf{v}| = \sqrt{355^2 + 710.14^2} \approx 794$ mi/h, and the true direction is $\theta = \tan^{-1}\left(-\frac{710.14}{355}\right) \approx 116.6°$ so θ is N $26.6°$ W.

63. **(a)** The velocity of the river is represented by the vector $\mathbf{r} = \langle 10, 0 \rangle$.

(b) Since the boater direction is $60°$ from the shore at 20 mi/h, the velocity of the boat is represented by the vector $\mathbf{b} = \langle 20 \cos 60°, 20 \sin 60° \rangle = \langle 10, 10\sqrt{3} \rangle \approx \langle 10, 17.32 \rangle$.

(c) $\mathbf{w} = \mathbf{r} + \mathbf{b} = \langle 10 + 10, 0 + 17.32 \rangle = \langle 20, 17.32 \rangle$

(d) The true speed of the boat is $|\mathbf{w}| = \sqrt{20^2 + 17.32^2} \approx 26.5$ mi/h, and the true direction is $\theta = \tan^{-1}\left(\frac{17.32}{20}\right) \approx 40.9° \approx$ N 49.1° E.

65. **(a)** Let $\mathbf{b} = \langle b_x, b_y \rangle$ represent the velocity of the boat relative to the water. Then $\mathbf{b} = \langle 24 \cos 18°, 24 \sin 18° \rangle \approx \langle 22.8, 7.4 \rangle$.

(b) Let $\mathbf{w} = \langle w_x, w_y \rangle$ represent the velocity of the water. Then $\mathbf{w} = \langle 0, w \rangle$ where w is the speed of the water. So the true velocity of the boat is $\mathbf{b} + \mathbf{w} = \langle 24 \cos 18°, 24 \sin 18° - w \rangle$. For the direction to be due east, we must have $24 \sin 18° - w = 0 \Leftrightarrow w = 7.42$ mi/h. Therefore, the true speed of the water is 7.4 mi/h. Since $\mathbf{b} + \mathbf{w} = \langle 24 \cos 18°, 0 \rangle$, the true speed of the boat is $|\mathbf{b} + \mathbf{w}| = 24 \cos 18° \approx 22.8$ mi/h.

67. $\mathbf{F}_1 = \langle 2, 5 \rangle$ and $\mathbf{F}_2 = \langle 3, -8 \rangle$.

(a) $\mathbf{F}_1 + \mathbf{F}_2 = \langle 2 + 3, 5 - 8 \rangle = \langle 5, -3 \rangle$

(b) The additional force required is $\mathbf{F}_3 = \langle 0, 0 \rangle - \langle 5, -3 \rangle = \langle -5, 3 \rangle$.

69. $\mathbf{F}_1 = 4\mathbf{i} - \mathbf{j}$, $\mathbf{F}_2 = 3\mathbf{i} - 7\mathbf{j}$, $\mathbf{F}_3 = -8\mathbf{i} + 3\mathbf{j}$, and $\mathbf{F}_4 = \mathbf{i} + \mathbf{j}$.

(a) $\mathbf{F}_1 + \mathbf{F}_2 + \mathbf{F}_3 + \mathbf{F}_4 = (4 + 3 - 8 + 1)\,\mathbf{i} + (-1 - 7 + 3 + 1)\,\mathbf{j} = 0\mathbf{i} - 4\mathbf{j} = -4\mathbf{j}$

(b) The additional force required is $\mathbf{F}_5 = 0\mathbf{i} + 0\mathbf{j} - (0\mathbf{i} - 4\mathbf{j}) = 4\mathbf{j}$.

71. $\mathbf{F}_1 = \langle 10 \cos 60°, 10 \sin 60° \rangle = \left\langle 5, 5\sqrt{3} \right\rangle$, $\mathbf{F}_2 = \langle -8 \cos 30°, 8 \sin 30° \rangle = \left\langle -4\sqrt{3}, 4 \right\rangle$, and $\mathbf{F}_3 = \langle -6 \cos 20°, -6 \sin 20° \rangle \approx \langle -5.638, -2.052 \rangle$.

(a) $\mathbf{F}_1 + \mathbf{F}_2 + \mathbf{F}_3 = \left\langle 5 - 4\sqrt{3} - 5.638, 5\sqrt{3} + 4 - 2.052 \right\rangle \approx \langle -7.57, 10.61 \rangle$.

(b) The additional force required is $\mathbf{F}_4 = \langle 0, 0 \rangle - \langle -7.57, 10.61 \rangle = \langle 7.57, -10.61 \rangle$.

73. From the figure we see that $\mathbf{T}_1 = -|\mathbf{T}_1| \cos 50° \mathbf{i} + |\mathbf{T}_1| \sin 50° \mathbf{j}$ and $\mathbf{T}_2 = |\mathbf{T}_2| \cos 30° \mathbf{i} + |\mathbf{T}_2| \sin 30° \mathbf{j}$. Since $\mathbf{T}_1 + \mathbf{T}_2 = 100\mathbf{j}$ we get $-|\mathbf{T}_1| \cos 50° + |\mathbf{T}_2| \cos 30° = 0$ and $|\mathbf{T}_1| \sin 50° + |\mathbf{T}_2| \sin 30° = 100$. From the first equation, $|\mathbf{T}_2| = |\mathbf{T}_1| \dfrac{\cos 50°}{\cos 30°}$, and substituting into the second equation gives $|\mathbf{T}_1| \sin 50° + |\mathbf{T}_1| \dfrac{\cos 50° \sin 30°}{\cos 30°} = 100$ $\Leftrightarrow |\mathbf{T}_1| (\sin 50° \cos 30° + \cos 50° \sin 30°) = 100 \cos 30° \Leftrightarrow |\mathbf{T}_1| \sin (50° + 30°) = 100 \cos 30° \Leftrightarrow$ $|\mathbf{T}_1| = 100 \dfrac{\cos 30°}{\sin 80°} \approx 87.9385$.

Similarly, solving for $|\mathbf{T}_1|$ in the first equation gives $|\mathbf{T}_1| = |\mathbf{T}_2| \dfrac{\cos 30°}{\cos 50°}$ and substituting gives $|\mathbf{T}_2| \dfrac{\cos 30° \sin 50°}{\cos 50°} + |\mathbf{T}_2| \sin 30° = 100 \Leftrightarrow |\mathbf{T}_2| (\cos 30° \sin 50° + \cos 50° \sin 30°) = 100 \cos 50° \Leftrightarrow$ $|\mathbf{T}_2| = \dfrac{100 \cos 50°}{\sin 80°} \approx 65.2704$. Thus, $\mathbf{T}_1 \approx (-87.9416 \cos 50°)\,\mathbf{i} + (87.9416 \sin 50°)\,\mathbf{j} \approx -56.5\mathbf{i} + 67.4\mathbf{j}$ and $\mathbf{T}_2 \approx (65.2704 \cos 30°)\,\mathbf{i} + (65.2704 \sin 30°)\,\mathbf{j} \approx 56.5\mathbf{i} + 32.6\mathbf{j}$.

75. When we add two (or more vectors), the resultant vector can be found by first placing the initial point of the second vector at the terminal point of the first vector. The resultant vector can then found by using the new terminal point of the second vector and the initial point of the first vector. When the n vectors are placed head to tail in the plane so that they form a polygon, the initial point and the terminal point are the same. Thus the sum of these n vectors is the zero vector.

9.2 THE DOT PRODUCT

1. The dot product of $\mathbf{u} = \langle a_1, a_2 \rangle$ and $\mathbf{v} = \langle b_1, b_2 \rangle$ is defined by $\mathbf{u} \cdot \mathbf{v} = a_1 a_2 + b_1 b_2$. The dot product of two vectors is a *real number, or scalar*, not a vector.

3. **(a)** The component of $\mathbf{u}$ along $\mathbf{v}$ is the scalar $|\mathbf{u}|\cos\theta$ and can be expressed in terms of the dot product as $\dfrac{\mathbf{u}\cdot\mathbf{v}}{|\mathbf{v}|}$.

(b) The projection of $\mathbf{u}$ onto $\mathbf{v}$ is the vector $\text{proj}_{\mathbf{v}}\mathbf{u} = \left(\dfrac{\mathbf{u}\cdot\mathbf{v}}{|\mathbf{v}|^2}\right)\mathbf{v}$.

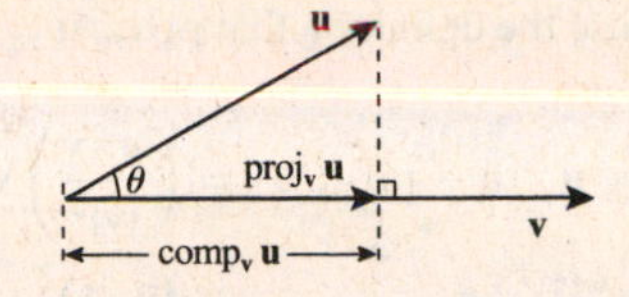

5. **(a)** $\mathbf{u}\cdot\mathbf{v} = \langle 2,0\rangle\cdot\langle 1,1\rangle = 2+0 = 2$

(b) $\cos\theta = \dfrac{\mathbf{u}\cdot\mathbf{v}}{|\mathbf{u}|\,|\mathbf{v}|} = \dfrac{2}{2\cdot\sqrt{2}} = \dfrac{1}{\sqrt{2}} \Rightarrow \theta = 45^\circ$

7. **(a)** $\mathbf{u}\cdot\mathbf{v} = \langle 2,7\rangle\cdot\langle 3,1\rangle = 6+7 = 13$

(b) $\cos\theta = \dfrac{\mathbf{u}\cdot\mathbf{v}}{|\mathbf{u}|\,|\mathbf{v}|} = \dfrac{13}{\sqrt{53}\cdot\sqrt{10}} \Rightarrow \theta \approx 56^\circ$

9. **(a)** $\mathbf{u}\cdot\mathbf{v} = \langle 3,-2\rangle\cdot\langle 1,2\rangle = 3+(-4) = -1$

(b) $\cos\theta = \dfrac{\mathbf{u}\cdot\mathbf{v}}{|\mathbf{u}|\,|\mathbf{v}|} = \dfrac{-1}{\sqrt{13}\cdot\sqrt{5}} \Rightarrow \theta \approx 97^\circ$

11. **(a)** $\mathbf{u}\cdot\mathbf{v} = \langle 0,-5\rangle\cdot\left\langle -1,-\sqrt{3}\right\rangle = 0+5\sqrt{3} = 5\sqrt{3}$

(b) $\cos\theta = \dfrac{\mathbf{u}\cdot\mathbf{v}}{|\mathbf{u}|\,|\mathbf{v}|} = \dfrac{5\sqrt{3}}{5\cdot 2} = \dfrac{\sqrt{3}}{2} \Rightarrow \theta = 30^\circ$

13. **(a)** $\mathbf{u}\cdot\mathbf{v} = (\mathbf{i}+3\mathbf{j})\cdot(4\mathbf{i}-\mathbf{j}) = 4-3 = 1$

(b) $\cos\theta = \dfrac{\mathbf{u}\cdot\mathbf{v}}{|\mathbf{u}|\,|\mathbf{v}|} = \dfrac{1}{\sqrt{10}\cdot\sqrt{17}} \Rightarrow \theta \approx 85.6^\circ$

15. $\mathbf{u}\cdot\mathbf{v} = -12+12 = 0 \Rightarrow$ vectors are orthogonal

17. $\mathbf{u}\cdot\mathbf{v} = -8+12 = 4 \neq 0 \Rightarrow$ vectors are not orthogonal

19. $\mathbf{u}\cdot\mathbf{v} = -24+24 = 0 \Rightarrow$ vectors are orthogonal

21. $\mathbf{u}\cdot\mathbf{v}+\mathbf{u}\cdot\mathbf{w} = \langle 2,1\rangle\cdot\langle 1,-3\rangle + \langle 2,1\rangle\cdot\langle 3,4\rangle$
$= 2-3+6+4 = 9$

23. $(\mathbf{u}+\mathbf{v})\cdot(\mathbf{u}-\mathbf{v}) = [\langle 2,1\rangle+\langle 1,-3\rangle]\cdot[\langle 2,1\rangle-\langle 1,-3\rangle]$
$= \langle 3,-2\rangle\cdot\langle 1,4\rangle = 3-8 = -5$

25. $x = \dfrac{\mathbf{u}\cdot\mathbf{v}}{|\mathbf{v}|} = \dfrac{12-24}{5} = -\dfrac{12}{5}$

27. $x = \dfrac{\mathbf{u}\cdot\mathbf{v}}{|\mathbf{v}|} = \dfrac{0-24}{1} = -24$

29. **(a)** $\mathbf{u}_1 = \text{proj}_{\mathbf{v}}\mathbf{u} = \left(\dfrac{\mathbf{u}\cdot\mathbf{v}}{|\mathbf{v}|^2}\right)\mathbf{v} = \left(\dfrac{\langle -2,4\rangle\cdot\langle 1,1\rangle}{1^2+1^2}\right)\langle 1,1\rangle = \langle 1,1\rangle$.

(b) $\mathbf{u}_2 = \mathbf{u}-\mathbf{u}_1 = \langle -2,4\rangle - \langle 1,1\rangle = \langle -3,3\rangle$. We resolve the vector $\mathbf{u}$ as $\mathbf{u}_1+\mathbf{u}_2$, where $\mathbf{u}_1 = \langle 1,1\rangle$ and $\mathbf{u}_2 = \langle -3,3\rangle$.

31. **(a)** $\mathbf{u}_1 = \text{proj}_{\mathbf{v}}\mathbf{u} = \left(\dfrac{\mathbf{u}\cdot\mathbf{v}}{|\mathbf{v}|^2}\right)\mathbf{v} = \left(\dfrac{\langle 1,2\rangle\cdot\langle 1,-3\rangle}{1^2+(-3)^2}\right)\langle 1,-3\rangle = -\frac{1}{2}\langle 1,-3\rangle = \left\langle -\frac{1}{2},\frac{3}{2}\right\rangle$

(b) $\mathbf{u}_2 = \mathbf{u}-\mathbf{u}_1 = \langle 1,2\rangle - \left\langle -\frac{1}{2},\frac{3}{2}\right\rangle = \left\langle \frac{3}{2},\frac{1}{2}\right\rangle$. We resolve the vector $\mathbf{u}$ as $\mathbf{u}_1+\mathbf{u}_2$, where $\mathbf{u}_1 = \left\langle -\frac{1}{2},\frac{3}{2}\right\rangle$ and $\mathbf{u}_2 = \left\langle \frac{3}{2},\frac{1}{2}\right\rangle$.

33. **(a)** $\mathbf{u}_1 = \text{proj}_{\mathbf{v}}\mathbf{u} = \left(\dfrac{\mathbf{u}\cdot\mathbf{v}}{|\mathbf{v}|^2}\right)\mathbf{v} = \left(\dfrac{\langle 2,9\rangle\cdot\langle -3,4\rangle}{(-3)^3+4^2}\right)\langle -3,4\rangle = \frac{6}{5}\langle -3,4\rangle = \left\langle -\frac{18}{5},\frac{24}{5}\right\rangle$

(b) $\mathbf{u}_2 = \mathbf{u}-\mathbf{u}_1 = \langle 2,9\rangle - \left\langle -\frac{18}{5},\frac{24}{5}\right\rangle = \left\langle \frac{28}{5},\frac{21}{5}\right\rangle$. We resolve the vector $\mathbf{u}$ as $\mathbf{u}_1+\mathbf{u}_2$, where $\mathbf{u}_1 = \left\langle -\frac{18}{5},\frac{24}{5}\right\rangle$ and $\mathbf{u}_2 = \left\langle \frac{28}{5},\frac{21}{5}\right\rangle$.

35. $W = \mathbf{F}\cdot\mathbf{d} = \langle 4,-5\rangle\cdot\langle 3,8\rangle = -28$

37. $W = \mathbf{F}\cdot\mathbf{d} = \langle 10,3\rangle\cdot\langle 4,-5\rangle = 25$

39. Let $\mathbf{u} = \langle \mathbf{u}_1,\mathbf{u}_2\rangle$ and $\mathbf{v} = \langle \mathbf{v}_1,\mathbf{v}_2\rangle$. Then
$\mathbf{u}\cdot\mathbf{v} = \langle u_1,u_2\rangle\cdot\langle v_1,v_2\rangle = u_1v_1+u_2v_2 = v_1u_1+v_2u_2 = \langle v_1,v_2\rangle\cdot\langle u_1,u_2\rangle = \mathbf{v}\cdot\mathbf{u}$

41. Let $\mathbf{u} = \langle u_1,u_2\rangle$, $\mathbf{v} = \langle v_1,v_2\rangle$, and $\mathbf{w} = \langle w_1,w_2\rangle$. Then

$$\begin{aligned}(\mathbf{u}+\mathbf{v})\cdot\mathbf{w} &= (\langle u_1,u_2\rangle+\langle v_1,v_2\rangle)\cdot\langle w_1,w_2\rangle = \langle u_1+v_1,u_2+v_2\rangle\cdot\langle w_1,w_2\rangle\\ &= u_1w_1+v_1w_1+u_2w_2+v_2w_2 = u_1w_1+u_2w_2+v_1w_1+v_2w_2\\ &= \langle u_1,u_2\rangle\cdot\langle w_1,w_2\rangle+\langle v_1,v_2\rangle\cdot\langle w_1,w_2\rangle = \mathbf{u}\cdot\mathbf{w}+\mathbf{v}\cdot\mathbf{w}\end{aligned}$$

43. We use the definition that $\text{proj}_{\mathbf{v}}\,\mathbf{u} = \left(\dfrac{\mathbf{u}\cdot\mathbf{v}}{|\mathbf{v}|^2}\right)\mathbf{v}$. Then

$$\text{proj}_{\mathbf{v}}\,\mathbf{u}\cdot(\mathbf{u}-\text{proj}_{\mathbf{v}}\,\mathbf{u}) = \left(\frac{\mathbf{u}\cdot\mathbf{v}}{|\mathbf{v}|^2}\right)\mathbf{v}\cdot\left[\mathbf{u}-\left(\frac{\mathbf{u}\cdot\mathbf{v}}{|\mathbf{v}|^2}\right)\mathbf{v}\right] = \left(\frac{\mathbf{u}\cdot\mathbf{v}}{|\mathbf{v}|^2}\right)(\mathbf{v}\cdot\mathbf{u}) - \left(\frac{\mathbf{u}\cdot\mathbf{v}}{|\mathbf{v}|^2}\right)\mathbf{v}\cdot\left(\frac{\mathbf{u}\cdot\mathbf{v}}{|\mathbf{v}|^2}\right)\mathbf{v}$$

$$= \frac{(\mathbf{u}\cdot\mathbf{v})^2}{|\mathbf{v}|^2} - \frac{(\mathbf{u}\cdot\mathbf{v})^2}{|\mathbf{v}|^4}|\mathbf{v}|^2 = \frac{(\mathbf{u}\cdot\mathbf{v})^2}{|\mathbf{v}|^2} - \frac{(\mathbf{u}\cdot\mathbf{v})^2}{|\mathbf{v}|^2} = 0$$

Thus $\mathbf{u}$ and $\mathbf{u}-\text{proj}_{\mathbf{v}}\,\mathbf{u}$ are orthogonal.

45. $W = \mathbf{F}\cdot\mathbf{d} = \langle 4,-7\rangle\cdot\langle 4,0\rangle = 16$ ft-lb

47. The distance vector is $\mathbf{D} = \langle 200, 0\rangle$ and the force vector is $\mathbf{F} = \langle 50\cos 30°, 50\sin 30°\rangle$. Hence, the work done is $W = \mathbf{F}\cdot\mathbf{D} = \langle 200,0\rangle\cdot\langle 50\cos 30°, 50\sin 30°\rangle = 200\cdot 50\cos 30° \approx 8660$ ft-lb.

49. **(a)** Since the force parallel to the driveway is $490 = |\mathbf{w}|\sin 10° \Leftrightarrow |\mathbf{w}| = \dfrac{490}{\sin 10°} \approx 2821.8$, and thus the weight of the car is about 2822 lb.

(b) The force exerted against the driveway is $2821.8\cos 10° \approx 2779$ lb.

51. Since the force required parallel to the plane is 80 lb and the weight of the package is 200 lb, it follows that $80 = 200\sin\theta$, where θ is the angle of inclination of the plane. Then $\theta = \sin^{-1}\left(\frac{80}{200}\right) \approx 23.58°$, and so the angle of inclination is approximately 23.6°.

53. **(a)** $2(0)+4(2) = 8$, so $Q(0,2)$ lies on L. $2(2)+4(1) = 4+4 = 8$, so $R(2,1)$ lies on L.

(b) $\mathbf{u} = \overrightarrow{QP} = \langle 0,2\rangle - \langle 3,4\rangle = \langle -3,-2\rangle$.

$\mathbf{v} = \overrightarrow{QR} = \langle 0,2\rangle - \langle 2,1\rangle = \langle -2,1\rangle$.

$$\mathbf{w} = \text{proj}_{\mathbf{v}}\,\mathbf{u} = \left(\frac{\mathbf{u}\cdot\mathbf{v}}{|\mathbf{v}|^2}\right)\mathbf{v} = \frac{\langle -3,-2\rangle\cdot\langle 2,1\rangle}{(-2)^2+1^2}\langle -2,1\rangle$$

$$= -\tfrac{8}{5}\langle -2,1\rangle = \left\langle \tfrac{16}{5}, -\tfrac{8}{5}\right\rangle$$

(c) From the graph, we can see that $\mathbf{u}-\mathbf{w}$ is orthogonal to $\mathbf{v}$ (and thus to L). Thus, the distance from P to L is $|\mathbf{u}-\mathbf{w}|$.

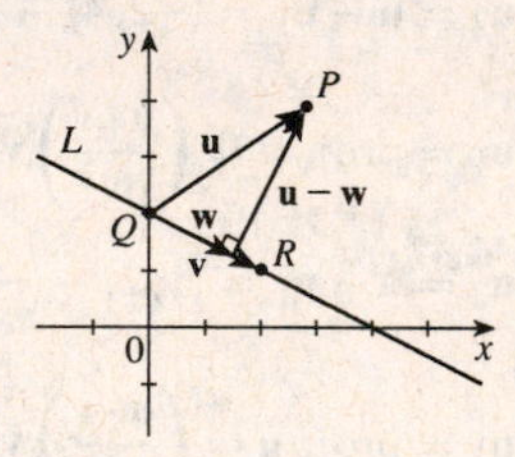

9.3 THREE-DIMENSIONAL COORDINATE GEOMETRY

1. In a three-dimensional coordinate system the three mutually perpendicular axes are called the x-axis, the y-axis, and the z-axis. The point P has coordinates $(5,2,3)$. The equation of the plane passing through P and parallel to the xz-plane is $y = 2$.

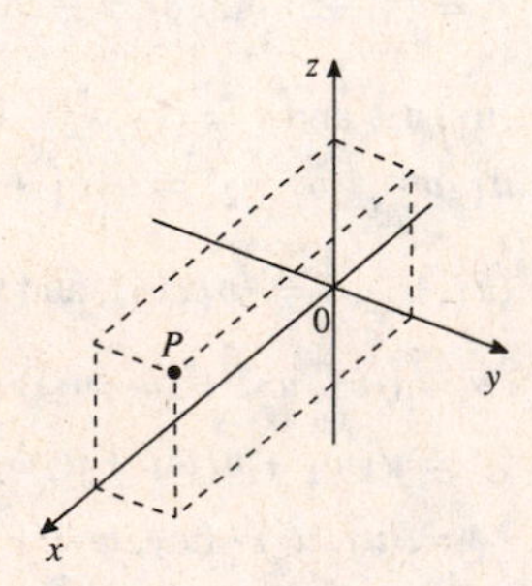

3. (a)

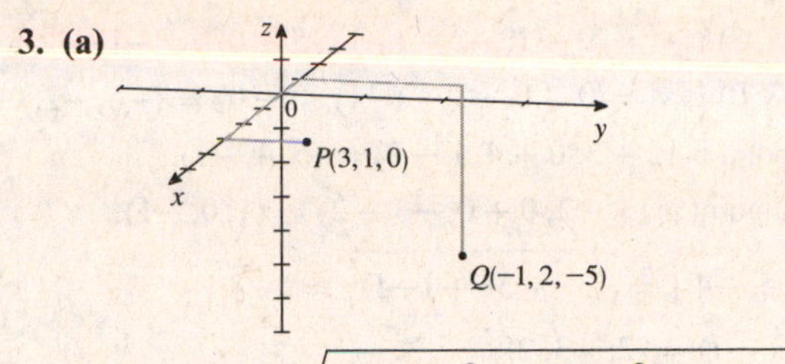

(b) $d(P, Q) = \sqrt{(-1-3)^2 + (2-1)^2 + (-5-0)^2} = \sqrt{42}$

5. (a)

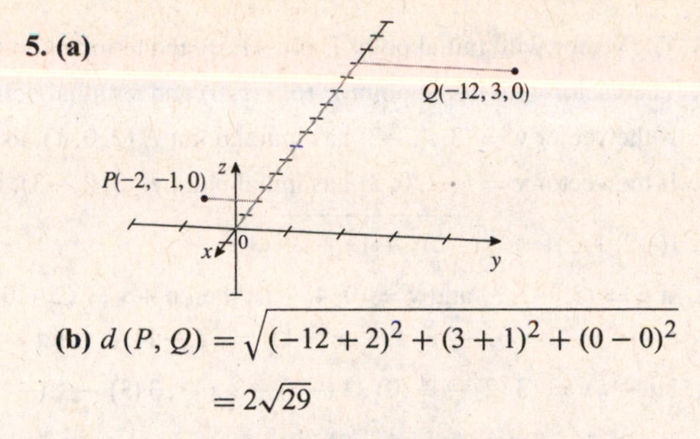

(b) $d(P, Q) = \sqrt{(-12+2)^2 + (3+1)^2 + (0-0)^2} = 2\sqrt{29}$

7. $x = 4$ is a plane parallel to the yz-plane.

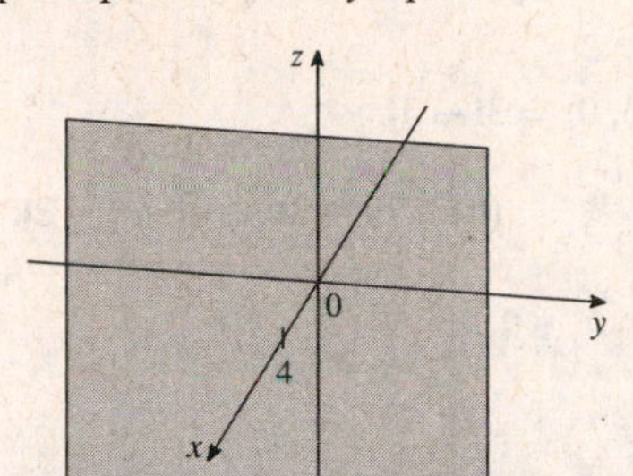

9. $z = 8$ is a plane parallel to the xy-plane.

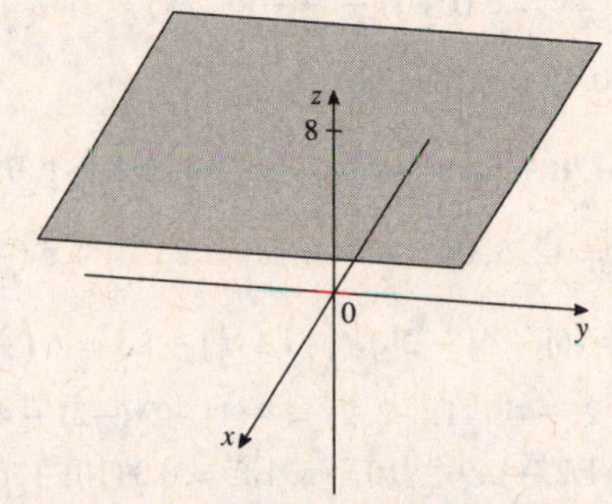

11. A sphere with radius $r = 5$ and center $C(2, -5, 3)$ has equation $(x-2)^2 + [y-(-5)]^2 + (z-3)^2 = 5^2$, or $(x-2)^2 + (y+5)^2 + (z-3)^2 = 25$.

13. A sphere with radius $r = \sqrt{6}$ and center $C(3, -1, 0)$ has equation $(x-3)^2 + (y+1)^2 + z^2 = 6$.

15. We complete the squares in x, y, and z: $x^2 + y^2 + z^2 - 10x + 2y + 8z = 9 \Leftrightarrow \left(x^2 - 10x + 25\right) + \left(y^2 + 2y + 1\right) + \left(z^2 + 8z + 16\right) = 9 + 25 + 1 + 16 \Leftrightarrow (x-5)^2 + (y+1)^2 + (z+4)^2 = 51$. This is an equation of a sphere with center $(5, -1, -4)$ and radius $\sqrt{51}$.

17. We complete the squares in x, y, and z: $x^2 + y^2 + z^2 = 12x + 2y \Leftrightarrow \left(x^2 - 12x + 36\right) + \left(y^2 - 2y + 1\right) + z^2 = 36 + 1 \Leftrightarrow (x-6)^2 + (y-1)^2 + z^2 = 37$. This is an equation of a sphere with center $(6, 1, 0)$ and radius $\sqrt{37}$.

19. (a) To find the trace in the yz-plane, we set $x = 0$: $(0+1)^2 + (y-2)^2 + (z+10)^2 = 100 \Leftrightarrow (y-2)^2 + (z+10)^2 = 99$. This represents a circle with center $(0, 2, -10)$ and radius $3\sqrt{11}$.

(b) We set $x = 4$ and find $(4+1)^2 + (y-2)^2 + (z+10)^2 = 100 \Leftrightarrow (y-2)^2 + (z+10)^2 = 75$. This represents a circle with center $(4, 2, -10)$ and radius $5\sqrt{3}$.

21. With the origin at its center, an equation of the tank is $x^2 + y^2 + z^2 = 25$. The metal circle is the trace in the plane $z = -4$, so its equation is $x^2 + y^2 + (-4)^2 = 25$ or $x^2 + y^2 = 9$. Therefore, its radius is 3.

23. We use the Distance Formula to write the condition that $X(x, y, z)$ is equidistant from $P(0, 0, 0)$ and $Q(0, 3, 0)$. (It is easier to equate the *squares* of the distances.) $|XP|^2 = |XQ|^2 \Leftrightarrow (x-0)^2 + (y-0)^2 + (z-0)^2 = (x-0)^2 + (y-3)^2 + (z-0)^2 \Leftrightarrow y^2 = (y-3)^2 \Leftrightarrow y^2 = y^2 - 6y + 9 \Leftrightarrow 6y = 9 \Leftrightarrow y = \frac{3}{2}$. This is an equation of a plane parallel to the xz-plane.

9.4 VECTORS IN THREE DIMENSIONS

1. A vector $\mathbf{v} = \langle a_1, a_2, a_3 \rangle$ in three dimensions can be written in terms of the *unit* vectors $\mathbf{i}$, $\mathbf{j}$, and $\mathbf{k}$ as $\mathbf{v} = a_1\mathbf{i} + a_2\mathbf{j} + a_3\mathbf{k}$. The magnitude of the vector $\mathbf{v}$ is $|\mathbf{v}| = \sqrt{a_1^2 + a_2^2 + a_3^2}$. So $\langle 4, -2, 4 \rangle = 4\mathbf{i} + (-2)\mathbf{j} + 4\mathbf{k}$ and $7\mathbf{j} - 24\mathbf{k} = \langle 0, 7, -24 \rangle$.

3. The vector with initial point $P(1,-1,0)$ and terminal point $Q(0,-2,5)$ is $\mathbf{v} = \langle 0-1, -2-(-1), 5-0\rangle = \langle -1,-1,5\rangle$.

5. The vector with initial point $P(6,-1,0)$ and terminal point $Q(0,-3,0)$ is $\mathbf{v} = \langle 0-6, -3-(-1), 0-0\rangle = \langle -6,-2,0\rangle$.

7. If the vector $\mathbf{v} = \langle 3,4,-2\rangle$ has initial point $P(2,0,1)$, its terminal point is $(2+3, 0+4, 1-2) = (5,4,-1)$.

9. If the vector $\mathbf{v} = \langle -2,0,2\rangle$ has initial point $P(3,0,-3)$, its terminal point is $(3-2, 0+0, -3+2) = (1,0,-1)$.

11. $|\langle -2,1,2\rangle| = \sqrt{(-2)^2+1^2+2^2} = 3$

13. $|\langle 3,5,-4\rangle| = \sqrt{3^2+5^2+(-4)^2} = 5\sqrt{2}$

15. If $\mathbf{u} = \langle 2,-7,3\rangle$ and $\mathbf{v} = \langle 0,4,-1\rangle$, then $\mathbf{u}+\mathbf{v} = \langle 2+0, -7+4, 3-1\rangle = \langle 2,-3,2\rangle$,
$\mathbf{u}-\mathbf{v} = \langle 2-0, -7-4, 3-(-1)\rangle = \langle 2,-11,4\rangle$, and
$3\mathbf{u}-\frac{1}{2}\mathbf{v} = \left\langle 3(2)-\frac{1}{2}(0), 3(-7)-\frac{1}{2}(4), 3(3)-\frac{1}{2}(-1)\right\rangle = \left\langle 6,-23,\frac{19}{2}\right\rangle$.

17. If $\mathbf{u} = \mathbf{i}+\mathbf{j}$ and $\mathbf{v} = -\mathbf{j}-2\mathbf{k}$, then $\mathbf{u}+\mathbf{v} = \mathbf{i}+\mathbf{j}-\mathbf{j}-2\mathbf{k} = \mathbf{i}-2\mathbf{k}$, $\mathbf{u}-\mathbf{v} = \mathbf{i}+\mathbf{j}-(-\mathbf{j}-2\mathbf{k}) = \mathbf{i}+2\mathbf{j}+2\mathbf{k}$, and
$3\mathbf{u}-\frac{1}{2}\mathbf{v} = 3(\mathbf{i}+\mathbf{j})-\frac{1}{2}(-\mathbf{j}-2\mathbf{k}) = 3\mathbf{i}+\frac{7}{2}\mathbf{j}+\mathbf{k}$.

19. $\langle 12,0,2\rangle = 12\mathbf{i}+2\mathbf{k}$

21. $\langle 3,-3,0\rangle = 3\mathbf{i}-3\mathbf{j}$

23. **(a)** $-2\mathbf{u}+3\mathbf{v} = -2\langle 0,-2,1\rangle + 3\langle 1,-1,0\rangle = \langle 3,1,-2\rangle$ **(b)** $-2\mathbf{u}+3\mathbf{v} = 3\mathbf{i}+\mathbf{j}-2\mathbf{k}$

25. $\mathbf{u}\cdot\mathbf{v} = \langle 2,5,0\rangle \cdot \left\langle \frac{1}{2},-1,10\right\rangle = 2\left(\frac{1}{2}\right)+5(-1)+0(10) = -4$

27. $\mathbf{u}\cdot\mathbf{v} = (6\mathbf{i}-4\mathbf{j}-2\mathbf{k})\cdot\left(\frac{5}{6}\mathbf{i}+\frac{3}{2}\mathbf{j}-\mathbf{k}\right) = 6\left(\frac{5}{6}\right)-4\left(\frac{3}{2}\right)-2(-1) = 1$

29. $\langle 4,-2,-4\rangle \cdot \langle 1,-2,2\rangle = 4(1)-2(-2)-4(2) = 0$, so the vectors are perpendicular.

31. $\langle 0.3,1.2,-0.9\rangle \cdot \langle 10,-5,10\rangle = 0.3(10)+1.2(-5)-0.9(10) = -12$, so the vectors are not perpendicular.

33. $\cos\theta = \dfrac{\mathbf{u}\cdot\mathbf{v}}{|\mathbf{u}|\,|\mathbf{v}|} = \dfrac{\langle 2,-2,-1\rangle\cdot\langle 1,2,2\rangle}{|\langle 2,-2,-1\rangle|\,|\langle 1,2,2\rangle|} = \dfrac{2(1)-2(2)-1(2)}{\sqrt{2^2+(-2)^2+(-1)^2}\sqrt{1^2+2^2+2^2}} = -\dfrac{4}{9}$, so
$\theta = \cos^{-1}\left(-\frac{4}{9}\right) \approx 116.4°$.

35. $\cos\theta = \dfrac{\mathbf{u}\cdot\mathbf{v}}{|\mathbf{u}|\,|\mathbf{v}|} = \dfrac{(\mathbf{j}+\mathbf{k})\cdot(\mathbf{i}+2\mathbf{j}-3\mathbf{k})}{|\mathbf{j}+\mathbf{k}|\,|\mathbf{i}+2\mathbf{j}-3\mathbf{k}|} = \dfrac{1(2)+1(-3)}{\sqrt{1^2+1^2}\sqrt{1^2+2^2+(-3)^2}} = -\dfrac{\sqrt{28}}{28}$, so $\theta = \cos^{-1}\left(-\frac{\sqrt{28}}{28}\right) \approx 100.9°$.

37. The length of the vector $3\mathbf{i}+4\mathbf{j}+5\mathbf{k}$ is $\sqrt{3^2+4^2+5^2} = 5\sqrt{2}$, so by definition, its direction angles satisfy $\cos\alpha = \frac{3}{5\sqrt{2}}$, $\cos\beta = \frac{4}{5\sqrt{2}}$, and $\cos\gamma = \frac{1}{\sqrt{2}}$. Thus, $\alpha = \cos^{-1}\frac{3}{5\sqrt{2}} \approx 65°$, $\beta = \cos^{-1}\frac{2\sqrt{2}}{5} \approx 56°$, and $\gamma = \cos^{-1}\frac{1}{\sqrt{2}} = 45°$.

39. $|\langle 2,3,-6\rangle| = \sqrt{2^2+3^2+(-6)^2} = 7$, so $\cos\alpha = \frac{2}{7}$, $\cos\beta = \frac{3}{7}$, and $\cos\gamma = \frac{-6}{7} \Leftrightarrow \alpha = \cos^{-1}\frac{2}{7} \approx 73°$, $\beta = \cos^{-1}\frac{3}{7} \approx 65°$, and $\gamma = \cos^{-1}\left(-\frac{6}{7}\right) \approx 149°$.

41. We are given that $\alpha = \frac{\pi}{3}$, $\gamma = \frac{2\pi}{3}$, and β is acute. Using the property of direction cosines $\cos^2\alpha+\cos^2\beta+\cos^2\gamma = 1$, we have $\left(\cos\frac{\pi}{3}\right)^2+\cos^2\beta+\left(\cos\frac{2\pi}{3}\right)^2 = 1 \Leftrightarrow \left(\frac{1}{2}\right)^2+\cos^2\beta+\left(-\frac{1}{2}\right)^2 = 1 \Leftrightarrow \cos^2\beta = \frac{1}{2} \Leftrightarrow \cos\beta = \pm\frac{1}{\sqrt{2}}$. Because β is acute, we have $\beta = \cos^{-1}\frac{1}{\sqrt{2}} = 45°$.

43. We are given that $\alpha = 60°$, $\beta = 50°$, and γ is obtuse, so $\cos^2 60° + \cos^2 50° + \cos^2\gamma = 1 \Leftrightarrow$ $\cos^2\gamma = 1-\cos^2 60°-\cos^2 50° \approx 0.337$. Because γ is obtuse, $\gamma \approx \cos^{-1}\left(-\sqrt{0.337}\right) \approx 125°$.

45. Here $\cos^2\alpha+\cos^2\beta = \cos^2 20°+\cos^2 45° \approx 1.38 > 1$, so there is no angle γ satisfying the property of direction cosines $\cos^2\alpha+\cos^2\beta+\cos^2\gamma = 1$.

47. **(a)** We solve $\mathbf{v} = a\mathbf{u} \Leftrightarrow \langle -6,4,-8\rangle = a\langle 3,-2,4\rangle \Leftrightarrow a = -2$. Therefore, the vectors are parallel and $\mathbf{v} = -2\mathbf{u}$.

(b) $\mathbf{v} = a\mathbf{u} \Leftrightarrow \langle 12,8,-16\rangle = a\langle -9,-6,12\rangle \Leftrightarrow a = -\frac{4}{3}$, so the vectors are parallel and $\mathbf{v} = -\frac{4}{3}\mathbf{u}$.

(c) $\mathbf{v} = a\mathbf{u} \Leftrightarrow 2\mathbf{i}+2\mathbf{j}-2\mathbf{k} = a(\mathbf{i}+\mathbf{j}+\mathbf{k})$ has no solution, so the vectors are not parallel.

49. (a) The second and third forces are $\mathbf{F}_2 = 24\mathbf{j}$ and $\mathbf{F}_3 = -25\mathbf{k}$. Therefore, $\mathbf{F}_1 + \mathbf{F}_2 + \mathbf{F}_3 + \mathbf{F}_4 = \mathbf{0} \Leftrightarrow$ $7\mathbf{i} + 24\mathbf{j} - 25\mathbf{k} + \mathbf{F}_4 = \mathbf{0} \Leftrightarrow \mathbf{F}_4 = -7\mathbf{i} - 24\mathbf{j} + 25\mathbf{k}$.

(b) $|\mathbf{F}_4| = \sqrt{(-7)^2 + (-24)^2 + (-25)^2} = 25\sqrt{2}$

51. (a) $(\mathbf{r} - \mathbf{u}) \cdot (\mathbf{r} - \mathbf{v}) = 0 \Leftrightarrow \langle x - 2, y - 2, z - 2\rangle \cdot \langle x - (-2), y - (-2), z - 0\rangle = 0 \Leftrightarrow (x - 2)(x + 2) +$ $(y - 2)(y + 2) + (z - 2)z = 0 \Leftrightarrow x^2 - 4 + y^2 - 4 + z^2 - 2z = 0 \Leftrightarrow x^2 + y^2 + (z - 1)^2 = 4 + 4 + 1 = 9$.

(b) The sphere with equation $x^2 + y^2 + (z - 1)^2 = 9$ has center $(0, 0, 1)$ and radius 3.

(c) The diagram shows the plane determined by **u**, **v**, and **r**, along with the trace of the sphere in that plane. We see that the equation $(\mathbf{r} - \mathbf{u}) \cdot (\mathbf{r} - \mathbf{v}) = 0$ states the fact that lines from the ends of a diameter of a circle to any point on its surface meet at right angles.

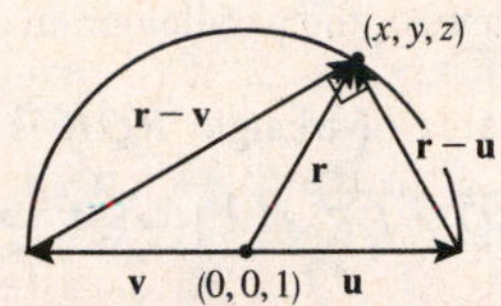

(d) Let $\mathbf{u} = \langle 0, 1, 3\rangle$ and $\mathbf{v} = \langle 2, -1, 4\rangle$. Then $(\mathbf{r} - \mathbf{u}) \cdot (\mathbf{r} - \mathbf{v}) = 0 \Leftrightarrow \langle x, y - 1, z - 3\rangle \cdot \langle x - 2, y + 1, z - 4\rangle = 0$ $\Leftrightarrow x(x - 2) + (y - 1)(y + 1) + (z - 3)(z - 4) = 0 \Leftrightarrow x^2 - 2x + y^2 - 1 + z^2 - 7z + 12 = 0 \Leftrightarrow$ $(x - 1)^2 + y^2 + \left(z - \frac{7}{2}\right)^2 = 1 - 12 - 1 + \frac{49}{4} = \frac{1}{4}$, an equation of a circle with center $\left(1, 0, \frac{7}{2}\right)$ and radius $\frac{1}{2}$.

9.5 THE CROSS PRODUCT

1. The cross product of the vectors $\mathbf{u} = \langle a_1, a_2, a_3\rangle$ and $\mathbf{v} = \langle b_1, b_2, b_3\rangle$ is the vector

$$\mathbf{u} \times \mathbf{v} = \begin{vmatrix} \mathbf{i} & \mathbf{j} & \mathbf{k} \\ a_1 & a_2 & a_3 \\ b_1 & b_2 & b_3 \end{vmatrix} = (a_2b_3 - a_3b_2)\,\mathbf{i} + (a_3b_1 - a_1b_3)\,\mathbf{j} + (a_1b_2 - a_2b_1)\,\mathbf{k}.$$ So the cross product of $\mathbf{u} = \langle 1, 0, 1\rangle$ and $\mathbf{v} = \langle 2, 3, 0\rangle$ is $\mathbf{u} \times \mathbf{v} = \begin{vmatrix} \mathbf{i} & \mathbf{j} & \mathbf{k} \\ 1 & 0 & 1 \\ 2 & 3 & 0 \end{vmatrix} = -3\mathbf{i} + 2\mathbf{j} + 3\mathbf{k}$.

3. $\mathbf{u} \times \mathbf{v} = \begin{vmatrix} \mathbf{i} & \mathbf{j} & \mathbf{k} \\ 1 & 0 & -3 \\ 2 & 3 & 0 \end{vmatrix} = 9\mathbf{i} - 6\mathbf{j} + 3\mathbf{k}$

5. $\mathbf{u} \times \mathbf{v} = \begin{vmatrix} \mathbf{i} & \mathbf{j} & \mathbf{k} \\ 6 & -2 & 8 \\ -9 & 3 & -12 \end{vmatrix} = \mathbf{0}$

7. $\mathbf{u} \times \mathbf{v} = \begin{vmatrix} \mathbf{i} & \mathbf{j} & \mathbf{k} \\ 1 & 1 & 1 \\ 3 & 0 & -4 \end{vmatrix} = -4\mathbf{i} + 7\mathbf{j} - 3\mathbf{k}$

9. (a) $\mathbf{u} \times \mathbf{v} = \begin{vmatrix} \mathbf{i} & \mathbf{j} & \mathbf{k} \\ 1 & 1 & -1 \\ -1 & 1 & -1 \end{vmatrix} = \langle 0, 2, 2\rangle$ is perpendicular to both **u** and **v**.

(b) $\dfrac{\mathbf{u} \times \mathbf{v}}{|\mathbf{u} \times \mathbf{v}|} = \dfrac{\langle 0, 2, 2\rangle}{\sqrt{2^2 + 2^2}} = \left\langle 0, \frac{\sqrt{2}}{2}, \frac{\sqrt{2}}{2}\right\rangle$ is a unit vector perpendicular to both **u** and **v**.

11. (a) $\mathbf{u} \times \mathbf{v} = \begin{vmatrix} \mathbf{i} & \mathbf{j} & \mathbf{k} \\ \frac{1}{2} & -1 & \frac{2}{3} \\ 6 & -12 & -6 \end{vmatrix} = \langle 14, 7, 0\rangle$ is perpendicular to both **u** and **v**.

(b) $\dfrac{\mathbf{u}\times\mathbf{v}}{|\mathbf{u}\times\mathbf{v}|}=\dfrac{\langle 14,7,0\rangle}{\sqrt{14^2+7^2}}=\left\langle \frac{2\sqrt{5}}{5},\frac{\sqrt{5}}{5},0\right\rangle$ is a unit vector perpendicular to both **u** and **v**.

13. $|\mathbf{u}\times\mathbf{v}|=|\mathbf{u}|\,|\mathbf{v}|\sin\theta=6\left(\frac{1}{2}\right)\sin 60^\circ=\frac{3\sqrt{3}}{2}$

15. $|\mathbf{u}\times\mathbf{v}|=|\mathbf{u}|\,|\mathbf{v}|\sin\theta=10\,(10)\sin 90^\circ=100$

17. $\overrightarrow{PQ}\times\overrightarrow{PR}=\langle 1,1,-1\rangle\times\langle -2,0,0\rangle=\langle 0,2,2\rangle$ is perpendicular to the plane passing through P, Q, and R.

19. $\overrightarrow{PQ}\times\overrightarrow{PR}=\langle 1,1,5\rangle\times\langle -1,-1,5\rangle=\langle 10,-10,0\rangle$ is perpendicular to the plane passing through P, Q, and R.

21. The area of the parallelogram determined by $\mathbf{u}=\langle 3,2,1\rangle$ and $\mathbf{v}=\langle 1,2,3\rangle$ is $|\mathbf{u}\times\mathbf{v}|=|\langle 4,-8,4\rangle|=4\sqrt{6}$.

23. The area of the parallelogram determined by $\mathbf{u}=2\mathbf{i}-\mathbf{j}+4\mathbf{k}$ and $\mathbf{v}=\frac{1}{2}\mathbf{i}+2\mathbf{j}-\frac{3}{2}\mathbf{k}$ is $|\mathbf{u}\times\mathbf{v}|=\left|\left\langle -\frac{13}{2},5,\frac{9}{2}\right\rangle\right|=\frac{5\sqrt{14}}{2}$.

25. The area of triangle PQR is one-half the area of the parallelogram determined by $\overrightarrow{PQ}$ and $\overrightarrow{PR}$, that is, $\frac{1}{2}\left|\overrightarrow{PQ}\times\overrightarrow{PR}\right|=\frac{1}{2}\left|\langle -1,1,-1\rangle\times\langle 1,3,3\rangle\right|=\frac{1}{2}\sqrt{6^2+2^2+(-4)^2}=\sqrt{14}$.

27. The area of triangle PQR is one-half the area of the parallelogram determined by $\overrightarrow{PQ}$ and $\overrightarrow{PR}$, that is, $\frac{1}{2}\left|\overrightarrow{PQ}\times\overrightarrow{PR}\right|=\frac{1}{2}\left|\langle -6,-6,0\rangle\times\langle -6,0,-6\rangle\right|=\frac{1}{2}\sqrt{36^2+(-36)^2+(-36)^2}=18\sqrt{3}$.

29. **(a)** $\mathbf{u}\cdot(\mathbf{v}\times\mathbf{w})=\langle 1,2,3\rangle\cdot(\langle -3,2,1\rangle\times\langle 0,8,10\rangle)=\langle 1,2,3\rangle\cdot\langle 12,30,-24\rangle=0$

(b) Because their scalar triple product is 0, the vectors are coplanar.

31. **(a)** $\mathbf{u}\cdot(\mathbf{v}\times\mathbf{w})=\langle 2,3,-2\rangle\cdot(\langle -1,4,0\rangle\times\langle 3,-1,3\rangle)=\langle 2,3,-2\rangle\cdot\langle 12,3,-11\rangle=55$

(b) Because their scalar triple product is nonzero, the vectors are not coplanar. The volume of the parallelepiped that they determine is $|\mathbf{u}\cdot(\mathbf{v}\times\mathbf{w})|=55$.

33. **(a)** $\mathbf{u}\cdot(\mathbf{v}\times\mathbf{w})=\langle 1,-1,1\rangle\cdot(\langle 0,-1,1\rangle\times\langle 1,1,1\rangle)=\langle 1,-1,1\rangle\cdot\langle -2,1,1\rangle=-2$

(b) Because their scalar triple product is nonzero, the vectors are not coplanar. The volume of the parallelepiped that they determine is $|\mathbf{u}\cdot(\mathbf{v}\times\mathbf{w})|=2$.

35. **(a)** We have $|\mathbf{u}|=120$ cm, $|\mathbf{v}|=150$ cm, $|\mathbf{w}|=300$ cm, the angle between **v** and **w** is $90^\circ-30^\circ=60^\circ$, and the angle between **u** and $\mathbf{v}\times\mathbf{w}$ is 0° (because **u** is perpendicular to both **v** and **w**). Therefore, $\mathbf{u}\cdot(\mathbf{v}\times\mathbf{w})=|\mathbf{u}|\,|\mathbf{v}\times\mathbf{w}|\cos 0^\circ=120\,(150\cdot 300\cdot\sin 60^\circ)=2{,}700{,}000\sqrt{3}\approx 4{,}676{,}537$.

(b) The capacity in liters is approximately $\dfrac{4{,}676{,}537\text{ cm}^3}{1000\text{ cm}^3/\text{L}}\approx 4677$ liters.

37. **(a)** $\mathbf{u}\cdot(\mathbf{v}\times\mathbf{w})=\langle 0,1,1\rangle\cdot(\langle 1,0,1\rangle\times\langle 1,1,0\rangle)=\langle 0,1,1\rangle\cdot\langle -1,1,1\rangle=2$, $\mathbf{u}\cdot(\mathbf{w}\times\mathbf{v})=\langle 0,1,1\rangle\cdot(\langle 1,1,0\rangle\times\langle 1,0,1\rangle)=\langle 0,1,1\rangle\cdot\langle 1,-1,-1\rangle=-2$, $\mathbf{v}\cdot(\mathbf{u}\times\mathbf{w})=\langle 1,0,1\rangle\cdot(\langle 0,1,1\rangle\times\langle 1,1,0\rangle)=\langle 1,0,1\rangle\cdot\langle -1,1,-1\rangle=-2$, $\mathbf{v}\cdot(\mathbf{w}\times\mathbf{u})=\langle 1,0,1\rangle\cdot(\langle 1,1,0\rangle\times\langle 0,1,1\rangle)=\langle 1,0,1\rangle\cdot\langle 1,-1,1\rangle=2$, $\mathbf{w}\cdot(\mathbf{u}\times\mathbf{v})=\langle 1,1,0\rangle\cdot(\langle 0,1,1\rangle\times\langle 1,0,1\rangle)=\langle 1,1,0\rangle\cdot\langle 1,1,-1\rangle=2$, and $\mathbf{w}\cdot(\mathbf{v}\times\mathbf{u})=\langle 1,1,0\rangle\cdot(\langle 1,0,1\rangle\times\langle 0,1,1\rangle)=\langle 1,1,0\rangle\cdot\langle -1,-1,1\rangle=-2$.

(b) It appears that $\mathbf{u}\cdot(\mathbf{v}\times\mathbf{w})=\mathbf{v}\cdot(\mathbf{w}\times\mathbf{u})=\mathbf{w}\cdot(\mathbf{u}\times\mathbf{v})=-\mathbf{u}\cdot(\mathbf{w}\times\mathbf{v})=-\mathbf{v}\cdot(\mathbf{u}\times\mathbf{w})=-\mathbf{w}\cdot(\mathbf{v}\times\mathbf{u})$.

(c) We know that the absolute values of the six scalar triple products must be equal because they all represent the volume of the parallelepiped determined by **u**, **v**, and **w**. The fact that $\mathbf{a}\times\mathbf{b}=-(\mathbf{b}\times\mathbf{a})$ completes the proof.

9.6 EQUATIONS OF LINES AND PLANES

1. A line in space is described algebraically by using *parametric* equations. The line that passes through the point $P(x_0,y_0,z_0)$ and is parallel to the vector $\mathbf{v}=\langle a,b,c\rangle$ is described by the equations $x=x_0+at$, $y=y_0+bt$, $z=z_0+ct$.

3. The line passing through $P(1,0,-2)$ parallel to $\mathbf{v}=\langle 3,2,-3\rangle$ has parametric equations $x=1+3t$, $y=2t$, $z=-2-3t$.

5. $x=3$, $y=2-4t$, $z=1+2t$

7. $x=1+2t$, $y=0$, $z=-2-5t$

9. We first find a vector determined by $P(1,-3,2)$ and $Q(2,1,-1)$: $\mathbf{v}=\langle 2-1,1-(-3),-1-2\rangle=\langle 1,4,-3\rangle$. Now we use **v** and the point $(1,-3,2)$ to find parametric equations: $x=1+t$, $y=-3+4t$, $z=2-3t$ where t is any real number.

11. A vector determined by $P(1, 1, 0)$ and $Q(0, 2, 2)$ is $\langle -1, 1, 2\rangle$, so parametric equations are $x = 1 - t$, $y = 1 + t$, $z = 2t$.

13. A vector determined by $P(3, 7, -5)$ and $Q(7, 3, -5)$ is $\langle 4, -4, 0\rangle$, so parametric equations are $x = 3 + 4t$, $y = 7 - 4t$, $z = -5$.

15. **(a)** An equation of the plane with normal vector $\mathbf{n} = \langle 1, 1, -1\rangle$ that passes through $P(0, 2, -3)$ is $1(x - 0) + 1(y - 2) + (-1)[z - (-3)] = 0$ or $x + y - z = 5$.

(b) Setting $y = z = 0$, we find $x = 5$, so the x-intercept is 5. Similarly, the y-intercept is 5 and the z-intercept is -5.

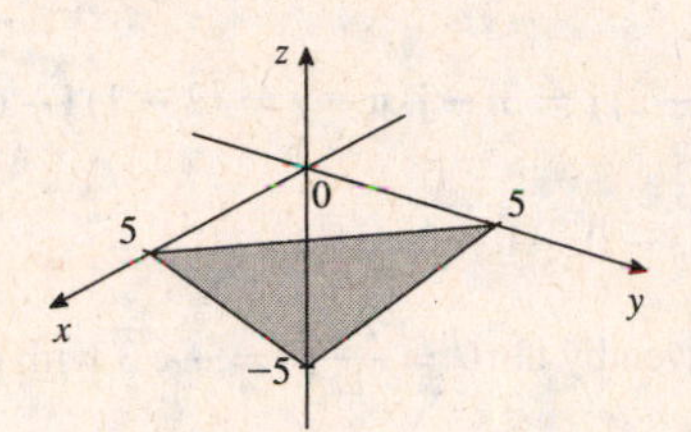

17. **(a)** $3(x - 2) - \frac{1}{2}(z - 8) = 0 \Leftrightarrow 6x - z = 4$

(b) x-intercept $\frac{2}{3}$, no y-intercept, z-intercept -4

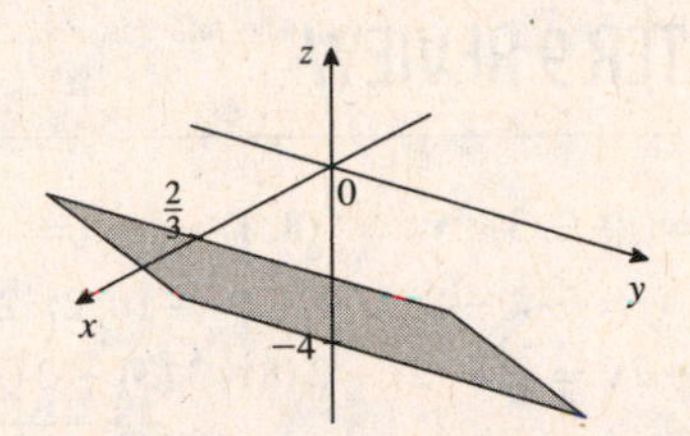

19. **(a)** $3x - (y - 2) + 2(z + 3) = 0 \Leftrightarrow 3x - y + 2z = -8$

(b) x-intercept $-\frac{8}{3}$, y-intercept 8, z-intercept -4

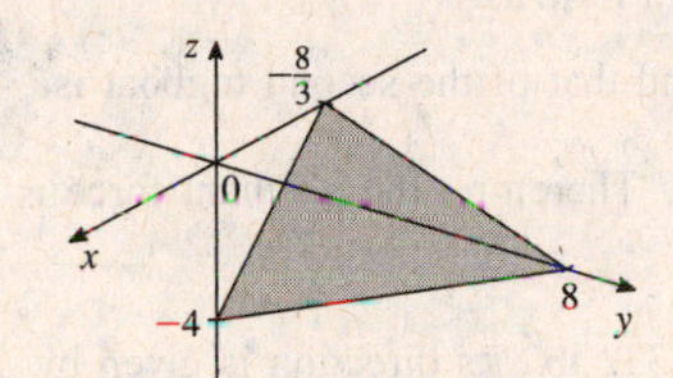

21. The vector $\overrightarrow{PQ} \times \overrightarrow{PR} = \langle -1, -1, -2\rangle \times \langle 1, 2, -1\rangle = \langle 5, -3, -1\rangle$ is perpendicular to both $\overrightarrow{PQ}$ and $\overrightarrow{PR}$ and is therefore perpendicular to the plane through P, Q, and R. Using the formula for an equation of a plane with the point P, we have $5(x - 6) - 3(y + 2) - (z - 1) = 0 \Leftrightarrow$ $5x - 3y - z = 35$.

23. $\overrightarrow{PQ} \times \overrightarrow{PR} = \left\langle 1, \frac{1}{3}, 2\right\rangle \times \left\langle -1, -\frac{1}{3}, 6\right\rangle = \left\langle \frac{8}{3}, -8, 0\right\rangle$, so an equation is $\frac{8}{3}(x - 3) - 8\left(y + \frac{1}{3}\right) = 0 \Leftrightarrow x - 3y = 2$.

25. $\overrightarrow{PQ} \times \overrightarrow{PR} = \langle -3, 1, -1\rangle \times \langle -6, -1, -1\rangle = \langle -2, 3, 9\rangle$, so an equation is $-2(x - 6) + 3(y - 1) + 9(z - 1) = 0 \Leftrightarrow$ $2x - 3y - 9z = 0$

27. The line passes through $(0, 0, 4)$ and $(2, 5, 0)$. A vector determined by these two points is $\mathbf{v} = \langle 2 - 0, 5 - 0, 0 - 4\rangle = \langle 2, 5, -4\rangle$. Now we use $\mathbf{v}$ and the point $(0, 0, 4)$ to find parametric equations: $x = 2t$, $y = 5t$, $z = 4 - 4t$, where t is any real number.

29. The line passes through $(2, -1, 5)$ and is parallel to $\mathbf{j}$, so equations are $x = 2$, $y = -1 + t$, $z = 5$, where t is any real number.

31. The plane passes through $P(1, 0, 0)$, $Q(0, 3, 0)$, and $R(0, 0, 4)$. The vector $\overrightarrow{PQ} \times \overrightarrow{PR} = \langle -1, 3, 0\rangle \times \langle -1, 0, 4\rangle = \langle 12, 4, 3\rangle$ is perpendicular to both $\overrightarrow{PQ}$ and $\overrightarrow{PR}$ and is therefore perpendicular to the plane through P, Q, and R. Using the formula for an equation of a plane, we have $12(x - 1) + 4y + 3z = 0 \Leftrightarrow 12x + 4y + 3z = 12$.

33. If the point (x, y, z) is equidistant from $P(-3, 2, 5)$ and $Q(1, -1, 4)$, then it satisfies $(x + 3)^2 + (y - 2)^2 + (z - 5)^2 = (x - 1)^2 + (y + 1)^2 + (z - 4)^2 \Leftrightarrow 6x + 9 - 4y + 4 - 10z + 25 = -2x + 1 + 2y + 1 - 8z + 16$ $\Leftrightarrow 8x - 6y - 2z = -20 \Leftrightarrow 4x - 3y - z = -10$.

35. **(a)** To find the point of intersection, we substitute the parametric equations of the line into the equation of the plane: $5(2 + t) - 2(3t) - 2(5 - t) = 1 \Leftrightarrow 10 + 5t - 6t - 10 + 2t = 1 \Leftrightarrow t = 1$.

(b) The parameter value $t = 1$ corresponds to the point $(3, 3, 4)$.

37. **(a)** Setting $t = 0$ in the equation for Line 1 gives the point $P(1, 0, -6)$. Setting $t = 1$ gives $Q(0, 3, -1)$. If we set $t = 1$ in the equation for Line 2, we have the point $P'(1, 0, -6) = P$, and if we set $t = \frac{1}{2}$, we get $Q'(0, 3, -1) = Q$.

(b) Setting $t = 0$ in the equation for Line 1 gives the point $(0, 3, -5)$. But if a point on Line 4 has x-coordinate 0, it must have $x = 8 - 2t = 0 \Leftrightarrow t = 4$. But this parameter value gives the point $(0, 3, 2)$ on Line 2, so the two lines are not the same.

CHAPTER 9 REVIEW

1. $\mathbf{u} = \langle -2, 3\rangle$, $\mathbf{v} = \langle 8, 1\rangle$. $|\mathbf{u}| = \sqrt{(-2)^2 + 3^2} = \sqrt{13}$, $\mathbf{u} + \mathbf{v} = \langle -2 + 8, 3 + 1\rangle = \langle 6, 4\rangle$, $\mathbf{u} - \mathbf{v} = \langle -2 - 8, 3 - 1\rangle = \langle -10, 2\rangle$, $2\mathbf{u} = \langle 2(-2), 2(3)\rangle = \langle -4, 6\rangle$, and $3\mathbf{u} - 2\mathbf{v} = \langle 3(-2) - 2(8), 3(3) - 2(1)\rangle = \langle -22, 7\rangle$.

3. $\mathbf{u} = 2\mathbf{i} + \mathbf{j}$, $\mathbf{v} = \mathbf{i} - 2\mathbf{j}$. $|\mathbf{u}| = \sqrt{2^2 + 1^2} = \sqrt{5}$, $\mathbf{u} + \mathbf{v} = (2 + 1)\mathbf{i} + (1 - 2)\mathbf{j} = 3\mathbf{i} - \mathbf{j}$, $\mathbf{u} - \mathbf{v} = (2 - 1)\mathbf{i} + (1 + 2)\mathbf{j} = \mathbf{i} + 3\mathbf{j}$, $2\mathbf{u} = 4\mathbf{i} + 2\mathbf{j}$, and $3\mathbf{u} - 2\mathbf{v} = 3(2\mathbf{i} + \mathbf{j}) - 2(\mathbf{i} - 2\mathbf{j}) = 4\mathbf{i} + 7\mathbf{j}$.

5. The vector with initial point $P(0, 3)$ and terminal point $Q(3, -1)$ is $\langle 3 - 0, -1 - 3\rangle = \langle 3, -4\rangle$.

7. $\mathbf{u} = \left\langle -2, 2\sqrt{3}\right\rangle$ has length $\sqrt{(-2)^2 + \left(2\sqrt{3}\right)^2} = 4$. Its direction is given by $\tan\theta = \dfrac{2\sqrt{3}}{-2} = -\sqrt{3}$ with θ in quadrant II, so $\theta = \pi + \tan^{-1}\left(-\sqrt{3}\right) = 120°$.

9. $\mathbf{u} = \langle |\mathbf{u}|\cos\theta, |\mathbf{u}|\sin\theta\rangle = \langle 20\cos 60°, 20\sin 60°\rangle = \left\langle 10, 10\sqrt{3}\right\rangle$.

11. **(a)** The force exerted by the first tugboat can be expressed in component form as $\mathbf{u} = \left\langle 2.0 \times 10^4 \cos 40°, 2.0 \times 10^4 \sin 40°\right\rangle \approx \langle 15321, 12856\rangle$, and that of the second tugboat is $\left\langle 3.4 \times 10^4 \cos(-15°), 3.4 \times 10^4 \sin(-15°)\right\rangle \approx \langle 32841, -8800\rangle$. Therefore, the resultant force is $\mathbf{w} = \mathbf{u} + \mathbf{v} \approx \langle 15321 + 32841, 12856 - 8800\rangle = \langle 48162, 4056\rangle$.

(b) The magnitude of the resultant force is $\sqrt{48162^2 + 4056^2} \approx 48{,}332$ lb. Its direction is given by $\tan\theta = \dfrac{4056}{48{,}162} \approx 0.084$, so $\theta \approx \tan^{-1} 0.084 \approx 4.8°$ or N 85.2° E.

13. $\mathbf{u} = \langle 4, -3\rangle$, $\mathbf{v} = \langle 9, -8\rangle$. $|\mathbf{u}| = \sqrt{4^2 + (-3)^2} = 5$, $\mathbf{u}\cdot\mathbf{u} = 4^2 + (-3)^2 = 25$, and $\mathbf{u}\cdot\mathbf{v} = 4(9) + (-3)(-8) = 60$.

15. $\mathbf{u} = -2\mathbf{i} + 2\mathbf{j}$, $\mathbf{v} = \mathbf{i} + \mathbf{j}$. $|\mathbf{u}| = \sqrt{(-2)^2 + 2^2} = 2\sqrt{2}$, $\mathbf{u}\cdot\mathbf{u} = (-2)^2 + 2^2 = 8$, and $\mathbf{u}\cdot\mathbf{v} = -2(1) + 2(1) = 0$.

17. $\mathbf{u}\cdot\mathbf{v} = \langle -4, 2\rangle \cdot \langle 3, 6\rangle = -4(3) + 2(6) = 0$, so the vectors are perpendicular.

19. $\mathbf{u}\cdot\mathbf{v} = (2\mathbf{i} + \mathbf{j})\cdot(\mathbf{i} + 3\mathbf{j}) = 2(1) + 1(3) = 5$, so the vectors are not perpendicular. The angle between them is given by $\cos\theta = \dfrac{\mathbf{u}\cdot\mathbf{v}}{|\mathbf{u}|\,|\mathbf{v}|} = \dfrac{5}{\sqrt{2^2 + 1^2}\sqrt{1^2 + 3^2}} = \dfrac{\sqrt{2}}{2}$, so $\theta = \cos^{-1}\frac{\sqrt{2}}{2} = 45°$.

21. **(a)** $\mathbf{u} = \langle 3, 1\rangle$, $\mathbf{v} = \langle 6, -1\rangle$. The component of $\mathbf{u}$ along $\mathbf{v}$ is $\dfrac{\mathbf{u}\cdot\mathbf{v}}{|\mathbf{v}|} = \dfrac{3(6) + 1(-1)}{\sqrt{6^2 + 1^2}} = \dfrac{17\sqrt{37}}{37}$.

(b) $\text{proj}_{\mathbf{v}}\mathbf{u} = \left(\dfrac{\mathbf{u}\cdot\mathbf{v}}{|\mathbf{v}|^2}\right)\mathbf{v} = \frac{17}{37}\langle 6, -1\rangle = \left\langle \frac{102}{37}, -\frac{17}{37}\right\rangle$

(c) $\mathbf{u}_1 = \text{proj}_{\mathbf{v}}\mathbf{u} = \left\langle \frac{102}{37}, -\frac{17}{37}\right\rangle$ and $\mathbf{u}_2 = \mathbf{u} - \text{proj}_{\mathbf{v}}\mathbf{u} = \langle 3, 1\rangle - \left\langle \frac{102}{37}, -\frac{17}{37}\right\rangle = \left\langle \frac{9}{37}, \frac{54}{37}\right\rangle$.

23. **(a)** $\mathbf{u} = \mathbf{i} + 2\mathbf{j}$, $\mathbf{v} = 4\mathbf{i} - 9\mathbf{j}$. The component of $\mathbf{u}$ along $\mathbf{v}$ is $\dfrac{\mathbf{u}\cdot\mathbf{v}}{|\mathbf{v}|} = \dfrac{1(4) + 2(-9)}{\sqrt{4^2 + (-9)^2}} = -\dfrac{14\sqrt{97}}{97}$.

(b) $\text{proj}_{\mathbf{v}}\mathbf{u} = \left(\dfrac{\mathbf{u}\cdot\mathbf{v}}{|\mathbf{v}|^2}\right)\mathbf{v} = \frac{-14}{97}(4\mathbf{i} - 9\mathbf{j}) = -\frac{56}{97}\mathbf{i} + \frac{126}{97}\mathbf{j}$

(c) $\mathbf{u}_1 = \text{proj}_\mathbf{v}\,\mathbf{u} = -\frac{56}{97}\mathbf{i} + \frac{126}{97}\mathbf{j}$ and $\mathbf{u}_2 = \mathbf{u} - \text{proj}_\mathbf{v}\,\mathbf{u} = \mathbf{i} + 2\mathbf{j} - \left(-\frac{56}{97}\mathbf{i} + \frac{126}{97}\mathbf{j}\right) = \frac{153}{97}\mathbf{i} + \frac{68}{97}\mathbf{j}$.

25.

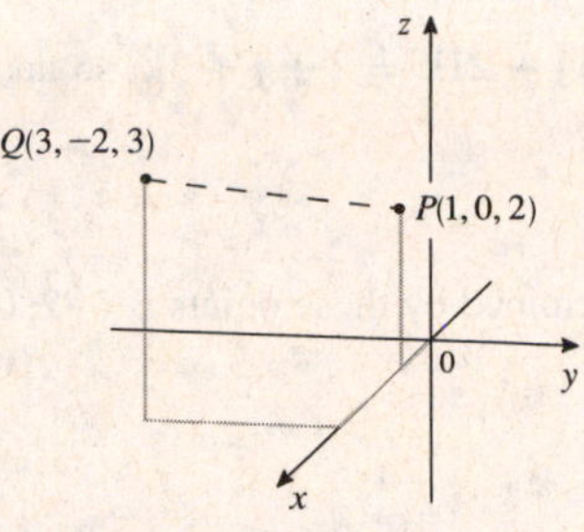

The distance between P and Q is $\sqrt{(3-1)^2 + (-2-0)^2 + (3-2)^2} = 3$.

27. The sphere with radius $r = 6$ and center $C\,(0, 0, 0)$ has equation $(x-0)^2 + (y-0)^2 + (z-0)^2 = 6^2 \Leftrightarrow x^2 + y^2 + z^2 = 36$.

29. We complete the squares to find

$$x^2 + y^2 + z^2 - 2x - 6y + 4z = 2 \Leftrightarrow$$

$$\left(x^2 - 2x + 1\right) + \left(y^2 - 6y + 9\right) + \left(z^2 + 4z + 4\right) = 2 + 1 + 9 + 4$$

$\Leftrightarrow (x-1)^2 + (y-3)^2 + (z+2)^2 = 16$, an equation of the sphere with center $(1, 3, -2)$ and radius 4.

31. $\mathbf{u} = \langle 4, -2, 4\rangle$ and $\mathbf{v} = \langle 2, 3, -1\rangle$, so $|\mathbf{u}| = \sqrt{4^2 + (-2)^2 + (4)^2} = 6$, $\mathbf{u} + \mathbf{v} = \langle 4+2, -2+3, 4+(-1)\rangle = \langle 6, 1, 3\rangle$, $\mathbf{u} - \mathbf{v} = \langle 4-2, -2-3, 4-(-1)\rangle = \langle 2, -5, 5\rangle$, and $\frac{3}{4}\mathbf{u} - 2\mathbf{v} = \frac{3}{4}\langle 4, -2, 4\rangle - 2\langle 2, 3, -1\rangle = \left\langle -1, -\frac{15}{2}, 5\right\rangle$.

33. (a) $\mathbf{u}\cdot\mathbf{v} = \langle 3, -2, 4\rangle \cdot \langle 3, 1, -2\rangle = 3\,(3) - 2\,(1) + 4\,(-2) = -1$

(b) $\mathbf{u}\cdot\mathbf{v} \neq 0$, so the vectors are not perpendicular. The angle between them is given by

$$\cos\theta = \frac{\mathbf{u}\cdot\mathbf{v}}{|\mathbf{u}|\,|\mathbf{v}|} = \frac{-1}{\sqrt{3^2 + (-2)^2 + 4^2}\sqrt{3^2 + 1^2 + (-2)^2}} = -\frac{\sqrt{406}}{406},\ \text{so}\ \theta = \cos^{-1}\left(-\frac{\sqrt{406}}{406}\right) \approx 92.8^\circ.$$

35. (a) $\mathbf{u}\cdot\mathbf{v} = (2\mathbf{i} - \mathbf{j} + 4\mathbf{k}) \cdot (3\mathbf{i} + 2\mathbf{j} - \mathbf{k}) = 2\,(3) - 1\,(2) + 4\,(-1) = 0$

(b) $\mathbf{u}\cdot\mathbf{v} = 0$, so the vectors are perpendicular.

37. (a) $\mathbf{u}\times\mathbf{v} = \langle 1, 1, 3\rangle \times \langle 5, 0, -2\rangle = (-2-0)\,\mathbf{i} - (-2-15)\,\mathbf{j} + (0-5)\,\mathbf{k} = \langle -2, 17, -5\rangle$

(b) A unit vector perpendicular to $\mathbf{u}$ and $\mathbf{v}$ is $\dfrac{\mathbf{u}\times\mathbf{v}}{|\mathbf{u}\times\mathbf{v}|} = \dfrac{\langle -2, 17, -5\rangle}{\sqrt{(-2)^2 + 17^2 + (-5)^2}} = \left\langle -\frac{\sqrt{318}}{159}, \frac{17\sqrt{318}}{318}, -\frac{5\sqrt{318}}{318}\right\rangle$.

39. (a) $\mathbf{u}\times\mathbf{v} = (\mathbf{i} - \mathbf{j}) \times (2\mathbf{j} - \mathbf{k}) = (1-0)\,\mathbf{i} - (-1-0)\,\mathbf{j} + (2-0)\,\mathbf{k} = \mathbf{i} + \mathbf{j} + 2\mathbf{k}$

(b) A unit vector perpendicular to $\mathbf{u}$ and $\mathbf{v}$ is $\dfrac{\mathbf{u}\times\mathbf{v}}{|\mathbf{u}\times\mathbf{v}|} = \dfrac{\mathbf{i} + \mathbf{j} + 2\mathbf{k}}{\sqrt{1^2 + 1^2 + 2^2}} = \frac{\sqrt{6}}{6}\mathbf{i} + \frac{\sqrt{6}}{6}\mathbf{j} + \frac{\sqrt{6}}{3}\mathbf{k}$.

41. The area of triangle PQR is one-half the area of the parallelogram determined by $\overrightarrow{PQ}$ and $\overrightarrow{PR}$, that is, $\frac{1}{2}\left|\overrightarrow{PQ} \times \overrightarrow{PR}\right| = \frac{1}{2}\left|(1-6)\,\mathbf{i} - (2+8)\,\mathbf{j} + (-6-4)\,\mathbf{k}\right| = \frac{1}{2}\sqrt{(-5)^2 + (-10)^2 + (-10)^2} = \frac{1}{2}\sqrt{225} = \frac{15}{2}$.

43. The volume of the parallelepiped determined by $\mathbf{u} = 2\mathbf{i} - \mathbf{j}$, $\mathbf{v} = 2\mathbf{j} + \mathbf{k}$, and $\mathbf{w} = 3\mathbf{i} + \mathbf{j} - \mathbf{k}$ is the absolute value of their scalar triple product: $V = |\mathbf{u}\cdot(\mathbf{v}\times\mathbf{w})| = |(2\mathbf{i} - \mathbf{j}) \cdot (-3\mathbf{i} + 3\mathbf{j} - 6\mathbf{k})| = |-6-3| = 9$.

45. The line that passes through $P\,(2, 0, -6)$ and is parallel to $\mathbf{v} = \langle 3, 1, 0\rangle$ has parametric equations $x = 2 + 3t$, $y = t$, $z = -6$.

47. A vector determined by $P\,(6, -2, -3)$ and $Q\,(4, 1, -2)$ is $\langle -2, 3, 1\rangle$, so parametric equations are $x = 6 - 2t$, $y = -2 + 3t$, $z = -3 + t$.

49. Using the formula for an equation of a plane, the plane with normal vector $\mathbf{n} = \langle 2, 3, -5\rangle$ passing through $P\,(2, 1, 1)$ has equation $2\,(x-2) + 3\,(y-1) - 5\,(z-1) = 0 \Leftrightarrow 2x + 3y - 5z = 2$.

51. The plane passes through $P\,(1,1,1)$, $Q\,(3,-4,2)$, and $R\,(6,-1,0)$.

$$\overrightarrow{PQ}\times\overrightarrow{PR}=\langle 2,-5,1\rangle\times\langle 5,-2,-1\rangle=\begin{vmatrix}\mathbf{i}&\mathbf{j}&\mathbf{k}\\2&-5&1\\5&-2&-1\end{vmatrix}=7\mathbf{i}-(-7)\,\mathbf{j}+21\mathbf{k}=\mathbf{i}+\mathbf{j}+3\mathbf{k}$$, so an equation is

$1\,(x-1)+1\,(y-1)+3\,(z-1)=0\Leftrightarrow x+y+3z=5$.

53. The line passes through the points $(2,0,0)$ and $(0,0,-4)$. A vector determined by these points is $\langle -2,0,-4\rangle$ or $\langle 1,0,2\rangle$, so parametric equations are $x=2+t$, $y=0$, $z=2t$.

CHAPTER 9 TEST

1. (a)

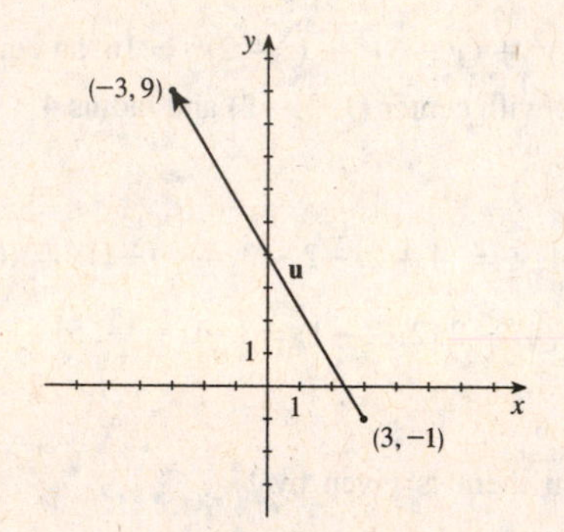

(b) $\mathbf{u}=(-3-3)\,\mathbf{i}+[9-(-1)]\,\mathbf{j}=-6\mathbf{i}+10\mathbf{j}$

(c) $|\mathbf{u}|=\sqrt{(-6)^2+10^2}=2\sqrt{34}$

3. (a)

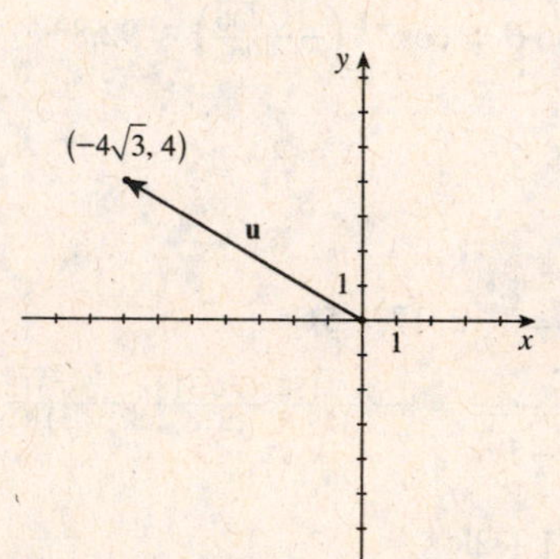

(b) The length of $\mathbf{u}$ is $|\mathbf{u}|=\sqrt{\left(-4\sqrt{3}\right)^2+4^2}=8$. Its direction is given by

$\tan\theta=\dfrac{4}{-4\sqrt{3}}=-\frac{\sqrt{3}}{3}$ with θ in quadrant II, so

$\theta=180^\circ-\tan^{-1}\left(\frac{\sqrt{3}}{3}\right)=150^\circ$.

5. (a) $\cos\theta=\dfrac{\mathbf{u}\cdot\mathbf{v}}{|\mathbf{u}|\,|\mathbf{v}|}=\dfrac{3\,(5)+2\,(-1)}{\sqrt{3^2+2^2}\sqrt{5^2+(-1)^2}}=\dfrac{\sqrt{338}}{26}$, so $\theta=\cos^{-1}\frac{\sqrt{338}}{26}=\cos^{-1}\frac{\sqrt{2}}{2}=45^\circ$.

(b) The component of $\mathbf{u}$ along $\mathbf{v}$ is $\dfrac{\mathbf{u}\cdot\mathbf{v}}{|\mathbf{v}|}=\dfrac{13}{\sqrt{26}}=\dfrac{\sqrt{26}}{2}$.

(c) $\operatorname{proj}_{\mathbf{v}}\mathbf{u}=\left(\dfrac{\mathbf{u}\cdot\mathbf{v}}{|\mathbf{v}|^2}\right)\mathbf{v}=\frac{13}{26}\,(5\mathbf{i}-\mathbf{j})=\frac{5}{2}\mathbf{i}-\frac{1}{2}\mathbf{j}$

7. (a) The distance between $P\,(4,3,-1)$ and $Q\,(6,-1,3)$ is $d=\sqrt{(6-4)^2+(-1-3)^2+[3-(-1)]^2}=6$.

(b) An equation is $(x-4)^2+(y-3)^2+[z-(-1)]^2=6^2\Leftrightarrow(x-4)^2+(y-3)^2+(z+1)^2=36$.

(c) $\mathbf{u}=\langle 6-4,-1-3,3-(-1)\rangle=\langle 2,-4,4\rangle=2\mathbf{i}-4\mathbf{j}+4\mathbf{k}$

9. A vector perpendicular to both $\mathbf{u}=\mathbf{j}+2\mathbf{k}$ and $\mathbf{v}=\mathbf{i}-2\mathbf{j}+3\mathbf{k}$ is $\mathbf{u}\times\mathbf{v}=\langle 0,1,2\rangle\times\langle 1,-2,3\rangle=\langle 7,2,-1\rangle$, so two unit vectors perpendicular to $\mathbf{u}$ and $\mathbf{v}$ are $\dfrac{\mathbf{u}\times\mathbf{v}}{|\mathbf{u}\times\mathbf{v}|}=\dfrac{\langle 7,2,-1\rangle}{\sqrt{7^2+2^2+(-1)^2}}=\left\langle\frac{7\sqrt{6}}{18},\frac{\sqrt{6}}{9},-\frac{\sqrt{6}}{18}\right\rangle$ and $\left\langle-\frac{7\sqrt{6}}{18},-\frac{\sqrt{6}}{9},\frac{\sqrt{6}}{18}\right\rangle$.

11. A vector determined by the two points is $\overrightarrow{PQ} = \langle -2, 1, -2 \rangle$, so parametric equations are $x = 2 - 2t$, $y = -4 + t$, $z = 7 - 2t$.

FOCUS ON MODELING Vector Fields

1. $\mathbf{F}(x, y) = \frac{1}{2}\mathbf{i} + \frac{1}{2}\mathbf{j}$

All vectors point in the same direction and have length $\frac{\sqrt{2}}{2}$.

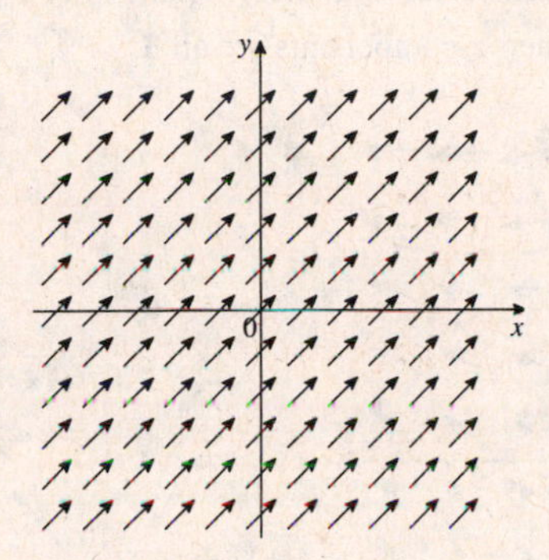

3. $\mathbf{F}(x, y) = y\mathbf{i} + \frac{1}{2}\mathbf{j}$

The vectors point to the left for $y < 0$ and to the right for $y > 0$.

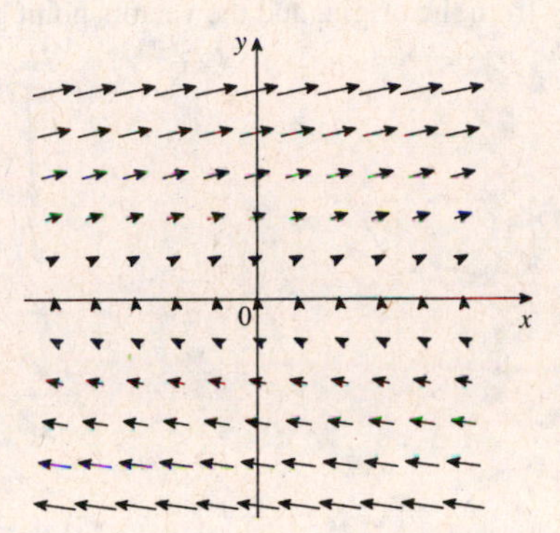

5. $\mathbf{F}(x, y) = \dfrac{y\mathbf{i} + x\mathbf{j}}{\sqrt{x^2 + y^2}}$

The length of the vector $\dfrac{y\mathbf{i} + x\mathbf{j}}{\sqrt{x^2 + y^2}}$ is 1.

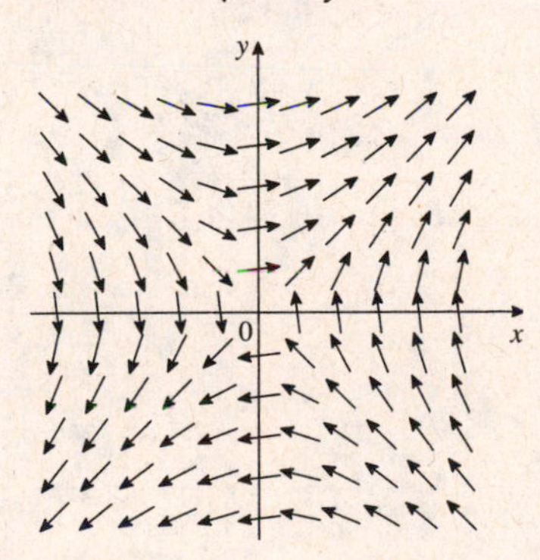

7. $\mathbf{F}(x, y, z) = \mathbf{j}$

All vectors in this field are parallel to the y-axis and have length 1.

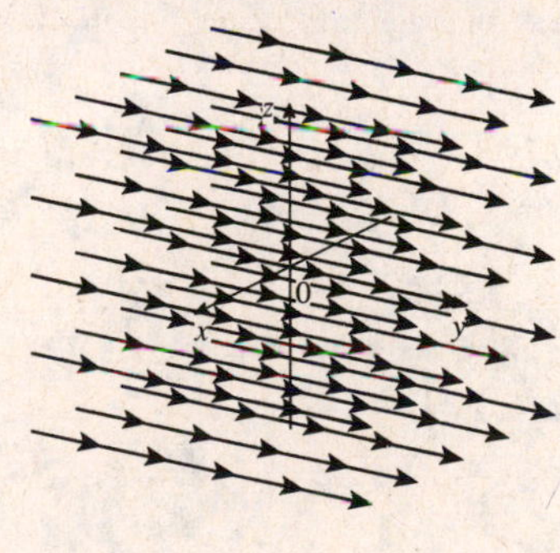

9. $\mathbf{F}(x, y, z) = z\mathbf{j}$

At each point (x, y, z), $\mathbf{F}(x, y, z)$ is a vector of length $|z|$. For $z > 0$, all point in the direction of the positive y-axis while for $z < 0$, all are in the direction of the negative y-axis.

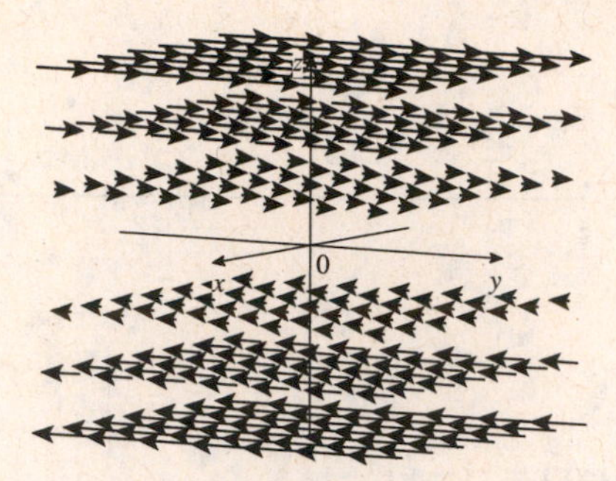

11. $\mathbf{F}(x, y) = \langle y, x \rangle$ corresponds to graph II, because in the first quadrant all the vectors have positive x- and y-components, in the second quadrant all vectors have positive x-components and negative y-components, in the third quadrant all vectors have negative x- and y-components, and in the fourth quadrant all vectors have negative x-components and positive y-components.

13. $\mathbf{F}(x, y) = \langle x - 2, x + 1 \rangle$ corresponds to graph I because the vectors are independent of y (vectors along vertical lines are identical) and, as we move to the right, both the x- and the y-components get larger.

15. $\mathbf{F}(x, y, z) = \mathbf{i} + 2\mathbf{j} + 3\mathbf{k}$ corresponds to graph IV, since all vectors have identical length and direction.

17. $\mathbf{F}(x, y, z) = x\mathbf{i} + y\mathbf{j} + 3\mathbf{k}$ corresponds to graph III; the projection of each vector onto the xy-plane is $x\mathbf{i} + y\mathbf{j}$, which points away from the origin, and the vectors point generally upward because their z-components are all 3.

19.

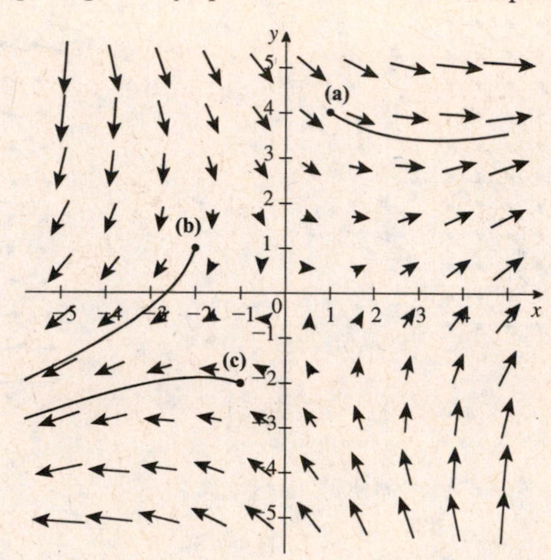

10 SYSTEMS OF EQUATIONS AND INEQUALITIES

10.1 SYSTEMS OF LINEAR EQUATIONS IN TWO VARIABLES

1. The given system is a system of two equations in the two variables x and y. To check if $(5, -1)$ is a solution of this system, we check if $x = 5$ and $y = -1$ satisfy each *equation* in the system. The only solution of the given system is $(2, 1)$.

3. A system of two linear equations in two variables can have one solution, *no* solution, or *infinitely many* solutions.

5. $\begin{cases} x - y = 1 \\ 4x + 3y = 18 \end{cases}$ Solving the first equation for x, we get $x = y + 1$, and substituting this into the second equation gives $4(y+1) + 3y = 18 \Leftrightarrow 7y + 4 = 18 \Leftrightarrow 7y = 14 \Leftrightarrow y = 2$. Substituting for y we get $x = y + 1 = 2 + 1 = 3$. Thus, the solution is $(3, 2)$.

7. $\begin{cases} x - y = 2 \\ 2x + 3y = 9 \end{cases}$ Solving the first equation for x, we get $x = y + 2$, and substituting this into the second equation gives $2(y+2) + 3y = 9 \Leftrightarrow 5y + 4 = 9 \Leftrightarrow 5y = 5 \Leftrightarrow y = 1$. Substituting for y we get $x = y + 2 = (1) + 2 = 3$. Thus, the solution is $(3, 1)$.

9. $\begin{cases} 3x + 4y = 10 \\ x - 4y = -2 \end{cases}$ Adding the two equations, we get $4x = 8 \Leftrightarrow x = 2$, and substituting into the first equation in the original system gives $3(2) + 4y = 10 \Leftrightarrow 4y = 4 \Leftrightarrow y = 1$. Thus, the solution is $(2, 1)$.

11. $\begin{cases} 3x - 2y = -13 \\ -6x + 5y = 28 \end{cases}$ Multiplying the first equation by 2 gives the system $\begin{cases} 6x - 4y = -26 \\ -6x + 5y = 28 \end{cases}$ Adding, we get $y = 2$, and substituting into the first equation in the original system gives $3x - 2(2) = -13 \Leftrightarrow 3x = -9 \Leftrightarrow x = -3$. The solution is $(-3, 2)$.

13. $\begin{cases} 2x + y = -1 \\ x - 2y = -8 \end{cases}$ By inspection of the graph, it appears that $(-2, 3)$ is the solution to the system. We check this in both equations to verify that it is a solution. $2(-2) + 3 = -4 + 3 = -1$ and $-2 - 2(3) = -2 - 6 = -8$. Since both equations are satisfied, the solution is $(-2, 3)$.

15. $\begin{cases} x - y = 4 \\ 2x + y = 2 \end{cases}$

The solution is $x = 2$, $y = -2$.

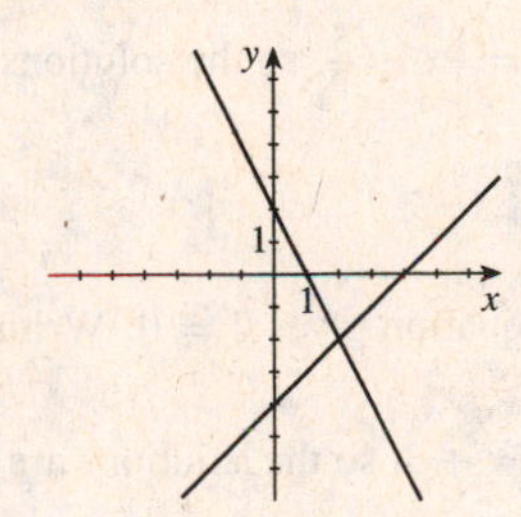

17. $\begin{cases} 2x - 3y = 12 \\ -x + \frac{3}{2}y = 4 \end{cases}$

The lines are parallel, so there is no intersection and hence no solution.

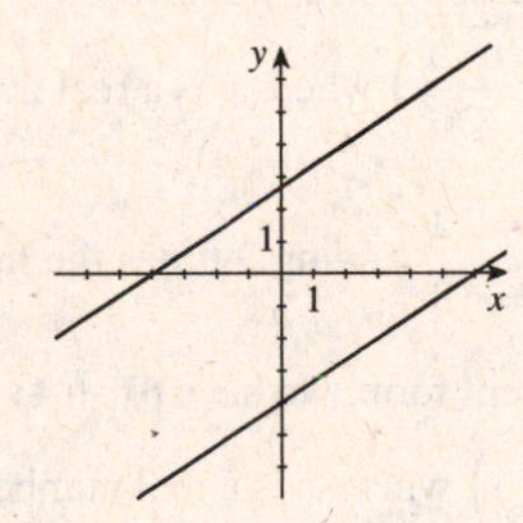

19. $\begin{cases} -x + \frac{1}{2}y = -5 \\ 2x - \quad y = 10 \end{cases}$

There are infinitely many solutions.

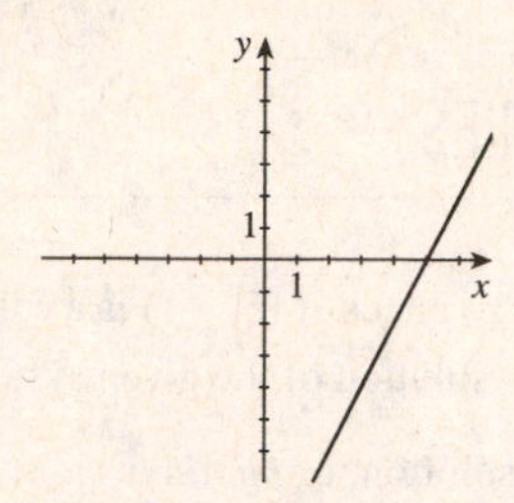

21. $\begin{cases} x + y = 4 \\ -x + y = 0 \end{cases}$ Adding the two equations gives $2y = 4 \Leftrightarrow y = 2$. Substituting for y in the first equation gives $x + 2 = 4 \Leftrightarrow x = 2$. Hence, the solution is $(2, 2)$.

23. $\begin{cases} 2x - 3y = 9 \\ 4x + 3y = 9 \end{cases}$ Adding the two equations gives $6x = 18 \Leftrightarrow x = 3$. Substituting for x in the second equation gives $4\,(3) + 3y = 9 \Leftrightarrow 12 + 3y = 9 \Leftrightarrow 3y = -3 \Leftrightarrow x = -1$. Hence, the solution is $(3, -1)$.

25. $\begin{cases} x + 3y = 5 \\ 2x - \quad y = 3 \end{cases}$ Solving the first equation for x gives $x = -3y + 5$. Substituting for x in the second equation gives $2\,(-3y + 5) - y = 3 \Leftrightarrow -6y + 10 - y = 3 \Leftrightarrow -7y = -7 \Leftrightarrow y = 1$. Then $x = -3\,(1) + 5 = 2$. Hence, the solution is $(2, 1)$.

27. $-x + y = 2 \Leftrightarrow y = x + 2$. Substituting for y into $4x - 3y = -3$ gives $4x - 3\,(x + 2) = -3 \Leftrightarrow 4x - 3x - 6 = -3 \Leftrightarrow x = 3$, and so $y = (3) + 2 = 5$. Hence, the solution is $(3, 5)$.

29. $x + 2y = 7 \Leftrightarrow x = 7 - 2y$. Substituting for x into $5x - y = 2$ gives $5\,(7 - 2y) - y = 2 \Leftrightarrow 35 - 10y - y = 2 \Leftrightarrow -11y = -33 \Leftrightarrow y = 3$, and so $x = 7 - 2\,(3) = 1$. Hence, the solution is $(1, 3)$.

31. $-\frac{1}{3}x - \frac{1}{6}y = -1 \Leftrightarrow -2x - y = -6 \Leftrightarrow y = -2x + 6$. Substituting for y into $\frac{2}{3}x + \frac{1}{6}y = 3$ gives $\frac{2}{3}x + \frac{1}{6}\,(-2x + 6) = 3 \Leftrightarrow 4x - 2x + 6 = 18 \Leftrightarrow 2x = 12 \Leftrightarrow x = 6$, and so $y = -2\,(6) + 6 = -6$. Hence, the solution is $(6, -6)$.

33. $\frac{1}{2}x + \frac{1}{3}y = 2 \Leftrightarrow x + \frac{2}{3}y = 4 \Leftrightarrow x = 4 - \frac{2}{3}y$. Substituting for x into $\frac{1}{5}x - \frac{2}{3}y = 8$ gives $\frac{1}{5}\left(4 - \frac{2}{3}y\right) - \frac{2}{3}y = 8 \Leftrightarrow \frac{4}{5} - \frac{2}{15}y - \frac{10}{15}y = 8 \Leftrightarrow 12 - 2y - 10y = 120 \Leftrightarrow y = -9$, and so $x = 4 - \frac{2}{3}\,(-9) = 10$. Hence, the solution is $(10, -9)$.

35. $\begin{cases} 3x + 2y = 8 \\ x - 2y = 0 \end{cases}$ Multiplying the second equation by 3 gives the system $\begin{cases} 3x + 2y = 8 \\ 3x - 6y = 0 \end{cases}$ Subtracting the second equation from the first gives $8y = 8 \Leftrightarrow y = 1$. Substituting into the first equation we get $3x + 2\,(1) = 8 \Leftrightarrow 3x = 6 \Leftrightarrow x = 2$. Thus, the solution is $(2, 1)$.

37. $\begin{cases} x + \ 4y = 8 \\ 3x + 12y = 2 \end{cases}$ Adding -3 times the first equation to the second equation gives $0 = -22$, which is never true. Thus, the system has no solution.

39. $\begin{cases} 2x - 6y = \ \ 10 \\ -3x + 9y = -15 \end{cases}$ Adding 3 times the first equation to 2 times the second equation gives $0 = 0$. Writing the equation in slope-intercept form, we have $2x - 6y = 10 \Leftrightarrow -6y = -2x + 10 \Leftrightarrow y = \frac{1}{3}x - \frac{5}{3}$, so the solutions are all pairs of the form $\left(x, \frac{1}{3}x - \frac{5}{3}\right)$ where x is a real number.

41. $\begin{cases} 6x + 4y = 12 \\ 9x + 6y = 18 \end{cases}$ Adding 3 times the first equation to -2 times the second equation gives $0 = 0$. Writing the equation in slope-intercept form, we have $6x + 4y = 12 \Leftrightarrow 4y = -6x + 12 \Leftrightarrow y = -\frac{3}{2}x + 3$, so the solutions are all pairs of the form $\left(x, -\frac{3}{2}x + 3\right)$ where x is a real number.

43. $\begin{cases} 8s - 3t = -3 \\ 5s - 2t = -1 \end{cases}$ Adding 2 times the first equation to 3 times the second equation gives $s = -3$, so $8(-3) - 3t = -3 \Leftrightarrow -24 - 3t = -3 \Leftrightarrow t = -7$. Thus, the solution is $(-3, -7)$.

45. $\begin{cases} \frac{1}{2}x + \frac{3}{5}y = 3 \\ \frac{5}{3}x + 2y = 10 \end{cases}$ Adding 10 times the first equation to -3 times the second equation gives $0 = 0$. Writing the equation in slope-intercept form, we have $\frac{1}{2}x + \frac{3}{5}y = 3 \Leftrightarrow \frac{3}{5}y = -\frac{1}{2}x + 3 \Leftrightarrow y = -\frac{5}{6}x + 5$, so the solutions are all pairs of the form $\left(x, -\frac{5}{6}x + 5\right)$ where x is a real number.

47. $\begin{cases} 0.4x + 1.2y = 14 \\ 12x - 5y = 10 \end{cases}$ Adding 30 times the first equation to -1 times the second equation gives $41y = 410 \Leftrightarrow y = 10$, so $12x - 5(10) = 10 \Leftrightarrow 12x = 60 \Leftrightarrow x = 5$. Thus, the solution is $(5, 10)$.

49. $\begin{cases} \frac{1}{3}x - \frac{1}{4}y = 2 \\ -8x + 6y = 10 \end{cases}$ Adding 24 times the first equation to the second equation gives $0 = 58$, which is never true. Thus, the system has no solution.

51. $\begin{cases} 0.21x + 3.17y = 9.51 \\ 2.35x - 1.17y = 5.89 \end{cases}$

The solution is approximately $(3.87, 2.74)$.

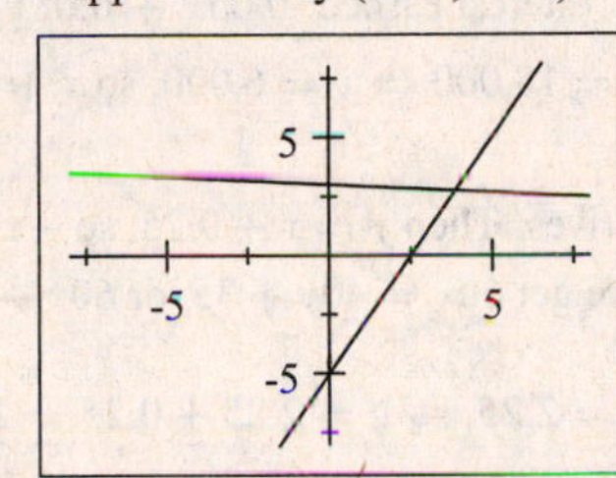

53. $\begin{cases} 2371x - 6552y = 13{,}591 \\ 9815x + 992y = 618{,}555 \end{cases}$

The solution is approximately $(61.00, 20.00)$.

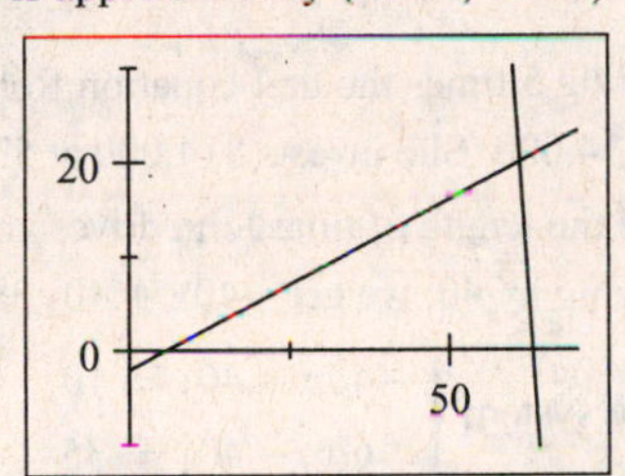

55. Subtracting the first equation from the second, we get $ay - y = 1 \Leftrightarrow y(a-1) = 1 \Leftrightarrow y = \frac{1}{a-1}$, $a \neq 1$. So $x + \left(\frac{1}{a-1}\right) = 0 \Leftrightarrow x = \frac{1}{1-a} = -\frac{1}{a-1}$. Thus, the solution is $\left(-\frac{1}{a-1}, \frac{1}{a-1}\right)$.

57. Subtracting b times the first equation from a times the second, we get $\left(a^2 - b^2\right)y = a - b \Leftrightarrow y = \frac{a-b}{a^2-b^2} = \frac{1}{a+b}$, $a^2 - b^2 \neq 0$. So $ax + \frac{b}{a+b} = 1 \Leftrightarrow ax = \frac{a}{a+b} \Leftrightarrow x = \frac{1}{a+b}$. Thus, the solution is $\left(\frac{1}{a+b}, \frac{1}{a+b}\right)$.

59. Let the two numbers be x and y. Then $\begin{cases} x + y = 34 \\ x - y = 10 \end{cases}$ Adding these two equations gives $2x = 44 \Leftrightarrow x = 22$. So $22 + y = 34 \Leftrightarrow y = 12$. Therefore, the two numbers are 22 and 12.

61. Let d be the number of dimes and q be the number of quarters. This gives $\begin{cases} d + q = 14 \\ 0.10d + 0.25q = 2.75 \end{cases}$ Subtracting the first equation from 10 times the second gives $1.5q = 13.5 \Leftrightarrow q = 9$. So $d + 9 = 14 \Leftrightarrow d = 5$. Thus, the number of dimes is 5 and the number of quarters is 9.

63. Let r be the amount of regular gas sold and p the amount of premium gas sold. Then $\begin{cases} r + p = 280 \\ 2.20r + 3.00p = 680 \end{cases}$ Subtracting the second equation from three times the first equation gives $3r - 2.2 = 3(280) - 680 \Leftrightarrow 0.8r = 160 \Leftrightarrow r = 200$. Substituting this value of r into the original first equation gives $200 + p = 280 \Leftrightarrow p = 80$. Thus, 200 gallons of regular gas and 80 gallons of premium were sold.

65. Let x be the speed of the plane in still air and y be the speed of the wind. This gives $\begin{cases} 2x - 2y = 180 \\ 1.2x + 1.2y = 180 \end{cases}$ Subtracting 6 times the first equation from 10 times the second gives $24x = 2880 \Leftrightarrow x = 120$, so $2(120) - 2y = 180 \Leftrightarrow -2y = -60 \Leftrightarrow y = 30$. Therefore, the speed of the plane is 120 mi/h and the wind speed is 30 mi/h.

67. Let a and b be the number of grams of food A and food B. Then $\begin{cases} 0.12a + 0.20b = 32 \\ 100a + 50b = 22{,}000 \end{cases}$ Subtracting 250 times the first equation from the second, we get $70a = 14{,}000 \Leftrightarrow a = 200$, so $0.12(200) + 0.20b = 32 \Leftrightarrow 0.20b = 8 \Leftrightarrow b = 40$. Thus, she should use 200 grams of food A and 40 grams of food B.

69. Let x and y be the sulfuric acid concentrations in the first and second containers. $\begin{cases} 300x + 600y = 900(0.15) \\ 100x + 500y = 600(0.125) \end{cases}$ Subtracting the first equation from 3 times the second gives $900y = 90 \Leftrightarrow y = 0.10$, so $100x + 500(0.10) = 75 \Leftrightarrow x = 0.25$. Thus, the concentrations of sulfuric acid are 25% in the first container and 10% in the second.

71. Let x be the amount invested at 5% and y the amount invested at 8%. $\begin{cases} \text{Total invested: } x + y = 20{,}000 \\ \text{Interest earned: } 0.05x + 0.08y = 1180 \end{cases}$ Subtracting 5 times the first equation from 100 times the second gives $3y = 18{,}000 \Leftrightarrow y = 6{,}000$, so $x + 6{,}000 = 20{,}000 \Leftrightarrow x = 14{,}000$. She invests \$14,000 at 5% and \$6,000 at 8%.

73. Let x be the length of time John drives and y be the length of time Mary drives. Then $y = x + 0.25$, so $-x + y = 0.25$, and multiplying by 40, we get $-40x + 40y = 10$. Comparing the distances, we get $60x = 40y + 35$, or $60x - 40y = 35$. This gives the system $\begin{cases} -40x + 40y = 10 \\ 60x - 40y = 35 \end{cases}$ Adding, we get $20x = 45 \Leftrightarrow x = 2.25$, so $y = 2.25 + 0.25 = 2.5$. Thus, John drives for $2\frac{1}{4}$ hours and Mary drives for $2\frac{1}{2}$ hours.

75. Let x be the tens digit and y be the ones digit of the number. $\begin{cases} x + y = 7 \\ 10y + x = 27 + 10x + y \end{cases}$ Adding 9 times the first equation to the second gives $18x = 36 \Leftrightarrow x = 2$, so $2 + y = 7 \Leftrightarrow y = 5$. Thus, the number is 25.

77. $n = 5$, so $\sum_{k=1}^{n} x_k = 1+2+3+5+7 = 18$, $\sum_{k=1}^{n} y_k = 3+5+6+6+9 = 29$, $\sum_{k=1}^{n} x_k y_k = 1(3) + 2(5) + 3(6) + 5(6) + 7(9) = 124$, and $\sum_{k=1}^{n} x_k^2 = 1^2 + 2^2 + 3^2 + 5^2 + 7^2 = 88$. Thus we get the system

$\begin{cases} 18a + 5b = 29 \\ 88a + 18b = 124 \end{cases}$ Subtracting 18 times the first equation from 5 times the second, we get $116a = 98 \Leftrightarrow a \approx 0.845$. Then $b = \frac{1}{5}[-18(0.845) + 29] \approx 2.758$. So the regression line is $y = 0.845x + 2.758$.

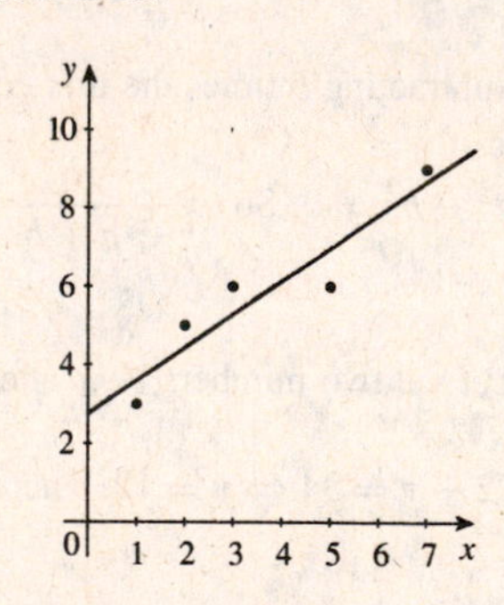

10.2 SYSTEMS OF LINEAR EQUATIONS IN SEVERAL VARIABLES

1. If we add 2 times the first equation to the second equation, the second equation becomes $x + 3z = 1$.

3. The equation $6x - \sqrt{3}y + \frac{1}{2}z = 0$ is linear.

5. The system $\begin{cases} xy - 3y + z = 5 \\ x - y^2 + 5z = 0 \\ 2x + yz = 3 \end{cases}$ is not a linear system, since the first equation contains a product of variables. In fact both the second and the third equation are not linear.

7. $\begin{cases} x - 3y + z = 0 \\ y - z = 3 \\ z = -2 \end{cases}$ Substituting $z = -2$ into the second equation gives $y - (-2) = 3 \Leftrightarrow y = 1$. Substituting $z = -2$ and $y = 1$ into the first equation gives $x - 3(1) + (-2) = 0 \Leftrightarrow x = 5$. Thus, the solution is $(5, 1, -2)$.

9. $\begin{cases} x + 2y + z = 7 \\ -y + 3z = 9 \\ 2z = 6 \end{cases}$ Solving we get $2z = 6 \Leftrightarrow z = 3$. Substituting $z = 3$ into the second equation gives $-y + 3(3) = 9 \Leftrightarrow y = 0$. Substituting $z = 3$ and $y = 0$ into the first equation gives $x + 2(0) + 3 = 7 \Leftrightarrow x = 4$. Thus, the solution is $(4, 0, 3)$.

11. $\begin{cases} 2x - y + 6z = 5 \\ y + 4z = 0 \\ -2z = 1 \end{cases}$ Solving we get $-2z = 1 \Leftrightarrow z = -\frac{1}{2}$. Substituting $z = -\frac{1}{2}$ into the second equation gives $y + 4\left(-\frac{1}{2}\right) = 0 \Leftrightarrow y = 2$. Substituting $z = -\frac{1}{2}$ and $y = 2$ into the first equation gives $2x - (2) + 6\left(-\frac{1}{2}\right) = 5 \Leftrightarrow x = 5$. Thus, the solution is $\left(5, 2, -\frac{1}{2}\right)$.

13. $\begin{cases} 3x + y + z = 4 \\ -x + y + 2z = 0 \\ x - 2y - z = -1 \end{cases}$ Add the third equation to the second equation: $\begin{cases} 3x + y + z = 4 \\ -y + z = -1 \\ x - 2y - z = -1 \end{cases}$

Or, add the first equation to three times the second equation: $\begin{cases} 3x + y + z = 4 \\ 4y + 7z = 4 \\ x - 2y - z = -1 \end{cases}$

15. $\begin{cases} 2x + y - 3z = 5 \\ 2x + 3y + z = 13 \\ 6x - 5y - z = 7 \end{cases}$ Add -3 times the first equation to the third equation: $\begin{cases} 2x + y - 3z = 5 \\ 2x + 3y + z = 13 \\ -8y + 8z = -8 \end{cases}$

Or, add -3 times the second equation to the third equation: $\begin{cases} 2x + y - 3z = 5 \\ 2x + 3y + z = 13 \\ -14y - 4z = -32 \end{cases}$

17. $\begin{cases} x - y - z = 4 \\ 2y + z = -1 \\ -x + y - 2z = 5 \end{cases} \Leftrightarrow \begin{cases} x - y - z = 4 \\ 2y + z = -1 \\ -3z = 9 \end{cases}$ Eq. 1 + Eq. 3

So $z = -3$ and $2y + (-3) = -1 \Leftrightarrow 2y = 2 \Leftrightarrow y = 1$. Thus, $x - 1 - (-3) = 4 \Leftrightarrow x = 2$. So the solution is $(2, 1, -3)$.

19. $\begin{cases} x + 2y - z = -6 \\ y - 3z = -16 \\ x - 3y + 2z = 14 \end{cases} \Leftrightarrow \begin{cases} x + 2y - z = -6 \\ y - 3z = -16 \\ 5y - 3z = -20 & \text{Eq. 1} + (-1) \times \text{Eq. 3} \end{cases} \Leftrightarrow$

$\begin{cases} x + 2y - z = -6 \\ y - 3z = -16 \\ 12z = 60 & (-5) \times \text{Eq. 2} + \text{Eq. 3} \end{cases}$

So $z = 5$ and $y - 3(5) = -16 \Leftrightarrow y = -1$. Then $x + 2(-1) - 5 = -6 \Leftrightarrow x = 1$. So the solution is $(1, -1, 5)$.

21. $\begin{cases} x + y + z = 4 \\ x + 3y + 3z = 10 \\ 2x + y - z = 3 \end{cases} \Leftrightarrow \begin{cases} x + y + z = 4 \\ 2y + 2z = 6 & (-1) \times \text{Eq. 1} + \text{Eq. 2} \\ y + 3z = 5 & 2 \times \text{Eq. 1} + (-1) \times \text{Eq. 3} \end{cases} \Leftrightarrow \begin{cases} x + y + z = 4 \\ y + 3z = 5 & \text{Eq. 3} \\ 2y + 2z = 6 & \text{Eq. 2} \end{cases} \Leftrightarrow$

$\begin{cases} x + y + z = 4 \\ y + 3z = 5 \\ -4z = -4 & (-2) \times \text{Eq. 2} + \text{Eq. 3} \end{cases}$

So $z = 1$ and $y + 3(1) = 8 \Leftrightarrow y = 2$. Then $x + 2 + 1 = 4 \Leftrightarrow x = 1$. So the solution is $(1, 2, 1)$.

23. $\begin{cases} x - 4z = 1 \\ 2x - y - 6z = 4 \\ 2x + 3y - 2z = 8 \end{cases} \Leftrightarrow \begin{cases} x - 4z = 1 \\ -y + 2z = 2 & (-2) \times \text{Eq. 1} + \text{Eq. 2} \\ 3y + 6z = 6 & (-2) \times \text{Eq. 1} + \text{Eq. 3} \end{cases} \Leftrightarrow \begin{cases} x - 4z = 1 \\ -y + 2z = 2 \\ 12z = 12 & 3 \times \text{Eq. 2} + \text{Eq. 3} \end{cases}$

So $z = 1$ and $-y + 2(1) = 2 \Leftrightarrow y = 0$. Then $x - 4(1) = 1 \Leftrightarrow x = 5$. So the solution is $(5, 0, 1)$.

25. $\begin{cases} 2x + 4y - z = 2 \\ x + 2y - 3z = -4 \\ 3x - y + z = 1 \end{cases} \Leftrightarrow \begin{cases} 2x + 4y - z = 2 \\ 5z = 10 & \text{Eq. 1} + (-2) \times \text{Eq. 2} \\ 14y - 5z = 4 & 3 \times \text{Eq. 1} + (-2) \times \text{Eq. 3} \end{cases} \Leftrightarrow \begin{cases} x + 2y - 3z = -4 \\ 14y - 5z = 4 & \text{Eq. 2} \leftrightarrow \text{Eq. 3} \\ 5z = 10 & \text{Eq. 2} \leftrightarrow \text{Eq. 3} \end{cases}$

So $z = 2$ and $14y - 5(2) = 4 \Leftrightarrow y = 1$. Then $x + 2(1) - 3(2) = -4 \Leftrightarrow x = 0$. So the solution is $(0, 1, 2)$.

27. $\begin{cases} 2y + 4z = -1 \\ -2x + y + 2z = -1 \\ 4x - 2y = 0 \end{cases} \Leftrightarrow \begin{cases} -2x + y + 2z = -1 & \text{Eq. 2} \\ 2y + 4z = -1 & \text{Eq. 1} \\ 4z = -2 & 2 \times \text{Eq. 2} + \text{Eq. 3} \end{cases}$

So $z = -\frac{1}{2}$ and $2y + 4\left(-\frac{1}{2}\right) = -1 \Leftrightarrow y = \frac{1}{2}$. Then $-2x + \frac{1}{2} + 2\left(-\frac{1}{2}\right) = -1 \Leftrightarrow x = \frac{1}{4}$. So the solution is $\left(\frac{1}{4}, \frac{1}{2}, -\frac{1}{2}\right)$.

29. $\begin{cases} x + 2y - z = 1 \\ 2x + 3y - 4z = -3 \\ 3x + 6y - 3z = 4 \end{cases} \Leftrightarrow \begin{cases} x + 2y - z = 1 \\ -y - 2z = -5 & (-2) \times \text{Eq. 1} + \text{Eq. 2} \\ 0 = 1 & (-3) \times \text{Eq. 1} + \text{Eq. 3} \end{cases}$ Since $0 = 1$ is false, this system is inconsistent.

31. $\begin{cases} 2x + 3y - z = 1 \\ x + 2y = 3 \\ x + 3y + z = 4 \end{cases} \Leftrightarrow \begin{cases} x + 2y = 3 & \text{Eq. 2} \\ 2x + 3y - z = 1 & \text{Eq. 1} \\ x + 3y + z = 4 \end{cases} \Leftrightarrow \begin{cases} x + 2y = 3 \\ -y - z = -5 & \text{Eq. 2} + (-2) \times \text{Eq. 1} \\ y + z = 1 & \text{Eq. 3} - \text{Eq. 1} \end{cases} \Leftrightarrow$

$\begin{cases} x + 2y = 3 \\ -y - z = -5 \\ 0 = -4 & \text{Eq. 2} + \text{Eq. 3} \end{cases}$ Since $0 = -4$ is false, this system is inconsistent.

33. $\begin{cases} x+y-z=0 \\ x+2y-3z=-3 \\ 2x+3y-4z=-3 \end{cases} \Leftrightarrow \begin{cases} x+y-z=0 \\ y-2z=-3 & \text{Eq. 2} - \text{Eq. 1} \\ y-2z=-3 & (-2)\times\text{Eq. 1}+\text{Eq. 3} \end{cases} \Leftrightarrow \begin{cases} x+y-z=0 \\ y-2z=-3 \\ 0=0 & \text{Eq. 2} - \text{Eq. 3} \end{cases}$

So $z=t$ and $y-2t=-3 \Leftrightarrow y=2t-3$. Then $x+(2t-3)-t=0 \Leftrightarrow x=-t+3$. So the solutions are $(-t+3, 2t-3, t)$, where t is any real number.

35. $\begin{cases} x+3y-2z=0 \\ 2x \quad\quad +4z=4 \\ 4x+6y \quad\quad =4 \end{cases} \Leftrightarrow \begin{cases} x+3y-2z=0 \\ -6y+8z=4 & \text{Eq. 2}+(-2)\times\text{Eq. 1} \\ -6y+8z=4 & \text{Eq. 3}+(-4)\times\text{Eq. 1} \end{cases} \Leftrightarrow \begin{cases} x+3y-2z=0 \\ -6y+8z=4 \\ 0=0 & \text{Eq. 2} - \text{Eq. 3} \end{cases}$

So $z=t$ and $-6y+8t=4 \Leftrightarrow -6y=-8t+4 \Leftrightarrow y=\frac{4}{3}t-\frac{2}{3}$. Then $x+3\left(\frac{4}{3}t-\frac{2}{3}\right)-2t=0 \Leftrightarrow x=-2t+2$. So the solutions are $\left(-2t+2, \frac{4}{3}t-\frac{2}{3}, t\right)$, where t is any real number.

37. $\begin{cases} x \quad\quad +z+2w=6 \\ y-2z \quad\quad =-3 \\ x+2y-z \quad\quad =-2 \\ 2x+y+3z-2w=0 \end{cases} \Leftrightarrow \begin{cases} x \quad\quad +z+2w=6 \\ y-2z \quad\quad =-3 \\ 2y-2z-2w=-8 & \text{Eq. 3} - \text{Eq. 1} \\ y+z-6w=-12 & \text{Eq. 4}+(-2)\times\text{Eq. 1} \end{cases} \Leftrightarrow$

$\begin{cases} x \quad\quad +z+2w=6 \\ y-2z \quad\quad =-3 \\ 2z-2w=-2 & \text{Eq. 3}+(-2)\times\text{Eq. 2} \\ 3z-6w=-9 & \text{Eq. 4} - \text{Eq. 2} \end{cases} \Leftrightarrow \begin{cases} x \quad\quad +z+2w=6 \\ y-2z \quad\quad =-3 \\ z-w=-1 & \frac{1}{2}\text{Eq. 3} \\ 6w=12 & 3\times\text{Eq. 3}+(-2)\times\text{Eq. 4} \end{cases}$

So $w=2$ and $z-2=-1 \Leftrightarrow z=1$. Then $y-2(1)=-3 \Leftrightarrow y=-1$ and $x+1+2(2)=6 \Leftrightarrow x=1$. Thus, the solution is $(1,-1,1,2)$.

39. Let x be the amount invested at 4%, y the amount invested at 5%, and z the amount invested at 6%. We set up a model and get the following equations: $\begin{cases} \text{Total money:} & x+y+z=100{,}000 \\ \text{Annual income:} & 0.04x+0.05y+0.06z=0.051\,(100{,}000) \\ \text{Equal amounts:} & x=y \end{cases}$

$\Leftrightarrow \begin{cases} x+y+z=100{,}000 \\ 4x+5y+6z=510{,}000 \\ x-y \quad\quad =0 \end{cases} \Leftrightarrow \begin{cases} x+y+z=100{,}000 \\ y+2z=110{,}000 & \text{Eq. 2}+(-4)\times\text{Eq. 1} \\ -2y-z=-100{,}000 & \text{Eq. 3} - \text{Eq. 1} \end{cases} \Leftrightarrow$

$\begin{cases} x+y+z=100{,}000 \\ y+2z=110{,}000 \\ 3z=120{,}000 & 2\times\text{Eq. 2}+\text{Eq. 3} \end{cases}$

So $z=40{,}000$ and $y+2(40{,}000)=110{,}000 \Leftrightarrow y=30{,}000$. Since $x=y$, $x=30{,}000$. Mark should invest \$30,000 in short-term bonds, \$30,000 in intermediate-term bonds, and \$40,000 in long-term bonds.

41. Let x, y, and z be the number of acres of land planted with corn, wheat, and soybeans. We set up a model and get the following equations: $\begin{cases} \text{Total acres:} & x+y+z=1200 \\ \text{Market demand:} & 2x=y \\ \text{Total cost:} & 45x+60y+50z=63{,}750 \end{cases}$ Substituting $2x$ for y, we get

$$\begin{cases} x+2x+z=1200 \\ 2x=y \\ 45x+60(2x)+50z=63{,}750 \end{cases} \Leftrightarrow \begin{cases} 3x \quad + \quad z = \quad 1200 \\ 2x-y \quad\quad = \quad 0 \\ 165x \quad +50z = 63{,}750 \end{cases} \Leftrightarrow \begin{cases} 3x \quad +z=1200 \\ 2x-y \quad = \quad 0 \\ 15x \quad\quad = 3750 & \text{Eq. } 3+(-50)\times\text{Eq. } 1 \end{cases}$$

So $15x=3{,}750 \Leftrightarrow x=250$ and $y=2(250)=500$. Substituting into the original equation, we have $250+500+z=1200$ $\Leftrightarrow z=450$. Thus the farmer should plant 250 acres of corn, 500 acres of wheat, and 450 acres of soybeans.

43. Let a, b, and c be the number of ounces of Type A, Type B, and Type C pellets used. The requirements for the different vitamins gives the following system: $\begin{cases} 2a+3b+\ c=\ 9 \\ 3a+\ b+3c=14 \\ 8a+5b+7c=32 \end{cases} \Leftrightarrow$

$$\begin{cases} 2a+3b+\ c=\ 9 \\ -7b+3c=\ 1 & 2\times\text{Eq. } 2+(-3)\times\text{Eq. } 1 \\ -7b+3c=-4 & \text{Eq. } 3+(-4)\times\text{Eq. } 1 \end{cases}$$

Equations 2 and 3 are inconsistent, so there is no solution.

45. Let a, b, and c represent the number of Midnight Mango, Tropical Torrent, and Pineapple Power smoothies sold. The given information leads to the system $\begin{cases} 8a+6b+2c=820 \\ 3a+5b+8c=690 \\ 3a+3b+4c=450 \end{cases} \Leftrightarrow \begin{cases} 8a+\ 6b+\ 2c=\ 820 \\ 22b+58c=3060 & 8\times\text{Eq. } 2+(-3)\times\text{Eq. } 1 \\ 2b+\ 4c=\ 240 & \text{Eq. } 2-\text{Eq. } 3 \end{cases} \Leftrightarrow$

$$\begin{cases} 8a+\ 6b+\ 2c=\ 820 \\ 22b+58c=3060 \\ 14c=\ 420 & \text{Eq. } 2+(-11)\times\text{Eq. } 3 \end{cases}$$

Thus, $c=30$, so $22b+58(30)=3060 \Leftrightarrow 22b=1320 \Leftrightarrow b=60$ and $8a+6(60)+2(30)=820 \Leftrightarrow a=50$. Thus, The Juice Company sold 50 Midnight Mango, 60 Tropical Torrent, and 30 Pineapple Power smoothies on that particular day.

47. Let a, b, and c be the number of shares of Stock A, Stock B, and Stock C in the investor's portfolio. Since the total value remains unchanged, we get the following system:

$$\begin{cases} 10a+25b+29c=74{,}000 \\ 12a+20b+32c=74{,}000 \\ 16a+15b+32c=74{,}000 \end{cases} \Leftrightarrow \begin{cases} 10a+25b+29c=\ 74{,}000 \\ 50b+14c=\ 74{,}000 & 6\times\text{Eq. } 1+(-5)\times\text{Eq. } 2 \\ 125b+72c=222{,}000 & 8\times\text{Eq. } 1+(-5)\times\text{Eq. } 3 \end{cases} \Leftrightarrow$$

$$\begin{cases} 10a+25b+29c=\ 74{,}000 \\ 50b+14c=\ 74{,}000 \\ 74c=74{,}000 & (-5)\times\text{Eq. } 2+2\times\text{Eq. } 3 \end{cases}$$

So $c=1{,}000$. Back-substituting we have $50b+14(1000)=74{,}000 \Leftrightarrow 50b=60{,}000 \Leftrightarrow b=1{,}200$. And finally $10a+25(1200)+29(1000)=74{,}000 \quad 10a+30{,}000+29{,}000=74{,}000 \Leftrightarrow 10a=15{,}000 \Leftrightarrow a=1{,}500$. Thus the portfolio consists of 1,500 shares of Stock A, 1,200 shares of Stock B, and 1,000 shares of Stock C.

49. (a) We begin by substituting $\dfrac{x_0+x_1}{2}$, $\dfrac{y_0+y_1}{2}$, and $\dfrac{z_0+z_1}{2}$ into the left-hand side of the first equation:

$$a_1\left(\frac{x_0+x_1}{2}\right)+b_1\left(\frac{y_0+y_1}{2}\right)+c_1\left(\frac{z_0+z_1}{2}\right)=\tfrac{1}{2}\left[(a_1x_0+b_1y_0+c_1z_0)+(a_1x_1+b_1y_1+c_1z_1)\right]$$
$$=\tfrac{1}{2}\left[d_1+d_1\right]=d_1$$

Thus the given ordered triple satisfies the first equation. We can show that it satisfies the second and the third in exactly the same way. Thus it is a solution of the system.

(b) We have shown in part (a) that if the system has two different solutions, we can find a third one by averaging the two solutions. But then we can find a fourth and a fifth solution by averaging the new one with each of the previous two. Then we can find four more by repeating this process with these new solutions, and so on. Clearly this process can continue indefinitely, so there are infinitely many solutions.

10.3 PARTIAL FRACTIONS

1. (iii): $r(x)=\dfrac{4}{x(x-2)^2}=\dfrac{A}{x}+\dfrac{B}{x-2}+\dfrac{C}{(x-2)^2}$

3. $\dfrac{1}{(x-1)(x+2)}=\dfrac{A}{x-1}+\dfrac{B}{x+2}$

5. $\dfrac{x^2-3x+5}{(x-2)^2(x+4)}=\dfrac{A}{x-2}+\dfrac{B}{(x-2)^2}+\dfrac{C}{x+4}$

7. $\dfrac{x^2}{(x-3)(x^2+4)}=\dfrac{A}{x-3}+\dfrac{Bx+C}{x^2+4}$

9. $\dfrac{x^3-4x^2+2}{(x^2+1)(x^2+2)}=\dfrac{Ax+B}{x^2+1}+\dfrac{Cx+D}{x^2+2}$

11. $\dfrac{x^3+x+1}{x(2x-5)^3(x^2+2x+5)^2}=\dfrac{A}{x}+\dfrac{B}{2x-5}+\dfrac{C}{(2x-5)^2}+\dfrac{D}{(2x-5)^3}+\dfrac{Ex+F}{x^2+2x+5}+\dfrac{Gx+H}{(x^2+2x+5)^2}$

13. $\dfrac{2}{(x-1)(x+1)}=\dfrac{A}{x-1}+\dfrac{B}{x+1}$. Multiplying by $(x-1)(x+1)$, we get $2=A(x+1)+B(x-1)\Leftrightarrow$ $2=Ax+A+Bx-B$. Thus $\begin{cases}A+B=0\\A-B=2\end{cases}$ Adding we get $2A=2\Leftrightarrow A=1$. Now $A+B=0\Leftrightarrow B=-A$, so $B=-1$. Thus, the required partial fraction decomposition is $\dfrac{2}{(x-1)(x+1)}=\dfrac{1}{x-1}-\dfrac{1}{x+1}$.

15. $\dfrac{5}{(x-1)(x+4)}=\dfrac{A}{x-1}+\dfrac{B}{x+4}$. Multiplying by $(x-1)(x+4)$, we get $5=A(x+4)+B(x-1)\Leftrightarrow$ $5=Ax+4A+Bx-B$. Thus $\begin{cases}A+B=0\\4A-B=5\end{cases}$ Now $A+B=0\Leftrightarrow B=-A$, so substituting,we get $4A-(-A)=5\Leftrightarrow$ $5A=5\Leftrightarrow A=1$and $B=-1$. The required partial fraction decomposition is $\dfrac{5}{(x-1)(x+4)}=\dfrac{1}{x-1}-\dfrac{1}{x+4}$.

17. $\dfrac{12}{x^2-9}=\dfrac{12}{(x-3)(x+3)}=\dfrac{A}{x-3}+\dfrac{B}{x+3}$. Multiplying by $(x-3)(x+3)$, we get $12=A(x+3)+B(x-3)\Leftrightarrow$ $12=Ax+3A+Bx-3B$. Thus $\begin{cases}A+\ B=\ 0\\3A-3B=12\end{cases}\Leftrightarrow\begin{cases}A+B=0\\A-B=4\end{cases}$ Adding, we get $2A=4\Leftrightarrow A=2$. So $2+B=0$ $\Leftrightarrow B=-2$. The required partial fraction decomposition is $\dfrac{12}{x^2-9}=\dfrac{2}{x-3}-\dfrac{2}{x+3}$.

19. $\dfrac{4}{x^2-4}=\dfrac{4}{(x-2)(x+2)}=\dfrac{A}{x-2}+\dfrac{B}{x+2}$. Multiplying by x^2-4, we get

$4=A(x+2)+B(x-2)=(A+B)x+(2A-2B)$, and so $\begin{cases}A+B=0\\2A-2B=4\end{cases}\Leftrightarrow\begin{cases}A+B=0\\A-B=2\end{cases}$ Adding we get $2A=2$

$\Leftrightarrow A=1$, and $B=-1$. Therefore, $\dfrac{4}{x^2-4}=\dfrac{1}{x-2}-\dfrac{1}{x+2}$.

21. $\dfrac{x+14}{x^2-2x-8}=\dfrac{x+14}{(x-4)(x+2)}=\dfrac{A}{x-4}+\dfrac{B}{x+2}$. Hence, $x+14=A(x+2)+B(x-4)=(A+B)x+(2A-4B)$,

and so $\begin{cases}A+B=1\\2A-4B=14\end{cases}\Leftrightarrow\begin{cases}2A+2B=2\\A-2B=7\end{cases}$ Adding, we get $3A=9\Leftrightarrow A=3$. So $(3)+B=1\Leftrightarrow B=-2$.

Therefore, $\dfrac{x+14}{x^2-2x-8}=\dfrac{3}{x-4}-\dfrac{2}{x+2}$.

23. $\dfrac{x}{8x^2-10x+3}=\dfrac{x}{(4x-3)(2x-1)}=\dfrac{A}{4x-3}+\dfrac{B}{2x-1}$. Hence,

$x=A(2x-1)+B(4x-3)=(2A+4B)x+(-A-3B)$, and so $\begin{cases}2A+4B=1\\-A-3B=0\end{cases}\Leftrightarrow\begin{cases}2A+4B=1\\-2A-6B=0\end{cases}$

Adding, we get $-2B=1\Leftrightarrow B=-\frac{1}{2}$, and $A=\frac{3}{2}$. Therefore, $\dfrac{x}{8x^2-10x+3}=\dfrac{\frac{3}{2}}{4x-3}-\dfrac{\frac{1}{2}}{2x-1}$.

25. $\dfrac{9x^2-9x+6}{2x^3-x^2-8x+4}=\dfrac{9x^2-9x+6}{(x-2)(x+2)(2x-1)}=\dfrac{A}{x-2}+\dfrac{B}{x+2}+\dfrac{C}{2x-1}$. Thus,

$$\begin{aligned}9x^2-9x+6&=A(x+2)(2x-1)+B(x-2)(2x-1)+C(x-2)(x+2)\\&=A\left(2x^2+3x-2\right)+B\left(2x^2-5x+2\right)+C\left(x^2-4\right)\\&=(2A+2B+C)x^2+(3A-5B)x+(-2A+2B-4C)\end{aligned}$$

This leads to the system $\begin{cases}2A+2B+C=9 & \text{Coefficients of } x^2\\3A-5B=-9 & \text{Coefficients of } x\\-2A+2B-4C=6 & \text{Constant terms}\end{cases}\Leftrightarrow\begin{cases}2A+2B+C=9\\16B+3C=45\\4B-3C=15\end{cases}\Leftrightarrow$

$\begin{cases}2A+2B+C=9\\16B+3C=45\\15C=-15\end{cases}$ Hence, $-15C=15\Leftrightarrow C=-1$; $16B-3=45\Leftrightarrow B=3$; and $2A+6-1=9\Leftrightarrow A=2$.

Therefore, $\dfrac{9x^2-9x+6}{2x^3-x^2-8x+4}=\dfrac{2}{x-2}+\dfrac{3}{x+2}-\dfrac{1}{2x-1}$.

27. $\dfrac{x^2+1}{x^3+x^2}=\dfrac{x^2+1}{x^2(x+1)}=\dfrac{A}{x}+\dfrac{B}{x^2}+\dfrac{C}{x+1}$. Hence,

$x^2+1=Ax(x+1)+B(x+1)+Cx^2=(A+C)x^2+(A+B)x+B$, and so $B=1$; $A+1=0\Leftrightarrow A=-1$; and

$-1+C=1\Leftrightarrow C=2$. Therefore, $\dfrac{x^2+1}{x^3+x^2}=\dfrac{-1}{x}+\dfrac{1}{x^2}+\dfrac{2}{x+1}$.

29. $\dfrac{2x}{4x^2+12x+9}=\dfrac{2x}{(2x+3)^2}=\dfrac{A}{2x+3}+\dfrac{B}{(2x+3)^2}$. Hence, $2x=A(2x+3)+B=2Ax+(3A+B)$. So $2A=2\Leftrightarrow$

$A=1$; and $3(1)+B=0\Leftrightarrow B=-3$. Therefore, $\dfrac{2x}{4x^2+12x+9}=\dfrac{1}{2x+3}-\dfrac{3}{(2x+3)^2}$.

31. $\dfrac{4x^2-x-2}{x^4+2x^3}=\dfrac{4x^2-x-2}{x^3(x+2)}=\dfrac{A}{x}+\dfrac{B}{x^2}+\dfrac{C}{x^3}+\dfrac{D}{x+2}$. Hence,

$$\begin{aligned}4x^2-x-2 &= Ax^2(x+2)+Bx(x+2)+C(x+2)+Dx^3\\ &= (A+D)x^3+(2A+B)x^2+(2B+C)x+2C\end{aligned}$$

So $2C=-2\Leftrightarrow C=-1$; $2B-1=-1\Leftrightarrow B=0$; $2A+0=4\Leftrightarrow A=2$; and $2+D=0\Leftrightarrow D=-2$. Therefore, $\dfrac{4x^2-x-2}{x^4+2x^3}=\dfrac{2}{x}-\dfrac{1}{x^3}-\dfrac{2}{x+2}$.

33. $\dfrac{-10x^2+27x-14}{(x-1)^3(x+2)}=\dfrac{A}{x+2}+\dfrac{B}{x-1}+\dfrac{C}{(x-1)^2}+\dfrac{D}{(x-1)^3}$. Thus,

$$\begin{aligned}-10x^2+27x-14 &= A(x-1)^3+B(x+2)(x-1)^2+C(x+2)(x-1)+D(x+2)\\ &= A\left(x^3-3x^2+3x-1\right)+B(x+2)\left(x^2-2x+1\right)+C\left(x^2+x-2\right)+D(x+2)\\ &= A\left(x^3-3x^2+3x-1\right)+B\left(x^3-3x+2\right)+C\left(x^2+x-2\right)+D(x+2)\\ &= (A+B)x^3+(-3A+C)x^2+(3A-3B+C+D)x+(-A+2B-2C+2D)\end{aligned}$$

which leads to the system

$$\begin{cases}A+B=0 & \text{Coefficients of } x^3\\ -3A+C=-10 & \text{Coefficients of } x^2\\ 3A-3B+C+D=27 & \text{Coefficients of } x\\ -A+2B-2C+2D=-14 & \text{Constant terms}\end{cases}\Leftrightarrow\begin{cases}A+B=0\\ 3B+C=-10\\ -3B+2C+D=17\\ 3B-5C+7D=-15\end{cases}\Leftrightarrow$$

$$\begin{cases}A+B=0\\ 3B+C=-10\\ 3C+D=7\\ -3C+8D=2\end{cases}\Leftrightarrow\begin{cases}A+B=0\\ 3B+C=-10\\ 3C+D=7\\ 9D=9\end{cases}$$

Hence, $9D=9\Leftrightarrow D=1$, $3C+1=7\Leftrightarrow C=2$, $3B+2=-10\Leftrightarrow B=-4$, and $A-4=0\Leftrightarrow A=4$. Therefore, $\dfrac{-10x^2+27x-14}{(x-1)^3(x+2)}=\dfrac{4}{x+2}-\dfrac{4}{x-1}+\dfrac{2}{(x-1)^2}+\dfrac{1}{(x-1)^3}$.

35. $\dfrac{3x^3+22x^2+53x+41}{(x+2)^2(x+3)^2}=\dfrac{A}{x+2}+\dfrac{B}{(x+2)^2}+\dfrac{C}{x+3}+\dfrac{D}{(x+3)^2}$. Thus,

$$\begin{aligned}3x^3+22x^2+53x+41 &= A(x+2)(x+3)^2+B(x+3)^2+C(x+2)^2(x+3)+D(x+2)^2\\ &= A\left(x^3+8x^2+21x+18\right)+B\left(x^2+6x+9\right)\\ &\qquad +C\left(x^3+7x^2+16x+12\right)+D\left(x^2+4x+4\right)\\ &= (A+C)x^3+(8A+B+7C+D)x^2\\ &\qquad +(21A+6B+16C+4D)x+(18A+9B+12C+4D)\end{aligned}$$

so we must solve the system $\begin{cases} A + C = 3 & \text{Coefficients of } x^3\\ 8A+B+7C+D=22 & \text{Coefficients of } x^2\\ 21A+6B+16C+4D=53 & \text{Coefficients of } x\\ 18A+9B+12C+4D=41 & \text{Constant terms}\end{cases}$ $\Leftrightarrow \begin{cases} A+C=3\\ B-C+D=-2\\ 6B-5C+4D=-10\\ 9B-6C+4D=-13\end{cases}$

$\Leftrightarrow \begin{cases} A+C=3\\ B-C+D=-2\\ C-2D=2\\ 3C-5D=5\end{cases} \Leftrightarrow \begin{cases} A+C=3\\ B-C+D=-2\\ C-2D=2\\ D=-1\end{cases}$ Hence, $D=-1$, $C+2=2\Leftrightarrow C=0$, $B-0-1=-2$

$\Leftrightarrow B=-1$, and $A+0=3\Leftrightarrow A=3$. Therefore, $\dfrac{3x^3+22x^2+53x+41}{(x+2)^2(x+3)^2}=\dfrac{3}{x+2}-\dfrac{1}{(x+2)^2}-\dfrac{1}{(x+3)^2}$.

37. $\dfrac{x-3}{x^3+3x}=\dfrac{x-3}{x(x^2+3)}=\dfrac{A}{x}+\dfrac{Bx+C}{x^2+3}$. Hence, $x-3=A\left(x^2+3\right)+Bx^2+Cx=(A+B)x^2+Cx+3A$. So

$3A=-3\Leftrightarrow A=-1$; $C=1$; and $-1+B=0\Leftrightarrow B=1$. Therefore, $\dfrac{x-3}{x^3+3x}=-\dfrac{1}{x}+\dfrac{x+1}{x^2+3}$.

39. $\dfrac{2x^3+7x+5}{(x^2+x+2)(x^2+1)}=\dfrac{Ax+B}{x^2+x+2}+\dfrac{Cx+D}{x^2+1}$. Thus,

$$\begin{aligned}2x^3+7x+5 &= (Ax+B)\left(x^2+1\right)+(Cx+D)\left(x^2+x+2\right)\\ &= Ax^3+Ax+Bx^2+B+Cx^3+Cx^2+2Cx+Dx^2+Dx+2D\\ &= (A+C)x^3+(B+C+D)x^2+(A+2C+D)x+(B+2D)\end{aligned}$$

We must solve the system

$\begin{cases} A+C=2 & \text{Coefficients of } x^3\\ B+C+D=0 & \text{Coefficients of } x^2\\ A+2C+D=7 & \text{Coefficients of } x\\ B+2D=5 & \text{Constant terms}\end{cases} \Leftrightarrow \begin{cases} A+C=2\\ B+C+D=0\\ C+D=5\\ C-D=-5\end{cases} \Leftrightarrow \begin{cases} A+C=2\\ B+C+D=0\\ C+D=5\\ 2D=10\end{cases}$

Hence, $2D = 10 \Leftrightarrow D = 5$, $C + 5 = 5 \Leftrightarrow C = 0$, $B + 0 + 5 = 0 \Leftrightarrow B = -5$, and $A + 0 = 2 \Leftrightarrow A = 2$. Therefore, $\dfrac{2x^3 + 7x + 5}{(x^2 + x + 2)(x^2 + 1)} = \dfrac{2x - 5}{x^2 + x + 2} + \dfrac{5}{x^2 + 1}$.

41. $\dfrac{x^4 + x^3 + x^2 - x + 1}{x(x^2 + 1)^2} = \dfrac{A}{x} + \dfrac{Bx + C}{x^2 + 1} + \dfrac{Dx + E}{(x^2 + 1)^2}$. Hence,

$$\begin{aligned} x^4 + x^3 + x^2 - x + 1 &= A\left(x^2 + 1\right)^2 + (Bx + C)x\left(x^2 + 1\right) + x(Dx + E) \\ &= A\left(x^4 + 2x^2 + 1\right) + \left(Bx^2 + Cx\right)\left(x^2 + 1\right) + Dx^2 + Ex \\ &= A\left(x^4 + 2x^2 + 1\right) + Bx^4 + Bx^2 + Cx^3 + Cx + Dx^2 + Ex \\ &= (A + B)x^4 + Cx^3 + (2A + B + D)x^2 + (C + E)x + A \end{aligned}$$

So $A = 1$, $1 + B = 1 \Leftrightarrow B = 0$; $C = 1$; $2 + 0 + D = 1 \Leftrightarrow D = -1$; and $1 + E = -1 \Leftrightarrow E = -2$. Therefore, $\dfrac{x^4 + x^3 + x^2 - x + 1}{x(x^2 + 1)^2} = \dfrac{1}{x} + \dfrac{1}{x^2 + 1} - \dfrac{x + 2}{(x^2 + 1)^2}$.

43. We must first get a proper rational function. Using long division, we find that $\dfrac{x^5 - 2x^4 + x^3 + x + 5}{x^3 - 2x^2 + x - 2} = x^2 + \dfrac{2x^2 + x + 5}{x^3 - 2x^2 + x - 2} = x^2 + \dfrac{2x^2 + x + 5}{(x - 2)(x^2 + 1)} = x^2 + \dfrac{A}{x - 2} + \dfrac{Bx + C}{x^2 + 1}$. Hence,

$$\begin{aligned} 2x^2 + x + 5 &= A\left(x^2 + 1\right) + (Bx + C)(x - 2) = Ax^2 + A + Bx^2 + Cx - 2Bx - 2C \\ &= (A + B)x^2 + (C - 2B)x + (A - 2C) \end{aligned}$$

Equating coefficients, we get the system

$$\begin{cases} A + B \quad\quad\; = 2 & \text{Coefficients of } x^2 \\ \;\; -2B + C = 1 & \text{Coefficients of } x \\ A \quad\quad - 2C = 5 & \text{Constant terms} \end{cases} \Leftrightarrow \begin{cases} A + B = 2 \\ -2B + C = 1 \\ B + 2C = -3 \end{cases} \Leftrightarrow \begin{cases} A + B = 2 \\ -2B + C = 1 \\ 5C = -5 \end{cases}$$

Therefore, $5C = -5 \Leftrightarrow C = -1$, $-2B - 1 = 1 \Leftrightarrow B = -1$, and $A - 1 = 2 \Leftrightarrow A = 3$, so $\dfrac{x^5 - 2x^4 + x^3 + x + 5}{x^3 - 2x^2 + x - 2} = x^2 + \dfrac{3}{x - 2} - \dfrac{x + 1}{x^2 + 1}$.

45. $\dfrac{ax + b}{x^2 - 1} = \dfrac{A}{x - 1} + \dfrac{B}{x + 1}$. Hence, $ax + b = A(x + 1) + B(x - 1) = (A + B)x + (A - B)$.

So $\begin{cases} A + B = a \\ A - B = b \end{cases}$ Adding, we get $2A = a + b \Leftrightarrow A = \dfrac{a + b}{2}$.

Substituting, we get $B = a - A = \dfrac{2a}{2} - \dfrac{a + b}{2} = \dfrac{a - b}{2}$. Therefore, $A = \dfrac{a + b}{2}$ and $B = \dfrac{a - b}{2}$.

47. **(a)** The expression $\dfrac{x}{x^2+1}+\dfrac{1}{x+1}$ is already a partial fraction decomposition. The denominator in the first term is a quadratic which cannot be factored and the degree of the numerator is less than 2. The denominator of the second term is linear and the numerator is a constant.

(b) The term $\dfrac{x}{(x+1)^2}$ can be decomposed further, since the numerator and denominator both have linear factors. $\dfrac{x}{(x+1)^2}=\dfrac{A}{x+1}+\dfrac{B}{(x+1)^2}$. Hence, $x=A(x+1)+B=Ax+(A+B)$. So $A=1$, $B=-1$, and $\dfrac{x}{(x+1)^2}=\dfrac{1}{x+1}+\dfrac{-1}{(x+1)^2}$.

(c) The expression $\dfrac{1}{x+1}+\dfrac{2}{(x+1)^2}$ is already a partial fraction decomposition, since each numerator is constant.

(d) The expression $\dfrac{x+2}{\left(x^2+1\right)^2}$ is already a partial fraction decomposition, since the denominator is the square of a quadratic which cannot be factored,and the degree of the numerator is less than 2.

10.4 SYSTEMS OF NONLINEAR EQUATIONS

1. The solutions of the system are the points of intersection of the two graphs, namely $(-2,2)$ and $(4,8)$.

3. $\begin{cases} y=x^2 \\ y=x+12 \end{cases}$ Substituting $y=x^2$ into the second equation gives $x^2=x+12 \Leftrightarrow 0=x^2-x-12=(x-4)(x+3) \Rightarrow x=4$ or $x=-3$. So since $y=x^2$, the solutions are $(-3,9)$ and $(4,16)$.

5. $\begin{cases} x^2+y^2=8 \\ x+y=0 \end{cases}$ Solving the second equation for y gives $y=-x$, and substituting this into the first equation gives $x^2+(-x)^2=8 \Leftrightarrow 2x^2=8 \Leftrightarrow x=\pm 2$. So since $y=-x$, the solutions are $(2,-2)$ and $(-2,2)$.

7. $\begin{cases} x+y^2=0 \\ 2x+5y^2=75 \end{cases}$ Solving the first equation for x gives $x=-y^2$, and substituting this into the second equation gives $2\left(-y^2\right)+5y^2=75 \Leftrightarrow 3y^2=75 \Leftrightarrow y^2=25 \Leftrightarrow y=\pm 5$. So since $x=-y^2$, the solutions are $(-25,-5)$ and $(-25,5)$.

9. $\begin{cases} x^2-2y=1 \\ x^2+5y=29 \end{cases}$ Subtracting the first equation from the second equation gives $7y=28 \Rightarrow y=4$. Substituting $y=4$ into the first equation of the original system gives $x^2-2(4)=1 \Leftrightarrow x^2=9 \Leftrightarrow x=\pm 3$. The solutions are $(3,4)$ and $(-3,4)$.

11. $\begin{cases} 3x^2-y^2=11 \\ x^2+4y^2=8 \end{cases}$ Multiplying the first equation by 4 gives the system $\begin{cases} 12x^2-4y^2=44 \\ x^2+4y^2=8 \end{cases}$ Adding the equations gives $13x^2=52 \Leftrightarrow x=\pm 2$. Substituting into the first equation we get $3(4)-y^2=11 \Leftrightarrow y=\pm 1$. Thus, the solutions are $(2,1)$, $(2,-1)$, $(-2,1)$, and $(-2,-1)$.

13. $\begin{cases} x-y^2+3=0 \\ 2x^2+y^2-4=0 \end{cases}$ Adding the two equations gives $2x^2+x-1=0$. Using the Quadratic Formula we have $x=\dfrac{-1\pm\sqrt{1-4(2)(-1)}}{2(2)}=\dfrac{-1\pm\sqrt{9}}{4}=\dfrac{-1\pm 3}{4}$. So $x=\dfrac{-1-3}{4}=-1$ or $x=\dfrac{-1+3}{4}=\frac{1}{2}$. Substituting $x=-1$ into the first equation gives $-1-y^2+3=0 \Leftrightarrow y^2=2 \Leftrightarrow y=\pm\sqrt{2}$. Substituting $x=\frac{1}{2}$ into the first equation gives $\frac{1}{2}-y^2+3=0 \Leftrightarrow y^2=\frac{7}{2} \Leftrightarrow y=\pm\sqrt{\frac{7}{2}}$. Thus the solutions are $\left(-1,\pm\sqrt{2}\right)$ and $\left(\frac{1}{2},\pm\sqrt{\frac{7}{2}}\right)$.

15. $\begin{cases} x^2 + y = 8 \\ x - 2y = -6 \end{cases}$ By inspection of the graph, it appears that $(2, 4)$ is a solution, but is difficult to get accurate values for the other point. Multiplying the first equation by 2 gives the system $\begin{cases} 2x^2 + 2y = 16 \\ x - 2y = -6 \end{cases}$ Adding the equations gives $2x^2 + x = 10 \Leftrightarrow 2x^2 + x - 10 = 0 \Leftrightarrow (2x+5)(x-2) = 0$. So $x = -\frac{5}{2}$ or $x = 2$. If $x = -\frac{5}{2}$, then $-\frac{5}{2} - 2y = -6 \Leftrightarrow -2y = -\frac{7}{2} \Leftrightarrow y = \frac{7}{4}$, and if $x = 2$, then $2 - 2y = -6 \Leftrightarrow -2y = -8 \Leftrightarrow y = 4$. Hence, the solutions are $\left(-\frac{5}{2}, \frac{7}{4}\right)$ and $(2, 4)$.

17. $\begin{cases} x^2 + y = 0 \\ x^3 - 2x - y = 0 \end{cases}$ By inspection of the graph, it appears that $(-2, -4)$, $(0, 0)$, and $(1, -1)$ are solutions to the system. We check each point in both equations to verify that it is a solution.

For $(-2, -4)$: $(-2)^2 + (-4) = 4 - 4 = 0$ and $(-2)^3 - 2(-2) - (-4) = -8 + 4 + 4 = 0$.

For $(0, 0)$: $(0)^2 + (0) = 0$ and $(0)^3 - 2(0) - (0) = 0$.

For $(1, -1)$: $(1)^2 + (-1) = 1 - 1 = 0$ and $(1)^3 - 2(1) - (-1) = 1 - 2 + 1 = 0$.

Thus, the solutions are $(-2, -4)$, $(0, 0)$, and $(1, -1)$.

19. $\begin{cases} y + x^2 = 4x \\ y + 4x = 16 \end{cases}$ Subtracting the second equation from the first equation gives $x^2 - 4x = 4x - 16 \Leftrightarrow x^2 - 8x + 16 = 0 \Leftrightarrow (x-4)^2 = 0 \Leftrightarrow x = 4$. Substituting this value for x into either of the original equations gives $y = 0$. Therefore, the solution is $(4, 0)$.

21. $\begin{cases} x - 2y = 2 \\ y^2 - x^2 = 2x + 4 \end{cases}$ Now $x - 2y = 2 \Leftrightarrow x = 2y + 2$. Substituting for x gives $y^2 - x^2 = 2x + 4 \Leftrightarrow y^2 - (2y+2)^2 = 2(2y+2) + 4 \Leftrightarrow y^2 - 4y^2 - 8y - 4 = 4y + 4 + 4 \Leftrightarrow y^2 + 4y + 4 = 0 \Leftrightarrow (y+2)^2 = 0 \Leftrightarrow y = -2$. Since $x = 2y + 2$, we have $x = 2(-2) + 2 = -2$. Thus, the solution is $(-2, -2)$.

23. $\begin{cases} x - y = 4 \\ xy = 12 \end{cases}$ Now $x - y = 4 \Leftrightarrow x = 4 + y$. Substituting for x gives $xy = 12 \Leftrightarrow (4+y)y = 12 \Leftrightarrow y^2 + 4y - 12 = 0 \Leftrightarrow (y+6)(y-2) = 0 \Leftrightarrow y = -6, y = 2$. Since $x = 4 + y$, the solutions are $(-2, -6)$ and $(6, 2)$.

25. $\begin{cases} x^2 y = 16 \\ x^2 + 4y + 16 = 0 \end{cases}$ Now $x^2 y = 16 \Leftrightarrow x^2 = \dfrac{16}{y}$. Substituting for x^2 gives $\dfrac{16}{y} + 4y + 16 = 0 \Rightarrow 4y^2 + 16y + 16 = 0 \Leftrightarrow y^2 + 4y + 4 = 0 \Leftrightarrow (y+2)^2 = 0 \Leftrightarrow y = -2$. Therefore, $x^2 = \dfrac{16}{-2} = -8$, which has no real solution, and so the system has no solution.

27. $\begin{cases} x^2 + y^2 = 9 \\ x^2 - y^2 = 1 \end{cases}$ Adding the equations gives $2x^2 = 10 \Leftrightarrow x^2 = 5 \Leftrightarrow x = \pm\sqrt{5}$. Now $x = \pm\sqrt{5} \Rightarrow y^2 = 9 - 5 = 4 \Leftrightarrow y = \pm 2$, and so the solutions are $\left(\sqrt{5}, 2\right)$, $\left(\sqrt{5}, -2\right)$, $\left(-\sqrt{5}, 2\right)$, and $\left(-\sqrt{5}, -2\right)$.

29. $\begin{cases} 2x^2 - 8y^3 = 19 \\ 4x^2 + 16y^3 = 34 \end{cases}$ Multiplying the first equation by 2 gives the system $\begin{cases} 4x^2 - 16y^3 = 38 \\ 4x^2 + 16y^3 = 34 \end{cases}$ Adding the two equations gives $8x^2 = 72 \Leftrightarrow x = \pm 3$, and then substituting into the first equation we have $2(9) - 8y^3 = 19 \Leftrightarrow y^3 = -\frac{1}{8} \Leftrightarrow y = -\frac{1}{2}$. Therefore, the solutions are $\left(3, -\frac{1}{2}\right)$ and $\left(-3, -\frac{1}{2}\right)$.

31. $\begin{cases} \dfrac{2}{x} - \dfrac{3}{y} = 1 \\ -\dfrac{4}{x} + \dfrac{7}{y} = 1 \end{cases}$ If we let $u = \dfrac{1}{x}$ and $v = \dfrac{1}{y}$, the system is equivalent to $\begin{cases} 2u - 3v = 1 \\ -4u + 7v = 1 \end{cases}$ Multiplying the first equation by 4 gives the system $\begin{cases} 4u - 6v = 2 \\ -4u + 7v = 1 \end{cases}$ Adding the equations gives $v = 3$, and then substituting into the first equation gives $2u - 9 = 1 \Leftrightarrow u = 5$. Thus, the solution is $\left(\frac{1}{5}, \frac{1}{3}\right)$.

33. $\begin{cases} y = x^2 + 8x \\ y = 2x + 16 \end{cases}$

The solutions are $(-8, 0)$ and $(2, 20)$.

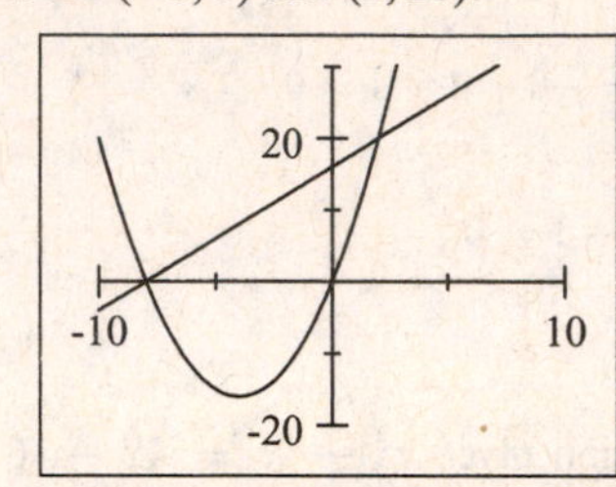

35. $\begin{cases} x^2 + y^2 = 25 \\ x + 3y = 2 \end{cases} \Leftrightarrow \begin{cases} y = \pm\sqrt{25 - x^2} \\ y = -\frac{1}{3}x + \frac{2}{3} \end{cases}$

The solutions are $(-4.51, 2.17)$ and $(4.91, -0.97)$.

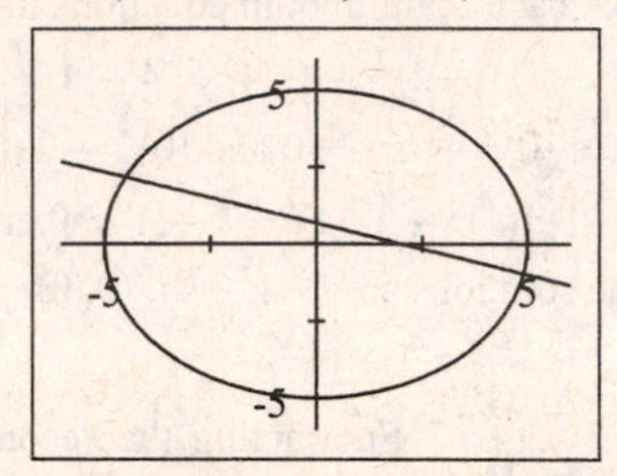

37. $\begin{cases} \dfrac{x^2}{9} + \dfrac{y^2}{18} = 1 \\ y = -x^2 + 6x - 2 \end{cases} \Leftrightarrow \begin{cases} y = \pm\sqrt{18 - 2x^2} \\ y = -x^2 + 6x - 2 \end{cases}$

The solutions are $(1.23, 3.87)$ and $(-0.35, -4.21)$.

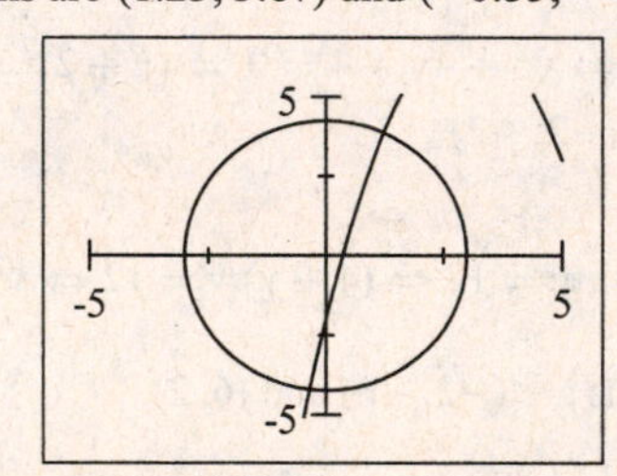

39. $\begin{cases} x^4 + 16y^4 = 32 \\ x^2 + 2x + y = 0 \end{cases} \Leftrightarrow \begin{cases} y = \pm\dfrac{\sqrt[4]{32 - x^4}}{2} \\ y = -x^2 - 2x \end{cases}$

The solutions are $(-2.30, -0.70)$ and $(0.48, -1.19)$.

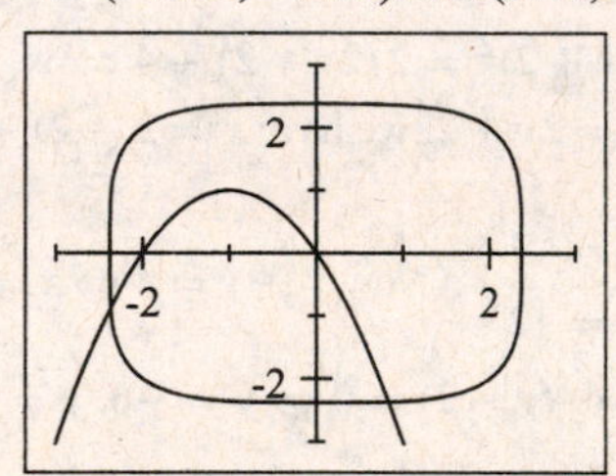

41. $\begin{cases} \log x + \log y = \frac{3}{2} \\ 2\log x - \log y = 0 \end{cases}$ Adding the two equations gives $3\log x = \frac{3}{2} \Leftrightarrow \log x = \frac{1}{2} \Leftrightarrow x = \sqrt{10}$. Substituting into the second equation we get $2\log 10^{1/2} - \log y = 0 \Leftrightarrow \log 10 - \log y = 0 \Leftrightarrow \log y = 1 \Leftrightarrow y = 10$. Thus, the solution is $\left(\sqrt{10}, 10\right)$.

43. $\begin{cases} x - y = 3 \\ x^3 - y^3 = 387 \end{cases}$ Solving the first equation for x gives $x = 3 + y$ and using the hint, $x^3 - y^3 = 387 \Leftrightarrow (x - y)\left(x^2 + xy + y^2\right) = 387$. Next, substituting for x, we get $3\left[(3 + y)^2 + y(3 + y) + y^2\right] = 387 \Leftrightarrow 9 + 6y + y^2 + 3y + y^2 + y^2 = 129 \Leftrightarrow 3y^2 + 9y + 9 = 129 \Leftrightarrow (y + 8)(y - 5) = 0 \Rightarrow y = -8$ or $y = 5$. If $y = -8$, then $x = 3 + (-8) = -5$, and if $y = 5$, then $x = 3 + 5 = 8$. Thus the solutions are $(-5, -8)$ and $(8, 5)$.

45. Let w and l be the lengths of the sides, in cm. Then we have the system $\begin{cases} lw = 180 \\ 2l + 2w = 54 \end{cases}$ We solve the second equation for w giving, $w = 27 - l$, and substitute into the first equation to get $l(27 - l) = 180 \Leftrightarrow l^2 - 27l + 180 = 0 \Leftrightarrow (l - 15)(l - 12) = 0 \Rightarrow l = 15$ or $l = 12$. If $l = 15$, then $w = 27 - 15 = 12$, and if $l = 12$, then $w = 27 - 12 = 15$. Therefore, the dimensions of the rectangle are 12 cm by 15 cm.

47. Let l and w be the length and width, respectively, of the rectangle. Then, the system of equations is $\begin{cases} 2l + 2w = 70 \\ \sqrt{l^2 + w^2} = 25 \end{cases}$ Solving the first equation for l, we have $l = 35 - w$, and substituting into the second gives $\sqrt{l^2 + w^2} = 25 \Leftrightarrow l^2 + w^2 = 625 \Leftrightarrow (35 - w)^2 + w^2 = 625 \Leftrightarrow 1225 - 70w + w^2 + w^2 = 625 \Leftrightarrow 2w^2 - 70w + 600 = 0 \Leftrightarrow (w - 15)(w - 20) = 0 \Rightarrow w = 15$ or $w = 20$. So the dimensions of the rectangle are 15 and 20.

49. At the points where the rocket path and the hillside meet, we have $\begin{cases} y = \frac{1}{2}x \\ y = -x^2 + 401x \end{cases}$ Substituting for y in the second equation gives $\frac{1}{2}x = -x^2 + 401x \Leftrightarrow x^2 - \frac{801}{2}x = 0 \Leftrightarrow x\left(x - \frac{801}{2}\right) = 0 \Rightarrow x = 0, x = \frac{801}{2}$. When $x = 0$, the rocket has not left the pad. When $x = \frac{801}{2}$, then $y = \frac{1}{2}\left(\frac{801}{2}\right) = \frac{801}{4}$. So the rocket lands at the point $\left(\frac{801}{2}, \frac{801}{4}\right)$. The distance from the base of the hill is $\sqrt{\left(\frac{801}{2}\right)^2 + \left(\frac{801}{4}\right)^2} \approx 447.77$ meters.

51. The point P is at an intersection of the circle of radius 26 centered at $A(22, 32)$ and the circle of radius 20 centered at $B(28, 20)$. We have the system

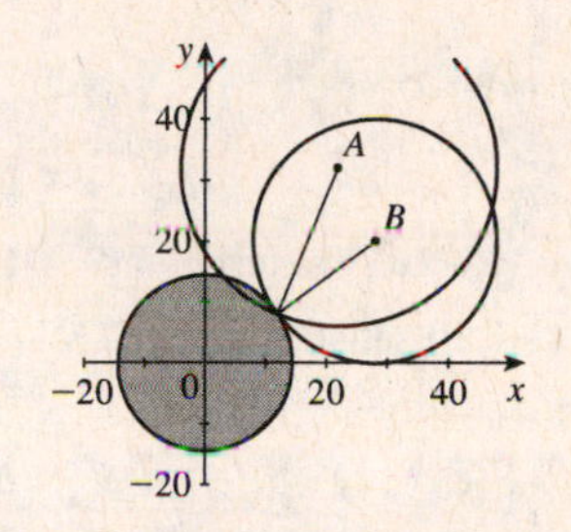

$$\begin{cases} (x - 22)^2 + (y - 32)^2 = 26^2 \\ (x - 28)^2 + (y - 20)^2 = 20^2 \end{cases} \Leftrightarrow$$

$$\begin{cases} x^2 - 44x + 484 + y^2 - 64y + 1024 = 676 \\ x^2 - 56x + 784 + y^2 - 40y + 400 = 400 \end{cases} \Leftrightarrow$$

$$\begin{cases} x^2 - 44x + y^2 - 64y = -832 \\ x^2 - 56x + y^2 - 40y = -784 \end{cases}$$

Subtracting the two equations, we get $12x - 24y = -48 \Leftrightarrow x - 2y = -4$, which is the equation of a line. Solving for x, we have $x = 2y - 4$. Substituting into the first equation gives $(2y - 4)^2 - 44(2y - 4) + y^2 - 64y = -832 \Leftrightarrow 4y^2 - 16y + 16 - 88y + 176 + y^2 - 64y = -832 \Leftrightarrow 5y^2 - 168y + 192 = -832 \Leftrightarrow 5y^2 - 168y + 1024 = 0$. Using the Quadratic Formula, we have $y = \frac{168 \pm \sqrt{168^2 - 4(5)(1024)}}{2(5)} = \frac{168 \pm \sqrt{7744}}{10} = \frac{168 \pm 88}{10} \Leftrightarrow y = 8$ or $y = 25.60$. Since the y-coordinate of the point P must be less than that of point A, we have $y = 8$. Then $x = 2(8) - 4 = 12$. So the coordinates of P are $(12, 8)$.

To solve graphically, we must solve each equation for y. This gives $(x - 22)^2 + (y - 32)^2 = 26^2 \Leftrightarrow (y - 32)^2 = 26^2 - (x - 22)^2 \Rightarrow y - 32 = \pm\sqrt{676 - (x - 22)^2} \Leftrightarrow y = 32 \pm \sqrt{676 - (x - 22)^2}$. We use the function $y = 32 - \sqrt{676 - (x - 22)^2}$ because the intersection we at interested in is below the point A. Likewise, solving the second equation for y, we would get the function $y = 20 - \sqrt{400 - (x - 28)^2}$. In a three-dimensional situation, you would need a minimum of three satellites, since a point on the earth can be uniquely specified as the intersection of three spheres centered at the satellites.

10.5 SYSTEMS OF INEQUALITIES

1. If the point $(2, 3)$ is a solution of an inequality in x and y, then the inequality is satisfied when we replace x by 2 and y by 3. Because $4(2) - 2(3) = 8 - 6 = 2 \geq 1$, the point $(2, 3)$ is a solution of the inequality $4x - 2y \geq 1$.

3. If the point $(2, 3)$ is a solution of a system of inequalities in x and y, then *each* inequality is satisfied when we replace x by 2 and y by 3. Because $2(2) + 4(3) = 16 \leq 17$ and $6(2) + 5(3) = 27 \leq 29$, the point $(2, 3)$ is a solution of the given system.

5.

Test Point	Inequality $x - 5y > 3$	Conclusion
$(-1, -2)$	$-1 - 5(-2) \stackrel{?}{>} 3$ ✓	Solution
$(1, -2)$	$1 - 5(-2) \stackrel{?}{>} 3$ ✓	Solution
$(1, 2)$	$-1 - 5(2) \stackrel{?}{>} 3$ ×	Not a solution
$(8, 1)$	$8 - 5(1) \stackrel{?}{>} 3$ ×	Not a solution

7.

Test Point	System $\begin{cases} 3x - 2y \leq 5 \\ 2x + y \geq 3 \end{cases}$	Conclusion
$(0, 0)$	$\begin{cases} 3(0) - 2(0) \stackrel{?}{\leq} 5 ✓ \\ 2(0) + 0 \stackrel{?}{\geq} 3 × \end{cases}$	Not a solution
$(1, 2)$	$\begin{cases} 3(1) - 2(2) \stackrel{?}{\leq} 5 ✓ \\ 2(1) + 2 \stackrel{?}{\geq} 3 ✓ \end{cases}$	Solution
$(1, 1)$	$\begin{cases} 3(1) - 2(1) \stackrel{?}{\leq} 5 ✓ \\ 2(1) + 1 \stackrel{?}{\geq} 3 ✓ \end{cases}$	Solution
$(3, 1)$	$\begin{cases} 3(3) - 2(1) \stackrel{?}{\leq} 5 × \\ 2(3) + 1 \stackrel{?}{\geq} 3 ✓ \end{cases}$	Not a solution

9. $y < -2x$. The test point $(-1, 0)$ satisfies the inequality.

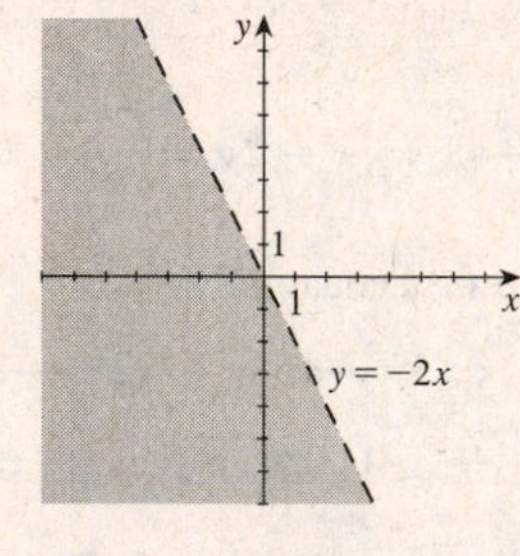

11. $y \geq 2$. The test point $(0, 3)$ satisfies the inequality.

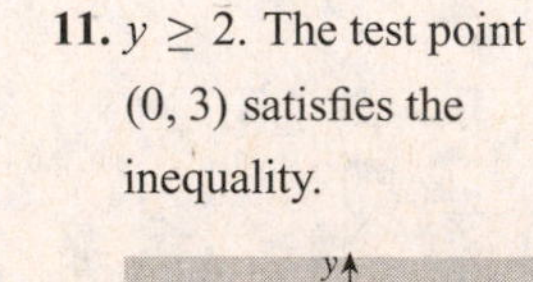
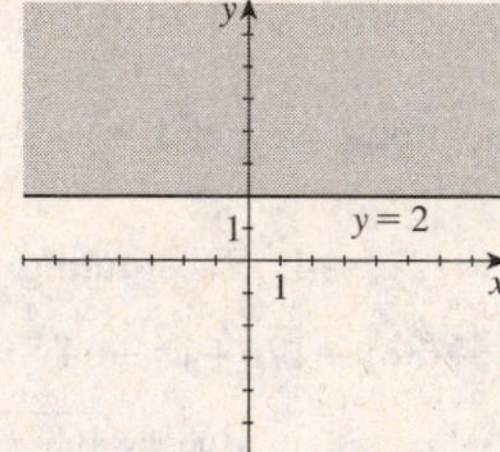

13. $x < 2$. The test point $(0, 0)$ satisfies the inequality.

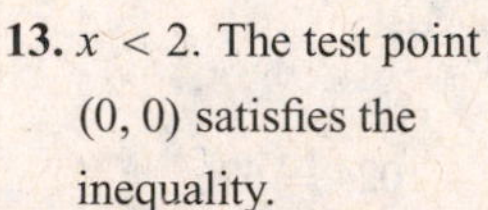
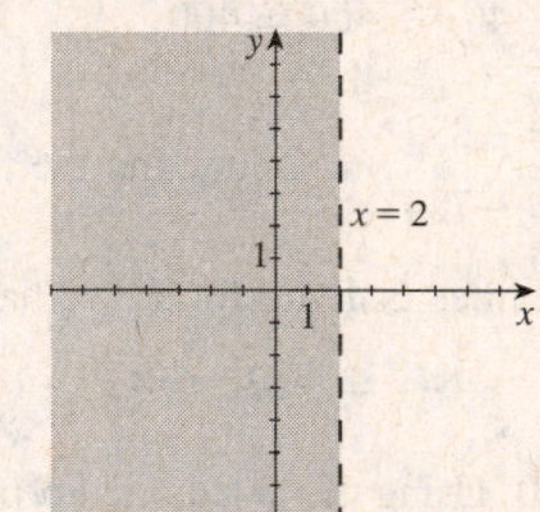

15. $y > x - 3$. The test point $(0, 0)$ satisfies the inequality.

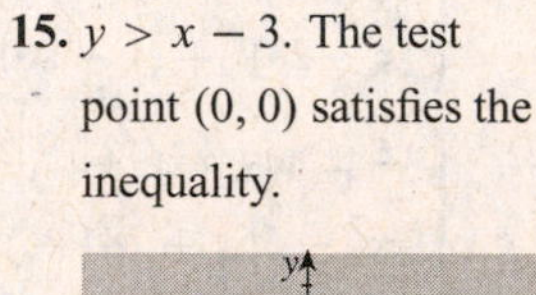
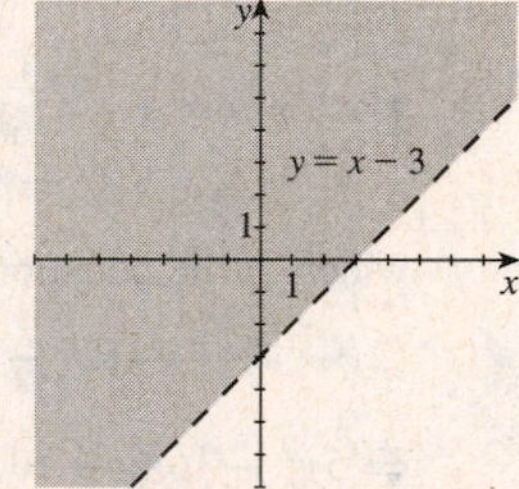

17. $2x - y \geq -4$. The test point $(0, 0)$ satisfies the inequality.

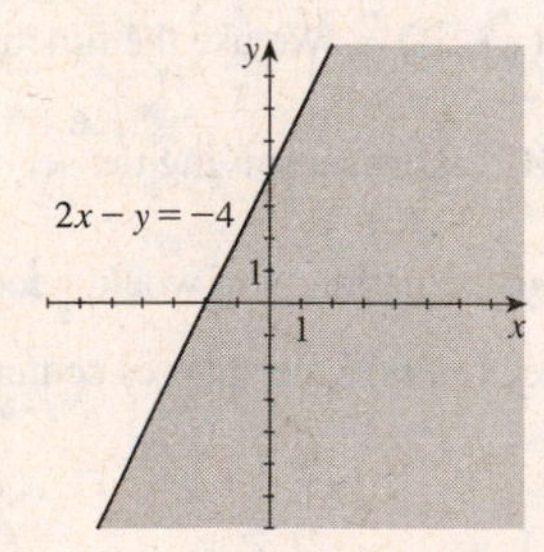

19. $-x^2 + y \geq 5$. The test point $(0, 6)$ satisfies the inequality.

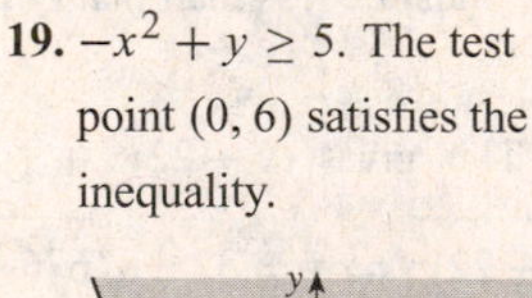
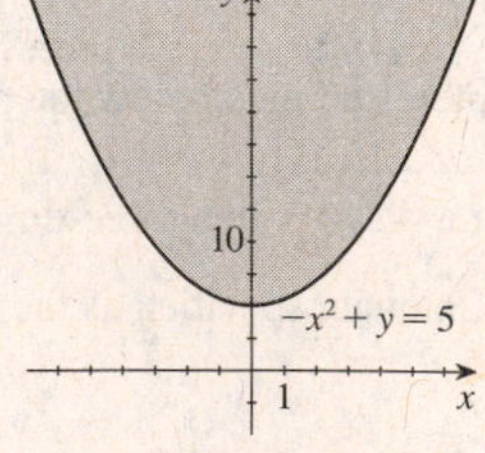

21. $x^2 + y^2 > 9$. The test point $(0, 4)$ satisfies the inequality.

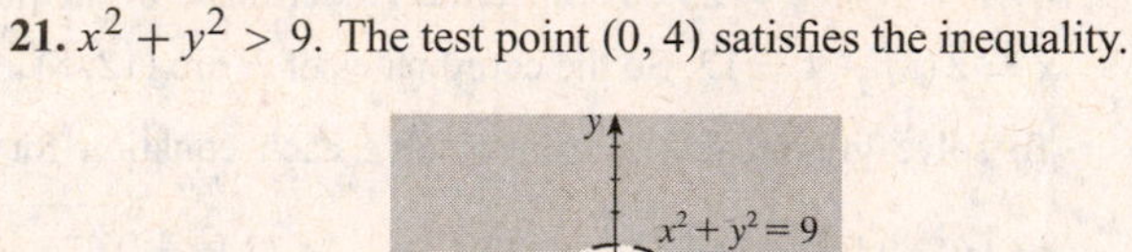

23. $3x - 2y \geq 18$

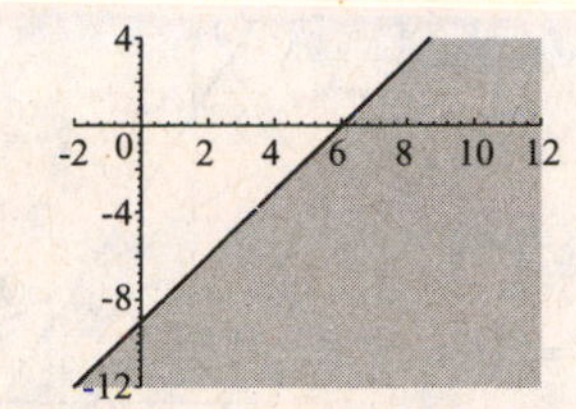

25. $5x + 2y > 8$

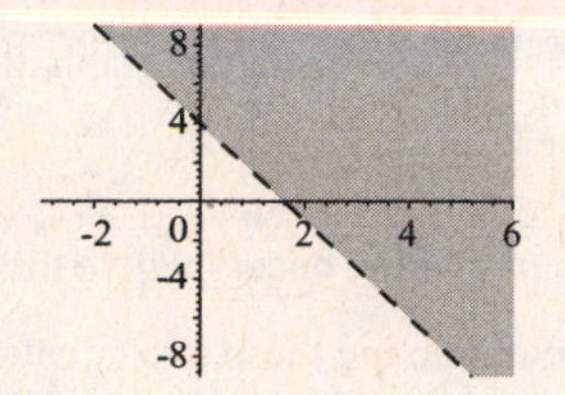

27. The boundary is a solid curve, so we have the inequality $y \leq \frac{1}{2}x - 1$. We take the test point $(0, -2)$ and verify that it satisfies the inequality: $-2 \leq \frac{1}{2}(0) - 1$.

29. The boundary is a broken curve, so we have the inequality $x^2 + y^2 > 4$. We take the test point $(0, 4)$ and verify that it satisfies the inequality: $0^2 + 4^2 > 4$.

31. $\begin{cases} x + y \leq 4 \\ y \geq x \end{cases}$ The vertices occur where $\begin{cases} x + y = 4 \\ y = x \end{cases}$ Substituting, we have $2x = 4 \Leftrightarrow x = 2$. Since $y = x$, the vertex is $(2, 2)$, and the solution set is not bounded. The test point $(0, 1)$ satisfies each inequality.

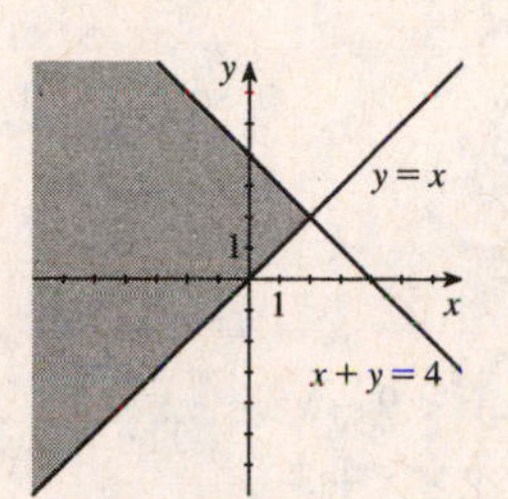

33. $\begin{cases} y < \frac{1}{4}x + 2 \\ y \geq 2x - 5 \end{cases}$ The vertex occurs where $\begin{cases} y = \frac{1}{4}x + 2 \\ y = 2x - 5 \end{cases}$ Substituting for y gives $\frac{1}{4}x + 2 = 2x - 5 \Leftrightarrow \frac{7}{4}x = 7 \Leftrightarrow x = 4$, so $y = 3$. Hence, the vertex is $(4, 3)$, and the solution is not bounded. The test point $(0, 1)$ satisfies each inequality.

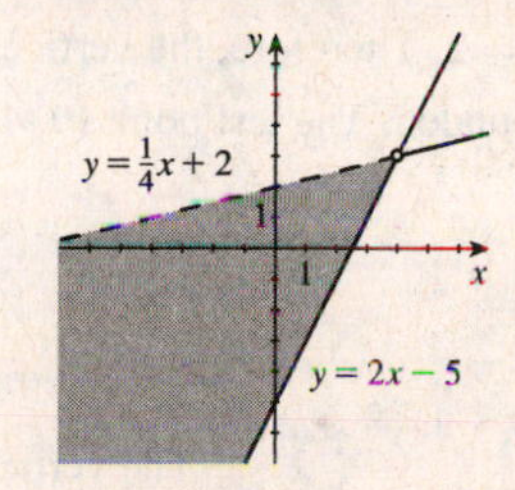

35. $\begin{cases} y \leq -2x + 8 \\ y \leq -\frac{1}{2}x + 5 \\ x \geq 0, y \geq 0 \end{cases}$ One vertex occurs where $\begin{cases} y \leq -2x + 8 \\ y \leq -\frac{1}{2}x + 5 \end{cases}$ Substituting for y gives $-2x + 8 = -\frac{1}{2}x + 5 \Leftrightarrow -\frac{3}{2}x = -3 \Leftrightarrow x = 2$, so $y = -2(2) + 8 = 4$.

Hence, this vertex is $(2, 4)$. Another vertex occurs where $\begin{cases} y = -2x + 8 \\ y = 0 \end{cases} \Leftrightarrow$ $-2x + 8 = 0 \Leftrightarrow x = 4$; this vertex is $(4, 0)$. Another occurs where $\begin{cases} y = -\frac{1}{2}x + 5 \\ x = 0 \end{cases} \Leftrightarrow y = 5$; this gives the vertex $(0, 5)$. The origin is another vertex, and the solution set is bounded. The test point $(1, 1)$ satisfies each inequality.

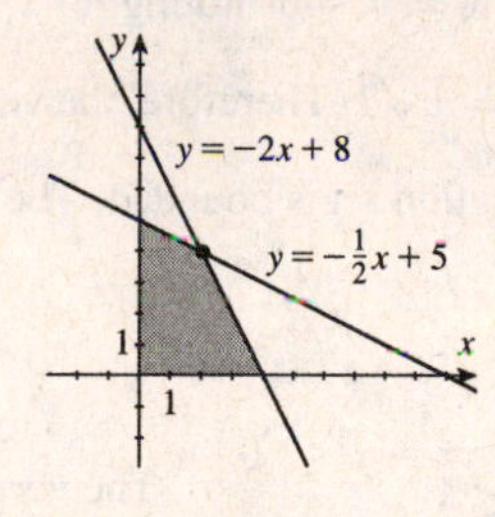

37. $\begin{cases} x \geq 0 \\ y \geq 0 \\ 3x + 5y \leq 15 \\ 3x + 2y \leq 9 \end{cases}$ From the graph, the points $(3, 0)$, $(0, 3)$ and $(0, 0)$ are vertices, and the fourth vertex occurs where the lines $3x + 5y = 15$ and $3x + 2y = 9$ intersect. Subtracting these two equations gives $3y = 6 \Leftrightarrow y = 2$, and so $x = \frac{5}{3}$. Thus, the fourth vertex is $\left(\frac{5}{3}, 2\right)$, and the solution set is bounded. The test point $(1, 1)$ satisfies each inequality.

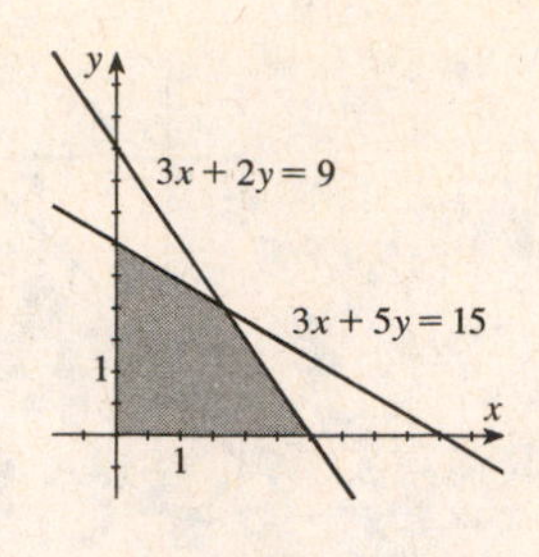

39. $\begin{cases} y \leq 9 - x^2 \\ x \geq 0, y \geq 0 \end{cases}$ From the graph, the vertices occur at $(0, 0)$, $(3, 0)$, and $(0, 9)$. The solution set is bounded. The test point $(1, 1)$ satisfies each inequality.

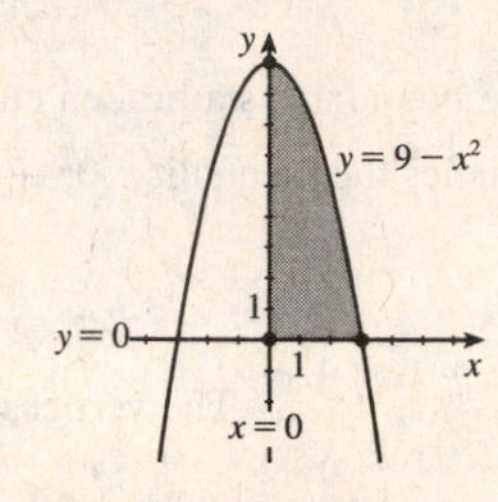

41. $\begin{cases} y < 9 - x^2 \\ y \geq x + 3 \end{cases}$ The vertices occur where $\begin{cases} y = 9 - x^2 \\ y = x + 3 \end{cases}$ Substituting for y gives $9 - x^2 = x + 3 \Leftrightarrow x^2 + x - 6 = 0 \Leftrightarrow (x - 2)(x + 3) = 0 \Rightarrow x = -3$, $x = 2$. Therefore, the vertices are $(-3, 0)$ and $(2, 5)$, and the solution set is bounded. The test point $(0, 4)$ satisfies each inequality.

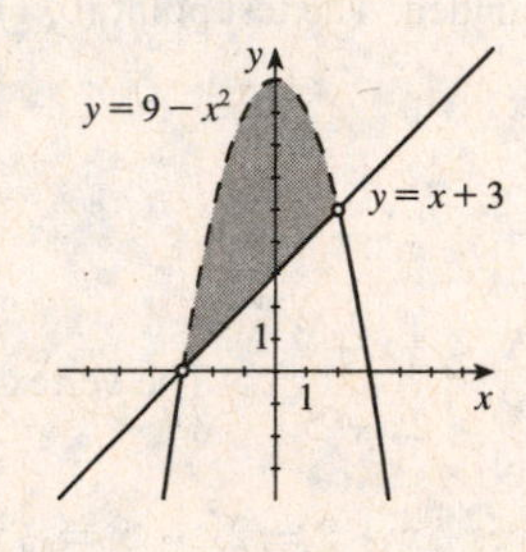

43. $\begin{cases} x^2 + y^2 \leq 4 \\ x - y > 0 \end{cases}$ The vertices occur where $\begin{cases} x^2 + y^2 = 4 \\ x - y = 0 \end{cases}$ Since $x - y = 0$ $\Leftrightarrow x = y$, substituting for x gives $y^2 + y^2 = 4 \Leftrightarrow y^2 = 2 \Rightarrow y = \pm\sqrt{2}$, and $x = \pm\sqrt{2}$. Therefore, the vertices are $\left(-\sqrt{2}, -\sqrt{2}\right)$ and $\left(\sqrt{2}, \sqrt{2}\right)$, and the solution set is bounded. The test point $(1, 0)$ satisfies each inequality.

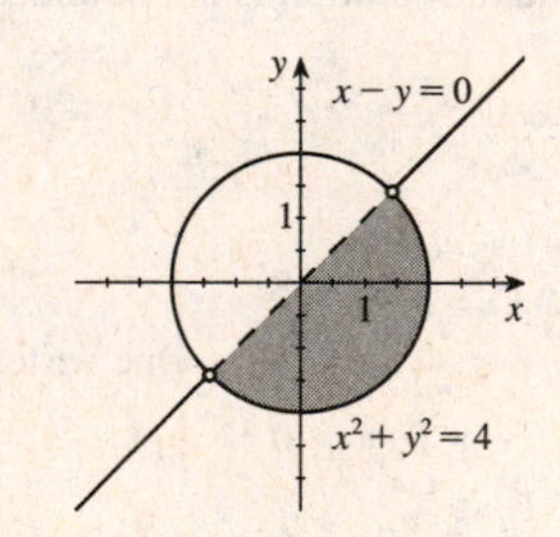

45. $\begin{cases} x^2 - y \leq 0 \\ 2x^2 + y \leq 12 \end{cases}$ The vertices occur where $\begin{cases} x^2 - y = 0 \\ 2x^2 + y = 12 \end{cases} \Leftrightarrow$ $\begin{cases} 2x^2 - 2y = 0 \\ 2x^2 + y = 12 \end{cases}$ Subtracting the equations gives $3y = 12 \Leftrightarrow y = 4$, and $x = \pm 2$. Thus, the vertices are $(2, 4)$ and $(-2, 4)$, and the solution set is bounded. The test point $(0, 1)$ satisfies each inequality.

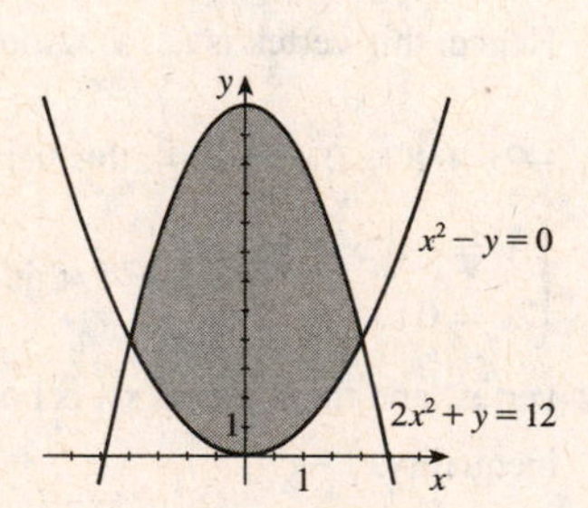

47. $\begin{cases} x^2 + y^2 \le 9 \\ 2x + y^2 \le 1 \end{cases}$ The vertices occur where $\begin{cases} x^2 + y^2 = 9 \\ 2x + y^2 = 1 \end{cases}$ Subtracting the equations gives $x^2 - 2x = 8 \Leftrightarrow x^2 - 2x - 8 = (x+2)(x-4) = 0$. Therefore, the vertices are $\left(-2, -\sqrt{5}\right)$ and $\left(-2, \sqrt{5}\right)$. The solution set is bounded. The test point $(0, 0)$ satisfies each inequality.

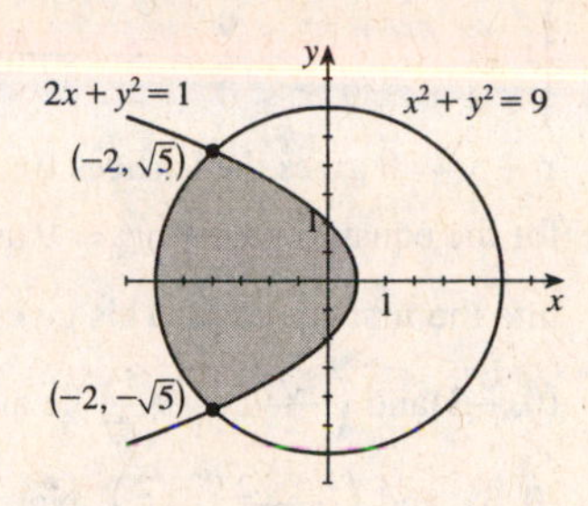

49. $\begin{cases} x + 2y \le 14 \\ 3x - \ \ y \ge \ 0 \\ \ \ x - \ \ y \ge \ 2 \end{cases}$ We find the vertices of the region by solving pairs of the

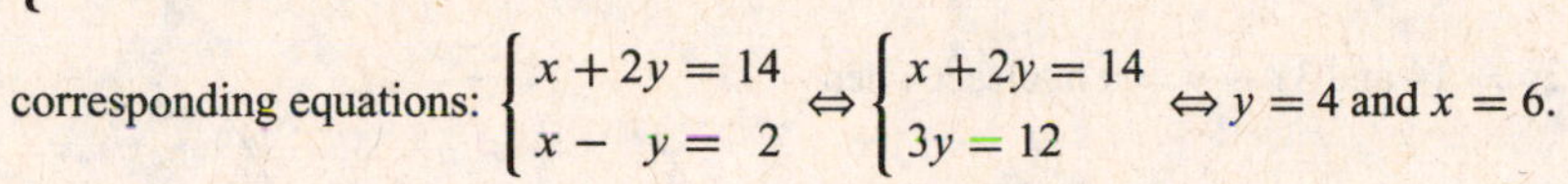

corresponding equations: $\begin{cases} x + 2y = 14 \\ x - \ \ y = \ 2 \end{cases} \Leftrightarrow \begin{cases} x + 2y = 14 \\ 3y = 12 \end{cases} \Leftrightarrow y = 4$ and $x = 6$.

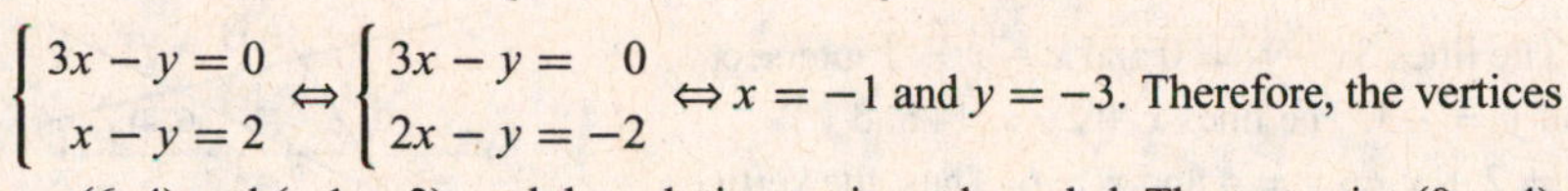

$\begin{cases} 3x - y = 0 \\ x - y = 2 \end{cases} \Leftrightarrow \begin{cases} 3x - y = \ \ 0 \\ 2x - y = -2 \end{cases} \Leftrightarrow x = -1$ and $y = -3$. Therefore, the vertices are $(6, 4)$ and $(-1, -3)$, and the solution set is not bounded. The test point $(0, -4)$ satisfies each inequality.

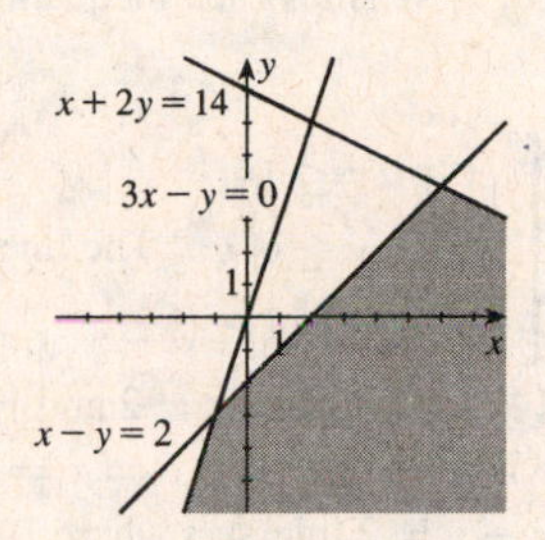

51. $\begin{cases} x \ge 0, y \ge 0 \\ x \le 5, x + y \le 7 \end{cases}$ The points of intersection are $(0, 7)$, $(0, 0)$, $(7, 0)$, $(5, 2)$, and $(5, 0)$. However, the point $(7, 0)$ is not in the solution set. Therefore, the vertices are $(0, 7)$, $(0, 0)$, $(5, 0)$, and $(5, 2)$, and the solution set is bounded. The test point $(1, 1)$ satisfies each inequality.

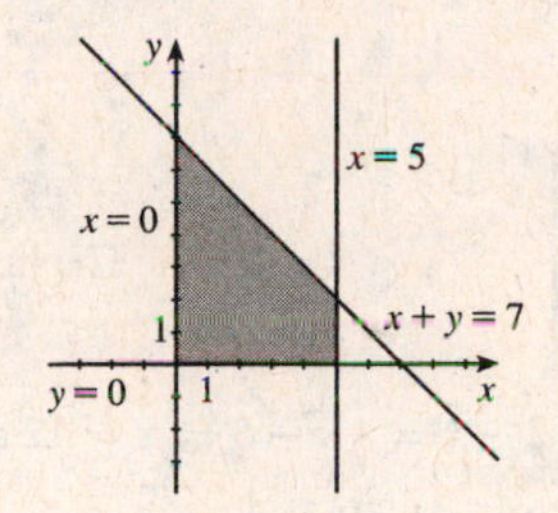

53. $\begin{cases} y > x + 1 \\ x + 2y \le 12 \\ x + 1 > \ \ 0 \end{cases}$ We find the vertices of the region by solving pairs of the corresponding equations. Using $x = -1$ and substituting for x in the line $y = x + 1$ gives the point $(-1, 0)$. Substituting for x in the line $x + 2y = 12$ gives the point $\left(-1, \frac{13}{2}\right)$. $\begin{cases} y = x + 1 \\ x + 2y = 12 \end{cases} \Leftrightarrow x = y - 1$ and $y - 1 + 2y = 12 \Leftrightarrow$ $3y = 13 \Leftrightarrow y = \frac{13}{3}$ and $x = \frac{10}{3}$. So the vertices are $(-1, 0)$, $\left(-1, \frac{13}{2}\right)$, and $\left(\frac{10}{3}, \frac{13}{3}\right)$, and none of these vertices is in the solution set. The solution set is bounded. The test point $(0, 2)$ satisfies each inequality.

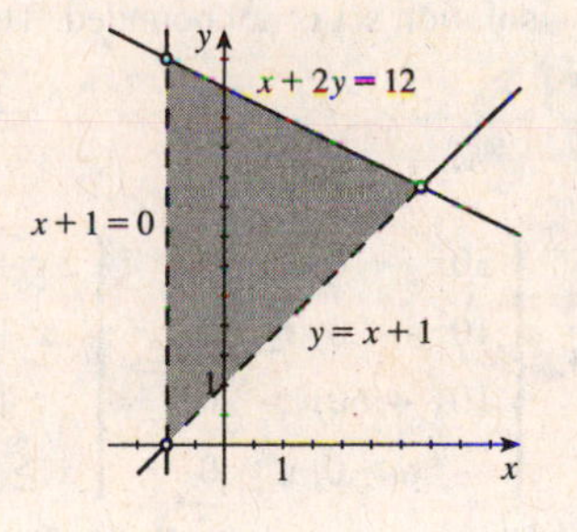

55. $\begin{cases} x^2 + y^2 \le 8 \\ x \ge 2, y \ge 0 \end{cases}$ The intersection points are $(2, \pm 2)$, $(2, 0)$, and $\left(2\sqrt{2}, 0\right)$. However, since $(2, -2)$ is not part of the solution set, the vertices are $(2, 2)$, $(2, 0)$, and $\left(2\sqrt{2}, 0\right)$. The solution set is bounded. The test point $(2.1, 1)$ satisfies each inequality.

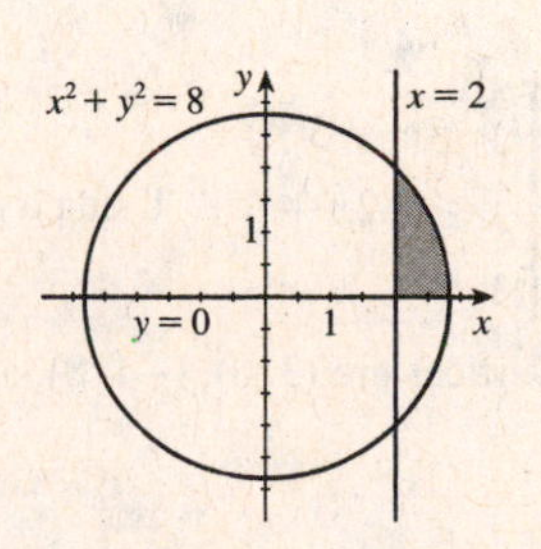

57. $\begin{cases} x^2 + y^2 < 9 \\ x + y > 0, x \le 0 \end{cases}$ Substituting $x = 0$ into the equations $x^2 + y^2 = 9$ and $x + y = 0$ gives the vertices $(0, \pm 3)$ and $(0, 0)$. To find the points of intersection for the equations $x^2 + y^2 = 9$ and $x + y = 0$, we solve for $x = -y$ and substitute into the first equation. This gives $(-y)^2 + y^2 = 9 \Rightarrow y = \pm\frac{3\sqrt{2}}{2}$. The points $(0, -3)$ and $\left(\frac{3\sqrt{2}}{2}, -\frac{3\sqrt{2}}{2}\right)$ lie away from the solution set, so the vertices are $(0, 0)$, $(0, 3)$, and $\left(-\frac{3\sqrt{2}}{2}, \frac{3\sqrt{2}}{2}\right)$. Note that the vertices are not solutions in this case. The solution set is bounded. The test point $(0, 1)$ satisfies each inequality.

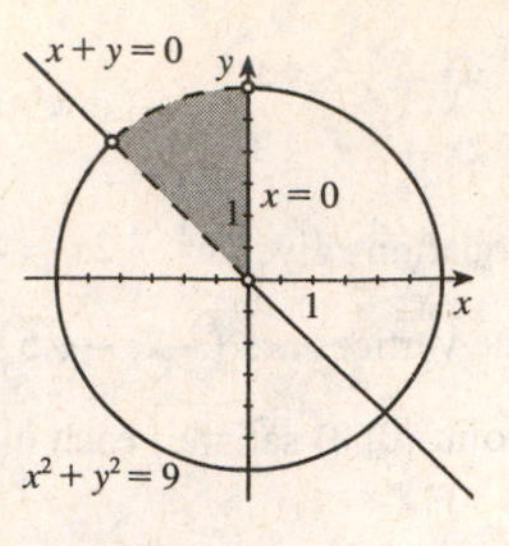

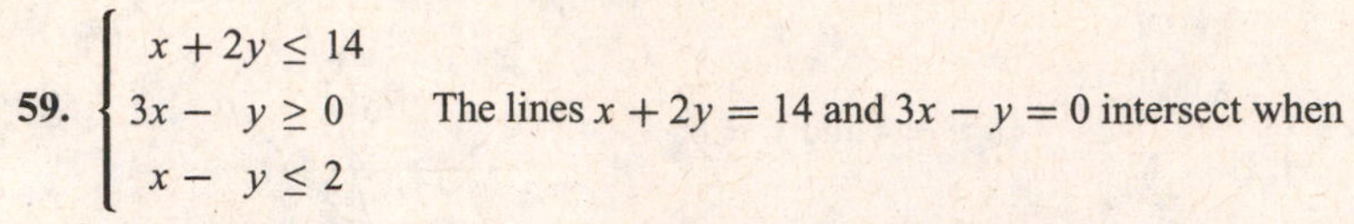

59. $\begin{cases} x + 2y \le 14 \\ 3x - \ \ y \ge 0 \\ x - \ \ y \le 2 \end{cases}$ The lines $x + 2y = 14$ and $3x - y = 0$ intersect when $14 - x = 6x \Leftrightarrow x = 2$ and $y = 6$. The lines $3x - y = 0$ and $x - y = 2$ intersect where $-3x = 2 - x \Leftrightarrow x = -1$ and $y = -3$. The lines $x + 2y = 14$ and $x - y = 2$ intersect where $14 - 2y = 2 + y \Leftrightarrow y = 4$ and $x = 6$. Thus, the vertices of the region are $(-1, -3)$, $(2, 6)$, and $(6, 4)$. The solution set is bounded. The test point $(1, 0)$ satisfies each inequality.

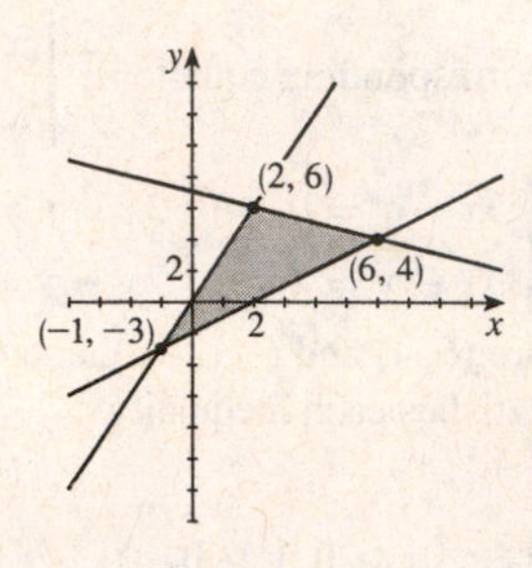

61. $\begin{cases} x + y \le 12 \\ y \le \frac{1}{2}x - 6 \\ y \le 2x + 6 \end{cases}$ The lines $x + y = 12$ and $y = \frac{1}{2}x - 6$ intersect where $12 - x = \frac{1}{2}x - 6 \Leftrightarrow x = 12$ and $y = 0$. The lines $y = \frac{1}{2}x - 6$ and $y = 2x + 6$ intersect where $\frac{1}{2}x - 6 = 2x + 6 \Leftrightarrow x = -8$ and $y = -10$. The lines $x + y = 12$ and $y = 2x + 6$ intersect where $12 - x = 2x + 6 \Leftrightarrow x = 2$ and $y = 12$. The solution set is not bounded. The test point $(0, -8)$ satisfies each inequality.

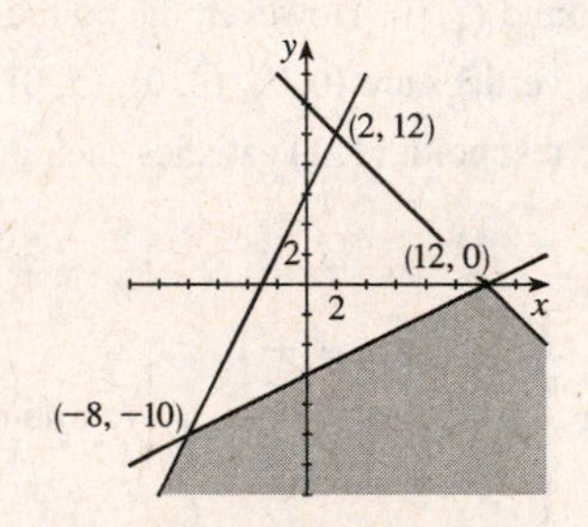

63. $\begin{cases} 30x + 10y \ge 50 \\ 10x + 20y \ge 50 \\ 10x + 60y \ge 90 \\ x \ge 0, y \ge 0 \end{cases} \Leftrightarrow \begin{cases} 3x + \ \ y \ge 5 \\ x + 2y \ge 5 \\ x + 6y \ge 9 \\ x \ge 0, y \ge 0 \end{cases}$

The vertices are $(0, 5)$, $(1, 2)$, $(3, 1)$, and $(9, 0)$. The solution set is not bounded. The test point $(1, 3)$ satisfies each inequality.

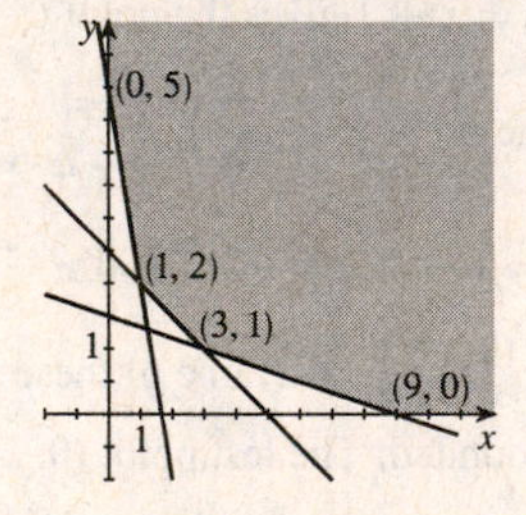

65. $\begin{cases} y \ge x - 3 \\ y \ge -2x + 6 \\ y \le 8 \end{cases}$ Using a graphing calculator, we find the region shown. The vertices are $(3, 0)$, $(-1, 8)$, and $(11, 8)$.

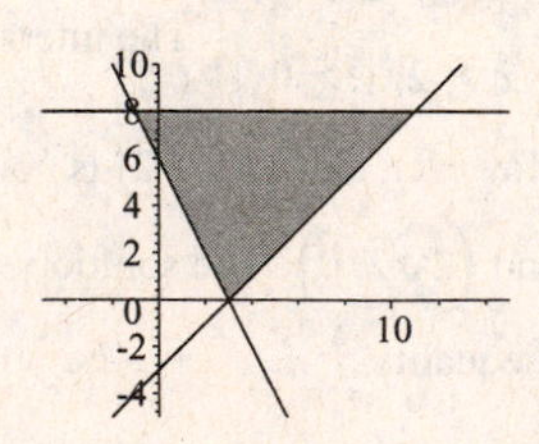

67. $\begin{cases} y \le 6x - x^2 \\ x + y \ge 4 \end{cases}$ Using a graphing calculator, we find the region shown. The vertices are (0.6, 3.4) and (6.4, −2.4).

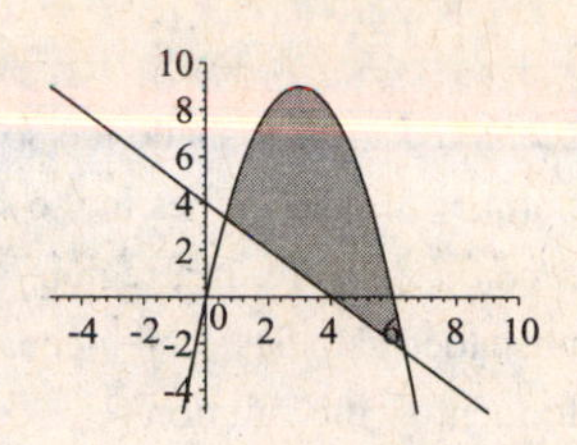

69. **(a)** Let x and y be the numbers of acres of potatoes and corn, respectively.

A system describing the possibilities is $\begin{cases} x + \quad y \le 500 \\ 90x + 50y \le 40{,}000 \\ 30x + 80y \le 30{,}000 \\ x \ge 0, y \ge 0 \end{cases}$

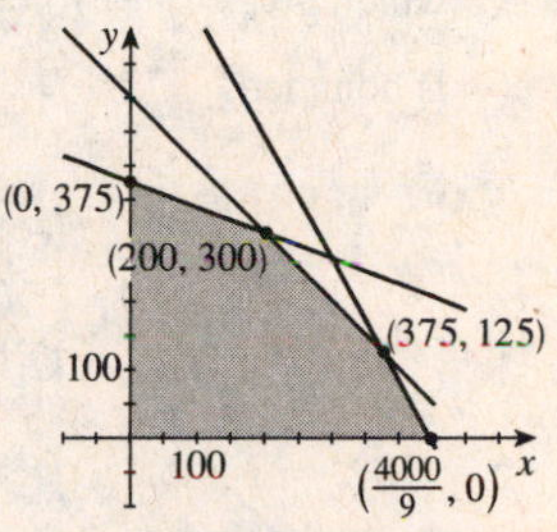

(b) Because the point (300, 180) lies in the feasible region, the farmer can plant 300 acres of potatoes and 180 acres of corn.

(c) Because the point (150, 325) lies outside the feasible region, the farmer cannot plant this combination of crops.

71. Let x be the number of fiction books published in a year and y the number of nonfiction books. Then the following system of inequalities holds:

$\begin{cases} x \ge 0, y \ge 0 \\ x + y \le 100 \\ y \ge 20, x \ge y \end{cases}$ From the graph, we see that the vertices are (50, 50), (80, 20) and (20, 20).

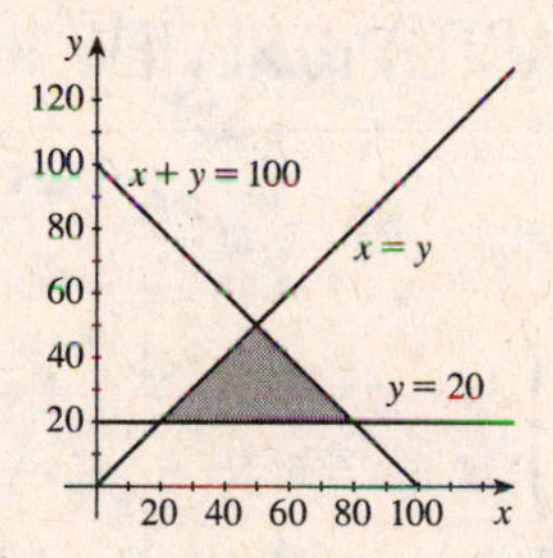

73. Let x be the number of Standard Blend packages and y be the number of Deluxe Blend packages. Since there are 16 ounces per pound, we get the following system of inequalities:

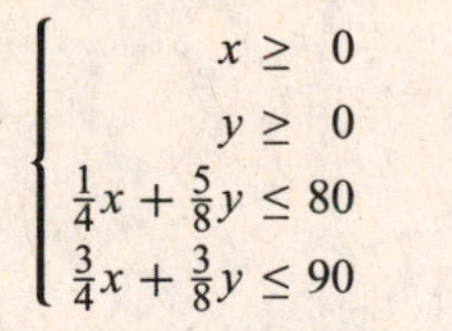

$$\begin{cases} x \ge 0 \\ y \ge 0 \\ \frac{1}{4}x + \frac{5}{8}y \le 80 \\ \frac{3}{4}x + \frac{3}{8}y \le 90 \end{cases}$$

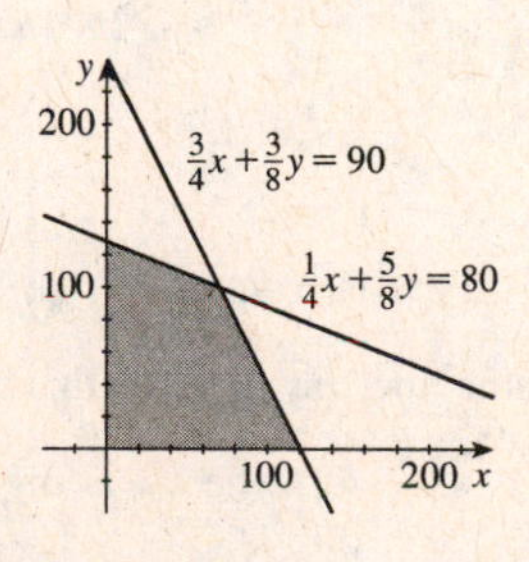

From the graph, we see that the vertices are (0, 0), (120, 0), (70, 100) and (0, 128).

75. $x + 2y > 4, -x + y < 1, x + 3y \le 9, x < 3.$

Method 1: We shade the solution to each inequality with lines perpendicular to the boundary. As you can see, as the number of inequalities in the system increases, it gets harder to locate the region where *all* of the shaded parts overlap.

Method 2: Here, if a region is shaded then it fails to satisfy at least one inequality. As a result, the region that is left unshaded satisfies each inequality, and is the solution to the system of inequalities. In this case, this method makes it easier to identify the solution set.

To finish, we find the vertices of the solution set. The line $x = 3$ intersects the line $x + 2y = 4$ at $\left(3, \frac{1}{2}\right)$ and the line $x + 3y = 9$ at $(3, 2)$. To find where the lines $-x + y = 1$ and $x + 2y = 4$ intersect, we add the two equations, which gives $3y = 5 \Leftrightarrow y = \frac{5}{3}$, and $x = \frac{2}{3}$. To find where the lines $-x + y = 1$ and $x + 3y = 9$ intersect, we add the two equations, which gives $4y = 10 \Leftrightarrow y = \frac{10}{4} = \frac{5}{2}$, and $x = \frac{3}{2}$. The vertices are $\left(3, \frac{1}{2}\right)$, $(3, 2)$, $\left(\frac{2}{3}, \frac{5}{3}\right)$, and $\left(\frac{3}{2}, \frac{5}{2}\right)$, and the solution set is bounded.

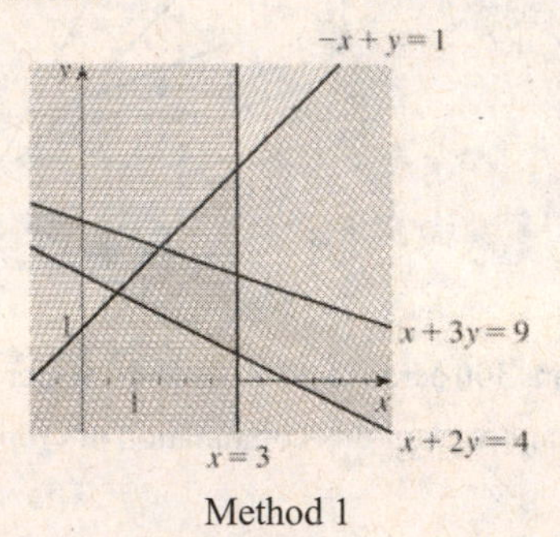

Method 1

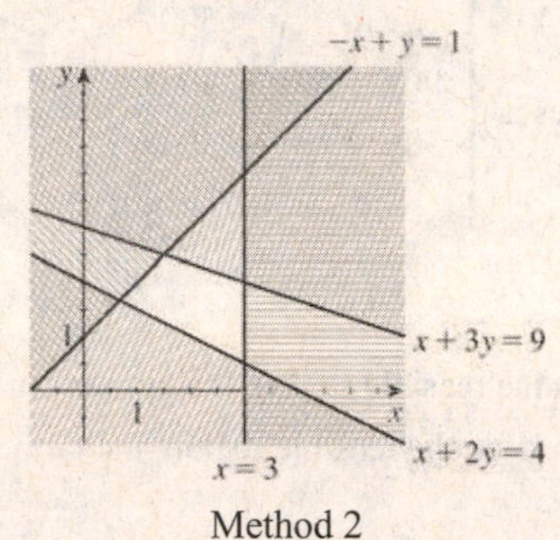

Method 2

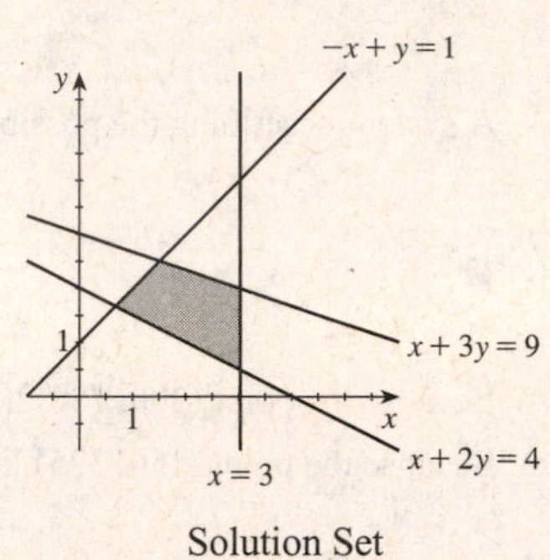

Solution Set

CHAPTER 10 REVIEW

1. $\begin{cases} 3x - y = 5 \\ 2x + y = 5 \end{cases}$ Adding, we get $5x = 10 \Leftrightarrow x = 2$. So $2(2) + y = 5 \Leftrightarrow y = 1$.

Thus, the solution is $(2, 1)$.

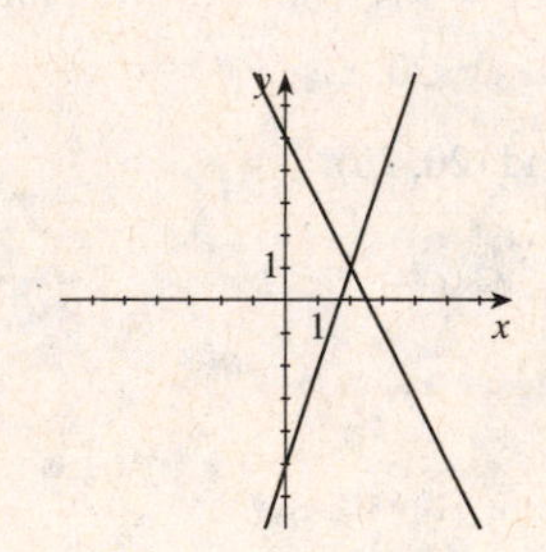

3. $\begin{cases} 2x - 7y = 28 \\ y = \frac{2}{7}x - 4 \end{cases} \Leftrightarrow \begin{cases} 2x - 7y = 28 \\ 2x - 7y = 28 \end{cases}$ Since these equations represent the same line, any point on this line will satisfy the system. Thus the solution are $\left(x, \frac{2}{7}x - 4\right)$, where x is any real number.

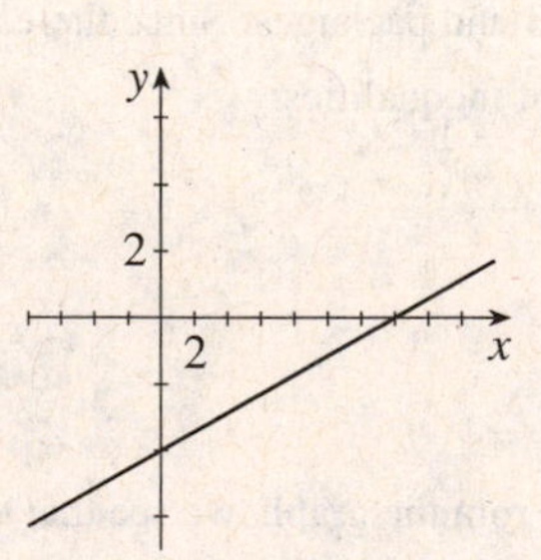

5. $\begin{cases} 2x - y = 1 \\ x + 3y = 10 \\ 3x + 4y = 15 \end{cases}$ Solving the first equation for y, we get $y = -2x + 1$.

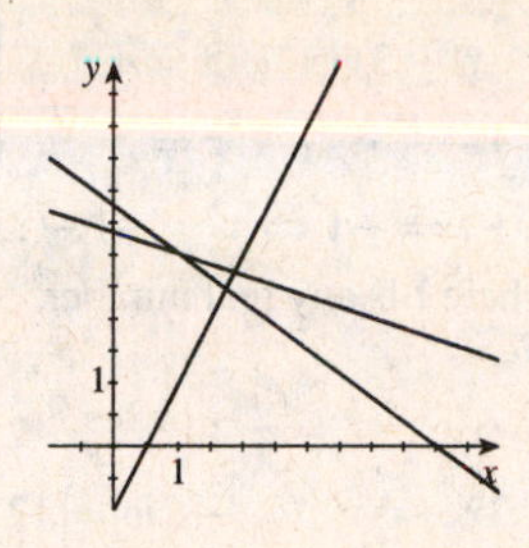

Substituting into the second equation gives $x + 3(-2x + 1) = 10 \Leftrightarrow -5x = 7 \Leftrightarrow$ $x = -\frac{7}{5}$. So $y = -\left(-\frac{7}{5}\right) + 1 = \frac{12}{5}$.Checking the point $\left(-\frac{7}{5}, \frac{12}{5}\right)$ in the third equation we have $3\left(-\frac{7}{5}\right) + 4\left(\frac{12}{5}\right) \stackrel{?}{=} 15$ but $-\frac{21}{5} + \frac{48}{5} \neq 15$. Thus, there is no solution, and the lines do not intersect at one point.

7. $\begin{cases} y = x^2 + 2x \\ y = 6 + x \end{cases}$ Substituting for y gives $6 + x = x^2 + 2x \Leftrightarrow x^2 + x - 6 = 0$. Factoring, we have $(x - 2)(x + 3) = 0$. Thus $x = 2$ or -3. If $x = 2$, then $y = 8$, and if $x = -3$, then $y = 3$. Thus the solutions are $(-3, 3)$ and $(2, 8)$.

9. $\begin{cases} 3x + \dfrac{4}{y} = 6 \\ x - \dfrac{8}{y} = 4 \end{cases}$ Adding twice the first equation to the second gives $7x = 16 \Leftrightarrow x = \frac{16}{7}$. So $\dfrac{16}{7} - \dfrac{8}{y} = 4 \Leftrightarrow$ $16y - 56 = 28y \Leftrightarrow -12y = 56 \Leftrightarrow y = -\frac{14}{3}$. Thus, the solution is $\left(\frac{16}{7}, -\frac{14}{3}\right)$.

11. $\begin{cases} 0.32x + 0.43y = 0 \\ 7x - 12y = 341 \end{cases} \Leftrightarrow \begin{cases} y = -\dfrac{32x}{43} \\ y = \dfrac{7x - 341}{12} \end{cases}$

The solution is approximately $(21.41, -15.93)$.

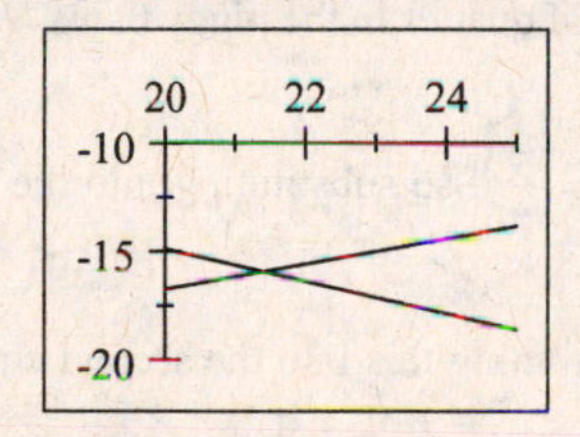

13. $\begin{cases} x - y^2 = 10 \\ x = \frac{1}{22}y + 12 \end{cases} \Leftrightarrow \begin{cases} y = \pm\sqrt{x - 10} \\ y = 22(x - 12) \end{cases}$

The solutions are $(11.94, -1.39)$ and $(12.07, 1.44)$.

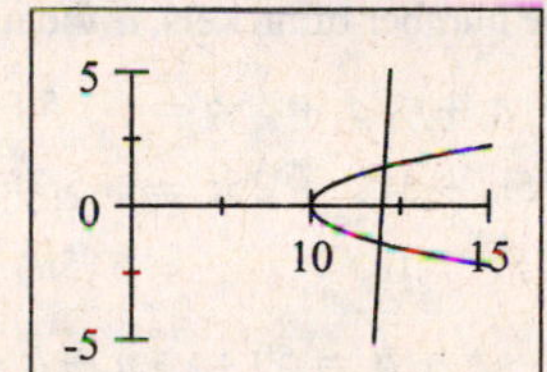

15. $\begin{cases} x - 2y + z = 8 \\ 4x + z = 9 \\ -2x + y - z = -8 \end{cases} \Leftrightarrow \begin{cases} x - 2y + z = 8 \\ 8y - 3z = -23 \\ -3y + z = 8 \end{cases} \Leftrightarrow \begin{cases} x - 2y + z = 8 \\ 8y - 3z = -23 \\ -z = -5 \end{cases}$ Therefore, $z = 5$, $8y - 3(5) = -23$ $\Leftrightarrow y = -1$, and $x - 2(-1) + 5 = 8 \Leftrightarrow x = 1$. Hence, the solution is $(1, -1, 5)$.

17. $\begin{cases} x + y + 2z = 6 \\ 2x + 5z = 12 \\ x + 2y + 3z = 9 \end{cases} \Leftrightarrow \begin{cases} x + y + 2z = 6 \\ 2y - z = 0 \\ 4y + z = 6 \end{cases} \Leftrightarrow \begin{cases} x + y + 2z = 6 \\ 2y - z = 0 \\ 3z = 6 \end{cases}$ Therefore, $3z = 6 \Leftrightarrow z = 2$, $2y - 2 = 0 \Leftrightarrow$ $y = 1$, and $x + 1 + 2(2) = 6 \Leftrightarrow x = 1$. Hence, the solution is $(1, 1, 2)$.

19. $\begin{cases} x - 2y + 3z = 1 \\ 2x - y + z = 3 \\ 2x - 7y + 11z = 2 \end{cases} \Leftrightarrow \begin{cases} x - 2y + 3z = 1 \\ 3y - 5z = 1 \\ 6y - 10z = 1 \end{cases} \Leftrightarrow \begin{cases} x - 2y + 3z = 1 \\ 3y - 5z = 1 \\ 0 = -1 \end{cases}$ which is impossible. Therefore, the system has no solution.

21. $\begin{cases} x - 3y + z = 4 \\ 4x - y + 15z = 5 \end{cases} \Leftrightarrow \begin{cases} x - 3y + z = 4 \\ y + z = -1 \end{cases}$ Thus, the system has infinitely many solutions given by $z = t$, $y + t = -1 \Leftrightarrow y = -1 - t$, and $x + 3(1 + t) + t = 4 \Leftrightarrow x = 1 - 4t$. Therefore, the solutions are $(1 - 4t, -1 - t, t)$, where t is any real number.

23. $\begin{cases} x - z + w = 2 \\ 2x + y - 2w = 12 \\ 3y + z + w = 4 \\ x + y - z = 10 \end{cases} \Leftrightarrow \begin{cases} x - z + w = 2 \\ y + 2z - 4w = 8 \\ 3y + z + w = 4 \\ y - w = 8 \end{cases} \Leftrightarrow \begin{cases} x - z + w = 2 \\ y + 2z - 4w = 8 \\ 5z - 13w = 20 \\ z + 4w = -20 \end{cases} \Leftrightarrow$

$\begin{cases} x - z + w = 2 \\ y + 2z - 4w = 8 \\ 5z - 13w = 20 \\ 33w = -120 \end{cases}$ Therefore, $33w = -120 \Leftrightarrow w = -\frac{40}{11}$, $5z - 13\left(-\frac{40}{11}\right) = 20 \Leftrightarrow z = -\frac{60}{11}$, $y + 2\left(-\frac{60}{11}\right) - 4\left(-\frac{40}{11}\right) = 8 \Leftrightarrow y = \frac{48}{11}$; and $x - \left(-\frac{60}{11}\right) + \left(-\frac{40}{11}\right) = 2 \Leftrightarrow x = \frac{2}{11}$. Hence, the solution is $\left(\frac{2}{11}, \frac{48}{11}, -\frac{60}{11}, -\frac{40}{11}\right)$.

25. Let k and s be be Kieran's and Siobhan's ages. From the given information, we have the system $\begin{cases} k = s + 4 \\ k + s = 22 \end{cases}$ Subtracting the first equation from the second gives $s = 22 - s - 4 \Leftrightarrow 2s = 18 \Leftrightarrow s = 9$, so $k = 22 - 9 = 13$. Thus, Kieran is 13 and Siobhan is 9.

27. Let n be the number of nickels, d the number of dimes, and q the number of quarter in the piggy bank. We get the following system: $\begin{cases} n + d + q = 50 \\ 5n + 10d + 25z = 560 \\ 10d = 5(5n) \end{cases}$ Since $10d = 25n$, we have $d = \frac{5}{2}n$, so substituting into the first equation we get $n + \frac{5}{2}n + q = 50 \Leftrightarrow \frac{7}{2}n + q = 50 \Leftrightarrow q = 50 - \frac{7}{2}n$. Now substituting this into the second equation we have $5n + 10\left(\frac{5}{2}n\right) + 25\left(50 - \frac{7}{2}n\right) = 560 \Leftrightarrow 5n + 25n + 1250 - \frac{175}{2}n = 560 \Leftrightarrow 1250 - \frac{115}{2}n = 560 \Leftrightarrow \frac{115}{2}n = 650 \Leftrightarrow$ $n = 12$. Then $d = \frac{5}{2}(12) = 30$ and $q = 50 - n - d = 50 - 12 - 30 = 8$. Thus the piggy bank contains 12 nickels, 30 dimes, and 8 quarters.

29. $\dfrac{3x + 1}{x^2 - 2x - 15} = \dfrac{3x + 1}{(x - 5)(x + 3)} = \dfrac{A}{x - 5} + \dfrac{B}{x + 3}$. Thus, $3x + 1 = A(x + 3) + B(x - 5) = x(A + B) + (3A - 5B)$, and so $\begin{cases} A + B = 3 \\ 3A - 5B = 1 \end{cases} \Leftrightarrow \begin{cases} -3A - 3B = -9 \\ 3A - 5B = 1 \end{cases}$ Adding, we have $-8B = -8 \Leftrightarrow B = 1$, and $A = 2$. Hence, $\dfrac{3x + 1}{x^2 - 2x - 15} = \dfrac{2}{x - 5} + \dfrac{1}{x + 3}$.

31. $\dfrac{2x - 4}{x(x - 1)^2} = \dfrac{A}{x} + \dfrac{B}{x - 1} + \dfrac{C}{(x - 1)^2}$. Then $2x - 4 = A(x - 1)^2 + Bx(x - 1) + Cx = Ax^2 - 2Ax + A + Bx^2 - Bx + Cx = x^2(A + B) + x(-2A - B + C) + A$. So $A = -4$, $-4 + B = 0 \Leftrightarrow B = 4$, and $8 - 4 + C = 2 \Leftrightarrow C = -2$. Therefore, $\dfrac{2x - 4}{x(x - 1)^2} = -\dfrac{4}{x} + \dfrac{4}{x - 1} - \dfrac{2}{(x - 1)^2}$.

33. $\dfrac{2x-1}{x^3+x} = \dfrac{2x-1}{x(x^2+1)} = \dfrac{A}{x} + \dfrac{Bx+C}{x^2+1}$. Then $2x-1 = A\left(x^2+1\right) + (Bx+C)x = Ax^2 + A + Bx^2 + Cx = (A+B)x^2 + Cx + A$. So $A=-1$, $C=2$, and $A+B=0$ gives us $B=1$. Thus $\dfrac{2x-1}{x^3+x} = -\dfrac{1}{x} + \dfrac{x+2}{x^2+1}$.

35. $\dfrac{3x^2-x+6}{(x^2+2)^2} = \dfrac{Ax+B}{x^2+2} + \dfrac{Cx+D}{(x^2+2)^2}$. Thus

$$\begin{aligned} 3x^2-x+6 &= \left(x^2+2\right)(Ax+B) + Cx + D = Ax^3 + Bx^2 + 2Ax + 2B + Cx + D \\ &= x^3A + x^2B + x(2A+C) + (2B+D) \end{aligned}$$

This leads to the system $\left\{\begin{array}{lllll} A & & & & = 0 \\ & B & & & = 3 \\ 2A & & +C & & = -1 \\ & 2B & & +D & = 6 \end{array}\right. \Leftrightarrow \left\{\begin{array}{llll} A & & & = 0 \\ & B & & = 3 \\ & & C & = -1 \\ & & & D = 0 \end{array}\right.$

Thus $\dfrac{3x^2-x+6}{(x^2+2)^2} = \dfrac{3}{x^2+2} - \dfrac{x}{(x^2+2)^2}$.

37. $\begin{cases} 2x+3y=7 \\ x-2y=0 \end{cases}$ By inspection of the graph, it appears that $(2, 1)$ is the solution to the system. We check this in both equations to verify that it is the solution. $2(2)+3(1) = 4+3 = 7$ and $2-2(1) = 2-2 = 0$. Since both equations are satisfied, the solution is indeed $(2, 1)$.

39. $\begin{cases} x^2+y=2 \\ x^2-3x-y=0 \end{cases}$ By inspection of the graph, it appears that $(2, -2)$ is a solution to the system, but is difficult to get accurate values for the other point. Adding the equations, we get $2x^2-3x=2 \Leftrightarrow 2x^2-3x-2=0 \Leftrightarrow (2x+1)(x-2)=0$. So $2x+1=0 \Leftrightarrow x=-\frac{1}{2}$ or $x=2$. If $x=-\frac{1}{2}$, then $\left(-\frac{1}{2}\right)^2 + y = 2 \Leftrightarrow y = \frac{7}{4}$. If $x=2$, then $2^2+y=2 \Leftrightarrow y=-2$. Thus, the solutions are $\left(-\frac{1}{2}, \frac{7}{4}\right)$ and $(2, -2)$.

41. The boundary is a solid curve, so we have the inequality $x+y^2 \le 4$. We take the test point $(0, 0)$ and verify that it satisfies the inequality: $0+0^2 \le 4$.

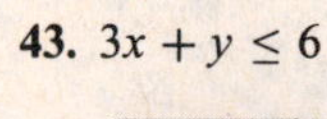

43. $3x+y \le 6$

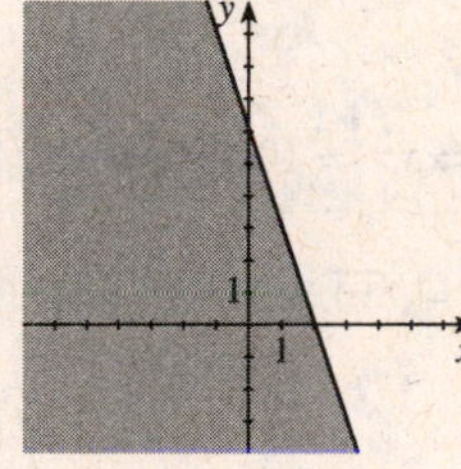

45. $x^2+y^2 > 9$

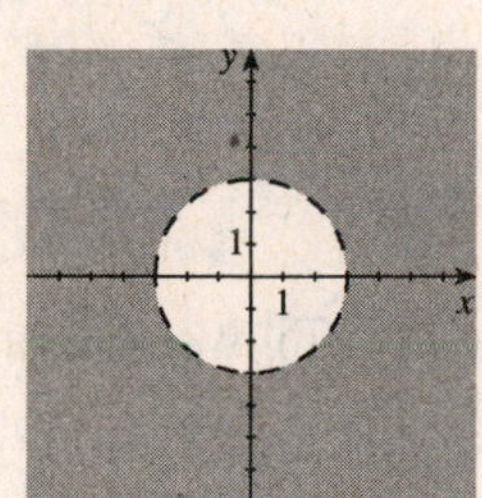

47. $\begin{cases} y \ge x^2-3x \\ y \le \frac{1}{3}x-1 \end{cases}$

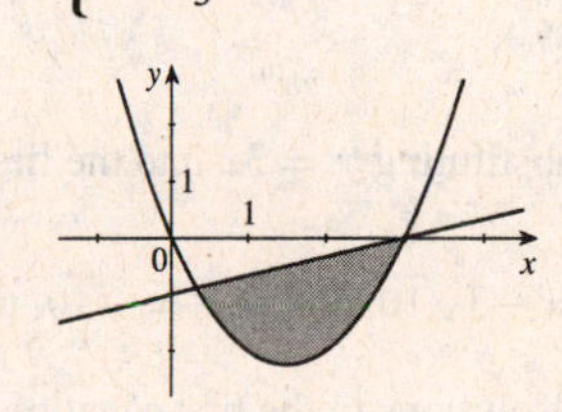

49. $\begin{cases} x+y \ge 2 \\ y-x \le 2 \\ x \le 3 \end{cases}$

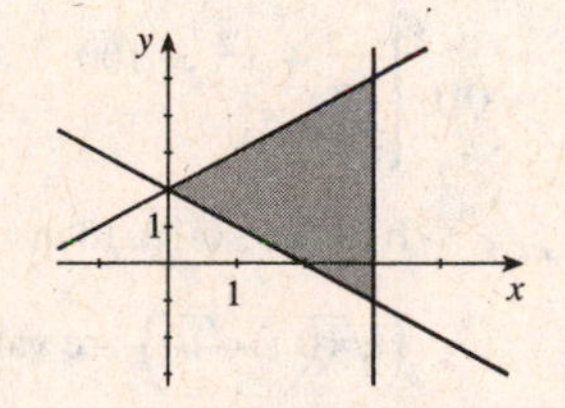

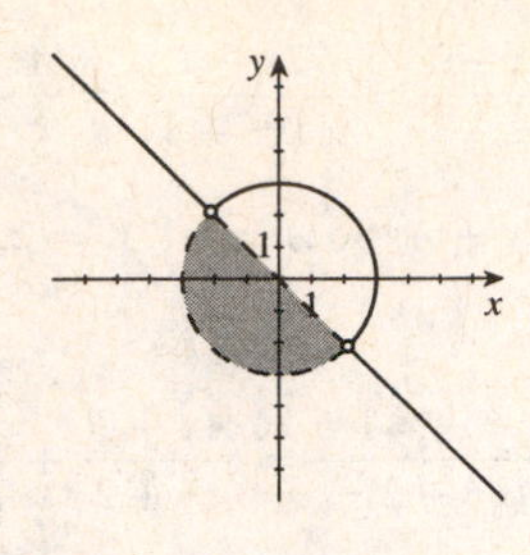

51. $\begin{cases} x^2 + y^2 < 9 \\ x + y < 0 \end{cases}$ The vertices occur where $y = -x$. By substitution, $x^2 + x^2 = 9 \Leftrightarrow x = \pm\frac{3}{\sqrt{2}}$, and so $y = \mp\frac{3}{\sqrt{2}}$. Therefore, the vertices are $\left(\frac{3}{\sqrt{2}}, -\frac{3}{\sqrt{2}}\right)$ and $\left(-\frac{3}{\sqrt{2}}, \frac{3}{\sqrt{2}}\right)$ and the solution set is bounded. The test point $(-1, -1)$ satisfies each inequality.

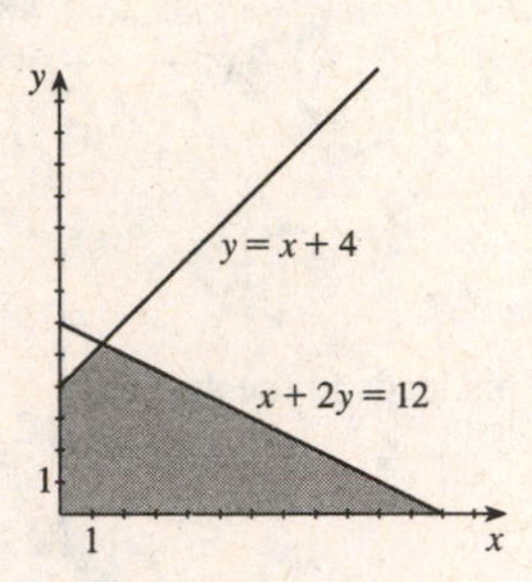

53. $\begin{cases} x \geq 0, y \geq 0 \\ x + 2y \leq 12 \\ y \leq x + 4 \end{cases}$ The intersection points are $(-4, 0)$, $(0, 4)$, $\left(\frac{4}{3}, \frac{16}{3}\right)$, $(0, 6)$, $(0, 0)$, and $(12, 0)$. Since the points $(-4, 0)$ and $(0, 6)$ are not in the solution set, the vertices are $(0, 4)$, $\left(\frac{4}{3}, \frac{16}{3}\right)$, $(12, 0)$, and $(0, 0)$. The solution set is bounded. The test point $(1, 1)$ satisfies each inequality.

55. $\begin{cases} -x + y + z = a \\ x - y + z = b \\ x + y - z = c \end{cases} \Leftrightarrow \begin{cases} -x + y + z = a \\ 2z = a + b \\ 2y = a + c \end{cases}$ Thus, $y = \dfrac{a+c}{2}$, $z = \dfrac{a+b}{2}$, and $-x + \dfrac{a+c}{2} + \dfrac{a+b}{2} = a \Leftrightarrow$ $x = \dfrac{b+c}{2}$. The solution is $\left(\dfrac{b+c}{2}, \dfrac{a+c}{2}, \dfrac{a+b}{2}\right)$.

57. Solving the second equation for y, we have $y = kx$. Substituting for $y$in the first equation gives us $x + kx = 12 \Leftrightarrow (1 + k)x = 12 \Leftrightarrow x = \dfrac{12}{k+1}$. Substituting for y in the third equation gives us $kx - x = 2k \Leftrightarrow$ $(k-1)x = 2k \Leftrightarrow x = \dfrac{2k}{k-1}$. These points of intersection are the same when the x-values are equal. Thus, $\dfrac{12}{k+1} = \dfrac{2k}{k-1}$ $\Leftrightarrow 12(k-1) = 2k(k+1) \Leftrightarrow 12k - 12 = 2k^2 + 2k \Leftrightarrow 0 = 2k^2 - 10k + 12 = 2\left(k^2 - 5k + 6\right) = 2(k-3)(k+2)$. Hence, $k = 2$ or $k = 3$.

CHAPTER 10 TEST

1. (a) The system is linear.

(b) $\begin{cases} x + 3y = 7 \\ 5x + 2y = -4 \end{cases}$ Multiplying the first equation by -5 and then adding gives $-13y = -39 \Leftrightarrow y = 3$. So $x + 3(3) = 7 \Leftrightarrow x = -2$. Thus, the solution is $(-2, 3)$.

3. (a) The system is nonlinear.

(b) $\begin{cases} x^2 + y^2 = 100 \\ y = 3x \end{cases}$ Substituting $y = 3x$ into the first equation gives $x^2 + (3x)^2 = 100 \Leftrightarrow x^2 = 10 \Leftrightarrow x = \pm\sqrt{10}$.

If $x = -\sqrt{10}$, then $y = -3\sqrt{10}$, and if $x = \sqrt{10}$, then $y = 3\sqrt{10}$. We can verify that $\left(-\sqrt{10}, -3\sqrt{10}\right)$ and $\left(\sqrt{10}, 3\sqrt{10}\right)$ are valid solutions to the first equation in the given system.

5. (a) $\begin{cases} x+2y+z=3 \\ x+3y+2z=3 \\ 2x+3y-z=8 \end{cases} \Leftrightarrow \begin{cases} x+2y+z=3 \\ y+z=0 & -\text{Eq. }1+\text{Eq. }2 \\ y+3z=-2 & 2\times\text{Eq. }1-\text{Eq. }3 \end{cases} \Leftrightarrow \begin{cases} x+2y+z=3 \\ y+z=0 \\ 2z=-2 & -\text{Eq. }2+\text{Eq. }3 \end{cases} \Rightarrow$

$z=-1$, so $y-1=0 \Leftrightarrow y=1$ and $z+2(1)+(-1)=3 \Leftrightarrow x=2$. Thus, the solution is $(2,1,-1)$.

(b) The system is neither inconsistent nor dependent.

7. (a) $\begin{cases} 2x-y+z=0 \\ 3x+2y-3z=1 \\ x-4y+5z=-1 \end{cases} \Leftrightarrow \begin{cases} 2x-y+z=0 \\ 7y-9z=2 & (-3)\times\text{Eq. }1+2\times\text{Eq. }2 \\ 7y-9z=2 & \text{Eq. }1+(-2)\times\text{Eq. }3 \end{cases} \Leftrightarrow \begin{cases} 2x-y+z=0 \\ 7y-9z=2 \\ 0=0 & \text{Eq. }2-\text{Eq. }3 \end{cases}$

Letting $z=t$, we have $7y-9t=2 \Leftrightarrow y=\frac{2}{7}+\frac{9}{7}t$, so $2x-\left(\frac{2}{7}+\frac{9}{7}t\right)+t=0 \Leftrightarrow x=\frac{1}{7}+\frac{1}{7}t$. The solutions are $\left(\frac{1}{7}+\frac{1}{7}t, \frac{2}{7}+\frac{9}{7}t, t\right)$ where t is any real number.

(b) The system is dependent.

9. Let w be the speed of the wind and a the speed of the airplane in still air, in kilometers per hour. Then the speed of the of the plane flying against the wind is $a-w$ and the speed of the plane flying with the wind is $a+w$. Using distance = rate × time, we get the system $\begin{cases} 600=2.5(a-w) \\ 300=\frac{50}{60}(a+w) \end{cases} \Leftrightarrow \begin{cases} 240=a-w \\ 360=a+w \end{cases}$ Adding the two equations, we get $600=2a \Leftrightarrow a=300$. So $360=300+w \Leftrightarrow w=60$.Thus the speed of the airplane in still air is 300 km/h and the speed of the wind is 60 km/h.

11. $3x+4y<6$

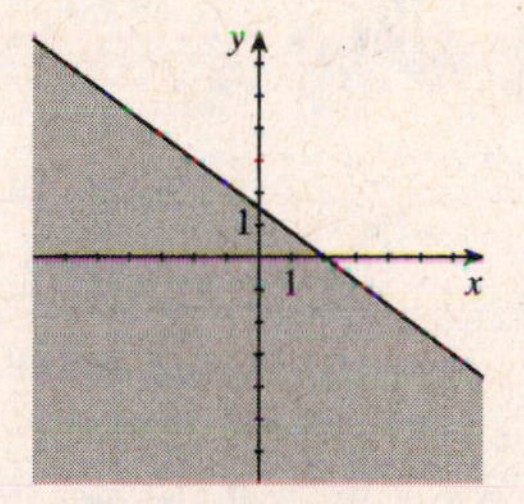

13. $\begin{cases} 2x+y\leq 8 \\ x-y\geq -2 \\ x+2y\geq 4 \end{cases}$ From the graph, the points $(4,0)$ and $(0,2)$ are vertices. The third vertex occurs where the lines $2x+y=8$ and $x-y=-2$ intersect. Adding these two equations gives $3x=6 \Leftrightarrow x=2$, and so $y=8-2(2)=4$. Thus, the third vertex is $(2,4)$.

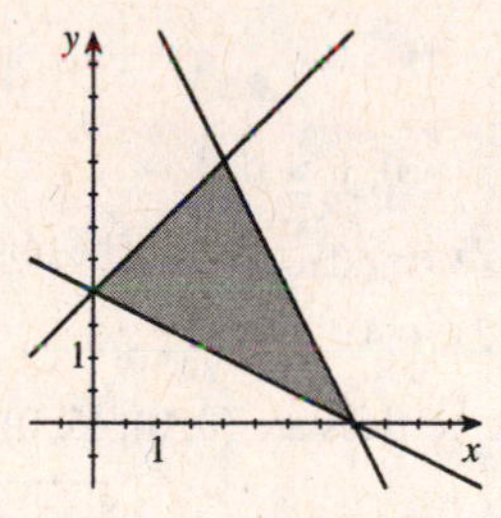

15. $\dfrac{4x-1}{(x-1)^2(x+2)}=\dfrac{A}{x-1}+\dfrac{B}{(x-1)^2}+\dfrac{C}{x+2}$. Thus,

$$\begin{aligned} 4x-1 &= A(x-1)(x+2)+B(x+2)+C(x-1)^2 = A\left(x^2+x-2\right)+B(x+2)+C\left(x^2-2x+1\right) \\ &= (A+C)x^2+(A+B-2C)x+(-2A+2B+C) \end{aligned}$$

which leads to the system of equations

$$\begin{cases} A \quad\quad + C = 0 \\ A + B - 2C = 4 \\ -2A + 2B + C = -1 \end{cases} \Leftrightarrow \begin{cases} A \quad + C = 0 \\ B - 3C = 4 \\ 2B + 3C = -1 \end{cases} \Leftrightarrow \begin{cases} A \quad + C = 0 \\ B - 3C = 4 \\ 9C = -9 \end{cases}$$

Therefore, $9C = -9 \Leftrightarrow C = -1$, $B - 3(-1) = 4 \Leftrightarrow B = 1$, and $A + (-1) = 0 \Leftrightarrow A = 1$. Therefore, $\dfrac{4x-1}{(x-1)^2(x+2)} = \dfrac{1}{x-1} + \dfrac{1}{(x-1)^2} - \dfrac{1}{x+2}$.

FOCUS ON MODELING Linear Programming

1.

Vertex	$M = 200 - x - y$
$(0,2)$	$200 - (0) - (2) = 198$
$(0,5)$	$200 - (0) - (5) = 195$
$(4,0)$	$200 - (4) - (0) = 196$

Thus, the maximum value is 198 and the minimum value is 195.

3. $\begin{cases} x \geq 0, y \geq 0 \\ 2x + y \leq 10 \\ 2x + 4y \leq 28 \end{cases}$ The objective function is $P = 140 - x + 3y$. From the graph, the vertices are $(0,0)$, $(5,0)$, $(2,6)$, and $(0,7)$.

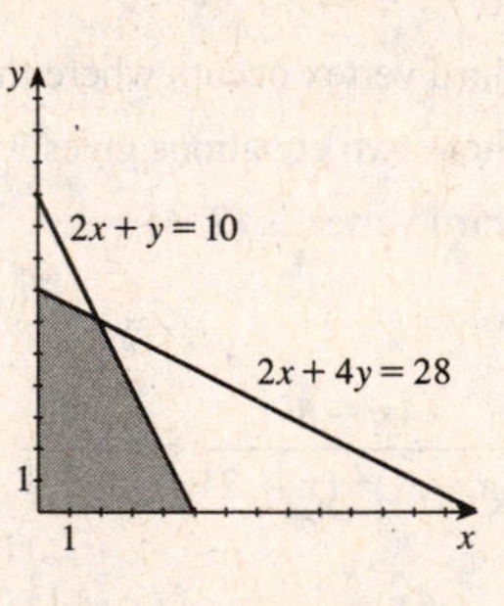

Vertex	$P = 140 - x + 3y$
$(0,0)$	$140 - (0) + 3(0) = 140$
$(5,0)$	$140 - (5) + 3(0) = 135$
$(2,6)$	$140 - (2) + 3(6) = 156$
$(0,7)$	$140 - (0) + 3(7) = 161$

Thus the maximum value is 161, and the minimum value is 135.

5. Let t be the number of tables made daily and c be the number of chairs made daily. Then the data given can be summarized by the following table:

	Tables t	Chairs c	Available time
Carpentry	2 h	3 h	108 h
Finishing	1 h	$\frac{1}{2}$ h	20 h
Profit	\$35	\$20	

Thus we wish to maximize the total profit $P = 35t + 20c$ subject to the constraints $\begin{cases} 2t + 3c \le 108 \\ t + \frac{1}{2}c \le 20 \\ t \ge 0, c \ge 0 \end{cases}$

From the graph, the vertices occur at (0, 0), (20, 0), (0, 36), and (3, 34).

Vertex	$P = 35t + 20c$
(0, 0)	$35(0) + 20(0) = 0$
(20, 0)	$35(20) + 20(0) = 700$
(0, 36)	$35(0) + 20(36) = 720$
(3, 34)	$35(3) + 20(34) = 785$

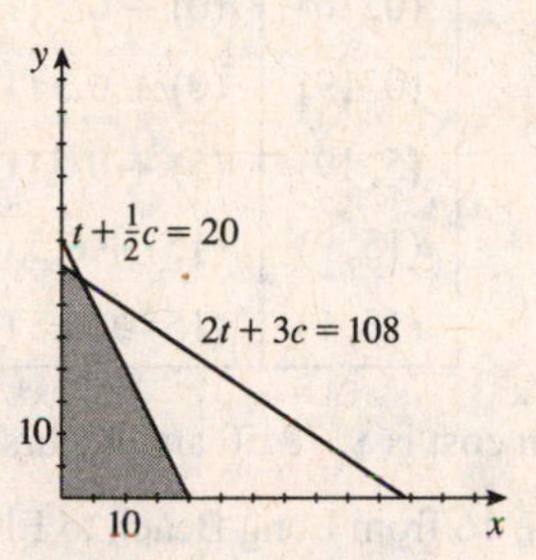

Hence, 3 tables and 34 chairs should be produced daily for a maximum profit of \$785.

7. Let x be the number of crates of oranges and y the number of crates of grapefruit. Then the data given can be summarized by the following table:

	Oranges	Grapefruit	Available
Volume	4 ft^3	6 ft^3	300 ft^3
Weight	80 lb	100 lb	5600 lb
Profit	\$2.50	\$4.00	

In addition, $x \ge y$. Thus we wish to maximize the total profit $P = 2.5x + 4y$ subject to the constraints $\begin{cases} x \ge 0, y \ge 0, x \ge y \\ 4x + 6y \le 300 \\ 80x + 100y \le 5600 \end{cases}$

From the graph, the vertices occur at (0, 0), (30, 30), (45, 20), and (70, 0).

Vertex	$P = 2.5x + 4y$
(0, 0)	$2.5(0) + 4(0) = 0$
(30, 30)	$2.5(30) + 4(30) = 195$
(45, 20)	$2.5(45) + 4(20) = 192.5$
(70, 0)	$2.5(70) + 4(0) = 175$

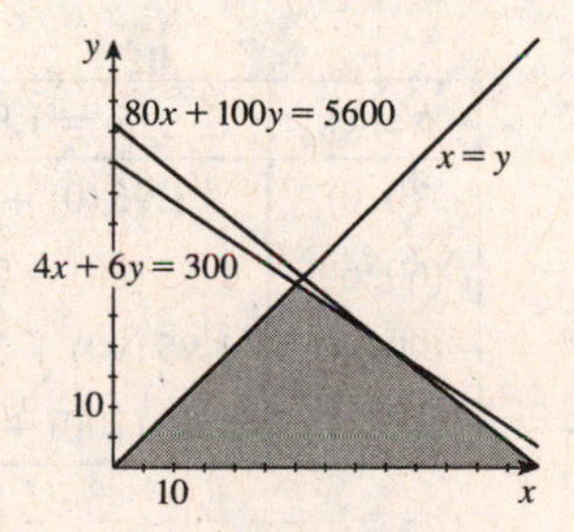

Thus, she should carry 30 crates of oranges and 30 crates of grapefruit for a maximum profit of \$195.

9. Let x be the number of stereo sets shipped from Long Beach to Santa Monica and y the number of stereo sets shipped from Long Beach to El Toro. Thus, $15 - x$ sets must be shipped to Santa Monica from Pasadena and $19 - y$ sets to El Toro from Pasadena. Thus, $x \geq 0$, $y \geq 0$, $15 - x \geq 0$, $19 - y \geq 0$, $x + y \leq 24$, and $(15 - x) + (19 - y) \leq 18$. Simplifying, we get

the constraints $\begin{cases} x \geq 0, y \geq 0 \\ x \leq 15, y \leq 19 \\ x + y \leq 24 \\ x + y \geq 16 \end{cases}$

The objective function is the cost $C = 5x + 6y + 4(15 - x) + 5.5(19 - y) = x + 0.5y + 164.5$, which we wish to minimize. From the graph, the vertices occur at (0, 16), (0, 19), (5, 19), (15, 9), and (15, 1).

Vertex	$C = x + 0.5y + 164.5$
(0, 16)	$(0) + 0.5(16) + 164.5 = 172.5$
(0, 19)	$(0) + 0.5(19) + 164.5 = 174$
(5, 19)	$(5) + 0.5(19) + 164.5 = 179$
(15, 9)	$(15) + 0.5(9) + 164.5 = 184$
(15, 1)	$(15) + 0.5(1) + 164.5 = 180$

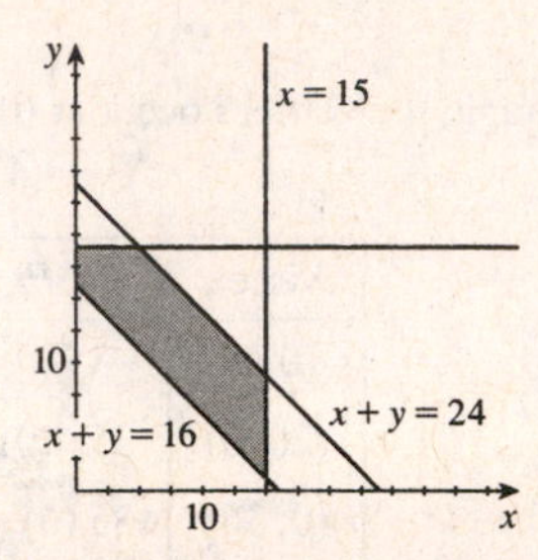

The minimum cost is \$172.50 and occurs when $x = 0$ and $y = 16$. Hence, no stereo should be shipped from Long Beach to Santa Monica, 16 from Long Beach to El Toro, 15 from Pasadena to Santa Monica, and 3 from Pasadena to El Toro.

11. Let x be the number of bags of standard mixtures and y be the number of bags of deluxe mixtures. Then the data can be summarized by the following table:

	Standard	Deluxe	Available
Cashews	100 g	150 g	15 kg
Peanuts	200 g	50 g	20 kg
Selling price	\$1.95	\$2.20	

Thus the total revenue, which we want to maximize, is given by $R = 1.95x + 2.25y$. We have the constraints

$\begin{cases} x \geq 0, y \geq 0, x \geq y \\ 0.1x + 0.15y \leq 15 \\ 0.2x + 0.05y \leq 20 \end{cases} \Leftrightarrow \begin{cases} x \geq 0, y \geq 0, x \geq y \\ 10x + 15y \leq 1500 \\ 20x + 5y \leq 2000 \end{cases}$

From the graph, the vertices occur at (0, 0), (60, 60), (90, 40), and (100, 0).

Vertex	$R = 1.95x + 2.25y$
(0, 0)	$1.96(0) + 2.25(0) = 0$
(60, 60)	$1.95(60) + 2.25(60) = 252$
(90, 40)	$1.95(90) + 2.25(40) = 265.5$
(100, 0)	$1.95(100) + 2.25(0) = 195$

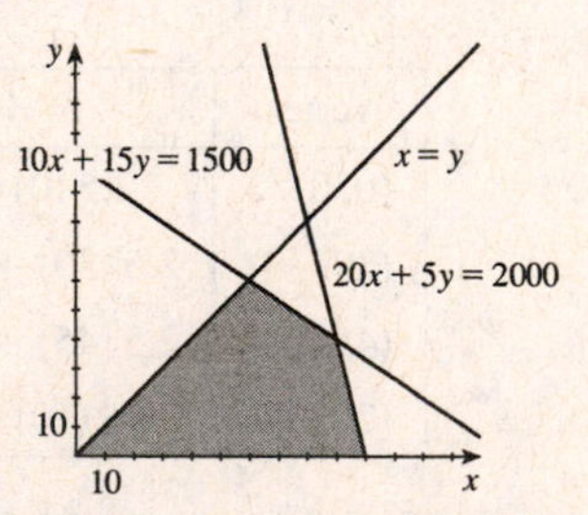

Hence, he should pack 90 bags of standard and 40 bags of deluxe mixture for a maximum revenue of \$265.50.

13. Let x be the amount in municipal bonds and y the amount in bank certificates, both in dollars. Then $12000 - x - y$ is the amount in high-risk bonds. So our constraints can be stated as

$$\begin{cases} x \geq 0, y \geq 0, x \geq 3y \\ 12{,}000 - x - y \geq 0 \\ 12{,}000 - x - y \leq 2000 \end{cases} \Leftrightarrow \begin{cases} x \geq 0, y \geq 0, x \geq 3y \\ x + y \leq 12{,}000 \\ x + y \geq 10{,}000 \end{cases}$$

From the graph, the vertices occur at $(7500, 2500)$, $(10000, 0)$, $(12000, 0)$, and $(9000, 3000)$. The objective function is $P = 0.07x + 0.08y + 0.12(12000 - x - y) = 1440 - 0.05x - 0.04y$, which we wish to maximize.

Vertex	$P = 1440 - 0.05x - 0.04y$
$(7500, 2500)$	$1440 - 0.05(7500) - 0.04(2500) = 965$
$(10000, 0)$	$1440 - 0.05(10{,}000) - 0.04(0) = 940$
$(12000, 0)$	$1440 - 0.05(12{,}000) - 0.04(0) = 840$
$(9000, 3000)$	$1440 - 0.05(9000) - 0.04(3000) = 870$

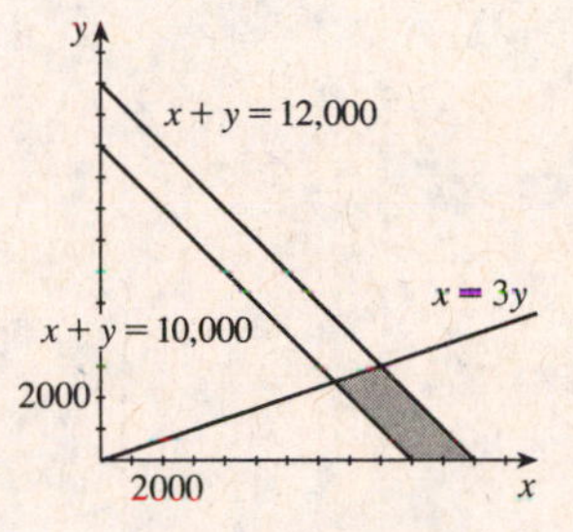

Hence, she should invest \$7500 in municipal bonds, \$2500 in bank certificates, and the remaining \$2000 in high-risk bonds for a maximum yield of \$965.

15. Let g be the number of games published and e be the number of educational programs published. Then the number of utility programs published is $36 - g - e$. Hence we wish to maximize profit, $P = 5000g + 8000e + 6000(36 - g - e) = 216{,}000 - 1000g + 2000e$, subject to the constraints

$$\begin{cases} g \geq 4, e \geq 0 \\ 36 - g - e \geq 0 \\ 36 - g - e \leq 2e \end{cases} \Leftrightarrow \begin{cases} g \geq 4, e \geq 0 \\ g + e \leq 36 \\ g + 3e \geq 36. \end{cases}$$

From the graph, the vertices are at $\left(4, \frac{32}{3}\right)$, $(4, 32)$, and $(36, 0)$. The objective function is $P = 216{,}000 - 1000g + 2000e$.

Vertex	$P = 216{,}000 - 1000g + 2000e$
$\left(4, \frac{32}{3}\right)$	$216{,}000 - 1000(4) + 2000\left(\frac{32}{3}\right) = 233{,}333.33$
$(4, 32)$	$216{,}000 - 1000(4) + 2000(32) = 276{,}000$
$(36, 0)$	$216{,}000 - 1000(36) + 2000(0) = 180{,}000$

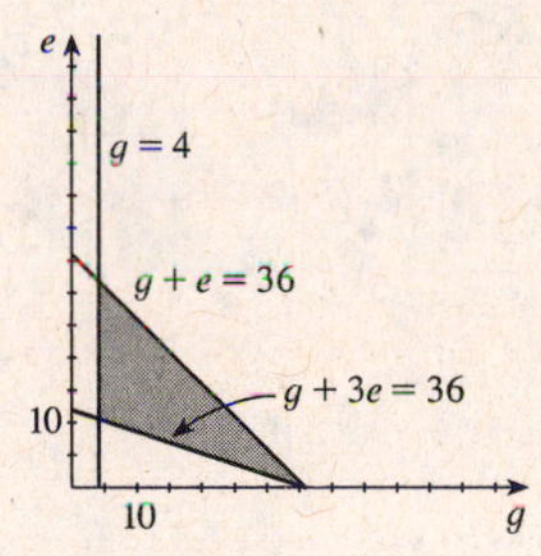

So, they should publish 4 games, 32 educational programs, and no utility program for a maximum profit of \$276,000 annually.

11 MATRICES AND DETERMINANTS

11.1 MATRICES AND SYSTEMS OF LINEAR EQUATIONS

1. A system of linear equations with infinitely many solutions is called *dependent*. A system of linear equations with no solution is called *inconsistent*.

3. **(a)** The leading variables are x *and* y.

(b) The system is dependent.

(c) The solution of the system is $x = 3 + t$, $y = 5 - 2t$, $z = t$.

5. 3×2

7. 2×1

9. 1×3

11. $\begin{bmatrix} 3 & 1 & -1 & 2 \\ 2 & -1 & 0 & 1 \\ 1 & 0 & -1 & 3 \end{bmatrix}$

13. **(a)** Yes, this matrix is in row-echelon form.

(b) Yes, this matrix is in reduced row-echelon form.

(c) $\begin{cases} x = -3 \\ y = 5 \end{cases}$

15. **(a)** Yes, this matrix is in row-echelon form.

(b) No, this matrix is not in reduced row-echelon form, since the leading 1 in the second row does not have a zero above it.

(c) $\begin{cases} x + 2y + 8z = 0 \\ \qquad y + 3z = 2 \\ \qquad\qquad 0 = 0 \end{cases}$

17. **(a)** No, this matrix is not in row-echelon form, since the row of zeros is not at the bottom.

(b) No, this matrix is not in reduced row-echelon form.

(c) $\begin{cases} x \qquad\quad = 0 \\ \qquad\quad 0 = 0 \\ \quad y + 5z = 1 \end{cases}$

19. **(a)** Yes, this matrix is in row-echelon form.

(b) Yes, this matrix is in reduced row-echelon form.

(c) $\begin{cases} x + 3y \qquad - w = 0 \\ \qquad\quad z + 2w = 0 \\ \qquad\qquad\quad 0 = 1 \\ \qquad\qquad\quad 0 = 0 \end{cases}$

Notice that this system has no solution.

21. $\begin{bmatrix} -1 & 1 & 2 & 0 \\ 3 & 1 & 1 & 4 \\ 1 & -2 & -1 & -1 \end{bmatrix} \xrightarrow{3R_1 + R_2 \to R_2} \begin{bmatrix} -1 & 1 & 2 & 0 \\ 0 & 4 & 7 & 4 \\ 1 & -2 & -1 & -1 \end{bmatrix}$

23. $\begin{bmatrix} 2 & 1 & -3 & 5 \\ 2 & 3 & 1 & 13 \\ 6 & -5 & -1 & 7 \end{bmatrix} \xrightarrow{-3R_1 + R_3 \to R_3} \begin{bmatrix} 2 & 1 & -3 & 5 \\ 2 & 3 & 1 & 13 \\ 0 & -8 & 8 & -8 \end{bmatrix}$

25. (a) $\begin{cases} x - 2y + 4z = 3 \\ y + 2z = 7 \\ z = 2 \end{cases}$

(b) $y + 2(2) = 7 \Leftrightarrow y = 3$, so $x - 2(3) + 4(2) = 3 \Leftrightarrow x = 1$. The solution is $(1, 3, 2)$.

27. (a) $\begin{cases} x + 2y + 3z - w = 7 \\ y - 2z = 5 \\ z + 2w = 5 \\ w = 3 \end{cases}$

(b) $z + 2(3) = 5 \Leftrightarrow z = -1$, so $y - 2(-1) = 5 \Leftrightarrow y = 3$ and $x + 2(3) + 3(-1) - 3 = 7 \Leftrightarrow x = 7$. The solution is $(7, 3, -1, 3)$.

29. $\begin{bmatrix} 1 & -2 & 1 & 1 \\ 0 & 1 & 2 & 5 \\ 1 & 1 & 3 & 8 \end{bmatrix} \xrightarrow{R_3 - R_1 \to R_3} \begin{bmatrix} 1 & -2 & 1 & 1 \\ 0 & 1 & 2 & 5 \\ 0 & 3 & 2 & 7 \end{bmatrix} \xrightarrow{R_3 - 3R_2 \to R_3} \begin{bmatrix} 1 & -2 & 1 & 1 \\ 0 & 1 & 2 & 5 \\ 0 & 0 & -4 & -8 \end{bmatrix}$. Thus, $-4z = -8 \Leftrightarrow z = 2$; $y + 2(2) = 5 \Leftrightarrow y = 1$; and $x - 2(1) + (2) = 1 \Leftrightarrow x = 1$. Therefore, the solution is $(1, 1, 2)$.

31. $\begin{bmatrix} 1 & 1 & 1 & 2 \\ 2 & -3 & 2 & 4 \\ 4 & 1 & -3 & 1 \end{bmatrix} \xrightarrow[R_3 - 4R_1 \to R_3]{R_2 - 2R_1 \to R_2} \begin{bmatrix} 1 & 1 & 1 & 2 \\ 0 & -5 & 0 & 0 \\ 0 & -3 & -7 & -7 \end{bmatrix} \xrightarrow{R_3 - \frac{3}{5}R_2 \to R_3} \begin{bmatrix} 1 & 1 & 1 & 2 \\ 0 & -5 & 0 & 0 \\ 0 & 0 & -7 & -7 \end{bmatrix}$.

Thus, $-7z = -7 \Leftrightarrow z = 1$; $-5y = 0 \Leftrightarrow y = 0$; and $x + 0 + 1 = 2 \Leftrightarrow x = 1$. Therefore, the solution is $(1, 0, 1)$.

33. $\begin{bmatrix} 1 & 2 & -1 & -2 \\ 1 & 0 & 1 & 0 \\ 2 & -1 & -1 & -3 \end{bmatrix} \xrightarrow[R_3 - 2R_1 \to R_3]{R_2 - R_1 \to R_2} \begin{bmatrix} 1 & 2 & -1 & -2 \\ 0 & -2 & 2 & 2 \\ 0 & -5 & 1 & 1 \end{bmatrix} \xrightarrow{-\frac{1}{2}R_2} \begin{bmatrix} 1 & 2 & -1 & -2 \\ 0 & 1 & 1 & 1 \\ 0 & -5 & 1 & 1 \end{bmatrix} \xrightarrow{R_3 + 5R_2 \to R_3}$

$\begin{bmatrix} 1 & 2 & -1 & -2 \\ 0 & 1 & 1 & 1 \\ 0 & 0 & 6 & 6 \end{bmatrix}$. Thus, $6z = 6 \Leftrightarrow z = 1$; $y + (1) = 1 \Leftrightarrow y = 0$; and $x + 2(0) - (1) = -2 \Leftrightarrow x = -1$. Therefore, the solution is $(-1, 0, 1)$.

35. $\begin{bmatrix} 1 & 2 & -1 & 9 \\ 2 & 0 & -1 & -2 \\ 3 & 5 & 2 & 22 \end{bmatrix} \xrightarrow[R_3 - 3R_1 \to R_3]{R_2 - 2R_1 \to R_2} \begin{bmatrix} 1 & 2 & -1 & 9 \\ 0 & -4 & 1 & -20 \\ 0 & -1 & 5 & -5 \end{bmatrix} \xrightarrow{4R_3 - R_2 \to R_3} \begin{bmatrix} 1 & 2 & -1 & 9 \\ 0 & -4 & 1 & -20 \\ 0 & 0 & 19 & 0 \end{bmatrix}$.

Thus, $19x_3 = 0 \Leftrightarrow x_3 = 0$; $-4x_2 = -20 \Leftrightarrow x_2 = 5$; and $x_1 + 2(5) = 9 \Leftrightarrow x_1 = -1$. Therefore, the solution is $(-1, 5, 0)$.

37. $\begin{bmatrix} 2 & -3 & -1 & 13 \\ -1 & 2 & -5 & 6 \\ 5 & -1 & -1 & 49 \end{bmatrix} \xrightarrow[2R_3 - 5R_1 \to R_3]{2R_2 + R_1 \to R_2} \begin{bmatrix} 2 & -3 & -1 & 13 \\ 0 & 1 & -11 & 25 \\ 0 & 13 & 3 & 33 \end{bmatrix} \xrightarrow{R_3 - 13R_2 \to R_3} \begin{bmatrix} 2 & -3 & -1 & 13 \\ 0 & 1 & -11 & 25 \\ 0 & 0 & 146 & -292 \end{bmatrix}$.

Thus, $146z = -292 \Leftrightarrow z = -2$; $y - 11(-2) = 25 \Leftrightarrow y = 3$; and $2x - 3 \cdot 3 + 2 = 13 \Leftrightarrow x = 10$. Therefore, the solution is $(10, 3, -2)$.

39. $\begin{bmatrix} 1 & 1 & 1 & 2 \\ 0 & 1 & -3 & 1 \\ 2 & 1 & 5 & 0 \end{bmatrix} \xrightarrow{R_3 - 2R_1 \to R_3} \begin{bmatrix} 1 & 1 & 1 & 2 \\ 0 & 1 & -3 & 1 \\ 0 & -1 & 3 & -4 \end{bmatrix} \xrightarrow{R_3 + R_2 \to R_3} \begin{bmatrix} 1 & 1 & 1 & 3 \\ 0 & 1 & -3 & 1 \\ 0 & 0 & 0 & -3 \end{bmatrix}$. The third row of the matrix states $0 = -3$, which is impossible. Hence, the system is inconsistent, and there is no solution.

41. $\begin{bmatrix} 2 & -3 & -9 & -5 \\ 1 & 0 & 3 & 2 \\ -3 & 1 & -4 & -3 \end{bmatrix} \xrightarrow{R_1 \leftrightarrow R_2} \begin{bmatrix} 1 & 0 & 3 & 2 \\ 2 & -3 & -9 & -5 \\ -3 & 1 & -4 & -3 \end{bmatrix} \xrightarrow[R_3+3R_1 \to R_3]{R_2-2R_1 \to R_2} \begin{bmatrix} 1 & 0 & 3 & 2 \\ 0 & -3 & -15 & -9 \\ 0 & 1 & 5 & 3 \end{bmatrix} \xrightarrow{-\frac{1}{3}R_2}$

$\begin{bmatrix} 1 & 0 & 3 & 2 \\ 0 & 1 & 5 & 3 \\ 0 & 1 & 5 & 3 \end{bmatrix} \xrightarrow{R_3-R_2 \to R_3} \begin{bmatrix} 1 & 0 & 3 & 2 \\ 0 & 1 & 5 & 3 \\ 0 & 0 & 0 & 0 \end{bmatrix}$. Therefore, this system has infinitely many solutions, given by $x + 3t = 2$ $\Leftrightarrow x = 2 - 3t$, and $y + 5t = 3 \Leftrightarrow y = 3 - 5t$. Hence, the solutions are $(2 - 3t, 3 - 5t, t)$, where t is any real number.

43. $\begin{bmatrix} 1 & -1 & 3 & 3 \\ 4 & -8 & 32 & 24 \\ 2 & -3 & 11 & 4 \end{bmatrix} \xrightarrow[R_3-2R_1 \to R_3]{R_2-4R_1 \to R_2} \begin{bmatrix} 1 & -1 & 3 & 3 \\ 0 & -4 & 20 & 12 \\ 0 & -1 & 5 & -2 \end{bmatrix} \xrightarrow[R_3+R_2 \to R_3]{-\frac{1}{4}R_2} \begin{bmatrix} 1 & -1 & 3 & 3 \\ 0 & 1 & -5 & -3 \\ 0 & 0 & 0 & -5 \end{bmatrix}$. The third row of the matrix states $0 = -5$, which is impossible. Hence, the system is inconsistent, and there is no solution.

45. $\begin{bmatrix} 1 & 4 & -2 & -3 \\ 2 & -1 & 5 & 12 \\ 8 & 5 & 11 & 30 \end{bmatrix} \xrightarrow[R_3-8R_1 \to R_3]{R_2-2R_1 \to R_2} \begin{bmatrix} 1 & 4 & -2 & -3 \\ 0 & -9 & 9 & 18 \\ 0 & -27 & 27 & 54 \end{bmatrix} \xrightarrow{R_3-3R_2 \to R_3} \begin{bmatrix} 1 & 4 & -2 & -3 \\ 0 & -9 & 9 & 18 \\ 0 & 0 & 0 & 0 \end{bmatrix}$.

Therefore, this system has infinitely many solutions, given by $-9y + 9t = 18 \Leftrightarrow y = -2 + t$, and $x + 4(-2 + t) - 2t = -3 \Leftrightarrow x = 5 - 2t$. Hence, the solutions are $(5 - 2t, -2 + t, t)$, where t is any real number.

47. $\begin{bmatrix} 2 & 1 & -2 & 12 \\ -1 & -\frac{1}{2} & 1 & -6 \\ 3 & \frac{3}{2} & -3 & 18 \end{bmatrix} \xrightarrow[-R_1]{R_1 \leftrightarrow R_2} \begin{bmatrix} 1 & \frac{1}{2} & -1 & 6 \\ 2 & 1 & -2 & 12 \\ 3 & \frac{3}{2} & -3 & 18 \end{bmatrix} \xrightarrow[R_3-3R_1 \to R_3]{R_2-2R_1 \to R_2} \begin{bmatrix} 1 & \frac{1}{2} & -1 & 6 \\ 0 & 0 & 0 & 0 \\ 0 & 0 & 0 & 0 \end{bmatrix}$.

Therefore, this system has infinitely many solutions, given by $x + \frac{1}{2}s - t = 6 \Leftrightarrow x = 6 - \frac{1}{2}s + t$. Hence, the solutions are $\left(6 - \frac{1}{2}s + t, s, t\right)$, where s and t are any real numbers.

49. $\begin{bmatrix} 4 & -3 & 1 & -8 \\ -2 & 1 & -3 & -4 \\ 1 & -1 & 2 & 3 \end{bmatrix} \xrightarrow{R_1 \leftrightarrow R_3} \begin{bmatrix} 1 & -1 & 2 & 3 \\ -2 & 1 & -3 & -4 \\ 4 & -3 & 1 & -8 \end{bmatrix} \xrightarrow[R_3-4R_1 \to R_3]{R_2+2R_1 \to R_2} \begin{bmatrix} 1 & -1 & 2 & 3 \\ 0 & -1 & 1 & 2 \\ 0 & 1 & -7 & -20 \end{bmatrix} \xrightarrow{-R_2}$

$\begin{bmatrix} 1 & -1 & 2 & 3 \\ 0 & 1 & -1 & -2 \\ 0 & 1 & -7 & -20 \end{bmatrix} \xrightarrow{R_3-R_2 \to R_3} \begin{bmatrix} 1 & -1 & 2 & 3 \\ 0 & 1 & -1 & -2 \\ 0 & 0 & -6 & -18 \end{bmatrix}$. Therefore, $-6z = -18 \Leftrightarrow z = 3$; $y - (3) = -2 \Leftrightarrow$ $y = 1$; and $x - (1) + 2(3) = 3 \Leftrightarrow x = -2$. Hence, the solution is $(-2, 1, 3)$.

51. $\begin{bmatrix} 2 & 1 & 3 & 9 \\ -1 & 0 & -7 & 10 \\ 3 & 2 & -1 & 4 \end{bmatrix} \xrightarrow[R_3+3R_2 \to R_3]{2R_2+R_1 \to R_2} \begin{bmatrix} 2 & 1 & 3 & 9 \\ 0 & 1 & -11 & 29 \\ 0 & 2 & -22 & 34 \end{bmatrix} \xrightarrow{2R_2-R_3 \to R_3} \begin{bmatrix} 2 & 1 & 3 & 9 \\ 0 & 1 & -11 & 29 \\ 0 & 0 & 0 & 24 \end{bmatrix}$.

Therefore, the system is inconsistent and there is no solution.

53. $\begin{bmatrix} 1 & 2 & -3 & -5 \\ -2 & -4 & -6 & 10 \\ 3 & 7 & -2 & -13 \end{bmatrix} \xrightarrow[R_3-3R_1 \to R_3]{R_2+2R_1 \to R_2} \begin{bmatrix} 1 & 2 & -3 & -5 \\ 0 & 0 & -12 & 0 \\ 0 & 1 & 7 & 2 \end{bmatrix} \xrightarrow{R_2 \leftrightarrow R_3} \begin{bmatrix} 1 & 2 & -3 & -5 \\ 0 & 1 & 7 & 2 \\ 0 & 0 & -12 & 0 \end{bmatrix}$. Therefore, $-12z = 0 \Leftrightarrow z = 0$; $y + 7(0) = 2 \Leftrightarrow y = 2$; and $x + 2(2) - 3(0) = -5 \Leftrightarrow x = -9$. Hence, the solution is $(-9, 2, 0)$.

55. $\begin{bmatrix} 1 & -1 & 6 & 8 \\ 1 & 0 & 1 & 5 \\ 1 & 3 & -14 & -4 \end{bmatrix} \xrightarrow[R_3 - R_1 \to R_3]{R_2 - R_1 \to R_2} \begin{bmatrix} 1 & -1 & 6 & 8 \\ 0 & 1 & -5 & -3 \\ 0 & 4 & -20 & -12 \end{bmatrix} \xrightarrow{R_3 - 4R_2 \to R_3} \begin{bmatrix} 1 & -1 & 6 & 8 \\ 0 & 1 & -5 & -3 \\ 0 & 0 & 0 & 0 \end{bmatrix}$. Therefore, the system is dependent. Let $z = t$. Then $y - 5t = -3 \Rightarrow y = -3 + 5t$ and $x - y + 6t = 8 \Leftrightarrow x = 8 + (-3 + 5t) - 6t = 5 - t$. The solutions are $(5 - t, -3 + 5t, t)$, where t is any real number.

57. $\begin{bmatrix} -1 & 2 & 1 & -3 & 3 \\ 3 & -4 & 1 & 1 & 9 \\ -1 & -1 & 1 & 1 & 0 \\ 2 & 1 & 4 & -2 & 3 \end{bmatrix} \xrightarrow{-R_1} \begin{bmatrix} 1 & -2 & -1 & 3 & -3 \\ 3 & -4 & 1 & 1 & 9 \\ -1 & -1 & 1 & 1 & 0 \\ 2 & 1 & 4 & -2 & 3 \end{bmatrix} \xrightarrow[\substack{R_3 + R_1 \to R_3 \\ R_4 - 2R_1 \to R_4}]{R_2 - 3R_1 \to R_2} \begin{bmatrix} 1 & -2 & -1 & 3 & -3 \\ 0 & 2 & 4 & -8 & 18 \\ 0 & -3 & 0 & 4 & -3 \\ 0 & 5 & 6 & -8 & 9 \end{bmatrix}$

$\xrightarrow{\frac{1}{2}R_2} \begin{bmatrix} 1 & -2 & -1 & 3 & -3 \\ 0 & 1 & 2 & -4 & 9 \\ 0 & -3 & 0 & 4 & -3 \\ 0 & 5 & 6 & -8 & 9 \end{bmatrix} \xrightarrow[R_4 - 5R_2 \to R_4]{R_3 + 3R_2 \to R_3} \begin{bmatrix} 1 & -2 & -1 & 3 & -3 \\ 0 & 1 & 2 & -4 & 9 \\ 0 & 0 & 6 & -8 & 24 \\ 0 & 0 & -4 & 12 & -36 \end{bmatrix} \xrightarrow{3R_4 + 2R_3 \to R_4}$

$\begin{bmatrix} 1 & -2 & -1 & 3 & -3 \\ 0 & 1 & 2 & -4 & 9 \\ 0 & 0 & 6 & -8 & 24 \\ 0 & 0 & 0 & 20 & -60 \end{bmatrix}$. Therefore, $20w = -60 \Leftrightarrow w = -3$; $6z + 24 = 24 \Leftrightarrow z = 0$. Then $y + 12 = 9 \Leftrightarrow y = -3$ and $x + 6 - 9 = -3 \Leftrightarrow x = 0$. Hence, the solution is $(0, -3, 0, -3)$.

59. $\begin{bmatrix} 1 & 1 & 2 & -1 & -2 \\ 0 & 3 & 1 & 2 & 2 \\ 1 & 1 & 0 & 3 & 2 \\ -3 & 0 & 1 & 2 & 5 \end{bmatrix} \xrightarrow[R_4 + 3R_1 \to R_4]{R_3 - R_1 \to R_3} \begin{bmatrix} 1 & 1 & 2 & -1 & -2 \\ 0 & 3 & 1 & 2 & 2 \\ 0 & 0 & -2 & 4 & 4 \\ 0 & 3 & 7 & -1 & -1 \end{bmatrix} \xrightarrow{R_4 - R_2 \to R_4} \begin{bmatrix} 1 & 1 & 2 & -1 & -2 \\ 0 & 3 & 1 & 2 & 2 \\ 0 & 0 & -2 & 4 & 4 \\ 0 & 0 & 6 & -3 & -3 \end{bmatrix}$

$\xrightarrow{R_4 + 3R_3 \to R_4} \begin{bmatrix} 1 & 1 & 2 & -1 & -2 \\ 0 & 3 & 1 & 2 & 2 \\ 0 & 0 & -2 & 4 & 4 \\ 0 & 0 & 0 & 9 & 9 \end{bmatrix}$. Therefore, $9w = 9 \Leftrightarrow w = 1$; $-2z + 4(1) = 4 \Leftrightarrow z = 0$. Then $3y + (0) + 2(1) = 2 \Leftrightarrow y = 0$ and $x + (0) + 2(0) - (1) = -2 \Leftrightarrow x = -1$. Hence, the solution is $(-1, 0, 0, 1)$.

61. $\begin{bmatrix} 1 & -1 & 0 & 1 & 0 \\ 3 & 0 & -1 & 2 & 0 \\ 1 & -4 & 1 & 2 & 0 \end{bmatrix} \xrightarrow[R_3 - R_1 \to R_3]{R_2 - 3R_1 \to R_2} \begin{bmatrix} 1 & -1 & 0 & 1 & 0 \\ 0 & 3 & -1 & -1 & 0 \\ 0 & -3 & 1 & 1 & 0 \end{bmatrix} \xrightarrow{R_3 + R_2 \to R_3} \begin{bmatrix} 1 & -1 & 0 & 1 & 0 \\ 0 & 3 & -1 & -1 & 0 \\ 0 & 0 & 0 & 0 & 0 \end{bmatrix}$.

Therefore, the system has infinitely many solutions, given by $3y - s - t = 0 \Leftrightarrow y = \frac{1}{3}(s + t)$ and $x - \frac{1}{3}(s + t) + t = 0 \Leftrightarrow x = \frac{1}{3}(s - 2t)$. So the solutions are $\left(\frac{1}{3}(s - 2t), \frac{1}{3}(s + t), s, t\right)$, where s and t are any real numbers.

63. $\begin{bmatrix} 1 & 0 & 1 & 1 & 4 \\ 0 & 1 & -1 & 0 & -4 \\ 1 & -2 & 3 & 1 & 12 \\ 2 & 0 & -2 & 5 & -1 \end{bmatrix} \xrightarrow[\text{R}_4 - 2\text{R}_1 \to \text{R}_4]{\text{R}_3 - \text{R}_1 \to \text{R}_3} \begin{bmatrix} 1 & 0 & 1 & 1 & 4 \\ 0 & 1 & -1 & 0 & -4 \\ 0 & -2 & 2 & 0 & 8 \\ 0 & 0 & -4 & 3 & -9 \end{bmatrix} \xrightarrow{\text{R}_3 + 2\text{R}_2 \to \text{R}_3} \begin{bmatrix} 1 & 0 & 1 & 1 & 4 \\ 0 & 1 & -1 & 0 & -4 \\ 0 & 0 & 0 & 0 & 0 \\ 0 & 0 & -4 & 3 & -9 \end{bmatrix}$

$\xrightarrow{\text{R}_3 \leftrightarrow -\text{R}_4} \begin{bmatrix} 1 & 0 & 1 & 1 & 4 \\ 0 & 1 & -1 & 0 & -4 \\ 0 & 0 & 4 & -3 & 9 \\ 0 & 0 & 0 & 0 & 0 \end{bmatrix}$. Therefore, $4z - 3t = 9 \Leftrightarrow 4z = 9 + 3t \Leftrightarrow z = \frac{9}{4} + \frac{3}{4}t$. Then we have $y - \left(\frac{9}{4} + \frac{3}{4}t\right) = -4 \Leftrightarrow y = \frac{-7}{4} + \frac{3}{4}t$ and $x + \left(\frac{9}{4} + \frac{3}{4}t\right) + t = 4 \Leftrightarrow x = \frac{7}{4} - \frac{7}{4}t$. Hence, the solutions are $\left(\frac{7}{4} - \frac{7}{4}t, -\frac{7}{4} + \frac{3}{4}t, \frac{9}{4} + \frac{3}{4}t, t\right)$, where t is any real number.

65. Using `rref` on the matrix $\begin{bmatrix} 0.75 & -3.75 & 2.95 & 4.0875 \\ 0.95 & -8.75 & 0 & 3.375 \\ 1.25 & -0.15 & 2.75 & 3.6625 \end{bmatrix}$, we find that the solution is $x = 1.25$, $y = -0.25$, $z = 0.75$.

67. Using `rref` on the matrix $\begin{bmatrix} 42 & -31 & 0 & -42 & -0.4 \\ -6 & 0 & 0 & -9 & 4.5 \\ 35 & 0 & -67 & 32 & 348.8 \\ 0 & 31 & 48 & -52 & -76.6 \end{bmatrix}$, we find that the solution is $x = 1.2$, $y = 3.4$, $z = -5.2$, $w = -1.3$.

69. Let x, y, z represent the number of VitaMax, Vitron, and VitaPlus pills taken daily. The matrix representation for the system of equations is

$\begin{bmatrix} 5 & 10 & 15 & 50 \\ 15 & 20 & 0 & 50 \\ 10 & 10 & 10 & 50 \end{bmatrix} \xrightarrow[\frac{1}{5}\text{R}_3]{\frac{1}{5}\text{R}_1,\ \frac{1}{5}\text{R}_2} \begin{bmatrix} 1 & 2 & 3 & 10 \\ 3 & 4 & 0 & 10 \\ 2 & 2 & 2 & 10 \end{bmatrix} \xrightarrow[\text{R}_3 - 2\text{R}_1 \to \text{R}_3]{\text{R}_2 - 3\text{R}_1 \to \text{R}_2} \begin{bmatrix} 1 & 2 & 3 & 10 \\ 0 & -2 & -9 & -20 \\ 0 & -2 & -4 & -10 \end{bmatrix} \xrightarrow{\text{R}_3 - \text{R}_2 \to \text{R}_3} \begin{bmatrix} 1 & 2 & 3 & 10 \\ 0 & -2 & -9 & -20 \\ 0 & 0 & 5 & 10 \end{bmatrix}$.

Thus, $5z = 10 \Leftrightarrow z = 2$; $-2y - 18 = -20 \Leftrightarrow y = 1$; and $x + 2 + 6 = 10 \Leftrightarrow x = 2$. Hence, he should take 2 VitaMax, 1 Vitron, and 2 VitaPlus pills daily.

71. Let x, y, and z represent the distance, in miles, of the run, swim, and cycle parts of the race respectively. Then, since $\text{time} = \dfrac{\text{distance}}{\text{speed}}$, we get the following equations from the three contestants' race times:

$\begin{cases} \left(\frac{x}{10}\right) + \left(\frac{y}{4}\right) + \left(\frac{z}{20}\right) = 2.5 \\ \left(\frac{x}{7.5}\right) + \left(\frac{y}{6}\right) + \left(\frac{z}{15}\right) = 3 \\ \left(\frac{x}{15}\right) + \left(\frac{y}{3}\right) + \left(\frac{z}{40}\right) = 1.75 \end{cases} \Leftrightarrow \begin{cases} 2x + 5y + z = 50 \\ 4x + 5y + 2z = 90 \\ 8x + 40y + 3z = 210 \end{cases}$ which has the following matrix representation:

$\begin{bmatrix} 2 & 5 & 1 & 50 \\ 4 & 5 & 2 & 90 \\ 8 & 40 & 3 & 210 \end{bmatrix} \xrightarrow[\text{R}_3 - 4\text{R}_1 \to \text{R}_3]{\text{R}_2 - 2\text{R}_1 \to \text{R}_2} \begin{bmatrix} 2 & 5 & 1 & 50 \\ 0 & -5 & 0 & -10 \\ 0 & 20 & -1 & 10 \end{bmatrix} \xrightarrow{\text{R}_3 + 4\text{R}_2 \to \text{R}_3} \begin{bmatrix} 2 & 5 & 1 & 50 \\ 0 & -5 & 0 & -10 \\ 0 & 0 & -1 & -30 \end{bmatrix}$.

Thus, $-z = -30 \Leftrightarrow z = 30$; $-5y = -10 \Leftrightarrow y = 2$; and $2x + 10 + 30 = 50 \Leftrightarrow x = 5$. So the race has a 5 mile run, 2 mile swim, and 30 mile cycle.

73. Let t be the number of tables produced, c the number of chairs, and a the number of armoires. Then, the system of equations is $\begin{cases} \frac{1}{2}t + c + a = 300 \\ \frac{1}{2}t + \frac{3}{2}c + a = 400 \\ t + \frac{3}{2}c + 2a = 590 \end{cases} \Leftrightarrow \begin{cases} t + 2c + 2a = 600 \\ t + 3c + 2a = 800 \\ 2t + 3c + 4a = 1180 \end{cases}$ and a matrix representation is

$$\begin{bmatrix} 1 & 2 & 2 & 600 \\ 1 & 3 & 2 & 800 \\ 2 & 3 & 4 & 1180 \end{bmatrix} \xrightarrow[R_3 - 2R_1 \to R_3]{R_2 - R_1 \to R_2} \begin{bmatrix} 1 & 2 & 2 & 600 \\ 0 & 1 & 0 & 200 \\ 0 & -1 & 0 & -20 \end{bmatrix} \xrightarrow{R_3 + R_2 \to R_3} \begin{bmatrix} 1 & 2 & 2 & 600 \\ 0 & 1 & 0 & 200 \\ 0 & 0 & 0 & 180 \end{bmatrix}.$$

The third row states $0 = 180$, which is impossible, and so the system is inconsistent. Therefore, it is impossible to use all of the available labor-hours.

75. *Line containing the points* $(0, 0)$ *and* $(1, 12)$: Using the general form of a line, $y = ax + b$, we substitute for x and y and solve for a and b. The point $(0, 0)$ gives $0 = a(0) + b \Rightarrow b = 0$; the point $(1, 12)$ gives $12 = a(1) + b \Rightarrow a = 12$. Since $a = 12$ and $b = 0$, the equation of the line is $y = 12x$.

Quadratic containing the points $(0, 0)$, $(1, 12)$, *and* $(3, 6)$: Using the general form of a quadratic, $y = ax^2 + bx + c$, we substitute for x and y and solve for a, b, and c. The point $(0, 0)$ gives $0 = a(0)^2 + b(0) + c \Rightarrow c = 0$; the point $(1, 12)$ gives $12 = a(1)^2 + b(1) + c \Rightarrow a + b = 12$; the point $(3, 6)$ gives $6 = a(3)^2 + b(3) + c \Rightarrow 9a + 3b = 6$. Subtracting the third equation from -3 times the third gives $6a = -30 \Leftrightarrow a = -5$. So $a + b = 12 \Leftrightarrow b = 12 - a \Rightarrow b = 17$. Since $a = -5$, $b = 17$, and $c = 0$, the equation of the quadratic is $y = -5x^2 + 17x$.

Cubic containing the points $(0, 0)$, $(1, 12)$, $(2, 40)$, *and* $(3, 6)$: Using the general form of a cubic, $y = ax^3 + bx^2 + cx + d$, we substitute for x and y and solve for a, b, c, and d. The point $(0, 0)$ gives $0 = a(0)^3 + b(0)^2 + c(0) + d \Rightarrow d = 0$; the point the point $(1, 12)$ gives $12 = a(1)^3 + b(1)^2 + c(1) + d \Rightarrow a + b + c + d = 12$; the point $(2, 40)$ gives $40 = a(2)^3 + b(2)^2 + c(2) + d \Rightarrow 8a + 4b + 2c + d = 40$; the point $(3, 6)$ gives $6 = a(3)^3 + b(3)^2 + c(3) + d \Rightarrow 27a + 9b + 3c + d = 6$. Since $d = 0$, the system reduces to $\begin{cases} a + b + c = 12 \\ 8a + 4b + 2c = 40 \\ 27a + 9b + 3c = 6 \end{cases}$ which has representation

$$\begin{bmatrix} 1 & 1 & 1 & 12 \\ 8 & 4 & 2 & 40 \\ 27 & 9 & 3 & 6 \end{bmatrix} \xrightarrow[R_3 - 27R_1 \to R_3]{R_2 - 8R_1 \to R_2} \begin{bmatrix} 1 & 1 & 1 & 12 \\ 0 & -4 & -6 & -56 \\ 0 & -18 & -24 & -318 \end{bmatrix} \xrightarrow[-\frac{1}{6}R_3]{-\frac{1}{2}R_2} \begin{bmatrix} 1 & 1 & 1 & 12 \\ 0 & 2 & 3 & 28 \\ 0 & 3 & 4 & 53 \end{bmatrix} \xrightarrow{2R_3 - 3R_2 \to R_3} \begin{bmatrix} 1 & 1 & 1 & 12 \\ 0 & 2 & 3 & 28 \\ 0 & 0 & -1 & 22 \end{bmatrix}.$$

So $c = -22$ and back-substituting we have $2b + 3(-22) = 28 \Leftrightarrow b = 47$ and $a + 47 + (-22) = 0 \Leftrightarrow a = -13$. So the cubic is $y = -13x^3 + 47x^2 - 22x$.

Fourth-degree polynomial containing the points $(0, 0)$, $(1, 12)$, $(2, 40)$, $(3, 6)$, *and* $(-1, -14)$: Using the general form of a fourth-degree polynomial, $y = ax^4 + bx^3 + cx^2 + dx + e$, we substitute for x and y and solve for a, b, c, d, and e. The point $(0, 0)$ gives $0 = a(0)^4 + b(0)^3 + c(0)^2 + d(0) + e \Rightarrow e = 0$; the point $(1, 12)$ gives $12 = a(1)^4 + b(1)^3 + c(1)^2 + d(1) + e$; the point $(2, 40)$ gives $40 = a(2)^4 + b(2)^3 + c(2)^2 + d(2) + e$; the point $(3, 6)$ gives $6 = a(3)^4 + b(3)^3 + c(3)^2 + d(3) + e$; the point $(-1, -14)$ gives $-14 = a(-1)^4 + b(-1)^3 + c(-1)^2 + d(-1) + e$.

Because the first equation is $e = 0$, we eliminate e from the other equations to get

$$\begin{cases} a + b + c + d = 12 \\ 16a + 8b + 4c + 2d = 40 \\ 81a + 27b + 9c + 3d = 6 \\ a - b + c - d = -14 \end{cases} \Leftrightarrow \begin{bmatrix} 1 & 1 & 1 & 1 & 12 \\ 16 & 8 & 4 & 2 & 40 \\ 81 & 27 & 9 & 3 & 6 \\ 1 & -1 & 1 & -1 & -14 \end{bmatrix} \xrightarrow[\substack{R_4 - R_1 \to R_4}]{\substack{R_2 - 16R_1 \to R_2 \\ R_3 - 81R_1 \to R_3}} \begin{bmatrix} 1 & 1 & 1 & 1 & 12 \\ 0 & -8 & -12 & -14 & -152 \\ 0 & -54 & -72 & -78 & -966 \\ 0 & -2 & 0 & -2 & -26 \end{bmatrix} \xrightarrow[\substack{R_3 \to R_4}]{\substack{-\frac{1}{2}R_4 \to R_2 \\ R_2 \to R_3}}$$

$$\begin{bmatrix} 1 & 1 & 1 & 1 & 12 \\ 0 & 1 & 0 & 1 & 13 \\ 0 & -8 & -12 & -14 & -152 \\ 0 & -54 & -72 & -78 & -966 \end{bmatrix} \xrightarrow[\substack{R_4 + 54R_2 \to R_4}]{\substack{R_3 + 8R_2 \to R_3}} \begin{bmatrix} 1 & 1 & 1 & 1 & 12 \\ 0 & 1 & 0 & 1 & 13 \\ 0 & 0 & -12 & -6 & -48 \\ 0 & 0 & -72 & -24 & -264 \end{bmatrix} \xrightarrow{R_4 - 6R_3 \to R_4} \begin{bmatrix} 1 & 1 & 1 & 1 & 12 \\ 0 & 1 & 0 & 1 & 13 \\ 0 & 0 & -12 & -6 & -48 \\ 0 & 0 & 0 & 12 & 24 \end{bmatrix}.$$

So $d = 2$. Then $-12c - 6(2) = -48 \Leftrightarrow c = 3$and $b + 2 = 13 \Leftrightarrow b = 11$. Finally, $a + 11 + 3 + 2 = 12 \Leftrightarrow a = -4$. So the fourth-degree polynomial containing these points is $y = -4x^4 + 11x^3 + 3x^2 + 2x$.

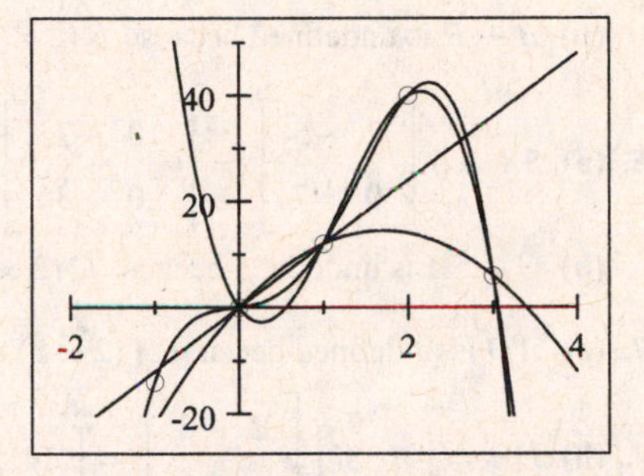

11.2 THE ALGEBRA OF MATRICES

1. We can add (or subtract) two matrices only if they have the same *dimension*.

3. (i) $A + A$ and (ii) $2A$ exist for all matrices A, but (iii) $A \cdot A$ is not defined when A is not square.

5. The matrices have different dimensions, so they cannot be equal.

7. All corresponding entries must be equal, so $a = -5$ and $b = 3$.

9. $\begin{bmatrix} 2 & 6 \\ -5 & 3 \end{bmatrix} + \begin{bmatrix} -1 & -3 \\ 6 & 2 \end{bmatrix} = \begin{bmatrix} 1 & 3 \\ 1 & 5 \end{bmatrix}$

11. $3 \begin{bmatrix} 1 & 2 \\ 4 & -1 \\ 1 & 0 \end{bmatrix} = \begin{bmatrix} 3 & 6 \\ 12 & -3 \\ 3 & 0 \end{bmatrix}$

13. $\begin{bmatrix} 2 & 6 \\ 1 & 3 \\ 2 & 4 \end{bmatrix} \begin{bmatrix} 1 & -2 \\ 3 & 6 \\ -2 & 0 \end{bmatrix}$ is undefined because these matrices have incompatible dimensions.

15. $\begin{bmatrix} 1 & 2 \\ -1 & 4 \end{bmatrix} \begin{bmatrix} 1 & -2 & 3 \\ 2 & 2 & -1 \end{bmatrix} = \begin{bmatrix} 5 & 2 & 1 \\ 7 & 10 & -7 \end{bmatrix}$

17. $2X + A = B \Leftrightarrow X = \frac{1}{2}(B - A) = \frac{1}{2}\left(\begin{bmatrix} 2 & 5 \\ 3 & 7 \end{bmatrix} - \begin{bmatrix} 4 & 6 \\ 1 & 3 \end{bmatrix}\right) = \frac{1}{2}\begin{bmatrix} -2 & -1 \\ 2 & 4 \end{bmatrix} = \begin{bmatrix} -1 & -\frac{1}{2} \\ 1 & 2 \end{bmatrix}$.

19. $2(B - X) = D$. Since B is a 2×2 matrix, $B - X$ is defined only when X is a 2×2 matrix, so $2(B - X)$ is a 2×2 matrix. But D is a 3×2 matrix. Thus, there is no solution.

21. $\frac{1}{5}(X + D) = C \Leftrightarrow X + D = 5C \Leftrightarrow$

$$X = 5C - D = 5\begin{bmatrix} 2 & 3 \\ 1 & 0 \\ 0 & 2 \end{bmatrix} - \begin{bmatrix} 10 & 20 \\ 30 & 20 \\ 10 & 0 \end{bmatrix} = \begin{bmatrix} 10 & 15 \\ 5 & 0 \\ 0 & 10 \end{bmatrix} - \begin{bmatrix} 10 & 20 \\ 30 & 20 \\ 10 & 0 \end{bmatrix} = \begin{bmatrix} 0 & -5 \\ -25 & -20 \\ -10 & 10 \end{bmatrix}.$$

In Solutions 23–35, the matrices A, B, C, D, E, F, G, and H are defined as follows:

$$A = \begin{bmatrix} 2 & -5 \\ 0 & 7 \end{bmatrix} \qquad B = \begin{bmatrix} 3 & \frac{1}{2} & 5 \\ 1 & -1 & 3 \end{bmatrix} \qquad C = \begin{bmatrix} 2 & -\frac{5}{2} & 0 \\ 0 & 2 & -3 \end{bmatrix} \qquad D = \begin{bmatrix} 7 & 3 \end{bmatrix}$$

$$E = \begin{bmatrix} 1 \\ 2 \\ 0 \end{bmatrix} \qquad F = \begin{bmatrix} 1 & 0 & 0 \\ 0 & 1 & 0 \\ 0 & 0 & 1 \end{bmatrix} \qquad G = \begin{bmatrix} 5 & -3 & 10 \\ 6 & 1 & 0 \\ -5 & 2 & 2 \end{bmatrix} \qquad H = \begin{bmatrix} 3 & 1 \\ 2 & -1 \end{bmatrix}$$

23. (a) $B + C = \begin{bmatrix} 3 & \frac{1}{2} & 5 \\ 1 & -1 & 3 \end{bmatrix} + \begin{bmatrix} 2 & -\frac{5}{2} & 0 \\ 0 & 2 & -3 \end{bmatrix} = \begin{bmatrix} 5 & -2 & 5 \\ 1 & 1 & 0 \end{bmatrix}$

(b) $B + F$ is undefined because B (2×3) and F (3×3) don't have the same dimensions.

25. (a) $5A = 5\begin{bmatrix} 2 & -5 \\ 0 & 7 \end{bmatrix} = \begin{bmatrix} 10 & -25 \\ 0 & 35 \end{bmatrix}$

(b) $C - 5A$ is undefined because C (2×3) and A (2×2) don't have the same dimensions.

27. (a) AD is undefined because A (2×2) and D (1×2) have incompatible dimensions.

(b) $DA = \begin{bmatrix} 7 & 3 \end{bmatrix}\begin{bmatrix} 2 & -5 \\ 0 & 7 \end{bmatrix} = \begin{bmatrix} 14 & -14 \end{bmatrix}$

29. (a) $AH = \begin{bmatrix} 2 & -5 \\ 0 & 7 \end{bmatrix}\begin{bmatrix} 3 & 1 \\ 2 & -1 \end{bmatrix} = \begin{bmatrix} -4 & 7 \\ 14 & -7 \end{bmatrix}$

(b) $\begin{bmatrix} 3 & 1 \\ 2 & -1 \end{bmatrix}\begin{bmatrix} 2 & -5 \\ 0 & 7 \end{bmatrix} = \begin{bmatrix} 6 & -8 \\ 4 & -17 \end{bmatrix}$

31. (a) $GF = \begin{bmatrix} 5 & -3 & 10 \\ 6 & 1 & 0 \\ -5 & 2 & 2 \end{bmatrix}\begin{bmatrix} 1 & 0 & 0 \\ 0 & 1 & 0 \\ 0 & 0 & 1 \end{bmatrix} = \begin{bmatrix} 5 & -3 & 10 \\ 6 & 1 & 0 \\ -5 & 2 & 2 \end{bmatrix}$

(b) $GE = \begin{bmatrix} 5 & -3 & 10 \\ 6 & 1 & 0 \\ -5 & 2 & 2 \end{bmatrix}\begin{bmatrix} 1 \\ 2 \\ 0 \end{bmatrix} = \begin{bmatrix} -1 \\ 8 \\ -1 \end{bmatrix}$

33. (a) $A^2 = \begin{bmatrix} 2 & -5 \\ 0 & 7 \end{bmatrix}\begin{bmatrix} 2 & -5 \\ 0 & 7 \end{bmatrix} = \begin{bmatrix} 4 & -45 \\ 0 & 49 \end{bmatrix}$

(b) $A^3 = \begin{bmatrix} 2 & -5 \\ 0 & 7 \end{bmatrix}\begin{bmatrix} 2 & -5 \\ 0 & 7 \end{bmatrix}\begin{bmatrix} 2 & -5 \\ 0 & 7 \end{bmatrix} = \begin{bmatrix} 4 & -45 \\ 0 & 49 \end{bmatrix}\begin{bmatrix} 2 & -5 \\ 0 & 7 \end{bmatrix} = \begin{bmatrix} 8 & -335 \\ 0 & 343 \end{bmatrix}$

35. (a) $ABE = \begin{bmatrix} 2 & -5 \\ 0 & 7 \end{bmatrix}\begin{bmatrix} 3 & \frac{1}{2} & 5 \\ 1 & -1 & 3 \end{bmatrix}\begin{bmatrix} 1 \\ 2 \\ 0 \end{bmatrix} = \begin{bmatrix} 1 & 6 & -5 \\ 7 & -7 & 21 \end{bmatrix}\begin{bmatrix} 1 \\ 2 \\ 0 \end{bmatrix} = \begin{bmatrix} 13 \\ -7 \end{bmatrix}$

(b) AHE is undefined because the dimensions of AH (2×2) and E (3×1) are incompatible.

In Solutions 37–41, the matrices A, B, and C are defined as follows:

$$A = \begin{bmatrix} 0.3 & 1.1 & 2.4 \\ 0.9 & -0.1 & 0.4 \\ -0.7 & 0.3 & -0.5 \end{bmatrix} \qquad B = \begin{bmatrix} 1.2 & -0.1 \\ 0 & -0.5 \\ 0.5 & -2.1 \end{bmatrix} \qquad C = \begin{bmatrix} -0.2 & 0.2 & 0.1 \\ 1.1 & 2.1 & -2.1 \end{bmatrix}$$

37. $AB = \begin{bmatrix} 1.56 & -5.62 \\ 1.28 & -0.88 \\ -1.09 & 0.97 \end{bmatrix}$ **39.** $BC = \begin{bmatrix} -0.35 & 0.03 & 0.33 \\ -0.55 & -1.05 & 1.05 \\ -2.41 & -4.31 & 4.46 \end{bmatrix}$

41. B and C have different dimensions, so $B + C$ is undefined.

43. $\begin{bmatrix} x & 2y \\ 4 & 6 \end{bmatrix} = \begin{bmatrix} 2 & -2 \\ 2x & -6y \end{bmatrix}$. Thus we must solve the system $\begin{cases} x = 2 \\ 2y = -2 \\ 4 = 2x \\ 6 = -6y \end{cases}$ So $x = 2$ and $2y = -2 \Leftrightarrow y = -1$. Since these values for x and y also satisfy the last two equations, the solution is $x = 2$, $y = -1$.

45. $2\begin{bmatrix} x & y \\ x + y & x - y \end{bmatrix} = \begin{bmatrix} 2 & -4 \\ -2 & 6 \end{bmatrix}$. Since $2\begin{bmatrix} x & y \\ x + y & x - y \end{bmatrix} = \begin{bmatrix} 2x & 2y \\ 2(x + y) & 2(x - y) \end{bmatrix}$, Thus we must solve the system $\begin{cases} 2x = 2 \\ 2y = -4 \\ 2(x + y) = -2 \\ 2(x - y) = 6 \end{cases}$ So $x = 1$ and $y = -2$. Since these values for x and y also satisfy the last two equations, the solution is $x = 1$, $y = -2$.

47. $\begin{cases} 2x - 5y = 7 \\ 3x + 2y = 4 \end{cases}$ written as a matrix equation is $\begin{bmatrix} 2 & -5 \\ 3 & 2 \end{bmatrix}\begin{bmatrix} x \\ y \end{bmatrix} = \begin{bmatrix} 7 \\ 4 \end{bmatrix}$.

49. $\begin{cases} 3x_1 + 2x_2 - x_3 + x_4 = 0 \\ x_1 - x_3 = 5 \\ 3x_2 + x_3 - x_4 = 4 \end{cases}$ written as a matrix equation is $\begin{bmatrix} 3 & 2 & -1 & 1 \\ 1 & 0 & -1 & 0 \\ 0 & 3 & 1 & -1 \end{bmatrix}\begin{bmatrix} x_1 \\ x_2 \\ x_3 \\ x_4 \end{bmatrix} = \begin{bmatrix} 0 \\ 5 \\ 4 \end{bmatrix}$.

51. $A = \begin{bmatrix} 1 & 0 & 6 & -1 \\ 2 & \frac{1}{2} & 4 & 0 \end{bmatrix}$, $B = \begin{bmatrix} 1 & 7 & -9 & 2 \end{bmatrix}$, and $C = \begin{bmatrix} 1 \\ 0 \\ -1 \\ -2 \end{bmatrix}$. ABC is undefined because the dimensions of A (2×4) and B (1×4) are not compatible. $ACB = \begin{bmatrix} -3 \\ -2 \end{bmatrix}\begin{bmatrix} 1 & 7 & -9 & 2 \end{bmatrix} = \begin{bmatrix} -3 & -21 & 27 & -6 \\ -2 & -14 & 18 & -4 \end{bmatrix}$. BAC is undefined because the dimensions of B (1×4) and A (2×4) are not compatible. BCA is undefined because the dimensions of C (4×1) and A (2×4) are not compatible. CAB is undefined because the dimensions of C (4×1) and A (2×4) are not compatible. CBA is undefined because the dimensions of B (1×4) and A (2×4) are not compatible.

53. (a) $AB = \begin{bmatrix} 0.75 & 0.10 & 0 \\ 0.25 & 0.70 & 0.70 \\ 0 & 0.20 & 0.30 \end{bmatrix} \begin{bmatrix} 4 \\ 20 \\ 10 \end{bmatrix} = \begin{bmatrix} 5 \\ 22 \\ 7 \end{bmatrix}$

(b) Five members of the group have no postsecondary education, 22 have 1 to 4 years, and seven have more than 4 years.

55. (a) $AB = \begin{bmatrix} 50 & 20 & 15 \\ 40 & 75 & 20 \\ 35 & 60 & 100 \end{bmatrix} \begin{bmatrix} 3.50 \\ 5.75 \\ 4.25 \end{bmatrix} = \begin{bmatrix} 353.75 \\ 656.25 \\ 892.50 \end{bmatrix}$

(b) The total revenue for Monday is the $(1, 1)$th entry of the product matrix, \$353.75.

(c) The total revenue is the sum of the three entries in the product matrix, \$1902.50.

57. (a) $AB = \begin{bmatrix} 12 & 10 & 0 \\ 4 & 4 & 20 \\ 8 & 9 & 12 \end{bmatrix} \begin{bmatrix} \$1000 & \$500 \\ \$2000 & \$1200 \\ \$1500 & \$1000 \end{bmatrix} = \begin{bmatrix} \$32{,}000 & \$18{,}000 \\ \$42{,}000 & \$26{,}800 \\ \$44{,}000 & \$26{,}800 \end{bmatrix}$

(b) The daily profit in January from the Biloxi plant is the $(2, 1)$ matrix entry, namely \$42,000.

(c) The total daily profit from all three plants in February was $\$18{,}000 + \$26{,}800 + \$26{,}800 = \$71{,}600$.

59. (a) $AC = \begin{bmatrix} 120 & 50 & 60 \\ 40 & 25 & 30 \\ 60 & 30 & 20 \end{bmatrix} \begin{bmatrix} 0.10 \\ 0.50 \\ 1.00 \end{bmatrix} = \begin{bmatrix} 97.00 \\ 46.50 \\ 41.00 \end{bmatrix}$ Amy's stand sold \$97 worth of produce on Saturday, Beth's stand sold \$46.50 worth, and Chad's stand sold \$41 worth.

(b) $BC = \begin{bmatrix} 100 & 60 & 30 \\ 35 & 20 & 20 \\ 60 & 25 & 30 \end{bmatrix} \begin{bmatrix} 0.10 \\ 0.50 \\ 1.00 \end{bmatrix} = \begin{bmatrix} 70.00 \\ 33.50 \\ 48.50 \end{bmatrix}$ Amy's stand sold \$70 worth of produce on Sunday, Beth's stand sold \$33.50 worth, and Chad's stand sold \$48.50 worth.

(c) $A + B = \begin{bmatrix} 120 & 50 & 60 \\ 40 & 25 & 30 \\ 60 & 30 & 20 \end{bmatrix} + \begin{bmatrix} 100 & 60 & 30 \\ 35 & 20 & 20 \\ 60 & 25 & 30 \end{bmatrix} = \begin{bmatrix} 220 & 110 & 90 \\ 75 & 45 & 50 \\ 120 & 55 & 50 \end{bmatrix}$ This represents the melons, squash, and tomatoes they sold during the weekend.

(d) $(A + B)C = \left(\begin{bmatrix} 120 & 50 & 60 \\ 40 & 25 & 30 \\ 60 & 30 & 20 \end{bmatrix} + \begin{bmatrix} 100 & 60 & 30 \\ 35 & 20 & 20 \\ 60 & 25 & 30 \end{bmatrix} \right) \begin{bmatrix} 0.10 \\ 0.50 \\ 1.00 \end{bmatrix} = \begin{bmatrix} 220 & 110 & 90 \\ 75 & 45 & 50 \\ 120 & 55 & 50 \end{bmatrix} \begin{bmatrix} 0.10 \\ 0.50 \\ 1.00 \end{bmatrix} = \begin{bmatrix} 167.00 \\ 80.00 \\ 89.50 \end{bmatrix}$

During the weekend, Amy's stand sold \$167 worth, Beth's stand sold \$80 worth, and Chad's stand sold \$89.50 worth of produce. Notice that $(A + B)C = AC + BC = \begin{bmatrix} 97.00 \\ 46.50 \\ 41.00 \end{bmatrix} + \begin{bmatrix} 70.00 \\ 33.50 \\ 48.50 \end{bmatrix} = \begin{bmatrix} 167.00 \\ 80.00 \\ 89.50 \end{bmatrix}$.

61. Suppose A is $n \times m$ and B is $i \times j$. If the product AB is defined, then $m = i$. If the product BA is defined, then $j = n$. Thus if both products are defined and if A is $n \times m$, then B must be $m \times n$.

63. $A = \begin{bmatrix} 1 & 1 \\ 1 & 1 \end{bmatrix}$; $A^2 = \begin{bmatrix} 1 & 1 \\ 1 & 1 \end{bmatrix} \begin{bmatrix} 1 & 1 \\ 1 & 1 \end{bmatrix} = \begin{bmatrix} 2 & 2 \\ 2 & 2 \end{bmatrix}$; $A^3 = A \cdot A^2 = \begin{bmatrix} 1 & 1 \\ 1 & 1 \end{bmatrix} \begin{bmatrix} 2 & 2 \\ 2 & 2 \end{bmatrix} = \begin{bmatrix} 4 & 4 \\ 4 & 4 \end{bmatrix}$;

$A^4 = A \cdot A^3 = \begin{bmatrix} 1 & 1 \\ 1 & 1 \end{bmatrix} \begin{bmatrix} 4 & 4 \\ 4 & 4 \end{bmatrix} = \begin{bmatrix} 8 & 8 \\ 8 & 8 \end{bmatrix}$. From this pattern, we see that $A^n = \begin{bmatrix} 2^{n-1} & 2^{n-1} \\ 2^{n-1} & 2^{n-1} \end{bmatrix}$.

11.3 INVERSES OF MATRICES AND MATRIX EQUATIONS

1. (a) The matrix $I = \begin{bmatrix} 1 & 0 \\ 0 & 1 \end{bmatrix}$ is called an *identity* matrix.

(b) If A is a 2×2 matrix then $A \times I = A$ and $I \times A = A$.

(c) If A and B are 2×2 matrices with $AB = I$ then B is the *inverse* of A.

3. $A = \begin{bmatrix} 4 & 1 \\ 7 & 2 \end{bmatrix}$; $B = \begin{bmatrix} 2 & -1 \\ -7 & 4 \end{bmatrix}$.

$AB = \begin{bmatrix} 4 & 1 \\ 7 & 2 \end{bmatrix}\begin{bmatrix} 2 & -1 \\ -7 & 4 \end{bmatrix} = \begin{bmatrix} 1 & 0 \\ 0 & 1 \end{bmatrix}$ and $BA = \begin{bmatrix} 2 & -1 \\ -7 & 4 \end{bmatrix}\begin{bmatrix} 4 & 1 \\ 7 & 2 \end{bmatrix} = \begin{bmatrix} 1 & 0 \\ 0 & 1 \end{bmatrix}$.

5. $A = \begin{bmatrix} 1 & 3 & -1 \\ 1 & 4 & 0 \\ -1 & -3 & 2 \end{bmatrix}$; $B = \begin{bmatrix} 8 & -3 & 4 \\ -2 & 1 & -1 \\ 1 & 0 & 1 \end{bmatrix}$. $AB = \begin{bmatrix} 1 & 3 & -1 \\ 1 & 4 & 0 \\ -1 & -3 & 2 \end{bmatrix}\begin{bmatrix} 8 & -3 & 4 \\ -2 & 1 & -1 \\ 1 & 0 & 1 \end{bmatrix} = \begin{bmatrix} 1 & 0 & 0 \\ 0 & 1 & 0 \\ 0 & 0 & 1 \end{bmatrix}$ and

$BA = \begin{bmatrix} 8 & -3 & 4 \\ -2 & 1 & -1 \\ 1 & 0 & 1 \end{bmatrix}\begin{bmatrix} 1 & 3 & -1 \\ 1 & 4 & 0 \\ -1 & -3 & 2 \end{bmatrix} = \begin{bmatrix} 1 & 0 & 0 \\ 0 & 1 & 0 \\ 0 & 0 & 1 \end{bmatrix}$

7. $A = \begin{bmatrix} 7 & 4 \\ 3 & 2 \end{bmatrix} \Leftrightarrow A^{-1} = \dfrac{1}{14-12}\begin{bmatrix} 2 & -4 \\ -3 & 7 \end{bmatrix} = \begin{bmatrix} 1 & -2 \\ -\frac{3}{2} & \frac{7}{2} \end{bmatrix}$. Then, $AA^{-1} = \begin{bmatrix} 7 & 4 \\ 3 & 2 \end{bmatrix}\begin{bmatrix} 1 & -2 \\ -\frac{3}{2} & \frac{7}{2} \end{bmatrix} = \begin{bmatrix} 1 & 0 \\ 0 & 1 \end{bmatrix}$

and $A^{-1}A = \begin{bmatrix} 1 & -2 \\ -\frac{3}{2} & \frac{7}{2} \end{bmatrix}\begin{bmatrix} 7 & 4 \\ 3 & 2 \end{bmatrix} = \begin{bmatrix} 1 & 0 \\ 0 & 1 \end{bmatrix}$.

9. Using a calculator, we find $A^{-1} = \begin{bmatrix} \frac{1}{3} & -\frac{1}{2} \\ 2 & 2 \end{bmatrix}$ and verify that $A^{-1}A = AA^{-1} = I_2$.

11. $\begin{bmatrix} -3 & -5 \\ 2 & 3 \end{bmatrix}^{-1} = \dfrac{1}{-9+10}\begin{bmatrix} 3 & 5 \\ -2 & -3 \end{bmatrix} = \begin{bmatrix} 3 & 5 \\ -2 & -3 \end{bmatrix}$

13. $\begin{bmatrix} 2 & 5 \\ -5 & -13 \end{bmatrix}^{-1} = \dfrac{1}{-26+25}\begin{bmatrix} -13 & -5 \\ 5 & 2 \end{bmatrix} = \begin{bmatrix} 13 & 5 \\ -5 & -2 \end{bmatrix}$

15. $\begin{bmatrix} 6 & -3 \\ -8 & 4 \end{bmatrix}^{-1} = \dfrac{1}{24-24}\begin{bmatrix} 4 & 3 \\ 8 & 6 \end{bmatrix}$, which is not defined, and so there is no inverse.

17. $\begin{bmatrix} 0.4 & -1.2 \\ 0.3 & 0.6 \end{bmatrix}^{-1} = \dfrac{1}{0.24+0.36}\begin{bmatrix} 0.6 & 1.2 \\ -0.3 & 0.4 \end{bmatrix} = \begin{bmatrix} 1 & 2 \\ -\frac{1}{2} & \frac{2}{3} \end{bmatrix}$

19. $\begin{bmatrix} 2 & 4 & 1 & 1 & 0 & 0 \\ -1 & 1 & -1 & 0 & 1 & 0 \\ 1 & 4 & 0 & 0 & 0 & 1 \end{bmatrix} \xrightarrow[2R_3 - R_1 \to R_3]{2R_2 + R_1 \to R_2} \begin{bmatrix} 2 & 4 & 1 & 1 & 0 & 0 \\ 0 & 6 & -1 & 1 & 2 & 0 \\ 0 & 4 & -1 & -1 & 0 & 2 \end{bmatrix} \xrightarrow[3R_1 - 2R_2 \to R_1]{3R_3 - 2R_2 \to R_3} \begin{bmatrix} 6 & 0 & 5 & 1 & -4 & 0 \\ 0 & 6 & -1 & 1 & 2 & 0 \\ 0 & 0 & -1 & -5 & -4 & 6 \end{bmatrix}$

$\xrightarrow[R_2 - R_3 \to R_2]{R_1 + 5R_3 \to R_1} \begin{bmatrix} 6 & 0 & 0 & -24 & -24 & 30 \\ 0 & 6 & 0 & 6 & 6 & -6 \\ 0 & 0 & -1 & -5 & -4 & 6 \end{bmatrix} \xrightarrow[\frac{1}{6}R_2,\, -R_3]{\frac{1}{6}R_1} \begin{bmatrix} 1 & 0 & 0 & -4 & -4 & 5 \\ 0 & 1 & 0 & 1 & 1 & -1 \\ 0 & 0 & 1 & 5 & 4 & -6 \end{bmatrix}$.

Therefore, the inverse matrix is $\begin{bmatrix} -4 & -4 & 5 \\ 1 & 1 & -1 \\ 5 & 4 & -6 \end{bmatrix}$.

21. $\begin{bmatrix} 1 & 2 & 3 & 1 & 0 & 0 \\ 4 & 5 & -1 & 0 & 1 & 0 \\ 1 & -1 & -10 & 0 & 0 & 1 \end{bmatrix} \xrightarrow[R_3 - R_1 \to R_3]{R_2 - 4R_1 \to R_2} \begin{bmatrix} 1 & 2 & 3 & 1 & 0 & 0 \\ 0 & -3 & -13 & -4 & 1 & 0 \\ 0 & -3 & -13 & -1 & 0 & 1 \end{bmatrix} \xrightarrow{R_3 - R_2 \to R_3} \begin{bmatrix} 1 & 2 & 3 & 1 & 0 & 0 \\ 0 & -3 & -13 & -4 & 1 & 0 \\ 0 & 0 & 0 & 3 & -1 & 1 \end{bmatrix}$.

Since the left half of the last row consists entirely of zeros, there is no inverse matrix.

23. $\begin{bmatrix} 0 & -2 & 2 & 1 & 0 & 0 \\ 3 & 1 & 3 & 0 & 1 & 0 \\ 1 & -2 & 3 & 0 & 0 & 1 \end{bmatrix} \xrightarrow{R_1 \leftrightarrow R_3} \begin{bmatrix} 1 & -2 & 3 & 0 & 0 & 1 \\ 3 & 1 & 3 & 0 & 1 & 0 \\ 0 & -2 & 2 & 1 & 0 & 0 \end{bmatrix} \xrightarrow{R_2 - 3R_1 \to R_2} \begin{bmatrix} 1 & -2 & 3 & 0 & 0 & 1 \\ 0 & 7 & -6 & 0 & 1 & -3 \\ 0 & -2 & 2 & 1 & 0 & 0 \end{bmatrix}$

$\xrightarrow[R_2 + 3R_3 \to R_2]{R_1 - R_3 \to R_1} \begin{bmatrix} 1 & 0 & 1 & -1 & 0 & 1 \\ 0 & 1 & 0 & 3 & 1 & -3 \\ 0 & -2 & 2 & 1 & 0 & 0 \end{bmatrix} \xrightarrow{R_3 + 2R_2 \to R_3} \begin{bmatrix} 1 & 0 & 1 & -1 & 0 & 1 \\ 0 & 1 & 0 & 3 & 1 & -3 \\ 0 & 0 & 2 & 7 & 2 & -6 \end{bmatrix} \xrightarrow{\frac{1}{2}R_3}$

$\begin{bmatrix} 1 & 0 & 1 & -1 & 0 & 1 \\ 0 & 1 & 0 & 3 & 1 & -3 \\ 0 & 0 & 1 & \frac{7}{2} & 1 & -3 \end{bmatrix} \xrightarrow{R_1 - R_3 \to R_1} \begin{bmatrix} 1 & 0 & 0 & -\frac{9}{2} & -1 & 4 \\ 0 & 1 & 0 & 3 & 1 & -3 \\ 0 & 0 & 1 & \frac{7}{2} & 1 & -3 \end{bmatrix}$. Therefore, the inverse matrix is

$\begin{bmatrix} -\frac{9}{2} & -1 & 4 \\ 3 & 1 & -3 \\ \frac{7}{2} & 1 & -3 \end{bmatrix}$.

25. $\left[\begin{array}{cccccccc} 1 & 2 & 0 & 3 & 1 & 0 & 0 & 0 \\ 0 & 1 & 1 & 1 & 0 & 1 & 0 & 0 \\ 0 & 1 & 0 & 1 & 0 & 0 & 1 & 0 \\ 1 & 2 & 0 & 2 & 0 & 0 & 0 & 1 \end{array}\right] \xrightarrow[R_4 - R_1 \to R_4]{R_3 - R_2 \to R_3} \left[\begin{array}{cccccccc} 1 & 2 & 0 & 3 & 1 & 0 & 0 & 0 \\ 0 & 1 & 1 & 1 & 0 & 1 & 0 & 0 \\ 0 & 0 & -1 & 0 & 0 & -1 & 1 & 0 \\ 0 & 0 & 0 & -1 & -1 & 0 & 0 & 1 \end{array}\right] \xrightarrow[-R_4]{-R_3}$

$\left[\begin{array}{cccccccc} 1 & 2 & 0 & 3 & 1 & 0 & 0 & 0 \\ 0 & 1 & 1 & 1 & 0 & 1 & 0 & 0 \\ 0 & 0 & 1 & 0 & 0 & 1 & -1 & 0 \\ 0 & 0 & 0 & 1 & 1 & 0 & 0 & -1 \end{array}\right] \xrightarrow[R_2 - R_3 \to R_2]{R_1 - 2R_2 \to R_1} \left[\begin{array}{cccccccc} 1 & 0 & -2 & 1 & 1 & -2 & 0 & 0 \\ 0 & 1 & 0 & 1 & 0 & 0 & 1 & 0 \\ 0 & 0 & 1 & 0 & 0 & 1 & -1 & 0 \\ 0 & 0 & 0 & 1 & 1 & 0 & 0 & -1 \end{array}\right] \xrightarrow[R_2 - R_4 \to R_2]{R_1 + 2R_3 \to R_1}$

$\left[\begin{array}{cccccccc} 1 & 0 & 0 & 1 & 1 & 0 & -2 & 0 \\ 0 & 1 & 0 & 0 & -1 & 0 & 1 & 1 \\ 0 & 0 & 1 & 0 & 0 & 1 & -1 & 0 \\ 0 & 0 & 0 & 1 & 1 & 0 & 0 & -1 \end{array}\right] \xrightarrow{R_1 \to R_1 - R_4} \left[\begin{array}{cccccccc} 1 & 0 & 0 & 0 & 0 & 0 & -2 & 1 \\ 0 & 1 & 0 & 0 & -1 & 0 & 1 & 1 \\ 0 & 0 & 1 & 0 & 0 & 1 & -1 & 0 \\ 0 & 0 & 0 & 1 & 1 & 0 & 0 & -1 \end{array}\right]$. Therefore, the inverse matrix is

$\begin{bmatrix} 0 & 0 & -2 & 1 \\ -1 & 0 & 1 & 1 \\ 0 & 1 & -1 & 0 \\ 1 & 0 & 0 & -1 \end{bmatrix}$.

27. $\begin{bmatrix} -3 & 2 & 3 \\ 0 & -1 & 3 \\ 1 & 0 & -2 \end{bmatrix}^{-1} = \begin{bmatrix} \frac{2}{3} & \frac{4}{3} & 3 \\ 1 & 1 & 3 \\ \frac{1}{3} & \frac{2}{3} & 1 \end{bmatrix}$

29. $\begin{bmatrix} -1 & -4 & 0 & 1 \\ 1 & 0 & -1 & 0 \\ 0 & 4 & 1 & -2 \\ 2 & 2 & -2 & 0 \end{bmatrix}^{-1} = \begin{bmatrix} -2 & 3 & -1 & -2 \\ 0 & -1 & 0 & \frac{1}{2} \\ -2 & 2 & -1 & -2 \\ -1 & -1 & -1 & 0 \end{bmatrix}$

31. $\begin{bmatrix} 1 & 7 & 3 \\ 0 & 2 & 1 \\ 0 & 0 & 3 \end{bmatrix}^{-1} = \begin{bmatrix} 1 & -\frac{7}{2} & \frac{1}{6} \\ 0 & \frac{1}{2} & -\frac{1}{6} \\ 0 & 0 & \frac{1}{3} \end{bmatrix}$

33. $\begin{bmatrix} 1 & 0 & 0 & 0 \\ 0 & 2 & 0 & 0 \\ 0 & 0 & 4 & 0 \\ 0 & 0 & 0 & 7 \end{bmatrix}^{-1} = \begin{bmatrix} 1 & 0 & 0 & 0 \\ 0 & \frac{1}{2} & 0 & 0 \\ 0 & 0 & \frac{1}{4} & 0 \\ 0 & 0 & 0 & \frac{1}{7} \end{bmatrix}$

In Solutions 35 and 37, the matrices A and B are defined as follows:

$$A = \begin{bmatrix} -1 & 0 & 2 \\ 0 & -2 & -1 \\ 4 & 2 & 1 \end{bmatrix} \qquad B = \begin{bmatrix} 2 & -1 & -2 \\ 0 & 3 & 1 \\ -1 & 0 & 2 \end{bmatrix}$$

35. $A^{-1}B = \begin{bmatrix} -\frac{1}{4} & \frac{3}{4} & \frac{3}{4} \\ -\frac{7}{16} & -\frac{23}{16} & -\frac{3}{16} \\ \frac{7}{8} & -\frac{1}{8} & -\frac{5}{8} \end{bmatrix}$

37. $BAB^{-1} = \begin{bmatrix} -7 & -3 & -4 \\ \frac{22}{7} & -\frac{2}{7} & \frac{16}{7} \\ \frac{50}{7} & \frac{26}{7} & \frac{37}{7} \end{bmatrix}$

39. $\begin{cases} -3x - 5y = 4 \\ 2x + 3y = 0 \end{cases}$ is equivalent to the matrix equation $\begin{bmatrix} -3 & -5 \\ 2 & 3 \end{bmatrix} \begin{bmatrix} x \\ y \end{bmatrix} = \begin{bmatrix} 4 \\ 0 \end{bmatrix}$. Using the inverse from Exercise 11,

$\begin{bmatrix} x \\ y \end{bmatrix} = \begin{bmatrix} 3 & 5 \\ -2 & -3 \end{bmatrix} \begin{bmatrix} 4 \\ 0 \end{bmatrix} = \begin{bmatrix} 12 \\ -8 \end{bmatrix}$. Therefore, $x = 12$ and $y = -8$.

41. $\begin{cases} 2x + 5y = 2 \\ -5x - 13y = 20 \end{cases}$ is equivalent to the matrix equation $\begin{bmatrix} 2 & 5 \\ -5 & -13 \end{bmatrix}\begin{bmatrix} x \\ y \end{bmatrix} = \begin{bmatrix} 2 \\ 20 \end{bmatrix}$. Using the inverse from Exercise 13, $\begin{bmatrix} x \\ y \end{bmatrix} = \begin{bmatrix} 13 & 5 \\ -5 & -2 \end{bmatrix}\begin{bmatrix} 2 \\ 20 \end{bmatrix} = \begin{bmatrix} 126 \\ -50 \end{bmatrix}$. Therefore, $x = 126$ and $y = -50$.

43. $\begin{cases} 2x + 4y + z = 7 \\ -x + y - z = 0 \\ x + 4y = -2 \end{cases}$ is equivalent to the matrix equation $\begin{bmatrix} 2 & 4 & 1 \\ -1 & 1 & -1 \\ 1 & 4 & 0 \end{bmatrix}\begin{bmatrix} x \\ y \\ z \end{bmatrix} = \begin{bmatrix} 7 \\ 0 \\ -2 \end{bmatrix}$. Using the inverse from Exercise 19, $\begin{bmatrix} x \\ y \\ z \end{bmatrix} = \begin{bmatrix} -4 & -4 & 5 \\ 1 & 1 & -1 \\ 5 & 4 & -6 \end{bmatrix}\begin{bmatrix} 7 \\ 0 \\ -2 \end{bmatrix} = \begin{bmatrix} -38 \\ 9 \\ 47 \end{bmatrix}$. Therefore, $x = -38$, $y = 9$, and $z = 47$.

45. $\begin{cases} -2y + 2z = 12 \\ 3x + y + 3z = -2 \\ x - 2y + 3z = 8 \end{cases}$ is equivalent to the matrix equation $\begin{bmatrix} 0 & -2 & 2 \\ 3 & 1 & 3 \\ 1 & -2 & 3 \end{bmatrix}\begin{bmatrix} x \\ y \\ z \end{bmatrix} = \begin{bmatrix} 12 \\ -2 \\ 8 \end{bmatrix}$. Using the inverse from Exercise 23, $\begin{bmatrix} x \\ y \\ z \end{bmatrix} = \begin{bmatrix} -\frac{9}{2} & -1 & 4 \\ 3 & 1 & -3 \\ \frac{7}{2} & 1 & -3 \end{bmatrix}\begin{bmatrix} 12 \\ -2 \\ 8 \end{bmatrix} = \begin{bmatrix} -20 \\ 10 \\ 16 \end{bmatrix}$. Therefore, $x = -20$, $y = 10$, and $z = 16$.

47. Using a calculator, we get the result $(3, 2, 1)$.

49. Using a calculator, we get the result $(3, -2, 2)$.

51. Using a calculator, we get the result $(8, 1, 0, 3)$.

53. This has the form $MX = C$, so $M^{-1}(MX) = M^{-1}C$ and $M^{-1}(MX) = \left(M^{-1}M\right)X = X$. Now $M^{-1} = \begin{bmatrix} 3 & -2 \\ -4 & 3 \end{bmatrix}^{-1} = \dfrac{1}{9-8}\begin{bmatrix} 3 & 2 \\ 4 & 3 \end{bmatrix} = \begin{bmatrix} 3 & 2 \\ 4 & 3 \end{bmatrix}$. Since $X = M^{-1}C$, we get $\begin{bmatrix} x & y & z \\ u & v & w \end{bmatrix} = \begin{bmatrix} 3 & 2 \\ 4 & 3 \end{bmatrix}\begin{bmatrix} 1 & 0 & -1 \\ 2 & 1 & 3 \end{bmatrix} = \begin{bmatrix} 7 & 2 & 3 \\ 10 & 3 & 5 \end{bmatrix}$.

55. $\begin{bmatrix} a & -a \\ a & a \end{bmatrix}^{-1} = \dfrac{1}{a^2 - (-a^2)}\begin{bmatrix} a & a \\ -a & a \end{bmatrix} = \dfrac{1}{2a^2}\begin{bmatrix} a & a \\ -a & a \end{bmatrix} = \dfrac{1}{2a}\begin{bmatrix} 1 & 1 \\ -1 & 1 \end{bmatrix}$

57. $\begin{bmatrix} 2 & x \\ x & x^2 \end{bmatrix}^{-1} = \dfrac{1}{2x^2 - x^2}\begin{bmatrix} x^2 & -x \\ -x & 2 \end{bmatrix} = \dfrac{1}{x^2}\begin{bmatrix} x^2 & -x \\ -x & 2 \end{bmatrix} = \begin{bmatrix} 1 & -1/x \\ -1/x & 2/x^2 \end{bmatrix}$.

The inverse does not exist when $x = 0$.

59. $\left[\begin{array}{ccc|ccc} 1 & e^x & 0 & 1 & 0 & 0 \\ e^x & -e^{2x} & 0 & 0 & 1 & 0 \\ 0 & 0 & 2 & 0 & 0 & 1 \end{array}\right] \xrightarrow{R_2 - e^x R_1 \to R_2} \left[\begin{array}{ccc|ccc} 1 & e^x & 0 & 1 & 0 & 0 \\ 0 & -2e^{2x} & 0 & -e^x & 1 & 0 \\ 0 & 0 & 2 & 0 & 0 & 1 \end{array}\right] \xrightarrow[\frac{1}{2}R_3]{-\frac{1}{2}e^{-2x}R_2}$

$\left[\begin{array}{ccc|ccc} 1 & e^x & 0 & 1 & 0 & 0 \\ 0 & 1 & 0 & \frac{1}{2}e^{-x} & -\frac{1}{2}e^{-2x} & 0 \\ 0 & 0 & 1 & 0 & 0 & \frac{1}{2} \end{array}\right] \xrightarrow{R_1 - e^x R_2 \to R_1} \left[\begin{array}{ccc|ccc} 1 & 0 & 0 & \frac{1}{2} & \frac{1}{2}e^{-x} & 0 \\ 0 & 1 & 0 & \frac{1}{2}e^{-x} & -\frac{1}{2}e^{-2x} & 0 \\ 0 & 0 & 1 & 0 & 0 & \frac{1}{2} \end{array}\right]$.

Therefore, the inverse matrix is $\begin{bmatrix} \frac{1}{2} & \frac{1}{2}e^{-x} & 0 \\ \frac{1}{2}e^{-x} & -\frac{1}{2}e^{-2x} & 0 \\ 0 & 0 & \frac{1}{2} \end{bmatrix}$. The inverse exists for all x.

61. (a) $\left[\begin{array}{ccc|ccc} 3 & 1 & 3 & 1 & 0 & 0 \\ 4 & 2 & 4 & 0 & 1 & 0 \\ 3 & 2 & 4 & 0 & 0 & 1 \end{array}\right] \xrightarrow[R_1 \leftrightarrow R_2]{R_3 - R_1 \to R_3} \left[\begin{array}{ccc|ccc} 4 & 2 & 4 & 0 & 1 & 0 \\ 3 & 1 & 3 & 1 & 0 & 0 \\ 0 & 1 & 1 & -1 & 0 & 1 \end{array}\right] \xrightarrow{R_1 - R_2 \to R_1} \left[\begin{array}{ccc|ccc} 1 & 1 & 1 & -1 & 1 & 0 \\ 3 & 1 & 3 & 1 & 0 & 0 \\ 0 & 1 & 1 & -1 & 0 & 1 \end{array}\right]$

$\xrightarrow{R_2 - 3R_1 \to R_2} \left[\begin{array}{ccc|ccc} 1 & 1 & 1 & -1 & 1 & 0 \\ 0 & -2 & 0 & 4 & -3 & 0 \\ 0 & 1 & 1 & -1 & 0 & 1 \end{array}\right] \xrightarrow[R_3 + \frac{1}{2}R_2 \to R_3]{R_1 + \frac{1}{2}R_2 \to R_1,\ -\frac{1}{2}R_2} \left[\begin{array}{ccc|ccc} 1 & 0 & 1 & 1 & -\frac{1}{2} & 0 \\ 0 & 1 & 0 & -2 & \frac{3}{2} & 0 \\ 0 & 0 & 1 & 1 & -\frac{3}{2} & 1 \end{array}\right] \xrightarrow{R_1 - R_3 \to R_1}$

$\left[\begin{array}{ccc|ccc} 1 & 0 & 0 & 0 & 1 & -1 \\ 0 & 1 & 0 & -2 & \frac{3}{2} & 0 \\ 0 & 0 & 1 & 1 & -\frac{3}{2} & 1 \end{array}\right]$. Therefore, the inverse of the matrix is $\begin{bmatrix} 0 & 1 & -1 \\ -2 & \frac{3}{2} & 0 \\ 1 & -\frac{3}{2} & 1 \end{bmatrix}$.

(b) $\begin{bmatrix} A \\ B \\ C \end{bmatrix} = \begin{bmatrix} 0 & 1 & -1 \\ -2 & \frac{3}{2} & 0 \\ 1 & -\frac{3}{2} & 1 \end{bmatrix} \begin{bmatrix} 10 \\ 14 \\ 13 \end{bmatrix} = \begin{bmatrix} 1 \\ 1 \\ 2 \end{bmatrix}$.

Therefore, he should feed the rats 1 oz of food A, 1 oz of food B, and 2 oz of food C.

(c) $\begin{bmatrix} A \\ B \\ C \end{bmatrix} = \begin{bmatrix} 0 & 1 & -1 \\ -2 & \frac{3}{2} & 0 \\ 1 & -\frac{3}{2} & 1 \end{bmatrix} \begin{bmatrix} 9 \\ 12 \\ 10 \end{bmatrix} = \begin{bmatrix} 2 \\ 0 \\ 1 \end{bmatrix}$.

Therefore, he should feed the rats 2 oz of food A, no food B, and 1 oz of food C.

(d) $\begin{bmatrix} A \\ B \\ C \end{bmatrix} = \begin{bmatrix} 0 & 1 & -1 \\ -2 & \frac{3}{2} & 0 \\ 1 & -\frac{3}{2} & 1 \end{bmatrix} \begin{bmatrix} 2 \\ 4 \\ 11 \end{bmatrix} = \begin{bmatrix} -7 \\ 2 \\ 7 \end{bmatrix}$.

Since $A < 0$, there is no combination of foods giving the required supply.

63. (a) $\begin{cases} 9x + 11y + 8z = 740 \\ 13x + 15y + 16z = 1204 \\ 8x + 7y + 14z = 828 \end{cases}$

(b) $\begin{bmatrix} 9 & 11 & 8 \\ 13 & 15 & 16 \\ 8 & 7 & 14 \end{bmatrix} \begin{bmatrix} x \\ y \\ z \end{bmatrix} = \begin{bmatrix} 740 \\ 1204 \\ 828 \end{bmatrix}$

(c) $\begin{bmatrix} 9 & 11 & 8 & 1 & 0 & 0 \\ 13 & 15 & 16 & 0 & 1 & 0 \\ 8 & 7 & 14 & 0 & 0 & 1 \end{bmatrix} \xrightarrow[9R_3 - 8R_1 \to R_3]{9R_2 - 13R_1 \to R_2} \begin{bmatrix} 9 & 11 & 8 & 1 & 0 & 0 \\ 0 & -8 & 40 & -13 & 9 & 0 \\ 0 & -25 & 62 & -8 & 0 & 9 \end{bmatrix} \xrightarrow{20R_3 - 31R_2 \to R_3}$

$\begin{bmatrix} 9 & 11 & 8 & 1 & 0 & 0 \\ 0 & -8 & 40 & -13 & 9 & 0 \\ 0 & -252 & 0 & 243 & -279 & 180 \end{bmatrix} \xrightarrow{R_2 \leftrightarrow R_3} \begin{bmatrix} 9 & 11 & 8 & 1 & 0 & 0 \\ 0 & -252 & 0 & 243 & -279 & 180 \\ 0 & -8 & 40 & -13 & 9 & 0 \end{bmatrix} \xrightarrow[63R_3 - 2R_2 \to R_3]{-\frac{1}{252}R_2}$

$\begin{bmatrix} 9 & 11 & 8 & 1 & 0 & 0 \\ 0 & 1 & 0 & -\frac{27}{28} & \frac{31}{28} & -\frac{5}{7} \\ 0 & 0 & 2520 & -1305 & 1125 & -360 \end{bmatrix} \xrightarrow{\frac{1}{2520}R_3} \begin{bmatrix} 9 & 11 & 8 & 1 & 0 & 0 \\ 0 & 1 & 0 & -\frac{27}{28} & \frac{31}{28} & -\frac{5}{7} \\ 0 & 0 & 1 & -\frac{29}{56} & \frac{25}{56} & -\frac{1}{7} \end{bmatrix} \xrightarrow{R_1 - 11R_2 - 8R_3 \to R_1}$

$\begin{bmatrix} 9 & 0 & 0 & \frac{63}{4} & -\frac{63}{4} & 9 \\ 0 & 1 & 0 & -\frac{27}{28} & \frac{31}{28} & -\frac{5}{7} \\ 0 & 0 & 1 & -\frac{29}{56} & \frac{25}{56} & -\frac{1}{7} \end{bmatrix} \xrightarrow{\frac{1}{9}R_1} \begin{bmatrix} 1 & 0 & 0 & \frac{7}{4} & -\frac{7}{4} & 1 \\ 0 & 1 & 0 & -\frac{27}{28} & \frac{31}{28} & -\frac{5}{7} \\ 0 & 0 & 1 & -\frac{29}{56} & \frac{25}{56} & -\frac{1}{7} \end{bmatrix}$. Thus, the inverse of the matrix is $\begin{bmatrix} \frac{7}{4} & -\frac{7}{4} & 1 \\ -\frac{27}{28} & \frac{31}{28} & -\frac{5}{7} \\ -\frac{29}{56} & \frac{25}{56} & -\frac{1}{7} \end{bmatrix}$. (You could also use a calculator to find the inverse.) Therefore,

$\begin{bmatrix} x \\ y \\ z \end{bmatrix} = \begin{bmatrix} \frac{7}{4} & -\frac{7}{4} & 1 \\ -\frac{27}{28} & \frac{31}{28} & -\frac{5}{7} \\ -\frac{29}{56} & \frac{25}{56} & -\frac{1}{7} \end{bmatrix} \begin{bmatrix} 740 \\ 1204 \\ 828 \end{bmatrix} = \begin{bmatrix} 16 \\ 28 \\ 36 \end{bmatrix}$. She earns \$16 on a standard phone, \$28 on a deluxe phone, and \$36 on a super deluxe phone.

11.4 DETERMINANTS AND CRAMER'S RULE

1. True. $\det(A)$ is defined only for a square matrix A.

3. True. If $\det(A) = 0$ then A is not invertible.

5. The matrix $\begin{bmatrix} 2 & 0 \\ 0 & 3 \end{bmatrix}$ has determinant $|D| = (2)(3) - (0)(0) = 6$.

7. The matrix $\begin{bmatrix} \frac{3}{2} & 1 \\ -1 & -\frac{2}{3} \end{bmatrix}$ has determinant $|D| = \frac{3}{2}\left(-\frac{2}{3}\right) - 1(-1) = 0$.

9. The matrix $\begin{bmatrix} 4 & 5 \\ 0 & -1 \end{bmatrix}$ has determinant $|D| = (4)(-1) - (5)(0) = -4$.

11. The matrix $\begin{bmatrix} 2 & 5 \end{bmatrix}$ does not have a determinant because it is not square.

13. The matrix $\begin{bmatrix} \frac{1}{2} & \frac{1}{8} \\ 1 & \frac{1}{2} \end{bmatrix}$ has determinant $|D| = \frac{1}{2} \cdot \frac{1}{2} - 1 \cdot \frac{1}{8} = \frac{1}{4} - \frac{1}{8} = \frac{1}{8}$.

In Solutions 15–19, $A = \begin{bmatrix} 1 & 0 & \frac{1}{2} \\ -3 & 5 & 2 \\ 0 & 0 & 4 \end{bmatrix}$.

15. $M_{11} = 5 \cdot 4 - 0 \cdot 2 = 20$, $A_{11} = (-1)^2 M_{11} = 20$

17. $M_{12} = -3 \cdot 4 - 0 \cdot 2 = -12$, $A_{12} = (-1)^3 M_{12} = 12$

19. $M_{23} = 1 \cdot 0 - 0 \cdot 0 = 0$, $A_{23} = (-1)^5 M_{23} = 0$

21. $M = \begin{bmatrix} 2 & 1 & 0 \\ 0 & -2 & 4 \\ 0 & 1 & -3 \end{bmatrix}$. Therefore, expanding by the first column, $|M| = 2\begin{vmatrix} -2 & 4 \\ 1 & -3 \end{vmatrix} = 2(6-4) = 4$. Since $|M| \neq 0$, the matrix has an inverse.

23. $M = \begin{bmatrix} 30 & 0 & 20 \\ 0 & -10 & -20 \\ 40 & 0 & 10 \end{bmatrix}$. Therefore, expanding by the first row,

$$|M| = 30\begin{vmatrix} -10 & -20 \\ 0 & 10 \end{vmatrix} + 20\begin{vmatrix} 0 & -10 \\ 40 & 0 \end{vmatrix} = 30(-100+0) + 20(0+400) = -3000 + 8000 = 5000$$, and so M^{-1} exists.

25. $M = \begin{bmatrix} 1 & 3 & 7 \\ 2 & 0 & 8 \\ 0 & 2 & 2 \end{bmatrix}$. Therefore, expanding by the second row, $|M| = -2\begin{vmatrix} 3 & 7 \\ 2 & 2 \end{vmatrix} - 8\begin{vmatrix} 1 & 3 \\ 0 & 2 \end{vmatrix} = -2(6-14) - 16 = 0$. Since $|M| = 0$, the matrix does not have an inverse.

27. $M = \begin{bmatrix} 1 & 3 & 3 & 0 \\ 0 & 2 & 0 & 1 \\ -1 & 0 & 0 & 2 \\ 1 & 6 & 4 & 1 \end{bmatrix}$. Therefore, expanding by the third row,

$$|M| = -1\begin{vmatrix} 3 & 3 & 0 \\ 2 & 0 & 1 \\ 6 & 4 & 1 \end{vmatrix} - 2\begin{vmatrix} 1 & 3 & 3 \\ 0 & 2 & 0 \\ 1 & 6 & 4 \end{vmatrix} = 1\begin{vmatrix} 3 & 3 \\ 6 & 4 \end{vmatrix} - 1\begin{vmatrix} 3 & 3 \\ 2 & 0 \end{vmatrix} - 4\begin{vmatrix} 1 & 3 \\ 1 & 4 \end{vmatrix} = -6 + 6 - 4 = -4$$, and so M^{-1} exists.

29. $\begin{vmatrix} 1 & 2 & -1 \\ 2 & 2 & 1 \\ 1 & 2 & 2 \end{vmatrix} = -6$. The matrix has an inverse.

31. $\begin{vmatrix} 1 & 10 & 2 & 7 \\ 2 & 18 & 18 & 13 \\ -3 & -30 & -4 & -24 \\ 1 & 10 & 2 & 10 \end{vmatrix} = -12$. The matrix has an inverse.

33. $\begin{vmatrix} 4 & 3 & -2 & 10 \\ -8 & -6 & 24 & -1 \\ 20 & 15 & 3 & 27 \\ 12 & 9 & -6 & -1 \end{vmatrix} = 0$. The matrix has no inverse.

35. $|M| = \begin{vmatrix} 0 & 0 & 4 & 6 \\ 2 & 1 & 1 & 3 \\ 2 & 1 & 2 & 3 \\ 3 & 0 & 1 & 7 \end{vmatrix} = \begin{vmatrix} 0 & 0 & 4 & 6 \\ 2 & 1 & 1 & 3 \\ 0 & 0 & 1 & 0 \\ 3 & 0 & 1 & 7 \end{vmatrix}$, by replacing R_3 with $R_3 - R_2$. Then, expanding by the third row,

$$|M| = 1\begin{vmatrix} 0 & 0 & 6 \\ 2 & 1 & 3 \\ 3 & 0 & 7 \end{vmatrix} = 6\begin{vmatrix} 2 & 1 \\ 3 & 0 \end{vmatrix} = 6(2 \cdot 0 - 3 \cdot 1) = -18.$$

37. $M = \begin{bmatrix} 1 & 2 & 3 & 4 & 5 \\ 0 & 2 & 4 & 6 & 8 \\ 0 & 0 & 3 & 6 & 9 \\ 0 & 0 & 0 & 4 & 8 \\ 0 & 0 & 0 & 0 & 5 \end{bmatrix}$, so $|M| = 5\begin{vmatrix} 1 & 2 & 3 & 4 \\ 0 & 2 & 4 & 6 \\ 0 & 0 & 3 & 6 \\ 0 & 0 & 0 & 4 \end{vmatrix} = 5 \cdot 4 \begin{vmatrix} 1 & 2 & 3 \\ 0 & 2 & 4 \\ 0 & 0 & 3 \end{vmatrix} = 20 \cdot 3 \begin{vmatrix} 1 & 2 \\ 0 & 2 \end{vmatrix} = 60 \cdot 2 = 120.$

39. $B = \begin{bmatrix} 4 & 1 & 0 \\ -2 & -1 & 1 \\ 4 & 0 & 3 \end{bmatrix}$

(a) $|B| = 2\begin{vmatrix} 1 & 0 \\ 0 & 3 \end{vmatrix} - 1\begin{vmatrix} 4 & 0 \\ 4 & 3 \end{vmatrix} - 1\begin{vmatrix} 4 & 1 \\ 4 & 0 \end{vmatrix} = 6 - 12 + 4 = -2$

(b) $|B| = -1\begin{vmatrix} 4 & 1 \\ 4 & 0 \end{vmatrix} + 3\begin{vmatrix} 4 & 1 \\ -2 & -1 \end{vmatrix} = 4 - 6 = -2$

(c) Yes, as expected, the results agree.

41. $\begin{cases} 2x - y = -9 \\ x + 2y = 8 \end{cases}$ Then $|D| = \begin{vmatrix} 2 & -1 \\ 1 & 2 \end{vmatrix} = 5$, $|D_x| = \begin{vmatrix} -9 & -1 \\ 8 & 2 \end{vmatrix} = -10$, and $|D_y| = \begin{vmatrix} 2 & -9 \\ 1 & 8 \end{vmatrix} = 25.$

Hence, $x = \dfrac{|D_x|}{|D|} = \dfrac{-10}{5} = -2$, $y = \dfrac{|D_y|}{|D|} = \dfrac{25}{5} = 5$, and so the solution is $(-2, 5)$.

43. $\begin{cases} x - 6y = 3 \\ 3x + 2y = 1 \end{cases}$ Then, $|D| = \begin{vmatrix} 1 & -6 \\ 3 & 2 \end{vmatrix} = 20$, $|D_x| = \begin{vmatrix} 3 & -6 \\ 1 & 2 \end{vmatrix} = 12$, and $|D_y| = \begin{vmatrix} 1 & 3 \\ 3 & 1 \end{vmatrix} = -8.$

Hence, $x = \dfrac{|D_x|}{|D|} = \frac{12}{20} = 0.6$, $y = \dfrac{|D_y|}{|D|} = \dfrac{-8}{20} = -0.4$,and so the solution is $(0.6, -0.4)$.

45. $\begin{cases} 0.4x + 1.2y = 0.4 \\ 1.2x + 1.6y = 3.2 \end{cases}$ Then, $|D| = \begin{vmatrix} 0.4 & 1.2 \\ 1.2 & 1.6 \end{vmatrix} = -0.8$, $|D_x| = \begin{vmatrix} 0.4 & 1.2 \\ 3.2 & 1.6 \end{vmatrix} = -3.2$, and $|D_y| = \begin{vmatrix} 0.4 & 0.4 \\ 1.2 & 3.2 \end{vmatrix} = 0.8.$

Hence, $x = \dfrac{|D_x|}{|D|} = \dfrac{-3.2}{-0.8} = 4$, $y = \dfrac{|D_y|}{|D|} = \dfrac{0.8}{-0.8} = -1$,and so the solution is $(4, -1)$.

47. $\begin{cases} x - y + 2z = 0 \\ 3x + z = 11 \\ -x + 2y = 0 \end{cases}$ Then expanding by the second row,

$|D| = \begin{vmatrix} 1 & -1 & 2 \\ 3 & 0 & 1 \\ -1 & 2 & 0 \end{vmatrix} = -3\begin{vmatrix} -1 & 2 \\ 2 & 0 \end{vmatrix} - 1\begin{vmatrix} 1 & -1 \\ -1 & 2 \end{vmatrix} = 12 - 1 = 11$, $|D_x| = \begin{vmatrix} 0 & -1 & 2 \\ 11 & 0 & 1 \\ 0 & 2 & 0 \end{vmatrix} = -11\begin{vmatrix} -1 & 2 \\ 2 & 0 \end{vmatrix} = 44,$

$|D_y| = \begin{vmatrix} 1 & 0 & 2 \\ 3 & 11 & 1 \\ -1 & 0 & 0 \end{vmatrix} = 11\begin{vmatrix} 1 & 2 \\ -1 & 0 \end{vmatrix} = 22$, and $|D_z| = \begin{vmatrix} 1 & -1 & 0 \\ 3 & 0 & 11 \\ -1 & 2 & 0 \end{vmatrix} = -11\begin{vmatrix} 1 & -1 \\ -1 & 2 \end{vmatrix} = -11.$

Therefore, $x = \frac{44}{11} = 4$, $y = \frac{22}{11} = 2$, $z = \frac{-11}{11} = -1$, and so the solution is $(4, 2, -1)$.

49. $\begin{cases} 2x_1 + 3x_2 - 5x_3 = 1 \\ x_1 + x_2 - x_3 = 2 \\ 2x_2 + x_3 = 8 \end{cases}$

Then, expanding by the third row,

$$|D| = \begin{vmatrix} 2 & 3 & -5 \\ 1 & 1 & -1 \\ 0 & 2 & 1 \end{vmatrix} = -2\begin{vmatrix} 2 & -5 \\ 1 & -1 \end{vmatrix} + \begin{vmatrix} 2 & 3 \\ 1 & 1 \end{vmatrix} = -6 - 1 = -7,$$

$$|D_{x_1}| = \begin{vmatrix} 1 & 3 & -5 \\ 2 & 1 & -1 \\ 8 & 2 & 1 \end{vmatrix} = \begin{vmatrix} 1 & -1 \\ 2 & 1 \end{vmatrix} - 3\begin{vmatrix} 2 & -1 \\ 8 & 1 \end{vmatrix} - 5\begin{vmatrix} 2 & 1 \\ 8 & 2 \end{vmatrix} = 3 - 30 + 20 = -7,$$

$$|D_{x_2}| = \begin{vmatrix} 2 & 1 & -5 \\ 1 & 2 & -1 \\ 0 & 8 & 1 \end{vmatrix} = -8\begin{vmatrix} 2 & -5 \\ 1 & -1 \end{vmatrix} + \begin{vmatrix} 2 & 1 \\ 1 & 2 \end{vmatrix} = -24 + 3 = -21, \text{ and}$$

$$|D_{x_3}| = \begin{vmatrix} 2 & 3 & 1 \\ 1 & 1 & 2 \\ 0 & 2 & 8 \end{vmatrix} = -2\begin{vmatrix} 2 & 1 \\ 1 & 2 \end{vmatrix} + 8\begin{vmatrix} 2 & 3 \\ 1 & 1 \end{vmatrix} = -6 - 8 = -14.$$

Thus, $x_1 = \frac{-7}{-7} = 1$, $x_2 = \frac{-21}{-7} = 3$, $x_3 = \frac{-14}{-7} = 2$, and so the solution is $(1, 3, 2)$.

51. $\begin{cases} \frac{1}{3}x - \frac{1}{5}y + \frac{1}{2}z = \frac{7}{10} \\ -\frac{2}{3}x + \frac{2}{5}y + \frac{3}{2}z = \frac{11}{10} \\ x - \frac{4}{5}y + z = \frac{9}{5} \end{cases} \Leftrightarrow \begin{cases} 10x - 6y + 15z = 21 \\ -20x + 12y + 45z = 33 \\ 5x - 4y + 5z = 9 \end{cases}$ Then

$$|D| = \begin{vmatrix} 10 & -6 & 15 \\ -20 & 12 & 45 \\ 5 & -4 & 5 \end{vmatrix} = 10\begin{vmatrix} 12 & 45 \\ -4 & 5 \end{vmatrix} + 6\begin{vmatrix} -20 & 45 \\ 5 & 5 \end{vmatrix} + 15\begin{vmatrix} -20 & 12 \\ 5 & -4 \end{vmatrix} = 2400 - 1950 + 300 = 750,$$

$$|D_x| = \begin{vmatrix} 21 & -6 & 15 \\ 33 & 12 & 45 \\ 9 & -4 & 5 \end{vmatrix} = 21\begin{vmatrix} 12 & 45 \\ -4 & 5 \end{vmatrix} + 6\begin{vmatrix} 33 & 45 \\ 9 & 5 \end{vmatrix} + 15\begin{vmatrix} 33 & 12 \\ 9 & -4 \end{vmatrix} = 5040 - 1440 - 3600 = 0,$$

$$|D_y| = \begin{vmatrix} 10 & 21 & 15 \\ -20 & 33 & 45 \\ 5 & 9 & 5 \end{vmatrix} = 10\begin{vmatrix} 33 & 45 \\ 9 & 5 \end{vmatrix} - 21\begin{vmatrix} -20 & 45 \\ 5 & 5 \end{vmatrix} + 15\begin{vmatrix} -20 & 33 \\ 5 & 9 \end{vmatrix} = -2400 + 6825 - 5175 = -750, \text{ and}$$

$$|D_z| = \begin{vmatrix} 10 & -6 & 21 \\ -20 & 12 & 33 \\ 5 & -4 & 9 \end{vmatrix} = 10\begin{vmatrix} 12 & 33 \\ -4 & 9 \end{vmatrix} + 6\begin{vmatrix} -20 & 33 \\ 5 & 9 \end{vmatrix} + 21\begin{vmatrix} -20 & 12 \\ 5 & -4 \end{vmatrix} = 2400 - 2070 + 420 = 750.$$

Therefore, $x = 0$, $y = -1$, $z = 1$, and so the solution is $(0, -1, 1)$.

53. $\begin{cases} 3y + 5z = 4 \\ 2x - z = 10 \\ 4x + 7y = 0 \end{cases}$ Then $|D| = \begin{vmatrix} 0 & 3 & 5 \\ 2 & 0 & -1 \\ 4 & 7 & 0 \end{vmatrix} = -3\begin{vmatrix} 2 & -1 \\ 4 & 0 \end{vmatrix} + 5\begin{vmatrix} 2 & 0 \\ 4 & 7 \end{vmatrix} = -12 + 70 = 58,$

$|D_x| = \begin{vmatrix} 4 & 3 & 5 \\ 10 & 0 & -1 \\ 0 & 7 & 0 \end{vmatrix} = -7\begin{vmatrix} 4 & 5 \\ 10 & -1 \end{vmatrix} = 378,$ $|D_y| = \begin{vmatrix} 0 & 4 & 5 \\ 2 & 10 & -1 \\ 4 & 0 & 0 \end{vmatrix} = 4\begin{vmatrix} 4 & 5 \\ 10 & -1 \end{vmatrix} = -216,$ and

$|D_z| = \begin{vmatrix} 0 & 3 & 4 \\ 2 & 0 & 10 \\ 4 & 7 & 0 \end{vmatrix} = 4\begin{vmatrix} 3 & 4 \\ 0 & 10 \end{vmatrix} - 7\begin{vmatrix} 0 & 4 \\ 2 & 10 \end{vmatrix} = 120 + 56 = 176.$

Thus, $x = \frac{189}{29}$, $y = -\frac{108}{29}$, and $z = \frac{88}{29}$, and so the solution is $\left(\frac{189}{29}, -\frac{108}{29}, \frac{88}{29}\right)$.

55. $\begin{cases} x + y + z + w = 0 \\ 2z + w = 0 \\ y - z = 0 \\ x + 2z = 1 \end{cases}$ Then

$|D| = \begin{vmatrix} 1 & 1 & 1 & 1 \\ 2 & 0 & 0 & 1 \\ 0 & 1 & -1 & 0 \\ 1 & 0 & 2 & 0 \end{vmatrix} = -1\begin{vmatrix} 2 & 0 & 1 \\ 0 & -1 & 0 \\ 1 & 2 & 0 \end{vmatrix} - 1\begin{vmatrix} 1 & 1 & 1 \\ 2 & 0 & 1 \\ 1 & 2 & 0 \end{vmatrix} = -\left(2\begin{vmatrix} -1 & 0 \\ 2 & 0 \end{vmatrix} + 1\begin{vmatrix} 0 & 1 \\ -1 & 0 \end{vmatrix}\right) - \left(-1\begin{vmatrix} 2 & 1 \\ 1 & 0 \end{vmatrix} - 2\begin{vmatrix} 1 & 1 \\ 2 & 1 \end{vmatrix}\right)$

$= -2(0) - 1(1) + 1(-1) + 2(-1) = -4,$

$|D_x| = \begin{vmatrix} 0 & 1 & 1 & 1 \\ 0 & 0 & 0 & 1 \\ 0 & 1 & -1 & 0 \\ 1 & 0 & 2 & 0 \end{vmatrix} = -1\begin{vmatrix} 1 & 1 & 1 \\ 0 & 0 & 1 \\ 1 & -1 & 0 \end{vmatrix} = -1(-1)\begin{vmatrix} 1 & 1 \\ 1 & -1 \end{vmatrix} = -2,$

$|D_y| = \begin{vmatrix} 1 & 0 & 1 & 1 \\ 2 & 0 & 0 & 1 \\ 0 & 0 & -1 & 0 \\ 1 & 1 & 2 & 0 \end{vmatrix} = 1\begin{vmatrix} 1 & 1 & 1 \\ 2 & 0 & 1 \\ 0 & -1 & 0 \end{vmatrix} = 1\begin{vmatrix} 0 & 1 \\ -1 & 0 \end{vmatrix} - 2\begin{vmatrix} 1 & 1 \\ -1 & 0 \end{vmatrix} = 1 - 2(1) = -1,$

$|D_z| = \begin{vmatrix} 1 & 1 & 0 & 1 \\ 2 & 0 & 0 & 1 \\ 0 & 1 & 0 & 0 \\ 1 & 0 & 1 & 0 \end{vmatrix} = -1\begin{vmatrix} 1 & 1 & 1 \\ 2 & 0 & 1 \\ 0 & 1 & 0 \end{vmatrix} = -1\begin{vmatrix} 0 & 1 \\ 1 & 0 \end{vmatrix} + 2\begin{vmatrix} 1 & 1 \\ 1 & 0 \end{vmatrix} = -1(-1) + 2(-1) = -1,$ and

$|D_w| = \begin{vmatrix} 1 & 1 & 1 & 0 \\ 2 & 0 & 0 & 0 \\ 0 & 1 & -1 & 0 \\ 1 & 0 & 2 & 1 \end{vmatrix} = 1\begin{vmatrix} 1 & 1 & 1 \\ 2 & 0 & 0 \\ 0 & 1 & -1 \end{vmatrix} = -2\begin{vmatrix} 1 & 1 \\ 1 & -1 \end{vmatrix} = -2(-2) = 4.$ Hence, we have $x = \frac{|D_x|}{|D|} = \frac{-2}{-4} = \frac{1}{2}$,

$y = \frac{|D_y|}{|D|} = \frac{-1}{-4} = \frac{1}{4}$, $z = \frac{|D_z|}{|D|} = \frac{-1}{-4} = \frac{1}{4}$, and $w = \frac{|D_w|}{|D|} = \frac{4}{-4} = -1$, and the solution is $\left(\frac{1}{2}, \frac{1}{4}, \frac{1}{4}, -1\right)$.

57. Area $= \pm\frac{1}{2}\begin{vmatrix} 0 & 0 & 1 \\ 6 & 2 & 1 \\ 3 & 8 & 1 \end{vmatrix} = \pm\frac{1}{2}\begin{vmatrix} 6 & 2 \\ 3 & 8 \end{vmatrix} = \pm\frac{1}{2}(48-6) = \frac{1}{2}(42) = 21$

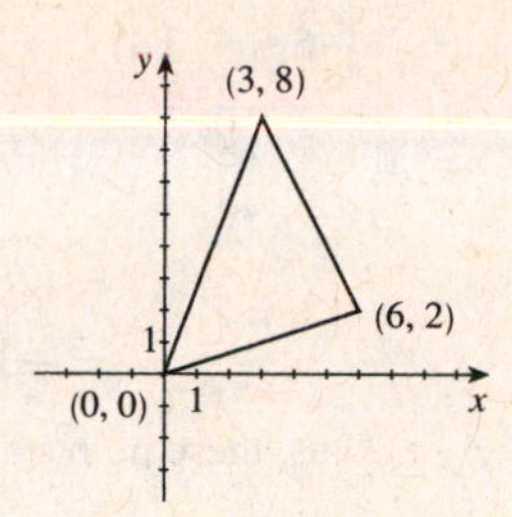

59. Area $= \pm\frac{1}{2}\begin{vmatrix} -1 & 3 & 1 \\ 2 & 9 & 1 \\ 5 & -6 & 1 \end{vmatrix} = \pm\frac{1}{2}\left[-1\begin{vmatrix} 9 & 1 \\ -6 & 1 \end{vmatrix} - 3\begin{vmatrix} 2 & 1 \\ 5 & 1 \end{vmatrix} + 1\begin{vmatrix} 2 & 9 \\ 5 & -6 \end{vmatrix}\right]$

$= \pm\frac{1}{2}[-1(9+6) - 3(2-5) + 1(-12-45)]$

$= \pm\frac{1}{2}[-15 - 3(-3) + (-57)] = \pm\frac{1}{2}(-63) = \frac{63}{2}$

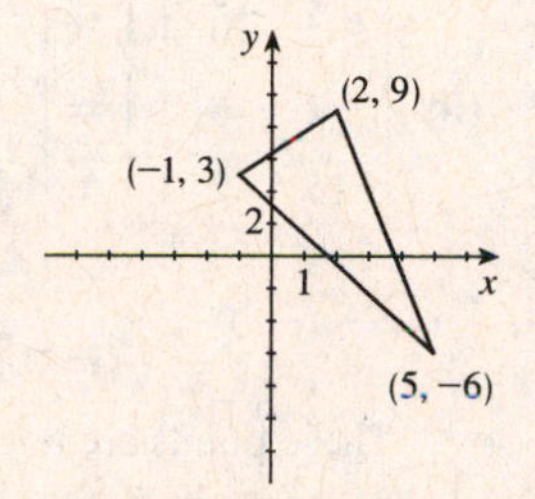

61. $\begin{vmatrix} a & 0 & 0 & 0 & 0 \\ 0 & b & 0 & 0 & 0 \\ 0 & 0 & c & 0 & 0 \\ 0 & 0 & 0 & d & 0 \\ 0 & 0 & 0 & 0 & e \end{vmatrix} = a\begin{vmatrix} b & 0 & 0 & 0 \\ 0 & c & 0 & 0 \\ 0 & 0 & d & 0 \\ 0 & 0 & 0 & e \end{vmatrix} = ab\begin{vmatrix} c & 0 & 0 \\ 0 & d & 0 \\ 0 & 0 & e \end{vmatrix} = abc\begin{vmatrix} d & 0 \\ 0 & e \end{vmatrix} = abcde$

63. $\begin{vmatrix} x & 12 & 13 \\ 0 & x-1 & 23 \\ 0 & 0 & x-2 \end{vmatrix} = 0 \Leftrightarrow (x-2)\begin{vmatrix} x & 12 \\ 0 & x-1 \end{vmatrix} = 0 \Leftrightarrow (x-2)\cdot x(x-1) = 0 \Leftrightarrow x = 0, 1, \text{ or } 2$

65. $\begin{vmatrix} 1 & 0 & x \\ x^2 & 1 & 0 \\ x & 0 & 1 \end{vmatrix} = 0 \Leftrightarrow 1\begin{vmatrix} 1 & 0 \\ 0 & 1 \end{vmatrix} + x\begin{vmatrix} x^2 & 1 \\ x & 0 \end{vmatrix} = 0 \Leftrightarrow 1 - x^2 = 0 \Leftrightarrow x^2 = 1 \Leftrightarrow x = \pm 1$

67. $\begin{vmatrix} 1 & x & x^2 \\ 1 & y & y^2 \\ 1 & z & z^2 \end{vmatrix} = 1\begin{vmatrix} y & y^2 \\ z & z^2 \end{vmatrix} - 1\begin{vmatrix} x & x^2 \\ z & z^2 \end{vmatrix} + 1\begin{vmatrix} x & x^2 \\ y & y^2 \end{vmatrix} = yz^2 - y^2z - \left(xz^2 - x^2z\right) + \left(xy^2 - xy^2\right)$

$= yz^2 - y^2z - xz^2 - x^2z + xy^2 - xy^2 + xyz - xyz = xyz - xz^2 - y^2z + yz^2 - x^2y + x^2z + zy^2 - xyz$

$= z\left(xy - xz - y^2 + yz\right) - x\left(xy - xz - y^2 + yz\right) = (z-x)\left(xy - xz - y^2 + yz\right)$

$= (z-x)\left[x(y-z) - y(y-z)\right] = (z-x)(x-y)(y-z)$

69. (a) If three points lie on a line then the area of the "triangle" they determine is 0, that is, $\pm\frac{1}{2}Q = \pm\frac{1}{2}\begin{vmatrix} a_1 & b_1 & 1 \\ a_2 & b_2 & 1 \\ a_3 & b_3 & 1 \end{vmatrix} = 0 \Leftrightarrow$ $Q = 0$. If the points are not collinear, then the point form a triangle, and the area of the triangle determined by these points is nonzero. If $Q = 0$, then $\pm\frac{1}{2}Q = \pm\frac{1}{2}(0) = 0$, so the "triangle" has no area, and the points are collinear.

(b) (i) $\begin{vmatrix} -6 & 4 & 1 \\ 2 & 10 & 1 \\ 6 & 13 & 1 \end{vmatrix} = \begin{vmatrix} 2 & 10 \\ 6 & 13 \end{vmatrix} - \begin{vmatrix} -6 & 4 \\ 6 & 13 \end{vmatrix} + \begin{vmatrix} -6 & 4 \\ 2 & 10 \end{vmatrix}$

$= (26 - 60) - (-78 - 24) + (-60 - 8)$

$= -34 + 104 - 68 = 0$

Thus, these points are collinear.

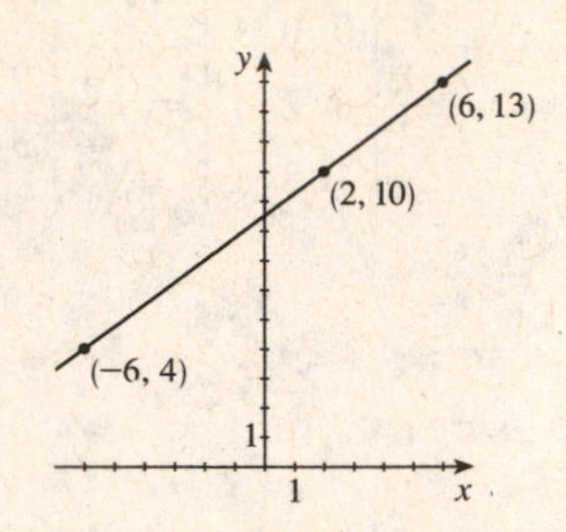

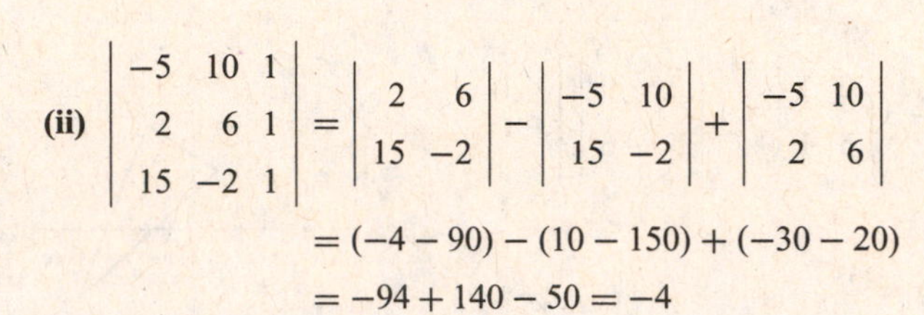

(ii) $\begin{vmatrix} -5 & 10 & 1 \\ 2 & 6 & 1 \\ 15 & -2 & 1 \end{vmatrix} = \begin{vmatrix} 2 & 6 \\ 15 & -2 \end{vmatrix} - \begin{vmatrix} -5 & 10 \\ 15 & -2 \end{vmatrix} + \begin{vmatrix} -5 & 10 \\ 2 & 6 \end{vmatrix}$

$= (-4 - 90) - (10 - 150) + (-30 - 20)$

$= -94 + 140 - 50 = -4$

These points are not collinear. Note that this is difficult to determine from the diagram.

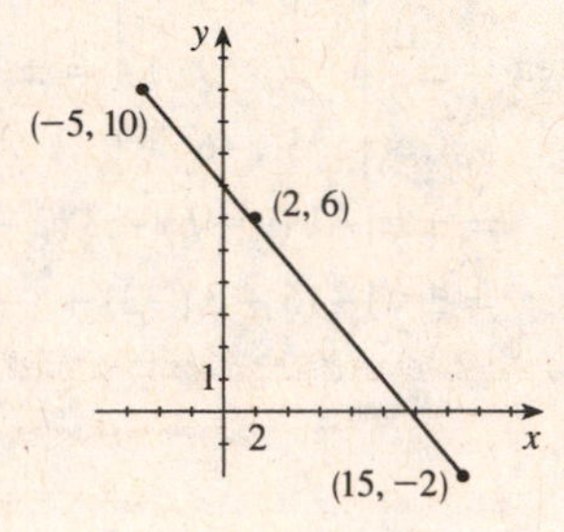

71. (a) Let x be the amount of apples, y the amount of peaches, and z the amount of pears (in pounds).

We get the model $\begin{cases} x + y + z = 18 \\ 0.75x + 0.90y + 0.60z = 13.80 \\ -0.75x + 0.90y + 0.60z = 1.80 \end{cases}$

(b) $|D| = \begin{vmatrix} 1 & 1 & 1 \\ 0.75 & 0.90 & 0.60 \\ -0.75 & 0.90 & 0.60 \end{vmatrix} = 1 \cdot \begin{vmatrix} 0.90 & 0.60 \\ 0.90 & 0.60 \end{vmatrix} - 1 \cdot \begin{vmatrix} 0.75 & 0.60 \\ -0.75 & 0.60 \end{vmatrix} + 1 \cdot \begin{vmatrix} 0.75 & 0.90 \\ -0.75 & 0.90 \end{vmatrix} = 0 - 0.90 + 1.35 = 0.45,$

$|D_x| = \begin{vmatrix} 18 & 1 & 1 \\ 13.80 & 0.90 & 0.60 \\ 1.80 & 0.90 & 0.60 \end{vmatrix} = 18 \cdot \begin{vmatrix} 0.90 & 0.60 \\ 0.90 & 0.60 \end{vmatrix} - 1 \cdot \begin{vmatrix} 13.80 & 0.60 \\ 1.80 & 0.60 \end{vmatrix} + 1 \cdot \begin{vmatrix} 13.80 & 0.90 \\ 1.80 & 0.90 \end{vmatrix} = 0 - 7.2 + 10.8 = 3.6,$

$|D_y| = \begin{vmatrix} 1 & 18 & 1 \\ 0.75 & 13.80 & 0.60 \\ -0.75 & 1.80 & 0.60 \end{vmatrix} = 1 \cdot \begin{vmatrix} 13.80 & 0.60 \\ 1.80 & 0.60 \end{vmatrix} - 18 \cdot \begin{vmatrix} 0.75 & 0.60 \\ -0.75 & 0.60 \end{vmatrix} + 1 \cdot \begin{vmatrix} 0.75 & 13.80 \\ -0.75 & 1.80 \end{vmatrix}$

$= 7.2 - 16.2 + 11.7 = 2.7$, and

$|D_z| = \begin{vmatrix} 1 & 1 & 18 \\ 0.75 & 0.90 & 13.80 \\ -0.75 & 0.90 & 1.80 \end{vmatrix} = 1 \cdot \begin{vmatrix} 0.90 & 13.80 \\ 0.90 & 1.80 \end{vmatrix} - 1 \cdot \begin{vmatrix} 0.75 & 13.80 \\ -0.75 & 1.80 \end{vmatrix} + 18 \cdot \begin{vmatrix} 0.75 & 0.90 \\ -0.75 & 0.90 \end{vmatrix}$

$= -10.8 - 11.7 + 24.3 = 1.8.$

So $x = \dfrac{|D_x|}{|D|} = \dfrac{3.6}{0.45} = 8$; $y = \dfrac{|D_y|}{|D|} = \dfrac{27}{0.45} = 6$; and $z = \dfrac{|D_z|}{|D|} = \dfrac{1.8}{0.45} = 4.$

Thus, Muriel buys 8 pounds of apples, 6 pounds of peaches, and 4 pounds of pears.

73. Using the determinant formula for the area of a triangle, we have

$$\text{Area} = \pm\frac{1}{2}\begin{vmatrix} 1000 & 2000 & 1 \\ 5000 & 4000 & 1 \\ 2000 & 6000 & 1 \end{vmatrix} = \pm\frac{1}{2}\left(1\cdot\begin{vmatrix} 5000 & 4000 \\ 2000 & 6000 \end{vmatrix} - 1\cdot\begin{vmatrix} 1000 & 2000 \\ 2000 & 6000 \end{vmatrix} + 1\cdot\begin{vmatrix} 1000 & 2000 \\ 5000 & 4000 \end{vmatrix}\right)$$

$$= \pm\tfrac{1}{2}\cdot(22{,}000{,}000 - 2{,}000{,}000 - 6{,}000{,}000) = \pm\tfrac{1}{2}\cdot 14{,}000{,}000 = \pm 7{,}000{,}000$$

Thus, the area is 7,000,000 ft^2.

75. **(a)** If A is a matrix with a row or column consisting entirely of zeros, then if we expand the determinant by this row or column, we get $|A| = 0\cdot|A_{1j}| - 0\cdot|A_{2j}| + \cdots - 0\cdot|A_{nj}| = 0$.

(b) Use the principle that if matrix B is a square matrix obtained from A by adding a multiple of one row to another, or a multiple of one column to another, then $|A| = |B|$. If we let B be the matrix obtained by subtracting the two rows (or columns) that are the same, then matrix B will have a row or column that consists entirely of zeros. So $|B| = 0 \Rightarrow |A| = 0$.

(c) Again use the principle that if matrix B is a square matrix obtained from A by adding a multiple of one row to another, or a multiple of one column to another, then $|A| = |B|$. If we let B be the matrix obtained by subtracting the proper multiple of the row (or column) from the other similar row (or column), then matrix B will have a row or column that consists entirely of zeros. So $|B| = 0 \Rightarrow |A| = 0$.

CHAPTER 11 REVIEW

1. **(a)** 2×3

(b) Yes, this matrix is in row-echelon form.

(c) No, this matrix is not in reduced row-echelon form, since the leading 1 in the second row does not have a 0 above it.

(d) $\begin{cases} x + 2y = -5 \\ \quad\;\; y = 3 \end{cases}$

3. **(a)** 3×4

(b) Yes, this matrix is in row-echelon form.

(c) Yes, this matrix is in reduced row-echelon form.

(d) $\begin{cases} x \quad\; + 8z = 0 \\ \quad y + 5z = -1 \\ \qquad\quad\; 0 = 0 \end{cases}$

5. **(a)** 3×4

(b) No, this matrix is not in row-echelon form. The leading 1 in the second row is not to the left of the one above it.

(c) No, this matrix is not in reduced row-echelon form.

(d) $\begin{cases} \quad\;\; y - 3z = 4 \\ x + y \qquad = 7 \\ x + 2y + z = 2 \end{cases}$

7. $\begin{bmatrix} 1 & 2 & 2 & 6 \\ 1 & -1 & 0 & -1 \\ 2 & 1 & 3 & 7 \end{bmatrix} \xrightarrow{R_2 \leftrightarrow R_1} \begin{bmatrix} 1 & -1 & 0 & -1 \\ 1 & 2 & 2 & 6 \\ 2 & 1 & 3 & 7 \end{bmatrix} \xrightarrow[R_3 - 2R_1 \to R_3]{R_2 - R_1 \to R_2} \begin{bmatrix} 1 & -1 & 0 & -1 \\ 0 & 3 & 2 & 7 \\ 0 & 3 & 3 & 9 \end{bmatrix} \xrightarrow{R_3 - R_2 \to R_3} \begin{bmatrix} 1 & -1 & 0 & -1 \\ 0 & 3 & 2 & 7 \\ 0 & 0 & 1 & 2 \end{bmatrix}$.

Thus, $z = 2$, $3y + 2(2) = 7 \Leftrightarrow 3y = 3 \Leftrightarrow y = 1$, and $x - (1) = -1 \Leftrightarrow x = 0$, and so the solution is $(0, 1, 2)$.

9. $\begin{bmatrix} 1 & -2 & 3 & -2 \\ 2 & -1 & 1 & 2 \\ 2 & -7 & 11 & -9 \end{bmatrix} \xrightarrow[R_3 - 2R_1 \to R_3]{R_2 - 2R_1 \to R_2} \begin{bmatrix} 1 & -2 & 3 & -2 \\ 0 & 3 & -5 & 6 \\ 0 & -3 & 5 & -5 \end{bmatrix} \xrightarrow{R_3 + R_2 \to R_3} \begin{bmatrix} 1 & -2 & 3 & -2 \\ 0 & 3 & -5 & 6 \\ 0 & 0 & 0 & 1 \end{bmatrix}$.

The last row corresponds to the equation $0 = 1$, which is always false. Thus, there is no solution.

11. $\begin{bmatrix} 1 & 1 & 1 & 1 & 0 \\ 1 & -1 & -4 & -1 & -1 \\ 1 & -2 & 0 & 4 & -7 \\ 2 & 2 & 3 & 4 & -3 \end{bmatrix} \xrightarrow[\substack{R_3 - R_1 \to R_3 \\ R_4 - 2R_1 \to R_4}]{R_2 - R_1 \to R_2} \begin{bmatrix} 1 & 1 & 1 & 1 & 0 \\ 0 & -2 & -5 & -2 & -1 \\ 0 & -3 & -1 & 3 & -7 \\ 0 & 0 & 1 & 2 & -3 \end{bmatrix} \xrightarrow{-R_3 + R_2 \to R_3} \begin{bmatrix} 1 & 1 & 1 & 1 & 0 \\ 0 & 1 & -4 & -5 & 6 \\ 0 & -3 & -1 & 3 & -7 \\ 0 & 0 & 1 & 2 & -3 \end{bmatrix}$

$\xrightarrow{R_3 + 3R_2 \to R_3} \begin{bmatrix} 1 & 1 & 1 & 1 & 0 \\ 0 & 1 & -4 & -5 & 6 \\ 0 & 0 & -13 & -12 & 11 \\ 0 & 0 & 1 & 2 & -3 \end{bmatrix} \xrightarrow{R_3 \leftrightarrow R_4} \begin{bmatrix} 1 & 1 & 1 & 1 & 0 \\ 0 & 1 & -4 & -5 & 6 \\ 0 & 0 & 1 & 2 & -3 \\ 0 & 0 & -13 & -12 & 11 \end{bmatrix} \xrightarrow{R_4 + 13R_3 \to R_4} \begin{bmatrix} 1 & 1 & 1 & 1 & 0 \\ 0 & 1 & -4 & -5 & 6 \\ 0 & 0 & 1 & 2 & -3 \\ 0 & 0 & 0 & 14 & -28 \end{bmatrix}$.

Therefore, $14w = -28 \Leftrightarrow w = -2$, $z + 2(-2) = -3 \Leftrightarrow z = 1$, $y - 4(1) - 5(-2) = 6 \Leftrightarrow y = 0$, and $x + 0 + 1 + (-2) = 0 \Leftrightarrow x = 1$. So the solution is $(1, 0, 1, -2)$.

13. $\begin{bmatrix} 1 & -1 & 3 & 2 \\ 2 & 1 & 1 & 2 \\ 3 & 0 & 4 & 4 \end{bmatrix} \xrightarrow[R_3 - 3R_1 \to R_3]{R_2 - 2R_1 \to R_2} \begin{bmatrix} 1 & -1 & 3 & 2 \\ 0 & 3 & -5 & -2 \\ 0 & 3 & -5 & -2 \end{bmatrix} \xrightarrow{R_3 - R_2 \to R_3} \begin{bmatrix} 1 & -1 & 3 & 2 \\ 0 & 3 & -5 & -2 \\ 0 & 0 & 0 & 0 \end{bmatrix} \xrightarrow{\frac{1}{3}R_2}$

$\begin{bmatrix} 1 & -1 & 3 & 2 \\ 0 & 1 & -\frac{5}{3} & -\frac{2}{3} \\ 0 & 0 & 0 & 0 \end{bmatrix} \xrightarrow{R_1 + R_2 \to R_1} \begin{bmatrix} 1 & 0 & \frac{4}{3} & \frac{4}{3} \\ 0 & 1 & -\frac{5}{3} & -\frac{2}{3} \\ 0 & 0 & 0 & 0 \end{bmatrix}$. The system is dependent, so let $z = t$: $y - \frac{5}{3}t = -\frac{2}{3} \Leftrightarrow$ $y = \frac{5}{3}t - \frac{2}{3}$ and $x + \frac{4}{3}t = \frac{4}{3} \Leftrightarrow x = -\frac{4}{3}t + \frac{4}{3}$. So the solution is $\left(-\frac{4}{3}t + \frac{4}{3}, \frac{5}{3}t - \frac{2}{3}, t\right)$, where t is any real number.

15. $\begin{bmatrix} 1 & -1 & 1 & -1 & 0 \\ 3 & -1 & -1 & -1 & 2 \end{bmatrix} \xrightarrow{R_2 - 3R_1 \to R_2} \begin{bmatrix} 1 & -1 & 1 & -1 & 0 \\ 0 & 2 & -4 & 2 & 2 \end{bmatrix} \xrightarrow{\frac{1}{2}R_4} \begin{bmatrix} 1 & -1 & 1 & -1 & 0 \\ 0 & 1 & -2 & 1 & 1 \end{bmatrix} \xrightarrow{R_1 + R_2 \to R_1}$

$\begin{bmatrix} 1 & 0 & -1 & 0 & 1 \\ 0 & 1 & -2 & 1 & 1 \end{bmatrix}$. Since the system is dependent, Let $z = s$ and $w = t$. Then $y - 2s + t = 1 \Leftrightarrow y = 2s - t + 1$ and $x - s = 1 \Leftrightarrow x = s + 1$. So the solution is $(s + 1, 2s - t + 1, s, t)$, where s and t are any real numbers.

17. $\begin{bmatrix} 1 & -1 & 1 & 0 \\ 3 & 2 & -1 & 6 \\ 1 & 4 & -3 & 3 \end{bmatrix} \xrightarrow[R_3 - R_1 \to R_3]{R_2 - 3R_1 \to R_2} \begin{bmatrix} 1 & -1 & 1 & 0 \\ 0 & 5 & -4 & 6 \\ 0 & 5 & -4 & 3 \end{bmatrix} \xrightarrow{R_3 - R_2 \to R_3} \begin{bmatrix} 1 & -1 & 1 & 0 \\ 0 & 5 & -4 & 6 \\ 0 & 0 & 0 & 3 \end{bmatrix}$. The last row of this matrix corresponds to the equation $0 = 3$, which is always false. Hence there is no solution.

19. $\begin{bmatrix} 1 & 1 & -1 & -1 & 2 \\ 1 & -1 & 1 & -1 & 0 \\ 2 & 0 & 0 & 2 & 2 \\ 2 & 4 & -4 & -2 & 6 \end{bmatrix} \xrightarrow{R_1 \leftrightarrow \frac{1}{2}R_3} \begin{bmatrix} 1 & 0 & 0 & 1 & 1 \\ 1 & -1 & 1 & -1 & 0 \\ 1 & 1 & -1 & -1 & 2 \\ 2 & 4 & -4 & -2 & 6 \end{bmatrix} \xrightarrow[R_4 - 2R_1 \to R_4]{\substack{R_2 - R_1 \to R_2 \\ R_3 - R_1 \to R_3}} \begin{bmatrix} 1 & 0 & 0 & 1 & 1 \\ 0 & -1 & 1 & -2 & -1 \\ 0 & 1 & -1 & -2 & 1 \\ 0 & 4 & -4 & -4 & 4 \end{bmatrix}$

$\xrightarrow[R_4 + 4R_2 \to R_4]{R_3 + R_2 \to R_3} \begin{bmatrix} 1 & 0 & 0 & 1 & 1 \\ 0 & -1 & 1 & -2 & -1 \\ 0 & 0 & 0 & -4 & 0 \\ 0 & 0 & 0 & -12 & 0 \end{bmatrix} \xrightarrow[-\frac{1}{12}R_4]{-\frac{1}{4}R_3} \begin{bmatrix} 1 & 0 & 0 & 1 & 1 \\ 0 & -1 & 1 & -2 & -1 \\ 0 & 0 & 0 & 0 & 0 \\ 0 & 0 & 0 & 1 & 0 \end{bmatrix} \xrightarrow[R_4 - R_3 \to R_4]{\substack{R_1 - R_3 \to R_1 \\ R_2 + 2R_3 \to R_2}}$

$\begin{bmatrix} 1 & 0 & 0 & 0 & 1 \\ 0 & -1 & 1 & 0 & -1 \\ 0 & 0 & 0 & 1 & 0 \\ 0 & 0 & 0 & 0 & 0 \end{bmatrix}$. This system is dependent. Let $z = t$, so $-y + t = -1 \Leftrightarrow y = t + 1$; $x = 1 \Leftrightarrow x = 1$. So the solution is $(1, t + 1, t, 0)$, where t is any real number.

21. A (3×3) and B (2×3) have different dimensions, so they are not equal.

In Solutions 23–33, the matrices A, B, C, D, E, F, and G are defined as follows:

$$A = \begin{bmatrix} 2 & 0 & -1 \end{bmatrix} \qquad B = \begin{bmatrix} 1 & 2 & 4 \\ -2 & 1 & 0 \end{bmatrix} \qquad C = \begin{bmatrix} \frac{1}{2} & 3 \\ 2 & \frac{3}{2} \\ -2 & 1 \end{bmatrix}$$

$$D = \begin{bmatrix} 1 & 4 \\ 0 & -1 \\ 2 & 0 \end{bmatrix} \qquad E = \begin{bmatrix} 2 & -1 \\ -\frac{1}{2} & 1 \end{bmatrix} \qquad F = \begin{bmatrix} 4 & 0 & 2 \\ -1 & 1 & 0 \\ 7 & 5 & 0 \end{bmatrix} \qquad G = \begin{bmatrix} 5 \end{bmatrix}$$

23. $A + B$ is not defined because the matrix dimensions 1×3 and 2×3 are not compatible.

25. $2C + 3D = 2\begin{bmatrix} \frac{1}{2} & 3 \\ 2 & \frac{3}{2} \\ -2 & 1 \end{bmatrix} + 3\begin{bmatrix} 1 & 4 \\ 0 & -1 \\ 2 & 0 \end{bmatrix} = \begin{bmatrix} 1 & 6 \\ 4 & 3 \\ -4 & 2 \end{bmatrix} + \begin{bmatrix} 3 & 12 \\ 0 & -3 \\ 6 & 0 \end{bmatrix} = \begin{bmatrix} 4 & 18 \\ 4 & 0 \\ 2 & 2 \end{bmatrix}$

27. $GA = \begin{bmatrix} 5 \end{bmatrix}\begin{bmatrix} 2 & 0 & -1 \end{bmatrix} = \begin{bmatrix} 10 & 0 & -5 \end{bmatrix}$

29. $BC = \begin{bmatrix} 1 & 2 & 4 \\ -2 & 1 & 0 \end{bmatrix}\begin{bmatrix} \frac{1}{2} & 3 \\ 2 & \frac{3}{2} \\ -2 & 1 \end{bmatrix} = \begin{bmatrix} -\frac{7}{2} & 10 \\ 1 & -\frac{9}{2} \end{bmatrix}$

31. $BF = \begin{bmatrix} 1 & 2 & 4 \\ -2 & 1 & 0 \end{bmatrix}\begin{bmatrix} 4 & 0 & 2 \\ -1 & 1 & 0 \\ 7 & 5 & 0 \end{bmatrix} = \begin{bmatrix} 30 & 22 & 2 \\ -9 & 1 & -4 \end{bmatrix}$

33. $(C + D)E = \left(\begin{bmatrix} \frac{1}{2} & 3 \\ 2 & \frac{3}{2} \\ -2 & 1 \end{bmatrix} + \begin{bmatrix} 1 & 4 \\ 0 & -1 \\ 2 & 0 \end{bmatrix}\right)\begin{bmatrix} 2 & -1 \\ -\frac{1}{2} & 1 \end{bmatrix} = \begin{bmatrix} \frac{3}{2} & 7 \\ 2 & \frac{1}{2} \\ 0 & 1 \end{bmatrix}\begin{bmatrix} 2 & -1 \\ -\frac{1}{2} & 1 \end{bmatrix} = \begin{bmatrix} -\frac{1}{2} & \frac{11}{2} \\ \frac{15}{4} & -\frac{3}{2} \\ -\frac{1}{2} & 1 \end{bmatrix}$

In Solutions 35–43, the matrices A and B are defined as follows:

$$A = \begin{bmatrix} 3 & 0 & -3 \\ -2 & 1 & 2 \\ 1 & 6 & 0 \end{bmatrix} \qquad B = \begin{bmatrix} -1 & 4 & -1 \\ 1 & -1 & 0 \\ -2 & 0 & 2 \end{bmatrix}$$

35. $AB^2 = \begin{bmatrix} 27 & 0 & -21 \\ -20 & 5 & 13 \\ -5 & 22 & -7 \end{bmatrix}$

37. $A^{-1}BA = \begin{bmatrix} 14 & 26 & -8 \\ -3 & -\frac{7}{3} & \frac{7}{3} \\ 18 & \frac{80}{3} & -\frac{35}{3} \end{bmatrix}$

39. $|AB| = -12$

41. $\left|A^{-1}\right| = \dfrac{1}{3}$

43. $\left|A^{-1}BA\right| = -4$

45. $AB = \begin{bmatrix} 2 & -5 \\ -2 & 6 \end{bmatrix}\begin{bmatrix} 3 & \frac{5}{2} \\ 1 & 1 \end{bmatrix} = \begin{bmatrix} 1 & 0 \\ 0 & 1 \end{bmatrix}$ and $BA = \begin{bmatrix} 3 & \frac{5}{2} \\ 1 & 1 \end{bmatrix}\begin{bmatrix} 2 & -5 \\ -2 & 6 \end{bmatrix} = \begin{bmatrix} 1 & 0 \\ 0 & 1 \end{bmatrix}$.

In Solutions 47–51, $A = \begin{bmatrix} 2 & 1 \\ 3 & 2 \end{bmatrix}$, $B = \begin{bmatrix} 1 & -2 \\ -2 & 4 \end{bmatrix}$, **and** $C = \begin{bmatrix} 0 & 1 & 3 \\ -2 & 4 & 0 \end{bmatrix}$.

47. $A + 3X = B \Leftrightarrow 3X = B - A \Leftrightarrow X = \frac{1}{3}(B - A)$. Thus, $X = \frac{1}{3}\left(\begin{bmatrix} 1 & -2 \\ -2 & 4 \end{bmatrix} - \begin{bmatrix} 2 & 1 \\ 3 & 2 \end{bmatrix}\right) = \frac{1}{3}\begin{bmatrix} -1 & -3 \\ -5 & 2 \end{bmatrix}$.

49. $2(X - A) = 3B \Leftrightarrow X - A = \frac{3}{2}B \Leftrightarrow X = A + \frac{3}{2}B$. Thus,

$$X = \begin{bmatrix} 2 & 1 \\ 3 & 2 \end{bmatrix} + \frac{3}{2}\begin{bmatrix} 1 & -2 \\ -2 & 4 \end{bmatrix} = \begin{bmatrix} 2 & 1 \\ 3 & 2 \end{bmatrix} + \begin{bmatrix} \frac{3}{2} & -3 \\ -3 & 6 \end{bmatrix} = \begin{bmatrix} \frac{7}{2} & -2 \\ 0 & 8 \end{bmatrix}.$$

51. $AX = C \Leftrightarrow A^{-1}AX = X = A^{-1}C$. Now

$$A^{-1} = \frac{1}{4-3}\begin{bmatrix} 2 & -1 \\ -3 & 2 \end{bmatrix} = \begin{bmatrix} 2 & -1 \\ -3 & 2 \end{bmatrix}. \text{ Thus, } X = A^{-1}C = \begin{bmatrix} 2 & -1 \\ -3 & 2 \end{bmatrix}\begin{bmatrix} 0 & 1 & 3 \\ -2 & 4 & 0 \end{bmatrix} = \begin{bmatrix} 2 & -2 & 6 \\ -4 & 5 & -9 \end{bmatrix}.$$

53. $D = \begin{bmatrix} 1 & 4 \\ 2 & 9 \end{bmatrix}$. Then $|D| = 1(9) - 2(4) = 1$, and so $D^{-1} = \begin{bmatrix} 9 & -4 \\ -2 & 1 \end{bmatrix}$.

55. $D = \begin{bmatrix} 4 & -12 \\ -2 & 6 \end{bmatrix}$. Then $|D| = 4(6) - 2(12) = 0$, and so D has no inverse.

57. $D = \begin{bmatrix} 3 & 0 & 1 \\ 2 & -3 & 0 \\ 4 & -2 & 1 \end{bmatrix}$. Then, $|D| = 1\begin{vmatrix} 2 & -3 \\ 4 & -2 \end{vmatrix} + 1\begin{vmatrix} 3 & 0 \\ 2 & -3 \end{vmatrix} = -4 + 12 - 9 = -1$. So D^{-1} exists.

$$\begin{bmatrix} 3 & 0 & 1 & 1 & 0 & 0 \\ 2 & -3 & 0 & 0 & 1 & 0 \\ 4 & -2 & 1 & 0 & 0 & 1 \end{bmatrix} \xrightarrow{R_1 - R_2 \to R_1} \begin{bmatrix} 1 & 3 & 1 & 1 & -1 & 0 \\ 2 & -3 & 0 & 0 & 1 & 0 \\ 4 & -2 & 1 & 0 & 0 & 1 \end{bmatrix} \xrightarrow[R_3 - 4R_1 \to R_3]{R_2 - 2R_1 \to R_2} \begin{bmatrix} 1 & 3 & 1 & 1 & -1 & 0 \\ 0 & -9 & -2 & -2 & 3 & 0 \\ 0 & -14 & -3 & -4 & 4 & 1 \end{bmatrix} \xrightarrow[-2R_3]{-3R_2}$$

$$\begin{bmatrix} 1 & 3 & 1 & 1 & -1 & 0 \\ 0 & 27 & 6 & 6 & -9 & 0 \\ 0 & 28 & 6 & 8 & -8 & -2 \end{bmatrix} \xrightarrow{R_3 - R_2 \to R_3} \begin{bmatrix} 1 & 3 & 1 & 1 & -1 & 0 \\ 0 & 27 & 6 & 6 & -9 & 0 \\ 0 & 1 & 0 & 2 & 1 & -2 \end{bmatrix} \xrightarrow[\frac{1}{3}R_3]{R_3 \leftrightarrow R_2} \begin{bmatrix} 1 & 3 & 1 & 1 & -1 & 0 \\ 0 & 1 & 0 & 2 & 1 & -2 \\ 0 & 9 & 2 & 2 & -3 & 0 \end{bmatrix} \xrightarrow[R_1 - 3R_2 \to R_1]{R_3 - 9R_2 \to R_3}$$

$$\begin{bmatrix} 1 & 0 & 1 & -5 & -4 & 6 \\ 0 & 1 & 0 & 2 & 1 & -2 \\ 0 & 0 & 2 & -16 & -12 & 18 \end{bmatrix} \xrightarrow[R_1 - R_3 \to R_1]{\frac{1}{2}R_3} \begin{bmatrix} 1 & 0 & 0 & 3 & 2 & -3 \\ 0 & 1 & 0 & 2 & 1 & -2 \\ 0 & 0 & 1 & -8 & -6 & 9 \end{bmatrix}. \text{ Thus, } D^{-1} = \begin{bmatrix} 3 & 2 & -3 \\ 2 & 1 & -2 \\ -8 & -6 & 9 \end{bmatrix}.$$

59. $D = \begin{bmatrix} 1 & 0 & 0 & 1 \\ 0 & 2 & 0 & 2 \\ 0 & 0 & 3 & 3 \\ 0 & 0 & 0 & 4 \end{bmatrix}$. Thus, $|D| = \begin{vmatrix} 2 & 0 & 2 \\ 0 & 3 & 3 \\ 0 & 0 & 4 \end{vmatrix} = 2\begin{vmatrix} 3 & 3 \\ 0 & 4 \end{vmatrix} = 24$ and D^{-1} exists. $\left[\begin{array}{cccc|cccc} 1 & 0 & 0 & 1 & 1 & 0 & 0 & 0 \\ 0 & 2 & 0 & 2 & 0 & 1 & 0 & 0 \\ 0 & 0 & 3 & 3 & 0 & 0 & 1 & 0 \\ 0 & 0 & 0 & 4 & 0 & 0 & 0 & 1 \end{array}\right] \xrightarrow[\frac{1}{4}R_4]{\frac{1}{2}R_2 \; \frac{1}{3}R_3}$

$\begin{bmatrix} 1 & 0 & 0 & 1 & 1 & 0 & 0 & 0 \\ 0 & 1 & 0 & 1 & 0 & \frac{1}{2} & 0 & 0 \\ 0 & 0 & 1 & 1 & 0 & 0 & \frac{1}{3} & 0 \\ 0 & 0 & 0 & 1 & 0 & 0 & 0 & \frac{1}{4} \end{bmatrix} \xrightarrow[R_3 - R_4 \to R_3]{R_1 - R_4 \to R_1 \;\; R_2 - R_4 \to R_2} \begin{bmatrix} 1 & 0 & 0 & 0 & 1 & 0 & 0 & -\frac{1}{4} \\ 0 & 1 & 0 & 0 & 0 & \frac{1}{2} & 0 & -\frac{1}{4} \\ 0 & 0 & 1 & 0 & 0 & 0 & \frac{1}{3} & -\frac{1}{4} \\ 0 & 0 & 0 & 1 & 0 & 0 & 0 & \frac{1}{4} \end{bmatrix}$. Therefore, $D^{-1} = \begin{bmatrix} 1 & 0 & 0 & -\frac{1}{4} \\ 0 & \frac{1}{2} & 0 & -\frac{1}{4} \\ 0 & 0 & \frac{1}{3} & -\frac{1}{4} \\ 0 & 0 & 0 & \frac{1}{4} \end{bmatrix}$.

61. $\begin{bmatrix} 12 & -5 \\ 5 & -2 \end{bmatrix}\begin{bmatrix} x \\ y \end{bmatrix} = \begin{bmatrix} 10 \\ 17 \end{bmatrix}$. If we let $A = \begin{bmatrix} 12 & -5 \\ 5 & -2 \end{bmatrix}$, then $A^{-1} = \frac{1}{-24+25}\begin{bmatrix} -2 & 5 \\ -5 & 12 \end{bmatrix} = \begin{bmatrix} -2 & 5 \\ -5 & 12 \end{bmatrix}$, and so

$\begin{bmatrix} x \\ y \end{bmatrix} = \begin{bmatrix} -2 & 5 \\ -5 & 12 \end{bmatrix}\begin{bmatrix} 10 \\ 17 \end{bmatrix} = \begin{bmatrix} 65 \\ 154 \end{bmatrix}$. Therefore, the solution is $(65, 154)$.

63. $\begin{bmatrix} 2 & 1 & 5 \\ 1 & 2 & 2 \\ 1 & 0 & 3 \end{bmatrix}\begin{bmatrix} x \\ y \\ z \end{bmatrix} = \begin{bmatrix} \frac{1}{3} \\ \frac{1}{4} \\ \frac{1}{6} \end{bmatrix}$. Let $A = \begin{bmatrix} 2 & 1 & 5 \\ 1 & 2 & 2 \\ 1 & 0 & 3 \end{bmatrix}$. Then $\begin{bmatrix} 2 & 1 & 5 & 1 & 0 & 0 \\ 1 & 2 & 2 & 0 & 1 & 0 \\ 1 & 0 & 3 & 0 & 0 & 1 \end{bmatrix} \xrightarrow{R_1 \leftrightarrow R_2} \begin{bmatrix} 1 & 2 & 2 & 0 & 1 & 0 \\ 2 & 1 & 5 & 1 & 0 & 0 \\ 1 & 0 & 3 & 0 & 0 & 1 \end{bmatrix}$

$\xrightarrow[R_3 - R_1 \to R_3]{R_2 - 2R_1 \to R_2} \begin{bmatrix} 1 & 2 & 2 & 0 & 1 & 0 \\ 0 & -3 & 1 & 1 & -2 & 0 \\ 0 & -2 & 1 & 0 & -1 & 1 \end{bmatrix} \xrightarrow{R_2 - 2R_3 \to R_2} \begin{bmatrix} 1 & 2 & 2 & 0 & 1 & 0 \\ 0 & 1 & -1 & 1 & 0 & -2 \\ 0 & -2 & 1 & 0 & -1 & 1 \end{bmatrix} \xrightarrow[R_3 \to R_3 + 2R_2]{R_1 - 2R_2 \to R_1}$

$\begin{bmatrix} 1 & 0 & 4 & -2 & 1 & 4 \\ 0 & 1 & -1 & 1 & 0 & -2 \\ 0 & 0 & -1 & 2 & -1 & -3 \end{bmatrix} \xrightarrow{-R_3} \begin{bmatrix} 1 & 0 & 4 & -2 & 1 & 4 \\ 0 & 1 & -1 & 1 & 0 & -2 \\ 0 & 0 & 1 & -2 & 1 & 3 \end{bmatrix} \xrightarrow[R_2 + R_3 \to R_2]{R_1 - 4R_3 \to R_1} \begin{bmatrix} 1 & 0 & 0 & 6 & -3 & -8 \\ 0 & 1 & 0 & -1 & 1 & 1 \\ 0 & 0 & 1 & -2 & 1 & 3 \end{bmatrix}$.

Hence, $A^{-1} = \begin{bmatrix} 6 & -3 & -8 \\ -1 & 1 & 1 \\ -2 & 1 & 3 \end{bmatrix}$ and $\begin{bmatrix} x \\ y \\ z \end{bmatrix} = \begin{bmatrix} 6 & -3 & -8 \\ -1 & 1 & 1 \\ -2 & 1 & 3 \end{bmatrix}\begin{bmatrix} \frac{1}{3} \\ \frac{1}{4} \\ \frac{1}{6} \end{bmatrix} = \begin{bmatrix} -\frac{1}{12} \\ \frac{1}{12} \\ \frac{1}{12} \end{bmatrix}$, and so the solution is $\left(-\frac{1}{12}, \frac{1}{12}, \frac{1}{12}\right)$.

65. (a) The (i, j)th entry of A represents how many pounds of vegetable j were sold on day i, and the ith entry of B represents the price of vegetable i.

(b) $AB = \begin{bmatrix} 25 & 16 & 30 \\ 14 & 12 & 16 \end{bmatrix}\begin{bmatrix} 1.50 \\ 1.00 \\ 0.50 \end{bmatrix} = \begin{bmatrix} 68.5 \\ 41.0 \end{bmatrix}$. The jth entry of AB represents the total revenue on day j.

67. $|D| = \begin{vmatrix} 2 & 7 \\ 6 & 16 \end{vmatrix} = 32 - 42 = -10$, $|D_x| = \begin{vmatrix} 13 & 7 \\ 30 & 16 \end{vmatrix} = 208 - 210 = -2$, and $|D_y| = \begin{vmatrix} 2 & 13 \\ 6 & 30 \end{vmatrix} = 60 - 78 = -18$.

Therefore, $x = \frac{-2}{-10} = \frac{1}{5}$ and $y = \frac{-18}{-10} = \frac{9}{5}$, and so the solution is $\left(\frac{1}{5}, \frac{9}{5}\right)$.

69. $|D| = \begin{vmatrix} 2 & -1 & 5 \\ -1 & 7 & 0 \\ 5 & 4 & 3 \end{vmatrix} = 5\begin{vmatrix} -1 & 7 \\ 5 & 4 \end{vmatrix} + 3\begin{vmatrix} 2 & -1 \\ -1 & 7 \end{vmatrix} = -195 + 39 = -156,$

$|D_x| = \begin{vmatrix} 0 & -1 & 5 \\ 9 & 7 & 0 \\ -9 & 4 & 3 \end{vmatrix} = 5\begin{vmatrix} 9 & 7 \\ -9 & 4 \end{vmatrix} + 3\begin{vmatrix} 0 & -1 \\ 9 & 7 \end{vmatrix} = 495 + 27 = 522,$

$|D_y| = \begin{vmatrix} 2 & 0 & 5 \\ -1 & 9 & 0 \\ 5 & -9 & 3 \end{vmatrix} = 5\begin{vmatrix} -1 & 9 \\ 5 & -9 \end{vmatrix} + 3\begin{vmatrix} 2 & 0 \\ -1 & 9 \end{vmatrix} = -180 + 54 = -126,$ and

$|D_z| = \begin{vmatrix} 2 & -1 & 0 \\ -1 & 7 & 9 \\ 5 & 4 & -9 \end{vmatrix} = -9\begin{vmatrix} 2 & -1 \\ 5 & 4 \end{vmatrix} - 9\begin{vmatrix} 2 & -1 \\ -1 & 7 \end{vmatrix} = -117 - 117 = -234.$

Therefore, $x = \frac{522}{-156} = -\frac{87}{26}$, $y = \frac{-126}{-156} = \frac{21}{26}$, and $z = \frac{-234}{-156} = \frac{3}{2}$, and so the solution is $\left(-\frac{87}{26}, \frac{21}{26}, \frac{3}{2}\right)$.

71. The area is $\pm\frac{1}{2}\begin{vmatrix} -1 & 3 & 1 \\ 3 & 1 & 1 \\ -2 & -2 & 1 \end{vmatrix} = \pm\frac{1}{2}\left(\begin{vmatrix} 3 & 1 \\ -2 & -2 \end{vmatrix} - \begin{vmatrix} -1 & 3 \\ -2 & -2 \end{vmatrix} + \begin{vmatrix} -1 & 3 \\ 3 & 1 \end{vmatrix}\right) = \pm\frac{1}{2}(-4 - 8 - 10) = 11.$

73. Let x be the amount invested in Bank A, y the amount invested in Bank B, and z the amount invested in Bank C.

We get the following system: $\begin{cases} x + y + z = 60{,}000 \\ 0.02x + 0.025y + 0.03z = 1575 \\ 2x + 2z = y \end{cases} \Leftrightarrow \begin{cases} x + y + z = 60{,}000 \\ 2x + 2.5y + 3z = 157{,}500 \\ 2x - y + 2z = 0 \end{cases}$ which

has matrix representation $\begin{bmatrix} 1 & 1 & 1 & 60{,}000 \\ 2 & 2.5 & 3 & 157{,}500 \\ 2 & -1 & 2 & 0 \end{bmatrix} \xrightarrow[R_3 - 2R_1 \to R_3]{R_2 - 2R_1 \to R_2} \begin{bmatrix} 1 & 1 & 1 & 60{,}000 \\ 0 & 0.5 & 1 & 37{,}500 \\ 0 & -3 & 0 & -120{,}000 \end{bmatrix} \quad R_2 \leftrightarrow -\frac{1}{3}R_3$

$\begin{bmatrix} 1 & 1 & 1 & 60{,}000 \\ 0 & 1 & 0 & 40{,}000 \\ 0 & 0.5 & 1 & 37{,}500 \end{bmatrix} \begin{matrix} R_1 - R_2 \to R_1 \\ R_3 - 0.5R_2 \to R_3 \end{matrix} \begin{bmatrix} 1 & 0 & 1 & 20{,}000 \\ 0 & 1 & 0 & 40{,}000 \\ 0 & 0 & 1 & 17{,}500 \end{bmatrix} \xrightarrow{R_1 - R_3 \to R_1} \begin{bmatrix} 1 & 0 & 1 & 2500 \\ 0 & 1 & 0 & 40{,}000 \\ 0 & 0 & 1 & 17{,}500 \end{bmatrix}$. Thus, she invests $2500 in Bank A, $40,000 in Bank B, and $17,500 in Bank C.

CHAPTER 11 TEST

1. $\begin{bmatrix} 1 & 8 & 0 & 0 \\ 0 & 1 & 7 & 10 \\ 0 & 0 & 0 & 0 \end{bmatrix}$ is in row-echelon form, but not reduced row-echelon form because the 1 in the second row does not have a 0 above it.

3. $\begin{bmatrix} 1 & 0 & 0 \\ 0 & 0 & 1 \end{bmatrix}$ is in reduced row-echelon form.

5. $\begin{cases} x - y + 2z = 0 \\ 2x - 4y + 5z = -5 \\ 2y - 3z = 5 \end{cases}$ has the matrix representation $\begin{bmatrix} 1 & -1 & 2 & 0 \\ 2 & -4 & 5 & -5 \\ 0 & 2 & -3 & 5 \end{bmatrix} \xrightarrow{R_2 - R_1 \to R_2} \begin{bmatrix} 1 & -1 & 2 & 0 \\ 0 & -2 & 1 & -5 \\ 0 & 2 & -3 & 5 \end{bmatrix}$

$\xrightarrow{R_3 + R_2 \to R_3} \begin{bmatrix} 1 & -1 & 2 & 0 \\ 0 & -2 & 1 & -5 \\ 0 & 0 & -2 & 0 \end{bmatrix} \xrightarrow{-\frac{1}{2}R_3} \begin{bmatrix} 1 & -1 & 2 & 0 \\ 0 & -2 & 1 & -5 \\ 0 & 0 & 1 & 0 \end{bmatrix}$. Thus $z = 0$, $-2y + 0 = -5 \Leftrightarrow y = \frac{5}{2}$, and $x - \frac{5}{2} + 2(0) = 0 \Leftrightarrow x = \frac{5}{2}$. Thus, the solution is $\left(\frac{5}{2}, \frac{5}{2}, 0\right)$.

7. $\begin{cases} x + 2y = 3 \\ 3y + z = 2 \\ x - 2y + z = -2 \end{cases}$ has the matrix representation $\begin{bmatrix} 1 & 2 & 0 & 3 \\ 0 & 3 & 1 & 2 \\ 1 & -2 & 1 & -2 \end{bmatrix} \xrightarrow{R_3 - R_1 \to R_3} \begin{bmatrix} 1 & 2 & 0 & 3 \\ 0 & 3 & 1 & 2 \\ 0 & -4 & 1 & -5 \end{bmatrix}$

$\xrightarrow{3R_3 + 4R_2 \to R_3} \begin{bmatrix} 1 & 2 & 0 & 3 \\ 0 & 3 & 1 & 2 \\ 0 & 0 & 7 & -7 \end{bmatrix} \xrightarrow{\frac{1}{7}R_3} \begin{bmatrix} 1 & 2 & 0 & 3 \\ 0 & 3 & 1 & 2 \\ 0 & 0 & 1 & -1 \end{bmatrix}$.

Thus $z = -1$, $3y + (-1) = 2 \Leftrightarrow y = 1$, and $x + 2(1) = 3 \Leftrightarrow x = 1$. Hence, the solution is $(1, 1, -1)$.

In Solutions 9–15, $A = \begin{bmatrix} 2 & 3 \\ 2 & 4 \end{bmatrix}$, $B = \begin{bmatrix} 2 & 4 \\ -1 & 1 \\ 3 & 0 \end{bmatrix}$, **and** $C = \begin{bmatrix} 1 & 0 & 4 \\ -1 & 1 & 2 \\ 0 & 1 & 3 \end{bmatrix}$.

9. $A + B$ is undefined because A is 2×2 and B is 3×2, so they have incompatible dimensions.

11. $BA - 3B = \begin{bmatrix} 2 & 4 \\ -1 & 1 \\ 3 & 0 \end{bmatrix} \begin{bmatrix} 2 & 3 \\ 2 & 4 \end{bmatrix} - 3\begin{bmatrix} 2 & 4 \\ -1 & 1 \\ 3 & 0 \end{bmatrix} = \begin{bmatrix} 12 & 22 \\ 0 & 1 \\ 6 & 9 \end{bmatrix} - \begin{bmatrix} 6 & 12 \\ -3 & 3 \\ 9 & 0 \end{bmatrix} = \begin{bmatrix} 6 & 10 \\ 3 & -2 \\ -3 & 9 \end{bmatrix}$

13. $A = \begin{bmatrix} 2 & 3 \\ 2 & 4 \end{bmatrix} \Leftrightarrow A^{-1} = \frac{1}{8 - 6}\begin{bmatrix} 4 & -3 \\ -2 & 2 \end{bmatrix} = \begin{bmatrix} 2 & -\frac{3}{2} \\ -1 & 1 \end{bmatrix}$

15. $\det(B)$ is not defined because B is not a square matrix.

17. (a) The system $\begin{cases} 4x - 3y = 10 \\ 3x - 2y = 30 \end{cases}$ is equivalent to the matrix equation $\begin{bmatrix} 4 & -3 \\ 3 & -2 \end{bmatrix}\begin{bmatrix} x \\ y \end{bmatrix} = \begin{bmatrix} 10 \\ 30 \end{bmatrix}$.

(b) We have $|D| = \begin{vmatrix} 4 & -3 \\ 3 & -2 \end{vmatrix} = 4(-2) - 3(-3) = 1$. So $D^{-1} = \begin{bmatrix} -2 & 3 \\ -3 & 4 \end{bmatrix}$ and $\begin{bmatrix} x \\ y \end{bmatrix} = \begin{bmatrix} -2 & 3 \\ -3 & 4 \end{bmatrix}\begin{bmatrix} 10 \\ 30 \end{bmatrix} = \begin{bmatrix} 70 \\ 90 \end{bmatrix}$.

Therefore, $x = 70$ and $y = 90$.

19. $\begin{cases} 2x \quad\quad - z = 14 \\ 3x - y + 5z = 0 \\ 4x + 2y + 3z = -2 \end{cases}$ Then $|D| = \begin{vmatrix} 2 & 0 & -1 \\ 3 & -1 & 5 \\ 4 & 2 & 3 \end{vmatrix} = 2\begin{vmatrix} -1 & 5 \\ 2 & 3 \end{vmatrix} - 1\begin{vmatrix} 3 & -1 \\ 4 & 2 \end{vmatrix} = -26 - 10 = -36$,

$$|D_x| = \begin{vmatrix} 14 & 0 & -1 \\ 0 & -1 & 5 \\ -2 & 2 & 3 \end{vmatrix} = 14\begin{vmatrix} -1 & 5 \\ 2 & 3 \end{vmatrix} - 1\begin{vmatrix} 0 & -1 \\ 4 & 2 \end{vmatrix} = -182 + 2 = -180,$$

$$|D_y| = \begin{vmatrix} 2 & 14 & -1 \\ 3 & 0 & 5 \\ 4 & -2 & 3 \end{vmatrix} = -3\begin{vmatrix} 14 & -1 \\ -2 & 3 \end{vmatrix} - 5\begin{vmatrix} 2 & 14 \\ 4 & -2 \end{vmatrix} = -120 + 300 = 180, \text{ and}$$

$$|D_z| = \begin{vmatrix} 2 & 0 & 14 \\ 3 & -1 & 0 \\ 4 & 2 & -2 \end{vmatrix} = 2\begin{vmatrix} -1 & 0 \\ 2 & -2 \end{vmatrix} + 14\begin{vmatrix} 3 & -1 \\ 2 & 2 \end{vmatrix} = 4 + 140 = 144.$$

Therefore, $x = \frac{-180}{-36} = 5$, $y = \frac{180}{-36} = -5$, $z = \frac{144}{-36} = -4$, and so the solution is $(5, -5, -4)$.

FOCUS ON MODELING Computer Graphics

1. The data matrix $D = \begin{bmatrix} 0 & 1 & 1 & 0 \\ 0 & 0 & 1 & 1 \end{bmatrix}$ represents the gray square.

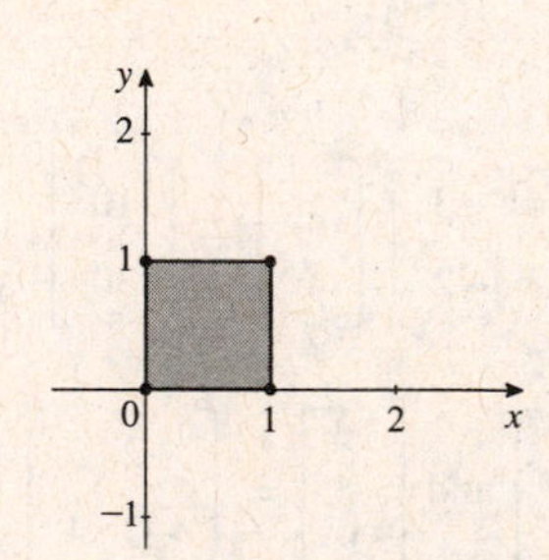

Reflection using $T = \begin{bmatrix} 1 & 0 \\ 0 & -1 \end{bmatrix}$:

$$TD = \begin{bmatrix} 1 & 0 \\ 0 & -1 \end{bmatrix}\begin{bmatrix} 0 & 1 & 1 & 0 \\ 0 & 0 & 1 & 1 \end{bmatrix} = \begin{bmatrix} 0 & 1 & 1 & 0 \\ 0 & 0 & -1 & -1 \end{bmatrix}$$

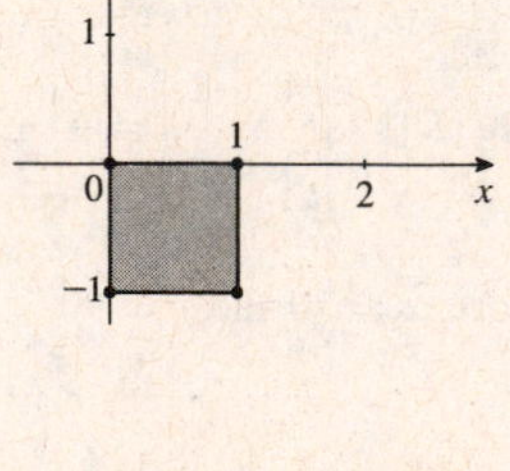

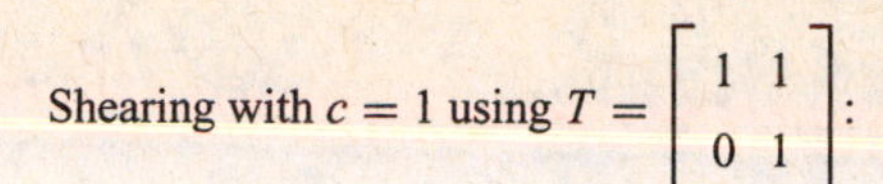

Expansion with $c = 2$ using $T = \begin{bmatrix} 2 & 0 \\ 0 & 1 \end{bmatrix}$:

$$TD = \begin{bmatrix} 2 & 0 \\ 0 & 1 \end{bmatrix} \begin{bmatrix} 0 & 1 & 1 & 0 \\ 0 & 0 & 1 & 1 \end{bmatrix} = \begin{bmatrix} 0 & 2 & 2 & 0 \\ 0 & 0 & 1 & 1 \end{bmatrix}$$

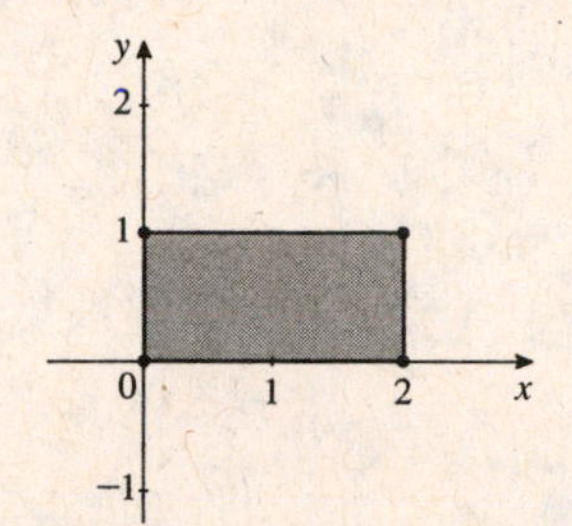

Shearing with $c = 1$ using $T = \begin{bmatrix} 1 & 1 \\ 0 & 1 \end{bmatrix}$:

$$TD = \begin{bmatrix} 1 & 1 \\ 0 & 1 \end{bmatrix} \begin{bmatrix} 0 & 1 & 1 & 0 \\ 0 & 0 & 1 & 1 \end{bmatrix} = \begin{bmatrix} 0 & 1 & 2 & 1 \\ 0 & 0 & 1 & 1 \end{bmatrix}$$

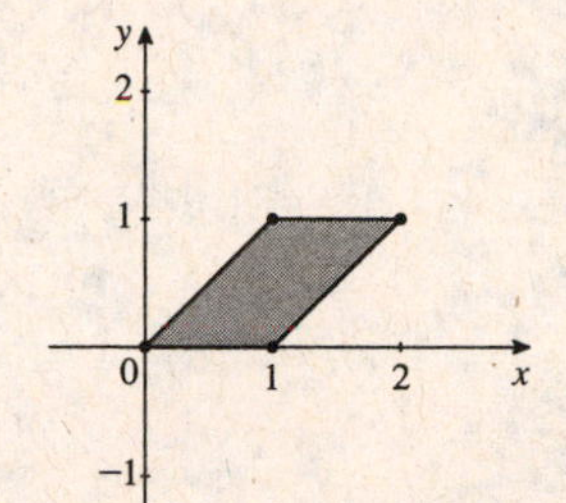

3. (a) $T = \begin{bmatrix} 1 & 1.5 \\ 0 & 1 \end{bmatrix}$ is a shear in the x-direction.

(b) $T^{-1} = \frac{1}{1}\begin{bmatrix} 1 & -1.5 \\ 0 & 1 \end{bmatrix} = \begin{bmatrix} 1 & -1.5 \\ 0 & 1 \end{bmatrix}$

(c) T^{-1} is a leftward shear in the x-direction.

(d) The result is the original matrix. Algebraically, $T^{-1}(TD) = \left(T^{-1}T\right)D = ID = D$ where I is the 2×2 identity matrix: $\begin{bmatrix} 1 & -1.5 \\ 0 & 1 \end{bmatrix}\left(\begin{bmatrix} 1 & 1.5 \\ 0 & 1 \end{bmatrix}\begin{bmatrix} 0 & 1 & 1 & 0 \\ 0 & 0 & 1 & 1 \end{bmatrix}\right) = \begin{bmatrix} 1 & -1.5 \\ 0 & 1 \end{bmatrix}\begin{bmatrix} 0 & 1 & 2.5 & 1.5 \\ 0 & 0 & 1 & 1 \end{bmatrix} = \begin{bmatrix} 0 & 1 & 1 & 0 \\ 0 & 0 & 1 & 1 \end{bmatrix}$

5. (a) $D = \begin{bmatrix} 0 & 1 & 1 & 4 & 4 & 1 & 1 & 6 & 6 & 0 & 0 \\ 0 & 0 & 4 & 4 & 5 & 5 & 7 & 7 & 8 & 8 & 0 \end{bmatrix}$

(b) $T = \begin{bmatrix} 0.75 & 0 \\ 0 & 1 \end{bmatrix}$,

$$TD = \begin{bmatrix} 0.75 & 0 \\ 0 & 1 \end{bmatrix}\begin{bmatrix} 0 & 1 & 1 & 4 & 4 & 1 & 1 & 6 & 6 & 0 & 0 \\ 0 & 0 & 4 & 4 & 5 & 5 & 7 & 7 & 8 & 8 & 0 \end{bmatrix} = \begin{bmatrix} 0 & 0.75 & 0.75 & 3 & 3 & 0.75 & 0.75 & 4.5 & 4.5 & 0 & 0 \\ 0 & 0 & 4 & 4 & 5 & 5 & 7 & 7 & 8 & 8 & 0 \end{bmatrix}$$

(c) $S = \begin{bmatrix} 1 & 0.25 \\ 0 & 1 \end{bmatrix}$,

$$SD = \begin{bmatrix} 1 & 0.25 \\ 0 & 1 \end{bmatrix}\begin{bmatrix} 0 & 1 & 1 & 4 & 4 & 1 & 1 & 6 & 6 & 0 & 0 \\ 0 & 0 & 4 & 4 & 5 & 5 & 7 & 7 & 8 & 8 & 0 \end{bmatrix} = \begin{bmatrix} 0 & 1 & 2 & 5 & 5.25 & 2.25 & 2.75 & 7.75 & 8 & 2 & 0 \\ 0 & 0 & 4 & 4 & 5 & 5 & 7 & 7 & 8 & 8 & 0 \end{bmatrix}$$

12 CONIC SECTIONS

12.1 PARABOLAS

1. A parabola is the set of all points in the plane equidistant from a fixed point called the *focus* and a fixed line called the *directrix* of the parabola.

3. The graph of the equation $y^2 = 4px$ is a parabola with focus $F(p, 0)$ and directrix $x = -p$. So the graph of $y^2 = 12x$ is a parabola with focus $F(3, 0)$ and directrix $x = -3$.

5. $y^2 = 2x$ is Graph III, which opens to the right and is not as wide as the graph for Exercise 5.

7. $x^2 = -6y$ is Graph II, which opens downward and is narrower than the graph for Exercise 6.

9. $y^2 - 8x = 0$ is Graph VI, which opens to the right and is wider than the graph for Exercise 1.

11. (a) $x^2 = 8y$, so $4p = 8 \Leftrightarrow p = 2$. The focus is $(0, 2)$, the directrix is $y = -2$, and the focal diameter is 8.

(b)

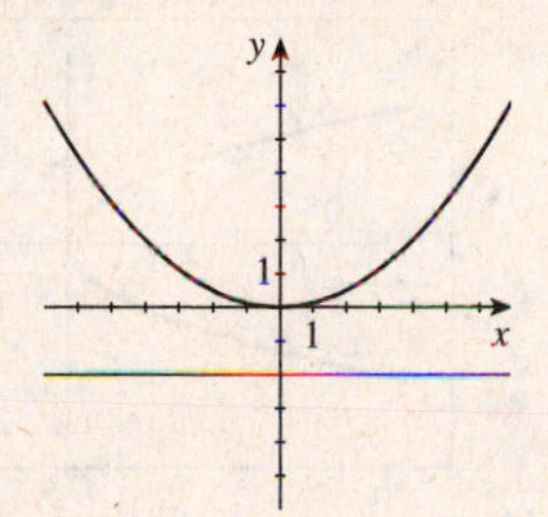

13. (a) $y^2 = -24x$, so $4p = -24 \Leftrightarrow p = -6$. The focus is $(-6, 0)$, the directrix is $x = 6$, and the focal diameter is 24.

(b)

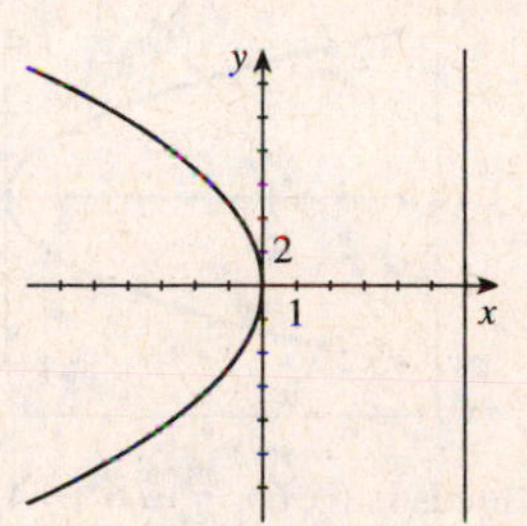

15. (a) $y = -\frac{1}{8}x^2 \Leftrightarrow x^2 = -8y$, so $4p = -8 \Leftrightarrow p = -2$.

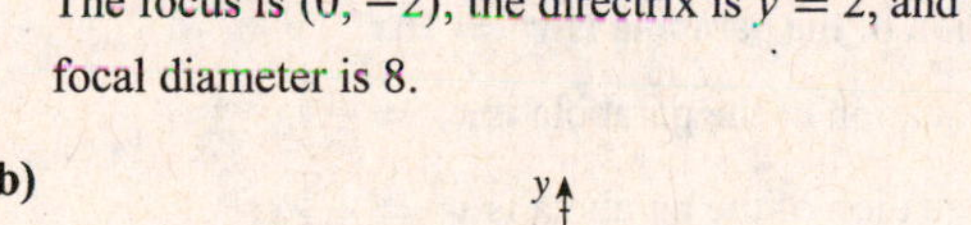

The focus is $(0, -2)$, the directrix is $y = 2$, and the focal diameter is 8.

(b)

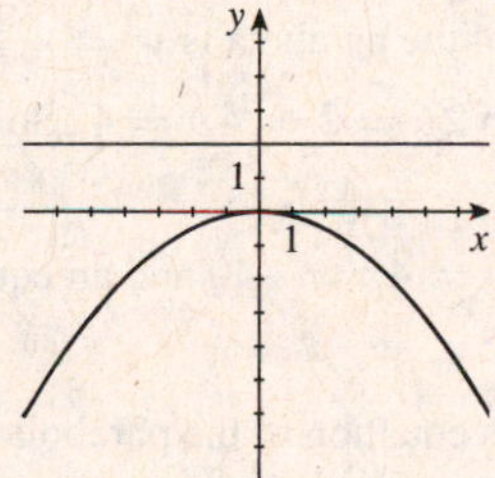

17. (a) $x = -2y^2 \Leftrightarrow y^2 = -\frac{1}{2}x$, so $4p = -\frac{1}{2} \Leftrightarrow p = -\frac{1}{8}$. The focus is $\left(-\frac{1}{8}, 0\right)$, the directrix is $x = \frac{1}{8}$, and the focal diameter is $\frac{1}{2}$.

(b)

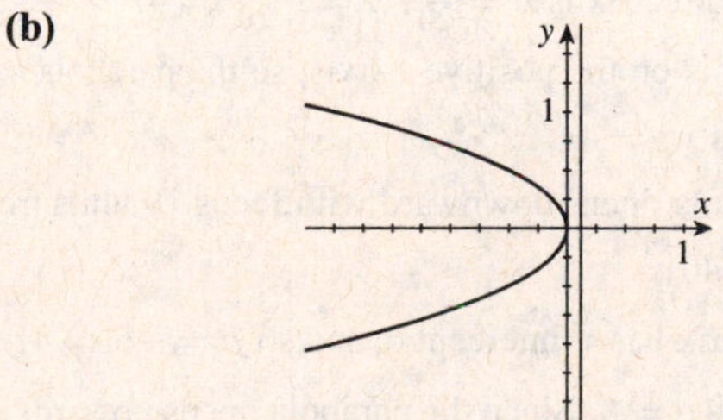

19. **(a)** $5y = x^2$, so $4p = 5 \Leftrightarrow p = \frac{5}{4}$. The focus is $\left(0, \frac{5}{4}\right)$, the directrix is $y = -\frac{5}{4}$, and the focal diameter is 5.

(b)

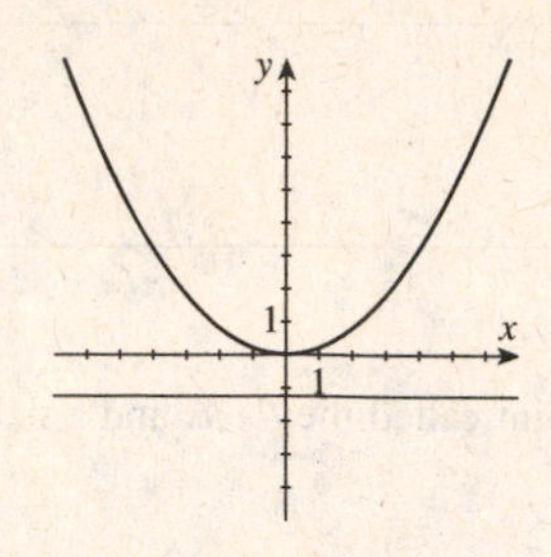

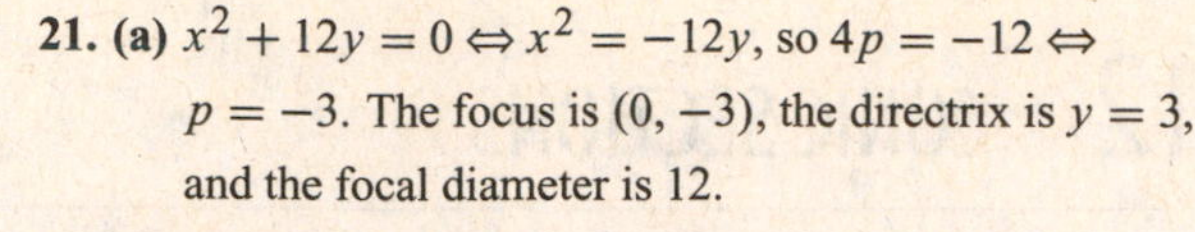

21. **(a)** $x^2 + 12y = 0 \Leftrightarrow x^2 = -12y$, so $4p = -12 \Leftrightarrow p = -3$. The focus is $(0, -3)$, the directrix is $y = 3$, and the focal diameter is 12.

(b)

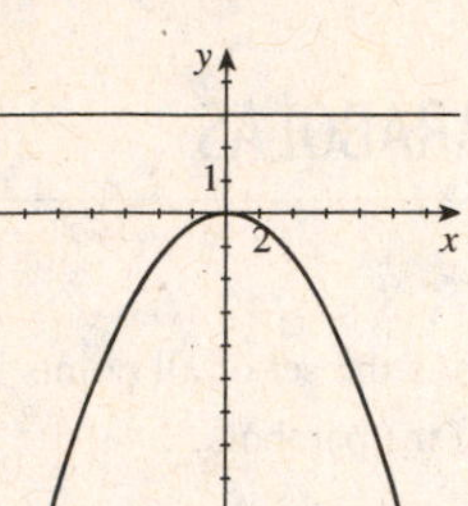

23. **(a)** $5x + 3y^2 = 0 \Leftrightarrow y^2 = -\frac{5}{3}x$. Then $4p = -\frac{5}{3} \Leftrightarrow p = -\frac{5}{12}$. The focus is $\left(-\frac{5}{12}, 0\right)$, the directrix is $x = \frac{5}{12}$, and the focal diameter is $\frac{5}{3}$.

(b)

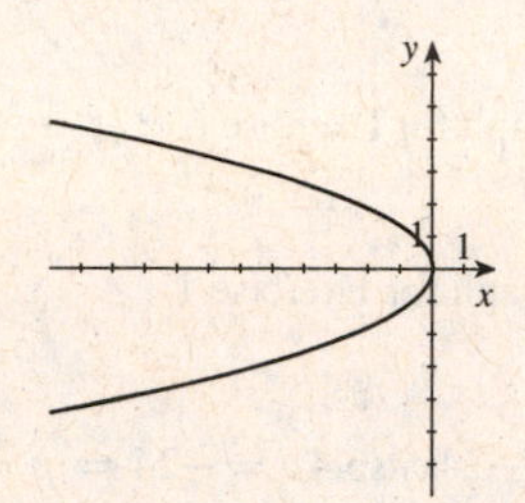

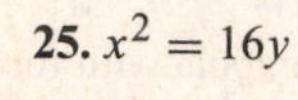

25. $x^2 = 16y$

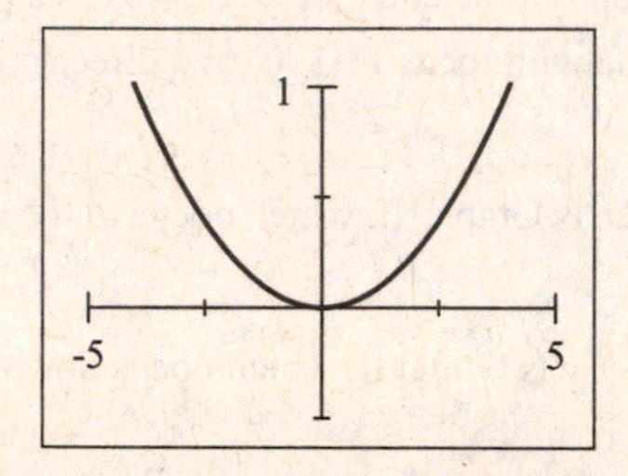

27. $y^2 = -\frac{1}{3}x$

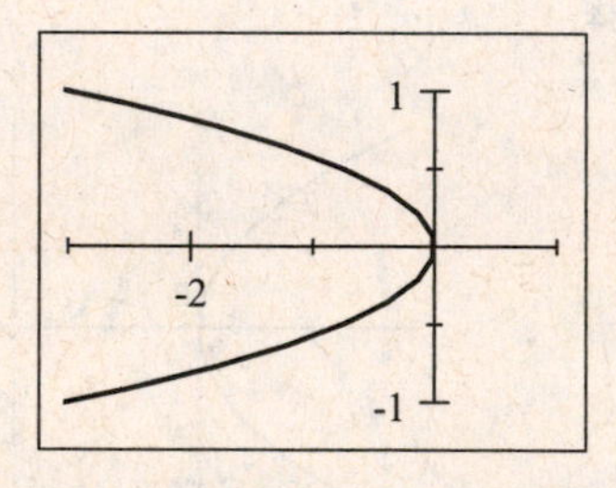

29. $4x + y^2 = 0$

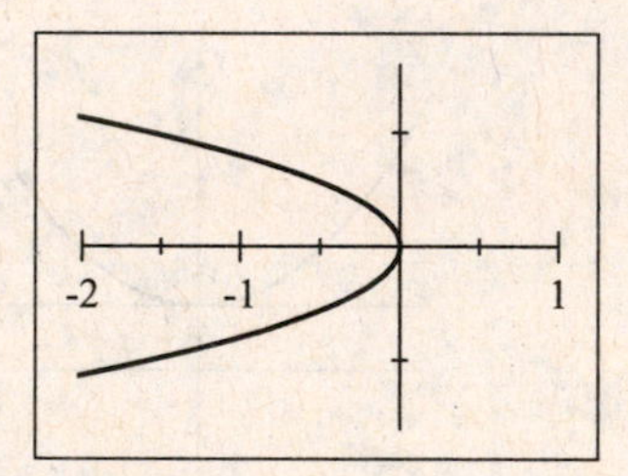

31. Since the focus is $(0, 6)$, $p = 6 \Leftrightarrow 4p = 24$. Hence, an equation of the parabola is $x^2 = 24y$.

33. Since the focus is $(-8, 0)$, $p = -8 \Leftrightarrow 4p = -32$. Hence, an equation of the parabola is $y^2 = -32x$.

35. Since the focus is $\left(0, -\frac{3}{4}\right)$, $p = -\frac{3}{4} \Leftrightarrow 4p = -3$. Hence, an equation of the parabola is $x^2 = -3y$.

37. Since the directrix is $x = -4$, $p = 4 \Leftrightarrow 4p = 16$. Hence, an equation of the parabola is $y^2 = 16x$.

39. Since the directrix is $y = \frac{1}{10}$, $p = -\frac{1}{10} \Leftrightarrow 4p = -\frac{2}{5}$. Hence, an equation of the parabola is $x^2 = -\frac{2}{5}y$.

41. Since the directrix is $x = \frac{1}{20}$, $p = -\frac{1}{20} \Leftrightarrow 4p = -\frac{1}{5}$. Hence, an equation of the parabola is $y^2 = -\frac{1}{5}x$.

43. The focus is on the positive x-axis, so the parabola opens horizontally with $2p = 2 \Leftrightarrow 4p = 4$. So an equation of the parabola is $y^2 = 4x$.

45. The parabola opens downward with focus 10 units from $(0, 0)$, so $p = -10 \Leftrightarrow 4p = -40$ and an equation of the parabola is $x^2 = -40y$.

47. The directrix has y-intercept 6, and so $p = -6 \Leftrightarrow 4p = -24$. Therefore, an equation of the parabola is $x^2 = -24y$.

49. $p = 2 \Leftrightarrow 4p = 8$. Since the parabola opens upward, its equation is $x^2 = 8y$.

51. $p = 4 \Leftrightarrow 4p = 16$. Since the parabola opens to the left, its equation is $y^2 = -16x$.

53. The focal diameter is $4p = \frac{3}{2} + \frac{3}{2} = 3$. Since the parabola opens to the left, its equation is $y^2 = -3x$.

55. The equation of the parabola has the form $y^2 = 4px$. Since the parabola passes through the point $(4, -2)$, $(-2)^2 = 4p(4)$ $\Leftrightarrow 4p = 1$, and so an equation is $y^2 = x$.

57. The area of the shaded region is width × height $= 4p \cdot p = 8$, and so $p^2 = 2 \Leftrightarrow p = -\sqrt{2}$ (because the parabola opens downward). Therefore, an equation is $x^2 = 4py = -4\sqrt{2}y \Leftrightarrow x^2 = -4\sqrt{2}y$.

59. (a) A parabola with directrix $y = -p$ has equation $x^2 = 4py$. If the directrix is $y = \frac{1}{2}$, then $p = -\frac{1}{2}$, so an equation is $x^2 = 4\left(-\frac{1}{2}\right)y \Leftrightarrow x^2 = -2y$. If the directrix is $y = 1$, then $p = -1$, so an equation is $x^2 = 4(-1)y \Leftrightarrow x^2 = -4y$. If the directrix is $y = 4$, then $p = -4$, so an equation is $x^2 = 4(-4)y \Leftrightarrow x^2 = -16y$. If the directrix is $y = 8$, then $p = -8$, so an equation is $x^2 = 4(-8)y \Leftrightarrow x^2 = -32y$.

(b)

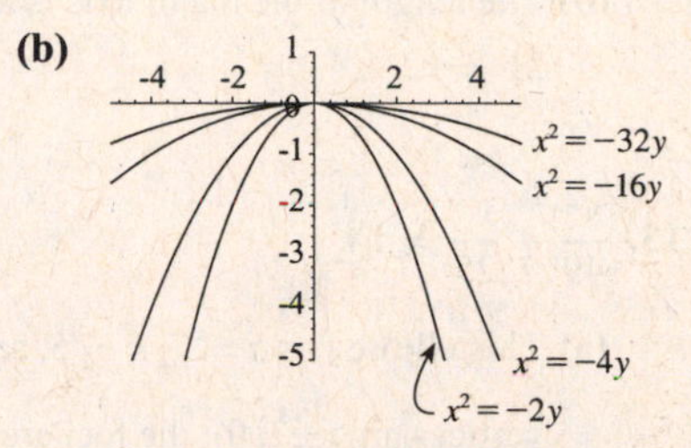

As the directrix moves further from the vertex, the parabolas get flatter.

61. (a) Since the focal diameter is 12 cm, $4p = 12$. Hence, the parabola has equation $y^2 = 12x$.

(b) At a point 20 cm horizontally from the vertex, the parabola passes through the point $(20, y)$, and hence from part (a), $y^2 = 12(20) \Leftrightarrow y^2 = 240 \Leftrightarrow y = \pm 4\sqrt{15}$. Thus, $|CD| = 8\sqrt{15} \approx 31$ cm.

63. With the vertex at the origin, the top of one tower will be at the point $(300, 150)$. Inserting this point into the equation $x^2 = 4py$ gives $(300)^2 = 4p(150) \Leftrightarrow 90000 = 600p \Leftrightarrow p = 150$. So an equation of the parabolic part of the cables is $x^2 = 4(150)y \Leftrightarrow x^2 = 600y$.

65. Many answers are possible: satellite dish TV antennas, sound surveillance equipment, solar collectors for hot water heating or electricity generation, bridge pillars, etc.

12.2 ELLIPSES

1. An ellipse is the set of all points in the plane for which the *sum* of the distances from two fixed points F_1 and F_2 is constant. The points F_1 and F_2 are called the *foci* of the ellipse.

3. The graph of the equation $\frac{x^2}{b^2} + \frac{y^2}{a^2} = 1$ with $a > b > 0$ is an ellipse with vertices $(0, a)$ and $(0, -a)$ and foci $(0, \pm c)$, where $c = \sqrt{a^2 - b^2}$. So the graph of $\frac{x^2}{4^2} + \frac{y^2}{5^2} = 1$ is an ellipse with vertices $(0, 5)$ and $(0, -5)$ and foci $(0, 3)$ and $(0, -3)$.

5. $\frac{x^2}{16} + \frac{y^2}{4} = 1$ is Graph II. The major axis is horizontal and the vertices are $(\pm 4, 0)$.

7. $4x^2 + y^2 = 4$ is Graph I. The major axis is vertical and the vertices are $(0, \pm 2)$.

9. $\frac{x^2}{25} + \frac{y^2}{9} = 1$.

(a) This ellipse has $a = 5$, $b = 3$, and so $c^2 = a^2 - b^2 = 16 \Leftrightarrow c = 4$. The vertices are $(\pm 5, 0)$, the foci are $(\pm 4, 0)$, and the eccentricity is $e = \frac{c}{a} = \frac{4}{5} = 0.8$.

(b) The length of the major axis is $2a = 10$, and the length of the minor axis is $2b = 6$.

(c)

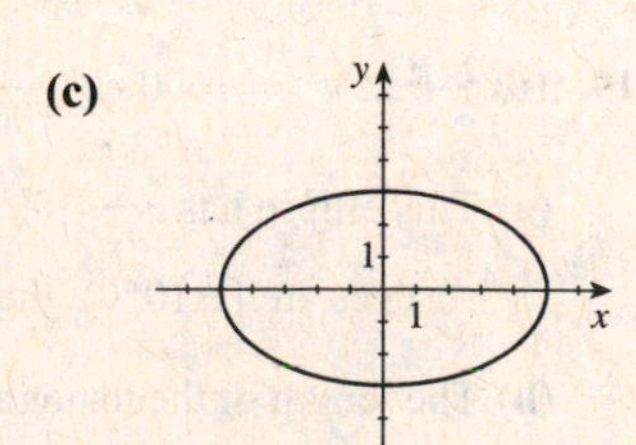

11. $\dfrac{x^2}{36}+\dfrac{y^2}{81}=1$

(a) This ellipse has $a=9$, $b=6$, and so $c^2=81-36=45 \Leftrightarrow c=3\sqrt{5}$. The vertices are $(0,\pm 9)$, the foci are $\left(0,\pm 3\sqrt{5}\right)$, and the eccentricity is $e=\frac{c}{a}=\frac{\sqrt{5}}{3}$.

(b) The length of the major axis is $2a=18$ and the length of the minor axis is $2b=12$.

(c)

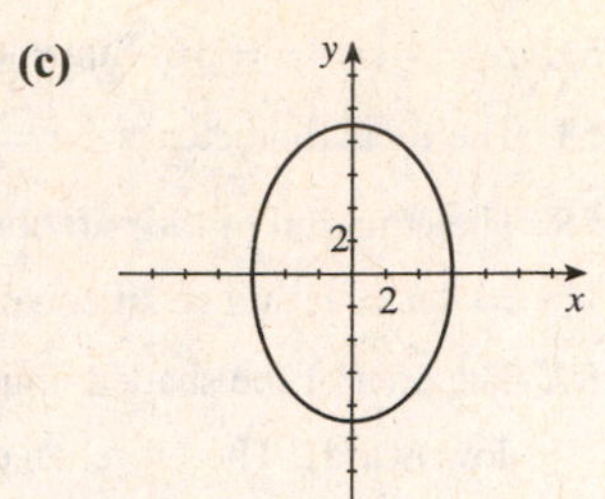

13. $\dfrac{x^2}{49}+\dfrac{y^2}{25}=1$

(a) This ellipse has $a=7$, $b=5$, and so $c^2=49-25=24 \Leftrightarrow c=2\sqrt{6}$. The vertices are $(\pm 7,0)$, the foci are $\left(\pm 2\sqrt{6},0\right)$, and the eccentricity is $e=\frac{c}{a}=\frac{2\sqrt{6}}{7}$.

(b) The length of the major axis is $2a=14$ and the length of the minor axis is $2b=10$.

(c)

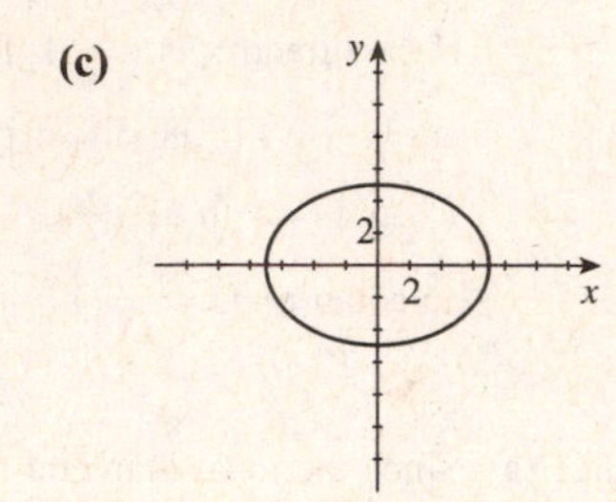

15. $9x^2+4y^2=36 \Leftrightarrow \dfrac{x^2}{4}+\dfrac{y^2}{9}=1$

(a) This ellipse has $a=3$, $b=2$, and so $c^2=9-4=5 \Leftrightarrow c=\sqrt{5}$. The vertices are $(0,\pm 3)$, the foci are $\left(0,\pm\sqrt{5}\right)$, and the eccentricity is $e=\frac{c}{a}=\frac{\sqrt{5}}{3}$.

(b) The length of the major axis is $2a=6$, and the length of the minor axis is $2b=4$.

(c)

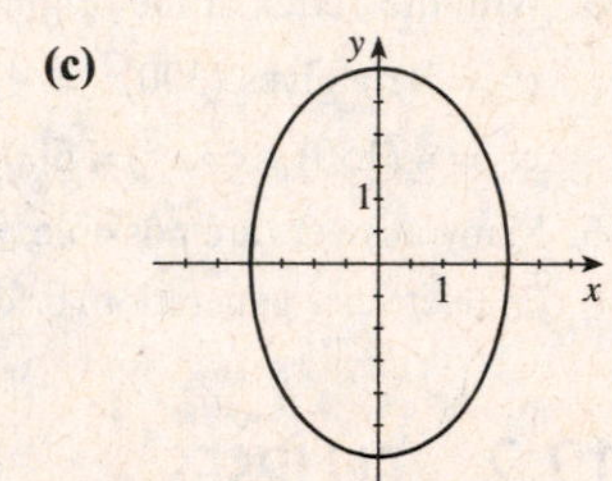

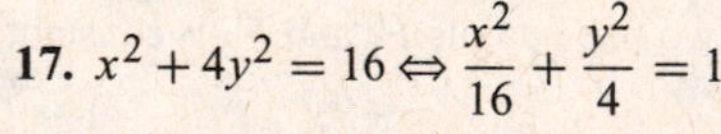

17. $x^2+4y^2=16 \Leftrightarrow \dfrac{x^2}{16}+\dfrac{y^2}{4}=1$

(a) This ellipse has $a=4$, $b=2$, and so $c^2=16-4=12 \Leftrightarrow c=2\sqrt{3}$.The vertices are $(\pm 4,0)$, the foci are $\left(\pm 2\sqrt{3},0\right)$, and the eccentricity is $e=\frac{c}{a}=\frac{2\sqrt{3}}{4}=\frac{\sqrt{3}}{2}$.

(b) The length of the major axis is $2a=8$, and the length of the minor axis is $2b=4$.

(c)

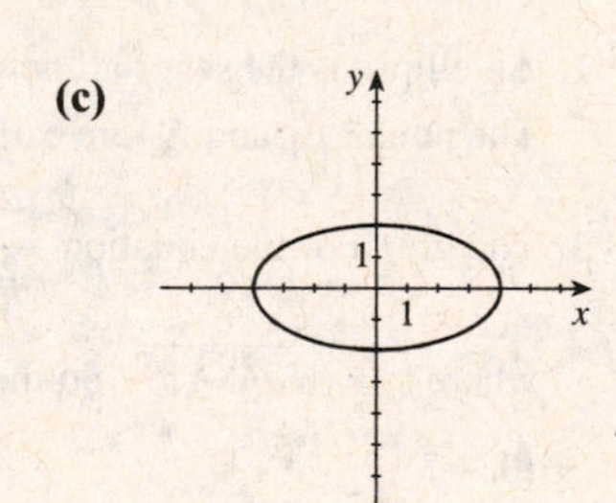

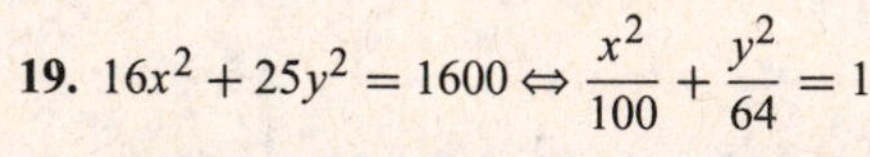

19. $16x^2+25y^2=1600 \Leftrightarrow \dfrac{x^2}{100}+\dfrac{y^2}{64}=1$

(a) This ellipse has $a=10$, $b=8$, and so $c^2=100-64=36 \Leftrightarrow c=6$. The vertices are $(\pm 10,0)$, the foci are $(\pm 6,0)$, and the eccentricity is $e=\frac{c}{a}=\frac{3}{5}$.

(b) The length of the major axis is $2a=20$ and the length of the minor axis is $2b=16$.

(c)

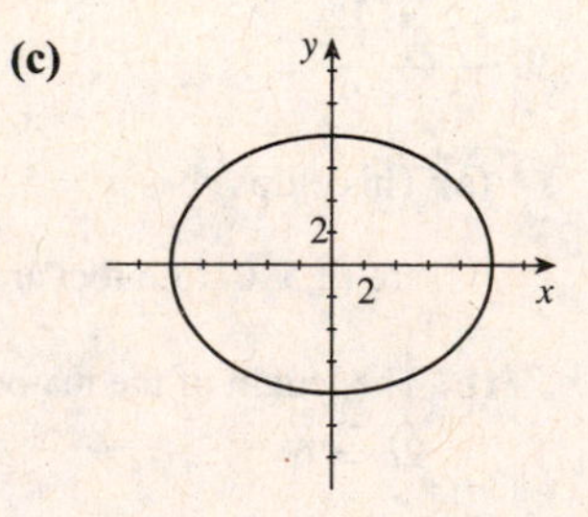

21. $3x^2 + y^2 = 9 \Leftrightarrow \frac{x^2}{3} + \frac{y^2}{9} = 1$

(a) This ellipse has $a = 3$, $b = \sqrt{3}$, and so $c^2 = 9 - 3 = 6 \Leftrightarrow c = \sqrt{6}$. The vertices are $(0, \pm 3)$, the foci are $\left(0, \pm\sqrt{6}\right)$, and the eccentricity is $e = \frac{c}{a} = \frac{\sqrt{6}}{3}$.

(b) The length of the major axis is $2a = 6$ and the length of the minor axis is $2b = 2\sqrt{3}$.

(c)

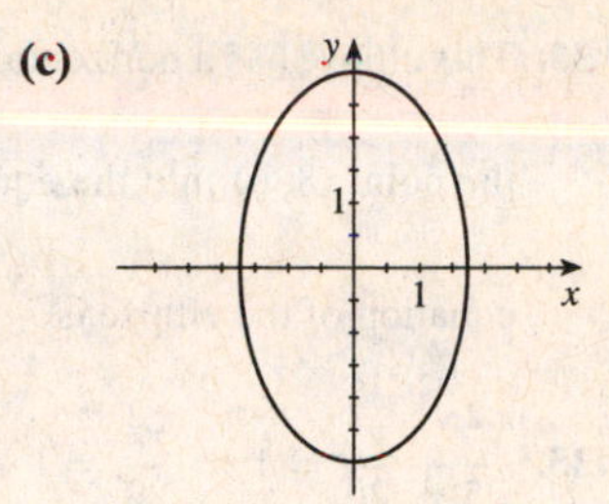

23. $2x^2 + y^2 = 4 \Leftrightarrow \frac{x^2}{2} + \frac{y^2}{4} = 1$

(a) This ellipse has $a = 2$, $b = \sqrt{2}$, and so $c^2 = 4 - 2 = 2 \Leftrightarrow c = \sqrt{2}$. The vertices are $(0, \pm 2)$, the foci are $\left(0, \pm\sqrt{2}\right)$, and the eccentricity is $e = \frac{c}{a} = \frac{\sqrt{2}}{2}$.

(b) The length of the major axis is $2a = 4$ and the length of the minor axis is $2b = 2\sqrt{2}$.

(c)

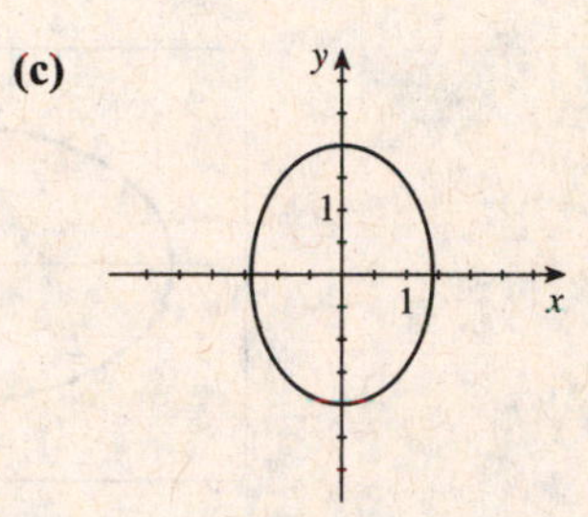

25. $x^2 + 4y^2 = 1 \Leftrightarrow \frac{x^2}{1} + \frac{y^2}{\frac{1}{4}} = 1$

(a) This ellipse has $a = 1$, $b = \frac{1}{2}$, and so $c^2 = 1 - \frac{1}{4} = \frac{3}{4} \Leftrightarrow c = \frac{\sqrt{3}}{2}$. The vertices are $(\pm 1, 0)$, the foci are $\left(\pm\frac{\sqrt{3}}{2}, 0\right)$, and the eccentricity is $e = \frac{c}{a} = \frac{\sqrt{3}/2}{1} = \frac{\sqrt{3}}{2}$.

(b) The length of the major axis is $2a = 2$, and the length of the minor axis is $2b = 1$.

(c)

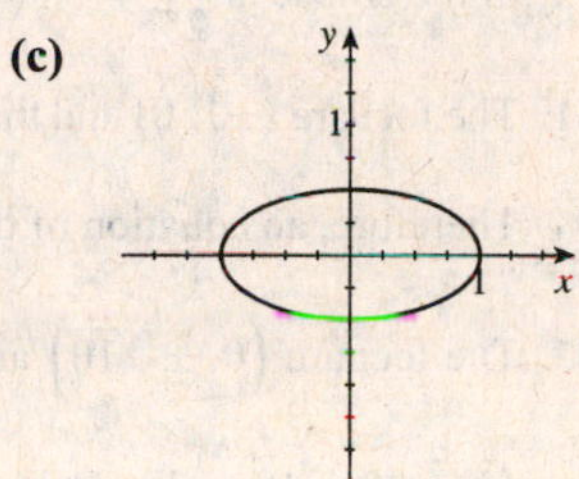

27. $x^2 = 4 - 2y^2 \Leftrightarrow x^2 + 2y^2 = 4 \Leftrightarrow \frac{x^2}{4} + \frac{y^2}{2} = 1$

(a) This ellipse has $a = 2$, $b = \sqrt{2}$, and so $c^2 = 4 - 2 = 2 \Leftrightarrow c = \sqrt{2}$. The vertices are $(\pm 2, 0)$, the foci are $\left(\pm\sqrt{2}, 0\right)$, and the eccentricity is $e = \frac{c}{a} = \frac{\sqrt{2}}{2}$.

(b) The length of the major axis is $2a = 4$, and the length of the minor axis is $2b = 2\sqrt{2}$.

(c)

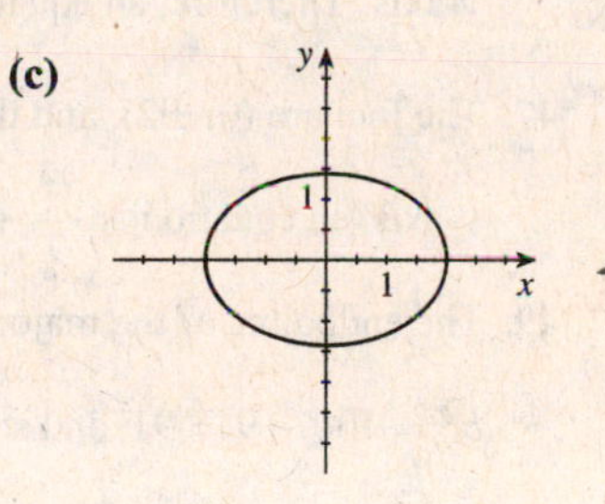

29. This ellipse has a horizontal major axis with $a = 5$ and $b = 4$, so an equation is $\frac{x^2}{(5)^2} + \frac{y^2}{(4)^2} = 1 \Leftrightarrow \frac{x^2}{25} + \frac{y^2}{16} = 1$.

31. This ellipse has a vertical major axis with $c = 2$ and $b = 2$. So $a^2 = c^2 + b^2 = 2^2 + 2^2 = 8 \Leftrightarrow a = 2\sqrt{2}$. So an equation is $\frac{x^2}{(2)^2} + \frac{y^2}{\left(2\sqrt{2}\right)^2} = 1 \Leftrightarrow \frac{x^2}{4} + \frac{y^2}{8} = 1$.

33. This ellipse has a horizontal major axis with $a = 16$, so an equation of the ellipse is of the form $\frac{x^2}{16^2} + \frac{y^2}{b^2} = 1$. Substituting the point $(8, 6)$ into the equation, we get $\frac{64}{256} + \frac{36}{b^2} = 1 \Leftrightarrow \frac{36}{b^2} = 1 - \frac{1}{4} \Leftrightarrow \frac{36}{b^2} = \frac{3}{4} \Leftrightarrow b^2 = \frac{4\,(36)}{3} = 48$. Thus, an equation of the ellipse is $\frac{x^2}{256} + \frac{y^2}{48} = 1$.

35. $\frac{x^2}{25} + \frac{y^2}{20} = 1 \Leftrightarrow \frac{y^2}{20} = 1 - \frac{x^2}{25} \Leftrightarrow y^2 = 20 - \frac{4x^2}{5} \Rightarrow$ $y = \pm\sqrt{20 - \frac{4x^2}{5}}$.

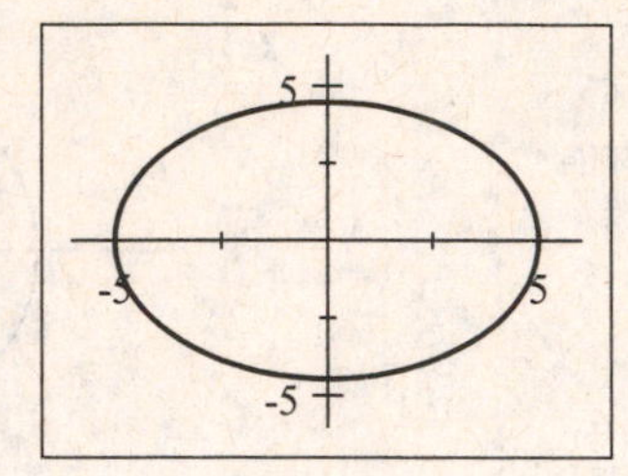

37. $6x^2 + y^2 = 36 \Leftrightarrow y^2 = 36 - 6x^2 \Rightarrow y = \pm\sqrt{36 - 6x^2}$.

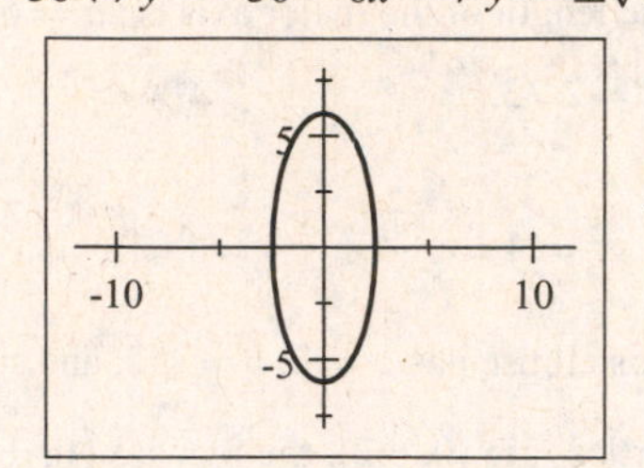

39. The foci are $(\pm 4, 0)$, and the vertices are $(\pm 5, 0)$. Thus, $c = 4$ and $a = 5$, and so $b^2 = 25 - 16 = 9$. Therefore, an equation of the ellipse is $\frac{x^2}{25} + \frac{y^2}{9} = 1$.

41. The foci are $(\pm 1, 0)$ and the vertices are $(\pm 2, 0)$. Thus, $c = 1$ and $a = 2$, so $c^2 = a^2 - b^2 \Leftrightarrow 1 = 4 - b^2 \Leftrightarrow b^2 = 4 - 1 = 3$. Therefore, an equation of the ellipse is $\frac{x^2}{4} + \frac{y^2}{3} = 1$.

43. The foci are $\left(0, \pm\sqrt{10}\right)$ and the vertices are $(0, \pm 7)$. Thus, $c = \sqrt{10}$ and $a = 7$, so $c^2 = a^2 - b^2 \Leftrightarrow 10 = 49 - b^2 \Leftrightarrow$ $b^2 = 49 - 10 = 39$. Therefore, an equation of the ellipse is $\frac{x^2}{39} + \frac{y^2}{49} = 1$.

45. The length of the major axis is $2a = 4 \Leftrightarrow a = 2$, the length of the minor axis is $2b = 2 \Leftrightarrow b = 1$, and the foci are on the y-axis. Therefore, an equation of the ellipse is $x^2 + \frac{y^2}{4} = 1$.

47. The foci are $(0, \pm 2)$, and the length of the minor axis is $2b = 6 \Leftrightarrow b = 3$. Thus, $a^2 = 4 + 9 = 13$. Since the foci are on the y-axis, an equation is $\frac{x^2}{9} + \frac{y^2}{13} = 1$.

49. The endpoints of the major axis are $(\pm 10, 0) \Leftrightarrow a = 10$, and the distance between the foci is $2c = 6 \Leftrightarrow c = 3$. Therefore, $b^2 = 100 - 9 = 91$, and so an equation of the ellipse is $\frac{x^2}{100} + \frac{y^2}{91} = 1$.

51. The length of the major axis is 10, so $2a = 10 \Leftrightarrow a = 5$, and the foci are on the x-axis, so the form of the equation is $\frac{x^2}{25} + \frac{y^2}{b^2} = 1$. Since the ellipse passes through $\left(\sqrt{5}, 2\right)$, we have $\frac{\left(\sqrt{5}\right)^2}{25} + \frac{(2)^2}{b^2} = 1 \Leftrightarrow \frac{5}{25} + \frac{4}{b^2} = 1 \Leftrightarrow \frac{4}{b^2} = \frac{4}{5} \Leftrightarrow$ $b^2 = 5$, and so an equation is $\frac{x^2}{25} + \frac{y^2}{5} = 1$.

53. The eccentricity is $\frac{1}{3}$, so $e = \frac{1}{3}$, and the foci are $(0, \pm 2)$, so $c = 2$. Thus, $e = \frac{c}{a} \Leftrightarrow a = \frac{c}{e} = \frac{2}{1/3} = 6$ and $b^2 = a^2 - c^2 = 36 - 4 = 32$. The major axis lies on the y-axis, so an equation is $\frac{x^2}{32} + \frac{y^2}{36} = 1$.

55. Since the length of the major axis is $2a = 4$, we have $a = 2$. The eccentricity is $\frac{\sqrt{3}}{2} = \frac{c}{a} = \frac{c}{2}$, so $c = \sqrt{3}$. Then $b^2 = a^2 - c^2 = 4 - 3 = 1$, and since the foci are on the y-axis, an equation of the ellipse is $x^2 + \dfrac{y^2}{4} = 1$.

57. $\begin{cases} 4x^2 + y^2 = 4 \\ 4x^2 + 9y^2 = 36 \end{cases}$ Subtracting the first equation from the second gives $8y^2 = 32 \Leftrightarrow y^2 = 4 \Leftrightarrow y = \pm 2$. Substituting $y = \pm 2$ in the first equation gives $4x^2 + (\pm 2)^2 = 4 \Leftrightarrow x = 0$, and so the points of intersection are $(0, \pm 2)$.

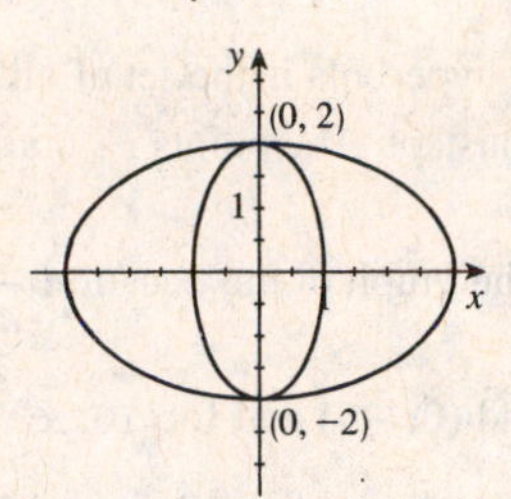

59. $\begin{cases} 100x^2 + 25y^2 = 100 \\ x^2 + \dfrac{y^2}{9} = 1 \end{cases}$ Dividing the first equation by 100 gives $x^2 + \dfrac{y^2}{4} = 1$.

Subtracting this equation from the second equation gives $\dfrac{y^2}{9} - \dfrac{y^2}{4} = 0 \Leftrightarrow \left(\frac{1}{9} - \frac{1}{4}\right) y^2 = 0 \Leftrightarrow y = 0$. Substituting $y = 0$ in the second equation gives $x^2 + (0)^2 = 1 \Leftrightarrow x = \pm 1$, and so the points of intersection are $(\pm 1, 0)$.

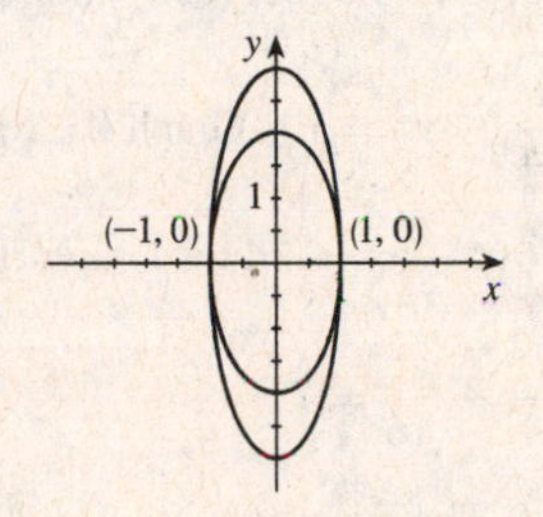

61. **(a)** The ellipse $x^2 + 4y^2 = 16 \Leftrightarrow \dfrac{x^2}{16} + \dfrac{y^2}{4} = 1$ has $a = 4$ and $b = 2$. Thus, an equation of the ancillary circle is $x^2 + y^2 = 4$.

(b) If (s, t) is a point on the ancillary circle, then $s^2 + t^2 = 4 \Leftrightarrow 4s^2 + 4t^2 = 16 \Leftrightarrow (2s)^2 + 4(t)^2 = 16$, which implies that $(2s, t)$ is a point on the ellipse.

63. $\dfrac{x^2}{k} + \dfrac{y^2}{4+k} = 1$ is an ellipse for $k > 0$. Then $a^2 = 4 + k$, $b^2 = k$, and so $c^2 = 4 + k - k = 4 \Leftrightarrow c = \pm 2$. Therefore, all of the ellipses' foci are $(0, \pm 2)$ regardless of the value of k.

65. Using the perihelion, $a - c = 147{,}000{,}000$, while using the aphelion, $a + c = 153{,}000{,}000$. Adding, we have $2a = 300{,}000{,}000 \Leftrightarrow a = 150{,}000{,}000$. So $b^2 = a^2 - c^2 = \left(150 \times 10^6\right)^2 - \left(3 \times 10^6\right)^2 = 22{,}491 \times 10^{12} = 2.2491 \times 10^{16}$.

Thus, an equation of the orbit is $\dfrac{x^2}{2.2500 \times 10^{16}} + \dfrac{y^2}{2.2491 \times 10^{16}} = 1$.

67. Using the perilune, $a - c = 1075 + 68 = 1143$, and using the apolune, $a + c = 1075 + 195 = 1270$. Adding, we get $2a = 2413 \Leftrightarrow a = 1206.5$. So $c = 1270 - 1206.5 \Leftrightarrow c = 63.5$. Therefore, $b^2 = (1206.5)^2 - (63.5)^2 = 1{,}451{,}610$. Since $a^2 \approx 1{,}455{,}642$, an equation of Apollo 11's orbit is $\dfrac{x^2}{1{,}455{,}642} + \dfrac{y^2}{1{,}451{,}610} = 1$.

69. From the diagram, $a = 40$ and $b = 20$, and so an equation of the ellipse whose top half is the window is $\dfrac{x^2}{1600} + \dfrac{y^2}{400} = 1$.

Since the ellipse passes through the point $(25, h)$, by substituting, we have $\dfrac{25^2}{1600} + \dfrac{h^2}{400} = 1 \Leftrightarrow 625 + 4y^2 = 1600 \Leftrightarrow$ $y = \dfrac{\sqrt{975}}{2} = \dfrac{5\sqrt{39}}{2} \approx 15.61$ in. Therefore, the window is approximately 15.6 inches high at the specified point.

71. We start with the flashlight perpendicular to the wall; this shape is a circle. As the angle of elevation increases, the shape of the light changes to an ellipse. When the flashlight is angled so that the outer edge of the light cone is parallel to the wall, the shape of the light is a parabola. Finally, as the angle of elevation increases further, the shape of the light is hyperbolic.

12.3 HYPERBOLAS

1. A hyperbola is the set of all points in the plane for which the *difference* of the distances from two fixed point F_1 and F_2 is constant. The points F_1 and F_2 are called the *foci* of the hyperbola.

3. The graph of the equation $\dfrac{y^2}{a^2} - \dfrac{x^2}{b^2} = 1$ with $a > 0$, $b > 0$ is a hyperbola with *vertical* transverse axis, vertices $(0, a)$ and $(0, -a)$ and foci $(0, \pm c)$, where $c = \sqrt{a^2 + b^2}$. So the graph of $\dfrac{y^2}{4^2} - \dfrac{x^2}{3^2} = 1$ is a hyperbola with vertices $(0, 4)$ and $(0, -4)$ and foci $(0, 5)$ and $(0, -5)$.

5. $\dfrac{x^2}{4} - y^2 = 1$ is Graph III, which opens horizontally and has vertices at $(\pm 2, 0)$.

7. $16y^2 - x^2 = 144$ is Graph II, which pens vertically and has vertices at $(0, \pm 3)$.

9. $\dfrac{x^2}{4} - \dfrac{y^2}{16} = 1$

(a) The hyperbola has $a = 2$, $b = 4$, and $c^2 = 16 + 4 \Rightarrow c = 2\sqrt{5}$. The vertices are $(\pm 2, 0)$, the foci are $\left(\pm 2\sqrt{5}, 0\right)$, and the asymptotes are $y = \pm \frac{4}{2}x \Leftrightarrow y = \pm 2x$.

(b) The transverse axis has length $2a = 4$.

(c)

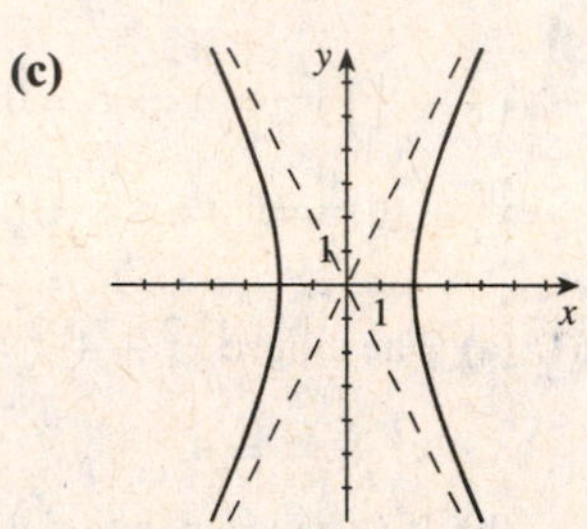

11. $\dfrac{y^2}{36} - \dfrac{x^2}{4} = 1$

(a) The hyperbola has $a = 6$, $b = 2$, and $c^2 = 36 + 4 = 40 \Rightarrow c = 2\sqrt{10}$. The vertices are $(0, \pm 6)$, the foci are $\left(0, \pm 2\sqrt{10}\right)$, and the asymptotes are $y = \pm 3x$.

(b) The transverse axis has length $2a = 12$.

(c)

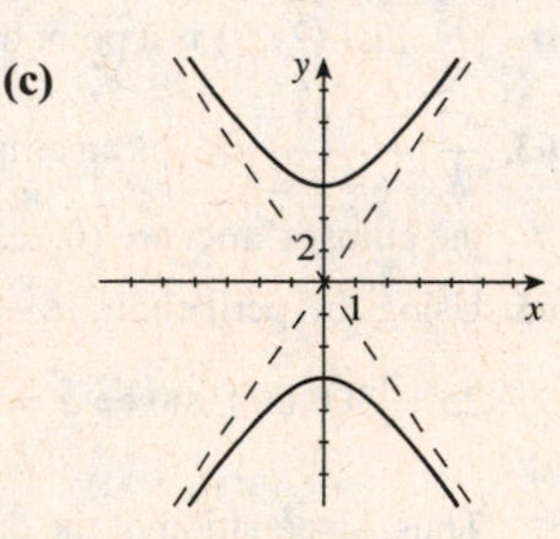

13. $\dfrac{y^2}{1} - \dfrac{x^2}{25} = 1$

(a) The hyperbola has $a = 1$, $b = 5$, and $c^2 = 1 + 25 = 26 \Rightarrow c = \sqrt{26}$. The vertices are $(0, \pm 1)$, the foci are $\left(0, \pm\sqrt{26}\right)$, and the asymptotes are $y = \pm \frac{1}{5}x$.

(b) The transverse axis has length $2a = 2$.

(c)

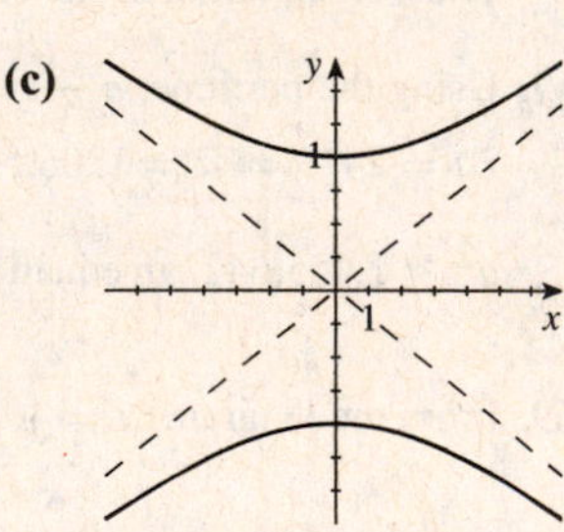

15. $x^2 - y^2 = 1$

(a) The hyperbola has $a = 1$, $b = 1$, and $c^2 = 1 + 1 = 2 \Rightarrow c = \sqrt{2}$. The vertices are $(\pm 1, 0)$, the foci are $\left(\pm\sqrt{2}, 0\right)$, and the asymptotes are $y = \pm x$.

(b) The transverse axis has length $2a = 2$.

(c)

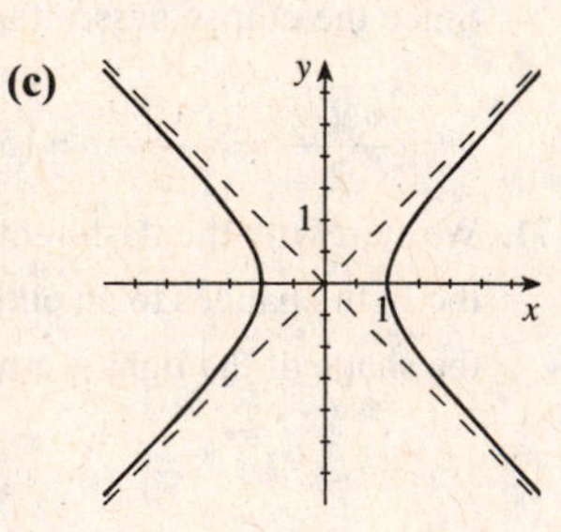

17. $9x^2 - 4y^2 = 36 \Leftrightarrow \dfrac{x^2}{4} - \dfrac{y^2}{9} = 1$

(c)

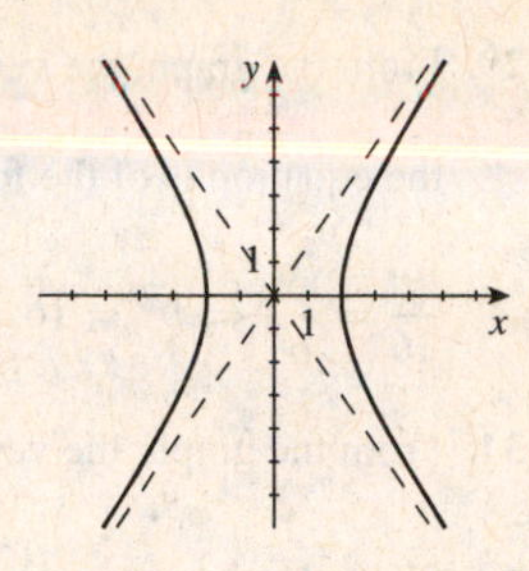

(a) The hyperbola has $a = 2$, $b = 3$, and $c^2 = 4 + 9 = 13 \Rightarrow c = \sqrt{13}$. The vertices are $(\pm 2, 0)$, the foci are $\left(\pm\sqrt{13}, 0\right)$, and the asymptotes are $y = \pm\frac{3}{2}x$.

(b) The transverse axis has length $2a = 4$.

19. $4y^2 - 9x^2 = 144 \Leftrightarrow \dfrac{y^2}{36} - \dfrac{x^2}{16} = 1$

(c)

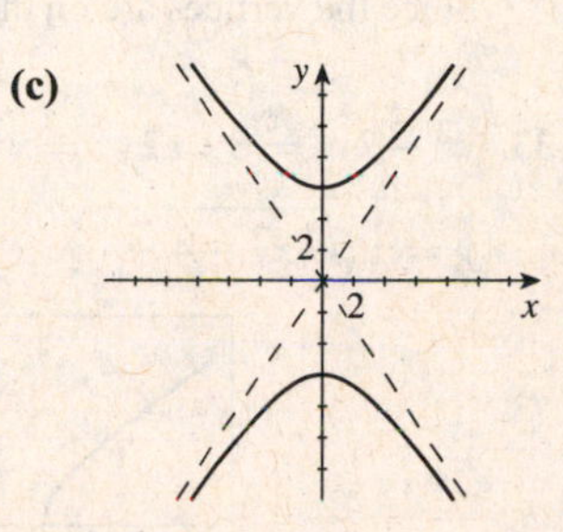

(a) The hyperbola has $a = 6$, $b = 4$, and $c^2 = a^2 + b^2 = 52 \Rightarrow c = 2\sqrt{13}$. The vertices are $(0, \pm 6)$, the foci are $\left(0, \pm 2\sqrt{13}\right)$, and the asymptotes are $y = \pm\frac{3}{2}x$.

(b) The transverse axis has length $2a = 12$.

21. $x^2 - 4y^2 - 8 = 0 \Leftrightarrow \dfrac{x^2}{8} - \dfrac{y^2}{2} = 1$

(c)

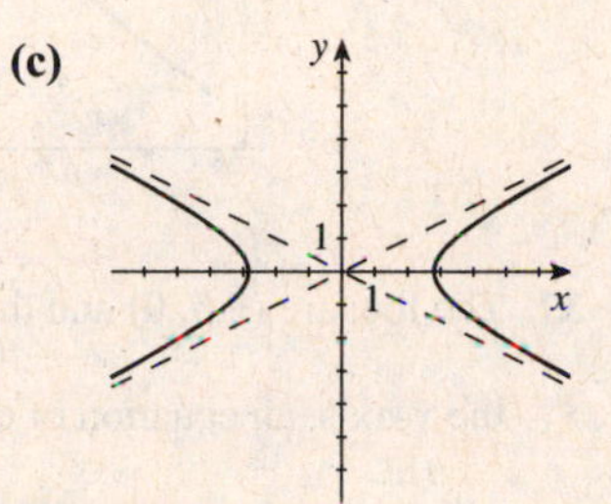

(a) The hyperbola has $a = 2\sqrt{2}$, $b = \sqrt{2}$, and $c^2 = 8 + 2 = 10 \Rightarrow c = \sqrt{10}$. The vertices are $\left(\pm 2\sqrt{2}, 0\right)$, the foci are $\left(\pm\sqrt{10}, 0\right)$, and the asymptotes are $y = \pm\frac{\sqrt{2}}{\sqrt{8}}x = \pm\frac{1}{2}x$.

(b) The transverse axis has length $2a = 4\sqrt{2}$.

23. $x^2 - y^2 + 4 = 0 \Leftrightarrow y^2 - x^2 = 4 \Leftrightarrow \dfrac{y^2}{4} - \dfrac{x^2}{4} = 1$

(c)

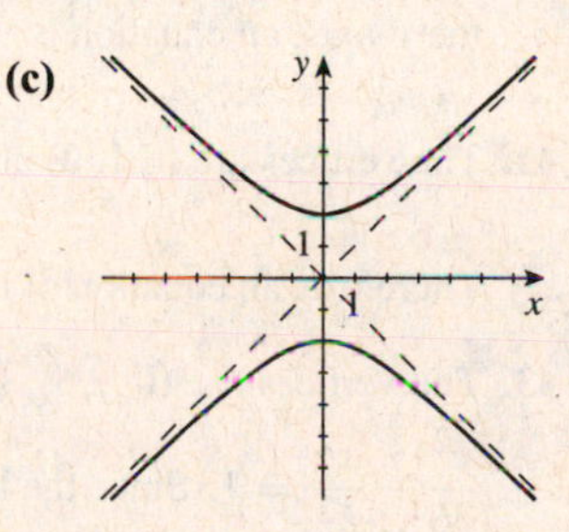

(a) The hyperbola has $a = 2$, $b = 2$, and $c^2 = 4 + 4 = 8 = 2\sqrt{2}$. The vertices are $(0, \pm 2)$, the foci are $\left(0, \pm 2\sqrt{2}\right)$, and the asymptotes are $y = \pm x$.

(b) The transverse axis has length $2a = 4$.

25. $4y^2 - x^2 = 1 \Leftrightarrow \dfrac{y^2}{\frac{1}{4}} - x^2 = 1$

(c)

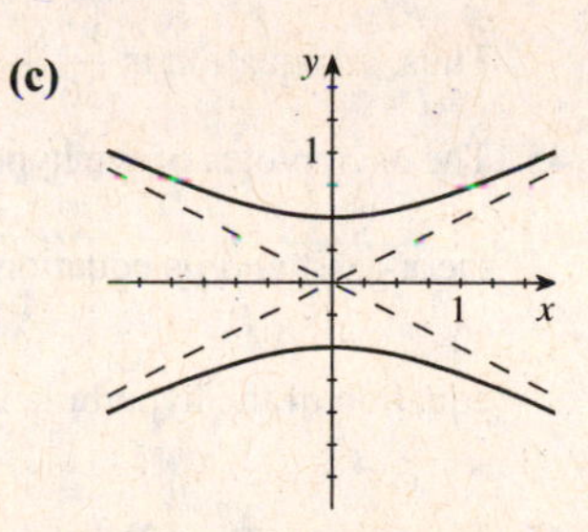

(a) The hyperbola has $a = \frac{1}{2}$, $b = 1$, and $c^2 = \frac{1}{4} + 1 = \frac{5}{4} \Rightarrow c = \frac{\sqrt{5}}{2}$. The vertices are $\left(0, \pm\frac{1}{2}\right)$, the foci are $\left(0, \pm\frac{\sqrt{5}}{2}\right)$, and the asymptotes are $y = \pm\frac{1/2}{1}x = \pm\frac{1}{2}x$.

(b) The transverse axis has length $2a = 1$.

27. From the graph, the foci are $(\pm 4, 0)$, and the vertices are $(\pm 2, 0)$, so $c = 4$ and $a = 2$. Thus, $b^2 = 16 - 4 = 12$, and since the vertices are on the x-axis, an equation of the hyperbola is $\dfrac{x^2}{4} - \dfrac{y^2}{12} = 1$.

29. From the graph, the vertices are $(0, \pm 4)$, the foci are on the y-axis, and the hyperbola passes through the point $(3, -5)$. So the equation is of the form $\frac{y^2}{16} - \frac{x^2}{b^2} = 1$. Substituting the point $(3, -5)$, we have $\frac{(-5)^2}{16} - \frac{(3)^2}{b^2} = 1 \Leftrightarrow \frac{25}{16} - 1 = \frac{9}{b^2} \Leftrightarrow \frac{9}{16} = \frac{9}{b^2} \Leftrightarrow b^2 = 16$. Thus, an equation of the hyperbola is $\frac{y^2}{16} - \frac{x^2}{16} = 1$.

31. From the graph, the vertices are $(0, \pm 3)$, so $a = 3$. Since the asymptotes are $y = \pm 3x = \pm \frac{a}{b} x$, we have $\frac{3}{b} = 3 \Leftrightarrow b = 1$. Since the vertices are on the x-axis, an equation is $\frac{y^2}{3^2} - \frac{x^2}{1^2} = 1 \Leftrightarrow \frac{y^2}{9} - x^2 = 1$.

33. $x^2 - 2y^2 = 8 \Leftrightarrow 2y^2 = x^2 - 8 \Leftrightarrow y^2 = \frac{1}{2}x^2 - 4 \Rightarrow$

$y = \pm\sqrt{\frac{1}{2}x^2 - 4}$

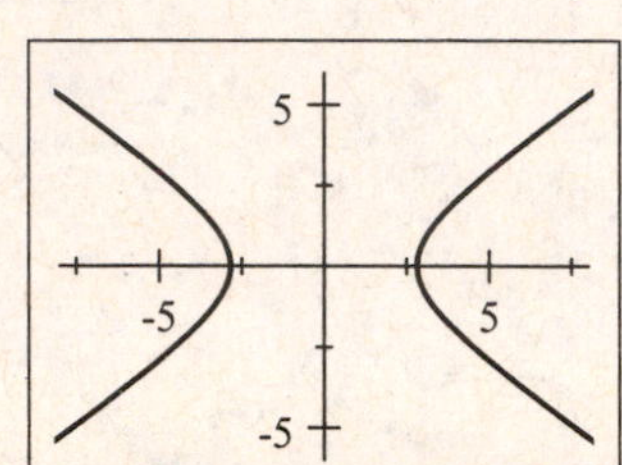

35. $\frac{y^2}{2} - \frac{x^2}{6} = 1 \Leftrightarrow \frac{y^2}{2} = \frac{x^2}{6} + 1 \Leftrightarrow y^2 = \frac{x^2}{3} + 2 \Rightarrow$

$y = \pm\sqrt{\frac{x^2}{3} + 2}$

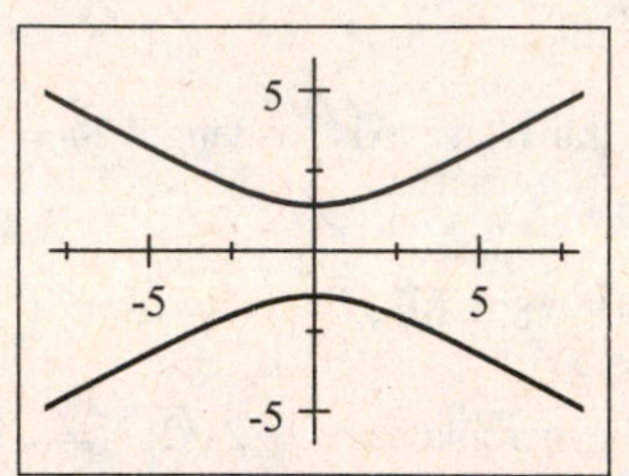

37. The foci are $(\pm 5, 0)$ and the vertices are $(\pm 3, 0)$, so $c = 5$ and $a = 3$. Then $b^2 = 25 - 9 = 16$, and since the vertices are on the x-axis, an equation of the hyperbola is $\frac{x^2}{9} - \frac{y^2}{16} = 1$.

39. The foci are $(0, \pm 2)$ and the vertices are $(0, \pm 1)$, so $c = 2$ and $a = 1$. Then $b^2 = 4 - 1 = 3$, and since the vertices are on the y-axis, an equation is $y^2 - \frac{x^2}{3} = 1$.

41. The vertices are $(\pm 1, 0)$ and the asymptotes are $y = \pm 5x$, so $a = 1$. The asymptotes are $y = \pm \frac{b}{a} x$, so $\frac{b}{1} = 5 \Leftrightarrow b = 5$. Therefore, an equation of the hyperbola is $x^2 - \frac{y^2}{25} = 1$.

43. The vertices are $(0, \pm 6)$, so $a = 6$. Since the vertices are on the y-axis, the hyperbola has an equation of the form $\frac{y^2}{36} - \frac{x^2}{b^2} = 1$. Since the hyperbola passes through the point $(-5, 9)$, we have $\frac{81}{36} - \frac{25}{b^2} = 1 \Leftrightarrow \frac{25}{b^2} = \frac{45}{36} \Leftrightarrow b^2 = 20$. Thus, an equation is $\frac{y^2}{36} - \frac{x^2}{20} = 1$.

45. The asymptotes of the hyperbola are $y = \pm x$, so $b = a$. Since the hyperbola passes through the point $(5, 3)$, its foci are on the x-axis, and its equation has the form, $\frac{x^2}{a^2} - \frac{y^2}{a^2} = 1$, so it follows that $\frac{25}{a^2} - \frac{9}{a^2} = 1 \Leftrightarrow a^2 = 16 = b^2$. Therefore, an equation of the hyperbola is $\frac{x^2}{16} - \frac{y^2}{16} = 1$.

47. The foci are $(0, \pm 3)$, so $c = 3$ and an equation is $\frac{y^2}{a^2} - \frac{x^2}{b^2} = 1$. The hyperbola passes through $(1, 4)$, so $\frac{4^2}{a^2} - \frac{1^2}{b^2} = 1 \Leftrightarrow 16b^2 - a^2 = a^2b^2 \Leftrightarrow 16b^2 - \left(9 - b^2\right) = \left(9 - b^2\right) b^2 \Leftrightarrow b^4 + 8b^2 - 9 = 0 \Leftrightarrow \left(b^2 - 1\right)\left(b^2 + 9\right) = 0$, Thus, $b^2 = 1$, $a^2 = 8$, and an equation is $\frac{y^2}{8} - x^2 = 1$.

49. The foci are $(\pm 5, 0)$, and the length of the transverse axis is 6, so $c = 5$ and $2a = 6 \Leftrightarrow a = 3$. Thus, $b^2 = 25 - 9 = 16$, and an equation is $\frac{x^2}{9} - \frac{y^2}{16} = 1$.

51. (a) The hyperbola $x^2 - y^2 = 5 \Leftrightarrow \frac{x^2}{5} - \frac{y^2}{5} = 1$ has $a = \sqrt{5}$ and $b = \sqrt{5}$. Thus, the asymptotes are $y = \pm x$, and their slopes are $m_1 = 1$ and $m_2 = -1$. Since $m_1 \cdot m_2 = -1$, the asymptotes are perpendicular.

(b) Since the asymptotes are perpendicular, they must have slopes ± 1, so $a = b$. Therefore, $c^2 = 2a^2 \Leftrightarrow a^2 = \frac{c^2}{2}$, and since the vertices are on the x-axis, an equation is $\frac{x^2}{\frac{1}{2}c^2} - \frac{y^2}{\frac{1}{2}c^2} = 1 \Leftrightarrow x^2 - y^2 = \frac{c^2}{2}$.

53. $\sqrt{(x+c)^2 + y^2} - \sqrt{(x-c)^2 + y^2} = \pm 2a$. Let us consider the positive case only. Then $\sqrt{(x+c)^2 + y^2} = 2a + \sqrt{(x-c)^2 + y^2}$, and squaring both sides gives $x^2 + 2cx + c^2 + y^2 = 4a^2 + 4a\sqrt{(x-c)^2 + y^2} + x^2 - 2cx + c^2 + y^2 \Leftrightarrow 4a\sqrt{(x-c)^2 + y^2} = 4cx - 4a^2$. Dividing by 4 and squaring both sides gives $a^2\left(x^2 - 2cx + c^2 + y^2\right) = c^2x^2 - 2a^2cx + a^4 \Leftrightarrow a^2x^2 - 2a^2cx + a^2c^2 + a^2y^2 = c^2x^2 - 2a^2cx + a^4$ $\Leftrightarrow a^2x^2 + a^2c^2 + a^2y^2 = c^2x^2 + a^4$. Rearranging the order, we have $c^2x^2 - a^2x^2 - a^2y^2 = a^2c^2 - a^4 \Leftrightarrow$ $\left(c^2 - a^2\right)x^2 - a^2y^2 = a^2\left(c^2 - a^2\right)$. The negative case gives the same result.

55. (a) From the equation, we have $a^2 = k$ and $b^2 = 16 - k$. Thus, $c^2 = a^2 + b^2 = k + 16 - k = 16 \Rightarrow c = \pm 4$. Thus the foci of the family of hyperbolas are $(0, \pm 4)$.

(b) $\frac{y^2}{k} - \frac{x^2}{16-k} = 1 \Leftrightarrow y^2 = k\left(1 + \frac{x^2}{16-k}\right) \Rightarrow y = \pm\sqrt{k + \frac{kx^2}{16-k}}$. For the top branch, we graph $y = \sqrt{k + \frac{kx^2}{16-k}}$, $k = 1, 4, 8, 12$. As k increases, the asymptotes get steeper and the vertices move further apart.

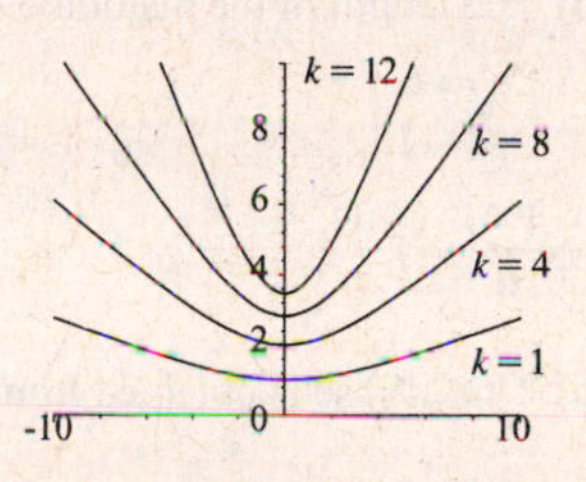

57. Since the asymptotes are perpendicular, $a = b$. Also, since the sun is a focus and the closest distance is 2×10^9, it follows that $c - a = 2 \times 10^9$. Now $c^2 = a^2 + b^2 = 2a^2$, and so $c = \sqrt{2}a$. Thus, $\sqrt{2}a - a = 2 \times 10^9 \Rightarrow a = \frac{2 \times 10^9}{\sqrt{2} - 1}$ and $a^2 = b^2 = \frac{4 \times 10^{18}}{3 - 2\sqrt{2}} \approx 2.3 \times 10^{19}$. Therefore, an equation of the hyperbola is $\frac{x^2}{2.3 \times 10^{19}} - \frac{y^2}{2.3 \times 10^{19}} = 1 \Leftrightarrow$ $x^2 - y^2 = 2.3 \times 10^{19}$.

59. Some possible answers are: as cross-sections of nuclear power plant cooling towers, or as reflectors for camouflaging the location of secret installations.

12.4 SHIFTED CONICS

1. (a) If we replace x by $x - 3$ the graph of the equation is shifted to the *right* by 3 units. If we replace x by $x + 3$ the graph is shifted to the *left* by 3 units.

(b) If we replace y by $y - 1$ the graph of the equation is shifted *upward* by 1 unit. If we replace y by $y + 1$ the graph is shifted *downward* by 1 unit.

3. $\dfrac{x^2}{5^2}+\dfrac{y^2}{4^2}=1$, from left to right: vertex $(-5,0)$, focus $(-3,0)$, focus $(3,0)$, vertex $(5,0)$. $\dfrac{(x-3)^2}{5^2}+\dfrac{(y-1)^2}{4^2}=1$, from left to right: vertex $(-2,1)$, focus $(0,1)$, focus $(6,1)$, vertex $(8,1)$.

5. 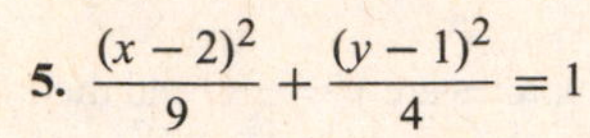
$\dfrac{(x-2)^2}{9}+\dfrac{(y-1)^2}{4}=1$

(c)

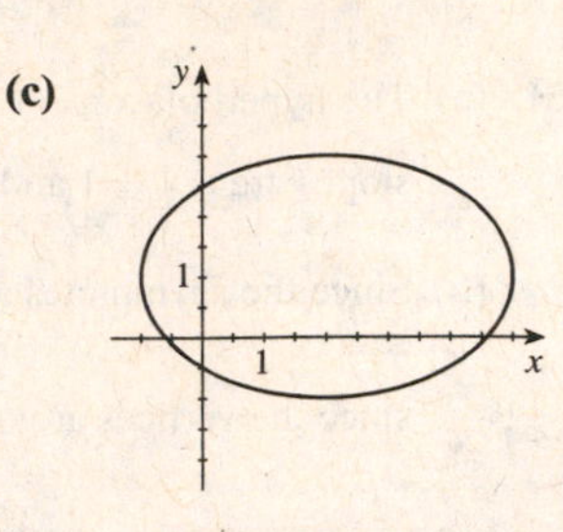

(a) This ellipse is obtained from the ellipse $\dfrac{x^2}{9}+\dfrac{y^2}{4}=1$ by shifting it 2 units to the right and 1 unit upward. So $a=3$, $b=2$, and $c=\sqrt{9-4}=\sqrt{5}$. The center is $(2,1)$, the vertices are $(2\pm 3,1)=(-1,1)$ and $(5,1)$, and the foci are $\left(2\pm\sqrt{5},1\right)$.

(b) The length of the major axis is $2a=6$ and the length of the minor axis is $2b=4$.

7. $\dfrac{x^2}{9}+\dfrac{(y+5)^2}{25}=1$

(c)

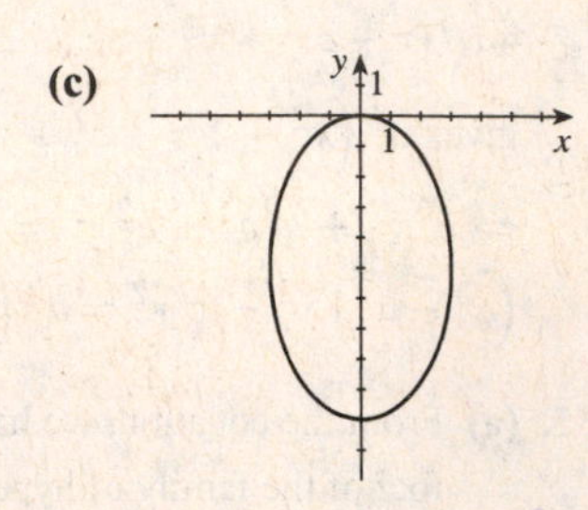

(a) This ellipse is obtained from the ellipse $\dfrac{x^2}{9}+\dfrac{y^2}{25}=1$ by shifting it 5 units downward. So $a=5$, $b=3$, and $c=\sqrt{25-9}=4$. The center is $(0,-5)$, the vertices are $(0,-5\pm 5)=(0,-10)$ and $(0,0)$, and the foci are $(0,-5\pm 4)=(0,-9)$ and $(0,-1)$.

(b) The length of the major axis is $2a=10$ and the length of the minor axis is $2b=6$.

9. $\dfrac{(x+5)^2}{16}+\dfrac{(y-1)^2}{4}=1$

(c)

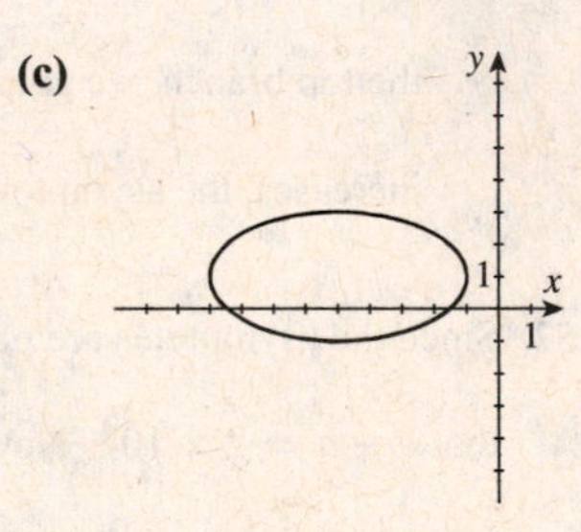

(a) This ellipse is obtained from the ellipse $\dfrac{x^2}{16}+\dfrac{y^2}{4}=1$ by shifting it 5 units to the left and 1 units upward. So $a=4$, $b=2$, and $c=\sqrt{16-4}=2\sqrt{3}$. The center is $(-5,1)$, the vertices are $(-5\pm 4,1)=(-9,1)$ and $(-1,1)$, and the foci are $\left(-5\pm 2\sqrt{3},1\right)=\left(-5-2\sqrt{3},1\right)$ and $\left(-5+2\sqrt{3},1\right)$.

(b) The length of the major axis is $2a=8$ and the length of the minor axis is $2b=4$.

11. $4x^2+25y^2-50y=75 \Leftrightarrow 4x^2+25(y-1)^2-25=75 \Leftrightarrow \dfrac{x^2}{25}+\dfrac{(y-1)^2}{4}=1$

(c)

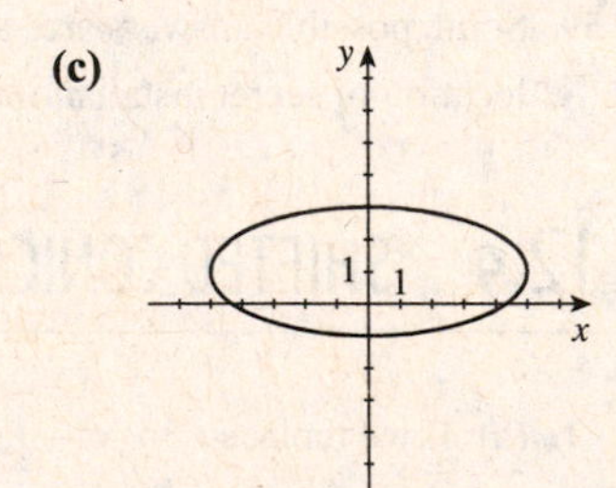

(a) This ellipse is obtained from the ellipse $\dfrac{x^2}{25}+\dfrac{y^2}{4}=1$ by shifting it 1 unit upward. So $a=5$, $b=2$, and $c=\sqrt{25-4}=\sqrt{21}$. The center is $(0,1)$, the vertices are $(\pm 5,1)=(-5,1)$ and $(5,1)$, and the foci are $\left(\pm\sqrt{21},1\right)=\left(-\sqrt{21},1\right)$ and $\left(\sqrt{21},1\right)$.

(b) The length of the major axis is $2a=10$ and the length of the minor axis is $2b=4$.

13. $(x-3)^2 = 8(y+1)$

(a) This parabola is obtained from the parabola $x^2 = 8y$ by shifting it 3 units to the right and 1 unit down. So $4p = 8 \Leftrightarrow p = 2$. The vertex is $(3, -1)$, the focus is $(3, -1+2) = (3, 1)$, and the directrix is $y = -1 - 2 = -3$.

(b)

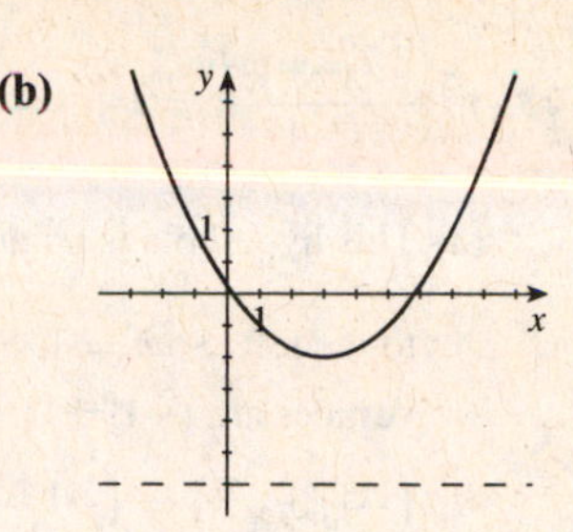

15. $(y+5)^2 = -6x + 12 = -6(x-2)$

(a) This parabola is obtained from the parabola $y^2 = -6x$ by shifting to the right 2 units and down 5 units. So $4p = -6 \Leftrightarrow p = -\frac{3}{2}$. The vertex is $(2, -5)$, the focus is $\left(2 - \frac{3}{2}, -5\right) = \left(\frac{1}{2}, -5\right)$, and the directrix is $x = 2 + \frac{3}{2} = \frac{7}{2}$.

(b)

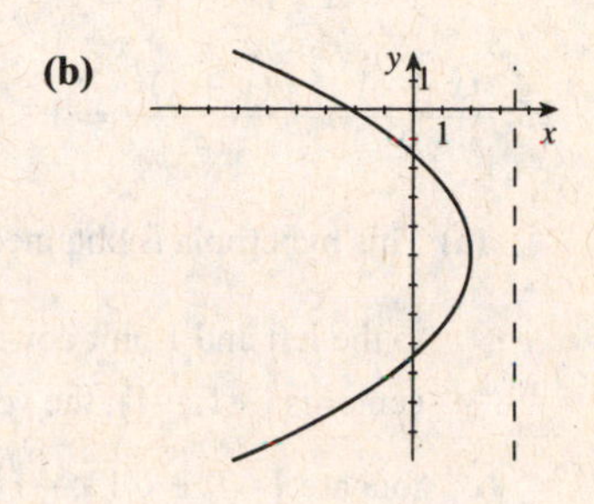

17. $2(x-1)^2 = y \Leftrightarrow (x-1)^2 = \frac{1}{2}y$

(a) This parabola is obtained from the parabola $x^2 = \frac{1}{2}y$ by shifting it 1 unit to the right. So $4p = \frac{1}{2} \Leftrightarrow p = \frac{1}{8}$. The vertex is $(1, 0)$, the focus is $\left(1, \frac{1}{8}\right)$, and the directrix is $y = -\frac{1}{8}$.

(b)

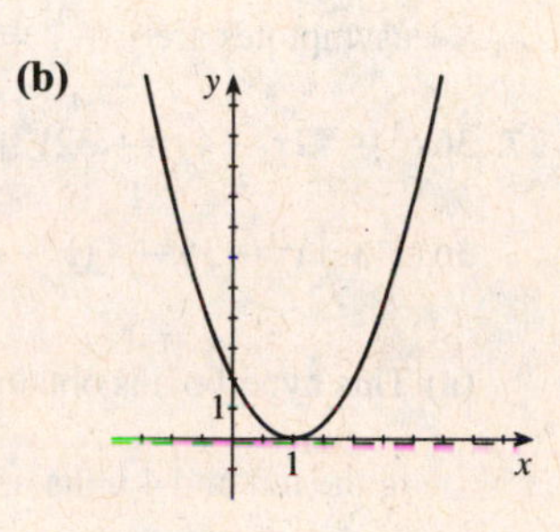

19. $y^2 - 6y - 12x + 33 = 0 \Leftrightarrow (y-3)^2 - 9 - 12x + 33 = 0 \Leftrightarrow (y-3)^2 = 12(x-2)$

(a) This parabola is obtained from the parabola $y^2 = 12x$ by shifting it 2 units to the right and 3 units upward. So $4p = 12 \Leftrightarrow p = 3$. The vertex is $(2, 3)$, the focus is $(2+3, 3) = (5, 3)$, and the directrix is $x = 2 - 3 = -1$.

(b)

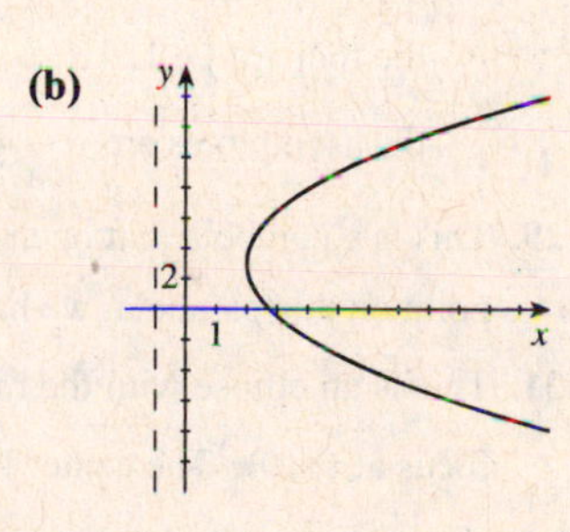

21. $\dfrac{(x+1)^2}{9} - \dfrac{(y-3)^2}{16} = 1$

(a) This hyperbola is obtained from the hyperbola $\dfrac{x^2}{9} - \dfrac{y^2}{16} = 1$ by shifting it 1 unit to the left and 3 units up. So $a = 3$, $b = 4$, and $c = \sqrt{9+16} = 5$. The center is $(-1, 3)$, the vertices are $(-1 \pm 3, 3) = (-4, 3)$ and $(2, 3)$, the foci are $(-1 \pm 5, 3) = (-6, 3)$ and $(4, 3)$, and the asymptotes are $(y-3) = \pm\frac{4}{3}(x+1) \Leftrightarrow y = \pm\frac{4}{3}(x+1) + 3 \Leftrightarrow y = \frac{4}{3}x + \frac{13}{3}$ and $y = -\frac{4}{3}x + \frac{5}{3}$.

(b)

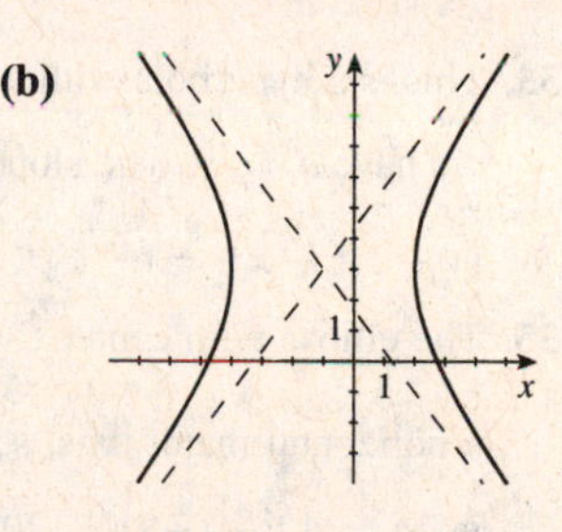

23. $y^2 - \dfrac{(x+1)^2}{4} = 1$

(a) This hyperbola is obtained from the hyperbola $y^2 - \dfrac{x^2}{4} = 1$ by shifting it 1 unit to the left. So $a = 1$, $b = 2$, and $c = \sqrt{1+4} = \sqrt{5}$. The center is $(-1, 0)$, the vertices are $(-1, \pm 1) = (-1, -1)$ and $(-1, 1)$, the foci are $\left(-1, \pm\sqrt{5}\right) = \left(-1, -\sqrt{5}\right)$ and $\left(-1, \sqrt{5}\right)$, and the asymptotes are $y = \pm\frac{1}{2}(x+1) \Leftrightarrow y = \frac{1}{2}x + \frac{1}{2}$ and $y = -\frac{1}{2}x - \frac{1}{2}$.

(b)

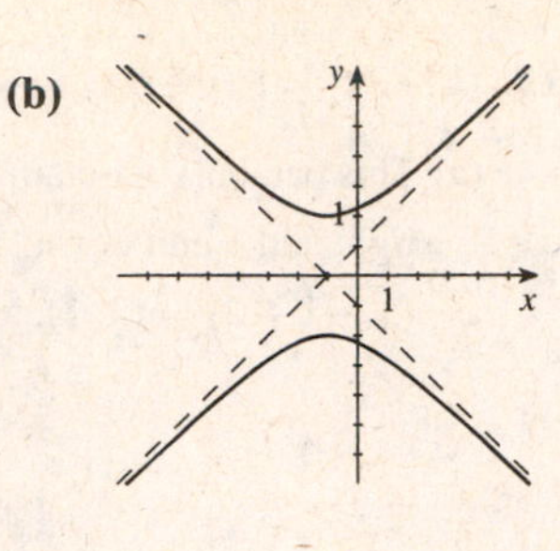

25. $\dfrac{(x+1)^2}{9} - \dfrac{(y+1)^2}{4} = 1$

(a) This hyperbola is obtained from the hyperbola $\dfrac{x^2}{9} - \dfrac{y^2}{4} = 1$ by shifting it 1 unit to the left and 1 unit downward, so $a = 3$, $b = 2$, and $c = \sqrt{9+4} = \sqrt{13}$. The center is $(-1, -1)$, the vertices are $(-1 \pm 3, -1) = (-4, -1)$ and $(2, -1)$, the foci are $\left(-1 \pm \sqrt{13}, -1\right) = \left(-1 - \sqrt{13}, -1\right)$ and $\left(\sqrt{13} - 1, -1\right)$, and the asymptotes are $y + 1 = \pm\frac{2}{3}(x+1) \Leftrightarrow y = \frac{2}{3}x - \frac{1}{3}$ and $y = -\frac{2}{3}x - \frac{5}{3}$.

(b)

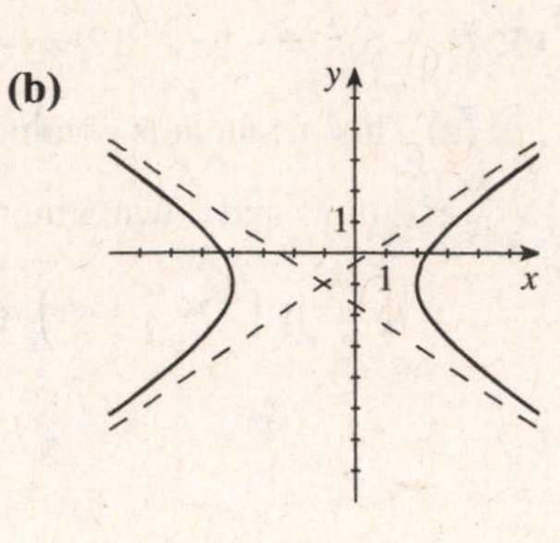

27. $36x^2 + 72x - 4y^2 + 32y + 116 = 0 \Leftrightarrow$

$36(x+1)^2 - 36 - 4(y-4)^2 + 64 + 116 = 0 \Leftrightarrow \dfrac{(y-4)^2}{36} - \dfrac{(x+1)^2}{4} = 1$

(a) This hyperbola is obtained from the hyperbola $\dfrac{y^2}{36} - \dfrac{x^2}{4} = 1$ by shifting it 1 unit to the left and 4 units upward. So $a = 6$, $b = 2$, and $c = \sqrt{36+4} = 2\sqrt{10}$. The center is $(-1, 4)$, the vertices are $(-1, 4 \pm 6) = (-1, -2)$ and $(-1, 10)$, the foci are $\left(-1, 4 \pm 2\sqrt{10}\right) = \left(-1, -2\sqrt{10} + 4\right)$ and $\left(-1, 4 + 2\sqrt{10}\right)$, and the asymptotes are $y - 4 = \pm 3(x+1) \Leftrightarrow y = 3x + 7$ and $y = -3x + 1$.

(b)

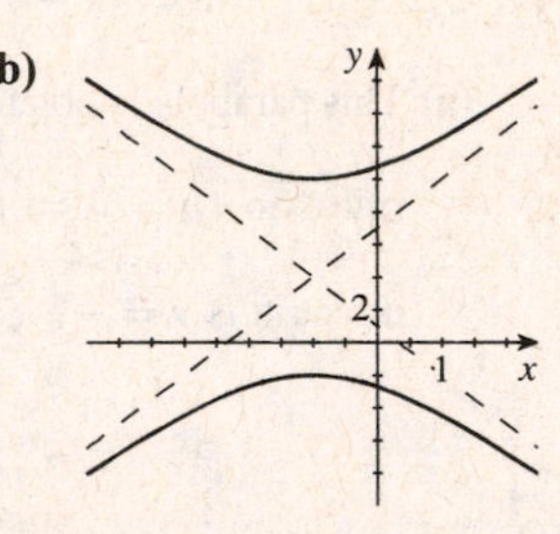

29. This is a parabola that opens down with its vertex at $(0, 4)$, so its equation is of the form $x^2 = a(y-4)$. Since $(1, 0)$ is a point on this parabola, we have $(1)^2 = a(0-4) \Leftrightarrow 1 = -4a \Leftrightarrow a = -\frac{1}{4}$. Thus, an equation is $x^2 = -\frac{1}{4}(y-4)$.

31. This is an ellipse with the major axis parallel to the x-axis, with one vertex at $(0, 0)$, the other vertex at $(10, 0)$, and one focus at $(8, 0)$. The center is at $\left(\frac{0+10}{2}, 0\right) = (5, 0)$, $a = 5$, and $c = 3$ (the distance from one focus to the center). So $b^2 = a^2 - c^2 = 25 - 9 = 16$. Thus, an equation is $\dfrac{(x-5)^2}{25} + \dfrac{y^2}{16} = 1$.

33. This is a hyperbola with center $(0, 1)$ and vertices $(0, 0)$ and $(0, 2)$. Since a is the distance form the center to a vertex, we have $a = 1$. The slope of the given asymptote is 1, so $\dfrac{a}{b} = 1 \Leftrightarrow b = 1$. Thus, an equation of the hyperbola is $(y-1)^2 - x^2 = 1$.

35. The ellipse with center $C(2, -3)$, vertices $V_1(-8, -3)$ and $V_2(12, -3)$, and foci $F_1(-4, -3)$ and $F_2(8, -3)$ has a horizontal major axis, so its equation has the form $\dfrac{(x-2)^2}{a^2} + \dfrac{(y+3)^2}{b^2} = 1$. The distance between the vertices is $2a = 12 - (-8) = 20$, so $a = 10$. Also, the distance from the center to each focus is $c = 2 - (-4) = 6$, so $b^2 = a^2 - c^2 = 100 - 36 = 64$. Thus, an equation is $\dfrac{(x-2)^2}{100} + \dfrac{(y+3)^2}{64} = 1$.

37. The hyperbola with center $C(-1,4)$, vertices $V_1(-1,-3)$ and $V_2(-1,11)$, and foci $F_1(-1,-5)$ and $F_2(-1,13)$ has a vertical transverse axis, so its equation has the form $\frac{(y-4)^2}{a^2}-\frac{(x+1)^2}{b^2}=1$. The distance between the vertices is $2a=11-(-3)=14$, so $a=7$. Also, the distance from the center to each focus is $c=4-(-5)=9$, so $b^2=c^2-a^2=81-49=32$. Thus, an equation is $\frac{(y-4)^2}{49}-\frac{(x+1)^2}{32}=1$.

39. The parabola with vertex $V(-3,5)$ and directrix $y=2$ has an equation of the form $(x+3)^2=4p(y-5)$. The distance from the vertex to the directrix is $p=5-2=3$, so an equation is $(x+3)^2=12(y-5)$.

41. The hyperbola with foci $F_1(1,-5)$ and $F_2(1,5)$ that passes through the point $(1,4)$ is centered midway between the foci; that is, it has $C\left(1,\frac{1}{2}(-5+5)\right)=(1,0)$. It has a vertical transverse axis, so its equation has the form $\frac{y^2}{a^2}-\frac{(x-1)^2}{b^2}=1$. The point $(1,4)$ lies on the transverse axis, so it is a vertex and we have $a=4-0=4$. Also, the distance from the center to each focus is $c=5-0=5$, so $b^2=c^2-a^2=25-16=9$. Thus, an equation is $\frac{y^2}{16}-\frac{(x-1)^2}{9}=1$.

43. The ellipse with foci $F_1(1,-4)$ and $F_2(5,-4)$ that passes through the point $(3,1)$ is centered midway between the foci; that is, it has $C\left(\frac{1}{2}(1+5),-4\right)=(3,-4)$, and so its equation has the form $\frac{(x-3)^2}{a^2}+\frac{(y+4)^2}{b^2}=1$. The distance from the center to each focus is $c=3-1=2$, so $c^2=a^2-b^2\Leftrightarrow a^2=b^2+4$. Substituting the point $(x,y)=(3,1)$ into the equation of the ellipse, we have $\frac{(3-3)^2}{a^2}+\frac{(1+4)^2}{b^2}=1\Leftrightarrow b^2=25$. From above, we have $a^2=b^2+4=29$. Thus, an equation of the ellipse is $\frac{(x-3)^2}{29}+\frac{(y+4)^2}{25}=1$.

45. The parabola that passes through the point $(6,1)$, with vertex $V(-1,2)$ and horizontal axis of symmetry has an equation of the form $(y-2)^2=4p(x+1)$. Substituting the point $(6,1)$ into this equation, we have $(1-2)^2=4p(6+1)\Leftrightarrow p=\frac{1}{28}$. Thus, an equation is $(y-2)^2=\frac{1}{7}(x+1)$.

47. $y^2=4(x+2y)\Leftrightarrow y^2-8y=4x\Leftrightarrow y^2-8y+16=4x+16\Leftrightarrow$ $(y-4)^2=4(x+4)$. This is a parabola with $4p=4\Leftrightarrow p=1$. The vertex is $(-4,4)$, the focus is $(-4+1,4)=(-3,4)$, and the directrix is $x=-4-1=-5$.

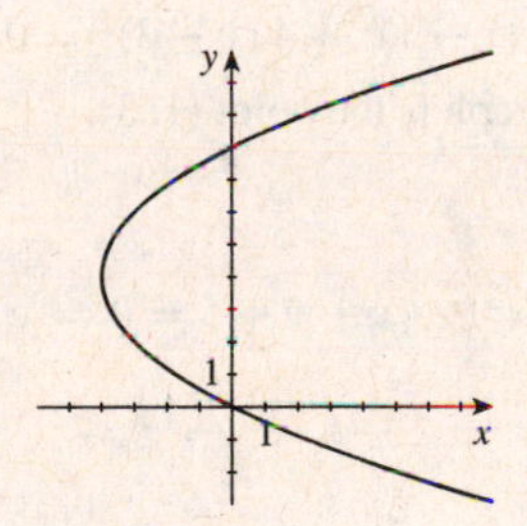

49. $x^2-5y^2-2x+20y=44\Leftrightarrow\left(x^2-2x+1\right)-5\left(y^2-4y+4\right)=44+1-20$ $\Leftrightarrow(x-1)^2-5(y-2)^2=25\Leftrightarrow\frac{(x-1)^2}{25}-\frac{(y-2)^2}{5}=1$. This is a hyperbola with $a=5$, $b=\sqrt{5}$, and $c=\sqrt{25+5}=\sqrt{30}$. The center is $(1,2)$, the foci are $\left(1\pm\sqrt{30},2\right)$, the vertices are $(1\pm5,2)=(-4,2)$ and $(6,2)$, and the asymptotes are $y-2=\pm\frac{\sqrt{5}}{5}(x-1)\Leftrightarrow y=\pm\frac{\sqrt{5}}{5}(x-1)+2\Leftrightarrow y=-\frac{\sqrt{5}}{5}x+2+\frac{\sqrt{5}}{5}$ and $y=\frac{\sqrt{5}}{5}x+2-\frac{\sqrt{5}}{5}$.

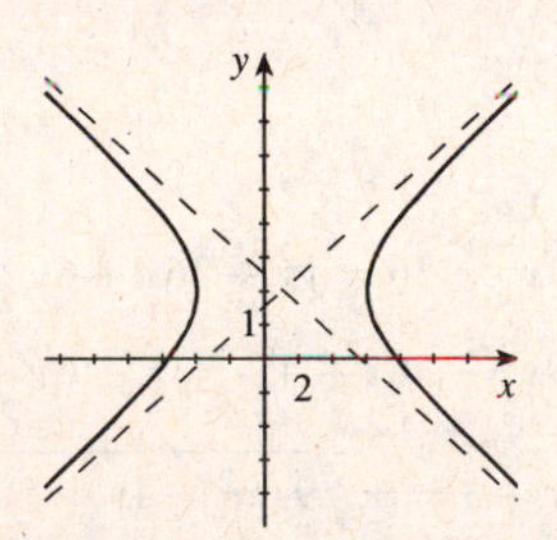

51. $4x^2 + 25y^2 - 24x + 250y + 561 = 0 \Leftrightarrow$

$4\left(x^2 - 6x + 9\right) + 25\left(y^2 + 10y + 25\right) = -561 + 36 + 625 \Leftrightarrow$

$4(x-3)^2 + 25(y+5)^2 = 100 \Leftrightarrow \dfrac{(x-3)^2}{25} + \dfrac{(y+5)^2}{4} = 1$. This is an ellipse with $a = 5$, $b = 2$, and $c = \sqrt{25-4} = \sqrt{21}$. The center is $(3, -5)$, the foci are $\left(3 \pm \sqrt{21}, -5\right)$, the vertices are $(3 \pm 5, -5) = (-2, -5)$ and $(8, -5)$, the length of the major axis is $2a = 10$, and the length of the minor axis is $2b = 4$.

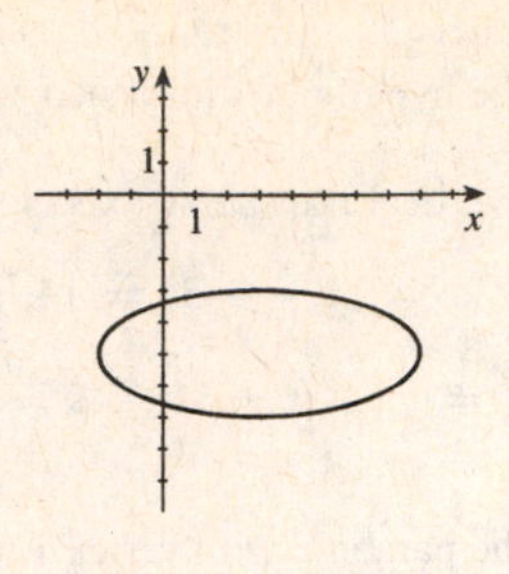

53. $16x^2 - 9y^2 - 96x + 288 = 0 \Leftrightarrow 16\left(x^2 - 6x\right) - 9y^2 + 288 = 0 \Leftrightarrow$

$16\left(x^2 - 6x + 9\right) - 9y^2 = 144 - 288 \Leftrightarrow 16(x-3)^2 - 9y^2 = -144 \Leftrightarrow$

$\dfrac{y^2}{16} - \dfrac{(x-3)^2}{9} = 1$. This is a hyperbola with $a = 4$, $b = 3$, and $c = \sqrt{16+9} = 5$. The center is $(3, 0)$, the foci are $(3, \pm 5)$, the vertices are $(3, \pm 4)$, and the asymptotes are $y = \pm\frac{4}{3}(x-3) \Leftrightarrow y = \frac{4}{3}x - 4$ and $y = 4 - \frac{4}{3}x$.

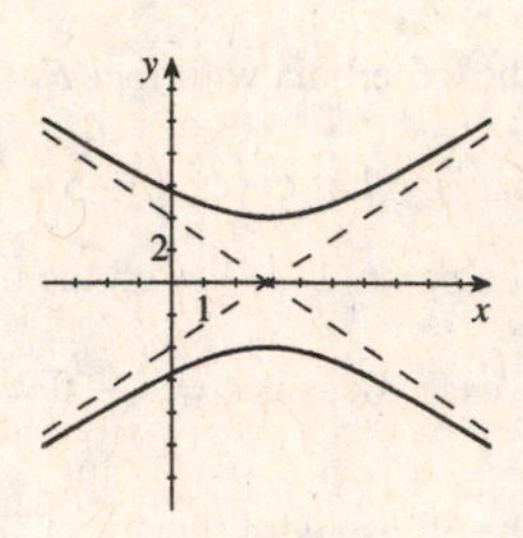

55. $x^2 + 16 = 4\left(y^2 + 2x\right) \Leftrightarrow x^2 - 8x - 4y^2 + 16 = 0 \Leftrightarrow$

$\left(x^2 - 8x + 16\right) - 4y^2 = -16 + 16 \Leftrightarrow 4y^2 = (x-4)^2 \Leftrightarrow y = \pm\frac{1}{2}(x-4)$.

Thus, the conic is degenerate, and its graph is the pair of lines $y = \frac{1}{2}(x-4)$ and $y = -\frac{1}{2}(x-4)$.

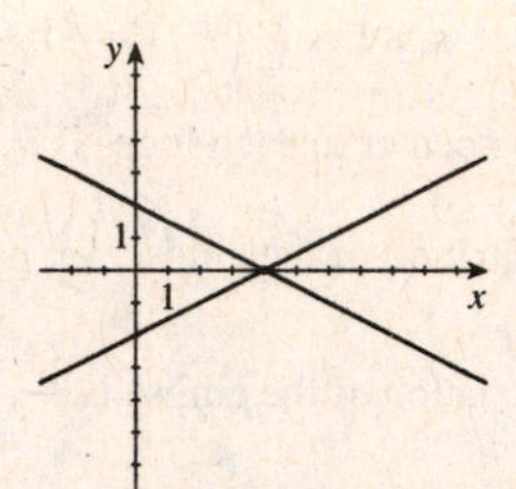

57. $3x^2 + 4y^2 - 6x - 24y + 39 = 0 \Leftrightarrow 3\left(x^2 - 2x\right) + 4\left(y^2 - 6y\right) = -39 \Leftrightarrow$

$3\left(x^2 - 2x + 1\right) + 4\left(y^2 - 6y + 9\right) = -39 + 3 + 36 \Leftrightarrow$

$3(x-1)^2 + 4(y-3)^2 = 0 \Leftrightarrow x = 1$ and $y = 3$. This is a degenerate conic whose graph is the point $(1, 3)$.

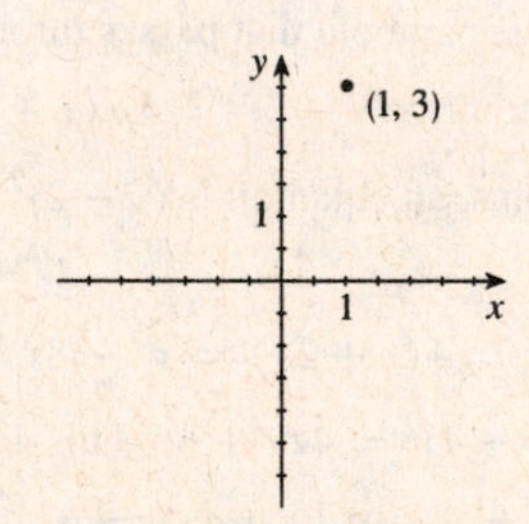

59. $2x^2 - 4x + y + 5 = 0 \Leftrightarrow y = -2x^2 + 4x - 5$.

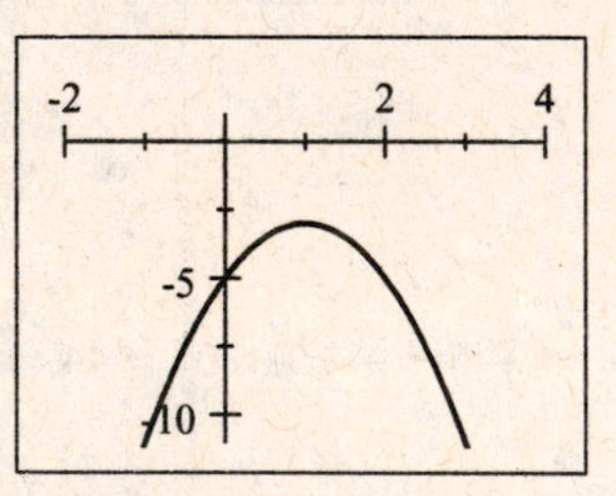

61. $9x^2 + 36 = y^2 + 36x + 6y \Leftrightarrow x^2 - 36x + 36 = y^2 + 6y \Leftrightarrow$

$9x^2 - 36x + 45 = y^2 + 6y + 9 \Leftrightarrow 9\left(x^2 - 4x + 5\right) = (y+3)^2 \Leftrightarrow$

$y + 3 = \pm\sqrt{9\left(x^2 - 4x + 5\right)} \Leftrightarrow y = -3 \pm 3\sqrt{x^2 - 4x + 5}$

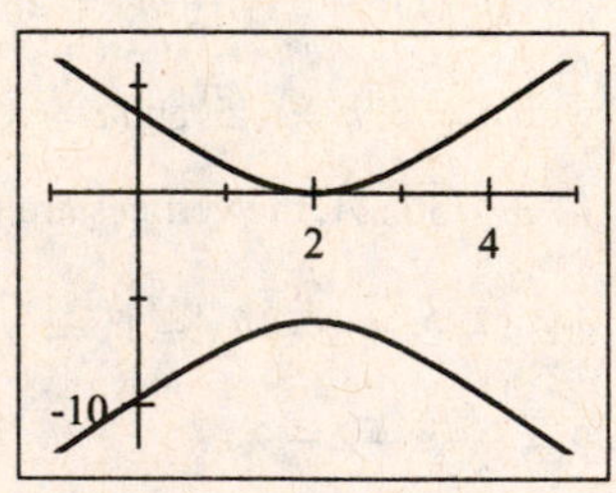

63. $4x^2 + y^2 + 4(x - 2y) + F = 0 \Leftrightarrow 4\left(x^2 + x\right) + \left(y^2 - 8y\right) = -F \Leftrightarrow$

$4\left(x^2 + x + \frac{1}{4}\right) + \left(y^2 - 8y + 16\right) = 16 + 1 - F \Leftrightarrow 4\left(x + \frac{1}{2}\right)^2 + (y - 1)^2 = 17 - F$

(a) For an ellipse, $17 - F > 0 \Leftrightarrow F < 17$.

(b) For a single point, $17 - F = 0 \Leftrightarrow F = 17$.

(c) For the empty set, $17 - F < 0 \Leftrightarrow F > 17$.

65. **(a)** $x^2 = 4p(y + p)$, for $p = -2, -\frac{3}{2}, -1, -\frac{1}{2}, \frac{1}{2}, 1, \frac{3}{2}, 2$.

(b) The graph of $x^2 = 4p(y + p)$ is obtained by shifting the graph of $x^2 = 4py$ vertically $-p$ units so that the vertex is at $(0, -p)$. The focus of $x^2 = 4py$ is at $(0, p)$, so this point is also shifted $-p$ units vertically to the point $(0, p - p) = (0, 0)$. Thus, the focus is located at the origin.

(c) The parabolas become narrower as the vertex moves toward the origin.

67. Since the height of the satellite above the earth varies between 140 and 440, the length of the major axis is $2a = 140 + 2(3960) + 440 = 8500 \Leftrightarrow a = 4250$. Since the center of the earth is at one focus, we have $a - c = (\text{earth radius}) + 140 = 3960 + 140 = 4100 \Leftrightarrow$ $c = a - 4100 = 4250 - 4100 = 150$. Thus, the center of the ellipse is $(-150, 0)$. So $b^2 = a^2 - c^2 = 4250^2 - 150^2 = 18{,}062{,}500 - 22500 = 18{,}040{,}000$. Hence, an equation is $\dfrac{(x + 150)^2}{18{,}062{,}500} + \dfrac{y^2}{18{,}040{,}000} = 1$.

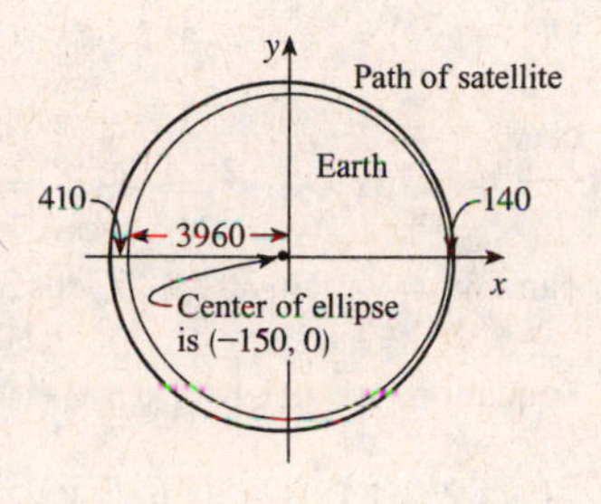

12.5 ROTATION OF AXES

1. If the x- and y-axes are rotated through an acute angle ϕ to produce the new X- and Y-axes, then the xy-coordinates (x, y) and the XY-coordinates (X, Y) of a point P in the plane are related by the formulas $x = X\cos\phi - Y\sin\phi$, $y = X\sin\phi + Y\cos\phi$, $X = x\cos\phi + y\sin\phi$, and $Y = -x\sin\phi + y\cos\phi$.

3. $(x, y) = (1, 1)$, $\phi = 45°$. Then $X = x\cos\phi + y\sin\phi = 1 \cdot \frac{1}{\sqrt{2}} + 1 \cdot \frac{1}{\sqrt{2}} = \sqrt{2}$ and $Y = -x\sin\phi + y\cos\phi = -1 \cdot \frac{1}{\sqrt{2}} + 1 \cdot \frac{1}{\sqrt{2}} = 0$. Therefore, the XY-coordinates of the given point are $(X, Y) = \left(\sqrt{2}, 0\right)$.

5. $(x, y) = \left(3, -\sqrt{3}\right)$, $\phi = 60°$. Then $X = x\cos\phi + y\sin\phi = 3 \cdot \frac{1}{2} - \sqrt{3} \cdot \frac{\sqrt{3}}{2} = 0$ and $Y = -x\sin\phi + y\cos\phi = -3 \cdot \frac{\sqrt{3}}{2} - \sqrt{3} \cdot \frac{1}{2} = -2\sqrt{3}$. Therefore, the XY-coordinates of the given point are $(X, Y) = \left(0, -2\sqrt{3}\right)$.

7. $(x, y) = (0, 2)$, $\phi = 55°$. Then $X = x\cos\phi + y\sin\phi = 0\cos 55° + 2\sin 55° \approx 1.6383$ and $Y = -x\sin\phi + y\cos\phi = -0\sin 55° + 2\cos 55° \approx 1.1472$. Therefore, the XY-coordinates of the given point are approximately $(X, Y) = (1.6383, 1.1472)$.

9. $x^2 - 3y^2 = 4$, $\phi = 60°$. Then $x = X\cos 60° - Y\sin 60° = \frac{1}{2}X - \frac{\sqrt{3}}{2}Y$ and $y = X\sin 60° + Y\cos 60° = \frac{\sqrt{3}}{2}X + \frac{1}{2}Y$.
Substituting these values into the equation, we get $\left(\frac{1}{2}X - \frac{\sqrt{3}}{2}Y\right)^2 - 3\left(\frac{\sqrt{3}}{2}X + \frac{1}{2}Y\right)^2 = 4 \Leftrightarrow$
$\frac{X^2}{4} - \frac{\sqrt{3}XY}{2} + \frac{3Y^2}{4} - 3\left(\frac{3X^2}{4} + \frac{\sqrt{3}XY}{2} + \frac{Y^2}{4}\right) = 4 \Leftrightarrow \frac{X^2}{4} - \frac{9}{4}X^2 + \frac{3Y^2}{4} - \frac{3Y^2}{4} - \frac{\sqrt{3}XY}{2} - \frac{3\sqrt{3}XY}{2} = 4 \Leftrightarrow$
$-2X^2 - 2\sqrt{3}XY = 4 \Leftrightarrow X^2 + \sqrt{3}XY = -2$.

11. $x^2 - y^2 = 2y$, $\phi = \cos^{-1}\left(\frac{3}{5}\right)$. So $\cos\phi = \frac{3}{5}$ and $\sin\phi = \frac{4}{5}$. Then
$(X\cos\phi - Y\sin\phi)^2 - (X\sin\phi + Y\cos\phi)^2 = 2(X\sin\phi + Y\cos\phi) \Leftrightarrow \left(\frac{3}{5}X - \frac{4}{5}Y\right)^2 - \left(\frac{4}{5}X + \frac{3}{5}Y\right)^2 = 2\left(\frac{4}{5}X + \frac{3}{5}Y\right)$
$\Leftrightarrow \frac{9X^2}{25} - \frac{24XY}{25} + \frac{16Y^2}{25} - \frac{16X^2}{25} - \frac{24XY}{25} - \frac{9Y^2}{25} = \frac{8X}{5} + \frac{6Y}{5} \Leftrightarrow -\frac{7X^2}{25} - \frac{48XY}{25} + \frac{7Y^2}{25} - \frac{8X}{5} - \frac{6Y}{5} = 0 \Leftrightarrow$
$7Y^2 - 48XY - 7X^2 - 40X - 30Y = 0$.

13. $x^2 + 2\sqrt{3}xy - y^2 = 4$, $\phi = 30°$. Then $x = X\cos 30° - Y\sin 30° = \frac{\sqrt{3}}{2}X - \frac{1}{2}Y = \frac{1}{2}\left(\sqrt{3}X - Y\right)$
and $y = X\sin 30° + Y\cos 30° = \frac{1}{2}X + \frac{\sqrt{3}}{2}Y = \frac{1}{2}\left(X + \sqrt{3}Y\right)$. Substituting these values into the
equation, we get $\left[\frac{1}{2}\left(\sqrt{3}X - Y\right)\right]^2 + 2\sqrt{3}\left[\frac{1}{2}\left(\sqrt{3}X - Y\right)\right]\left[\frac{1}{2}\left(X + \sqrt{3}Y\right)\right] - \left[\frac{1}{2}\left(X + \sqrt{3}Y\right)\right]^2 = 4 \Leftrightarrow$
$\left(\sqrt{3}X - Y\right)^2 + 2\sqrt{3}\left(\sqrt{3}X - Y\right)\left(X + \sqrt{3}Y\right) - \left(X + \sqrt{3}Y\right)^2 = 16 \Leftrightarrow$
$\left(3X^2 - 2\sqrt{3}XY + Y^2\right) + \left(6X^2 + 4\sqrt{3}XY - 6Y^2\right) - \left(X^2 + 2\sqrt{3}XY + 3Y^2\right) = 16 \Leftrightarrow 8X^2 - 8Y^2 = 16 \Leftrightarrow$
$\frac{X^2}{2} - \frac{Y^2}{2} = 1$. This is a hyperbola.

15. **(a)** $xy = 8 \Leftrightarrow 0x^2 + xy + 0y^2 = 8$. So $A = 0$, $B = 1$, and $C = 0$, and so the discriminant is $B^2 - 4AC = 1^2 - 4(0)(0) = 1$. Since the discriminant is positive, the equation represents a hyperbola.

(b) $\cot 2\phi = \frac{A - C}{B} = 0 \Rightarrow 2\phi = 90° \Leftrightarrow \phi = 45°$. Therefore,
$x = \frac{\sqrt{2}}{2}X - \frac{\sqrt{2}}{2}Y$ and $y = \frac{\sqrt{2}}{2}X + \frac{\sqrt{2}}{2}Y$. After substitution, the original equation becomes $\left(\frac{\sqrt{2}}{2}X - \frac{\sqrt{2}}{2}Y\right)\left(\frac{\sqrt{2}}{2}X + \frac{\sqrt{2}}{2}Y\right) = 8 \Leftrightarrow$
$\frac{(X - Y)(X + Y)}{2} = 8 \Leftrightarrow \frac{X^2}{16} - \frac{Y^2}{16} = 1$. This is a hyperbola with $a = 4$, $b = 4$, and $c = 4\sqrt{2}$. Hence, the vertices are $V(\pm 4, 0)$ and the foci are $F\left(\pm 4\sqrt{2}, 0\right)$.

(c)

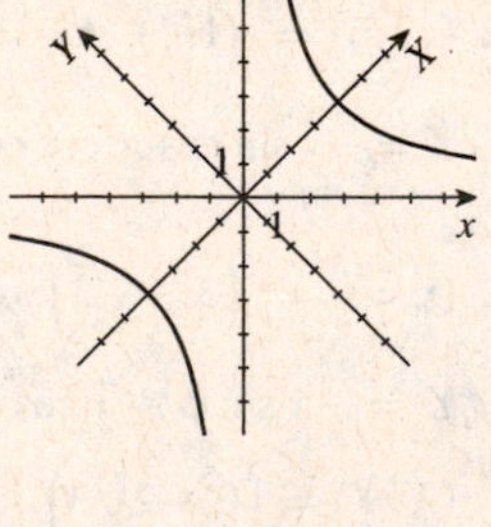

17. (a) $x^2 + 2\sqrt{3}xy - y^2 + 2 = 0$. So $A = 1$, $B = 2\sqrt{3}$, and $C = -1$, and so the discriminant is $B^2 - 4AC = \left(2\sqrt{3}\right)^2 - 4\,(1)\,(-1) > 0$. Since the discriminant is positive, the equation represents a hyperbola.

(b) $\cot 2\phi = \dfrac{A - C}{B} = \dfrac{1+1}{2\sqrt{3}} = \dfrac{1}{\sqrt{3}} \Rightarrow 2\phi = 60° \Leftrightarrow \phi = 30°$. Therefore, $x = \frac{\sqrt{3}}{2}X - \frac{1}{2}Y$ and $y = \frac{1}{2}X + \frac{\sqrt{3}}{2}Y$. After substitution, the original equation becomes

$$\left(\tfrac{\sqrt{3}}{2}X - \tfrac{1}{2}Y\right)^2 + 2\sqrt{3}\left(\tfrac{\sqrt{3}}{2}X - \tfrac{1}{2}Y\right)\left(\tfrac{1}{2}X + \tfrac{\sqrt{3}}{2}Y\right) - \left(\tfrac{1}{2}X + \tfrac{\sqrt{3}}{2}Y\right)^2 + 2 = 0 \Leftrightarrow$$

$$\tfrac{3}{4}X^2 - \tfrac{\sqrt{3}}{2}XY + \tfrac{1}{4}Y^2 + \tfrac{\sqrt{3}}{2}\left(\sqrt{3}X^2 + 2XY - \sqrt{3}Y^2\right) - \tfrac{1}{4}X^2 - \tfrac{\sqrt{3}}{2}XY - \tfrac{3}{4}Y^2 + 2 = 0 \Leftrightarrow$$

$$X^2\left(\tfrac{3}{4} + \tfrac{3}{2} - \tfrac{1}{4}\right) + XY\left(-\tfrac{\sqrt{3}}{2} + \sqrt{3} - \tfrac{\sqrt{3}}{2}\right) + Y^2\left(\tfrac{1}{4} - \tfrac{3}{2} - \tfrac{3}{4}\right) = -2 \Leftrightarrow 2X^2 - 2Y^2 = -2 \Leftrightarrow Y^2 - X^2 = 1.$$

(c)

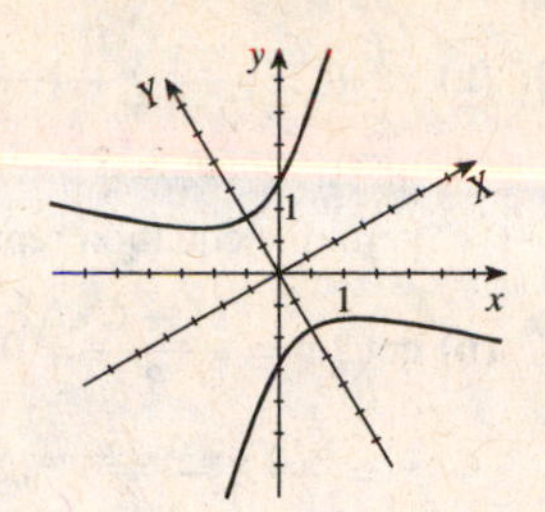

19. (a) $11x^2 - 24xy + 4y^2 + 20 = 0$. So $A = 11$, $B = -24$, and $C = 4$, and so the discriminant is $B^2 - 4AC = (-24)^2 - 4\,(11)\,(4) > 0$. Since the discriminant is positive, the equation represents a hyperbola.

(b) $\cot 2\phi = \dfrac{A - C}{B} = \dfrac{11-4}{-24} = -\dfrac{7}{24} \Rightarrow \cos 2\phi = -\frac{7}{25}$. Therefore, $\cos\phi = \sqrt{\frac{1+(-7/25)}{2}} = \frac{3}{5}$ and $\sin\phi = \sqrt{\frac{1-(-7/25)}{2}} = \frac{4}{5}$. Hence, $x = \dfrac{3X}{5} - \frac{4}{5}Y$ and $y = \frac{4}{5}X + \frac{3}{5}Y$. After substitution, the original equation becomes

$$11\left(\tfrac{3}{5}X - \tfrac{4}{5}Y\right)^2 - 24\left(\tfrac{3}{5}X - \tfrac{4}{5}Y\right)\left(\tfrac{4}{5}X + \tfrac{3}{5}Y\right) + 4\left(\tfrac{4}{5}X + \tfrac{3}{5}Y\right)^2 + 20 = 0 \Leftrightarrow$$

$$\tfrac{11}{25}\left(9X^2 - 24XY + 16Y^2\right) - \tfrac{24}{25}\left(12X^2 - 7XY - 12Y^2\right) + \tfrac{4}{25}\left(16X^2 + 24XY + 9Y^2\right) + 20 = 0 \Leftrightarrow$$

$$X^2\,(99 - 288 + 64) + XY\,(-264 + 168 + 96) + Y^2\,(176 + 288 + 36) = -500 \Leftrightarrow -125X^2 + 500Y^2 = -500 \Leftrightarrow$$

$$\tfrac{1}{4}X^2 - Y^2 = 1.$$

(c)

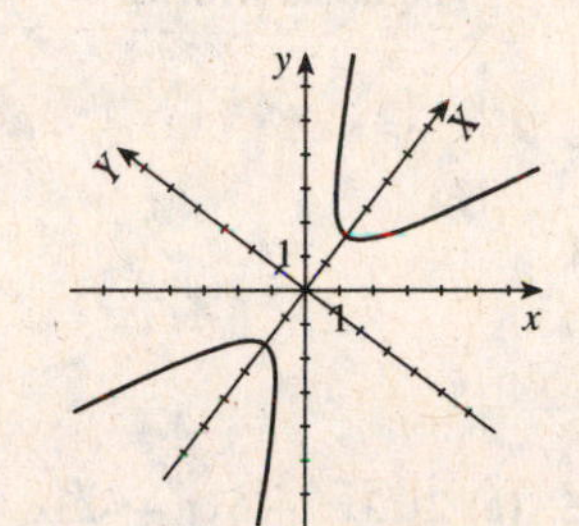

Since $\cos 2\phi = -\frac{7}{25}$, we have $2\phi \approx 106.26°$, so $\phi \approx 53°$.

21. (a) $\sqrt{3}x^2 + 3xy = 3$. So $A = \sqrt{3}$, $B = 3$, and $C = 0$, and so the discriminant is $B^2 - 4AC = (3)^2 - 4\left(\sqrt{3}\right)(0) = 9$. Since the discriminant is positive, the equation represents a hyperbola.

(b) $\cot 2\phi = \dfrac{A - C}{B} = \dfrac{1}{\sqrt{3}} \Rightarrow 2\phi = 60° \Leftrightarrow \phi = 30°$. Therefore, $x = \frac{\sqrt{3}}{2}X - \frac{1}{2}Y$ and $y = \frac{1}{2}X + \frac{\sqrt{3}}{2}Y$. After substitution, the equation becomes $\sqrt{3}\left(\frac{\sqrt{3}}{2}X - \frac{1}{2}Y\right)^2 + 3\left(\frac{\sqrt{3}}{2}X - \frac{1}{2}Y\right)\left(\frac{1}{2}X + \frac{\sqrt{3}}{2}Y\right) = 3$

$$\Leftrightarrow \tfrac{\sqrt{3}}{4}\left(3X^2 - 2\sqrt{3}XY + Y^2\right) + \tfrac{3}{4}\left(\sqrt{3}X^2 + 2XY - \sqrt{3}Y^2\right) = 3 \Leftrightarrow$$

$$X^2\left(\tfrac{3\sqrt{3}}{4} + \tfrac{3\sqrt{3}}{4}\right) + XY\left(\tfrac{-6}{4} + \tfrac{6}{4}\right) + Y^2\left(\tfrac{\sqrt{3}}{4} - \tfrac{3\sqrt{3}}{4}\right) = 3 \Leftrightarrow \tfrac{3\sqrt{3}}{2}X^2 - \tfrac{\sqrt{3}}{2}Y^2 = 3 \Leftrightarrow \tfrac{\sqrt{3}}{2}X^2 - \tfrac{1}{2\sqrt{3}}Y^2 = 1.$$

This is a hyperbola with $a = \sqrt{\frac{2}{\sqrt{3}}}$ and $b = \sqrt{2\sqrt{3}}$.

(c)

23. **(a)** $x^2+2xy+y^2+x-y=0$. So $A=1$, $B=2$, and $C=1$, and so the discriminant is $B^2-4AC=2^2-4(1)(1)=0$. Since the discriminant is zero, the equation represents a parabola.

(b) $\cot 2\phi=\dfrac{A-C}{B}=0 \Rightarrow 2\phi=90^\circ \Leftrightarrow \phi=45^\circ$. Therefore, $x=\frac{\sqrt{2}}{2}X-\frac{\sqrt{2}}{2}Y$ and $y=\frac{\sqrt{2}}{2}X+\frac{\sqrt{2}}{2}Y$. After substitution, the original equation becomes

$$\left(\tfrac{\sqrt{2}}{2}X-\tfrac{\sqrt{2}}{2}Y\right)^2+2\left(\tfrac{\sqrt{2}}{2}X-\tfrac{\sqrt{2}}{2}Y\right)\left(\tfrac{\sqrt{2}}{2}X+\tfrac{\sqrt{2}}{2}Y\right)+\left(\tfrac{\sqrt{2}}{2}X+\tfrac{\sqrt{2}}{2}Y\right)^2+\left(\tfrac{\sqrt{2}}{2}X-\tfrac{\sqrt{2}}{2}Y\right)-\left(\tfrac{\sqrt{2}}{2}X+\tfrac{\sqrt{2}}{2}Y\right)=0 \Leftrightarrow$$

$\frac{1}{2}X^2-XY+\frac{1}{2}Y^2+X^2-Y^2+\frac{1}{2}X^2+XY+Y^2-\sqrt{2}Y=0 \Leftrightarrow 2X^2-\sqrt{2}Y=0 \Leftrightarrow X^2=\frac{\sqrt{2}}{2}Y$. This is a parabola with $4p=\frac{1}{\sqrt{2}}$ and hence the focus is $F\left(0,\frac{1}{4\sqrt{2}}\right)$.

(c)

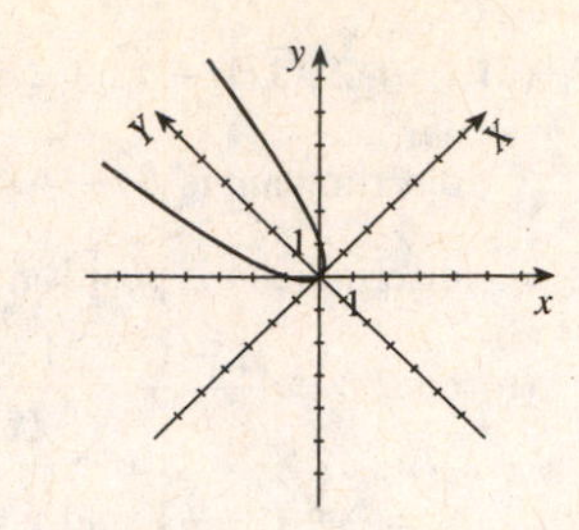

25. **(a)** $2\sqrt{3}x^2-6xy+\sqrt{3}x+3y=0$. So $A=2\sqrt{3}$, $B=-6$, and $C=0$, and so the discriminant is $B^2-4AC=(-6)^2-4\left(2\sqrt{3}\right)(0)=36$. Since the discriminant is positive, the equation represents a hyperbola.

(b) $\cot 2\phi=\dfrac{A-C}{B}=\dfrac{2\sqrt{3}}{-6}=-\dfrac{1}{\sqrt{3}} \Rightarrow 2\phi=120^\circ \Leftrightarrow \phi=60^\circ$. Therefore, $x=\frac{1}{2}X-\frac{\sqrt{3}}{2}Y$ and $y=\frac{\sqrt{3}}{2}X+\frac{1}{2}Y$, and substituting gives

$$2\sqrt{3}\left(\tfrac{1}{2}X-\tfrac{\sqrt{3}}{2}Y\right)^2-6\left(\tfrac{1}{2}X-\tfrac{\sqrt{3}}{2}Y\right)\left(\tfrac{\sqrt{3}}{2}X+\tfrac{1}{2}Y\right)+\sqrt{3}\left(\tfrac{1}{2}X-\tfrac{\sqrt{3}}{2}Y\right)+3\left(\tfrac{\sqrt{3}}{2}X+\tfrac{1}{2}Y\right)=0 \Leftrightarrow$$

$$\tfrac{\sqrt{3}}{2}\left(X^2-2\sqrt{3}XY+3Y^2\right)-\tfrac{3}{2}\left(\sqrt{3}X^2-2XY-\sqrt{3}Y^2\right)+\tfrac{\sqrt{3}}{2}\left(X-\sqrt{3}Y\right)+\tfrac{3}{2}\left(\sqrt{3}X+Y\right)=0 \Leftrightarrow$$

$$X^2\left(\tfrac{\sqrt{3}}{2}-\tfrac{3\sqrt{3}}{2}\right)+X\left(\tfrac{\sqrt{3}}{2}+\tfrac{3\sqrt{3}}{2}\right)+XY(-3+3)+Y^2\left(\tfrac{3\sqrt{3}}{2}+\tfrac{3\sqrt{3}}{2}\right)+Y\left(-\tfrac{3}{2}+\tfrac{3}{2}\right)=0 \Leftrightarrow$$

$$-\sqrt{3}X^2+2\sqrt{3}X+3\sqrt{3}Y^2=0 \Leftrightarrow -X^2+2X+3Y^2=0 \Leftrightarrow 3Y^2-\left(X^2-2X+1\right)=-1 \Leftrightarrow$$

$(X-1)^2-3Y^2=1$. This is a hyperbola with $a=1$, $b=\frac{\sqrt{3}}{3}$, $c=\sqrt{1+\frac{1}{3}}=\frac{2}{\sqrt{3}}$, and $C(1,0)$.

(c)

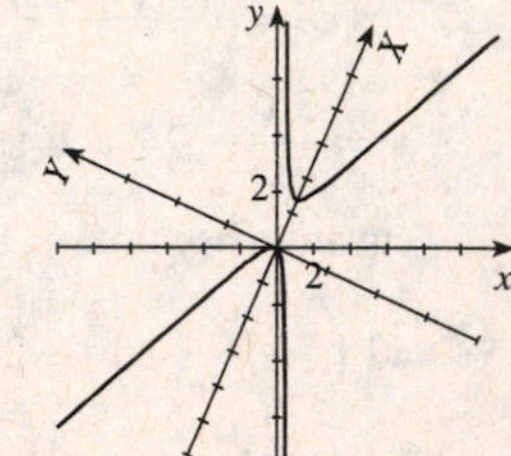

27. **(a)** $52x^2+72xy+73y^2=40x-30y+75$. So $A=52$, $B=72$, and $C=73$, and so the discriminant is $B^2-4AC=(72)^2-4(52)(73)=-10{,}000$. Since the discriminant is decidedly negative, the equation represents an ellipse.

(b) $\cot 2\phi = \dfrac{A-C}{B} = \dfrac{52-73}{72} = -\dfrac{7}{24}$. Therefore, as in Exercise 19(b), we get $\cos\phi = \frac{3}{5}$, $\sin\phi = \frac{4}{5}$, and $x = \frac{3}{5}X - \frac{4}{5}Y$, $y = \frac{4}{5}X + \frac{3}{5}Y$. By substitution,

$$52\left(\tfrac{3}{5}X - \tfrac{4}{5}Y\right)^2 + 72\left(\tfrac{3}{5}X - \tfrac{4}{5}Y\right)\left(\tfrac{4}{5}X + \tfrac{3}{5}Y\right) + 73\left(\tfrac{4}{5}X + \tfrac{3}{5}Y\right)^2$$
$$= 40\left(\tfrac{3}{5}X - \tfrac{4}{5}Y\right) - 30\left(\tfrac{4}{5}X + \tfrac{3}{5}Y\right) + 75 \Leftrightarrow$$
$$\tfrac{52}{25}\left(9X^2 - 24XY + 16Y^2\right) + \tfrac{72}{25}\left(12X^2 - 7XY - 12Y^2\right) + \tfrac{73}{25}\left(16X^2 + 24XY + 9Y^2\right)$$
$$= 24X - 32Y - 24X - 18Y + 75 \Leftrightarrow$$

$468X^2 + 832Y^2 + 864X^2 - 864Y^2 + 1168X^2 + 657Y^2 = -1250Y + 1875 \Leftrightarrow 2500X^2 + 625Y^2 + 1250Y = 1875 \Leftrightarrow$ $100X^2 + 25Y^2 + 50Y = 75 \Leftrightarrow X^2 + \frac{1}{4}(Y+1)^2 = 1$. This is an ellipse with $a = 2$, $b = 1$, $c = \sqrt{4-1} = \sqrt{3}$, and center $C(0,-1)$.

(c)

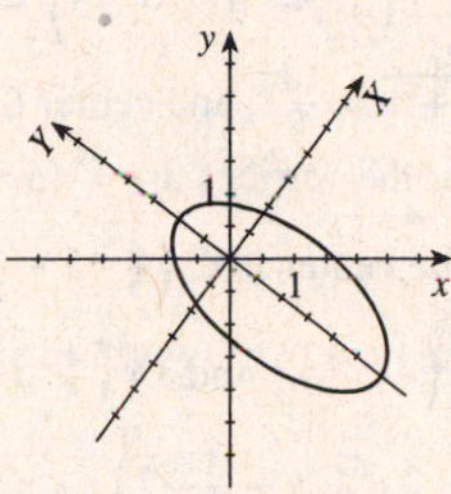

Since $\cos 2\phi = -\frac{7}{25}$, we have $2\phi = \cos^{-1}\left(-\frac{7}{25}\right) \approx 106.26°$ and so $\phi \approx 53°$.

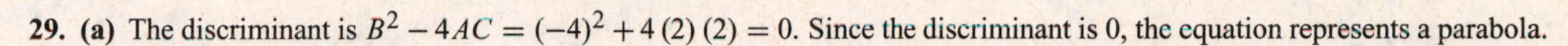
29. (a) The discriminant is $B^2 - 4AC = (-4)^2 + 4(2)(2) = 0$. Since the discriminant is 0, the equation represents a parabola.

(b) $2x^2 - 4xy + 2y^2 - 5x - 5 = 0 \Leftrightarrow 2y^2 - 4xy = -2x^2 + 5x + 5 \Leftrightarrow$
$2\left(y^2 - 2xy\right) = -2x^2 + 5x + 5 \Leftrightarrow$
$2\left(y^2 - 2xy + x^2\right) = -2x^2 + 5x + 5 + 2x^2 \Leftrightarrow 2(y-x)^2 = 5x + 5 \Leftrightarrow$
$(y-x)^2 = \frac{5}{2}x + \frac{5}{2} \Rightarrow y - x = \pm\sqrt{\frac{5}{2}x + \frac{5}{2}} \Leftrightarrow y = x \pm \sqrt{\frac{5}{2}x + \frac{5}{2}}$

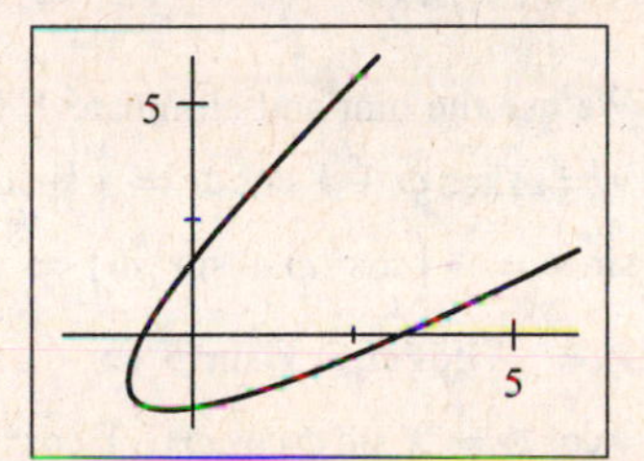

31. (a) The discriminant is $B^2 - 4AC = 10^2 + 4(6)(3) = 28 > 0$. Since the discriminant is positive, the equation represents a hyperbola.

(b) $6x^2 + 10xy + 3y^2 - 6y = 36 \Leftrightarrow 3y^2 + 10xy - 6y = 36 - 6x^2 \Leftrightarrow$
$3y^2 + 2(5x - 3)y = 36 - 6x^2 \Leftrightarrow y^2 + 2\left(\frac{5}{3}x - 1\right)y = 12 - 2x^2 \Leftrightarrow$
$y^2 + 2\left(\frac{5}{3}x - 1\right)y + \left(\frac{5}{3}x - 1\right)^2 = \left(\frac{5}{3}x - 1\right)^2 + 12 - 2x^2 \Leftrightarrow$
$\left[y + \left(\frac{5}{3}x - 1\right)\right]^2 = \frac{25}{9}x^2 - \frac{10}{3}x + 1 + 12 - 2x^2 \Leftrightarrow$
$\left[y + \left(\frac{5}{3}x - 1\right)\right]^2 = \frac{7}{9}x^2 - \frac{10}{3}x + 13 \Leftrightarrow$
$y + \left(\frac{5}{3}x - 1\right) = \pm\sqrt{\frac{7}{9}x^2 - \frac{10}{3}x + 13} \Leftrightarrow$
$y = -\frac{5}{3}x + 1 \pm \sqrt{\frac{7}{9}x^2 - \frac{10}{3}x + 13}$

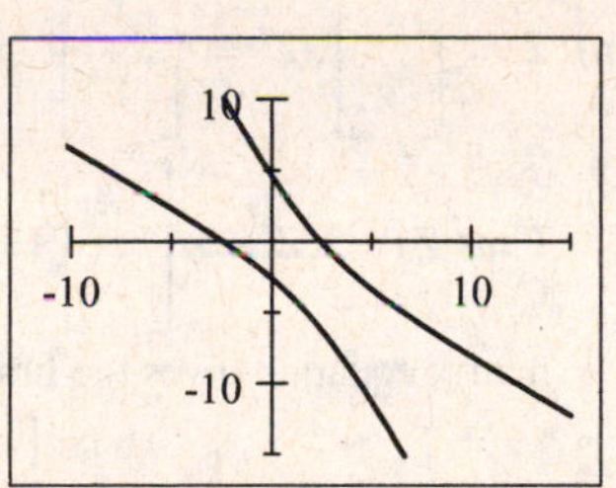

33. **(a)** $7x^2 + 48xy - 7y^2 - 200x - 150y + 600 = 0$. Then $A = 7$, $B = 48$, and $C = -7$, and so the discriminant is $B^2 - 4AC = (48)^2 - 4(7)(7) > 0$. Since the discriminant is positive, the equation represents a hyperbola. We now find the equation in terms of XY-coordinates. We have $\cot 2\phi = \dfrac{A-C}{B} = \dfrac{7}{24} \Rightarrow \cos\phi = \frac{4}{5}$ and $\sin\phi = \frac{3}{5}$. Therefore, $x = \frac{4}{5}X - \frac{3}{5}Y$ and $y = \frac{3}{5}X + \frac{4}{5}Y$, and substitution gives

$$7\left(\tfrac{4}{5}X - \tfrac{3}{5}Y\right)^2 + 48\left(\tfrac{4}{5}X - \tfrac{3}{5}Y\right)\left(\tfrac{3}{5}X + \tfrac{4}{5}Y\right) - 7\left(\tfrac{3}{5}X + \tfrac{4}{5}Y\right)^2 - 200\left(\tfrac{4}{5}X - \tfrac{3}{5}Y\right) - 150\left(\tfrac{3}{5}X + \tfrac{4}{5}Y\right) + 600 = 0$$

$$\Leftrightarrow \tfrac{7}{25}\left(16X^2 - 24XY + 9Y^2\right) + \tfrac{48}{25}\left(12X^2 + 7XY - 12Y^2\right) - \tfrac{7}{25}\left(9X^2 + 24XY + 16Y^2\right) - 160X + 120Y - 90X - 120Y + 600 = 0$$

$\Leftrightarrow 112X^2 - 168XY + 63Y^2 + 576X^2 + 336XY - 576Y^2 - 63X^2 - 168XY - 112Y^2 - 6250X + 15{,}000 = 0 \Leftrightarrow$ $25X^2 - 25Y^2 - 250X + 600 = 0 \Leftrightarrow 25\left(X^2 - 10X + 25\right) - 25Y^2 = -600 + 625 \Leftrightarrow (X-5)^2 - Y^2 = 1$. This is a hyperbola with $a = 1$, $b = 1$, $c = \sqrt{1+1} = \sqrt{2}$, and center $C(5, 0)$.

(b) In the XY-plane, the center is $C(5, 0)$, the vertices are $V(5 \pm 1, 0) = V_1(4, 0)$ and $V_2(6, 0)$, and the foci are $F\left(5 \pm \sqrt{2}, 0\right)$. In the xy-plane, the center is $C\left(\frac{4}{5}\cdot 5 - \frac{3}{5}\cdot 0, \frac{3}{5}\cdot 5 + \frac{4}{5}\cdot 0\right) = C(4, 3)$, the vertices are $V_1\left(\frac{4}{5}\cdot 4 - \frac{3}{5}\cdot 0, \frac{3}{5}\cdot 4 + \frac{4}{5}\cdot 0\right) = V_1\left(\frac{16}{5}, \frac{12}{5}\right)$ and $V_2\left(\frac{4}{5}\cdot 6 - \frac{3}{5}\cdot 0, \frac{3}{5}\cdot 6 + \frac{4}{5}\cdot 0\right) = V_2\left(\frac{24}{5}, \frac{18}{5}\right)$, and the foci are $F_1\left(4 + \frac{4}{5}\sqrt{2}, 3 + \frac{3}{5}\sqrt{2}\right)$ and $F_2\left(4 - \frac{4}{5}\sqrt{2}, 3 - \frac{3}{5}\sqrt{2}\right)$.

(c) In the XY-plane, the equations of the asymptotes are $Y = X - 5$ and $Y = -X + 5$. In the xy-plane, these equations become $-x\cdot\frac{3}{5} + y\cdot\frac{4}{5} = x\cdot\frac{4}{5} + y\cdot\frac{3}{5} - 5 \Leftrightarrow 7x - y - 25 = 0$. Similarly, $-x\cdot\frac{3}{5} + y\cdot\frac{4}{5} = -x\cdot\frac{4}{5} - y\cdot\frac{3}{5} + 5 \Leftrightarrow$ $x + 7y - 25 = 0$.

35. We use the hint and eliminate Y by adding: $x = X\cos\phi - Y\sin\phi \Leftrightarrow x\cos\phi = X\cos^2\phi - Y\sin\phi\cos\phi$ and $y = X\sin\phi + Y\cos\phi \Leftrightarrow y\sin\phi = X\sin^2\phi + Y\sin\phi\cos\phi$, and adding these two equations gives $x\cos\phi + y\sin\phi = X\left(\cos^2\phi + \sin^2\phi\right) \Leftrightarrow x\cos\phi + y\sin\phi = X$. In a similar manner, we eliminate X by subtracting: $x = X\cos\phi - Y\sin\phi \Leftrightarrow -x\sin\phi = -X\cos\phi\sin\phi + Y\sin^2\phi$ and $y = X\sin\phi + Y\cos\phi \Leftrightarrow$ $y\cos\phi = X\sin\phi\cos\phi + Y\cos^2\phi$, so $-x\sin\phi + y\cos\phi = Y\left(\cos^2\phi + \sin^2\phi\right) \Leftrightarrow -x\sin\phi + y\cos\phi = Y$. Thus, $X = x\cos\phi + y\sin\phi$ and $Y = -x\sin\phi + y\cos\phi$.

37. **(a)** $Z = \begin{bmatrix} x \\ y \end{bmatrix}$, $Z' = \begin{bmatrix} X \\ Y \end{bmatrix}$, and $R = \begin{bmatrix} \cos\phi & -\sin\phi \\ \sin\phi & \cos\phi \end{bmatrix}$.

Thus $Z = RZ' \Leftrightarrow \begin{bmatrix} x \\ y \end{bmatrix} = \begin{bmatrix} \cos\phi & -\sin\phi \\ \sin\phi & \cos\phi \end{bmatrix}\begin{bmatrix} X \\ Y \end{bmatrix} = \begin{bmatrix} X\cos\phi - Y\sin\phi Y \\ X\sin\phi + Y\cos\phi \end{bmatrix}$. Equating the entries in this matrix equation gives the first pair of rotation of axes formulas. Now

$R^{-1} = \dfrac{1}{\cos^2\phi + \sin^2\phi}\begin{bmatrix} \cos\phi & \sin\phi \\ -\sin\phi & \cos\phi \end{bmatrix} = \begin{bmatrix} \cos\phi & \sin\phi \\ -\sin\phi & \cos\phi \end{bmatrix}$ and so $Z' = R^{-1}Z \Leftrightarrow$

$\begin{bmatrix} X \\ Y \end{bmatrix} = \begin{bmatrix} \cos\phi & \sin\phi \\ -\sin\phi & \cos\phi \end{bmatrix}\begin{bmatrix} x \\ y \end{bmatrix} = \begin{bmatrix} x\cos\phi + y\sin\phi \\ -x\sin\phi + y\cos\phi \end{bmatrix}$. Equating the entries in this matrix equation gives the second pair of rotation of axes formulas.

(b) $R_1R_2 = \begin{bmatrix} \cos\phi_1 & -\sin\phi_1 \\ \sin\phi_1 & \cos\phi_1 \end{bmatrix} \begin{bmatrix} \cos\phi_2 & -\sin\phi_2 \\ \sin\phi_2 & \cos\phi_2 \end{bmatrix}$

$$= \begin{bmatrix} \cos\phi_1\cos\phi_2 - \sin\phi_1\sin\phi_2 & -\cos\phi_1\sin\phi_2 - \sin\phi_1\cos\phi_2 \\ \sin\phi_1\cos\phi_2 + \cos\phi_1\sin\phi_2 & -\sin\phi_1\sin\phi_2 + \cos\phi_1\cos\phi_2 \end{bmatrix} = \begin{bmatrix} \cos(\phi_1+\phi_2) & -\sin(\phi_1+\phi_2) \\ \sin(\phi_1+\phi_2) & \cos(\phi_1+\phi_2) \end{bmatrix}$$

39. Let P be the point (x_1, y_1) and Q be the point (x_2, y_2) and let $P'(X_1, Y_1)$ and $Q'(X_2, Y_2)$ be the images of P and Q under the rotation of ϕ. So $X_1 = x_1\cos\phi + y_1\sin\phi$, $Y_1 = -x_1\sin\phi + y_1\cos\phi$, $X_2 = x_2\cos\phi + y_2\sin\phi$,and $Y_2 = -x_2\sin\phi + y_2\cos\phi$. Thus $d(P', Q') = \sqrt{(X_2 - X_1)^2 + (Y_2 - Y_1)^2}$, where

$$\begin{aligned} (X_2 - X_1)^2 &= [(x_2\cos\phi + y_2\sin\phi) - (x_1\cos\phi + y_1\sin\phi)]^2 = [(x_2 - x_1)\cos\phi + (y_2 - y_1)\sin\phi]^2 \\ &= (x_2 - x_1)^2\cos^2\phi + (x_2 - x_1)(y_2 - y_1)\sin\phi\cos\phi + (y_2 - y_1)^2\sin^2\phi \end{aligned}$$

and

$$\begin{aligned} (Y_2 - Y_1)^2 &= [(-x_2\sin\phi + y_2\cos\phi) - (-x_1\sin\phi + y_1\cos\phi)]^2 = [-(x_2 - x_1)\sin\phi + (y_2 - y_1)\cos\phi]^2 \\ &= (x_2 - x_1)^2\sin^2\phi - (x_2 - x_1)(y_2 - y_1)\sin\phi\cos\phi + (y_2 - y_1)^2\cos^2\phi \end{aligned}$$

So

$$\begin{aligned} (X_2 - X_1)^2 + (Y_2 - Y_1)^2 &= (x_2 - x_1)^2\cos^2\phi + (x_2 - x_1)(y_2 - y_1)\sin\phi\cos\phi + (y_2 - y_1)^2\sin^2\phi \\ &\quad + (x_2 - x_1)^2\sin^2\phi - (x_2 - x_1)(y_2 - y_1)\sin\phi\cos\phi + (y_2 - y_1)^2\cos^2\phi \\ &= (x_2 - x_1)^2\cos^2\phi + (y_2 - y_1)^2\sin^2\phi + (x_2 - x_1)^2\sin^2\phi + (y_2 - y_1)^2\cos^2\phi \\ &= (x_2 - x_1)^2\left(\cos^2\phi + \sin^2\phi\right) + (y_2 - y_1)^2\left(\sin^2\phi + \cos^2\phi\right) = (x_2 - x_1)^2 + (y_2 - y_1)^2 \end{aligned}$$

Putting these equations together gives $d(P', Q') = \sqrt{(X_2 - X_1)^2 + (Y_2 - Y_1)^2} = \sqrt{(x_2 - x_1)^2 + (y_2 - y_1)^2} = d(P, Q)$.

12.6 POLAR EQUATIONS OF CONICS

1. All conics can be described geometrically using a fixed point F called the *focus* and a fixed line ℓ called the *directrix*. For a fixed positive number e the set of all points P satisfying $\dfrac{\text{distance from } P \text{ to } F}{\text{distance from } P \text{ to } \ell} = e$ is a *conic section*. If $e = 1$ the conic is a *parabola*, if $e < 1$ the conic is an *ellipse*, and if $e > 1$ the conic is a *hyperbola*. The number e is called the *eccentricity* of the conic.

3. Substituting $e = \frac{2}{3}$ and $d = 3$ into the general equation of a conic with vertical directrix, we get $r = \dfrac{\frac{2}{3}\cdot 3}{1 + \frac{2}{3}\cos\theta} \Leftrightarrow$ $r = \dfrac{6}{3 + 2\cos\theta}$.

5. Substituting $e = 1$ and $d = 2$ into the general equation of a conic with horizontal directrix, we get $r = \dfrac{1\cdot 2}{1 + \sin\theta} \Leftrightarrow$ $r = \dfrac{2}{1 + \sin\theta}$.

7. $r = 5\sec\theta \Leftrightarrow r\cos\theta = 5 \Leftrightarrow x = 5$. So $d = 5$ and $e = 4$ gives $r = \dfrac{4\cdot 5}{1 + 4\cos\theta} \Leftrightarrow r = \dfrac{20}{1 + 4\cos\theta}$.

9. Since this is a parabola whose focus is at the origin and vertex at $(5, \pi/2)$, the directrix must be $y = 10$. So $d = 10$ and $e = 1$ gives $r = \dfrac{1\cdot 10}{1 + \sin\theta} = \dfrac{10}{1 + \sin\theta}$.

11. $r = \dfrac{6}{1 + \cos\theta}$ is Graph II. The eccentricity is 1, so this is a parabola. When $\theta = 0$, we have $r = 3$ and when $\theta = \frac{\pi}{2}$, we have $r = 6$.

13. $r = \dfrac{3}{1 - 2\sin\theta}$ is Graph VI. $e = 2$, so this is a hyperbola. When $\theta = 0$, $r = 3$, and when $\theta = \pi$, $r = 3$.

15. $r = \dfrac{12}{3 + 2\sin\theta}$ is Graph IV. $r = \dfrac{4}{1 + \frac{2}{3}\sin\theta}$, so $e = \frac{2}{3}$ and this is an ellipse. When $\theta = 0$, $r = 4$, and when $\theta = \pi$, $r = 4$.

17. (a) The equation $r = \dfrac{4}{1 - \sin\theta}$ has $e = 1$ and $d = 4$, so it represents a parabola.

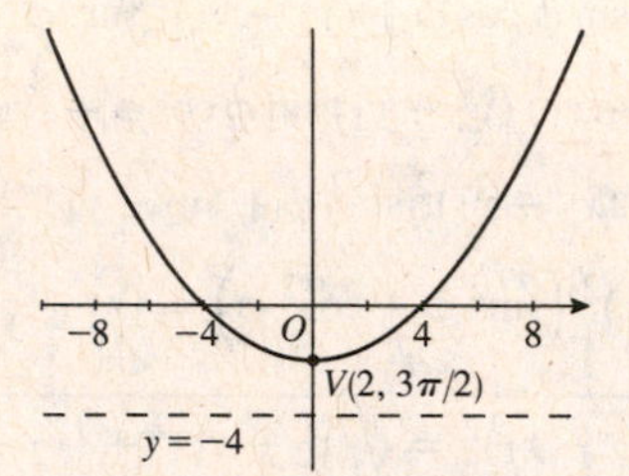

(b) Because the equation is of the form $r = \dfrac{ed}{1 - e\sin\theta}$, the directrix is parallel to the polar axis and has equation $y = -4$. The vertex is $\left(2, \frac{3\pi}{2}\right)$.

19. (a) The equation $r = \dfrac{5}{3 + 3\cos\theta} = \dfrac{\frac{5}{3}}{1 + \cos\theta}$ has $e = 1$ and $d = \frac{5}{3}$, so it represents a parabola.

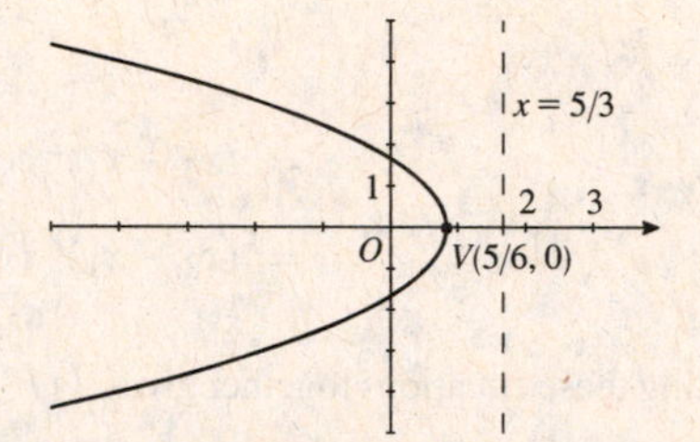

(b) Because the equation is of the form $r = \dfrac{ed}{1 + e\cos\theta}$, the directrix is parallel to the polar axis and has equation $x = d = \frac{5}{3}$. The vertex is $\left(\frac{5}{6}, 0\right)$.

21. (a) The equation $r = \dfrac{4}{2 - \cos\theta} = \dfrac{2}{1 - \frac{1}{2}\cos\theta}$ has $e = \frac{1}{2} < 1$, so it represents an ellipse.

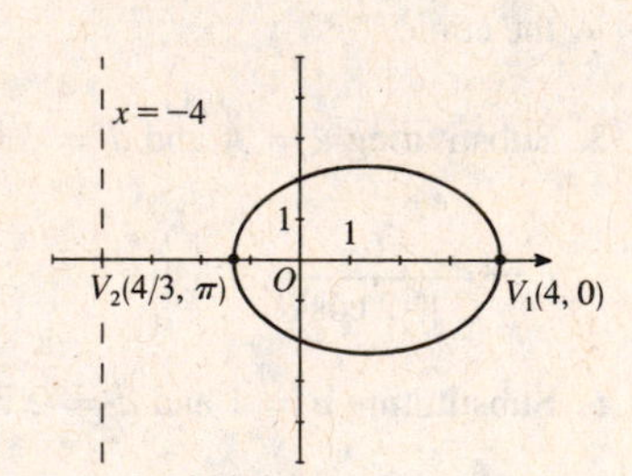

(b) Because the equation is of the form $r = \dfrac{ed}{1 - e\cos\theta}$ with $d = 4$, the directrix is vertical and has equation $x = -4$. Thus, the vertices are $V_1\ (4, 0)$ and $V_2\left(\frac{4}{3}, \pi\right)$.

(c) The length of the major axis is $2a = |V_1V_2| = 4 + \frac{4}{3} = \frac{16}{3}$ and the center is at the midpoint of V_1V_2, $\left(\frac{4}{3}, 0\right)$. The minor axis has length $2b$ where $b^2 = a^2 - c^2 = a^2 - (ae)^2 = \left(\frac{8}{3}\right)^2 - \left(\frac{8}{3} \cdot \frac{1}{2}\right)^2 = \frac{16}{3}$, so $2b = 2 \cdot \sqrt{\frac{16}{3}} = \frac{8\sqrt{3}}{3} \approx 4.62$.

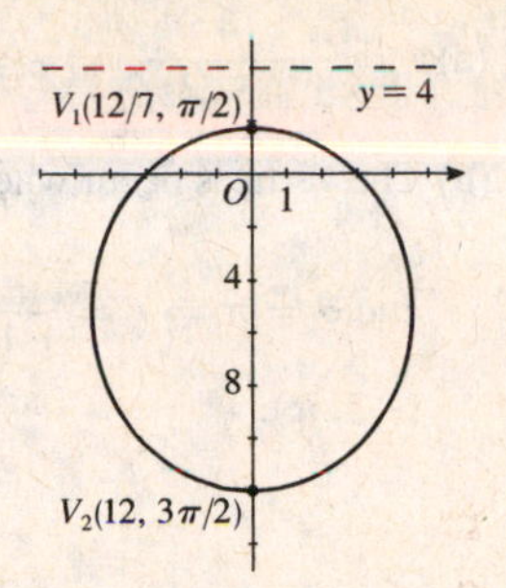

23. **(a)** The equation $r = \dfrac{12}{4 + 3\sin\theta} = \dfrac{3}{1 + \frac{3}{4}\sin\theta}$ has $e = \frac{3}{4} < 1$, so it represents an ellipse.

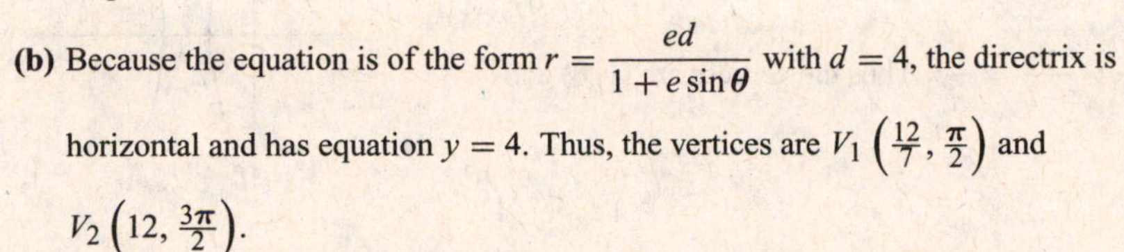

(b) Because the equation is of the form $r = \dfrac{ed}{1 + e\sin\theta}$ with $d = 4$, the directrix is horizontal and has equation $y = 4$. Thus, the vertices are $V_1\left(\frac{12}{7}, \frac{\pi}{2}\right)$ and $V_2\left(12, \frac{3\pi}{2}\right)$.

(c) The length of the major axis is $2a = |V_1V_2| = \frac{12}{7} + 12 = \frac{96}{7}$ and the center is at the midpoint of V_1V_2, $\left(\frac{36}{7}, \frac{3\pi}{2}\right)$. The minor axis has length $2b$ where

$b^2 = a^2 - c^2 = a^2 - (ae)^2 = \left(\frac{48}{7}\right)^2 - \left(\frac{48}{7} \cdot \frac{3}{4}\right)^2 = \frac{144}{7}$, so

$2b = 2 \cdot \sqrt{\frac{144}{7}} = \frac{24\sqrt{7}}{7} \approx 9.07$.

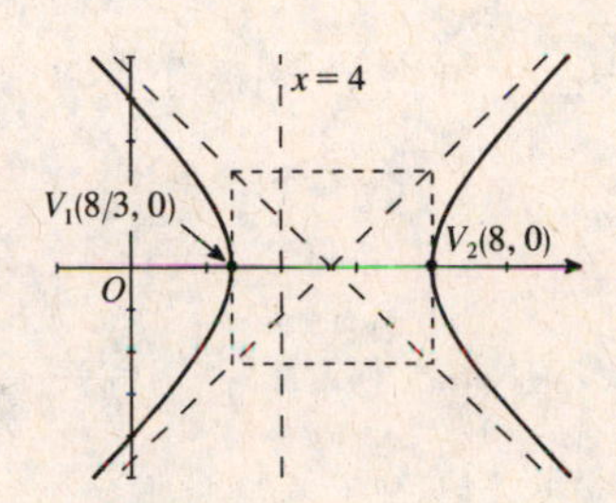

25. **(a)** The equation $r = \dfrac{8}{1 + 2\cos\theta}$ has $e = 2 > 1$, so it represents a hyperbola.

(b) Because the equation has the form $r = \dfrac{ed}{1 + \cos\theta}$ with $d = 4$, the transverse axis is horizontal and the directrix has equation $x = 4$. The vertices are $V_1\left(\frac{8}{3}, 0\right)$ and $V_2(-8, \pi) = (8, 0)$.

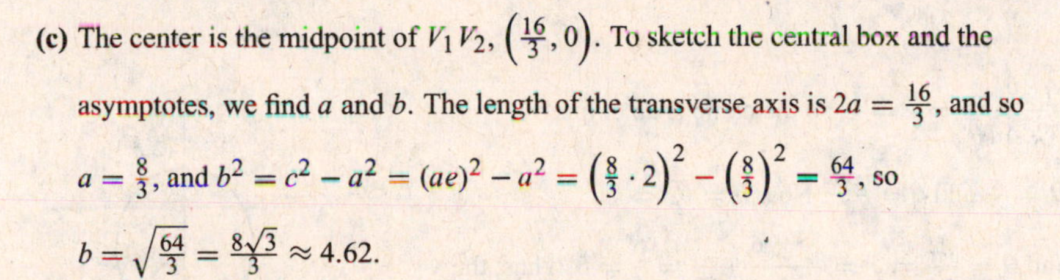

(c) The center is the midpoint of V_1V_2, $\left(\frac{16}{3}, 0\right)$. To sketch the central box and the asymptotes, we find a and b. The length of the transverse axis is $2a = \frac{16}{3}$, and so

$a = \frac{8}{3}$, and $b^2 = c^2 - a^2 = (ae)^2 - a^2 = \left(\frac{8}{3} \cdot 2\right)^2 - \left(\frac{8}{3}\right)^2 = \frac{64}{3}$, so

$b = \sqrt{\frac{64}{3}} = \frac{8\sqrt{3}}{3} \approx 4.62$.

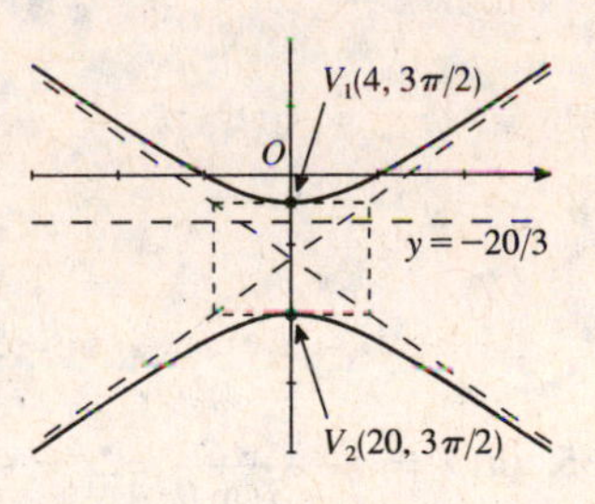

27. **(a)** The equation $r = \dfrac{20}{2 - 3\sin\theta} = \dfrac{10}{1 - \frac{3}{2}\sin\theta}$ has $e = \frac{3}{2} > 1$, so it represents a hyperbola.

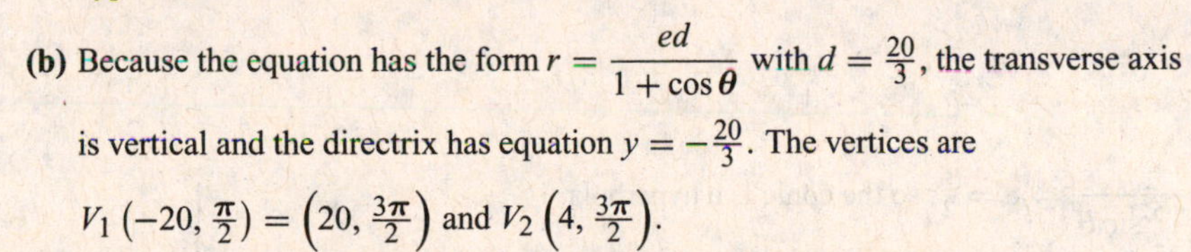

(b) Because the equation has the form $r = \dfrac{ed}{1 + \cos\theta}$ with $d = \frac{20}{3}$, the transverse axis is vertical and the directrix has equation $y = -\frac{20}{3}$. The vertices are $V_1\left(-20, \frac{\pi}{2}\right) = \left(20, \frac{3\pi}{2}\right)$ and $V_2\left(4, \frac{3\pi}{2}\right)$.

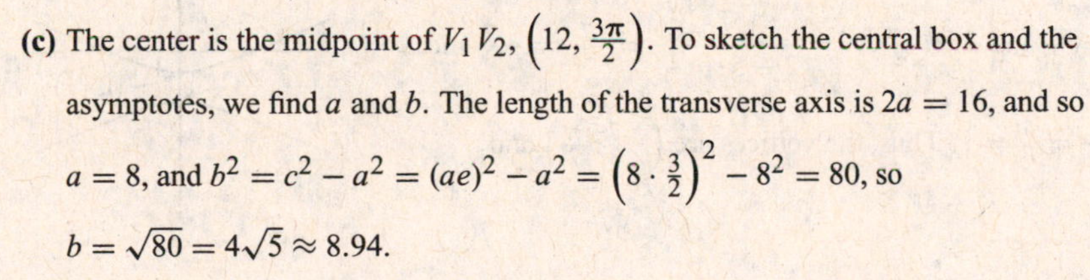

(c) The center is the midpoint of V_1V_2, $\left(12, \frac{3\pi}{2}\right)$. To sketch the central box and the asymptotes, we find a and b. The length of the transverse axis is $2a = 16$, and so

$a = 8$, and $b^2 = c^2 - a^2 = (ae)^2 - a^2 = \left(8 \cdot \frac{3}{2}\right)^2 - 8^2 = 80$, so

$b = \sqrt{80} = 4\sqrt{5} \approx 8.94$.

29. **(a)** $r = \dfrac{4}{1 + 3\cos\theta} \Rightarrow e = 3$, so the conic is a hyperbola.

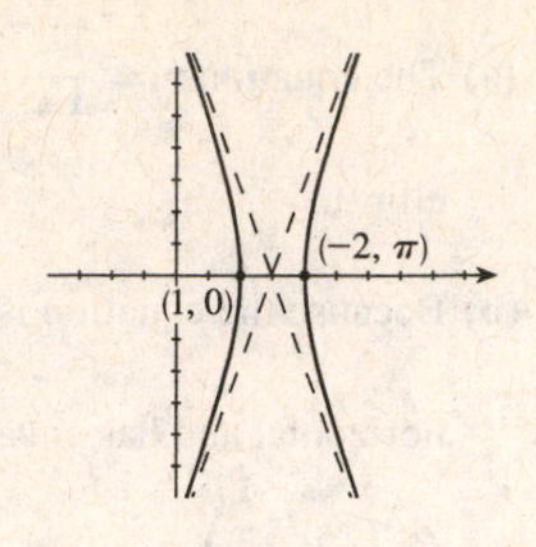

(b) The vertices occur where $\theta = 0$ and $\theta = \pi$. Now $\theta = 0 \Rightarrow r = \dfrac{4}{1 + 3\cos 0} = 1$, and $\theta = \pi \Rightarrow r = \dfrac{4}{1 + 3\cos\pi} = \dfrac{4}{-2} = -2$. Thus the vertices are $(1, 0)$ and $(-2, \pi)$.

31. **(a)** $r = \dfrac{2}{1 - \cos\theta} \Rightarrow e = 1$, so the conic is a parabola.

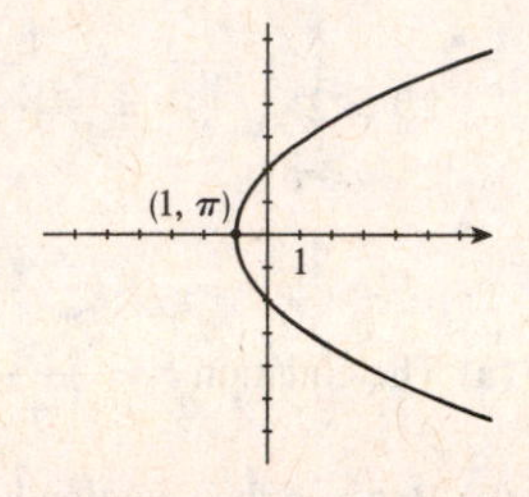

(b) Substituting $\theta = \pi$, we have $r = \dfrac{2}{1 - \cos\pi} = \frac{2}{2} = 1$. Thus the vertex is $(1, \pi)$.

33. **(a)** $r = \dfrac{6}{2 + \sin\theta} \Leftrightarrow r = \dfrac{\frac{1}{2} \cdot 6}{1 + \frac{1}{2}\sin\theta} \Rightarrow e = \frac{1}{2}$, so the conic is an ellipse.

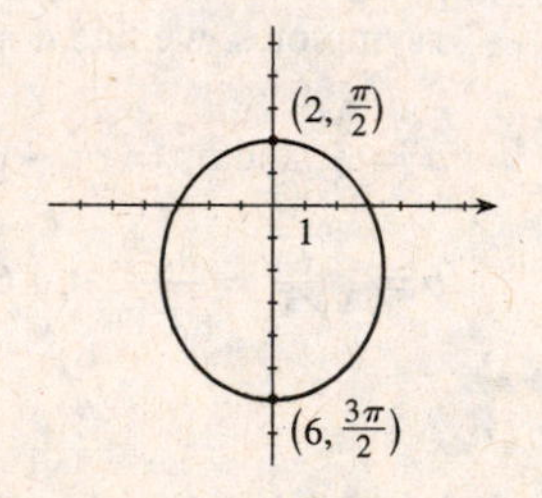

(b) The vertices occur where $\theta = \frac{\pi}{2}$ and $\theta = \frac{3\pi}{2}$. Now $\theta = \frac{\pi}{2} \Rightarrow r = \dfrac{6}{2 + \sin\frac{\pi}{2}} = \dfrac{6}{3} = 2$ and $\theta = \frac{3\pi}{2} \Rightarrow r = \dfrac{6}{2 + \sin\frac{3\pi}{2}} = \dfrac{6}{1} = 6$. Thus, the vertices are $\left(2, \frac{\pi}{2}\right)$ and $\left(6, \frac{3\pi}{2}\right)$.

35. **(a)** $r = \dfrac{7}{2 - 5\sin\theta} \Leftrightarrow r = \dfrac{\frac{7}{2}}{1 - \frac{5}{2}\sin\theta} \Rightarrow e = \frac{5}{2}$, so the conic is a hyperbola.

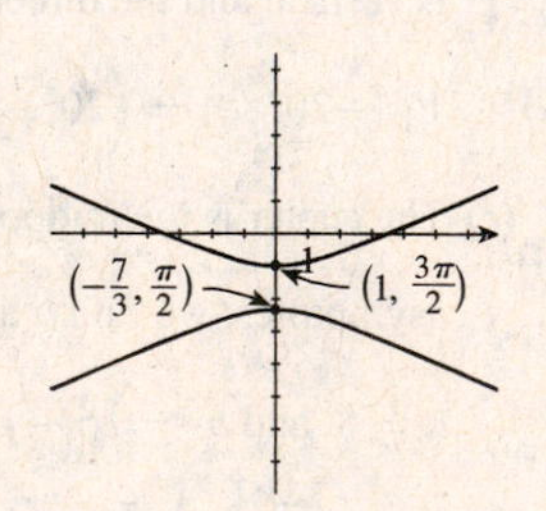

(b) The vertices occur where $\theta = \frac{\pi}{2}$ and $\theta = \frac{3\pi}{2}$. $r = \dfrac{7}{2 - 5\sin\frac{\pi}{2}} = \frac{7}{-3} = -\frac{7}{3}$ and $\theta = \frac{3\pi}{2} \Rightarrow r = \dfrac{7}{2 - 5\sin\frac{3\pi}{2}} = \frac{7}{7} = 1$. Thus, the vertices are $\left(-\frac{7}{3}, \frac{\pi}{2}\right)$ and $\left(1, \frac{3\pi}{2}\right)$.

37. **(a)** $r = \dfrac{1}{4 - 3\cos\theta} = \dfrac{\frac{1}{4}}{1 - \frac{3}{4}\cos\theta} \Rightarrow e = \frac{3}{4}$, so the conic is an ellipse. The vertices occur where $\theta = 0$ and $\theta = \pi$. Now $\theta = 0 \Rightarrow r = \dfrac{1}{4 - 3\cos 0} = 1$ and $\theta = \pi \Rightarrow r = \dfrac{1}{4 - 3\cos\pi} = \frac{1}{7}$. Thus, the vertices are $(1, 0)$ and $\left(\frac{1}{7}, \pi\right)$. We have $d = \frac{1}{3}$, so the directrix is $x = -\frac{1}{3}$.

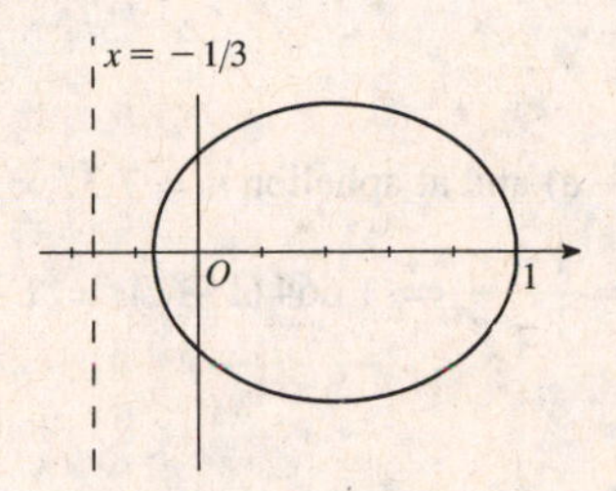

(b) If the ellipse is rotated through $\frac{\pi}{3}$, the equation of the resulting conic is $r = \dfrac{1}{4 - 3\cos\left(\theta - \frac{\pi}{3}\right)}$.

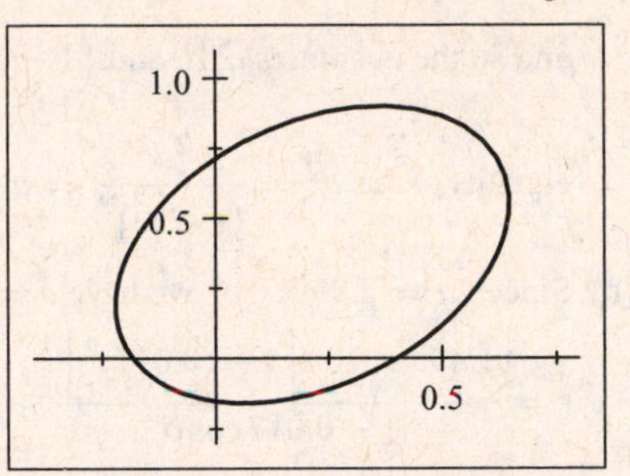

39. **(a)** $r = \dfrac{2}{1 + \sin\theta} \Rightarrow e = 1$, so the conic is a parabola. Substituting $\theta = \frac{\pi}{2}$, we have $r = t\dfrac{2}{1 + \sin\frac{\pi}{2}} = 1$, so the vertex is $\left(1, \frac{\pi}{2}\right)$. Because $d = 2$, the directrix is $y = 2$.

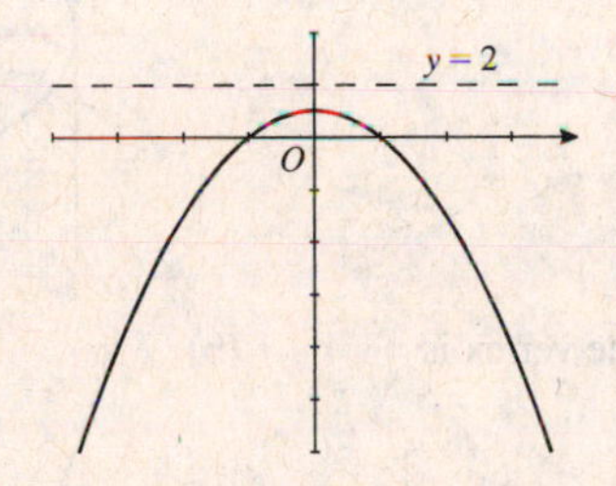

(b) If the ellipse is rotated through $\theta = -\frac{\pi}{4}$, the equation of the resulting conic is
$$r = \frac{2}{1 + \sin\left(\theta + \frac{\pi}{4}\right)}.$$

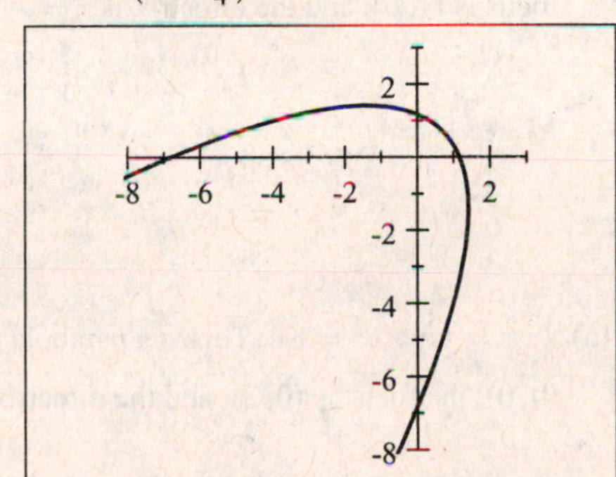

41. The ellipse is nearly circular when e is close to 0 and becomes more elongated as $e \to 1^-$. At $e = 1$, the curve becomes a parabola.

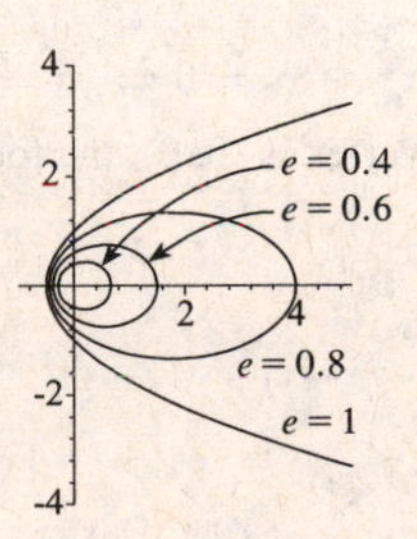

43. **(a)** Since the polar form of an ellipse with directrix $x = -d$ is $r = \dfrac{ed}{1 - e\cos\theta}$ we need to show that $ed = a\left(1 - e^2\right)$.

From the proof of the Equivalent Description of Conics we have $a^2 = \dfrac{e^2d^2}{\left(1 - e^2\right)^2}$. Since the conic is an ellipse, $e < 1$ and so the quantities a, d, and $\left(1 - e^2\right)$ are all positive. Thus we can take the square roots of both sides and maintain equality. Thus $a^2 = \dfrac{e^2d^2}{\left(1 - e^2\right)^2} \Leftrightarrow a = \dfrac{ed}{1 - e^2} \Leftrightarrow ed = a\left(1 - e^2\right)$. As a result, $r = \dfrac{ed}{1 - e\cos\theta} \Leftrightarrow r = \dfrac{a\left(1 - e^2\right)}{1 - e\cos\theta}$.

(b) Since $2a = 2.99 \times 10^8$ we have $a = 1.495 \times 10^8$, so a polar equation for the earth's orbit (using $e \approx 0.017$) is $r = \dfrac{1.495 \times 10^8\left[1 - (0.017)^2\right]}{1 - 0.017\cos\theta} \approx \dfrac{1.49 \times 10^8}{1 - 0.017\cos\theta}$.

45. From Exercise 44, we know that at perihelion $r = 4.43 \times 10^9 = a(1 - e)$ and at aphelion $r = 7.37 \times 10^9 = a(1 + e)$. Dividing these equations gives $\dfrac{7.37 \times 10^9}{4.43 \times 10^9} = \dfrac{a(1 + e)}{a(1 - e)} \Leftrightarrow 1.664 = \dfrac{1 + e}{1 - e} \Leftrightarrow 1.664(1 - e) = 1 + e \Leftrightarrow$ $1.664 - 1 = e + 1.664 \Leftrightarrow 0.664 = 2.664e \Leftrightarrow e = \dfrac{0.664}{2.664} \approx 0.25$.

47. The r-coordinate of the satellite will be its distance from the focus (the center of the earth). From the r-coordinate we can easily calculate the height of the satellite.

CHAPTER 12 REVIEW

1. **(a)** $y^2 = 4x$. This is a parabola with $4p = 4 \Leftrightarrow p = 1$. The vertex is $(0, 0)$, the focus is $(1, 0)$, and the directrix is $x = -1$.

(b)

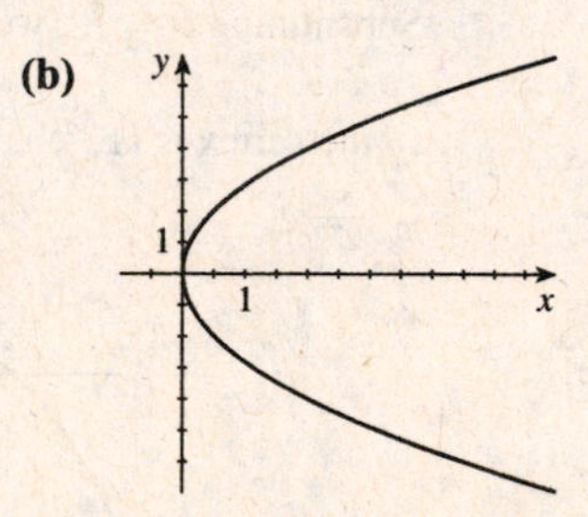

3. **(a)** $\frac{1}{8}x^2 = y \Leftrightarrow x^2 = 8y$. This is a parabola with $4p = 8 \Leftrightarrow p = 2$. The vertex is $(0, 0)$, the focus is $(0, 2)$, and the directrix is $y = -2$.

(b)

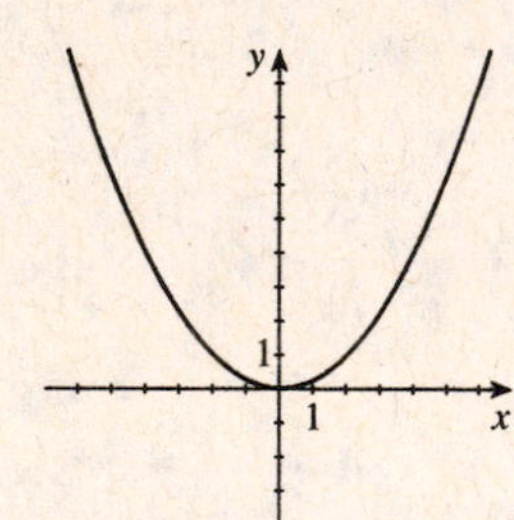

5. **(a)** $x^2 + 8y = 0 \Leftrightarrow x^2 = -8y$. This is a parabola with $4p = -8 \Leftrightarrow p = -2$. The vertex is $(0, 0)$, the focus is $(0, -2)$, and the directrix is $y = 2$.

(b)

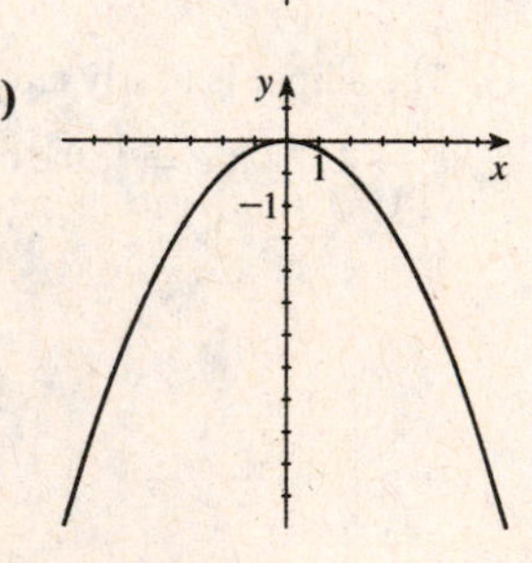

7. (a) $(y-2)^2 = 4(x+2)$. This is a parabola with $4p = 4 \Leftrightarrow p = 1$. The vertex is $(-2, 2)$, the focus is $(-1, 2)$, and the directrix is $x = -3$.

(b)

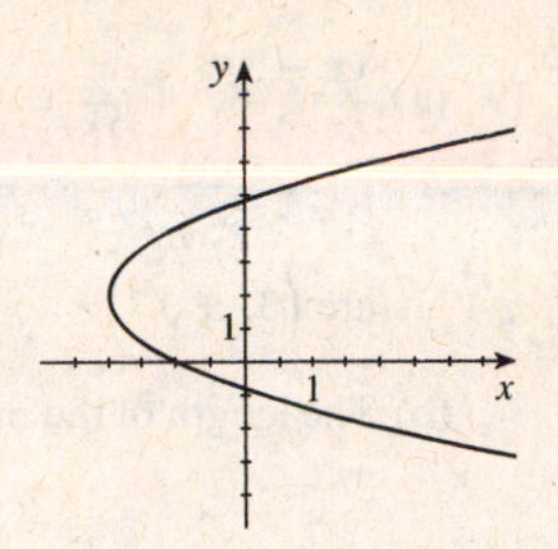

9. (a) $\frac{1}{2}(y-3)^2 + x = 0 \Leftrightarrow (y-3)^2 = -2x$. This is a parabola with $4p = -2 \Leftrightarrow p = -\frac{1}{2}$. The vertex is $(0, 3)$, the focus is $\left(-\frac{1}{2}, 3\right)$, and the directrix is $x = \frac{1}{2}$.

(b)

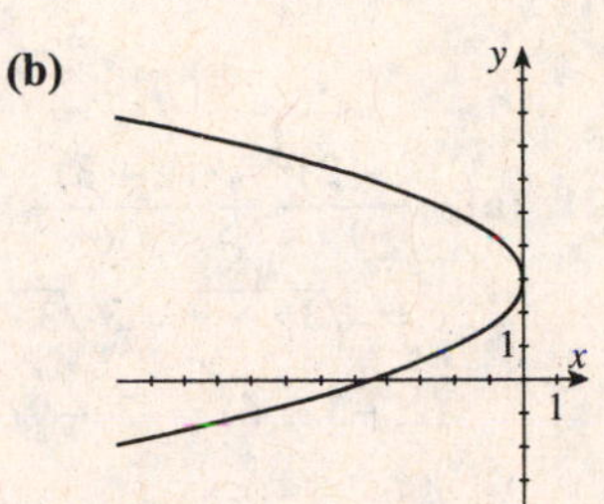

11. (a) $\frac{1}{2}x^2 + 2x = 2y + 4 \Leftrightarrow x^2 + 4x = 4y + 8 \Leftrightarrow x^2 + 4x + 4 = 4y + 8 \Leftrightarrow (x+2)^2 = 4(y+3)$. This is a parabola with $4p = 4 \Leftrightarrow p = 1$. The vertex is $(-2, -3)$, the focus is $(-2, -3+1) = (-2, -2)$, and the directrix is $y = -3 - 1 = -4$.

(b)

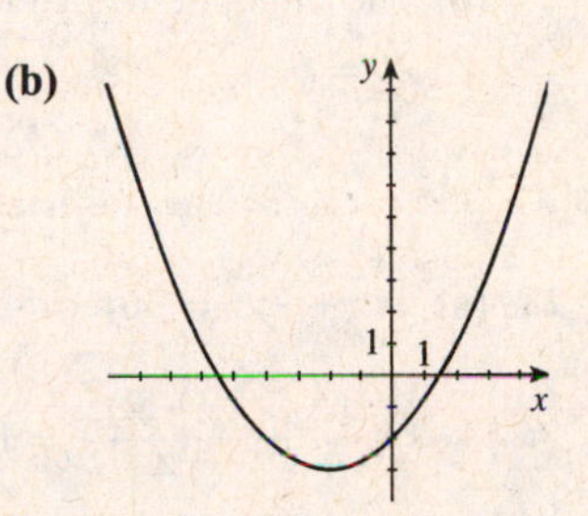

13. (a) $\frac{x^2}{9} + \frac{y^2}{25} = 1$. This is an ellipse with $a = 5$, $b = 3$, and $c = \sqrt{25 - 9} = 4$. The center is $(0, 0)$, the vertices are $(0, \pm 5)$, and the foci are $(0, \pm 4)$.

(b) The length of the major axis is $2a = 10$ and the length of the minor axis is $2b = 6$.

(c)

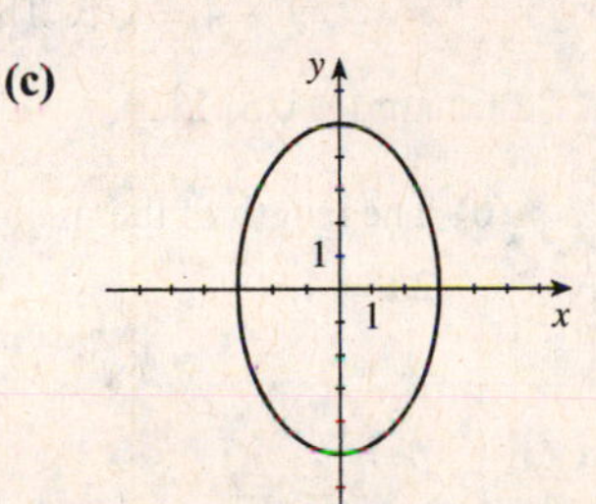

15. (a) $\frac{x^2}{49} + \frac{y^2}{4} = 1$. This is an ellipse with $a = 7$, $b = 2$, and $c = \sqrt{49 - 4} = 3\sqrt{5}$. The center is $(0, 0)$, the vertices are $(\pm 7, 0)$, and the foci are $\left(\pm 3\sqrt{5}, 0\right)$.

(b) The length of the major axis is $2a = 14$ and the length of the minor axis is $2b = 4$.

(c)

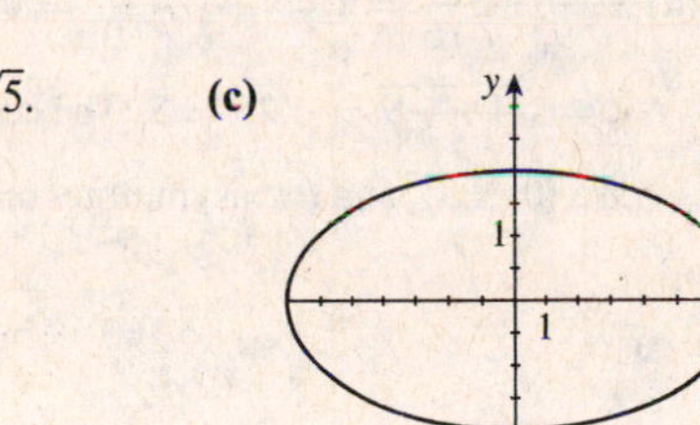

17. (a) $x^2 + 4y^2 = 16 \Leftrightarrow \frac{x^2}{16} + \frac{y^2}{4} = 1$. This is an ellipse with $a = 4$, $b = 2$, and $c = \sqrt{16 - 4} = 2\sqrt{3}$. The center is $(0, 0)$, the vertices are $(\pm 4, 0)$, and the foci are $\left(\pm 2\sqrt{3}, 0\right)$.

(b) The length of the major axis is $2a = 8$ and the length of the minor axis is $2b = 4$.

(c)

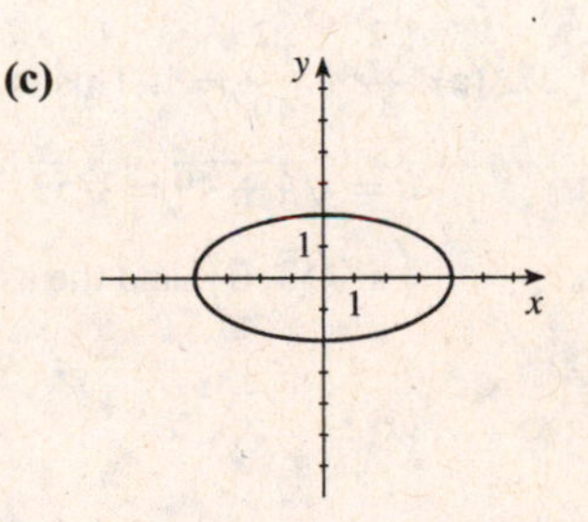

19. **(a)** $\dfrac{(x-3)^2}{9}+\dfrac{y^2}{16}=1$.This is an ellipse with $a=4$, $b=3$, and $c=\sqrt{16-9}=\sqrt{7}$. The center is $(3,0)$, the vertices are $(3,\pm 4)$, and the foci are $\left(3,\pm\sqrt{7}\right)$.

(b) The length of the major axis is $2a=8$ and the length of the minor axis is $2b=6$.

(c)

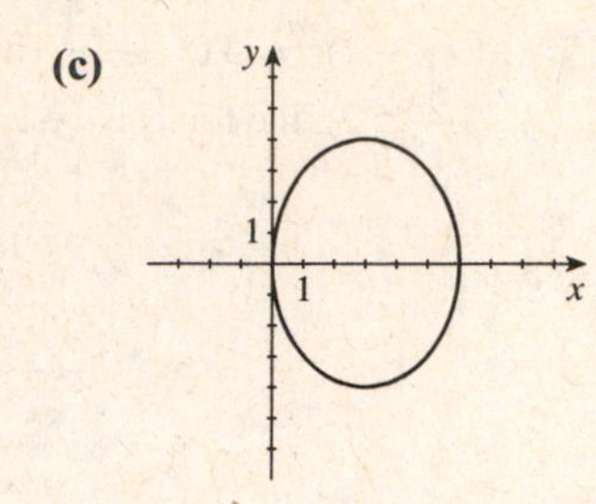

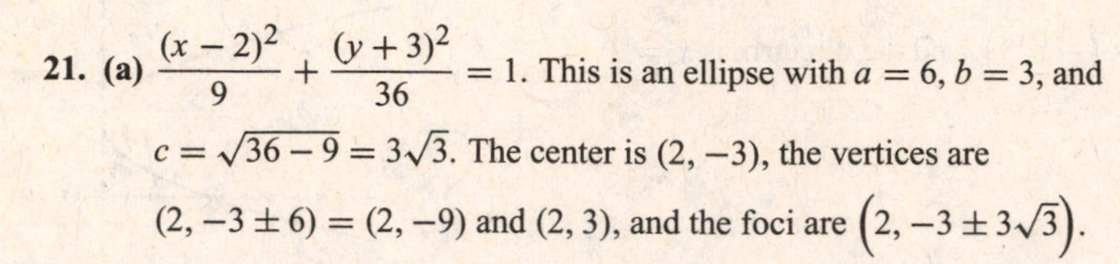

21. **(a)** $\dfrac{(x-2)^2}{9}+\dfrac{(y+3)^2}{36}=1$. This is an ellipse with $a=6$, $b=3$, and $c=\sqrt{36-9}=3\sqrt{3}$. The center is $(2,-3)$, the vertices are $(2,-3\pm 6)=(2,-9)$ and $(2,3)$, and the foci are $\left(2,-3\pm 3\sqrt{3}\right)$.

(b) The length of the major axis is $2a=12$ and the length of the minor axis is $2b=6$.

(c)

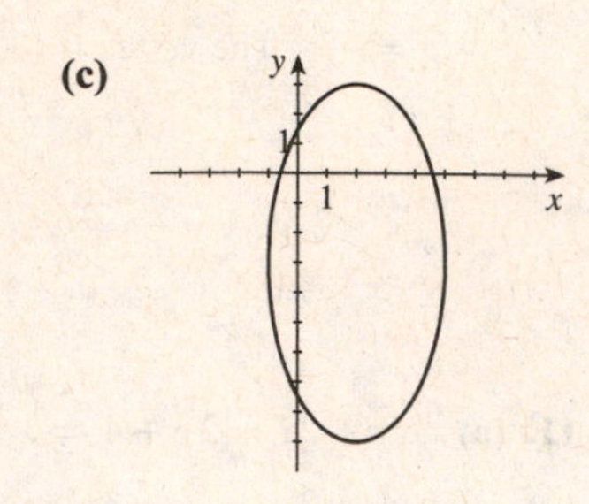

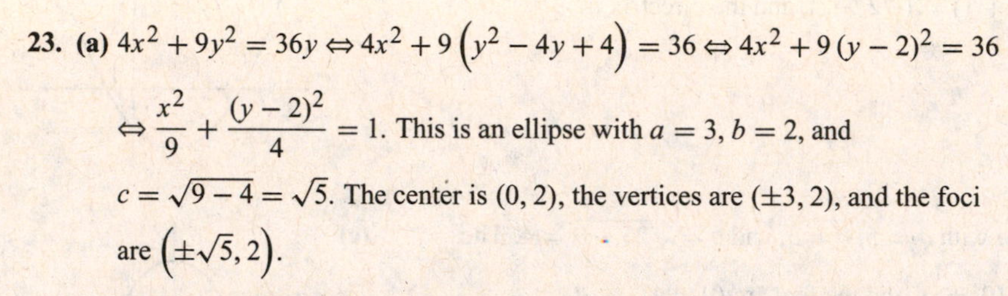

23. **(a)** $4x^2+9y^2=36y \Leftrightarrow 4x^2+9\left(y^2-4y+4\right)=36 \Leftrightarrow 4x^2+9(y-2)^2=36$

$\Leftrightarrow \dfrac{x^2}{9}+\dfrac{(y-2)^2}{4}=1$. This is an ellipse with $a=3$, $b=2$, and $c=\sqrt{9-4}=\sqrt{5}$. The center is $(0,2)$, the vertices are $(\pm 3,2)$, and the foci are $\left(\pm\sqrt{5},2\right)$.

(b) The length of the major axis is $2a=6$ and the length of the minor axis is $2b=4$.

(c)

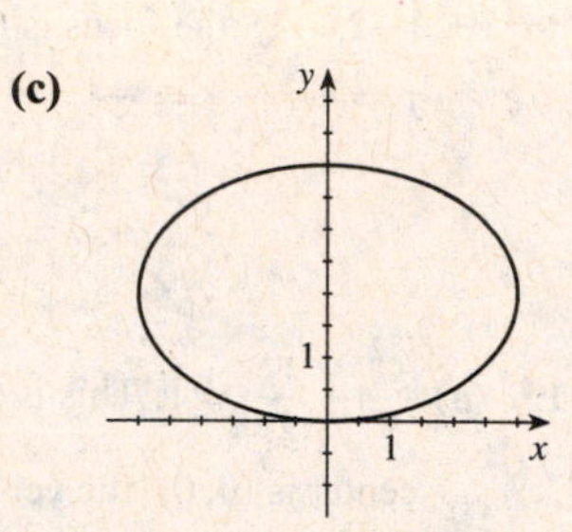

25. **(a)** $-\dfrac{x^2}{9}+\dfrac{y^2}{16}=1 \Leftrightarrow \dfrac{y^2}{16}-\dfrac{x^2}{9}=0$. This is a hyperbola with $a=4$, $b=3$, and $c=\sqrt{16+9}=\sqrt{25}=5$. The center is $(0,0)$, the vertices are $(0,\pm 4)$, the foci are $(0,\pm 5)$, and the asymptotes are $y=\pm\frac{4}{3}x$.

(b)

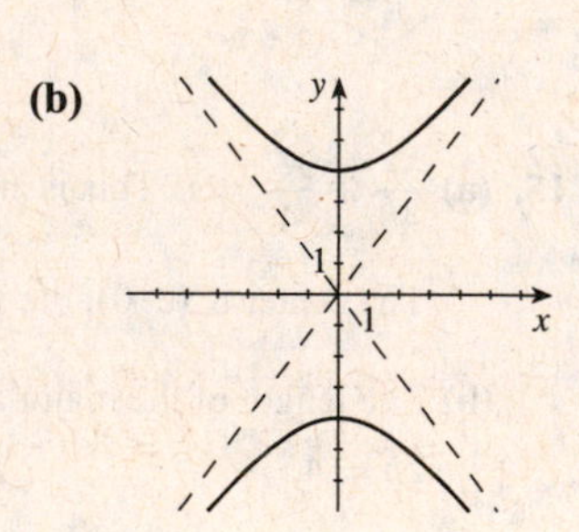

27. **(a)** $\dfrac{x^2}{4}-\dfrac{y^2}{49}=1$. This is a hyperbola with $a=2$, $b=7$, and $c=\sqrt{4+49}=\sqrt{53}$. The center is $(0,0)$, the vertices are $(\pm 2,0)$, the foci are $\left(\pm\sqrt{53},0\right)$, and the asymptotes are $y=\pm\frac{7}{2}x$.

(b)

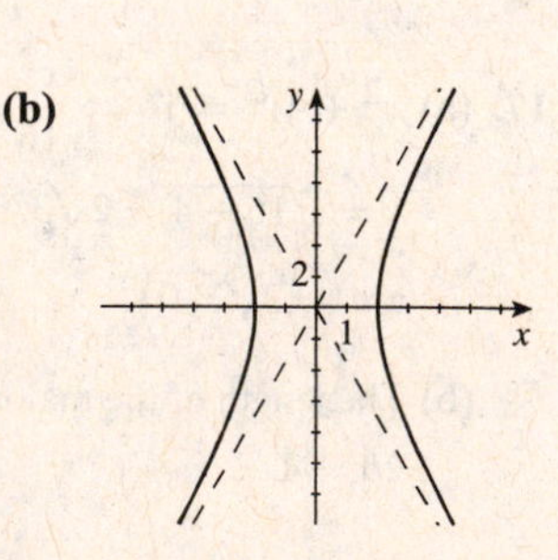

29. (a) $x^2 - 2y^2 = 16 \Leftrightarrow \dfrac{x^2}{16} - \dfrac{y^2}{8} = 1$. This is a hyperbola with $a = 4$, $b = 2\sqrt{2}$, and $c = \sqrt{16+8} = \sqrt{24} = 2\sqrt{6}$. The center is $(0,0)$, the vertices are $(\pm 4, 0)$, the foci are $\left(\pm 2\sqrt{6}, 0\right)$, and the asymptotes are $y = \pm \frac{2\sqrt{2}}{4} x \Leftrightarrow y = \pm \frac{1}{\sqrt{2}} x$.

(b)

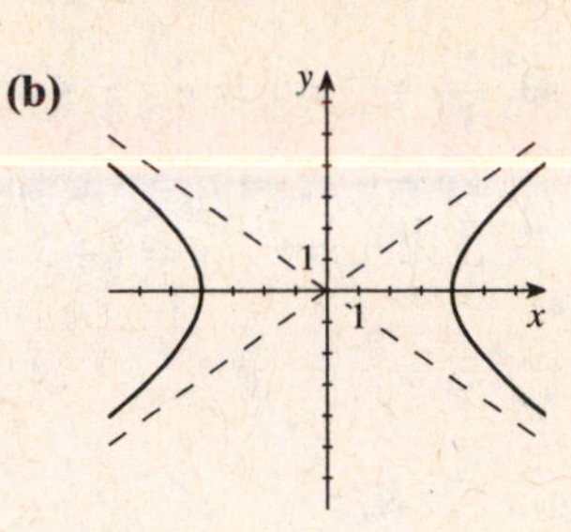

31. (a) $\dfrac{(x+4)^2}{16} - \dfrac{y^2}{16} = 1$. This is a hyperbola with $a = 4$, $b = 4$ and $c = \sqrt{16+16} = 4\sqrt{2}$. The center is $(-4, 0)$, the vertices are $(-4 \pm 4, 0)$ which are $(-8, 0)$ and $(0, 0)$, the foci are $\left(-4 \pm 4\sqrt{2}, 0\right)$, and the asymptotes are $y = \pm (x+4)$.

(b)

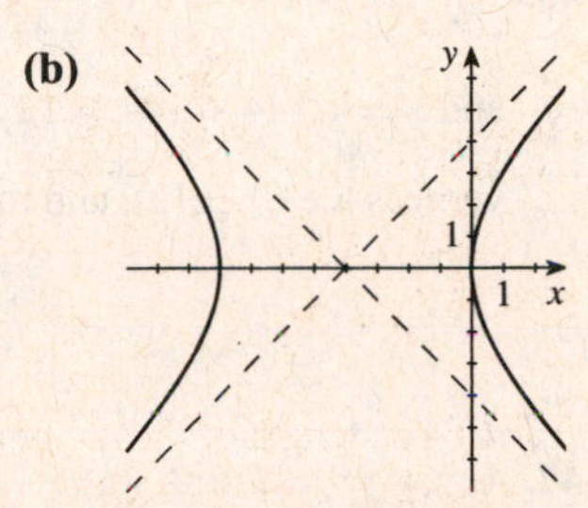

33. (a) $\dfrac{(y-3)^2}{4} - \dfrac{(x+1)^2}{36} = 1$. This is a hyperbola with $a = 2$, $b = 6$, and $c = \sqrt{4+36} = 2\sqrt{10}$. The center is $(-1, 3)$, the vertices are $(-1, 3 \pm 2) = (-1, 1)$ and $(-1, 5)$, the foci are $\left(-1, 3 \pm 2\sqrt{10}\right)$, and the asymptotes are $y - 3 = \pm \frac{1}{3}(x+1) \Leftrightarrow y = \frac{1}{3}x + \frac{10}{3}$ and $y = -\frac{1}{3}x + \frac{8}{3}$.

(b)

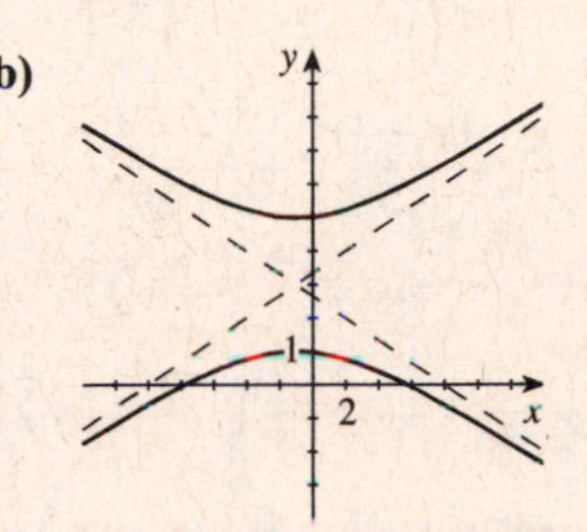

35. (a) $9y^2 + 18y = x^2 + 6x + 18 \Leftrightarrow 9\left(y^2 + 2y + 1\right) = \left(x^2 + 6x + 9\right) + 9 - 9 + 18$ $\Leftrightarrow 9(y+1)^2 - (x+3)^2 = 18 \Leftrightarrow \dfrac{(y+1)^2}{2} - \dfrac{(x+3)^2}{18} = 1$. This is a hyperbola with $a = \sqrt{2}$, $b = 3\sqrt{2}$, and $c = \sqrt{2+18} = 2\sqrt{5}$. The center is $(-3, -1)$, the vertices are $\left(-3, -1 \pm \sqrt{2}\right)$, the foci are $\left(-3, -1 \pm 2\sqrt{5}\right)$, and the asymptotes are $y + 1 = \pm \frac{1}{3}(x+3) \Leftrightarrow y = \frac{1}{3}x$ and $y = -\frac{1}{3}x - 2$.

(b)

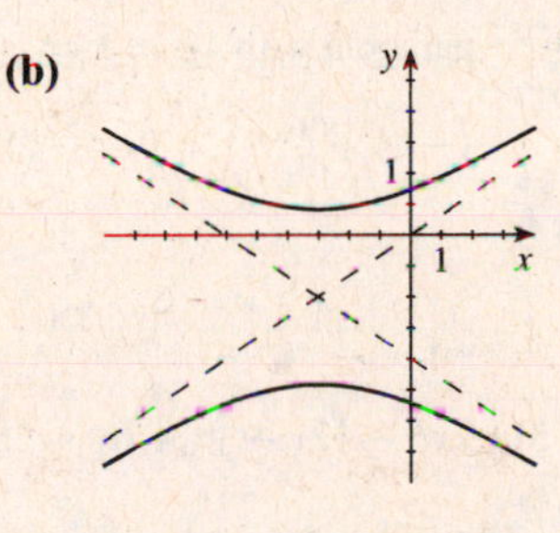

37. This is a parabola that opens to the right with its vertex at $(0,0)$ and the focus at $(2,0)$. So $p = 2$, and the equation is $y^2 = 4(2)x \Leftrightarrow y^2 = 8x$.

39. From the graph, the center is $(0,0)$, and the vertices are $(0,-4)$ and $(0,4)$. Since a is the distance from the center to a vertex, we have $a = 4$. Because one focus is $(0,5)$, we have $c = 5$, and since $c^2 = a^2 + b^2$, we have $25 = 16 + b^2 \Leftrightarrow b^2 = 9$. Thus an equation of the hyperbola is $\dfrac{y^2}{16} - \dfrac{x^2}{9} = 1$.

41. From the graph, the center of the ellipse is $(4,2)$, and so $a = 4$ and $b = 2$. The equation is $\dfrac{(x-4)^2}{4^2} + \dfrac{(y-2)^2}{2^2} = 1 \Leftrightarrow \dfrac{(x-4)^2}{16} + \dfrac{(y-2)^2}{4} = 1$.

43. $\frac{x^2}{12}+y=1 \Leftrightarrow \frac{x^2}{12}=-(y-1) \Leftrightarrow x^2=-12(y-1)$. This is a parabola with $4p=-12 \Leftrightarrow p=-3$. The vertex is $(0,1)$, the focus is $(0,1-3)=(0,-2)$, and the directrix is $y=1+3=4$.

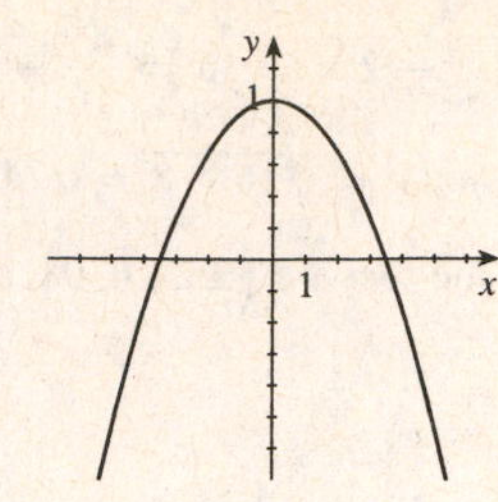

45. $x^2-y^2+144=0 \Leftrightarrow \frac{y^2}{144}-\frac{x^2}{144}=1$. This is a hyperbola with $a=12$, $b=12$, and $c=\sqrt{144+144}=12\sqrt{2}$. The center is $(0,0)$, the foci are $\left(0,\pm 12\sqrt{2}\right)$, the vertices are $(0,\pm 12)$, and the asymptotes are $y=\pm x$.

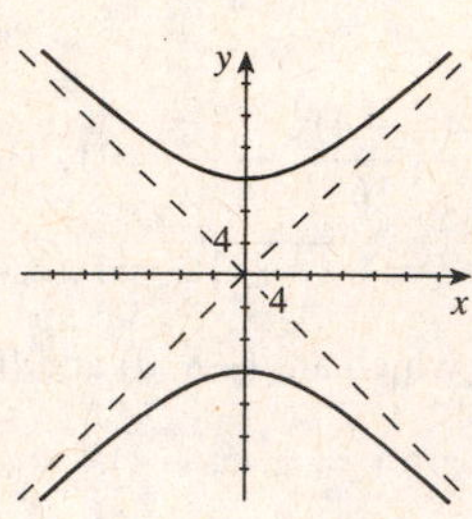

47. $4x^2+y^2=8(x+y) \Leftrightarrow 4\left(x^2-2x\right)+\left(y^2-8y\right)=0 \Leftrightarrow$
$4\left(x^2-2x+1\right)+\left(y^2-8y+16\right)=4+16 \Leftrightarrow 4(x-1)^2+(y-4)^2=20 \Leftrightarrow$
$\frac{(x-1)^2}{5}+\frac{(y-4)^2}{20}=1$. This is an ellipse with $a=2\sqrt{5}$, $b=\sqrt{5}$, and $c=\sqrt{20-5}=\sqrt{15}$. The center is $(1,4)$, the foci are $\left(1,4\pm\sqrt{15}\right)$, and the vertices are $\left(1,4\pm 2\sqrt{5}\right)$.

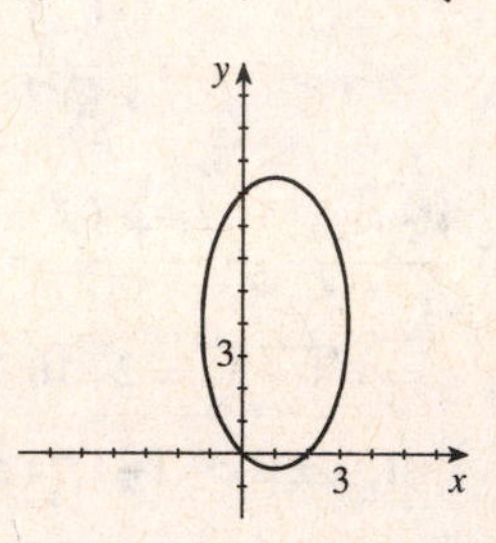

49. $x=y^2-16y \Leftrightarrow x+64=y^2-16y+64 \Leftrightarrow (y-8)^2=x+64$. This is a parabola with $4p=1 \Leftrightarrow p=\frac{1}{4}$. The vertex is $(-64,8)$, the focus is $\left(-64+\frac{1}{4},8\right)=\left(-\frac{255}{4},8\right)$, and the directrix is $x=-64-\frac{1}{4}=-\frac{257}{4}$.

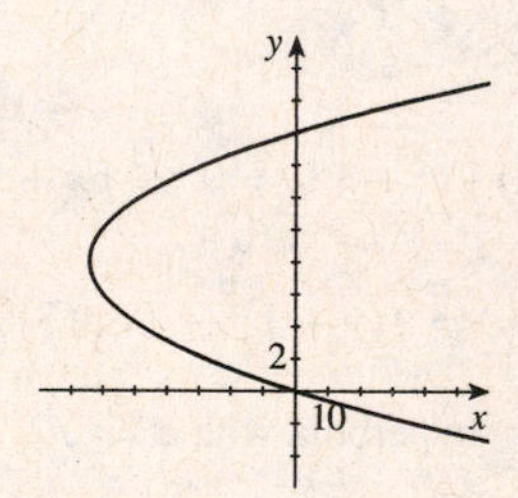

51. $2x^2-12x+y^2+6y+26=0 \Leftrightarrow 2\left(x^2-6x\right)+\left(y^2+6y\right)=-26 \Leftrightarrow$
$2\left(x^2-6x+9\right)+\left(y^2+6y+9\right)=-26+18+9 \Leftrightarrow 2(x-3)^2+(y+3)^2=1$
$\Leftrightarrow \frac{(x-3)^2}{\frac{1}{2}}+(y+3)^2=1$. This is an ellipse with $a=1$, $b=\frac{\sqrt{2}}{2}$, and $c=\sqrt{1-\frac{1}{2}}=\frac{\sqrt{2}}{2}$. The center is $(3,-3)$, the foci are $\left(3,-3\pm\frac{\sqrt{2}}{2}\right)$, and the vertices are $(3,-3\pm 1)=(3,-4)$ and $(3,-2)$.

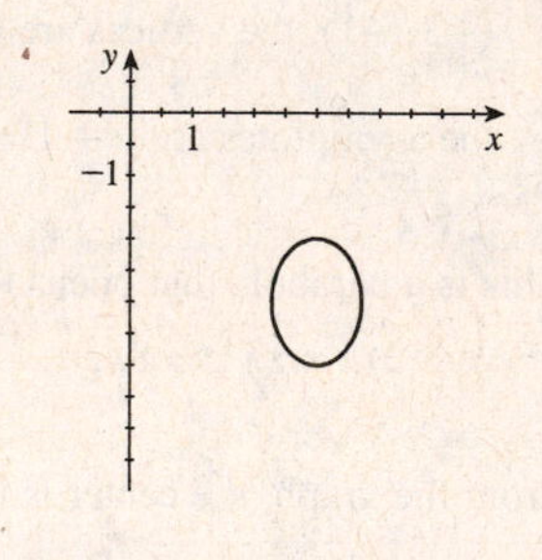

53. $9x^2+8y^2-15x+8y+27=0 \Leftrightarrow 9\left(x^2-\frac{5}{3}x+\frac{25}{36}\right)+8\left(y^2+y+\frac{1}{4}\right)=-27+\frac{25}{4}+2 \Leftrightarrow 9\left(x-\frac{5}{6}\right)^2+8\left(y+\frac{1}{2}\right)^2=-\frac{75}{4}$. However, since the left-hand side of the equation is greater than or equal to 0, there is no point that satisfies this equation. The graph is empty.

55. The parabola has focus $(0,1)$ and directrix $y=-1$. Therefore, $p=1$ and so $4p=4$. Since the focus is on the y-axis and the vertex is $(0,0)$, an equation of the parabola is $x^2=4y$.

57. The ellipse with center at the origin and with x-intercepts ± 2 and y-intercepts ± 5 has a vertical major axis, $a = 5$, and $b = 2$, so an equation is $\frac{x^2}{4} + \frac{y^2}{25} = 1$.

59. The ellipse has center $C(0, 4)$, foci $F_1(0, 0)$ and $F_2(0, 8)$, and major axis of length 10. Then $2c = 8 - 0 \Leftrightarrow c = 4$. Also, since the length of the major axis is 10, $2a = 10 \Leftrightarrow a = 5$. Therefore, $b^2 = a^2 - c^2 = 25 - 16 = 9$. Since the foci are on the y-axis, the vertices are on the y-axis, and an equation of the ellipse is $\frac{x^2}{9} + \frac{(y-4)^2}{25} = 1$.

61. The ellipse has foci $F_1(1, 1)$ and $F_2(1, 3)$, and one vertex is on the x-axis. Thus, $2c = 3 - 1 = 2 \Leftrightarrow c = 1$, and so the center of the ellipse is $C(1, 2)$. Also, since one vertex is on the x-axis, $a = 2 - 0 = 2$, and thus $b^2 = 4 - 1 = 3$. So an equation of the ellipse is $\frac{(x-1)^2}{3} + \frac{(y-2)^2}{4} = 1$.

63. The ellipse has vertices $V_1(7, 12)$ and $V_2(7, -8)$ and passes through the point $P(1, 8)$. Thus, $2a = 12 - (-8) = 20 \Leftrightarrow a = 10$, and the center is $\left(7, \frac{-8+12}{2}\right) = (7, 2)$. Thus an equation of the ellipse has the form $\frac{(x-7)^2}{b^2} + \frac{(y-2)^2}{100} = 1$.
Since the point $P(1, 8)$ is on the ellipse, $\frac{(1-7)^2}{b^2} + \frac{(8-2)^2}{100} = 1 \Leftrightarrow 3600 + 36b^2 = 100b^2 \Leftrightarrow 64b^2 = 3600 \Leftrightarrow b^2 = \frac{225}{4}$.
Therefore, an equation of the ellipse is $\frac{(x-7)^2}{225/4} + \frac{(y-2)^2}{100} = 1 \Leftrightarrow \frac{4(x-7)^2}{225} + \frac{(y-2)^2}{100} = 1$.

65. The length of the major axis is $2a = 186{,}000{,}000 \Leftrightarrow a = 93{,}000{,}000$. The eccentricity is $e = c/a = 0.017$, and so $c = 0.017(93{,}000{,}000) = 1{,}581{,}000$.

(a) The earth is closest to the sun when the distance is $a - c = 93{,}000{,}000 - 1{,}581{,}000 = 91{,}419{,}000$.

(b) The earth is furthest from the sun when the distance is $a + c = 93{,}000{,}000 + 1{,}581{,}000 = 94{,}581{,}000$.

67. (a) The graphs of $\frac{x^2}{16+k^2} + \frac{y^2}{k^2} = 1$ for $k = 1, 2, 4$, and 8 are shown in the figure.

(b) $c^2 = \left(16 + k^2\right) - k^2 = 16 \Rightarrow \quad c = \pm 4$. Since the center is $(0, 0)$, the foci of each of the ellipses are $(\pm 4, 0)$.

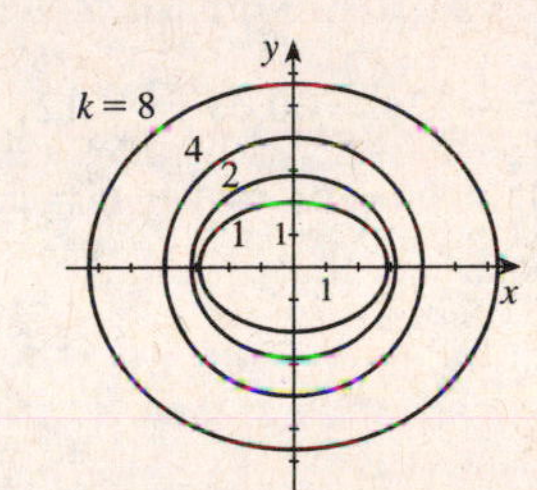

69. (a) $x^2 + 4xy + y^2 = 1$. Then $A = 1$, $B = 4$, and $C = 1$, so the discriminant is $4^2 - 4(1)(1) = 12$. Since the discriminant is positive, the equation represents a hyperbola.

(b) $\cot 2\phi = \frac{A - C}{B} = \frac{1-1}{4} = 0 \Rightarrow 2\phi = 90° \Leftrightarrow \phi = 45°$. Therefore, $x = \frac{\sqrt{2}}{2}X - \frac{\sqrt{2}}{2}Y$ and $y = \frac{\sqrt{2}}{2}X + \frac{\sqrt{2}}{2}Y$.
Substituting into the original equation gives

$$\left(\frac{\sqrt{2}}{2}X - \frac{\sqrt{2}}{2}Y\right)^2 + 4\left(\frac{\sqrt{2}}{2}X - \frac{\sqrt{2}}{2}Y\right)\left(\frac{\sqrt{2}}{2}X + \frac{\sqrt{2}}{2}Y\right) + \left(\frac{\sqrt{2}}{2}X + \frac{\sqrt{2}}{2}Y\right)^2 = 1 \Leftrightarrow$$

$$\frac{1}{2}\left(X^2 - 2XY + Y^2\right) + 2\left(X^2 + XY - XY - Y^2\right) + \frac{1}{2}\left(X^2 + 2XY + Y^2\right) = 1 \Leftrightarrow$$

$3X^2 - Y^2 = 1 \Leftrightarrow 3X^2 - Y^2 = 1$. This is a hyperbola with $a = \frac{1}{\sqrt{3}}$, $b = 1$, and $c = \sqrt{\frac{1}{3} + 1} = \frac{2}{\sqrt{3}}$. Therefore, the hyperbola has vertices $V\left(\pm\frac{1}{\sqrt{3}}, 0\right)$ and foci $F\left(\pm\frac{2}{\sqrt{3}}, 0\right)$, in XY-coordinates.

(c)

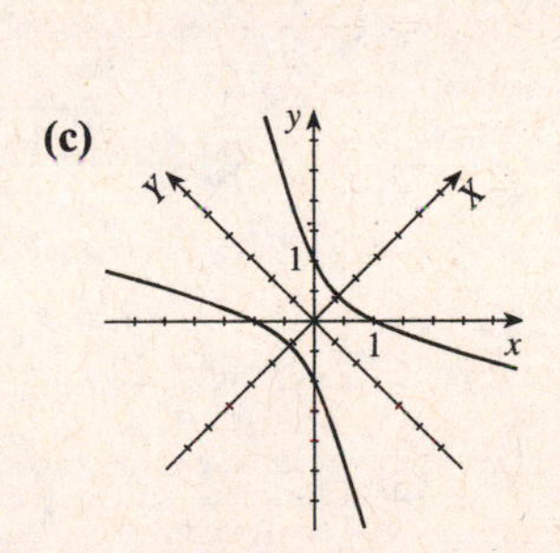

71. (a) $7x^2 - 6\sqrt{3}xy + 13y^2 - 4\sqrt{3}x - 4y = 0$. Then $A = 7$, $B = -6\sqrt{3}$, and $C = 13$, so the discriminant is $\left(-6\sqrt{3}\right)^2 - 4(7)(13) = -256$. Since the discriminant is negative, the equation represents an ellipse.

(b) $\cot 2\phi = \dfrac{A-C}{B} = \dfrac{7-13}{-6\sqrt{3}} = \dfrac{1}{\sqrt{3}} \Rightarrow 2\phi = 60^\circ \Leftrightarrow \phi = 30^\circ$. Therefore, $x = \frac{\sqrt{3}}{2}X - \frac{1}{2}Y$ and $y = \frac{1}{2}X + \frac{\sqrt{3}}{2}Y$.

Substituting into the original equation gives

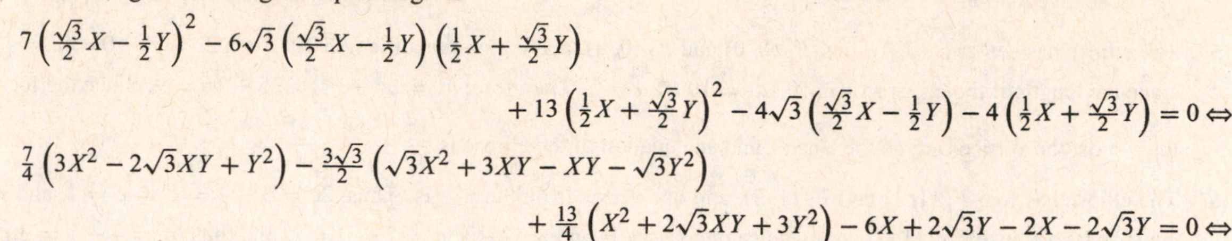

$$7\left(\tfrac{\sqrt{3}}{2}X - \tfrac{1}{2}Y\right)^2 - 6\sqrt{3}\left(\tfrac{\sqrt{3}}{2}X - \tfrac{1}{2}Y\right)\left(\tfrac{1}{2}X + \tfrac{\sqrt{3}}{2}Y\right)$$
$$+ 13\left(\tfrac{1}{2}X + \tfrac{\sqrt{3}}{2}Y\right)^2 - 4\sqrt{3}\left(\tfrac{\sqrt{3}}{2}X - \tfrac{1}{2}Y\right) - 4\left(\tfrac{1}{2}X + \tfrac{\sqrt{3}}{2}Y\right) = 0 \Leftrightarrow$$
$$\tfrac{7}{4}\left(3X^2 - 2\sqrt{3}XY + Y^2\right) - \tfrac{3\sqrt{3}}{2}\left(\sqrt{3}X^2 + 3XY - XY - \sqrt{3}Y^2\right)$$
$$+ \tfrac{13}{4}\left(X^2 + 2\sqrt{3}XY + 3Y^2\right) - 6X + 2\sqrt{3}Y - 2X - 2\sqrt{3}Y = 0 \Leftrightarrow$$

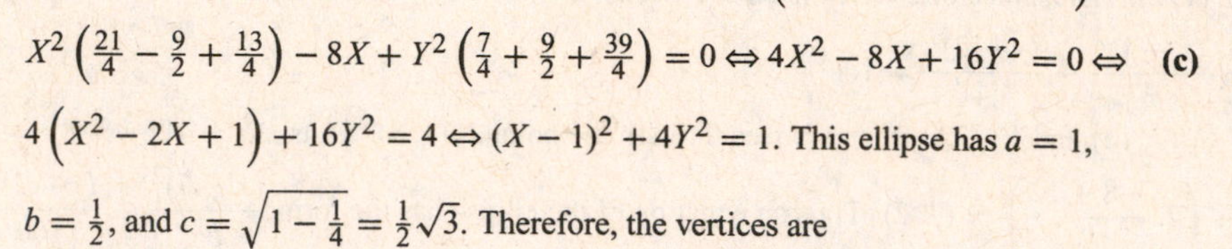

$X^2\left(\frac{21}{4} - \frac{9}{2} + \frac{13}{4}\right) - 8X + Y^2\left(\frac{7}{4} + \frac{9}{2} + \frac{39}{4}\right) = 0 \Leftrightarrow 4X^2 - 8X + 16Y^2 = 0 \Leftrightarrow$ **(c)**

$4\left(X^2 - 2X + 1\right) + 16Y^2 = 4 \Leftrightarrow (X-1)^2 + 4Y^2 = 1$. This ellipse has $a = 1$,

$b = \frac{1}{2}$, and $c = \sqrt{1 - \frac{1}{4}} = \frac{1}{2}\sqrt{3}$. Therefore, the vertices are

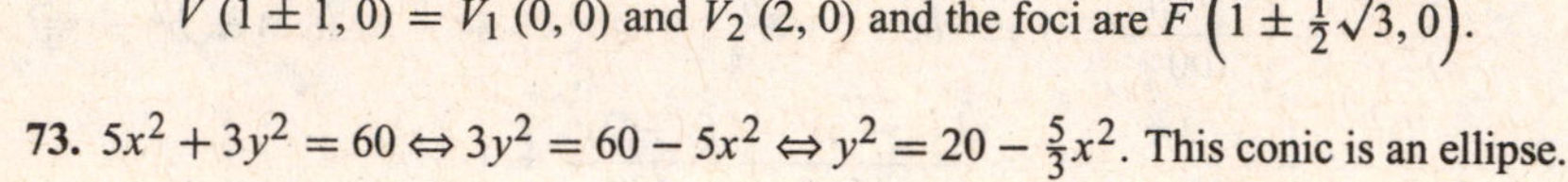

$V(1 \pm 1, 0) = V_1(0,0)$ and $V_2(2,0)$ and the foci are $F\left(1 \pm \frac{1}{2}\sqrt{3}, 0\right)$.

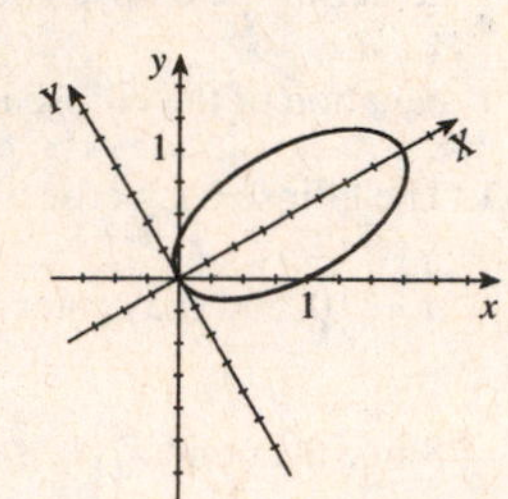

73. $5x^2 + 3y^2 = 60 \Leftrightarrow 3y^2 = 60 - 5x^2 \Leftrightarrow y^2 = 20 - \frac{5}{3}x^2$. This conic is an ellipse.

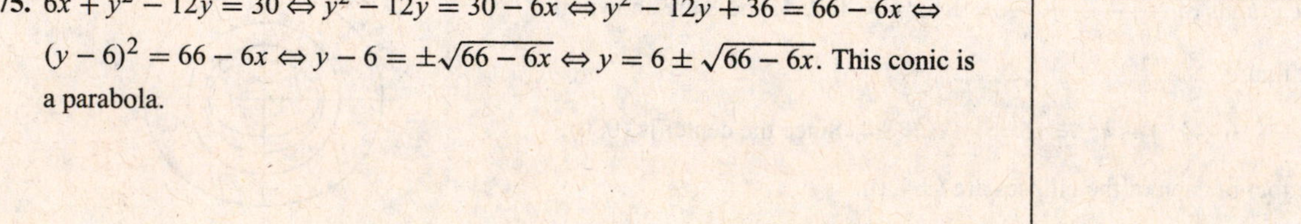

75. $6x + y^2 - 12y = 30 \Leftrightarrow y^2 - 12y = 30 - 6x \Leftrightarrow y^2 - 12y + 36 = 66 - 6x \Leftrightarrow$ $(y-6)^2 = 66 - 6x \Leftrightarrow y - 6 = \pm\sqrt{66-6x} \Leftrightarrow y = 6 \pm \sqrt{66-6x}$. This conic is a parabola.

77. (a) $r = \dfrac{1}{1 - \cos\theta} \Rightarrow e = 1$. Therefore, this is a parabola.

(b)

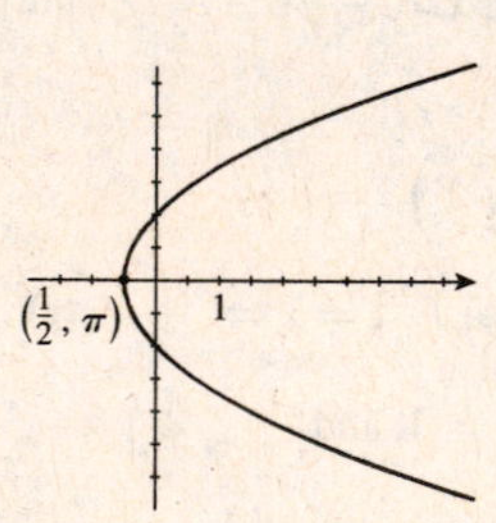

79. (a) $r = \dfrac{4}{1 + 2\sin\theta} \Rightarrow e = 2$. Therefore, this is a hyperbola.

(b)

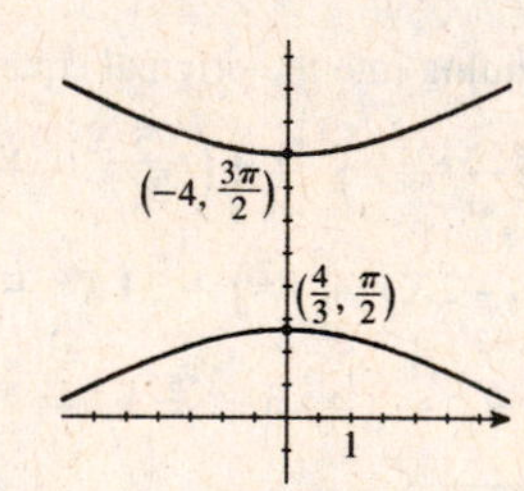

(b) $\cot 2\phi = \dfrac{A-C}{B} = \dfrac{7-13}{-6\sqrt{3}} = \dfrac{1}{\sqrt{3}} \Rightarrow 2\phi = 60^\circ \Leftrightarrow \phi = 30^\circ$. Therefore, $x = \frac{\sqrt{3}}{2}X - \frac{1}{2}Y$ and $y = \frac{1}{2}X + \frac{\sqrt{3}}{2}Y$.

Substituting into the original equation gives

$$7\left(\tfrac{\sqrt{3}}{2}X - \tfrac{1}{2}Y\right)^2 - 6\sqrt{3}\left(\tfrac{\sqrt{3}}{2}X - \tfrac{1}{2}Y\right)\left(\tfrac{1}{2}X + \tfrac{\sqrt{3}}{2}Y\right)$$
$$+ 13\left(\tfrac{1}{2}X + \tfrac{\sqrt{3}}{2}Y\right)^2 - 4\sqrt{3}\left(\tfrac{\sqrt{3}}{2}X - \tfrac{1}{2}Y\right) - 4\left(\tfrac{1}{2}X + \tfrac{\sqrt{3}}{2}Y\right) = 0 \Leftrightarrow$$
$$\tfrac{7}{4}\left(3X^2 - 2\sqrt{3}XY + Y^2\right) - \tfrac{3\sqrt{3}}{2}\left(\sqrt{3}X^2 + 3XY - XY - \sqrt{3}Y^2\right)$$
$$+ \tfrac{13}{4}\left(X^2 + 2\sqrt{3}XY + 3Y^2\right) - 6X + 2\sqrt{3}Y - 2X - 2\sqrt{3}Y = 0 \Leftrightarrow$$

$X^2\left(\frac{21}{4} - \frac{9}{2} + \frac{13}{4}\right) - 8X + Y^2\left(\frac{7}{4} + \frac{9}{2} + \frac{39}{4}\right) = 0 \Leftrightarrow 4X^2 - 8X + 16Y^2 = 0 \Leftrightarrow$ **(c)**

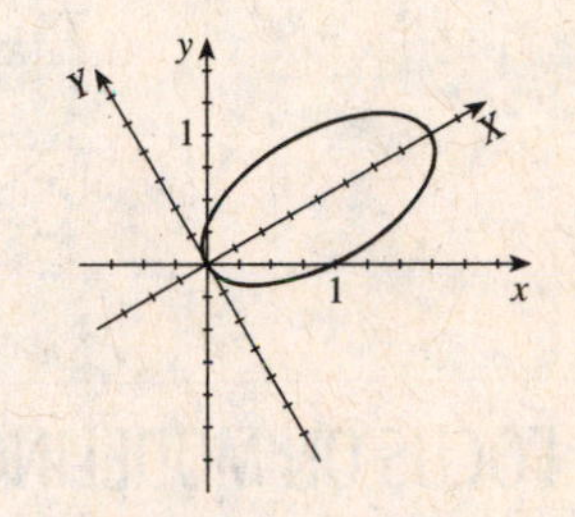

$4\left(X^2 - 2X + 1\right) + 16Y^2 = 4 \Leftrightarrow (X-1)^2 + 4Y^2 = 1$. This ellipse has $a = 1$, $b = \frac{1}{2}$, and $c = \sqrt{1 - \frac{1}{4}} = \frac{1}{2}\sqrt{3}$. Therefore, the vertices are $V(1 \pm 1, 0) = V_1(0,0)$ and $V_2(2,0)$ and the foci are $F\left(1 \pm \frac{1}{2}\sqrt{3}, 0\right)$.

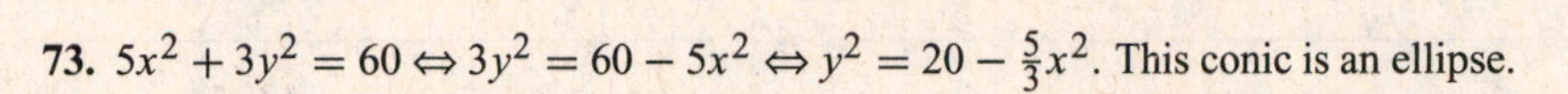

73. $5x^2 + 3y^2 = 60 \Leftrightarrow 3y^2 = 60 - 5x^2 \Leftrightarrow y^2 = 20 - \frac{5}{3}x^2$. This conic is an ellipse.

75. $6x + y^2 - 12y = 30 \Leftrightarrow y^2 - 12y = 30 - 6x \Leftrightarrow y^2 - 12y + 36 = 66 - 6x \Leftrightarrow (y-6)^2 = 66 - 6x \Leftrightarrow y - 6 = \pm\sqrt{66 - 6x} \Leftrightarrow y = 6 \pm \sqrt{66 - 6x}$. This conic is a parabola.

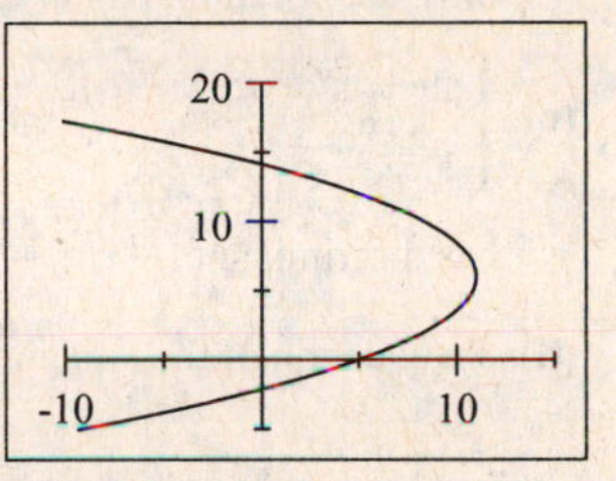

77. (a) $r = \dfrac{1}{1 - \cos\theta} \Rightarrow e = 1$. Therefore, this is a parabola.

(b)

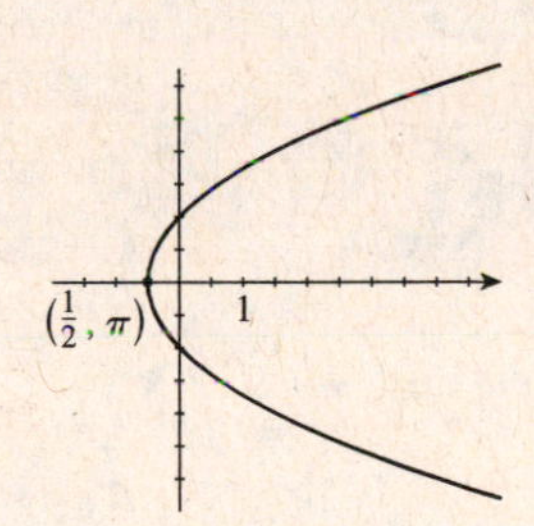

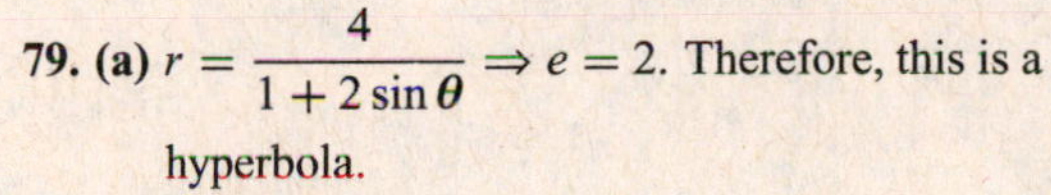

79. (a) $r = \dfrac{4}{1 + 2\sin\theta} \Rightarrow e = 2$. Therefore, this is a hyperbola.

(b)

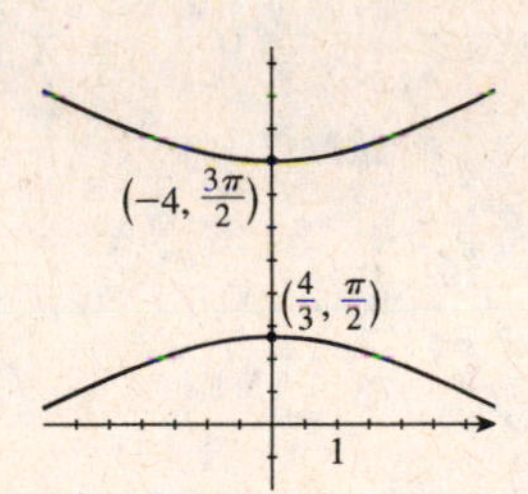

17. (a) Since the focus of this conic is the origin and the directrix is $x = 2$, the equation has the form $r = \dfrac{ed}{1 + e\cos\theta}$. Subsituting $e = \frac{1}{2}$ and $d = 2$ we get $r = \dfrac{1}{1 + \frac{1}{2}\cos\theta} \Leftrightarrow r = \dfrac{2}{2 + \cos\theta}$.

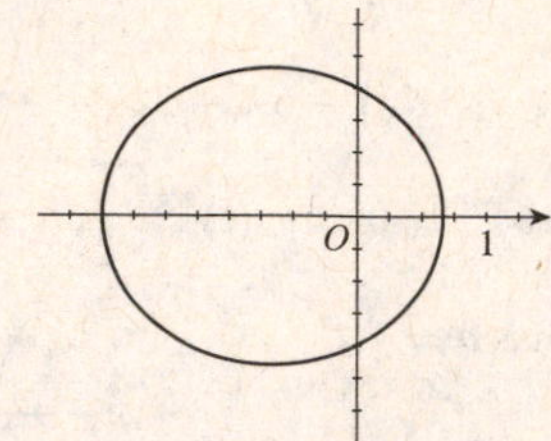

(b) $r = \dfrac{3}{2 - \sin\theta} \Leftrightarrow r = \dfrac{\frac{3}{2}}{1 - \frac{1}{2}\sin\theta}$. So $e = \frac{1}{2}$ and the conic is an ellipse.

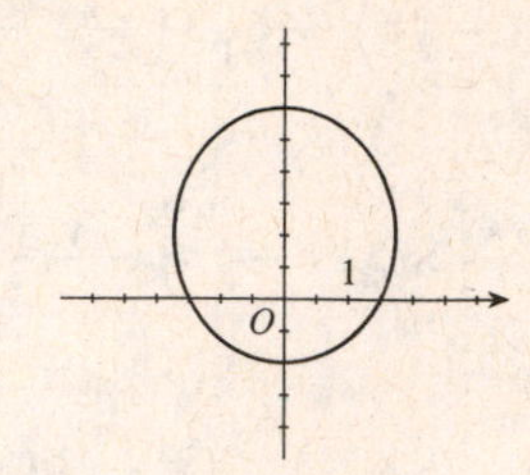

FOCUS ON MODELING Conics in Architecture

1. Answers will vary.

5. (a) The tangent line passes though the point $\left(a, a^2\right)$, so an equation is $y - a^2 = m(x - a)$.

(b) Because the tangent line intersects the parabola at only the one point $\left(a, a^2\right)$, the system $\begin{cases} y - a^2 = m(x - a) \\ y = x^2 \end{cases}$ has only one solution, namely $x = a$, $y = a^2$.

(c) $\begin{cases} y - a^2 = m(x - a) \\ y = x^2 \end{cases} \Leftrightarrow \begin{cases} y = a^2 + m(x - a) \\ y = x^2 \end{cases} \Leftrightarrow a^2 + m(x - a) = x^2 \Leftrightarrow x^2 - mx + am - a^2 = 0$. This quadratic has discriminant $(-m)^2 - 4(1)\left(am - a^2\right) = m^2 - 4am + 4a^2 = (m - 2a)^2$. Setting this equal to 0, we find $m = 2a$.

(d) An equation of the tangent line is $y - a^2 = 2a(x - a) \Leftrightarrow y = a^2 + 2ax - 2a^2 \Leftrightarrow y = 2ax - a^2$.

13 SEQUENCES AND SERIES

13.1 SEQUENCES AND SUMMATION NOTATION

1. A sequence is a function whose domain is *the natural numbers*.

3. $a_n = n - 3$. Then $a_1 = 1 - 3 = -2$, $a_2 = 2 - 3 = -1$, $a_3 = 3 - 3 = 0$, $a_4 = 4 - 3 = 1$, and $a_{100} = 100 - 3 = 97$.

5. $a_n = \dfrac{1}{2n+1}$. Then $a_1 = \dfrac{1}{2(1)+1} = \dfrac{1}{3}$, $a_2 = \dfrac{1}{2(2)+1} = \dfrac{1}{5}$, $a_3 = \dfrac{1}{2(3)+1} = \dfrac{1}{7}$, $a_4 = \dfrac{1}{2(4)+1} = \dfrac{1}{9}$, and $a_{100} = \dfrac{1}{2(100)+1} = \dfrac{1}{201}$.

7. $a_n = 5^n$. Then $a_1 = 5^1 = 5$, $a_2 = 5^2 = 25$, $a_3 = 5^3 = 125$, $a_4 = 5^4 = 625$, and $a_{100} = 5^{100} \approx 7.9 \times 10^{69}$.

9. $a_n = \dfrac{(-1)^n}{n^2}$. Then $a_1 = \dfrac{(-1)^1}{1^2} = -1$, $a_2 = \dfrac{(-1)^2}{2^2} = \dfrac{1}{4}$, $a_3 = \dfrac{(-1)^3}{3^2} = -\dfrac{1}{9}$, $a_4 = \dfrac{(-1)^4}{4^2} = \dfrac{1}{16}$, and $a_{100} = \dfrac{(-1)^{100}}{100^2} = \dfrac{1}{10{,}000}$.

11. $a_n = 1 + (-1)^n$. Then $a_1 = 1 + (-1)^1 = 0$, $a_2 = 1 + (-1)^2 = 2$, $a_3 = 1 + (-1)^3 = 0$, $a_4 = 1 + (-1)^4 = 2$, and $a_{100} = 1 + (-1)^{100} = 2$.

13. $a_n = n^n$. Then $a_1 = 1^1 = 1$, $a_2 = 2^2 = 4$, $a_3 = 3^3 = 27$, $a_4 = 4^4 = 256$, and $a_{100} = 100^{100} = 10^{200}$.

15. $a_n = 2(a_{n-1} + 3)$ and $a_1 = 4$. Then $a_2 = 2(4+3) = 14$, $a_3 = 2(14+3) = 34$, $a_4 = 2(34+3) = 74$, and $a_5 = 2(74+3) = 154$.

17. $a_n = 2a_{n-1} + 1$ and $a_1 = 1$. Then $a_2 = 2(1)+1 = 3$, $a_3 = 2(3)+1 = 7$, $a_4 = 2(7)+1 = 15$, and $a_5 = 2(15)+1 = 31$.

19. $a_n = a_{n-1} + a_{n-2}$, $a_1 = 1$, and $a_2 = 2$. Then $a_3 = 2 + 1 = 3$, $a_4 = 3 + 2 = 5$, and $a_5 = 5 + 3 = 8$.

21. **(a)** $a_1 = 7$, $a_2 = 11$, $a_3 = 15$, $a_4 = 19$, $a_5 = 23$, $a_6 = 27$, $a_7 = 31$, $a_8 = 35$, $a_9 = 39$, $a_{10} = 43$

(b)

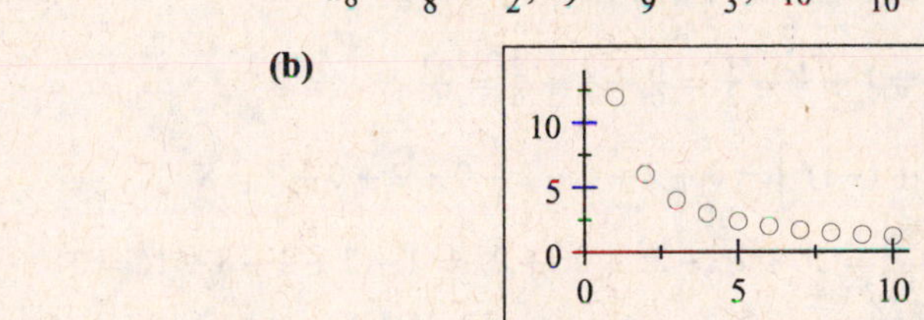

23. **(a)** $a_1 = \frac{12}{1} = 12$, $a_2 = \frac{12}{2} = 6$, $a_3 = \frac{12}{3} = 4$, $a_4 = \frac{12}{4} = 3$, $a_5 = \frac{12}{5}$, $a_6 = \frac{12}{6} = 2$, $a_7 = \frac{12}{7}$, $a_8 = \frac{12}{8} = \frac{3}{2}$, $a_9 = \frac{12}{9} = \frac{4}{3}$, $a_{10} = \frac{12}{10} = \frac{6}{5}$

(b)

25. **(a)** $a_1 = 2$, $a_2 = 0.5$, $a_3 = 2$, $a_4 = 0.5$, $a_5 = 2$, $a_6 = 0.5$, $a_7 = 2$, $a_8 = 0.5$, $a_9 = 2$, $a_{10} = 0.5$

(b)

27. 2, 4, 6, 8, ... All are multiples of 2, so $a_1 = 2$, $a_2 = 2 \cdot 2$, $a_3 = 3 \cdot 2$, $a_4 = 4 \cdot 2$, ... Thus $a_n = 2n$.

29. 2, 4, 8, 16, ... All are powers of 2, so $a_1 = 2$, $a_2 = 2^2$, $a_3 = 2^3$, $a_4 = 2^4$, ... Thus $a_n = 2^n$.

31. $-2, 3, 8, 13, \ldots$. The difference between any two consecutive terms is 5, so $a_1 = 5(1) - 7$, $a_2 = 5(2) - 7$, $a_3 = 5(3) - 7$, $a_4 = 5(4) - 7$, ... Thus, $a_n = 5n - 7$.

33. $5, -25, 125, -625, \ldots$. These terms are powers of 5, and the terms alternate in sign. So $a_1 = (-1)^2 \cdot 5^1$, $a_2 = (-1)^3 \cdot 5^2$, $a_3 = (-1)^4 \cdot 5^3$, $a_4 = (-1)^5 \cdot 5^4, \ldots$. Thus $a_n = (-1)^{n+1} \cdot 5^n$.

35. $1, \frac{3}{4}, \frac{5}{9}, \frac{7}{16}, \frac{9}{25}, \ldots$. We consider the numerator separately from the denominator. The numerators of the terms differ by 2, and the denominators are perfect squares. So $a_1 = \dfrac{2(1)-1}{1^2}$, $a_2 = \dfrac{2(2)-1}{2^2}$, $a_3 = \dfrac{2(3)-1}{3^2}$, $a_4 = \dfrac{2(4)-1}{4^2}$, $a_5 = \dfrac{2(5)-1}{5^2}, \ldots$. Thus $a_n = \dfrac{2n-1}{n^2}$.

37. $0, 2, 0, 2, 0, 2, \ldots$. These terms alternate between 0 and 2. So $a_1 = 1-1$, $a_2 = 1+1$, $a_3 = 1-1$, $a_4 = 1+1$, $a_5 = 1-1$, $a_6 = 1+1, \ldots$ Thus $a_n = 1 + (-1)^n$.

39. $a_1 = 1$, $a_2 = 3$, $a_3 = 5$, $a_4 = 7, \ldots$. Therefore, $a_n = 2n - 1$. So $S_1 = 1$, $S_2 = 1 + 3 = 4$, $S_3 = 1 + 3 + 5 + = 9$, $S_4 = 1 + 3 + 5 + 7 = 16$, $S_5 = 1 + 3 + 5 + 7 + 9 = 25$, and $S_6 = 1 + 3 + 5 + 7 + 9 + 11 = 36$.

41. $a_1 = \frac{1}{3}$, $a_2 = \dfrac{1}{3^2}$, $a_3 = \dfrac{1}{3^3}$, $a_4 = \dfrac{1}{3^4}, \ldots$. Therefore, $a_n = \dfrac{1}{3^n}$. So $S_1 = \frac{1}{3}$, $S_2 = \frac{1}{3} + \dfrac{1}{3^2} = \dfrac{4}{9}$, $S_3 = \frac{1}{3} + \dfrac{1}{3^2} + \dfrac{1}{3^3} = \frac{13}{27}$, $S_4 = \frac{1}{3} + \dfrac{1}{3^2} + \dfrac{1}{3^3} + \dfrac{1}{3^4} = \frac{40}{81}$, and $S_5 = \frac{1}{3} + \dfrac{1}{3^2} + \dfrac{1}{3^3} + \dfrac{1}{3^4} + \dfrac{1}{3^5} = \frac{121}{243}$, $S_6 = \frac{1}{3} + \dfrac{1}{3^2} + \dfrac{1}{3^3} + \dfrac{1}{3^4} + \dfrac{1}{3^5} + \dfrac{1}{3^6} = \frac{364}{729}$.

43. $a_n = \dfrac{2}{3^n}$. So $S_1 = \frac{2}{3}$, $S_2 = \dfrac{2}{3} + \dfrac{2}{3^2} = \frac{8}{9}$, $S_3 = \frac{2}{3} + \dfrac{2}{3^2} + \dfrac{2}{3^3} = \frac{26}{27}$, and $S_4 = \frac{2}{3} + \dfrac{2}{3^2} + \dfrac{2}{3^3} + \dfrac{2}{3^4} = \frac{80}{81}$. Therefore, $S_n = \dfrac{3^n - 1}{3^n}$.

45. $a_n = \sqrt{n} - \sqrt{n+1}$. So $S_1 = \sqrt{1} - \sqrt{2} = 1 - \sqrt{2}$, $S_2 = \left(\sqrt{1} - \sqrt{2}\right) + \left(\sqrt{2} - \sqrt{3}\right) = 1 + \left(-\sqrt{2} + \sqrt{2}\right) - \sqrt{3} = 1 - \sqrt{3}$,
$S_3 = \left(\sqrt{1} - \sqrt{2}\right) + \left(\sqrt{2} - \sqrt{3}\right) + \left(\sqrt{3} - \sqrt{4}\right) = 1 + \left(-\sqrt{2} + \sqrt{2}\right) + \left(-\sqrt{3} + \sqrt{3}\right) - \sqrt{4} = 1 - \sqrt{4}$,

$$\begin{aligned} S_4 &= \left(\sqrt{1} - \sqrt{2}\right) + \left(\sqrt{2} - \sqrt{3}\right) + \left(\sqrt{3} - \sqrt{4}\right) + \left(\sqrt{4} - \sqrt{5}\right) \\ &= 1 + \left(-\sqrt{2} + \sqrt{2}\right) + \left(-\sqrt{3} + \sqrt{3}\right) + \left(-\sqrt{4} + \sqrt{4}\right) - \sqrt{5} = 1 - \sqrt{5} \end{aligned}$$

Therefore,

$$\begin{aligned} S_n &= \left(\sqrt{1} - \sqrt{2}\right) + \left(\sqrt{2} - \sqrt{3}\right) + \cdots + \left(\sqrt{n} - \sqrt{n+1}\right) \\ &= 1 + \left(-\sqrt{2} + \sqrt{2}\right) + \left(-\sqrt{3} + \sqrt{3}\right) + \cdots + \left(-\sqrt{n} + \sqrt{n}\right) - \sqrt{n+1} = 1 - \sqrt{n+1} \end{aligned}$$

47. $\sum_{k=1}^{4} k = 1 + 2 + 3 + 4 = 10$

49. $\sum_{k=1}^{3} \dfrac{1}{k} = 1 + \frac{1}{2} + \frac{1}{3} = \frac{6}{6} + \frac{3}{6} + \frac{2}{6} = \frac{11}{6}$

51. $\sum_{i=1}^{8} \left[1 + (-1)^i\right] = 0 + 2 + 0 + 2 + 0 + 2 + 0 + 2 = 8$

53. $\sum_{k=1}^{5} 2^{k-1} = 2^0 + 2^1 + 2^2 + 2^3 + 2^4 = 1 + 2 + 4 + 8 + 16 = 31$

55. 385

57. 46,438

59. 22

61. $\sum_{k=1}^{4} k^3 = 1^3 + 2^3 + 3^3 + 4^3 = 1 + 8 + 27 + 64$

63. $\sum_{k=0}^{6} \sqrt{k+4} = \sqrt{4} + \sqrt{5} + \sqrt{6} + \sqrt{7} + \sqrt{8} + \sqrt{9} + \sqrt{10}$

65. $\sum_{k=3}^{100} x^k = x^3 + x^4 + x^5 + \cdots + x^{100}$

67. $2 + 4 + 6 + \cdots + 50 = \sum_{k=1}^{25} 2k$

69. $1^2 + 2^2 + 3^2 + \cdots + 10^2 = \sum_{k=1}^{10} k^2$

71. $\dfrac{1}{1 \cdot 2} + \dfrac{1}{2 \cdot 3} + \dfrac{1}{3 \cdot 4} + \cdots + \dfrac{1}{999 \cdot 1000} = \sum_{k=1}^{999} \dfrac{1}{k(k+1)}$

73. $1 + x + x^2 + x^3 + \cdots + x^{100} = \sum_{k=0}^{100} x^k$

75. $\sqrt{2}, \sqrt{2\sqrt{2}}, \sqrt{2\sqrt{2\sqrt{2}}}, \sqrt{2\sqrt{2\sqrt{2\sqrt{2}}}}, \ldots$. We simplify each term in an attempt to determine a formula for a_n. So $a_1 = 2^{1/2}$, $a_2 = \sqrt{2 \cdot 2^{1/2}} = \sqrt{2^{3/2}} = 2^{3/4}$, $a_3 = \sqrt{2 \cdot 2^{3/4}} = \sqrt{2^{7/4}} = 2^{7/8}$, $a_4 = \sqrt{2 \cdot 2^{7/8}} = \sqrt{2^{15/8}} = 2^{15/16}, \ldots$. Thus $a_n = 2^{(2^n-1)/2^n}$.

77. **(a)** $A_1 = \$2004$, $A_2 = \$2008.01$, $A_3 = \$2012.02$, $A_4 = \$2016.05$, $A_5 = \$2020.08$, $A_6 = \$2024.12$

(b) Since 3 years is 36 months, we get $A_{36} = \$2149.16$.

79. **(a)** $P_1 = 35{,}700$, $P_2 = 36{,}414$, $P_3 = 37{,}142$, $P_4 = 37{,}885$, $P_5 = 38{,}643$

(b) Since 2014 is 10 years after 2004, $P_{10} = 42{,}665$.

81. **(a)** The number of catfish at the end of the month, P_n, is the population at the start of the month, P_{n-1}, plus the increase in population, $0.08P_{n-1}$, minus the 300 catfish harvested. Thus $P_n = P_{n-1} + 0.08P_{n-1} - 300 \Leftrightarrow P_n = 1.08P_{n-1} - 300$.

(b) $P_1 = 5100$, $P_2 = 5208$, $P_3 = 5325$, $P_4 = 5451$, $P_5 = 5587$, $P_6 = 5734$, $P_7 = 5892$, $P_8 = 6064$, $P_9 = 6249$, $P_{10} = 6449$, $P_{11} = 6665$, $P_{12} = 6898$. Thus there should be 6898 catfish in the pond at the end of 12 months.

83. **(a)** Let S_n be his salary in the nth year. Then $S_1 = \$30{,}000$. Since his salary increase by 2000 each year, $S_n = S_{n-1} + 2000$. Thus $S_1 = \$30{,}000$ and $S_n = S_{n-1} + 2000$.

(b) $S_5 = S_4 + 2000 = (S_3 + 2000) + 2000 = (S_2 + 2000) + 4000 = (S_1 + 2000) + 6000 = \$38{,}000$.

85. Let F_n be the number of pairs of rabbits in the nth month. Clearly $F_1 = F_2 = 1$. In the nth month each pair that is two or more months old (that is, F_{n-2} pairs) will add a pair of offspring to the F_{n-1} pairs already present. Thus $F_n = F_{n-1} + F_{n-2}$. So F_n is the Fibonacci sequence.

87. $a_{n+1} = \begin{cases} \dfrac{a_n}{2} & \text{if } a_n \text{ is even} \\ 3a_n + 1 & \text{if } a_n \text{ is odd} \end{cases}$ With $a_1 = 11$, we have $a_2 = 34$, $a_3 = 17$, $a_4 = 52$, $a_5 = 26$, $a_6 = 13$, $a_7 = 40$, $a_8 = 20$, $a_9 = 10$, $a_{10} = 5$, $a_{11} = 16$, $a_{12} = 8$, $a_{13} = 4$, $a_{14} = 2$, $a_{15} = 1$, $a_{16} = 4$, $a_{17} = 2$, $a_{18} = 1, \ldots$ (with 4, 2, 1 repeating). So $a_{3n+1} = 4$, $a_{3n+2} = 2$, and $a_{3n} = 1$, for $n \geq 5$. With $a_1 = 25$, we have $a_2 = 76$, $a_3 = 38$, $a_4 = 19$, $a_5 = 58$, $a_6 = 29$, $a_7 = 88$, $a_8 = 44$, $a_9 = 22$, $a_{10} = 11$, $a_{11} = 34$, $a_{12} = 17$, $a_{13} = 52$, $a_{14} = 26$, $a_{15} = 13$, $a_{16} = 40$, $a_{17} = 20$, $a_{18} = 10$, $a_{19} = 5$, $a_{20} = 16$, $a_{21} = 8$, $a_{22} = 4$, $a_{23} = 2$, $a_{24} = 1$, $a_{25} = 4$, $a_{26} = 2$, $a_{27} = 1, \ldots$ (with 4, 2, 1 repeating). So $a_{3n+1} = 4$, $a_{3n+2} = 2$, and $a_{3n+3} = 1$ for $n \geq 7$.

We conjecture that the sequence will always return to the numbers 4, 2, 1 repeating.

13.2 ARITHMETIC SEQUENCES

1. An arithmetic sequence is sequence where the *difference* between successive terms is constant.

3. True. The nth partial sum of an arithmetic sequence is the average of the first and last terms times n.

5. **(a)** $a_1 = 7 + 3(1 - 1) = 7$, $a_2 = 7 + 3(2 - 1) = 10$, $a_3 = 7 + 3(3 - 1) = 13$, $a_4 = 7 + 3(4 - 1) = 16$, $a_5 = 7 + 3(5 - 1) = 19$

(b) The common difference is 3.

(c)

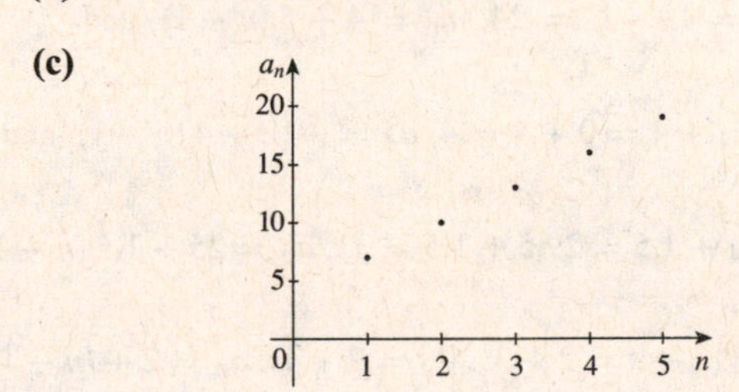

7. **(a)** $a_1 = -6 - 4(1 - 1) = -6$, $a_2 = -6 - 4(2 - 1) - 10$, $a_3 = -6 - 4(3 - 1) - 14$, $a_4 = -6 - 4(4 - 1) - 18$, $a_5 = -6 - 4(5 - 1) - 22$

(b) The common difference is -4.

(c)

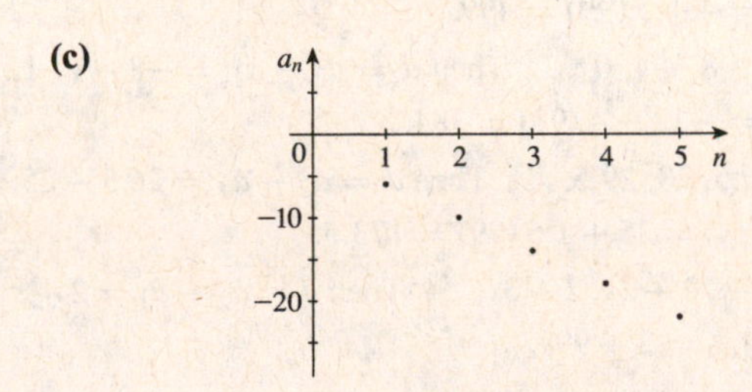

9. (a) $a_1 = \frac{5}{2} - (1-1) = \frac{5}{2}, a_2 = \frac{5}{2} - (2-1) = \frac{3}{2},$
$a_3 = \frac{5}{2} - (3-1) = \frac{1}{2}, a_4 = \frac{5}{2} - (4-1) = -\frac{1}{2},$
$a_5 = \frac{5}{2} - (5-1) = -\frac{3}{2}$

(b) The common difference is -1.

(c)

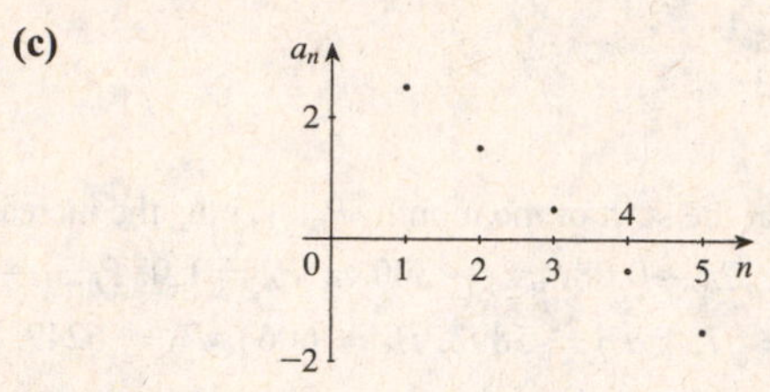

11. $a = 9, d = 4, a_n = a + d(n-1) = 9 + 4(n-1)$. So $a_{10} = 9 + 4(10-1) = 45$.

13. $a = -0.7, d = -0.2, a_n = a + d(n-1) = -0.7 - 0.2(n-1)$. So $a_{10} = -0.7 - 0.2(10-1) = -2.5$.

15. $a = \frac{5}{2}, d = -\frac{1}{2}, a_n = a + d(n-1) = \frac{5}{2} - \frac{1}{2}(n-1)$. So $a_{10} = \frac{5}{2} - \frac{1}{2}(10-1) = -2$.

17. $a_4 - a_3 = a_3 - a_2 = a_2 - a_1 = 6$. The sequence is arithmetic with common difference 6.

19. Since $a_3 - a_2 = -7$ and $a_4 - a_3 = 6$, the terms of the sequence do not have a common difference. This sequence is not arithmetic.

21. Since $a_2 - a_1 = 4 - 2 = 2$ and $a_4 - a_3 = 16 - 8 = 8$, the terms of the sequence do not have a common difference. This sequence is not arithmetic.

23. $a_4 - a_3 = -\frac{3}{2} - 0 = -\frac{3}{2}, a_3 - a_2 = 0 - \dfrac{3}{2} = -\frac{3}{2}, a_2 - a_1 = \frac{3}{2} - 3 = -\frac{3}{2}$. This sequence is arithmetic with common difference $-\frac{3}{2}$.

25. $a_4 - a_3 = 7.7 - 6.0 = 1.7, a_3 - a_2 = 6.0 - 4.3 = 1.7, 4. - a_1 = 4.3 - 2.6 = 1.7$. This sequence is arithmetic with common difference 1.7.

27. $a_1 = 4 + 7(1) = 11, a_2 = 4 + 7(2) = 18, a_3 = 4 + 7(3) = 25, a_4 = 4 + 7(4) = 32, a_5 = 4 + 7(5) = 39$. This sequence is arithmetic, the common difference is $d = 7$ and $a_n = 4 + 7n = 4 + 7n - 7 + 7 = 11 + 7(n-1)$.

29. $a_1 = \dfrac{1}{1+2(1)} = \dfrac{1}{3}, a_2 = \dfrac{1}{1+2(2)} = \dfrac{1}{5}, a_3 = \dfrac{1}{1+2(3)} = \dfrac{1}{7}, a_4 = \dfrac{1}{1+2(4)} = \dfrac{1}{9}, a_5 = \dfrac{1}{1+2(5)} = \dfrac{1}{11}$. Since $a_4 - a_3 = \frac{1}{9} - \frac{1}{7} = -\frac{2}{63}$ and $a_3 - a_2 = \frac{1}{7} - \frac{1}{3} = -\frac{2}{21}$, the terms of the sequence do not have a common difference. This sequence is not arithmetic.

31. $a_1 = 6(1) - 10 = -4, a_2 = 6(2) - 10 = 2, a_3 = 6(3) - 10 = 8, a_4 = 6(4) - 10 = 14, a_5 = 6(5) - 10 = 20$. This sequence is arithmetic, the common difference is $d = 6$ and $a_n = 6n - 10 = 6n - 6 + 6 - 10 = -4 + 6(n-1)$.

33. 4, 10, 16, 22, ... Then $d = a_2 - a_1 = 10 - 4 = 6, a_5 = a_4 + 6 = 22 + 6 = 28, a_n = 4 + 6(n-1)$, and $a_{100} = 4 + 6(99) = 598$.

35. $29, 11, -7, -25, \ldots$. Then $d = a_2 - a_1 = 11 - 29 = -18, a_5 = a_4 - 18 = -25 - 18 = -43, a_n = 29 - 18(n-1)$, and $a_{100} = 29 - 18(99) = -1753$.

37. 4, 9, 14, 19, ... Then $d = a_2 - a_1 = 9 - 4 = 5, a_5 = a_4 + 5 = 19 + 5 = 24, a_n = 4 + 5(n-1)$, and $a_{100} = 4 + 5(99) = 499$.

39. $-12, -8, -4, 0, \ldots$. Then $d = a_2 - a_1 = -8 - (-12) = 4, a_5 = a_4 + 4 = 0 + 4 = 4, a_n = -12 + 4(n-1)$, and $a_{100} = -12 + 4(99) = 384$.

41. 25, 26.5, 28, 29.5, ... Then $d = a_2 - a_1 = 26.5 - 25 = 1.5, a_5 = a_4 + 1.5 = 29.5 + 1.5 = 31, a_n = 25 + 1.5(n-1)$, and $a_{100} = 25 + 1.5(99) = 173.5$.

43. $2, 2+s, 2+2s, 2+3s, \ldots$. Then $d = a_2 - a_1 = 2 + s - 2 = s, a_5 = a_4 + s = 2 + 3s + s = 2 + 4s, a_n = 2 + (n-1)s$, and $a_{100} = 2 + 99s$.

45. $a_{50} = 1000$ and $d = 6$. Thus, $a_{50} = a_1 + d(50-1) \Leftrightarrow 1000 = a_1 + 6(50-1) \Leftrightarrow a_1 = 1000 - 294 = 706$ and $a_2 = 706 + 6 = 712$.

47. $a_{14} = \frac{2}{3}$ and $a_9 = \frac{1}{4}$, so $\frac{2}{3} = a_1 + 13d$ and $\frac{1}{4} = a_1 + 8d \Leftrightarrow \frac{2}{3} - \frac{1}{4} = 5d \Leftrightarrow d = \frac{1}{12}$. Thus, $a_1 = \frac{1}{4} - 8\left(\frac{1}{12}\right) = -\frac{5}{12}$ and $a_n = -\frac{5}{12} + \frac{1}{12}(n-1)$.

49. $a_1 = 25$ and $d = 18$, so $a_n = 601 \Leftrightarrow 25 + 18(n-1) = 601 \Leftrightarrow 33$. Thus, 601 is the 33rd term of the sequence.

51. $a = 3, d = 5, n = 20$. Then $S_{20} = \frac{20}{2}(2 \cdot 3 + 19 \cdot 5) = 1010$.

53. $a = -40, d = 14, n = 15$. Then $S_{15} = \frac{15}{2}[2(-40) + 14 \cdot 14] = 870$.

55. $a_1 = 55, d = 12, n = 10$. Then $S_{10} = \frac{10}{2}(2 \cdot 55 + 9 \cdot 12) = 1090$.

57. $1 + 5 + 9 + \cdots + 401$ is a partial sum of an arithmetic series with $a = 1$ and $d = 5 - 1 = 4$. The last term is $401 = a_n = 1 + 4(n-1)$, so $n - 1 = 100 \Leftrightarrow n = 101$. So the partial sum is $S_{101} = \frac{101}{2}(1 + 401) = 101 \cdot 201 = 20{,}301$.

59. $250 + 233 + 216 + \cdots + 97$ is a partial sum of an arithmetic sequence with $a = 250$ and $d = 233 - 250 = -17$. The last term is $97 = a_n = 250 - 17(n-1)$, so $n - 1 = \frac{97-250}{-17} = 9 \Leftrightarrow n = 10$. So the partial sum is $S_{10} = \frac{10}{2}(250 + 97) = 1735$.

61. $0.7 + 2.7 + 4.7 + \cdots + 56.7$ is a partial sum of an arithmetic sequence with $a = 0.7$ and $d = 2.7 - 0.7 = 2$. The last term is $56.7 = a_n = 0.7 + 2(n-1) \Leftrightarrow 28 = n - 1 \Leftrightarrow n = 29$. So the partial sum is $S_{29} = \frac{29}{2}(0.7 + 56.7) = 832.3$.

63. $\sum_{k=0}^{10}(3 + 0.25k)$ is a partial sum of an arithmetic sequence with $a = 3 + 0.25 \cdot 0 = 3$ and $d = 0.25$. The last term is $a_{11} = 3 + 0.25 \cdot 10 = 5.5$. So the partial sum is $S_{11} = \frac{11}{2}(3 + 5.5) = 46.75$.

65. We have an arithmetic sequence with $a = 5$ and $d = 2$. We seek n such that $2700 = S_n = \dfrac{n}{2}[2a + (n-1)d]$. Solving for n, we have $2700 = \dfrac{n}{2}[10 + 2(n-1)] \Leftrightarrow 5400 = 10n + 2n^2 - 2n \Leftrightarrow n^2 + 4n - 2700 = 0 \Leftrightarrow (n-50)(n+54) = 0 \Leftrightarrow$ $n = 50$ or $n = -54$. Since n is a positive integer, 50 terms of the sequence must be added to get 2700.

67. Let x denote the length of the side between the length of the other two sides. Then the lengths of the three sides of the triangle are $x - a$, x, and $x + a$, for some $a > 0$. Since $x + a$ is the longest side, it is the hypotenuse, and by the Pythagorean Theorem, we know that $(x-a)^2 + x^2 = (x+a)^2 \Leftrightarrow x^2 - 2ax + a^2 + x^2 = x^2 + 2ax + a^2 \Leftrightarrow x^2 - 4ax = 0 \Leftrightarrow$ $x(x - 4a) = 0 \Rightarrow x = 4a$ ($x = 0$ is not a possible solution). Thus, the lengths of the three sides are $x - a = 4a - a = 3a$, $x = 4a$, and $x + a = 4a + a = 5a$. The lengths $3a, 4a, 5a$ are proportional to 3, 4, 5, and so the triangle is similar to a 3-4-5 triangle.

69. The sequence $1, \frac{3}{5}, \frac{3}{7}, \frac{1}{3}, \ldots$ is harmonic if $1, \frac{5}{3}, \frac{7}{3}, 3, \ldots$ forms an arithmetic sequence. Since $\frac{5}{3} - 1 = \frac{7}{3} - \frac{5}{3} = 3 - \frac{7}{3} = \frac{2}{3}$, the sequence of reciprocals is arithmetic and thus the original sequence is harmonic.

71. The diminishing values of the computer form an arithmetic sequence with $a_1 = 12{,}500$ and common difference $d = -1875$. Thus the value of the computer after 6 years is $a_7 = 12{,}500 + (7-1)(-1875) = \$1250$.

73. The increasing values of the man's salary form an arithmetic sequence with $a_1 = 30{,}000$ and common difference $d = 2300$. Then his total earnings for a ten-year period are $S_{10} = \frac{10}{2}[2(30{,}000) + 9(2300)] = 403{,}500$. Thus his total earnings for the 10 year period are \$403,500.

75. The number of seats in the nth row is given by the nth term of an arithmetic sequence with $a_1 = 15$ and common difference $d = 3$. We need to find n such that $S_n = 870$. So we solve $870 = S_n = \dfrac{n}{2}[2(15) + (n-1)3]$ for n. We have $870 = \dfrac{n}{2}(27 + 3n) \Leftrightarrow 1740 = 3n^2 + 27n \Leftrightarrow 3n^2 + 27n - 1740 = 0 \Leftrightarrow n^2 + 9n - 580 = 0 \Leftrightarrow (x-20)(x+29) = 0 \Rightarrow$ $n = 20$ or $n = -29$. Since the number of rows is positive, the theater must have 20 rows.

77. The number of gifts on the 12th day is $1 + 2 + 3 + 4 + \cdots + 12$. Since $a_2 - a_1 = a_3 - a_2 = a_4 - a_3 = \cdots = 1$, the number of gifts on the 12th day is the partial sum of an arithmetic sequence with $a = 1$ and $d = 1$. So the sum is $S_{12} = 12\left(\dfrac{1+12}{2}\right) = 6 \cdot 13 = 78$.

13.3 GEOMETRIC SEQUENCES

1. A geometric sequence is a sequence where the *ratio* between successive terms is constant.

3. True. If we know the first and second terms of a geometric sequence then we can find all other terms.

5. (a) $a_1 = 7(3)^0 = 7, a_2 = 7(3)^1 = 21,$

$a_3 = 7(3)^2 = 63, a_4 = 7(3)^3 = 189,$

$a_5 = 7(3)^4 = 567$

(b) The common ratio is 3.

(c)

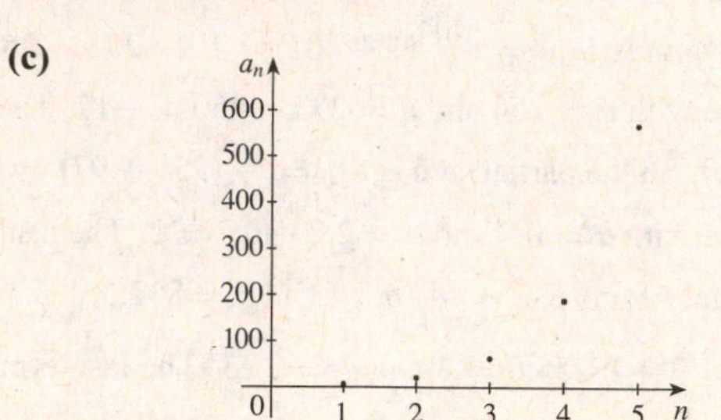

7. (a) $a_1 = \frac{5}{2}\left(-\frac{1}{2}\right)^0 = \frac{5}{2}, a_2 = \frac{5}{2}\left(-\frac{1}{2}\right)^1 = -\frac{5}{4},$

$a_3 = \frac{5}{2}\left(-\frac{1}{2}\right)^2 = \frac{5}{8}, a_4 = \frac{5}{2}\left(-\frac{1}{2}\right)^3 = -\frac{5}{16},$

$a_5 = \frac{5}{2}\left(-\frac{1}{2}\right)^4 = \frac{5}{32}$

(b) The common ratio is $-\frac{1}{2}$.

(c)

9. $a = 7, r = 4$. So $a_n = ar^{n-1} = 7(4)^{n-1}$ and $a_4 = 7 \cdot 4^3 = 448$.

11. $a = \frac{5}{2}, r = -\frac{1}{2}$. So $a_n = ar^{n-1} = \frac{5}{2}\left(-\frac{1}{2}\right)^{n-1}$ and $a_4 = \frac{5}{2} \cdot \left(-\frac{1}{2}\right)^3 = -\frac{5}{16}$.

13. $\frac{a_2}{a_1} = \frac{6}{3} = 2, \frac{a_3}{a_2} = \frac{12}{6} = 2$, and $\frac{a_4}{a_3} = \frac{24}{12} = 2$. Since these ratios are the same, the sequence is geometric with the common ratio 2.

15. $\frac{a_2}{a_1} = \frac{1536}{3072} = \frac{1}{2}, \frac{a_3}{a_2} = \frac{768}{1536} = \frac{1}{2}$, and $\frac{a_4}{a_3} = \frac{384}{768} = \frac{1}{2}$. Since these ratios are the same, the sequence is geometric with the common ratio $\frac{1}{2}$.

17. $\frac{a_2}{a_1} = \frac{3/2}{3} = \frac{1}{2}, \frac{a_3}{a_2} = \frac{3/4}{3/2} = \frac{1}{2}$, and $\frac{a_4}{a_3} = \frac{3/8}{3/4} = \frac{1}{2}$. Since these ratios are the same, the sequence is geometric with the common ratio $\frac{1}{2}$.

19. $\frac{a_2}{a_1} = \frac{1/3}{1/2} = \frac{2}{3}$ and $\frac{a_4}{a_3} = \frac{1/5}{1/4} = \frac{4}{5}$. Since these ratios are not the same, this is not a geometric sequence.

21. $\frac{a_2}{a_1} = \frac{1.1}{1.0} = 1.1, \frac{a_3}{a_2} = \frac{1.21}{1.1} = 1.1$, and $\frac{a_4}{a_3} = \frac{1.331}{1.21} = 1.1$. Since these ratios are the same, the sequence is geometric with the common ratio 1.1.

23. $a_1 = 2(3)^1 = 6, a_2 = 2(3)^2 = 18, a_3 = 2(3)^3 = 54, a_4 = 2(3)^4 = 162$, and $a_5 = 2(3)^5 = 486$. This sequence is geometric, the common ratio is $r = 3$, and $a_n = a_1 r^{n-1} = 6(3)^{n-1}$.

25. $a_1 = \frac{1}{4}, a_2 = \frac{1}{4^2} = \frac{1}{16}, a_3 = \frac{1}{4^3} = \frac{1}{64}, a_4 = \frac{1}{4^4} = \frac{1}{256}$, and $a_5 = \frac{1}{4^5} = \frac{1}{1024}$. This sequence is geometric, the common ratio is $r = \frac{1}{4}$ and $a_n = a_1 r^{n-1} = \frac{1}{4}\left(\frac{1}{4}\right)^{n-1}$.

27. Since $\ln a^b = b \ln a$, we have $a_1 = \ln\left(5^0\right) = \ln 1 = 0, a_2 = \ln\left(5^1\right) = \ln 5, a_3 = \ln\left(5^2\right) = 2\ln 5, a_4 = \ln\left(5^3\right) = 3\ln 5,$ $a_5 = \ln\left(5^4\right) = 4\ln 5$. Since $a_1 = 0$ and $a_2 \neq 0$, this sequence is not geometric.

29. 2, 6, 18, 54, ... Then $r = \frac{a_2}{a_1} = \frac{6}{2} = 3, a_5 = a_4 \cdot 3 = 54(3) = 162$, and $a_n = 2 \cdot 3^{n-1}$.

31. $0.3, -0.09, 0.027, -0.0081, \ldots$. Then $r = \dfrac{a_2}{a_1} = \dfrac{-0.09}{0.3} = -0.3$, $a_5 = a_4 \cdot (-0.3) = -0.0081\,(-0.3) = 0.00243$, and $a_n = 0.3\,(-0.3)^{n-1}$.

33. $144, -12, 1, -\frac{1}{12}, \ldots$. Then $r = \dfrac{a_2}{a_1} = \dfrac{-12}{144} = -\frac{1}{12}$, $a_5 = a_4 \cdot \left(-\frac{1}{12}\right) = -\frac{1}{12}\left(-\frac{1}{12}\right) = \frac{1}{144}$, $a_n = 144\left(-\frac{1}{12}\right)^{n-1}$.

35. $3, 3^{5/3}, 3^{7/3}, 27, \ldots$. Then $r = \dfrac{a_2}{a_1} = \dfrac{3^{5/3}}{3} = 3^{2/3}$, $a_5 = a_4 \cdot \left(3^{2/3}\right) = 27 \cdot 3^{2/3} = 3^{11/3}$, and $a_n = 3\left(3^{2/3}\right)^{n-1} = 3 \cdot 3^{(2n-2)/3} = 3^{(2n+1)/3}$.

37. $1, s^{2/7}, s^{4/7}, s^{6/7}, \ldots$. Then $r = \dfrac{a_2}{a_1} = \dfrac{s^{2/7}}{1} = s^{2/7}$, $a_5 = a_4 \cdot s^{2/7} = s^{6/7} \cdot s^{2/7} = s^{8/7}$, and $a_n = \left(s^{2/7}\right)^{n-1} = s^{(2n-2)/7}$.

39. $a_1 = 15$, $a_2 = 6$. Thus $r = \dfrac{a_2}{a_1} = \dfrac{6}{15} = \dfrac{2}{5}$ and $a_4 = a_1 r^{4-1} = 15\left(\frac{2}{5}\right)^3 = \frac{24}{25}$.

41. $a_3 = -\dfrac{1}{3}$ and $a_6 = 9$. Thus, $\dfrac{a_6}{a_3} = \dfrac{a_1 r^5}{a_1 r^2} = r^3 \Leftrightarrow r^3 = \dfrac{9}{-1/3} = -27$, so $r = -3$. Therefore, $a_1 (-3)^2 = -\frac{1}{3} \Leftrightarrow a_1 = -\frac{1}{27}$ and $a_2 = -\frac{1}{27}(-3) = \frac{1}{9}$.

43. $a_3 = -18$ and $a_6 = 9216$. Thus, $r^3 = \dfrac{a_6}{a_3} = \dfrac{9216}{-18} = -512 \Leftrightarrow r = -8$. Therefore, $a_1 = \dfrac{a_3}{r^2} = \dfrac{-18}{64} = -\dfrac{9}{32}$ and $a_n = -\dfrac{9}{32}(-8)^{n-1}$.

45. $r = 0.75$ and $a_4 = 729$, so $a_1 = \dfrac{a_4}{r^3} = \dfrac{729}{(0.75)^3} = 1728$, $a_2 = 1728\,(0.75) = 1296$, and $a_3 = 1296\,(0.75) = 972$.

47. $a = 1536$ and $r = \frac{1}{2}$, so $a_n = ar^{n-1} \Leftrightarrow 6 = 1536\left(\frac{1}{2}\right)^{n-1} = 1536\,(2)^{1-n} \Leftrightarrow \log_2 6 = \log_2 1536 + 1 - n \Leftrightarrow n = 1 + \log_2 \frac{1536}{6} = 1 + \log_2 256 = 9$. Thus, 6 is the ninth term.

49. $a = 5$, $r = 2$, $n = 6$. Then $S_6 = 5\dfrac{1 - 2^6}{1 - 2} = (-5)(-63) = 315$.

51. $a_3 = 28$, $a_6 = 224$, $n = 6$. So $\dfrac{a_6}{a_3} = \dfrac{ar^5}{ar^2} = r^3$. So we have $r^3 = \dfrac{a_6}{a_3} = \frac{224}{28} = 8$, and hence $r = 2$. Since $a_3 = a \cdot r^2$, we get $a = \dfrac{a_3}{r^2} = \dfrac{28}{2^2} = 7$. So $S_6 = 7\dfrac{1 - 2^6}{1 - 2} = (-7)(-63) = 441$.

53. $1 + 3 + 9 + \cdots + 2187$ is a partial sum of a geometric sequence, where $a = 1$ and $r = \dfrac{a_2}{a_1} = \frac{3}{1} = 3$. Then the last term is $2187 = a_n = 1 \cdot 3^{n-1} \Leftrightarrow n - 1 = \log_3 2187 = 7 \Leftrightarrow n = 8$. So the partial sum is $S_8 = (1)\dfrac{1 - 3^8}{1 - 3} = 3280$.

55. $-15 + 30 - 60 + \cdots - 960$ is a partial sum of a geometric sequence for which $a = -15$ and $r = \dfrac{a_2}{a_1} = \dfrac{30}{-15} = -2$. The last term is $a_n = -960 = -15\,(-2)^{n-1} \Leftrightarrow n = 7$, so the partial sum is $S_7 = -15\dfrac{1 - (-2)^7}{1 - (-2)} = -645$.

57. $1.25 + 12.5 + 125 + \cdots + 12{,}500{,}000$ is a partial sum of a geometric sequence for which $a = 1.25$ and $r = \dfrac{a_2}{a_1} = 10$. The last term is $a_n = 12{,}500{,}000 = 1.25 \cdot 10^{n-1} \Leftrightarrow n = 8$, so the partial sum is $S_8 = 1.25\dfrac{1 - 10^8}{1 - 10} = 13{,}888{,}888.75$.

59. $\displaystyle\sum_{k=1}^{5} 3\left(\tfrac{1}{2}\right)^{k-1} = 3\frac{1 - \left(\frac{1}{2}\right)^5}{1 - \frac{1}{2}} = \frac{93}{16}$

61. $\displaystyle\sum_{k=1}^{6} 5\,(-2)^{k-1} = 5\frac{1 - (-2)^6}{1 - (-2)} = -105$

63. $\sum_{k=1}^{5} 3\left(\frac{2}{3}\right)^{k-1} = 3\frac{1-\left(\frac{2}{3}\right)^5}{1-\frac{2}{3}} = \frac{211}{27}$

65. $1 + \frac{1}{3} + \frac{1}{9} + \frac{1}{27} + \cdots$ is an infinite geometric series with $a = 1$ and $r = \frac{1}{3}$. Therefore, it is convergent with sum $S = \frac{a}{1-r} = \frac{1}{1-\left(\frac{1}{3}\right)} = \frac{3}{2}$.

67. $1 - \frac{1}{3} + \frac{1}{9} - \frac{1}{27} + \cdots$ is an infinite geometric series with $a = 1$ and $r = -\frac{1}{3}$. Therefore, it is convergent with sum $S = \frac{a}{1-r} = \frac{1}{1-\left(-\frac{1}{3}\right)} = \frac{3}{4}$.

69. $1 + \frac{3}{2} + \left(\frac{3}{2}\right)^2 + \left(\frac{3}{2}\right)^3 + \cdots$ is an infinite geometric series with $a = 1$ and $r = \frac{3}{2} > 1$. Therefore, the series diverges.

71. $3 - \frac{3}{2} + \frac{3}{4} - \frac{3}{8} + \cdots$ is an infinite geometric series with $a = 3$ and $r = -\frac{1}{2}$. Therefore, it is convergent with sum $S = \frac{3}{1-\left(-\frac{1}{2}\right)} = 2$.

73. $3 - 3(1.1) + 3(1.1)^2 - 3(1.1)^3 + \cdots$ is an infinite geometric series with $a = 3$ and $r = 1.1 > 1$. Therefore, the series diverges.

75. $\frac{1}{\sqrt{2}} + \frac{1}{2} + \frac{1}{2\sqrt{2}} + \frac{1}{4} + \cdots$ is an infinite geometric series with $a = \frac{1}{\sqrt{2}}$ and $r = \frac{1}{\sqrt{2}}$. Therefore, the sum of the series is $S = \frac{\frac{1}{\sqrt{2}}}{1-\frac{1}{\sqrt{2}}} = \frac{1}{\sqrt{2}-1} = \sqrt{2}+1$.

77. $0.777\ldots = \frac{7}{10} + \frac{7}{100} + \frac{7}{1000} + \cdots$ is an infinite geometric series with $a = \frac{7}{10}$ and $r = \frac{1}{10}$. Thus $0.777\ldots = \frac{a}{1-r} = \frac{\frac{7}{10}}{1-\frac{1}{10}} = \frac{7}{9}$.

79. $0.030303\ldots = \frac{3}{100} + \frac{3}{10{,}000} + \frac{3}{1{,}000{,}000} + \cdots$ is an infinite geometric series with $a = \frac{3}{100}$ and $r = \frac{1}{100}$. Thus $0.030303\ldots = \frac{a}{1-r} = \frac{\frac{3}{100}}{1-\frac{1}{100}} = \frac{3}{99} = \frac{1}{33}$.

81. $0.\overline{112} = 0.112112112\ldots = \frac{112}{1000} + \frac{112}{1{,}000{,}000} + \frac{112}{1{,}000{,}000{,}000} + \cdots$ is an infinite geometric series with $a = \frac{112}{1000}$ and $r = \frac{1}{1000}$. Thus $0.112112112\ldots = \frac{a}{1-r} = \frac{\frac{112}{1000}}{1-\frac{1}{1000}} = \frac{112}{999}$.

83. Since we have 5 terms, let us denote $a_1 = 5$ and $a_5 = 80$. Also, $\frac{a_5}{a_1} = r^4$ because the sequence is geometric, and so $r^4 = \frac{80}{5} = 16 \Leftrightarrow r = \pm 2$. If $r = 2$, the three geometric means are $a_2 = 10$, $a_3 = 20$, and $a_4 = 40$. (If $r = -2$, the three geometric means are $a_2 = -10$, $a_3 = 20$, and $a_4 = -40$, but these are not between 5 and 80.)

85. **(a)** $5, -3, 5, -3, \ldots$ is neither arithmetic nor geometric because $a_2 - a_1 \neq a_3 - a_2$ and $\frac{a_2}{a_1} \neq \frac{a_3}{a_2}$.

(b) $\frac{1}{3}, 1, \frac{5}{3}, \frac{7}{3}, \ldots$ is arithmetic with $d = \frac{2}{3}$, so the next term is $a_5 = \frac{7}{3} + \frac{2}{3} = 3$.

(c) $\sqrt{3}, 3, 3\sqrt{3}, 9, \ldots$ is geometric with $r = \sqrt{3}$, so the next term is $a_5 = 9\sqrt{3}$.

(d) $-3, -\frac{3}{2}, 0, \frac{3}{2}, \ldots$ is arithmetic with $d = \frac{3}{2}$, so the next term is $a_5 = \frac{3}{2} + \frac{3}{2} = 3$.

87. **(a)** The value at the end of the year is equal to the value at beginning less the depreciation, so $V_n = V_{n-1} - 0.2V_{n-1} = 0.8V_{n-1}$ with $V_1 = 160{,}000$. Thus $V_n = 160{,}000 \cdot 0.8^{n-1}$.

(b) $V_n < 100{,}000 \Leftrightarrow 0.8^{n-1} \cdot 160{,}000 < 100{,}000 \Leftrightarrow 0.8^{n-1} < 0.625 \Leftrightarrow (n-1)\log 0.8 < \log 0.625 \Leftrightarrow$

$n - 1 > \dfrac{\log 0.625}{\log 0.8} = 2.11$. Thus it will depreciate to below \$100,000 during the fourth year.

89. Since the ball is dropped from a height of 80 feet, $a = 80$. Also since the ball rebounds three-fourths of the distance fallen, $r = \frac{3}{4}$. So on the nth bounce, the ball attains a height of $a_n = 80\left(\frac{3}{4}\right)^n$. Hence, on the fifth bounce, the ball goes $a_5 = 80\left(\frac{3}{4}\right)^5 = \frac{80 \cdot 243}{1024} \approx 19$ ft high.

91. Let a_n be the amount of water remaining at the nth stage. We start with 5 gallons, so $a = 5$. When 1 gallon (that is, $\frac{1}{5}$ of the mixture) is removed, $\frac{4}{5}$ of the mixture (and hence $\frac{4}{5}$ of the water in the mixture) remains. Thus, $a_1 = 5 \cdot \frac{4}{5}$, $a_2 = 5 \cdot \frac{4}{5} \cdot \frac{4}{5}$, ..., and in general, $a_n = 5\left(\frac{4}{5}\right)^n$. The amount of water remaining after 3 repetitions is $a_3 = 5\left(\frac{4}{5}\right)^3 = \frac{64}{25}$, and after 5 repetitions it is $a_5 = 5\left(\frac{4}{5}\right)^5 = \frac{1024}{625}$.

93. Let a_n be the height the ball reaches on the nth bounce. From the given information, a_n is the geometric sequence $a_n = 9 \cdot \left(\frac{1}{3}\right)^n$. (Notice that the ball hits the ground for the fifth time after the fourth bounce.)

(a) $a_0 = 9$, $a_1 = 9 \cdot \frac{1}{3} = 3$, $a_2 = 9 \cdot \left(\frac{1}{3}\right)^2 = 1$, $a_3 = 9 \cdot \left(\frac{1}{3}\right)^3 = \frac{1}{3}$, and $a_4 = 9 \cdot \left(\frac{1}{3}\right)^4 = \frac{1}{9}$. The total distance traveled is

$a_0 + 2a_1 + 2a_2 + 2a_3 + 2a_4 = 9 + 2 \cdot 3 + 2 \cdot 1 + 2 \cdot \frac{1}{3} + 2 \cdot \frac{1}{9} = \frac{161}{9} = 17\frac{8}{9}$ ft.

(b) The total distance traveled at the instant the ball hits the ground for the nth time is

$$\begin{aligned} D_n &= 9 + 2 \cdot 9 \cdot \tfrac{1}{3} + 2 \cdot 9 \cdot \left(\tfrac{1}{3}\right)^2 + 2 \cdot 9 \cdot \left(\tfrac{1}{3}\right)^3 + 2 \cdot 9 \cdot \left(\tfrac{1}{3}\right)^4 + \cdots + 2 \cdot 9 \cdot \left(\tfrac{1}{3}\right)^{n-1} \\ &= 2\left[9 + 9 \cdot \tfrac{1}{3} + 9 \cdot \left(\tfrac{1}{3}\right)^2 + 9 \cdot \left(\tfrac{1}{3}\right)^3 + 9 \cdot \left(\tfrac{1}{3}\right)^4 + \cdots + 9 \cdot \left(\tfrac{1}{3}\right)^{n-1}\right] - 9 \\ &= 2\left[9 \cdot \frac{1 - \left(\frac{1}{3}\right)^n}{1 - \frac{1}{3}}\right] - 9 = 27\left[1 - \left(\tfrac{1}{3}\right)^n\right] - 9 = 18 - \left(\tfrac{1}{3}\right)^{n-3} \end{aligned}$$

95. Let $a_1 = 1$ be the man with 7 wives. Also, let $a_2 = 7$ (the wives), $a_3 = 7a_2 = 7^2$ (the sacks), $a_4 = 7a_3 = 7^3$ (the cats), and $a_5 = 7a_4 = 7^4$ (the kits). The total is $a_1 + a_2 + a_3 + a_4 + a_5 = 1 + 7 + 7^2 + 7^3 + 7^4$, which is a partial sum of a geometric sequence with $a = 1$ and $r = 7$. Thus, the number in the party is $S_5 = 1 \cdot \dfrac{1 - 7^5}{1 - 7} = 2801$.

97. Let a_n be the height the ball reaches on the nth bounce. We have $a_0 = 1$ and $a_n = \frac{1}{2}a_{n-1}$. Since the total distance d traveled includes the bounce up as well and the distance down, we have

$$\begin{aligned} d &= a_0 + 2 \cdot a_1 + 2 \cdot a_2 + \cdots = 1 + 2\left(\tfrac{1}{2}\right) + 2\left(\tfrac{1}{2}\right)^2 + 2\left(\tfrac{1}{2}\right)^3 + 2\left(\tfrac{1}{2}\right)^4 + \cdots \\ &= 1 + 1 + \tfrac{1}{2} + \left(\tfrac{1}{2}\right)^2 + \left(\tfrac{1}{2}\right)^3 + \cdots = 1 + \sum_{i=0}^{\infty} \left(\tfrac{1}{2}\right)^i = 1 + \frac{1}{1 - \frac{1}{2}} = 3 \end{aligned}$$

Thus the total distance traveled is about 3 m.

99. (a) If a square has side x, then by the Pythagorean Theorem the length of the side of the square formed by joining the midpoints is, $\sqrt{\left(\frac{x}{2}\right)^2 + \left(\frac{x}{2}\right)^2} = \sqrt{\frac{x^2}{4} + \frac{x^2}{4}} = \frac{x}{\sqrt{2}}$. In our case, $x = 1$ and the side of the first inscribed square is $\frac{1}{\sqrt{2}}$, the side of the second inscribed square is $\frac{1}{\sqrt{2}} \cdot \frac{1}{\sqrt{2}} = \left(\frac{1}{\sqrt{2}}\right)^2$, the side of the third inscribed square is $\left(\frac{1}{\sqrt{2}}\right)^3$, and so on. Since this pattern continues, the total area of all the squares is $A = 1^2 + \left(\frac{1}{\sqrt{2}}\right)^2 + \left(\frac{1}{\sqrt{2}}\right)^4 + \left(\frac{1}{\sqrt{2}}\right)^6 + \cdots = 1 + \frac{1}{2} + \left(\frac{1}{2}\right)^2 + \left(\frac{1}{2}\right)^3 + \cdots = \frac{1}{1 - \frac{1}{2}} = 2$.

(b) As in part (a), the sides of the squares are $1, \frac{1}{\sqrt{2}}, \left(\frac{1}{\sqrt{2}}\right)^2, \left(\frac{1}{\sqrt{2}}\right)^3, \ldots$. Thus the sum of the perimeters is $S = 4 \cdot 1 + 4 \cdot \frac{1}{\sqrt{2}} + 4 \cdot \left(\frac{1}{\sqrt{2}}\right)^2 + 4 \cdot \left(\frac{1}{\sqrt{2}}\right)^3 + \cdots$, which is an infinite geometric series with $a = 4$ and $r = \frac{1}{\sqrt{2}}$. Thus the sum of the perimeters is $S = \frac{4}{1 - \frac{1}{\sqrt{2}}} = \frac{4\sqrt{2}}{\sqrt{2} - 1} = \frac{4\sqrt{2}}{\sqrt{2} - 1} \cdot \frac{\sqrt{2} + 1}{\sqrt{2} + 1} = \frac{4 \cdot 2 + 4\sqrt{2}}{2 - 1} = 8 + 4\sqrt{2}$.

101. Let a_n denote the area colored blue at nth stage. Since only the middle squares are colored blue, $a_n = \frac{1}{9} \times$ (area remaining yellow at the $(n-1)$ th stage). Also, the area remaining yellow at the nth stage is $\frac{8}{9}$ of the area remaining yellow at the preceding stage. So $a_1 = \frac{1}{9}$, $a_2 = \frac{1}{9}\left(\frac{8}{9}\right)$, $a_3 = \frac{1}{9}\left(\frac{8}{9}\right)^2$, $a_4 = \frac{1}{9}\left(\frac{8}{9}\right)^3$, ... Thus the total area colored blue $A = \frac{1}{9} + \frac{1}{9}\left(\frac{8}{9}\right) + \frac{1}{9}\left(\frac{8}{9}\right)^2 + \frac{1}{9}\left(\frac{8}{9}\right)^3 + \cdots$ is an infinite geometric series with $a = \frac{1}{9}$ and $r = \frac{8}{9}$. So the total area is $A = \frac{\frac{1}{9}}{1 - \frac{8}{9}} = 1$.

103. $a_1, a_2, a_3, \ldots$ is a geometric sequence with common ratio r. Thus $a_2 = a_1 r$, $a_3 = a_1 \cdot r^2, \ldots, a_n = a_1 \cdot r^{n-1}$. Hence $\log a_2 = \log(a_1 r) = \log a_1 + \log r$, $\log a_3 = \log\left(a_1 \cdot r^2\right) = \log a_1 + \log\left(r^2\right) = \log a_1 + 2\log r, \ldots$, $\log a_n = \log\left(a_1 \cdot r^{n-1}\right) = \log a_1 + \log\left(r^{n-1}\right) = \log a_1 + (n-1)\log r$, and so $\log a_1, \log a_2, \log a_3, \ldots$ is an arithmetic sequence with common difference $\log r$.

13.4 MATHEMATICS OF FINANCE

1. An annuity is a sum of money that is paid in regular equal payments. The *amount* of an annuity is the sum of all the individual payments together with all the interest.

3. $n = 10$, $R = \$1000$, $i = 0.06$. So $A_f = R\frac{(1+i)^n - 1}{i} = 1000\frac{(1+0.06)^{10} - 1}{0.06} = \$13{,}180.79$.

5. $n = 20$, $R = \$5000$, $i = 0.12$. So $A_f = R\frac{(1+i)^n - 1}{i} = 5000\frac{(1+0.12)^{20} - 1}{0.12} = \$360{,}262.21$.

7. $n = 16$, $R = \$300$, $i = \frac{0.08}{4} = 0.02$. So $A_f = R\frac{(1+i)^n - 1}{i} = 300\frac{(1+0.02)^{16} - 1}{0.02} = \$5{,}591.79$.

9. $A_f = 5000$, $n = 4 \cdot 2 = 8$, $i = \frac{0.10}{4} = 0.025$. So $R = \frac{iA_f}{(1+i)^n - 1} = \frac{(0.025)(5000)}{(1.025)^8 - 1} = \$572.34$.

11. $R = 1000$, $n = 20$, $i = \frac{0.09}{2} = 0.045$. So $A_p = R\frac{1 - (1+i)^{-n}}{i} = 1000\frac{1 - (1.045)^{-20}}{0.045} = \$13{,}007.94$.

13. $R = \$200$, $n = 20$, $i = \frac{0.09}{2} = 0.045$. So $A_p = R\frac{1 - (1+i)^{-n}}{i} = (200)\frac{1 - (1+0.045)^{-20}}{0.045} = \$2601.59$.

15. $A_P = \$12,000$, $i = \dfrac{0.105}{12} = 0.00875$, $n = 48$. Then $R = \dfrac{iA_P}{1-(1+i)^{-n}} = \dfrac{(0.00875)(12000)}{1-(1+0.00875)^{-48}} = \$307.24$.

17. $A_P = \$100{,}000$, $i = \dfrac{0.08}{12} \approx 0.006667$, $n = 360$. Then $R = \dfrac{iA_P}{1-(1+i)^{-n}} = \dfrac{(0.006667)(100{,}000)}{1-(1+0.006667)^{-360}} = \$733.76$.
Therefore, the total amount paid on this loan over the 30 year period is $(360)(733.76) = \$264{,}153.60$.

19. $R = 3500$, $n = 12(30) = 360$, $i = \dfrac{0.06}{12} = 0.005$. So $A_P = R\dfrac{1-(1+i)^{-n}}{i} = 3500\dfrac{1-(1.005)^{-360}}{0.005} = \$583{,}770.65$.
Therefore, Dr. Gupta can afford a loan of \$583,770.65.

21. $R = 220$, $n = 12(3) = 36$, $i = \dfrac{0.08}{12} \approx 0.00667$. The amount borrowed is $A_P = R\dfrac{1-(1+i)^{-n}}{i} = 220\dfrac{1-(1.00667)^{-36}}{0.00667} = \$7{,}020.60$. So she purchased the car for $\$7{,}020.60 + \$2000 = \$9020.60$.

23. $A_P = 100{,}000$, $n = 360$, $i = \dfrac{0.0975}{12} = 0.008125$.

(a) $R = \dfrac{iA_P}{1-(1+i)^{-n}} = \dfrac{(0.008125)(100{,}000)}{1-(1+0.008125)^{-360}} = \$859.15$.

(b) The total amount that will be paid over the 30 year period is $(360)(859.15) = \$309{,}294.00$.

(c) $R = \$859.15$, $i = \dfrac{0.0975}{12} = 0.008125$, $n = 360$. So $A_f = 859.15\dfrac{(1+0.008125)^{360}-1}{0.008125} = \$1{,}841{,}519.29$.

25. $A_P = \$640$, $R = \$32$, $n = 24$. We want to solve the equation $R = \dfrac{iA_P}{1-(1+i)^n}$ for the interest rate i. Let x be the interest rate, then $i = \dfrac{x}{12}$. So we can express R as a function of x by $R(x) = \dfrac{\frac{x}{12}\cdot 640}{1-\left(1+\frac{x}{12}\right)^{-24}}$. We graph $R(x)$ and $y = 32$ in the rectangle $[0.12, 0.22] \times [30, 34]$. The x-coordinate of the intersection is about 0.1816, which corresponds to an interest rate of 18.16%.

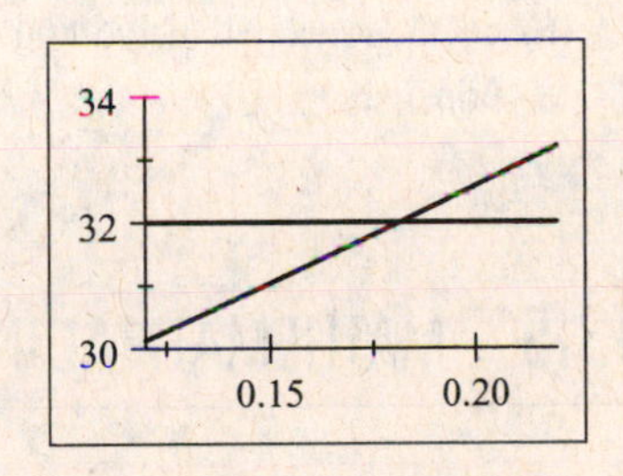

27. $A_P = \$189.99$, $R = \$10.50$, $n = 20$. We want to solve the equation $R = \dfrac{iA_P}{1-(1+i)^n}$ for the interest rate i. Let x be the interest rate, then $i = \dfrac{x}{12}$. So we can express R as a function of x by $R(x) = \dfrac{\frac{x}{12}\cdot 189.99}{1-\left(1+\frac{x}{12}\right)^{-20}}$. We graph $R(x)$ and $y = 10.50$ in the rectangle $[0.10, 0.18] \times [10, 11]$. The x-coordinate of the intersection is about 0.1168, which corresponds to an interest rate of 11.68%.

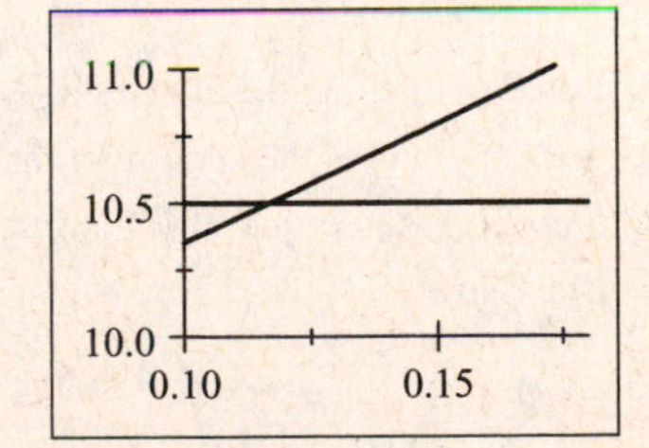

29. **(a)** The present value of the kth payment is $PV = R(1+i)^{-k} = \dfrac{R}{(1+i)^k}$. The present value of an annuity is the sum of the present values of each of the payments of R dollars, as shown in the time line.

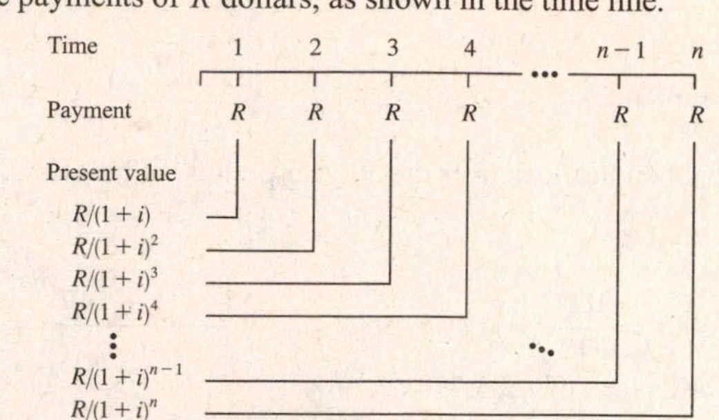

(b)
$$\begin{aligned} A_p &= \frac{R}{1+i} + \frac{R}{(1+i)^2} + \frac{R}{(1+i)^3} + \cdots + \frac{R}{(1+i)^n} \\ &= \frac{R}{1+i} + \left(\frac{R}{1+i}\right)\left(\frac{1}{1+i}\right) + \left(\frac{R}{1+i}\right)\left(\frac{1}{1+i}\right)^2 + \cdots + \left(\frac{R}{1+i}\right)\left(\frac{1}{1+i}\right)^{n-1} \end{aligned}$$

This is a geometric series with $a = \dfrac{R}{1+i}$ and $r = \dfrac{1}{1+i}$. Since $S_n = a\dfrac{1-r^n}{1-r}$, we have

$$A_p = \left(\frac{R}{1+i}\right)\frac{1-\left[\dfrac{1}{(1+i)}\right]^n}{1-\left(\dfrac{1}{1+i}\right)} = R\frac{1-(1+i)^{-n}}{(1+i)\left[1-\left(\dfrac{1}{1+i}\right)\right]} = R\frac{1-(1+i)^{-n}}{(1+i)-1} = R\frac{1-(1+i)^{-n}}{i}.$$

31. **(a)** Using the hint, we calculate the present value of the remaining 240 payments with $R = 724.17$, $i = 0.0075$, and $n = 240$. Since $A_p = R\dfrac{1-(1+i)^{-n}}{i} = (724.17)\dfrac{1-(1.0075)^{-240}}{0.0075} = 80{,}487.84$, they still owe \$80,487.84 on their mortgage.

(b) On their next payment, $0.0075\,(80{,}487.84) = \$603.66$ is interest and $\$724.17 - 603.66 = \$120.51$ goes toward the principal.

13.5 MATHEMATICAL INDUCTION

1. Mathematical induction is a method of proving that a statement $P(n)$ is true for all *natural* numbers n. In Step 1 we prove that $P(1)$ is true.

3. Let $P(n)$ denote the statement $2+4+6+\cdots+2n = n(n+1)$.
Step 1: $P(1)$ is the statement that $2 = 1(1+1)$, which is true.
Step 2: Assume that $P(k)$ is true; that is, $2+4+6+\cdots+2k = k(k+1)$. We want to use this to show that $P(k+1)$ is true. Now

$$\begin{aligned} 2+4+6+\cdots+2k+2(k+1) &= k(k+1)+2(k+1) && \text{induction hypothesis} \\ &= (k+1)(k+2) = (k+1)[(k+1)+1] \end{aligned}$$

Thus, $P(k+1)$ follows from $P(k)$. So by the Principle of Mathematical Induction, $P(n)$ is true for all n.

5. Let $P(n)$ denote the statement $5+8+11+\cdots+(3n+2)=\frac{n(3n+7)}{2}$.

Step 1: We need to show that $P(1)$ is true. But $P(1)$ says that $5=\frac{1\cdot(3\cdot1+7)}{2}$, which is true.

Step 2: Assume that $P(k)$ is true; that is, $5+8+11+\cdots+(3k+2)=\frac{k(3k+7)}{2}$. We want to use this to show that $P(k+1)$ is true. Now

$$\begin{aligned}5+8+11+\cdots+(3k+2)+[3(k+1)+2] &= \frac{k(3k+7)}{2}+(3k+5) \qquad \text{induction hypothesis}\\ &= \frac{3k^2+7k}{2}+\frac{6k+10}{2}=\frac{3k^2+13k+10}{2}\\ &= \frac{(3k+10)(k+1)}{2}=\frac{(k+1)[3(k+1)+7]}{2}\end{aligned}$$

Thus, $P(k+1)$ follows from $P(k)$. So by the Principle of Mathematical Induction, $P(n)$ is true for all n.

7. Let $P(n)$ denote the statement $1\cdot2+2\cdot3+3\cdot4+\cdots+n(n+1)=\frac{n(n+1)(n+2)}{3}$.

Step 1: $P(1)$ is the statement that $1\cdot2=\frac{1\cdot(1+1)\cdot(1+2)}{3}$, which is true.

Step 2: Assume that $P(k)$ is true; that is, $1\cdot2+2\cdot3+3\cdot4+\cdots+k(k+1)=\frac{k(k+1)(k+2)}{3}$. We want to use this to show that $P(k+1)$ is true. Now

$$\begin{aligned}&1\cdot2+2\cdot3+3\cdot4+\cdots+k(k+1)+(k+1)[(k+1)+1]\\ &\quad= \frac{k(k+1)(k+2)}{3}+(k+1)(k+2) \qquad \text{induction hypothesis}\\ &\quad= \frac{k(k+1)(k+2)}{3}+\frac{3(k+1)(k+2)}{3}=\frac{(k+1)(k+2)(k+3)}{3}\end{aligned}$$

Thus, $P(k+1)$ follows from $P(k)$. So by the Principle of Mathematical Induction, $P(n)$ is true for all n.

9. Let $P(n)$ denote the statement $1^3+2^3+3^3+\cdots+n^3=\frac{n^2(n+1)^2}{4}$.

Step 1: $P(1)$ is the statement that $1^3=\frac{1^2\cdot(1+1)^2}{4}$, which is clearly true.

Step 2: Assume that $P(k)$ is true; that is, $1^3+2^3+3^3+\cdots+k^3=\frac{k^2(k+1)^2}{4}$. We want to use this to show that $P(k+1)$ is true. Now

$$\begin{aligned}1^3+2^3+3^3+\cdots+k^3+(k+1)^3 &= \frac{k^2(k+1)^2}{4}+(k+1)^3 \qquad \text{induction hypothesis}\\ &= \frac{(k+1)^2\left[k^2+4(k+1)\right]}{4}=\frac{(k+1)^2\left[k^2+4k+4\right]}{4}\\ &= \frac{(k+1)^2(k+2)^2}{4}=\frac{(k+1)^2[(k+1)+1]^2}{4}\end{aligned}$$

Thus, $P(k+1)$ follows from $P(k)$. So by the Principle of Mathematical Induction, $P(n)$ is true for all n.

11. Let $P(n)$ denote the statement $2^3 + 4^3 + 6^3 + \cdots + (2n)^3 = 2n^2(n+1)^2$.

Step 1: $P(1)$ is true since $2^3 = 2(1)^2(1+1)^2 = 2 \cdot 4 = 8$.

Step 2: Assume that $P(k)$ is true; that is, $2^3 + 4^3 + 6^3 + \cdots + (2k)^3 = 2k^2(k+1)^2$. We want to use this to show that $P(k+1)$ is true. Now

$$\begin{aligned} 2^3 + 4^3 + 6^3 + \cdots + (2k)^3 + [2(k+1)]^3 &= 2k^2(k+1)^2 + [2(k+1)]^3 && \text{induction hypothesis} \\ &= 2k^2(k+1)^2 + 8(k+1)(k+1)^2 = (k+1)^2\left(2k^2 + 8k + 8\right) \\ &= 2(k+1)^2(k+2)^2 = 2(k+1)^2[(k+1)+1]^2 \end{aligned}$$

Thus, $P(k+1)$ follows from $P(k)$. So by the Principle of Mathematical Induction, $P(n)$ is true for all n.

13. Let $P(n)$ denote the statement $1 \cdot 2 + 2 \cdot 2^2 + 3 \cdot 2^3 + 4 \cdot 2^4 + \cdots + n \cdot 2^n = 2[1 + (n-1)2^n]$.

Step 1: $P(1)$ is the statement that $1 \cdot 2 = 2[1+0]$, which is clearly true.

Step 2: Assume that $P(k)$ is true; that is, $1 \cdot 2 + 2 \cdot 2^2 + 3 \cdot 2^3 + 4 \cdot 2^4 + \cdots + k \cdot 2^k = 2\left[1 + (k-1)2^k\right]$. We want to use this to show that $P(k+1)$ is true. Now

$$\begin{aligned} &1 \cdot 2 + 2 \cdot 2^2 + 3 \cdot 2^3 + 4 \cdot 2^4 + \cdots + k \cdot 2^k + (k+1) \cdot 2^{(k+1)} \\ &\quad = 2\left[1 + (k-1)2^k\right] + (k+1) \cdot 2^{k+1} && \text{induction hypothesis} \\ &\quad = 2\left[1 + (k-1) \cdot 2^k + (k+1) \cdot 2^k\right] = 2\left[1 + 2k \cdot 2^k\right] \\ &\quad = 2\left[1 + k \cdot 2^{k+1}\right] = 2\left\{1 + [(k+1) - 1]2^{k+1}\right\} \end{aligned}$$

Thus $P(k+1)$ follows from $P(k)$. So by the Principle of Mathematical Induction, $P(n)$ is true for all n.

15. Let $P(n)$ denote the statement $n^2 + n$ is divisible by 2.

Step 1: $P(1)$ is the statement that $1^2 + 1 = 2$ is divisible by 2, which is clearly true.

Step 2: Assume that $P(k)$ is true; that is, $k^2 + k$ is divisible by 2. Now $(k+1)^2 + (k+1) = k^2 + 2k + 1 + k + 1 = \left(k^2 + k\right) + 2k + 2 = \left(k^2 + k\right) + 2(k+1)$. By the induction hypothesis, $k^2 + k$ is divisible by 2, and clearly $2(k+1)$ is divisible by 2. Thus, the sum is divisible by 2, so $P(k+1)$ is true. Therefore, $P(k+1)$ follows from $P(k)$. So by the Principle of Mathematical Induction, $P(n)$ is true for all n.

17. Let $P(n)$ denote the statement that $n^2 - n + 41$ is odd.

Step 1: $P(1)$ is the statement that $1^2 - 1 + 41 = 41$ is odd, which is clearly true.

Step 2: Assume that $P(k)$ is true; that is, $k^2 - k + 41$ is odd. We want to use this to show that $P(k+1)$ is true. Now, $(k+1)^2 - (k+1) + 41 = k^2 + 2k + 1 - k - 1 + 41 = \left(k^2 - k + 41\right) + 2k$, which is also odd because $k^2 - k + 41$ is odd by the induction hypothesis, $2k$ is always even, and an odd number plus an even number is always odd. Therefore, $P(k+1)$ follows from $P(k)$. So by the Principle of Mathematical Induction, $P(n)$ is true for all n.

19. Let $P(n)$ denote the statement that $8^n - 3^n$ is divisible by 5.

Step 1: $P(1)$ is the statement that $8^1 - 3^1 = 5$ is divisible by 5, which is clearly true.

Step 2: Assume that $P(k)$ is true; that is, $8^k - 3^k$ is divisible by 5. We want to use this to show that $P(k+1)$ is true. Now, $8^{k+1} - 3^{k+1} = 8 \cdot 8^k - 3 \cdot 3^k = 8 \cdot 8^k - (8-5) \cdot 3^k = 8 \cdot \left(8^k - 3^k\right) + 5 \cdot 3^k$, which is divisible by 5 because $8^k - 3^k$ is divisible by 5 by our induction hypothesis, and $5 \cdot 3^k$ is divisible by 5. Thus $P(k+1)$ follows from $P(k)$. So by the Principle of Mathematical Induction, $P(n)$ is true for all n.

21. Let $P(n)$ denote the statement $n < 2^n$.

Step 1: $P(1)$ is the statement that $1 < 2^1 = 2$, which is clearly true.

Step 2: Assume that $P(k)$ is true; that is, $k < 2^k$. We want to use this to show that $P(k+1)$ is true. Adding 1 to both sides of $P(k)$ we have $k + 1 < 2^k + 1$. Since $1 < 2^k$ for $k \geq 1$, we have $2^k + 1 < 2^k + 2^k = 2 \cdot 2^k = 2^{k+1}$. Thus $k + 1 < 2^{k+1}$, which is exactly $P(k+1)$. Therefore, $P(k+1)$ follows from $P(k)$. So by the Principle of Mathematical Induction, $P(n)$ is true for all n.

23. Let $P(n)$ denote the statement $(1+x)^n \geq 1 + nx$, if $x > -1$.

Step 1: $P(1)$ is the statement that $(1+x)^1 \geq 1 + 1x$, which is clearly true.

Step 2: Assume that $P(k)$ is true; that is, $(1+x)^k \geq 1 + kx$. Now, $(1+x)^{k+1} = (1+x)(1+x)^k \geq (1+x)(1+kx)$, by the induction hypothesis. Since $(1+x)(1+kx) = 1 + (k+1)x + kx^2 \geq 1 + (k+1)x$ (since $kx^2 \geq 0$), we have $(1+x)^{k+1} \geq 1 + (k+1)x$, which is $P(k+1)$. Thus $P(k+1)$ follows from $P(k)$. So the Principle of Mathematical Induction, $P(n)$ is true for all n.

25. Let $P(n)$ be the statement that $a_n = 5 \cdot 3^{n-1}$.

Step 1: $P(1)$ is the statement that $a_1 = 5 \cdot 3^0 = 5$, which is true.

Step 2: Assume that $P(k)$ is true; that is, $a_k = 5 \cdot 3^{k-1}$. We want to use this to show that $P(k+1)$ is true. Now, $a_{k+1} = 3a_k = 3 \cdot \left(5 \cdot 3^{k-1}\right)$, by the induction hypothesis. Therefore, $a_{k+1} = 3 \cdot \left(5 \cdot 3^{k-1}\right) = 5 \cdot 3^k$, which is exactly $P(k+1)$. Thus, $P(k+1)$ follows from $P(k)$. So by the Principle of Mathematical Induction, $P(n)$ is true for all n.

27. Let $P(n)$ be the statement that $x - y$ is a factor of $x^n - y^n$ for all natural numbers n.

Step 1: $P(1)$ is the statement that $x - y$ is a factor of $x^1 - y^1$, which is clearly true.

Step 2: Assume that $P(k)$ is true; that is, $x - y$ is a factor of $x^k - y^k$. We want to use this to show that $P(k+1)$ is true. Now, $x^{k+1} - y^{k+1} = x^{k+1} - x^k y + x^k y - y^{k+1} = x^k(x-y) + \left(x^k - y^k\right)y$, for which $x - y$ is a factor because $x - y$ is a factor of $x^k(x-y)$, and $x - y$ is a factor of $\left(x^k - y^k\right)y$, by the induction hypothesis. Thus $P(k+1)$ follows from $P(k)$. So by the Principle of Mathematical Induction, $P(n)$ is true for all n.

29. Let $P(n)$ denote the statement that F_{3n} is even for all natural numbers n.

Step 1: $P(1)$ is the statement that F_3 is even. Since $F_3 = F_2 + F_1 = 1 + 1 = 2$, this statement is true.

Step 2: Assume that $P(k)$ is true; that is, F_{3k} is even. We want to use this to show that $P(k+1)$ is true. Now, $F_{3(k+1)} = F_{3k+3} = F_{3k+2} + F_{3k+1} = F_{3k+1} + F_{3k} + F_{3k+1} = F_{3k} + 2 \cdot F_{3k+1}$, which is even because F_{3k} is even by the induction hypothesis, and $2 \cdot F_{3k+1}$ is even. Thus $P(k+1)$ follows from $P(k)$. So by the Principle of Mathematical Induction, $P(n)$ is true for all n.

31. Let $P(n)$ denote the statement that $F_1^2 + F_2^2 + F_3^2 + \cdots + F_n^2 = F_n \cdot F_{n+1}$.

Step 1: $P(1)$ is the statement that $F_1^2 = F_1 \cdot F_2$ or $1^2 = 1 \cdot 1$, which is true.

Step 2: Assume that $P(k)$ is true, that is, $F_1^2 + F_2^2 + F_3^2 + \cdots + F_k^2 = F_k \cdot F_{k+1}$. We want to use this to show that $P(k+1)$ is true. Now

$$\begin{aligned} F_1^2 + F_2^2 + F_3^2 + \cdots + F_k^2 + F_{k+1}^2 &= F_k \cdot F_{k+1} + F_{k+1}^2 && \text{induction hypothesis} \\ &= F_{k+1}\left(F_k + F_{k+1}\right) \\ &= F_{k+1} \cdot F_{k+2} && \text{by definition of the Fibonacci sequence} \end{aligned}$$

Thus $P(k+1)$ follows from $P(k)$. So by the Principle of Mathematical Induction, $P(n)$ is true for all n.

33. Let $P(n)$ denote the statement $\begin{bmatrix} 1 & 1 \\ 1 & 0 \end{bmatrix}^n = \begin{bmatrix} F_{n+1} & F_n \\ F_n & F_{n-1} \end{bmatrix}$.

Step 1: Since $\begin{bmatrix} 1 & 1 \\ 1 & 0 \end{bmatrix}^2 = \begin{bmatrix} 1 & 1 \\ 1 & 0 \end{bmatrix}\begin{bmatrix} 1 & 1 \\ 1 & 0 \end{bmatrix} = \begin{bmatrix} 2 & 1 \\ 1 & 1 \end{bmatrix} = \begin{bmatrix} F_3 & F_2 \\ F_2 & F_1 \end{bmatrix}$, it follows that $P(2)$ is true.

Step 2: Assume that $P(k)$ is true; that is, $\begin{bmatrix} 1 & 1 \\ 1 & 0 \end{bmatrix}^k = \begin{bmatrix} F_{k+1} & F_k \\ F_k & F_{k-1} \end{bmatrix}$. We show that $P(k+1)$ follows from this. Now,

$$\begin{bmatrix} 1 & 1 \\ 1 & 0 \end{bmatrix}^{k+1} = \begin{bmatrix} 1 & 1 \\ 1 & 0 \end{bmatrix}^k \begin{bmatrix} 1 & 1 \\ 1 & 0 \end{bmatrix} = \begin{bmatrix} F_{k+1} & F_k \\ F_k & F_{k-1} \end{bmatrix}\begin{bmatrix} 1 & 1 \\ 1 & 0 \end{bmatrix} \quad \text{induction hypothesis}$$

$$= \begin{bmatrix} F_{k+1}+F_k & F_{k+1} \\ F_k+F_{k-1} & F_k \end{bmatrix} = \begin{bmatrix} F_{k+2} & F_{k+1} \\ F_{k+1} & F_k \end{bmatrix} \quad \text{by definition of the Fibonacci sequence}$$

Thus $P(k+1)$ follows from $P(k)$. So by the Principle of Mathematical Induction, $P(n)$ is true for all $n \geq 2$.

35. Since $F_1 = 1$, $F_2 = 1$, $F_3 = 2$, $F_4 = 3$, $F_5 = 5$, $F_6 = 8$, $F_7 = 13, \ldots$ our conjecture is that $F_n \geq n$, for all $n \geq 5$. Let $P(n)$ denote the statement that $F_n \geq n$.

Step 1: $P(5)$ is the statement that $F_5 = 5 \geq 5$, which is clearly true.

Step 2: Assume that $P(k)$ is true; that is, $F_k \geq k$, for some $k \geq 5$. We want to use this to show that $P(k+1)$ is true. Now, $F_{k+1} = F_k + F_{k-1} \geq k + F_{k-1}$ (by the induction hypothesis)$\geq k+1$ (because $F_{k-1} \geq 1$). Thus $P(k+1)$ follows from $P(k)$. So by the Principle of Mathematical Induction, $P(n)$ is true for all $n \geq 5$.

37. (a) $P(n) = n^2 - n + 11$ is prime for all n. This is false as the case for $n = 11$ demonstrates: $P(11) = 11^2 - 11 + 11 = 121$, which is not prime since $11^2 = 121$.

(b) $n^2 > n$, for all $n \geq 2$. This is true. Let $P(n)$ denote the statement that $n^2 > n$.

Step 1: $P(2)$ is the statement that $2^2 = 4 > 2$, which is clearly true.

Step 2: Assume that $P(k)$ is true; that is, $k^2 > k$. We want to use this to show that $P(k+1)$ is true. Now $(k+1)^2 = k^2 + 2k + 1$. Using the induction hypothesis (to replace k^2), we have $k^2 + 2k + 1 > k + 2k + 1 = 3k + 1 > k + 1$, since $k \geq 2$. Therefore, $(k+1)^2 > k+1$, which is exactly $P(k+1)$. Thus $P(k+1)$ follows from $P(k)$. So by the Principle of Mathematical Induction, $P(n)$ is true for all n.

(c) $2^{2n+1} + 1$ is divisible by 3, for all $n \geq 1$. This is true. Let $P(n)$ denote the statement that $2^{2n+1} + 1$ is divisible by 3.

Step 1: $P(1)$ is the statement that $2^3 + 1 = 9$ is divisible by 3, which is clearly true.

Step 2: Assume that $P(k)$ is true; that is, $2^{2k+1} + 1$ is divisible by 3. We want to use this to show that $P(k+1)$ is true. Now, $2^{2(k+1)+1} + 1 = 2^{2k+3} + 1 = 4 \cdot 2^{2k+1} + 1 = (3+1)\,2^{2k+1} + 1 = 3 \cdot 2^{2k+1} + \left(2^{2k+1} + 1\right)$, which is divisible by 3 since $2^{2k+1} + 1$ is divisible by 3 by the induction hypothesis, and $3 \cdot 2^{2k+1}$ is clearly divisible by 3. Thus $P(k+1)$ follows from $P(k)$. So by the Principle of Mathematical Induction, $P(n)$ is true for all n.

(d) The statement $n^3 \geq (n+1)^2$ for all $n \geq 2$ is false. The statement fails when $n = 2$: $2^3 = 8 < (2+1)^2 = 9$.

(e) $n^3 - n$ is divisible by 3, for all $n \geq 2$. This is true. Let $P(n)$ denote the statement that $n^3 - n$ is divisible by 3.

Step 1: $P(2)$ is the statement that $2^3 - 2 = 6$ is divisible by 3, which is clearly true.

Step 2: Assume that $P(k)$ is true; that is, $k^3 - k$ is divisible by 3. We want to use this to show that $P(k+1)$ is true. Now $(k+1)^3 - (k+1) = k^3 + 3k^2 + 3k + 1 - (k+1) = k^3 + 3k^2 + 2k = k^3 - k + 3k^2 + 2k + k = \left(k^3 - k\right) + 3\left(k^2 + k\right)$.

The term $k^3 - k$ is divisible by 3 by our induction hypothesis, and the term $3\left(k^2 + k\right)$ is clearly divisible by 3. Thus $(k+1)^3 - (k+1)$ is divisible by 3, which is exactly $P(k+1)$. So by the Principle of Mathematical Induction, $P(n)$ is true for all n.

(f) $n^3 - 6n^2 + 11n$ is divisible by 6, for all $n \geq 1$. This is true. Let $P(n)$ denote the statement that $n^3 - 6n^2 + 11n$ is divisible by 6.

Step 1: $P(1)$ is the statement that $(1)^3 - 6(1)^2 + 11(1) = 6$ is divisible by 6, which is clearly true.

Step 2: Assume that $P(k)$ is true; that is, $k^3 - 6k^2 + 11k$ is divisible by 6. We show that $P(k+1)$ is then also true. Now

$$\begin{aligned}
(k+1)^3 - 6(k+1)^2 + 11(k+1) &= k^3 + 3k^2 + 3k + 1 - 6k^2 - 12k - 6 + 11k + 11 \\
&= k^3 - 3k^2 + 2k + 6 = k^3 - 6k^2 + 11k + \left(3k^2 - 9k + 6\right) \\
&= \left(k^3 - 6k^2 + 11k\right) + 3\left(k^2 - 3k + 2\right) = \left(k^3 - 6k^2 + 11k\right) + 3(k-1)(k-2)
\end{aligned}$$

In this last expression, the first term is divisible by 6 by our induction hypothesis. The second term is also divisible by 6. To see this, notice that $k - 1$ and $k - 2$ are consecutive natural numbers, and so one of them must be even (divisible by 2). Since 3 also appears in this second term, it follows that this term is divisible by 2 and 3 and so is divisible by 6. Thus $P(k+1)$ follows from $P(k)$. So by the Principle of Mathematical Induction, $P(n)$ is true for all n.

13.6 THE BINOMIAL THEOREM

1. An algebraic expression of the form $a + b$, which consists of a sum of two terms, is called a *binomial*.

3. The binomial coefficients can be calculated directly using the formula $\binom{n}{k} = \dfrac{n!}{k!\,(n-k)!}$. So $\binom{4}{3} = \dfrac{4!}{3!1!} = 4$.

5. $(x+y)^6 = x^6 + 6x^5y + 15x^4y^2 + 20x^3y^3 + 15x^2y^4 + 6xy^5 + y^6$

7. $\left(x + \dfrac{1}{x}\right)^4 = x^4 + 4x^3 \cdot \dfrac{1}{x} + 6x^2\left(\dfrac{1}{x}\right)^2 + 4x\left(\dfrac{1}{x}\right)^3 + \left(\dfrac{1}{x}\right)^4 = x^4 + 4x^2 + 6 + \dfrac{4}{x^2} + \dfrac{1}{x^4}$

9. $(x-1)^5 = x^5 - 5x^4 + 10x^3 - 10x^2 + 5x - 1$

11. $\left(x^2y - 1\right)^5 = \left(x^2y\right)^5 - 5\left(x^2y\right)^4 + 10\left(x^2y\right)^3 - 10\left(x^2y\right)^2 + 5x^2y - 1 = x^{10}y^5 - 5x^8y^4 + 10x^6y^3 - 10x^4y^2 + 5x^2y - 1$

13. $(2x - 3y)^3 = (2x)^3 - 3(2x)^2 3y + 3 \cdot 2x(3y)^2 - (3y)^3 = 8x^3 - 36x^2y + 54xy^2 - 27y^3$

15. $\left(\dfrac{1}{x} - \sqrt{x}\right)^5 = \left(\dfrac{1}{x}\right)^5 - 5\left(\dfrac{1}{x}\right)^4\sqrt{x} + 10\left(\dfrac{1}{x}\right)^3 x - 10\left(\dfrac{1}{x}\right)^2 x\sqrt{x} + 5\left(\dfrac{1}{x}\right)x^2 - x^2\sqrt{x}$

$$= \frac{1}{x^5} - \frac{5}{x^{7/2}} + \frac{10}{x^2} - \frac{10}{x^{1/2}} + 5x - x^{5/2}$$

17. $\binom{6}{4} = \dfrac{6!}{4!\,2!} = \dfrac{6 \cdot 5 \cdot 4!}{2 \cdot 1 \cdot 4!} = 15$

19. $\binom{100}{98} = \dfrac{100!}{98!\,2!} = \dfrac{100 \cdot 99 \cdot 98!}{98! \cdot 2 \cdot 1} = 4950$

21. $\binom{3}{1}\binom{4}{2} = \dfrac{3!}{1!\,2!}\dfrac{4!}{2!\,2!} = \dfrac{3 \cdot 2! \cdot 4 \cdot 3 \cdot 2!}{1 \cdot 2! \cdot 2 \cdot 1 \cdot 2!} = 18$

23. $\binom{5}{0}+\binom{5}{1}+\binom{5}{2}+\binom{5}{3}+\binom{5}{4}+\binom{5}{5}=(1+1)^5=2^5=32$

25. $(x+2y)^4=\binom{4}{0}x^4+\binom{4}{1}x^3\cdot 2y+\binom{4}{2}x^2\cdot 4y^2+\binom{4}{3}x\cdot 8y^3+\binom{4}{4}16y^4=x^4+8x^3y+24x^2y^2+32xy^3+16y^4$

27. $\left(1+\frac{1}{x}\right)^6=\binom{6}{0}1^6+\binom{6}{1}1^5\left(\frac{1}{x}\right)+\binom{6}{2}1^4\left(\frac{1}{x}\right)^2+\binom{6}{3}1^3\left(\frac{1}{x}\right)^3+\binom{6}{4}1^2\left(\frac{1}{x}\right)^4+\binom{6}{5}1\left(\frac{1}{x}\right)^5+\binom{6}{6}\left(\frac{1}{x}\right)^6$
$=1+\frac{6}{x}+\frac{15}{x^2}+\frac{20}{x^3}+\frac{15}{x^4}+\frac{6}{x^5}+\frac{1}{x^6}$

29. The first three terms in the expansion of $(x+2y)^{20}$ are $\binom{20}{0}x^{20}=x^{20}$, $\binom{20}{1}x^{19}\cdot 2y=40x^{19}y$, and $\binom{20}{2}x^{18}\cdot(2y)^2=760x^{18}y^2$.

31. The last two terms in the expansion of $\left(a^{2/3}+a^{1/3}\right)^{25}$ are $\binom{25}{24}a^{2/3}\cdot\left(a^{1/3}\right)^{24}=25a^{26/3}$, and $\binom{25}{25}a^{25/3}=a^{25/3}$.

33. The middle term in the expansion of $\left(x^2+1\right)^{18}$ occurs when both terms are raised to the 9th power. So this term is $\binom{18}{9}\left(x^2\right)^9 1^9=48{,}620x^{18}$.

35. The 24th term in the expansion of $(a+b)^{25}$ is $\binom{25}{23}a^2b^{23}=300a^2b^{23}$.

37. The 100th term in the expansion of $(1+y)^{100}$ is $\binom{100}{99}1^1\cdot y^{99}=100y^{99}$.

39. The term that contains x^4 in the expansion of $(x+2y)^{10}$ has exponent $r=4$. So this term is $\binom{10}{4}x^4\cdot(2y)^{10-4}=13{,}440x^4y^6$.

41. The rth term is $\binom{12}{r}a^r\left(b^2\right)^{12-r}=\binom{12}{r}a^rb^{24-2r}$. Thus the term that contains b^8 occurs where $24-2r=8\Leftrightarrow r=8$. So the term is $\binom{12}{8}a^8b^8=495a^8b^8$.

43. $x^4+4x^3y+6x^2y^2+4xy^3+y^4=(x+y)^4$

45. $8a^3+12a^2b+6ab^2+b^3=\binom{3}{0}(2a)^3+\binom{3}{1}(2a)^2b+\binom{3}{2}2ab^2+\binom{3}{3}b^3=(2a+b)^3$

47. $\dfrac{(x+h)^3-x^3}{h}=\dfrac{x^3+3x^2h+3xh^2+h^3-x^3}{h}=\dfrac{3x^2h+3xh^2+h^3}{h}=\dfrac{h\left(3x^2+3xh+h^2\right)}{h}=3x^2+3xh+h^2$

49. $(1.01)^{100}=(1+0.01)^{100}$. Now the first term in the expansion is $\binom{100}{0}1^{100}=1$, the second term is $\binom{100}{1}1^{99}(0.01)=1$, and the third term is $\binom{100}{2}1^{98}(0.01)^2=0.495$. Now each term is nonnegative, so $(1.01)^{100}=(1+0.01)^{100}>1+1+.0.495>2$. Thus $(1.01)^{100}>2$.

51. $\binom{n}{1}=\dfrac{n!}{1!\,(n-1)!}=\dfrac{n\,(n-1)!}{1\,(n-1)!}=\dfrac{n}{1}=n$. $\binom{n}{n-1}=\dfrac{n!}{(n-1)!\,1!}=\dfrac{n\,(n-1)!}{(n-1)!\,1}=n$. Therefore, $\binom{n}{1}=\binom{n}{n-1}=n$.

53. (a) $\binom{n}{r-1}+\binom{n}{r}=\dfrac{n!}{(r-1)!\,[n-(r-1)]!}+\dfrac{n!}{r!\,(n-r)!}$.

(b) $\dfrac{n!}{(r-1)!\,[n-(r-1)]!}+\dfrac{n!}{r!\,(n-r)!}=\dfrac{r\cdot n!}{r\cdot(r-1)!\,(n-r+1)!}+\dfrac{(n-r+1)\cdot n!}{r!\,(n-r+1)\,(n-r)!}$
$=\dfrac{r\cdot n!}{r!\,(n-r+1)!}+\dfrac{(n-r+1)\cdot n!}{r!\,(n-r+1)!}$.

Thus a common denominator is $r!\,(n-r+1)!$.

(c) Therefore, using the results of parts (a) and (b),

$$\binom{n}{r-1}+\binom{n}{r}=\frac{n!}{(r-1)!\,[n-(r-1)]!}+\frac{n!}{r!\,(n-r)!}=\frac{r\cdot n!}{r!\,(n-r+1)!}+\frac{(n-r+1)\cdot n!}{r!\,(n-r+1)!}$$
$$=\frac{r\cdot n!+(n-r+1)\cdot n!}{r!\,(n-r+1)!}=\frac{n!\,(r+n-r+1)}{r!\,(n-r+1)!}=\frac{n!\,(n+1)}{r!\,(n+1-r)!}=\frac{(n+1)!}{r!\,(n+1-r)!}=\binom{n+1}{r}$$

55. By the Binomial Theorem, the volume of a cube of side $x+2$ inches is $(x+2)^3 = \binom{3}{0}x^3 + \binom{3}{1}x^2(2) + \binom{3}{2}x(2)^2 + \binom{3}{3}2^3 = x^3 + 3\cdot 2x^2 + 3\cdot 4x + 8 = x^3 + 6x^2 + 12x + 8$. The volume of a cube of side x inches is x^3, so the difference in volumes is $x^3 + 6x^2 + 12x + 8 - x^3 = 6x^2 + 12x + 8$ cubic inches.

57. Notice that $(100!)^{101} = (100!)^{100}\cdot 100!$ and $(101!)^{100} = (101\cdot 100!)^{100} = 101^{100}\cdot(100!)^{100}$. Now $100! = 1\cdot 2\cdot 3\cdot 4\cdot\cdots\cdot 99\cdot 100$ and $101^{100} = 101\cdot 101\cdot 101\cdot\cdots\cdot 101$. Thus each of these last two expressions consists of 100 factors multiplied together, and since each factor in the product for 101^{100} is larger than each factor in the product for $100!$, it follows that $100! < 101^{100}$. Thus $(100!)^{100}\cdot 100! < (100!)^{100}\cdot 101^{100}$. So $(100!)^{101} < (101!)^{100}$.

59. $0 = 0^n = (-1+1)^n = \binom{n}{0}(-1)^0(1)^n + \binom{n}{1}(-1)^1(1)^{n-1} + \binom{n}{2}(-1)^2(1)^{n-2} + \cdots + \binom{n}{n}(-1)^n(1)^0$
$= \binom{n}{0} - \binom{n}{1} + \binom{n}{2} - \cdots + (-1)^k\binom{n}{k} + \cdots + (-1)^n\binom{n}{n}$

CHAPTER 13 REVIEW

1. $a_n = \frac{n^2}{n+1}$. Then $a_1 = \frac{1^2}{1+1} = \frac{1}{2}$, $a_2 = \frac{2^2}{2+1} = \frac{4}{3}$, $a_3 = \frac{3^2}{3+1} = \frac{9}{4}$, $a_4 = \frac{4^2}{4+1} = \frac{16}{5}$, and $a_{10} = \frac{10^2}{10+1} = \frac{100}{11}$.

3. $a_n = \frac{(-1)^n + 1}{n^3}$. Then $a_1 = \frac{(-1)^1+1}{1^3} = 0$, $a_2 = \frac{(-1)^2+1}{2^3} = \frac{2}{8} = \frac{1}{4}$, $a_3 = \frac{(-1)^3+1}{3^3} = 0$, $a_4 = \frac{(-1)^4+1}{4^3} = \frac{2}{64} = \frac{1}{32}$, and $a_{10} = \frac{(-1)^{10}+1}{10^3} = \frac{1}{500}$.

5. $a_n = \frac{(2n)!}{2^n n!}$. Then $a_1 = \frac{(2\cdot 1)!}{2^1\cdot 1!} = 1$, $a_2 = \frac{(2\cdot 2)!}{2^2\cdot 2!} = 3$, $a_3 = \frac{(2\cdot 3)!}{2^3\cdot 3!} = \frac{6\cdot 5\cdot 4}{8} = 15$, $a_4 = \frac{(2\cdot 4)!}{2^4\cdot 4!} = \frac{8\cdot 7\cdot 6\cdot 5}{16} = 105$, and $a_{10} = \frac{(2\cdot 10)!}{2^{10}\cdot 10!} = 654{,}729{,}075$.

7. $a_n = a_{n-1} + 2n - 1$ and $a_1 = 1$. Then $a_2 = a_1 + 4 - 1 = 4$, $a_3 = a_2 + 6 - 1 = 9$, $a_4 = a_3 + 8 - 1 = 16$, $a_5 = a_4 + 10 - 1 = 25$, $a_6 = a_5 + 12 - 1 = 36$, and $a_7 = a_6 + 14 - 1 = 49$.

9. $a_n = a_{n-1} + 2a_{n-2}$, $a_1 = 1$ and $a_2 = 3$. Then $a_3 = a_2 + 2a_1 = 5$, $a_4 = a_3 + 2a_2 = 11$, $a_5 = a_4 + 2a_3 = 21$, $a_6 = a_5 + 2a_4 = 43$, and $a_7 = a_6 + 2a_5 = 85$.

11. **(a)** $a_1 = 2(1) + 5 = 7$, $a_2 = 2(2) + 5 = 9$, $a_3 = 2(3) + 5 = 11$, $a_4 = 2(4) + 5 = 13$, $a_5 = 2(5) + 5 = 15$

(b)

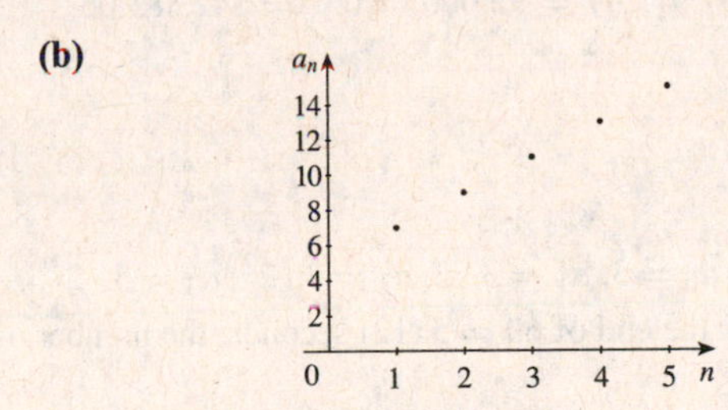

(c) $S_5 = 7 + 9 + 11 + 13 + 15 = 55$

(d) This sequence is arithmetic with common difference 2.

13. **(a)** $a_1 = \frac{3^1}{2^2} = \frac{3}{4}$, $a_2 = \frac{3^2}{2^3} = \frac{9}{8}$, $a_3 = \frac{3^3}{2^4} = \frac{27}{16}$, $a_4 = \frac{3^4}{2^5} = \frac{81}{32}$, $a_5 = \frac{3^5}{2^6} = \frac{243}{64}$

(b)

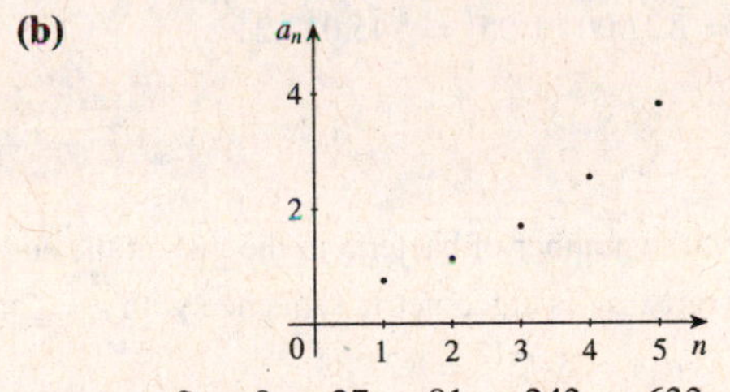

(c) $S_5 = \frac{3}{4} + \frac{9}{8} + \frac{27}{16} + \frac{81}{32} + \frac{243}{64} = \frac{633}{64}$

(d) This sequence is geometric with common ratio $\frac{3}{2}$.

15. $5, 5.5, 6, 6.5, \ldots$. Since $5.5 - 5 = 6 - 5.5 = 6.5 - 6 = 0.5$, this is an arithmetic sequence with $a_1 = 5$ and $d = 0.5$. Then $a_5 = a_4 + 0.5 = 7$.

17. $t-3, t-2, t-1, t, \ldots$. Since $(t-2)-(t-3)=(t-1)-(t-2)=t-(t-1)=1$, this is an arithmetic sequence with $a_1=t-3$ and $d=1$. Then $a_5=a_4+1=t+1$.

19. $t^3, t^2, t, 1, \ldots$. Since $\frac{t^2}{t^3}=\frac{t}{t^2}=\frac{1}{t}$, this is a geometric sequence with $a_1=t^3$ and $r=\frac{1}{t}$. Then $a_5=a_4\cdot r=1\cdot\frac{1}{t}=\frac{1}{t}$.

21. $\frac{3}{4}, \frac{1}{2}, \frac{1}{3}, \frac{2}{9}, \ldots$. Since $\frac{\frac{1}{2}}{\frac{3}{4}}=\frac{\frac{1}{3}}{\frac{1}{2}}=\frac{\frac{2}{9}}{\frac{1}{3}}=\frac{2}{3}$, this is a geometric sequence with $a_1=\frac{3}{4}$ and $r=\frac{2}{3}$. Then $a_5=a_4\cdot r=\frac{2}{9}\cdot\frac{2}{3}=\frac{4}{27}$. .

23. $3, 6i, -12, -24i, \ldots$. Since $\frac{6i}{3}=2i$, $\frac{-12}{6i}=\frac{-2}{i}=\frac{-2i}{i^2}=2i$, $\frac{-24i}{-12}=2i$, this is a geometric sequence with common ratio $r=2i$.

25. $a_6=17=a+5d$ and $a_4=11=a+3d$. Then, $a_6-a_4=17-11 \Leftrightarrow (a+5d)-(a+3d)=6 \Leftrightarrow 6=2d \Leftrightarrow d=3$. Substituting into $11=a+3d$ gives $11=a+3\cdot 3$, and so $a=2$. Thus $a_2=a+(2-1)\,d=2+3=5$.

27. $a_3=9$ and $r=\frac{3}{2}$. Then $a_5=a_3\cdot r^2=9\cdot\left(\frac{3}{2}\right)^2=\frac{81}{4}$.

29. **(a)** $A_n=32{,}000\cdot 1.05^{n-1}$

(b) $A_1=\$32{,}000$, $A_2=32{,}000\cdot 1.05^1=\$33{,}600$, $A_3=32{,}000\cdot 1.05^2=\$35{,}280$, $A_4=32{,}000\cdot 1.05^3=\$37{,}044$, $A_5=32{,}000\cdot 1.05^4=\$38{,}896.20$, $A_6=32{,}000\cdot 1.05^5=\$40{,}841.01$, $A_7=32{,}000\cdot 1.05^6=\$42{,}883.06$, $A_8=32{,}000\cdot 1.05^7=\$45{,}027.21$

31. Let a_n be the number of bacteria in the dish at the end of $5n$ seconds. So $a_0=3$, $a_1=3\cdot 2$, $a_2=3\cdot 2^2$, $a_3=3\cdot 2^3, \ldots$. Then, clearly, a_n is a geometric sequence with $r=2$ and $a=3$. Thus at the end of $60=5\,(12)$ seconds, the number of bacteria is $a_{12}=3\cdot 2^{12}=12{,}288$.

33. Suppose that the common ratio in the sequence $a_1, a_2, a_3, \ldots$ is r. Also, suppose that the common ratio in the sequence $b_1, b_2, b_3, \ldots$ is s. Then $a_n=a_1r^{n-1}$ and $b_n=b_1s^{n-1}$, $n=1,2,3,\ldots$. Thus $a_nb_n=a_1r^{n-1}\cdot b_1s^{n-1}=(a_1b_1)(rs)^{n-1}$. So the sequence $a_1b_1, a_2b_2, a_3b_3, \ldots$ is geometric with first term a_1b_1 and common ratio rs.

35. (a) $6, x, 12, \ldots$ is arithmetic if $x - 6 = 12 - x \Leftrightarrow 2x = 18 \Leftrightarrow x = 9$.

(b) $6, x, 12, \ldots$ is geometric if $\dfrac{x}{6} = \dfrac{12}{x} \Leftrightarrow x^2 = 72 \Leftrightarrow x = \pm 6\sqrt{2}$.

37. $\sum_{k=3}^{6} (k+1)^2 = (3+1)^2 + (4+1)^2 + (5+1)^2 + (6+1)^2 = 16 + 25 + 36 + 49 = 126$

39. $\sum_{k=1}^{6} (k+1)\, 2^{k-1} = 2 \cdot 2^0 + 3 \cdot 2^1 + 4 \cdot 2^2 + 5 \cdot 2^3 + 6 \cdot 2^4 + 7 \cdot 2^5 = 2 + 6 + 16 + 40 + 96 + 224 = 384$

41. $\sum_{k=1}^{10} (k-1)^2 = 0^2 + 1^2 + 2^2 + 3^2 + 4^2 + 5^2 + 6^2 + 7^2 + 8^2 + 9^2$

43. $\sum_{k=1}^{50} \dfrac{3^k}{2^{k+1}} = \dfrac{3}{2^2} + \dfrac{3^2}{2^3} + \dfrac{3^3}{2^4} + \dfrac{3^4}{2^5} + \cdots + \dfrac{3^{49}}{2^{50}} + \dfrac{3^{50}}{2^{51}}$

45. $3 + 6 + 9 + 12 + \cdots + 99 = 3\,(1) + 3\,(2) + 3\,(3) + \cdots + 3\,(33) = \sum_{k=1}^{33} 3k$

47. $1 \cdot 2^3 + 2 \cdot 2^4 + 3 \cdot 2^5 + 4 \cdot 2^6 + \cdots + 100 \cdot 2^{102}$

$$= (1)\,2^{(1)+2} + (2)\,2^{(2)+2} + (3)\,2^{(3)+2} + (4)\,2^{(4)+2} + \cdots + (100)\,2^{(100)+2}$$

$$= \sum_{k=1}^{100} k \cdot 2^{k+2}$$

49. $1 + 0.9 + (0.9)^2 + \cdots + (0.9)^5$ is a geometric series with $a = 1$ and $r = \dfrac{0.9}{1} = 0.9$. Thus, the sum of the series is

$$S_6 = \frac{1 - (0.9)^6}{1 - 0.9} = \frac{1 - 0.531441}{0.1} = 4.68559.$$

51. $\sqrt{5} + 2\sqrt{5} + 3\sqrt{5} + \cdots + 100\sqrt{5}$ is an arithmetic series with $a = \sqrt{5}$ and $d = \sqrt{5}$. Then $100\sqrt{5} = a_n = \sqrt{5} + \sqrt{5}\,(n-1)$ $\Leftrightarrow n = 100$. So the sum is $S_{100} = \frac{100}{2}\left(\sqrt{5} + 100\sqrt{5}\right) = 50\left(101\sqrt{5}\right) = 5050\sqrt{5}$.

53. $\sum_{n=0}^{6} 3 \cdot (-4)^n$ is a geometric series with $a = 3$, $r = -4$, and $n = 7$. Therefore, the sum of the series is $S_7 = 3 \cdot \dfrac{1 - (-4)^7}{1 - (-4)} = \frac{3}{5}\left(1 + 4^7\right) = 9831$.

55. $1 - \frac{2}{5} + \frac{4}{25} - \frac{8}{125} + \cdots$ is a geometric series with $a = 1$ and $r = -\frac{2}{5}$. Therefore, it is convergent with sum $S = \dfrac{a}{1-r} = \dfrac{1}{1 - \left(-\frac{2}{5}\right)} = \dfrac{5}{7}$.

57. $5 - 5\,(1.01) + 5\,(1.01)^2 - 5\,(1.01)^3 + \cdots$ is an infinite geometric series with $a = 5$ and $r = 1.01$. Because $|r| = 1.01 > 1$, the series diverges.

59. $-1 + \frac{9}{8} - \left(\frac{9}{8}\right)^2 + \left(\frac{9}{8}\right)^3 - \cdots$ is an infinite geometric series with $a = -1$ and $r = -\frac{9}{8}$. Because $|r| = \frac{9}{8} > 1$, the series diverges.

61. We have an arithmetic sequence with $a = 7$ and $d = 3$. Then $S_n = 325 = \dfrac{n}{2}\,[2a + (n-1)\,d] = \dfrac{n}{2}\,[14 + 3\,(n-1)] = \dfrac{n}{2}\,(11 + 3n) \Leftrightarrow 650 = 3n^2 + 11n \Leftrightarrow (3n + 50)\,(n - 13) = 0 \Leftrightarrow$ $n = 13$ (because $n = -\frac{50}{3}$ is inadmissible). Thus, 13 terms must be added.

63. This is a geometric sequence with $a = 2$ and $r = 2$. Then $S_{15} = 2 \cdot \dfrac{1 - 2^{15}}{1 - 2} = 2\left(2^{15} - 1\right) = 65{,}534$, and so the total number of ancestors is 65,534.

65. $A = 10{,}000$, $i = 0.03$, and $n = 4$. Thus, $10{,}000 = R\dfrac{(1.03)^4 - 1}{0.03} \Leftrightarrow R = \dfrac{10{,}000 \cdot 0.03}{(1.03)^4 - 1} = \$2390.27$.

67. Let $P(n)$ denote the statement that $1+4+7+\cdots+(3n-2)=\frac{n(3n-1)}{2}$.

Step 1: $P(1)$ is the statement that $1=\frac{1[3(1)-1]}{2}=\frac{1\cdot 2}{2}$, which is true.

Step 2: Assume that $P(k)$ is true; that is, $1+4+7+\cdots+(3k-2)=\frac{k(3k-1)}{2}$. We want to use this to show that $P(k+1)$ is true. Now

$$\begin{aligned}1+4+7+10+\cdots+(3k-2)+[3(k+1)-2] &= \frac{k(3k-1)}{2}+3k+1 && \text{induction hypothesis}\\ &= \frac{k(3k-1)}{2}+\frac{6k+2}{2}=\frac{3k^2-k+6k+2}{2}\\ &= \frac{3k^2+5k+2}{2}=\frac{(k+1)(3k+2)}{2}\\ &= \frac{(k+1)[3(k+1)-1]}{2}\end{aligned}$$

Thus, $P(k+1)$ follows from $P(k)$. So by the Principle of Mathematical Induction, $P(n)$ is true for all n.

69. Let $P(n)$ denote the statement that $\left(1+\frac{1}{1}\right)\left(1+\frac{1}{2}\right)\left(1+\frac{1}{3}\right)\cdots\cdots\left(1+\frac{1}{n}\right)=n+1$.

Step 1: $P(1)$ is the statement that $1+\frac{1}{1}=1+1$, which is clearly true.

Step 2: Assume that $P(k)$ is true; that is, $\left(1+\frac{1}{1}\right)\left(1+\frac{1}{2}\right)\left(1+\frac{1}{3}\right)\cdots\cdots\left(1+\frac{1}{k}\right)=k+1$. We want to use this to show that $P(k+1)$ is true. Now

$$\begin{aligned}&\left(1+\frac{1}{1}\right)\left(1+\frac{1}{2}\right)\left(1+\frac{1}{3}\right)\cdots\cdots\left(1+\frac{1}{k}\right)\left(1+\frac{1}{k+1}\right)\\ &\quad= \left[\left(1+\frac{1}{1}\right)\left(1+\frac{1}{2}\right)\left(1+\frac{1}{3}\right)\cdots\cdots\left(1+\frac{1}{k}\right)\right]\left(1+\frac{1}{k+1}\right)\\ &\quad= (k+1)\left(1+\frac{1}{k+1}\right) && \text{induction hypothesis}\\ &\quad= (k+1)+1\end{aligned}$$

Thus, $P(k+1)$ follows from $P(k)$. So by the Principle of Mathematical Induction, $P(n)$ is true for all n.

71. $a_{n+1}=3a_n+4$ and $a_1=4$. Let $P(n)$ denote the statement that $a_n=2\cdot 3^n-2$.

Step 1: $P(1)$ is the statement that $a_1=2\cdot 3^1-2=4$, which is clearly true.

Step 2: Assume that $P(k)$ is true; that is, $a_k=2\cdot 3^k-2$. We want to use this to show that $P(k+1)$ is true. Now

$$\begin{aligned}a_{k+1} &= 3a_k+4 && \text{definition of } a_{k+1}\\ &= 3\left(2\cdot 3^k-2\right)+4 && \text{induction hypothesis}\\ &= 2\cdot 3^{k+1}-6+4=2\cdot 3^{k+1}-2\end{aligned}$$

Thus $P(k+1)$ follows from $P(k)$. So by the Principle of Mathematical Induction, $P(n)$ is true for all n.

73. $\binom{5}{2}\binom{5}{3}=\frac{5!}{2!\,3!}\cdot\frac{5!}{3!\,2!}=\frac{5\cdot 4}{2}\cdot\frac{5\cdot 4}{2}=10\cdot 10=100$

75. $\sum_{k=0}^{5}\binom{5}{k}=\binom{5}{0}+\binom{5}{1}+\binom{5}{2}+\binom{5}{3}+\binom{5}{4}+\binom{5}{5}=2\left(\frac{5!}{0!\,5!}+\frac{5!}{1!\,4!}+\frac{5!}{2!\,3!}\right)=2(1+5+10)=32$

77. $(A-B)^3=\binom{3}{0}A^3-\binom{3}{1}A^2B+\binom{3}{2}AB^2-\binom{3}{3}B^3=A^3-3A^2B+3AB^2-B^3$

79. $\left(1-x^2\right)^6 = \binom{6}{0}1^6 - \binom{6}{1}1^5x^2 + \binom{6}{2}1^4x^4 - \binom{6}{3}1^3x^6 + \binom{6}{4}1^2x^8 - \binom{6}{5}x^{10} + \binom{6}{6}x^{12}$
$= 1 - 6x^2 + 15x^4 - 20x^6 + 15x^8 - 6x^{10} + x^{12}$

81. The 20th term is $\binom{22}{19}a^3b^{19} = 1540a^3b^{19}$.

83. The rth term in the expansion of $(A+3B)^{10}$ is $\binom{10}{r}A^r(3B)^{10-r}$. The term that contains A^6 occurs when $r=6$. Thus, the term is $\binom{10}{6}A^6(3B)^4 = 210A^6 81B^4 = 17{,}010A^6B^4$.

CHAPTER 13 TEST

1. $a_n = 2n^2 - n \Rightarrow a_1 = 1, a_2 = 6, a_3 = 15, a_4 = 28, a_5 = 45, a_6 = 66$, and $S_6 = 1+6+15+28+45+66 = 161$.

3. (a) The common difference is $d = 5-2 = 3$.

(b) $a_n = 2 + (n-1)3$

(c) $a_{35} = 2 + 3(35-1) = 104$

5. (a) $a_1 = 25, a_4 = \frac{1}{5}$. Then $r^3 = \dfrac{\frac{1}{5}}{25} = \dfrac{1}{125} \Leftrightarrow r = \frac{1}{5}$, so $a_5 = ra_4 = \frac{1}{25}$.

(b) $S_8 = 25\dfrac{1-\left(\frac{1}{5}\right)^8}{1-\frac{1}{5}} = \dfrac{5^8-1}{12{,}500} = \dfrac{97{,}656}{3125}$

7. Let the common ratio for the geometric series $a_1, a_2, a_3, \ldots$ be r, so that $a_n = a_1r^{n-1}$, $n = 1, 2, 3, \ldots$. Then $a_n^2 = \left(a_1r^{n-1}\right)^2 = \left(a_1^2\right)\left(r^2\right)^{n-1}$. Therefore, the sequence $a_1^2, a_2^2, a_3^2, \ldots$ is geometric with common ratio r^2.

9. (a) The geometric sum $\frac{1}{3} + \frac{2}{3^2} + \frac{2^2}{3^3} + \frac{2^3}{3^4} + \cdots + \frac{2^9}{3^{10}}$ has $a = \frac{1}{3}$, $r = \frac{2}{3}$, and $n = 10$. So $S_{10} = \frac{1}{3}\cdot\frac{1-(2/3)^{10}}{1-(2/3)} = \frac{1}{3}\cdot 3\left(1-\frac{1024}{59{,}049}\right) = \frac{58{,}025}{59{,}049}$.

(b) The infinite geometric series $1 + \frac{1}{2^{1/2}} + \frac{1}{2} + \frac{1}{2^{3/2}} + \cdots$ has $a = 1$ and $r = 2^{-1/2} = \frac{1}{\sqrt{2}}$. Thus, $S = \frac{1}{1-1/\sqrt{2}} = \frac{\sqrt{2}}{\sqrt{2}-1} = \frac{\sqrt{2}}{\sqrt{2}-1}\cdot\frac{\sqrt{2}+1}{\sqrt{2}+1} = 2+\sqrt{2}$.

11. $\left(2x+y^2\right)^5 = \binom{5}{0}(2x)^5 + \binom{5}{1}(2x)^4y^2 + \binom{5}{2}(2x)^3\left(y^2\right)^2 + \binom{5}{3}(2x)^2\left(y^2\right)^3 + \binom{5}{4}(2x)\left(y^2\right)^4 + \binom{5}{5}\left(y^2\right)^5$
$= 32x^5 + 80x^4y^2 + 80x^3y^4 + 40x^2y^6 + 10xy^8 + y^{10}$

13. (a) Each week he gains 24% in weight, that is, $0.24a_n$. Thus, $a_{n+1} = a_n + 0.24a_n = 1.24a_n$ for $n \geq 1$. a_0 is given to be 0.85 lb. Then $a_0 = 0.85$, $a_1 = 1.24(0.85)$, $a_2 = 1.24(1.24(0.85)) = 1.24^2(0.85)$, $a_3 = 1.24\left(1.24^2(0.85)\right) = 1.24^3(0.85)$, and so on. So we can see that $a_n = 0.85(1.24)^n$.

(b) $a_6 = 1.24a_5 = 1.24(1.24a_4) = \cdots = 1.24^6a_0 = 1.24^6(0.85) \approx 3.1$ lb

(c) The sequence $a_1, a_2, a_3, \ldots$ is geometric with common ratio 1.24.

FOCUS ON MODELING Modeling with Recursive Sequences

1. (a) Since there are 365 days in a year, the interest earned per day is $\frac{0.0365}{365} = 0.0001$. Thus the amount in the account at the end of the nth day is $A_n = 1.0001A_{n-1}$ with $A_0 = \$275{,}000$.

(b) $A_0 = \$275{,}000$, $A_1 = 1.0001A_0 = 1.0001 \cdot 275{,}000 = \$275{,}027.50$,

$A_2 = 1.0001A_1 = 1.0001\,(1.0001A_0) = 1.0001^2 A_0 = \$275{,}055.00$,

$A_3 = 1.0001A_2 = 1.0001^3 A_0 = \$275{,}082.51$, $A_4 = 1.0001^4 A_0 = \$275{,}110.02$, $A_5 = 1.0001^5 A_0 = \$275{,}137.53$,

$A_6 = 1.0001^6 A_0 = \$275{,}165.04$, $A_7 = 1.0001^7 A_0 = \$275{,}192.56$

(c) $A_n = 1.0001^n \cdot 275{,}000$

3. (a) Since there are 12 months in a year, the interest earned per day is $\frac{0.03}{12} = 0.0025$. Thus the amount in the account at the end of the nth month is $A_n = 1.0025A_{n-1} + 100$ with $A_0 = \$100$.

(b) $A_0 = \$100$, $A_1 = 1.0025A_0 + 100 = 1.0025 \cdot 100 + 100 = \$200.25$,

$A_2 = 1.0025A_1 + 100 = 1.0025\,(1.0025 \cdot 100 + 100) + 100 = 1.0025^2 \cdot 100 + 1.0025 \cdot 100 + 100 = \$300.75$,

$$A_3 = 1.0025A_2 + 100 = 1.0025\left(1.0025^2 \cdot 100 + 1.0025 \cdot 100 + 100\right) + 100$$
$$= 1.0025^3 \cdot 100 + 1.0025^2 \cdot 100 + 1.0025 \cdot 100 + 100 = \$401.50,$$

$$A_4 = 1.0025A_3 + 100 = 1.0025\left(1.0025^3 \cdot 100 + 1.0025^2 \cdot 100 + 1.0025 \cdot 100 + 100\right) + 100$$
$$= 1.0025^4 \cdot 100 + 1.0025^3 \cdot 100 + 1.0025^2 \cdot 100 + 1.0025 \cdot 100 + 100 = \$502.51$$

(c) $A_n = 1.0025^n \cdot 100 + \cdots + 1.0025^2 \cdot 100 + 1.0025 \cdot 100 + 100$, the partial sum of a geometric series, so

$$A_n = 100 \cdot \frac{1 - 1.0025^{n+1}}{1 - 1.0025} = 100 \cdot \frac{1.0025^{n+1} - 1}{0.0025}.$$

(d) Since 5 years is 60 months, we have $A_{60} = 100 \cdot \frac{1.0025^{61} - 1}{0.0025} \approx \$6580.83$.

5. (a) $U_n = 1.05U_{n-1} + 0.10\left(1.05U_{n-1}\right) = 1.10\,(1.05)\,U_{n-1} = 1.155U_{n-1}$ with $U_0 = 5000$.

(b) $U_0 = \$5000$, $U_1 = 1.155U_0 = \$5775$,

$U_2 = 1.155U_1 = 1.155\,(1.155 \cdot 5000) = 1.155^2 \cdot 5000 = \$6670.13$,

$U_3 = 1.155U_2 = 1.155\left(1.155^2 \cdot 5000\right) = 1.155^3 \cdot 5000 = \$7703.99$,

$U_4 = 1.155U_3 = 1.155\left(1.155^3 \cdot 5000\right) = 1.155^4 \cdot 5000 = \$8898.11$.

(c) Using the pattern we found in part (b), we have $U_n = 1.155^n \cdot 5000$.

(d) $U_{10} = 1.155^{10} \cdot 5000 = \$21{,}124.67$.

14 COUNTING AND PROBABILITY

14.1 COUNTING

1. The Fundamental Counting Principle says that if one event can occur in m ways and a second event can occurs in n ways, then the two events can occur in order in $m \times n$ ways. So if you have two choices for shoes and three choices for hats, then the number of different shoe-hat combinations you can wear is $2 \times 3 = 6$.

3. The number of ways of choosing r objects from n objects is called the number of *combinations* of n objects taken r at a time, and is given by the formula $C(n,r) = \dfrac{n!}{r!\,(n-r)!}$.

5. $P(8,3) = \dfrac{8!}{(8-3)!} = \dfrac{8!}{5!} = 8 \cdot 7 \cdot 6 = 336$

7. $P(11,4) = \dfrac{11!}{(11-4)!} = \dfrac{11!}{7!} = 11 \cdot 10 \cdot 9 \cdot 8 = 7920$

9. $P(100,1) = \dfrac{100!}{(100-1)!} = \dfrac{100!}{99!} = 100$

11. $C(8,3) = \dfrac{8!}{3!\,(8-3)!} = \dfrac{8!}{3!\,5!} = \dfrac{8 \cdot 7 \cdot 6}{3 \cdot 2 \cdot 1} = 56$

13. $C(11,4) = \dfrac{11!}{4!\,7!} = \dfrac{11 \cdot 10 \cdot 9 \cdot 8}{4 \cdot 3 \cdot 2 \cdot 1} = 330$

15. $C(100,1) = \dfrac{100!}{1!\,99!} = \frac{100}{1} = 100$

17. By the Fundamental Counting Principle, the number of possible single-scoop ice cream cones is $\begin{pmatrix}\text{number of ways to}\\ \text{choose the flavor}\end{pmatrix} \cdot \begin{pmatrix}\text{number of ways to}\\ \text{choose the type of cone}\end{pmatrix} = 4 \cdot 3 = 12.$

19. (a) By the Fundamental Counting Principle, the possible number of ways 8 horses can complete a race, assuming no ties in any position, is

$$\begin{pmatrix}\text{number of ways to}\\ \text{choose the 1st finisher}\end{pmatrix} \cdot \begin{pmatrix}\text{number of ways to}\\ \text{choose the 2nd finisher}\end{pmatrix} \cdot \cdots \cdot \begin{pmatrix}\text{number of ways to}\\ \text{choose the 8th finisher}\end{pmatrix} = 8 \cdot 7 \cdot 6 \cdot 5 \cdot 4 \cdot 3 \cdot 2 \cdot 1 = 8! = 40{,}320$$

(b) By the Fundamental Counting Principle, the possible number of ways the first, second, and third place can be decided, assuming no ties, is $\begin{pmatrix}\text{number of ways to}\\ \text{choose the 1st finisher}\end{pmatrix} \cdot \begin{pmatrix}\text{number of ways to}\\ \text{choose the 2nd finisher}\end{pmatrix} \cdot \begin{pmatrix}\text{number of ways to}\\ \text{choose the 3th finisher}\end{pmatrix} = 8 \cdot 7 \cdot 6 = 336.$

21. The number of possible seven-digit phone numbers is $\begin{pmatrix}\text{number of ways to}\\ \text{choose the 1st digit}\end{pmatrix} \cdot \begin{pmatrix}\text{number of ways to}\\ \text{choose the 2nd digit}\end{pmatrix} \cdot \cdots \cdot \begin{pmatrix}\text{number of ways to}\\ \text{choose the 7th digit}\end{pmatrix}$. Since the first digit cannot be a 0 or a 1, there are only 8 digits to choose from, while there are 10 digits to choose from for the other six digits in the phone number. Thus the number of possible seven-digit phone numbers is $8 \cdot 10 \cdot 10 \cdot 10 \cdot 10 \cdot 10 \cdot 10 = 8{,}000{,}000$.

23. Since there are 4 main courses, there are 6 ways to choose a main course. Likewise, there are 5 drinks and 3 desserts so there are 5 ways to choose a drink and 3 ways to choose a dessert. So the number of different meals consisting of a main course, a drink, and a dessert is $\begin{pmatrix}\text{number of ways to}\\ \text{choose the main course}\end{pmatrix} \cdot \begin{pmatrix}\text{number of ways to}\\ \text{choose a drink}\end{pmatrix} \cdot \begin{pmatrix}\text{number of ways to}\\ \text{choose a dessert}\end{pmatrix} = (4)(5)(3) = 60.$

25. The number of possible sequences of heads and tails when a coin is flipped 5 times is

$$\begin{pmatrix}\text{number of possible}\\ \text{outcomes on the 1st flip}\end{pmatrix} \cdot \begin{pmatrix}\text{number of possible}\\ \text{outcomes on the 2nd flip}\end{pmatrix} \cdot \cdots \cdot \begin{pmatrix}\text{number of possible}\\ \text{outcomes on the 5th flip}\end{pmatrix} = (2)(2)(2)(2)(2) = 2^5 = 32$$

Here there are only two choices, heads or tails, for each flip.

27. Since there are six different faces on each die, the number of possible outcomes when a red die and a blue die and a white die are rolled is
$$\begin{pmatrix}\text{number of possible}\\ \text{outcomes on the red die}\end{pmatrix}\cdot\begin{pmatrix}\text{number of possible}\\ \text{outcomes on the blue die}\end{pmatrix}\cdot\begin{pmatrix}\text{number of possible}\\ \text{outcomes on the white die}\end{pmatrix}=(6)(6)(6)=6^3=216.$$

29. The number of different California license plates possible is
$$\begin{pmatrix}\text{number of ways to}\\ \text{choose a nonzero digit}\end{pmatrix}\cdot\begin{pmatrix}\text{number of ways}\\ \text{to choose 3 letters}\end{pmatrix}\cdot\begin{pmatrix}\text{number of ways}\\ \text{to choose 3 digits}\end{pmatrix}=(9)\left(26^3\right)\left(10^3\right)=158{,}184{,}000.$$

31. Since successive numbers cannot be the same, the number of possible choices for the second number in the combination is only 59. The third number in the combination cannot be the same as the second in the combination, but it can be the same as the first number, so the number of possible choices for the third number in the combination is also 59. So the number of possible combinations consisting of a number in the clockwise direction, a number in the counterclockwise direction, and then a number in the clockwise direction is $(60)(59)(59)=208{,}860$.

33. Since a student can hold only one office, the number of ways that a president, a vice-president and a secretary can be chosen from a class of 30 students is
$$\begin{pmatrix}\text{number of ways}\\ \text{to choose a president}\end{pmatrix}\cdot\begin{pmatrix}\text{number of ways to}\\ \text{choose a vice-president}\end{pmatrix}\cdot\begin{pmatrix}\text{number of ways}\\ \text{to choose a secretary}\end{pmatrix}=(30)(29)(28)=24{,}360.$$

35. We have seven choices for the first digit and 10 choices for each of the other 8 digits. Thus, the number of Social Security numbers is $7\cdot 10^8=700{,}000{,}000$.

37. (a) The number of ways to select 5 of the 8 objects is $C(8,5)=\dfrac{8!}{5!\,3!}=56$.

(b) A set with 8 elements has $2^8=256$ subsets.

39. Each subset of toppings constitutes a different way a hamburger can be ordered. Since a set with 10 elements has $2^{10}=1024$ subsets, there are 1024 different ways to order a hamburger.

41. (a) The number of ways to seat ten people in a row of ten chairs is $10!=3{,}628{,}800$.

(b) The number of ways to choose six out of ten people and seat them in six chairs is $C(10,6)\cdot 6!=\dfrac{10!}{6!\,4!}\cdot 6!=\dfrac{10!}{4!}=151{,}200$.

43. In selecting these officers, order is important and repetition is not allowed, so the number of ways of choosing 3 officers from 15 students is $P(15,3)=2730$.

45. Since the order of finish is important, we want the number of permutations of 8 objects (the contestants) taken three at a time, which is $P(8,3)=\dfrac{8!}{(8-3)!}=\dfrac{8!}{5!}=8\cdot 7\cdot 6=336$.

47. The number of ways of ordering 9 distinct objects (the contestants) is $P(9,9)=9!=362{,}880$. Here a runner cannot finish more than once, so no repetitions are allowed, and order is important.

49. The number of ways of ordering 1000 distinct objects (the contestants) taking 3 at a time is $P(1000,3)=1000\cdot 999\cdot 998=997{,}002{,}000$. We are assuming that a person cannot win more than once, that is, there are no repetitions.

51. We first place Jack in the first seat, and then seat the remaining four students. Thus the number of these arrangements is
$$\begin{pmatrix}\text{number of ways to}\\ \text{seat Jack in the first seat}\end{pmatrix}\cdot\begin{pmatrix}\text{number of ways to seat}\\ \text{the remaining four students}\end{pmatrix}=P(1,1)\cdot P(4,4)=1!\,4!=24.$$

53. Here we have 6 objects, of which 2 are blue marbles and 4 are red marbles. Thus the number of distinguishable permutations is $\dfrac{6!}{2!\,4!}=\dfrac{6\cdot 5\cdot 4!}{2\cdot 4!}=15$.

55. The number of distinguishable permutations of 12 objects (the 12 coins), from like groups of size 4 (the pennies), of size 3 (the nickels), of size 2 (the dimes) and of size 3 (the quarters) is $\dfrac{12!}{4!\,3!\,2!\,3!}=277{,}200$.

57. The number of distinguishable permutations of 12 objects (the 12 ice cream cones) from like groups of size 3 (the vanilla cones), of size 2 (the chocolate cones), of size 4 (the strawberry cones), and of size 5 (the butterscotch cones) is $\dfrac{14!}{3!\,2!\,4!\,5!} = 2{,}522{,}520$.

59. The number of distinguishable permutations of 8 objects (the 8 cleaning tasks) from like groups of size 5, 2, and 1 workers, respectively is $\dfrac{8!}{5!\,2!\,1!} = 168$.

61. Here we are interested in the number of ways of choosing three objects (the three members of the committee) from a set of 25 objects (the 25 members). The number of combinations of 25 objects taken three at a time is $C(25,3) = \dfrac{25!}{3!\,22!} = 2300$.

63. We want the number of ways of choosing a group of three from a group of 12. This number is $C(12,3) = \dfrac{12!}{3!\,9!} = 220$.

65. We want the number of ways of choosing a group (the 5-card hand) where order of selection is not important. The number of combinations of 52 objects (the 52 cards) taken 5 at a time is $C(52,5) = \dfrac{52!}{5!\,47!} = 2{,}598{,}960$.

67. The order of selection is not important, hence we must calculate the number of combinations of 10 objects (the 10 questions) taken 7 at a time, this gives $C(10,7) = \dfrac{10!}{7!\,3!} = 120$.

69. We assume that the order in which he plays the pieces in the recital is not important, so the number of combinations of 12 objects (the 12 pieces) taken 8 at a time is $C(12,8) = \dfrac{12!}{8!\,4!} = 495$.

71. The order in which the pants are selected is not important and no pair of pants is repeated, so the number of combinations of ten pairs of pants taken three at a time is $C(10,3) = \dfrac{10!}{3!\,7!} = 120$.

73. Since the order in which the numbers are selected is not important, the number of combinations of 49 numbers taken 6 at a time is $C(49,6) = \dfrac{49!}{6!\,43!} = 13{,}983{,}816$.

75. **(a)** The number of ways of choosing 5 students from the 20 students is $C(20,5) = \dfrac{20!}{5!\,15!} = 15{,}504$.

(b) The number of ways of choosing 5 students for the committee from the 12 females is $C(12,5) = \dfrac{12!}{5!\,7!} = 792$.

(c) We use the Fundamental Counting Principle to count the number of possible committees with 3 females and 2 males. Thus, we get

$$\begin{pmatrix}\text{number of ways to choose the}\\ \text{3 females from the 12 females}\end{pmatrix}\cdot\begin{pmatrix}\text{number of ways to choose the}\\ \text{2 males from the 8 males}\end{pmatrix} = C(12,3)\cdot C(8,2) = (220)(28) = 6160.$$

77. The number of ways the committee can be chosen is

$$\begin{aligned}\begin{pmatrix}\text{number of ways to}\\ \text{choose 1 president}\end{pmatrix}\cdot\begin{pmatrix}\text{number of ways to}\\ \text{choose 1 vice-president}\end{pmatrix}\cdot\begin{pmatrix}\text{number of ways to}\\ \text{choose 4 other members}\end{pmatrix} &= C(20,1)\cdot C(19,1)\cdot C(18,4)\\ &= 20\cdot 19\cdot 3060 = 1{,}162{,}800\end{aligned}$$

79. The number of ways the committee can be chosen is

$$\begin{aligned}&\begin{pmatrix}\text{number of ways to}\\ \text{choose 2 of 6 freshmen}\end{pmatrix}\cdot\begin{pmatrix}\text{number of ways to}\\ \text{choose 3 of 8 sophomores}\end{pmatrix}\cdot\begin{pmatrix}\text{number of ways to}\\ \text{choose 4 of 12 juniors}\end{pmatrix}\cdot\begin{pmatrix}\text{number of ways to}\\ \text{choose 5 of 10 seniors}\end{pmatrix}\\ &\quad = C(6,2)\cdot C(8,3)\cdot C(12,4)\cdot C(10,5) = 15\cdot 56\cdot 495\cdot 252 = 104{,}781{,}600\end{aligned}$$

81. We choose 3 forwards from the forwards, 2 defensemen from the defensemen, and the goalie from the two goalies. Thus the number of ways to pick the 6 starting players is

$$\begin{aligned}\begin{pmatrix}\text{number of ways to}\\ \text{pick 3 of 12 forwards}\end{pmatrix}\cdot\begin{pmatrix}\text{number of ways to}\\ \text{pick 2 of 6 defensemen}\end{pmatrix}\cdot\begin{pmatrix}\text{number of ways to}\\ \text{pick 1 of 2 goalies}\end{pmatrix} &= C(12,3)\cdot C(6,2)\cdot C(2,1)\\ &= (220)(15)(2) = 6600\end{aligned}$$

83. We count the total number of committees and subtract the number that contain both Barry and Harry. The total number of committees possible is $C(10,4)$ and the number that contain both Barry and Harry is $C(8,2)$, so the number of possible committees is $C(10,4) - C(8,2) = 210 - 28 = 182$.

85. Since the two algebra books must be next to each other, we first consider them as one object. So we now have four objects to arrange and there are 4! ways to arrange these four objects. Now there are 2 ways to arrange the two algebra books. Thus the number of ways that 5 mathematics books may be placed on a shelf if the two algebra books are to be next to each other is $2 \cdot 4! = 48$.

87. (a) To find the number of ways the men and women can be seated we first select and place a man in the first seat and then arrange the other 7 people. Thus we get $\begin{pmatrix}\text{select 1 of} \\ \text{the 4 men}\end{pmatrix} \cdot \begin{pmatrix}\text{arrange the} \\ \text{remaining 7 people}\end{pmatrix} = C(4,1) \cdot P(7,7) = 4 \cdot 7! = 20{,}160$.

(b) To find the number of ways the men and women can be seated we first select and place a woman in the first and last seats and then arrange the other 6 people. Thus we get

$\begin{pmatrix}\text{arrange 2 of} \\ \text{the 4 women}\end{pmatrix} \cdot \begin{pmatrix}\text{arrange the} \\ \text{remaining 6 people}\end{pmatrix} = P(4,2) \cdot P(6,6) = 12 \cdot 6! = 8{,}640$.

89. The number of ways the top finalist can be chosen is

$$\begin{pmatrix}\text{number of ways} \\ \text{to choose the 6} \\ \text{semifinalists from the 30}\end{pmatrix} \cdot \begin{pmatrix}\text{number of ways} \\ \text{to choose the 2} \\ \text{finalists from the 6}\end{pmatrix} \cdot \begin{pmatrix}\text{number of ways} \\ \text{to choose the winner} \\ \text{from the 2 finalists}\end{pmatrix} = C(30,6) \cdot C(6,2) \cdot C(2,1)$$

$$= (593{,}775)(15)(2) = 17{,}813{,}250$$

91. Since there are 26 letters, the possible number of combinations of the first and the last initials is $(26)(26) = 676$. Since $677 > 676$, there must be at least two people that have the same first and last initials in any group of 677 people.

93. We are only interested in selecting a set of three marbles to give to Luke and a set of two marbles to give to Mark, not the order in which we hand out the marbles. Since both $C(10,3) \cdot C(7,2)$ and $C(10,2) \cdot C(8,3)$ count the number of ways this can be done, these numbers must be equal. (Calculating these values shows that they are indeed equal.) In general, if we wish to find two distinct sets of k and r objects selected from n objects ($k + r \leq n$), then we can either first select the k objects from the n objects and then select the r objects from the $n - k$ remaining objects, or we can first select the r objects from the n objects and then the k objects from the $n - r$ remaining objects. Thus $\binom{n}{r} \cdot \binom{n-r}{k} = \binom{n}{k} \cdot \binom{n-k}{r}$.

14.2 PROBABILITY

1. The set of all possible outcomes of an experiment is called the *sample space*. A subset of the sample space is called an *event*. The sample space for the experiment of tossing two coins is $S = \{HH, HT, TH, TT\}$, and the event "getting at least one head" is $E = \{HH, HT, TH\}$. The probability of getting at least one head is $P(E) = \dfrac{n(E)}{n(S)} = \dfrac{3}{4}$.

3. The conditional probability of E given that F occurs is $P(E \mid F) = \dfrac{n(E \cap F)}{n(F)}$. So in tossing a die, the conditional probability of the event E "getting a six" given that that the event F "getting an even number" has occurred is $P(E \mid F) = \dfrac{1}{3}$.

5. (a) $S = \{1, 2, 3, 4, 5, 6\}$

(b) $E = \{2, 4, 6\}$

(c) $E = \{5, 6\}$

7. Let H stand for head and T for tails.

(a) The sample space is $S = \{HH, HT, TH, TT\}$.

(b) Let E be the event of getting exactly two heads, so $E = \{HH\}$. Then $P(E) = \frac{n(E)}{n(S)} = \frac{1}{4}$.

(c) Let F be the event of getting at least one head. Then $F = \{HH, HT, TH\}$, and $P(F) = \frac{n(F)}{n(S)} = \frac{3}{4}$.

(d) Let G be the event of getting exactly one head, that is, $G = \{HT, TH\}$. Then $P(G) = \frac{n(G)}{n(S)} = \frac{2}{4} = \frac{1}{2}$.

9. (a) Let E be the event of rolling a six. Then $P(E) = \frac{n(E)}{n(S)} = \frac{1}{6}$.

(b) Let F be the event of rolling an even number. Then $F = \{2, 4, 6\}$. So $P(F) = \frac{n(F)}{n(S)} = \frac{3}{6} = \frac{1}{2}$.

(c) Let G be the event of rolling a number greater than 5. Since 6 is the only face greater than 5, $P(G) = \frac{n(G)}{n(S)} = \frac{1}{6}$.

11. (a) Let E be the event of choosing a king. Since a deck has four kings, $P(E) = \frac{n(E)}{n(S)} = \frac{4}{52} = \frac{1}{13}$.

(b) Let F be the event of choosing a face card. Since there are three face cards per suit and four suits, $P(F) = \frac{n(F)}{n(S)} = \frac{12}{52} = \frac{3}{13}$.

(c) Let F be the event of choosing a face card. Then $P(F') = 1 - P(F) = 1 - \frac{3}{13} = \frac{10}{13}$.

13. (a) Let E be the event of selecting a red ball. Since the jar contains five red balls, $P(E) = \frac{n(E)}{n(S)} = \frac{5}{8}$.

(b) Let F be the event of selecting a yellow ball. Since there is only one yellow ball, $P(F') = 1 - P(F) = 1 - \frac{n(F)}{n(S)} = 1 - \frac{1}{8} = \frac{7}{8}$.

(c) Let G be the event of selecting a black ball. Since there are no black balls in the jar, $P(G) = \frac{n(G)}{n(S)} = \frac{0}{8} = 0$.

15. (a) Let E be the event of dealing five hearts. Since there are 13 hearts, $P(E) = \dfrac{C(13, 5)}{C(52, 5)} = \dfrac{1287}{2{,}598{,}960} \approx 0.000495$.

(b) Let E be the event of choosing five cards of the same suit. Since there are four suits and 13 cards in each suit, $n(E) = 4 \cdot C(13, 5)$. Also, $n(S) = C(52, 5)$. Therefore, $P(E) = \dfrac{4 \cdot C(13, 5)}{C(52, 5)} = \dfrac{5{,}148}{2{,}598{,}960} \approx 0.00198$.

(c) Let E be the event of dealing five face cards. Since there are 3 face cards for each suit and 4 suits, $P(E) = \dfrac{C(12, 5)}{C(52, 5)} = \dfrac{792}{2{,}598{,}960} \approx 0.000305$.

(d) Let E be the event of dealing a royal flush (ace, king, queen, jack, and 10 of the same suit). Since there is only one such sequence for each suit, there are only 4 royal flushes, so $P(E) = \dfrac{4}{C(52, 5)} = \dfrac{4}{2{,}598{,}960} \approx 1.53908 \times 10^{-6}$.

17. (a) Let E be the event of choosing two red balls. Since there are three red balls, $n(E) = C(3, 2)$. Also, $n(S) = C(8, 2)$. Therefore, $P(E) = \dfrac{n(E)}{n(S)} = \dfrac{C(3, 2)}{C(8, 2)} = \dfrac{3}{28} \approx 0.107$.

(b) Let E be the event of choosing two white balls. Since there are five white balls, $n(E) = C(5, 2)$, so $P(E) = \dfrac{n(E)}{n(S)} = \dfrac{C(5, 2)}{C(8, 2)} = \dfrac{10}{28} = \dfrac{5}{14} \approx 0.357$.

19. (a) Let E be the probability that at least one card is a spade. The number of hands that do not contain a spade is the number of possible 5-card hands using the other three suits, that is, $C(39, 5)$. Thus, $P(E) = 1 - \dfrac{C(39, 5)}{C(52, 5)} = \dfrac{7411}{9520} \approx 0.778$.

(b) Let E be the probability that at least one card is a face card. The number of hands that *do not* contain a face card is the number of possible 5-card hands using the cards of the deck that are not face cards, that is, $C(40, 5)$. Thus, $P(E) = 1 - \dfrac{C(40, 5)}{C(52, 5)} = \dfrac{6221}{8330} \approx 0.747$.

21. **(a)** Let E be the event that the spinner stops on red. Since 12 of the regions are red, $P(E) = \frac{12}{16} = \frac{3}{4}$.

(b) Let F be the event that the spinner stops on an even number. Since 8 of the regions are even-numbered, $P(F) = \frac{8}{16} = \frac{1}{2}$.

(c) Since 4 of the even-numbered regions are red, $P(E \cup F) = P(E) + P(F) - P(E \cap F) = \frac{3}{4} + \frac{1}{2} - \frac{4}{16} = 1$.

23. **(a)** Yes, the events are mutually exclusive since the number cannot be both even and odd. So $P(E \cup F) = P(E) + P(F) = \frac{3}{6} + \frac{3}{6} = 1$.

(b) No, the events are not mutually exclusive since 6 is both even and greater than 4. So $P(E \cup F) = P(E) + P(F) - P(E \cap F) = \frac{3}{6} + \frac{2}{6} - \frac{1}{6} = \frac{2}{3}$.

25. **(a)** No, the events E and F are not mutually exclusive since the jack, queen, and king of spades are both face cards and spades. So $P(E \cup F) = P(E) + P(F) - P(E \cap F) = \frac{13}{52} + \frac{12}{52} - \frac{3}{52} = \frac{11}{26}$.

(b) Yes, the events E and F are mutually exclusive since the card cannot be both a heart and a spade. So $P(E \cup F) = P(E) + P(F) = \frac{13}{52} + \frac{13}{52} = \frac{1}{2}$.

27. **(a)** Let E denote a roll of five and F a roll greater than three. Using the formula for conditional probability, we have $P(E \mid F) = \dfrac{n(E \cap F)}{n(F)} = \dfrac{1}{3}$.

(b) Let E denote a roll of three and F an odd roll. Using the formula for conditional probability, we have $P(E \mid F) = \dfrac{n(E \cap F)}{n(F)} = \dfrac{1}{3}$.

29. Let E denote the spinner stopping on an even number and F the spinner stopping on red. Then $P(E \mid F) = \dfrac{n(E \cap F)}{n(F)} = \dfrac{4}{12} = \dfrac{1}{3}$.

31. **(a)** There is only one red ball numbered 3 and one green ball numbered 3. If the ball drawn is numbered 3, then the probability it is red is $\frac{1}{2}$.

(b) There is only one ball numbered 7 and it is green, so if the ball drawn is numbered 7, then the probability it is green is 1.

(c) There are two even-numbered red balls and three even-numbered green balls, so if the ball is even-numbered, then the probability it is a red ball is $\frac{2}{5}$.

(d) There are five red balls, and two are even-numbered, so if the ball drawn is red, then the probability it is even-numbered is $\frac{2}{5}$.

33. **(a)** Let E be the event of drawing a black ball first. Because the jar contains seven black balls and three white balls, the probability of the first ball being black is $P(E) = \frac{7}{10}$. The probability of the second ball being white is $\frac{3}{9} = \frac{1}{3}$, so the probability of the intersection is $\frac{7}{10} \cdot \frac{1}{3} = \frac{7}{30}$.

(b) Here the probability of the second ball being black is $\frac{6}{9} = \frac{2}{3}$, so the probability of the intersection is $\frac{7}{10} \cdot \frac{2}{3} = \frac{7}{15}$.

35. **(a)** The probability of the first card being an ace is $\frac{4}{52}$ and the probability of the second being a king is $\frac{4}{51}$, so the probability of the intersection is $\frac{4}{52} \cdot \frac{4}{51} = \frac{4}{663}$.

(b) The probability of the first card being an ace is $\frac{4}{52}$ and the probability of the second being an ace is $\frac{3}{51}$, so the probability of the intersection is $\frac{4}{52} \cdot \frac{3}{51} = \frac{1}{221}$.

37. Let E be the event of getting a 1 on the first roll, and let F be the event of getting an even number on the second roll. Since these events are independent, $P(E \cap F) = P(E) \cdot P(F) = \frac{1}{6} \cdot \frac{3}{6} = \frac{1}{12}$.

39. **(a)** Yes. What happens on spinner A does not influence what happens on spinner B.

(b) The probability that A stops on red and B stops on yellow is $P(E \cap F) = P(E) \cdot P(F) = \left(\frac{2}{4}\right)\left(\frac{2}{8}\right) = \frac{1}{8}$.

41. **(a)** Let B and G stand for "boy"and "girl". Then
$S = \{BBBB, GBBB, BGBB, BBGB, BBBG, GGBB, GBGB, GBBG,$
$BGGB, BGBG, BBGG, BGGG, GBGG, GGBG, GGGB, GGGG\}$

(b) Let E be the event that the couple has only boys. Then $E = \{BBBB\}$ and $P(E) = \frac{1}{16}$.

(c) Let F be the event that the couple has 2 boys and 2 girls. Then
$F = \{GGBB, GBGB, GBBG, BGGB, BGBG, BBGG\}$, so $P(F) = \frac{6}{16} = \frac{3}{8}$.

(d) Let G be the event that the couple has 4 children of the same sex. Then $G = \{BBBB, GGGG\}$, and $P(G) = \frac{2}{16} = \frac{1}{8}$.

(e) Let H be the event that the couple has at least 2 girls. Then H' is the event that the couple has fewer than two girls. Thus, $H' = \{BBBB, GBBB, BGBB, BBGB, BBBG\}$, so $n(H') = 5$, and $P(H) = 1 - P(H') = 1 - \frac{5}{16} = \frac{11}{16}$.

43. Let E be the event that the ball lands in an odd numbered slot. Since there are 18 odd numbers between 1 and 36, $P(E) = \frac{18}{38} = \frac{9}{19}$.

45. Let E be the event of picking the 6 winning numbers. Since there is only one way to pick these, $P(E) = \dfrac{1}{C(49,6)} = \dfrac{1}{13{,}983{,}816} \approx 7.15 \times 10^{-8}$.

47. The sample space consist of all possible True-False combinations, so $n(S) = 2^{10}$. Let E be the event that the student answers all 10 questions correctly. Since there is only one way to answer all 10 questions correctly, $P(E) = \dfrac{1}{2^{10}} = \dfrac{1}{1024}$.

49. **(a)** Let E be the event that the monkey types "Hamlet" as his first word. Since "Hamlet" contains 6 letters and there are 48 typewriter keys, $P(E) = \dfrac{1}{48^6} \approx 8.18 \times 10^{-11}$.

(b) Let F be the event that the monkey types "to be or not to be" as his first words. Since this phrase has 18 characters (including the blanks), $P(F) = \dfrac{1}{48^{18}} \approx 5.47 \times 10^{-31}$.

51. Let E be the event that the toddler will arrange the 8 blocks to spell *TRIANGLE* or *INTEGRAL*. The number of ways of arranging these blocks is the number of distinguishable permutations of 8 blocks. Since no two blocks are the same, the number of distinguishable permutations is 8!. Two of these arrangements result in event E, so $P(E) = \dfrac{2}{8!} \approx 4.96 \times 10^{-5}$ or 0.0000496.

53. **(a)** Let E be the event that the pea is tall. Since tall is dominant, $E = \{TT, Tt, tT\}$. So $P(E) = \frac{3}{4}$.

(b) E' is the event that the pea is short. So $P(E') = 1 - P(E) = 1 - \frac{3}{4} = \frac{1}{4}$.

55. Let E be the event that the player wins on spin 1, and let F be the event that the player wins on spin 2. What happens on the first spin does not influence what happens on the second spin, so the events are independent. Thus, $P(E \cap F) = P(E) \cdot P(F) = \frac{1}{38} \cdot \frac{1}{38} = \frac{1}{1444}$.

57. Let E, F and G denote the events of rolling two ones on the first, second, and third rolls, respectively, of a pair of dice. The events are independent, so $P(E \cap F \cap G) = P(E) \cdot P(F) \cdot P(G) = \frac{1}{36} \cdot \frac{1}{36} \cdot \frac{1}{36} = \frac{1}{36^3} \approx 2.14 \times 10^{-5}$.

59. Let E be the event that the marble is red and F be the event that the number is odd-numbered. Then E' is the event that the marble is blue, and F' is the event that the marble is even-numbered.

(a) $P(E) = \frac{6}{16} = \frac{3}{8}$

(b) $P(F) = \frac{8}{16} = \frac{1}{2}$

(c) $P(E \cup F) = P(E) + P(F) - P(E \cap F) = \frac{6}{16} + \frac{8}{16} - \frac{3}{16} = \frac{11}{16}$

(d) $P(E' \cup F') = P(E') + P(F') - P(E' \cap F') = \frac{10}{16} + \frac{8}{16} - \frac{5}{16} = \frac{13}{16}$.

61. The probability of getting 2 red balls by picking from jar B is $\left(\frac{5}{7}\right)\left(\frac{4}{6}\right) = \frac{10}{21}$. The probability of getting 2 red balls by picking one ball from each jar is $\left(\frac{3}{7}\right)\left(\frac{5}{7}\right) = \frac{15}{49}$. The probability of getting 2 red balls after putting all balls in one jar is $\left(\frac{8}{14}\right)\left(\frac{7}{13}\right) = \frac{4}{13}$. Hence, picking both balls from jar B gives the greatest probability.

63. Let E be the event that she opens the lock within an hour. The number of combinations she can try in one hour is $10 \cdot 60 = 600$. The number of possible combinations is $P(40, 3)$ assuming that no number can be repeated. Thus $P(E) = \dfrac{600}{P(40,3)} = \dfrac{600}{59{,}280} = \dfrac{5}{494} \approx 0.010$.

65. Let E be the event that Paul stands next to Phyllis. To find $n(E)$ we treat Paul and Phyllis as one object and find the number of ways to arrange the 19 objects and then multiply the result by the number of ways to arrange Paul and Phyllis. So $n(E) = 19! \cdot 2!$. The sample space is all the ways that 20 people can be arranged. Thus $P(E) = \dfrac{19! \cdot 2!}{20!} = \dfrac{2}{20} = 0.10$.

67. Let E be the event that the monkey arranges the 11 blocks to spell *PROBABILITY*. The number of ways of arranging these blocks is the number of distinguishable permutations of 11 blocks. Since there are two blocks labeled B and two blocks labeled I, the number of distinguishable permutations is $\dfrac{11!}{2!\,2!}$. Only one of these arrangements spells the word *PROBABILITY*. Thus $P(E) = \dfrac{1}{\frac{11!}{2!\,2!}} = \dfrac{2!\,2!}{11!} = \dfrac{1}{9{,}979{,}200}$.

14.3 BINOMIAL PROBABILITY

1. A binomial experiment is one in which there are exactly *two* outcomes. One outcome is called *success* and the other is called *failure*.

3. $P(\text{2 successes in 5}) = C(5,2) \cdot \left(0.7^2\right)\left(0.3^3\right) = 0.13230$

5. $P(\text{0 success in 5}) = C(5,0) \cdot \left(0.7^0\right)\left(0.3^5\right) = 0.00243$

7. $P(\text{1 success in 5}) = C(5,1) \cdot \left(0.7^1\right)\left(0.3^4\right) = 0.02835$

9. $P(\text{at least 4 successes}) = P(\text{4 successes}) + P(\text{5 successes}) = C(5,4)(0.7)^4(0.3)^1 + C(5,5)(0.7)^5(0.3)^0$
$= 0.36015 + 0.16807 = 0.52822$

11. $P(\text{at most 1 failure}) = P(\text{0 failure}) + P(\text{1 failure}) = P(\text{5 successes}) + P(\text{4 successes})$
$= C(5,5)(0.7)^5(0.3)^0 + C(5,4)(0.7)^4(0.3)^1$
$= 0.16807 + 0.36015 = 0.52822$

13. $P(\text{at least 2 successes}) = P(\text{2 successes}) + P(\text{3 successes}) + P(\text{4 successes}) + P(\text{5 successes})$
$= 0.13230 + 0.30870 + 0.36015 + 0.16807 = 0.96922$

15. (a)

Outcome	Probability
1	0.2
2	0.2
3	0.2
4	0.2
5	0.2

(b)

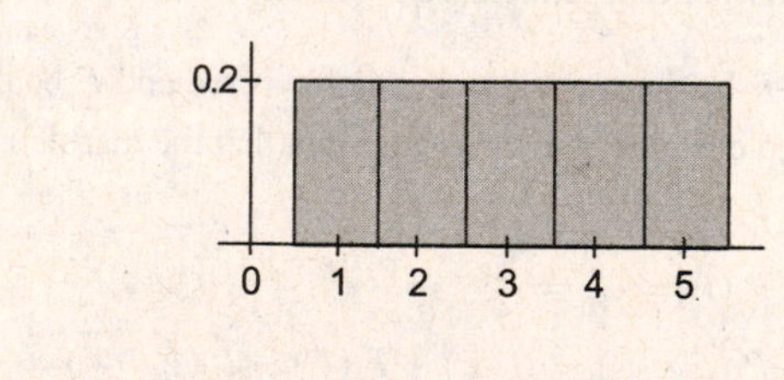

17. (a)

r	Probability
0	$\frac{1}{16}$
1	$\frac{1}{4}$
2	$\frac{3}{8}$
3	$\frac{1}{4}$
4	$\frac{1}{16}$

(b)

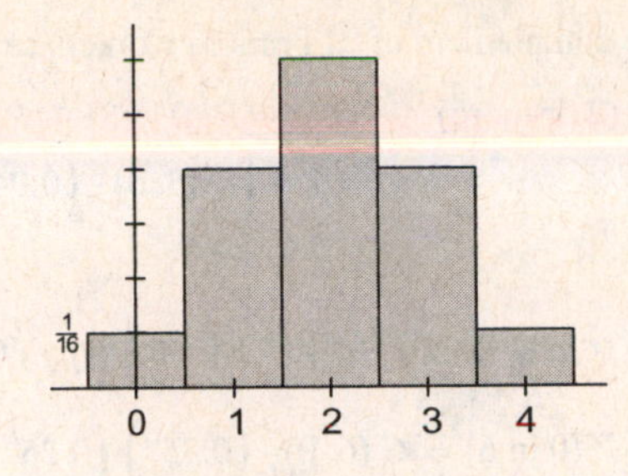

19. (a)

r	Probability
0	0.2097
1	0.3670
2	0.2753
3	0.1147
4	0.0287
5	0.0043
6	0.00036
7	0.000013

(b)

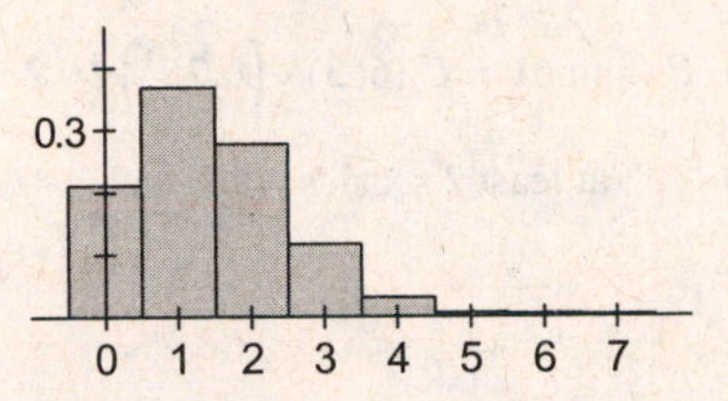

21. Here "success" is "face is 4" and P (face is 4) $= \frac{1}{6}$. Then $P\,(2 \text{ successes in } 6) = C\,(6,2) \cdot \left(\frac{1}{6}\right)^2 \left(\frac{5}{6}\right)^4 = 0.20094$.

23. $P\,(4 \text{ successes in } 10) = C\,(10,4) \cdot \left(0.4^4\right)\left(0.6^6\right) \approx 0.25082$

25. (a) $P\,(5 \text{ in } 10) = C\,(10,5) \cdot \left(0.45^5\right)\left(0.55^5\right) \approx 0.23403$

(b) P (at least 3) $= 1 - P\,(\text{at most } 2) = P\,(0 \text{ in } 10) + P\,(1 \text{ in } 10) + P\,(2 \text{ in } 10)$

$= 1 - \left[C\,(10,0) \cdot \left(0.45^0\right)\left(0.55^{10}\right) + C\,(10,1) \cdot \left(0.45^1\right)\left(0.55^9\right) + C\,(10,2) \cdot \left(0.45^2\right)\left(0.55^8\right)\right]$

≈ 0.90044

27. (a) The complement of at least 1 germinating is no seed germinating, so

$P\,(\text{at least 1 germinates}) = 1 - P\,(0 \text{ germinates}) = 1 - C\,(4,0) \cdot \left(0.75^0\right)\left(0.25^4\right) \approx 0.99609$.

(b) P (at least 2 germinate) $= P\,(2 \text{ germinates}) + P\,(3 \text{ germinates}) + P\,(4 \text{ germinates})$

$= C\,(4,2) \cdot \left(0.75^2\right)\left(0.25^2\right) + C\,(4,3) \cdot \left(0.75^3\right)\left(0.25^1\right) + C\,(4,4) \cdot \left(0.75^4\right)\left(0.25^0\right)$

≈ 0.94922

(c) $P\,(4 \text{ germinates}) = C\,(4,4) \cdot \left(0.75^4\right)\left(0.25^0\right) \approx 0.31641$

29. (a) $P\,(\text{all 10 are boys}) = C\,(10,10) \cdot \left(0.52^{10}\right)\left(0.48^0\right) \approx 0.0014456$.

(b) $P\,(\text{all 10 are girls}) = C\,(10,0) \cdot \left(0.52^0\right)\left(0.48^{10}\right) \approx 0.00064925$.

(c) $P\,(5 \text{ in 10 are boys}) = C\,(10,5) \cdot \left(0.52^5\right)\left(0.48^5\right) \approx 0.24413$.

31. (a) $P\,(3 \text{ in } 3) = C\,(3,3) \cdot \left(0.005^3\right)\left(0.995^0\right) \approx 0.000000125$.

(b) The complement of "one or more bulbs is defective" is "none of the bulbs is defective." So

$P\,(\text{at least 1 defective}) = 1 - P\,(\text{none is defective}) = 1 - C\,(3,0) \cdot \left(0.005^0\right)\left(0.995^3\right) \approx 0.014925$.

33. The complement of "2 or more workers call in sick" is "0 or 1 worker calls in sick." So

$$\begin{aligned} P\,(2 \text{ or more}) &= 1 - [P\,(0 \text{ in } 8) + P\,(1 \text{ in } 8)] \\ &= 1 - \left[C\,(8,0)\cdot\left(0.04^0\right)\left(0.96^8\right) + C\,(8,1)\cdot\left(0.04^1\right)\left(0.96^7\right)\right] \approx 0.038147 \end{aligned}$$

35. (a) $P\,(6 \text{ in } 6) = C\,(6,6)\cdot\left(0.75^6\right)\left(0.25^0\right) \approx 0.17798$

(b) $P\,(0 \text{ in } 6) = C\,(6,0)\cdot\left(0.75^0\right)\left(0.25^6\right) \approx 0.00024414$

(c) $P\,(3 \text{ in } 6) = C\,(6,3)\cdot\left(0.75^3\right)\left(0.25^3\right) \approx 0.13184$

(d) $$\begin{aligned} P\,(\text{at least 2 seasick}) &= 1 - P\,(\text{at most 1 seasick}) = 1 - [P\,(6 \text{ in } 6 \text{ OK}) + P\,(5 \text{ in } 6 \text{ OK})] \\ &= 1 - \left[C\,(6,6)\cdot\left(0.75^6\right)\left(0.25^0\right) + C\,(6,5)\cdot\left(0.75^5\right)\left(0.25^1\right)\right] \approx 0.46606 \end{aligned}$$

37. (a) The complement of "at least one gets the disease" is "none gets the disease." Then

$P\,(\text{at least 1 gets the disease}) = 1 - P\,(0 \text{ gets the disease}) = 1 - C\,(4,0)\cdot\left(0.25^0\right)\left(0.75^4\right) \approx 0.68359.$

(b) $$\begin{aligned} P\,(\text{at least 3 get the disease}) &= P\,(3 \text{ get the disease}) + P\,(4 \text{ get the disease}) \\ &= C\,(4,3)\cdot\left(0.25^3\right)\left(0.75^1\right) + C\,(4,4)\cdot\left(0.25^4\right)\left(0.75^0\right) \approx 0.05078 \end{aligned}$$

39. Fred (a nonsmoker) is already in the room, so this exercise concerns the remaining 4 participants assigned to the room.

(a) $P\,(1 \text{ in } 4 \text{ is a smoker}) = C\,(4,1)\cdot\left(0.3^1\right)\left(0.7^3\right) = 0.4116$

(b) $P\,(\text{at least 1 smoker}) = 1 - P\,(0 \text{ in } 4 \text{ is a smoker}) = 1 - C\,(4,0)\cdot\left(0.3^0\right)\left(0.7^4\right) = 0.7599$

41. (a) $$\begin{aligned} P\,(8 \text{ or more recover}) &= P\,(0 \text{ dies}) + P\,(1 \text{ dies}) + P\,(2 \text{ die}) \\ &= C\,(10,0)\,(0.6)^0\,(0.4)^{10} + C\,(10,1)\,(0.6)^1\,(0.4)^9 + C\,(10,2)\,(0.6)^2\,(0.4)^8 \approx 0.0123 \end{aligned}$$

(b) Yes, the drug appears to be effective.

43. (a)

Number of heads	Probability
0	0.003906
1	0.03125
2	0.109375
3	0.21875
4	0.273475
5	0.21875
6	0.109375
7	0.03125
8	0.003906

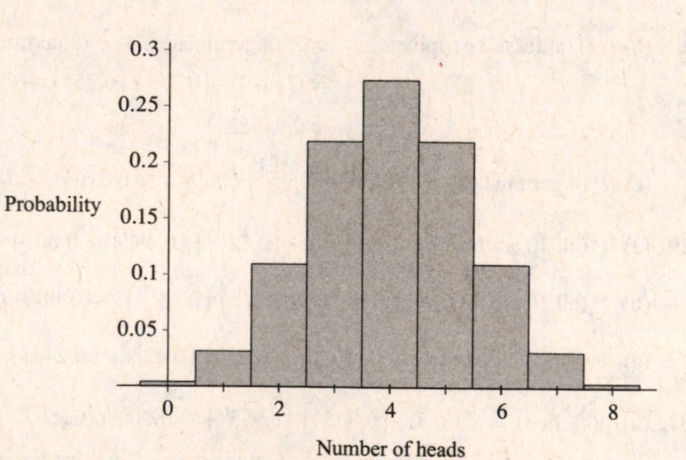

If $n = 8$, then 4 heads has the greatest probability of occurring. If the coin is flipped 100 times, then 50 heads has the greatest probability of occurring.

(b)

Number of heads	Probability
0	0.001953
1	0.017578
2	0.070313
3	0.164063
4	0.246094
5	0.246094
6	0.164063
7	0.070313
8	0.017578
9	0.001953

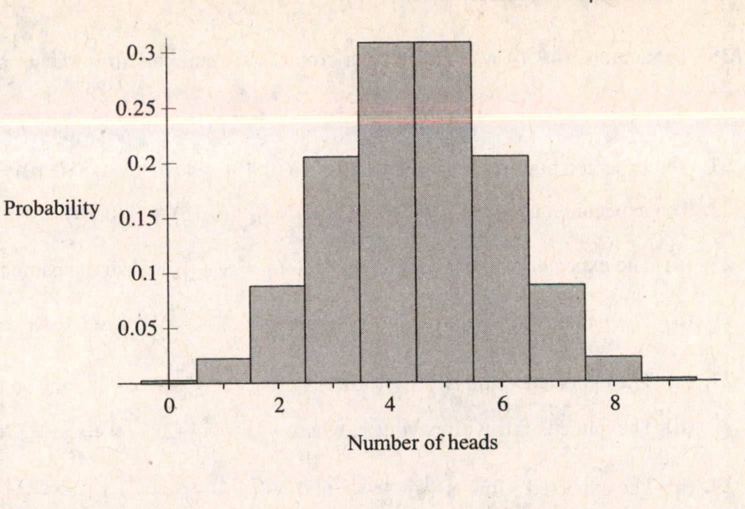

If $n = 9$, then 4 and 5 heads are the most likely outcomes. If the coin is flipped 101 times, then 50 and 51 heads are the most likely outcomes.

14.4 EXPECTED VALUE

1. If a game gives payoffs of \$10 and \$100 with probabilities 0.9 and 0.1, respectively, then the expected value of this game is $E = 10 \times 0.9 + 100 \times 0.1 = \$19$.

3. Mike gets \$2 with probability $\frac{1}{2}$ and \$1 with probability $\frac{1}{2}$. Thus, $E = (2)\left(\frac{1}{2}\right) + (1)\left(\frac{1}{2}\right) = 1.5$, and so his expected winnings are \$1.50 per game.

5. Since the probability of drawing the ace of spades is $\frac{1}{52}$, the expected value of this game is $E = (100)\left(\frac{1}{52}\right) + (-1)\left(\frac{51}{52}\right) = \frac{49}{52} \approx 0.94$. So your expected winnings are \$0.94 per game.

7. Since the probability that Carol rolls a six is $\frac{1}{6}$, the expected value of this game is $E = (3)\left(\frac{1}{6}\right) + (0.50)\left(\frac{5}{6}\right) = \frac{5.5}{6} \approx 0.9167$. So Carol expects to win \$0.92 per game.

9. Since the probability that the die shows an even number equals the probability that that die shows an odd number, the expected value of this game is $E = (2)\left(\frac{1}{2}\right) + (-2)\left(\frac{1}{2}\right) = 0$. So Tom should expect to break even after playing this game many times.

11. Since it costs \$0.50 to play, if you get a silver dollar, you win only $1 - 0.50 = \$0.50$. Thus the expected value of this game is $E = (0.50)\left(\frac{2}{10}\right) + (-0.50)\left(\frac{8}{10}\right) = -0.30$. So your expected winnings are $-\$0.30$ per game. In other words, you should expect to lose \$0.30 per game.

13. You can either win \$35 or lose \$1, so the expected value of this game is $E = (35)\left(\frac{1}{38}\right) + (-1)\left(\frac{37}{38}\right) = -\frac{2}{38} = -0.0526$. Thus the expected value is $-\$0.0526$ per game.

15. By the rules of the game, a player can win \$10 or \$5, break even, or lose \$100. Thus the expected value of this game is $E = (10)\left(\frac{10}{100}\right) + (5)\left(\frac{10}{100}\right) + (-100)\left(\frac{2}{100}\right) + (0)\left(\frac{78}{100}\right) = -0.50$. So the expected winnings per game are $-\$0.50$.

17. If the stock goes up to \$20, she expects to make $\$20 - \$5 = \$15$. And if the stock falls to \$1, then she has lost $\$5 - \$1 = \$4$. So the expected value of her profit is $E = (15)(0.1) - (4)(0.9) = -2.1$. Thus, her expected profit per share is $-\$2.10$, that is, she should expect to lose \$2.10 per share. She did not make a wise investment.

19. There are $C(49,6)$ ways to select a group of six numbers from the group of 49 numbers, of which only one is a winning set. Thus the expected value of this game is $E = \left(10^6 - 1\right)\left(\frac{1}{C(49,6)}\right) + (-1)\left(1 - \frac{1}{C(49,6)}\right) \approx -\$0.93$.

21. The expected number is $E = 2(0.15) + 3(0.45) + 4(0.30) + 5(0.10) = 3.35$ hours of TV.

23. The expected number is $3(0.30) + 2(0.45) + 1(0.15) + 0(0.10) = 1.95$ times in any given week.

25. **(a)** The expected value is $\frac{1}{52}(12) - \frac{51}{52}\left(\frac{1}{2}\right) = -\frac{27}{104}$, and so the game is not fair.

(b) The game is fair with payout x, where $\frac{1}{52}(x) - \frac{51}{52}\left(\frac{1}{2}\right) = 0 \Leftrightarrow x = \frac{51}{2} = \$25.50$.

27. **(a)** The expected value is $\frac{1}{6} \cdot \frac{1}{6}(30) - \frac{35}{36}(2) = \frac{30-70}{36} = -\frac{10}{9}$, and so the game is not fair.

(b) The game is fair with payout x, where $\frac{1}{36}x - \frac{35}{36}(2) = 0 \Leftrightarrow x = \$70$.

29. **(a)** The expected value is $\frac{1}{52} \cdot \frac{1}{6} \cdot \frac{1}{2}(600) - \left(1 - \frac{1}{52} \cdot \frac{1}{6} \cdot \frac{1}{2}\right)1 = -\frac{23}{624}$, and so the game is not fair.

(b) The game is fair with payout x, where $\frac{1}{52} \cdot \frac{1}{6} \cdot \frac{1}{2}x - \left(1 - \frac{1}{52} \cdot \frac{1}{6} \cdot \frac{1}{2}\right) = 0 \Leftrightarrow x = \$623$.

31. If you win, you win \$1 million minus the price of the stamp. If you lose, you lose only the price of the stamp (currently 44 cents). So the expected value of this game is $(999{,}999.56) \cdot \frac{1}{20 \times 10^6} + (-0.44) \cdot \frac{20 \times 10^6 - 1}{20 \times 10^6} = -0.39$. Thus, you expect to lose 39 cents on each entry, and so it's not worth it.

CHAPTER 14 REVIEW

1. The number of possible outcomes is $\left(\begin{array}{c}\text{number of outcomes}\\ \text{when a coin is tossed}\end{array}\right) \cdot \left(\begin{array}{c}\text{number of outcomes}\\ \text{a die is rolled}\end{array}\right) \cdot \left(\begin{array}{c}\text{number of ways}\\ \text{to draw a card}\end{array}\right) = (2)(6)(52) = 624$.

3. **(a)** Order is not important, and there are no repetitions, so the number of different two-element subsets is $C(5,2) = \frac{5!}{2!\,3!} = \frac{5 \cdot 4}{2} = 10$.

(b) Order is important, and there are no repetitions, so the number of different two-letter words is $P(5,2) = \frac{5!}{3!} = 20$.

5. You earn a score of 70% by answering exactly 7 of the 10 questions correctly. The number of different ways to answer the questions correctly is $C(10,7) = \frac{10!}{7!\,3!} = 120$.

7. You must choose two of the ten questions to omit, and the number of ways of choosing these two questions is $C(10,2) = \frac{10!}{2!\,8!} = 45$.

9. The maximum number of employees using this security system is $\left(\begin{array}{c}\text{number of choices}\\ \text{for the first letter}\end{array}\right) \cdot \left(\begin{array}{c}\text{number of choices}\\ \text{for the second letter}\end{array}\right) \cdot \left(\begin{array}{c}\text{number of choices}\\ \text{for the third letter}\end{array}\right) = (26)(26)(26) = 17{,}576$.

11. We could count the number of ways of choosing 7 of the flips to be heads; equivalently we could count the number of ways of choosing 3 of the flips to be tails. Thus, the number of different ways this can occur is $C(10,7) = C(10,3) = \frac{10!}{3!\,7!} = 120$.

13. Let x be the number of people in the group. Then $C(x,2) = 10 \Leftrightarrow \frac{x!}{2!\,(x-2)!} = 10 \Leftrightarrow \frac{x!}{(x-2)!} = 20 \Leftrightarrow x(x-1) = 20$ $\Leftrightarrow x^2 - x - 20 = 0 \Leftrightarrow (x-5)(x+4) = 0 \Leftrightarrow x = 5$ or $x = -4$. So there are 5 people in this group.

15. A letter can be represented by a sequence of length 1, a sequence of length 2, or a sequence of length 3. Since each symbol is either a dot or a dash, the possible number of letters is $\binom{\text{number of letters}}{\text{using 3 symbols}} + \binom{\text{number of letters}}{\text{using 2 symbols}} + \binom{\text{number of letters}}{\text{using 1 symbol}} = 2^3 + 2^2 + 2 = 14$.

17. (a) Since we cannot choose a major and a minor in the same subject, the number of ways a student can select a major and a minor is $P(16, 2) = 16 \cdot 15 = 240$.

(b) Again, since we cannot have repetitions and the order of selection is important, the number of ways to select a major, a first minor, and a second minor is $P(16, 3) = 16 \cdot 15 \cdot 14 = 3360$.

(c) When we select a major and 2 minors, the order in which we choose the minors is not important. Thus the number of ways to select a major and 2 minors is $\binom{\text{number of ways}}{\text{to select a major}} \cdot \binom{\text{number of ways to}}{\text{select two minors}} = 16 \cdot C(15, 2) = 16 \cdot 105 = 1680$.

19. Because the letters are distinct, the number of anagrams of the word *RANDOM* is $6! = 720$.

21. Because three letters are the same, the number of anagrams of the word *BUBBLE* is $\frac{6!}{3!} = 120$.

23. (a) The possible number of committees is $C(18, 7) = 31{,}824$.

(b) Since we must select the 4 men from the group of 10 men and the 3 women from the group of 8 women, the possible number of committees is $\binom{\text{number of ways to}}{\text{choose 4 of 10 men}} \cdot \binom{\text{number of ways to}}{\text{choose 3 of 8 women}} = C(10, 4) \cdot C(8, 3) = 210 \cdot 56 = 11{,}760$.

(c) We remove Susie from the group of 18, so the possible number of committees is $C(17, 7) = 19{,}448$.

(d) The possible number of committees is

$$\binom{\text{possible number of}}{\text{committees with 5 women}} + \binom{\text{possible number of}}{\text{committees with 6 women}} + \binom{\text{possible number of}}{\text{committees with 7 women}}$$
$$= C(8, 5) \cdot C(10, 2) + C(8, 6) \cdot C(10, 1) + C(8, 7) \cdot C(10, 0) = 56 \cdot 45 + 28 \cdot 10 + 8 \cdot 1 = 2808$$

(e) Since the committee is to have 7 members, "at most two men" is the same as "at least five women," which we found in part (d). So the number is also 2808.

(f) We select the specific offices first, then complete the committee from the remaining members of the group. So the number of possible committees is

$$\left(\begin{array}{c}\text{number of ways to choose}\\ \text{a chairman, a vice-chairman,}\\ \text{and a secretary}\end{array}\right) \cdot \left(\begin{array}{c}\text{number of ways to choose}\\ \text{4 other members}\end{array}\right) = P(18, 3) \cdot C(15, 4) = 4896 \cdot 1365 = 6{,}683{,}040.$$

25. (a) The probability that the ball is red is $\frac{10}{15} = \frac{2}{3}$.

(b) The probability that the ball is even numbered is $\frac{8}{15}$.

(c) The probability that the ball is white and an odd number is $\frac{2}{15}$.

(d) The probability that the ball is red or odd numbered is $P(\text{red}) + P(\text{odd}) - P(\text{red} \cap \text{odd}) = \frac{10}{15} + \frac{7}{15} - \frac{5}{15} = \frac{12}{15} = \frac{4}{5}$.

27. (a) $S = \{HHH, HHT, HTH, HTT, THH, THT, TTH, TTT\}$.

(b) $P(HHH) = \frac{1}{8}$

(c) $P(\text{2 or more heads}) = P(\text{exactly 2 heads}) + P(\text{3 heads}) = \frac{3}{8} + \frac{1}{8} = \frac{4}{8} = \frac{1}{2}$

(d) $P(\text{tails on the first toss}) = \frac{4}{8} = \frac{1}{2}$

29. Since rolling a die and selecting a card is independent,
$P(\text{both show a six}) = P(\text{die shows a six}) \cdot P(\text{card is a six}) = \frac{1}{6} \cdot \frac{4}{52} = \frac{1}{78}$.

31. (a) Since these events are independent, the probability of getting the ace of spades, a six, and a head is $\frac{1}{52} \cdot \frac{1}{6} \cdot \frac{1}{2} = \frac{1}{624}$.

(b) The probability of getting a spade, a six, and a head is $\frac{13}{52} \cdot \frac{1}{6} \cdot \frac{1}{2} = \frac{1}{48}$.

(c) The probability of getting a face card, a number greater than 3, and a head is $\frac{12}{52} \cdot \frac{3}{6} \cdot \frac{1}{2} = \frac{3}{52}$.

33. (a) Since there are four kings in a standard deck, $P(4 \text{ kings}) = \dfrac{C(4,4)}{C(52,4)} = \dfrac{1}{\frac{52 \cdot 51 \cdot 50 \cdot 49}{4 \cdot 3 \cdot 2 \cdot 1}} = \dfrac{1}{270{,}725} \approx 3.69 \times 10^{-6}$.

(b) Since there are 13 spades in a standard deck, $P(4 \text{ spades}) = \dfrac{C(13,4)}{C(52,4)} = \dfrac{\frac{13 \cdot 12 \cdot 11 \cdot 10}{4 \cdot 3 \cdot 2 \cdot 1}}{\frac{52 \cdot 51 \cdot 50 \cdot 49}{4 \cdot 3 \cdot 2 \cdot 1}} = \frac{11}{4165} \approx 0.00264$.

(c) Since there are 26 red cards and 26 black cards, $P(\text{all same color}) = \dfrac{2 \cdot C(26,4)}{C(52,4)} = \dfrac{2 \cdot \frac{26 \cdot 25 \cdot 24 \cdot 23}{4 \cdot 3 \cdot 2 \cdot 1}}{\frac{52 \cdot 51 \cdot 50 \cdot 49}{4 \cdot 3 \cdot 2 \cdot 1}} = \frac{92}{833} \approx 0.11044$.

35. She knows the first digit and must arrange the other four digits. Since only one of the $P(4,4) = 24$ arrangements is correct, the probability that she guesses correctly is $\frac{1}{24}$.

37. (a) Since there are only two colors of socks, any 3 socks must contain a matching pair.

(b) *Method 1:* If the two socks drawn form a matching pair then they are either both red or both blue. So

$$P\left(\begin{array}{c}\text{choosing a}\\ \text{matching pair}\end{array}\right) = P\left(\begin{array}{c}\text{both red or}\\ \text{both blue}\end{array}\right) = P(\text{both red}) + P(\text{both blue}) = \frac{C(20,2)}{C(50,2)} + \frac{C(30,2)}{C(50,2)} \approx 0.51.$$

Method 2: The complement of choosing a matching pair is choosing one sock of each color. So

$$P(\text{choosing a matching pair}) = 1 - P(\text{different colors}) = 1 - \frac{C(20,1) \cdot C(30,1)}{C(50,2)} \approx 1 - 0.49 = 0.51.$$

39. (a) Order is important, and repeats are possible. Thus there are 10 choices for each digit. So the number of different Zip+4 codes is $10 \cdot 10 \cdot \cdots \cdot 10 = 10^9$.

(b) If a Zip+4 code is to be a palindrome, the first 5 digits can be chosen arbitrarily. But once chosen, the last 4 digits are determined. Since there are 10 ways to choose each of the first 5 digits, there are 10^5 palindromes.

(c) By parts (a) and (b), the probability that a randomly chosen Zip+4 code is a palindrome is $\frac{10^5}{10^9} = 10^{-4}$.

41. (a) $P(\text{king}) = \frac{4}{52} = \frac{1}{13}$

(b) $P(\text{king or ace}) = \frac{8}{52} = \frac{2}{13}$

(c) The probability that the card is a king given that it is a face card is $\dfrac{\text{number of kings}}{\text{number of face cards}} = \dfrac{4}{12} = \dfrac{1}{3}$.

(d) The probability that the card is a king given that it is not an ace is $\dfrac{\text{number of kings}}{\text{number of non-aces}} = \dfrac{4}{48} = \dfrac{1}{12}$.

43. (a) $P(4 \text{ sixes in 8 rolls}) = C(8,4) \cdot \left(\frac{1}{6}\right)^4 \left(\frac{5}{6}\right)^4 \approx 0.026048$.

(b) There are three even numbers on a die and three odds numbers, so $P(\text{even}) = P(\text{odd}) = 0.5$. Thus

$$\begin{aligned} P(2 \text{ or more evens in 8 rolls}) &= 1 - P(\text{fewer than 2 evens}) = 1 - [P(0 \text{ evens}) + P(1 \text{ even})] \\ &= 1 - \left[C(8,0) \cdot \left(0.5^0\right)\left(0.5^8\right) + C(8,1) \cdot \left(0.5^1\right)\left(0.5^7\right)\right] \approx 0.96484 \end{aligned}$$

45. (a) The probability that nine or more patients would have recovered without the drug is $C(12,9)(0.65)^9(0.35)^3 + C(12,10)(0.65)^{10}(0.35)^2 + C(12,11)(0.65)^{11}(0.35)^1 + C(12,12)(0.65)^{12}(0.35)^0 \approx 0.347$.

(b) No, the drug does not appear to be effective.

47. There are 36 possible outcomes in rolling two dice and 6 ways in which both dice show the same numbers, namely, (1, 1), (2, 2), (3, 3), (4, 4), (5, 5), and (6, 6). So the expected value of this game is $E = (5)\left(\frac{6}{36}\right) + (-1)\left(\frac{30}{36}\right) = 0$.

49. Since Mary makes a guess as to the order of ratification of the 13 original states, the number of such guesses is $P(13,13) = 13!$, while the probability that she guesses the correct order is $\dfrac{1}{13!}$. Thus the expected value is $E = (1{,}000{,}000)\left(\dfrac{1}{13!}\right) + (0)\left(\dfrac{13! - 1}{13!}\right) = 0.00016$. So Mary's expected winnings are \$0.00016.

CHAPTER 14 TEST

1. The order is fixed, but for each grandchild they have three choices of pictures. Thus, the number of possibilities is $3 \times 3 \times 3 \times 3 = 81$.

3. **(a)** If repetition is allowed, then each letter can be chosen in 26 ways and each digit in 10 ways, so the number of possible passwords is $26^4 \cdot 10^3 = 456{,}976{,}000$.

(b) If repetition is not allowed, then the first letter can be chosen in 26 ways, the second in 25 ways, the third in 24 ways, and the fourth in 23 ways. The first digit can be chosen in 10 ways, the second in 9 ways, and the third in 8 ways. Thus, in this case the total number of possible passwords is $26 \cdot 25 \cdot 24 \cdot 23 \cdot 10 \cdot 9 \cdot 8 = 258{,}336{,}000$.

5. There are two choices to be made: choose a road to travel from Ajax to Barrie, and then choose a different road from Barrie to Ajax. Since there are 4 roads joining the two cities, we need the number of permutations of 4 objects (the roads) taken 2 at a time (the road there and the road back). This number is $P(4, 2) = 4 \cdot 3 = 12$.

7. **(a)** We want the number of ways of arranging 4 distinct objects (the letters L, O, V, E). This is the number of permutations of 4 objects taken 4 at a time. Therefore, the number of anagrams of the word LOVE is $P(4, 4) = 4! = 24$.

(b) We want the number of distinguishable permutations of 6 objects (the letters K, I, S, S, E, S) consisting of three like groups of size 1 and a like group of size 3 (the S's). Therefore, the number of different anagrams of the word KISSES is $\dfrac{6!}{1!\,1!\,1!\,3!} = \dfrac{6!}{3!} = 120$.

9. One card is drawn from a deck.

(a) Since there are 26 red cards, the probability that the card is red is $\frac{26}{52} = \frac{1}{2}$.

(b) Since there are 4 kings, the probability that the card is a king is $\frac{4}{52} = \frac{1}{13}$.

(c) Since there are 2 red kings, the probability that the card is a red king is $\frac{2}{52} = \frac{1}{26}$.

11. Let E be the event of choosing 3 men. Then $P(E) = \dfrac{n(E)}{n(S)} = \dfrac{\text{number of ways to choose 3 men}}{\text{number of ways to choose 3 people}} = \dfrac{C(5, 3)}{C(15, 3)} \approx 0.022$.

13. There are 4 students and 12 astrological signs. Let E be the event that at least 2 have the same astrological sign. Then E' is the event that no 2 have the same astrological sign. It is easier to find E'. So $P(E') = \dfrac{\text{number of ways to assign 4 different astrological signs}}{\text{number of ways to assign 4 astrological signs}} = \dfrac{P(12, 4)}{12^4} = \dfrac{12 \cdot 11 \cdot 10 \cdot 9}{12 \cdot 12 \cdot 12 \cdot 12} = \dfrac{55}{96}$.
Therefore, $P(E) = 1 - P(E') = 1 - \frac{55}{96} = \frac{41}{96} \approx 0.427$.

15. A deck of cards contains 4 aces, 12 face cards, and 36 other cards. So the probability of an ace is $\frac{4}{52} = \frac{1}{13}$, the probability of a face card is $\frac{12}{52} = \frac{3}{13}$, and the probability of a non-ace, non-face card is $\frac{36}{52} = \frac{9}{13}$. Thus the expected value of this game is $E = (10)\left(\frac{1}{13}\right) + (1)\left(\frac{3}{13}\right) + (-.5)\left(\frac{9}{13}\right) = \frac{8.5}{13} \approx 0.654$, that is, about \$0.65.

FOCUS ON MODELING The Monte Carlo Method

1. **(a)** You should find that with the switching strategy, you win about 90% of the time. The more games you play, the closer to 90% your winning ratio will be.

(b) The probability that the contestant has selected the winning door to begin with is $\frac{1}{10}$, since there are ten doors and only one is a winner. So the probability that he has selected a losing door is $\frac{9}{10}$. If the contestant switches, he exchanges a losing door for a winning door (and vice versa), so the probability that he loses is now $\frac{1}{10}$, and the probability that he wins is now $\frac{9}{10}$.

3. (a) You should find that player A wins about $\frac{7}{8}$ of the time. That is, if you play this game 80 times, player A should win approximately 70 times.

(b) The game will end when either player A gets one more head or player B gets three more tails. Each toss is independent, and both heads and tails have probability $\frac{1}{2}$, so we obtain the following probabilities.

Outcome	Probability
H	$\frac{1}{2}$
TH	$\frac{1}{2} \cdot \frac{1}{2} = \frac{1}{4}$
TTH	$\frac{1}{2} \cdot \frac{1}{2} \cdot \frac{1}{2} = \frac{1}{8}$
TTT	$\frac{1}{2} \cdot \frac{1}{2} \cdot \frac{1}{2} = \frac{1}{8}$

Since Player A wins for any outcome that ends in heads, the probability that he wins is $\frac{1}{2} + \frac{1}{4} + \frac{1}{8} = \frac{7}{8}$.

5. With 1000 trials, you are likely to obtain an estimate for π that is between 3.1 and 3.2.

7. (a) We can use the following TI-83 program to model this experiment. It is a minor modification of the one given in Problem 5.

```
PROGRAM:PROB7
:0→P
:For(N,1,1000)
:rand→X:rand→Y
:P+((X+Y)<1)→P
:End
:Disp "PROBABILITY IS APPROX",P/1000
```

You should find that the probability is very close to $\frac{1}{2}$.

(b) Following the hint, the points in the square for which $x + y < 1$ are the ones that lie below the line $x + y = 1$. This triangle has area $\frac{1}{2}$ (it takes up half the square), so the probability that $x + y < 1$ is $\frac{1}{2}$.

APPENDIXES

A GEOMETRY REVIEW

1. Congruent by ASA

3. Not necessarily congruent

5. Similar

7. Similar

9. $\frac{5}{6} = \frac{x}{150} \Leftrightarrow x = 125$

11. $\frac{7}{2} = \frac{y}{3/2} \Leftrightarrow y = \frac{21}{4}$, so $\frac{x}{7} = \frac{9/2}{21/4} \Leftrightarrow x = 6.$

13. $\frac{x}{a} = \frac{c}{a+b} \Leftrightarrow x = \frac{ac}{a+b}$

15. From the diagram, we see that because $AC \parallel EF$ and $BC \parallel ED$, $a = \alpha$ and $b = \beta$. Thus, $\triangle ABC \sim \triangle AED \sim \triangle EBF$.

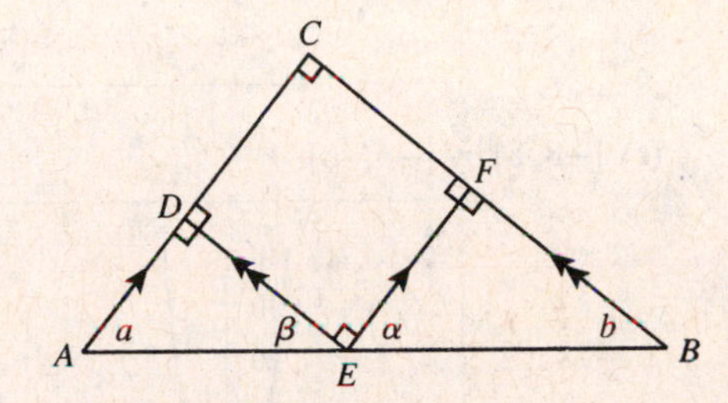

17. $x = \sqrt{8^2 + 6^2} = 10$

19. $x = \sqrt{2^2 - 1^2} = \sqrt{3}$

21. $x^2 + (x+2)^2 = 58^2 \Leftrightarrow 2x^2 + 4x + 4 = 3364 \Leftrightarrow 2x^2 + 4x - 3360 = 2(x+42)(x-40) = 0 \Leftrightarrow x = 40$

23. $5^2 + 12^2 = 169 = 13^2$, so the triangle is a right triangle.

25. $8^2 + 10^2 = 164 \neq 12^2$, so the triangle is not a right triangle.

27. $48^2 + 55^2 = 5329 = 73^2$, so the triangle is a right triangle.

29. Let the other leg have length x. Then $11^2 + x^2 = (x+1)^2 \Leftrightarrow 121 + x^2 = x^2 + 2x + 1 \Leftrightarrow 2x = 122 \Leftrightarrow x = 61$ cm.

31. If the quadrilateral were a rectangle, we would have $17^2 + 21^2 = 27^2$. But this is false, so it is not a rectangle.

33. The diagonal of the left face of the box is $\sqrt{3^2 + 4^2} = 5$, so the length of the desired diagonal is $\sqrt{5^2 + 12^2} = 13$.

35.

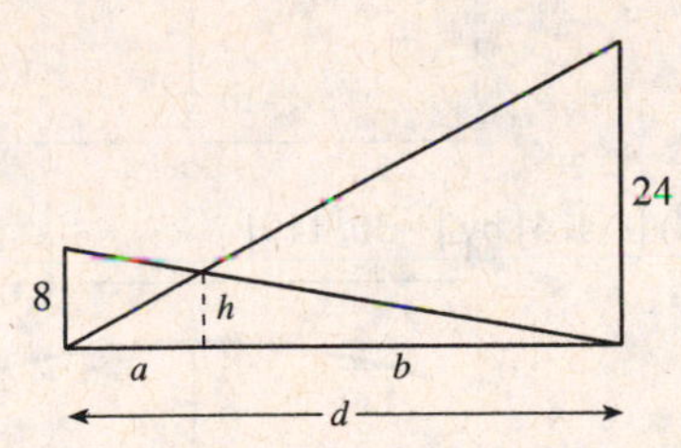

By similarity, $\frac{h}{a} = \frac{24}{d}$ and $\frac{h}{b} = \frac{8}{d}$. Thus, $24a = 8b \Leftrightarrow b = 3a$ $\Leftrightarrow 4a = d$, and so $\frac{h}{a} = \frac{24}{4a} \Leftrightarrow h = 6.$

B CALCULATIONS AND SIGNIFICANT FIGURES

1. $3.27 - 0.1834 \approx 3.09$

3. $28.36 \times 501.375 \approx 14{,}220$

5. $(1.36)^3 \approx 2.52$

7. $3.3\,(642.75 + 66.787) \approx 3.3\,(709.54) \approx 2300$

9. $\left(5.10 \times 10^{-3}\right)\left(12.4 \times 10^{7}\right)\left(6.007 \times 10^{-6}\right) \approx 3.80$

11. The circumference is $2\pi r \approx 2\pi\,(5.27) \approx 33.1$ ft and the area is $\pi r^2 \approx \pi\,(5.27)^2 \approx 87.3$ ft^2.

13. The force is $F = G\dfrac{m_1 m_2}{r^2} \approx 6.67428 \times 10^{-11} \cdot \dfrac{(11{,}426)^2}{(57{,}200)^2} \approx 2.66 \times 10^{-12}$ N.

C GRAPHING WITH A GRAPHING CALCULATOR

1. $y = x^4 + 2$

(a) $[-2, 2]$ by $[-2, 2]$

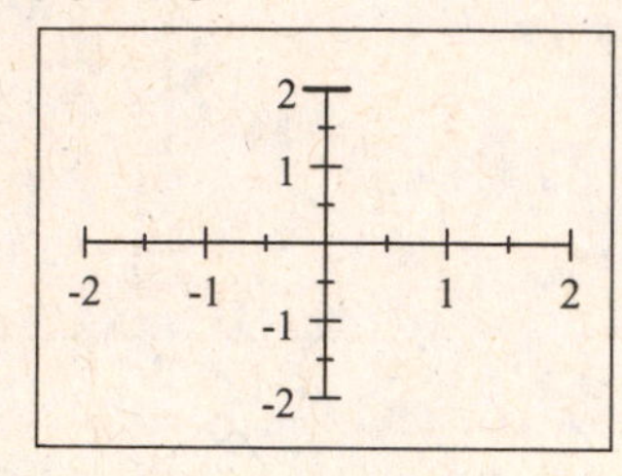

(b) $[0, 4]$ by $[0, 4]$

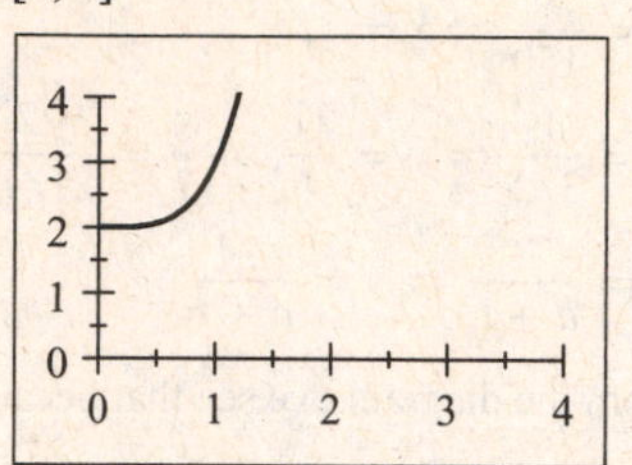

(c) $[-8, 8]$ by $[-4, 40]$

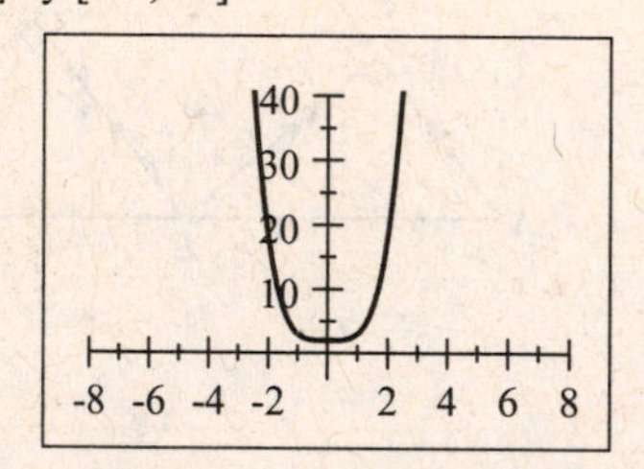

(d) $[-40, 40]$ by $[-80, 800]$

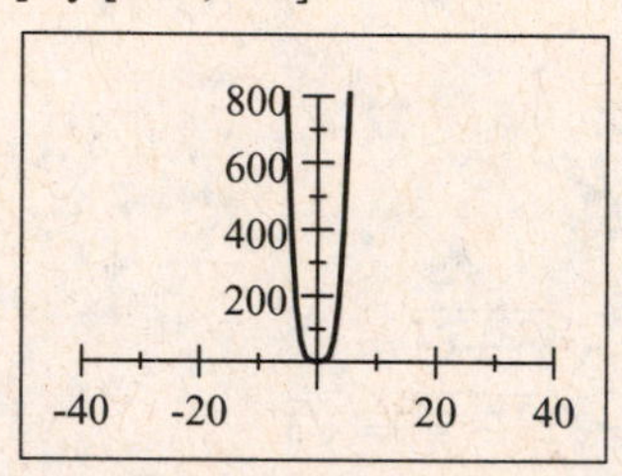

The viewing rectangle in part (c) produces the most appropriate graph of the equation.

3. $y = 100 - x^2$

(a) $[-4, 4]$ by $[-4, 4]$

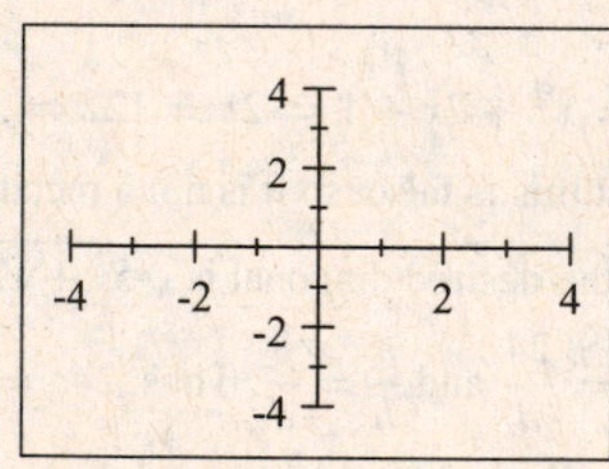

(b) $[-10, 10]$ by $[-10, 10]$

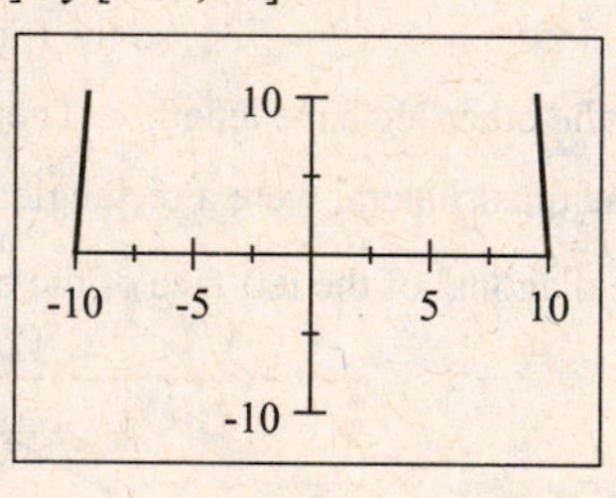

(c) $[-15, 15]$ by $[-30, 110]$

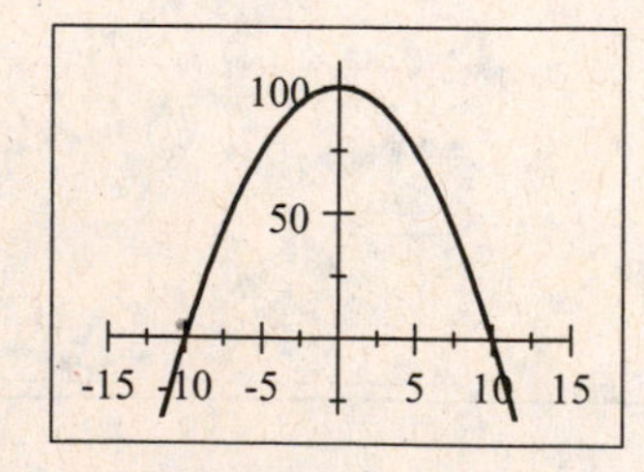

(d) $[-4, 4]$ by $[-30, 110]$

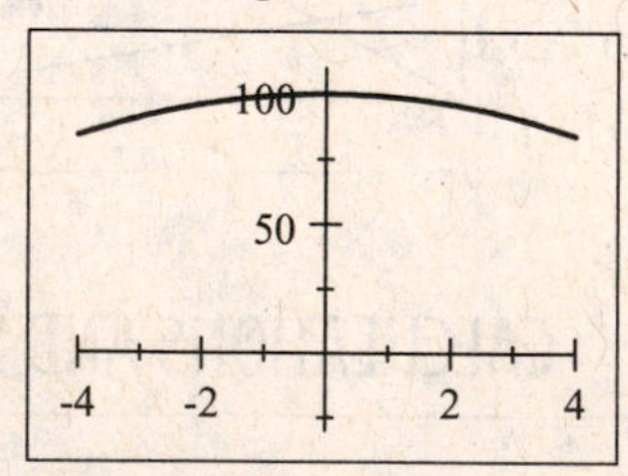

The viewing rectangle in part (c) produces the most appropriate graph of the equation.

5. $y = 10 + 25x - x^3$

(a) $[-4, 4]$ by $[-4, 4]$

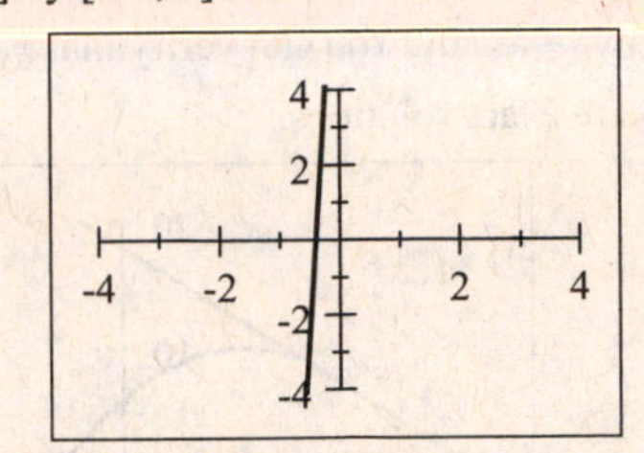

(b) $[-10, 10]$ by $[-10, 10]$

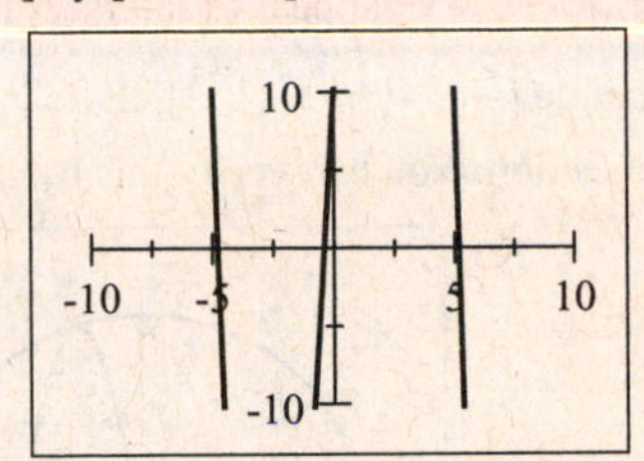

(c) $[-20, 20]$ by $[-100, 100]$

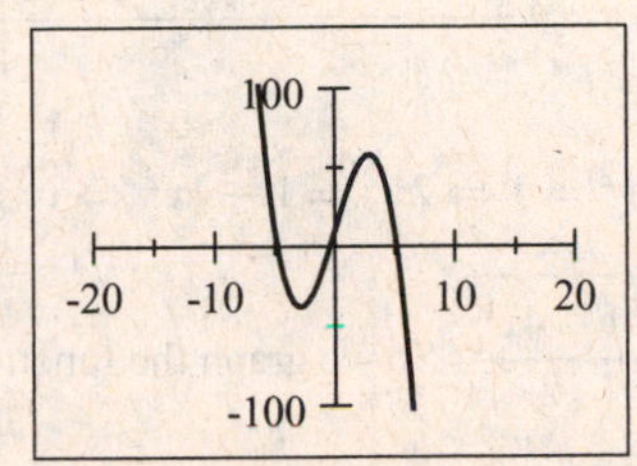

(d) $[-100, 100]$ by $[-200, 200]$

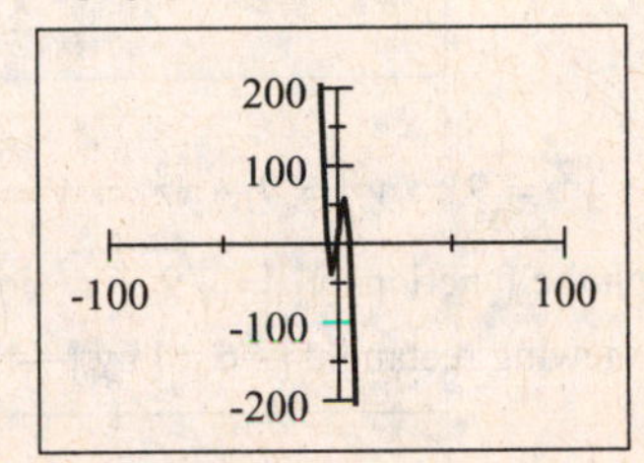

The viewing rectangle in part (c) produces the most appropriate graph of the equation.

7. $y = 100x^2$, $[-2, 2]$ by $[-10, 400]$

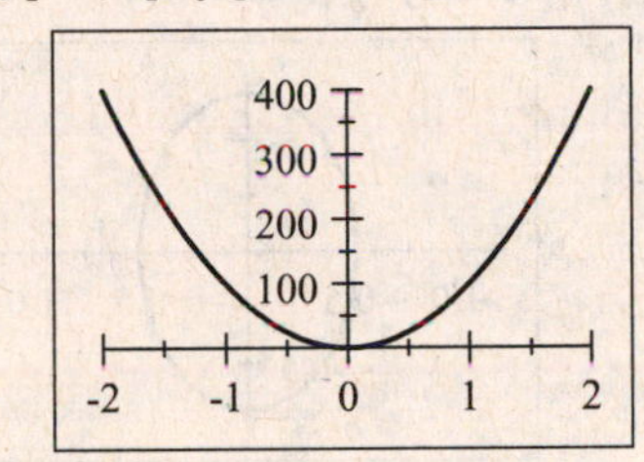

9. $y = 4 + 6x - x^2$, $[-4, 10]$ by $[-10, 20]$

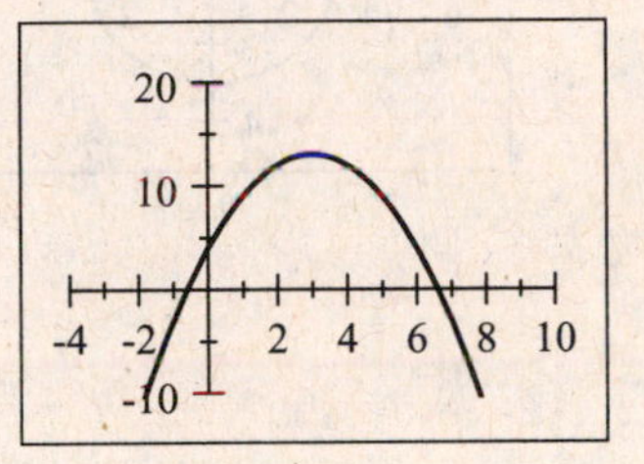

11. $y = \sqrt[4]{256 - x^2}$. We require that $256 - x^2 \geq 0 \Rightarrow -16 \leq x \leq 16$, so we graph $y = \sqrt[4]{256 - x^2}$ in the viewing rectangle $[-20, 20]$ by $[-1, 5]$.

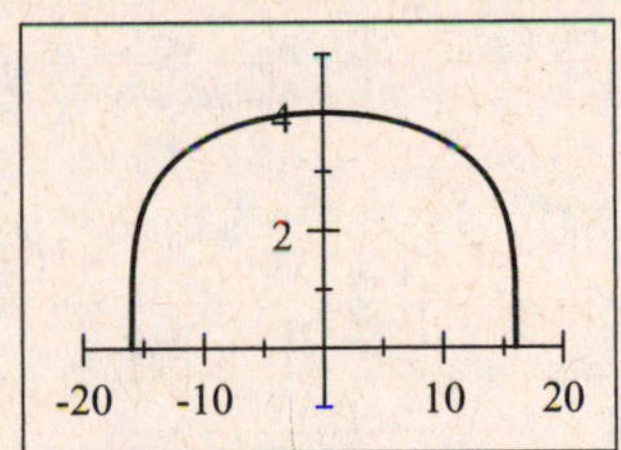

13. $y = 0.01x^3 - x^2 + 5$, $[-50, 150]$ by $[-2000, 2000]$

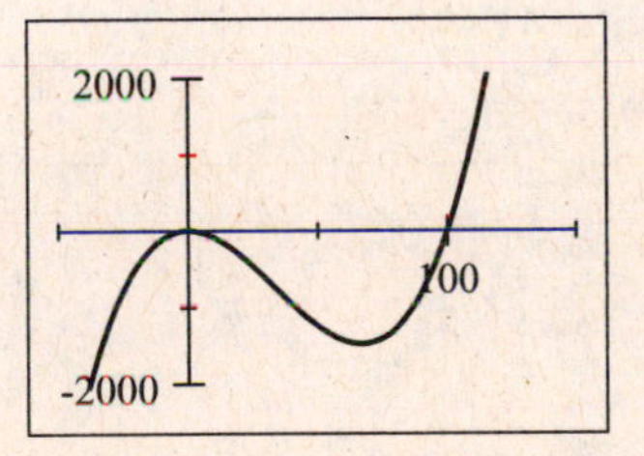

15. $y = \dfrac{1}{x^2 - 2x}$, $[-2, 4]$ by $[-8, 8]$

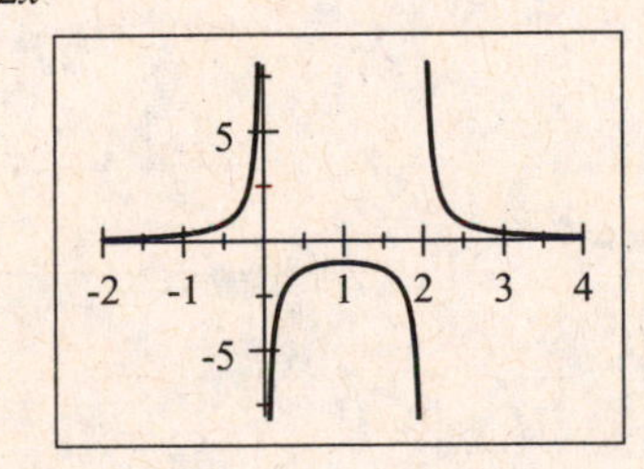

17. $y = 1 + |x - 1|$, $[-3, 5]$ by $[-1, 5]$

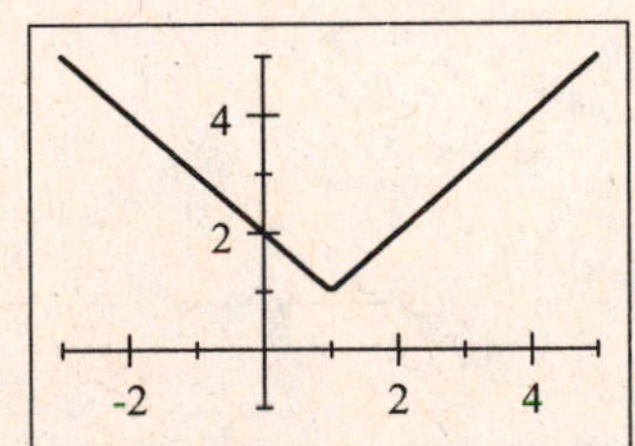

19. Although the graphs of $y = -3x^2 + 6x - \frac{1}{2}$ and $y = \sqrt{7 - \frac{7}{12}x^2}$ appear to intersect in the viewing rectangle $[-4, 4]$ by $[-1, 3]$, there is no point of intersection. You can verify this by zooming in.

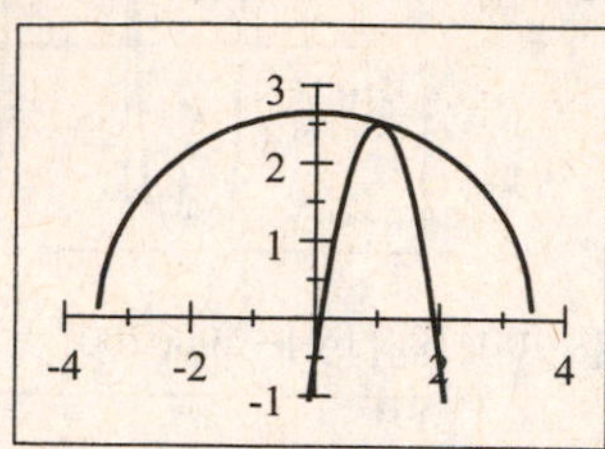

21. The graphs of $y = 6 - 4x - x^2$ and $y = 3x + 18$ appear to have two points of intersection in the viewing rectangle $[-6, 2]$ by $[-5, 20]$. You can verify that $x = -4$ and $x = -3$ are exact solutions.

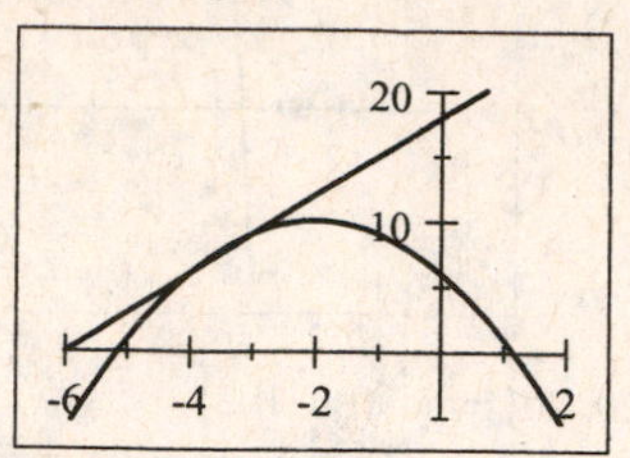

23. $x^2 + y^2 = 9 \Leftrightarrow y^2 = 9 - x^2 \Rightarrow y = \pm\sqrt{9 - x^2}$. So we graph the functions $y_1 = \sqrt{9 - x^2}$ and $y_2 = -\sqrt{9 - x^2}$ in the viewing rectangle $[-6, 6]$ by $[-4, 4]$.

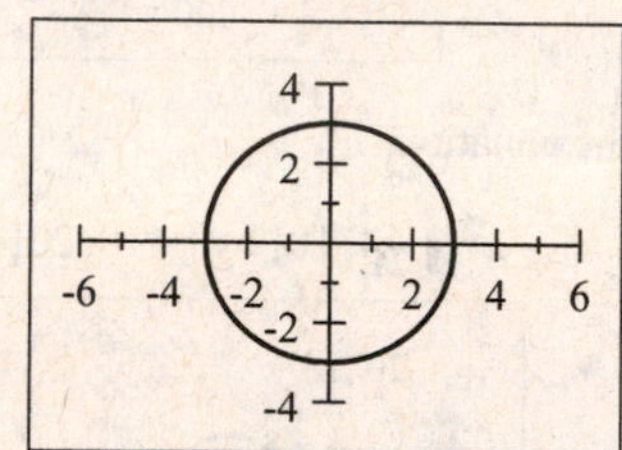

25. $4x^2 + 2y^2 = 1 \Leftrightarrow 2y^2 = 1 - 4x^2 \Leftrightarrow y^2 = \dfrac{1 - 4x^2}{2} \Rightarrow$ $y = \pm\sqrt{\dfrac{1 - 4x^2}{2}}$. So we graph the functions $y_1 = \sqrt{\dfrac{1 - 4x^2}{2}}$ and $y_2 = -\sqrt{\dfrac{1 - 4x^2}{2}}$ in the viewing rectangle $[-1.2, 1.2]$ by $[-0.8, 0.8]$.

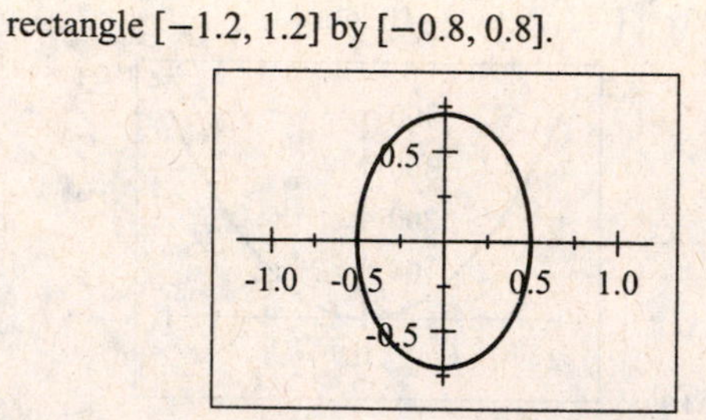